Gerhard Philipp

I. S. GRADSTEIN • I. M. RYSHIK
TAFELN • TABLES

GRADSTEIN
RYSHIK

SUMMEN-, PRODUKT- UND INTEGRAL-

TAFELN

*

TABLES

OF SERIES, PRODUCTS, AND INTEGRALS

BAND 1/VOLUME 1

deutsch und englisch

German and English

1981

Verlag Harri Deutsch

Thun · Frankfurt/M

Diese Ausgabe wurde übersetzt und zum Druck vorbereitet von Dr. h. c. LUDWIG BOLL, DDR Berlin. Mit freundlicher Genehmigung des VEB Deutscher Verlag der Wissenschaften, DDR 108 Berlin, wurde die 1963 erschienene 2., berichtigte Auflage mit der deutschen Übersetzung von CHRISTA BERG und Dr. LOTHAR BERG und der englischen Übersetzung von Dr. MARTIN STRAUSS benutzt.

This edition was translated and prepared for press by Dr. h. c. LUDWIG BOLL, Berlin, GDR. The second corrected edition of VEB Deutscher Verlag der Wissenschaften, 108 Berlin GDR, (1963), which was translated into German by CHRISTA BERG and Dr. LOTHAR BERG and into English by Dr. MARTIN STRAUSS, has been used by the kind permission of VEB Deutscher Verlag der Wissenschaften.

Diese Ausgabe wurde auf der Basis der 5. russischen Auflage, die von J. GERONIMUS und M. ZEITLIN überarbeitet wurde, übersetzt.

This edition has been translated incorporating the fifth russian edition, which was prepared by YU. GERONIMUS and M. TSEITLIN.

Titel der Originalausgabe:
И. С. Градштейн и И. М. Рыжик
Таблицы интегралов, сумм, рядов и произведений
Издательство „Наука", Москва 1971

ISBN 3 87144 350 6
© Verlag MIR, Moskau, 1981
Lizenzausgabe für den Verlag Harri Deutsch, Thun 1981
1. Auflage
Einband: A. W. Schipow

Aus dem Vorwort zur ersten russischen Auflage

Die in den existierenden mathematischen Tafelwerken enthaltenen Formeln für Integrale, Summen, Reihen und Produkte reichen in keiner Weise aus, um die Bedürfnisse von wissenschaftlich arbeitenden Mathematikern und theoretisch oder forschend tätigen Ingenieuren zu befriedigen. Das vorliegende Tafelwerk soll diese Lücke schließen. Es enthält mehr als 5000 Formeln, die verschiedenen Quellen entnommen wurden.

Da das Buch für Wissenschaftler und Forschungsingenieure auf dem Gebiet der physikalisch-mathematischen Wissenschaften bestimmt ist, nehmen Erläuterungen nur geringen Raum ein. Das Buch ist im wesentlichen eine Formelsammlung. Ausführlich wird auf spezielle Funktionen, insbesondere auf elliptische Funktionen, Zylinder- und Kugelfunktionen eingegangen. Viele Formeln des Buches beziehen sich auf diese Funktionen.

Ich möchte die Gelegenheit benutzen, den Professoren W. W. STEPANOW, A. I. MARKUSCHEWITSCH und I. N. BRONSTEIN für ihre wertvollen Ratschläge und Hinweise bei der Zusammenstellung dieser Tafeln herzlich zu danken.

I. RYSHIK

Aus dem Vorwort zur dritten russischen Auflage

I. M. RYSHIK, der Herausgeber der ersten zwei Auflagen dieses Werkes, starb während des zweiten Weltkrieges. Auf Vorschlag des Verlages habe ich diese Tabellen überarbeitet.

Im Abschnitt über bestimmte Integrale wurden folgende Änderungen durchgeführt: Alle „Fakultäten" wie $2^{n-2/3}$ wurden durch die Gammafunktion ersetzt, und wo möglich, durch gewöhnliche Produkte und Fakultäten, da wir der Meinung waren, daß die „Fakultäten" dem modernen Leser wenig vertraut sind und nur überflüssige Schwierigkeiten mit sich bringen. Wo die rechten Seiten von Formeln durch spezielle Funktionen oder Zahlen ersetzt werden konnten, wurde das getan. Es wurde eine Reihe von Integralen hinzugefügt, die auf spezielle Funktionen führen. Integrale über Ausdrücke, die komplexe Größen enthalten, wurden weggelassen, ebenso einige andere Integrale. Außerdem wurde die Reihenfolge der Formeln geändert.

Auch die Art der Numerierung der Formeln wurde geändert. Alle Formeln, Definitionen und Sätze wurden auf numerierte Abschnitte verteilt. Die neue Art der Numerierung ähnelt der Dezimalklassifikation und kann leicht aus dem Inhaltsverzeichnis ersehen werden. Im Inhaltsverzeichnis sind nur die Abschnitte erwähnt, deren Nummern eine, zwei oder drei Ziffern umfassen. Die kleinsten Abschnitte des Buches tragen vier Ziffern und enthalten numeriert eine oder mehrere Formeln (Definitionen oder Sätze). Die Ziffer „Null" ist Abschnitten allgemeinen Charakters vorbehalten: Einführungen, Definitionen usw. Das erste Kapitel des Buches, das eine Reihe von Sätzen allgemeiner Art enthält und einführenden Charakter trägt, ist ebenfalls mit einer Null versehen.

Neu in dieser Auflage sind Verweise auf die Literatur, der die Formeln entnommen sind*. Ich habe mich bemüht, in erster Linie auf sowjetische Ausgaben und besonders auf die Originalausgaben zu verweisen, an zweiter Stelle auf ausländische Bücher und erst an dritter Stelle auf Handbücher. Auf Zeitschriftenartikel wird nicht verwiesen. Formeln, die irgendeinem Buch entnommen sind, wurden manchmal umgeformt. In diesen Fällen steht am Ende des Literaturhinweises der Buchstabe $I\!I$. Insbesondere kann dieser Buchstabe auf eine Druckfehlerberichtigung hinweisen.

I. GRADSTEIN

Aus dem Vorwort zur vierten russischen Auflage

Bei der Vorbereitung der vierten Auflage plante I. S. GRADSTEIN eine beträchtliche Erweiterung des Nachschlagewerkes. Der Tod hinderte ihn leider daran, sein Vorhaben zu realisieren. Der Verlag hat uns beauftragt, das von I. S. GRADSTEIN hinterlassene Manuskript zum Druck vorzubereiten und die nicht fertiggestellten Abschnitte zu vervollständigen. Bei der Ausführung dieser Arbeit waren wir bemüht, dem Gerüst des Manuskripts und der dritten Auflage zu folgen und die wichtigsten Besonderheiten beizubehalten, nämlich die Reihenfolge der Formeln und die Quellenhinweise.

Wir möchten A. F. LAPKO für die sorgfältige Durchsicht des Manuskripts und eine Reihe nützlicher Bemerkungen sehr herzlich danken.

J. GERONIMUS und M. ZEITLIN

Vorwort zur fünften russischen Auflage

Die fünfte Auflage ist ein verbesserter Nachdruck der vierten Auflage.
Wir benutzten Mitteilungen von S. I. ANISSIMOW, I. E. DSJALOSCHINSKI, I. B. LEWINSON, W. L. LOBYSSEW, B. E. MEJEROWITSCH, A. A. OWTSCHINNIKOW und J. T. SOMOW.

Wir benutzten auch Mitteilungen über Fehler in Nachschlagewerken, die in der Zeitschrift „Mathematics of Computation" erschienen, insbesondere Mitteilungen von J. R. BLAKE, E. R. HANSEN, M. L. PATRICK und H. E. FETTIS. Wertvolle Hilfe leisteten E. M. LIFSCHITZ, K. A. SEMENDJAJEW und J. S. JAKOWLEW.

Allen genannten Personen, sowie allen Lesern, die uns schrieben, danken wir sehr.

J. GERONIMUS und M. ZEITLIN

---

* Das Literaturverzeichnis ist am Buchende aufgeführt. Die Literaturhinweise werden in Form von kleingedruckten Zahlen hinter den Formeln oder Definitionen gegeben. Dabei bedeuten: das Kurzzeichen — die Bezeichnung des Buches im Literaturverzeichnis, die Zahl ohne Klammern — die Seite in dem betreffenden Werk, die Zahl in runden Klammern — die Formelnummer, die Zahl in eckigen Klammern — die Nummer der Tafel in dem Quellenwerk. Das Symbol + bedeutet, daß die Formel geändert wurde (Bezeichnungen, Faktoren, Druckfehler).

From the Preface to the First Russian Edition

The formulae for integrals, sums, series, and products included in mathematical tables published to date in the Soviet Union and abroad do not meet the needs of research not only in mathematics but in engineering as well. This collection undertakes to fill the gap. The more than 5000 formulae have been selected from various sources.
The book is intended primarily for theoretical and applied research in the physico-mathematical sciences. Therefore the book is largely a collection of formulae with few explanations.
Attention is focussed on special functions, notably elliptic, cylinder, and spherical functions. Many formulae in this book relate to these functions.
I wish to take this opportunity to express my sincere thanks to professors V. V. STEPANOV, A. I. MARKUSHEVICH, and I. N. BRONSHTEIN for their valuable advice.

I. RYZHIK

From the Preface to the Third Russian Edition

I. M. RYZHIK, to whom we owe the first two editions of this book, lost his life in World War II. The publishers have asked me to update the tables.
The chapter on definite integrals contains several changes. All "factorials", such as $2^{n-2/3}$, are expressed in terms of the gamma function and, where possible, by ordinary products and factorials. There was the fear that the old notation would not be familiar to most readers and would present unnecessary difficulty. Where the right-hand side of a formula can be expressed in terms of a special function or number, this has been done. Many new integrals leading to special functions have been added. On the other hand, integrals with complex integrands and a few other integrals have been omitted. The arrangement of the formulae has also been changed, and so has the numbering. All the formulae, definitions, and theorems are arranged in numbered chapters and sections. The numbering corresponds roughly to a decimal classification, and its principles may easily be recognized from the table of contents. Only the larger subsections with reference numbers up to three figures are mentioned in the table. The smallest subsections have reference numbers with four figures; they contain definitions (formulae or some theorems) numbered in the ordinary way. The figure "0" has been reserved for sections of general character: introductions, definitions, etc. The first chapter, which serves as a kind of introduction and contains theorems of a rather general nature, has also been assigned the reference number 0.

A new feature of this present edition is the references to the literature from which the formulae have been taken *. I have quoted largely from Soviet sources, where possible original works, foreign books, and other reference works. I have not used periodicals as sources. Where the letter *Π* appears after a reference, it means that the formula in the text differs from that in the source. For one, that may mean that a misprint has been corrected.

I. GRADSHTEIN

From the Preface to the Fourth Russian Edition

When preparing the fourth edition, I. S. GRADSHTEIN had in mind to greatly expand this handbook. When his untimely death made this impossible, we were asked to complete the unfinished parts and prepare his manuscript for press. We endeavored to follow the plan of the manuscript and of the third edition and to preserve the main features, namely, arrangement of formulae and references to sources.
We wish to thank A. F. LAPKO for reading the manuscript and making helpful comments.

YU. GERONIMUS and M. TSEITLIN

Preface to the Fifth Russian Edition

The fifth edition is a corrected reprint. The errors were noted by S. I. ANISIMOV, I. E. DZYALOSHINSKII, I. B. LEVINSON, V. L. LOBYSEV, B. E. MEYEROVICH, A. A. OVCHINNIKOV, and E. T. SOMOV, and also by the journal *Mathematics of Computation* (USA) in reports from J. R. BLAKE, E. R. HANSEN, M. L. PATRICK, and H. E. FETTIS.
We received invaluable assistance from E. M. LIFSHITZ, K. A. SEMENDYAEV, and YU. S. YAKOVLEV.
We extend our thanks to all the above mentioned and all who sent us their comments.

YU. GERONIMUS and M. TSEITLIN

---

* In this German-English edition references are given at the end of each volume. The references in the text consist of letters and numbers. Letters indicate the sources, numbers without parentheses indicate the pages in the sources, numbers inside parentheses give the number of the formulae, while numbers in brackets refer to tables. The plus sign shows that the formula is changed compared to the source (new variables, constants, and correction of misprints).

Zur Anordnung der Formeln
Arrangement of Formulae

Die zweckmäßigste Anordnung der Formeln erwies sich insbesondere im Abschnitt „Bestimmte Integrale" als ein recht schwieriges Problem. Naturgemäß schwebt einem eine lexikographische Anordnung vor, jedoch läßt sich eine solche gerade bei den Formeln der Integralrechnung kaum verwirklichen. Man kann nämlich in einer Formel

$$\int_a^b f(x)\, dx = A$$

eine Reihe von Substitutionen der Form $x = \varphi(t)$ durchführen und so „Synonyme" der ursprünglichen Formel erhalten. Sowohl die Tafeln bestimmter Integrale von BIERENS DE HAAN als auch die ersten Auflagen dieser Sammlung enthalten zuviele solcher Synonyme. Wir waren bestrebt, in dieser Auflage nur die einfachsten dieser „Formel-Synonyme" beizubehalten. Dabei schätzten wir eine Formel als einfach ein, wenn die Argumente der „äußeren" Funktion im Integranden einfach sind. Nach Möglichkeit wurden komplizierte Funktionen durch einfachere ersetzt. Manchmal ließen sich mehrere kompliziertere Formeln auf eine einzige einfachere reduzieren; dann wurde nur diese aufgenommen. Gelegentlich ergab sich nach solchen vereinfachenden Substitutionen ein Integral, das aufgrund der Formeln des Abschnitts 2 und der NEWTON-LEIBNIZ-schen Formel auswertbar ist, oder ein Integral der Gestalt

$$\int_{-a}^{a} f(x)\, dx,$$

The most suitable arrangement of the formulae proved to be a rather difficult problem, especially of those referring to definite integrals. One might be tempted to use a lexicographical order. It is almost impossible, however, to arrive in this way at anything resembling a natural order of the formulae for the integrals because substitutions of the form $x = \varphi(t)$ lead to "synonymous" formulae for any given formula

It must be said that both the table of definite integrals by BIERENS DE HAAN and the first editions of this book have far too many such synonyms and complicated formulae. We have tried in this edition to retain only the simplest synonyms. By "simple" we mean that the arguments of what are called the "outer" function are simple. Where possible we have replaced a complicated formula with a simpler one. Sometimes in the process several complicated formulae could have been reduced to one simpler formula. Then we have retained the simpler one. At times as a result of such substitutions we have arrived at an integral that can be evaluated with the help of the formula of Chapter 2 and the NEWTON-LEIBNIZ formula or at an integral of the form

wobei $f(x)$ eine ungerade Funktion ist. Ein solches Integral wurde nicht aufgenommen.

Wir führen Beispiele dafür an, wie Integrale vor dem Aufsuchen in den Tafeln umgeformt werden sollten (Nr. 26 auf S. 159 der zweiten russischen Auflage):

with $f(x)$ an odd function. We have omitted such integrals.

Here is one example (p. 159, No. 26 of the second Russian edition):

$$\int_0^{\frac{\pi}{4}} \frac{(\operatorname{ctg} x - 1)^{p-1}}{\sin^2 x} \ln \operatorname{tg} x \, dx = -\frac{\pi}{p} \operatorname{cosec} p\pi. \tag{1}$$

Als naturgemäße Substitution bietet sich $\operatorname{ctg} x - 1 = u$ an; mit ihrer Hilfe erhalten wir

The obvious substitution $\operatorname{ctg} x - 1 = u$ leads to

$$\int_0^\infty u^{p-1} \ln(1+u) \, du = \frac{\pi}{p} \operatorname{cosec} p\pi. \tag{2}$$

Dieses Integral war nicht direkt in der Sammlung enthalten; es konnte aus anderen, komplizierteren Formeln der Sammlung hergeleitet werden. Ferner sind Nr. 59 und 60 Spezialfälle der Formel Nr. 26 auf S. 159. Alle diese Integrale sind in der neuen Auflage nicht enthalten. An ihrer Stelle steht jetzt Formel (2) und die sich aus dem Integral (1) durch die Substitution $\operatorname{ctg} x = v$ ergebende Formel.

Ein zweites Beispiel ist (Nr. 24 auf S. 172 der zweiten russischen Auflage)

This integral was not included in the former editions; it could be derived from the more complicated formulae. Furthermore, Nos. 59 and 60 are special cases of formula 26. This edition omits all such integrals. Instead it contains formula (2) and the formula derived from (1) by the substitution $\operatorname{ctg} x = v$.

A second example is (No. 24 on p. 172 of the second Russian edition)

$$I = \int_0^{\frac{\pi}{2}} \ln(\operatorname{tg}^p x + \operatorname{ctg}^p x) \ln \operatorname{tg} x \, dx = 0.$$

Die Substitution $\operatorname{tg} x = u$ liefert

The substitution $\operatorname{tg} x = u$ leads to

$$I = \int_0^\infty \frac{\ln(u^p + u^{-p}) \ln u}{1 + u^2} \, du.$$

Nun setzen wir $v = \ln u$. Dann folgt

A further substitution $v = \ln u$ yields

$$I = \int_{-\infty}^\infty \frac{v e^v}{1 + e^{2v}} \ln(e^{pv} + e^{-pv}) \, dv = \int_{-\infty}^\infty v \frac{\ln 2 \operatorname{ch} pv}{2 \operatorname{ch} v} \, dv.$$

Da der Integrand ungerade ist, ist das Integral gleich Null.

Ehe man also ein Integral in den Tabellen sucht, muß man den Inte-

Since the integrand is an odd function, the integral is zero.

Thus before an integral is looked up in the tables, the function to be inte-

grated must be simplified in such a way that the arguments of the integrand (the "inner" functions) become as simple as possible.
The functions are arranged in the following manner.
First we have the elementary functions:
(1) $f(x) = x$
(2) exponential function
(3) hyperbolic functions
(4) trigonometric functions
(5) logarithmic function
(6) inverse hyperbolic functions (replaced by the corresponding logarithms in formulae containing definite integrals)
(7) inverse trigonometric functions
Then the special functions follow:
(8) elliptic integrals
(8) elliptic functions
(10) logarithmic, exponential, sine, and cosine integrals
(11) error functions and FRESNEL integrals
(12) the gamma function and related functions
(13) cylinder functions
(14) MATHIEU functions
(15) spherical harmonics
(16) orthogonal polynomials
(17) hypergeometric function
(18) confluent hypergeometric function
(19) parabolic cylinder function
(20) MEIJER function and MACROBERT function
(21) RIEMANN zeta function
The higher the "rank" of a function in the list, the later it appears in the tables. Expressions built with the same outer functions are listed according to the rank of the next, inner, function. In this way the functions

$$\sin e^x, \quad \sin x, \quad \sin \ln x$$

(with the common outer function sine) will appear in the following order:

$$\sin x, \quad \sin e^x, \quad \sin \ln x.$$

11

In der obigen Liste fehlen folgende Funktionen: Polynome, rationale Funktionen, algebraische Funktionen und Potenzen. Eine in den Tafeln der bestimmten Integrale vorkommende algebraische Funktion läßt sich aber gewöhnlich auf eine endliche Kombination von Wurzeln mit rationalen Exponenten zurückführen. Daher können wir bei der Klassifikation unserer Formeln die Potenzfunktion als Verallgemeinerung einer algebraischen und somit auch einer rationalen Funktion ansehen.* Alle diese Funktionen fassen wir zum Unterschied von den oben angegebenen als gewisse Operatoren (Rechenvorschriften) auf.

Beispielsweise nehmen wir an, daß in dem Ausdruck $\sin^2 e^x$ auf die äußere Funktion sin der Quadrierungsoperator angewendet wurde. Bei $\dfrac{\sin x + \cos x}{\sin x - \cos x}$ nehmen wir an, daß auf die Funktionen sin und cos ein rationaler Operator angewendet wird.

Die Operatoren ordnen wir folgendermaßen an:
1. Polynome (nach wachsendem Grad geordnet)
2. Rationale Operatoren
3. Algebraische Operatoren (im wesentlichen $A^{p/q}$, wobei $q > 0$, q und p rationale Zahlen sind, nach wachsendem q geordnet)
4. Potenzoperatoren

Ausdrücke mit gleichen äußeren und inneren Funktionen werden nach den Operatoren geordnet. Als Beispiel sei die Anordnung gegeben:

The list does not include polynomials and rational, algebraic, and power functions. An algebraic function included in the tables of definite integrals can usually be reduced to a finite combination of roots with rational exponents. In this way a power function may be considered a generalization of the algebraic and hence of the rational function.* The functions just mentioned have been treated as operators, in contrast to those listed above.

To give an example: in the expression $\sin^2 e^x$ the squaring operator is applied to the outer function, the sine. The expression $\dfrac{\sin x + \cos x}{\sin x - \cos x}$ is the result of a rational operator applied to the trigonometric functions sine and cosine.

We will arrange such operators as follows:
(1) polynomial operators (in the order of their degree)
(2) rational operators
(3) algebraic operators (essentially, $A^{p/q}$ with $q > 0$, where p and q are rational numbers; the larger the value of q the higher the rank)
(4) power operators

Expressions with the same outer and inner functions are arranged according to the rank of the operators, for instance:

$$\sin x, \; \sin x \cos x, \; \frac{1}{\cos x} = \sec x, \; \frac{\sin x}{\cos x} = \operatorname{tg} x,$$

$$\frac{\sin x + \cos x}{\sin x - \cos x}, \; \sin^m x, \; \sin^m x \cos x.$$

Kommen in einem Integranden zwei äußere Funktionen $\varphi_1(x)$ und $\varphi_2(x)$ vor, auf welche einer der erwähnten Operatoren angewendet wurde, und

Integrals with integrands resulting from the application of one of the above operators to two outer functions $\varphi_1(x)$ and $\varphi_2(x)$ have been placed after all

---

* Für natürliches n ist die Potenz $(a+bx)^n$ des Binoms $a+bx$ ein Polynom, für ganzes negatives n eine rationale Funktion, für irrationales n keine algebraische Funktion mehr.

* $(a+bx)^n$ is a polynomial in x if n is a natural number, a rational function if n is a negative integer, and a nonalgebraic function if n is irrational.

hat $\varphi_1(x)$ den höheren Rang, so steht das entsprechende Integral später als alle Integrale, die nur die Funktion $\varphi_1(x)$ enthalten, und ist gemäß dem Rang von $\varphi_2(x)$ angeordnet. So werden die trigonometrischen Funktionen von trigonometrischen Funktionen und Potenzen (d.h. $\varphi_2(x) = x$) gefolgt; dann kommen
Kombinationen von trigonometrischen Funktionen und Exponentialfunktionen,
Kombinationen von trigonometrischen Funktionen, Exponentialfunktionen und Potenzen,
Kombinationen von trigonometrischen Funktionen und Hyperbelfunktionen, usw.
Die Reihenfolge der Integrale, die zwei Funktionen $\varphi_1(x)$ und $\varphi_2(x)$ enthalten, wird von der ranghöheren Funktion bestimmt. Bei gleicher Rangnummer bezüglich der ersten Funktion ist die Rangnummer der zweiten maßgebend.

Zu diesen allgemeinen Regeln kommen noch einige spezielle, die man unmittelbar aus den Tafeln ersieht. So hat zwar $e^{1/x}$ nach unseren Ausführungen einen höheren Rang als e^x, doch haben $\ln x$ und $\ln \frac{1}{x}$ denselben Rang; denn es ist $\ln \frac{1}{x} = -\ln x$. Im Abschnitt „Potenzen und algebraische Funktionen" lassen sich aus Potenzen der Gestalt $(a + bx)^n$, $(\alpha + \beta x)^\nu$ Polynome, rationale Funktionen und sogar Potenzen von Potenzen bilden.

integrals containing only φ_1, where φ_1 is the function with the higher rank, and they are arranged according to the rank of φ_2. Thus the trigonometric functions are followed by the trigonometric and power functions (for instance, $\varphi_2(x) = x$); these are followed by

combinations of trigonometric and exponential functions;
combinations of trigonometric, exponential, and power functions;

combinations of trigonometric and hyperbolic functions; etc.

Integrals with two functions, φ_1 and φ_2, in the integrand have been arranged according to the function with the higher rank. But if the arrangement of a group with respect to the rank of this function is ambiguous, the rank of the second function is used.
In addition to these general rules we must note some cases, which are evident in the tables. The function $e^{1/x}$ is given a higher rank than e^x, while $\ln x$ and $\ln \frac{1}{x}$ have the same rank because $\ln \frac{1}{x} = -\ln x$. The section "Power and Algebraic Functions" can be made to cover polynomials, rational functions, and even power functions of power functions by applying operators to power functions of the form $(a + bx)^n$ and $(\alpha + \beta x)^\nu$.

Inhalt

0. Einführung

0.1 Endliche Summen
    0.11 Progressionen   23
    0.12 Summen von Potenzen natürlicher Zahlen   23
    0.13 Summen von reziproken natürlichen Zahlen   24
    0.14 Summen von Produkten reziproker natürlicher Zahlen .   25
    0.15 Summen von Binomialkoeffizienten.   25
0.2 Zahlenreihen und unendliche Produkte
    0.21 Konvergenz von Zahlenreihen   27
    0.22 Konvergenzkriterien   27
    0.23–0.24 Beispiele für Zahlenreihen   29
    0.25 Unendliche Produkte   33
    0.26 Beispiele für unendliche Produkte   34
0.3 Funktionenreihen
    0.30 Definitionen und Sätze . . .   35
    0.31 Potenzreihen   37
    0.32 Trigonometrische Reihen . .   40
    0.33 Asymptotische Reihen. . . .   43
0.4 Einige Formeln aus der Differentialrechnung
    0.41 Differentiation eines bestimmten Integrals nach einem Parameter   44
    0.42 Die n-te Ableitung eines Produkts   45
    0.43 Die n-te Ableitung einer zusammengesetzten Funktion .   45

1. Elementare Funktionen

1.1 Potenzen von Binomen
    1.11 Potenzreihen   47
    1.12 Reihen rationaler Brüche . .   48
1.2 Die Exponentialfunktion
    1.21 Reihendarstellung   48
    1.22 Funktionalgleichungen. . . .   49
    1.23 Reihen von Exponentialfunktionen.   49
1.3–1.4 Trigonometrische und hyperbolische Funktionen
    1.30 Einführung   50
    1.31 Die grundlegenden Funktionalgleichungen   51
    1.32 Darstellung von Potenzen trigonometrischer und hyper-

Contents

0. Introduction

0.1 Finite Series
    0.11 Progressions   23
    0.12 Series of powers of natural numbers   23
    0.13 Series of reciprocals of natural numbers   24
    0.14 Series of products of reciprocals of natural numbers . .   25
    0.15 Series of the binomial coefficients   25
0.2 Numerical Series and Infinite Products
    0.21 The convergence of numerical series   27
    0.22 Convergence tests.   27
    0.23–0.24 Examples of numerical series   29
    0.25 Infinite products   33
    0.26 Examples of infinite products   34
0.3 Functional Series
    0.30 Definitions and theorems . .   35
    0.31 Power series   37
    0.32 FOURIER series   40
    0.33 Asymptotic series   43
0.4 Formulae from Differential Calculus
    0.41 Differentiation of a definite integral with respect to a parameter   44
    0.42 The nth derivative of a product   45
    0.43 The nth derivative of a composite function   45

1. Elementary Functions

1.1 Powers of Binomials
    1.11 Power series   47
    1.12 Series of rational fractions .   48
1.2 The Exponential Function
    1.21 Series representation   48
    1.22 Functional relations.   49
    1.23 Series of exponentials . . .   49
1.3–1.4 Trigonometric and Hyperbolic Functions
    1.30 Introduction   50
    1.31 The fundamental functional relations   51
    1.32 The representation of powers of trigonometric and hyperbo-

bolischer Funktionen durch Funktionen von Vielfachen der Argumente (Bogen)	52	lic functions in terms of functions of multiples of the argument (angle)	52
1.33 Darstellung trigonometrischer und hyperbolischer Funktionen von Vielfachen der Argumente (Bogen) durch Potenzen dieser Funktionen.	55	1.33 The representation of trigonometric and hyperbolic functions of multiples of the argument (angle) in terms of powers of these functions	55
1.34 Einige Summen trigonometrischer und hyperbolischer Funktionen.	57	1.34 Finite series of trigonometric and hyperbolic functions . .	57
1.35 Summen von Potenzen trigonometrischer Funktionen von Vielfachen des Bogens . . .	59	1.35 Finite series of powers of trigonometric functions of multiple angles	59
1.36 Summen von Produkten trigonometrischer Funktionen von Vielfachen der Bögen . .	60	1.36 Finite series of products of trigonometric functions of multiple angles	60
1.37 Summen von Tangensfunktionen von Vielfachen des Bogens	60	1.37 Finite series of tangents of multiple angles	60
1.38 Summen, die auf hyperbolische Tangens- und Cotangensfunktionen führen	61	1.38 Finite series leading to hyperbolic tangents and cotangents	61
1.39 Darstellung von Cosinus und Sinus von Vielfachen des Bogens als endliche Produkte	62	1.39 Cosine and sine of multiples of the angle expressed in terms of finite products	62
1.41 Potenzreihenentwicklung der trigonometrischen und hyperbolischen Funktionen	63	1.41 The expansion of trigonometric and hyperbolic functions in power series	63
1.42 Partialbruchzerlegung	64	1.42 Expansion in series of simple fractions.	64
1.43 Darstellung durch ein unendliches Produkt	65	1.43 Representation in the form of an infinite product	65
1.44−1.45 Trigonometrische Reihen	66	1.44−1.45 Fourier series . . .	66
1.46 Reihen von Produkten von Exponentialfunktionen mit trigonometrischen Funktionen .	70	1.46 Series of products of exponential and trigonometric functions.	70
1.47 Reihen von Hyperbelfunktionen	70	1.47 Series of hyperbolic functions	70
1.48 Der Lobatschewskische „Parallelenwinkel" $\Pi(x)$	70	1.48 Lobachevskiy's "Angle of parallelism" $\Pi(x)$.	70
1.49 Die Hyperbelamplitude (der Gudermannsche Winkel) gd x	71	1.49 The Gudermannian gd x . .	71
1.5 Der Logarithmus		1.5 The Logarithmic Function	
1.51 Reihendarstellungen . . .	72	1.51 Series representation . . .	72
1.52 Reihen von Logarithmen . .	74	1.52 Series of logarithms	74
1.6 Zyklometrische und Area-Funktionen		1.6 The Inverse Trigonometric and Hyperbolic Functions	
1.61 Definitionsbereiche	74	1.61 The domain of definition. . .	74
1.62−1.63 Funktionalgleichungen	75	1.62−1.63 Functional equations	75
1.64 Reihendarstellungen	79	1.64 Series representations . . .	79

2. Unbestimmte Integrale elementarer Funktionen

2. Indefinite Integrals of Elementary Functions

2.0 Einführung		2.0 Introduction	
2.00 Bemerkungen allgemeiner Art.	81	2.00 General remarks	81
2.01 Die Grundintegrale	82	2.01 The basic integrals	82
2.02 Allgemeine Formeln	83	2.02 General formulae	83
2.1 Rationale Funktionen		2.1 Rational Functions	
2.10 Allgemeine Integrationsregeln	85	2.10 General rules of integration .	85
2.11−2.13 Funktionen von $a+bx^k$	88	2.11−2.13 Functions of $a+bx^k$	88
2.14 Funktionen von $1 \pm x^n$ . . .	93	2.14 Functions of $1 \pm x^n$ . . .	93
2.15 Funktionen von $a+bx$ und $\alpha+\beta x$	97	2.15 Functions of $a+bx$ and $\alpha+\beta x$	97

2.16 Funktionen von $a+bx^k+cx^{2k}$ 98	2.16 Functions of $a+bx^k+cx^{2k}$. 98
2.17 Funktionen von $a+bx+cx^2$ und Potenzen von x 99	2.17 Functions of $a+bx+cx^2$ and powers of x 99
2.18 Funktionen von $a+bx+cx^2$ und $\alpha+\beta x$ 101	2.18 Functions of $a+bx+cx^2$ and $\alpha+\beta x$ 101
2.2 Algebraische Funktionen	2.2 Algebraic Functions
2.20 Einführung 102	2.20 Introduction 102
2.21 Funktionen von $a+bx^k$ und $\sqrt{x}$ 103	2.21 Functions of $a+bx^k$ and $\sqrt{x}$ 103
2.22—2.23 Funktionen von $\sqrt[n]{(a+bx)^k}$ 105	2.22—2.23 Functions of $\sqrt[n]{(a+bx)^k}$ 105
2.24 Funktionen von $\sqrt{a+bx}$ und $\alpha+\beta x$ 108	2.24 Functions of $\sqrt{a+bx}$ and $\alpha+\beta x$ 108
2.25 Funktionen von $\sqrt{a+bx+cx^2}$ 112	2.25 Functions of $\sqrt{a+bx+cx^2}$ 112
2.26 Funktionen von $\sqrt{a+bx+cx^2}$ und ganzen Potenzen von x . 115	2.26 Functions of $\sqrt{a+bx+cx^2}$ and integral powers of x . . 115
2.27 Funktionen von $\sqrt{a+cx^2}$ und ganzen Potenzen von x . . 120	2.27 Functions of $\sqrt{a+cx^2}$ and integral powers of x . . . 120
2.28 Funktionen von $\sqrt{a+bx+cx^2}$ und Polynomen ersten und zweiten Grades 123	2.28 Functions of $\sqrt{a+bx+cx^2}$ and of first- and second-degree polynomials 123
2.29 Integrale, die auf elliptische und pseudoelliptische Integrale führen 124	2.29 Integrals that can be reduced to elliptic and pseudo-elliptic integrals 124
2.3 Die Exponentialfunktion	2.3 The Exponential Function
2.31 Funktionen von e^{ax} 125	2.31 Functions of e^{ax} 125
2.32 Die Exponentialfunktion und rationale Funktionen von x . 126	2.32 The exponential and rational functions of x 126
2.4 Hyperbelfunktionen	2.4 Hyperbolic Functions
2.41—2.43 Potenzen von sh x, ch x, th x und cth x 127	2.41—2.43 Powers of sh x, ch x, th x, and cth x. 127
2.44—2.45 Rationale Funktionen von Hyperbelfunktionen . . 142	2.44—2.45 Rational functions of hyperbolic functions. . . . 142
2.46 Algebraische Funktionen von Hyperbelfunktionen 149	2.46 Algebraic functions of hyperbolic functions 149
2.47 Hyperbelfunktionen und Potenzen 155	2.47 Combinations of hyperbolic functions and powers . . . 155
2.48 Hyperbelfunktionen, Exponentialfunktionen und Potenzen 164	2.48 Combinations of hyperbolic functions, exponentials, and powers 164
2.5—2.6 Trigonometrische Funktionen	2.5—2.6 Trigonometric Functions
2.50 Einführung 166	2.50 Introduction 166
2.51—2.52 Potenzen trigonometrischer Funktionen 166	2.51—2.52 Powers of trigonometric functions. 166
2.53—2.54 Sinus und Cosinus von Vielfachen des Bogens, von linearen und komplizierteren Funktionen des Arguments . 177	2.53—2.54 Sines and cosines of multiple angles and of linear and more complicated functions of the argument . . . 177
2.55—2.56 Rationale Funktionen von Sinus und Cosinus . . . 185	2.55—2.56 Rational functions of the sine and cosine. 185
2.57 Funktionen von $\sqrt{a\pm b\sin x}$, $\sqrt{a\pm b\cos x}$ und darauf zurückführbare Funktionen 192	2.57 Functions of $\sqrt{a\pm b\sin x}$ or $\sqrt{a\pm b\cos x}$ and functions reducible to such roots 192
2.58—2.62 Integrale, die auf elliptische und pseudoelliptische Integrale führen 196	2.58—2.62 Integrals reducible to elliptic and pseudo-elliptic integrals 196
2.63—2.65 Trigonometrische Funktionen und Potenzen 225	2.63—2.65 Trigonometric functions and powers 225
2.66 Trigonometrische Funktionen und Exponentialfunktion . . 237	2.66 Trigonometric functions and exponential functions 237
2.67 Trigonometrische Funktionen und Hyperbelfunktionen . . 242	2.67 Trigonometric and hyperbolic functions. 242

2.7 Der Logarithmus; Area-Funktionen	2.7 The Logarithmic and Inverse Hyperbolic Functions
2.71 Der Logarithmus 247	2.71 The logarithmic function . . 247
2.72–2.73 Logarithmus und algebraische Funktionen 247	2.72–2.73 Logarithmic and algebraic functions 247
2.74 Area-Funktionen 250	2.74 Inverse hyperbolic functions 250
2.8 Zyklometrische Funktionen	2.8 Inverse Trigonometric Functions
2.81 Arkussinus und Arkuscosinus 251	2.81 Arcsine and arccosine . . . 251
2.82 Arkussekans und Arkuscosekans, Arkustangens und Arkuscotangens 251	2.82 The arcsecant, the arccosecant, the arctangent and the arccotangent 251
2.83 Arkussinus, Arkuscosinus und algebraische Funktionen . . 252	2.83 Arcsine or arccosine and algebraic functions 252
2.84 Arkussekans, Arkuscosekans und Potenzen von x 253	2.84 Arcsecant and arccosecant with powers of x 253
2.85 Arkustangens, Arkuscotangens und algebraische Funktionen 253	2.85 Arctangent and arccotangent with algebraic functions . . . 253

3–4. Bestimmte Integrale elementarer Funktionen

3–4. Definite Integrals of Elementary Functions

3.0 Einführung	3.0 Introduction
3.01 Allgemeine Sätze 255	3.01 General theorems 255
3.02 Variablentransformation in einem bestimmten Integral . 257	3.02 Change of variable in a definite integral 257
3.03 Allgemeine Formeln 258	3.03 General formulae 258
3.04 Uneigentliche Integrale . . . 260	3.04 Improper integrals 260
3.05 Hauptwerte uneigentlicher Integrale 262	3.05 Principle values of improper integrals 262
3.1–3.2 Potenzfunktionen und algebraische Funktionen	3.1–3.2 Power and Algebraic Functions
3.11 Rationale Funktionen . . . 264	3.11 Rational functions 264
3.12 Produkte rationaler Funktionen mit Ausdrücken, die auf Quadratwurzeln aus linearen und quadratischen Polynomen führen 265	3.12 Products of rational functions with expressions reducible to square roots of first- and second-degree polynomials 265
3.13–3.17 Ausdrücke, die auf Quadratwurzeln aus Polynomen dritten und vierten Grades führen und ihre Produkte mit rationalen Funktionen . . . 266	3.13–3.17 Expressions reducible to square roots of third- and fourth-degree polynomials and their products with rational functions 266
3.18 Ausdrücke, die auf vierte Wurzeln aus Polynomen zweiten Grades führen, und ihre Produkte mit rationalen Funktionen 335	3.18 Expressions reducible to fourth roots of second-degree polynomials and their products with rational functions 335
3.19–3.23 Potenzen von x und von Binomen $\alpha + \beta x$ 337	3.19–3.23 Powers of x and powers of binomials of the form $(\alpha + \beta x)$ 337
3.24–3.27 Potenzen von x, von Binomen $\alpha + \beta x^p$ und von Polynomen in x 345	3.24–3.27 Powers of x, of binomials of the form $(\alpha + \beta x^p)$ and of polynomials in x . . 345
3.3–3.4 Die Exponentialfunktion	3.3–3.4 The Exponential Function
3.31 Die Exponentialfunktion. . . 357	3.31 The exponential function . . 357
3.32–3.34 Die Exponentialfunktion von komplizierteren Argumenten 359	3.32–3.34 Exponentials of compound arguments 359
3.35 Die Exponentialfunktion und rationale Funktionen 362	3.35 Exponentials and rational functions 362
3.36–3.37 Die Exponentialfunktion und algebraische Funktionen 367	3.36–3.37 Exponentials and algebraic functions 367
3.38–3.39 Die Exponentialfunktion und Potenzen mit beliebigen Exponenten. 369	3.38–3.39 Exponentials and arbitrary powers 369

3.41–3.44 Rationale Funktionen von Potenzen und der Exponentialfunktion	376	
3.45 Algebraische Funktionen von der Exponentialfunktion und Potenzen	386	
3.46–3.48 Exponentialfunktionen mit komplizierteren Argumenten und Potenzen	387	
3.5 Hyperbelfunktionen		
3.51 Hyperbelfunktionen	394	
3.52–3.53 Hyperbelfunktionen und algebraische Funktionen	398	
3.54 Hyperbelfunktionen und die Exponentialfunktion	407	
3.55–3.56 Hyperbel-, Exponential- und Potenzfunktionen	412	
3.6–4.1 Trigonometrische Funktionen		
3.61 Rationale Funktionen von Sinus und Cosinus und trigonometrische Funktionen von Vielfachen des Bogens	417	
3.62 Potenzen trigonometrischer Funktionen	421	
3.63 Potenzen trigonometrischer Funktionen und trigonometrische Funktionen linearer Argumentsfunktionen	423	
3.64–3.65 Potenzen trigonometrischer Funktionen und rationale Funktionen trigonometrischer Funktionen	428	
3.66 Potenzen linearer Funktionen von trigonometrischen Funktionen	433	
3.67 Quadratwurzeln aus Ausdrücken, die trigonometrische Funktionen enthalten	437	
3.68 Funktionen von Potenzen trigonometrischer Funktionen	439	
3.69–3.71 Trigonometrische Funktionen mit komplizierteren Argumenten	444	
3.72–3.74 Trigonometrische und rationale Funktionen	454	
3.75 Trigonometrische und algebraische Funktionen	466	
3.76–3.77 Produkte von trigonometrischen Funktionen mit Potenzen	467	
3.78–3.81 Rationale Funktionen von x und von trigonometrischen Funktionen	479	
3.82–3.83 Potenzen trigonometrischer Funktionen und Potenzen	492	
3.84 Integrale, die $\sqrt{1-k^2\sin^2 x}$, $\sqrt{1-k^2\cos^2 x}$ und ähnliche Ausdrücke enthalten	506	
3.85–3.88 Trigonometrische Funktionen mit komplizierteren Argumenten und Potenzen	510	
3.89–3.91 Trigonometrische Funktionen und die Exponentialfunktion	522	

3.41–3.44 Rational functions of powers and exponentials	376	
3.45 Algebraic functions of exponentials and powers	386	
3.46–3.48 Exponentials of compound arguments and powers	387	
3.5 Hyperbolic Functions		
3.51 Hyperbolic functions	394	
3.52–3.53 Hyperbolic functions and algebraic functions	398	
3.54 Hyperbolic functions and exponentials	407	
3.55–3.56 Hyperbolic functions, exponentials and powers	412	
3.6–4.1 Trigonometric Functions		
3.61 Rational functions of sines and cosines and trigonometric functions of multiple angles	417	
3.62 Powers of trigonometric functions	421	
3.63 Powers of trigonometric functions and trigonometric functions of linear functions	423	
3.64–3.65 Powers of trigonometric functions and rational functions of trigonometric functions	428	
3.66 Powers of linear functions of trigonometric functions	433	
3.67 Square roots of expressions containing trigonometric functions	437	
3.68 Functions of powers of trigonometric functions	439	
3.69–3.71 Trigonometric functions of compound arguments	444	
3.72–3.74 Trigonometric and rational functions	454	
3.75 Trigonometric and algebraic functions	466	
3.76–3.77 Trigonometric functions and powers	467	
3.78–3.81 Rational functions of x and of trigonometric functions	479	
3.82–3.83 Powers of trigonometric functions combined with other powers	492	
3.84 Integrals containing $\sqrt{1-k^2\sin^2 x}$, $\sqrt{1-k^2\cos^2 x}$, and similar expressions	506	
3.85–3.88 Trigonometric functions of compound arguments combined with powers	510	
3.89–3.91 Trigonometric functions and exponentials	522	

3.92 Trigonometrische Funktionen mit komplizierteren Argumenten und die Exponentialfunktion	530	3.92 Trigonometric functions of compound arguments combined with exponentials	530
3.93 Trigonometrische Funktionen und Exponentialfunktionen von trigonometrischen Funktionen	532	3.93 Trigonometric and exponential functions of trigonometric functions	532
3.94–3.97 Trigonometrische Funktionen, Exponentialfunktionen und Potenzen	535	3.94–3.97 Trigonometric functions, exponentials, and powers	535
3.98–3.99 Trigonometrische und hyperbolische Funktionen	548	3.98–3.99 Trigonometric and hyperbolic functions	548
4.11–4.12 Trigonometrische Funktionen, Hyperbelfunktionen und Potenzen	556	4.11–4.12 Trigonometric and hyperbolic functions and powers	556
4.13 Trigonometrische Funktionen, Hyperbelfunktionen und die Exponentialfunktion	564	4.13 Trigonometric and hyperbolic functions and exponentials	564
4.14 Trigonometrische Funktionen, Hyperbelfunktionen, die Exponentialfunktion und Potenzen	567	4.14 Trigonometric and hyperbolic functions, exponentials, and powers	567
4.2–4.4 Der Logarithmus		4.2–4.4 The Logarithmic Function	
4.21 Der Logarithmus	568	4.21 The logarithmic function	568
4.22 Logarithmen von zusammengesetzten Argumenten	570	4.22 Logarithms of compound arguments	570
4.23 Logarithmus und rationale Funktionen	577	4.23 Logarithms and rational functions	577
4.24 Logarithmus und algebraische Funktionen	580	4.24 Logarithms and algebraic functions	580
4.25 Logarithmus und Potenzen	582	4.25 Logarithms and powers	582
4.26–4.27 Potenzen des Logarithmus und Potenzen	585	4.26–4.27 Powers of the logarithm and other powers	585
4.28 Rationale Funktionen von $\ln x$ und Potenzen	599	4.28 Rational functions of $\ln x$ and powers	599
4.29–4.32 Der Logarithmus von zusammengesetzten Argumenten und Potenzen	602	4.29–4.32 Logarithmic functions of compound arguments and powers	602
4.33–4.34 Logarithmus und Exponentialfunktion	621	4.33–4.34 The logarithmic and the exponential function	621
4.35–4.36 Logarithmus, Exponentialfunktion und Potenzen	623	4.35–4.36 The logarithmic and the exponential function and powers	623
4.37 Logarithmus und Hyperbelfunktionen	628	4.37 The logarithmic and the hyperbolic functions	628
4.38–4.41 Logarithmus und trigonometrische Funktionen	631	4.38–4.41 The logarithmic and the trigonometric function	631
4.42–4.43 Logarithmus, trigonometrische Funktionen und Potenzen	646	4.42–4.43 The logarithmic and the trigonometric function and powers	646
4.44 Logarithmus, trigonometrische Funktionen und Exponentialfunktion	652	4.44 The logarithmic and the trigonometric function and exponentials	652
4.5 Zyklometrische Funktionen		4.5 Inverse Trigonometric Functions	
4.51 Zyklometrische Funktionen	653	4.51 Inverse trigonometric functions	653
4.52 Arkussinus, Arkuscosinus und Potenzen	653	4.52 Arcsines, arccosines, and powers	653
4.53–4.54 Arkustangens, Arkuscotangens und Potenzen	655	4.53–4.54 Arctangents, arccotangents, and powers	655
4.55 Zyklometrische Funktionen und Potenzen	659	4.55 Inverse trigonometric functions and exponentials	659
4.56 Arkustangens und Hyperbelfunktion	659	4.56 The arctangent and the hyperbolic function	659
4.57 Zyklometrische und trigonometrische Funktionen	659	4.57 Inverse and direct trigonometric functions	659

4.58 Zyklometrische und trigonometrische Funktionen und Potenzen	661
4.59 Zyklometrische Funktionen und Logarithmus	661
4.6 Mehrfache Integrale	
4.60 Variablentransformation in Mehrfachintegralen	662
4.61 Vertauschung der Integrationsfolge und Variablentransformation	663
4.62 Zwei- und dreifache Integrale mit konstanten Grenzen . .	666
4.63—4.64 Mehrfache Integrale .	668
Bezeichnungen	674
Zitierte Literatur	676

4.58 An inverse and a direct trigonometric function and a power	661
4.59 Inverse trigonometric functions and logarithms	661
4.6 Multiple Integrals	
4.60 Change of variables in multiple integrals	662
4.61 Change of the order of integration and change of variables	663
4.62 Double and triple integrals with constant limits	666
4.63—4.64 Multiple integrals. . .	668
Notations	674
References	676

Einführung
Introduction

Endliche Summen
Finite Series

0.1

Progressionen
Progressions

0.11

Arithmetische Progression: | Arithmetic progression:

0.111

$$\sum_{k=0}^{n-1}(a+kr)=\frac{n}{2}[2a+(n-1)r]=\frac{n}{2}(a+l)$$

[l ist das letzte Glied]. | [l is the last term].

Geometrische Progression: | Geometric progression:

0.112

$$\sum_{k=1}^{n} aq^{k-1} = \frac{a(q^n-1)}{q-1}.$$

Arithmetisch-geometrische Progression: | Arithmetic-geometric progression:

0.113

$$\sum_{k=0}^{n-1}(a+kr)q^k = \frac{a-[a+(n-1)r]q^n}{1-q} + \frac{rq(1-q^{n-1})}{(1-q)^2}.$$

Jo(5)

Summen von Potenzen natürlicher Zahlen
Series of powers of natural numbers

0.12

$$\sum_{k=1}^{n} k^q = \frac{n^{q+1}}{q+1}+\frac{n^q}{2}+\frac{1}{2}\binom{q}{1}B_2 n^{q-1}+\frac{1}{4}\binom{q}{3}B_4 n^{q-3}+\frac{1}{6}\binom{q}{5}B_6 n^{q-5}+\ldots =$$

$$= \frac{n^{q+1}}{q+1}+\frac{n^q}{2}+\frac{qn^{q-1}}{12}-\frac{q(q-1)(q-2)}{720}n^{q-3}+\frac{q(q-1)(q-2)(q-3)(q-4)}{30\,240}n^{q-5}-\ldots$$

0.121

C 300

[das letzte Glied enthält n oder n^2]. | [the last term contains n or n^2].

1. $\sum_{k=1}^{n} k = \frac{n(n+1)}{2}.$ C 300

2. $\sum_{k=1}^{n} k^2 = \frac{n(n+1)(2n+1)}{6}.$ C 300

3. $\sum_{k=1}^{n} k^3 = \left[\frac{n(n+1)}{2}\right]^2.$ C 300

0.121　　4.　　$\sum_{k=1}^{n} k^4 = \dfrac{1}{30} n(n+1)(2n+1)(3n^2+3n-1).$ 　　　C 300

5.　　$\sum_{k=1}^{n} k^5 = \dfrac{1}{12} n^2(n+1)^2 (2n^2+2n-1).$ 　　　C 300

6.　　$\sum_{k=1}^{n} k^6 = \dfrac{1}{42} n(n+1)(2n+1)(3n^4+6n^3-3n+1).$ 　　　C 301

7.　　$\sum_{k=1}^{n} k^7 = \dfrac{1}{24} n^2(n+1)^2 (3n^4+6n^3-n^2-4n+2).$ 　　　C 301

0.122　　$\sum_{k=1}^{n} (2k-1)^q = \dfrac{2^q}{q+1} n^{q+1} - \dfrac{1}{2}\binom{q}{1} 2^{q-1} B_2 n^{q-1} -$

$\qquad\qquad - \dfrac{1}{4}\binom{q}{3} 2^{q-3}(2^3-1) B_4 n^{q-3} - \ldots$

[das letzte Glied enthält n oder n^2]. | [the last term contains n or n^2].

1.　　$\sum_{k=1}^{n} (2k-1) = n^2.$

2.　　$\sum_{k=1}^{n} (2k-1)^2 = \dfrac{1}{3} n(4n^2-1).$ 　　　Jo(32a)

3.　　$\sum_{k=1}^{n} (2k-1)^3 = n^2 (2n^2-1).$ 　　　Jo(32b)

0.123　　$\sum_{k=1}^{n} k(k+1)^2 = \dfrac{1}{12} n(n+1)(n+2)(3n+5).$

0.124　　$\sum_{k=1}^{q} k(n^2-k^2) = \dfrac{1}{4} q(q+1)(2n^2-q^2-q).$

0.125　　$\sum_{k=1}^{n} k! \cdot k = (n+1)! - 1.$ 　　　A(188.1)

0.126　　$\sum_{k=0}^{n} \dfrac{(n+k)!}{k!(n-k)!} = \sqrt{\dfrac{e}{\pi}}\, K_{n+\frac{1}{2}}\!\left(\dfrac{1}{2}\right).$ 　　　W80(12)

Summen von reziproken natürlichen Zahlen
Series of reciprocals of natural numbers

0.13

0.131　　$\sum_{k=1}^{n} \dfrac{1}{k} = C + \ln n + \dfrac{1}{2n} - \sum_{k=2}^{\infty} \dfrac{A_k}{n(n+1)\ldots(n+k-1)},$

wobei | where

$A_k = \dfrac{1}{k} \int_{0}^{1} x(1-x)(2-x)(3-x)\ldots(k-1-x)\, dx.$

$A_2 = \dfrac{1}{12}, \quad A_3 = \dfrac{1}{12},$

$A_4 = \dfrac{19}{120}, \quad A_5 = \dfrac{9}{20}.$ 　　　Jo(59), A(1876)

$$\sum_{k=1}^{n} \frac{1}{2k-1} = \frac{1}{2}(C + \ln n) + \ln 2 + \frac{B_2}{8n^2} + \frac{(2^3-1)B_4}{64 n^4} + \ldots \qquad \text{Jo(71a)+} \qquad 0.132$$

$$\sum_{k=2}^{n} \frac{1}{k^2-1} = \frac{3}{4} - \frac{2n+1}{2n(n+1)}. \qquad \text{Jo(184f)} \qquad 0.133$$

Summen von Produkten reziproker natürlicher Zahlen
Series of products of reciprocals of natural numbers 0.14

1. $$\sum_{k=1}^{n} \frac{1}{[p+(k-1)q](p+kq)} = \frac{n}{p(p+nq)}. \qquad \text{GK II 19(64)+} \qquad 0.141$$

2. $$\sum_{k=1}^{n} \frac{1}{[p+(k-1)q](p+kq)[p+(k+1)q]} = \frac{n(2p+nq+q)}{2p(p+q)(p+nq)[p+(n+1)q]}.$$
$$\text{GK II 19(65)+}$$

3. $$\sum_{k=1}^{n} \frac{1}{[p+(k-1)q](p+kq)\ldots[p+(k+l)q]} = \frac{1}{(l+1)q} \left\{ \frac{1}{p(p+q)\ldots(p+lq)} - \frac{1}{(p+nq)[p+(n+1)q]\ldots[p+(n+l)q]} \right\}. \qquad \text{A(1856)+}$$

4. $$\sum_{k=1}^{n} \frac{1}{[1+(k-1)q][1+(k-1)q+p]} =$$
$$= \frac{1}{p} \left[\sum_{k=1}^{n} \frac{1}{1+(k-1)q} - \sum_{k=1}^{n} \frac{1}{1+(k-1)q+p} \right]. \qquad \text{GK II 20(66)+}$$

$$\sum_{k=1}^{n} \frac{k^2+k-1}{(k+2)!} = \frac{1}{2} - \frac{n+1}{(n+2)!}. \qquad \text{Jo(157)} \qquad 0.142$$

Summen von Binomialkoeffizienten (n ist eine natürliche Zahl)
Series of the binomial coefficients (n is a natural number) 0.15

1. $$\sum_{k=0}^{m} \binom{n+k}{n} = \binom{n+m+1}{n+1}. \qquad \text{Kr 64(70.1)} \qquad 0.151$$

2. $$1 + \binom{n}{2} + \binom{n}{4} + \ldots = 2^{n-1}. \qquad \text{Kr 62(58.1)}$$

3. $$\binom{n}{1} + \binom{n}{3} + \binom{n}{5} + \ldots = 2^{n-1}. \qquad \text{Kr 62(58.1)}$$

4. $$\sum_{k=0}^{m} (-1)^k \binom{n}{k} = (-1)^m \binom{n-1}{m}. \qquad \text{Kr 64(70.2)}$$

1. $$\binom{n}{0} + \binom{n}{3} + \binom{n}{6} + \ldots = \frac{1}{3}\left(2^n + 2\cos\frac{n\pi}{3}\right). \qquad \text{Kr 62(59.1)} \qquad 0.152$$

2. $$\binom{n}{1} + \binom{n}{4} + \binom{n}{7} + \ldots = \frac{1}{3}\left(2^n + 2\cos\frac{(n-2)\pi}{3}\right). \qquad \text{Kr 62(59.2)}$$

0.152 3. $\binom{n}{2} + \binom{n}{5} + \binom{n}{8} + \ldots = \frac{1}{3}\left(2^n + 2\cos\frac{(n-4)\pi}{3}\right).$ Kr 62(59.3)

0.153 1. $\binom{n}{0} + \binom{n}{4} + \binom{n}{8} + \ldots = \frac{1}{2}\left(2^{n-1} + 2^{\frac{n}{2}}\cos\frac{n\pi}{4}\right).$ Kr 63(60.1)

2. $\binom{n}{1} + \binom{n}{5} + \binom{n}{9} + \ldots = \frac{1}{2}\left(2^{n-1} + 2^{\frac{n}{2}}\sin\frac{n\pi}{4}\right).$ Kr 63(60.2)

3. $\binom{n}{2} + \binom{n}{6} + \binom{n}{10} + \ldots = \frac{1}{2}\left(2^{n-1} - 2^{\frac{n}{2}}\cos\frac{n\pi}{4}\right).$ Kr 63(60.3)

4. $\binom{n}{3} + \binom{n}{7} + \binom{n}{11} + \ldots = \frac{1}{2}\left(2^{n-1} - 2^{\frac{n}{2}}\sin\frac{n\pi}{4}\right).$ Kr 63(60.4)

0.154 1. $\sum_{k=0}^{n}(k+1)\binom{n}{k} = 2^{n-1}(n+2).$ Kr 63(66.1)

2. $\sum_{k=1}^{n}(-1)^{k+1}k\binom{n}{k} = 0.$ Kr 63(66.2)

0.155 1. $\sum_{k=1}^{n}\frac{(-1)^{k+1}}{k+1}\binom{n}{k} = \frac{n}{n+1}.$ Kr 63(67)

2. $\sum_{k=0}^{n}\frac{1}{k+1}\binom{n}{k} = \frac{2^{n+1}-1}{n+1}.$ Kr 63(68.1)

3. $\sum_{k=0}^{n}\frac{\alpha^{k+1}}{k+1}\binom{n}{k} = \frac{(\alpha+1)^{n+1}-1}{n+1}.$ Kr 63(68.2)

4. $\sum_{k=1}^{n}\frac{(-1)^{k+1}}{k}\binom{n}{k} = \sum_{m=1}^{n}\frac{1}{m}.$ Kr 64(69)

0.156 1. $\sum_{k=0}^{p}\binom{n}{k}\binom{m}{p-k} = \binom{n+m}{p}$ Kr 64(71.1)

2. $\sum_{k=0}^{n-p}\binom{n}{k}\binom{n}{p+k} = \frac{(2n)!}{(n-p)!(n+p)!}.$ Kr 64(71.2)

0.157 1. $\sum_{k=0}^{n}\binom{n}{k}^2 = \binom{2n}{n}.$ Kr 64(72.1)

2. $\sum_{k=0}^{2n}(-1)^k\binom{2n}{k}^2 = (-1)^n\binom{2n}{n}.$ Kr 64(72.2)

3. $\sum_{k=0}^{2n+1}(-1)^k\binom{2n+1}{k}^2 = 0.$ Kr 64(72.3)

4. $\sum_{k=1}^{n}k\binom{n}{k}^2 = \frac{(2n-1)!}{[(n-1)!]^2}.$ Kr 64(72.4)

Zahlenreihen und unendliche Produkte
Numerical Series and Infinite Products

0.2

Konvergenz von Zahlenreihen
The convergence of numerical series

0.21

| Die Reihe | The series |

$$\sum_{k=1}^{\infty} u_k = u_1 + u_2 + u_3 + \ldots$$

0.211

| wird *absolut konvergent* genannt, wenn die Reihe der absoluten Beträge ihrer Glieder | is called *absolutely convergent* if and only if the series of the absolute values of its terms, |

$$\sum_{k=1}^{\infty} |u_k| = |u_1| + |u_2| + |u_3| + \ldots$$

0.212

| konvergiert. Wenn die Reihe 0.211 konvergiert und die Reihe 0.212 divergiert, so wird die Reihe 0.211 als *bedingt konvergent* bezeichnet. Jede absolut konvergente Reihe *konvergiert*. | converges. If the series 0.211 converges while the series 0.212 diverges, the former is called *conditionally convergent*. Every absolutely convergent series *converges*. |

Konvergenzkriterien
Convergence tests

0.22

| Es sei | Let |

0.221

$$\lim_{k \to \infty} |u_k|^{\frac{1}{k}} = q.$$

| Im Fall $q < 1$ konvergiert die Reihe 0.211 absolut; ist jedoch $q > 1$, so divergiert die Reihe 0.211 (Cauchy). | The series 0.211 converges absolutely if $q < 1$ and diverges if $q > 1$ (Cauchy). |
| Es sei | Let |

0.222

$$\lim_{k \to \infty} \left| \frac{u_{k+1}}{u_k} \right| = q.$$

| Die Reihe 0.211 konvergiert absolut für $q < 1$ und divergiert für $q > 1$. Strebt $|u_{k+1}/u_k|$ von oben gegen 1, so divergiert die Reihe 0.211 (D'Alembert). | The series 0.211 converges absolutely if $q < 1$ and diverges if $q > 1$. If for every k the quantity $|u_{k+1}/u_k|$ tends to 1 and remains greater than 1, the series 0.211 diverges (D'Alembert). |
| Es sei | Let |

0.223

$$\lim_{k \to \infty} k \left\{ \left| \frac{u_k}{u_{k+1}} \right| - 1 \right\} = q.$$

| Die Reihe 0.211 konvergiert absolut für $q > 1$ und divergiert für $q < 1$ (Raabe). | The series 0.211 converges absolutely if $q > 1$ and diverges if $q < 1$ (Raabe). |

0.224 Es sei $f(x)$ eine positive abnehmende Funktion, und es gelte für natürliche k

Let $f(x)$ be a positive-valued decreasing function and for any natural number k

$$\lim_{k\to\infty} \frac{e^k f(e^k)}{f(k)} = q.$$

Die Reihe $\sum_{k=1}^{\infty} f(k)$ konvergiert (absolut) für $q < 1$ und divergiert für $q > 1$ (ERMAKOW).

The series $\sum_{k=1}^{\infty} f(k)$ (absolutely) converges if $q < 1$ and diverges if $q > 1$ (ERMAKOV).

0.225 Es sei

Let

$$\left|\frac{u_k}{u_{k+1}}\right| = 1 + \frac{q}{k} + \frac{|v_k|}{k^p},$$

wobei $p > 1$ ist und die $|v_k|$ beschränkt sind, d.h. kleiner als eine gewisse Zahl M, die nicht von k abhängt. Für $q > 1$ konvergiert die Reihe 0.211 absolut, für $q \leqslant 1$ divergiert sie (GAUSS).

where $p > 1$ and all the $|v_k|$ are bounded, that is, smaller than a certain number M that does not depend on k. The series 0.211 converges absolutely if $q > 1$ and diverges if $q \leqslant 1$ (GAUSS).

0.226 Die für $x \geqslant q \geqslant 1$ definierte Funktion $f(x)$ sei stetig, positiv und monoton abnehmend. Dann konvergiert oder divergiert die Reihe

Let $f(x)$ for $x \geqslant q \geqslant 1$ be continuous, positive-valued, and decreasing monotonically. Then the series

$$\sum_{k=1}^{\infty} f(k)$$

je nachdem, ob das Integral

converges or diverges according to whether the integral

$$\int_{q}^{\infty} f(x)\,dx$$

konvergiert oder divergiert (CAUCHY-sches Integralkriterium).

converges or diverges (CAUCHY's integral test).

0.227 Sind alle Glieder der Folge $u_1, u_2, \ldots, u_n, \ldots$ p o s i t i v, so wird die Reihe

Let all the members of the sequence $u_1, u_2, \ldots, u_n, \ldots$ be p o s i t i v e. Then the series

1. $$\sum_{k=1}^{\infty} (-1)^{k+1} u_k = u_1 - u_2 + u_3 - \cdots$$

als *alternierend* bezeichnet.
Nehmen die absoluten Beträge der Glieder einer alternierenden Reihe monoton ab und streben sie gegen Null, d.h., gilt

is called *alternating*.
If the absolute values of the terms of an alternating series monotonically decrease and, moreover, tend to zero, that is, if

2. $$u_{k+1} < u_k, \lim_{k\to\infty} u_k = 0,$$

so konvergiert die Reihe 0.227 1. Hierbei gilt für den **Reihenrest**

then the series 0.227 1. converges. For the **remainder of the series** we have

3. $\sum_{k=n+1}^{\infty}(-1)^{k-n+1}u_k = \left|\sum_{k=1}^{\infty}(-1)^{k+1}u_k - \sum_{k=1}^{n}(-1)^{k+1}u_k\right| < u_{n+1}$ (LEIBNIZ).

Konvergiert die Reihe

If the series

0.228

1. $\sum_{k=1}^{\infty} v_k = v_1 + v_2 + \ldots + v_k + \ldots$

und bilden die Zahlen u_k eine monotone und beschränkte Folge, d.h., ist für eine gewisse Zahl M und für alle k die Ungleichung $|u_k| < M$ erfüllt, so konvergiert die Reihe

converges and if the sequence of the numbers u_k is monotonic and bounded (that is, if $|u_k| < M$ for a certain M and all k), then the series

2. $\sum_{k=1}^{\infty} u_k v_k = u_1 v_1 + u_2 v_2 + \ldots + u_k v_k + \ldots$ F II 319

(ABEL). Sind die Partialsummen der Reihe 0.228 1. gleichmäßig beschränkt und bilden die Zahlen u_k eine monotone, gegen Null strebende Folge, d.h., gilt

converges (ABEL). If the partial sums of the series 0.228 1. are uniformly bounded and if the sequence of the numbers u_k tends monotonically to zero, that is, if

0.229

$\left|\sum_{k=1}^{n} v_k\right| < M \ [n = 1, 2, 3, \ldots]$ und and $\lim_{k \to \infty} u_k = 0,$ F II 319—320

so konvergiert die Reihe 0.228 2. (DIRICHLET).

then the series 0.228 2. converges (DIRICHLET).

Beispiele für Zahlenreihen
Examples of numerical series

0.23—0.24

Progressionen.

Progressions:

0.231

1. $\sum_{k=0}^{\infty} aq^k = \frac{a}{1-q}$ $[|q| < 1]$.

2. $\sum_{k=0}^{\infty}(a+kr)q^k = \frac{a}{1-q} + \frac{rq}{(1-q)^2}$ $[|q| < 1]$ (cf. 0.113).

1. $\sum_{k=1}^{\infty}(-1)^{k+1}\frac{1}{k} = \ln 2$ (cf. 1.511).   0.232

2. $\sum_{k=1}^{\infty}(-1)^{k+1}\frac{1}{2k-1} = 1 - 2\sum_{k=1}^{\infty}\frac{1}{(4k-1)(4k+1)} = \frac{\pi}{4}$ (cf. 1.643).

1. $\sum_{k=1}^{\infty}\frac{1}{k^p} = 1 + \frac{1}{2^p} + \frac{1}{3^p} + \ldots = \zeta(p)$ $[\operatorname{Re} p > 1]$.   WW 265 0.233

0.233

2. $\sum_{k=1}^{\infty} (-1)^{k+1} \dfrac{1}{k^p} = (1 - 2^{1-p}) \zeta(p)$ [Re $p > 0$].   WW 267

3. $\sum_{k=1}^{\infty} \dfrac{1}{k^{2n}} = \dfrac{2^{2n-1} \pi^{2n}}{(2n)!} |B_{2n}|.$ F II 514

4. $\sum_{k=1}^{\infty} (-1)^{k+1} \dfrac{1}{k^{2n}} = \dfrac{(2^{2n-1} - 1) \pi^{2n}}{(2n)!} |B_{2n}|.$ Jo(165)

5. $\sum_{k=1}^{\infty} \dfrac{1}{(2k-1)^{2n}} = \dfrac{(2^{2n} - 1) \pi^{2n}}{2 \cdot (2n)!} |B_{2n}|.$ Jo(184b)

6. $\sum_{k=1}^{\infty} (-1)^{k+1} \dfrac{1}{(2k-1)^{2n+1}} = \dfrac{\pi^{2n+1}}{2^{2n+2} (2n)!} |E_{2n}|.$ Jo(184d)

0.234

1. $\sum_{k=1}^{\infty} (-1)^{k+1} \dfrac{1}{k^2} = \dfrac{\pi^2}{12}.$ Kn 247+

2. $\sum_{k=1}^{\infty} \dfrac{1}{(2k-1)^2} = \dfrac{\pi^2}{8}.$ Kn 247

3. $\sum_{k=0}^{\infty} \dfrac{(-1)^k}{(2k+1)^2} = \mathbf{G}.$ F II 476

4. $\sum_{k=1}^{\infty} \dfrac{(-1)^{k+1}}{(2k-1)^3} = \dfrac{\pi^3}{32}.$ Kn 248

5. $\sum_{k=1}^{\infty} \dfrac{1}{(2k-1)^4} = \dfrac{\pi^4}{96}.$ Kn 247

6. $\sum_{k=1}^{\infty} \dfrac{(-1)^{k+1}}{(2k-1)^5} = \dfrac{5\pi^5}{1536}.$ Kn 248

7. $\sum_{k=1}^{\infty} (-1)^{k+1} \dfrac{k}{(k+1)^2} = \dfrac{\pi^2}{12} - \ln 2.$

0.235 $S_n = \sum_{k=1}^{\infty} \dfrac{1}{(4k^2 - 1)^n},$

$S_1 = \dfrac{1}{2}, \quad S_2 = \dfrac{\pi^2 - 8}{16}, \quad S_3 = \dfrac{32 - 3\pi^2}{64}, \quad S_4 = \dfrac{\pi^4 + 30\pi^2 - 384}{768}.$ Jo(186)

0.236

1. $\sum_{k=1}^{\infty} \dfrac{1}{k(4k^2 - 1)} = 2 \ln 2 - 1.$ Br$_{26}$ 51+

2. $\sum_{k=1}^{\infty} \dfrac{1}{k(9k^2 - 1)} = \dfrac{3}{2} (\ln 3 - 1).$ Br$_{26}$ 51+

3. $\sum_{k=1}^{\infty} \dfrac{1}{k(36k^2 - 1)} = -3 + \dfrac{3}{2} \ln 3 + 2 \ln 2.$ Br$_{26}$ 52, A(6913.3)

4. $\sum_{k=1}^{\infty} \dfrac{k}{(4k^2 - 1)^2} = \dfrac{1}{8}.$ Br$_{26}$ 52

5. $$\sum_{k=1}^{\infty} \frac{1}{k(4k^2-1)^2} = \frac{3}{2} - 2\ln 2.$$ Br$_{26}$ 52   0.236

6. $$\sum_{k=1}^{\infty} \frac{12k^2-1}{k(4k^2-1)^2} = 2\ln 2.$$ A(6917.3), Br$_{26}$ 52

1. $$\sum_{k=1}^{\infty} \frac{1}{(2k-1)(2k+1)} = \frac{1}{2}.$$ A(6917.2), Br$_{26}$ 52   0.237

2. $$\sum_{k=1}^{\infty} \frac{1}{(4k-1)(4k+1)} = \frac{1}{2} - \frac{\pi}{8}.$$

3. $$\sum_{k=2}^{\infty} \frac{1}{(k-1)(k+1)} = \frac{3}{4} \quad (\text{cf. } 0.133).$$

4. $$\sideset{}{'}\sum_{k=1,\,k\neq m}^{\infty} \frac{1}{(m+k)(m-k)} = -\frac{3}{4m^2} \quad \begin{bmatrix} m & \text{ganzzahlig} \\ & \text{integer} \end{bmatrix}.$$ A(6916.1)

5. $$\sideset{}{'}\sum_{k=1,\,k\neq m}^{\infty} \frac{(-1)^{k-1}}{(m-k)(m+k)} = \frac{3}{4m^2} \quad \begin{bmatrix} m & \text{geradzahlig} \\ & \text{even integer} \end{bmatrix}.$$ A(6916.2)

1. $$\sum_{k=1}^{\infty} \frac{1}{(2k-1)\,2k\,(2k+1)} = \ln 2 - \frac{1}{2}.$$ GK II 21(93)   0.238

2. $$\sum_{k=1}^{\infty} \frac{(-1)^{k+1}}{(2k-1)\,2k(2k+1)} = \frac{1}{2}(1-\ln 2).$$ GK II 21(94)+

3. $$\sum_{k=0}^{\infty} \frac{1}{(3k+1)(3k+2)(3k+3)(3k+4)} = \frac{1}{6} - \frac{1}{4}\ln 3 + \frac{\pi}{12\sqrt{3}}.$$ GK II 21(95)

1. $$\sum_{k=1}^{\infty} (-1)^{k+1} \frac{1}{3k-2} = \frac{1}{3}\left(\frac{\pi}{\sqrt{3}} + \ln 2\right).$$ GK II 21(85), Br$_{08}$ 161(1)   0.239

2. $$\sum_{k=1}^{\infty} (-1)^{k+1} \frac{1}{3k-1} = \frac{1}{3}\left(\frac{\pi}{\sqrt{3}} - \ln 2\right).$$ Br$_{08}$ 161(1)

3. $$\sum_{k=1}^{\infty} (-1)^{k+1} \frac{1}{4k-3} = \frac{1}{4\sqrt{2}}[\pi + 2\ln(\sqrt{2}+1)].$$ Br$_{08}$ 161(1)

4. $$\sum_{k=1}^{\infty} (-1)^{E\left(\frac{k+3}{2}\right)} \frac{1}{k} = \frac{\pi}{4} + \frac{1}{2}\ln 2.$$ GK II 21(87)

5. $$\sum_{k=1}^{\infty} (-1)^{E\left(\frac{k+3}{2}\right)} \frac{1}{2k-1} = \frac{\pi}{2\sqrt{2}}.$$

6. $$\sum_{k=1}^{\infty} (-1)^{E\left(\frac{k+5}{3}\right)} \frac{1}{2k-1} = \frac{5\pi}{12}.$$ GK II 21(88)

7. $$\sum_{k=1}^{\infty} \frac{1}{(8k-1)(8k+1)} = \frac{1}{2} - \frac{\pi}{16}(\sqrt{2}+1).$$

0.241   1. $$\sum_{k=1}^{\infty} \frac{1}{2^k k} = \ln 2.$$ Jo(172g)

2. $$\sum_{k=1}^{\infty} \frac{1}{2^k k^2} = \frac{\pi^2}{12} - \frac{1}{2}(\ln 2)^2.$$ Jo(174)

0.242 $$\sum_{k=0}^{\infty} (-1)^k \frac{1}{n^{2k}} = \frac{n^2}{n^2+1}.$$

0.243   1. $$\sum_{k=1}^{\infty} \frac{1}{[p+(k-1)q](p+kq)\ldots[p+(k+l)q]} = \frac{1}{(l+1)q} \frac{1}{p(p+q)\ldots(p+lq)}$$
(cf. 0.141 3.).

2. $$\sum_{k=1}^{\infty} \frac{x^{k-1}}{[p+(k-1)q][p+(k-1)q+1][p+(k-1)q+2]\ldots[p+(k-1)q+l]} =$$
$$= \frac{1}{l!} \int_0^1 \frac{t^{p-1}(1-t)^l}{1-xt^q} dt \quad [q>0, \ x^2 \leq 1].$$ Br$_{08}$ 161(2)+, A(6.704)

0.244   1. $$\sum_{k=1}^{\infty} \frac{1}{(k+p)(k+q)} = \frac{1}{q-p} \int_0^1 \frac{x^p - x^q}{1-x} dx \quad [p>-1, \ q>-1, \ p \neq q].$$
GK II 21(90)

2. $$\sum_{k=1}^{\infty} (-1)^{k+1} \frac{1}{p+(k-1)q} = \int_0^1 \frac{t^{p-1}}{1+t^q} dt \quad [p>0, \ q>0].$$ Br$_{08}$ 161(1)

Summen reziproker Fakultäten
Series of reciprocal factorials

0.245   1. $$\sum_{k=0}^{\infty} \frac{1}{k!} = e = 2{,}718\,28\ldots$$

2. $$\sum_{k=0}^{\infty} \frac{(-1)^k}{k!} = \frac{1}{e} = 0{,}367\,88\ldots$$

3. $$2\sum_{k=1}^{\infty} \frac{k}{(2k+1)!} = \frac{1}{e} = 0{,}367\,88\ldots$$

4. $$\sum_{k=1}^{\infty} \frac{k}{(k+1)!} = 1.$$

5. $$\sum_{k=0}^{\infty} \frac{1}{(2k)!} = \frac{1}{2}\left(e + \frac{1}{e}\right) = 1{,}543\,08\ldots$$

6. $$\sum_{k=0}^{\infty} \frac{1}{(2k+1)!} = \frac{1}{2}\left(e - \frac{1}{e}\right) = 1{,}175\,20\ldots$$

7. $$\sum_{k=0}^{\infty} \frac{(-1)^k}{(2k)!} = \cos 1 = \cos 57°17'45'' = 0{,}540\,30\ldots$$

8. $$\sum_{k=1}^{\infty} \frac{(-1)^{k-1}}{(2k-1)!} = \sin 1 = \sin 57°17'45'' = 0{,}841\,47\ldots$$ 0.245

1. $$\sum_{k=0}^{\infty} \frac{1}{(k!)^2} = I_0(2) = 2{,}279\,585\,30\ldots$$ 0.246

2. $$\sum_{k=0}^{\infty} \frac{1}{k!\,(k+1)!} = I_1(2) = 1{,}590\,636\,855\ldots$$

3. $$\sum_{k=0}^{\infty} \frac{1}{k!\,(k+n)!} = I_n(2).$$

4. $$\sum_{k=0}^{\infty} \frac{(-1)^k}{(k!)^2} = J_0(2) = 0{,}223\,890\,78\ldots$$

5. $$\sum_{k=0}^{\infty} \frac{(-1)^k}{k!\,(k+1)!} = J_1(2) = 0{,}576\,724\,81\ldots$$

6. $$\sum_{k=0}^{\infty} \frac{(-1)^k}{k!\,(k+n)!} = J_n(2).$$

$$\sum_{k=1}^{\infty} \frac{k!}{(n+k-1)!} = \frac{1}{(n-2)\cdot(n-1)!}.$$ Jo(159) 0.247

$$\sum_{k=1}^{\infty} \frac{k^n}{k!} = S_n,$$ 0.248

$S_1 = e,\quad S_2 = 2e,\quad S_3 = 5e,\quad S_4 = 15e,$

$S_5 = 52e,\quad S_6 = 203e,\quad S_7 = 877e,\quad S_8 = 4140e.$ Jo(185)

$$\sum_{k=1}^{\infty} \frac{(k+1)^3}{k!} = 15e.$$ Jo(76) 0.249

Unendliche Produkte
Infinite products

0.25

Gegeben sei die Zahlenfolge $a_1, a_2, \ldots, a_k, \ldots$ Existiert der Grenzwert

Let $a_1, a_2, \ldots, a_k, \ldots$ be an infinite sequence of numbers. If the limit

0.250

$$\lim_{n\to\infty} \prod_{k=1}^{n} (1+a_k)$$

(er kann endlich, $+\infty$ oder $-\infty$ sein), so nennt man ihn den Wert des unendlichen Produktes $\prod_{k=1}^{\infty}(1+a_k)$ und schreibt:

exists (its value may be finite or $+\infty$ or $-\infty$), it is called the value of the infinite product $\prod_{k=1}^{\infty}(1+a_k)$:

1. $$\lim_{n\to\infty} \prod_{k=1}^{\infty}(1+a_k) = \prod_{k=1}^{\infty}(1+a_k).$$

If the value is finite and n o n z e r o, the infinite product is called *convergent*; otherwise it is called *divergent*.
F II 363

0.251 A necessary condition for the convergence of the infinite product 0.250 1. is $\lim_{k \to \infty} a_k = 0$.
F II 366

0.252 If for all values of $k \geqslant k_0$ either $a_k > 0$ or $a_k < 0$, then the convergence of the series $\sum_{k=1}^{\infty} a_k$ is necessary and sufficient for the convergence of the product 0.250 1.

0.253 The product $\prod_{k=1}^{\infty}(1 + a_k)$ is called *absolutely convergent* if $\prod_{k=1}^{\infty}(1 + |a_k|)$ converges.
F II 368

0.254 If an infinite product is absolutely convergent, it is also convergent.

0.255 The product $\prod_{k=1}^{\infty}(1 + a_k)$ converges absolutely if and only if the series $\sum_{k=1}^{\infty} a_k$ converges absolutely.
F II 368

0.26 **Beispiele für unendliche Produkte**
Examples of infinite products

0.261 $$\prod_{k=1}^{\infty}\left(1 + \frac{(-1)^{k+1}}{2k-1}\right) = \sqrt{2}.$$ Kn 235–236

0.262 1. $$\prod_{k=2}^{\infty}\left(1 - \frac{1}{k^2}\right) = \frac{1}{2}.$$ F II 363

2. $$\prod_{k=1}^{\infty}\left(1 - \frac{1}{(2k)^2}\right) = \frac{2}{\pi}.$$ F II 364

3. $$\prod_{k=1}^{\infty}\left(1 - \frac{1}{(2k+1)^2}\right) = \frac{\pi}{4}.$$ F II 364

0.263 $$\frac{2}{1} \cdot \left(\frac{4}{3}\right)^{\frac{1}{2}} \left(\frac{6 \cdot 8}{5 \cdot 7}\right)^{\frac{1}{4}} \left(\frac{10 \cdot 12 \cdot 14 \cdot 16}{9 \cdot 11 \cdot 13 \cdot 15}\right)^{\frac{1}{8}} \ldots = e.$$

0.264 $$\prod_{k=1}^{\infty} \frac{\sqrt[k]{e}}{1 + \frac{1}{k}} = e^{C}.$$ F II 365

$$\sqrt{\frac{1}{2}} \cdot \sqrt{\frac{1}{2}+\frac{1}{2}\sqrt{\frac{1}{2}}} \cdot \sqrt{\frac{1}{2}+\frac{1}{2}\sqrt{\frac{1}{2}+\frac{1}{2}\sqrt{\frac{1}{2}}}} \ldots = \frac{2}{\pi}.$$
F II 364   0.265

$$\prod_{k=0}^{\infty} (1 + x^{2k}) = \frac{1}{1-x} \quad [|x| < 1].$$
F II 364   0.266

Funktionenreihen
Functional Series

0.3

Definitionen und Sätze
Definitions and theorems

0.30

| Die Reihe | The series | 0.301 |

1.
$$\sum_{k=1}^{\infty} f_k(x),$$

deren Glieder Funktionen sind, wird als *Funktionenreihe* bezeichnet. Die Menge der Werte der unabhängigen Veränderlichen x, für welche die Reihe 0.301 1. konvergiert, bildet den *Konvergenzbereich* dieser Reihe.

where the terms are functions of x, is called a series of functions or a *functional series*. The set of values of the independent variable x for which the series 0.301 1. converges is called the *domain of convergence* of this series.

Eine Reihe, die für alle Werte von x aus einem Bereich M konvergiert, wird als in diesem Bereich *gleichmäßig konvergent* bezeichnet, wenn zu jedem $\varepsilon > 0$ eine Zahl N existiert derart, daß für $n > N$ die Ungleichung

A functional series which converges for all values of x in a domain M is said to be *uniformly convergent* in M if for every $\varepsilon > 0$ there exists a number N such that for all $n > N$ the inequality

0.302

$$\left| \sum_{k=n+1}^{\infty} f_k(x) \right| < \varepsilon$$

für alle x aus M erfüllt ist.

holds for every value of x pertaining to M.

Genügen die Glieder der Funktionenreihe 0.301 1. in einem Bereich M den Ungleichungen

If the terms of the functional series 0.301 1. satisfy, in a certain domain M, the inequalities

0.303

$$|f_k(x)| < u_k \quad (k = 1, 2, 3, \ldots),$$

wobei die u_k Glieder einer gewissen konvergenten Zahlenreihe

where the u_k form a convergent number series

$$\sum_{k=1}^{\infty} u_k = u_1 + u_2 + \ldots + u_k + \ldots$$

sind, so konvergiert die Reihe 0.301 1. in M gleichmäßig (WEIERSTRASS).
F II 443

then the series 0.301 1. converges uniformly in M (WEIERSTRASS).
F II 443

Wir setzen voraus, die Reihe 0.301 1. konvergiere in einem Bereich M gleich-

Let the series 0.301 1. be uniformly convergent in M and let the functions

0.304

mäßig und die Funktionen $g_k(x)$ bilden (für jedes x) eine monotone und gleichmäßig beschränkte Folge, d.h., für eine gewisse Zahl L sowie für alle n und x seien die Ungleichungen

$g_k(x)$ be the elements of a monotonic and uniformly bounded sequence, that is, for a certain number L and for all n and x let the inequalities

1. $$|g_n(x)| \leqslant L$$

erfüllt. Dann konvergiert die Reihe | be satisfied. Then the series

2. $$\sum_{k=1}^{\infty} f_k(x)\, g_k(x)$$

im Bereich M gleichmäßig (ABEL).

converges uniformly in M (ABEL).

F II 445

0.305 Die Partialsummen der Reihe 0.301 1. seien gleichmäßig beschränkt, d.h., für ein gewisses L und für alle n und alle x aus M seien die Ungleichungen

Let the partial sums of the series 0.301 1. be uniformly bounded, that is, let the inequalities

$$\left|\sum_{k=1}^{\infty} f_k(x)\right| < L$$

erfüllt. Bilden außerdem die Funktionen $g_n(x)$ (für jedes x) eine monotone Folge, die im Bereich M gleichmäßig gegen Null konvergiert, so konvergiert die Reihe 0.304 2. im Bereich M gleichmäßig (DIRICHLET).

be satisfied for a certain L and all n, and all values of x in M, and let the values of the functions $g_n(x)$ (for every x) form a monotonic sequence that converges uniformly in M towards zero. Then the series 0.304 2. converges uniformly in M (DIRICHLET).

F II 445

0.306 Sind die Funktionen $f_k(x)$ ($k=1,2,3,\ldots$) auf dem Intervall $[a,b]$ integrierbar und konvergiert die aus ihnen gebildete Reihe 0.301 1. auf diesem Intervall gleichmäßig, so kann man sie **gliedweise integrieren**, d.h., es gilt

If the functions $f_k(x)$ ($k=1,2,3,\ldots$) are integrable over the interval $[a,b]$ and if the series 0.301 1. converges uniformly in $[a,b]$, then the series may be **integrated termwise**, that is

$$\int_a^x \left(\sum_{k=1}^{\infty} f_k(x)\right) dx = \sum_{k=1}^{\infty} \int_a^x f_k(x)\, dx \quad [a \leqslant x \leqslant b].$$

F II 452

0.307 Die Funktionen $f_k(x)$ ($k=1,2,3,\ldots$) mögen auf dem Intervall $[a,b]$ stetige Ableitungen $f'_k(x)$ besitzen. Konvergiert auf diesem Intervall die Reihe 0.301 1. und konvergiert die aus den Ableitungen gebildete Reihe $\sum_{k=1}^{\infty} f'_k(x)$ dort gleichmäßig, so kann man die Reihe 0.301 1. **gliedweise differenzieren**, d.h., es ist

Let the functions $f_k(x)$ ($k=1,2,3,\ldots$) have continuous first derivations $f'_k(x)$ in the interval $[a,b]$. If the series 0.301 1. converges in $[a,b]$ and if the series of the derivatives, $\sum_{k=1}^{\infty} f'_k(x)$, converges uniformly within this interval, then the series 0.301 1. may be **differentiated termwise**, that is

$$\left\{\sum_{k=1}^{\infty} f_k(x)\right\}' = \sum_{k=1}^{\infty} f'_k(x).$$

F II 454

Potenzreihen
Power series

0.31

A functional series of the form

0.311

$$\sum_{k=0}^{\infty} a_k(x-\xi)^k = a_0 + a_1(x-\xi) + a_2(x-\xi)^2 + \ldots$$

is called a *power series*. The domain of convergence of the power series 0.311 1., not everywhere divergent, is a circle with centre ξ and radius R; at each point of the interior of this circle the power series 0.311 1. is absolutely convergent, and outside of this circle it diverges. This circle is called the *circle of convergence*, its radius the *radius of convergence*. If the series converges at all points in the complex plane, one says its *radius of convergence is infinite* ($R = +\infty$).

In the interior of its convergence circle a power series may be integrated and differentiated termwise, that is

0.312

$$\int_{\xi}^{x} \left\{ \sum_{k=0}^{\infty} a_k(x-\xi)^k \right\} dx = \sum_{k=0}^{\infty} \frac{a_k}{k+1}(x-\xi)^{k+1},$$

$$\frac{d}{dx} \left\{ \sum_{k=0}^{\infty} a_k(x-\xi)^k \right\} = \sum_{k=1}^{\infty} k a_k(x-\xi)^{k-1}.$$

The radius of convergence of the series obtained from the original one by way of termwise integration or differentiation is equal to the radius of convergence of the original series.

Operations with power series

Division of power series:

0.313

$$\frac{\sum_{k=0}^{\infty} b_k x^k}{\sum_{k=0}^{\infty} a_k x^k} = \frac{1}{a_0} \sum_{k=0}^{\infty} c_k x^k$$

mit | with

$$c_n + \frac{1}{a_0} \sum_{k=1}^{n} c_{n-k} a_k - b_n = 0$$

oder | or

$$c_n = \frac{(-1)^n}{a_0^n} \begin{vmatrix} a_1 b_0 - a_0 b_1 & a_0 & 0 & \ldots & 0 \\ a_2 b_0 - a_0 b_2 & a_1 & a_0 & \ldots & 0 \\ a_3 b_0 - a_0 b_3 & a_2 & a_1 & \ldots & 0 \\ \cdot & \cdot & \cdot & \ldots & \cdot \\ \cdot & \cdot & \cdot & \ldots & \cdot \\ a_{n-1} b_0 - a_0 b_{n-1} & a_{n-2} & a_{n-3} & \ldots & a_0 \\ a_n b_0 - a_0 b_n & a_{n-1} & a_{n-2} & \ldots & a_1 \end{vmatrix}$$

A(6360)

0.314 Potenzieren von Potenzreihen: | Powers of power series:

$$\left(\sum_{k=0}^{\infty} a_k x^k \right)^n = \sum_{k=0}^{\infty} c_k x^k$$

mit | with

$$c_0 = a_0^n, \quad c_m = \frac{1}{m a_0} \sum_{k=1}^{m} (kn - m + k) a_k c_{m-k}$$

A(6361)

für $m \geqslant 1$ [n eine natürliche Zahl]. | for $m \geqslant 1$ [n natural number].

0.315 Einsetzen einer Reihe in eine andere: | Substitution of one series into another:

$$\sum_{k=1}^{\infty} b_k y^k = \sum_{k=1}^{\infty} c_k x^k, \quad y = \sum_{k=1}^{\infty} a_k x^k;$$

$$c_1 = a_1 b_1, \quad c_2 = a_2 b_1 + a_1^2 b_2, \quad c_3 = a_3 b_1 + 2 a_1 a_2 b_2 + a_1^3 b_3,$$

$$c_4 = a_4 b_1 + a_2^2 b_2 + 2 a_1 a_3 b_2 + 3 a_1^2 a_2 b_3 + a_1^4 b_4, \ldots$$

A(6362)

0.316 Multiplikation von Potenzreihen: | Multiplication of power series:

$$\sum_{k=0}^{\infty} a_k x^k \sum_{k=0}^{\infty} b_k x^k = \sum_{k=0}^{\infty} c_k x^k; \quad c_n = \sum_{k=0}^{n} a_k b_{n-k}.$$

F II 335

Die Taylorsche Reihe
The Taylor series

0.317 Besitzt eine Funktion $f(x)$ in einer Umgebung eines Punktes ξ Ableitungen beliebiger Ordnung, so kann man folgende Reihe bilden: | If in the neighbourhood of point ξ the function $f(x)$ possesses derivatives of every order n, we may form the series

1. $$f(\xi) + \frac{(x-\xi)}{1!} f'(\xi) + \frac{(x-\xi)^2}{2!} f''(\xi) + \frac{(x-\xi)^3}{3!} f'''(\xi) + \ldots,$$

die sogenannte TAYLORsche Reihe der Funktion $f(x)$.
Die TAYLORsche Reihe strebt gegen die Funktion $f(x)$, wenn das *Restglied*

called the TAYLOR *series* of the function $f(x)$.
The TAYLOR series converges towards $f(x)$ if the *remainder*

2. $$R_n(x) = f(x) - f(\xi) - \sum_{k=1}^{n} \frac{(x-\xi)^k}{k!} f^{(k)}(\xi)$$

für $n \to \infty$ gegen Null strebt.
Formen des R e s t g l i e d e s einer TAYLORschen Reihe:

tends to zero as $n \to \infty$.
Expressions for the r e m a i n d e r of a TAYLOR series:

3. $$R_n(x) = \frac{(x-\xi)^{n+1}}{(n+1)!} f^{(n+1)}(\xi + \theta(x-\xi)) \quad [0 < \theta < 1] \quad \text{(LAGRANGE)}.$$

4. $$R_n(x) = \frac{(x-\xi)^{n+1}}{n!} (1-\theta)^n f^{(n+1)}(\xi + \theta(x-\xi)) \quad [0 < \theta < 1] \quad \text{(CAUCHY)}.$$

5. $$R_n(x) = \frac{\psi(x-\xi) - \psi(0)}{\psi'[(x-\xi)(1-\theta)]} \frac{(x-\xi)^n (1-\theta)^n}{n!} f^{(n+1)}(\xi + \theta(x-\xi)) \quad [0 < \theta < 1]$$
$$\text{(SCHLÖMILCH)},$$

wobei $\psi(x)$ eine willkürliche Funktion ist, die den beiden folgenden Bedingungen genügt: 1. Sie ist nebst ihrer Ableitung $\psi'(x)$ im Intervall $(0, x-\xi)$ stetig; 2. die Ableitung $\psi'(x)$ ändert in diesem Intervall ihr Vorzeichen nicht. Setzt man $\psi(x) = x^{p+1}$, so ergibt sich folgende Gestalt des Restgliedes:

where $\psi(x)$ is any function satisfying the following two conditions: (1) both $\psi(x)$ and $\psi'(x)$ are continuous in $(0, x-\xi)$ and (2) in the same interval the derivative $\psi'(x)$ does not change its sign. If we put $\psi(x) = x^{p+1}$, we obtain for the remainder the expression

$$R_n(x) = \frac{(x-\xi)^{n+1} (1-\theta)^{n-p-1}}{(p+1) n!} f^{(n+1)}(\xi + \theta(x-\xi)) \quad [0 < p \leq n; \ 0 < \theta < 1]$$
$$\text{(ROUCHÉ)}.$$

6. $$R_n(x) = \frac{1}{n!} \int_{\xi}^{x} f^{(n+1)}(t) (x-t)^n \, dt.$$

Andere Schreibweisen der T A Y L O Rschen R e i h e:

Another way of writing the T A Y L O R series:

0.318

1. $$f(a+x) = \sum_{k=0}^{\infty} \frac{x^k}{k!} f^{(k)}(a) = f(a) + \frac{x}{1!} f'(a) + \frac{x^2}{2!} f''(a) + \cdots$$

2. $$f(x) = \sum_{k=0}^{\infty} \frac{x^k}{k!} f^{(k)}(0) = f(0) + \frac{x}{1!} f'(0) + \frac{x^2}{2!} f''(0) + \cdots$$

(MACLAURINsche R e i h e).
Die T A Y L O R s c h e R e i h e f ü r F u n k t i o n e n m e h r e r e r V e r ä n d e r l i c h e r:

(MACLAURIN s e r i e s).
The TAYLOR series for functions of several variables:

0.319

$$f(x, y) = f(\xi, \eta) + (x - \xi) \frac{\partial f(\xi, \eta)}{\partial x} + (y - \eta) \frac{\partial f(\xi, \eta)}{\partial y} +$$
$$+ \frac{1}{2!} \left\{ (x-\xi)^2 \frac{\partial^2 f(\xi, \eta)}{\partial x^2} + 2(x-\xi)(y-\eta) \frac{\partial^2 f(\xi, \eta)}{\partial x \partial y} + (y-\eta)^2 \frac{\partial^2 f(\xi, \eta)}{\partial y^2} \right\} + \cdots$$

0.32 Trigonometrische Reihen / FOURIER series

0.320 Es sei $f(x)$ eine periodische Funktion mit der Periode $2l$, die im Intervall $(-l, l)$ (zumindest im uneigentlichen Sinne) absolut integrierbar ist.
Die trigonometrische Reihe

Let $f(x)$ be any periodic function with period $2l$ which is absolutely integrable (at least in the sense of an improper integral) in the interval $(-l, l)$.
The FOURIER series

1. $$\frac{a_0}{2} + \sum_{k=1}^{\infty} a_k \cos \frac{k\pi x}{l} + b_k \sin \frac{k\pi x}{l}$$

mit den Koeffizienten | whose coefficients are given by

2. $$a_k = \frac{1}{l} \int_{-l}^{l} f(t) \cos \frac{k\pi t}{l} dt = \frac{1}{l} \int_{\alpha}^{\alpha+2l} f(t) \cos \frac{k\pi t}{l} dt \quad (k = 0, 1, 2, \ldots),$$

3. $$b_k = \frac{1}{l} \int_{-l}^{l} f(t) \sin \frac{k\pi t}{l} dt = \frac{1}{l} \int_{\alpha}^{\alpha+2l} f(t) \sin \frac{k\pi t}{l} dt \quad (k = 1, 2, \ldots)$$

wird FOURIER-Reihe der Funktion $f(x)$ genannt, die Koeffizienten werden als FOURIER-Koeffizienten bezeichnet.

is called the FOURIER series of $f(x)$; the coefficients are called FOURIER coefficients.

Konvergenzkriterien
Conditions of convergence

0.321 Die FOURIER-Reihe einer Funktion $f(x)$ im Punkt x_0 konvergiert gegen die Zahl

The FOURIER series of $f(x)$ converges at point x_0 towards the number

$$\frac{f(x_0 + 0) + f(x_0 - 0)}{2},$$

wenn für ein gewisses $h > 0$ das Integral | if for a certain $h > 0$ the integral

$$\int_0^h \frac{|f(x_0 + t) + f(x_0 - t) - f(x_0 + 0) - f(x_0 - 0)|}{t} dt$$

existiert. Hierbei wird vorausgesetzt, daß die Funktion $f(x)$ im Punkt x_0 stetig ist oder von beiden Seiten her eine Unstetigkeitsstelle erster Art (einen Sprung) besitzt und daß die beiden Grenzwerte $f(x_0 + 0)$ und $f(x_0 - 0)$ existieren (DINI). F III 427

exists. Here the function $f(x)$ is supposed to be either continuous at x_0 or to have a discontinuity at x_0 of the first kind with finite limits $f(x_0 + 0)$ and $f(x_0 - 0)$ (DINI). F III 427

Die FOURIER-Reihe einer periodischen Funktion $f(x)$, die im Intervall $[a, b]$ den DIRICHLETschen Bedingungen genügt, konvergiert in jedem Punkt x_0 gegen den Wert	The FOURIER series of a periodic function $f(x)$, which satisfies the DIRICHLET conditions in the interval $[a, b]$, converges at every point of this interval towards

0.322

$$\frac{1}{2}\{f(x_0 + 0) + f(x_0 - 0)\} \quad \text{(DIRICHLET)}.$$

Von einer Funktion $f(x)$ sagt man, sie *erfülle im Intervall $[a, b]$ die DIRICHLETschen Bedingungen*, wenn sie in diesem Intervall beschränkt ist und man das Intervall $[a, b]$ in endlich viele Teilintervalle zerlegen kann, in denen die Funktion $f(x)$ stetig und monoton ist.	A function $f(x)$ is said to *satisfy the DIRICHLET conditions in the interval $[a, b]$* if it is bounded in the whole interval $[a, b]$ and if the latter can be partitioned into a finite number of subintervals so that in each the function $f(x)$ is continuous and monotonic.
Die FOURIER-Reihe einer Funktion $f(x)$ im Punkt x_0 konvergiert gegen	The FOURIER series of $f(x)$ converges at x_0 towards

0.323

$$\frac{1}{2}\{f(x_0 + 0) + f(x_0 - 0)\},$$

wenn in einem Intervall $(x_0 - h, x_0 + h)$ mit dem Mittelpunkt x_0 die Funktion $f(x)$ von *endlicher Schwankung (beschränkter Variation)* ist (JORDAN-DIRICHLET).	if in an interval $(x_0 - h, x_0 + h)$ there exists a number $h < 0$ such that the function $f(x)$ is of *bounded variation* (JORDAN-DIRICHLET).

F III 431

Definition einer Funktion von beschränkter Variation. Die Funktion $f(x)$ sei auf einem Intervall $[a, b]$ mit $a < b$ definiert. Wir zerlegen dieses Intervall durch beliebig gewählte Punkte

Definition of a function of bounded variation. Let the function $f(x)$ be defined in the interval $[a, b]$ with $a < b$. Let this interval be partitioned into subintervals by the arbitrarily chosen points

$$a = x_0 < x_1 < x_2 < \ldots < x_{n-1} < x_n = b$$

und bilden die Summen	and the sum

$$\sum_{k=1}^{n} |f(x_k) - f(x_{k-1})|$$

Verschiedenen Zerlegungen des Intervalls $[a, b]$ (d. h. verschiedenen n und Punkten x_i) entsprechen im allgemeinen verschiedene Summen. Sind diese Summen **gleichmäßig nach oben beschränkt**, so sagt man, $f(x)$ besitze im Intervall $[a, b]$ eine *beschränkte Schwankung*. Die obere Grenze dieser Summen wird *vollständige Schwankung* (oder *Totalvariation*) der Funktion $f(x)$ im Intervall $[a, b]$ genannt.	In general, the sums obtained for different partitions of the interval $[a, b]$ (that is, for different n and different choices of the points x_i) will not be the same. If these sums are **uniformly bounded from above**, the functions $f(x)$ is said to be of *bounded variation* in the interval $[a, b]$. The least upper bound of these sums is called the *total variation* of $f(x)$ in the interval $[a, b]$.

F III 73–81

Die Funktion $f(x)$ sei im Intervall $[a, b]$ stückweise stetig und besitze	Let the functions $f(x)$ be piecewise continuous in the interval $[a, b]$ and have,

0.324

| in jedem Stetigkeitsintervall eine stückweise stetige Ableitung. Dann konvergiert in jedem Punkt x_0 des Intervalls $[a, b]$ die FOURIER-Reihe von $f(x)$ gegen | in each subinterval of continuity, a piecewise continuous derivative. Then the FOURIER series of $f(x)$ converges at every point x_0 of $[a, b]$ towards |

$$\frac{1}{2}\{f(x_0 + 0) + f(x_0 - 0)\}.$$

0.325 | Eine im Intervall $(0, l)$ definierte Funktion $f(x)$ läßt sich in eine **Cosinusreihe** entwickeln: | A function $f(x)$ defined in an interval $(0, l)$ may be expanded into a **cosine series**

1.
$$\frac{a_0}{2} + \sum_{k=1}^{\infty} a_k \cos \frac{k\pi x}{l}$$

mit | with

2.
$$a_k = \frac{2}{l} \int_0^l f(t) \cos \frac{k\pi t}{l}\, dt.$$

0.326 | Eine im Intervall $(0, l)$ definierte Funktion $f(x)$ läßt sich in eine **Sinusreihe** entwickeln: | A function $f(x)$ defined in an interval $(0, l)$ may be expanded into a **sine series**

1.
$$\sum_{k=1}^{\infty} b_k \sin \frac{k\pi x}{l}$$

mit | with

2.
$$b_k = \frac{2}{l} \int_0^l f(t) \sin \frac{k\pi t}{l}\, dt.$$

0.327 | Die Konvergenzkriterien für die Reihen 0.325 1. und 0.326 1. sind den Konvergenzkriterien für die Reihe 0.320 1. analog (vgl. 0.321–0.324). Die durch die Formeln 0.320 2. und 0.320 3. definierten FOURIER-Koeffizienten a_k und b_k einer absolut integrierbaren Funktion streben für $k \to \infty$ gegen Null. Für eine im Intervall $(-l, l)$ quadratisch integrierbare Funktion $f(x)$ gilt die *Vollständigkeitsrelation* | The conditions of convergence for the series 0.325 1. and 0.326 2. correspond exactly to those for the series 0.320 1. (see 0.321–0.324). The FOURIER **coefficients** a_k and b_k defined by 0.320 2. and 0.320 3., respectively, tend to zero for $k \to \infty$ if the function $f(x)$ is absolutely integrable. A function $f(x)$ which is quadratically integrable over the interval $(-l, l)$ satisfies the *completeness relation*

$$\frac{a_0^2}{2} + \sum_{k=1}^{\infty} (a_k^2 + b_k^2) = \frac{1}{l} \int_{-l}^{l} f^2(x)\, dx \quad \text{(LYAPUNOV)}. \qquad \text{F III 574}$$

0.328 | Es seien $f(x)$ und $\varphi(x)$ auf dem Intervall $(-l, l)$ quadratisch integrierbare Funktionen und a_k, b_k bzw. α_k, β_k ihre FOURIER-Koeffizienten. Für solche | Let $f(x)$ and $\varphi(x)$ be any two functions that are square integrable over the interval $(-l, l)$ and let a_k, b_k and α_k, β_k, respectively, be their FOURIER

Funktionen gilt die *verallgemeinerte Vollständigkeitsrelation (die* PARSEVAL*sche Gleichung)*

coefficients. Then $f(x)$ and $\varphi(x)$ satisfy the *generalized relation of completeness* (PARSEVAL *equation*)

$$\frac{a_0\alpha_0}{2} + \sum_{k=1}^{\infty}(a_k\alpha_k + b_k\beta_k) = \frac{1}{l}\int_{-l}^{l} f(x)\varphi(x)\,dx.$$

F III 577

Beispiele für trigonometrische Reihen findet man in 1.44, 1.45.

Examples of FOURIER series may be found in 1.44 and 1.45.

Asymptotische Reihen
Asymptotic series

0.33

Unter den divergenten Reihen läßt sich eine umfangreiche Klasse von Reihen aussondern, die als *asymptotische* oder *halbkonvergente Reihen* bezeichnet werden. Obgleich diese Reihen divergieren, kann man die Werte der durch sie dargestellten Funktionen mit großer Genauigkeit berechnen, wenn man die Summe einer geeigneten Anzahl von Gliedern dieser Reihen nimmt. Bei a l t e r n i e r e n d e n asymptotischen Reihen erhält man die größte Genauigkeit, wenn man die Reihe bei demjenigen Glied abbricht, das einem Glied kleinsten absoluten Betrages vorangeht. In diesem Fall übersteigt der Fehler (absolut genommen) den absoluten Betrag des ersten vernachlässigten Gliedes nicht (vgl. 0.227 3.).

From the class of divergent series a large subclass of series may be singled out which are called *asymptotic* or *semiconvergent series*. Although these series diverge, the values of the function represented by them may by calculated with a high accuracy by taking the sum of a suitable number of terms. For a l t e r n a t i n g asymptotic series the highest accuracy is obtained by breaking off the sum at the term that is followed by the one with the smallest modulus. In this case the absolute value of the error is not greater than the absolute value of the first neglected term (cf. 0.227 3.).

0.330

Die asymptotischen Reihen haben sehr viele Eigenschaften, die denen der konvergenten Reihen entsprechen. Daher spielen sie eine große Rolle in der Analysis.

Asymptotic series have many properties similar to those of convergent series. This is the reason why they are so important in analysis.

Die a s y m p t o t i s c h e E n t w i c k l u n g einer Funktion $f(z)$ bezeichnen wir mit

The a s y m p t o t i c e x p a n s i o n of a function $f(z)$ is denoted by

$$f(z) \sim \sum_{n=0}^{\infty} A_n z^{-n}.$$

Definition der *asymptotischen Entwicklung*. Eine divergente Reihe $\sum_{n=0}^{\infty} \frac{A_n}{z^n}$ stellt die asymptotische Entwicklung einer Funktion $f(z)$ in einem vorgegebenen Gebiet für die Werte von arg z

Definition of *asymptotic expansion:* A divergent series $\sum_{n=0}^{\infty} \frac{A_n}{z^n}$ is said to be the asymptotic expansion of a function $f(z)$ in a given domain of values of arg z if the expression $R_n(z) = z^n[f(z) - S_n(z)]$

dar, wenn der Ausdruck $R_n(z) = z^n[f(z) - S_n(z)]$ mit $S_n(z) = \sum_{k=0}^{n} \frac{A_k}{z^k}$ der Bedingung $\lim_{|z| \to \infty} R_n(z) = 0$ bei festem n genügt. F II 555

Eine divergente Reihe, welche eine asymptotische Entwicklung einer Funktion darstellt, wird eine *asymptotische Reihe* genannt.

0.331 Eigenschaften asymptotischer Reihen:
1. Mit asymptotischen Reihen lassen sich die Operationen Addition, Subtraktion, Multiplikation und Potenzieren genau so ausführen wie mit absolut konvergenten Reihen; die Reihen, die sich in dieser Weise ergeben, sind wieder asymptotische Reihen.
2. Eine asymptotische Reihe kann durch eine asymptotische Reihe dividiert werden unter der einzigen Bedingung, daß das erste Glied A_0 des Divisors nicht Null ist. Die sich durch Division ergebende Reihe ist wieder eine asymptotische Reihe. F II 556–559
3. Eine asymptotische Reihe läßt sich gliedweise **integrieren**; die sich dadurch ergebende Reihe ist wieder eine asymptotische Reihe. Dagegen ist die **Differentiation** einer asymptotischen Reihe im allgemeinen unzulässig. F II 559–560
4. Ein und dieselbe asymptotische Reihe kann verschiedene Funktionen darstellen. Dagegen kann eine gegebene Funktion nur auf eine einzige Weise in eine asymptotische Reihe entwickelt werden. Kn 556

with $S_n(z) = \sum_{k=0}^{n} \frac{A_k}{z^k}$ satisfies the condition $\lim_{|z| \to \infty} R_n(z) = 0$ for fixed n. F II 555

A diverging series that represents an asymptotic expansion of a function is called an *asymptotic series*.

Properties of asymptotic series:
(1) The operations of addition, subtraction, multiplication, and raising to a power can be carried out with asymptotic series in the same way as with absolutely convergent series. The series obtained will be asymptotic series.
(2) The division of an asymptotic series by another one can be carried out provided that the first term of the divisor series A_0 is nonzero. The series obtained by division is again an asymptotic series. F II 556–559
(3) An asymptotic series may be **integrated** termwise, and the series obtained is again asymptotic. Termwise **differentiation** of an asymptotic series is not, generally speaking, correct. F II 559–560
(4) The same asymptotic series may represent different functions; but a given function has only one asymptotic expansion. Kn 556

Einige Formeln aus der Differentialrechnung
Formulae from Differential Calculus

0.4

Differentiation eines bestimmten Integrals nach einem Parameter
Differentiation of a definite integral with respect to a parameter

0.41

0.410 $\quad \dfrac{d}{da} \displaystyle\int_{\psi(a)}^{\varphi(a)} f(x, a)\, dx = f(\varphi(a), a)\, \dfrac{d\varphi(a)}{da} - f(\psi(a), a)\, \dfrac{d\psi(a)}{da} + \displaystyle\int_{\psi(a)}^{\varphi(a)} \dfrac{d}{da} f(x, a)\, dx.$

F II 689

Insbesondere ist | In particular, 0.411

1. $$\frac{d}{da}\int_b^a f(x)\,dx = f(a).$$

2. $$\frac{d}{db}\int_b^a f(x)\,dx = -f(b).$$

Die n-te Ableitung eines Produkts (LEIBNIZsche Regel)
The nth derivative of a product (LEIBNIZ' formula) 0.42

Es seien u und v zwei n-mal differenzierbare Funktionen von x. Dann gilt | Let u and v be two functions of x of which the nth derivatives exist. Then

$$\frac{d^n(uv)}{dx^n} = u\frac{d^n v}{dx^n} + \binom{n}{1}\frac{du}{dx}\frac{d^{n-1}v}{dx^{n-1}} + \binom{n}{2}\frac{d^2 u}{dx^2}\frac{d^{n-2}v}{dx^{n-2}} + \binom{n}{3}\frac{d^3 u}{dx^3}\frac{d^{n-3}v}{dx^{n-3}} + \ldots + v\frac{d^n u}{dx^n}$$

oder symbolisch | or symbolically

$$\frac{d^n(uv)}{dx^n} = (u+v)^{(n)}.$$

F I 217

Die n-te Ableitung einer zusammengesetzten Funktion
The nth derivative of a composite function 0.43

Ist $f(x) = F(y)$ und $y = \varphi(x)$, so gilt | Let $f(x) = F(y)$ and $y = \varphi(x)$, then 0.430

1. $$\frac{d^n}{dx^n} f(x) = \frac{U_1}{1!} F'(y) + \frac{U_2}{2!} F''(y) + \frac{U_3}{3!} F'''(y) + \ldots + \frac{U_n}{n!} F^{(n)}(y)$$

mit | with

$$U_k = \frac{d^n}{dx^n} y^k - \frac{k}{1!} y \frac{d^n}{dx^n} y^{k-1} + \frac{k(k-1)}{2!} y^2 \frac{d^n}{dx^n} y^{k-2} - \ldots + (-1)^{k-1} k y^{k-1} \frac{d^n y}{dx^n}.$$

A(7361), Gou I 76

2. $$\frac{d^n}{dx^n} f(x) = \sum \frac{n!}{i!\,j!\,h!\,\ldots\,k!} \frac{d^m F}{dy^m} \left(\frac{y'}{1!}\right)^i \left(\frac{y''}{2!}\right)^j \left(\frac{y'''}{3!}\right)^h \ldots \left(\frac{y^{(l)}}{l!}\right)^k,$$

wobei sich die Summation über alle positiven ganzzahligen Lösungen der Gleichung $i + 2j + 3h + \ldots + lk = n$ erstreckt und $m = i + j + h + \ldots + k$ gilt. | where the summation is over all positive integer solutions of the equation $i + 2j + 3h + \ldots + lk = n$; m is given by $m = i + j + h + \ldots + k$.

Gou I 78

1. $$(-1)^n \frac{d^n}{dx^n} F\left(\frac{1}{x}\right) = \frac{1}{x^{2n}} F^{(n)}\left(\frac{1}{x}\right) + \frac{n-1}{x^{2n-1}} \frac{n}{1!} F^{(n-1)}\left(\frac{1}{x}\right) +$$

0.431

$$+ \frac{(n-1)(n-2)}{x^{2n-2}} \frac{n(n-1)}{2!} F^{(n-2)}\left(\frac{1}{x}\right) + \ldots$$

A(7362.1)

0.431 2. $$(-1)^n \frac{d^n}{dx^n} e^{\frac{a}{x}} = \frac{1}{x^n} e^{\frac{a}{x}} \left\{ \left(\frac{a}{x}\right)^n + (n-1)\binom{n}{1}\left(\frac{a}{x}\right)^{n-1} + \right.$$
$$\left. + (n-1)(n-2)\binom{n}{2}\left(\frac{a}{x}\right)^{n-2} + (n-1)(n-2)(n-3)\binom{n}{3}\left(\frac{a}{x}\right)^{n-3} + \ldots \right\}.$$
A(7362.2)

0.432 1. $$\frac{d^n}{dx^n} F(x^2) = (2x)^n F^{(n)}(x^2) + \frac{n(n-1)}{1!} (2x)^{n-2} F^{(n-1)}(x^2) +$$
$$+ \frac{n(n-1)(n-2)(n-3)}{2!} (2x)^{n-4} F^{(n-2)}(x^2) +$$
$$+ \frac{n(n-1)(n-2)(n-3)(n-4)(n-5)}{3!} (2x)^{n-6} F^{(n-3)}(x^2) + \ldots$$
A(7363.1)

2. $$\frac{d^n}{dx^n} e^{ax^2} = (2ax)^n e^{ax^2} \left\{ 1 + \frac{n(n-1)}{1!(4ax^2)} + \frac{n(n-1)(n-2)(n-3)}{2!(4ax^2)^2} + \right.$$
$$\left. + \frac{n(n-1)(n-2)(n-3)(n-4)(n-5)}{3!(4ax^2)^3} + \ldots \right\}.$$
A(7363.2)

3. $$\frac{d^n}{dx^n} (1+ax^2)^p = \frac{p(p-1)(p-2)\ldots(p-n+1)(2ax)^n}{(1+ax^2)^{n-p}} \times$$
$$\times \left\{ 1 + \frac{n(n-1)}{1!(p-n+1)} \frac{1+ax^2}{4ax^2} + \frac{n(n-1)(n-2)(n-3)}{2!(p-n+1)(p-n+2)} \left(\frac{1+ax^2}{4ax^2}\right)^2 + \ldots \right\}.$$
A(7363.3)

4. $$\frac{d^{m-1}}{dx^{m-1}} (1-x^2)^{m-\frac{1}{2}} = (-1)^{m-1} \frac{(2m-1)!!}{m} \sin(m \arccos x).$$
A(7363.4)

0.433 1. $$\frac{d^n}{dx^n} F(\sqrt{x}) = \frac{F^{(n)}(\sqrt{x})}{(2\sqrt{x})^n} - \frac{n(n-1)}{1!} \frac{F^{(n-1)}(\sqrt{x})}{(2\sqrt{x})^{n+1}} +$$
$$+ \frac{(n+1)n(n-1)(n-2)}{2!} \frac{F^{(n-2)}(\sqrt{x})}{(2\sqrt{x})^{n+2}} - \ldots$$
A(7364.1)

2. $$\frac{d^n}{dx^n} (1 + a\sqrt{x})^{2n-1} = \frac{(2n-1)!!}{2^n} \frac{a}{\sqrt{x}} \left(a^2 - \frac{1}{x}\right)^{n-1}.$$
A(7364.2)

0.434 $$\frac{d^n}{dx^n} y^p = p \binom{n-p}{n} \left\{ -\binom{n}{1} \frac{1}{p-1} y^{p-1} \frac{d^n y}{dx^n} + \binom{n}{2} \frac{1}{p-2} y^{p-2} \frac{d^n y^2}{dx^n} - \ldots \right\}.$$
A(737.1)

0.435 $$\frac{d^n}{dx^n} \ln y = \binom{n}{1} \left\{ \frac{1}{1 \cdot y} \frac{d^n y}{dx^n} - \binom{n}{2} \frac{1}{2 \cdot y^2} \frac{d^n y^2}{dx^n} + \binom{n}{3} \frac{1}{3 \cdot y^3} \frac{d^n y^3}{dx^n} - \ldots \right\}.$$
A(737.2)

Elementare Funktionen
Elementary Functions

Potenzen von Binomen
Powers of Binomials

1.1

Potenzreihen
Power series

1.11

$$(1+x)^q = 1 + qx + \frac{q(q-1)}{2!}x^2 + \ldots + \frac{q(q-1)\ldots(q-k+1)}{k!}x^k + \ldots$$ 1.110

Ist q keine nichtnegative ganze Zahl, so konvergiert die Reihe für $|x|<1$ absolut und divergiert für $|x|>1$; im Punkt $x=1$ konvergiert die Reihe für $q>-1$ und divergiert für $q\leqslant-1$; dort konvergiert sie für $q>0$ absolut. Im Punkt $x=-1$ konvergiert sie für $q>0$ absolut, während sie für $q<0$ divergiert. Ist q eine natürliche Zahl n, so geht die Reihe 1.110 in die endliche Summe 1.111 über. F II 385

For q nonzero and different from any natural number the series converges absolutely for $|x|<1$ and diverges for $|x|>1$. If $x=1$, the series converges for $q>-1$ and diverges for $q\leqslant-1$; if $x=1$, it converges absolutely for $q>0$; and if $x=-1$, it converges absolutely for $q>0$ and diverges for $q<0$. If q is a natural number, n, the series 1.110 reduces to the finite series 1.111. F II 385

$$(a+x)^n = \sum_{k=0}^{n} \binom{n}{k} x^k a^{n-k}.$$ 1.111

1. $(1+x)^{-1} = 1 - x + x^2 - x^3 + \ldots = \sum_{k=1}^{\infty}(-1)^{k-1} x^{k-1}$ (cf. 1.121 2.). 1.112

2. $(1+x)^{-2} = 1 - 2x + 3x^2 - 4x^3 + \ldots = \sum_{k=1}^{\infty}(-1)^{k-1} kx^{k-1}.$

3. $(1+x)^{\frac{1}{2}} = 1 + \frac{1}{2}x - \frac{1\cdot 1}{2\cdot 4}x^2 + \frac{1\cdot 1\cdot 3}{2\cdot 4\cdot 6}x^3 - \frac{1\cdot 1\cdot 3\cdot 5}{2\cdot 4\cdot 6\cdot 8}x^4 + \ldots$

4. $(1+x)^{-\frac{1}{2}} = 1 - \frac{1}{2}x + \frac{1\cdot 3}{2\cdot 4}x^2 - \frac{1\cdot 3\cdot 5}{2\cdot 4\cdot 6}x^3 + \ldots$

$$\frac{x}{(1-x)^2} = \sum_{k=1}^{\infty} kx^k \quad [x^2 < 1].$$ 1.113

1.114 1. $$(1+\sqrt{1+x})^q = 2^q\left\{1 + \frac{q}{1!}\left(\frac{x}{4}\right) + \frac{q(q-3)}{2!}\left(\frac{x}{4}\right)^2 + \right.$$
$$\left. + \frac{q(q-4)(q-5)}{3!}\left(\frac{x}{4}\right)^3 + \cdots\right\}$$ A(6351.1)

$[x^2 < 1,\ q$ eine reelle Zahl$]$. $\quad|\quad [x^2 < 1,\ q$ real$]$.

2. $$(x+\sqrt{1+x^2})^q = 1 + \sum_{k=0}^{\infty} \frac{q^2(q^2-2^2)(q^2-4^2)\cdots[q^2-(2k)^2]x^{2k+2}}{(2k+2)!} +$$
$$+ qx + q\sum_{k=1}^{\infty} \frac{(q^2-1^2)(q^2-3^2)\cdots[q^2-(2k-1)^2]}{(2k+1)!}x^{2k+1}$$ A(6351.2)

$[x^2 < 1,\ q$ eine reelle Zahl$]$. $\quad|\quad [x^2 < 1,\ q$ real$]$.

Reihen rationaler Brüche
Series of rational fractions

1.12

1.121 1. $$\frac{x}{1-x} = \sum_{k=1}^{\infty} \frac{2^{k-1} x^{2^{k-1}}}{1+x^{2^{k-1}}} = \sum_{k=1}^{\infty} \frac{x^{2^{k-1}}}{1-x^{2^k}} \quad [x^2 < 1].$$ A(6350.3)

2. $$\frac{1}{x-1} = \sum_{k=1}^{\infty} \frac{2^{k-1}}{x^{2^{k-1}}+1} \quad [x^2 > 1].$$ A(6350.3)

Die Exponentialfunktion
The Exponential Function

1.2

Reihendarstellung
Series representation

1.21

1.211 1. $$e^x = \sum_{k=0}^{\infty} \frac{x^k}{k!}.$$

2. $$a^x = \sum_{k=0}^{\infty} \frac{(x \ln a)^k}{k!}.$$

3. $$e^{-x^2} = \sum_{k=0}^{\infty} (-1)^k \frac{x^{2k}}{k!}.$$

1.212 $$e^x(1+x) = \sum_{k=0}^{\infty} \frac{x^k(k+1)}{k!}.$$

Die Exponentialfunktion / The Exponential Function

$$\frac{x}{e^x - 1} = 1 - \frac{x}{2} + \sum_{k=1}^{\infty} \frac{B_{2k} x^{2k}}{(2k)!} \quad [x < 2\pi].$$
F II 512   1.213

$$e^{e^x} = e\left(1 + x + \frac{2x^2}{2!} + \frac{5x^3}{3!} + \frac{15x^4}{4!} + \ldots\right).$$
A(6460.3)   1.214

1. $$e^{\sin x} = 1 + x + \frac{x^2}{2!} - \frac{3x^4}{4!} - \frac{8x^5}{5!} - \frac{3x^6}{6!} + \frac{56x^7}{7!} + \ldots$$
A(6460.4)   1.215

2. $$e^{\cos x} = e\left(1 - \frac{x^2}{2!} + \frac{4x^4}{4!} - \frac{31x^6}{6!} + \ldots\right).$$
A(6460.5)

3. $$e^{\operatorname{tg} x} = 1 + x + \frac{x^2}{2!} + \frac{3x^3}{3!} + \frac{9x^4}{4!} + \frac{37x^5}{5!} + \ldots$$
A(6460.6)

1. $$e^{\arcsin x} = 1 + x + \frac{x^2}{2!} + \frac{2x^3}{3!} + \frac{5x^4}{4!} + \ldots$$
A(6460.7)   1.216

2. $$e^{\operatorname{arctg} x} = 1 + x + \frac{x^2}{2!} - \frac{x^3}{3!} - \frac{7x^4}{4!} - \ldots$$
A(6460.8)

1. $$\pi \frac{e^{\pi x} + e^{-\pi x}}{e^{\pi x} - e^{-\pi x}} = \frac{1}{x} + 2x \sum_{k=1}^{\infty} \frac{1}{x^2 + k^2} \quad (\text{cf. } 1.421\ 3.).$$
A(6707.1)   1.217

2. $$\frac{2\pi}{e^{\pi x} - e^{-\pi x}} = \frac{1}{x} + 2x \sum_{k=1}^{\infty} (-1)^k \frac{1}{x^2 + k^2} \quad (\text{cf. } 1.422\ 3.).$$
A(6707.2)

Funktionalgleichungen
Functional relations
1.22

1. $$a^x = e^{x \ln a}.$$
1.221

2. $$a^{\log_a x} = a^{\frac{1}{\log_x a}} = x.$$

1. $$e^x = \operatorname{ch} x + \operatorname{sh} x.$$
1.222

2. $$e^{ix} = \cos x + i \sin x.$$

$$e^{ax} - e^{bx} = (a - b) x \exp\left[\frac{1}{2}(a + b)x\right] \prod_{k=1}^{\infty} \left[1 + \frac{(a - b)^2 x^2}{4k^2 \pi^2}\right].$$
MO 216   1.223

Reihen von Exponentialfunktionen
Series of exponentials
1.23

$$\sum_{k=0}^{\infty} a^{kx} = \frac{1}{1 - a^x}$$
1.231

$[a>1$ und $x<0$ oder $0<a<1$ und $x>0]$.	$[a>1$ and $x<0$ or $0<a<1$ and $x>0]$.

1.232

1. $$\operatorname{th} x = 1 + 2\sum_{k=1}^{\infty}(-1)^k e^{-2kx} \quad [x>0].$$

2. $$\operatorname{sech} x = 2\sum_{k=0}^{\infty}(-1)^k e^{-(2k+1)x} \quad [x>0].$$

3. $$\operatorname{cosech} x = 2\sum_{k=0}^{\infty} e^{-(2k+1)x} \quad [x>0].$$

1.3–1.4

Trigonometrische und hyperbolische Funktionen
Trigonometric and Hyperbolic Functions

1.30

Einführung
Introduction

Zwischen dem trigonometrischen Sinus und dem hyperbolischen Sinus bestehen die Beziehungen	The trigonometric and the hyperbolic sine are related to one another as follows:

$$\operatorname{sh} x = \frac{1}{i}\sin ix, \quad \sin x = \frac{1}{i}\operatorname{sh} ix.$$

Für den trigonometrischen und den hyperbolischen Cosinus gilt	For the trigonometric and the hyperbolic cosine we have

$$\operatorname{ch} x = \cos ix, \quad \cos x = \operatorname{ch} ix.$$

Aufgrund dieser Dualität kann man jeder Beziehung, in welcher trigonometrische Funktionen vorkommen, formal eine Beziehung zuordnen, in welche die hyperbolischen Funktionen eingehen, und umgekehrt läßt sich jeder Beziehung, in welcher hyperbolische Funktionen auftreten, formal eine Beziehung gegenüberstellen, in welcher trigonometrische Funktionen vorkommen. In vielen (wenn auch nicht in allen) Fällen haben tatsächlich beide Paare von Beziehungen einen Sinn. In den nachstehenden Beziehungen liegt Dualität vor, jedoch sind nicht alle sinnvollen „Paare" aufgeführt.	This duality shows that every relation containing trigonometric functions has corresponding to it an analogous relation containing hyperbolic functions, and vice versa. In many (though not in all) cases both pairs of relations have sense. The dual nature of such formulae will be clarified through the material that follows. However, not all possible "dualities" are stated.

Die grundlegenden Funktionalgleichungen
The fundamental functional relations

1.31

1. $\sin x = \dfrac{1}{2i}(e^{ix} - e^{-ix});$ 2. $\operatorname{sh} x = \dfrac{1}{2}(e^{x} - e^{-x});$ 1.311

        $= -i \operatorname{sh} ix.$ $= -i \sin(ix).$

3. $\cos x = \dfrac{1}{2}(e^{ix} + e^{-ix});$ 4. $\operatorname{ch} x = \dfrac{1}{2}(e^{x} + e^{-x});$

        $= \operatorname{ch} ix.$ $= \cos ix.$

5. $\operatorname{tg} x = \dfrac{\sin x}{\cos x} = \dfrac{1}{i}\operatorname{th} ix.$ 6. $\operatorname{th} x = \dfrac{\operatorname{sh} x}{\operatorname{ch} x} = \dfrac{1}{i}\operatorname{tg} ix.$

7. $\operatorname{ctg} x = \dfrac{\cos x}{\sin x} = \dfrac{1}{\operatorname{tg} x} = i\operatorname{cth} ix.$ 8. $\operatorname{cth} x = \dfrac{\operatorname{ch} x}{\operatorname{sh} x} = \dfrac{1}{\operatorname{th} x} = i\operatorname{ctg} ix.$

1. $\cos^2 x + \sin^2 x = 1.$ 2. $\operatorname{ch}^2 x - \operatorname{sh}^2 x = 1.$ 1.312

1. $\sin(x \pm y) = \sin x \cos y \pm \sin y \cos x.$ 1.313

2. $\operatorname{sh}(x \pm y) = \operatorname{sh} x \operatorname{ch} y \pm \operatorname{sh} y \operatorname{ch} x.$

3. $\sin(x \pm iy) = \sin x \operatorname{ch} y \pm i \operatorname{sh} y \cos x.$

4. $\operatorname{sh}(x \pm iy) = \operatorname{sh} x \cos y \pm i \sin y \operatorname{ch} x.$

5. $\cos(x \pm y) = \cos x \cos y \mp \sin x \sin y.$

6. $\operatorname{ch}(x \pm y) = \operatorname{ch} x \operatorname{ch} y \pm \operatorname{sh} x \operatorname{sh} y.$

7. $\cos(x \pm iy) = \cos x \operatorname{ch} y \mp i \sin x \operatorname{sh} y.$

8. $\operatorname{ch}(x \pm iy) = \operatorname{ch} x \cos y \pm i \operatorname{sh} x \sin y.$

9. $\operatorname{tg}(x \pm y) = \dfrac{\operatorname{tg} x \pm \operatorname{tg} y}{1 \mp \operatorname{tg} x \operatorname{tg} y}.$

10. $\operatorname{th}(x \pm y) = \dfrac{\operatorname{th} x \pm \operatorname{th} y}{1 \pm \operatorname{th} x \operatorname{th} y}.$

11. $\operatorname{tg}(x \pm iy) = \dfrac{\operatorname{tg} x \pm i \operatorname{th} y}{1 \mp i \operatorname{tg} x \operatorname{th} y}.$

12. $\operatorname{th}(x \pm iy) = \dfrac{\operatorname{th} x \pm i \operatorname{tg} y}{1 \pm i \operatorname{th} x \operatorname{tg} y}.$

1. $\sin x \pm \sin y = 2 \sin \dfrac{1}{2}(x \pm y) \cos \dfrac{1}{2}(x \mp y).$ 1.314

2. $\operatorname{sh} x \pm \operatorname{sh} y = 2 \operatorname{sh} \dfrac{1}{2}(x \pm y) \operatorname{ch} \dfrac{1}{2}(x \mp y).$

1.314 3. $\cos x + \cos y = 2\cos\frac{1}{2}(x+y)\cos\frac{1}{2}(x-y).$

4. $\operatorname{ch} x + \operatorname{ch} y = 2\operatorname{ch}\frac{1}{2}(x+y)\operatorname{ch}\frac{1}{2}(x-y).$

5. $\cos x - \cos y = 2\sin\frac{1}{2}(x+y)\sin\frac{1}{2}(y-x).$

6. $\operatorname{ch} x - \operatorname{ch} y = 2\operatorname{sh}\frac{1}{2}(x+y)\operatorname{sh}\frac{1}{2}(x-y).$

7. $\operatorname{tg} x \pm \operatorname{tg} y = \dfrac{\sin(x\pm y)}{\cos x \cos y}.$ 8. $\operatorname{th} x \pm \operatorname{th} y = \dfrac{\operatorname{sh}(x\pm y)}{\operatorname{ch} x \operatorname{ch} y}.$

1.315 1. $\sin^2 x - \sin^2 y = \sin(x+y)\sin(x-y) = \cos^2 y - \cos^2 x.$

2. $\operatorname{sh}^2 x - \operatorname{sh}^2 y = \operatorname{sh}(x+y)\operatorname{sh}(x-y) = \operatorname{ch}^2 x - \operatorname{ch}^2 y.$

3. $\cos^2 x - \sin^2 y = \cos(x+y)\cos(x-y) = \cos^2 y - \sin^2 x.$

4. $\operatorname{sh}^2 x + \operatorname{ch}^2 y = \operatorname{ch}(x+y)\operatorname{ch}(x-y) = \operatorname{ch}^2 x + \operatorname{sh}^2 y.$

1.316 1. $(\cos x + i\sin x)^n = \cos nx + i\sin nx.$ 2. $(\operatorname{ch} x + \operatorname{sh} x)^n = \operatorname{sh} nx + \operatorname{ch} ny$

$$\left[n\ \begin{matrix}\text{ganzzahlig}\\ \text{integer}\end{matrix}\right].$$

1.317 1. $\sin\dfrac{x}{2} = \pm\sqrt{\dfrac{1}{2}(1-\cos x)}.$ 2. $\operatorname{sh}\dfrac{x}{2} = \pm\sqrt{\dfrac{1}{2}(\operatorname{ch} x - 1)}.$

3. $\cos\dfrac{x}{2} = \pm\sqrt{\dfrac{1}{2}(1+\cos x)}.$ 4. $\operatorname{ch}\dfrac{x}{2} = \sqrt{\dfrac{1}{2}(\operatorname{ch} x + 1)}.$

5. $\operatorname{tg}\dfrac{x}{2} = \dfrac{1-\cos x}{\sin x} = \dfrac{\sin x}{1+\cos x}.$ 6. $\operatorname{th}\dfrac{x}{2} = \dfrac{\operatorname{ch} x - 1}{\operatorname{sh} x} = \dfrac{\operatorname{sh} x}{\operatorname{ch} x + 1}.$

Das Vorzeichen der Wurzel in den Formeln 1.317 1., 1.317 2. und 1.317 3. muß in Übereinstimmung mit dem Vorzeichen der linken Seite gewählt werden; dieses hängt von dem Wert x ab.

The sign of the root in formulae 1.317 1., 1.317 2., and 1.317 3. has to be chosen in accordance with the sign of the left-hand side; the latter depends on x.

Darstellung von Potenzen trigonometrischer und hyperbolischer Funktionen durch Funktionen von Vielfachen der Argumente (Bogen)

1.32

The representation of powers of trigonometric and hyperbolic functions in terms of functions of multiples of the argument (angle)

1.320 1. $\sin^{2n} x = \dfrac{1}{2^{2n}}\left\{\displaystyle\sum_{k=0}^{n-1}(-1)^{n-k} 2\binom{2n}{k}\cos 2(n-k)x + \binom{2n}{n}\right\}.$ Kr 56(10,2)

2. $\operatorname{sh}^{2n} x = \dfrac{(-1)^n}{2^{2n}}\left\{\displaystyle\sum_{k=0}^{n-1}(-1)^{n-k} 2\binom{2n}{k}\operatorname{ch} 2(n-k)x + \binom{2n}{n}\right\}.$

3. $\sin^{2n-1} x = \dfrac{1}{2^{2n-2}} \sum_{k=0}^{n-1} (-1)^{n+k-1} \binom{2n-1}{k} \sin(2n-2k-1)x.$ Kr 56(10,4) 1.320

4. $\text{sh}^{2n-1} x = \dfrac{(-1)^{n-1}}{2^{2n-2}} \sum_{k=0}^{n-1} (-1)^{n+k-1} \binom{2n-1}{k} \text{sh}(2n-2k-1)x.$

5. $\cos^{2n} x = \dfrac{1}{2^{2n}} \left\{ \sum_{k=0}^{n-1} 2\binom{2n}{k} \cos 2(n-k)x + \binom{2n}{n} \right\}.$ Kr 56(10,1)

6. $\text{ch}^{2n} x = \dfrac{1}{2^{2n}} \left\{ \sum_{k=0}^{n-1} 2\binom{2n}{k} \text{ch}\, 2(n-k)x + \binom{2n}{n} \right\}.$

7. $\cos^{2n-1} x = \dfrac{1}{2^{2n-2}} \sum_{k=0}^{n-1} \binom{2n-1}{k} \cos(2n-2k-1)x.$ Kr 56(10,3)

8. $\text{ch}^{2n-1} x = \dfrac{1}{2^{2n-2}} \sum_{k=0}^{n-1} \binom{2n-1}{k} \text{ch}(2n-2k-1)x.$

Spezialfälle
Special cases

1. $\sin^2 x = \dfrac{1}{2}(-\cos 2x + 1).$ 1.321

2. $\sin^3 x = \dfrac{1}{4}(-\sin 3x + 3\sin x).$

3. $\sin^4 x = \dfrac{1}{8}(\cos 4x - 4\cos 2x + 3).$

4. $\sin^5 x = \dfrac{1}{16}(\sin 5x - 5\sin 3x + 10\sin x).$

5. $\sin^6 x = \dfrac{1}{32}(-\cos 6x + 6\cos 4x - 15\cos 2x + 10).$

6. $\sin^7 x = \dfrac{1}{64}(-\sin 7x + 7\sin 5x - 21\sin 3x + 35\sin x).$

1. $\text{sh}^2 x = \dfrac{1}{2}(\text{ch}\, 2x - 1).$ 1.322

2. $\text{sh}^3 x = \dfrac{1}{4}(\text{sh}\, 3x - 3\,\text{sh}\, x).$

3. $\text{sh}^4 x = \dfrac{1}{8}(\text{ch}\, 4x - 4\,\text{ch}\, 2x + 3).$

1.322

4. $\quad \operatorname{sh}^5 x = \dfrac{1}{16}(\operatorname{sh} 5x - 5\operatorname{sh} 3x + 10\operatorname{sh} x).$

5. $\quad \operatorname{sh}^6 x = \dfrac{1}{32}(\operatorname{ch} 6x - 6\operatorname{ch} 4x + 15\operatorname{ch} 2x - 10).$

6. $\quad \operatorname{sh}^7 x = \dfrac{1}{64}(\operatorname{sh} 7x - 7\operatorname{sh} 5x + 21\operatorname{sh} 3x - 35\operatorname{sh} x).$

1.323

1. $\quad \cos^2 x = \dfrac{1}{2}(\cos 2x + 1).$

2. $\quad \cos^3 x = \dfrac{1}{4}(\cos 3x + 3\cos x).$

3. $\quad \cos^4 x = \dfrac{1}{8}(\cos 4x + 4\cos 2x + 3).$

4. $\quad \cos^5 x = \dfrac{1}{16}(\cos 5x + 5\cos 3x + 10\cos x).$

5. $\quad \cos^6 x = \dfrac{1}{32}(\cos 6x + 6\cos 4x + 15\cos 2x + 10).$

6. $\quad \cos^7 x = \dfrac{1}{64}(\cos 7x + 7\cos 5x + 21\cos 3x + 35\cos x).$

1.324

1. $\quad \operatorname{ch}^2 x = \dfrac{1}{2}(\operatorname{ch} 2x + 1).$

2. $\quad \operatorname{ch}^3 x = \dfrac{1}{4}(\operatorname{ch} 3x + 3\operatorname{ch} x).$

3. $\quad \operatorname{ch}^4 x = \dfrac{1}{8}(\operatorname{ch} 4x + 4\operatorname{ch} 2x + 3).$

4. $\quad \operatorname{ch}^5 x = \dfrac{1}{16}(\operatorname{ch} 5x + 5\operatorname{ch} 3x + 10\operatorname{ch} x).$

5. $\quad \operatorname{ch}^6 x = \dfrac{1}{32}(\operatorname{ch} 6x + 6\operatorname{ch} 4x + 15\operatorname{ch} 2x + 10).$

6. $\quad \operatorname{ch}^7 x = \dfrac{1}{64}(\operatorname{ch} 7x + 7\operatorname{ch} 5x + 21\operatorname{ch} 3x + 35\operatorname{ch} x).$

Darstellung trigonometrischer und hyperbolischer Funktionen von Vielfachen der Argumente (Bogen) durch Potenzen dieser Funktionen
The representation of trigonometric and hyperbolic functions of multiples of the argument (angle) in terms of powers of these functions

1.33

1. $\sin nx = n \cos^{n-1} x \sin x - \binom{n}{3} \cos^{n-3} x \sin^3 x + \binom{n}{5} \cos^{n-5} x \sin^5 x - \ldots;$ 1.331

$$= \sin x \left\{ 2^{n-1} \cos^{n-1} x - \binom{n-2}{1} 2^{n-3} \cos^{n-3} x + \right.$$

$$\left. + \binom{n-3}{2} 2^{n-5} \cos^{n-5} x - \binom{n-4}{3} 2^{n-7} \cos^{n-7} x + \ldots \right\}. \quad \text{A(3.175)}$$

2. $\operatorname{sh} nx = \operatorname{sh} x \sum_{k=1}^{E\left(\frac{n+1}{2}\right)} \binom{n}{2k-1} \operatorname{sh}^{2k-2} x \operatorname{ch}^{n-2k+1} x;$

$$= \operatorname{sh} x \sum_{k=0}^{E\left(\frac{n-1}{2}\right)} (-1)^k \binom{n-k-1}{k} 2^{n-2k-1} \operatorname{ch}^{n-2k-1} x.$$

3. $\cos nx = \cos^n x - \binom{n}{2} \cos^{n-2} x \sin^2 x + \binom{n}{4} \cos^{n-4} x \sin^4 x - \ldots;$

$$= 2^{n-1} \cos^n x - \frac{n}{1} 2^{n-3} \cos^{n-2} x +$$

$$+ \frac{n}{2} \binom{n-3}{1} 2^{n-5} \cos^{n-4} x - \frac{n}{3} \binom{n-4}{2} 2^{n-7} \cos^{n-6} x + \ldots \quad \text{A(3.175)}$$

4. $\operatorname{ch} nx = \sum_{k=0}^{E\left(\frac{n}{2}\right)} \binom{n}{2k} \operatorname{sh}^{2k} x \operatorname{ch}^{n-2k} x =$

$$= 2^{n-1} \operatorname{ch}^n x + n \sum_{k=1}^{E\left(\frac{n}{2}\right)} (-1)^k \frac{1}{k} \binom{n-k-1}{k-1} 2^{n-2k-1} \operatorname{ch}^{n-2k} x.$$

1. $\sin 2nx = 2n \cos x \left\{ \sin x - \frac{4n^2 - 2^2}{3!} \sin^3 x + \frac{(4n^2 - 2^2)(4n^2 - 4^2)}{5!} \sin^5 x - \ldots \right\};$ 1.332

A(3.171)

$$= (-1)^{n-1} \cos x \left\{ 2^{2n-1} \sin^{2n-1} x - \frac{2n-2}{1!} 2^{2n-3} \sin^{2n-3} x + \right.$$

$$\left. + \frac{(2n-3)(2n-4)}{2!} 2^{2n-5} \sin^{2n-5} x - \frac{(2n-4)(2n-5)(2n-6)}{3!} 2^{2n-7} \sin^{2n-7} x + \ldots \right\}.$$

A(3.173)

1.332

2. $$\sin(2n-1)x = (2n-1)\left\{\sin x - \frac{(2n-1)^2 - 1^2}{3!}\sin^3 x + \right.$$
$$\left. + \frac{[(2n-1)^2 - 1^2][(2n-1)^2 - 3^2]}{5!}\sin^5 x - \ldots\right\};\quad \text{A(3.172)}$$

$$= (-1)^{n-1}\left\{2^{2n-2}\sin^{2n-1}x - \frac{2n-1}{1!}2^{2n-4}\sin^{2n-3}x + \right.$$
$$\left. + \frac{(2n-1)(2n-4)}{2!}2^{2n-6}\sin^{2n-5}x - \frac{(2n-1)(2n-5)(2n-6)}{3!}2^{2n-8}\sin^{2n-7}x + \ldots\right\}.$$

A(3.174)+

3. $$\cos 2nx = 1 - \frac{4n^2}{2!}\sin^2 x +$$
$$+ \frac{4n^2(4n^2 - 2^2)}{4!}\sin^4 x - \frac{4n^2(4n^2 - 2^2)(4n^2 - 4^2)}{6!}\sin^6 x + \ldots;\quad \text{A(3.171)}$$

$$= (-1)^n\left\{2^{2n-1}\sin^{2n}x - \frac{2n}{1!}2^{2n-3}\sin^{2n-2}x + \right.$$
$$\left. + \frac{2n(2n-3)}{2!}2^{2n-5}\sin^{2n-4} - \frac{2n(2n-4)(2n-5)}{3!}2^{2n-7}\sin^{2n-6} + \ldots\right\}.$$

A(3.173)+

4. $$\cos(2n-1)x = \cos x\left\{1 - \frac{(2n-1)^2 - 1^2}{2!}\sin^2 x + \right.$$
$$\left. + \frac{[(2n-1)^2 - 1^2][(2n-1)^2 - 3^2]}{4!}\sin^4 x - \ldots\right\};\quad \text{A(3.172)}$$

$$= (-1)^{n-1}\cos x\left\{2^{2n-2}\sin^{2n-2}x - \frac{2n-3}{1!}2^{2n-4}\sin^{2n-4}x + \right.$$
$$+ \frac{(2n-4)(2n-5)}{2!}2^{2n-6}\sin^{2n-6}x -$$
$$\left. - \frac{(2n-5)(2n-6)(2n-7)}{3!}2^{2n-8}\sin^{2n-8}x + \ldots\right\}.\quad \text{A(3.174)}$$

Mit Hilfe der Formeln und Bemerkungen in 1.30 kann man zu Formeln für sh $2nx$, sh $(2n-1)x$, ch $2nx$, ch $(2n-1)x$ gelangen, die den Formeln 1.332 entsprechen, indem man wie in 1.331 verfährt.

Analogous formulae for sh $2nx$, sh$(2n-1)x$, ch $2nx$, and ch $(2n-1)x$ can be obtained with the help of the formulae and remarks under 1.30, proceeding like in 1.331.

Spezialfälle
Special cases

1.333

1. $\sin 2x = 2 \sin x \cos x.$
2. $\sin 3x = 3 \sin x - 4 \sin^3 x.$
3. $\sin 4x = \cos x\, (4 \sin x - 8 \sin^3 x).$
4. $\sin 5x = 5 \sin x - 20 \sin^3 x + 16 \sin^5 x.$
5. $\sin 6x = \cos x\, (6 \sin x - 32 \sin^3 x + 32 \sin^5 x).$
6. $\sin 7x = 7 \sin x - 56 \sin^3 x + 112 \sin^5 x - 64 \sin^7 x.$

1.334

1. $\operatorname{sh} 2x = 2 \operatorname{sh} x \operatorname{ch} x.$
2. $\operatorname{sh} 3x = 3 \operatorname{sh} x + 4 \operatorname{sh}^3 x.$
3. $\operatorname{sh} 4x = \operatorname{ch} x\, (4 \operatorname{sh} x + 8 \operatorname{sh}^3 x).$
4. $\operatorname{sh} 5x = 5 \operatorname{sh} x + 20 \operatorname{sh}^3 x + 16 \operatorname{sh}^5 x.$
5. $\operatorname{sh} 6x = \operatorname{ch} x\, (6 \operatorname{sh} x + 32 \operatorname{sh}^3 x + 32 \operatorname{sh}^5 x).$
6. $\operatorname{sh} 7x = 7 \operatorname{sh} x + 56 \operatorname{sh}^3 x + 112 \operatorname{sh}^5 x + 64 \operatorname{sh}^7 x.$

1.335

1. $\cos 2x = 2 \cos^2 x - 1.$
2. $\cos 3x = 4 \cos^3 x - 3 \cos x.$
3. $\cos 4x = 8 \cos^4 x - 8 \cos^2 x + 1.$
4. $\cos 5x = 16 \cos^5 x - 20 \cos^3 x + 5 \cos x.$
5. $\cos 6x = 32 \cos^6 x - 48 \cos^4 x + 18 \cos^2 x - 1.$
6. $\cos 7x = 64 \cos^7 x - 112 \cos^5 x + 56 \cos^3 x - 7 \cos x.$

1.336

1. $\operatorname{ch} 2x = 2 \operatorname{ch}^2 x - 1.$
2. $\operatorname{ch} 3x = 4 \operatorname{ch}^3 x - 3 \operatorname{ch} x.$
3. $\operatorname{ch} 4x = 8 \operatorname{ch}^4 x - 8 \operatorname{ch}^2 x + 1.$
4. $\operatorname{ch} 5x = 16 \operatorname{ch}^5 x - 20 \operatorname{ch}^3 x + 5 \operatorname{ch} x.$
5. $\operatorname{ch} 6x = 32 \operatorname{ch}^6 x - 48 \operatorname{ch}^4 x + 18 \operatorname{ch}^2 x - 1.$
6. $\operatorname{ch} 7x = 64 \operatorname{ch}^7 x - 112 \operatorname{ch}^5 x + 56 \operatorname{ch}^3 x - 7 \operatorname{ch} x.$

Einige Summen trigonometrischer und hyperbolischer Funktionen
Finite series of trigonometric and hyperbolic functions

1.34

1. $\displaystyle\sum_{k=0}^{n-1} \sin(x + ky) = \sin\left(x + \frac{n-1}{2} y\right) \sin \frac{ny}{2} \operatorname{cosec} \frac{y}{2}.$ A(361.8)   **1.341**

1.341

2. $\sum_{k=0}^{n-1} \operatorname{sh}(x+ky) = \operatorname{sh}\left(x+\frac{n-1}{2}y\right)\operatorname{sh}\frac{ny}{2}\frac{1}{\operatorname{sh}\frac{y}{2}}.$

3. $\sum_{k=0}^{n-1} \cos(x+ky) = \cos\left(x+\frac{n-1}{2}y\right)\sin\frac{ny}{2}\operatorname{cosec}\frac{y}{2}.$ A(361.9)

4. $\sum_{k=0}^{n-1} \operatorname{ch}(x+ky) = \operatorname{ch}\left(x+\frac{n-1}{2}y\right)\operatorname{sh}\frac{ny}{2}\frac{1}{\operatorname{sh}\frac{y}{2}}.$

5. $\sum_{k=0}^{2n-1} (-1)^k \cos(x+ky) = \sin\left(x+\frac{2n-1}{2}y\right)\sin ny \sec\frac{y}{2}.$ Jo(202)

6. $\sum_{k=0}^{n-1} (-1)^k \sin(x+ky) = \sin\left\{x+\frac{n-1}{2}(y+\pi)\right\}\sin\frac{n(y+\pi)}{2}\sec\frac{y}{2}.$

Jo(202a)

Spezialfälle
Special cases

1.342

1. $\sum_{k=1}^{n} \sin kx = \sin\frac{n+1}{2}x \sin\frac{nx}{2}\operatorname{cosec}\frac{x}{2}.$ A(361.1)

2. $\sum_{k=0}^{n} \cos kx = \cos\frac{n+1}{2}x \sin\frac{nx}{2}\operatorname{cosec}\frac{x}{2} + 1 = \cos\frac{nx}{2}\sin\frac{n+1}{2}x\operatorname{cosec}\frac{x}{2}.$

A(361.2)

3. $\sum_{k=1}^{n} \sin(2k-1)x = \sin^2 nx \operatorname{cosec} x.$ A(361.7)

4. $\sum_{k=1}^{n} \cos(2k-1)x = \frac{1}{2}\sin 2nx \operatorname{cosec} x.$ Jo(207)

1.343

1. $\sum_{k=1}^{n} (-1)^k \cos kx = -\frac{1}{2} + \frac{(-1)^n \cos\left(\frac{2n+1}{2}x\right)}{2\cos\frac{x}{2}}.$ A(361.11)

2. $\sum_{k=1}^{n} (-1)^{k+1} \sin(2k-1)x = (-1)^{n+1}\frac{\sin 2nx}{2\cos x}.$ A(361.10)

3. $\sum_{k=1}^{n} \cos(4k-3)x + \sum_{k=1}^{n} \sin(4k-1)x =$

$= \sin 2nx (\cos 2nx + \sin 2nx)(\cos x + \sin x) \operatorname{cosec} 2x.$ Jo(208)

1.344

1. $\sum_{k=1}^{n-1} \sin\frac{\pi k}{n} = \operatorname{ctg}\frac{\pi}{2n}.$ A(361.19)

2. $\sum_{k=1}^{n-1} \sin \frac{2\pi k^2}{n} = \frac{\sqrt{n}}{2}\left(1 + \cos \frac{n\pi}{2} - \sin \frac{n\pi}{2}\right).$ A(361.18) 1.344

3. $\sum_{k=0}^{n-1} \cos \frac{2\pi k^2}{n} = \frac{\sqrt{n}}{2}\left(1 + \cos \frac{n\pi}{2} + \sin \frac{n\pi}{2}\right).$ A(361.17)

Summen von Potenzen trigonometrischer Funktionen von Vielfachen des Bogens
Finite series of powers of trigonometric functions of multiple angles 1.35

1. $\sum_{k=1}^{n} \sin^2 kx = \frac{1}{4}[(2n+1)\sin x - \sin(2n+1)x]\,\operatorname{cosec} x;$ 1.351

$$= \frac{n}{2} - \frac{\cos(n+1)x \sin nx}{2\sin x}.$$ A(361.3)

2. $\sum_{k=1}^{n} \cos^2 kx = \frac{n-1}{2} + \frac{1}{2}\cos nx \sin(n+1)x \operatorname{cosec} x;$

$$= \frac{n}{2} + \frac{\cos(n+1)x \sin nx}{2\sin x}.$$ A(361.4)+

3. $\sum_{k=1}^{n} \sin^3 kx = \frac{3}{4} \sin \frac{n+1}{2} x \sin \frac{nx}{2} \operatorname{cosec} \frac{x}{2} -$

$$- \frac{1}{4} \sin \frac{3(n+1)x}{2} \sin \frac{3nx}{2} \operatorname{cosec} \frac{3x}{2}.$$ Jo(210)

4. $\sum_{k=1}^{n} \cos^3 kx = \frac{3}{4} \cos \frac{n+1}{2} x \sin \frac{nx}{2} \operatorname{cosec} \frac{x}{2} +$

$$+ \frac{1}{4} \cos \frac{3(n+1)}{2} x \sin \frac{3nx}{2} \operatorname{cosec} \frac{3x}{2}.$$ Jo(211)+

5. $\sum_{k=1}^{n} \sin^4 kx = \frac{1}{8}[3n - 4\cos(n+1)x \sin nx \operatorname{cosec} x +$

$$+ \cos 2(n+1)x \sin 2nx \operatorname{cosec} 2x].$$ Jo(212)

6. $\sum_{k=1}^{n} \cos^4 kx = \frac{1}{8}[3n + 4\cos(n+1)x \sin nx \operatorname{cosec} x +$

$$+ \cos 2(n+1)x \sin 2nx \operatorname{cosec} 2x].$$ Jo(213)

1. $\sum_{k=1}^{n-1} k \sin kx = \frac{\sin nx}{4\sin^2 \frac{x}{2}} - \frac{n \cos \frac{2n-1}{2} x}{2 \sin \frac{x}{2}}.$ A(361.5) 1.352

1.352 2. $$\sum_{k=1}^{n-1} k \cos kx = \frac{n \sin \frac{2n-1}{2} x}{2 \sin \frac{x}{2}} - \frac{1-\cos nx}{4 \sin^2 \frac{x}{2}}.$$ A(361.6)

1.353 1. $$\sum_{k=1}^{n-1} p^k \sin kx = \frac{p \sin x - p^n \sin nx + p^{n+1} \sin (n-1) x}{1 - 2p \cos x + p^2}.$$ A(361.12)+

2. $$\sum_{k=1}^{n-1} p^k \operatorname{sh} kx = \frac{p \operatorname{sh} x - p^n \operatorname{sh} nx + p^{n+1} \operatorname{sh}(n-1)x}{1 - 2p \operatorname{ch} x + p^2}.$$

3. $$\sum_{k=0}^{n-1} p^k \cos kx = \frac{1 - p \cos x - p^n \cos nx + p^{n+1} \cos (n-1) x}{1 - 2p \cos x + p^2}.$$ A(361.13)+

4. $$\sum_{k=0}^{n-1} p^k \operatorname{ch} kx = \frac{1 - p \operatorname{ch} x - p^n \operatorname{ch} nx + p^{n+1} \operatorname{ch}(n-1) x}{1 - 2p \operatorname{ch} x + p^2}.$$ Jo(396)

1.36 Summen von Produkten trigonometrischer Funktionen von Vielfachen der Bögen
Finite series of products of trigonometric functions of multiple angles

1.361 1. $$\sum_{k=1}^{n} \sin kx \sin (k+1) x = \frac{1}{4}[(n+1) \sin 2x - \sin 2(n+1) x] \operatorname{cosec} x.$$ Jo(214)

2. $$\sum_{k=1}^{n} \sin kx \sin (k+2) x = \frac{n}{2} \cos 2x - \frac{1}{2} \cos (n+3) x \sin nx \operatorname{cosec} x.$$ Jo(216)

3. $$\sum_{k=1}^{n} \sin kx \cos (2k-1) y = \frac{1}{2}\left[\sin\left(ny + \frac{n+1}{2} x\right) \sin\frac{n(x+2y)}{2} \operatorname{cosec} \frac{x+2y}{2} \right.$$
$$\left. - \sin\left(ny - \frac{n+1}{2} x\right) \sin\frac{n(2y-x)}{2} \operatorname{cosec} \frac{2y-x}{2} \right].$$ Jo(217)

1.362 1. $$\sum_{k=1}^{n} \left(2^k \sin^2 \frac{x}{2^k} \right)^2 = \left(2^n \sin \frac{x}{2^n} \right)^2 - \sin^2 x.$$ A(361.15)

2. $$\sum_{k=1}^{n} \left(\frac{1}{2^k} \sec \frac{x}{2^k} \right)^2 = \operatorname{cosec}^2 x - \left(\frac{1}{2^n} \operatorname{cosec} \frac{x}{2^n} \right)^2.$$ A(361.14)

1.37 Summen von Tangensfunktionen von Vielfachen des Bogens
Finite series of tangents of multiple angles

1.371 1. $$\sum_{k=0}^{n} \frac{1}{2^k} \operatorname{tg} \frac{x}{2^k} = \frac{1}{2^n} \operatorname{ctg} \frac{x}{2^n} - 2 \operatorname{ctg} 2x.$$ A(361.16)

2. $$\sum_{k=0}^{n} \frac{1}{2^{2k}} \operatorname{tg}^2 \frac{x}{2^k} = \frac{2^{2n+2}-1}{3 \cdot 2^{2n-1}} + 4 \operatorname{ctg}^2 2x - \frac{1}{2^{2n}} \operatorname{ctg}^2 \frac{x}{2^n}.$$ A(361.20)

Summen, die auf hyperbolische Tangens- und Cotangensfunktionen führen
Finite series leading to hyperbolic tangents and cotangents

1.38

1. $$\sum_{k=0}^{n-1} \frac{\operatorname{th} x \dfrac{1}{n \sin^2 \dfrac{2k+1}{4n}\pi}}{1 + \dfrac{\operatorname{th}^2 x}{\operatorname{tg}^2 \dfrac{2k+1}{4n}\pi}} = \operatorname{th} 2nx.$$ Jo(402)+ 1.381

2. $$\sum_{k=1}^{n-1} \frac{\operatorname{th} x \dfrac{1}{n \sin^2 \dfrac{k\pi}{2n}}}{1 + \dfrac{\operatorname{th}^2 x}{\operatorname{tg}^2 \dfrac{k\pi}{2n}}} = \operatorname{cth} 2nx - \dfrac{1}{2n}(\operatorname{th} x + \operatorname{cth} x).$$ Jo(403)

3. $$\sum_{k=0}^{n-1} \frac{\operatorname{th} x \dfrac{2}{(2n+1) \sin^2 \dfrac{2k+1}{2(2n+1)}\pi}}{1 + \dfrac{\operatorname{th}^2 x}{\operatorname{tg}^2 \dfrac{2k+1}{2(2n+1)}\pi}} = \operatorname{th}(2n+1)x - \dfrac{\operatorname{th} x}{2n+1}.$$ Jo(404)

4. $$\sum_{k=1}^{n} \frac{\operatorname{th} x \dfrac{2}{(2n+1) \sin^2 \dfrac{k\pi}{(2n+1)}}}{1 + \dfrac{\operatorname{th}^2 x}{\operatorname{tg}^2 \dfrac{k\pi}{(2n+1)}}} = \operatorname{cth}(2n+1)x - \dfrac{\operatorname{cth} x}{2n+1}.$$ Jo(405)

1. $$\sum_{k=0}^{n-1} \frac{1}{\dfrac{\sin^2 \dfrac{2k+1}{4n}\pi}{\operatorname{sh} x} + \dfrac{1}{2}\operatorname{th}\dfrac{x}{2}} = 2n \operatorname{th} nx.$$ Jo(406) 1.382

2. $$\sum_{k=1}^{n-1} \frac{1}{\dfrac{\sin^2 \dfrac{k\pi}{2n}}{\operatorname{sh} x} + \dfrac{1}{2}\operatorname{th}\dfrac{x}{2}} = 2n \operatorname{cth} nx - 2\operatorname{cth} x.$$ Jo(407)

3. $$\sum_{k=0}^{n-1} \frac{1}{\dfrac{\sin^2 \dfrac{2k+1}{2(2n+1)}\pi}{\operatorname{sh} x} + \dfrac{1}{2}\operatorname{th}\dfrac{x}{2}} = (2n+1)\operatorname{th}\dfrac{(2n+1)x}{2} - \operatorname{th}\dfrac{x}{2}.$$ Jo(408)

4. $$\sum_{k=1}^{n} \frac{1}{\dfrac{\sin^2 \dfrac{k\pi}{2n+1}}{\operatorname{sh} x} + \dfrac{1}{2}\operatorname{th}\dfrac{x}{2}} = (2n+1)\operatorname{cth}\dfrac{(2n+1)x}{2} - \operatorname{cth}\dfrac{x}{2}.$$ Jo(409)

Elementare Funktionen/Elementary Functions

1.39 **Darstellung von Cosinus und Sinus von Vielfachen des Bogens als endliche Produkte**
Cosine and sine of multiples of the angle expressed in terms of finite products

1.391

1. $\sin nx = n \sin x \cos x \prod_{k=1}^{\frac{n-2}{2}} \left(1 - \dfrac{\sin^2 x}{\sin^2 \dfrac{k\pi}{n}}\right) \quad \left[n \; \begin{matrix}\text{geradzahlig}\\ \text{even}\end{matrix}\right].$ Jo(568)

2. $\cos nx = \prod_{k=1}^{\frac{n}{2}} \left(1 - \dfrac{\sin^2 x}{\sin^2 \dfrac{(2k-1)\pi}{2n}}\right) \quad \left[n \; \begin{matrix}\text{geradzahlig}\\ \text{even}\end{matrix}\right].$ Jo(569)

3. $\sin nx = n \sin x \prod_{k=1}^{\frac{n-1}{2}} \left(1 - \dfrac{\sin^2 x}{\sin^2 \dfrac{k\pi}{n}}\right) \quad \left[n \; \begin{matrix}\text{ungeradzahlig}\\ \text{odd}\end{matrix}\right].$ Jo(570)

4. $\cos nx = \cos x \prod_{k=1}^{\frac{n-1}{2}} \left(1 - \dfrac{\sin^2 x}{\sin^2 \dfrac{(2k-1)\pi}{2n}}\right) \quad \left[n \; \begin{matrix}\text{ungeradzahlig}\\ \text{odd}\end{matrix}\right].$ Jo(571)+

1.392

1. $\sin nx = 2^{n-1} \prod_{k=0}^{n-1} \sin\left(x + \dfrac{k\pi}{n}\right).$ Jo(548)

2. $\cos nx = 2^{n-1} \prod_{k=1}^{n} \sin\left(x + \dfrac{2k-1}{2n}\pi\right).$ Jo(549)

1.393

1. $\prod_{k=0}^{n-1} \cos\left(x + \dfrac{2k}{n}\pi\right) = \dfrac{1}{2^{n-1}} \cos nx \quad \left[n \; \begin{matrix}\text{ungeradzahlig}\\ \text{odd}\end{matrix}\right];$

   $= \dfrac{1}{2^{n-1}} [(-1)^{\frac{n}{2}} - \cos nx] \quad \left[n \; \begin{matrix}\text{geradzahlig}\\ \text{even}\end{matrix}\right].$ Jo(543)

2. $\prod_{k=0}^{n-1} \sin\left(x + \dfrac{2k}{n}\pi\right) = \dfrac{(-1)^{\frac{n-1}{2}}}{2^{n-1}} \sin nx \quad \left[n \; \begin{matrix}\text{ungeradzahlig}\\ \text{odd}\end{matrix}\right];$

   $= \dfrac{(-1)^{\frac{n}{2}}}{2^{n-1}} (1 - \cos nx) \quad \left[n \; \begin{matrix}\text{geradzahlig}\\ \text{even}\end{matrix}\right].$ Jo(544)

1.394 $\prod_{k=0}^{n-1}\left\{x^2 - 2xy \cos\left(\alpha + \dfrac{2k\pi}{n}\right) + y^2\right\} = x^{2n} - 2x^n y^n \cos n\alpha + y^{2n}.$ Jo(573)

1.395

1. $\cos nx - \cos ny = 2^{n-1} \prod_{k=0}^{n-1} \left\{\cos x - \cos\left(y + \dfrac{2k\pi}{n}\right)\right\}.$ Jo(572)

2. $\operatorname{ch} nx - \cos ny = 2^{n-1} \prod_{k=0}^{n-1} \left\{\operatorname{ch} x - \cos\left(y + \dfrac{2k\pi}{n}\right)\right\}.$ Jo(538)

Trigonometrische und Hyperbelfunktionen/Trigonometric and Hyperbolic Functions

1. $\displaystyle\prod_{k=1}^{n-1}\left(x^2 - 2x\cos\frac{k\pi}{n} + 1\right) = \frac{x^{2n}-1}{x^2-1}.$ Kr 58(28.1)   1.396

2. $\displaystyle\prod_{k=1}^{n}\left(x^2 - 2x\cos\frac{2k\pi}{2n+1} + 1\right) = \frac{x^{2n+1}-1}{x-1}.$ Kr 58(28.2)

3. $\displaystyle\prod_{k=1}^{n}\left(x^2 + 2x\cos\frac{2k\pi}{2n+1} + 1\right) = \frac{x^{2n+1}-1}{x+1}.$ Kr 58(28.3)

4. $\displaystyle\prod_{k=0}^{n-1}\left(x^2 - 2x\cos\frac{(2k+1)\pi}{2n} + 1\right) = x^{2n}+1.$ Kr 58(28.4)

Potenzreihenentwicklung der trigonometrischen und hyperbolischen Funktionen

The expansion of trigonometric and hyperbolic functions in power series   1.41

1. $\displaystyle\sin x = \sum_{k=0}^{\infty}(-1)^k\frac{x^{2k+1}}{(2k+1)!}.$ 2. $\displaystyle\operatorname{sh} x = \sum_{k=0}^{\infty}\frac{x^{2k+1}}{(2k+1)!}.$ 1.411

3. $\displaystyle\cos x = \sum_{k=0}^{\infty}(-1)^k\frac{x^{2k}}{(2k)!}.$ 4. $\displaystyle\operatorname{ch} x = \sum_{k=0}^{\infty}\frac{x^{2k}}{(2k)!}.$

5. $\displaystyle\operatorname{tg} x = \sum_{k=1}^{\infty}\frac{2^{2k}(2^{2k}-1)}{(2k)!}|B_{2k}|x^{2k-1}\quad\left[x^2<\frac{\pi^2}{4}\right].$ F II 515

6. $\displaystyle\operatorname{th} x = x - \frac{x^3}{3}+\frac{2x^5}{15}-\frac{17}{315}x^7+\ldots = \sum_{k=1}^{\infty}\frac{2^{2k}(2^{2k}-1)}{(2k)!}B_{2k}x^{2k-1}\quad\left[x^2<\frac{\pi^2}{4}\right].$

7. $\displaystyle\operatorname{ctg} x = \frac{1}{x}-\sum_{k=1}^{\infty}\frac{2^{2k}|B_{2k}|}{(2k)!}x^{2k-1}\quad[x^2<\pi^2].$ F II 515+

8. $\displaystyle\operatorname{cth} x = \frac{1}{x}+\frac{x}{3}-\frac{x^3}{45}+\frac{2x^5}{945}-\ldots = \frac{1}{x}+\sum_{k=1}^{\infty}\frac{2^{2k}B_{2k}}{(2k)!}x^{2k-1}\quad[x^2<\pi^2].$ F II 514+

9. $\displaystyle\sec x = \sum_{k=0}^{\infty}\frac{|E_{2k}|}{(2k)!}x^{2k}\quad\left[x^2<\frac{\pi^2}{4}\right].$ C 297+

10. $\displaystyle\operatorname{sech} x = 1 - \frac{x^2}{2}+\frac{5x^4}{24}-\frac{61x^6}{720}+\ldots = 1+\sum_{k=1}^{\infty}\frac{E_{2k}}{(2k)!}x^{2k}\quad\left[x^2<\frac{\pi^2}{4}\right].$ C 297

11. $\displaystyle\operatorname{cosec} x = \frac{1}{x}+\sum_{k=1}^{\infty}\frac{2(2^{2k-1}-1)|B_{2k}|x^{2k-1}}{(2k)!}\quad[x^2<\pi^2].$ C 297+

12. $\displaystyle\operatorname{cosech} x = \frac{1}{x}-\frac{1}{6}x+\frac{7x^3}{360}-\frac{31x^5}{15120}+\ldots = \frac{1}{x}-\sum_{k=1}^{\infty}\frac{2(2^{2k-1}-1)B_{2k}}{(2k)!}x^{2k-1}$
$[x^2<\pi^2].$ Jo(418)

1. $\displaystyle\sin^2 x = \sum_{k=1}^{\infty}(-1)^{k+1}\frac{2^{2k-1}x^{2k}}{(2k)!}.$ Jo(452)+   1.412

2. $\displaystyle\cos^2 x = 1 - \sum_{k=1}^{\infty}(-1)^{k+1}\frac{2^{2k-1}x^{2k}}{(2k)!}.$ Jo(443)

1.412 3.
$$\sin^3 x = \frac{1}{4}\sum_{k=1}^{\infty}(-1)^{k+1}\frac{3^{2k+1}-3}{(2k+1)!}x^{2k+1}.$$
Jo(452a)+

4.
$$\cos^3 x = \frac{1}{4}\sum_{k=0}^{\infty}(-1)^{k}\frac{(3^{2k}+3)x^{2k}}{(2k)!}.$$
Jo(443a)

1.413 1.
$$\operatorname{sh} x = \operatorname{cosec} x \sum_{k=1}^{\infty}(-1)^{k+1}\frac{2^{2k-1}x^{4k-2}}{(4k-2)!}.$$
Jo(508)

2.
$$\operatorname{ch} x = \sec x + \sec x \sum_{k=1}^{\infty}(-1)^{k}\frac{2^{2k}x^{4k}}{(4k)!}.$$
Jo(507)

3.
$$\operatorname{sh} x = \sec x \sum_{k=1}^{\infty}(-1)^{E\left(\frac{k}{2}\right)}\frac{2^{k-1}x^{2k-1}}{(2k-1)!}.$$
Jo (510)

4.
$$\operatorname{ch} x = \operatorname{cosec} x \sum_{k=1}^{\infty}(-1)^{E\left(\frac{k-1}{2}\right)}\frac{2^{k-1}x^{2k-1}}{(2k-1)!}.$$
Jo(509)

1.414 1.
$$\cos[n \ln(x+\sqrt{1+x^2})] =$$
$$= 1 - \sum_{k=0}^{\infty}(-1)^{k}\frac{(n^2+0^2)(n^2+2^2)\cdots[n^2+(2k)^2]}{(2k+2)!}x^{2k+2} \quad [x^2 < 1].$$
A(6456.1)

2.
$$\sin[n \ln(x+\sqrt{1+x^2})] =$$
$$= nx - n\sum_{k=1}^{\infty}(-1)^{k+1}\frac{(n^2+1^2)(n^2+3^2)\cdots[n^2+(2k-1)^2]x^{2k+1}}{(2k+1)!} \quad [x^2 < 1].$$
A(6456.2)

Potenzreihen für ln sin x, ln cos x und ln tg x siehe 1.518. | For the expansions of ln sin x, ln cos x, and ln tg x into power series see 1.518.

Partialbruchzerlegung
Expansion in series of simple fractions

1.42

1.421 1.
$$\operatorname{t}\frac{\pi x}{2} = \frac{4x}{\pi}\sum_{k=1}^{\infty}\frac{1}{(2k-1)^2-x^2}.$$
Br$_{03}$ (191), A(6495.1)

2.
$$\operatorname{th}\frac{\pi x}{2} = \frac{4x}{\pi}\sum_{k=1}^{\infty}\frac{1}{(2k-1)^2+x^2}.$$

3.
$$\operatorname{ctg}\pi x = \frac{1}{\pi x} + \frac{2x}{\pi}\sum_{k=1}^{\infty}\frac{1}{x^2-k^2} = \frac{1}{\pi x} + \frac{x}{\pi}\sum_{\substack{k=-\infty \\ k\neq 0}}^{\infty}\frac{1}{k(x-k)}.$$
A(6495.2), Jo(450a)

4.
$$\operatorname{cth}\pi x = \frac{1}{\pi x} + \frac{2x}{\pi}\sum_{k=1}^{\infty}\frac{1}{x^2+k^2} \quad (\text{cf. } 1.2171.).$$

5.
$$\operatorname{tg}^2\frac{\pi x}{2} = x^2\sum_{k=1}^{\infty}\frac{2(2k-1)^2-x^2}{(1^2-x^2)^2(3^2-x^2)^2\cdots[(2k-1)^2-x^2]^2}.$$
Jo(450)

Trigonometrische und Hyperbelfunktionen/Trigonometric and Hyperbolic Functions

1. $$\sec\frac{\pi x}{2} = \frac{4}{\pi}\sum_{k=1}^{\infty}(-1)^{k+1}\frac{2k-1}{(2k-1)^2 - x^2}.$$ A(6495.3)+ 1.422

2. $$\sec^2\frac{\pi x}{2} = \frac{4}{\pi^2}\sum_{k=1}^{\infty}\left\{\frac{1}{(2k-1-x)^2} + \frac{1}{(2k-1+x)^2}\right\}.$$ Jo(451)+

3. $$\operatorname{cosec}\pi x = \frac{1}{\pi x} + \frac{2x}{\pi}\sum_{k=1}^{\infty}\frac{(-1)^k}{x^2 - k^2} \quad (\text{cf. 1.217 2.}).$$ A(6495.4)+

4. $$\operatorname{cosec}^2 \pi x = \frac{1}{\pi^2}\sum_{k=-\infty}^{\infty}\frac{1}{(x-k)^2} = \frac{1}{\pi^2 x^2} + \frac{2}{\pi^2}\sum_{k=1}^{\infty}\frac{x^2 + k^2}{(x^2 - k^2)^2}.$$ Jo(446)

5. $$\frac{1 + x\operatorname{cosec} x}{2x^2} = \frac{1}{x^2} - \sum_{k=1}^{\infty}\frac{(-1)^{k+1}}{(x^2 - k^2\pi^2)}.$$ Jo(449)

6. $$\operatorname{cosec}\pi x = \frac{1}{\pi x} + \frac{1}{\pi}\sum_{k=-\infty}^{\infty}(-1)^k\left(\frac{1}{x-k} + \frac{1}{k}\right).$$ Jo(450b)

$$\frac{\pi^2}{4m^2}\operatorname{cosec}^2\frac{\pi}{m} + \frac{\pi}{4m}\operatorname{ctg}\frac{\pi}{m} - \frac{1}{2} = \sum_{k=1}^{\infty}\frac{1}{(1-k^2 m^2)^2}.$$ Jo(477) 1.423

Darstellung durch ein unendliches Produkt
Representation in the form of an infinite product   1.43

1. $$\sin x = x\prod_{k=1}^{\infty}\left(1 - \frac{x^2}{k^2\pi^2}\right).$$ F II 389 1.431

2. $$\operatorname{sh} x = x\prod_{k=1}^{\infty}\left(1 + \frac{x^2}{k^2\pi^2}\right).$$ F II 390

3. $$\cos x = \prod_{k=0}^{\infty}\left(1 - \frac{4x^2}{(2k+1)^2\pi^2}\right).$$ F II 390

4. $$\operatorname{ch} x = \prod_{k=0}^{\infty}\left(1 + \frac{4x^2}{(2k+1)^2\pi^2}\right).$$ F II 390

1. $$\cos x - \cos y = 2\left(1 - \frac{x^2}{y^2}\right)\sin^2\frac{y}{2}\prod_{k=1}^{\infty}\left(1 - \frac{x^2}{(2k\pi + y)^2}\right)\left(1 - \frac{x^2}{(2k\pi - y)^2}\right).$$ A(653.2) 1.432

2. $$\operatorname{ch} x - \cos y = 2\left(1 + \frac{x^2}{y^2}\right)\sin^2\frac{y}{2}\prod_{k=1}^{\infty}\left(1 + \frac{x^2}{(2k\pi + y)^2}\right)\left(1 + \frac{x^2}{(2k\pi - y)^2}\right).$$ A(653.1)

$$\cos\frac{\pi x}{4} - \sin\frac{\pi x}{4} = \prod_{k=1}^{\infty}\left[1 + \frac{(-1)^k x}{2k-1}\right].$$ Br$_{08}$ 189 1.433

$$\cos^2 x = \frac{1}{4}(\pi + 2x)^2\prod_{k=1}^{\infty}\left[1 - \left(\frac{\pi + 2x}{2k\pi}\right)^2\right]^2.$$ MO 216 1.434

$$\frac{\sin\pi(x+a)}{\sin\pi a} = \frac{x+a}{a}\prod_{k=1}^{\infty}\left(1 - \frac{x}{k-a}\right)\left(1 + \frac{x}{k+a}\right).$$ MO 216 1.435

1.436
$$1 - \frac{\sin^2 \pi x}{\sin^2 \pi a} = \prod_{k=-\infty}^{\infty} \left[1 - \left(\frac{x}{k-a}\right)^2\right].$$
MO 216

1.437
$$\frac{\sin 3x}{\sin x} = -\prod_{k=-\infty}^{\infty} \left[1 - \left(\frac{2x}{x+k\pi}\right)^2\right].$$
MO 216

1.438
$$\frac{\operatorname{ch} x - \cos a}{1 - \cos a} = \prod_{k=-\infty}^{\infty} \left[1 + \left(\frac{x}{2k\pi + a}\right)^2\right].$$
MO 216

1.439
1.
$$\sin x = x \prod_{k=1}^{\infty} \cos \frac{x}{2^k} \quad [|x| < 1].$$
A(651), MO 216

2.
$$\frac{\sin x}{x} = \prod_{k=1}^{\infty} \left[1 - \frac{4}{3} \sin^2\left(\frac{x}{3^k}\right)\right].$$
MO 216

1.44–1.45

Trigonometrische Reihen
FOURIER series

1.441
1.
$$\sum_{k=1}^{\infty} \frac{\sin kx}{k} = \frac{\pi - x}{2} \quad [0 < x < 2\pi].$$
F III 439

2.
$$\sum_{k=1}^{\infty} \frac{\cos kx}{k} = \frac{1}{2} \ln \frac{1}{2(1 - \cos x)} \quad [0 < x < 2\pi].$$
F III 448+, A(6814)

3.
$$\sum_{k=1}^{\infty} \frac{(-1)^{k-1} \sin kx}{k} = \frac{x}{2} \quad [-\pi < x < \pi].$$
F III 442

4.
$$\sum_{k=1}^{\infty} (-1)^{k-1} \frac{\cos kx}{k} = \ln\left(2 \cos \frac{x}{2}\right) \quad [-\pi < x < \pi].$$
F III 448

1.442
1.
$$\sum_{k=1}^{\infty} \frac{\sin(2k-1)x}{2k-1} = \frac{\pi}{4} \quad [0 < x < \pi].$$
F III 440

2.
$$\sum_{k=1}^{\infty} \frac{\cos(2k-1)x}{2k-1} = \frac{1}{2} \ln \operatorname{ctg} \frac{x}{2} \quad [0 < x < \pi].$$
Br$_{08}$ 168, Jo(266), GK II 31(195)

3.
$$\sum_{k=1}^{\infty} (-1)^{k-1} \frac{\sin(2k-1)x}{2k-1} = \frac{1}{2} \ln \operatorname{tg}\left(\frac{\pi}{4} + \frac{x}{2}\right) \quad \left[-\frac{\pi}{2} < x < \frac{\pi}{2}\right].$$
Br$_{08}$ 168, Jo(268)+

4.
$$\sum_{k=1}^{\infty} (-1)^{k-1} \frac{\cos(2k-1)x}{2k-1} = \frac{\pi}{4} \quad \left[0 < x < \frac{\pi}{2}\right]$$
$$= -\frac{\pi}{4} \quad \left[\frac{\pi}{2} < x < \pi\right].$$
Br$_{08}$ 168, Jo(269)

1.443
1.
$$\sum_{k=1}^{\infty} \frac{\cos k\pi x}{k^{2n}} = (-1)^{n-1} 2^{2n-1} \frac{\pi^{2n}}{(2n)!} \sum_{k=0}^{2n} \binom{2n}{k} B_{2n-k} \rho^k;$$
$$= (-1)^{n-1} \frac{1}{2} \frac{(2\pi)^{2n}}{(2n)!} B_{2n}\left(\frac{x}{2}\right)$$
$$\left[0 < x < 1, \quad \rho = \frac{x}{2} - E\left(\frac{x}{2}\right)\right].$$
C 307, Ge 237

2. $\sum_{k=1}^{\infty} \frac{\sin k\pi x}{k^{2n+1}} = (-1)^{n-1} 2^{2n} \frac{\pi^{2n+1}}{(2n+1)!} \sum_{k=0}^{2n+1} \binom{2n+1}{k} B_{2n-k+1} \rho^k;$ 1.443

$$= (-1)^{n-1} \frac{1}{2} \frac{(2\pi)^{2n+1}}{(2n+1)!} B_{2n+1}\left(\frac{x}{2}\right)$$

$$\left[0 < x < 1; \; \rho = \frac{x}{2} - E\left(\frac{x}{2}\right)\right].$$ C 308

3. $\sum_{k=1}^{\infty} \frac{\cos kx}{k^2} = \frac{\pi^2}{6} - \frac{\pi x}{2} + \frac{x^2}{4}$ $[0 \leqslant x \leqslant 2\pi]$.   F III 446

4. $\sum_{k=1}^{\infty} (-1)^{k-1} \frac{\cos kx}{k^2} = \frac{\pi^2}{12} - \frac{x^2}{4}$ $[-\pi \leqslant x \leqslant \pi]$.   F III 443

5. $\sum_{k=1}^{\infty} \frac{\sin kx}{k^3} = \frac{\pi^2 x}{6} - \frac{\pi x^2}{4} + \frac{x^3}{12}$ A(6816)

6. $\sum_{k=1}^{\infty} \frac{\cos kx}{k^4} = \frac{\pi^4}{90} - \frac{\pi^2 x^2}{12} + \frac{\pi x^3}{12} - \frac{x^4}{48}$ $[0 \leqslant x \leqslant 2\pi]$.   A(6817)

7. $\sum_{k=1}^{\infty} \frac{\sin kx}{k^5} = \frac{\pi^4 x}{90} - \frac{\pi^2 x^3}{36} + \frac{\pi x^4}{48} - \frac{x^5}{240}$ A(6818)

1. $\sum_{k=1}^{\infty} \frac{\sin 2(k+1)x}{k(k+1)} = \sin 2x - (\pi - 2x) \sin^2 x - \sin x \cos x \ln(4 \sin^2 x)$ 1.444

$[0 \leqslant x \leqslant \pi]$.   Br$_{08}$ 168, GK II 30(189)

2. $\sum_{k=1}^{\infty} \frac{\cos 2(k+1)x}{k(k+1)} = \cos 2x - \left(\frac{\pi}{2} - x\right) \sin 2x + \sin^2 x \ln(4 \sin^2 x)$

$[0 \leqslant x \leqslant \pi]$.   Br$_{08}$ 168

3. $\sum_{k=1}^{\infty} (-1)^k \frac{\sin(k+1)x}{k(k+1)} = \sin x - \frac{x}{2}(1 + \cos x) - \sin x \ln\left|2 \cos \frac{x}{2}\right|$.   MO 213

4. $\sum_{k=1}^{\infty} (-1)^k \frac{\cos(k+1)x}{k(k+1)} = \cos x - \frac{x}{2} \sin x - (1 + \cos x) \ln\left|2 \cos \frac{x}{2}\right|$.   MO 213

5. $\sum_{k=0}^{\infty} (-1)^k \frac{\sin(2k+1)x}{(2k+1)^2} = \frac{\pi}{4} x$ $\left[-\frac{\pi}{2} \leqslant x \leqslant \frac{\pi}{2}\right]$;

$= \frac{\pi}{4}(\pi - x)$ $\left[\frac{\pi}{2} \leqslant x \leqslant \frac{3}{2}\pi\right]$.   MO 213

6. $\sum_{k=0}^{\infty} \frac{\cos(2k-1)x}{(2k-1)^2} = \frac{\pi}{4}\left(\frac{\pi}{2} - |x|\right)$ $[-\pi \leqslant x \leqslant \pi]$.   F III 445

7. $\sum_{k=1}^{\infty} \frac{\cos 2kx}{(2k-1)(2k+1)} = \frac{1}{2} - \frac{\pi}{4} \sin x$ $\left[0 \leqslant x \leqslant \frac{\pi}{2}\right]$.   Jo(591)

1.445

1. $$\sum_{k=1}^{\infty} \frac{k \sin kx}{k^2 + \alpha^2} = \frac{\pi}{2} \frac{\operatorname{sh} \alpha(\pi - x)}{\operatorname{sh} \alpha\pi} \quad [0 < x < 2\pi].$$ Br$_{08}$ 257, Jo(411)

2. $$\sum_{k=1}^{\infty} \frac{\cos kx}{k^2 + \alpha^2} = \frac{\pi}{2\alpha} \frac{\operatorname{ch} \alpha (\pi - x)}{\operatorname{sh} \alpha\pi} - \frac{1}{2\alpha^2} \quad [0 \leqslant x \leqslant 2\pi].$$ Br$_{08}$ 257, Jo(410)

3. $$\sum_{k=1}^{\infty} \frac{(-1)^k \cos kx}{k^2 + \alpha^2} = \frac{\pi}{2\alpha} \frac{\operatorname{ch} \alpha x}{\operatorname{sh} \alpha\pi} - \frac{1}{2\alpha^2} \quad [-\pi \leqslant x \leqslant \pi].$$ F III 444

4. $$\sum_{k=1}^{\infty} (-1)^{k-1} \frac{k \sin kx}{k^2 + \alpha^2} = \frac{\pi}{2} \frac{\operatorname{sh} \alpha x}{\operatorname{sh} \alpha\pi} \quad [-\pi < x < \pi].$$ F III 444

5. $$\sum_{k=0}^{\infty} \frac{k \sin kx}{k^2 - \alpha^2} = \pi \frac{\sin \{\alpha[(2m + 1) \pi - x]\}}{2 \sin \alpha\pi}$$

$$\left[2m\pi < x < (2m + 2) \pi, \alpha \text{ nicht ganzzahlig / noninteger} \right].$$ MO 213

6. $$\sum_{k=1}^{\infty} \frac{\cos kx}{k^2 - \alpha^2} = \frac{1}{2\alpha^2} - \frac{\pi}{2} \frac{\cos [\alpha\{(2m + 1) \pi - x\}]}{\alpha \sin \alpha\pi}$$

$$\left[2m\pi \leqslant x \leqslant (2m + 2) \pi, \alpha \text{ nicht ganzzahlig / noninteger} \right].$$ MO 213

7. $$\sum_{k=0}^{\infty} (-1)^k \frac{k \sin kx}{k^2 - \alpha^2} = \pi \frac{\sin [\alpha(2m\pi - x)]}{2 \sin \alpha\pi}$$

$$\left[(2m - 1) \pi < x < (2m + 1) \pi, \alpha \text{ nicht ganzzahlig / noninteger} \right].$$ F III 444+

8. $$\sum_{k=1}^{\infty} (-1)^k \frac{\cos kx}{k^2 - \alpha^2} = \frac{1}{2\alpha^2} - \frac{\pi}{2} \frac{\cos [\alpha (2m\pi - x)]}{\alpha \sin \alpha\pi}$$

$$\left[(2m - 1) \pi \leqslant x \leqslant (2m + 1) \pi, \alpha \text{ nicht ganzzahlig / noninteger} \right].$$ F III 444+

1.446 $$\sum_{k=1}^{\infty} \frac{(-1)^{k+1} \cos (2k + 1) x}{(2k - 1) (2k + 1) (2k + 3)} = \frac{\pi}{8} \cos^2 x - \frac{1}{3} \cos x \quad \left[-\frac{\pi}{2} \leqslant x \leqslant \frac{\pi}{2}\right].$$

Br$_{08}$ 256, GK II 30(191)

1.447

1. $$\sum_{k=1}^{\infty} p^k \sin kx = \frac{p \sin x}{1 - 2p \cos x + p^2}$$ F II 479

2. $$\sum_{k=0}^{\infty} p^k \cos kx = \frac{1 - p \cos x}{1 - 2p \cos x + p^2} \quad [|p| < 1].$$ F II 479+

3. $$1 + 2 \sum_{k=1}^{\infty} p^k \cos kx = \frac{1 - p^2}{1 - 2p \cos x + p^2}$$ F II 479+, MO 213

1.448

1. $\displaystyle\sum_{k=1}^{\infty}\frac{p^k\sin kx}{k}=\operatorname{arctg}\frac{p\sin x}{1-p\cos x}$ F II 479

2. $\displaystyle\sum_{k=1}^{\infty}\frac{p^k\cos kx}{k}=\ln\frac{1}{\sqrt{1-2p\cos x+p^2}}$ F II 479

$[0<x<2\pi,\ p^2\leqslant 1]$.

3. $\displaystyle\sum_{k=1}^{\infty}\frac{p^{2k-1}\sin(2k-1)x}{2k-1}=\frac{1}{2}\operatorname{arctg}\frac{2p\sin x}{1-p^2}$ Jo(594)

4. $\displaystyle\sum_{k=1}^{\infty}\frac{p^{2k-1}\cos(2k-1)x}{2k-1}=\frac{1}{4}\ln\frac{1+2p\cos x+p^2}{1-2p\cos x+p^2}$ Jo(259)

5. $\displaystyle\sum_{k=1}^{\infty}\frac{(-1)^{k-1}p^{2k-1}\sin(2k-1)x}{2k-1}=$

$\displaystyle =\frac{1}{4}\ln\frac{1+2p\sin x+p^2}{1-2p\sin x+p^2}$ Jo(261)

$[0<x<\pi,\ p^2\leqslant 1]$.

6. $\displaystyle\sum_{k=1}^{\infty}\frac{(-1)^{k-1}p^{2k-1}\cos(2k-1)x}{2k-1}=$

$\displaystyle =\frac{1}{2}\operatorname{arctg}\frac{2p\cos x}{1-p^2}$ Jo(597)

1.449

1. $\displaystyle\sum_{k=1}^{\infty}\frac{p^k\sin kx}{k!}=e^{p\cos x}\sin(p\sin x)$ Jo(486)

2. $\displaystyle\sum_{k=0}^{\infty}\frac{p^k\cos kx}{k!}=e^{p\cos x}\cos(p\sin x)$ Jo(485)

$[p^2\leqslant 1]$.

Entwicklung hyperbolischer Funktionen in trigonometrische Reihen
Expansions of hyperbolic functions in Fourier series

1.451

1. $\operatorname{sh} x=\cos x\displaystyle\sum_{k=0}^{\infty}\frac{(1^2+0^2)(1^2+2^2)\ldots[1^2+(2k)^2]}{(2k+1)!}\sin^{2k+1}x.$ Jo(504)

2. $\operatorname{ch} x=\cos x+\cos x\displaystyle\sum_{k=1}^{\infty}\frac{(1^2+1^2)(1^2+3^2)\ldots[1^2+(2k-1)^2]}{(2k)!}\sin^{2k}x.$ Jo(503)

1.452

1. $\operatorname{sh}(x\cos\theta)=\sec(x\sin\theta)\displaystyle\sum_{k=0}^{\infty}\frac{x^{2k+1}\cos(2k+1)\theta}{(2k+1)!}$ Jo(391)

2. $\operatorname{ch}(x\cos\theta)=\sec(x\sin\theta)\displaystyle\sum_{k=0}^{\infty}\frac{x^{2k}\cos 2k\theta}{(2k)!}$ Jo(390)

3. $\operatorname{sh}(x\cos\theta)=\operatorname{cosec}(x\sin\theta)\displaystyle\sum_{k=1}^{\infty}\frac{x^{2k}\sin 2k\theta}{(2k)!}$ Jo(393)

4. $\operatorname{ch}(x\cos\theta)=\operatorname{cosec}(x\sin\theta)\displaystyle\sum_{k=0}^{\infty}\frac{x^{2k+1}\sin(2k+1)\theta}{(2k+1)!}$ Jo(392)

$[x^2<1]$.

Reihen von Produkten von Exponentialfunktionen mit trigonometrischen Funktionen
Series of products of exponential and trigonometric functions

1.46

1.461

1. $$\sum_{k=0}^{\infty} e^{-kt} \sin kx = \frac{1}{2} \frac{\sin x}{\operatorname{ch} t - \cos x} \quad [t > 0].$$ MO 213

2. $$1 + 2 \sum_{k=1}^{\infty} e^{-kt} \cos kx = \frac{\operatorname{sh} t}{\operatorname{ch} t - \cos x} \quad [t > 0].$$ MO 213

1.462

$$\sum_{k=1}^{\infty} \frac{\sin kx \sin ky}{k} e^{-2k|t|} = \frac{1}{4} \ln \left[\frac{\sin^2 \frac{x+y}{2} + \operatorname{sh}^2 t}{\sin^2 \frac{x-y}{2} + \operatorname{sh}^2 t} \right].$$ MO 214

1.463

1. $$e^{x \cos \varphi} \cos (x \sin \varphi) = \sum_{n=0}^{\infty} \frac{x^n \cos n\varphi}{n!} \quad [x^2 < 1].$$ A(6476.1)

2. $$e^{x \cos \varphi} \sin (x \sin \varphi) = \sum_{n=1}^{\infty} \frac{x^n \sin n\varphi}{n!} \quad [x^2 < 1].$$ A(6476.2)

Reihen von Hyperbelfunktionen
Series of hyperbolic functions

1.47

1.471

1. $$\sum_{k=1}^{\infty} \frac{\operatorname{sh} kx}{k!} = e^{\operatorname{ch} x} \operatorname{sh} (\operatorname{sh} x).$$ Jo(395)

2. $$\sum_{k=0}^{\infty} \frac{\operatorname{ch} kx}{k!} = e^{\operatorname{ch} x} \operatorname{ch} (\operatorname{sh} x).$$ Jo(394)

1.472

1. $$\sum_{k=1}^{\infty} p^k \operatorname{sh} kx = \frac{p \operatorname{sh} x}{1 - 2p \operatorname{ch} x + p^2} \quad [p^2 < 1].$$ Jo(396)

2. $$\sum_{k=0}^{\infty} p^k \operatorname{ch} kx = \frac{1 - p \operatorname{ch} x}{1 - 2p \operatorname{ch} x + p^2} \quad [p^2 < 1].$$ Jo(397)+

Der LOBATSCHEWSKI sche „Parallelenwinkel" $\Pi(x)$
LOBACHEVSKIY's "Angle of parallelism" $\Pi(x)$

1.48

1.480 Definition. | Definition.

1. $\Pi(x) = 2 \operatorname{arcctg} e^x = 2 \operatorname{arctg} e^{-x} \quad [x \geqslant 0].$ Lo III 297, Lo I 120

2. $\Pi(x) = \pi - \Pi(-x) \quad [x < 0].$ Lo III 183, Lo I 93

1.481 Funktionalgleichungen. | Functional relations.

1. $$\sin \Pi(x) = \frac{1}{\operatorname{ch} x}.$$ Lo III 297

2. $$\cos \Pi(x) = \operatorname{th} x.$$ Lo III 297

3. $\qquad \operatorname{tg} \Pi(x) = \dfrac{1}{\operatorname{sh} x}.$ \hfill Lo III 297 \quad 1.481

4. $\qquad \operatorname{ctg} \Pi(x) = \operatorname{sh} x.$ \hfill Lo III 297

5. $\qquad \sin \Pi(x+y) = \dfrac{\sin \Pi(x)\,\sin \Pi(y)}{1 + \cos \Pi(x)\cos \Pi(y)}.$ \hfill Lo III 297

6. $\qquad \cos \Pi(x+y) = \dfrac{\cos \Pi(x) + \cos \Pi(y)}{1 + \cos \Pi(x)\cos \Pi(y)}.$ \hfill Lo III 183

Der Zusammenhang mit der (GUDERMANNschen) Hyperbelamplitude.	Relation to the Gudermannian.	1.482

$$\operatorname{gd}(-x) = \Pi(x) - \frac{\pi}{2}.$$

Ein (bestimmtes) Integral des Parallelenwinkels findet sich in 4.561.	For a (definite) integral of $\Pi(x)$ see 4.561.

Die Hyperbelamplitude (der Gudermannsche Winkel) gd x
The Gudermannian gd x 1.49

Definition. | Definition. \hfill 1.490

1. $\qquad \operatorname{gd} x = \displaystyle\int_0^x \dfrac{dt}{\operatorname{ch} t} = 2 \operatorname{arctg} e^x - \dfrac{\pi}{2}.$ \hfill JE 58

2. $\qquad x = \displaystyle\int_0^{\operatorname{gd} x} \dfrac{dt}{\cos t} = \ln \operatorname{tg}\left(\dfrac{\operatorname{gd} x}{2} + \dfrac{\pi}{4}\right).$ \hfill JE 58

Funktionalgleichungen. | Functional relations: \hfill 1.491

1. $\qquad \operatorname{ch} x = \sec(\operatorname{gd} x).$ \hfill A(343.1), JE 58

2. $\qquad \operatorname{sh} x = \operatorname{tg}(\operatorname{gd} x).$ \hfill A(343.2), JE 58

3. $\qquad e^x = \sec(\operatorname{gd} x) + \operatorname{tg}(\operatorname{gd} x) = \operatorname{tg}\left(\dfrac{\pi}{4} + \dfrac{\operatorname{gd} x}{2}\right) = \dfrac{1 + \sin(\operatorname{gd} x)}{\cos(\operatorname{gd} x)}.$
\hfill A(343.3), A(344.5), JE 58+

4. $\qquad \operatorname{th} x = \sin(\operatorname{gd} x).$ \hfill A(344.3), JE 58

5. $\qquad \operatorname{th} \dfrac{x}{2} = \operatorname{tg}\left(\dfrac{1}{2} \operatorname{gd} x\right).$ \hfill A(344.4), JE 58

6. $\qquad \operatorname{arctg}(\operatorname{th} x) = \dfrac{1}{2} \operatorname{gd} 2x.$ \hfill A(344.6)+

Für $\gamma = \operatorname{gd} x$ ist $ix = \operatorname{gd} i\gamma$. Reihenentwicklung.	If $\gamma = \operatorname{gd} x$, then $ix = \operatorname{gd} i\gamma$. Expansion in series.	JE 58 \quad 1.492 \newline 1.493

1. $\qquad \dfrac{\operatorname{gd} x}{2} = \displaystyle\sum_{k=0}^{\infty} \dfrac{(-1)^k}{2k+1} \operatorname{th}^{2k+1} \dfrac{x}{2}.$ \hfill JE 58

2. $\qquad \dfrac{x}{2} = \displaystyle\sum_{k=0}^{\infty} \dfrac{1}{2k+1} \operatorname{tg}^{2k+1}\left(\dfrac{1}{2} \operatorname{gd} x\right).$ \hfill JE 58

1.493 3. $$\operatorname{gd} x = x - \frac{x^3}{6} + \frac{x^5}{24} - \frac{61 x^7}{5040} + \ldots$$ JE 58

4. $$x = \operatorname{gd} x + \frac{(\operatorname{gd} x)^3}{6} + \frac{(\operatorname{gd} x)^5}{24} + \frac{61(\operatorname{gd} x)^7}{5040} + \ldots \quad \left[\operatorname{gd} x < \frac{\pi}{2}\right].$$ JE 58

Der Logarithmus
The Logarithmic Function

1.5

Reihendarstellungen
Series representation

1.51

1.511 $$\ln(1+x) = x - \frac{1}{2}x^2 + \frac{1}{3}x^3 - \frac{1}{4}x^4 + \ldots = \sum_{k=1}^{\infty} (-1)^{k+1} \frac{x^k}{k} \quad [-1 < x \leq 1].$$

1.512 1. $$\ln x = (x-1) - \frac{1}{2}(x-1)^2 + \frac{1}{3}(x-1)^3 - \ldots =$$
$$= \sum_{k=1}^{\infty} (-1)^{k+1} \frac{(x-1)^k}{k} \quad [0 < x \leq 2].$$

2. $$\ln x = 2\left[\frac{x-1}{x+1} + \frac{1}{3}\left(\frac{x-1}{x+1}\right)^3 + \frac{1}{5}\left(\frac{x-1}{x+1}\right)^5 + \ldots\right] =$$
$$= 2\sum_{k=1}^{\infty} \frac{1}{2k-1}\left(\frac{x-1}{x+1}\right)^{2k-1} \quad [0 < x].$$

3. $$\ln x = \frac{x-1}{x} + \frac{1}{2}\left(\frac{x-1}{x}\right)^2 + \frac{1}{3}\left(\frac{x-1}{x}\right)^3 + \ldots = \sum_{k=1}^{\infty} \frac{1}{k}\left(\frac{x-1}{x}\right)^k$$
$$\left[x \geq \frac{1}{2}\right].$$ A(644.6)

1.513 1. $$\ln \frac{1+x}{1-x} = 2\sum_{k=1}^{\infty} \frac{1}{2k-1} x^{2k-1} \quad [x^2 < 1].$$ F II 382

2. $$\ln \frac{x+1}{x-1} = 2\sum_{k=1}^{\infty} \frac{1}{(2k-1)x^{2k-1}} \quad [x^2 > 1].$$ A(644.9)

3. $$\ln \frac{x}{x-1} = \sum_{k=1}^{\infty} \frac{1}{kx^k} \quad [x^2 > 1].$$ Jo(88a)

4. $$\ln \frac{1}{1-x} = \sum_{k=1}^{\infty} \frac{x^k}{k} \quad [x^2 < 1].$$ Jo(88b)

5. $$\frac{1-x}{x} \ln \frac{1}{1-x} = 1 - \sum_{k=1}^{\infty} \frac{x^k}{k(k+1)} \quad [x^2 < 1].$$ Jo(102)

6. $$\frac{1}{1-x} \ln \frac{1}{1-x} = \sum_{k=1}^{\infty} x^k \sum_{n=1}^{k} \frac{1}{n} \quad [x^2 < 1].$$ Jo(88e)

7. $\dfrac{(1-x)^2}{2x^3} \ln \dfrac{1}{1-x} = \dfrac{1}{2x^2} - \dfrac{3}{4x} + \sum_{k=1}^{\infty} \dfrac{x^{k-1}}{k(k+1)(k+2)}$ $\quad [x^2 < 1]$. $\qquad$ A(6445.1) $\quad$ 1.513

$$\ln(1 - 2x\cos\varphi + x^2) = -2 \sum_{k=1}^{\infty} \dfrac{\cos k\varphi}{k} x^k. \qquad \text{MO 98, F II 480} \quad 1.514$$

$\ln(x + \sqrt{1 + x^2}) = \operatorname{Arsh} x$ cf. 1.631, 1.641, 1.642, 1.646.

1. $\ln(1 + \sqrt{1 + x^2}) = \ln 2 + \dfrac{1 \cdot 1}{2 \cdot 2} x^2 - \dfrac{1 \cdot 1 \cdot 3}{2 \cdot 4 \cdot 4} x^4 + \dfrac{1 \cdot 1 \cdot 3 \cdot 5}{2 \cdot 4 \cdot 6 \cdot 6} x^6 - \cdots;$ $\qquad$ 1.515

$$= \ln 2 - \sum_{k=1}^{\infty} (-1)^k \dfrac{(2k-1)!}{2^{2k} (k!)^2} x^{2k} \quad [x^2 \leq 1]. \qquad \text{Jo(91)}$$

2. $\ln(1 + \sqrt{1 + x^2}) = \ln x + \dfrac{1}{x} - \dfrac{1}{2 \cdot 3x^3} + \dfrac{1 \cdot 3}{2 \cdot 4 \cdot 5x^5} - \cdots;$

$$= \ln x + \dfrac{1}{x} + \sum_{k=1}^{\infty} (-1)^k \dfrac{(2k-1)!}{2^{2k-1} \cdot k! \, (k-1)! \, (2k+1) \, x^{2k+1}}$$

$$[x^2 \geq 1]. \qquad \text{A(644.4)}$$

3. $\sqrt{1 + x^2} \ln(x + \sqrt{1 + x^2}) = x - \sum_{k=1}^{\infty} (-1)^k \dfrac{2^{2k-1} (k-1)! \, k!}{(2k+1)!} x^{2k+1}$ $[x^2 < 1]$. Jo(93)

4. $\dfrac{\ln(x + \sqrt{1 + x^2})}{\sqrt{1 + x^2}} = \sum_{k=0}^{\infty} (-1)^k \dfrac{2^{2k} (k!)^2}{(2k+1)!} x^{2k+1}$ $\quad [x^2 < 1]$. $\qquad$ Jo(94)

1. $\dfrac{1}{2} \{\ln(1 \pm x)\}^2 = \sum_{k=1}^{\infty} \dfrac{(\mp 1)^{k+1} x^{k+1}}{k+1} \sum_{n=1}^{k} \dfrac{1}{n}$ $\quad [x^2 < 1]$. $\qquad$ Jo(86), Jo(85) $\quad$ 1.516

2. $\dfrac{1}{6} \{\ln(1 + x)\}^3 = \sum_{k=1}^{\infty} \dfrac{(-1)^{k+1} x^{k+2}}{k+2} \sum_{n=1}^{k} \dfrac{1}{n+1} \sum_{m=1}^{n} \dfrac{1}{m}$ $\quad [x^2 < 1]$. $\qquad$ A(644.14)

3. $-\ln(1 + x) \cdot \ln(1 - x) = \sum_{k=1}^{\infty} \dfrac{x^{2k}}{k} \sum_{n=1}^{2k-1} \dfrac{(-1)^{n+1}}{n}$ $\quad [x^2 < 1]$. $\qquad$ Jo(87)

4. $\dfrac{1}{4x} \left\{ \dfrac{1+x}{\sqrt{x}} \ln \dfrac{1+\sqrt{x}}{1-\sqrt{x}} + 2\ln(1-x) \right\} = \dfrac{1}{2x} + \sum_{k=1}^{\infty} \dfrac{x^{k-1}}{(2k-1) \, 2k \, (2k+1)}$

$$[0 < x < 1]. \qquad \text{A(6445.2)}$$

1. $\dfrac{1}{2x} \left\{ 1 - \ln(1+x) - \dfrac{1-x}{\sqrt{x}} \operatorname{arctg} x \right\} = \sum_{k=1}^{\infty} \dfrac{(-1)^{k+1} x^{k-1}}{(2k-1) \, 2k \, (2k+1)}$ $[0 < x \leq 1]$. $\quad$ 1.517

A(6445.3)

2. $\dfrac{1}{2} \operatorname{arctg} x \ln \dfrac{1+x}{1-x} = \sum_{k=1}^{\infty} \dfrac{x^{4k-2}}{2k-1} \sum_{n=1}^{2k-1} \dfrac{(-1)^{n-1}}{2n-1}$ $\quad [x^2 < 1]$. $\qquad$ Br$_{08}$ 163

3. $\dfrac{1}{2} \operatorname{arctg} x \ln(1 + x^2) = \sum_{k=1}^{\infty} \dfrac{(-1)^{k+1} x^{2k+1}}{2k+1} \sum_{n=1}^{2k} \dfrac{1}{n}$ $\quad [x^2 < 1]$. $\qquad$ A(6455.3)

1.518

1. $$\ln \sin x = \ln x - \frac{x^2}{6} - \frac{x^4}{180} - \frac{x^6}{2835} - \ldots;$$

$$= \ln x + \sum_{k=1}^{\infty} \frac{(-1)^k \, 2^{2k-1} B_{2k} x^{2k}}{k \, (2k)!} \quad [x^2 < \pi^2].$$ A(643.1)+

2. $\ln \cos x = -\dfrac{x^2}{2} - \dfrac{x^4}{12} - \dfrac{x^6}{45} - \dfrac{17 x^8}{2520} - \ldots;$

$$= \sum_{k=1}^{\infty} (-1)^k \frac{2^{2k-1}(2^{2k}-1) B_{2k}}{k \, (2k)!} x^{2k} = -\frac{1}{2} \sum_{k=1}^{\infty} \frac{\sin^{2k} x}{k} \left[x^2 < \frac{\pi^2}{4} \right].$$ F II 515

3. $\ln \operatorname{tg} x = \ln x + \dfrac{x^2}{3} + \dfrac{7}{90} x^4 + \dfrac{62}{2835} x^6 + \dfrac{127}{18\,900} x^8 + \ldots;$

$$= \ln x + \sum_{k=1}^{\infty} (-1)^{k+1} \frac{(2^{2k-1}-1) \, 2^{2k} \, B_{2k} \, x^{2k}}{k \, (2k)!} \left[x^2 < \frac{\pi^2}{4} \right].$$ A(643.3)+

Die Potenzreihen für $\genfrac{}{}{0pt}{}{\cos}{\sin} \{n \ln(x + \sqrt{1+x^2})\}$ findet man in 1.414.

For the power expansions of $\genfrac{}{}{0pt}{}{\cos}{\sin} \{n \ln(x + \sqrt{1+x^2})\}$ see 1.414.

1.52

Reihen von Logarithmen (vgl. 1.431)

Series of logarithms (cf. 1.431)

1.521

1. $$\sum_{k=1}^{\infty} \ln \left(1 - \frac{4x^2}{(2k-1)^2 \pi^2} \right) = \ln \cos x.$$

2. $$\sum_{k=1}^{\infty} \ln \left(1 - \frac{x^2}{k^2 \pi^2} \right) = \ln \sin x - \ln x.$$

1.6

Zyklometrische und Area-Funktionen

The Inverse Trigonometric and Hyperbolic Functions

1.61

Definitionsbereiche

The domain of definition

Die Hauptwerte der zyklometrischen Funktionen sind durch die folgenden Ungleichungen bestimmt:

The principal values of the inverse trigonometric functions are determined by the following inequalities:

$$-\frac{\pi}{2} \leqslant \arcsin x \leqslant \frac{\pi}{2}; \quad 0 \leqslant \arccos x \leqslant \pi \quad [-1 \leqslant x \leqslant 1].$$ F II 545

$$-\frac{\pi}{2} < \operatorname{arctg} x < \frac{\pi}{2}; \quad 0 < \operatorname{arcctg} x < \pi \quad [-\infty < x < +\infty].$$ F II 543

Funktionalgleichungen
Functional equations 1.62–1.63

Verknüpfung zyklometrischer Funktionen mit den gleichnamigen trigonometrischen Funktionen: | Relations between the inverse trigonometric and the corresponding trigonometric functions: 1.621

1. $\arcsin(\sin x) = x - 2n\pi \quad \left[2n\pi - \frac{\pi}{2} \leqslant x \leqslant 2n\pi + \frac{\pi}{2}\right];$

   $= -x + (2n+1)\pi \quad \left[(2n+1)\pi - \frac{\pi}{2} \leqslant x \leqslant (2n+1)\pi + \frac{\pi}{2}\right].$

2. $\arccos(\cos x) = x - 2n\pi \quad [2n\pi \leqslant x \leqslant (2n+1)\pi];$

   $= -x + 2(n+1)\pi \quad [(2n+1)\pi \leqslant x \leqslant 2(n+1)\pi].$

3. $\operatorname{arctg}(\operatorname{tg} x) = x - n\pi \quad \left[n\pi - \frac{\pi}{2} < x < n\pi + \frac{\pi}{2}\right].$

4. $\operatorname{arcctg}(\operatorname{ctg} x) = x - n\pi \quad [n\pi < x < (n+1)\pi].$

Zusammenhang zwischen zyklometrischen Funktionen, Area-Funktionen und dem Logarithmus. | Relations between inverse trigonometric, inverse hyperbolic, and the logarithmic functions. 1.622

1. $\arcsin z = \frac{1}{i}\ln(iz + \sqrt{1-z^2}) = \frac{1}{i}\operatorname{Arsh}(iz).$

2. $\arccos z = \frac{1}{i}\ln(z + \sqrt{z^2-1}) = \frac{1}{i}\operatorname{Arch} z.$

3. $\operatorname{arctg} z = \frac{1}{2i}\ln\frac{1+iz}{1-iz} = \frac{1}{i}\operatorname{Arth}(iz).$

4. $\operatorname{arcctg} z = \frac{1}{2i}\ln\frac{iz-1}{iz+1} = i\operatorname{Arcth}(iz).$

5. $\operatorname{Arsh} z = \ln(z + \sqrt{z^2+1}) = \frac{1}{i}\arcsin(iz).$

6. $\operatorname{Arch} z = \ln(z + \sqrt{z^2-1}) = i\arccos z.$

7. $\operatorname{Arth} z = \frac{1}{2}\ln\frac{1+z}{1-z} = \frac{1}{i}\operatorname{arctg} iz.$

8. $\operatorname{Arcth} z = \frac{1}{2}\ln\frac{z+1}{z-1} = \frac{1}{i}\operatorname{arcctg}(-iz).$

Beziehungen zwischen verschiedenen zyklometrischen Funktionen
Relations between different inverse trigonometric functions

1. $\arcsin x + \arccos x = \frac{\pi}{2}.$ No 43 1.623

2. $\operatorname{arctg} x + \operatorname{arcctg} x = \frac{\pi}{2}.$ No 43

1.624

1. $\arcsin x = \arccos \sqrt{1 - x^2}$ $[0 \leqslant x \leqslant 1]$;

$= -\arccos \sqrt{1 - x^2}$ $[-1 \leqslant x \leqslant 0]$. No 47(5)

2. $\arcsin x = \operatorname{arctg} \dfrac{x}{\sqrt{1 - x^2}}$ $[x^2 < 1]$. No 46(2)

3. $\arcsin x = \operatorname{arcctg} \dfrac{\sqrt{1 - x^2}}{x}$ $[0 < x \leqslant 1]$;

$= \operatorname{arcctg} \dfrac{\sqrt{1 - x^2}}{x} - \pi$ $[-1 \leqslant x < 0]$. No 49(10)

4. $\arccos x = \arcsin \sqrt{1 - x^2}$ $[0 \leqslant x \leqslant 1]$;

$= \pi - \arcsin \sqrt{1 - x^2}$ $[-1 \leqslant x \leqslant 0]$. No 48(6)

5. $\arccos x = \operatorname{arctg} \dfrac{\sqrt{1 - x^2}}{x}$ $[0 < x \leqslant 1]$;

$= \pi + \operatorname{arctg} \dfrac{\sqrt{1 - x^2}}{x}$ $[-1 \leqslant x < 0]$. No 48(8)

6. $\arccos x = \operatorname{arcctg} \dfrac{x}{\sqrt{1 - x^2}}$ $[-1 \leqslant x < 1]$. No 46(4)

7. $\operatorname{arctg} x = \arcsin \dfrac{x}{\sqrt{1 + x^2}}$. No 6(3)

8. $\operatorname{arctg} x = \arccos \dfrac{1}{\sqrt{1 + x^2}}$ $[x \geqslant 0]$;

$= -\arccos \dfrac{1}{\sqrt{1 + x^2}}$ $[x \leqslant 0]$. No 48(7)

9. $\operatorname{arctg} x = \operatorname{arcctg} \dfrac{1}{x}$ $[x > 0]$;

$= \operatorname{arcctg} \dfrac{1}{x} - \pi$ $[x < 0]$. No 49(9)

10. $\operatorname{arcctg} x = \arcsin \dfrac{1}{\sqrt{1 + x^2}}$ $[x > 0]$;

$= \pi - \arcsin \dfrac{1}{\sqrt{1 + x^2}}$ $[x < 0]$. No 49(11)

11. $\operatorname{arcctg} x = \arccos \dfrac{x}{\sqrt{1 + x^2}}$. No 46(4)

12. $\operatorname{arcctg} x = \operatorname{arctg} \dfrac{1}{x}$ $[x > 0]$;

$= \pi + \operatorname{arctg} \dfrac{1}{x}$ $[x < 0]$. No 49(12)

Zyklometrische Funktionen/The Inverse Trigonometric Functions

1. $\arcsin x + \arcsin y = \arcsin (x\sqrt{1-y^2} + y\sqrt{1-x^2})$ 1.625

$$\left[xy \leqslant 0 \;\;\substack{\text{oder}\\\text{or}}\;\; x^2 + y^2 \leqslant 1\right];$$

$$= \pi - \arcsin (x\sqrt{1-y^2} + y\sqrt{1-x^2})$$

$$\left[x > 0,\; y > 0 \;\;\substack{\text{und}\\\text{and}}\;\; x^2 + y^2 > 1\right];$$

$$= -\pi - \arcsin (x\sqrt{1-y^2} + y\sqrt{1-x^2})$$

$$\left[x < 0,\; y < 0 \;\;\substack{\text{und}\\\text{and}}\;\; x^2 + y^2 > 1\right]. \quad \text{No 54(1), GK I 71(880)}$$

2. $\arcsin x + \arcsin y = \arccos (\sqrt{1-x^2}\sqrt{1-y^2} - xy) \quad [x \geqslant 0,\; y \geqslant 0]$;

$$= -\arccos (\sqrt{1-x^2}\sqrt{1-y^2} - xy) \quad [x < 0,\; y < 0]. \quad \text{No 55}$$

3. $\arcsin x + \arcsin y = \operatorname{arctg} \dfrac{x\sqrt{1-y^2} + y\sqrt{1-x^2}}{\sqrt{1-x^2}\sqrt{1-y^2} - xy}$

$$\left[xy \leqslant 0 \;\;\substack{\text{oder}\\\text{or}}\;\; x^2 + y^2 < 1\right];$$

$$= \operatorname{arctg} \dfrac{x\sqrt{1-y^2} + y\sqrt{1-x^2}}{\sqrt{1-x^2}\sqrt{1-y^2} - xy} + \pi$$

$$\left[x > 0,\; y > 0 \;\;\substack{\text{und}\\\text{and}}\;\; x^2 + y^2 > 1\right];$$

$$= \operatorname{arctg} \dfrac{x\sqrt{1-y^2} + y\sqrt{1-x^2}}{\sqrt{1-x^2}\sqrt{1-y^2} - xy} - \pi$$

$$\left[x < 0,\; y < 0 \;\;\substack{\text{und}\\\text{and}}\;\; x^2 + y^2 > 1\right]. \quad \text{No 56}$$

4. $\arcsin x - \arcsin y = \arcsin (x\sqrt{1-y^2} - y\sqrt{1-x^2})$

$$\left[xy \geqslant 0 \;\;\substack{\text{oder}\\\text{or}}\;\; x^2 + y^2 \leqslant 1\right];$$

$$= \pi - \arcsin (x\sqrt{1-y^2} - y\sqrt{1-x^2})$$

$$\left[x > 0,\; y < 0 \;\;\substack{\text{und}\\\text{and}}\;\; x^2 + y^2 > 1\right];$$

$$= -\pi - \arcsin (x\sqrt{1-y^2} - y\sqrt{1-x^2})$$

$$\left[x < 0,\; y > 0 \;\;\substack{\text{und}\\\text{and}}\;\; x^2 + y^2 > 1\right]. \quad \text{No 55(2)}$$

5. $\arcsin x - \arcsin y = \arccos (\sqrt{1-x^2}\sqrt{1-y^2} + xy) \quad [x > y]$;

$$= -\arccos (\sqrt{1-x^2}\sqrt{1-y^2} + xy) \quad [x < y]. \quad \text{No 56}$$

6. $\arccos x + \arccos y = \arccos (xy - \sqrt{1-x^2}\sqrt{1-y^2}) \quad [x+y \geqslant 0]$;

$$= 2\pi - \arccos (xy - \sqrt{1-x^2}\sqrt{1-y^2}) \quad [x+y < 0].$$

No 57(3)

1.625   7. $\arccos x - \arccos y = -\arccos(xy + \sqrt{1-x^2}\sqrt{1-y^2})$ $[x \geqslant y]$;

$$= \arccos(xy + \sqrt{1-x^2}\sqrt{1-y^2}) \quad [x < y].$$

No 57(4)

8. $\arctg x + \arctg y = \arctg \dfrac{x+y}{1-xy}$ $[xy < 1]$;

$$= \pi + \arctg \dfrac{x+y}{1-xy} \quad [x > 0,\ xy > 1];$$

$$= -\pi + \arctg \dfrac{x+y}{1-xy} \quad [x < 0,\ xy > 1].$$

No 59(5), GK I 71(879)

9. $\arctg x - \arctg y = \arctg \dfrac{x-y}{1+xy}$ $[xy > -1]$;

$$= \pi + \arctg \dfrac{x-y}{1+xy} \quad [x > 0,\ xy < -1];$$

$$= -\pi + \arctg \dfrac{x-y}{1+xy} \quad [x < 0,\ xy < -1].$$

No 59(6)

1.626   1. $2\arcsin x = \arcsin(2x\sqrt{1-x^2})$ $\left[|x| \leqslant \dfrac{1}{\sqrt{2}}\right]$;

$$= \pi - \arcsin(2x\sqrt{1-x^2}) \quad \left[\dfrac{1}{\sqrt{2}} < x \leqslant 1\right];$$

$$= -\pi - \arcsin(2x\sqrt{1-x^2}) \quad \left[-1 \leqslant x < -\dfrac{1}{\sqrt{2}}\right].$$

No 61(7)

2. $2\arccos x = \arccos(2x^2 - 1)$ $[0 \leqslant x \leqslant 1]$;

$$= 2\pi - \arccos(2x^2 - 1) \quad [-1 \leqslant x < 0].$$

No 61(8)

3. $2\arctg x = \arctg \dfrac{2x}{1-x^2}$ $[|x| < 1]$;

$$= \arctg \dfrac{2x}{1-x^2} + \pi \quad [x > 1];$$

$$= \arctg \dfrac{2x}{1-x^2} - \pi \quad [x < -1].$$

No 61(9)

1.627   1. $\arctg x + \arctg \dfrac{1}{x} = \dfrac{\pi}{2}$ $[x > 0]$;

$$= -\dfrac{\pi}{2} \quad [x < 0].$$

GK I 71(878)

2. $\arctg x + \arctg \dfrac{1-x}{1+x} = \dfrac{\pi}{4}$ $[x > -1]$;

$$= -\dfrac{3}{4}\pi \quad [x < -1]. \quad \text{No 62, GK I 72(881)}$$

Zyklometrische Funktionen/The Inverse Trigonometric Functions

1. $\arcsin \dfrac{2x}{1+x^2} = -\pi - 2\arctg x \quad [x < -1];$ 1.628

$= 2\arctg x \quad [-1 \leqslant x \leqslant 1];$

$= \pi - 2\arctg x \quad [x > 1].$ No 65

2. $\arccos \dfrac{1-x^2}{1+x^2} = 2\arctg x \quad [x \geqslant 0];$

$= -2\arctg x \quad [x \leqslant 0].$ No 66

$\dfrac{2x-1}{2} - \dfrac{1}{\pi}\arctg\left(\tg\dfrac{2x-1}{2}\pi\right) = E(x).$ GK I 72(886) 1.629

Beziehungen zwischen Area-Funktionen. | Relations between inverse hyperbolic functions. 1.631

1. $\operatorname{Arsh} x = \operatorname{Arch}\sqrt{x^2+1} = \operatorname{Arth}\dfrac{x}{\sqrt{x^2+1}}.$ JE 52

2. $\operatorname{Arch} x = \operatorname{Arsh}\sqrt{x^2-1} = \operatorname{Arth}\dfrac{\sqrt{x^2-1}}{x}.$ JE 52

3. $\operatorname{Arth} x = \operatorname{Arsh}\dfrac{x}{\sqrt{1-x^2}} = \operatorname{Arch}\dfrac{1}{\sqrt{1-x^2}} = \operatorname{Arcth}\dfrac{1}{x}.$ JE 53

4. $\operatorname{Arsh} x \pm \operatorname{Arsh} y = \operatorname{Arsh}(x\sqrt{1+y^2} \pm y\sqrt{1+x^2}).$ JE 54

5. $\operatorname{Arch} x \pm \operatorname{Arch} y = \operatorname{Arch}(xy \pm \sqrt{(x^2-1)(y^2-1)}).$ JE 54

6. $\operatorname{Arth} x \pm \operatorname{Arth} y = \operatorname{Arth}\dfrac{x \pm y}{1 \pm xy}.$ JE 54

Reihendarstellungen
Series representations 1.64

1. $\arcsin x = \dfrac{\pi}{2} - \arccos x = x + \dfrac{1}{2\cdot 3}x^3 + \dfrac{1\cdot 3}{2\cdot 4\cdot 5}x^5 + \dfrac{1\cdot 3\cdot 5}{2\cdot 4\cdot 6\cdot 7}x^7 + \ldots;$ 1.641

$= \displaystyle\sum_{k=0}^{\infty}\dfrac{(2k)!}{2^{2k}(k!)^2(2k+1)}x^{2k+1} = xF\left(\dfrac{1}{2},\dfrac{1}{2};\dfrac{3}{2};x^2\right) \quad [x^2 < 1].$ F II 545

2. $\operatorname{Arsh} x = x - \dfrac{1}{2\cdot 3}x^3 + \dfrac{1\cdot 3}{2\cdot 4\cdot 5}x^5 - \ldots;$

$= \displaystyle\sum_{k=0}^{\infty}(-1)^k\dfrac{(2k)!}{2^{2k}(k!)^2(2k+1)}x^{2k+1};$

$= xF\left(\dfrac{1}{2},\dfrac{1}{2};\dfrac{3}{2};-x^2\right) \quad [x^2 < 1].$ F II 474

1. $\operatorname{Arsh} x = \ln 2x + \dfrac{1}{2}\dfrac{1}{2x^2} - \dfrac{1\cdot 3}{2\cdot 4}\dfrac{1}{4x^4} + \ldots;$ 1.642

$= \ln 2x + \displaystyle\sum_{k=1}^{\infty}(-1)^{k+1}\dfrac{(2k)!\, x^{-2k}}{2^{2k}(k!)^2\, 2k} \quad [x^2 > 1].$ A(6480.2)+

1.642 2. $$\text{Arch } x = \ln 2x - \sum_{k=1}^{\infty} \frac{(2k)!}{2^{2k}(k!)^2 \, 2k} x^{-2k} \quad [x^2 > 1].$$ A(6480.3)+

1.643 1. $$\text{arctg } x = x - \frac{x^3}{3} + \frac{x^5}{5} - \frac{x^7}{7} + \ldots;$$

$$= \sum_{k=0}^{\infty} \frac{(-1)^k x^{2k+1}}{2k+1} \quad [x^2 \leqslant 1].$$ F II 544

2. $$\text{Arth } x = x + \frac{x^3}{3} + \frac{x^5}{5} + \ldots = \sum_{k=0}^{\infty} \frac{x^{2k+1}}{2k+1} \quad [x^2 < 1].$$ A(6480.4)

1.644 1. $$\text{arctg } x = \frac{x}{\sqrt{1+x^2}} \sum_{k=0}^{\infty} \frac{(2k)!}{2^{2k}(k!)^2 (2k+1)} \left(\frac{x^2}{1+x^2}\right)^k;$$

$$= \frac{x}{\sqrt{1+x^2}} F\left(\frac{1}{2}, \frac{1}{2}; \frac{3}{2}; \frac{x^2}{1+x^2}\right) \quad [x^2 < \infty].$$ A(641.3)

2. $$\text{arctg } x = \frac{\pi}{2} - \frac{1}{x} + \frac{1}{3x^3} - \frac{1}{5x^5} + \frac{1}{7x^7} - \ldots;$$

$$= \frac{\pi}{2} - \sum_{k=0}^{\infty} (-1)^k \frac{1}{(2k+1) x^{2k+1}} \quad [x^2 \geqslant 1] \quad (\text{cf. } 1.643).$$

A(641.4)

1.645 1. $$\text{arcsec } x = \frac{\pi}{2} - \frac{1}{x} - \frac{1}{2 \cdot 3x^3} - \frac{1 \cdot 3}{2 \cdot 4 \cdot 5x^5} - \ldots;$$

$$= \frac{\pi}{2} - \sum_{k=0}^{\infty} \frac{(2k)! \, x^{-(2k+1)}}{(k!)^2 2^{2k}(2k+1)};$$

$$= \frac{\pi}{2} - \frac{1}{x} F\left(\frac{1}{2}, \frac{1}{2}; \frac{3}{2}; \frac{1}{x^2}\right) \quad [x^2 > 1].$$ A(641.5)

2. $$(\arcsin x)^2 = \sum_{k=0}^{\infty} \frac{2^{2k}(k!)^2 x^{2k+2}}{(2k+1)! \, (k+1)} \quad [x^2 \leqslant 1].$$ A(642.1), GK II 27(152)+

3. $$(\arcsin x)^3 = x^3 + \frac{3!}{5!} 3^2 \left(1 + \frac{1}{3^2}\right) x^5 + \frac{3!}{7!} 3^2 \cdot 5^2 \left(1 + \frac{1}{3^2} + \frac{1}{5^2}\right) x^7 + \ldots$$
$$[x^2 \leqslant 1].$$ Br$_{68}$ 188, A(642.2), GK II 28(153)+

1.646 1. $$\text{Arsh } \frac{1}{x} = \text{Arcosech } x = \sum_{k=0}^{\infty} \frac{(-1)^k (2k)!}{2^{2k}(k!)^2(2k+1)} x^{-2k-1} \quad [x^2 > 1].$$ A(6480.5)

2. $$\text{Arch } \frac{1}{x} = \text{Arsech } x = \ln \frac{2}{x} - \sum_{k=1}^{\infty} \frac{(2k)!}{2^{2k}(k!)^2 \, 2k} x^{2k} \quad [0 < x < 1].$$ A(6480.6)

3. $$\text{Arsh } \frac{1}{x} = \text{Arcosech } x = \ln \frac{2}{x} + \sum_{k=1}^{\infty} \frac{(-1)^{k+1}(2k)!}{2^{2k}(k!)^2 \, 2k} x^{2k} \quad [0 < x < 1].$$ A(6480.7)+

4. $$\text{Arth } \frac{1}{x} = \text{Arcth } x = \sum_{k=0}^{\infty} \frac{x^{-(2k+1)}}{2k+1} \quad [x^2 > 1].$$ A(6480.8)

Unbestimmte Integrale elementarer Funktionen
Indefinite Integrals of Elementary Functions

Einführung
Introduction

Bemerkungen allgemeiner Art
General remarks

In allen Formeln dieses Abschnitts ist die Integrationskonstante weggelassen worden. Daher ist das Gleichheitszeichen (=) hier so zu verstehen, daß die links bzw. rechts von diesem Zeichen stehenden Funktionen sich um eine Konstante unterscheiden können. Beispielsweise (vgl. 2.01 15.) schreiben wir	In the formulae of this chapter the integration constant will be omitted. Hence the sign of equality means that the functions standing on the left-hand side and the right-hand side may differ by a constant. Thus (see 2.01 15.) we write, for instance,

$$\int \frac{dx}{1+x^2} = \operatorname{arctg} x = -\operatorname{arcctg} x$$

anstelle von	though

$$\operatorname{arctg} x = -\operatorname{arcctg} x + \frac{\pi}{2}.$$

Bei der Integration einiger Funktionen ergibt sich der Logarithmus eines absoluten Betrages (z. B. ist	Integration of certain functions yields the logarithm of a modulus (for instance

$$\int \frac{dx}{\sqrt{1+x^2}} = \ln\left| x + \sqrt{1+x^2} \right|).$$

In solchen Formeln haben wir zur Vereinfachung der Schreibweise die Betragsstriche weggelassen.	For the sake of simplicity the bars will be omitted.
In einigen Fällen ist es wesentlich, eine wohlbestimmte Stammfunktion anzugeben. Solche Stammfunktionen in Gestalt von bestimmten Integralen findet man in anderen Abschnitten aber nicht in diesem.	Sometimes it will be significant to introduce a well defined antiderivative. These antiderivatives defined by definite integrals are to be found in other chapters rather than in this.
An diese Formeln schließen sich weitere an, bei denen die Integrations-	These formulae are closely related to formulae in which the integrand and

grenzen und der Integrand von ein und demselben Parameter abhängen.

Manche Formeln verlieren für gewisse Werte der Konstanten (Parameter) oder für gewisse Beziehungen zwischen diesen Konstanten ihren Sinn (so etwa Formel 2.02 8. für $n = -1$ oder Formel 2.02 15. für $a = b$). Diese Werte der Konstanten bzw. diese Beziehungen zwischen ihnen sind in den meisten Fällen aus der Struktur der rechten Seite der Formel (die kein Integralzeichen enthält) klar ersichtlich. Deshalb haben wir in diesem Abschnitt auf die entsprechenden Hinweise verzichtet. Wenn sich jedoch für diejenigen Werte der Parameter, für welche eine Formel ihren Sinn verliert, der Wert des Integrals mit Hilfe einer anderen Formel berechnen läßt, haben wir diese zweite Formel mit einer entsprechenden Erläuterung versehen.

Die Buchstaben $x, y, t, \ldots$ bezeichnen unabhängige Veränderliche, $f, g, \varphi, \ldots$ Funktionen von $x, y, t, \ldots$. Mit $f', g', \varphi', \ldots$ bzw. $f'', g'', \varphi'', \ldots$ werden die Ableitungen von $f, g, \varphi, \ldots$ erster bzw. zweiter Ordnung usw. bezeichnet, während $a, b, m, p, \ldots$ Konstanten bedeuten, unter denen im allgemeinen beliebige reelle Zahlen zu verstehen sind. Ist eine Formel nur für bestimmte Werte dieser Konstanten gültig (z.B. nur für positive bzw. nur für ganze Zahlen), so wird darauf hingewiesen, sofern sich diese Einschränkung nicht unmittelbar ergibt (wie in 2.148 4. und 2.442 6., da aus ihrer Gestalt klar hervorgeht, daß n darin eine natürliche Zahl sein muß).

the limit (or limits) of integration depend on the same parameter (or parameters).

Some formulae lose their meaning for certain values of the parameters or for certain relations between them (for instance, formula 2.02 8. for $n = -1$ and formula 2.02 15. for $a = b$). These values or relations will not be stated here, since they are evident from the inspection of the right-hand side (which does not contain the integral sign). For this matter we will not make any stipulations concerning such cases. However, where an integral can be calculated by another formula for such values of the parameters or such relations between them, the second formula will be accompanied by appropriate explanations.

The letters $x, y, t, \ldots$ are used to denote indepent variables; $f, g, \varphi, \ldots$ are functions of these variables; $f', g', \varphi', \ldots, f'', g'', \varphi'', \ldots$ are the first, second, etc. derivatives; $a, b, m, p, \ldots$ are constants, in general arbitrary real numbers. If a formula holds only for some values of these constants (for instance, only for positive numbers or only for integers), its domain of validity is stated unless this is evident by inspection (as in the case of the formulae 2.148 4. and 2.442 6., where it is clear that n is a positive integer).

Die Grundintegrale
The basic integrals

2.01

1. $$\int x^n \, dx = \frac{x^{n+1}}{n+1} \quad (n \neq -1).$$

Für $n = -1$ gilt | For $n = -1$:

2. $$\int \frac{dx}{x} = \ln x.$$

Einführung/Introduction

2.01

3. $\int e^x \, dx = e^x.$

4. $\int a^x \, dx = \dfrac{a^x}{\ln a}.$

5. $\int \sin x \, dx = -\cos x.$

6. $\int \cos x \, dx = \sin x.$

7. $\int \dfrac{dx}{\sin^2 x} = -\operatorname{ctg} x.$

8. $\int \dfrac{dx}{\cos^2 x} = \operatorname{tg} x.$

9. $\int \dfrac{\sin x}{\cos^2 x} \, dx = \sec x.$

10. $\int \dfrac{\cos x}{\sin^2 x} \, dx = -\operatorname{cosec} x.$

11. $\int \operatorname{tg} x \, dx = -\ln \cos x.$

12. $\int \operatorname{ctg} x \, dx = \ln \sin x.$

13. $\int \dfrac{dx}{\sin x} = \ln \operatorname{tg} \dfrac{x}{2}.$

14. $\int \dfrac{dx}{\cos x} = \ln \operatorname{tg}\left(\dfrac{\pi}{4} + \dfrac{x}{2}\right) = \ln(\sec x + \operatorname{tg} x).$

15. $\int \dfrac{dx}{1 + x^2} = \operatorname{arctg} x = -\operatorname{arcctg} x.$

16. $\int \dfrac{dx}{1 - x^2} = \operatorname{Arth} x = \dfrac{1}{2} \ln \dfrac{1 + x}{1 - x}.$

17. $\int \dfrac{dx}{\sqrt{1 - x^2}} = \arcsin x = -\arccos x.$

18. $\int \dfrac{dx}{\sqrt{x^2 + 1}} = \operatorname{Arsh} x = \ln(x + \sqrt{x^2 + 1}).$

19. $\int \dfrac{dx}{\sqrt{x^2 - 1}} = \operatorname{Arch} x = \ln(x + \sqrt{x^2 - 1}).$

20. $\int \operatorname{sh} x \, dx = \operatorname{ch} x.$

21. $\int \operatorname{ch} x \, dx = \operatorname{sh} x.$

22. $\int \dfrac{dx}{\operatorname{sh}^2 x} = -\operatorname{cth} x.$

23. $\int \dfrac{dx}{\operatorname{ch}^2 x} = \operatorname{th} x.$

24. $\int \operatorname{th} x \, dx = \ln \operatorname{ch} x.$

25. $\int \operatorname{cth} x \, dx = \ln \operatorname{sh} x.$

26. $\int \dfrac{dx}{\operatorname{sh} x} = \ln \operatorname{th} \dfrac{x}{2}.$

Allgemeine Formeln
General formulae

2.02

1. $\int af \, dx = a \int f \, dx.$

2. $\int [af \pm b\varphi \pm c\psi \pm \ldots] \, dx = a \int f \, dx \pm b \int \varphi \, dx \pm c \int \psi \, dx \pm \ldots$

2.02

3. $$\frac{d}{dx}\int f\,dx = f.$$

4. $$\int f'\,dx = f.$$

5. $$\int f'\varphi\,dx = f\varphi - \int f\varphi'\,dx \quad \begin{bmatrix}\text{partielle Integration}\\ \text{integration by parts}\end{bmatrix}.$$

6. $$\int f^{(n+1)}\varphi\,dx = \varphi f^{(n)} - \varphi' f^{(n-1)} + \varphi'' f^{(n-2)} - \ldots +(-1)^n \varphi^{(n)} f +$$
$$+ (-1)^{n+1}\int \varphi^{(n+1)} f\,dx.$$

7. $$\int f(x)\,dx = \int f[\varphi(y)]\,\varphi'(y)\,dy \quad [x = \varphi(y)] \quad \begin{bmatrix}\text{Substitutionsregel}\\ \text{rule of substitution}\end{bmatrix}.$$

8. $$\int (f)^n f'\,dx = \frac{(f)^{n+1}}{n+1}.$$

Für $n = -1$ gilt | For $n = -1$:

$$\int \frac{f'\,dx}{f} = \ln f.$$

9. $$\int (af+b)^n f'\,dx = \frac{(af+b)^{n+1}}{a(n+1)}.$$

10. $$\int \frac{f'\,dx}{\sqrt{af+b}} = \frac{2\sqrt{af+b}}{a}.$$

11. $$\int \frac{f'\varphi - \varphi' f}{\varphi^2}\,dx = \frac{f}{\varphi}.$$

12. $$\int \frac{f'\varphi - \varphi' f}{f\varphi}\,dx = \ln \frac{f}{\varphi}.$$

13. $$\int \frac{dx}{f(f\pm\varphi)} = \pm\int \frac{dx}{f\varphi} \mp \int \frac{dx}{\varphi(f\pm\varphi)}.$$

14. $$\int \frac{f'\,dx}{\sqrt{f^2+a}} = \ln(f+\sqrt{f^2+a}).$$

15. $$\int \frac{f\,dx}{(f+a)(f+b)} = \frac{a}{a-b}\int \frac{dx}{(f+a)} - \frac{b}{a-b}\int \frac{dx}{(f+b)}.$$

Für $a = b$ gilt | For $a = b$:

$$\int \frac{f\,dx}{(f+a)^2} = \int \frac{dx}{f+a} - a\int \frac{dx}{(f+a)^2}.$$

16. $$\int \frac{f\,dx}{(f+\varphi)^n} = \int \frac{dx}{(f+\varphi)^{n-1}} - \int \frac{\varphi\,dx}{(f+\varphi)^n}.$$

17. $$\int \frac{f'\,dx}{p^2+q^2 f^2} = \frac{1}{pq}\operatorname{arctg}\frac{qf}{p}.$$

18. $$\int \frac{f'\,dx}{q^2 f^2 - p^2} = \frac{1}{2pq} \ln \frac{qf - p}{qf + p}.$$

19. $$\int \frac{f\,dx}{1 - f} = -x + \int \frac{dx}{1 - f}.$$

20. $$\int \frac{f^2\,dx}{f^2 - a^2} = \frac{1}{2} \int \frac{f\,dx}{f - a} + \frac{1}{2} \int \frac{f\,dx}{f + a}.$$

21. $$\int \frac{f'\,dx}{\sqrt{a^2 - f^2}} = \arcsin \frac{f}{a}.$$

22. $$\int \frac{f'\,dx}{af^2 + bf} = \frac{1}{b} \ln \frac{f}{af + b}.$$

23. $$\int \frac{f'\,dx}{f\sqrt{f^2 - a^2}} = \frac{1}{a} \operatorname{arcsec} \frac{f}{a}.$$

24. $$\int \frac{(f'\varphi - f\varphi')\,dx}{f^2 + \varphi^2} = \operatorname{arctg} \frac{f}{\varphi}.$$

25. $$\int \frac{(f'\varphi - f\varphi')\,dx}{f^2 - \varphi^2} = \frac{1}{2} \ln \frac{f - \varphi}{f + \varphi}.$$

Rationale Funktionen
Rational Functions   2.1

Allgemeine Integrationsregeln
General rules of integration   2.10

Um eine beliebige (gebrochene) rationale Funktion $\frac{F(x)}{f(x)}$, wobei $F(x)$ und $f(x)$ Polynome ohne gemeinsamen Faktor sind, zu integrieren, muß man zunächst den eventuell vorhandenen ganzen Teil $E(x)$ abtrennen ($E(x)$ ein Polynom) und das Integral über diesen und das Integral über den Rest bilden:

Let the integrand be a rational function $\frac{F(x)}{f(x)}$, where $F(x)$ and $f(x)$ are polynomials without a common factor. Then in the integral of this function we must first separate out the integral part $E(x)$ ($E(x)$ is a polynomial) and then evaluate two integrals, namely   2.101

$$\int \frac{F(x)\,dx}{f(x)} = \int E(x)\,dx + \int \frac{\varphi(x)}{f(x)}\,dx.$$

Die Integration des Restes, der eine echte gebrochene rationale Funktion ist (der Grad des Zählers ist kleiner als der Grad des Nenners), beruht auf seiner Zerlegung in Partialbrüche. Sind $a, b, c, \ldots, m$ die Nullstellen von $f(x)$ und $\alpha, \beta, \gamma, \ldots, \mu$ die zugehörigen Vielfachheiten, so daß also $(fx) = (x - a)^\alpha (x - b)^\beta \ldots (x - m)^\mu$

where the degree of $\varphi(x)$ is less than the degree of $f(x)$. The second integrand is to be written as a sum of partial fractions.

Let $a, b, c, \ldots, m$ be the zeros of $f(x)$ and $\alpha, \beta, \gamma, \ldots, \mu$ their multiplicities, that is, $f(x) = (x - a)^\alpha (x - b)^\beta \ldots (x - m)^\mu$. Then the decom-   2.102

ist, so kann $\dfrac{\varphi(x)}{f(x)}$ in Partialbrüche zerlegt werden: | position of $\dfrac{\varphi(x)}{f(x)}$ into partial fractions is

$$\frac{\varphi(x)}{f(x)} = \frac{A_\alpha}{(x-a)^\alpha} + \frac{A_{\alpha-1}}{(x-a)^{\alpha-1}} + \cdots + \frac{A_1}{x-a} + \frac{B_\beta}{(x-b)^\beta} +$$

$$+ \frac{B_{\beta-1}}{(x-b)^{\beta-1}} + \cdots + \frac{B_1}{x-b} + \cdots + \frac{M_\mu}{(x-m)^\mu} + \frac{M_{\mu-1}}{(x-m)^{\mu-1}} + \cdots + \frac{M_1}{x-m},$$

wobei die Zähler der einzelnen Brüche durch die folgenden Formeln bestimmt sind: | where the numerators are given by the following formulas:

$$A_{\alpha-k+1} = \frac{\psi_1^{(k-1)}(a)}{(k-1)!}, \quad B_{\beta-k+1} = \frac{\psi_2^{(k-1)}(b)}{(k-1)!}, \quad \ldots, \quad M_{\mu-k+1} = \frac{\psi_m^{(k-1)}(m)}{(k-1)!},$$

$$\psi_1(x) = \frac{\varphi(x)(x-a)^\alpha}{f(x)}, \quad \psi_2(x) = \frac{\varphi(x)(x-b)^\beta}{f(x)}, \quad \ldots, \quad \psi_m(x) = \frac{\varphi(x)(x-m)^\mu}{f(x)}.$$

T 51+

Sind $a, b, \ldots, m$ einfache Nullstellen, also $\alpha = \beta = \gamma = \cdots = \mu = 1$, so gilt | If $a, b, \ldots, m$ are simple zeros, that is, if $\alpha = \beta = \gamma = \cdots = \mu = 1$, we have

$$\frac{\varphi(x)}{f(x)} = \frac{A}{x-a} + \frac{B}{x-b} + \cdots + \frac{M}{x-m}$$

mit | with

$$A = \frac{\varphi(a)}{f'(a)}, \quad B = \frac{\varphi(b)}{f'(b)}, \quad \ldots, \quad M = \frac{\varphi(m)}{f'(m)}.$$

Sind einige Wurzeln der Gleichung $f(x) = 0$ komplex, so kann man durch Zusammenfassung der Partialbrüche, die zu konjugierten Wurzeln gehören, nach einigen Umformungen die entsprechenden Paare von Brüchen in Form von reellen Brüchen darstellen: | If some of the zeros of the equation $f(x) = 0$ are complex the partial fractions pertaining to complex conjugate roots can be grouped together. This leads, after some simple manipulations, to real fractions of the form

$$\frac{M_1 x + N_1}{x^2 + 2Bx + C} + \frac{M_2 x + N_2}{(x^2 + 2Bx + C)^2} + \cdots + \frac{M_p x + N_p}{(x^2 + 2Bx + C)^p}.$$

2.103 Somit reduziert sich die Integration eines echten Bruches $\dfrac{\varphi(x)}{f(x)}$ auf die Auswertung von Integralen der Gestalt | Thus the integration of a proper fraction $\dfrac{\varphi(x)}{f(x)}$ may be reduced to evaluating integrals of the form

$$\int \frac{g\, dx}{(x-a)^\alpha} \quad \text{und} \quad \int \frac{Mx + N}{(A + 2Bx + Cx^2)^p}\, dx.$$

Erstere ergeben für $\alpha > 1$ rationale Funktionen, für $\alpha = 1$ Logarithmen; letztere liefern rationale Funktionen, | The integrals of the first form are rational functions if $\alpha > 1$ and logarithms if $\alpha = 1$; the integrals of the

Logarithmen oder Arkustangens: | second form are rational functions and logarithms or arc tangents:

1. $$\int \frac{g\,dx}{(x-a)^\alpha} = g \int \frac{d(x-a)}{(x-a)^\alpha} = -\frac{g}{(\alpha-1)(x-a)^{\alpha-1}}.$$

2. $$\int \frac{g\,dx}{x-a} = g \int \frac{d(x-a)}{x-a} = g \ln|x-a|.$$

3. $$\int \frac{Mx+N}{(A+2Bx+Cx^2)^p}\,dx = \frac{NB - MA + (NC - MB)x}{2(p-1)(AC - B^2)(A+2Bx+Cx^2)^{p-1}} +$$
$$+ \frac{(2p-3)(NC - MB)}{2(p-1)(AC - B^2)} \int \frac{dx}{(A+2Bx+Cx^2)^{p-1}}.$$

4. $$\int \frac{dx}{A+2Bx+Cx^2} = \frac{1}{\sqrt{AC-B^2}} \operatorname{arctg} \frac{Cx+B}{\sqrt{AC-B^2}} \quad [AC > B^2];$$
$$= \frac{1}{2\sqrt{B^2-AC}} \ln \left| \frac{Cx+B-\sqrt{B^2-AC}}{Cx+B+\sqrt{B^2-AC}} \right| \quad [AC < B^2].$$

5. $$\int \frac{(Mx+N)\,dx}{A+2Bx+Cx^2} = \frac{M}{2C} \ln|A+2Bx+Cx^2| +$$
$$+ \frac{NC - MB}{C\sqrt{AC-B^2}} \operatorname{arctg} \frac{Cx+B}{\sqrt{AC-B^2}} \quad [AC > B^2];$$
$$= \frac{M}{2C} \ln|A+2Bx+Cx^2| +$$
$$+ \frac{NC - MB}{2C\sqrt{B^2-AC}} \ln \left| \frac{Cx+B-\sqrt{B^2-AC}}{Cx+B+\sqrt{B^2-AC}} \right| \quad [AC < B^2].$$

Die Methode von OSTROGRADSKI-HERMITE
The OSTROGRADSKI-HERMITE method

Mit Hilfe der Methode von OSTROGRADSKI-HERMITE kann man den (echten) rationalen Teil von $\int \frac{\varphi(x)}{f(x)} dx$ ohne die Bestimmung der Nullstellen von $f(x)$ und ohne Zerlegung in Partialbrüche finden:

With the help of a method due to OSTROGRADSKI and HERMITE the rational part of the integral $\int \frac{\varphi(x)}{f(x)} dx$ can be found without finding the zeros of $f(x)$ and decomposing it in partial functions. This can be done by writing

2.104

$$\int \frac{\varphi(x)}{f(x)} dx = \frac{M}{D} + \int \frac{N\,dx}{Q}.$$

F II 46, 47

Dabei sind M, N, D, Q ganze rationale Funktionen von x, wobei D der größte gemeinsame Teiler der Funktion $f(x)$ und ihrer Ableitung $f'(x)$ ist und $Q = \frac{f(x)}{D}$ gilt. Ferner ist M ein Polynom höchstens $(m-1)$-ten Grades,

Here M, N, D, Q are polynomials in x. In particular, D is the greatest common divisor of $f(x)$ and its derivative $f'(x)$, while $Q = \frac{f(x)}{D}$. Moreover, the degree of the polynomial M is not higher than $(m-1)$, where

| wenn m der Grad von D ist, und N ein Polynom höchstens $(n-1)$-ten Grades, wenn n der Grad von Q ist. Die Koeffizienten von M und N ergeben sich durch Koeffizientenvergleich aus der Identität | m is the degree of D; similarly, if n is the degree of Q, then N is a polynomial of a degree not higher than $(n-1)$. The coefficients of polynomials M and N are to be determined by comparing the coefficients of like powers in the identity |

$$\varphi(x) = M'Q - M(T - Q') + ND,$$

| wobei $T = \dfrac{f'(x)}{D}$ ist. M' und Q' sind die Ableitungen von M bzw. Q. | where $T = \dfrac{f'(x)}{D}$, while M' and Q' are the derivatives of M and Q, respectively. |

2.11–2.13

Funktionen von $a + bx^k$

Functions of $a + bx^k$

2.110 Rekursionsformeln für $z_k = a + bx^k$. | Recurrence formulae for $z_k = a + bx^k$.

1. $\displaystyle\int x^n z_k^m\, dx = \frac{x^{n+1} z_k^m}{km+n+1} + \frac{amk}{km+n+1} \int x^n z_k^{m-1}\, dx =$ La 126(4)

$$= \frac{x^{n+1}}{m+1} \sum_{s=0}^{p} \frac{(ak)^s (m+1) m (m-1) \ldots (m-s+1) z_k^{m-s}}{[mk+n+1][(m-1)k+n+1]\ldots[(m-s)k+n+1]} +$$

$$+ \frac{(ak)^{p+1} m(m-1)\ldots(m-p+1)(m-p)}{[mk+n+1][(m-1)k+n+1]\ldots[(m-p)k+n+1]} \int x^n z_k^{m-p-1}\, dx.$$

2. $\displaystyle\int x^n z_k^m\, dx = \frac{-x^{n+1} z_k^{m+1}}{ak(m+1)} + \frac{km+k+n+1}{ak(m+1)} \int x^n z_k^{m+1}\, dx.$ La 126(6)

3. $\displaystyle\int x^n z_k^m\, dx = \frac{x^{n+1} z_k^m}{n+1} - \frac{bkm}{n+1} \int x^{n+k} z_k^{m-1}\, dx.$ La 125(1)

4. $\displaystyle\int x^n z_k^m\, dx = \frac{x^{n+1-k} z_k^{m+1}}{bk(m+1)} - \frac{n+1-k}{bk(m+1)} \int x^{n-k} z_k^{m+1}\, dx.$ La 125(2)

5. $\displaystyle\int x^n z_k^m\, dx = \frac{x^{n+1-k} z_k^{m+1}}{b(km+n+1)} - \frac{a(n+1-k)}{b(km+n+1)} \int x^{n-k} z_k^m\, dx.$ La 126(3)

6. $\displaystyle\int x^n z_k^m\, dx = \frac{x^{n+1} z_k^{m+1}}{a(n+1)} - \frac{b(km+k+n+1)}{a(n+1)} \int x^{n+k} z_k^m\, dx.$ La 126(5)

Funktionen von $z_1 = a + bx$

Functions of $z_1 = a + bx$

2.111   1. $\displaystyle\int z_1^m\, dx = \frac{z_1^{m+1}}{b(m+1)}.$

Rationale Funktionen/Rational Functions

Für $m = -1$ gilt | For $m = -1$: 2.111

$$\int \frac{dx}{z_1} = \frac{1}{b} \ln z_1.$$

2. $$\int \frac{x^n \, dx}{z_1^m} = \frac{x^n}{z_1^{m-1}(n+1-m)b} - \frac{na}{(n+1-m)b} \int \frac{x^{n-1} \, dx}{z_1^m}.$$

Für $n = m - 1$ kann man folgende Formel benutzen: | For $n = m - 1$ the following formula can be used:

3. $$\int \frac{x^{m-1} \, dx}{z_1^m} = -\frac{x^{m-1}}{z_1^{m-1}(m-1)b} + \frac{1}{b} \int \frac{x^{m-2} \, dx}{z_1^{m-1}}.$$

Für $m = 1$ gilt | For $m = 1$:

$$\int \frac{x^n \, dx}{z_1} = \frac{x^n}{nb} - \frac{ax^{n-1}}{(n-1)b^2} + \frac{a^2 x^{n-2}}{(n-2)b^3} - \ldots + (-1)^{n-1} \frac{a^{n-1} x}{1 \cdot b^n} + \frac{(-1)^n a^n}{b^{n+1}} \ln z_1.$$

4. $$\int \frac{x^n \, dx}{z_1^2} = \sum_{k=1}^{n-1} (-1)^{k-1} \frac{k a^{k-1} x^{n-k}}{(n-k) b^{k+1}} + (-1)^{n-1} \frac{a^n}{b^{n+1} z_1} + (-1)^{n+1} \frac{n a^{n-1}}{b^{n+1}} \ln z_1.$$

1. $$\int \frac{x \, dx}{z_1} = \frac{x}{b} - \frac{a}{b^2} \ln z_1.$$ 2.112

2. $$\int \frac{x^2 \, dx}{z_1} = \frac{x^2}{2b} - \frac{ax}{b^2} + \frac{a^2}{b^3} \ln z_1.$$

1. $$\int \frac{dx}{z_1^2} = -\frac{1}{bz_1}.$$ 2.113

2. $$\int \frac{x \, dx}{z_1^2} = -\frac{x}{bz_1} + \frac{1}{b^2} \ln z_1 = \frac{a}{b^2 z_1} + \frac{1}{b^2} \ln z_1.$$

3. $$\int \frac{x^2 \, dx}{z_1^2} = \frac{x}{b^2} - \frac{a^2}{b^3 z_1} - \frac{2a}{b^3} \ln z_1$$

1. $$\int \frac{dx}{z_1^3} = -\frac{1}{2b z_1^2}.$$ 2.114

2. $$\int \frac{x \, dx}{z_1^3} = -\left[\frac{x}{b} + \frac{a}{2b^2}\right] \frac{1}{z_1^2}.$$

3. $$\int \frac{x^2 \, dx}{z_1^3} = \left[\frac{2ax}{b^2} + \frac{3a^2}{2b^3}\right] \frac{1}{z_1^2} + \frac{1}{b^3} \ln z_1.$$

4. $$\int \frac{x^3 \, dx}{z_1^3} = \left[\frac{x^3}{b} + 2\frac{a}{b^2} x^2 - 2\frac{a^2}{b^3} x - \frac{5}{2} \frac{a^3}{b^4}\right] \frac{1}{z_1^2} - 3\frac{a}{b^4} \ln z_1.$$

1. $$\int \frac{dx}{z_1^4} = -\frac{1}{3b z_1^3}.$$ 2.115

2. $$\int \frac{x \, dx}{z_1^4} = -\left[\frac{x}{2b} + \frac{a}{6b^2}\right] \frac{1}{z_1^3}.$$

2.115 3. $\int \dfrac{x^2\, dx}{z_1^4} = -\left[\dfrac{x^2}{b} + \dfrac{ax}{b^2} + \dfrac{a^2}{3b^3}\right]\dfrac{1}{z_1^3}$.

4. $\int \dfrac{x^3\, dx}{z_1^4} = \left[\dfrac{3ax^2}{b^2} + \dfrac{9a^2 x}{2b^3} + \dfrac{11a^3}{6b^4}\right]\dfrac{1}{z_1^3} + \dfrac{1}{b^4}\ln z_1$.

2.116 1. $\int \dfrac{dx}{z_1^5} = -\dfrac{1}{4bz_1^4}$.

2. $\int \dfrac{x\, dx}{z_1^5} = -\left[\dfrac{x}{3b} + \dfrac{a}{12b^2}\right]\dfrac{1}{z_1^4}$.

3. $\int \dfrac{x^2\, dx}{z_1^5} = -\left[\dfrac{x^2}{2b} + \dfrac{ax}{3b^2} + \dfrac{a^2}{12b^3}\right]\dfrac{1}{z_1^4}$.

4. $\int \dfrac{x^3\, dx}{z_1^5} = -\left[\dfrac{x^3}{b} + \dfrac{3ax^2}{2b^2} + \dfrac{a^2 x}{b^3} + \dfrac{a^3}{4b^4}\right]\dfrac{1}{z_1^4}$.

2.117 1. $\int \dfrac{dx}{x^n z_1^m} = \dfrac{-1}{(n-1)\, ax^{n-1} z_1^{m-1}} + \dfrac{b(2-n-m)}{a(n-1)}\int \dfrac{dx}{x^{n-1} z_1^m}$.

2. $\int \dfrac{dx}{z_1^m} = -\dfrac{1}{(m-1)\, bz_1^{m-1}}$.

3. $\int \dfrac{dx}{x z_1^m} = \dfrac{1}{z_1^{m-1}\, a(m-1)} + \dfrac{1}{a}\int \dfrac{dx}{x z_1^{m-1}}$.

4. $\int \dfrac{dx}{x^n z_1} = \sum_{k=1}^{n-1} \dfrac{(-1)^k\, b^{k-1}}{(n-k)\, a^k\, x^{n-k}} + \dfrac{(-1)^n\, b^{n-1}}{a^n}\ln\dfrac{z_1}{x}$.

2.118 1. $\int \dfrac{dx}{x z_1} = -\dfrac{1}{a}\ln\dfrac{z_1}{x}$.

2. $\int \dfrac{dx}{x^2 z_1} = -\dfrac{1}{ax} + \dfrac{b}{a^2}\ln\dfrac{z_1}{x}$.

3. $\int \dfrac{dx}{x^3 z_1} = -\dfrac{1}{2ax^2} + \dfrac{b}{a^2 x} - \dfrac{b^2}{a^3}\ln\dfrac{z_1}{x}$.

2.119 1. $\int \dfrac{dx}{x z_1^2} = \dfrac{1}{az_1} - \dfrac{1}{a^2}\ln\dfrac{z_1}{x}$.

2. $\int \dfrac{dx}{x^2 z_1^2} = -\left[\dfrac{1}{ax} + \dfrac{2b}{a^2}\right]\dfrac{1}{z_1} + \dfrac{2b}{a^3}\ln\dfrac{z_1}{x}$.

3. $\int \dfrac{dx}{x^3 z_1^2} = \left[-\dfrac{1}{2ax^2} + \dfrac{3b}{2a^2 x} + \dfrac{3b^2}{a^3}\right]\dfrac{1}{z_1} - \dfrac{3b^2}{a^4}\ln\dfrac{z_1}{x}$.

2.121 1. $\int \dfrac{dx}{x z_1^3} = \left[\dfrac{3}{2a} + \dfrac{bx}{a^2}\right]\dfrac{1}{z_1^2} - \dfrac{1}{a^3}\ln\dfrac{z_1}{x}$.

2. $\int \dfrac{dx}{x^2 z_1^3} = -\left[\dfrac{1}{ax} + \dfrac{9b}{2a^2} + \dfrac{3b^2 x}{a^3}\right]\dfrac{1}{z_1^2} + \dfrac{3b}{a^4}\ln\dfrac{z_1}{x}$.

3. $\int \dfrac{dx}{x^3 z_1^3} = \left[-\dfrac{1}{2ax^2} + \dfrac{2b}{a^2 x} + \dfrac{9b^2}{a^3} + \dfrac{6b^3 x}{a^4} \right] \dfrac{1}{z_1^2} - \dfrac{6b^2}{a^5} \ln \dfrac{z_1}{x}$.   2.121

1. $\int \dfrac{dx}{x z_1^4} = \left[\dfrac{11}{6a} + \dfrac{5bx}{2a^2} + \dfrac{b^2 x^2}{a^3} \right] \dfrac{1}{z_1^3} - \dfrac{1}{a^4} \ln \dfrac{z_1}{x}$.   2.122

2. $\int \dfrac{dx}{x^2 z_1^4} = -\left[\dfrac{1}{ax} + \dfrac{22b}{3a^2} + \dfrac{10b^2 x}{a^3} + \dfrac{4b^3 x^2}{a^4} \right] \dfrac{1}{z_1^3} + \dfrac{4b}{a^5} \ln \dfrac{z_1}{x}$.

3. $\int \dfrac{dx}{x^3 z_1^4} = \left[-\dfrac{1}{2ax^2} + \dfrac{5b}{2a^2 x} + \dfrac{55b^2}{3a^3} + \dfrac{25b^3 x}{a^4} + \dfrac{10b^4 x^2}{a^5} \right] \dfrac{1}{z_1^2} - \dfrac{10b^2}{a^6} \ln \dfrac{z_1}{x}$.

1. $\int \dfrac{dx}{x z_1^5} = \left[\dfrac{25}{12a} + \dfrac{13bx}{3a^2} + \dfrac{7b^2 x^2}{2a^3} + \dfrac{b^3 x^3}{a^4} \right] \dfrac{1}{z_1^4} - \dfrac{1}{a^5} \ln \dfrac{z_1}{x}$.   2.123

2. $\int \dfrac{dx}{x^2 z_1^5} = \left[-\dfrac{1}{ax} - \dfrac{125b}{12a^2} - \dfrac{65b^2 x}{3a^3} - \dfrac{35b^3 x^2}{2a^4} - \dfrac{5b^4 x^3}{a^5} \right] \dfrac{1}{z_1^4} + \dfrac{5b}{a^6} \ln \dfrac{z_1}{x}$.

3. $\int \dfrac{dx}{x^3 z_1^5} = \left[-\dfrac{1}{2ax^2} + \dfrac{3b}{a^2 x} + \dfrac{125b^2}{4a^3} + \dfrac{65b^3 x}{a^4} + \dfrac{105b^4 x^2}{2a^5} + \dfrac{15b^5 x^3}{a^6} \right] \dfrac{1}{z_1^4} - \dfrac{15b^2}{a^7} \ln \dfrac{z_1}{x}$.

Funktionen von $z_2 = a + bx^2$. | Functions of $z_2 = a + bx^2$.   2.124

1. $\int \dfrac{dx}{z_2} = \dfrac{1}{\sqrt{ab}} \operatorname{arctg} x \sqrt{\dfrac{b}{a}}$ $[ab > 0]$ (cf. 2.141 2.);

$\qquad = \dfrac{1}{2i \sqrt{ab}} \ln \dfrac{a + xi \sqrt{ab}}{a - xi \sqrt{ab}}$ $[ab < 0]$ (cf. 2.143 2., 2.143 3.).

2. $\int \dfrac{x \, dx}{z_2^m} = -\dfrac{1}{2b(m-1) z_2^{m-1}}$ (cf. 2.145 2., 2.145 6., 2.18).

Funktionen von $z_3 = a + bx^3$
Functions of $z_3 = a + bx^3$

Abkürzung: $\alpha = \sqrt[3]{\dfrac{a}{b}}$.
Abbreviation:

1. $\int \dfrac{x^n \, dx}{z_3^m} = \dfrac{x^{n-2}}{z_3^{m-1}(n + 1 - 3m) b} - \dfrac{(n-2) a}{b(n + 1 - 3m)} \int \dfrac{x^{n-3} \, dx}{z_3^m}$.   2.125

2. $\int \dfrac{x^n \, dx}{z_3^m} = \dfrac{x^{n+1}}{3a(m-1) z_3^{m-1}} - \dfrac{n + 4 - 3m}{3a(m-1)} \int \dfrac{x^n \, dx}{z_3^{m-1}}$.   La 133(1)

1. $\int \dfrac{dx}{z_3} = \dfrac{\alpha}{3a} \left\{ \dfrac{1}{2} \ln \dfrac{(x + \alpha)^2}{x^2 - \alpha x + \alpha^2} + \sqrt{3} \operatorname{arctg} \dfrac{x \sqrt{3}}{2\alpha - x} \right\}$;   2.126

$\qquad = \dfrac{\alpha}{3a} \left\{ \dfrac{1}{2} \ln \dfrac{(x + \alpha)^2}{x^2 - \alpha x + \alpha^2} + \sqrt{3} \operatorname{arctg} \dfrac{2x - \alpha}{\alpha \sqrt{3}} \right\}$ (cf. 2.141 3., 2.143 4.).

2. $\int \dfrac{x \, dx}{z_3} = -\dfrac{1}{3b\alpha} \left\{ \dfrac{1}{2} \ln \dfrac{(x + \alpha)^2}{x^2 - \alpha x + \alpha^2} - \sqrt{3} \operatorname{arctg} \dfrac{2x - \alpha}{\alpha \sqrt{3}} \right\}$

(cf. 2.145 3., 2.145 7.).

2.126　　3.　　$$\int \frac{x^2\,dx}{z_3} = \frac{1}{3b}\ln(1+x^3\alpha^{-3}) = \frac{1}{3b}\ln z_3.$$

4.　　$$\int \frac{x^3\,dx}{z_3} = \frac{x}{b} - \frac{a}{b}\int \frac{dx}{z_3} \quad \text{(cf. 2.126 1.).}$$

5.　　$$\int \frac{x^4\,dx}{z_3} = \frac{x^2}{2b} - \frac{a}{b}\int \frac{x\,dx}{z_3} \quad \text{(cf. 2.126 2.).}$$

2.127　　1.　　$$\int \frac{dx}{z_3^2} = \frac{x}{3az_3} + \frac{2}{3a}\int \frac{dx}{z_3} \quad \text{(cf. 2.126 1.).}$$

2.　　$$\int \frac{x\,dx}{z_3^2} = \frac{x^2}{3az_3} + \frac{1}{3a}\int \frac{x\,dx}{z_3} \quad \text{(cf. 2.126 2.).}$$

3.　　$$\int \frac{x^2\,dx}{z_3^2} = -\frac{1}{3bz_3}.$$

4.　　$$\int \frac{x^3\,dx}{z_3^2} = -\frac{x}{3bz_3} + \frac{1}{3b}\int \frac{dx}{z_3} \quad \text{(cf. 2.126 1.).}$$

2.128　　1.　　$$\int \frac{dx}{x^n z_3^m} = -\frac{1}{(n-1)ax^{n-1}z_3^{m-1}} - \frac{b(3m+n-4)}{a(n-1)}\int \frac{dx}{x^{n-3}z_3^m}.$$

2.　　$$\int \frac{dx}{x^n z_3^m} = \frac{1}{3a(m-1)x^{n-1}z_3^{m-1}} + \frac{n+3m-4}{3a(m-1)}\int \frac{dx}{x^n z_3^{m-1}}. \qquad \text{La 133(2)}$$

2.129　　1.　　$$\int \frac{dx}{xz_3} = \frac{1}{3a}\ln\frac{x^3}{z_3}.$$

2.　　$$\int \frac{dx}{x^2 z_3} = -\frac{1}{ax} - \frac{b}{a}\int \frac{x\,dx}{z_3} \quad \text{(cf. 2.126 2.).}$$

3.　　$$\int \frac{dx}{x^3 z_3} = -\frac{1}{2ax^2} - \frac{b}{a}\int \frac{dx}{z_3} \quad \text{(cf. 2.126 1.).}$$

2.131　　1.　　$$\int \frac{dx}{xz_3^2} = \frac{1}{3az_3} + \frac{1}{3a^2}\ln\frac{x^3}{z_3}.$$

2.　　$$\int \frac{dx}{x^2 z_3^2} = -\left[\frac{1}{ax} + \frac{4bx^2}{3a^2}\right]\frac{1}{z_3} - \frac{4b}{3a^2}\int \frac{x\,dx}{z_3} \quad \text{(cf. 2.126 2.).}$$

3.　　$$\int \frac{dx}{x^3 z_3^2} = -\left[\frac{1}{2ax^2} + \frac{5bx}{6a^2}\right]\frac{1}{z_3} - \frac{5b}{3a^2}\int \frac{dx}{z_3} \quad \text{(cf. 2.126 1.).}$$

Funktionen von $z_4 = a + bx^4$

Functions of $z_4 = a + bx^4$

Abkürzungen:　　$\alpha = \sqrt[4]{\dfrac{a}{b}}, \quad \alpha' = \sqrt[4]{\dfrac{-a}{b}}.$
Abbreviations:

2.132　　1.　　$$\int \frac{dx}{z_4} = \frac{\alpha}{4a\sqrt{2}}\left\{\ln\frac{x^2+\alpha x\sqrt{2}+\alpha^2}{x^2-\alpha x\sqrt{2}+\alpha^2} + 2\,\text{arctg}\,\frac{\alpha x\sqrt{2}}{\alpha^2-x^2}\right\} \quad [ab > 0]$$
(cf. 2.141 4.);

$$= \frac{\alpha'}{4a}\left\{\ln\frac{x+\alpha'}{x-\alpha'} + 2\,\text{arctg}\,\frac{x}{\alpha'}\right\} \quad [ab < 0] \quad \text{(cf. 2.143 5.).}$$

2. $\int \frac{x\,dx}{z_4} = \frac{1}{2\sqrt{ab}} \operatorname{arctg} x^2 \sqrt{\frac{b}{a}}$ $[ab > 0]$ (cf. 2.145 4.);   2.132

$\qquad = \frac{1}{4i\sqrt{ab}} \ln \frac{a + x^2 i\sqrt{ab}}{a - x^2 i\sqrt{ab}}$ $[ab < 0]$ (cf. 2.145 8.).

3. $\int \frac{x^2\,dx}{z_4} = \frac{1}{4b\alpha\sqrt{2}} \left\{ \ln \frac{x^2 - \alpha x\sqrt{2} + \alpha^2}{x^2 + \alpha x\sqrt{2} + \alpha^2} + 2 \operatorname{arctg} \frac{\alpha x\sqrt{2}}{\alpha^2 - x^2} \right\}$ $[ab > 0];$

$\qquad = -\frac{1}{4b\alpha'} \left\{ \ln \frac{x + \alpha'}{x - \alpha'} - 2 \operatorname{arctg} \frac{x}{\alpha'} \right\}$ $[ab < 0].$

4. $\int \frac{x^3\,dx}{z_4} = \frac{1}{4b} \ln z_4.$

1. $\int \frac{x^n\,dx}{z_4^m} = \frac{x^{n+1}}{4a(m-1)z_4^{m-1}} + \frac{4m - n - 5}{4a(m-1)} \int \frac{x^n\,dx}{z_4^{m-1}}.$ La 134 1   2.133

2. $\int \frac{x^n\,dx}{z_4^m} = \frac{x^{n-3}}{z_4^{m-1}(n+1-4m)b} - \frac{(n-3)a}{b(n+1-4m)} \int \frac{x^{n-4}\,dx}{z_4^m}.$

1. $\int \frac{dx}{z_4^2} = \frac{x}{4az_4} + \frac{3}{4a} \int \frac{dx}{z_4}$ (cf. 2.132 1.).   2.134

2. $\int \frac{x\,dx}{z_4^2} = \frac{x^2}{4az_4} + \frac{1}{2a} \int \frac{x\,dx}{z_4}$ (cf. 2.132 2.).

3. $\int \frac{x^2\,dx}{z_4^2} = \frac{x^3}{4az_4} + \frac{1}{4a} \int \frac{x^2\,dx}{z_4}$ (cf. 2.132 3.).

4. $\int \frac{x^3\,dx}{z_4^2} = \frac{x^4}{4az_4} = -\frac{1}{4bz_4}.$

$\int \frac{dx}{x^n z_4^m} = -\frac{1}{(n-1)ax^{n-1}z_4^{m-1}} - \frac{b(4m + n - 5)}{(n-1)a} \int \frac{dx}{x^{n-4}z_4^m}.$ 2.135

Für $n = 1$ gilt | For $n = 1$:

$\int \frac{dx}{xz_4^m} = \frac{1}{a} \int \frac{dx}{xz_4^{m-1}} - \frac{b}{a} \int \frac{dx}{x^{-3}z_4^m}.$

1. $\int \frac{dx}{xz_4} = \frac{\ln x}{a} - \frac{\ln z_4}{4a} = \frac{1}{4a} \ln \frac{x^4}{z_4}.$ 2.136

2. $\int \frac{dx}{x^2 z_4} = -\frac{1}{ax} - \frac{b}{a} \int \frac{x^2\,dx}{z_4}$ (cf. 2.132 3.).

Funktionen von $1 \pm x^n$
Functions of $1 \pm x^n$ 2.14

1. $\int \frac{dx}{1 + x} = \ln(1 \pm x).$ 2.141

2.141 2. $\int \dfrac{dx}{1+x^2} = \operatorname{arctg} x = -\operatorname{arcctg} x$ (cf. 2.124 1.).

3. $\int \dfrac{dx}{1+x^3} = \dfrac{1}{3}\ln\dfrac{1+x}{\sqrt{1-x+x^2}} + \dfrac{1}{\sqrt{3}}\operatorname{arctg}\dfrac{x\sqrt{3}}{2-x}$ (cf. 2.126 1.).

4. $\int \dfrac{dx}{1+x^4} = \dfrac{1}{4\sqrt{2}}\ln\dfrac{1+x\sqrt{2}+x^2}{1-x\sqrt{2}+x^2} + \dfrac{1}{2\sqrt{2}}\operatorname{arctg}\dfrac{x\sqrt{2}}{1-x^2}$ (cf. 2.132 1.).

2.142 $\int \dfrac{dx}{1+x^n} = -\dfrac{2}{n}\sum_{k=0}^{\frac{n}{2}-1} P_k \cos\dfrac{2k+1}{n}\pi + \dfrac{2}{n}\sum_{k=0}^{\frac{n}{2}-1} Q_k \sin\dfrac{2k+1}{n}\pi$

$\left[n \begin{array}{l} \text{positiv geradzahlig} \\ \text{any positive, even integer} \end{array}\right]$; T(43)+

$= \dfrac{1}{n}\ln(1+x) - \dfrac{2}{n}\sum_{k=0}^{\frac{n-3}{2}} P_k \cos\dfrac{2k+1}{n}\pi + \dfrac{2}{n}\sum_{k=0}^{\frac{n-3}{2}} Q_k \sin\dfrac{2k+1}{n}\pi$

$\left[n \begin{array}{l} \text{positiv ungeradzahlig} \\ \text{any positive, odd integer} \end{array}\right]$. T(45)

$P_k = \dfrac{1}{2}\ln\left(x^2 - 2x\cos\dfrac{2k+1}{n}\pi + 1\right),$

$Q_k = \operatorname{arctg}\dfrac{x\sin\dfrac{2k+1}{n}\pi}{1-x\cos\dfrac{2k+1}{n}\pi} = \operatorname{arctg}\dfrac{x-\cos\dfrac{2k+1}{n}\pi}{\sin\dfrac{2k+1}{n}\pi}.$

2.143 1. $\int \dfrac{dx}{1-x} = -\ln(1-x).$

2. $\int \dfrac{dx}{1-x^2} = \dfrac{1}{2}\ln\dfrac{1+x}{1-x} = \operatorname{Arth} x$ $[-1 < x < 1]$ (cf. 2.141 1.).

3. $\int \dfrac{dx}{x^2-1} = \dfrac{1}{2}\ln\dfrac{x-1}{x+1} = -\operatorname{Arth} x$ $[x>1, x<-1]$.

4. $\int \dfrac{dx}{1-x^3} = \dfrac{1}{3}\ln\dfrac{\sqrt{1+x+x^2}}{1-x} + \dfrac{1}{\sqrt{3}}\operatorname{arctg}\dfrac{x\sqrt{3}}{2+x}$ (cf. 2.126 1.).

5. $\int \dfrac{dx}{1-x^4} = \dfrac{1}{4}\ln\dfrac{1+x}{1-x} + \dfrac{1}{2}\operatorname{arctg} x = \dfrac{1}{2}(\operatorname{Arth} x + \operatorname{arctg} x)$ (cf. 2.132 1.).

2.144 1. $\int \dfrac{dx}{1-x^n} = \dfrac{1}{n}\ln\dfrac{1+x}{1-x} - \dfrac{2}{n}\sum_{k=1}^{\frac{n}{2}-1} P_k \cos\dfrac{2k}{n}\pi + \dfrac{2}{n}\sum_{k=1}^{\frac{n}{2}-1} Q_k \sin\dfrac{2k}{n}\pi$

$\left[n \begin{array}{l} \text{positiv geradzahlig} \\ \text{any positive, even integer} \end{array}\right].$

$P_k = \dfrac{1}{2}\ln\left(x^2 - 2x\cos\dfrac{2k}{n}\pi + 1\right), \quad Q_k = \operatorname{arctg}\dfrac{x-\cos\dfrac{2k}{n}\pi}{\sin\dfrac{2k}{n}\pi}.$ T(47)

2. $\displaystyle\int\frac{dx}{1-x^n} = -\frac{1}{n}\ln(1-x) + \frac{2}{n}\sum_{k=0}^{\frac{n-3}{2}} P_k \cos\frac{2k+1}{n}\pi +$ 2.144

$$+\frac{2}{n}\sum_{k=0}^{\frac{n-3}{2}} Q_k \sin\frac{2k+1}{n}\pi$$

$$\begin{bmatrix} n \text{ positiv ungeradzahlig} \\ \text{any positive, odd integer} \end{bmatrix}.$$ T(49)

$$P_k = \frac{1}{2}\ln\left(x^2 + 2x\cos\frac{2k+1}{n}\pi + 1\right), \quad Q_k = \operatorname{arctg}\frac{x+\cos\dfrac{2k+1}{n}\pi}{\sin\dfrac{2k+1}{n}\pi}.$$

1. $\displaystyle\int\frac{x\,dx}{1+x} = x - \ln(1+x).$ 2.145

2. $\displaystyle\int\frac{x\,dx}{1+x^2} = \frac{1}{2}\ln(1+x^2).$

3. $\displaystyle\int\frac{x\,dx}{1+x^3} = -\frac{1}{6}\ln\frac{(1+x)^2}{1-x+x^2} + \frac{1}{\sqrt{3}}\operatorname{arctg}\frac{2x-1}{\sqrt{3}}$ (cf. 2.126 2.).

4. $\displaystyle\int\frac{x\,dx}{1+x^4} = \frac{1}{2}\operatorname{arctg} x^2.$

5. $\displaystyle\int\frac{x\,dx}{1-x} = -\ln(1-x) - x.$

6. $\displaystyle\int\frac{x\,dx}{1-x^2} = -\frac{1}{2}\ln(1-x^2).$

7. $\displaystyle\int\frac{x\,dx}{1-x^3} = -\frac{1}{6}\ln\frac{(1-x)^2}{1+x+x^2} - \frac{1}{\sqrt{3}}\operatorname{arctg}\frac{2x+1}{\sqrt{3}}$ (cf. 2.126 2.).

8. $\displaystyle\int\frac{x\,dx}{1-x^4} = \frac{1}{4}\ln\frac{1+x^2}{1-x^2}$ (cf. 2.132 2.).

Sind m und n natürliche Zahlen, so gilt | For m and n positive integers: 2.146

1. $\displaystyle\int\frac{x^{m-1}\,dx}{1+x^{2n}} = -\frac{1}{2n}\sum_{k=1}^{n}\cos\frac{m\pi(2k-1)}{2n}\ln\left\{1 - 2x\cos\frac{2k-1}{2n}\pi + x^2\right\} +$

$$+\frac{1}{n}\sum_{k=1}^{n}\sin\frac{m\pi(2k-1)}{2n}\operatorname{arctg}\frac{x-\cos\dfrac{2k-1}{2n}\pi}{\sin\dfrac{2k-1}{2n}\pi} \quad [m<2n].$$ T(44)+

2. $\displaystyle\int\frac{x^{m-1}\,dx}{1+x^{2n+1}} = (-1)^{m+1}\frac{\ln(1+x)}{2n+1} -$

$$-\frac{1}{2n+1}\sum_{k=1}^{n}\cos\frac{m\pi(2k-1)}{2n+1}\ln\left\{1 - 2x\cos\frac{2k-1}{2n+1}\pi + x^2\right\} +$$

$$+\frac{2}{2n+1}\sum_{k=1}^{n}\sin\frac{m\pi(2k-1)}{2n+1}\operatorname{arctg}\frac{x-\cos\dfrac{2k-1}{2n+1}\pi}{\sin\dfrac{2k-1}{2n+1}\pi} \quad [m\leqslant 2n].$$ T(46)+

2.146

3. $\int \frac{x^{m-1}\,dx}{1-x^{2n}} = \frac{1}{2n}\{(-1)^{m+1}\ln(1+x) - \ln(1-x)\} -$

$$- \frac{1}{2n}\sum_{k=1}^{n-1}\cos\frac{km\pi}{n}\ln\left(1 - 2x\cos\frac{k\pi}{n} + x^2\right) +$$

$$+ \frac{1}{n}\sum_{k=1}^{n-1}\sin\frac{km\pi}{n}\,\text{arctg}\,\frac{x - \cos\frac{k\pi}{n}}{\sin\frac{k\pi}{n}} \qquad [m < 2n]. \qquad \text{T(43)}$$

4. $\int \frac{x^{m-1}\,dx}{1-x^{2n+1}} = -\frac{1}{2n+1}\ln(1-x) +$

$$+ (-1)^{m+1}\frac{1}{2n+1}\sum_{k=1}^{n}\cos\frac{m\pi(2k-1)}{2n+1}\ln\left(1 + 2x\cos\frac{2k-1}{2n+1}\pi + x^2\right) +$$

$$+ (-1)^{m+1}\frac{2}{2n+1}\sum_{k=1}^{n}\sin\frac{m\pi(2k-1)}{2n+1}\,\text{arctg}\,\frac{x + \cos\frac{2k-1}{2n+1}\pi}{\sin\frac{2k-1}{2n+1}\pi} \qquad [m \leqslant 2n]. \qquad \text{T(50)}$$

2.147

1. $\int \frac{x^m\,dx}{1-x^{2n}} = \frac{1}{2}\int \frac{x^m\,dx}{1-x^n} + \frac{1}{2}\int \frac{x^m\,dx}{1+x^n}.$

2. $\int \frac{x^m\,dx}{(1+x^2)^n} = -\frac{1}{2n-m-1}\cdot\frac{x^{m-1}}{(1+x^2)^{n-1}} + \frac{m-1}{2n-m-1}\int \frac{x^{m-2}\,dx}{(1+x^2)^n}.$

La 139(28)

3. $\int \frac{x^m}{1+x^2}\,dx = \frac{x^{m-1}}{m-1} - \int \frac{x^{m-2}}{1+x^2}\,dx.$

4. $\int \frac{x^m\,dx}{(1-x^2)^n} = \frac{1}{2n-m-1}\frac{x^{m-1}}{(1-x^2)^{n-1}} - \frac{m-1}{2n-m-1}\int \frac{x^{m-2}\,dx}{(1-x^2)^n};$

$$= \frac{1}{2n-2}\frac{x^{m-1}}{(1-x^2)^{n-1}} - \frac{m-1}{2n-2}\int \frac{x^{m-2}\,dx}{(1-x^2)^{n-1}}. \qquad \text{La 139(33)}$$

5. $\int \frac{x^m\,dx}{1-x^2} = -\frac{x^{m-1}}{m-1} + \int \frac{x^{m-2}\,dx}{1-x^2}.$

2.148

1. $\int \frac{dx}{x^m(1+x^2)^n} = -\frac{1}{m-1}\frac{1}{x^{m-1}(1+x^2)^{n-1}} - \frac{2n+m-3}{m-1}\int \frac{dx}{x^{m-2}(1+x^2)^n}.$

La 139(29)

Für $m = 1$ gilt | For $m = 1$:

$$\int \frac{dx}{x(1+x^2)^n} = \frac{1}{2n-2}\frac{1}{(1+x^2)^{n-1}} + \int \frac{dx}{x(1+x^2)^{n-1}}. \qquad \text{La 139(31)}$$

Für $m = 1$ und $n = 1$ gilt | For $m = 1$ and $n = 1$:

$$\int \frac{dx}{x(1+x^2)} = \ln \frac{x}{\sqrt{1+x^2}}.$$

2. $\int \frac{dx}{x^m(1+x^2)} = -\frac{1}{(m-1)x^{m-1}} - \int \frac{dx}{x^{m-2}(1+x^2)}.$

3. $\int \frac{dx}{(1+x^2)^n} = \frac{1}{2n-2}\frac{x}{(1+x^2)^{n-1}} + \frac{2n-3}{2n-2}\int \frac{dx}{(1+x^2)^{n-1}}.$

F II 37

4. $\int \frac{dx}{(1+x^2)^n} = \frac{x}{2n-1} \sum_{k=1}^{n-1} \frac{(2n-1)(2n-3)(2n-5)\ldots(2n-2k+1)}{2^k(n-1)(n-2)\ldots(n-k)(1+x^2)^{n-k}} +$ 2.148

$$+ \frac{(2n-3)!!}{2^{n-1}(n-1)!} \operatorname{arctg} x. \quad \text{T(91)}$$

1. $\int \frac{dx}{x^m(1-x^2)^n} = -\frac{1}{(m-1)x^{m-1}(1-x^2)^{n-1}} + \frac{2n+m-3}{m-1} \int \frac{dx}{x^{m-2}(1-x^2)^n}.$ 2.149

La 139(34)

Für $m = 1$ gilt | For $m = 1$:

$$\int \frac{dx}{x(1-x^2)^n} = \frac{1}{2(n-1)(1-x^2)^{n-1}} + \int \frac{dx}{x(1-x^2)^{n-1}}. \quad \text{La 139(36)}$$

Für $m = 1$ und $n = 1$ gilt | For $m = 1$ and $n = 1$:

$$\int \frac{dx}{x(1-x^2)} = \ln \frac{x}{\sqrt{1-x^2}}.$$

2. $\int \frac{dx}{(1-x^2)^n} = \frac{1}{2n-2} \frac{x}{(1-x^2)^{n-1}} + \frac{2n-3}{2n-2} \int \frac{dx}{(1-x^2)^{n-1}}.$ La 139(35)

3. $\int \frac{dx}{(1-x^2)^n} = \frac{x}{2n-1} \sum_{k=1}^{n-1} \frac{(2n-1)(2n-3)(2n-5)\ldots(2n-2k+1)}{2^k(n-1)(n-2)\ldots(n-k)(1-x^2)^{n-k}} +$

$$+ \frac{(2n-3)!!}{2^n \cdot (n-1)!} \ln \frac{1+x}{1-x}. \quad \text{T(91)}$$

Funktionen von $a + bx$ und $\alpha + \beta x$
Functions of $a + bx$ and $\alpha + \beta x$ 2.15

Abkürzungen:
Abbreviations:
$$z = a + bx; \quad t = \alpha + \beta x; \quad \Delta = a\beta - \alpha b.$$

$$\int z^n t^m \, dx = \frac{z^{n+1} t^m}{(m+n+1)b} - \frac{m\Delta}{(m+n+1)b} \int z^n t^{m-1} \, dx. \quad 2.151$$

1. $\int \frac{z}{t} \, dx = \frac{bx}{\beta} + \frac{\Delta}{\beta^2} \ln t.$ 2.152

2. $\int \frac{t}{z} \, dx = \frac{\beta x}{b} - \frac{\Delta}{b^2} \ln z.$

$$\int \frac{t^m \, dx}{z^n} = \frac{1}{(m-n+1)b} \frac{t^m}{z^{n-1}} - \frac{m\Delta}{(m-n+1)b} \int \frac{t^{m-1} \, dx}{z^n}; \quad 2.153$$

$$= \frac{1}{(n-1)\Delta} \frac{t^{m+1}}{z^{n-1}} - \frac{(m-n+2)\beta}{(n-1)\Delta} \int \frac{t^m \, dx}{z^{n-1}};$$

$$= -\frac{1}{(n-1)b} \frac{t^m}{z^{n-1}} + \frac{m\beta}{(n-1)b} \int \frac{t^{m-1}}{z^{n-1}} \, dx.$$

$$\int \frac{dx}{zt} = \frac{1}{\Delta} \ln \frac{t}{z}. \quad 2.154$$

2.155 $$\int \frac{dx}{z^n t^m} = -\frac{1}{(m-1)\Delta} \frac{1}{t^{m-1} z^{n-1}} - \frac{(m+n-2)b}{(m-1)\Delta} \int \frac{dx}{t^{m-1} z^n};$$

$$= \frac{1}{(n-1)\Delta} \frac{1}{t^{m-1} z^{n-1}} + \frac{(m+n-2)\beta}{(n-1)\Delta} \int \frac{dx}{t^m z^{n-1}}.$$

2.156 $$\int \frac{x\,dx}{zt} = \frac{1}{\Delta}\left(\frac{a}{b}\ln z - \frac{\alpha}{\beta}\ln t\right).$$

Funktionen von $a + bx^k + cx^{2k}$
2.16
Functions of $a + bx^k + cx^{2k}$

2.160 Rekursionsformeln für | Recurrence formulae for

$$R_k = a + bx^k + cx^{2k}.$$

1. $$\int x^{m-1} R_k^n \, dx = \frac{x^m R_k^{n+1}}{ma} - \frac{(m+k+nk)b}{ma} \int x^{m+k-1} R_k^n \, dx -$$

$$- \frac{(m+2k+2kn)c}{ma} \int x^{m+2k-1} R_k^n \, dx.$$

2. $$\int x^{m-1} R_k^n \, dx = \frac{x^m R_k^n}{m} - \frac{bkn}{m} \int x^{m+k-1} R_k^{n-1} \, dx - \frac{2ckn}{m} \int x^{m+2k-1} R_k^{n-1} \, dx.$$

3. $$\int x^{m-1} R_k^n \, dx = \frac{x^{m-2k} R_k^{n+1}}{(m+2kn)c} - \frac{(m-2k)a}{(m+2kn)c} \int x^{m-2k-1} R_k^n \, dx -$$

$$- \frac{(m-k+kn)b}{(m+2kn)c} \int x^{m-k-1} R_k^n \, dx;$$

$$= \frac{x^m R_k^n}{m+2kn} + \frac{2kna}{m+2kn} \int x^{m-1} R_k^{n-1} \, dx + \frac{bkn}{m+2kn} \int x^{m+k-1} R_k^{n-1} \, dx.$$

2.161 Funktionen von | Functions of
$R_2 = a + bx^2 + cx^4.$ | $R_2 = a + bx^2 + cx^4.$

Abkürzungen: $f = \dfrac{b}{2} - \dfrac{1}{2}\sqrt{b^2 - 4ac}, \quad g = \dfrac{b}{2} + \dfrac{1}{2}\sqrt{b^2 - 4ac},$
Abbreviations:

$$h = \sqrt{b^2 - 4ac}, \quad q = \sqrt[4]{\frac{a}{c}}, \quad l = 2a(n-1)(b^2 - 4ac), \quad \cos\alpha = -\frac{b}{2\sqrt{ac}}.$$

1. $$\int \frac{dx}{R_2} = \frac{c}{h}\left\{\int \frac{dx}{cx^2 + f} - \int \frac{dx}{cx^2 + g}\right\} \quad [h^2 > 0]; \qquad \text{La 146(5)}$$

$$= \frac{1}{4cq^3 \sin\alpha}\left\{\sin\frac{\alpha}{2} \ln \frac{x^2 + 2qx\cos\dfrac{\alpha}{2} + q^2}{x^2 - 2qx\cos\dfrac{\alpha}{2} + q^2} + 2\cos\frac{\alpha}{2} \operatorname{arctg} \frac{x^2 - q^2}{2qx\sin\dfrac{\alpha}{2}}\right\} \quad [h^2 < 0].$$

La 146(8)+

2. $$\int \frac{x\,dx}{R_2} = \frac{1}{2h} \ln \frac{cx^2 + f}{cx^2 + g} \quad [h^2 > 0]; \qquad \text{La 146(6)}$$

$$= \frac{1}{2cq^2 \sin\alpha} \operatorname{arctg} \frac{x^2 - q^2 \cos\alpha}{q^2 \sin\alpha} \quad [h^2 < 0]. \qquad \text{La 146(9)+}$$

3. $\int \frac{x^2 dx}{R_2} = \frac{g}{h} \int \frac{dx}{cx^2 + g} - \frac{f}{h} \int \frac{dx}{cx^2 + f}$ $[h^2 > 0]$. La 146(7)   2.161

4. $\int \frac{dx}{R_2^2} = \frac{bcx^3 + (b^2 - 2ac)x}{lR_2} + \frac{b^2 - 6ac}{l} \int \frac{dx}{R_2} + \frac{bc}{l} \int \frac{x^2 dx}{R_2}$.

5. $\int \frac{dx}{R_2^n} = \frac{bcx^3 + (b^2 - 2ac)x}{lR_2^{n-1}} + \frac{(4n-7)bc}{l} \int \frac{x^2 dx}{R_2^{n-1}} +$

$\qquad + \frac{2(n-1)h^2 + 2ac - b^2}{l} \int \frac{dx}{R_2^{n-1}}$ $[n > 1]$. La 146(10)

6. $\int \frac{dx}{x^m R_2^n} = -\frac{1}{(m-1)ax^{m-1} R_2^{n-1}} - \frac{(m+2n-3)b}{(m-1)a} \int \frac{dx}{x^{m-2} R^n} -$

$\qquad - \frac{(m+4n-5)b}{(m-1)a} \int \frac{dx}{x^{m-4} R_2^n}$. La 147(12)+

Funktionen von $a + bx + cx^2$ und Potenzen von x
Functions of $a + bx + cx^2$ and powers of x   2.17

Abkürzungen:
Abbreviations: $R = a + bx + cx^2$; $\Delta = 4ac - b^2$.

1. $\int x^{m+1} R^n dx = \frac{x^m R^{n+1}}{c(m+2n+2)} - \frac{am}{c(m+2n+2)} \int x^{m-1} R^n dx -$ 2.171

$\qquad - \frac{b(m+n+1)}{c(m+2n+2)} \int x^m R^n dx$. T(97)

2. $\int \frac{R^n dx}{x^{m+1}} = -\frac{R^{n+1}}{amx^m} + \frac{b(n-m+1)}{am} \int \frac{R^n dx}{x^m} + \frac{c(2n-m+2)}{am} \int \frac{R^n dx}{x^{m-}}$.

La 142(3), T(98)+

3. $\int \frac{dx}{R^{n+1}} = \frac{b + 2cx}{n\Delta R^n} + \frac{(4n-2)c}{n\Delta} \int \frac{dx}{R^n}$. T(94)+

4. $\int \frac{dx}{R^{n+1}} = \frac{(2cx+b)}{2n+1} \sum_{k=0}^{n-1} \frac{2^k (2n+1)(2n-1)(2n-3) \ldots (2n-2k+1) c^k}{n(n-1) \ldots (n-k) \Delta^{k+1} R^{n-k}} +$

$\qquad + 2^n \frac{(2n-1)!! c^n}{n! \Delta^n} \int \frac{dx}{R}$. T(96)+

$\int \frac{dx}{R} = \frac{1}{\sqrt{-\Delta}} \ln \frac{b + 2cx - \sqrt{-\Delta}}{b + 2cx + \sqrt{-\Delta}} = \frac{-2}{\sqrt{-\Delta}} \text{Arth} \frac{b+2cx}{\sqrt{-\Delta}}$ $[\Delta < 0]$;   2.172

$\qquad\qquad\qquad = \frac{-2}{b + 2cx}$ $[\Delta = 0]$;

$\qquad\qquad\qquad = \frac{2}{\sqrt{\Delta}} \arctg \frac{b+2cx}{\sqrt{\Delta}}$ $[\Delta > 0]$.

1. $\int \frac{dx}{R^2} = \frac{b + 2cx}{\Delta R} + \frac{2c}{\Delta} \int \frac{dx}{R}$ (cf. 2.172).   2.173

2. $\int \frac{dx}{R^3} = \frac{b + 2cx}{\Delta} \left\{ \frac{1}{2R^2} + \frac{3c}{\Delta R} \right\} + \frac{6c^2}{\Delta^2} \int \frac{dx}{R}$ (cf. 2.172).

2.174 1. $\int \dfrac{x^m\,dx}{R^n} = -\dfrac{x^{m-1}}{(2n-m-1)c R^{n-1}} - \dfrac{(n-m)b}{(2n-m-1)c}\int \dfrac{x^{m-1}\,dx}{R^n} +$

$\qquad\qquad\qquad\qquad + \dfrac{(m-1)a}{(2n-m-1)c}\int \dfrac{x^{m-2}\,dx}{R^n}.$

Für $m = 2n-1$ kann man diese Formel nicht benutzen; an ihre Stelle tritt die Formel

For $m = 2n-1$, when the above formula cannot be used, its place is taken by

2. $\int \dfrac{x^{2n-1}\,dx}{R^n} = \dfrac{1}{c}\int \dfrac{x^{2n-3}\,dx}{R^{n-1}} - \dfrac{a}{c}\int \dfrac{x^{2n-3}\,dx}{R^n} - \dfrac{b}{c}\int \dfrac{x^{2n-2}\,dx}{R^n}.$

2.175 1. $\int \dfrac{x\,dx}{R} = \dfrac{1}{2c}\ln R - \dfrac{b}{2c}\int \dfrac{dx}{R}$ (cf. 2.172).

2. $\int \dfrac{x\,dx}{R^2} = -\dfrac{2a+bx}{\Delta R} - \dfrac{b}{\Delta}\int \dfrac{dx}{R}$ (cf. 2.172).

3. $\int \dfrac{x\,dx}{R^3} = -\dfrac{2a+bx}{2\Delta R^2} - \dfrac{3b(b+2cx)}{2\Delta^2 R} - \dfrac{3bc}{\Delta^2}\int \dfrac{dx}{R}$ (cf. 2.172).

4. $\int \dfrac{x^2\,dx}{R} = \dfrac{x}{c} - \dfrac{b}{2c^2}\ln R + \dfrac{b^2-2ac}{2c^2}\int \dfrac{dx}{R}$ (cf. 2.172).

5. $\int \dfrac{x^2\,dx}{R^2} = \dfrac{ab+(b^2-2ac)x}{c\Delta R} + \dfrac{2a}{\Delta}\int \dfrac{dx}{R}$ (cf. 2.172).

6. $\int \dfrac{x^2\,dx}{R^3} = \dfrac{ab+(b^2-2ac)x}{2c\Delta R^2} + \dfrac{(2ac+b^2)(b+2cx)}{2c\Delta^2 R} + \dfrac{2ac+b^2}{\Delta^2}\int \dfrac{dx}{R}$

(cf. 2.172).

7. $\int \dfrac{x^3\,dx}{R} = \dfrac{x^2}{2c} - \dfrac{bx}{c^2} + \dfrac{b^2-ac}{2c^3}\ln R - \dfrac{b(b^2-3ac)}{2c^3}\int \dfrac{dx}{R}$ (cf. 2.172).

8. $\int \dfrac{x^3\,dx}{R^2} = \dfrac{1}{2c^2}\ln R + \dfrac{a(2ac-b^2)+b(3ac-b^2)x}{c^2\Delta R} - \dfrac{b(6ac-b^2)}{2c^2\Delta}\int \dfrac{dx}{R}$

(cf. 2.172).

9. $\int \dfrac{x^3\,dx}{R^3} = -\left(\dfrac{x^2}{c} + \dfrac{abx}{c\Delta} + \dfrac{2a^2}{c\Delta}\right)\dfrac{1}{2R^2} - \dfrac{3ab}{2c\Delta}\int \dfrac{dx}{R^2}$ (cf. 2.173 1.).

2.176 $\int \dfrac{dx}{x^m R^n} = \dfrac{-1}{(m-1)ax^{m-1}R^{n-1}} - \dfrac{b(m+n-2)}{a(m-1)}\int \dfrac{dx}{x^{m-1}R^n} -$

$\qquad\qquad\qquad\qquad - \dfrac{c(m+2n-3)}{a(m-1)}\int \dfrac{dx}{x^{m-2}R^n}.$

2.177 1. $\int \dfrac{dx}{xR} = \dfrac{1}{2a}\ln \dfrac{x^2}{R} - \dfrac{b}{2a}\int \dfrac{dx}{R}$ (cf. 2.172).

2. $\int \dfrac{dx}{xR^2} = \dfrac{1}{2a^2}\ln \dfrac{x^2}{R} + \dfrac{1}{2aR}\left\{1 - \dfrac{b(b+2cx)}{\Delta}\right\} - \dfrac{b}{2a^2}\left(1 + \dfrac{2ac}{\Delta}\right)\int \dfrac{dx}{R}$

(cf. 2.172).

3. $\int \dfrac{dx}{xR^3} = \dfrac{1}{4aR^2} + \dfrac{1}{2a^2 R} + \dfrac{1}{2a^3}\ln \dfrac{x^2}{R} - \dfrac{b}{2a}\int \dfrac{dx}{R^3} - \dfrac{b}{2a^2}\int \dfrac{dx}{R^2} - \dfrac{b}{2a^3}\int \dfrac{dx}{R}$

(cf. 2.172, 2.173).

4. $\int \frac{dx}{x^2 R} = -\frac{b}{2a^2} \ln \frac{x^2}{R} - \frac{1}{ax} + \frac{b^2 - 2ac}{2a^2} \int \frac{dx}{R}$ (cf. 2.172). 2.177

5. $\int \frac{dx}{x^2 R^2} = -\frac{b}{a^3} \ln \frac{x^2}{R} - \frac{a+bx}{a^2 xR} + \frac{(b^2 - 3ac)(b + 2cx)}{a^2 \Delta R} -$

$$- \frac{1}{\Delta}\left(\frac{b^4}{a^3} - \frac{6b^2 c}{a^2} + \frac{6c^2}{a}\right) \int \frac{dx}{R} \quad \text{(cf. 2.172).}$$

6. $\int \frac{dx}{x^2 R^3} = -\frac{1}{axR^2} - \frac{3b}{a} \int \frac{dx}{xR^3} - \frac{5c}{a} \int \frac{dx}{R^3}$ (cf. 2.173, 2.177 3.).

7. $\int \frac{dx}{x^3 R} = -\frac{ac - b^2}{2a^3} \ln \frac{x^2}{R} + \frac{b}{a^2 x} - \frac{1}{2ax^2} + \frac{b(3ac - b^2)}{2a^3} \int \frac{dx}{R}$ (cf. 2.172).

8. $\int \frac{dx}{x^3 R^2} = \left(-\frac{1}{2ax^2} + \frac{3b}{2a^2 x}\right) \frac{1}{R} + \left(\frac{3b^2}{a^2} - \frac{2c}{a}\right) \int \frac{dx}{xR^2} + \frac{9bc}{2a^2} \int \frac{dx}{R^2}$

(cf. 2.173 1., 2.177 2.).

9. $\int \frac{dx}{x^3 R^3} = \left(\frac{-1}{2ax^2} + \frac{2b}{a^2 x}\right) \frac{1}{R^2} + \left(\frac{6b^2}{a^2} - \frac{3c}{a}\right) \int \frac{dx}{xR^3} + \frac{10bc}{a^2} \int \frac{dx}{R^3}$

(cf. 2.173 2., 2.177 3.).

Funktionen von $a + bx + cx^2$ und $\alpha + \beta x$
Functions of $a + bx + cx^2$ and $\alpha + \beta x$

2.18

Abkürzungen: $\begin{cases} R = a + bx + cx^2; \ z = \alpha + \beta x; \ A = a\beta^2 - \alpha b\beta + c\alpha^2; \\ B = b\beta - 2c\alpha; \ \Delta = 4ac - b^2. \end{cases}$
Abbreviations:

1. $\int z^m R^n dx = \frac{\beta z^{m-1} R^{n+1}}{(m + 2n + 1)c} - \frac{(m+n)B}{(m + 2n + 1)c} \int z^{m-1} R^n dx -$

$$- \frac{(m-1)A}{(m + 2n + 1)c} \int z^{m-2} R^n dx.$$

2. $\int \frac{R^n dx}{z^m} = -\frac{1}{(m - 2n - 1)\beta} \frac{R^n}{z^{m-1}} - \frac{2nA}{(m - 2n - 1)\beta^2} \int \frac{R^{n-1} dx}{z^m} -$

$$- \frac{nB}{(m - 2n - 1)\beta^2} \int \frac{R^{n-1} dx}{z^{m-1}}; \quad \text{La 148(4)}+$$

$= -\frac{\beta}{(m-1)A} \frac{R^{n+1}}{z^{m-1}} - \frac{(m-n-2)B}{(m-1)A} \int \frac{R^n dx}{z^{m-1}} - \frac{(m-2n-3)c}{(m-1)A} \int \frac{R^n dx}{z^{m-2}};$ La 148(5)

$= -\frac{1}{(m-1)\beta} \frac{R^n}{z^{m-1}} + \frac{nB}{(m-1)\beta^2} \int \frac{R^{n-1} dx}{z^{m-1}} + \frac{2nc}{(m-1)\beta^2} \int \frac{R^{n-1} dx}{z^{m-2}}.$ La 148(6)

3. $\int \frac{z^m dx}{R^n} = \frac{\beta}{(m - 2n + 1)c} \frac{z^{m-1}}{R^{n-1}} - \frac{(m-n)B}{(m - 2n + 1)c} \int \frac{z^{m-1} dx}{R^n} -$

$$- \frac{(m-1)A}{(m - 2n + 1)c} \int \frac{z^{m-2} dx}{R^n}; \quad \text{La 147(1)}$$

$= \frac{b + 2cx}{(n-1)\Delta} \frac{z^m}{R^{n-1}} - \frac{2(m - 2n + 3)c}{(n-1)\Delta} \int \frac{z^m dx}{R^n} - \frac{Bm}{(n-1)\Delta} \int \frac{z^{m-1} dx}{R^{n-1}}.$

La 148(3)

2.18   4. $\int \dfrac{dx}{z^m R^n} = -\dfrac{\beta}{(m-1)A} \dfrac{1}{z^{m-1}R^{n-1}} - \dfrac{(m+n-2)B}{(m-1)A} \int \dfrac{dx}{z^{m-1}R^n} -$

$$- \dfrac{(m+2n-3)c}{(m-1)A} \int \dfrac{dx}{z^{m-2}R^n} ; \qquad \text{La 148(7)}$$

$$= \dfrac{\beta}{2(n-1)A} \dfrac{1}{z^{m-1}R^{n-1}} - \dfrac{B}{2A} \int \dfrac{dx}{z^{m-1}R^n} + \dfrac{(m+2n-3)\beta^2}{2(n-1)A} \int \dfrac{dx}{z^m R^{n-1}}.$$

$$\text{La 148(8)}$$

Für $m = 1$ und $n = 1$ gilt | For $m = 1$ and $n = 1$:

$$\int \dfrac{dx}{zR} = \dfrac{\beta}{2A} \ln \dfrac{z^2}{R} - \dfrac{B}{2A} \int \dfrac{dx}{R}.$$

Für $A = 0$ gilt | For $A = 0$:

$$\int \dfrac{dx}{z^m R^n} = -\dfrac{\beta}{(m+n-1)B} \dfrac{1}{z^m R^{n-1}} - \dfrac{(m+2n-2)c}{(m+n-1)B} \int \dfrac{dx}{z^{m-1}R^n}. \qquad \text{La 148(9)}$$

Algebraische Funktionen
Algebraic Functions

2.2

Einführung
Introduction

2.20

2.201   Integrale der Gestalt | Integrals of the form

$$\int R\left(x, \left(\dfrac{\alpha x + \beta}{\gamma x + \delta}\right)^r, \left(\dfrac{\alpha x + \beta}{\gamma x + \delta}\right)^s, \ldots\right) dx$$

mit rationalen Zahlen r, s, ... gehen durch die Substitution | where r, s, ... are rational numbers, can be reduced to integrals of rational functions by the substitution

$$\dfrac{\alpha x + \beta}{\gamma x + \delta} = t^m, \qquad \text{F II 53}$$

in der m der Hauptnenner der Brüche r, s, ... ist, in Integrale rationaler Funktionen über. | where m is the common denominator of the fractions r, s, ...

2.202   Integrale der Gestalt $\int x^m(a+bx^n)^p dx$ (Integrale von Binomialausdrücken), wobei m, n und p rationale Zahlen sind, lassen sich nur in den folgenden Fällen durch elementare Funktionen ausdrücken: | Integrals of binomial expressions of the form $\int x^m(a+bx^n)^p dx$, where m, n, and p are rational, lead to elementary functions only in the following cases:

a) p ist ganzzahlig; dann besteht dieses Integral aus einer Summe von Integralen, wie sie in 2.201 behandelt wurden; | (a) p is an integer—the integral reduces to a sum of integrals that can be treated according to 2.201;

b) $\dfrac{m+1}{n}$ ist ganzzahlig; dann führt | (b) $\dfrac{m+1}{n}$ is an integer—by means of die Substitution $x^n = z$ dieses Integral in die in 2.201 behandelte Gestalt

the substitution $x^n = z$ the integral is transformed into

$$\frac{1}{n} \int (a + bz)^p z^{\frac{m+1}{n} - 1} dz$$

über;

which can be treated according to 2.201;

c) $\frac{m+1}{n} + p$ ist ganzzahlig; dieselbe Substitution $x^n = z$ bringt dann das vorgegebene Integral auf die in 2.201 behandelte Gestalt

(c) $\frac{m+1}{n} + p$ is an integer—the same substitution $x^n = z$ transforms the integral into

$$\frac{1}{n} \int \left(\frac{a + bz}{z}\right)^p z^{\frac{m+1}{n} + p - 1} dz,$$

which may be treated according to 2.201.

Rekursionsformeln für Integrale von Binomialausdrücken siehe 2.110.

For recurrence formulae for integrals of binomials see 2.110.

Funktionen von $a + bx^k$ und $\sqrt{x}$

Functions of $a + bx^k$ and $\sqrt{x}$

2.21

Abkürzung:
Abbreviation: $z_1 = a + bx$.

$$\int \frac{dx}{z_1 \sqrt{x}} = \frac{2}{\sqrt{ab}} \operatorname{arctg} \sqrt{\frac{bx}{a}} \quad [ab > 0];$$ 2.211

$$= \frac{1}{i\sqrt{ab}} \ln \frac{a - bx + 2i\sqrt{xab}}{z_1} \quad [ab < 0].$$

$$\int \frac{x^m \sqrt{x}}{z_1} dx = 2\sqrt{x} \sum_{k=0}^{m} \frac{(-1)^k a^k x^{m-k}}{(2m - 2k + 1) b^{k+1}} + (-1)^{m+1} \frac{a^{m+1}}{b^{m+1}} \int \frac{dx}{z_1 \sqrt{x}} \quad \text{(cf. 2.211).} \quad 2.212$$

1. $\int \frac{\sqrt{x}\, dx}{z_1} = \frac{2\sqrt{x}}{b} - \frac{a}{b} \int \frac{dx}{z_1 \sqrt{x}}$ (cf. 2.211). 2.213

2. $\int \frac{x\sqrt{x}\, dx}{z_1} = \left(\frac{x}{3b} - \frac{a}{b^2}\right) 2\sqrt{x} + \frac{a^2}{b^2} \int \frac{dx}{z_1 \sqrt{x}}$ (cf. 2.211).

3. $\int \frac{x^2 \sqrt{x}\, dx}{z_1} = \left(\frac{x^2}{5b} - \frac{xa}{3b^2} + \frac{a^2}{b^3}\right) 2\sqrt{x} - \frac{a^3}{b^3} \int \frac{dx}{z_1 \sqrt{x}}$ (cf. 2.211).

4. $\int \frac{dx}{z_1^2 \sqrt{x}} = \frac{\sqrt{x}}{az_1} + \frac{1}{2a} \int \frac{dx}{z_1 \sqrt{x}}$ (cf. 2.211).

5. $\int \frac{\sqrt{x}\, dx}{z_1^2} = -\frac{\sqrt{x}}{bz_1} + \frac{1}{2b} \int \frac{dx}{z_1 \sqrt{x}}$ (cf. 2.211).

6. $\int \frac{x \sqrt{x}\, dx}{z_1^2} = \frac{2x\sqrt{x}}{bz_1} - \frac{3a}{b} \int \frac{\sqrt{x}\, dx}{z_1^2}$ (cf. 2.213 5.).

Unbestimmte Integrale elem. Funktionen/Indefinite Integrals of Elem. Functions

2.213

7. $\int \dfrac{x^2 \sqrt{x}\, dx}{z_1^2} = \left(\dfrac{x^2}{3b} - \dfrac{5ax}{3b^2} \right) \dfrac{2\sqrt{x}}{z_1} + \dfrac{5a^2}{b^2} \int \dfrac{\sqrt{x}\, dx}{z_1^2}$ (cf. 2.213 5.).

8. $\int \dfrac{dx}{z_1^3 \sqrt{x}} = \left(\dfrac{1}{2az_1^2} + \dfrac{3}{4a^2 z_1} \right) \sqrt{x} + \dfrac{3}{8a^2} \int \dfrac{dx}{z_1 \sqrt{x}}$ (cf. 2.211).

9. $\int \dfrac{\sqrt{x}\, dx}{z_1^3} = \left(-\dfrac{1}{2bz_1^2} + \dfrac{1}{4abz_1} \right) \sqrt{x} + \dfrac{1}{8ab} \int \dfrac{dx}{z_1 \sqrt{x}}$ (cf. 2.211).

10. $\int \dfrac{x \sqrt{x}\, dx}{z_1^3} = -\dfrac{2x\sqrt{x}}{bz_1^2} + \dfrac{3a}{b} \int \dfrac{\sqrt{x}\, dx}{z_1^3}$ (cf. 2.213 9.).

11. $\int \dfrac{x^2 \sqrt{x}\, dx}{z_1^3} = \left(\dfrac{x^2}{b} + \dfrac{5ax}{b^2} \right) \dfrac{2\sqrt{x}}{z_1^2} - \dfrac{15a^2}{b^2} \int \dfrac{\sqrt{x}\, dx}{z_1^3}$ (cf. 2.213 9.).

Abkürzungen:
Abbreviations: $z_2 = a + bx^2$, $\alpha = \sqrt[4]{\dfrac{a}{b}}$; $\alpha' = \sqrt[4]{-\dfrac{a}{b}}$.

2.214

$\int \dfrac{dx}{z_2 \sqrt{x}} = \dfrac{1}{b\alpha^3 \sqrt{2}} \left[\ln \dfrac{x + \alpha\sqrt{2x} + \alpha^2}{\sqrt{z_2}} + \operatorname{arctg} \dfrac{\alpha\sqrt{2x}}{\alpha^2 - x} \right]$ $\left[\dfrac{a}{b} > 0 \right]$;

$= \dfrac{1}{2b\alpha'^3} \left(\ln \dfrac{\alpha' - \sqrt{x}}{\alpha' + \sqrt{x}} - 2 \operatorname{arctg} \dfrac{\sqrt{x}}{\alpha'} \right)$ $\left[\dfrac{a}{b} < 0 \right]$.

2.215

$\int \dfrac{\sqrt{x}\, dx}{z_2} = \dfrac{1}{b\alpha \sqrt{2}} \left[-\ln \dfrac{x + \alpha\sqrt{2x} + \alpha^2}{\sqrt{z_2}} + \operatorname{arctg} \dfrac{\alpha\sqrt{2x}}{\alpha^2 - x} \right]$ $\left[\dfrac{a}{b} > 0 \right]$;

$= \dfrac{1}{2b\alpha'} \left[\ln \dfrac{\alpha' - \sqrt{x}}{\alpha' + \sqrt{x}} + 2 \operatorname{arctg} \dfrac{\sqrt{x}}{\alpha'} \right]$ $\left[\dfrac{a}{b} < 0 \right]$.

2.216

1. $\int \dfrac{x \sqrt{x}\, dx}{z_2} = \dfrac{2\sqrt{x}}{b} - \dfrac{a}{b} \int \dfrac{dx}{z_2 \sqrt{x}}$ (cf. 2.214).

2. $\int \dfrac{x^2 \sqrt{x}\, dx}{z_2} = \dfrac{2x\sqrt{x}}{3b} - \dfrac{a}{b} \int \dfrac{\sqrt{x}\, dx}{z_2}$ (cf. 2.215).

3. $\int \dfrac{dx}{z_2^2 \sqrt{x}} = \dfrac{\sqrt{x}}{2az_2} + \dfrac{1}{4a} \int \dfrac{dx}{z_2 \sqrt{x}}$ (cf. 2.214).

4. $\int \dfrac{\sqrt{x}\, dx}{z_2^2} = \dfrac{x\sqrt{x}}{2az_2} + \dfrac{1}{4a} \int \dfrac{\sqrt{x}\, dx}{z_2}$ (cf. 2.215).

5. $\int \dfrac{x \sqrt{x}\, dx}{z_2^2} = -\dfrac{\sqrt{x}}{2bz_2} + \dfrac{1}{4b} \int \dfrac{dx}{z_2 \sqrt{x}}$ (cf. 2.214).

6. $\int \dfrac{x^2 \sqrt{x}\, dx}{z_2^2} = -\dfrac{x\sqrt{x}}{2bz_2} + \dfrac{3}{4b} \int \dfrac{\sqrt{x}\, dx}{z_2}$ (cf. 2.215).

7. $\int \dfrac{dx}{z_2^3 \sqrt{x}} = \left(\dfrac{1}{4az_2^2} + \dfrac{7}{16a^2 z_2} \right) \sqrt{x} + \dfrac{21}{32a^2} \int \dfrac{dx}{z_2 \sqrt{x}}$ (cf. 2.214).

8. $\int \dfrac{\sqrt{x}\, dx}{z_2^3} = \left(\dfrac{1}{4az_2^2} + \dfrac{5}{16a^2 z_2} \right) x\sqrt{x} + \dfrac{5}{32a^2} \int \dfrac{\sqrt{x}\, dx}{z_2}$ (cf. 2.215).

9. $\int \dfrac{x \sqrt{x}\, dx}{z_2^3} = \dfrac{(bx^2 - 3a)\sqrt{x}}{16abz_2^2} + \dfrac{3}{32ab} \int \dfrac{dx}{z_2 \sqrt{x}}$ (cf. 2.214).

10. $\quad\int\frac{x^2\sqrt{x}\,dx}{z_2^3} = -\frac{2x\sqrt{x}}{5bz_2^2} + \frac{3a}{5b}\int\frac{\sqrt{x}\,dx}{z_2^3}\quad$ (cf. 2.216 8.). $\qquad$ 2.216

Funktionen von $\sqrt[n]{(a+bx)^k}$

Functions of $\sqrt[n]{(a+bx)^k}$ $\qquad$ 2.22–2.23

Abkürzung:
Abbreviation: $\quad z = a + bx.$

$$\int x^n \sqrt[l]{z^{lm+f}}\,dx = \left\{\sum_{k=0}^{n}\frac{(-1)^k\binom{n}{k}z^{n-k}a^k}{ln - lk + l(m+1) + f}\right\}\frac{l\sqrt[l]{z^{l(m+1)+f}}}{b^{n+1}}.\qquad 2.220$$

Quadratwurzeln

Square roots

$$\int x^n \sqrt{z^{2m-1}}\,dx = \left\{\sum_{k=0}^{n}\frac{(-1)^k\binom{n}{k}z^{n-k}a^k}{2n - 2k + 2m + 1}\right\}\frac{2\sqrt{z^{2m+1}}}{b^{n+1}}.\qquad 2.221$$

1. $\quad\int\frac{dx}{\sqrt{z}} = \frac{2}{b}\sqrt{z}.\qquad$ 2.222

2. $\quad\int\frac{x\,dx}{\sqrt{z}} = \left(\frac{1}{3}z - a\right)\frac{2\sqrt{z}}{b^2}.$

3. $\quad\int\frac{x^2\,dx}{\sqrt{z}} = \left(\frac{1}{5}z^2 - \frac{2}{3}az + a^2\right)\frac{2\sqrt{z}}{b^3}.$

1. $\quad\int\frac{dx}{\sqrt{z^3}} = -\frac{2}{b\sqrt{z}}.\qquad$ 2.223

2. $\quad\int\frac{x\,dx}{\sqrt{z^3}} = (z+a)\frac{2}{b^2\sqrt{z}}.$

3. $\quad\int\frac{x^2\,dx}{\sqrt{z^3}} = \left(\frac{z^2}{3} - 2az - a^2\right)\frac{2}{b^3\sqrt{z}}.$

1. $\quad\int\frac{z^m\,dx}{x^n\sqrt{z}} = -\frac{z^m\sqrt{z}}{(n-1)ax^{n-1}} + \frac{2m - 2n + 3}{2(n-1)}\frac{b}{a}\int\frac{z^m\,dx}{x^{n-1}\sqrt{z}}.\qquad$ 2.224

2. $\quad\int\frac{z^m\,dx}{x^n\sqrt{z}} = -z^m\sqrt{z}\left\{\frac{1}{(n-1)ax^{n-1}} + \right.$

$\qquad + \sum_{k=1}^{n-2}\frac{(2m-2n+3)(2m-2n+5)\ldots(2m-2n+2k+1)}{2^k(n-1)(n-2)\ldots(n-k-1)x^{n-k-1}}\frac{b^k}{a^{k+1}}\bigg\} +$

$\qquad + \frac{(2m-2n+3)(2m-2n+5)\ldots(2m-3)(2m-1)}{2^{n-1}(n-1)!\,x}\frac{b^{n-1}}{a^{n-1}}\int\frac{z^m\,dx}{x\sqrt{z}}.$

Für $n=1$ gilt $\qquad\qquad\big|\qquad$ For $n=1$:

3. $\quad\int\frac{z^m}{x\sqrt{z}}\,dx = \frac{2z^m}{(2m-1)\sqrt{z}} + a\int\frac{z^{m-1}}{x\sqrt{z}}\,dx.$

2.224   4. $\quad \int \dfrac{z^m}{x\sqrt{z}}\,dx = \sum\limits_{k=1}^{m} \dfrac{2a^{m-k}z^k}{(2k-1)\sqrt{z}} + a^m \int \dfrac{dx}{x\sqrt{z}}$.

5. $\quad \int \dfrac{dx}{x\sqrt{z}} = \dfrac{1}{\sqrt{a}} \ln \dfrac{\sqrt{z}-\sqrt{a}}{\sqrt{z}+\sqrt{a}} \quad [a>0];$

$\qquad\qquad\quad = \dfrac{2}{\sqrt{-a}} \operatorname{arctg} \dfrac{\sqrt{z}}{\sqrt{-a}} \quad [a<0].$

2.225   1. $\quad \int \dfrac{\sqrt{z}\,dx}{x} = 2\sqrt{z} + a\int \dfrac{dx}{x\sqrt{z}} \quad$ (cf. 2.224 4.).

2. $\quad \int \dfrac{\sqrt{z}\,dx}{x^2} = -\dfrac{\sqrt{z}}{x} + \dfrac{b}{2}\int \dfrac{dx}{x\sqrt{z}} \quad$ (cf. 2.224 4.).

3. $\quad \int \dfrac{\sqrt{z}\,dx}{x^3} = -\dfrac{\sqrt{z^3}}{2ax^2} + \dfrac{b\sqrt{z}}{4ax} - \dfrac{b^2}{8a}\int \dfrac{dx}{x\sqrt{z}} \quad$ (cf. 2.224 4.).

2.226   1. $\quad \int \dfrac{\sqrt{z^3}\,dx}{x} = \left(\dfrac{z}{3}+a\right) 2\sqrt{z} + a^2 \int \dfrac{dx}{x\sqrt{z}} \quad$ (cf. 2.224 4.).

2. $\quad \int \dfrac{\sqrt{z^3}\,dx}{x^2} = -\dfrac{\sqrt{z^5}}{ax} + \dfrac{3b}{2a}\int \dfrac{\sqrt{z^3}\,dx}{x} \quad$ (cf. 2.226 1.).

3. $\quad \int \dfrac{\sqrt{z^3}\,dx}{x^3} = -\left(\dfrac{1}{2ax^2}+\dfrac{b}{4a^2 x}\right)\sqrt{z^5} + \dfrac{3b^2}{8a^2}\int \dfrac{\sqrt{z^3}\,dx}{x} \quad$ (cf. 2.226 1.).

2.227 $\quad \int \dfrac{dx}{xz^m\sqrt{z}} = \sum\limits_{k=0}^{m-1} \dfrac{2}{(2k+1)a^{m-k}z^k\sqrt{z}} + \dfrac{1}{a^m}\int \dfrac{dx}{x\sqrt{z}} \quad$ (cf. 2.224 4.).

2.228   1. $\quad \int \dfrac{dx}{x^2\sqrt{z}} = -\dfrac{\sqrt{z}}{ax} - \dfrac{b}{2a}\int \dfrac{dx}{x\sqrt{z}} \quad$ (cf. 2.224 4.).

2. $\quad \int \dfrac{dx}{x^3\sqrt{z}} = \left(-\dfrac{1}{2ax^2}+\dfrac{3b}{4a^2 x}\right)\sqrt{z} + \dfrac{3b^2}{8a^2}\int \dfrac{dx}{x\sqrt{z}} \quad$ (cf. 2.224 4.).

2.229   1. $\quad \int \dfrac{dx}{x\sqrt{z^3}} = \dfrac{2}{a\sqrt{z}} + \dfrac{1}{a}\int \dfrac{dx}{x\sqrt{z}} \quad$ (cf. 2.224 4.).

2. $\quad \int \dfrac{dx}{x^2\sqrt{z^3}} = \left(-\dfrac{1}{ax} - \dfrac{3b}{a^2}\right)\dfrac{1}{\sqrt{z}} - \dfrac{3b}{2a^2}\int \dfrac{dx}{x\sqrt{z}} \quad$ (cf. 2.224 4.).

3. $\quad \int \dfrac{dx}{x^3\sqrt{z^3}} = \left(-\dfrac{1}{2ax^2} + \dfrac{5b}{4a^2 x} + \dfrac{15b^2}{4a^3}\right)\dfrac{1}{\sqrt{z}} + \dfrac{15b^2}{8a^3}\int \dfrac{dx}{x\sqrt{z}} \quad$ (cf. 2.224 4.).

Kubikwurzeln
Cube roots

2.231   1. $\quad \int \sqrt[3]{z^{3m+1}}\, x^n\, dx = \left\{\sum\limits_{k=0}^{n} \dfrac{(-1)^k \binom{n}{k} z^{n-k} a^k}{3n-3k+3(m+1)+1}\right\} \dfrac{3\sqrt[3]{z^{3(m+1)+1}}}{b^{n+1}}$.

2. $\quad \int \dfrac{x^n\,dx}{\sqrt[3]{z^{3m+2}}} = \left\{\sum\limits_{k=0}^{n} \dfrac{(-1)^k \binom{n}{k} z^{n-k} a^k}{3n-3k-3(m-1)-2}\right\} \dfrac{3}{b^{n+1}\sqrt[3]{z^{3(m-1)+2}}}$.

Algebraische Funktionen/Algebraic Functions

3. $\int \sqrt[3]{z^{3m+2}}\, x^n dx = \left\{ \sum_{k=0}^{n} \frac{(-1)^k \binom{n}{k} z^{n-k} a^k}{3n - 3k + 3(m+1) + 2} \right\} \frac{3\sqrt[3]{z^{3(m+1)+2}}}{b^{n+1}}.$ 2.231

4. $\int \frac{x^n dx}{\sqrt[3]{z^{3m+1}}} = \left\{ \sum_{k=0}^{n} \frac{(-1)^k \binom{n}{k} z^{n-k} a^k}{3n - 3k - 3(m-1) - 1} \right\} \frac{3}{b^{n+1} \sqrt[3]{z^{3(m-1)+1}}}.$

5. $\int \frac{z^n dx}{x^m \sqrt[3]{x^2}} = -\frac{z^{n+\frac{1}{3}}}{(m-1) a x^{m-1}} + \frac{3n - 3m + 4}{3(m-1)} \frac{b}{a} \int \frac{z^n dx}{x^{m-1} \sqrt[3]{z^2}}.$

Für $m = 1$ gilt | For $m = 1$:

$$\int \frac{z^n dx}{x\sqrt[3]{z^2}} = \frac{3z^n}{(3n-2)\sqrt[3]{z^2}} + a \int \frac{z^{n-1} dx}{x\sqrt[3]{z^2}}.$$

6. $\int \frac{dx}{xz^n \sqrt[3]{z^2}} = \frac{3\sqrt[3]{z}}{(3n-1) az^n} + \frac{1}{a} \int \frac{\sqrt[3]{z}\, dx}{xz^n}.$

$$\int \frac{dx}{x\sqrt[3]{z^2}} = \frac{1}{\sqrt[3]{a^2}} \left\{ \frac{3}{2} \ln \frac{\sqrt[3]{z} - \sqrt[3]{a}}{\sqrt[3]{x}} - \sqrt{3} \arctg \frac{\sqrt{3}\sqrt[3]{z}}{\sqrt[3]{z} + 2\sqrt[3]{a}} \right\}.$$ 2.232

1. $\int \frac{\sqrt[3]{z}\, dx}{x} = 3\sqrt[3]{z} + a \int \frac{dx}{x\sqrt[3]{z^2}}$ (cf. 2.232).   2.233

2. $\int \frac{\sqrt[3]{z}\, dx}{x^2} = -\frac{z\sqrt[3]{z}}{ax} + \frac{b}{a} \sqrt[3]{z} + \frac{b}{3} \int \frac{dx}{x\sqrt[3]{z^2}}$ (cf. 2.232).

3. $\int \frac{\sqrt[3]{z}\, dx}{x^3} = \left(-\frac{1}{2ax^2} + \frac{b}{3a^2 x} \right) z\sqrt[3]{z} - \frac{b^2}{3a^2} \sqrt[3]{z} - \frac{b^2}{9a} \int \frac{dx}{x\sqrt[3]{z^2}}$ (cf. 2.232).

4. $\int \frac{dx}{x^2 \sqrt[3]{z^2}} = -\frac{\sqrt[3]{z}}{ax} - \frac{2b}{3a} \int \frac{dx}{x\sqrt[3]{z^2}}$ (cf. 2.232).

5. $\int \frac{dx}{x^3 \sqrt[3]{z^2}} = \left[-\frac{1}{2ax^2} + \frac{5b}{6a^2 x} \right] \sqrt[3]{z} + \frac{5b^2}{9a^2} \int \frac{dx}{x\sqrt[3]{z^2}}$ (cf. 2.232).

1. $\int \frac{z^n dx}{x^m \sqrt[3]{z}} = -\frac{z^n \sqrt[3]{z}}{(m-1) ax^{m-1}} + \frac{3n - 3m + 5}{3(m-1)} \frac{b}{a} \int \frac{z^n dx}{x^{m-1} \sqrt[3]{z}}.$ 2.234

Für $m = 1$ gilt | For $m = 1$:

2. $\int \frac{z^n dx}{x\sqrt[3]{z}} = \frac{3z^n}{(3n-1)\sqrt[3]{z}} + a \int \frac{z^{n-1} dx}{x\sqrt[3]{z}}.$

2.234 3.
$$\int \frac{dx}{xz^n \sqrt[3]{z}} = \frac{3\sqrt[3]{z^2}}{(3n-2)az^n} + \frac{1}{a}\int \frac{\sqrt[3]{z^2}dx}{xz^n}.$$

2.235
$$\int \frac{dx}{x\sqrt[3]{z}} = \frac{1}{\sqrt[3]{a^2}}\left\{ \frac{3}{2}\ln\frac{\sqrt[3]{z}-\sqrt[3]{a}}{\sqrt[3]{x}} + \sqrt{3}\operatorname{arctg}\frac{\sqrt{3}\sqrt[3]{z}}{\sqrt[3]{z}+2\sqrt[3]{a}}\right\}.$$

2.236 1.
$$\int \frac{\sqrt[3]{z^2}dx}{x} = \frac{3}{2}\sqrt[3]{z^2} + a\int \frac{dx}{x\sqrt[3]{z}} \quad \text{(cf. 2.235)}.$$

2.
$$\int \frac{\sqrt[3]{z^2}dx}{x^2} = -\frac{\sqrt[3]{z^5}}{ax} + \frac{b}{a}\sqrt[3]{z^2} + \frac{2b}{3}\int \frac{dx}{x\sqrt[3]{z}} \quad \text{(cf. 2.235)}.$$

3.
$$\int \frac{\sqrt[3]{z^2}dx}{x^3} = \left[-\frac{1}{2ax^2} + \frac{b}{6a^2x}\right]z^{\frac{5}{3}} - \frac{b^2}{6a^2}\sqrt[3]{z^2} - \frac{b^2}{9a}\int \frac{dx}{x\sqrt[3]{z}} \quad \text{(cf. 2.235)}.$$

4.
$$\int \frac{dx}{x^2\sqrt[3]{z}} = -\frac{\sqrt[3]{z^2}}{ax} - \frac{b}{3a}\int \frac{dx}{x\sqrt[3]{z}} \quad \text{(cf. 2.235)}.$$

5.
$$\int \frac{dx}{x^3\sqrt[3]{z}} = \left[-\frac{1}{2ax^2} + \frac{2b}{3a^2x}\right]\sqrt[3]{z} + \frac{2b^2}{9a^2}\int \frac{dx}{x\sqrt[3]{z}} \quad \text{(cf. 2.235)}.$$

2.24 **Funktionen von $\sqrt{a+bx}$ und $\alpha+\beta x$**
Functions of $\sqrt{a+bx}$ and $\alpha+\beta x$

Abkürzungen: $z = a+bx$, $t = \alpha+\beta x$, $\Delta = a\beta - b\alpha$.
Abbreviations:

2.241 1.
$$\int \frac{z^m t^n dx}{\sqrt{z}} = \frac{2}{(2n+2m+1)\beta}t^{n+1}z^{m-1}\sqrt{z} + \frac{(2m-1)\Delta}{(2n+2m+1)\beta}\int \frac{z^{m-1}t^n dx}{\sqrt{z}}.$$

La 176(1)

2.
$$\int \frac{t^n z^m dx}{\sqrt{z}} = 2\sqrt{z^{2m+1}}\sum_{k=0}^{n}\binom{n}{k}\frac{\alpha^{n-k}\beta^k}{b^{k+1}}\sum_{p=0}^{k}(-1)^p\binom{k}{p}\frac{z^{k-p}a^p}{2k-2p+2m+1}.$$

2.242 1.
$$\int \frac{t\,dx}{\sqrt{z}} = \frac{2\alpha\sqrt{z}}{b} + \beta\left(\frac{z}{3}-a\right)\frac{2\sqrt{z}}{b^2}.$$

2.
$$\int \frac{t^2\,dx}{\sqrt{z}} = \frac{2\alpha^2\sqrt{z}}{b} + 2\alpha\beta\left(\frac{z}{3}-a\right)\frac{2\sqrt{z}}{b^2} + \beta^2\left(\frac{z^2}{5}-\frac{2}{3}za+a^2\right)\frac{2\sqrt{z}}{b^3}.$$

3.
$$\int \frac{t^3\,dx}{\sqrt{z}} = \frac{2\alpha^3\sqrt{z}}{b} + 3\alpha^2\beta\left(\frac{z}{3}-a\right)\frac{2\sqrt{z}}{b^2} +$$
$$+ 3\alpha\beta^2\left(\frac{z^2}{5}-\frac{2}{3}za+a^2\right)\frac{2\sqrt{z}}{b^3} + \beta^3\left(\frac{z^3}{7}-\frac{3z^2a}{5}+za^2-a^3\right)\frac{2\sqrt{z}}{b^4}.$$

Algebraische Funktionen/Algebraic Functions

4. $\int \dfrac{tz\,dx}{\sqrt{z}} = \dfrac{2\alpha\sqrt{z^3}}{3b} + \beta\left(\dfrac{z}{5} - \dfrac{a}{3}\right)\dfrac{2\sqrt{z^3}}{b^2}.$ 2.242

5. $\int \dfrac{t^2 z\,dx}{\sqrt{z}} = \dfrac{2\alpha^2\sqrt{z^3}}{3b} + 2\alpha\beta\left(\dfrac{z}{5} - \dfrac{a}{3}\right)\dfrac{2\sqrt{z^3}}{b^2} + \beta^2\left(\dfrac{z^2}{7} - \dfrac{2za}{5} + \dfrac{a^2}{3}\right)\dfrac{2\sqrt{z^3}}{b^3}.$

6. $\int \dfrac{t^3 z\,dx}{\sqrt{z}} = \dfrac{2\alpha^3\sqrt{z^3}}{3b} + 3\alpha^2\beta\left(\dfrac{z}{5} - \dfrac{a}{3}\right)\dfrac{2\sqrt{z^3}}{b^2} +$

$+ 3\alpha\beta^2\left(\dfrac{z^2}{7} - \dfrac{2za}{5} + \dfrac{a^2}{3}\right)\dfrac{2\sqrt{z^3}}{b^3} + \beta^3\left(\dfrac{z^3}{9} - \dfrac{3z^2 a}{7} + \dfrac{3za^2}{5} - \dfrac{a^3}{3}\right)\dfrac{2\sqrt{z^3}}{b^4}.$

7. $\int \dfrac{tz^2\,dx}{\sqrt{z}} = \dfrac{2\alpha\sqrt{z^5}}{5b} + \beta\left(\dfrac{z}{7} - \dfrac{a}{5}\right)\dfrac{2\sqrt{z^5}}{b^2}.$

8. $\int \dfrac{t^2 z^2\,dx}{\sqrt{z}} = \dfrac{2\alpha^2\sqrt{z^5}}{5b} + 2\alpha\beta\left(\dfrac{z}{7} - \dfrac{a}{5}\right)\dfrac{2\sqrt{z^5}}{b^2} + \beta^2\left(\dfrac{z^2}{9} - \dfrac{2za}{7} + \dfrac{a^2}{5}\right)\dfrac{2\sqrt{z^5}}{b^3}.$

9. $\int \dfrac{t^3 z^2\,dx}{\sqrt{z}} = \dfrac{2\alpha^3\sqrt{z^5}}{5b} + 3\alpha^2\beta\left(\dfrac{z}{7} - \dfrac{a}{5}\right)\dfrac{2\sqrt{z^5}}{b^2} +$

$+ 3\alpha\beta^2\left(\dfrac{z^2}{9} - \dfrac{2za}{7} + \dfrac{a^2}{5}\right)\dfrac{2\sqrt{z^5}}{b^3} + \beta^3\left(\dfrac{z^3}{11} - \dfrac{3z^2 a}{9} + \dfrac{3za^2}{7} - \dfrac{a^3}{5}\right)\dfrac{2\sqrt{z^5}}{b^4}.$

10. $\int \dfrac{tz^3\,dx}{\sqrt{z}} = \dfrac{2\alpha\sqrt{z^7}}{7b} + \beta\left(\dfrac{z}{9} - \dfrac{a}{7}\right)\dfrac{2\sqrt{z^7}}{b^2}.$

11. $\int \dfrac{t^2 z^3\,dx}{\sqrt{z}} = \dfrac{2\alpha^2\sqrt{z^7}}{7b} + 2\alpha\beta\left(\dfrac{z}{9} - \dfrac{a}{7}\right)\dfrac{2\sqrt{z^7}}{b^2} + \beta^2\left(\dfrac{z^2}{11} - \dfrac{2za}{9} + \dfrac{a^2}{7}\right)\dfrac{2\sqrt{z^7}}{b^3}.$

12. $\int \dfrac{t^3 z^3\,dx}{\sqrt{z}} = \dfrac{2\alpha^3\sqrt{z^7}}{7b} + 3\alpha^2\beta\left(\dfrac{z}{9} - \dfrac{a}{7}\right)\dfrac{2\sqrt{z^7}}{b^2} +$

$+ 3\alpha\beta^2\left(\dfrac{z^2}{11} - \dfrac{2za}{9} + \dfrac{a^2}{7}\right)\dfrac{2\sqrt{z^7}}{b^3} + \beta^3\left(\dfrac{z^3}{13} - \dfrac{3z^2 a}{11} + \dfrac{3za^2}{9} - \dfrac{a^3}{7}\right)\dfrac{2\sqrt{z^7}}{b^4}.$

1. $\int \dfrac{t^n\,dx}{z^m\sqrt{z}} = \dfrac{2}{(2m-1)\Delta}\dfrac{t^{n+1}}{z^m}\sqrt{z} - \dfrac{(2n-2m+3)\beta}{(2m-1)\Delta}\int \dfrac{t^n\,dx}{z^{m-1}\sqrt{z}};$ 2.243

$= -\dfrac{2}{(2m-1)b}\dfrac{t^n}{z^m}\sqrt{z} + \dfrac{2n\beta}{(2m-1)b}\int \dfrac{t^{n-1}\,dx}{z^{m-1}\sqrt{z}}.$ La 176(2)

2. $\int \dfrac{t^n\,dx}{z^m\sqrt{z}} = \dfrac{2}{\sqrt{z^{2m-1}}}\sum_{k=0}^{n}\binom{n}{k}\dfrac{\alpha^{n-k}\beta^k}{b^{k+1}}\sum_{p=0}^{k}(-1)^p\binom{k}{p}\dfrac{z^{k-p}a^p}{2k-2p-2m+1}.$

1. $\int \dfrac{t\,dx}{z\sqrt{z}} = -\dfrac{2\alpha}{b\sqrt{z}} + \dfrac{2\beta(z+a)}{b^2\sqrt{z}}.$ 2.244

2. $\int \dfrac{t^2\,dx}{z\sqrt{z}} = -\dfrac{2\alpha^2}{b\sqrt{z}} + \dfrac{4\alpha\beta(z+a)}{b^2\sqrt{z}} + \dfrac{2\beta^2\left(\dfrac{z^2}{3} - 2za - a^2\right)}{b^3\sqrt{z}}.$

3. $\int \dfrac{t^3\,dx}{z\sqrt{z}} = -\dfrac{2\alpha^3}{b\sqrt{z}} + \dfrac{6\alpha^2\beta(z+a)}{b^2\sqrt{z}} + \dfrac{6\alpha\beta^2\left(\dfrac{z^2}{3} - 2za - a^2\right)}{b^3\sqrt{z}} +$

$+ \dfrac{2\beta^3\left(\dfrac{z^3}{5} - z^2 a + 3za^2 + a^3\right)}{b^4\sqrt{z}}.$

2.244

4. $$\int \frac{t\,dx}{z^2\sqrt{z}} = -\frac{2\alpha}{3b\sqrt{z^3}} - \frac{2\beta\left(z-\dfrac{a}{3}\right)}{b^2\sqrt{z^3}}.$$

5. $$\int \frac{t^2\,dx}{z^2\sqrt{z}} = -\frac{2\alpha^2}{3b\sqrt{z^3}} - \frac{4\alpha\beta\left(z-\dfrac{a}{3}\right)}{b^2\sqrt{z^3}} + \frac{2\beta^2\left(z^2+2az-\dfrac{a^2}{3}\right)}{b^3\sqrt{z^3}}.$$

6. $$\int \frac{t^3\,dx}{z^2\sqrt{z}} = -\frac{2\alpha^3}{3b\sqrt{z^3}} - \frac{6\alpha^2\beta\left(z-\dfrac{a}{3}\right)}{b^2\sqrt{z^3}} + \frac{6\alpha\beta^2\left(z^2+2za-\dfrac{a^2}{3}\right)}{b^3\sqrt{z^3}} +$$
$$+ \frac{2\beta^3\left(\dfrac{z^3}{3} - 3z^2a - 3za^2 + \dfrac{a^3}{3}\right)}{b^4\sqrt{z^3}}.$$

7. $$\int \frac{t\,dx}{z^3\sqrt{z}} = -\frac{2\alpha}{5b\sqrt{z^5}} - \frac{2\beta\left(\dfrac{z}{3}-\dfrac{a}{5}\right)}{b^2\sqrt{z^5}}.$$

8. $$\int \frac{t^2\,dx}{z^3\sqrt{z}} = -\frac{2\alpha^2}{5b\sqrt{z^5}} - \frac{4\alpha\beta\left(\dfrac{z}{3}-\dfrac{a}{5}\right)}{b^2\sqrt{z^5}} - \frac{2\beta^2\left(z^2-\dfrac{2za}{3}+\dfrac{a^2}{5}\right)}{b^3\sqrt{z^5}}.$$

9. $$\int \frac{t^3\,dx}{z^3\sqrt{z}} = -\frac{2\alpha^3}{5b\sqrt{z^5}} - \frac{6\alpha^2\beta\left(\dfrac{z}{3}-\dfrac{a}{5}\right)}{b^2\sqrt{z^5}} - \frac{6\alpha\beta^2\left(z^2-\dfrac{2za}{3}+\dfrac{a^2}{5}\right)}{b^3\sqrt{z^5}} +$$
$$+ \frac{2\beta^3\left(z^3 + 3z^2a - za^2 + \dfrac{a^3}{5}\right)}{b^4\sqrt{z^5}}.$$

2.245

1. $$\int \frac{z^m\,dx}{t^n\sqrt{z}} = -\frac{2}{(2n-2m-1)\beta} \frac{z^{m-1}}{t^{n-1}}\sqrt{z} - \frac{(2m-1)\Delta}{(2n-2m-1)\beta}\int \frac{z^{m-1}\,dx}{t^n\sqrt{z}};$$

La 176(3)

$$= -\frac{1}{(n-1)\beta}\frac{z^{m-1}}{t^{n-1}}\sqrt{z} + \frac{(2m-1)b}{2(n-1)\beta}\int \frac{z^{m-1}}{t^{n-1}\sqrt{z}}\,dx;$$

$$= -\frac{1}{(n-1)\Delta}\frac{z^m}{t^{n-1}}\sqrt{z} - \frac{(2n-2m-3)b}{2(n-1)\Delta}\int \frac{z^m\,dx}{t^{n-1}\sqrt{z}}.$$

2. $$\int \frac{z^m\,dx}{t^n\sqrt{z}} = -z^m\sqrt{z}\left\{\frac{1}{(n-1)\Delta}\frac{1}{t^{n-1}} + \right.$$
$$\left. + \sum_{k=2}^{n-1} \frac{(2n-2m-3)(2n-2m-5)\ldots(2n-2m-2k+1)b^{k-1}}{2^{k-1}(n-1)(n-2)\ldots(n-k)\Delta^k} \frac{1}{t^{n-k}}\right\} -$$
$$- \frac{(2n-2m-3)(2n-2m-5)\ldots(-2m+3)(-2m+1)b^{n-1}}{2^{n-1}\cdot(n-1)!\,\Delta^n}\int \frac{z^m\,dx}{t\sqrt{z}}.$$

Für $n=1$ gilt | For $n=1$:

3. $$\int \frac{z^m\,dx}{t\sqrt{z}} = \frac{2}{(2m-1)\beta}\frac{z^m}{\sqrt{z}} + \frac{\Delta}{\beta}\int \frac{z^{m-1}\,dx}{t\sqrt{z}}.$$

4. $$\int \frac{z^m\,dx}{t\sqrt{z}} = 2\sum_{k=0}^{m-1}\frac{\Delta^k}{(2m-2k-1)\beta^{k+1}}\frac{z^{m-k}}{\sqrt{z}} + \frac{\Delta^m}{\beta^m}\int \frac{dx}{t\sqrt{z}}.$$

$$\int \frac{dx}{t\sqrt{z}} = \frac{1}{\sqrt{\beta\Delta}} \ln \frac{\beta\sqrt{z} - \sqrt{\beta\Delta}}{\beta\sqrt{z} + \sqrt{\beta\Delta}} \quad [\beta\Delta > 0]; \qquad 2.246$$

$$= \frac{2}{\sqrt{-\beta\Delta}} \operatorname{arctg} \frac{\beta\sqrt{z}}{\sqrt{-\beta\Delta}} \quad [\beta\Delta < 0];$$

$$= -\frac{2\sqrt{z}}{bt} \qquad [\Delta = 0].$$

$$\int \frac{dx}{tz^m\sqrt{z}} = \frac{2}{z^{m-1}\sqrt{z}} + \sum_{k=1}^{m} \frac{\beta^{k-1}z^k}{\Delta^k(2m-2k+1)} + \frac{\beta^m}{\Delta^m} \int \frac{dx}{t\sqrt{z}} \quad \text{(cf. 2.246)}. \qquad 2.247$$

1. $$\int \frac{dx}{tz\sqrt{z}} = \frac{2}{\Delta\sqrt{z}} + \frac{\beta}{\Delta} \int \frac{dx}{t\sqrt{z}} \quad \text{(cf. 2.246)}. \qquad 2.248$$

2. $$\int \frac{dx}{tz^2\sqrt{z}} = \frac{2}{3\Delta z\sqrt{z}} + \frac{2\beta}{\Delta^2\sqrt{z}} + \frac{\beta^2}{\Delta^2} \int \frac{dx}{t\sqrt{z}} \quad \text{(cf. 2.246)}.$$

3. $$\int \frac{dx}{tz^3\sqrt{z}} = \frac{2}{5\Delta z^2\sqrt{z}} + \frac{2\beta}{3\Delta^2 z\sqrt{z}} + \frac{2\beta^2}{\Delta^3\sqrt{z}} + \frac{\beta^3}{\Delta^3} \int \frac{dx}{t\sqrt{z}} \quad \text{(cf. 2.246)}.$$

4. $$\int \frac{dx}{t^2\sqrt{z}} = -\frac{\sqrt{z}}{\Delta t} - \frac{b}{2\Delta} \int \frac{dx}{t\sqrt{z}} \quad \text{(cf. 2.246)}.$$

5. $$\int \frac{dx}{t^2 z\sqrt{z}} = -\frac{1}{\Delta t\sqrt{z}} - \frac{3b}{\Delta^2\sqrt{z}} - \frac{3b\beta}{2\Delta^2} \int \frac{dx}{t\sqrt{z}} \quad \text{(cf. 2.246)}.$$

6. $$\int \frac{dx}{t^2 z^2\sqrt{z}} = -\frac{1}{\Delta t z^2\sqrt{z}} - \frac{5b}{3\Delta^2 z\sqrt{z}} - \frac{5b\beta}{\Delta^3\sqrt{z}} - \frac{5b\beta^2}{2\Delta^3} \int \frac{dx}{t\sqrt{z}} \quad \text{(cf. 2.246)}.$$

7. $$\int \frac{dx}{t^2 z^3\sqrt{z}} = -\frac{1}{\Delta t z^2\sqrt{z}} - \frac{7b}{5\Delta^2 z^2\sqrt{z}} - \frac{7b\beta}{3\Delta^3 z\sqrt{z}} - \frac{7b\beta^2}{\Delta^4\sqrt{z}} -$$
$$-\frac{7b\beta^3}{2\Delta^4} \int \frac{dx}{t\sqrt{z}} \quad \text{(cf. 2.246)}.$$

8. $$\int \frac{dx}{t^3\sqrt{z}} = -\frac{\sqrt{z}}{2\Delta t^2} + \frac{3b\sqrt{z}}{4\Delta^2 t} + \frac{3b^2}{8\Delta^2} \int \frac{dx}{t\sqrt{z}} \quad \text{(cf. 2.246)}.$$

9. $$\int \frac{dx}{t^3 z\sqrt{z}} = -\frac{1}{2\Delta t^2\sqrt{z}} + \frac{5b}{4\Delta^2 t\sqrt{z}} + \frac{15b^2}{4\Delta^3\sqrt{z}} + \frac{15b^2\beta}{8\Delta^3} \int \frac{dx}{t\sqrt{z}} \quad \text{(cf. 2.246)}.$$

10. $$\int \frac{dx}{t^3 z^2\sqrt{z}} = -\frac{1}{2\Delta t^2 z\sqrt{z}} + \frac{7b\sqrt{z}}{4\Delta^3 tz\sqrt{z}} + \frac{35b^2}{12\Delta^3 z\sqrt{z}} +$$
$$+ \frac{35b^2\beta}{4\Delta^4\sqrt{z}} + \frac{35b^2\beta^2}{8\Delta^4} \int \frac{dx}{t\sqrt{z}} \quad \text{(cf. 2.246)}.$$

11. $$\int \frac{dx}{t^3 z^3\sqrt{z}} = -\frac{1}{2\Delta t^2 z^2\sqrt{z}} + \frac{9b}{4\Delta^2 t z^2\sqrt{z}} + \frac{63b^2}{20\Delta^3 z^2\sqrt{z}} +$$
$$+ \frac{21b^2\beta}{4\Delta^4 z\sqrt{z}} + \frac{63b^2\beta^2}{4\Delta^5\sqrt{z}} + \frac{63b^2\beta^3}{8\Delta^5} \int \frac{dx}{t\sqrt{z}} \quad \text{(cf. 2.246)}.$$

12. $$\int \frac{z\,dx}{t\sqrt{z}} = \frac{2\sqrt{z}}{\beta} + \frac{\Delta}{\beta} \int \frac{dx}{t\sqrt{z}} \quad \text{(cf. 2.246)}.$$

13. $$\int \frac{z^2\,dx}{t\sqrt{z}} = \frac{2z\sqrt{z}}{3\beta} + \frac{2\Delta\sqrt{z}}{\beta^2} + \frac{\Delta^2}{\beta^2} \int \frac{dx}{t\sqrt{z}} \quad \text{(cf. 2.246)}.$$

2.248 14. $\int \frac{z^3 \, dx}{t \sqrt{z}} = \frac{2z^2 \sqrt{z}}{5\beta} + \frac{2\Delta z \sqrt{z}}{3\beta^2} + \frac{2\Delta^2 \sqrt{z}}{\beta^3} + \frac{\Delta^3}{\beta^3} \int \frac{dx}{t \sqrt{z}}$ (cf. 2.246).

15. $\int \frac{z \, dx}{t^2 \sqrt{z}} = -\frac{z \sqrt{z}}{\Delta t} + \frac{b \sqrt{z}}{\beta \Delta} + \frac{b}{2\beta} \int \frac{dx}{t \sqrt{z}}$ (cf. 2.246).

16. $\int \frac{z^2 \, dx}{t^2 \sqrt{z}} = -\frac{z^2 \sqrt{z}}{\Delta t} + \frac{bz \sqrt{z}}{\beta \Delta} + \frac{3b \sqrt{z}}{\beta^2} + \frac{3b\Delta}{2\beta^2} \int \frac{dx}{t \sqrt{z}}$ (cf. 2.246).

17. $\int \frac{z^3 \, dx}{t^2 \sqrt{z}} = -\frac{z^3 \sqrt{z}}{\Delta t} + \frac{bz^2 \sqrt{z}}{\beta \Delta} + \frac{5bz \sqrt{z}}{3\beta^2} + \frac{5b\Delta \sqrt{z}}{\beta^3} + \frac{5\Delta^2 b}{2\beta^3} \int \frac{dx}{t \sqrt{z}}$

(cf. 2.246).

18. $\int \frac{z \, dx}{t^3 \sqrt{z}} = -\frac{z \sqrt{z}}{2\Delta t^2} - \frac{bz \sqrt{z}}{4\Delta^2 t} + \frac{b^2 \sqrt{z}}{4\beta \Delta^2} + \frac{b^2}{8\beta \Delta} \int \frac{dx}{t \sqrt{z}}$ (cf. 2.246).

19. $\int \frac{z^2 \, dx}{t^3 \sqrt{z}} = -\frac{z^2 \sqrt{z}}{2\Delta t^2} + \frac{bz^2 \sqrt{z}}{4\Delta^2 t} + \frac{b^2 z \sqrt{z}}{4\beta \Delta^2} + \frac{3b^2 \sqrt{z}}{4\beta^2 \Delta} + \frac{3b^2}{8\beta^2} \int \frac{dx}{t \sqrt{z}}$

(cf. 2.246).

20. $\int \frac{z^3 \, dx}{t^3 \sqrt{z}} = -\frac{z^3 \sqrt{z}}{2\Delta t^2} + \frac{3bz^3 \sqrt{z}}{\Delta^2 t} + \frac{3b^2 z^2 \sqrt{z}}{4\beta \Delta^2} + \frac{5b^2 z \sqrt{z}}{4\beta^2 \Delta} +$

$+ \frac{15 b^2 \sqrt{z}}{4\beta^3} + \frac{15 b^2 \Delta}{8\beta^3} \int \frac{dx}{t \sqrt{z}}$ (cf. 2.246).

2.249 1. $\int \frac{dx}{z^m t^n \sqrt{z}} = \frac{2}{(2m-1)\Delta} \frac{\sqrt{z}}{t^{n-1} z^m} + \frac{(2n+2m-3)\beta}{(2m-1)\Delta} \int \frac{dx}{t^n z^{m-1} \sqrt{z}};$ La 177(4)

$= -\frac{1}{(n-1)\Delta} \frac{\sqrt{z}}{z^m t^{n-1}} - \frac{(2n+2m-3)b}{2(n-1)\Delta} \int \frac{dx}{t^{n-1} z^m \sqrt{z}}.$

2. $\int \frac{dx}{z^m t^n \sqrt{z}} = \frac{\sqrt{z}}{z^m} \left\{ \frac{-1}{(n-1)\Delta} \frac{1}{t^{n-1}} + \right.$

$+ \sum_{k=2}^{n-1} (-1)^k \frac{(2n+2m-3)(2n+2m-5)\ldots(2n+2m-2k+1) b^{k-1}}{2^{k-1}(n-1)(n-2)\ldots(n-k)\Delta^k} \cdot \frac{1}{t^{n-k}} \Bigg\} +$

$+ (-1)^{n-1} \frac{(2n+2m-3)(2n+2m-5)\ldots(-2m+3)(-2m+1) b^{n-1}}{2^{n-1}(n-1)!\Delta^{n-1}} \int \frac{dx}{tz^m \sqrt{z}}.$

Für $n = 1$ gilt | For $n = 1$:

$$\int \frac{dx}{z^m t \sqrt{z}} = \frac{2}{(2m-1)\Delta} \frac{1}{z^{m-1} \sqrt{z}} + \frac{\beta}{\Delta} \int \frac{dx}{tz^{m-1} \sqrt{z}}.$$

2.25

Funktionen von $\sqrt{a + bx + cx^2}$

Functions of $\sqrt{a + bx + cx^2}$

Integrationsmethoden

Integration methods

2.251 Der Integrand von Integralen der Gestalt | The integrand of the integral

$$\int R(x, \sqrt{a + bx + cx^2}) \, dx$$

can be made rational by the use of at least one of the following EULER substitutions:

1. $$\sqrt{a+bx+cx^2} = xt \pm \sqrt{a} \quad \text{for } a>0;$$

2. $$\sqrt{a+bx+cx^2} = t \pm x\sqrt{c} \quad \text{for } c>0;$$

3. $$\sqrt{c(x-x_1)(x-x_2)} = t(x-x_1),$$

if x_1 and x_2, that is, the roots of the equation $a+bx+cx^2 = 0$, are real.

Apart from the EULER substitutions, to calculate such integrals the following method can be used. By removing the irrationality in the denominator and by other simple algebraic operations the integrand is first transformed into a sum of a certain rational function of x and expressions of the form

$$\frac{P_1(x)}{P_2(x)\sqrt{a+bx+cx^2}}$$

where $P_1(x)$ and $P_2(x)$ are polynomials. Then, by taking the integer part of the function $\frac{P_1(x)}{P_2(x)}$ and using partial fractions for the remainder, the integral of any of the second expressions is transformed into a sum of integrals of one of the following three forms:

2.252

I. $$\int \frac{P(x)\,dx}{\sqrt{a+bx+cx^2}},$$

where $P(x)$ is a polynomial of degree r,

II. $$\int \frac{dx}{(x+p)^k \sqrt{a+bx+cx^2}};$$

III. $$\int \frac{(Mx+N)\,dx}{(\alpha+\beta x+x^2)^m \sqrt{c(a_1+b_1 x+x^2)}}, \quad \left(a_1 = \frac{a}{c},\ b_1 = \frac{b}{c}\right).$$

I. $$\int \frac{P(x)\,dx}{\sqrt{a+bx+cx^2}} = Q(x)\sqrt{a+bx+cx^2} + \lambda \int \frac{dx}{\sqrt{a+bx+cx^2}},$$

where $Q(x)$ is a polynomial of degree $r-1$. Its coefficients and the number λ can be calculated in the usual way by comparing coefficients in the identity

$$P(x) = Q'(x)(a+bx+cx^2) + \frac{1}{2} Q(x)(b+2cx) + \lambda.$$

F II 70

Im Fall $r \leqslant 3$ lassen sich Integrale der Gestalt

$$\int \frac{P(x)\,dx}{\sqrt{a+bx+cx^2}}$$

auch unter Benutzung der Formeln 2.26 berechnen.

II. Ist der Grad n des Polynoms $P(x)$ kleiner als k, so lassen sich Integrale der Gestalt

$$\int \frac{P(x)\,dx}{(x+p)^k \sqrt{a+bx+cx^2}}$$

mit Hilfe der Substitution $t = \dfrac{1}{x+p}$ in Integrale der Gestalt

$$\int \frac{P(t)\,dt}{\sqrt{\alpha+\beta t+\gamma t^2}}$$

überführen (vgl. auch 2.281).

III. Integrale der Gestalt

$$\int \frac{(Mx+N)\,dx}{(\alpha+\beta x+x^2)^m \sqrt{c(a_1+b_1 x+x^2)}}$$

lassen sich folgendermaßen berechnen: Im Fall $b_1 \neq \beta$ geht dieses Integral mit Hilfe der Substitution

$$x = \frac{a_1 - \alpha}{\beta - b_1} + \frac{t-1}{t+1}\,\frac{\sqrt{(a_1-\alpha)^2 - (\alpha b_1 - a_1\beta)(\beta-b_1)}}{\beta - b_1}$$

in ein Integral der Gestalt

$$\int \frac{P(t)\,dt}{(t^2+p)^m \sqrt{c(t^2+q)}}$$

über, wobei $P(t)$ ein Polynom höchstens $(2m-1)$-ten Grades ist. Das Integral

$$\int \frac{P(t)\,dt}{(t^2+p)^m \sqrt{t^2+q}}$$

reduziert sich dann auf eine Summe von Integralen der Gestalt

$$\int \frac{t\,dt}{(t^2+p)^k \sqrt{t^2+q}} \quad \text{und} \quad \int \frac{dt}{(t^2+p)^k \sqrt{t^2+q}}.$$

Im Fall $b_1 = \beta$ führt die Substitution $t = x + \dfrac{b_1}{2}$ auf ein Integral der Gestalt

$$\int \frac{P(t)\,dt}{(t^2+p)^m \sqrt{c(t^2+q)}}.$$

For $r \leqslant 3$ the integrals of the form

$$\int \frac{P(x)\,dx}{\sqrt{a+bx+cx^2}}$$

can be evaluated with the help of formulae 2.26.

II. If the degree n of the polynomial $P(x)$ is less than k, the integrals of the form

$$\int \frac{P(x)\,dx}{(x+p)^k \sqrt{a+bx+cx^2}}$$

can, by means of the substitution $t = \dfrac{1}{x+p}$, be reduced to

$$\int \frac{P(t)\,dt}{\sqrt{\alpha+\beta t+\gamma t^2}}$$

(see also 2.281).

III. The integrals of the form

$$\int \frac{(Mx+N)\,dx}{(\alpha+\beta x+x^2)^m \sqrt{c(a_1+b_1 x+x^2)}}$$

may be treated as follows: If $b_1 \neq \beta$, the substitution

$$x = \frac{a_1 - \alpha}{\beta - b_1} + \frac{t-1}{t+1}\,\frac{\sqrt{(a_1-\alpha)^2 - (\alpha b_1 - a_1\beta)(\beta-b_1)}}{\beta - b_1}$$

is used to reduce the integral to the form

$$\int \frac{P(t)\,dt}{(t^2+p)^m \sqrt{c(t^2+q)}}$$

where $P(t)$ is a polynomial of a degree less or equal to $2m-1$. The integral

$$\int \frac{P(t)\,dt}{(t^2+p)^m \sqrt{t^2+q}}$$

can then be reduced to a sum of integrals of the form

$$\int \frac{t\,dt}{(t^2+p)^k \sqrt{t^2+q}} \quad \text{and} \quad \int \frac{dt}{(t^2+p)^k \sqrt{t^2+q}}.$$

If $b_1 = \beta$, the substitution $t = x + \dfrac{b_1}{2}$ leads to

$$\int \frac{P(t)\,dt}{(t^2+p)^m \sqrt{c(t^2+q)}}.$$

Das Integral / For the integral

$$\int \frac{t\,dt}{(t^2+p)^k \sqrt{c(t^2+q)}}$$

läßt sich mit Hilfe der Substitution $t^2 + q = u^2$ berechnen. / the substitution $t^2 + q = u^2$ is used.

Das Integral / For the integral

$$\int \frac{dt}{(t^2+p)^k \sqrt{c(t^2+q)}}$$

findet man mit Hilfe der Substitution $\frac{t}{\sqrt{t^2+q}} = v$ (vgl. auch 2.283). / the substitution $\frac{t}{\sqrt{t^2+q}} = v$ is used (see also 2.283).

F II 74–76

Funktionen von $\sqrt{a+bx+cx^2}$ und ganzen Potenzen von x
Functions of $\sqrt{a+bx+cx^2}$ and integral powers of x 2.26

Abkürzungen: / Abbreviations: $R = a + bx + cx^2$, $\Delta = 4ac - b^2$.

Vereinfachte Formeln für den Fall $b = 0$ siehe 2.27. / Simplified formulae for $b = 0$: see 2.27.

1. $\int x^m \sqrt{R^{2n+1}}\,dx = \dfrac{x^{m-1}\sqrt{R^{2n+3}}}{(m+2n+2)c} - \dfrac{(2m+2n+1)b}{2(m+2n+2)c}\int x^{m-1}\sqrt{R^{2n+1}}\,dx -$ 2.260

   $\qquad - \dfrac{(m-1)a}{(m+2n+2)c}\int x^{m-2}\sqrt{R^{2n+1}}\,dx.$ T(192)+

2. $\int \sqrt{R^{2n+1}}\,dx = \dfrac{2cx+b}{4(n+1)c}\sqrt{R^{2n+1}} + \dfrac{2n+1}{8(n+1)}\dfrac{\Delta}{c}\int \sqrt{R^{2n-1}}\,dx.$ T(188)

3. $\int \sqrt{R^{2n+1}}\,dx = \dfrac{(2cx+b)\sqrt{R}}{4(n+1)c}\left\{R^n + \right.$

   $\qquad + \sum_{k=0}^{n-1} \dfrac{(2n+1)(2n-1)\ldots(2n-2k+1)}{8^{k+1} n(n-1)\ldots(n-k)}\left(\dfrac{\Delta}{c}\right)^{k+1} R^{n-k-1}\biggr\} +$

   $\qquad + \dfrac{(2n+1)!!}{8^{n+1}(n+1)!}\left(\dfrac{\Delta}{c}\right)^{n+1}\int \dfrac{dx}{\sqrt{R}}.$ T(190)

Für $n = -1$ gilt / For $n = -1$: 2.261

$$\int \frac{dx}{\sqrt{R}} = \frac{1}{\sqrt{c}} \ln(2\sqrt{cR} + 2cx + b) \quad [c > 0];$$ T(127)

$$= \frac{1}{\sqrt{c}} \operatorname{Arsh} \frac{2cx+b}{\sqrt{\Delta}} \quad [c > 0,\ \Delta > 0];$$ D(380 001)

$$= \frac{-1}{\sqrt{-c}} \arcsin \frac{2cx+b}{\sqrt{-\Delta}} \quad [c < 0,\ \Delta < 0];$$ T(128)

$$= \frac{1}{\sqrt{c}} \ln(2cx + b) \quad [c > 0,\ \Delta = 0].$$ D(380 001)

2.262

1. $\int \sqrt{R}\, dx = \dfrac{(2cx+b)\sqrt{R}}{4c} + \dfrac{\Delta}{8c}\int \dfrac{dx}{\sqrt{R}}$ (cf. 2.261).

2. $\int x\sqrt{R}\, dx = \dfrac{\sqrt{R^3}}{3c} - \dfrac{(2cx+b)\,b}{8c^2}\sqrt{R} - \dfrac{b\Delta}{16c^2}\int \dfrac{dx}{\sqrt{R}}$ (cf. 2.261).

3. $\int x^2\sqrt{R}\, dx = \left(\dfrac{x}{4c} - \dfrac{5b}{24c^2}\right)\sqrt{R^3} + \left(\dfrac{5b^2}{16c^2} - \dfrac{a}{4c}\right)\dfrac{(2cx+b)\sqrt{R}}{4c} +$
$\qquad + \left(\dfrac{5b^2}{16c^2} - \dfrac{a}{4c}\right)\dfrac{\Delta}{8c}\int \dfrac{dx}{\sqrt{R}}$ (cf. 2.261).

4. $\int x^3\sqrt{R}\, dx = \left(\dfrac{x^2}{5c} - \dfrac{7bx}{40c^2} + \dfrac{7b^2}{48c^3} - \dfrac{2a}{15c^2}\right)\sqrt{R^3} -$
$\qquad - \left(\dfrac{7b^3}{32c^3} - \dfrac{3ab}{8c^2}\right)\dfrac{(2cx+b)\sqrt{R}}{4c} - \left(\dfrac{7b^3}{32c^3} - \dfrac{3ab}{8c^2}\right)\dfrac{\Delta}{8c}\int \dfrac{dx}{\sqrt{R}}$ (cf. 2.261).

5. $\int \sqrt{R^3}\, dx = \left(\dfrac{R}{8c} + \dfrac{3\Delta}{64c^2}\right)(2cx+b)\sqrt{R} + \dfrac{3\Delta^2}{128c^2}\int \dfrac{dx}{\sqrt{R}}$ (cf. 2.261).

6. $\int x\sqrt{R^3}\, dx = \dfrac{\sqrt{R^5}}{5c} - (2cx+b)\left(\dfrac{b}{16c^2}\sqrt{R^3} + \dfrac{3\Delta b}{128c^3}\sqrt{R}\right) -$
$\qquad - \dfrac{3\Delta^2 b}{256c^3}\int \dfrac{dx}{\sqrt{R}}$ (cf. 2.261).

7. $\int x^2\sqrt{R^3}\, dx = \left(\dfrac{x}{6c} + \dfrac{7b}{60c^2}\right)\sqrt{R^5} +$
$\qquad + \left(\dfrac{7b^2}{24c^2} - \dfrac{a}{6c}\right)\left(2x + \dfrac{b}{c}\right)\left(\dfrac{\sqrt{R^3}}{8} + \dfrac{3\Delta}{64c}\sqrt{R}\right) +$
$\qquad + \left(\dfrac{7b^3}{4c} - a\right)\dfrac{\Delta^2}{256c^3}\int \dfrac{dx}{\sqrt{R}}$ (cf. 2.261).

8. $\int x^3\sqrt{R^3}\, dx = \left(\dfrac{x^2}{7c} - \dfrac{3bx}{28c^2} + \dfrac{3b^2}{40c^3} - \dfrac{2a}{35c^2}\right)\sqrt{R^5} -$
$\qquad - \left(\dfrac{3b^3}{16c^3} - \dfrac{ab}{4c^2}\right)\left(2x + \dfrac{b}{c}\right)\left(\dfrac{\sqrt{R^3}}{8} + \dfrac{3\Delta}{64c}\sqrt{R}\right) -$
$\qquad - \left(\dfrac{3b^2}{4c} - a\right)\dfrac{3\Delta^2 b}{512c^4}\int \dfrac{dx}{\sqrt{R}}$ (cf. 2.261).

2.263

1. $\int \dfrac{x^m\, dx}{\sqrt{R^{2n+1}}} = \dfrac{x^{m-1}}{(m-2n)\,c\sqrt{R^{2n-1}}} - \dfrac{(2m-2n-1)\,b}{2(m-2n)\,c}\int \dfrac{x^{m-1}\, dx}{\sqrt{R^{2n+1}}} -$
$\qquad - \dfrac{(m-1)\,a}{(m-2n)\,c}\int \dfrac{x^{m-2}\, dx}{\sqrt{R^{2n+1}}}.$ T(193)+

Für $m = 2n$ gilt | For $m = 2n$:

2. $\int \dfrac{x^{2n}\, dx}{\sqrt{R^{2n+1}}} = -\dfrac{x^{2n-1}}{(2n-1)\,c\sqrt{R^{2n-1}}} - \dfrac{b}{2c}\int \dfrac{x^{2n-1}}{\sqrt{R^{2n+1}}}\, dx + \dfrac{1}{c}\int \dfrac{x^{2n-2}}{\sqrt{R^{2n-1}}}\, dx.$
T(194)+

3. $\int \dfrac{dx}{\sqrt{R^{2n+1}}} = \dfrac{2(2cx+b)}{(2n-1)\,\Delta\sqrt{R^{2n-1}}} + \dfrac{8(n-1)\,c}{(2n-1)\,\Delta}\int \dfrac{dx}{\sqrt{R^{2n-1}}}.$ T(189)

4. $\int \dfrac{dx}{\sqrt{R^{2n+1}}} = \dfrac{2(2cx+b)}{(2n-1)\Delta\sqrt{R^{2n-1}}} \times$ 2.263

$\times \left\{ 1 + \sum\limits_{k=1}^{n-1} \dfrac{8^k(n-1)(n-2)\ldots(n-k)}{(2n-3)(2n-5)\ldots(2n-2k-1)} \dfrac{c^k}{\Delta^k} R^k \right\}$ $[n \geqslant 1]$. T(191)

1. $\int \dfrac{dx}{\sqrt{R}}$ (cf. 2.261). 2.264

2. $\int \dfrac{x\,dx}{\sqrt{R}} = \dfrac{\sqrt{R}}{c} - \dfrac{b}{2c} \int \dfrac{dx}{\sqrt{R}}$ (cf. 2.261).

3. $\int \dfrac{x^2\,dx}{\sqrt{R}} = \left(\dfrac{x}{2c} - \dfrac{3b}{4c^2}\right)\sqrt{R} + \left(\dfrac{3b^2}{8c^2} - \dfrac{a}{2c}\right)\int \dfrac{dx}{\sqrt{R}}$ (cf. 2.261).

4. $\int \dfrac{x^3\,dx}{\sqrt{R}} = \left(\dfrac{x^2}{3c} - \dfrac{5bx}{12c^2} + \dfrac{5b^2}{8c^3} - \dfrac{2a}{3c^2}\right)\sqrt{R} - \left(\dfrac{5b^3}{16c^3} - \dfrac{3ab}{4c^2}\right)\int \dfrac{dx}{\sqrt{R}}$

(cf. 2.261).

5. $\int \dfrac{dx}{\sqrt{R^3}} = \dfrac{2(2cx+b)}{\Delta\sqrt{R}}$.

6. $\int \dfrac{x\,dx}{\sqrt{R^3}} = -\dfrac{2(2a+bx)}{\Delta\sqrt{R}}$.

7. $\int \dfrac{x^2\,dx}{\sqrt{R^3}} = -\dfrac{(\Delta-b^2)x - 2ab}{c\Delta\sqrt{R}} + \dfrac{1}{c}\int \dfrac{dx}{\sqrt{R}}$ (cf. 2.261).

8. $\int \dfrac{x^3\,dx}{\sqrt{R^3}} = \dfrac{c\Delta x^2 + b(10ac - 3b^2)x + a(8ac - 3b^2)}{c^2\Delta\sqrt{R}} - \dfrac{3b}{2c^2}\int \dfrac{dx}{\sqrt{R}}$ (cf. 2.261).

$\int \dfrac{\sqrt{R^{2n+1}}}{x^m}\,dx = -\dfrac{\sqrt{R^{2n+3}}}{(m-1)ax^{m-1}} + \dfrac{(2n - 2m + 5)b}{2(m-1)a} \int \dfrac{\sqrt{R^{2n+1}}}{x^{m-1}}\,dx +$ 2.265

$+ \dfrac{(2n - m + 4)c}{(m-1)a} \int \dfrac{\sqrt{R^{2n+1}}}{x^{m-2}}\,dx.$ T(195)

Für $m = 1$ gilt | For $m = 1$:

$\int \dfrac{\sqrt{R^{2n+1}}}{x}\,dx = \dfrac{\sqrt{R^{2n+1}}}{2n+1} + \dfrac{b}{2}\int \sqrt{R^{2n-1}}\,dx + a\int \dfrac{\sqrt{R^{2n-1}}}{x}\,dx.$ T(198)

Im Fall $a = 0$ gilt | For $a = 0$:

$\int \dfrac{\sqrt{(bx + cx^2)^{2n+1}}}{x^m}\,dx = \dfrac{2\sqrt{(bx + cx^2)^{2n+3}}}{(2n - 2m + 3)bx^m} + \dfrac{2(m - 2n - 3)c}{(2n - 2m + 3)b} \int \dfrac{\sqrt{(bx + cx^2)^{2n+1}}}{x^{m-1}}\,dx.$

La 169(3)

Für $m = 0$ vgl. 2.260 2. und 2.260 3. | For $m = 0$ see 2.260 2. and 2.260 3.
Für $n = -1$ und $m = 1$ gilt | For $n = -1$ and $m = 1$:

$\int \dfrac{dx}{x\sqrt{R}} = -\dfrac{1}{\sqrt{a}} \ln \dfrac{2a + bx + 2\sqrt{aR}}{x}$ $[a > 0]$; T(137) 2.266

$= \dfrac{1}{\sqrt{-a}} \arcsin \dfrac{2a + bx}{x\sqrt{b^2 - 4ac}}$ $[a < 0, \Delta < 0]$; T(138)

$= \dfrac{1}{\sqrt{-a}} \operatorname{arctg} \dfrac{2a + bx}{2\sqrt{-a}\sqrt{R}}$ $[a < 0]$; La 178(6)+

2.266
$$= -\frac{1}{\sqrt{a}} \operatorname{Arsh} \frac{2a+bx}{x\sqrt{\Delta}} \quad [a>0,\ \Delta>0];$$ D(380 111)

$$= -\frac{1}{\sqrt{a}} \operatorname{Arth} \frac{2a+bx}{2\sqrt{a}\sqrt{R}} \quad [a>0];$$

$$= \frac{1}{\sqrt{a}} \ln \frac{x}{2a+bx} \quad [a>0,\ \Delta=0];$$

$$= -\frac{2\sqrt{bx+cx^2}}{bx} \quad [a=0,\ b\ne 0].$$ La 170(16)

2.267 1. $\int \frac{\sqrt{R}\,dx}{x} = \sqrt{R} + a\int \frac{dx}{x\sqrt{R}} + \frac{b}{2}\int \frac{dx}{\sqrt{R}}$ (cf. 2.261, 2.266).

2. $\int \frac{\sqrt{R}\,dx}{x^2} = -\frac{\sqrt{R}}{x} + \frac{b}{2}\int \frac{dx}{x\sqrt{R}} + c\int \frac{dx}{\sqrt{R}}$ (cf. 2.261, 2.266).

Im Fall $a=0$ gilt | For $a=0$:

$$\int \frac{\sqrt{bx+cx^2}}{x^2}\,dx = -\frac{2\sqrt{bx+cx^2}}{x} + c\int \frac{dx}{\sqrt{bx+cx^2}} \quad \text{(cf. 2.261)}.$$

3. $\int \frac{\sqrt{R}\,dx}{x^3} = -\left(\frac{1}{2x^2} + \frac{b}{4ax}\right)\sqrt{R} - \left(\frac{b^2}{8a} - \frac{c}{2}\right)\int \frac{dx}{x\sqrt{R}}$ (cf. 2.266).

Im Fall $a=0$ gilt | For $a=0$:

$$\int \frac{\sqrt{bx+cx^2}}{x^3}\,dx = -\frac{2\sqrt{(bx+cx^2)^3}}{3bx^3}.$$

4. $\int \frac{\sqrt{R^3}}{x}\,dx = \frac{\sqrt{R^3}}{3} + \frac{2bcx+b^2+8ac}{8c}\sqrt{R} + a^2\int \frac{dx}{x\sqrt{R}} +$

$\qquad + \frac{b(12ac-b^2)}{16c}\int \frac{dx}{\sqrt{R}}$ (cf. 2.261, 2.266).

5. $\int \frac{\sqrt{R^3}}{x^2}\,dx = -\frac{\sqrt{R^5}}{ax} + \frac{cx+b}{a}\sqrt{R^3} + \frac{3}{4}(2cx+3b)\sqrt{R} +$

$\qquad + \frac{3}{2}ab\int \frac{dx}{x\sqrt{R}} + \frac{3(4ac+b^2)}{8}\int \frac{dx}{\sqrt{R}}$ (cf. 2.261, 2.266).

Im Fall $a=0$ gilt | For $a=0$:

$$\int \frac{\sqrt{(bx+cx^2)^3}}{x^2} = \frac{\sqrt{(bx+cx^2)^3}}{2x} + \frac{3b}{4}\sqrt{bx+cx^2} + \frac{3b^2}{8}\int \frac{dx}{\sqrt{bx+cx^2}} \quad \text{(cf. 2.261)}.$$

6. $\int \frac{\sqrt{R^3}}{x^3}\,dx = -\left(\frac{1}{2ax^2} + \frac{b}{4a^2x}\right)\sqrt{R^5} + \frac{bcx+2ac+b^2}{4a^2}\sqrt{R^3} +$

$\qquad + \frac{3(bcx+2ac+b^2)}{4a}\sqrt{R} + \frac{3}{8}(4ac+b^2)\int \frac{dx}{x\sqrt{R}} + \frac{3}{2}bc\int \frac{dx}{\sqrt{R}}$

(cf. 2.261, 2.266).

Im Fall $a=0$ gilt | For $a=0$:

$$\int \frac{\sqrt{(bx+cx^2)^3}}{x^3}\,dx = \left(c - \frac{2b}{x}\right)\sqrt{bx+cx^2} + \frac{3bc}{2}\int \frac{dx}{\sqrt{bx+cx^2}} \quad \text{(cf. 2.261)}.$$

$$\int \frac{dx}{x^m \sqrt{R^{2n+1}}} = -\frac{1}{(m-1)ax^{m-1}\sqrt{R^{2n-1}}} -$$ 2.268

$$-\frac{(2n+2m-3)b}{2(m-1)a}\int\frac{dx}{x^{m-1}\sqrt{R^{2n+1}}} - \frac{(2n+m-2)c}{(m-1)a}\int\frac{dx}{x^{m-2}\sqrt{R^{2n+1}}}.$$ T(196)

Für $m=1$ gilt | For $m=1$:

$$\int\frac{dx}{x\sqrt{R^{2n+1}}} = \frac{1}{(2n-1)a\sqrt{R^{2n-1}}} - \frac{b}{2a}\int\frac{dx}{\sqrt{R^{2n+1}}} + \frac{1}{a}\int\frac{dx}{x\sqrt{R^{2n-1}}}.$$ T(199)

Im Fall $a=0$ gilt | For $a=0$:

$$\int\frac{dx}{x^m\sqrt{(bx+cx^2)^{2n+1}}} = -\frac{2}{(2n+2m-1)bx^m\sqrt{(bx+cx^2)^{2n-1}}} -$$

$$-\frac{(4n+2m-2)c}{(2n+2m-1)b}\int\frac{dx}{x^{m-1}\sqrt{(bx+cx^2)^{2n+1}}} \quad \text{(cf. 2.265).}$$

1. $\quad \int\frac{dx}{x\sqrt{R}} \quad$ (cf. 2.266). 2.269

2. $\quad \int\frac{dx}{x^2\sqrt{R}} = -\frac{\sqrt{R}}{ax} - \frac{b}{2a}\int\frac{dx}{x\sqrt{R}} \quad$ (cf. 2.266).

Im Fall $a=0$ gilt | For $a=0$:

$$\int\frac{dx}{x^2\sqrt{bx+cx^2}} = \frac{2}{3}\left(-\frac{1}{bx^2}+\frac{2c}{b^2x}\right)\sqrt{bx+cx^2}.$$

3. $\quad \int\frac{dx}{x^3\sqrt{R}} = \left(-\frac{1}{2ax^2}+\frac{3b}{4a^2x}\right)\sqrt{R} + \left(\frac{3b^2}{8a^2}-\frac{c}{2a}\right)\int\frac{dx}{x\sqrt{R}} \quad$ (cf. 2.266).

Im Fall $a=0$ gilt | For $a=0$:

$$\int\frac{dx}{x^3\sqrt{bx+cx^2}} = \frac{2}{5}\left(-\frac{1}{bx^3}+\frac{4c}{3b^2x^2}-\frac{8c^2}{3b^3x}\right)\sqrt{bx+cx^2}.$$

4. $\quad \int\frac{dx}{x\sqrt{R^3}} = -\frac{2(bcx-2ac+b^2)}{a\Delta\sqrt{R}} + \frac{1}{a}\int\frac{dx}{x\sqrt{R}} \quad$ (cf. 2.266).

Im Fall $a=0$ gilt | For $a=0$:

$$\int\frac{dx}{x\sqrt{(bx+cx^2)^3}} = \frac{2}{3}\left(-\frac{1}{bx}+\frac{4c}{b^2}+\frac{8c^2x}{b^3}\right)\frac{1}{\sqrt{bx+cx^2}}.$$

5. $\quad \int\frac{dx}{x^2\sqrt{R^3}} = \left(-\frac{1}{ax} - \frac{b(10ac-3b^2)}{a^2\Delta} - \frac{c(8ac-3b^2)x}{a^2\Delta}\right)\frac{1}{\sqrt{R}} - \frac{3b}{2a^2}\int\frac{dx}{x\sqrt{R}}$

(cf. 2.266).

Im Fall $a=0$ gilt | For $a=0$:

$$\int\frac{dx}{x^2\sqrt{(bx+cx^2)^3}} = \frac{2}{5}\left(-\frac{1}{bx^2}+\frac{2c}{b^2x}-\frac{8c^2}{b^3}-\frac{16c^3x}{b^4}\right)\frac{1}{\sqrt{bx+cx^2}}.$$

6. $\quad \int\frac{dx}{x^3\sqrt{R^3}} = \left(-\frac{1}{ax^3}+\frac{5b}{2a^2x} - \frac{15b^4-62acb^2+24a^2c^2}{2a^3\Delta} - \right.$

$$\left. -\frac{bc(15b^2-52ac)x}{2a^3\Delta}\right)\frac{1}{2\sqrt{R}} + \frac{15b^3-12ac}{8a^3}\int\frac{dx}{x\sqrt{R}} \quad \text{(cf. 2.266).}$$

2.269 Im Fall $a = 0$ gilt | For $a = 0$:

$$\int \frac{dx}{x^3 \sqrt{(bx + cx^2)^3}} = \frac{2}{7}\left(-\frac{1}{bx^3} + \frac{8c}{5b^2x^2} - \frac{16c^2}{5b^3x} + \frac{64c^3}{5b^4} + \frac{128c^4x}{5b^5}\right)\frac{1}{\sqrt{bx + cx^2}}.$$

2.27 Funktionen von $\sqrt{a + cx^2}$ und ganzen Potenzen von x
Functions of $\sqrt{a + cx^2}$ and integral powers of x

Abkürzung:
Abbreviation: $u = \sqrt{a + cx^2}$.

$$I_1 = \frac{1}{\sqrt{c}} \ln(x\sqrt{c} + u) \quad [c > 0];$$

$$= \frac{1}{\sqrt{-c}} \arcsin x \sqrt{-\frac{c}{a}} \quad [c < 0, a > 0].$$

$$I_2 = \frac{1}{2\sqrt{a}} \ln \frac{u - \sqrt{a}}{u + \sqrt{a}} \quad [a > 0, c > 0];$$

$$= \frac{1}{2\sqrt{a}} \ln \frac{\sqrt{a} - u}{\sqrt{a} + u} \quad [a > 0, c < 0];$$

$$= \frac{1}{\sqrt{-a}} \operatorname{arcsec} x \sqrt{-\frac{c}{a}} = \frac{1}{\sqrt{-a}} \arccos \frac{1}{x} \sqrt{-\frac{a}{c}} \quad [a < 0, c > 0].$$

2.271

1. $\int u^5 dx = \frac{1}{6} xu^5 + \frac{5}{24} axu^3 + \frac{5}{16} a^2 xu + \frac{5}{16} a^3 I_1.$ D(230.05)+

2. $\int u^3 dx = \frac{1}{4} xu^3 + \frac{3}{8} axu + \frac{3}{8} a^2 I_1.$ D(230.03)+

3. $\int u\, dx = \frac{1}{2} xu + \frac{1}{2} aI_1.$ D(230.01)+

4. $\int \frac{dx}{u} = I_1.$ D(200.01)+

5. $\int \frac{dx}{u^3} = \frac{1}{a}\frac{x}{u}.$ D(200.03)+

6. $\int \frac{dx}{u^{2n+1}} = \frac{1}{a^n} \sum_{k=0}^{n-1} \frac{(-1)^k}{2k+1} \binom{n-1}{k} \frac{c^k x^{2k+1}}{u^{2k+1}}.$

7. $\int \frac{x\, dx}{u^{2n+1}} = -\frac{1}{(2n-1)cu^{2n-1}}$ D(201.9)+

2.272

1. $\int x^2 u^3 dx = \frac{1}{6}\frac{xu^5}{c} - \frac{1}{24}\frac{axu^3}{c} - \frac{1}{16}\frac{a^2 xu}{c} - \frac{1}{16}\frac{a^3}{c} I_1.$ D(232.03)+

2. $\int x^2 u\, dx = \frac{1}{4}\frac{xu^3}{c} - \frac{1}{8}\frac{axu}{c} - \frac{1}{8}\frac{a^2}{c} I_1.$ D(232.01)+

3. $\int \frac{x^2}{u} dx = \frac{1}{2}\frac{xu}{c} - \frac{1}{2}\frac{a}{c} I_1.$ D(202.01)+

4. $\int \frac{x^2}{u^3} dx = -\frac{x}{cu} + \frac{1}{c} I_1.$ D(202.03)+

5. $\int \dfrac{x^2}{u^5} dx = \dfrac{1}{3} \dfrac{x^3}{au^3}.$ D(202.05)+   2.272

6. $\int \dfrac{x^2 dx}{u^{2n+1}} = \dfrac{1}{a^{n-1}} \sum_{k=0}^{n-2} \dfrac{(-1)^k}{2k+3} \binom{n-2}{k} \dfrac{c^k x^{2k+3}}{u^{2k+3}}.$

7. $\int \dfrac{x^3 dx}{u^{2n+1}} = -\dfrac{1}{(2n-3)c^2 u^{2n-3}} + \dfrac{a}{(2n-1)c^2 u^{2n-1}}.$ D(203.09)+

1. $\int x^4 u^3 dx = \dfrac{1}{8} \dfrac{x^3 u^5}{c} - \dfrac{axu^5}{16c^2} + \dfrac{a^2 x u^3}{64c^2} + \dfrac{3a^3 xu}{128c^2} + \dfrac{3a^4}{128c^2} I_1.$ D(234.03)+   2.273

2. $\int x^4 u\, dx = \dfrac{1}{6} \dfrac{x^3 u^3}{c} - \dfrac{axu^3}{8c^2} + \dfrac{a^2 xu}{16c^2} + \dfrac{a^3}{16c^2} I_1.$ D(234.01)+

3. $\int \dfrac{x^4}{u} dx = \dfrac{1}{4} \dfrac{x^3 u}{c} - \dfrac{3}{8} \dfrac{axu}{c^2} + \dfrac{3}{8} \dfrac{a^2}{c^2} I_1.$ D(204.01)+

4. $\int \dfrac{x^4}{u^3} dx = \dfrac{1}{2} \dfrac{xu}{c^2} + \dfrac{ax}{c^2 u} - \dfrac{3}{2} \dfrac{a}{c^2} I_1.$ D(204.03)+

5. $\int \dfrac{x^4}{u^5} dx = -\dfrac{x}{c^2 u} - \dfrac{1}{3} \dfrac{x^3}{cu^3} + \dfrac{1}{c^2} I_1.$ D(204.05)+

6. $\int \dfrac{x^4}{u^7} dx = \dfrac{1}{5} \dfrac{x^5}{au^5}.$ D(204.07)+

7. $\int \dfrac{x^4 dx}{u^{2n+1}} = \dfrac{1}{a^{n-2}} \sum_{k=0}^{n-3} \dfrac{(-1)^k}{2k+5} \binom{n-3}{k} \dfrac{c^k x^{2k+5}}{u^{2k+5}}.$

8. $\int \dfrac{x^5 dx}{u^{2n+1}} = -\dfrac{1}{(2n-5)c^3 u^{2n-5}} + \dfrac{2a}{(2n-3)c^3 u^{2n-3}} - \dfrac{a^2}{(2n-1)c^3 u^{2n-1}}.$ D(205.9)+

1. $\int x^6 u^3 dx = \dfrac{1}{10} \dfrac{x^5 u^5}{c} - \dfrac{ax^3 u^5}{16c^2} + \dfrac{a^2 xu^5}{32c^3} - \dfrac{a^3 xu^3}{128c^3} - \dfrac{3a^4 xu}{256c^3} - \dfrac{3}{256} \dfrac{a^5}{c^3} I_1.$ 2.274

2. $\int x^6 u\, dx = \dfrac{1}{8} \dfrac{x^5 u^3}{c} - \dfrac{5}{48} \dfrac{ax^3 u^3}{c^2} + \dfrac{5a^2 xu^3}{64c^3} - \dfrac{5a^3 xu}{128c^3} - \dfrac{5}{128} \dfrac{a^4}{c^3} I_1.$

3. $\int \dfrac{x^6}{u} dx = \dfrac{1}{6} \dfrac{x^5 u}{c} - \dfrac{5}{24} \dfrac{ax^3 u}{c^2} + \dfrac{5}{16} \dfrac{a^2 xu}{c^3} - \dfrac{5}{16} \dfrac{a^3}{c^3} I_1.$ D(206.01)+

4. $\int \dfrac{x^6}{u^3} dx = \dfrac{1}{4} \dfrac{x^5}{cu} - \dfrac{5}{8} \dfrac{ax^3}{c^2 u} - \dfrac{15}{8} \dfrac{a^2 x}{c^3 u} + \dfrac{15}{8} \dfrac{a^2}{c^3} I_1.$ D(206.03)+

5. $\int \dfrac{x^6}{u^5} dx = \dfrac{1}{2} \dfrac{x^5}{cu^3} + \dfrac{10}{3} \dfrac{ax^3}{c^2 u^3} + \dfrac{5}{2} \dfrac{a^2 x}{c^3 u^3} - \dfrac{5}{2} \dfrac{a}{c^3} I_1.$ D(206.05)+

6. $\int \dfrac{x^6}{u^7} dx = -\dfrac{23}{15} \dfrac{x^5}{cu^5} - \dfrac{7}{3} \dfrac{ax^3}{c^2 u^5} - \dfrac{a^2 x}{c^3 u^5} + \dfrac{1}{c^3} I_1.$ D(206.07)+

7. $\int \dfrac{x^6}{u^9} dx = \dfrac{1}{7} \dfrac{x^7}{au^7}.$ D(206.09)+

8. $\int \dfrac{x^6 dx}{u^{2n+1}} = \dfrac{1}{a^{n-3}} \sum_{k=0}^{n-4} \dfrac{(-1)^k}{2k+7} \binom{n-4}{k} \dfrac{c^k x^{2k+7}}{u^{2k+7}}.$

9. $\int \dfrac{x^7 dx}{u^{2n+1}} = -\dfrac{1}{(2n-7)c^4 u^{2n-7}} + \dfrac{3a}{(2n-5)c^4 u^{2n-5}} - \dfrac{3a^2}{(2n-3)c^4 u^{2n-3}} +$

$$+ \dfrac{a^3}{(2n-1)c^4 u^{2n-1}}.$$ D(207.9)+

2.275

1. $\int \dfrac{u^5}{x}\,dx = \dfrac{u^5}{5} + \dfrac{1}{3}au^3 + a^2u + a^3 I_2.$ D(241.05)+

2. $\int \dfrac{u^3}{x}\,dx = \dfrac{u^3}{3} + au + a^2 I_2.$ D(241.03)+

3. $\int \dfrac{u}{x}\,dx = u + a I_2.$ D(241.01)+

4. $\int \dfrac{dx}{xu} = I_2.$ D(221.01)+

5. $\int \dfrac{dx}{xu^{2n+1}} = \dfrac{1}{a^n} I_2 + \sum_{k=0}^{n-1} \dfrac{1}{(2k+1)\,a^{n-k}\,u^{2k+1}}.$

6. $\int \dfrac{u^5}{x^2}\,dx = -\dfrac{u^5}{x} + \dfrac{5}{4} cxu^3 + \dfrac{15}{8} acxu + \dfrac{15}{8} a^2 I_1.$ D(242.05)+

7. $\int \dfrac{u^3}{x^2}\,dx = -\dfrac{u^3}{x} + \dfrac{3}{2} cxu + \dfrac{3}{2} a I_1.$ D(242.03)+

8. $\int \dfrac{u}{x^2}\,dx = -\dfrac{u}{x} + I_1.$ D(242.01)+

9. $\int \dfrac{dx}{x^2 u^{2n+1}} = -\dfrac{1}{a^{n+1}} \left\{ \dfrac{u}{x} + \sum_{k=1}^{n} \dfrac{(-1)^{k+1}}{2k-1} \binom{n}{k} c^k \left(\dfrac{x}{u}\right)^{2k-1} \right\}.$

2.276

1. $\int \dfrac{u^5}{x^3}\,dx = -\dfrac{u^5}{2x^2} + \dfrac{5}{6} cu^3 + \dfrac{5}{2} acu + \dfrac{5}{2} a^2 c I_2.$ D(243.05)+

2. $\int \dfrac{u^3}{x^3}\,dx = -\dfrac{u^3}{2x^2} + \dfrac{3}{2} cu + \dfrac{3}{2} ac I_2.$ D(243.03)+

3. $\int \dfrac{u}{x^3}\,dx = -\dfrac{u}{2x^2} + \dfrac{c}{2} \cdot I_2.$ D(243.01)+

4. $\int \dfrac{dx}{x^3 u} = -\dfrac{u}{2ax^2} - \dfrac{c}{2a} I_2.$ D(223.01)+

5. $\int \dfrac{dx}{x^3 u^3} = -\dfrac{1}{2ax^2 u} - \dfrac{3c}{2a^2 u} - \dfrac{3c}{2a^2} I_2.$ D(223.03)+

6. $\int \dfrac{dx}{x^3 u^5} = -\dfrac{1}{2ax^2 u^3} - \dfrac{5}{6}\dfrac{c}{a^2 u^3} - \dfrac{5}{2}\dfrac{c}{a^3 u} - \dfrac{5}{2}\dfrac{c}{a^3} I_2.$ D(223.05)+

7. $\int \dfrac{u^5}{x^4}\,dx = -\dfrac{au^3}{3x^3} - \dfrac{2acu}{x} + \dfrac{c^2 xu}{2} + \dfrac{5}{2} ac I_1.$ D(244.05)+

8. $\int \dfrac{u^3}{x^4}\,dx = -\dfrac{u^3}{3x^3} - \dfrac{cu}{x} + c I_1.$ D(244.03)+

9. $\int \dfrac{u}{x^4}\,dx = -\dfrac{u^3}{3ax^3}.$ D(244.01)+

10. $\int \dfrac{dx}{x^4 u^{2n+1}} = \dfrac{1}{a^{n+2}} \left\{ -\dfrac{u^3}{3x^3} + (n+1)\dfrac{cu}{x} + \sum_{k=2}^{n+1} \dfrac{(-1)^k}{2k-3} \binom{n+1}{k} c^k \left(\dfrac{x}{u}\right)^{2k-3} \right\}.$

2.277

1. $\int \dfrac{u^3}{x^5}\,dx = -\dfrac{u^3}{4x^4} - \dfrac{3}{8}\dfrac{cu^3}{ax^2} + \dfrac{3}{8}\dfrac{c^2 u}{a} + \dfrac{3}{8} c^2 I_2.$ D(245.03)+

2. $\int \dfrac{u}{x^5}\,dx = -\dfrac{u}{4x^4} - \dfrac{1}{8}\dfrac{cu}{ax^2} - \dfrac{1}{8}\dfrac{c^2}{a} I_2.$ D(245.01)+

3. $\int \dfrac{dx}{x^5 u} = -\dfrac{u}{4ax^4} + \dfrac{3}{8}\dfrac{cu}{a^2 x^2} + \dfrac{3}{8}\dfrac{c^2}{a^2} I_2.$ D(225.01)+   2.277

4. $\int \dfrac{dx}{x^5 u^3} = -\dfrac{1}{4ax^4 u} + \dfrac{5}{8}\dfrac{c}{a^2 x^2 u} + \dfrac{15}{8}\dfrac{c^2}{a^3 u} + \dfrac{15}{8}\dfrac{c^2}{a^3} I_2.$ D(225.03)+

1. $\int \dfrac{u^3}{x^6} dx = -\dfrac{u^5}{5ax^5}.$ D(246.03)+   2.278

2. $\int \dfrac{u}{x^6} dx = -\dfrac{u^3}{5ax^5} + \dfrac{2}{15}\dfrac{cu^3}{a^2 x^3}.$ D(246.01)+

3. $\int \dfrac{dx}{x^6 u} = \dfrac{1}{a^3}\left(-\dfrac{u^5}{5x^5} + \dfrac{2}{3}\dfrac{cu^3}{x^3} - \dfrac{c^2 u}{x}\right).$ D(226.01)+

4. $\int \dfrac{dx}{x^6 u^{2n+1}} = \dfrac{1}{a^{n+3}}\left\{-\dfrac{u^5}{5x^5} + \dfrac{1}{3}\binom{n+2}{1}\dfrac{cu^3}{x^3} - \binom{n+2}{2}\dfrac{c^2 u}{x} + \right.$

$$\left. + \sum_{k=3}^{n+2} \dfrac{(-1)^k}{2k-5} \binom{n+2}{k} c^k \left(\dfrac{x}{u}\right)^{2k-5}\right\}.$$

Funktionen von $\sqrt{a+bx+cx^2}$ und Polynomen ersten und zweiten Grades
Functions of $\sqrt{a+bx+cx^2}$ and of first- and second-degree polynomials   2.28

Abkürzung:
Abbreviation: $R = a + bx + cx^2.$

cf. 2.252

$\int \dfrac{dx}{(x+p)^n \sqrt{R}} = -\int \dfrac{t^{n-1} dt}{\sqrt{c+(b-2pc)t+(a-bp+cp^2)t^2}} \quad \left[t = \dfrac{1}{x+p}\right].$ 2.281

1. $\int \dfrac{\sqrt{R}\, dx}{x+p} = c\int \dfrac{x\,dx}{\sqrt{R}} + (b-cp)\int \dfrac{dx}{\sqrt{R}} + (a-bp+cp^2)\int \dfrac{dx}{(x+p)\sqrt{R}}.$ 2.282

2. $\int \dfrac{dx}{(x+p)(x+q)\sqrt{R}} = \dfrac{1}{q-p}\int \dfrac{dx}{(x+p)\sqrt{R}} + \dfrac{1}{p-q}\int \dfrac{dx}{(x+q)\sqrt{R}}.$

3. $\int \dfrac{\sqrt{R}\, dx}{(x+p)(x+q)} = \dfrac{1}{q-p}\int \dfrac{\sqrt{R}\, dx}{x+p} + \dfrac{1}{p-q}\int \dfrac{\sqrt{R}\, dx}{x+q}.$

4. $\int \dfrac{(x+p)\sqrt{R}\, dx}{x+q} = \int \sqrt{R}\, dx + (p-q)\int \dfrac{\sqrt{R}\, dx}{x+q}.$

5. $\int \dfrac{(rx+s)\, dx}{(x+p)(x+q)\sqrt{R}} = \dfrac{s-pr}{q-p}\int \dfrac{dx}{(x+p)\sqrt{R}} + \dfrac{s-qr}{p-q}\int \dfrac{dx}{(x+q)\sqrt{R}}.$

$\int \dfrac{(Ax+B)\, dx}{(p+R)^n \sqrt{R}} = \dfrac{A}{c}\int \dfrac{du}{(p+u^2)^n} + \dfrac{2Bc-Ab}{2c}\int \dfrac{(1-cv^2)^{n-1}\, dv}{\left[p+a-\dfrac{b^2}{4c}-cpv^2\right]^n}$ 2.283

mit $u = \sqrt{R}$ und $v = \dfrac{b+2cx}{2c\sqrt{R}}.$ | where $u = \sqrt{R}$ and $v = \dfrac{b+2cx}{2c\sqrt{R}}.$

$\int \dfrac{Ax+B}{(p+R)\sqrt{R}}\, dx = \dfrac{A}{c} I_1 + \dfrac{2Bc-Ab}{\sqrt{c^2 p[b^2-4(a+p)c]}} I_2$ 2.284

2.284 mit | where

$$I_1 = \frac{1}{\sqrt{p}} \operatorname{arctg} \sqrt{\frac{R}{p}} \quad [p > 0];$$

$$= \frac{1}{2\sqrt{-p}} \ln \frac{\sqrt{-p} - \sqrt{R}}{\sqrt{-p} + \sqrt{R}} \quad [p < 0].$$

$$I_2 = -\operatorname{arctg} \sqrt{\frac{p}{b^2 - 4(a+p)c}} \frac{b + 2cx}{\sqrt{R}} \quad [p\{b^2 - 4(a+p)c\} > 0];$$

$$= \frac{1}{2i} \ln \frac{\sqrt{4(a+p)c - b^2}\sqrt{R} + \sqrt{p}\,(b+2cx)}{\sqrt{4(a+p)c - b^2}\sqrt{R} - \sqrt{p}\,(b+2cx)} \quad [p\{b^2 - 4(a+p)c\} < 0,\, p > 0];$$

$$= \frac{1}{2i} \ln \frac{\sqrt{b^2 - 4(a+p)c}\sqrt{R} - \sqrt{-p}\,(b+2cx)}{\sqrt{b^2 - 4(a+p)c}\sqrt{R} + \sqrt{-p}\,(b+2cx)} \quad [p\{b^2 - 4(a+p)c\} < 0,\, p < 0].$$

2.29 Integrale, die auf elliptische und pseudoelliptische Integrale führen
Integrals that can be reduced to elliptic and pseudo-elliptic integrals

2.290 Die Integrale $\int R(x, \sqrt{P(x)})\,dx$, wobei $P(x)$ ein Polynom dritten oder vierten Grades ist, lassen sich durch algebraische Umformungen in eine Summe von Integralen, die durch elementare Funktionen ausgedrückt werden könen, und von elliptischen Integralen verwandeln (vgl. 8.11). Da die Substitutionen, die ein gegebenes Integral in ein elliptisches Integral in der LEGENDREschen Normalform überführen, sich für verschiedene Integrationsintervalle voneinander unterscheiden, werden die entsprechenden Formeln im Abschnitt „Bestimmte Integrale" aufgeführt (siehe 3.13, 3.17).

The integrals $\int R(x, \sqrt{P(x)})\,dx$, where $P(x)$ is a polynomial of third or fourth degree, can algebraically be transformed into a sum of integrals that can be expressed in terms of elementary functions and elliptic integrals (see 8.11). Since the substitutions that transform a given integral into the LEGENDRE standard form of an elliptic integral depend on the integration interval, the formulae for these integrals are given in the chapter on definite integrals (see 3.13 and 3.17).

2.291 Auf Integrale der Gestalt $\int R(x, \sqrt{P(x)})\,dx$ führen einige Integrale der Gestalt $\int R(x, \sqrt[k]{P_n(x)})\,dx$, wobei $k \geq 2$ und $P_n(x)$ ein Polynom mindestens fünften Grades ist. Dafür seien einige Beispiele angegeben:

Some integrals of the form $\int R(x, \sqrt[k]{P_n(x)})\,dx$, where $k \geq 2$ and $P_n(x)$ is a polynomial of a degree greater than four, also lead to integrals of the form $\int R(x, \sqrt{P(x)})\,dx$. The examples of such integrals are

1. $$\int \frac{dx}{\sqrt{1 - x^6}} = -\int \frac{dz}{\sqrt{3 + 3z^2 + z^4}} \quad \left[x^2 = \frac{1}{1+z^2}\right].$$

2. $$\int \frac{dx}{\sqrt{a + bx^2 + cx^4 + dx^6}} = \frac{1}{2} \int \frac{dz}{\sqrt{az + bz^2 + cz^3 + dz^4}} \quad [x^2 = z].$$

3. $\int (a + 2bx + cx^2 + gx^3)^{\pm\frac{1}{3}} dx = \frac{3}{2} \int \frac{z^2 A^{\pm\frac{1}{3}} dz}{B}$ 2.291

$$[a + 2bx + cx^2 = z^3,\ A = g\left(\frac{-b + \sqrt{b^2 + (z^3 - a)c}}{c}\right)^3 + z^3,$$

$$B = \sqrt{b^2 + (z^3 - a)c}].$$

In einigen Fällen lassen sich Integrale der Gestalt $\int R(x, \sqrt{P(x)})\,dx$, wobei $P(x)$ ein Polynom dritten oder vierten Grades ist, durch elementare Funktionen ausdrücken. Solche Integrale werden *pseudoelliptisch* genannt. Gelten beispielsweise die Beziehungen	In some cases integrals of the form $\int R(x, \sqrt{P(x)})\,dx$, where $P(x)$ is a polynomial of degree three or four, can be expressed in terms of elementary functions. Such integrals are called *pseudoelliptic*. For example, if the following relations hold:

2.292

$$f_1(x) = -f_1\left(\frac{1}{k^2 x}\right),\quad f_2(x) = -f_2\left(\frac{1 - k^2 x}{k^2(1 - x)}\right),\quad f_3(x) = -f_3\left(\frac{1 - x}{1 - k^2 x}\right),$$

so ist | then

1. $\int \frac{f_1(x)\,dx}{\sqrt{x(1-x)(1-k^2 x)}} = \int R_1(z)\,dz \quad [zx = \sqrt{x(1-x)(1-k^2 x)}];$

2. $\int \frac{f_2(x)\,dx}{\sqrt{x(1-x)(1-k^2 x)}} = \int R_2(z)\,dz \quad \left[z = \frac{\sqrt{x(1-k^2 x)}}{\sqrt{1-x}}\right];$

3. $\int \frac{f_3(x)\,dx}{\sqrt{x(1-x)(1-k^2 x)}} = \int R_3(z)\,dz \quad \left[z = \frac{\sqrt{x(1-x)}}{\sqrt{1-k^2 x}}\right],$

wobei $R_1(z)$, $R_2(z)$, $R_3(z)$ rationale Funktionen von z sind. | where $R_1(z)$, $R_2(z)$, and $R_3(z)$ are rational functions of z.

Die Exponentialfunktion
The Exponential Function

2.3

Funktionen von e^{ax}
Functions of e^{ax}

2.31

$$\int e^{ax}\,dx = \frac{e^{ax}}{a}.$$ 2.311

a^x in Integranden ersetze man durch $e^{x \ln a}$. | Function a^x has to be replaced by $e^{x \ln a}$. 2.312

1. $\int \frac{dx}{a + be^{mx}} = \frac{1}{am}[mx - \ln(a + be^{mx})].$ P(410) 2.313

2. $\int \frac{dx}{1 + e^x} = \ln \frac{e^x}{1 + e^x} = x - \ln(1 + e^x).$ P(409)

2.314 $$\int \frac{dx}{ae^{mx}+be^{-mx}} = \frac{1}{m\sqrt{ab}} \operatorname{arctg}\left(e^{mx}\sqrt{\frac{a}{b}}\right) \quad [ab>0];$$ P(411)

$$= \frac{1}{2m\sqrt{-ab}} \ln \frac{b+e^{mx}\sqrt{-ab}}{b-e^{mx}\sqrt{-ab}} \quad [ab<0].$$

2.315 $$\int \frac{dx}{\sqrt{a+be^{mx}}} = \frac{1}{m\sqrt{a}} \ln \frac{\sqrt{a+be^{mx}}-\sqrt{a}}{\sqrt{a+be^{mx}}+\sqrt{a}} \quad [a>0];$$

$$= \frac{2}{m\sqrt{-a}} \operatorname{arctg} \frac{\sqrt{a+be^{mx}}}{\sqrt{-a}} \quad [a<0].$$

2.32 Die Exponentialfunktion und rationale Funktionen von x
The exponential and rational functions of x

2.321 1. $$\int x^m e^{ax}\,dx = \frac{x^m e^{ax}}{a} - \frac{m}{a}\int x^{m-1} e^{ax}\,dx.$$

2. $$\int x^n e^{ax}\,dx = e^{ax}\left(\frac{x^n}{a} + \sum_{k=1}^{n}(-1)^k \frac{n(n-1)\dots(n-k+1)}{a^{k+1}} x^{n-k}\right).$$

2.322 1. $$\int x e^{ax}\,dx = e^{ax}\left(\frac{x}{a} - \frac{1}{a^2}\right).$$

2. $$\int x^2 e^{ax}\,dx = e^{ax}\left(\frac{x^2}{a} - \frac{2x}{a^2} + \frac{2}{a^3}\right).$$

3. $$\int x^3 e^{ax}\,dx = e^{ax}\left(\frac{x^3}{a} - \frac{3x^2}{a^2} + \frac{6x}{a^3} - \frac{6}{a^4}\right).$$

2.323 $$\int P_m(x) e^{ax}\,dx = \frac{e^{ax}}{a} \sum_{k=0}^{m} (-1)^k \frac{P^{(k)}(x)}{a^k},$$

wobei $P_m(x)$ ein Polynom m-ten Grades in x und $P^{(k)}(x)$ die k-te Ableitung von $P_m(x)$ nach x ist. | where $P_m(x)$ is a polynomial in x of degree m, while $P^{(k)}(x)$ is its kth derivative with respect to x.

2.324 1. $$\int \frac{e^{ax}\,dx}{x^m} = \frac{1}{m-1}\left[-\frac{e^{ax}}{x^{m-1}} + a\int \frac{e^{ax}\,dx}{x^{m-1}}\right].$$

2. $$\int \frac{e^{ax}}{x^n}\,dx = -e^{ax}\sum_{k=1}^{n-1} \frac{a^{k-1}}{(n-1)(n-2)\dots(n-k)\,x^{n-k}} + \frac{a^{n-1}}{(n-1)!}\operatorname{Ei}(ax).$$

2.325 1. $$\int \frac{e^{ax}}{x}\,dx = \operatorname{Ei}(ax).$$

2. $$\int \frac{e^{ax}}{x^2}\,dx = -\frac{e^{ax}}{x} + a\operatorname{Ei}(ax).$$

3. $$\int \frac{e^{ax}}{x^3}\,dx = -\frac{e^{ax}}{2x^2} - \frac{ae^{ax}}{2x} + \frac{a^2}{2}\operatorname{Ei}(ax).$$

2.326 $$\int \frac{x e^{ax}\,dx}{(1+ax)^2} = \frac{e^{ax}}{a^2(1+ax)}.$$

Hyperbelfunktionen
Hyperbolic Functions

2.4

Potenzen von sh x, ch x, th x **und** cth x
Powers of sh x, ch x, th x, **and** cth x

2.41–2.43

$$\int \text{sh}^p x \, \text{ch}^q x \, dx = \frac{\text{sh}^{p+1}x \, \text{ch}^{q-1}x}{p+q} + \frac{q-1}{p+q} \int \text{sh}^p x \, \text{ch}^{q-2} x \, dx;$$

2.411

$$= \frac{\text{sh}^{p-1}x \, \text{ch}^{q+1}x}{p+q} - \frac{p-1}{p+q} \int \text{sh}^{p-2} x \, \text{ch}^q x \, dx;$$

$$= \frac{\text{sh}^{p-1}x \, \text{ch}^{q+1}x}{q+1} - \frac{p-1}{q+1} \int \text{sh}^{p-2} x \, \text{ch}^{q+2} x \, dx;$$

$$= \frac{\text{sh}^{p+1}x \, \text{ch}^{q-1}x}{p+1} - \frac{q-1}{p+1} \int \text{sh}^{p+2} x \, \text{ch}^{q-2} x \, dx;$$

$$= \frac{\text{sh}^{p+1}x \, \text{ch}^{q+1}x}{p+1} - \frac{p+q+2}{p+1} \int \text{sh}^{p+2} x \, \text{ch}^q x \, dx;$$

$$= -\frac{\text{sh}^{p+1}x \, \text{ch}^{q+1}x}{q+1} + \frac{p+q+2}{q+1} \int \text{sh}^p x \, \text{ch}^{q+2} x \, dx.$$

1. $$\int \text{sh}^p x \, \text{ch}^{2n} x \, dx = \frac{\text{sh}^{p+1} x}{2n+p} \Bigg\{ \text{ch}^{2n-1} x +$$

2.412

$$+ \sum_{k=1}^{n-1} \frac{(2n-1)(2n-3) \ldots (2n-2k+1)}{(2n+p-2)(2n+p-4) \ldots (2n+p-2k)} \text{ch}^{2n-2k-1} x \Bigg\} +$$

$$+ \frac{(2n-1)!!}{(2n+p)(2n+p-2) \ldots (p+2)} \int \text{sh}^p x \, dx.$$

Diese Formel gilt für alle reellen p mit Ausnahme der negativen geraden Zahlen $-2, -4, \ldots, -2n$. Ist p eine natürliche Zahl und $n = 0$, so gilt

This formula holds for all real p with the exception of negative even integers $-2, -4, \ldots, -2n$. If p is a positive integer and $n = 0$, then

2. $$\int \text{sh}^{2m} x \, dx = (-1)^m \binom{2m}{m} \frac{x}{2^{2m}} + \frac{1}{2^{2m-1}} \sum_{k=0}^{m-1} (-1)^k \binom{2m}{k} \frac{\text{sh}(2m-2k)x}{2m-2k}.$$ T(543)

3. $$\int \text{sh}^{2m+1} x \, dx = \frac{1}{2^{2m}} \sum_{k=0}^m (-1)^k \binom{2m+1}{k} \frac{\text{ch}(2m-2k+1)x}{2m-2k+1};$$ T(544)

$$= (-1)^n \sum_{k=0}^m (-1)^k \binom{m}{k} \frac{\text{ch}^{2k+1} x}{2k+1}.$$ GHH[351](5)

4. $$\int \text{sh}^p x \, \text{ch}^{2n+1} x \, dx = \frac{\text{sh}^{p+1} x}{2n+p+1} \Bigg\{ \text{ch}^{2n} x +$$

$$+ \sum_{k=1}^n \frac{2^k n(n-1) \ldots (n-k+1) \text{ch}^{2n-2k} x}{(2n+p-1)(2n+p-3) \ldots (2n+p-2k+1)} \Bigg\}.$$

Diese Formel gilt für alle reellen p mit Ausnahme der negativen ungeraden Zahlen $-1, -3, \ldots, -(2n+1)$. | This formula holds for all real p with the exception of negative odd integers $-1, -3, \ldots, -(2n+1)$.

2.413

1. $\int \operatorname{ch}^p x \operatorname{sh}^{2n} x \, dx = \dfrac{\operatorname{ch}^{p+1} x}{2n+p} \Big\{ \operatorname{sh}^{2n-1} x +$

$+ \sum\limits_{k=1}^{n-1} (-1)^k \dfrac{(2n-1)(2n-3) \ldots (2n-2k+1) \operatorname{sh}^{2n-2k-1} x}{(2n+p-2)(2n+p-4) \ldots (2n+p-2k)} \Big\} +$

$+ (-1)^n \dfrac{(2n-1)!!}{(2n+p)(2n+p-2) \ldots (p+2)} \int \operatorname{ch}^p x \, dx.$

Diese Formel gilt für alle reellen p mit Ausnahme der negativen geraden Zahlen $-2, -4, \ldots, -2n$. Ist p eine natürliche Zahl und $n = 0$, so gilt | This formula holds for all real p with the exception of negative even integers $-2, -4, \ldots, -2n$. If p is a positive integer and $n = 0$, then

2. $\int \operatorname{ch}^{2m} x \, dx = \binom{2m}{m} \dfrac{x}{2^{2m}} + \dfrac{1}{2^{2m-1}} \sum\limits_{k=0}^{m-1} \binom{2m}{k} \dfrac{\operatorname{sh}(2m-2k)x}{2m-2k} .$ T(541)

3. $\int \operatorname{ch}^{2m+1} x \, dx = \dfrac{1}{2^{2m}} \sum\limits_{k=0}^{m} \binom{2m+1}{k} \dfrac{\operatorname{sh}(2m-2k+1)x}{2m-2k+1} ;$ T(542)

$= \sum\limits_{k=0}^{m} \binom{m}{k} \dfrac{\operatorname{sh}^{2k+1} x}{2k+1} .$ GHH[351](8)

4. $\int \operatorname{ch}^p x \operatorname{sh}^{2n+1} x \, dx = \dfrac{\operatorname{ch}^{p+1} x}{2n+p+1} \Big\{ \operatorname{sh}^{2n} x +$

$+ \sum\limits_{k=1}^{n} (-1)^k \dfrac{2^k n(n-1) \ldots (n-k+1) \operatorname{sh}^{2n-2k} x}{(2n+p-1)(2n+p-3) \ldots (2n+p-2k+1)} \Big\} .$

Diese Formel gilt für alle reellen p mit Ausnahme der negativen ungeraden Zahlen $-1, -3, \ldots, -(2n+1)$. | This formula holds for all real p with the exception of negative odd integers $-1, -3, \ldots, -(2n+1)$.

2.414

1. $\int \operatorname{sh} ax \, dx = \dfrac{1}{a} \operatorname{ch} ax.$

2. $\int \operatorname{sh}^2 ax \, dx = \dfrac{1}{4a} \operatorname{sh} 2ax - \dfrac{x}{2}.$

3. $\int \operatorname{sh}^3 x \, dx = -\dfrac{3}{4} \operatorname{ch} x + \dfrac{1}{12} \operatorname{ch} 3x = \dfrac{1}{3} \operatorname{ch}^3 x - \operatorname{ch} x.$

4. $\int \operatorname{sh}^4 x \, dx = \dfrac{3}{8} x - \dfrac{1}{4} \operatorname{sh} 2x + \dfrac{1}{32} \operatorname{sh} 4x = \dfrac{3}{8} x - \dfrac{3}{8} \operatorname{sh} x \operatorname{ch} x + \dfrac{1}{4} \operatorname{sh}^3 x \operatorname{ch} x.$

5. $\int \operatorname{sh}^5 x \, dx = \dfrac{5}{8} \operatorname{ch} x - \dfrac{5}{48} \operatorname{ch} 3x + \dfrac{1}{80} \operatorname{ch} 5x;$

$= \dfrac{4}{5} \operatorname{ch} x + \dfrac{1}{5} \operatorname{sh}^4 x \operatorname{ch} x - \dfrac{4}{15} \operatorname{ch}^3 x.$

6. $\int \operatorname{sh}^6 x \, dx = -\dfrac{5}{16} x + \dfrac{15}{64} \operatorname{sh} 2x - \dfrac{3}{64} \operatorname{sh} 4x + \dfrac{1}{192} \operatorname{sh} 6x;$

$= -\dfrac{5}{16} x + \dfrac{1}{6} \operatorname{sh}^5 x \operatorname{ch} x - \dfrac{5}{24} \operatorname{sh}^3 x \operatorname{ch} x + \dfrac{5}{16} \operatorname{sh} x \operatorname{ch} x.$

7. $\int \text{sh}^7 x \, dx = -\frac{35}{64} \text{ch} \, x + \frac{7}{64} \text{ch} \, 3x - \frac{7}{320} \text{ch} \, 5x + \frac{1}{448} \text{ch} \, 7x;$ 2.414

$$= -\frac{24}{35} \text{ch} \, x + \frac{8}{35} \text{ch}^3 x - \frac{6}{35} \text{ch} \, x \, \text{sh}^4 x + \frac{1}{7} \text{ch} \, x \, \text{sh}^6 x.$$

8. $$\int \text{ch} \, ax \, dx = \frac{1}{a} \text{sh} \, ax.$$

9. $$\int \text{ch}^2 ax \, dx = \frac{x}{2} + \frac{1}{4a} \text{sh} \, 2ax.$$

10. $$\int \text{ch}^3 x \, dx = \frac{3}{4} \text{sh} \, x + \frac{1}{12} \text{sh} \, 3x = \text{sh} \, x + \frac{1}{3} \text{sh}^3 x.$$

11. $\int \text{ch}^4 x \, dx = \frac{3}{8} x + \frac{1}{4} \text{sh} \, 2x + \frac{1}{32} \text{sh} \, 4x = \frac{3}{8} x + \frac{3}{8} \text{sh} \, x \, \text{ch} \, x + \frac{1}{4} \text{sh} \, x \, \text{ch}^3 x.$

12. $\int \text{ch}^5 x \, dx = \frac{5}{8} \text{sh} \, x + \frac{5}{48} \text{sh} \, 3x + \frac{1}{80} \text{sh} \, 5x;$

$$= \frac{4}{5} \text{sh} \, x + \frac{1}{5} \text{ch}^4 x \, \text{sh} \, x + \frac{4}{15} \text{sh}^3 x.$$

13. $\int \text{ch}^6 x \, dx = \frac{5}{16} x + \frac{15}{64} \text{sh} \, 2x + \frac{3}{64} \text{sh} \, 4x + \frac{1}{192} \text{sh} \, 6x;$

$$= \frac{5}{16} x + \frac{5}{16} \text{sh} \, x \, \text{ch} x + \frac{5}{24} \text{sh} \, x \, \text{ch}^3 x + \frac{1}{6} \text{sh} \, x \, \text{ch}^5 x.$$

14. $\int \text{ch}^7 x \, dx = \frac{35}{64} \text{sh} \, x + \frac{7}{64} \text{sh} \, 3x + \frac{7}{320} \text{sh} \, 5x + \frac{1}{448} \text{sh} \, 7x;$

$$= \frac{24}{35} \text{sh} \, x + \frac{8}{35} \text{sh}^3 x + \frac{6}{35} \text{sh} \, x \, \text{ch}^4 x + \frac{1}{7} \text{sh} \, x \, \text{ch}^6 x.$$

1. $$\int \text{sh} \, ax \, \text{ch} \, bx \, dx = \frac{\text{ch}(a+b)x}{2(a+b)} + \frac{\text{ch}(a-b)x}{2(a-b)}.$$ 2.415

2. $$\int \text{sh} \, ax \, \text{ch} \, ax \, dx = \frac{1}{4a} \text{ch} \, 2ax.$$

3. $$\int \text{sh}^2 x \, \text{ch} \, x \, dx = \frac{1}{3} \text{sh}^3 x.$$

4. $$\int \text{sh}^3 x \, \text{ch} x \, dx = \frac{1}{4} \text{sh}^4 x.$$

5. $$\int \text{sh}^4 x \, \text{ch} \, x \, dx = \frac{1}{5} \text{sh}^5 x.$$

6. $$\int \text{sh} \, x \, \text{ch}^2 x \, dx = \frac{1}{3} \text{ch}^3 x.$$

7. $$\int \text{sh}^2 x \, \text{ch}^2 x \, dx = -\frac{x}{8} + \frac{1}{32} \text{sh} \, 4x.$$

8. $$\int \text{sh}^3 x \, \text{ch}^2 x \, dx = \frac{1}{5} \left(\text{sh}^2 x - \frac{2}{3} \right) \text{ch}^3 x.$$

9. $$\int \text{sh}^4 x \, \text{ch}^2 x \, dx = \frac{x}{16} - \frac{1}{64} \text{sh} \, 2x - \frac{1}{64} \text{sh} \, 4x + \frac{1}{192} \text{sh} \, 6x.$$

2.415

10. $\int \operatorname{sh} x \operatorname{ch}^3 x \, dx = \frac{1}{4} \operatorname{ch}^4 x.$

11. $\int \operatorname{sh}^2 x \operatorname{ch}^3 x \, dx = \frac{1}{5}\left(\operatorname{ch}^2 x + \frac{2}{3}\right) \operatorname{sh}^3 x.$

12. $\int \operatorname{sh}^3 x \operatorname{ch}^3 x \, dx = -\frac{3}{64} \operatorname{ch} 2x + \frac{1}{192} \operatorname{ch} 6x = \frac{1}{48} \operatorname{ch}^3 2x - \frac{1}{16} \operatorname{ch} 2x;$

$$= \frac{\operatorname{sh}^6 x}{6} + \frac{\operatorname{sh}^4 x}{4} = \frac{\operatorname{ch}^6 x}{6} - \frac{\operatorname{ch}^4 x}{4}.$$

13. $\int \operatorname{sh}^4 x \operatorname{ch}^3 x \, dx = \frac{1}{7} \operatorname{sh}^3 x \left(\operatorname{ch}^4 x - \frac{3}{5} \operatorname{ch}^2 x - \frac{2}{5}\right) = \frac{1}{7}\left(\operatorname{ch}^2 x + \frac{2}{5}\right) \operatorname{sh}^5 x.$

14. $\int \operatorname{sh} x \operatorname{ch}^4 x \, dx = \frac{1}{5} \operatorname{ch}^5 x.$

15. $\int \operatorname{sh}^2 x \operatorname{ch}^4 x \, dx = -\frac{x}{16} - \frac{1}{64} \operatorname{sh} 2x + \frac{1}{64} \operatorname{sh} 4x + \frac{1}{192} \operatorname{sh} 6x.$

16. $\int \operatorname{sh}^3 x \operatorname{ch}^4 x \, dx = \frac{1}{7} \operatorname{ch}^3 x \left(\operatorname{sh}^4 x + \frac{3}{5} \operatorname{sh}^2 x - \frac{2}{5}\right) = \frac{1}{7}\left(\operatorname{sh}^2 x - \frac{2}{5}\right) \operatorname{ch}^5 x.$

17. $\int \operatorname{sh}^4 x \operatorname{ch}^4 x \, dx = \frac{3x}{128} - \frac{1}{128} \operatorname{sh} 4x + \frac{1}{1024} \operatorname{sh} 8x.$

2.416

1. $\int \frac{\operatorname{sh}^p x}{\operatorname{ch}^{2n} x} dx = \frac{\operatorname{sh}^{p+1} x}{2n-1} \Big\{ \operatorname{sech}^{2n-1} x +$

$$+ \sum_{k=1}^{n-1} \frac{(2n-p-2)(2n-p-4)\ldots(2n-p-2k)}{(2n-3)(2n-5)\ldots(2n-2k-1)} \operatorname{sech}^{2n-2k-1} x \Big\} +$$

$$+ \frac{(2n-p-2)(2n-p-4)\ldots(-p+2)(-p)}{(2n-1)!!} \int \operatorname{sh}^p x \, dx.$$

Diese Formel gilt für alle reellen p. Die Formeln für $\int \operatorname{sh}^p x \, dx$ bei natürlichem p findet man unter 2.412 2. und 2.412 3. Für $n=0$ und negatives ganzzahliges p gelten die Formeln	This formula holds for all real p. For $\int \operatorname{sh}^p x \, dx$, where p is a positive integer, see 2.412 2. and 2.412 3. If $n=0$ and p a negative integer, we have

2. $\int \frac{dx}{\operatorname{sh}^{2m} x} = \frac{\operatorname{ch} x}{2m-1} \Big\{ -\operatorname{cosech}^{2m-1} x +$

$$+ \sum_{k=1}^{m-1} (-1)^{k-1} \cdot \frac{2^k (m-1)(m-2)\ldots(m-k)}{(2m-3)(2m-5)\ldots(2m-2k-1)} \operatorname{cosech}^{2m-2k-1} x \Big\}.$$

3. $\int \frac{dx}{\operatorname{sh}^{2m+1} x} = \frac{\operatorname{ch} x}{2m} \Big\{ -\operatorname{cosech}^{2m} x +$

$$+ \sum_{k=1}^{m-1} (-1)^{k-1} \cdot \frac{(2m-1)(2m-3)\ldots(2m-2k+1)}{2^k (m-1)(m-2)\ldots(m-k)} \operatorname{cosech}^{2m-2k} x \Big\} +$$

$$+ (-1)^m \frac{(2m-1)!!}{(2m)!!} \ln \operatorname{th} \frac{x}{2}.$$

1. $\int \dfrac{\operatorname{sh}^p x}{\operatorname{ch}^{2n+1} x} dx = \dfrac{\operatorname{sh}^{p+1} x}{2n} \Big\{ \operatorname{sech}^{2n} x +$ 2.417

$+ \sum_{k=1}^{n-1} \dfrac{(2n-p-1)(2n-p-3)\ldots(2n-p-2k+1)}{2^k(n-1)(n-2)\ldots(n-k)} \operatorname{sech}^{2n-2k} x \Big\} +$

$+ \dfrac{(2n-p-1)(2n-p-3)\ldots(3-p)(1-p)}{2^n n!} \int \dfrac{\operatorname{sh}^p x}{\operatorname{ch} x} dx.$

| Diese Formel gilt für alle reellen p. | This formula holds for all real p. |
| Ist $n = 0$ und p ganzzahlig, so gilt | For $n = 0$ and integral p we have |

2. $\int \dfrac{\operatorname{sh}^{2m+1} x}{\operatorname{ch} x} dx = \sum_{k=1}^{m} \dfrac{(-1)^{m+k}}{2k} \operatorname{sh}^{2k} x + (-1)^m \ln \operatorname{ch} x;$

$= \sum_{k=1}^{m} \dfrac{(-1)^{m+k}}{2k} \binom{m}{k} \operatorname{ch}^{2k} x + (-1)^m \ln \operatorname{ch} x \quad [m \geq 1].$

3. $\int \dfrac{\operatorname{sh}^{2m} x}{\operatorname{ch} x} dx = \sum_{k=1}^{m} \dfrac{(-1)^{m+k}}{2k-1} \operatorname{sh}^{2k-1} x + (-1)^m \operatorname{arctg}(\operatorname{sh} x) \quad [m \geq 1].$

4. $\int \dfrac{dx}{\operatorname{sh}^{2m+1} x \operatorname{ch} x} = \sum_{k=1}^{m} \dfrac{(-1)^k \operatorname{cosech}^{2m-2k+2} x}{2m-2k+2} + (-1)^m \ln \operatorname{th} x.$

5. $\int \dfrac{dx}{\operatorname{sh}^{2m} x \operatorname{ch} x} = \sum_{k=1}^{m} \dfrac{(-1)^k \operatorname{cosech}^{2m-2k+1} x}{2m-2k+1} + (-1)^m \operatorname{arctg} \operatorname{sh} x.$

1. $\int \dfrac{\operatorname{ch}^p x}{\operatorname{sh}^{2n} x} dx = -\dfrac{\operatorname{ch}^{p+1} x}{2n-1} \Big\{ \operatorname{cosech}^{2n-1} x +$ 2.418

$+ \sum_{k=1}^{n-1} \dfrac{(-1)^k (2n-p-2)(2n-p-4)\ldots(2n-p-2k)}{(2n-3)(2n-5)\ldots(2n-2k-1)} \operatorname{cosech}^{2n-2k-1} x \Big\} +$

$+ \dfrac{(-1)^n (2n-p-2)(2n-p-4)\ldots(-p+2)(-p)}{(2n-1)!!} \int \operatorname{ch}^p x \, dx.$

| Diese Formel gilt für alle reellen p. | This formula holds for all real p. |
| Die Formeln für $\int \operatorname{ch}^p x \, dx$ bei natürlichem p findet man unter 2.413 2. und 2.413 3. Für negatives ganzzahliges p gelten die Formeln | For $\int \operatorname{ch}^p x \, dx$, where p is a positive integer, see 2.413 2. and 2.413 3. If p is a negative integer, we have |

2. $\int \dfrac{dx}{\operatorname{ch}^{2m} x} = \dfrac{\operatorname{sh} x}{2m-1} \Big\{ \operatorname{sech}^{2m-1} x +$

$+ \sum_{k=1}^{m-1} \dfrac{2^k (m-1)(m-2)\ldots(m-k)}{(2m-3)(2m-5)\ldots(2m-2k-1)} \operatorname{sech}^{2m-2k-1} x \Big\}.$

3. $\int \dfrac{dx}{\operatorname{ch}^{2m+1} x} = \dfrac{\operatorname{sh} x}{2m} \Big\{ \operatorname{sech}^{2m} x +$

$+ \sum_{k=1}^{m-1} \dfrac{(2m-1)(2m-3)\ldots(2m-2k+1)}{2^k (m-1)(m-2)\ldots(m-k)} \operatorname{sech}^{2m-2k} x \Big\} + \dfrac{(2m-1)!!}{(2m)!!} \operatorname{arctg} \operatorname{sh} x.$

2.419

1. $\int \dfrac{\operatorname{ch}^p x}{\operatorname{sh}^{2n+1} x} dx = -\dfrac{\operatorname{ch}^{p+1} x}{2n} \Big\{ \operatorname{cosech}^{2n} x +$

$+ \sum_{k=1}^{n-1} \dfrac{(-1)^k (2n-p-1)(2n-p-3)\ldots(2n-p-2k+1)}{2^k (n-1)(n-2)\ldots(n-k)} \operatorname{cosech}^{2n-2k} x \Big\} +$

$+ \dfrac{(-1)^n (2n-p-1)(2n-p-3)\ldots(3-p)(1-p)}{2^n n!} \int \dfrac{\operatorname{ch}^p x}{\operatorname{sh} x} dx.$

Diese Formel gilt für alle reellen p. Für $n=0$ und ganzzahliges p gelten die Formeln | This formula holds for all real p. For $n=0$ and integral p we have

2. $\int \dfrac{\operatorname{ch}^{2m} x}{\operatorname{sh} x} dx = \sum_{k=1}^{m} \dfrac{\operatorname{ch}^{2k-1} x}{2k-1} + \ln \operatorname{th} \dfrac{x}{2}.$

3. $\int \dfrac{\operatorname{ch}^{2m+1} x}{\operatorname{sh} x} dx = \sum_{k=1}^{m} \dfrac{\operatorname{ch}^{2k} x}{2k} + \ln \operatorname{sh} x;$

$= \sum_{k=1}^{m} \binom{m}{k} \dfrac{\operatorname{sh}^{2k} x}{2k} + \ln \operatorname{sh} x.$

4. $\int \dfrac{dx}{\operatorname{sh} x \operatorname{ch}^{2m} x} = \sum_{k=1}^{m} \dfrac{\operatorname{sech}^{2m-2k+1} x}{2m-2k+1} + \ln \operatorname{th} \dfrac{x}{2}.$

5. $\int \dfrac{dx}{\operatorname{sh} x \operatorname{ch}^{2m+1} x} = \sum_{k=1}^{m} \dfrac{\operatorname{sech}^{2m-2k+2} x}{2m-2k+2} + \ln \operatorname{th} x.$

2.421

1. $\int \dfrac{\operatorname{sh}^{2n+1} x}{\operatorname{ch}^m x} dx = \sum_{\substack{k=0 \\ k \neq \frac{m-1}{2}}}^{n} (-1)^{n+k} \binom{n}{k} \dfrac{\operatorname{ch}^{2k-m+1} x}{2k-m+1} +$

$+ s(-1)^{n+\frac{m-1}{2}} \binom{n}{\frac{m-1}{2}} \ln \operatorname{ch} x.$

2. $\int \dfrac{\operatorname{ch}^{2n+1} x}{\operatorname{sh}^m x} dx = \sum_{\substack{k=0 \\ k \neq \frac{m-1}{2}}}^{n} \binom{n}{k} \dfrac{\operatorname{sh}^{2k-m+1} x}{2k-m+1} + s \binom{n}{\frac{m-1}{2}} \ln \operatorname{sh} x.$

[In den Formeln 2.421 1. und 2.421 2. gilt für ungerades m und $m < 2n+1$, daß $s=1$, aber $s=0$ sonst.] | [In 2.421 1. and 2.421 2. $s=1$ if m is an odd integer and $m < 2n+1$; otherwise $s=0$.]

GHH[351](11, 13)

2.422

1. $\int \dfrac{dx}{\operatorname{sh}^{2m} x \operatorname{ch}^{2n} x} = \sum_{k=0}^{m+n-1} \dfrac{(-1)^{k+1}}{2m-2k-1} \binom{m+n-1}{k} \operatorname{th}^{2k-2m+1} x.$

2. $\int \dfrac{dx}{\operatorname{sh}^{2m+1} x \operatorname{ch}^{2n+1} x} = \sum_{\substack{k=0 \\ k \neq m}}^{m+n} \dfrac{(-1)^{k+1}}{2m-2k} \binom{m+n}{k} \operatorname{th}^{2k-2m} x +$

$+ (-1)^m \binom{m+n}{m} \ln \operatorname{th} x.$ GHH[351](15)

Hyperbelfunktionen/Hyperbolic Functions

2.423

1. $\int \dfrac{dx}{\operatorname{sh} x} = \ln \operatorname{th} \dfrac{x}{2} = \dfrac{1}{2} \ln \dfrac{\operatorname{ch} x - 1}{\operatorname{ch} x + 1}$.

2. $\int \dfrac{dx}{\operatorname{sh}^2 x} = -\operatorname{cth} x$.

3. $\int \dfrac{dx}{\operatorname{sh}^3 x} = -\dfrac{\operatorname{ch} x}{2\operatorname{sh}^2 x} - \dfrac{1}{2} \ln \operatorname{th} \dfrac{x}{2}$.

4. $\int \dfrac{dx}{\operatorname{sh}^4 x} = -\dfrac{\operatorname{ch} x}{3\operatorname{sh}^3 x} + \dfrac{2}{3} \operatorname{cth} x = -\dfrac{1}{3} \operatorname{cth}^3 x + \operatorname{cth} x$.

5. $\int \dfrac{dx}{\operatorname{sh}^5 x} = -\dfrac{\operatorname{ch} x}{4\operatorname{sh}^4 x} + \dfrac{3}{8} \dfrac{\operatorname{ch} x}{\operatorname{sh}^2 x} + \dfrac{3}{8} \ln \operatorname{th} \dfrac{x}{2}$.

6. $\int \dfrac{dx}{\operatorname{sh}^6 x} = -\dfrac{\operatorname{ch} x}{5\operatorname{sh}^5 x} + \dfrac{4}{15} \operatorname{cth}^3 x - \dfrac{4}{5} \operatorname{cth} x$;

   $= -\dfrac{1}{5} \operatorname{cth}^5 x + \dfrac{2}{3} \operatorname{cth}^3 x - \operatorname{cth} x$.

7. $\int \dfrac{dx}{\operatorname{sh}^7 x} = -\dfrac{\operatorname{ch} x}{6\operatorname{sh}^2 x} \left(\dfrac{1}{\operatorname{sh}^4 x} - \dfrac{5}{4\operatorname{sh}^2 x} + \dfrac{15}{8} \right) - \dfrac{5}{16} \ln \operatorname{th} \dfrac{x}{2}$.

8. $\int \dfrac{dx}{\operatorname{sh}^8 x} = \operatorname{cth} x - \operatorname{cth}^3 x + \dfrac{3}{5} \operatorname{cth}^5 x - \dfrac{1}{7} \operatorname{cth}^7 x$.

9. $\int \dfrac{dx}{\operatorname{ch} x} = \operatorname{arctg}(\operatorname{sh} x) = 2\operatorname{arctg}(e^x)$;

   $= \arcsin(\operatorname{th} x)$;
   $= \operatorname{gd} x$.

10. $\int \dfrac{dx}{\operatorname{ch}^2 x} = \operatorname{th} x$.

11. $\int \dfrac{dx}{\operatorname{ch}^3 x} = \dfrac{\operatorname{sh} x}{2\operatorname{ch}^2 x} + \dfrac{1}{2} \operatorname{arctg}(\operatorname{sh} x)$.

12. $\int \dfrac{dx}{\operatorname{ch}^4 x} = \dfrac{\operatorname{sh} x}{3\operatorname{ch}^3 x} + \dfrac{2}{3} \operatorname{th} x$;

    $= -\dfrac{1}{3} \operatorname{th}^3 x + \operatorname{th} x$.

13. $\int \dfrac{dx}{\operatorname{ch}^5 x} = \dfrac{\operatorname{sh} x}{4\operatorname{ch}^4 x} + \dfrac{3}{8} \dfrac{\operatorname{sh} x}{\operatorname{ch}^2 x} + \dfrac{3}{8} \operatorname{arctg}(\operatorname{sh} x)$.

14. $\int \dfrac{dx}{\operatorname{ch}^6 x} = \dfrac{\operatorname{sh} x}{5\operatorname{ch}^5 x} - \dfrac{4}{15} \operatorname{th}^3 x + \dfrac{4}{5} \operatorname{th} x$;

    $= \dfrac{1}{5} \operatorname{th}^5 x - \dfrac{2}{3} \operatorname{th}^3 x + \operatorname{th} x$.

15. $\int \dfrac{dx}{\operatorname{ch}^7 x} = \dfrac{\operatorname{sh} x}{6\operatorname{ch}^6 x} \left(\dfrac{1}{\operatorname{ch}^4 x} + \dfrac{5}{4\operatorname{ch}^2 x} + \dfrac{15}{8} \right) + \dfrac{5}{16} \operatorname{arctg}(\operatorname{sh} x)$.

16. $\int \dfrac{dx}{\operatorname{ch}^8 x} = -\dfrac{1}{7} \operatorname{th}^7 x + \dfrac{3}{5} \operatorname{th}^5 x - \operatorname{th}^3 x + \operatorname{th} x$.

17. $\int \dfrac{\operatorname{sh} x}{\operatorname{ch} x} dx = \ln \operatorname{ch} x$.

2.423

18. $\int \dfrac{\mathrm{sh}^2 x}{\mathrm{ch}\, x}\, dx = \mathrm{sh}\, x - \mathrm{arctg}\,(\mathrm{sh}\, x)$.

19. $\int \dfrac{\mathrm{sh}^3 x}{\mathrm{ch}\, x}\, dx = \dfrac{1}{2}\mathrm{sh}^2 x - \ln \mathrm{ch}\, x;$

$\qquad = \dfrac{1}{2}\mathrm{ch}^2 x - \ln \mathrm{ch}\, x$.

20. $\int \dfrac{\mathrm{sh}^4 x}{\mathrm{ch}\, x}\, dx = \dfrac{1}{3}\mathrm{sh}^3 x - \mathrm{sh}\, x + \mathrm{arctg}\,(\mathrm{sh}\, x)$.

21. $\int \dfrac{\mathrm{sh}\, x}{\mathrm{ch}^2 x}\, dx = -\dfrac{1}{\mathrm{ch}\, x}$.

22. $\int \dfrac{\mathrm{sh}^2 x}{\mathrm{ch}^2 x}\, dx = x - \mathrm{th}\, x$.

23. $\int \dfrac{\mathrm{sh}^3 x}{\mathrm{ch}^2 x}\, dx = \mathrm{ch}\, x + \dfrac{1}{\mathrm{ch}\, x}$.

24. $\int \dfrac{\mathrm{sh}^4 x}{\mathrm{ch}^2 x}\, dx = -\dfrac{3}{2}x + \dfrac{1}{4}\mathrm{sh}\, 2x + \mathrm{th}\, x$.

25. $\int \dfrac{\mathrm{sh}\, x}{\mathrm{ch}^3 x}\, dx = -\dfrac{1}{2\mathrm{ch}^2 x};$

$\qquad = \dfrac{1}{2}\mathrm{th}^2 x$.

26. $\int \dfrac{\mathrm{sh}^2 x}{\mathrm{ch}^3 x}\, dx = -\dfrac{\mathrm{sh}\, x}{2\mathrm{ch}^2 x} + \dfrac{1}{2}\mathrm{arctg}\,(\mathrm{sh}\, x)$.

27. $\int \dfrac{\mathrm{sh}^3 x}{\mathrm{ch}^3 x}\, dx = -\dfrac{1}{2}\mathrm{th}^2 x + \ln \mathrm{ch}\, x;$

$\qquad = \dfrac{1}{2\mathrm{ch}^2 x} + \ln \mathrm{ch}\, x$.

28. $\int \dfrac{\mathrm{sh}^4 x}{\mathrm{ch}^3 x}\, dx = \dfrac{\mathrm{sh}\, x}{2\mathrm{ch}\, x} + \mathrm{sh}\, x - \dfrac{3}{2}\mathrm{arctg}\,(\mathrm{sh}\, x)$.

29. $\int \dfrac{\mathrm{sh}\, x}{\mathrm{ch}^4 x}\, dx = -\dfrac{1}{3\mathrm{ch}^3 x}$.

30. $\int \dfrac{\mathrm{sh}^2 x}{\mathrm{ch}^4 x}\, dx = \dfrac{1}{3}\mathrm{th}^3 x$.

31. $\int \dfrac{\mathrm{sh}^3 x}{\mathrm{ch}^4 x}\, dx = -\dfrac{1}{\mathrm{ch}\, x} + \dfrac{1}{3\mathrm{ch}^3 x}$.

32. $\int \dfrac{\mathrm{sh}^4 x}{\mathrm{ch}^4 x}\, dx = -\dfrac{1}{3}\mathrm{th}^3 x - \mathrm{th}\, x + x$.

33. $\int \dfrac{\mathrm{ch}\, x}{\mathrm{sh}\, x}\, dx = \ln \mathrm{sh}\, x$.

34. $\int \dfrac{\mathrm{ch}^2 x}{\mathrm{sh}\, x}\, dx = \mathrm{ch}\, x + \ln \mathrm{th}\,\dfrac{x}{2}$.

35. $\int \dfrac{\mathrm{ch}^3 x}{\mathrm{sh}\, x}\, dx = \dfrac{1}{2}\mathrm{ch}^2 x + \ln \mathrm{sh}\, x$.

36. $\int \frac{\operatorname{ch}^4 x}{\operatorname{sh} x} dx = \frac{1}{3} \operatorname{ch}^3 x + \operatorname{ch} x + \ln \operatorname{th} \frac{x}{2}$.

37. $\int \frac{\operatorname{ch} x}{\operatorname{sh}^2 x} dx = -\frac{1}{\operatorname{sh} x}$.

38. $\int \frac{\operatorname{ch}^2 x}{\operatorname{sh}^2 x} dx = x - \operatorname{cth} x$.

39. $\int \frac{\operatorname{ch}^3 x}{\operatorname{sh}^2 x} dx = \operatorname{sh} x - \frac{1}{\operatorname{sh} x}$.

40. $\int \frac{\operatorname{ch}^4 x}{\operatorname{sh}^2 x} dx = \frac{3}{2} x + \frac{1}{4} \operatorname{sh} 2x - \operatorname{cth} x$.

41. $\int \frac{\operatorname{ch} x}{\operatorname{sh}^3 x} dx = -\frac{1}{2 \operatorname{sh}^2 x}$;

$\qquad = -\frac{1}{2} \operatorname{cth}^2 x$.

42. $\int \frac{\operatorname{ch}^2 x}{\operatorname{sh}^3 x} dx = -\frac{\operatorname{ch} x}{2 \operatorname{sh}^2 x} + \ln \operatorname{th} \frac{x}{2}$.

43. $\int \frac{\operatorname{ch}^3 x}{\operatorname{sh}^3 x} dx = -\frac{1}{2 \operatorname{sh}^2 x} + \ln \operatorname{sh} x$;

$\qquad = -\frac{1}{2} \operatorname{cth}^2 x + \ln \operatorname{sh} x$.

44. $\int \frac{\operatorname{ch}^4 x}{\operatorname{sh}^3 x} dx = -\frac{\operatorname{ch} x}{2 \operatorname{sh}^2 x} + \operatorname{ch} x + \frac{3}{2} \ln \operatorname{th} \frac{x}{2}$.

45. $\int \frac{\operatorname{ch} x}{\operatorname{sh}^4 x} dx = -\frac{1}{3 \operatorname{sh}^3 x}$.

46. $\int \frac{\operatorname{ch}^2 x}{\operatorname{sh}^4 x} dx = -\frac{1}{3} \operatorname{cth}^3 x$.

47. $\int \frac{\operatorname{ch}^3 x}{\operatorname{sh}^4 x} dx = -\frac{1}{\operatorname{sh} x} - \frac{1}{3 \operatorname{sh}^3 x}$.

48. $\int \frac{\operatorname{ch}^4 x}{\operatorname{sh}^4 x} dx = -\frac{1}{3} \operatorname{cth}^3 x - \operatorname{cth} x + x$.

49. $\int \frac{dx}{\operatorname{sh} x \operatorname{ch} x} = \ln \operatorname{th} x$.

50. $\int \frac{dx}{\operatorname{sh} x \operatorname{ch}^2 x} = \frac{1}{\operatorname{ch} x} + \ln \operatorname{th} \frac{x}{2}$.

51. $\int \frac{dx}{\operatorname{sh} x \operatorname{ch}^3 x} = \frac{1}{2 \operatorname{ch}^2 x} + \ln \operatorname{th} x$;

$\qquad = -\frac{1}{2} \operatorname{th}^2 x + \ln \operatorname{th} x$.

52. $\int \frac{dx}{\operatorname{sh} x \operatorname{ch}^4 x} = \frac{1}{\operatorname{ch} x} + \frac{1}{3 \operatorname{ch}^3 x} + \ln \operatorname{th} \frac{x}{2}$.

53. $\int \frac{dx}{\operatorname{sh}^2 x \operatorname{ch} x} = -\frac{1}{\operatorname{sh} x} - \operatorname{arctg} \operatorname{sh} x$.

2.423

54. $\displaystyle\int\frac{dx}{\operatorname{sh}^2 x\,\operatorname{ch}^2 x} = -2\operatorname{cth}2x.$

55. $\displaystyle\int\frac{dx}{\operatorname{sh}^2 x\,\operatorname{ch}^3 x} = -\frac{\operatorname{sh}x}{2\operatorname{ch}^2 x} - \frac{1}{\operatorname{sh}x} - \frac{3}{2}\operatorname{arctg}\operatorname{sh}x.$

56. $\displaystyle\int\frac{dx}{\operatorname{sh}^2 x\,\operatorname{ch}^4 x} = \frac{1}{3\operatorname{sh}x\,\operatorname{ch}^3 x} - \frac{8}{3}\operatorname{cth}2x.$

57. $\displaystyle\int\frac{dx}{\operatorname{sh}^3 x\,\operatorname{ch}x} = -\frac{1}{2\operatorname{sh}^2 x} - \ln\operatorname{th}x;$

$\displaystyle\phantom{\int\frac{dx}{\operatorname{sh}^3 x\,\operatorname{ch}x}} = -\frac{1}{2}\operatorname{cth}^2 x + \ln\operatorname{cth}x.$

58. $\displaystyle\int\frac{dx}{\operatorname{sh}^3 x\,\operatorname{ch}^2 x} = -\frac{1}{\operatorname{ch}x} - \frac{\operatorname{ch}x}{2\operatorname{sh}^2 x} - \frac{3}{2}\ln\operatorname{th}\frac{x}{2}.$

59. $\displaystyle\int\frac{dx}{\operatorname{sh}^3 x\,\operatorname{ch}^3 x} = -\frac{2\operatorname{ch}2x}{\operatorname{sh}^2 2x} - 2\ln\operatorname{th}x;$

$\displaystyle\phantom{\int\frac{dx}{\operatorname{sh}^3 x\,\operatorname{ch}^3 x}} = \frac{1}{2}\operatorname{th}^2 x - \frac{1}{2}\operatorname{cth}^2 x - 2\ln\operatorname{th}x.$

60. $\displaystyle\int\frac{dx}{\operatorname{sh}^3 x\,\operatorname{ch}^4 x} = -\frac{2}{\operatorname{ch}x} - \frac{1}{3\operatorname{ch}^3 x} - \frac{\operatorname{ch}x}{2\operatorname{sh}^2 x} - \frac{5}{2}\ln\operatorname{th}\frac{x}{2}.$

61. $\displaystyle\int\frac{dx}{\operatorname{sh}^4 x\,\operatorname{ch}x} = \frac{1}{\operatorname{sh}x} - \frac{1}{3\operatorname{sh}^3 x} + \operatorname{arctg}\operatorname{sh}x.$

62. $\displaystyle\int\frac{dx}{\operatorname{sh}^4 x\,\operatorname{ch}^2 x} = -\frac{1}{3\operatorname{ch}x\,\operatorname{sh}^3 x} + \frac{8}{3}\operatorname{cth}2x.$

63. $\displaystyle\int\frac{dx}{\operatorname{sh}^4 x\,\operatorname{ch}^3 x} = \frac{2}{\operatorname{sh}x} - \frac{1}{3\operatorname{sh}^3 x} + \frac{\operatorname{sh}x}{2\operatorname{ch}^2 x} + \frac{5}{2}\operatorname{arctg}\operatorname{sh}x.$

64. $\displaystyle\int\frac{dx}{\operatorname{sh}^4 x\,\operatorname{ch}^4 x} = 8\operatorname{cth}2x - \frac{8}{3}\operatorname{cth}^3 2x.$

2.424

1. $\displaystyle\int\operatorname{th}^p x\,dx = -\frac{\operatorname{th}^{p-1}x}{p-1} + \int\operatorname{th}^{p-2}x\,dx \quad [p\neq 1].$

2. $\displaystyle\int\operatorname{th}^{2n+1}x\,dx = \sum_{k=1}^{n}\frac{(-1)^{k-1}}{2k}\binom{n}{k}\frac{1}{\operatorname{ch}^{2k}x} + \ln\operatorname{ch}x;$

$\displaystyle\phantom{\int\operatorname{th}^{2n+1}x\,dx} = -\sum_{k=1}^{n}\frac{\operatorname{th}^{2n-2k+2}x}{2n-2k+2} + \ln\operatorname{ch}x.$

3. $\displaystyle\int\operatorname{th}^{2n}x\,dx = -\sum_{k=1}^{n}\frac{\operatorname{th}^{2n-2k+1}x}{2n-2k+1} + x.$ GHH[351](12)

4. $\displaystyle\int\operatorname{cth}^p x\,dx = -\frac{\operatorname{cth}^{p-1}x}{p-1} + \int\operatorname{cth}^{p-2}x\,dx \quad [p\neq 1].$

5. $\displaystyle\int\operatorname{cth}^{2n+1}x\,dx = -\sum_{k=1}^{n}\frac{1}{2n}\binom{n}{k}\frac{1}{\operatorname{sh}^{2k}x} + \ln\operatorname{sh}x;$

$\displaystyle\phantom{\int\operatorname{cth}^{2n+1}x\,dx} = -\sum_{k=1}^{n}\frac{\operatorname{cth}^{2n-2k+2}x}{2n-2k+2} + \ln\operatorname{sh}x.$

6. $\displaystyle\int\operatorname{cth}^{2n}x\,dx = -\sum_{k=1}^{n}\frac{\operatorname{cth}^{2n-2k+1}x}{2n-2k+1} + x.$ GHH[351](14)

Die Formeln für Potenzen von th x und cth x mit den Exponenten $n = 1, 2, 3, 4$ findet man unter 2.423 17., 2.423 22., 2.423 27., 2.423 32., 2.423 33., 2.423 38., 2.423 43., 2.423 48.

The formulae for powers of th x and cth x with the exponents $n = 1, 2, 3, 4$ are given in 2.423 17., 2.423 22., 2.423 27., 2.423 32., 2.423 33., 2.423 38., 2.423 43., 2.423 48.

Potenzen von Hyperbelfunktionen und Hyperbelfunktionen von linearen Funktionen des Arguments
Powers of hyperbolic functions and hyperbolic functions of linear functions of the argument

2.425

1. $\int \text{sh}\,(ax+b)\,\text{sh}\,(cx+d)\,dx = \dfrac{1}{2(a+c)}\,\text{sh}\,[(a+c)x+b+d] - \dfrac{1}{2(a-c)}\,\text{sh}\,[(a-c)x+b-d] \quad [a^2 \neq c^2].$ GHH[352](2a)

2. $\int \text{sh}\,(ax+b)\,\text{ch}\,(cx+d)\,dx = \dfrac{1}{2(a+c)}\,\text{ch}\,[(a+c)x+b+d] + \dfrac{1}{2(a-c)}\,\text{ch}\,[(a-c)x+b-d] \quad [a^2 \neq c^2].$ GHH[352](2c)

3. $\int \text{ch}\,(ax+b)\,\text{ch}\,(cx+d)\,dx = \dfrac{1}{2(a+c)}\,\text{sh}\,[(a+c)x+b+d] + \dfrac{1}{2(a-c)}\,\text{sh}\,[(a-c)x+b-d] \quad [a^2 \neq c^2]$ GHH[352](2b)

Für $a = c$ gelten die Formeln | For $a = c$:

4. $\int \text{sh}\,(ax+b)\,\text{sh}\,(ax+d)\,dx = -\dfrac{x}{2}\,\text{ch}\,(b-d) + \dfrac{1}{4a}\,\text{sh}\,(2ax+b+d).$ GHH[352](3a)

5. $\int \text{sh}\,(ax+b)\,\text{ch}\,(ax+d)\,dx = \dfrac{x}{2}\,\text{sh}\,(b-d) + \dfrac{1}{4a}\,\text{ch}\,(2ax+b+d).$ GHH[352](3c)

6. $\int \text{ch}\,(ax+b)\,\text{ch}\,(ax+d)\,dx = \dfrac{x}{2}\,\text{ch}\,(b-d) + \dfrac{1}{4a}\,\text{sh}\,(2ax+b+d).$ GHH[352](3b)

2.426

1. $\int \text{sh}\,ax\,\text{sh}\,bx\,\text{sh}\,cx\,dx = \dfrac{\text{ch}\,(a+b+c)x}{4(a+b+c)} - \dfrac{\text{ch}\,(-a+b+c)x}{4(-a+b+c)} - \dfrac{\text{ch}\,(a-b+c)x}{4(a-b+c)} - \dfrac{\text{ch}\,(a+b-c)x}{4(a+b-c)}.$ GHH[352](4a)

2. $\int \text{sh}\,ax\,\text{sh}\,bx\,\text{ch}\,cx\,dx = \dfrac{\text{sh}\,(a+b+c)x}{4(a+b+c)} - \dfrac{\text{sh}\,(-a+b+c)x}{4(-a+b+c)} - \dfrac{\text{sh}\,(a-b+c)x}{4(a-b+c)} + \dfrac{\text{sh}\,(a+b-c)x}{4(a+b-c)}.$ GHH[352](4b)

3. $\int \text{sh}\,ax\,\text{ch}\,bx\,\text{ch}\,cx\,dx = \dfrac{\text{ch}\,(a+b+c)x}{4(a+b+c)} - \dfrac{\text{ch}\,(-a+b+c)x}{4(-a+b+c)} + \dfrac{\text{ch}\,(a-b+c)x}{4(a-b+c)} + \dfrac{\text{ch}\,(a+b-c)x}{4(a+b-c)}.$ GHH[352](4c)

2.426 4. $\int \operatorname{ch} ax \operatorname{ch} bx \operatorname{ch} cx \, dx = \dfrac{\operatorname{sh}(a+b+c)x}{4(a+b+c)} + \dfrac{\operatorname{sh}(-a+b+c)x}{4(-a+b+c)} +$

$+ \dfrac{\operatorname{sh}(a-b+c)x}{4(a-b+c)} + \dfrac{\operatorname{sh}(a+b-c)x}{4(a+b-c)}$.   GHH[352](4d)

2.427 1. $\int \operatorname{sh}^p x \operatorname{sh} ax \, dx = \dfrac{1}{p+a} \left\{ \operatorname{sh}^p x \operatorname{ch} ax - p \int \operatorname{sh}^{p-1} x \operatorname{ch} (a-1) x \, dx \right\}$.

2. $\int \operatorname{sh}^p x \operatorname{sh}(2n+1) x \, dx = \dfrac{\Gamma(p+1)}{\Gamma\left(\dfrac{p+3}{2}+n\right)} \times$

$\times \left\{ \sum\limits_{k=0}^{n-1} \left[\dfrac{\Gamma\left(\dfrac{p+1}{2}+n-2k\right)}{2^{2k+1}\Gamma(p-2k+1)} \operatorname{sh}^{p-2k} x \operatorname{ch}(2n-2k+1)x - \right.\right.$

$\left. - \dfrac{\Gamma\left(\dfrac{p-1}{2}+n-2k\right)}{2^{2k+2}\Gamma(p-2k)} \operatorname{sh}^{p-2k-1} x \operatorname{sh}(2n-2k)x \right] +$

$\left. + \dfrac{\Gamma\left(\dfrac{p+3}{2}-n\right)}{2^{2n}\Gamma(p+1-2n)} \int \operatorname{sh}^{p-2n} x \operatorname{sh} x \, dx \right\}$

[p ist keine negative ganze Zahl]. | [p not a negative integer].

3. $\int \operatorname{sh}^p x \operatorname{sh} 2n x \, dx = \dfrac{\Gamma(p+1)}{\Gamma\left(\dfrac{p}{2}+n+1\right)} \times$

$\times \sum\limits_{k=0}^{n-1} \left[\dfrac{\Gamma\left(\dfrac{p}{2}+n-2k\right)}{2^{2k+1}\Gamma(p-2k+1)} \operatorname{sh}^{p-2k} x \operatorname{ch}(2n-2k)x - \right.$

$\left. - \dfrac{\Gamma\left(\dfrac{p}{2}+n-2k-1\right)}{2^{2k+2}\Gamma(p-2k)} \operatorname{sh}^{p-2k-1} x \operatorname{sh}(2n-2k-1)x \right]$ GHH[352](5)+

[p ist keine negative ganze Zahl]. | [p not a negative integer].

2.428 1. $\int \operatorname{sh}^p x \operatorname{ch} ax \, dx = \dfrac{1}{p+a} \left\{ \operatorname{sh}^p x \operatorname{sh} ax - p \int \operatorname{sh}^{p-1} x \operatorname{sh}(a-1) x \, dx \right\}$.

2. $\int \operatorname{sh}^p x \operatorname{ch}(2n+1) x \, dx = \dfrac{\Gamma(p+1)}{\Gamma\left(\dfrac{p+3}{2}+n\right)} \times$

$\times \left\{ \sum\limits_{k=0}^{n-1} \left[\dfrac{\Gamma\left(\dfrac{p+1}{2}+n-2k\right)}{2^{2k+1}\Gamma(p-2k+1)} \operatorname{sh}^{p-2k} x \operatorname{sh}(2n-2k+1)x - \right.\right.$

$\left. - \dfrac{\Gamma\left(\dfrac{p-1}{2}+n-2k\right)}{2^{2k+2}\Gamma(p-2k)} \operatorname{sh}^{p-2k-1} x \operatorname{ch}(2n-2k)x \right] +$

$\left. + \dfrac{\Gamma\left(\dfrac{p+3}{2}-n\right)}{2^{2n}\Gamma(p+1-2n)} \int \operatorname{sh}^{p-2n} x \operatorname{ch} x \, dx \right\}$

[p ist keine negative ganze Zahl]. | [p not a negative integer]. 2.428

3. $\int \operatorname{sh}^p x \operatorname{ch} 2nx\, dx = \dfrac{\Gamma(p+1)}{\Gamma\left(\dfrac{p}{2}+n+1\right)} \times$

$\times \left\{ \sum\limits_{k=0}^{n-1}\left[\dfrac{\Gamma\left(\dfrac{p}{2}+n-2k\right)}{2^{2k+1}\Gamma(p-2k+1)} \operatorname{sh}^{p-2k} x \operatorname{sh}(2n-2k)x - \right.\right.$

$\left.- \dfrac{\Gamma\left(\dfrac{p}{2}+n-2k-1\right)}{2^{2k+2}\Gamma(p-2k)} \operatorname{sh}^{p-2k-1} x \operatorname{ch}(2n-2k-1)x \right] + \dfrac{\Gamma\left(\dfrac{p}{2}-n+1\right)}{2^{2n}\Gamma(p+1-2n)} \int \operatorname{sh}^{p-2n} x\, dx \bigg\}$

GHH[352](6)+

[p ist keine negative ganze Zahl]. | [p not a negative integer]. 2.429

1. $\int \operatorname{ch}^p x \operatorname{sh} ax\, dx = \dfrac{1}{p+a}\left\{\operatorname{ch}^p x \operatorname{ch} ax + p\int \operatorname{ch}^{p-1} x \operatorname{sh}(a-1)x\, dx\right\}.$

2. $\int \operatorname{ch}^p x \operatorname{sh}(2n+1)x\, dx = \dfrac{\Gamma(p+1)}{\Gamma\left(\dfrac{p+3}{2}+n\right)} \left\{ \sum\limits_{k=0}^{n-1} \dfrac{\Gamma\left(\dfrac{p+1}{2}+n-k\right)}{2^{k+1}\Gamma(p-k+1)} \times \right.$

$\left.\times \operatorname{ch}^{p-k} x \operatorname{ch}(2n-k+1)x + \dfrac{\Gamma\left(\dfrac{p+3}{2}\right)}{2^n \Gamma(p-n+1)} \int \operatorname{ch}^{p-n} x \operatorname{sh}(n+1)x\, dx \right\}$

[p ist keine negative ganze Zahl]. | [p not a negative integer].

3. $\int \operatorname{ch}^p x \operatorname{sh} 2nx\, dx = \dfrac{\Gamma(p+1)}{\Gamma\left(\dfrac{p}{2}+n+1\right)} \left\{ \sum\limits_{k=0}^{n-1} \dfrac{\Gamma\left(\dfrac{p}{2}+n-k\right)}{2^{k+1}\Gamma(p-k+1)} \times \right.$

$\left.\times \operatorname{ch}^{p-k} x \operatorname{ch}(2n-k)x + \dfrac{\Gamma\left(\dfrac{p}{2}+1\right)}{2^n \Gamma(p-n+1)} \int \operatorname{ch}^{p-n} x \operatorname{sh} nx\, dx \right\}$

GHH[352](7)+

[p ist keine negative ganze Zahl]. | [p not a negative integer]. 2.431

1. $\int \operatorname{ch}^p x \operatorname{ch} ax\, dx = \dfrac{1}{p+a}\left\{\operatorname{ch}^p x \operatorname{sh} ax + p\int \operatorname{ch}^{p-1} x \operatorname{ch}(a-1)x\, dx\right\}.$

2. $\int \operatorname{ch}^p x \operatorname{ch}(2n+1)x\, dx = \dfrac{\Gamma(p+1)}{\Gamma\left(\dfrac{p+3}{2}+n\right)} \left\{ \sum\limits_{k=0}^{n-1} \dfrac{\Gamma\left(\dfrac{p+1}{2}+n-k\right)}{2^{k+1}\Gamma(p-k+1)} \times \right.$

$\left.\times \operatorname{ch}^{p-k} x \operatorname{sh}(2n-k+1)x + \dfrac{\Gamma\left(\dfrac{p+3}{2}\right)}{2^n \Gamma(p-n+1)} \int \operatorname{ch}^{p-n} x \operatorname{ch}(n+1)x\, dx \right\}$

[p ist keine negative ganze Zahl]. | [p not a negative integer].

2.431

3. $\int \operatorname{ch}^p x \operatorname{ch} 2nx \, dx = \dfrac{\Gamma(p+1)}{\Gamma\left(\dfrac{p}{2}+n+1\right)} \left\{ \sum_{k=0}^{n-1} \dfrac{\Gamma\left(\dfrac{p}{2}+n-k\right)}{2^{k+1}\Gamma(p-k+1)} \times \right.$

$\left. \times \operatorname{ch}^{p-k} x \operatorname{sh}(2n-k)x + \dfrac{\Gamma\left(\dfrac{p}{2}+1\right)}{2^n \Gamma(p-n+1)} \int \operatorname{ch}^{p-n} x \operatorname{ch} nx \, dx \right\}$ GHH[352](8)+

[p ist keine negative ganze Zahl]. | [p not a negative integer].

2.432

1. $\int \operatorname{sh}(n+1)x \operatorname{sh}^{n-1} x \, dx = \dfrac{1}{n} \operatorname{sh}^n x \operatorname{sh} nx.$

2. $\int \operatorname{sh}(n+1)x \operatorname{ch}^{n-1} x \, dx = \dfrac{1}{n} \operatorname{ch}^n x \operatorname{ch} nx.$

3. $\int \operatorname{ch}(n+1)x \operatorname{sh}^{n-1} x \, dx = \dfrac{1}{n} \operatorname{sh}^n x \operatorname{ch} nx.$

4. $\int \operatorname{ch}(n+1)x \operatorname{ch}^{n-1} x \, dx = \dfrac{1}{n} \operatorname{ch}^n x \operatorname{sh} nx.$

2.433

1. $\int \dfrac{\operatorname{sh}(2n+1)x}{\operatorname{sh} x} dx = 2 \sum_{k=0}^{n-1} \dfrac{\operatorname{sh}(2n-2k)x}{2n-2k} + x.$

2. $\int \dfrac{\operatorname{sh} 2nx}{\operatorname{sh} x} dx = 2 \sum_{k=0}^{n-1} \dfrac{\operatorname{sh}(2n-2k-1)x}{2n-2k-1}.$ GHH[352](5d)

3. $\int \dfrac{\operatorname{ch}(2n+1)x}{\operatorname{sh} x} dx = 2 \sum_{k=0}^{n-1} \dfrac{\operatorname{ch}(2n-2k)x}{2n-2k} + \ln \operatorname{sh} x.$

4. $\int \dfrac{\operatorname{ch} 2nx}{\operatorname{sh} x} dx = 2 \sum_{k=0}^{n-1} \dfrac{\operatorname{ch}(2n-2k-1)x}{2n-2k-1} + \ln \operatorname{th} \dfrac{x}{2}.$ GHH[352](6d)

5. $\int \dfrac{\operatorname{sh}(2n+1)x}{\operatorname{ch} x} dx = 2 \sum_{k=0}^{n-1} (-1)^k \dfrac{\operatorname{ch}(2n-2k)x}{2n-2k} + (-1)^n \ln \operatorname{ch} x.$

6. $\int \dfrac{\operatorname{sh} 2nx}{\operatorname{ch} x} dx = 2 \sum_{k=0}^{n-1} (-1)^k \dfrac{\operatorname{ch}(2n-2k-1)x}{2n-2k-1}.$ GHH[352](7d)

7. $\int \dfrac{\operatorname{ch}(2n+1)x}{\operatorname{ch} x} dx = 2 \sum_{k=0}^{n-1} (-1)^k \dfrac{\operatorname{sh}(2n-2k)x}{2n-2k} + (-1)^n x.$

8. $\int \dfrac{\operatorname{ch} 2nx}{\operatorname{ch} x} dx = 2 \sum_{k=0}^{n-1} (-1)^k \dfrac{\operatorname{sh}(2n-2k-1)x}{2n-2k-1} + (-1)^n \arcsin(\operatorname{th} x).$

GHH[352](8d)

9. $\int \dfrac{\operatorname{sh} 2x}{\operatorname{sh}^n x} dx = -\dfrac{2}{(n-2)\operatorname{sh}^{n-2} x}.$

Für $n = 2$ gilt | For $n = 2$:

10. $\displaystyle\int \frac{\operatorname{sh} 2x}{\operatorname{sh}^2 x} dx = 2 \ln \operatorname{sh} x.$

11. $\displaystyle\int \frac{\operatorname{sh} 2x \, dx}{\operatorname{ch}^n x} = \frac{2}{(2-n) \operatorname{ch}^{n-2} x}.$

Für $n = 2$ gilt | For $n = 2$:

12. $\displaystyle\int \frac{\operatorname{sh} 2x}{\operatorname{ch}^2 x} dx = 2 \ln \operatorname{ch} x.$

13. $\displaystyle\int \frac{\operatorname{ch} 2x}{\operatorname{sh} x} dx = 2 \operatorname{ch} x + \ln \operatorname{th} \frac{x}{2}.$

14. $\displaystyle\int \frac{\operatorname{ch} 2x}{\operatorname{sh}^2 x} dx = -\operatorname{cth} x + 2x.$

15. $\displaystyle\int \frac{\operatorname{ch} 2x}{\operatorname{sh}^3 x} dx = -\frac{\operatorname{ch} x}{2 \operatorname{sh}^2 x} + \frac{3}{2} \ln \operatorname{th} \frac{x}{2}.$

16. $\displaystyle\int \frac{\operatorname{ch} 2x}{\operatorname{ch} x} dx = 2 \operatorname{sh} x - \arcsin(\operatorname{th} x).$

17. $\displaystyle\int \frac{\operatorname{ch} 2x}{\operatorname{ch}^2 x} dx = -\operatorname{th} x + 2x.$

18. $\displaystyle\int \frac{\operatorname{ch} 2x}{\operatorname{ch}^3 x} dx = -\frac{\operatorname{sh} x}{2 \operatorname{ch}^2 x} + \frac{3}{2} \arcsin(\operatorname{th} x).$

19. $\displaystyle\int \frac{\operatorname{sh} 3x}{\operatorname{sh} x} dx = x + \operatorname{sh} 2x.$

20. $\displaystyle\int \frac{\operatorname{sh} 3x}{\operatorname{sh}^2 x} dx = 3 \ln \operatorname{th} \frac{x}{2} + 4 \operatorname{ch} x.$

21. $\displaystyle\int \frac{\operatorname{sh} 3x}{\operatorname{sh}^3 x} dx = -3 \operatorname{cth} x + 4x.$

22. $\displaystyle\int \frac{\operatorname{sh} 3x}{\operatorname{ch}^n x} dx = \frac{4}{(3-n) \operatorname{ch}^{n-3} x} - \frac{1}{(1-n) \operatorname{ch}^{n-1} x}.$

Für $n = 1$ und $n = 3$ gilt | For $n = 1$ and $n = 3$:

23. $\displaystyle\int \frac{\operatorname{sh} 3x}{\operatorname{ch} x} dx = 2 \operatorname{sh}^2 x - \ln \operatorname{ch} x.$

24. $\displaystyle\int \frac{\operatorname{sh} 3x}{\operatorname{ch}^3 x} dx = \frac{1}{2 \operatorname{ch}^2 x} + 4 \ln \operatorname{ch} x.$

25. $\displaystyle\int \frac{\operatorname{ch} 3x}{\operatorname{sh}^n x} dx = \frac{4}{(3-n) \operatorname{sh}^{n-3} x} + \frac{1}{(1-n) \operatorname{sh}^{n-1} x}.$

Für $n = 1$ und $n = 3$ gilt | For $n = 1$ and $n = 3$:

26. $\displaystyle\int \frac{\operatorname{ch} 3x}{\operatorname{sh} x} dx = 2 \operatorname{sh}^2 x + \ln \operatorname{sh} x.$

27. $\displaystyle\int \frac{\operatorname{ch} 3x}{\operatorname{sh}^3 x} dx = -\frac{1}{2 \operatorname{sh}^2 x} + 4 \ln \operatorname{sh} x.$

2.433 28. $\int \dfrac{\operatorname{ch} 3x}{\operatorname{ch} x}\,dx = \operatorname{sh} 2x - x.$

29. $\int \dfrac{\operatorname{ch} 3x}{\operatorname{ch}^2 x}\,dx = 4\operatorname{sh} x - 3\arcsin(\operatorname{th} x).$

30. $\int \dfrac{\operatorname{ch} 3x}{\operatorname{ch}^3 x}\,dx = 4x - 3\operatorname{th} x.$

Rationale Funktionen von Hyperbelfunktionen
Rational functions of hyperbolic functions

2.44–2.45

2.441 1. $\int \dfrac{A + B\operatorname{sh} x}{(a + b\operatorname{sh} x)^n}\,dx = \dfrac{aB - bA}{(n-1)(a^2 + b^2)} \cdot \dfrac{\operatorname{ch} x}{(a + b\operatorname{sh} x)^{n-1}} +$
$+ \dfrac{1}{(n-1)(a^2+b^2)} \int \dfrac{(n-1)(aA+bB) + (n-2)(aB-bA)\operatorname{sh} x}{(a+b\operatorname{sh} x)^{n-1}}\,dx.$

Für $n = 1$ gilt | For $n = 1$:

2. $\int \dfrac{A + B\operatorname{sh} x}{a + b\operatorname{sh} x}\,dx = \dfrac{B}{b}x - \dfrac{aB - bA}{b}\int \dfrac{dx}{a + b\operatorname{sh} x}$ (cf. 2.441 3.).

3. $\int \dfrac{dx}{a + b\operatorname{sh} x} = \dfrac{1}{\sqrt{a^2 + b^2}} \ln \dfrac{a\operatorname{th}\tfrac{x}{2} - b + \sqrt{a^2 + b^2}}{a\operatorname{th}\tfrac{x}{2} - b - \sqrt{a^2 + b^2}};$

$= \dfrac{2}{\sqrt{a^2 + b^2}} \operatorname{Arth} \dfrac{a\operatorname{th}\tfrac{x}{2} - b}{\sqrt{a^2 + b^2}}.$

2.442 1. $\int \dfrac{A + B\operatorname{ch} x}{(a + b\operatorname{sh} x)^n}\,dx = -\dfrac{B}{(n-1)b(a+b\operatorname{sh} x)^{n-1}} + A\int \dfrac{dx}{(a+b\operatorname{sh} x)^n}.$

Für $n = 1$ gilt | For $n = 1$:

2. $\int \dfrac{A + B\operatorname{ch} x}{a + b\operatorname{sh} x}\,dx = \dfrac{B}{b}\ln(a + b\operatorname{sh} x) + A\int \dfrac{dx}{a + b\operatorname{sh} x}$ (cf. 2.441 3.).

2.443 1. $\int \dfrac{A + B\operatorname{ch} x}{(a + b\operatorname{ch} x)^n}\,dx = \dfrac{aB - bA}{(n-1)(a^2 - b^2)} \cdot \dfrac{\operatorname{sh} x}{(a + b\operatorname{ch} x)^{n-1}} +$
$+ \dfrac{1}{(n-1)(a^2 - b^2)} \int \dfrac{(n-1)(aA - bB) + (n-2)(aB - bA)\operatorname{ch} x}{(a+b\operatorname{ch} x)^{n-1}}\,dx.$

Für $n = 1$ gilt | For $n = 1$:

2. $\int \dfrac{A + B\operatorname{ch} x}{a + b\operatorname{ch} x}\,dx = \dfrac{B}{b}x - \dfrac{aB - bA}{b}\int \dfrac{dx}{a + b\operatorname{ch} x}$ (cf. 2.443 3.).

3. $\int \dfrac{dx}{a + b\operatorname{ch} x} = \dfrac{1}{\sqrt{b^2 - a^2}} \arcsin \dfrac{b + a\operatorname{ch} x}{a + b\operatorname{ch} x}$ $[b^2 > a^2,\ x < 0];$

$= -\dfrac{1}{\sqrt{b^2 - a^2}} \arcsin \dfrac{b + a\operatorname{ch} x}{a + b\operatorname{ch} x}$ $[b^2 > a^2,\ x > 0];$

$= \dfrac{1}{\sqrt{a^2 - b^2}} \ln \dfrac{a + b + \sqrt{a^2 - b^2}\operatorname{th}\tfrac{x}{2}}{a + b - \sqrt{a^2 - b^2}\operatorname{th}\tfrac{x}{2}}$ $[a^2 > b^2].$

1. $\int \dfrac{dx}{\operatorname{ch} a + \operatorname{ch} x} = \operatorname{cosech} a \left[\ln \operatorname{ch} \dfrac{x+a}{2} - \ln \operatorname{ch} \dfrac{x-a}{2} \right];$ 2.444

$\qquad\qquad = 2 \operatorname{cosech} a \operatorname{Arth}\left(\operatorname{th} \dfrac{x}{2} \operatorname{th} \dfrac{a}{2} \right).$

2. $\int \dfrac{dx}{\cos a + \operatorname{ch} x} = 2 \operatorname{cosec} a \operatorname{arctg}\left(\operatorname{th} \dfrac{x}{2} \operatorname{tg} \dfrac{a}{2} \right).$

1. $\int \dfrac{A + B \operatorname{sh} x}{(a + b \operatorname{ch} x)^n} dx = -\dfrac{B}{(n-1) b (a + b \operatorname{ch} x)^{n-1}} + A \int \dfrac{dx}{(a + b \operatorname{ch} x)^n}.$ 2.445

Für $n = 1$ gilt | For $n = 1$:

2. $\int \dfrac{A + B \operatorname{sh} x}{a + b \operatorname{ch} x} dx = \dfrac{B}{b} \ln (a + b \operatorname{ch} x) + A \int \dfrac{dx}{a + b \operatorname{ch} x}$ (cf. 2.443 3.).

Bei der Berechnung bestimmter Integrale mit Hilfe der Formeln unter 2.441–2.443 und 2.445 muß man die Punkte umgehen, in denen der Integrand unendlich wird, d.h. die Punkte | In evaluating definite integrals with the help of the formulae 2.441– –2.443 and 2.445 it is necessary to avoid the points in which the integrand becomes infinite, that is, the points

$$x = \operatorname{Arsh}\left(-\dfrac{a}{b}\right)$$

in den Formeln 2.441, 2.442 und die Punkte | in 2.441 and 2.442, and the points

$$x = \operatorname{Arch}\left(-\dfrac{a}{b}\right)$$

in den Formeln 2.443, 2.445. Die Formel 2.443 ist für $a^2 = b^2$ nicht anwendbar. In diesen Fällen benutze man statt dessen die folgenden Formeln | in 2.443 and 2.445. Formula 2.443 is not valid if $a^2 = b^2$. In these cases we have

1. $\int \dfrac{A + B \operatorname{ch} x}{(\varepsilon + \operatorname{ch} x)^n} dx = \dfrac{B \operatorname{sh} x}{(1-n)(\varepsilon + \operatorname{ch} x)^n} +$ 2.446

$+ \left(\varepsilon A + \dfrac{n}{n-1} B \right) \dfrac{(n-1)!}{(2n-1)!!} \operatorname{sh} x \sum_{k=0}^{n-1} \dfrac{(2n-2k-3)!!}{(n-k-1)!} \cdot \dfrac{\varepsilon^k}{(\varepsilon + \operatorname{ch} x)^{n-k}} \quad [\varepsilon = \pm 1, n > 1].$

Für $n = 1$ gilt | For $n=1$:

2. $\int \dfrac{A + B \operatorname{ch} x}{\varepsilon + \operatorname{ch} x} dx = Bx + (\varepsilon A - B) \dfrac{\operatorname{ch} x - \varepsilon}{\operatorname{sh} x} \quad [\varepsilon = \pm 1].$

1. $\int \dfrac{\operatorname{sh} x\, dx}{a \operatorname{ch} x + b \operatorname{sh} x} = \dfrac{a \ln \operatorname{ch}\left(x + \operatorname{Arth} \dfrac{b}{a} \right) - bx}{a^2 - b^2} \quad [a > |b|];$ 2.447

$\qquad\qquad\qquad = \dfrac{bx - a \ln \operatorname{sh}\left(x + \operatorname{Arth} \dfrac{a}{b} \right)}{b^2 - a^2} \quad [b > |a|].$ MC 215+

Für $a = b = 1$ gilt | For $a = b = 1$:

2. $\int \dfrac{\operatorname{sh} x\, dx}{\operatorname{ch} x + \operatorname{sh} x} = \dfrac{x}{2} + \dfrac{1}{4} e^{-2x}.$

2.447   Für $a = -b = 1$ gilt | For $a = -b = 1$:

3. $\displaystyle\int \frac{\operatorname{sh} x\, dx}{\operatorname{ch} x - \operatorname{sh} x} = -\frac{x}{2} + \frac{1}{4} e^{2x}.$ MC 215

2.448   1. $\displaystyle\int \frac{\operatorname{ch} x\, dx}{a \operatorname{ch} x + b \operatorname{sh} x} = \frac{ax - b \ln \operatorname{ch}\left(x + \operatorname{Arth} \frac{b}{a}\right)}{a^2 - b^2}$ $[a > |b|];$

$\displaystyle = \frac{-ax + b \ln \operatorname{sh}\left(x + \operatorname{Arth} \frac{a}{b}\right)}{b^2 - a^2}$ $[b > |a|].$

Für $a = b = 1$ gilt | For $a = b = 1$:

2. $\displaystyle\int \frac{\operatorname{ch} x\, dx}{\operatorname{ch} x + \operatorname{sh} x} = \frac{x}{2} - \frac{1}{4} e^{-2x}.$

Im Fall $a = -b = 1$ gilt | For $a = -b = 1$:

3. $\displaystyle\int \frac{\operatorname{ch} x\, dx}{\operatorname{ch} x - \operatorname{sh} x} = \frac{x}{2} + \frac{1}{4} e^{2x}.$ MC 214, 215

2.449   1. $\displaystyle\int \frac{dx}{(a \operatorname{ch} x + b \operatorname{sh} x)^n} = \frac{1}{\sqrt{(a^2 - b^2)^n}} \int \frac{dx}{\operatorname{ch}^n\left(x + \operatorname{Arth} \frac{b}{a}\right)}$ $[a > |b|];$

$\displaystyle = \frac{1}{\sqrt{(b^2 - a^2)^n}} \int \frac{dx}{\operatorname{ch}^n\left(x + \operatorname{Arth} \frac{a}{b}\right)}$ $[b > |a|].$

Für $n = 1$ gilt | For $n = 1$:

2. $\displaystyle\int \frac{dx}{a \operatorname{ch} x + b \operatorname{sh} x} = \frac{1}{\sqrt{a^2 - b^2}} \operatorname{arctg} \left|\operatorname{sh}\left(x + \operatorname{Arth} \frac{b}{a}\right)\right|$ $[a > |b|];$

$\displaystyle = \frac{1}{\sqrt{b^2 - a^2}} \ln \left|\operatorname{th} \frac{x + \operatorname{Arth} \frac{a}{b}}{2}\right|$ $[b > |a|].$

Im Fall $a = b = 1$ ist | For $a = b = 1$:

3. $\displaystyle\int \frac{dx}{\operatorname{ch} x + \operatorname{sh} x} = -e^{-x} = \operatorname{sh} x - \operatorname{ch} x.$

Im Fall $a = -b = 1$ ist | For $a = -b = 1$:

4. $\displaystyle\int \frac{dx}{\operatorname{ch} x - \operatorname{sh} x} = e^x = \operatorname{sh} x + \operatorname{ch} x.$ MC 214

2.451   1. $\displaystyle\int \frac{A + B \operatorname{ch} x + C \operatorname{sh} x}{(a + b \operatorname{ch} x + c \operatorname{sh} x)^n} dx = \frac{Bc - Cb + (Ac - Ca) \operatorname{ch} x + (Ab - Ba) \operatorname{sh} x}{(1 - n)(a^2 - b^2 + c^2)(a + b \operatorname{ch} x + c \operatorname{sh} x)^{n-1}} +$

$\displaystyle + \frac{1}{(n - 1)(a^2 - b^2 + c^2)} \times$

$\displaystyle \times \int \frac{(n - 1)(Aa - Bb + Cc) - (n - 2)(Ab - Ba) \operatorname{ch} x - (n - 2)(Ac - Ca) \operatorname{sh} x}{(a + b \operatorname{ch} x + c \operatorname{sh} x)^{n-1}} dx$

$[a^2 + c^2 \neq b^2];$

$$= \frac{Bc - Cb - Ca \operatorname{ch} x - Ba \operatorname{sh} x}{(n-1) a (a + b \operatorname{ch} x + c \operatorname{sh} x)^n} +$$

2.451

$$+ \left[\frac{A}{a} + \frac{n(Bb - Cc)}{(n-1) a^2} \right] (c \operatorname{ch} x + b \operatorname{sh} x) \frac{(n-1)!}{(2n-1)!!} \times$$

$$\times \sum_{k=0}^{n-1} \frac{(2n - 2k - 3)!!}{(n-k-1)! \, a^k} \frac{1}{(a + b \operatorname{ch} x + c \operatorname{sh} x)^{n-k}} \quad [a^2 + c^2 = b^2].$$

2. $\displaystyle\int \frac{A + B \operatorname{ch} x + C \operatorname{sh} x}{a + b \operatorname{ch} x + c \operatorname{sh} x} dx = \frac{Cb - Bc}{b^2 - c^2} \ln(a + b \operatorname{ch} x + c \operatorname{sh} x) +$

$$+ \frac{Bb - Cc}{b^2 - c^2} x + \left(A - a \frac{Bb - Cc}{b^2 - c^2} \right) \int \frac{dx}{a + b \operatorname{ch} x + c \operatorname{sh} x} \quad [b^2 \neq c^2]$$

(cf. 2.451 4.).

3. $\displaystyle\int \frac{A + B \operatorname{ch} x + C \operatorname{sh} x}{a + b \operatorname{ch} x \pm b \operatorname{sh} x} dx = \frac{C \mp B}{2a} (\operatorname{ch} x \mp \operatorname{sh} x) + \left[\frac{A}{a} - \frac{(B \mp C) b}{2a^2} \right] x +$

$$+ \left[\frac{C \pm B}{2b} \pm \frac{A}{a} - \frac{(C \mp B) b}{2a^2} \right] \ln(a + b \operatorname{ch} x \pm b \operatorname{sh} x) \, [ab \neq 0].$$

4. $\displaystyle\int \frac{dx}{a + b \operatorname{ch} x + c \operatorname{sh} x} = \frac{2}{\sqrt{b^2 - a^2 - c^2}} \operatorname{arctg} \frac{(b-a) \operatorname{th} \frac{x}{2} + c}{\sqrt{b^2 - a^2 - c^2}}$

$$[b^2 > a^2 + c^2, \, a \neq b];$$

$$= \frac{1}{\sqrt{a^2 - b^2 + c^2}} \ln \frac{(a-b) \operatorname{th} \frac{x}{2} - c + \sqrt{a^2 - b^2 + c^2}}{(a-b) \operatorname{th} \frac{x}{2} - c - \sqrt{a^2 - b^2 + c^2}} \quad [b^2 < a^2 + c^2, \, a \neq b];$$

$$= \frac{1}{c} \ln \left(a + c \operatorname{th} \frac{x}{2} \right) \quad [a = b, \, c \neq 0];$$

$$= \frac{2}{(a-b) \operatorname{th} \frac{x}{2} + c} \quad [b^2 = a^2 + c^2].$$

GHH[351](18)

1. $\displaystyle\int \frac{A + B \operatorname{ch} x + C \operatorname{sh} x}{(a_1 + b_1 \operatorname{ch} x + c_1 \operatorname{sh} x)(a_2 + b_2 \operatorname{ch} x + c_2 \operatorname{sh} x)} dx = A_0 \ln \frac{a_1 + b_1 \operatorname{ch} x + c_1 \operatorname{sh} x}{a_2 + b_2 \operatorname{ch} x + c_2 \operatorname{sh} x} +$

2.452

$$+ A_1 \int \frac{dx}{a_1 + b_1 \operatorname{ch} x + c_1 \operatorname{sh} x} + A_2 \int \frac{dx}{a_2 + b_2 \operatorname{ch} x + c_2 \operatorname{sh} x}$$

mit | where

$$A_0 = \frac{\begin{vmatrix} a_1 & b_1 & c_1 \\ A & B & C \\ a_2 & b_2 & c_2 \end{vmatrix}}{\begin{vmatrix} a_1 & b_1 \\ a_2 & b_2 \end{vmatrix}^2 + \begin{vmatrix} b_1 & c_1 \\ b_2 & c_2 \end{vmatrix}^2 - \begin{vmatrix} c_1 & a_1 \\ c_2 & a_2 \end{vmatrix}^2}, \quad A_1 = \frac{\begin{vmatrix} a_1 & b_1 & c_1 \\ b_1 & c_1 \\ B & C \end{vmatrix} \begin{vmatrix} c_1 & a_1 \\ C & A \end{vmatrix} \begin{vmatrix} a_1 & b_1 \\ A & B \end{vmatrix}}{\begin{vmatrix} a_1 & b_1 \\ a_2 & b_2 \end{vmatrix}^2 + \begin{vmatrix} b_1 & c_1 \\ b_2 & c_2 \end{vmatrix}^2 - \begin{vmatrix} c_1 & a_1 \\ c_2 & a_2 \end{vmatrix}^2},$$

2.452
$$A_2 = \frac{\begin{vmatrix} a_1 & b_1 & c_1 \\ \begin{vmatrix} C & B \\ c_2 & b_2 \end{vmatrix} & \begin{vmatrix} C & A \\ c_2 & a_2 \end{vmatrix} & \begin{vmatrix} B & A \\ b_2 & a_2 \end{vmatrix} \\ a_2 & b_2 & c_2 \end{vmatrix}}{\begin{vmatrix} a_1 & b_1 \\ a_2 & b_2 \end{vmatrix}^2 + \begin{vmatrix} b_1 & c_1 \\ b_2 & c_2 \end{vmatrix}^2 - \begin{vmatrix} c_1 & a_1 \\ c_2 & a_2 \end{vmatrix}^2},$$

$$\left[\begin{vmatrix} a_1 & b_1 \\ a_2 & b_2 \end{vmatrix}^2 + \begin{vmatrix} b_1 & c_1 \\ b_2 & c_2 \end{vmatrix}^2 \neq \begin{vmatrix} c_1 & a_1 \\ c_2 & a_2 \end{vmatrix}^2 \right].$$

GHH[351](19)

2. $\displaystyle\int \frac{A \operatorname{ch}^2 x + 2B \operatorname{sh} x \operatorname{ch} x + C \operatorname{sh}^2 x}{a \operatorname{ch}^2 x + 2b \operatorname{sh} x \operatorname{ch} x + c \operatorname{sh}^2 x} dx =$

$$= \frac{1}{4b^2 - (a+c)^2} \{[4Bb - (A+C)(a+c)]x +$$

$$+ [(A+C)b - B(a+c)] \ln (a \operatorname{ch}^2 x + 2b \operatorname{sh} x \operatorname{ch} x + c \operatorname{sh}^2 x) +$$

$$+ [2(A-C) b^2 + 2Bb(a-c) + (Ca - Ac)(a+c)] f(x)\}$$

mit | where

$$f(x) = \frac{1}{2\sqrt{b^2 - ac}} \ln \frac{c \operatorname{th} x + b - \sqrt{b^2 - ac}}{c \operatorname{th} x + b + \sqrt{b^2 - ac}} \quad [b^2 > ac];$$

$$= \frac{1}{\sqrt{ac - b^2}} \operatorname{arctg} \frac{c \operatorname{th} x + b}{\sqrt{ac - b^2}} \quad [b^2 < ac];$$

$$= -\frac{1}{c \operatorname{th} x + b} \quad [b^2 = ac].$$

GHH[351](24)

2.453
1. $\displaystyle\int \frac{(A + B \operatorname{sh} x) dx}{\operatorname{sh} x (a + b \operatorname{sh} x)} = \frac{1}{a} \left[A \ln \left| \operatorname{th} \frac{x}{2} \right| + (aB - bA) \int \frac{dx}{a + b \operatorname{sh} x} \right]$

(cf. 2.441 3.).

2. $\displaystyle\int \frac{(A + B \operatorname{sh} x) dx}{\operatorname{sh} x (a + b \operatorname{ch} x)} = \frac{A}{a^2 - b^2} \left(a \ln \left| \operatorname{th} \frac{x}{2} \right| + b \ln \left| \frac{a + b \operatorname{ch} x}{\operatorname{sh} x} \right| \right) + B \int \frac{dx}{a + b \operatorname{ch} x}$

(cf. 2.443 3.).

Im Fall $a^2 = b^2 (= 1)$ gilt | For $a^2 = b^2 (= 1)$:

3. $\displaystyle\int \frac{(A + B \operatorname{sh} x) dx}{\operatorname{sh} x (1 + \operatorname{ch} x)} = \frac{A}{2} \left(\ln \left| \operatorname{th} \frac{x}{2} \right| - \frac{1}{2} \operatorname{th}^2 \frac{x}{2} \right) + B \operatorname{th} \frac{x}{2}.$

4. $\displaystyle\int \frac{(A + B \operatorname{sh} x) dx}{\operatorname{sh} x (1 - \operatorname{ch} x)} = \frac{A}{2} \left(-\ln \left| \operatorname{cth} \frac{x}{2} \right| + \frac{1}{2} \operatorname{cth}^2 \frac{x}{2} \right) + B \operatorname{cth} \frac{x}{2}.$

2.454
1. $\displaystyle\int \frac{(A + B \operatorname{sh} x) dx}{\operatorname{ch} x (a + b \operatorname{sh} x)} = \frac{1}{a^2 + b^2} \Big[(Aa + Bb) \operatorname{arctg} (\operatorname{sh} x) +$

$$+ (Ab - Ba) \ln \left| \frac{a + b \operatorname{sh} x}{\operatorname{ch} x} \right| \Big].$$

2. $\displaystyle\int \frac{(A + B \operatorname{ch} x) dx}{\operatorname{sh} x (a + b \operatorname{sh} x)} = \frac{1}{a} \left(A \ln \left| \operatorname{th} \frac{x}{2} \right| + B \ln \left| \frac{\operatorname{sh} x}{a + b \operatorname{sh} x} \right| - Ab \int \frac{dx}{a + b \operatorname{sh} x} \right)$

(cf. 2.441 3.).

1. $\int \dfrac{(A + B\,\text{ch}\,x)\,dx}{\text{sh}\,x\,(a + b\,\text{ch}\,x)} = \dfrac{1}{a^2 - b^2}\left[(Aa + Bb)\ln\left|\text{th}\,\dfrac{x}{2}\right| + (Ab - Ba)\ln\left|\dfrac{a + b\,\text{ch}\,x}{\text{sh}\,x}\right|\right].$ 2.455

Im Fall $a^2 = b^2\,(= 1)$ gelten die Formeln | For $a^2 = b^2\,(= 1)$:

2. $\int \dfrac{(A + B\,\text{ch}\,x)\,dx}{\text{sh}\,x\,(1 + \text{ch}\,x)} = \dfrac{A + B}{2}\ln\left|\text{th}\,\dfrac{x}{2}\right| - \dfrac{A - B}{4}\text{th}^2\dfrac{x}{2}.$

3. $\int \dfrac{(A + B\,\text{ch}\,x)\,dx}{\text{sh}\,x\,(1 - \text{ch}\,x)} = \dfrac{A + B}{4}\text{cth}^2\dfrac{x}{2} - \dfrac{A - B}{2}\ln\,\text{cth}\,\dfrac{x}{2}.$

$\int \dfrac{(A + B\,\text{ch}\,x)\,dx}{\text{ch}\,x\,(a + b\,\text{sh}\,x)} = \dfrac{A}{a^2 + b^2}\left[a\,\text{arctg}\,(\text{sh}\,x) + b\ln\left|\dfrac{a + b\,\text{sh}\,x}{\text{ch}\,x}\right|\right] + B\int \dfrac{dx}{a + b\,\text{sh}\,x}$ 2.456

(cf. 2.441 3.).

$\int \dfrac{(A + B\,\text{ch}\,x)\,dx}{\text{ch}\,x\,(a + b\,\text{ch}\,x)} = \dfrac{1}{a}\left[A\,\text{arctg}\,\text{sh}\,x - (Ab - Ba)\int \dfrac{dx}{a + b\,\text{ch}\,x}\right]$ (cf. 2.443 3.).   2.457

1. $\int \dfrac{dx}{a + b\,\text{sh}^2 x} = \dfrac{1}{\sqrt{a(b-a)}}\,\text{arctg}\left(\sqrt{\dfrac{b}{a} - 1}\,\text{th}\,x\right)\quad\left[\dfrac{b}{a} > 1\right]$ 2.458

$= \dfrac{1}{\sqrt{a(a-b)}}\,\text{Arth}\left(\sqrt{1 - \dfrac{b}{a}}\,\text{th}\,x\right)$

$\left[0 < \dfrac{b}{a} < 1\quad\text{oder}\atop\text{or}\quad \dfrac{b}{a} < 0\quad\text{und}\atop\text{and}\quad \text{sh}^2 x < -\dfrac{a}{b}\right];$

$= \dfrac{1}{\sqrt{a(a-b)}}\,\text{Arcth}\left(\sqrt{1 - \dfrac{b}{a}}\,\text{th}\,x\right)$

$\left[\dfrac{b}{a} < 0\quad\text{und}\atop\text{and}\quad \text{sh}^2 x > -\dfrac{a}{b}\right].$ MC 195

2. $\int \dfrac{dx}{a + b\,\text{ch}^2 x} = \dfrac{1}{\sqrt{-a(a+b)}}\,\text{arctg}\left(\sqrt{-\left(1 + \dfrac{b}{a}\right)}\,\text{cth}\,x\right)\left[\dfrac{b}{a} < -1\right];$

$= \dfrac{1}{\sqrt{a(a+b)}}\,\text{Arth}\left(\sqrt{1 + \dfrac{b}{a}}\,\text{cth}\,x\right)$

$\left[-1 < \dfrac{b}{a} < 0\quad\text{und}\atop\text{and}\quad \text{ch}^2 x > -\dfrac{a}{b}\right];$

$= \dfrac{1}{\sqrt{a(a+b)}}\,\text{Arcth}\left(\sqrt{1 + \dfrac{b}{a}}\,\text{cth}\,x\right)$

$\left[\dfrac{b}{a} > 0\quad\text{oder}\atop\text{or}\quad -1 < \dfrac{b}{a} < 0\quad\text{und}\atop\text{and}\quad \text{ch}^2 x < -\dfrac{a}{b}\right].$ MC 202

Im Fall $a^2 = b^2 = 1$ gelten die Formeln | For $a^2 = b^2 = 1$:

3. $\int \dfrac{dx}{1 + \text{sh}^2 x} = \text{th}\,x.$

2.458   4. $$\int \frac{dx}{1 - \operatorname{sh}^2 x} = \frac{1}{\sqrt{2}} \operatorname{Arth}(\sqrt{2}\,\operatorname{th} x) \quad [\operatorname{sh}^2 x < 1];$$

$$= \frac{1}{\sqrt{2}} \operatorname{Arcth}(\sqrt{2}\,\operatorname{th} x) \quad [\operatorname{sh}^2 x > 1].$$

5. $$\int \frac{dx}{1 + \operatorname{ch}^2 x} = \frac{1}{\sqrt{2}} \operatorname{Arcth}(\sqrt{2}\,\operatorname{cth} x).$$

6. $$\int \frac{dx}{1 - \operatorname{ch}^2 x} = \operatorname{cth} x.$$

2.459 1. $$\int \frac{dx}{(a + b \operatorname{sh}^2 x)^2} = \frac{1}{2a(b-a)} \left[\frac{b \operatorname{sh} x \operatorname{ch} x}{a + b \operatorname{sh}^2 x} + (b - 2a) \int \frac{dx}{a + b \operatorname{sh}^2 x} \right]$$
(cf. 2.458 1.).   MC 196

2. $$\int \frac{dx}{(a + b \operatorname{ch}^2 x)^2} = \frac{1}{2a(a+b)} \left[-\frac{b \operatorname{sh} x \operatorname{ch} x}{a + b \operatorname{ch}^2 x} + (2a + b) \int \frac{dx}{a + b \operatorname{ch}^2 x} \right]$$
(cf. 2.458 2.).   MC 203

3. $$\int \frac{dx}{(a + b \operatorname{sh}^2 x)^3} = \frac{1}{8pa^3} \left[\left(3 - \frac{2}{p^2} + \frac{3}{p^4}\right) \operatorname{arctg}(p \operatorname{th} x) + \right.$$
$$\left. + \left(3 - \frac{2}{p^2} - \frac{3}{p^4}\right) \frac{p \operatorname{th} x}{1 + p^2 \operatorname{th}^2 x} + \left(1 + \frac{2}{p^2} - \frac{1}{p^2} \operatorname{th}^2 x\right) \frac{2p \operatorname{th} x}{(1 + p^2 \operatorname{th}^2 x)^2} \right]$$
$$\left[p^2 = \frac{b}{a} - 1 > 0\right];$$

$$= \frac{1}{8qa^3} \left[\left(3 + \frac{2}{q^2} + \frac{3}{q^4}\right) \operatorname{Arth}(q \operatorname{th} x) + \right.$$
$$\left. + \left(3 + \frac{2}{q^2} - \frac{3}{q^4}\right) \frac{q \operatorname{th} x}{1 - q^2 \operatorname{th}^2 x} + \left(1 - \frac{2}{q^2} + \frac{1}{q^2} \operatorname{th}^2 x\right) \frac{2q \operatorname{th} x}{(1 - q^2 \operatorname{th}^2 x)^2} \right]$$
$$\left[q^2 = 1 - \frac{b}{a} > 0\right]. \quad \text{MC 196}$$

4. $$\int \frac{dx}{(a + b \operatorname{ch}^2 x)^3} = \frac{1}{8pa^3} \left[\left(3 - \frac{2}{p^2} + \frac{3}{p^4}\right) \operatorname{arctg}(p \operatorname{cth} x) + \right.$$
$$\left. + \left(3 - \frac{2}{p^2} - \frac{3}{p^4}\right) \frac{p \operatorname{cth} x}{1 + p^2 \operatorname{cth}^2 x} + \left(1 + \frac{2}{p^2} - \frac{1}{p^2} \operatorname{cth}^2 x\right) \frac{2p \operatorname{cth} x}{(1 + p^2 \operatorname{cth}^2 x)^2} \right]$$
$$\left[p^2 = -1 - \frac{b}{a} > 0\right];$$

$$= \frac{1}{8qa^3} \left[\left(3 + \frac{2}{q^2} + \frac{3}{q^4}\right) \varphi(x)^* + \right.$$
$$\left. + \left(3 + \frac{2}{q^2} - \frac{3}{q^4}\right) \frac{q \operatorname{cth} x}{1 - q^2 \operatorname{cth}^2 x} + \left(1 - \frac{2}{q^2} + \frac{1}{q^2} \operatorname{cth}^2 x\right) \frac{2q \operatorname{cth} x}{(1 - q^2 \operatorname{cth}^2 x)^2} \right]$$
$$\left[q^2 = 1 + \frac{b}{a} > 0\right].$$

---

* Ist $\dfrac{b}{a} < 0$ und $\operatorname{ch}^2 x > -\dfrac{a}{b}$, so ist $\varphi(x) = \operatorname{Arth}(q \operatorname{cth} x)$. Ist $\dfrac{b}{a} < 0$, aber $\operatorname{ch}^2 x < -\dfrac{a}{b}$, oder ist $\dfrac{b}{a} > 0$, so ist $\varphi(x) = \operatorname{Arcth}(q \operatorname{cth} x)$.   MC 203

* If $\dfrac{b}{a} < 0$ and $\operatorname{ch}^2 x > -\dfrac{a}{b}$, then $\varphi(x) = \operatorname{Arth}(q \operatorname{cth} x)$. If $\dfrac{b}{a} < 0$ and $\operatorname{ch}^2 x < -\dfrac{a}{b}$ or if $\dfrac{b}{a} > 0$, then $\varphi(x) = \operatorname{Arcth}(q \operatorname{cth} x)$.   MC 203

Algebraische Funktionen von Hyperbelfunktionen
Algebraic functions of hyperbolic functions

2.46

1. $\int \sqrt{\operatorname{th} x}\, dx = \operatorname{Arth} \sqrt{\operatorname{th} x} - \operatorname{arctg} \sqrt{\operatorname{th} x}.$ MC 221 2.461

2. $\int \sqrt{\operatorname{cth} x}\, dx = \operatorname{Arcth} \sqrt{\operatorname{cth} x} - \operatorname{arctg} \sqrt{\operatorname{cth} x}.$ MC 222

1. $\int \dfrac{\operatorname{sh} x\, dx}{\sqrt{a^2 + \operatorname{sh}^2 x}} = \operatorname{Arsh} \dfrac{\operatorname{ch} x}{\sqrt{a^2 - 1}} = \ln(\operatorname{ch} x + \sqrt{a^2 + \operatorname{sh}^2 x}) \quad [a^2 > 1];$ 2.462

$\qquad\qquad = \operatorname{Arch} \dfrac{\operatorname{ch} x}{\sqrt{1-a^2}} = \ln(\operatorname{ch} x + \sqrt{a^2 + \operatorname{sh}^2 x}) \quad [a^2 < 1];$

$\qquad\qquad = \ln \operatorname{ch} x \qquad [a^2 = 1].$

2. $\int \dfrac{\operatorname{sh} x\, dx}{\sqrt{a^2 - \operatorname{sh}^2 x}} = \arcsin \dfrac{\operatorname{ch} x}{\sqrt{a^2 + 1}} \quad [\operatorname{sh}^2 x < a^2].$

3. $\int \dfrac{\operatorname{sh} x\, dx}{\sqrt{\operatorname{sh}^2 x - a^2}} = \operatorname{Arch} \dfrac{\operatorname{ch} x}{\sqrt{a^2 + 1}} = \ln(\operatorname{ch} x + \sqrt{\operatorname{sh}^2 x - a^2}) \quad [\operatorname{sh}^2 x > a^2].$

MC 199

4. $\int \dfrac{\operatorname{ch} x\, dx}{\sqrt{a^2 + \operatorname{sh}^2 x}} = \operatorname{Arsh} \dfrac{\operatorname{sh} x}{a} = \ln(\operatorname{sh} x + \sqrt{a^2 + \operatorname{sh}^2 x}).$

5. $\int \dfrac{\operatorname{ch} x\, dx}{\sqrt{a^2 - \operatorname{sh}^2 x}} = \arcsin \dfrac{\operatorname{sh} x}{a} \quad [\operatorname{sh}^2 x < a^2].$

6. $\int \dfrac{\operatorname{ch} x\, dx}{\sqrt{\operatorname{sh}^2 x - a^2}} = \operatorname{Arch} \dfrac{\operatorname{sh} x}{a} = \ln(\operatorname{sh} x + \sqrt{\operatorname{sh}^2 x - a^2}) \quad [\operatorname{sh}^2 x > a^2].$

7. $\int \dfrac{\operatorname{sh} x\, dx}{\sqrt{a^2 + \operatorname{ch}^2 x}} = \operatorname{Arsh} \dfrac{\operatorname{ch} x}{a} = \ln(\operatorname{ch} x + \sqrt{a^2 + \operatorname{ch}^2 x}).$

8. $\int \dfrac{\operatorname{sh} x\, dx}{\sqrt{a^2 - \operatorname{ch}^2 x}} = \arcsin \dfrac{\operatorname{ch} x}{a} \quad [\operatorname{ch}^2 x < a^2].$

9. $\int \dfrac{\operatorname{sh} x\, dx}{\sqrt{\operatorname{ch}^2 x - a^2}} = \operatorname{Arch} \dfrac{\operatorname{ch} x}{a} = \ln(\operatorname{ch} x + \sqrt{\operatorname{ch}^2 x - a^2}) \quad [\operatorname{ch}^2 x > a^2].$

MC 215–216

10. $\int \dfrac{\operatorname{ch} x\, dx}{\sqrt{a^2 + \operatorname{ch}^2 x}} = \operatorname{Arsh} \dfrac{\operatorname{sh} x}{\sqrt{a^2 + 1}} = \ln(\operatorname{sh} x + \sqrt{a^2 + \operatorname{ch}^2 x}).$

11. $\int \dfrac{\operatorname{ch} x\, dx}{\sqrt{a^2 - \operatorname{ch}^2 x}} = \arcsin \dfrac{\operatorname{sh} x}{\sqrt{a^2 - 1}} \quad [\operatorname{ch}^2 x < a^2].$

12. $\int \dfrac{\operatorname{ch} x\, dx}{\sqrt{\operatorname{ch}^2 x - a^2}} = \operatorname{Arch} \dfrac{\operatorname{sh} x}{\sqrt{a^2 - 1}} \quad [a^2 > 1];$

$\qquad\qquad = \ln \operatorname{sh} x \qquad [a^2 = 1].$ MC 206

13. $\int \dfrac{\operatorname{cth} x\, dx}{\sqrt{a + b \operatorname{sh} x}} = 2\sqrt{a}\, \operatorname{Arcth} \sqrt{1 + \dfrac{b}{a} \operatorname{sh} x} \quad [b \operatorname{sh} x > 0,\ a > 0];$

$\qquad\qquad = 2\sqrt{a}\, \operatorname{Arth} \sqrt{1 + \dfrac{b}{a} \operatorname{sh} x} \quad [b \operatorname{sh} x < 0,\ a > 0];$

$\qquad\qquad = 2\sqrt{-a}\, \operatorname{Arth} \sqrt{-\left(1 + \dfrac{b}{a} \operatorname{sh} x\right)} \quad a < 0.$

2.462

14. $\int \dfrac{\operatorname{th} x\, dx}{\sqrt{a + b\operatorname{ch} x}} = 2\sqrt{a}\operatorname{Arcth} \sqrt{1 + \dfrac{b}{a}\operatorname{ch} x}$ $\quad [b\operatorname{ch} x > 0,\ a > 0];$

$\hspace{4.5em} = 2\sqrt{a}\operatorname{Arth} \sqrt{1 + \dfrac{b}{a}\operatorname{ch} x}$ $\quad [b\operatorname{ch} x < 0,\ a > 0];$

$\hspace{4.5em} = 2\sqrt{-a}\operatorname{Arth} \sqrt{-\left(1 + \dfrac{b}{a}\operatorname{ch} x\right)}$ $\quad [a < 0].$ $\hspace{2em}$ MC 220, 221

2.463

1. $\int \dfrac{\operatorname{sh} x \sqrt{a + b\operatorname{ch} x}}{p + q\operatorname{ch} x}\, dx = 2\sqrt{\dfrac{aq - bp}{q}}\operatorname{Arcth} \sqrt{\dfrac{q(a + b\operatorname{ch} x)}{aq - bp}}$

$\hspace{12em} \left[b\operatorname{ch} x > 0,\ \dfrac{aq - bp}{q} > 0\right];$

$\hspace{6em} = 2\sqrt{\dfrac{aq - bp}{q}}\operatorname{Arth} \sqrt{\dfrac{q(a + b\operatorname{ch} x)}{aq - bp}}$

$\hspace{12em} \left[b\operatorname{ch} x < 0,\ \dfrac{aq - bp}{q} > 0\right];$

$\hspace{6em} = 2\sqrt{\dfrac{bp - aq}{q}}\operatorname{Arth} \sqrt{\dfrac{q(a + b\operatorname{ch} x)}{bp - aq}}$

$\hspace{12em} \left[\dfrac{aq - bp}{q} < 0\right].$ $\hspace{2em}$ MC 220

2. $\int \dfrac{\operatorname{ch} x \sqrt{a + b\operatorname{sh} x}}{p + q\operatorname{sh} x}\, dx = 2\sqrt{\dfrac{aq - bp}{q}}\operatorname{Arcth} \sqrt{\dfrac{q(a + b\operatorname{sh} x)}{aq - bp}}$

$\hspace{12em} \left[b\operatorname{sh} x > 0,\ \dfrac{aq - bp}{q} > 0\right];$

$\hspace{6em} = 2\sqrt{\dfrac{aq - bp}{q}}\operatorname{Arth} \sqrt{\dfrac{q(a + b\operatorname{sh} x)}{aq - bp}}$

$\hspace{12em} \left[b\operatorname{sh} x < 0,\ \dfrac{aq - bp}{q} > 0\right];$

$\hspace{6em} = 2\sqrt{\dfrac{bp - aq}{q}}\operatorname{Arth} \sqrt{\dfrac{q(a + b\operatorname{sh} x)}{bp - aq}}$

$\hspace{12em} \left[\dfrac{aq - bp}{q} < 0\right].$ $\hspace{2em}$ MC 221

2.464

1. $\int \dfrac{dx}{\sqrt{k^2 + k'^2 \operatorname{ch}^2 x}} = \int \dfrac{dx}{\sqrt{1 + k'^2 \operatorname{sh}^2 x}} = F(\arcsin(\operatorname{th} x), k)$ $\quad [x > 0].$

$\hspace{20em}$ BF(295.00), BF(295.10)

2. $\int \dfrac{dx}{\sqrt{\operatorname{ch}^2 x - k^2}} = \int \dfrac{dx}{\sqrt{\operatorname{sh}^2 x + k'^2}} = F\left(\arcsin\left(\dfrac{1}{\operatorname{ch} x}\right), k\right)$ $\quad [x > 0].$

$\hspace{20em}$ BF(295.40), BF(295.30)

3. $\int \dfrac{dx}{\sqrt{1 - k'^2 \operatorname{ch}^2 x}} = F\left(\arcsin\left(\dfrac{\operatorname{th} x}{k}\right), k\right)$ $\quad \left[0 < x < \operatorname{Arch}\dfrac{1}{k'}\right].$ $\hspace{1em}$ BF(295.20)

In 2.464 4.−2.464 8. ist | In 2.464 4.−2.464 8. we put

$$\alpha = \arccos\frac{1-\operatorname{sh} 2ax}{1+\operatorname{sh} 2ax},\ r = \frac{1}{\sqrt{2}}\quad [ax > 0]:$$

4. $\displaystyle\int\frac{dx}{\sqrt{\operatorname{sh} 2ax}} = \frac{1}{2a}F(\alpha,r).$ \hfill BF(296.50)

5. $\displaystyle\int\sqrt{\operatorname{sh} 2ax}\,dx = \frac{1}{2a}[F(\alpha,r) - 2E(\alpha,r)] + \frac{1}{a}\frac{\sqrt{\operatorname{sh} 2ax(1+\operatorname{sh}^2 2ax)}}{1+\operatorname{sh} 2ax}.$ \hfill BF(296.53)

6. $\displaystyle\int\frac{\operatorname{ch}^2 2ax\,dx}{(1+\operatorname{sh} 2ax)^2\sqrt{\operatorname{sh} 2ax}} = \frac{1}{2a}E(\alpha,r).$ \hfill BF(296.51)

7. $\displaystyle\int\frac{(1-\operatorname{sh} 2ax)^2\,dx}{(1+\operatorname{sh} 2ax)^2\sqrt{\operatorname{sh} 2ax}} = \frac{1}{2a}[2E(\alpha,r) - F(\alpha,r)].$ \hfill BF(296.55)

8. $\displaystyle\int\frac{\sqrt{\operatorname{sh} 2ax}\,dx}{(1+\operatorname{sh} 2ax)^2} = \frac{1}{4a}[F(\alpha,r) - E(\alpha,r)].$ \hfill BF(296.54)

In 2.464 9.−2.464 15. ist | In 2.464 9.−2.464 15. we put

$$\alpha = \arcsin\sqrt{\frac{\operatorname{ch} 2ax - 1}{\operatorname{ch} 2ax}},\ r = \frac{1}{\sqrt{2}}\quad [x \neq 0]:$$

9. $\displaystyle\int\frac{dx}{\sqrt{\operatorname{ch} 2ax}} = \frac{1}{a\sqrt{2}}F(\alpha,r).$ \hfill BF(296.00)

10. $\displaystyle\int\sqrt{\operatorname{ch} 2ax}\,dx = \frac{1}{a\sqrt{2}}[F(\alpha,r) - 2E(\alpha,r)] + \frac{\operatorname{sh} 2ax}{a\sqrt{\operatorname{ch} 2ax}}.$ \hfill BF(296.03)

11. $\displaystyle\int\frac{dx}{\sqrt{\operatorname{ch}^3 2ax}} = \frac{1}{a\sqrt{2}}[2E(\alpha,r) - F(\alpha,r)].$ \hfill BF(296.04)

12. $\displaystyle\int\frac{dx}{\sqrt{\operatorname{ch}^5 2ax}} = \frac{1}{3\sqrt{2a}}F(\alpha,r) + \frac{\operatorname{th} 2ax}{3a\sqrt{\operatorname{ch} 2ax}}.$ \hfill BF(296.04)

13. $\displaystyle\int\frac{\operatorname{sh}^2 2ax\,dx}{\sqrt{\operatorname{ch} 2ax}} = -\frac{\sqrt{2}}{3a}F(\alpha,r) + \frac{1}{3a}\operatorname{sh} 2ax\sqrt{\operatorname{ch} 2ax}.$ \hfill BF(296.07)

14. $\displaystyle\int\frac{\operatorname{th}^2 2ax\,dx}{\sqrt{\operatorname{ch} 2ax}} = \frac{\sqrt{2}}{3a}F(\alpha,r) - \frac{\operatorname{th} 2ax}{3a\sqrt{\operatorname{ch} 2ax}}.$ \hfill BF(296.05)

15. $\displaystyle\int\frac{\sqrt{\operatorname{ch} 2ax}\,dx}{p^2 + (1-p^2)\operatorname{ch} 2ax} = \frac{1}{a\sqrt{2}}\Pi(\alpha, p^2, r).$ \hfill BF(296.02)

In 2.464 16.−2.464 20. ist | In 2.464 16.−2.464 20. we put

$$\alpha = \arccos\frac{\sqrt{a^2+b^2} - a - b\operatorname{sh} x}{\sqrt{a^2+b^2} + a + b\operatorname{sh} x},$$

$$r = \sqrt{\frac{a + \sqrt{a^2+b^2}}{2\sqrt{a^2+b^2}}}\quad \left[a > 0,\ b > 0,\ x > -\operatorname{Arsh}\frac{a}{b}\right]:$$

16. $\displaystyle\int\frac{dx}{\sqrt{a+b\operatorname{sh} x}} = \frac{1}{\sqrt[4]{a^2+b^2}}F(\alpha,r).$ \hfill F(98.00)

2.464

17. $$\int \sqrt{a+b\,\mathrm{sh}\,x}\,dx = \sqrt[4]{a^2+b^2}\,[F(\alpha,r) - 2E(\alpha,r)] +$$
$$+ \frac{2b\,\mathrm{ch}\,x\,\sqrt{a+b\,\mathrm{sh}\,x}}{\sqrt{a^2+b^2}+a+b\,\mathrm{sh}\,x}.\qquad \text{BF(298.02)}$$

18. $$\int \frac{\sqrt{a+b\,\mathrm{sh}\,x}}{\mathrm{ch}^2 x}\,dx = \sqrt[4]{a^2+b^2}\,E(\alpha,r) - \frac{\sqrt{a^2+b^2}-a}{2\sqrt[4]{a^2+b^2}}\,F(\alpha,r) -$$
$$- \frac{a+\sqrt{a^2+b^2}}{b} \cdot \frac{\sqrt{a^2+b^2}-a-b\,\mathrm{sh}\,x}{\sqrt{a^2+b^2}+a+b\,\mathrm{sh}\,x} \cdot \frac{\sqrt{a+b\,\mathrm{sh}\,x}}{\mathrm{ch}\,x}.\qquad \text{BF(298.03)}$$

19. $$\int \frac{\mathrm{ch}^2 x\,dx}{[\sqrt{a^2+b^2}+a+b\,\mathrm{sh}\,x]^2 \sqrt{a+b\,\mathrm{sh}\,x}} = \frac{1}{b^2\sqrt[4]{a^2+b^2}}\,E(\alpha,r).\qquad \text{BF(298.01)}$$

20. $$\int \frac{\sqrt{a+b\,\mathrm{sh}\,x}\,dx}{[\sqrt{a^2+b^2}-a-b\,\mathrm{sh}\,x]^2} = -\frac{1}{\sqrt[4]{a^2+b^2}(\sqrt{a^2+b^2}-a)}\,E(\alpha,r) +$$
$$+ \frac{b}{\sqrt{a^2+b^2}-a} \cdot \frac{\mathrm{ch}\,x\,\sqrt{a+b\,\mathrm{sh}\,x}}{a^2+b^2-(a+b\,\mathrm{sh}\,x)^2}.\qquad \text{BF(298.04)}$$

In 2.464 21.–2.464 31. ist | In 2.464 21.–2.464 31. we put

$$\alpha = \arcsin\left(\mathrm{th}\,\frac{x}{2}\right),\ r = \sqrt{\frac{a-b}{a+b}}\quad [0 < b < a,\ x > 0]:$$

21. $$\int \frac{dx}{\sqrt{a+b\,\mathrm{ch}\,x}} = \frac{2}{\sqrt{a+b}}\,F(\alpha,r).\qquad \text{BF(297.25)}$$

22. $$\int \sqrt{a+b\,\mathrm{ch}\,x}\,dx = 2\sqrt{a+b}\,[F(\alpha,r) - E(\alpha,r)] + 2\,\mathrm{th}\,\frac{x}{2}\,\sqrt{a+b\,\mathrm{ch}\,x}.\qquad \text{BF(297.29)}$$

23. $$\int \frac{\mathrm{ch}\,x\,dx}{\sqrt{a+b\,\mathrm{ch}\,x}} = \frac{2}{\sqrt{a+b}}\,F(\alpha,r) - \frac{2\sqrt{a+b}}{b}\,E(\alpha,r) +$$
$$+ \frac{2}{b}\,\mathrm{th}\,\frac{x}{2}\,\sqrt{a+b\,\mathrm{ch}\,x}.\qquad \text{BF(297.33)}$$

24. $$\int \frac{\mathrm{th}^2\frac{x}{2}}{\sqrt{a+b\,\mathrm{ch}\,x}}\,dx = \frac{2\sqrt{a+b}}{a-b}\,[F(\alpha,r) - E(\alpha,r)].\qquad \text{BF(297.28)}$$

25. $$\int \frac{\mathrm{th}^4\frac{x}{2}}{\sqrt{a+b\,\mathrm{ch}\,x}}\,dx = \frac{2\sqrt{a+b}}{3(a-b)^2}\,[(3a+b)F(\alpha,r) - 4aE(\alpha,r)] +$$
$$+ \frac{2}{3(a-b)} \cdot \frac{\mathrm{sh}\,\frac{x}{2}\sqrt{a+b\,\mathrm{ch}\,x}}{\mathrm{ch}^3\,\frac{x}{2}}.\qquad \text{BF(297.28)}$$

26. $$\int \frac{\mathrm{ch}\,x - 1}{\sqrt{a+b\,\mathrm{ch}\,x}}\,dx = \frac{2}{b}\left[\mathrm{th}\,\frac{x}{2}\,\sqrt{a+b\,\mathrm{ch}\,x} - \sqrt{a+b}\,E(\alpha,r)\right].\qquad \text{BF(297.31)}$$

27. $\int \dfrac{(\mathrm{ch}\, x - 1)^2}{\sqrt{a + b\, \mathrm{ch}\, x}}\, dx = \dfrac{4\sqrt{a+b}}{3b^2}\,[(a + 3b)\, E(\alpha, r) - bF(\alpha, r)] +$ 2.464

$\qquad + \dfrac{4}{3b^2}\left[b\, \mathrm{ch}^2\, \dfrac{x}{2} - (a + 3b)\right]\mathrm{th}\,\dfrac{x}{2}\sqrt{a + b\, \mathrm{ch}\, x}.$ BF(297.31)

28. $\int \dfrac{\sqrt{a + b\, \mathrm{ch}\, x}}{\mathrm{ch}\, x + 1}\, dx = \sqrt{a + b}\, E(\alpha, r).$ BF(297.26)

29. $\int \dfrac{dx}{(\mathrm{ch}\, x + 1)\sqrt{a + b\, \mathrm{ch}\, x}} = \dfrac{\sqrt{a+b}}{a - b}\, E(\alpha, r) - \dfrac{2b}{(a-b)\sqrt{a+b}}\, F(\alpha, r).$

BF(297.30)

30. $\int \dfrac{dx}{(\mathrm{ch}\, x + 1)^2 \sqrt{a + b\, \mathrm{ch}\, x}} = \dfrac{1}{3(a-b)^2 \sqrt{a+b}}\,[b(5b - a)\, F(\alpha, r) +$

$\qquad + (a - 3b)(a + b)\, E(\alpha, r)] + \dfrac{1}{6(a-b)} \cdot \dfrac{\mathrm{sh}\,\dfrac{x}{2}}{\mathrm{ch}^3\,\dfrac{x}{2}}\sqrt{a + b\, \mathrm{ch}\, x}.$ BF(297.30)

31. $\int \dfrac{(1 + \mathrm{ch}\, x)\, dx}{[1 + p^2 + (1 - p^2)\,\mathrm{ch}\, x]\sqrt{a + b\, \mathrm{ch}\, x}} = \dfrac{2}{\sqrt{a+b}}\, \Pi(\alpha, p^2, r).$ BF(297.27)

In 2.464 32.–2.464 40. ist | In 2.464 32.–2.464 40. we put

$\alpha = \arcsin \sqrt{\dfrac{a - b\, \mathrm{ch}\, x}{a - b}},\quad r = \sqrt{\dfrac{a - b}{a + b}}\quad \left[0 < b < a,\ 0 < x < \mathrm{Arch}\,\dfrac{a}{b}\right]:$

32. $\int \dfrac{dx}{\sqrt{a - b\, \mathrm{ch}\, x}} = \dfrac{2}{\sqrt{a+b}}\, F(\alpha, r).$ BF(297.50)

33. $\int \sqrt{a - b\, \mathrm{ch}\, x}\, dx = 2\sqrt{a+b}\,[F(\alpha, r) - E(\alpha, r)].$ BF(297.54)

34. $\int \dfrac{\mathrm{ch}\, x\, dx}{\sqrt{a - b\, \mathrm{ch}\, x}} = \dfrac{2\sqrt{a+b}}{b}\, E(\alpha, r) - \dfrac{2}{\sqrt{a+b}}\, F(\alpha, r).$ BF(297.56)

35. $\int \dfrac{\mathrm{ch}^2\, x\, dx}{\sqrt{a - b\, \mathrm{ch}\, x}} = \dfrac{2(b - 2a)}{3b\sqrt{a+b}}\, F(\alpha, r) + \dfrac{4a\sqrt{a+b}}{3b^2}\, E(\alpha, r) +$

$\qquad + \dfrac{2}{3b}\,\mathrm{sh}\, x\, \sqrt{a - b\, \mathrm{ch}\, x}.$ BF(297.56)

36. $\int \dfrac{(1 + \mathrm{ch}\, x)\, dx}{\sqrt{a - b\, \mathrm{ch}\, x}} = \dfrac{2\sqrt{a+b}}{b}\, E(\alpha, r).$ BF(297.51)

37. $\int \dfrac{dx}{\mathrm{ch}\, x\, \sqrt{a - b\, \mathrm{ch}\, x}} = \dfrac{2b}{a\sqrt{a+b}}\, \Pi\!\left(\alpha, \dfrac{a-b}{a}, r\right).$ BF(297.57)

38. $\int \dfrac{dx}{(1 + \mathrm{ch}\, x)\sqrt{a - b\, \mathrm{ch}\, x}} = \dfrac{1}{\sqrt{a+b}}\, E(\alpha, r) - \dfrac{1}{a+b}\,\mathrm{th}\,\dfrac{x}{2}\,\sqrt{a - b\, \mathrm{ch}\, x}.$

BF(297.58)

39. $\int \dfrac{dx}{(1 + \mathrm{ch}\, x)^2 \sqrt{a - b\, \mathrm{ch}\, x}} = \dfrac{1}{3\sqrt{(a+b)^3}}\,[(a + 3b)\, E(\alpha, r) - bF(\alpha, r)] -$

$\qquad - \dfrac{1}{3(a+b)^2} \cdot \dfrac{\mathrm{th}\,\dfrac{x}{2}\sqrt{a - b\, \mathrm{ch}\, x}}{\mathrm{ch}\, x + 1}\,[2a + 4b + (a + 3b)\,\mathrm{ch}\, x].$ BF(297.58)

2.464

40. $\displaystyle\int\frac{dx}{(a-b-ap^2+bp^2\operatorname{ch} x)\sqrt{a-b\operatorname{ch} x}}=\frac{2}{(a-b)\sqrt{a+b}}\Pi(\alpha,p^2,r).$ BF(297.52)

In 2.464 41.—2.464 47. ist | In 2.464 41.—2.464 47. we put

$$\alpha=\arcsin\sqrt{\frac{b(\operatorname{ch} x-1)}{b\operatorname{ch} x-a}},\quad r=\sqrt{\frac{a+b}{2b}}\quad [0<a<b,\ x>0]:$$

41. $\displaystyle\int\frac{dx}{\sqrt{b\operatorname{ch} x-a}}=\sqrt{\frac{2}{b}}\,F(\alpha,r).$ BF(297.00)

42. $\displaystyle\int\sqrt{b\operatorname{ch} x-a}\,dx=(b-a)\sqrt{\frac{2}{b}}\,F(\alpha,r)-2\sqrt{2b}\,E(\alpha,r)+\frac{2b\operatorname{sh} x}{\sqrt{b\operatorname{ch} x-a}}.$ BF(297.05)

43. $\displaystyle\int\frac{dx}{\sqrt{(b\operatorname{ch} x-a)^3}}=\frac{1}{b^2-a^2}\cdot\sqrt{\frac{2}{b}}\,[2bE(\alpha,r)-(b-a)F(\alpha,r)].$ BF(297.06)

44. $\displaystyle\int\frac{dx}{\sqrt{(b\operatorname{ch} x-a)^5}}=\frac{1}{3(b^2-a^2)^2}\sqrt{\frac{2}{b}}\,[(b-3a)(b-a)F(\alpha,r)+$
$+8abE(\alpha,r)]+\frac{2b}{3(b^2-a^2)}\cdot\frac{\operatorname{sh} x}{\sqrt{(b\operatorname{ch} x-a)^3}}.$ BF(297.06)

45. $\displaystyle\int\frac{\operatorname{ch} x\,dx}{\sqrt{b\operatorname{ch} x-a}}=\sqrt{\frac{2}{b}}\,[F(\alpha,r)-2E(\alpha,r)]+\frac{2\operatorname{sh} x}{\sqrt{b\operatorname{ch} x-a}}.$ BF(297.03)

46. $\displaystyle\int\frac{(\operatorname{ch} x+1)\,dx}{\sqrt{(b\operatorname{ch} x-a)^3}}=\frac{2}{b-a}\sqrt{\frac{2}{b}}\,E(\alpha,r).$ BF(297.01)

47. $\displaystyle\int\frac{\sqrt{b\operatorname{ch} x-a}\,dx}{p^2b-a+b(1-p^2)\operatorname{ch} x}=\sqrt{\frac{2}{b}}\,\Pi(\alpha,p^2,r).$ BF(297.02)

In 2.464 48.—2.464 55. ist | In 2.464 48.—2.464 55. we put

$$\alpha=\arcsin\sqrt{\frac{b\operatorname{ch} x-a}{b(\operatorname{ch} x-1)}},\quad r=\sqrt{\frac{2b}{a+b}}\quad \left[0<b<a,\ x>\operatorname{Arch}\frac{a}{b}\right]:$$

48. $\displaystyle\int\frac{dx}{\sqrt{b\operatorname{ch} x-a}}=\frac{2}{\sqrt{a+b}}\,F(\alpha,r).$ BF(297.75)

49. $\displaystyle\int\sqrt{b\operatorname{ch} x-a}\,dx=-2\sqrt{a+b}\,E(\alpha,r)+2\operatorname{cth}\frac{x}{2}\sqrt{b\operatorname{ch} x-a}.$ BF(297.79)

50. $\displaystyle\int\frac{\operatorname{cth}^2\frac{x}{2}\,dx}{\sqrt{b\operatorname{ch} x-a}}=\frac{2\sqrt{a+b}}{a-b}\,E(\alpha,r).$ BF(297.76)

51. $\displaystyle\int\frac{\sqrt{b\operatorname{ch} x-a}}{\operatorname{ch} x-1}\,dx=\sqrt{a+b}\,[F(\alpha,r)-E(\alpha,r)].$ BF(297.77)

52. $\displaystyle\int\frac{dx}{(\operatorname{ch} x-1)\sqrt{b\operatorname{ch} x-a}}=\frac{\sqrt{a+b}}{a-b}\,E(\alpha,r)-\frac{1}{\sqrt{a+b}}\,F(\alpha,r).$ BF(297.78)

53. $\displaystyle\int\frac{dx}{(\operatorname{ch} x-1)^2\sqrt{b\operatorname{ch} x-a}}=\frac{1}{3(a-b)^2\sqrt{a+b}}\,[(a-2b)(a-b)F(\alpha,r)+$
$+(3a-b)(a+b)E(\alpha,r)]+\frac{a+b}{6b(a-b)}\cdot\frac{\operatorname{ch}\frac{x}{2}}{\operatorname{sh}^3\frac{x}{2}}\sqrt{b\operatorname{ch} x-a}.$ BF(297.78)

54. $\int \dfrac{dx}{(\operatorname{ch} x + 1)\sqrt{b\operatorname{ch} x - a}} = \dfrac{1}{\sqrt{a+b}}[F(\alpha, r) - E(\alpha, r)] + \dfrac{2\sqrt{b\operatorname{ch} x - a}}{(a+b)\operatorname{sh} x}.$ 2.464

BF(297.80)

55. $\int \dfrac{dx}{(\operatorname{ch} x + 1)^2 \sqrt{b\operatorname{ch} x - a}} = \dfrac{1}{3\sqrt{(a+b)^3}}\Big[(a + 2b)\, F(\alpha, r) -$

$- (a + 3b)\, E(\alpha, r)\Big] + \dfrac{\sqrt{b\operatorname{ch} x - a}}{3(a+b)\operatorname{sh} x}\left(2\dfrac{a+3b}{a+b} - \operatorname{th}^2 \dfrac{x}{2}\right).$ BF(297.80)

In 2.464 56.–2.464 60. ist | In 2.464 56.–2.464 60. we put

$$\alpha = \arccos \dfrac{\sqrt[4]{b^2 - a^2}}{\sqrt{a\operatorname{sh} x + b\operatorname{ch} x}},\quad r = \dfrac{1}{\sqrt{2}}\ \left[0 < a < b,\ -\operatorname{Arsh} \dfrac{a}{\sqrt{b^2 - a^2}} < x\right]:$$

56. $\int \dfrac{dx}{\sqrt{a\operatorname{sh} x + b\operatorname{ch} x}} = \sqrt[4]{\dfrac{4}{b^2 - a^2}}\, F(\alpha, r).$ BF(299.00)

57. $\int \sqrt{a\operatorname{sh} x + b\operatorname{ch} x}\, dx = \sqrt[4]{4(b^2 - a^2)}\, [F(\alpha, r) - 2E(\alpha, r)] +$

$+ \dfrac{2(a\operatorname{ch} x + b\operatorname{sh} x)}{\sqrt{a\operatorname{sh} x + b\operatorname{ch} x}}.$ BF(299.02)

58. $\int \dfrac{dx}{\sqrt{(a\operatorname{sh} x + b\operatorname{ch} x)^3}} = \sqrt[4]{\dfrac{4}{(b^2 - a^2)^3}}\, [2E(\alpha, r) - F(\alpha, r)].$ BF(299.03)

59. $\int \dfrac{dx}{\sqrt{(a\operatorname{sh} x + b\operatorname{ch} x)^5}} = \dfrac{1}{3}\sqrt[4]{\dfrac{4}{(b^2 - a^2)^5}}\, F(\alpha, r) +$

$+ \dfrac{2}{3(b^2 - a^2)} \cdot \dfrac{a\operatorname{ch} x + b\operatorname{sh} x}{\sqrt{(a\operatorname{sh} x + b\operatorname{ch} x)^3}}.$ BF(299.03)

60. $\int \dfrac{(\sqrt{b^2 - a^2} + a\operatorname{sh} x + b\operatorname{ch} x)\, dx}{\sqrt{(a\operatorname{sh} x + b\operatorname{ch} x)^3}} = 2\sqrt[4]{\dfrac{4}{b^2 - a^2}}\, E(\alpha, r).$ BF(299.01)

Hyperbelfunktionen und Potenzen
Combinations of hyperbolic functions and powers 2.47

1. $\int x^r \operatorname{sh}^p x \operatorname{ch}^q x\, dx = \dfrac{1}{(p+q)^2}\Big[(p+q)\, x^r \operatorname{sh}^{p+1} x \operatorname{ch}^{q-1} x -$ 2.471

$- rx^{r-1} \operatorname{sh}^p x \operatorname{ch}^q x + r(r+1)\int x^{r-2} \operatorname{sh}^p x \operatorname{ch}^q x\, dx +$

$+ rp \int x^{r-1} \operatorname{sh}^{p-1} x \operatorname{ch}^{q-1} x\, dx + (q-1)(p+q)\int x^r \operatorname{sh}^p x \operatorname{ch}^{q-2} x\, dx\Big];$

$= \dfrac{1}{(p+q)^2}\Big[(p+q)\, x^r \operatorname{sh}^{p-1} x \operatorname{ch}^{q+1} x -$

$- rx^{r-1} \operatorname{sh}^p x \operatorname{ch}^q x + r(r-1)\int x^{r-2} \operatorname{sh}^p x \operatorname{ch}^q x\, dx -$

$- rq \int x^{r-1} \operatorname{sh}^{p-1} x \operatorname{ch}^{q-1} x\, dx - (p-1)(p+q)\int x^r \operatorname{sh}^{p-2} x \operatorname{ch}^q x\, dx\Big].$

GHH[353](1)

2.471

2. $\int x^n \operatorname{sh}^{2m} x \, dx = (-1)^m \binom{2m}{m} \frac{x^{n+1}}{2^{2m}(n+1)} +$
$$+ \frac{1}{2^{2m-1}} \sum_{k=0}^{m-1} (-1)^k \binom{2m}{k} \int x^n \operatorname{ch}(2m - 2k) x \, dx.$$

3. $\int x^n \operatorname{sh}^{2m+1} x \, dx = \frac{1}{2^{2m}} \sum_{k=0}^{m} (-1)^k \binom{2m+1}{k} \int x^n \operatorname{sh}(2m - 2k + 1) x \, dx.$

4. $\int x^n \operatorname{ch}^{2m} x \, dx = \binom{2m}{m} \frac{x^{n+1}}{2^{2m}(n+1)} +$
$$+ \frac{1}{2^{2m-1}} \sum_{k=0}^{m-1} \binom{2m}{k} \int x^n \operatorname{ch}(2m - 2k) x \, dx.$$

5. $\int x^n \operatorname{ch}^{2m+1} x \, dx = \frac{1}{2^{2m}} \sum_{k=0}^{m} \binom{2m+1}{k} \int x^n \operatorname{ch}(2m - 2k + 1) x \, dx.$

2.472

1. $\int x^n \operatorname{sh} x \, dx = x^n \operatorname{ch} x - n \int x^{n-1} \operatorname{ch} x \, dx =$
$$= x^n \operatorname{ch} x - n x^{n-1} \operatorname{sh} x + n(n-1) \int x^{n-2} \operatorname{sh} x \, dx.$$

2. $\int x^n \operatorname{ch} x \, dx = x^n \operatorname{sh} x - n \int x^{n-1} \operatorname{sh} x \, dx =$
$$= x^n \operatorname{sh} x - n x^{n-1} \operatorname{ch} x + n(n-1) \int x^{n-2} \operatorname{ch} x \, dx.$$

3. $\int x^{2n} \operatorname{sh} x \, dx = (2n)! \left\{ \sum_{k=0}^{n} \frac{x^{2k}}{(2k)!} \operatorname{ch} x - \sum_{k=1}^{n} \frac{x^{2k-1}}{(2k-1)!} \operatorname{sh} x \right\}.$

4. $\int x^{2n+1} \operatorname{sh} x \, dx = (2n+1)! \sum_{k=0}^{n} \left\{ \frac{x^{2k+1}}{(2k+1)!} \operatorname{ch} x - \frac{x^{2k}}{(2k)!} \operatorname{sh} x \right\}.$

5. $\int x^{2n} \operatorname{ch} x \, dx = (2n)! \left\{ \sum_{k=1}^{n} \frac{x^{2k}}{(2k)!} \operatorname{sh} x - \sum_{k=1}^{n} \frac{x^{2k-1}}{(2k-1)!} \operatorname{ch} x \right\}.$

6. $\int x^{2n+1} \operatorname{ch} x \, dx = (2n+1)! \sum_{k=0}^{n} \left\{ \frac{x^{2k+1}}{(2k+1)!} \operatorname{sh} x - \frac{x^{2k}}{(2k)!} \operatorname{ch} x \right\}.$

7. $\int x \operatorname{sh} x \, dx = x \operatorname{ch} x - \operatorname{sh} x.$

8. $\int x^2 \operatorname{sh} x \, dx = (x^2 + 2) \operatorname{ch} x - 2x \operatorname{sh} x.$

9. $\int x \operatorname{ch} x \, dx = x \operatorname{sh} x - \operatorname{ch} x.$

10. $\int x^2 \operatorname{ch} x \, dx = (x^2 + 2) \operatorname{sh} x - 2x \operatorname{ch} x.$

2.473 Abkürzung:
Abbreviation: $z_1 = a + bx.$

1. $\int z_1 \operatorname{sh} kx \, dx = \frac{1}{k} z_1 \operatorname{ch} kx - \frac{b}{k^2} \operatorname{sh} kx.$

2. $\quad \int z_1 \operatorname{ch} kx \, dx = \frac{1}{k} z_1 \operatorname{sh} kx - \frac{b}{k^2} \operatorname{ch} kx.$ 2.473

3. $\quad \int z_1^2 \operatorname{sh} kx \, dx = \frac{1}{k} \left(z_1^2 + \frac{2b^2}{k^2} \right) \operatorname{ch} kx - \frac{2bz_1}{k^2} \operatorname{sh} kx.$

4. $\quad \int z_1^2 \operatorname{ch} kx \, dx = \frac{1}{k} \left(z_1^2 + \frac{2b^2}{k^2} \right) \operatorname{sh} kx - \frac{2bz_1}{k^2} \operatorname{ch} kx.$

5. $\quad \int z_1^3 \operatorname{sh} kx \, dx = \frac{z_1}{k} \left(z_1^2 + \frac{6b^2}{k^2} \right) \operatorname{ch} kx - \frac{3b}{k^2} \left(z_1^2 + \frac{2b^2}{k^2} \right) \operatorname{sh} kx.$

6. $\quad \int z_1^3 \operatorname{ch} kx \, dx = \frac{z_1}{k} \left(z_1^2 + \frac{6b^2}{k^2} \right) \operatorname{sh} kx - \frac{3b}{k^2} \left(z_1^2 + \frac{2b^2}{k^2} \right) \operatorname{ch} kx.$

7. $\quad \int z_1^4 \operatorname{sh} kx \, dx = \frac{1}{k} \left(z_1^4 + \frac{12b^2}{k^2} z_1^2 + \frac{24b^4}{k^4} \right) \operatorname{ch} kx - \frac{4bz_1}{k^2} \left(z_1^2 + \frac{6b^2}{k^2} \right) \operatorname{sh} kx.$

8. $\quad \int z_1^4 \operatorname{ch} kx \, dx = \frac{1}{k} \left(z_1^4 + \frac{12b^2}{k^2} z_1^2 + \frac{24b^4}{k^4} \right) \operatorname{sh} kx - \frac{4bz_1}{k^2} \left(z_1^2 + \frac{6b^2}{k^2} \right) \operatorname{ch} kx.$

9. $\quad \int z_1^5 \operatorname{sh} kx \, dx = \frac{z_1}{k} \left(z_1^4 + \frac{20b^2}{k^2} z_1^2 + 120 \frac{b^4}{k^4} \right) \operatorname{ch} kx -$
$$- \frac{5b}{k^2} \left(z_1^4 + 12 \frac{b^2}{k^2} z_1^2 + 24 \frac{b^4}{k^4} \right) \operatorname{sh} kx.$$

10. $\quad \int z_1^5 \operatorname{ch} kx \, dx = \frac{z_1}{k} \left(z_1^4 + 20 \frac{b^2}{k^2} z_1^2 + 120 \frac{b^4}{k^4} \right) \operatorname{sh} kx -$
$$- \frac{5b}{k^2} \left(z_1^4 + 12 \frac{b^2}{k^2} z_1^2 + 24 \frac{b^4}{k^4} \right) \operatorname{ch} kx.$$

11. $\quad \int z_1^6 \operatorname{sh} kx \, dx = \frac{1}{k} \left(z_1^6 + 30 \frac{b^2}{k^2} z_1^4 + 360 \frac{b^4}{k^4} z_1^2 + 720 \frac{b^6}{k^6} \right) \operatorname{ch} kx -$
$$- \frac{6bz_1}{k^2} \left(z_1^4 + 20 \frac{b^2}{k^2} z_1^2 + 120 \frac{b^4}{k^4} \right) \operatorname{sh} kx.$$

12. $\quad \int z_1^6 \operatorname{ch} kx \, dx = \frac{1}{k} \left(z_1^6 + 30 \frac{b^2}{k^2} z_1^4 + 360 \frac{b^4}{k^4} z_1^2 + 720 \frac{b^6}{k^6} \right) \operatorname{sh} kx -$
$$- \frac{6bz_1}{k^2} \left(z_1^4 + 20 \frac{b^2}{k^2} z_1^2 + 120 \frac{b^4}{k^4} \right) \operatorname{ch} kx.$$

1. $\quad \int x^n \operatorname{sh}^2 x \, dx = -\frac{x^{n+1}}{2(n+1)} +$ 2.474
$$+ \frac{n!}{4} \sum_{k=0}^{E\left(\frac{n}{2}\right)} \left\{ \frac{x^{n-2k}}{2^{2k} (n-2k)!} \operatorname{sh} 2x - \frac{x^{n-2k-1}}{2^{2k+1} (n-2k-1)!} \operatorname{ch} 2x \right\}. \quad \text{GHH[353](2b)}$$

2. $\quad \int x^n \operatorname{ch}^2 x \, dx = \frac{x^{n+1}}{2(n+1)} +$
$$+ \frac{n!}{4} \sum_{k=0}^{E\left(\frac{n}{2}\right)} \left\{ \frac{x^{n-2k}}{2^{2k} (n-2k)!} \operatorname{sh} 2x - \frac{x^{n-2k-1}}{2^{2k+1} (n-2k-1)!} \operatorname{ch} 2x \right\}. \quad \text{GHH[353](3e)}$$

3. $\quad \int x \operatorname{sh}^2 x \, dx = \frac{1}{4} x \operatorname{sh} 2x - \frac{1}{8} \operatorname{ch} 2x - \frac{x^2}{4}.$

2.474

4. $\int x^2 \operatorname{sh}^2 x \, dx = \frac{1}{4}\left(x^2 + \frac{1}{2}\right) \operatorname{sh} 2x - \frac{x}{4} \operatorname{ch} 2x - \frac{x^3}{6}$. MC 257

5. $\int x \operatorname{ch}^2 x \, dx = \frac{x}{4} \operatorname{sh} 2x - \frac{1}{8} \operatorname{ch} 2x + \frac{x^2}{4}$.

6. $\int x^2 \operatorname{ch}^2 x \, dx = \frac{1}{4}\left(x^2 + \frac{1}{2}\right) \operatorname{sh} 2x - \frac{x}{4} \operatorname{ch} 2x + \frac{x^3}{6}$. MC 261

7. $\int x^n \operatorname{sh}^3 x \, dx = \frac{n!}{4} \sum_{k=0}^{E\left(\frac{n}{2}\right)} \left\{ \frac{x^{n-2k}}{(n-2k)!}\left(\frac{\operatorname{ch} 3x}{3^{2k+1}} - 3 \operatorname{ch} x\right) - \right.$
$\left. - \frac{x^{n-2k-1}}{(n-2k-1)!}\left(\frac{\operatorname{sh} 3x}{3^{2k+2}} - 3 \operatorname{sh} x\right)\right\}$. GHH[353](2f)

8. $\int x^n \operatorname{ch}^3 x \, dx = \frac{n!}{4} \sum_{k=0}^{E\left(\frac{n}{2}\right)} \left\{ \frac{x^{n-2k}}{(n-2k)!}\left(\frac{\operatorname{sh} 3x}{3^{2k+1}} + 3 \operatorname{sh} x\right) - \right.$
$\left. - \frac{x^{n-2k-1}}{(n-2k-1)!}\left(\frac{\operatorname{ch} 3x}{3^{2k+2}} + 3 \operatorname{ch} x\right)\right\}$. GHH[353](3f)

9. $\int x \operatorname{sh}^3 x \, dx = \frac{3}{4} \operatorname{sh} x - \frac{1}{36} \operatorname{sh} 3x - \frac{3}{4} x \operatorname{ch} x - \frac{x}{12} \operatorname{ch} 3x$.

10. $\int x^2 \operatorname{sh}^3 x \, dx = -\left(\frac{3x^2}{4} + \frac{3}{2}\right) \operatorname{ch} x + \left(\frac{x^2}{12} + \frac{1}{54}\right) \operatorname{ch} 3x +$
$+ \frac{3x}{2} \operatorname{sh} x - \frac{x}{18} \operatorname{sh} 3x$. MC 257

11. $\int x \operatorname{ch}^3 x \, dx = -\frac{3}{4} \operatorname{ch} x - \frac{1}{36} \operatorname{ch} 3x + \frac{3}{4} x \operatorname{sh} x + \frac{x}{12} \operatorname{sh} 3x$.

12. $\int x^2 \operatorname{ch}^3 x \, dx = \left(\frac{3}{4} x^2 + \frac{3}{2}\right) \operatorname{sh} x + \left(\frac{x^2}{12} + \frac{1}{54}\right) \operatorname{sh} 3x -$
$- \frac{3}{2} x \operatorname{ch} x - \frac{x}{18} \operatorname{ch} 3x$. MC 262

2.475

1. $\int \frac{\operatorname{sh}^q x}{x^p} dx = -\frac{(p-2) \operatorname{sh}^q x + qx \operatorname{sh}^{q-1} x \operatorname{ch} x}{(p-1)(p-2) x^{p-1}} +$
$+ \frac{q(q-1)}{(p-1)(p-2)} \int \frac{\operatorname{sh}^{q-2} x}{x^{p-2}} dx + \frac{q^2}{(p-1)(p-2)} \int \frac{\operatorname{sh}^q x}{x^{p-2}} dx \quad [p > 2]$. GHH[353](6a)

2. $\int \frac{\operatorname{ch}^q x}{x^p} dx = -\frac{(p-2) \operatorname{ch}^q x + qx \operatorname{ch}^{q-1} x \operatorname{sh} x}{(p-1)(p-2) x^{p-1}} -$
$- \frac{q(q-1)}{(p-1)(p-2)} \int \frac{\operatorname{ch}^{q-2} x}{x^{p-2}} dx + \frac{q^2}{(p-1)(p-2)} \int \frac{\operatorname{ch}^q x}{x^{p-2}} dx \quad [p > 2]$. GHH[353](7a)

3. $\int \frac{\operatorname{sh} x}{x^{2n}} dx = -\frac{1}{x(2n-1)!}\left\{\sum_{k=0}^{n-2} \frac{(2k+1)!}{x^{2k+1}} \operatorname{ch} x + \right.$
$\left. + \sum_{k=0}^{n-1} \frac{(2k)!}{x^{2k}} \operatorname{sh} x\right\} + \frac{1}{(2n-1)!} \operatorname{chi}(x)$. GHH[353](6b)

4. $\int \dfrac{\operatorname{sh} x}{x^{2n+1}} dx = -\dfrac{1}{x(2n)!} \left\{ \sum_{k=0}^{n-1} \dfrac{(2k)!}{x^{2k}} \operatorname{ch} x + \right.$

$\left. + \sum_{k=0}^{n-1} \dfrac{(2k+1)!}{x^{2k+1}} \operatorname{sh} x \right\} + \dfrac{1}{(2n)!} \operatorname{shi}(x).$ GHH[353](6b) 2.475

5. $\int \dfrac{\operatorname{ch} x}{x^{2n}} dx = -\dfrac{1}{x(2n-1)!} \left\{ \sum_{k=0}^{n-2} \dfrac{(2k+1)!}{x^{2k+1}} \operatorname{sh} x + \right.$

$\left. + \sum_{k=0}^{n-1} \dfrac{(2k)!}{x^{2k}} \operatorname{ch} x \right\} + \dfrac{1}{(2n-1)!} \operatorname{shi}(x).$ GHH[353](7b)

6. $\int \dfrac{\operatorname{ch} x}{x^{2n+1}} dx = -\dfrac{1}{(2n)!x} \left\{ \sum_{k=0}^{n-1} \dfrac{(2k)!}{x^{2k}} \operatorname{sh} x + \right.$

$\left. + \sum_{k=0}^{n-1} \dfrac{(2k+1)!}{x^{2k+1}} \operatorname{ch} x \right\} + \dfrac{1}{(2n)!} \operatorname{chi}(x).$ GHH[353](7b)

7. $\int \dfrac{\operatorname{sh}^{2m} x}{x} dx = \dfrac{1}{2^{2m-1}} \sum_{k=0}^{m-1} (-1)^k \binom{2m}{k} \operatorname{chi}(2m-2k)x +$

$+ \dfrac{(-1)^m}{2^{2m}} \binom{2m}{m} \ln x.$ GHH[353](6c)

8. $\int \dfrac{\operatorname{sh}^{2m+1} x}{x} dx = \dfrac{1}{2^{2m}} \sum_{k=0}^{m} (-1)^k \binom{2m+1}{k} \operatorname{shi}(2m-2k+1)x.$ GHH[353](6d)

9. $\int \dfrac{\operatorname{ch}^{2m} x}{x} dx = \dfrac{1}{2^{2m-1}} \sum_{k=0}^{m-1} \binom{2m}{k} \operatorname{chi}(2m-2k)x + \dfrac{1}{2^{2m}} \binom{2m}{m} \ln x.$ GHH[353](7c)

10. $\int \dfrac{\operatorname{ch}^{2m+1} x}{x} dx = \dfrac{1}{2^{2m}} \sum_{k=0}^{m} \binom{2m+1}{k} \operatorname{chi}(2m-2k+1)x.$ GHH[353](7c)

11. $\int \dfrac{\operatorname{sh}^{2m} x}{x^2} dx = \dfrac{(-1)^{m-1}}{2^{2m} x} \binom{2m}{m} +$

$+ \dfrac{1}{2^{2m-1}} \sum_{k=0}^{m-1} (-1)^{k+1} \binom{2m}{k} \left\{ \dfrac{\operatorname{ch}(2m-2k)x}{x} - (2m-2k) \operatorname{shi}(2m-2k)x \right\}.$

12. $\int \dfrac{\operatorname{sh}^{2m+1} x}{x^2} dx = \dfrac{1}{2^{2m}} \sum_{k=0}^{m} (-1)^{k+1} \binom{2m+1}{k} \times$

$\times \left\{ \dfrac{\operatorname{sh}(2m-2k+1)x}{x} - (2m-2k+1) \operatorname{chi}(2m-2k+1)x \right\}.$

13. $\int \dfrac{\operatorname{ch}^{2m} x}{x^2} dx = -\dfrac{1}{2^{2m} x} \binom{2m}{m} -$

$- \dfrac{1}{2^{2m-1}} \sum_{k=0}^{m-1} \binom{2m}{k} \left\{ \dfrac{\operatorname{ch}(2m-2k)x}{x} - (2m-2k) \operatorname{shi}(2m-2k)x \right\}.$

14. $\int \dfrac{\operatorname{ch}^{2m+1} x}{x^2} dx = -\dfrac{1}{2^{2m}} \sum_{k=0}^{m} \binom{2m+1}{k} \times$

$\times \left\{ \dfrac{\operatorname{ch}(2m-2k+1)x}{x} - (2m-2k+1) \operatorname{shi}(2m-2k+1)x \right\}.$

2.476

1. $\int \frac{\operatorname{sh} kx}{a+bx} dx = \frac{1}{b}\left[\operatorname{ch}\frac{ka}{b}\operatorname{shi}(u) - \operatorname{sh}\frac{ka}{b}\operatorname{chi}(u)\right];$

$= \frac{1}{2b}\left[\exp\left(-\frac{ka}{b}\right)\operatorname{Ei}(u) - \exp\left(\frac{ka}{b}\right)\operatorname{Ei}(-u)\right]$

$\left[u = \frac{k}{b}(a+bx)\right].$

2. $\int \frac{\operatorname{ch} kx}{a+bx} dx = \frac{1}{b}\left[\operatorname{ch}\frac{ka}{b}\operatorname{chi}(u) - \operatorname{sh}\frac{ka}{b}\operatorname{shi}(u)\right];$

$= \frac{1}{2b}\left[\exp\left(-\frac{ka}{b}\right)\operatorname{Ei}(u) + \exp\left(\frac{ka}{b}\right)\operatorname{Ei}(-u)\right]$

$\left[u = \frac{k}{b}(a+bx)\right].$

3. $\int \frac{\operatorname{sh} kx}{(a+bx)^2} dx = -\frac{1}{b}\cdot\frac{\operatorname{sh} kx}{a+bx} + \frac{k}{b}\int \frac{\operatorname{ch} kx}{a+bx} dx \quad \text{(cf. 2.476 2.)}.$

4. $\int \frac{\operatorname{ch} kx}{(a+bx)^2} dx = -\frac{1}{b}\cdot\frac{\operatorname{ch} kx}{a+bx} + \frac{k}{b}\int \frac{\operatorname{sh} kx}{a+bx} dx \quad \text{(cf. 2.476 1.)}.$

5. $\int \frac{\operatorname{sh} kx}{(a+bx)^3} dx = -\frac{\operatorname{sh} kx}{2b(a+bx)^2} - \frac{k\operatorname{ch} kx}{2b^2(a+bx)} +$

$+ \frac{k^2}{2b^2}\int \frac{\operatorname{sh} kx}{a+bx} dx \quad \text{(cf. 2.476 1.)}.$

6. $\int \frac{\operatorname{ch} kx}{(a+bx)^3} dx = -\frac{\operatorname{ch} kx}{2b(a+bx)^2} - \frac{k\operatorname{sh} kx}{2b^2(a+bx)} +$

$+ \frac{k^2}{2b^2}\int \frac{\operatorname{ch} kx}{a+bx} dx \quad \text{(cf. 2.476 2.)}.$

7. $\int \frac{\operatorname{sh} kx}{(a+bx)^4} dx = -\frac{\operatorname{sh} kx}{3b(a+bx)^3} - \frac{k\operatorname{ch} kx}{6b^2(a+bx)^2} - \frac{k^2\operatorname{sh} kx}{6b^3(a+bx)} +$

$+ \frac{k^3}{6b^3}\int \frac{\operatorname{ch} kx}{a+bx} dx \quad \text{(cf. 2.476 2.)}.$

8. $\int \frac{\operatorname{ch} kx}{(a+bx)^4} dx = -\frac{\operatorname{ch} kx}{3b(a+bx)^3} - \frac{k\operatorname{sh} kx}{6b^2(a+bx)^2} - \frac{k^2\operatorname{ch} kx}{6b^3(a+bx)} +$

$+ \frac{k^3}{6b^3}\int \frac{\operatorname{sh} kx}{a+bx} dx \quad \text{(cf. 2.476 1.)}.$

9. $\int \frac{\operatorname{sh} kx}{(a+bx)^5} dx = -\frac{\operatorname{sh} kx}{4b(a+bx)^4} - \frac{k\operatorname{ch} kx}{12b^2(a+bx)^3} -$

$- \frac{k^2\operatorname{sh} kx}{24b^3(a+bx)^2} - \frac{k^3\operatorname{ch} kx}{24b^4(a+bx)} + \frac{k^4}{24b^4}\int \frac{\operatorname{sh} kx}{a+bx} dx \quad \text{(cf. 2.476 1.)}.$

10. $\int \frac{\operatorname{ch} kx}{(a+bx)^5} dx = -\frac{\operatorname{ch} kx}{4b(a+bx)^4} - \frac{k\operatorname{sh} kx}{12b^2(a+bx)^3} -$

$- \frac{k^2\operatorname{ch} kx}{24b^3(a+bx)^2} - \frac{k^3\operatorname{sh} kx}{24b^4(a+bx)} + \frac{k^4}{24b^4}\int \frac{\operatorname{ch} kx}{a+bx} dx \quad \text{(cf. 2.476 2.)}.$

11. $\int \frac{\operatorname{sh} kx}{(a+bx)^6} dx = -\frac{\operatorname{sh} kx}{5b(a+bx)^5} - \frac{k\operatorname{ch} kx}{20b^2(a+bx)^4} -$

$- \frac{k^2\operatorname{sh} kx}{60b^3(a+bx)^3} - \frac{k^3\operatorname{ch} kx}{120b^4(a+bx)^2} - \frac{k^4\operatorname{sh} kx}{120b^5(a+bx)} + \frac{k^5}{120b^5}\int \frac{\operatorname{ch} kx}{a+bx} dx \quad \text{(cf. 2.476 2.)}.$

12. $\int \frac{\operatorname{ch} kx}{(a+bx)^6} dx = -\frac{\operatorname{ch} kx}{5b(a+bx)^5} - \frac{k \operatorname{sh} kx}{20b^2(a+bx)^4} -$ 2.476

$- \frac{k^2 \operatorname{ch} kx}{60 b^3 (a+bx)^3} - \frac{k^3 \operatorname{sh} kx}{120 b^4 (a+bx)^2} - \frac{k^4 \operatorname{ch} kx}{120 b^5 (a+bx)} + \frac{k^5}{120 b^5} \int \frac{\operatorname{sh} kx}{a+bx} dx$ (cf. 2.476 1.).

1. $\int \frac{x^p dx}{\operatorname{sh}^q x} = \frac{-px^{p-1} \operatorname{sh} x - (q-2) x^p \operatorname{ch} x}{(q-1)(q-2) \operatorname{sh}^{q-1} x} +$ 2.477

$+ \frac{p(p-1)}{(q-1)(q-2)} \int \frac{x^{p-2}}{\operatorname{sh}^{q-2} x} dx - \frac{q-2}{q-1} \int \frac{x^p dx}{\operatorname{sh}^{q-2} x}$ $[q > 2]$. GHH[353](8a)

2. $\int \frac{x^p dx}{\operatorname{ch}^q x} = \frac{px^{p-1} \operatorname{ch} x + (q-2) x^p \operatorname{sh} x}{(q-1)(q-2) \operatorname{ch}^{q-1} x} -$

$- \frac{p(p-1)}{(q-1)(q-2)} \int \frac{x^{p-2} dx}{\operatorname{ch}^{q-2} x} + \frac{q-2}{q-1} \int \frac{x^p dx}{\operatorname{ch}^{q-2} x}$ $[q > 2]$. GHH[353](10a)

3. $\int \frac{x^n}{\operatorname{sh} x} dx = \sum_{k=0}^{\infty} \frac{(2-2^{2k}) B_{2k}}{(n+2k)(2k)!} x^{n+2k}$ $[|x| < \pi, n > 0]$. GHH[353](8b)

4. $\int \frac{x^n}{\operatorname{ch} x} dx = \sum_{k=0}^{\infty} \frac{E_{2k} x^{n+2k+1}}{(n+2k+1)(2k)!}$ $\left[|x| < \frac{\pi}{2}, n \geq 0\right]$. GHH[353](10b)

5. $\int \frac{dx}{x^n \operatorname{sh} x} = -[1+(-1)^n] \frac{2^{n-1}-1}{n!} B_n \ln x +$

$+ \sum_{\substack{k=0 \\ k \neq \frac{n}{2}}}^{\infty} \frac{2-2^{2k}}{(2k-n)(2k)!} B_{2k} x^{2k-n}$ $[|x| < \pi, n \geq 1]$. GHH[353](9b)

6. $\int \frac{dx}{x^n \operatorname{ch} x} = \sum_{\substack{k=0 \\ k \neq \frac{n-1}{2}}}^{\infty} \frac{E_{2k}}{(2k-n+1)(2k)!} x^{2k-n+1} +$

$+ \frac{1}{2}[1-(-1)^{n-1}] + \frac{E_{n-1}}{(n-1)!} \ln x$ $\left[|x| < \frac{\pi}{2}\right]$. GHH[353](11b)

7. $\int \frac{x^n}{\operatorname{sh}^2 x} dx = -x^n \operatorname{cth} x + n \sum_{k=0}^{\infty} \frac{2^{2k} B_{2k}}{(n+2k-1)(2k)!} x^{n+2k-1}$ $[n>1, |x| < \pi]$. GHH[353](8c)

8. $\int \frac{x^n}{\operatorname{ch}^2 x} dx = x^n \operatorname{th} x - n \sum_{k=1}^{\infty} \frac{2^{2k}(2^{2k}-1) B_{2k}}{(n+2k-1)(2k)!} x^{n+2k-1}$ $\left[n>1, |x| < \frac{\pi}{2}\right]$. GHH[353](10c)

9. $\int \frac{dx}{x^n \operatorname{sh}^2 x} = -\frac{\operatorname{cth} x}{x^n} - [1-(-1)^n] \frac{2^n n}{(n+1)!} B_{n+1} \ln x -$

$- \frac{n}{x^{n+1}} \sum_{\substack{k=0 \\ k \neq \frac{n+1}{2}}}^{\infty} \frac{B_{2k}}{(2k-n-1)(2k)!} (2x)^{2k}$ $[|x| < \pi]$. GHH[353](9c)

10. $\int \frac{dx}{x^n \operatorname{ch}^2 x} = \frac{\operatorname{th} x}{x^n} + [1-(-1)^n] \frac{2^n (2^{n+1}-1) n}{(n+1)!} B_{n+1} \ln x +$

$+ \frac{n}{x^{n+1}} \sum_{\substack{k=1 \\ k \neq \frac{n+1}{2}}}^{\infty} \frac{(2^{2k}-1) B_{2k}}{(2k-n-1)(2k)!} (2x)^{2k}$ $\left[|x| < \frac{\pi}{2}\right]$. GHH[353](11c)

2.477

11. $\int \dfrac{x}{\operatorname{sh}^{2n} x} dx = \sum_{k=1}^{n-1} (-1)^k \dfrac{(2n-2)(2n-4)\ldots(2n-2k+2)}{(2n-1)(2n-3)\ldots(2n-2k+1)} \times$

$\times \left\{ \dfrac{x \operatorname{ch} x}{\operatorname{sh}^{2n-2k+1} x} + \dfrac{1}{(2n-2k)\operatorname{sh}^{2n-2k} x} \right\} + (-1)^{n-1} \dfrac{(2n-2)!!}{(2n-1)!!} \int \dfrac{x\, dx}{\operatorname{sh}^2 x}$

(cf. 2.477 17.). GHH[353](8e)

12. $\int \dfrac{x}{\operatorname{sh}^{2n-1} x} dx = \sum_{k=1}^{n-1} (-1)^k \dfrac{(2n-3)(2n-5)\ldots(2n-2k+1)}{(2n-2)(2n-4)\ldots(2n-2k)} \times$

$\times \left\{ \dfrac{x \operatorname{ch} x}{\operatorname{sh}^{2n-2k} x} + \dfrac{1}{(2n-2k-1)\operatorname{sh}^{2n-2k-1} x} \right\} + (-1)^{n-1} \dfrac{(2n-3)!!}{(2n-2)!!} \int \dfrac{x\, dx}{\operatorname{sh} x}$

(cf. 2.477 15.). GHH[353](8e

13. $\int \dfrac{x}{\operatorname{ch}^{2n} x} dx = \sum_{k=1}^{n-1} \dfrac{(2n-2)(2n-4)\ldots(2n-2k+2)}{(2n-1)(2n-3)\ldots(2n-2k+1)} \times$

$\times \left\{ \dfrac{x \operatorname{sh} x}{\operatorname{ch}^{2n-2k+1} x} + \dfrac{1}{(2n-2k)\operatorname{ch}^{2n-2k} x} \right\} + \dfrac{(2n-2)!!}{(2n-1)!!} \int \dfrac{x\, dx}{\operatorname{ch}^2 x}$

(cf. 2.477 18.). GHH[353](10e)

14. $\int \dfrac{x}{\operatorname{ch}^{2n-1} x} dx = \sum_{k=1}^{n-1} \dfrac{(2n-3)(2n-5)\ldots(2n-2k+1)}{(2n-2)(2n-4)\ldots(2n-2k)} \times$

$\times \left\{ \dfrac{x \operatorname{sh} x}{\operatorname{ch}^{2n-2k} x} + \dfrac{1}{(2n-2k-1)\operatorname{ch}^{2n-2k-1} x} \right\} + \dfrac{(2n-3)!!}{(2n-2)!!} \int \dfrac{x\, dx}{\operatorname{ch} x}$

(cf. 2.477 16.). GHH[353](10e)

15. $\int \dfrac{x\, dx}{\operatorname{sh} x} = \sum_{k=0}^{\infty} \dfrac{2 - 2^{2k}}{(2k+1)(2k)!} B_{2k} x^{2k+1} \quad |x| < \pi.$ GHH[353](8b)+

16. $\int \dfrac{x\, dx}{\operatorname{ch} x} = \sum_{k=0}^{\infty} \dfrac{E_{2k} x^{2k+2}}{(2k+2)(2k)!} \quad |x| < \dfrac{\pi}{2}.$ GHH[353](10b)+

17. $\int \dfrac{x\, dx}{\operatorname{sh}^2 x} = -x \operatorname{cth} x + \ln \operatorname{sh} x.$ MC 257

18. $\int \dfrac{x\, dx}{\operatorname{ch}^2 x} = x \operatorname{th} x - \ln \operatorname{ch} x.$ MC 262

19. $\int \dfrac{x\, dx}{\operatorname{sh}^3 x} = -\dfrac{x \operatorname{ch} x}{2 \operatorname{sh}^2 x} - \dfrac{1}{2 \operatorname{sh} x} - \dfrac{1}{2} \int \dfrac{x\, dx}{\operatorname{sh} x}$ (cf. 2.477 15.). MC 257

20. $\int \dfrac{x\, dx}{\operatorname{ch}^3 x} = \dfrac{x \operatorname{sh} x}{2 \operatorname{ch}^2 x} + \dfrac{1}{2 \operatorname{ch} x} + \dfrac{1}{2} \int \dfrac{x\, dx}{\operatorname{ch} x}$ (cf. 2.477 16.). MC 262

21. $\int \dfrac{x\, dx}{\operatorname{sh}^4 x} = -\dfrac{x \operatorname{ch} x}{3 \operatorname{sh}^3 x} - \dfrac{1}{6 \operatorname{sh}^2 x} + \dfrac{2}{3} x \operatorname{cth} x - \dfrac{2}{3} \ln \operatorname{sh} x.$ MC 258

22. $\int \dfrac{x\, dx}{\operatorname{ch}^4 x} = \dfrac{x \operatorname{sh} x}{3 \operatorname{ch}^3 x} + \dfrac{1}{6 \operatorname{ch}^2 x} + \dfrac{2}{3} x \operatorname{th} x - \dfrac{2}{3} \ln \operatorname{ch} x.$ MC 262

23. $\int \dfrac{x\, dx}{\operatorname{sh}^5 x} = -\dfrac{x \operatorname{ch} x}{4 \operatorname{sh}^4 x} - \dfrac{1}{12 \operatorname{sh}^3 x} + \dfrac{3x \operatorname{ch} x}{8 \operatorname{sh}^2 x} + \dfrac{3}{8 \operatorname{sh} x} + \dfrac{3}{8} \int \dfrac{x\, dx}{\operatorname{sh} x}$ (cf. 2.477 15.).

MC 258

24. $\int \dfrac{x\, dx}{\operatorname{ch}^5 x} = \dfrac{x \operatorname{sh} x}{4 \operatorname{ch}^4 x} + \dfrac{1}{12 \operatorname{ch}^3 x} + \dfrac{3x \operatorname{sh} x}{8 \operatorname{ch}^2 x} + \dfrac{3}{8 \operatorname{ch} x} + \dfrac{3}{8} \int \dfrac{x\, dx}{\operatorname{ch} x}$ (cf. 2.477 16.). MC 262

2.478

1. $\int \dfrac{x^n \operatorname{ch} x\, dx}{(a+b \operatorname{sh} x)^m} = -\dfrac{x^n}{(m-1)\,b\,(a+b \operatorname{sh} x)^{m-1}} +$

$\qquad + \dfrac{n}{(m-1)\,b} \int \dfrac{x^{n-1}\,dx}{(a+b \operatorname{sh} x)^{m-1}} \quad [m \neq 1].$ MC 263+

2. $\int \dfrac{x^n \operatorname{sh} x\, dx}{(a+b \operatorname{ch} x)^m} = -\dfrac{x^n}{(m-1)\,b\,(a+b \operatorname{ch} x)^{m-1}} +$

$\qquad + \dfrac{n}{(m-1)\,b} \int \dfrac{x^{n-1}\,dx}{(a+b \operatorname{ch} x)^{m-1}} \quad [m \neq 1].$ MC 263

3. $\int \dfrac{x\,dx}{1+\operatorname{ch} x} = x \operatorname{th} \dfrac{x}{2} - 2 \ln \operatorname{ch} \dfrac{x}{2}.$

4. $\int \dfrac{x\,dx}{1-\operatorname{ch} x} = x \operatorname{cth} \dfrac{x}{2} - 2 \ln \operatorname{sh} \dfrac{x}{2}.$

5. $\int \dfrac{x \operatorname{sh} x\, dx}{(1+\operatorname{ch} x)^2} = -\dfrac{x}{1+\operatorname{ch} x} + \operatorname{th} \dfrac{x}{2}.$

6. $\int \dfrac{x \operatorname{sh} x\, dx}{(1-\operatorname{ch} x)^2} = \dfrac{x}{1-\operatorname{ch} x} - \operatorname{cth} \dfrac{x}{2}.$ MC 262–264

7. $\int \dfrac{x\,dx}{\operatorname{ch} 2x - \cos 2t} = \dfrac{1}{2 \sin 2t}[L(u+t) - L(u-t) - 2L(t)]$

$\qquad [u = \operatorname{arctg}(\operatorname{th} x \operatorname{ctg} t),\ t \neq \pm n\pi].$ Lo III 402

8. $\int \dfrac{x \operatorname{ch} x\,dx}{\operatorname{ch} 2x - \cos 2t} = \dfrac{1}{2 \sin t}\left[L\left(\dfrac{u+t}{2}\right) - L\left(\dfrac{u-t}{2}\right) +\right.$

$\qquad \left. + L\left(\pi - \dfrac{v+t}{2}\right) + L\left(\dfrac{v-t}{2}\right) - 2L\left(\dfrac{t}{2}\right) - 2L\left(\dfrac{\pi-t}{2}\right)\right]$

$\left[u = 2 \operatorname{arctg}\left(\operatorname{th}\dfrac{x}{2} \cdot \operatorname{ctg}\dfrac{t}{2}\right),\ v = 2 \operatorname{arctg}\left(\operatorname{cth}\dfrac{x}{2} \cdot \operatorname{ctg}\dfrac{t}{2}\right);\ t \neq \pm n\pi\right].$ Lo III 403

2.479

1. $\int x^p \dfrac{\operatorname{sh}^{2m} x}{\operatorname{ch}^n x}\,dx = \sum_{k=0}^{m} (-1)^{m+k} \binom{m}{k} \int \dfrac{x^p\,dx}{\operatorname{ch}^{n-2k} x} \quad (\text{cf. 2.477 2.}).$

2. $\int x^p \dfrac{\operatorname{sh}^{2m+1} x}{\operatorname{ch}^n x}\,dx = \sum_{k=0}^{m} (-1)^{m+k} \binom{m}{k} \int x^p \dfrac{\operatorname{sh} x}{\operatorname{ch}^{n-2k} x}\,dx$

$\qquad [n > 1] \quad (\text{cf. 2.479 3.}).$

3. $\int x^p \dfrac{\operatorname{sh} x}{\operatorname{ch}^n x}\,dx = -\dfrac{x^p}{(n-1)\operatorname{ch}^{n-1} x} + \dfrac{p}{n-1} \int \dfrac{x^{p-1}\,dx}{\operatorname{ch}^{n-1} x}$

$\qquad [n > 1] \quad (\text{cf. 2.477 2.}).$ GHH[353](12)

4. $\int x^p \dfrac{\operatorname{ch}^{2m} x}{\operatorname{sh}^n x}\,dx = \sum_{k=0}^{m} \binom{m}{k} \int \dfrac{x^p\,dx}{\operatorname{sh}^{n-2k} x} \quad (\text{cf. 2.477 1.}).$

5. $\int x^p \dfrac{\operatorname{ch}^{2m+1} x}{\operatorname{sh}^n x}\,dx = \sum_{k=0}^{m} \binom{m}{k} \int \dfrac{x^p \operatorname{ch} x}{\operatorname{sh}^{n-2k} x}\,dx \quad (\text{cf. 2.479 6.}).$

6. $\int x^p \dfrac{\operatorname{ch} x}{\operatorname{sh}^n x}\,dx = -\dfrac{x^p}{(n-1)\operatorname{sh}^{n-1} x} + \dfrac{p}{n-1} \int \dfrac{x^{p-1}\,dx}{\operatorname{sh}^{n-1} x}$

$\qquad [n > 1] \quad (\text{cf. 2.477 1.}).$ GHH[353](13c)

2.479

7. $\int x^p \operatorname{th} x \, dx = \sum_{k=1}^{\infty} \frac{2^{2k}(2^{2k}-1) B_{2k}}{(2k+p)(2k)!} x^{p+2k} \quad \left[p \geqslant -1, \ |x| < \frac{\pi}{2}\right].$ GHH[353](12d)

8. $\int x^p \operatorname{cth} x \, dx = \sum_{k=0}^{\infty} \frac{2^{2k} B_{2k}}{(p+2k)(2k)!} x^{p+2k} \quad [p \geqslant +1, \ |x| < \pi].$ GHH[353](13d)

9. $\int \frac{x \operatorname{ch} x}{\operatorname{sh}^2 x} dx = \ln \operatorname{th} \frac{x}{2} - \frac{x}{\operatorname{sh} x}.$

10. $\int \frac{x \operatorname{sh} x}{\operatorname{ch}^2 x} dx = -\frac{x}{\operatorname{ch} x} + \operatorname{arctg}(\operatorname{sh} x).$ MC 263

Hyperbelfunktionen, Exponentialfunktionen und Potenzen

2.48

Combinations of hyperbolic functions, exponentials, and powers

2.481

1. $\int e^{ax} \operatorname{sh}(bx+c) \, dx = \frac{e^{ax}}{a^2-b^2}[a \operatorname{sh}(bx+c) - b \operatorname{ch}(bx+c)] \quad [a^2 \neq b^2].$

2. $\int e^{ax} \operatorname{ch}(bx+c) \, dx = \frac{e^{ax}}{a^2-b^2}[a \operatorname{ch}(bx+c) - b \operatorname{sh}(bx+c)] \quad [a^2 \neq b^2].$

Im Fall $a^2 = b^2$ gelten die Formeln $\ |\ $ For $a^2 = b^2$:

3. $\int e^{ax} \operatorname{sh}(ax+c) \, dx = -\frac{1}{2} x e^{-c} + \frac{1}{4a} e^{2ax+c}.$

4. $\int e^{-ax} \operatorname{sh}(ax+c) \, dx = \frac{1}{2} x e^{c} + \frac{1}{4a} e^{-(2ax+c)}.$

5. $\int e^{ax} \operatorname{ch}(ax+c) \, dx = \frac{1}{2} x e^{-c} + \frac{1}{4a} e^{2ax+c}.$

6. $\int e^{-ax} \operatorname{ch}(ax+c) \, dx = \frac{1}{2} x e^{c} - \frac{1}{4a} e^{-(2ax+c)}.$ MC 275-277

2.482

1. $\int x^p e^{ax} \operatorname{sh} bx \, dx = \frac{1}{2}\left\{\int x^p e^{(a+b)x} dx - \int x^p e^{(a-b)x} dx\right\}$

$[a^2 \neq b^2]$ (cf. 2.321).

2. $\int x^p e^{ax} \operatorname{ch} bx \, dx = \frac{1}{2}\left\{\int x^p e^{(a+b)x} dx + \int x^p e^{(a-b)x} dx\right\}$

$[a^2 \neq b^2]$ (cf. 2.321).

Im Fall $a^2 = b^2$ gelten die Formeln $\ |\ $ For $a^2 = b^2$:

3. $\int x^p e^{ax} \operatorname{sh} ax \, dx = \frac{1}{2} \int x^p e^{2ax} dx - \frac{x^{p+1}}{2(p+1)}$ (cf. 2.321).

4. $\int x^p e^{-ax} \operatorname{sh} ax \, dx = \frac{x^{p+1}}{2(p+1)} - \frac{1}{2} \int x^p e^{-2ax} dx$ (cf. 2.321).

5. $\int x^p e^{ax} \operatorname{ch} ax \, dx = \frac{x^{p+1}}{2(p+1)} + \frac{1}{2} \int x^p e^{2ax} dx$ (cf. 2.321). MC 276, 278

2.483

1. $\int x e^{ax} \operatorname{sh} bx \, dx = \frac{e^{ax}}{a^2-b^2}\left[\left(ax - \frac{a^2+b^2}{a^2-b^2}\right) \operatorname{sh} bx - \left(bx - \frac{2ab}{a^2-b^2}\right) \operatorname{ch} bx\right] \quad [a^2 \neq b^2].$

2.483

2. $\int xe^{ax} \operatorname{ch} bx\, dx = \dfrac{e^{ax}}{a^2-b^2}\left[\left(ax - \dfrac{a^2+b^2}{a^2-b^2}\right)\operatorname{ch} bx - \left(bx - \dfrac{2ab}{a^2-b^2}\right)\operatorname{sh} bx\right]\quad [a^2 \ne b^2].$

3. $\int x^2 e^{ax} \operatorname{sh} bx\, dx = \dfrac{e^{ax}}{a^2-b^2}\left\{\left[ax^2 - \dfrac{2(a^2+b^2)}{a^2-b^2}x + \dfrac{2a(a^2+3b^2)}{(a^2-b^2)^2}\right]\operatorname{sh} bx - \left[bx^2 - \dfrac{4ab}{a^2-b^2}x + \dfrac{2b(3a^2+b^2)}{(a^2-b^2)^2}\right]\operatorname{ch} x\right\}\quad [a^2 \ne b^2].$

4. $\int x^2 e^{ax} \operatorname{ch} bx\, dx = \dfrac{e^{ax}}{a^2-b^2}\left\{\left[ax^2 - \dfrac{2(a^2+b^2)}{a^2-b^2}x + \dfrac{2a(a^2+3b^2)}{(a^2-b^2)^2}\right]\operatorname{ch} bx - \left[bx^2 - \dfrac{4ab}{a^2-b^2}x + \dfrac{2b(3a^2+b^2)}{(a^2-b^2)^2}\right]\operatorname{sh} x\right\}\quad [a^2 \ne b^2].$

Im Fall $a^2 = b^2$ gelten die Formeln | For $a^2 = b^2$:

5. $\int xe^{ax}\operatorname{sh} ax\, dx = \dfrac{e^{2ax}}{4a}\left(x - \dfrac{1}{2a}\right) - \dfrac{x^2}{4}.$

6. $\int xe^{-ax}\operatorname{sh} ax\, dx = \dfrac{e^{-2ax}}{4a}\left(x + \dfrac{1}{2a}\right) + \dfrac{x^2}{4}.$ \quad MC 276, 278

7. $\int xe^{ax}\operatorname{ch} ax\, dx = \dfrac{x^2}{4} + \dfrac{e^{2ax}}{4a}\left(x - \dfrac{1}{2a}\right).$

8. $\int xe^{-ax}\operatorname{ch} ax\, dx = \dfrac{x^2}{4} - \dfrac{e^{-2ax}}{4a}\left(x + \dfrac{1}{2a}\right).$

9. $\int x^2 e^{ax}\operatorname{sh} ax\, dx = \dfrac{e^{2ax}}{4a}\left(x^2 - \dfrac{x}{a} + \dfrac{1}{2a^2}\right) - \dfrac{x^3}{6}.$

10. $\int x^2 e^{-ax}\operatorname{sh} ax\, dx = \dfrac{e^{-2ax}}{4a}\left(x^2 + \dfrac{x}{a} + \dfrac{1}{2a^2}\right) + \dfrac{x^3}{6}.$

11. $\int x^2 e^{ax}\operatorname{ch} ax\, dx = \dfrac{x^3}{6} + \dfrac{e^{2ax}}{4a}\left(x^2 - \dfrac{x}{a} + \dfrac{1}{2a^2}\right).$

2.484

1. $\int e^{ax}\operatorname{sh} bx\,\dfrac{dx}{x} = \dfrac{1}{2}\{\operatorname{Ei}[(a+b)x] - \operatorname{Ei}[(a-b)x]\}\quad [a^2 \ne b^2].$

2. $\int e^{ax}\operatorname{ch} bx\,\dfrac{dx}{x} = \dfrac{1}{2}\{\operatorname{Ei}[(a+b)x] + \operatorname{Ei}[(a-b)x]\}\quad [a^2 \ne b^2].$

3. $\int e^{ax}\operatorname{sh} bx\,\dfrac{dx}{x^2} = -\dfrac{e^{ax}\operatorname{sh} bx}{2x} + \dfrac{1}{2}\{(a+b)\operatorname{Ei}[(a+b)x] - (a-b)\operatorname{Ei}[(a-b)x]\}\quad [a^2 \ne b^2].$

4. $\int e^{ax}\operatorname{ch} bx\,\dfrac{dx}{x^2} = -\dfrac{e^{ax}\operatorname{ch} bx}{2x} + \dfrac{1}{2}\{(a+b)\operatorname{Ei}[(a+b)x] + (a-b)\operatorname{Ei}[(a-b)x]\}\quad [a^2 \ne b^2].$

Im Fall $a^2 = b^2$ gelten die Formeln | For $a^2 = b^2$:

5. $\int e^{ax}\operatorname{sh} ax\,\dfrac{dx}{x} = \dfrac{1}{2}[\operatorname{Ei}(2ax) - \ln x].$

6. $\int e^{-ax}\operatorname{sh} ax\,\dfrac{dx}{x} = \dfrac{1}{2}[\ln x - \operatorname{Ei}(-2ax)].$

2.484

7. $\int e^{ax} \operatorname{ch} ax \frac{dx}{x} = \frac{1}{2}[\ln x + \operatorname{Ei}(2ax)]$.

8. $\int e^{ax} \operatorname{sh} ax \frac{dx}{x^2} = -\frac{1}{2x}(e^{2ax} - 1) + a\operatorname{Ei}(2ax)$.

9. $\int e^{-ax} \operatorname{sh} ax \frac{dx}{x^2} = -\frac{1}{2x}(1 - e^{-2ax}) + a\operatorname{Ei}(-2ax)$.

10. $\int e^{ax} \operatorname{ch} ax \frac{dx}{x^2} = -\frac{1}{2x}(e^{2ax} + 1) + a\operatorname{Ei}(2ax)$.   MC 276−278

Trigonometrische Funktionen
Trigonometric Functions

2.5−2.6

Einführung
Introduction

2.50

2.501 Integrale der Gestalt $\int R(\sin x, \cos x)dx$ lassen sich mit Hilfe der Substitution $t = \operatorname{tg} \frac{x}{2}$ stets auf Integrale rationaler Funktionen zurückführen.

Integrals of the form $\int R(\sin x, \cos x)dx$ can always be reduced to integrals of rational functions by the substitution $t = \operatorname{tg} \frac{x}{2}$.

2.502 Genügen dabei die Funktionen $R(\sin x, \cos x)$ der Relation

If the function $R(\sin x, \cos x)$ satisfies the relation

$$R(\sin x, \cos x) = -R(-\sin x, \cos x),$$

so ist die Substitution $t = \cos x$ geeignet.

the substitution $t = \cos x$ may be used with advantage.

2.503 Genügt die Funktion $R(\sin x, \cos x)$ der Beziehung

If the function $R(\sin x, \cos x)$ satisfies the relation

$$R(\sin x, \cos x) = -R(\sin x, -\cos x),$$

so ist die Substitution $t = \sin x$ zu verwenden.

the substitution $t = \sin x$ may be used with advantage.

2.504 Gilt aber

If the function $R(\sin x, \cos x)$ satisfies the relation

$$R(\sin x, \cos x) = R(-\sin x, -\cos x),$$

so verwende man die Substitution $t = \operatorname{tg} x$.

the substitution $t = \operatorname{tg} x$ may be used with advantage.

Potenzen trigonometrischer Funktionen
Powers of trigonometric functions

2.51−2.52

2.510 $\int \sin^p x \cos^q x \, dx = -\frac{\sin^{p-1} x \cos^{q+1} x}{q+1} + \frac{p-1}{q+1} \int \sin^{p-2} x \cos^{q+2} x \, dx;$

$= -\frac{\sin^{p-1} x \cos^{q+1} x}{p+q} + \frac{p-1}{p+q} \int \sin^{p-2} x \cos^q x \, dx;$

$$= \frac{\sin^{p+1} x \cos^{q+1} x}{p+1} + \frac{p+q+2}{p+1} \int \sin^{p+2} x \cos^q x \, dx;$$

$$= \frac{\sin^{p+1} x \cos^{q-1} x}{p+1} + \frac{q-1}{p+1} \int \sin^{p+2} x \cos^{q-2} x \, dx;$$

$$= \frac{\sin^{p+1} x \cos^{q-1} x}{p+q} + \frac{q-1}{p+q} \int \sin^p x \cos^{q-2} x \, dx;$$

$$= -\frac{\sin^{p+1} x \cos^{q+1} x}{q+1} + \frac{p+q+2}{q+1} \int \sin^p x \cos^{q+2} x \, dx;$$

$$= \frac{\sin^{p-1} x \cos^{q-1} x}{p+q} \left\{ \sin^2 x - \frac{q-1}{p+q-2} \right\} +$$

$$+ \frac{(p-1)(q-1)}{(p+q)(p+q-2)} \int \sin^{p-2} x \cos^{q-2} x \, dx. \qquad \text{F II 79, T(214)}$$

2.510

1. $\int \sin^p x \cos^{2n} x \, dx = \frac{\sin^{p+1} x}{2n+p} \left\{ \cos^{2n-1} x + \right.$

2.511

$$\left. + \sum_{k=1}^{n-1} \frac{(2n-1)(2n-3)\ldots(2n-2k+1)\cos^{2n-2k-1} x}{(2n+p-2)(2n+p-4)\ldots(2n+p-2k)} \right\} +$$

$$+ \frac{(2n-1)!!}{(2n+p)(2n+p-2)\ldots(p+2)} \int \sin^p x \, dx.$$

Diese Formel gilt für alle reellen p mit Ausnahme der negativen geraden Zahlen $-2, -4, \ldots, -2n$. Für natürliches p und $n=0$ gelten die Formeln	This formula can be used for any real p except for $p = -2, -4, \ldots, -2n$. For p a positive integer and $n = 0$:

2. $\int \sin^{2l} x \, dx = -\frac{\cos x}{2l} \left\{ \sin^{2l-1} x + \right.$

$$\left. + \sum_{k=1}^{l-1} \frac{(2l-1)(2l-3)\ldots(2l-2k+1)}{2^k (l-1)(l-2)\ldots(l-k)} \sin^{2l-2k-1} x \right\} + \frac{(2l-1)!!}{2^l \, l!} x$$

(cf. 2.513 1.). T(232)

3. $\int \sin^{2l+1} x \, dx = -\frac{\cos x}{2l+1} \left\{ \sin^{2l} x + \right.$

$$\left. + \sum_{k=0}^{l-1} \frac{2^{k+1} l(l-1)\ldots(l-k)}{(2l-1)(2l-3)\ldots(2l-2k-1)} \sin^{2l-2k-2} x \right\}$$

(cf. 2.513 2.). T(233)

4. $\int \sin^p x \cos^{2n+1} x \, dx = \frac{\sin^{p+1} x}{2n+p+1} \left\{ \cos^{2n} x + \right.$

$$\left. + \sum_{k=1}^{n} \frac{2^k n(n-1)\ldots(n-k+1) \cos^{2n-2k} x}{(2n+p-1)(2n+p-3)\ldots(2n+p-2k+1)} \right\}.$$

Diese Formel gilt für alle reellen p mit Ausnahme der negativen ungeraden Zahlen $-1, -3, \ldots, -(2n+1)$.

This formula can be used for any real p except for $p = -1, -3, \ldots, -(2n+1)$.

2.512

1. $\int \cos^p x \sin^{2n} x \, dx = -\dfrac{\cos^{p+1} x}{2n+p} \Big\{ \sin^{2n-1} x +$

$+ \sum\limits_{k=1}^{n-1} \dfrac{(2n-1)(2n-3) \ldots (2n-2k+1) \sin^{2n-2k-1} x}{(2n+p-2)(2n+p-4) \ldots (2n+p-2k)} \Big\} +$

$+ \dfrac{(2n-1)!!}{(2n+p)(2n+p-2) \ldots (p+2)} \int \cos^p x \, dx.$

Diese Formel gilt für alle reellen p mit Ausnahme der negativen geraden Zahlen $-2, -4, \ldots, -2n$. Für natürliches p und $n = 0$ gelten die Formeln

This formula can be used for any real p except for $p = -2, -4, \ldots, -2n$. For p a positive integer and $n = 0$:

2. $\int \cos^{2l} x \, dx = \dfrac{\sin x}{2l} \Big\{ \cos^{2l-1} x +$

$+ \sum\limits_{k=1}^{l-1} \dfrac{(2l-1)(2l-3) \ldots (2l-2k+1)}{2^k (l-1)(l-2) \ldots (l-k)} \cos^{2l-2k-1} x \Big\} + \dfrac{(2l-1)!!}{2^l \, l!} x$

(cf. 2.513 3.).   T(230)

3. $\int \cos^{2l+1} x \, dx = \dfrac{\sin x}{2l+1} \Big\{ \cos^{2l} x +$

$+ \sum\limits_{k=0}^{l-1} \dfrac{2^{k+1} l (l-1) \ldots (l-k)}{(2l-1)(2l-3) \ldots (2l-2k-1)} \cos^{2l-2k-2} x \Big\}$

(cf. 2.513 4.).   T(231)

4. $\int \cos^p x \sin^{2n+1} x \, dx = -\dfrac{\cos^{p+1} x}{2n+p+1} \Big\{ \sin^{2n} x +$

$+ \sum\limits_{k=1}^{n} \dfrac{2^k n (n-1) \ldots (n-k+1) \sin^{2n-2k} x}{(2n+p-1)(2n+p-3) \ldots (2n+p-2k+1)} \Big\}.$

Diese Formel gilt für alle reellen p mit Ausnahme der negativen ungeraden Zahlen $-1, -3, \ldots, -(2n+1)$.

This formula can be used for any real p except for $p = -1, -3, \ldots, -(2n+1)$.

2.513

1. $\int \sin^{2n} x \, dx = \dfrac{1}{2^{2n}} \binom{2n}{n} x + \dfrac{(-1)^n}{2^{2n-1}} \sum\limits_{k=0}^{n-1} (-1)^k \binom{2n}{k} \dfrac{\sin(2n-2k)x}{2n-2k}$

(cf. 2.511 2.).   T(226)

2. $\int \sin^{2n+1} x \, dx = \dfrac{1}{2^{2n}} (-1)^{n+1} \sum\limits_{k=0}^{n} (-1)^k \binom{2n+1}{k} \dfrac{\cos(2n+1-2k)x}{2n+1-2k}$

(cf. 2.511 3.).   T(227)

3. $\int \cos^{2n} x \, dx = \dfrac{1}{2^{2n}} \binom{2n}{n} x + \dfrac{1}{2^{2n-1}} \sum\limits_{k=0}^{n-1} \binom{2n}{k} \dfrac{\sin(2n-2k)x}{2n-2k}$

(cf. 2.512 2.).   T(224)

4. $\int \cos^{2n+1} x \, dx = \dfrac{1}{2^{2n}} \sum\limits_{k=0}^{n} \binom{2n+1}{k} \dfrac{\sin(2n-2k+1)x}{2n-2k+1}$ 2.513

(cf. 2.512 3.). T(225)

5. $\int \sin^2 x \, dx = -\dfrac{1}{4} \sin 2x + \dfrac{1}{2} x = -\dfrac{1}{2} \sin x \cos x + \dfrac{1}{2} x.$

6. $\int \sin^3 x \, dx = \dfrac{1}{12} \cos 3x - \dfrac{3}{4} \cos x = \dfrac{1}{3} \cos^3 x - \cos x.$

7. $\int \sin^4 x \, dx = \dfrac{3x}{8} - \dfrac{\sin 2x}{4} + \dfrac{\sin 4x}{32} =$

$$= -\dfrac{3}{8} \sin x \cos x - \dfrac{1}{4} \sin^3 x \cos x + \dfrac{3}{8} x.$$

8. $\int \sin^5 x \, dx = -\dfrac{5}{8} \cos x + \dfrac{5}{48} \cos 3x - \dfrac{1}{80} \cos 5x =$

$$= -\dfrac{1}{5} \sin^4 x \cos x + \dfrac{4}{15} \cos^3 x - \dfrac{4}{5} \cos x.$$

9. $\int \sin^6 x \, dx = \dfrac{5}{16} x - \dfrac{15}{64} \sin 2x + \dfrac{3}{64} \sin 4x - \dfrac{1}{192} \sin 6x =$

$$= -\dfrac{1}{6} \sin^5 x \cos x - \dfrac{5}{24} \sin^3 x \cos x - \dfrac{5}{16} \sin x \cos x + \dfrac{5}{16} x.$$

10. $\int \sin^7 x \, dx = -\dfrac{35}{64} \cos x + \dfrac{7}{64} \cos 3x - \dfrac{7}{320} \cos 5x + \dfrac{1}{448} \cos 7x =$

$$= -\dfrac{1}{7} \sin^6 x \cos x - \dfrac{6}{35} \sin^4 x \cos x + \dfrac{8}{35} \cos^3 x - \dfrac{24}{35} \cos x.$$

11. $\int \cos^2 x \, dx = \dfrac{1}{4} \sin 2x + \dfrac{x}{2} = \dfrac{1}{2} \sin x \cos x + \dfrac{1}{2} x.$

12. $\int \cos^3 x \, dx = \dfrac{1}{12} \sin 3x + \dfrac{3}{4} \sin x = \sin x - \dfrac{1}{3} \sin^3 x.$

13. $\int \cos^4 x \, dx = \dfrac{3}{8} x + \dfrac{1}{4} \sin 2x + \dfrac{1}{32} \sin 4x =$

$$= \dfrac{3}{8} x + \dfrac{3}{8} \sin x \cos x + \dfrac{1}{4} \sin x \cos^3 x.$$

14. $\int \cos^5 x \, dx = \dfrac{5}{8} \sin x + \dfrac{5}{48} \sin 3x + \dfrac{1}{80} \sin 5x =$

$$= \dfrac{4}{5} \sin x - \dfrac{4}{15} \sin^3 x + \dfrac{1}{5} \cos^4 x \sin x.$$

15. $\int \cos^6 x \, dx = \dfrac{5}{16} x + \dfrac{15}{64} \sin 2x + \dfrac{3}{64} \sin 4x + \dfrac{1}{192} \sin 6x =$

$$= \dfrac{5}{16} x + \dfrac{5}{16} \sin x \cos x + \dfrac{5}{24} \sin x \cos^3 x + \dfrac{1}{6} \sin x \cos^5 x.$$

16. $\int \cos^7 x \, dx = \dfrac{35}{64} \sin x + \dfrac{7}{64} \sin 3x + \dfrac{7}{320} \sin 5x + \dfrac{1}{448} \sin 7x =$

$$= \dfrac{24}{35} \sin x - \dfrac{8}{35} \sin^3 x + \dfrac{6}{35} \sin x \cos^4 x + \dfrac{1}{7} \sin x \cos^6 x.$$

2.513

17. $\int \sin x \cos^2 x \, dx = -\frac{1}{4}\left\{\frac{1}{3}\cos 3x + \cos x\right\} = -\frac{\cos^3 x}{3}$.

18. $\int \sin x \cos^3 x \, dx = -\frac{\cos^4 x}{4}$.

19. $\int \sin x \cos^4 x \, dx = -\frac{\cos^5 x}{5}$.

20. $\int \sin^2 x \cos x \, dx = -\frac{1}{4}\left\{\frac{1}{3}\sin 3x - \sin x\right\} = \frac{\sin^3 x}{3}$.

21. $\int \sin^2 x \cos^2 x \, dx = -\frac{1}{8}\left\{\frac{1}{4}\sin 4x - x\right\}$.

22. $\int \sin^2 x \cos^3 x \, dx = -\frac{1}{16}\left\{\frac{1}{5}\sin 5x + \frac{1}{3}\sin 3x - 2\sin x\right\} =$
$= \frac{\sin^3 x}{5}\left(\cos^2 x + \frac{2}{3}\right) = \frac{\sin^3 x}{5}\left(\frac{5}{3} - \sin^2 x\right)$.

23. $\int \sin^2 x \cos^4 x \, dx = \frac{x}{16} + \frac{1}{64}\sin 2x - \frac{1}{64}\sin 4x - \frac{1}{192}\sin 6x$.

24. $\int \sin^3 x \cos x \, dx = \frac{1}{8}\left(\frac{1}{4}\cos 4x - \cos 2x\right) = \frac{\sin^4 x}{4}$.

25. $\int \sin^3 x \cos^2 x \, dx = \frac{1}{16}\left(\frac{1}{5}\cos 5x - \frac{1}{3}\cos 3x - 2\cos x\right) =$
$= \frac{1}{5}\cos^5 x - \frac{1}{3}\cos^3 x$.

26. $\int \sin^3 x \cos^3 x \, dx = \frac{1}{32}\left(\frac{1}{6}\cos 6x - \frac{3}{2}\cos 2x\right)$.

27. $\int \sin^3 x \cos^4 x \, dx = \frac{1}{7}\cos^3 x\left(-\frac{2}{5} - \frac{3}{5}\sin^2 x + \sin^4 x\right)$.

28. $\int \sin^4 x \cos x \, dx = \frac{\sin^5 x}{5}$.

29. $\int \sin^4 x \cos^2 x \, dx = \frac{1}{16}x - \frac{1}{64}\sin 2x - \frac{1}{64}\sin 4x + \frac{1}{192}\sin 6x$.

30. $\int \sin^4 x \cos^3 x \, dx = \frac{1}{7}\sin^3 x\left(\frac{2}{5} + \frac{3}{5}\cos^2 x - \cos^4 x\right)$.

31. $\int \sin^4 x \cos^4 x \, dx = \frac{3}{128}x - \frac{1}{128}\sin 4x + \frac{1}{1024}\sin 8x$.

2.514 $\int \frac{\sin^p x}{\cos^{2n} x} dx = \frac{\sin^{p+1} x}{2n-1}\left\{\sec^{2n-1} x + \right.$
$\left. + \sum_{k=1}^{n-1} \frac{(2n-p-2)(2n-p-4)\ldots(2n-p-2k)}{(2n-3)(2n-5)\ldots(2n-2k-1)} \sec^{2n-2k-1} x\right\} +$
$+ \frac{(2n-p-2)(2n-p-4)\ldots(-p+2)(-p)}{(2n-1)!!}\int \sin^p x \, dx$.

Diese Formel gilt für jedes reelle p. Formeln für $\int \sin^p x\, dx$ mit natürlichem p findet man unter 2.511 2., 2.511 3. und 2.513 1., 2.513 2. Ist $n = 0$ und p negativ ganzzahlig, so gelten die Formeln

This formula can be used for any real p. For $\int \sin^p x\, dx$, where p is a positive integer, see 2.511 2., 2.511 3., 2.513 1., and 2.513 2. If $n = 0$ and p is a negative integer, we have

2.515

1. $\int \dfrac{dx}{\sin^{2l} x} = -\dfrac{\cos x}{2l-1}\left\{\operatorname{cosec}^{2l-1} x + \sum_{k=1}^{l-1} \dfrac{2^k(l-1)(l-2)\ldots(l-k)}{(2l-3)(2l-5)\ldots(2l-2k-1)} \operatorname{cosec}^{2l-2k-1} x\right\}.$ T(242)

2. $\int \dfrac{dx}{\sin^{2l+1} x} = -\dfrac{\cos x}{2l}\left\{\operatorname{cosec}^{2l} x + \sum_{k=1}^{l-1} \dfrac{(2l-1)(2l-3)\ldots(2l-2k+1)}{2^k(l-1)(l-2)\ldots(l-k)} \operatorname{cosec}^{2l-2k} x\right\} + \dfrac{(2l-1)!!}{2^l\, l!} \ln \operatorname{tg} \dfrac{x}{2}.$ T(243)

2.516

1. $\int \dfrac{\sin^p x\, dx}{\cos^{2n+1} x} = \dfrac{\sin^{p+1}}{2n}\left\{\sec^{2n} x + \sum_{k=1}^{n-1} \dfrac{(2n-p-1)(2n-p-3)\ldots(2n-p-2k+1)}{2^k(n-1)(n-2)\ldots(n-k)} \sec^{2n-2k} x\right\} + \dfrac{(2n-p-1)(2n-p-3)\ldots(3-p)(1-p)}{2^n\, n!} \int \dfrac{\sin^p x}{\cos x}\, dx.$

Diese Formel gilt für alle reellen p. Ist $n = 0$ und p eine natürliche Zahl, so gelten die Formeln

This formula holds for any real p. For $n = 0$ and p a positive integer we have

2. $\int \dfrac{\sin^{2l+1} x\, dx}{\cos x} = -\sum_{k=1}^{l} \dfrac{\sin^{2k} x}{2k} - \ln \cos x.$

3. $\int \dfrac{\sin^{2l} x\, dx}{\cos x} = -\sum_{k=1}^{l} \dfrac{\sin^{2k-1} x}{2k-1} + \ln \operatorname{tg}\left(\dfrac{\pi}{4} + \dfrac{x}{2}\right).$

2.517

1. $\int \dfrac{dx}{\sin^{2m+1} x \cos x} = -\sum_{k=1}^{m} \dfrac{1}{(2m-2k+2)\sin^{2m-2k+2} x} + \ln \operatorname{tg} x.$

2. $\int \dfrac{dx}{\sin^{2m} x \cos x} = -\sum_{k=1}^{m} \dfrac{1}{(2m-2k+1)\sin^{2m-2k+1} x} + \ln \operatorname{tg}\left(\dfrac{\pi}{4} - \dfrac{x}{2}\right).$

2.518

1. $\int \dfrac{\sin^p x}{\cos^2 x}\, dx = \dfrac{\sin^{p-1} x}{\cos x} - (p-1) \int \sin^{p-2} x\, dx.$

2. $\int \dfrac{\cos^p x\, dx}{\sin^{2n} x} = -\dfrac{\cos^{p+1} x}{2n-1}\left\{\operatorname{cosec}^{2n-1} x + \sum_{k=1}^{n-1} \dfrac{(2n-p-2)(2n-p-4)\ldots(2n-p-2k)}{(2n-3)(2n-5)\ldots(2n-2k-1)} \operatorname{cosec}^{2n-2k-1} x\right\} + \dfrac{(2n-p-2)(2n-p-4)\ldots(2-p)(-p)}{(2n-1)!!} \int \cos^p x\, dx.$

Diese Formel gilt für alle reellen p. Ist p eine natürliche Zahl, so gelten für $\int \cos^p x \, dx$ die Formeln 2.512 2., 2.512 3. und 2.513 3., 2.513 4. Ist $n=0$ und p negativ ganzzahlig, so gilt

This formula can be used for any real p. For $\int \cos^p x \, dx$ with p a positive integer see 2.512 2., 2.512 3., 2.513 3., and 2.513 4. If $n=0$ and p is a negative integer, we have

2.519 1. $\int \dfrac{dx}{\cos^{2l} x} = \dfrac{\sin x}{2l-1} \Big\{ \sec^{2l-1} x +$

$+ \sum\limits_{k=1}^{l-1} \dfrac{2^k (l-1)(l-2) \ldots (l-k)}{(2l-3)(2l-5) \ldots (2l-2k-1)} \sec^{2l-2k-1} x \Big\}.$ T(240)

2. $\int \dfrac{dx}{\cos^{2l+1} x} = \dfrac{\sin x}{2l} \Big\{ \sec^{2l} x +$

$+ \sum\limits_{k=1}^{l-1} \dfrac{(2l-1)(2l-3) \ldots (2l-2k+1)}{2^k (l-1)(l-2) \ldots (l-k)} \sec^{2l-2k} x \Big\} + \dfrac{(2l-1)!!}{2^l \, l!} \ln \operatorname{tg}\left(\dfrac{\pi}{4} + \dfrac{x}{2}\right).$ T(241)

2.521 1. $\int \dfrac{\cos^p x \, dx}{\sin^{2n+1} x} = - \dfrac{\cos^{p+1} x}{2n} \Big\{ \operatorname{cosec}^{2n} x +$

$+ \sum\limits_{k=1}^{n-1} \dfrac{(2n-p-1)(2n-p-3) \ldots (2n-p-2k+1)}{2^k (n-1)(n-2) \ldots (n-k)} \operatorname{cosec}^{2n-2k} x \Big\} +$

$+ \dfrac{(2n-p-1)(2n-p-3) \ldots (3-p)(1-p)}{2^n \cdot n!} \int \dfrac{\cos^p x}{\sin x} \, dx.$

Diese Formel gilt für alle reellen p. Für $n=0$ und natürliches p ergibt sich

This formula can be used for any real p. For $n=0$ and p a positive integer we have

2. $\int \dfrac{\cos^{2l+1} x \, dx}{\sin x} = \sum\limits_{k=1}^{l} \dfrac{\cos^{2k} x}{2k} + \ln \sin x.$

3. $\int \dfrac{\cos^{2l} x \, dx}{\sin x} = \sum\limits_{k=1}^{l} \dfrac{\cos^{2k-1} x}{2k-1} + \ln \operatorname{tg} \dfrac{x}{2}.$

2.522 1. $\int \dfrac{dx}{\sin x \cos^{2m+1} x} = \sum\limits_{k=1}^{m} \dfrac{1}{(2m-2k+2) \cos^{2m-2k+2} x} + \ln \operatorname{tg} x.$

2. $\int \dfrac{dx}{\sin x \cos^{2m} x} = \sum\limits_{k=1}^{m} \dfrac{1}{(2m-2k+1) \cos^{2m-2k+1} x} + \ln \operatorname{tg} \dfrac{x}{2}.$ GHH[331](15)

2.523 $\int \dfrac{\cos^m x}{\sin^2 x} \, dx = - \dfrac{\cos^{m-1} x}{\sin x} - (m-1) \int \cos^{m-2} x \, dx.$

2.524 1. $\int \dfrac{\sin^{2n+1} x}{\cos^m x} \, dx = \sum\limits_{\substack{k=0 \\ k \neq \frac{m-1}{2}}}^{n} (-1)^{k+1} \binom{n}{k} \dfrac{\cos^{2k-m+1} x}{2k-m+1} +$

$+ s(-1)^{\frac{m+1}{2}} \dbinom{n}{\frac{m-1}{2}} \ln \cos x.$ GHH[331](11d)

2. $\int \frac{\cos^{2n+1} x}{\sin^m x} dx = \sum_{\substack{k=0 \\ k \neq \frac{m-1}{2}}}^{n} (-1)^k \binom{n}{k} \frac{\sin^{2k-m+1} x}{2k-m+1} +$ 2.524

$$+ s(-1)^{\frac{m-1}{2}} \binom{n}{\frac{m-1}{2}} \ln \sin x.$$

[In den Formeln 2.524 1., 2.524 2. ist für ungerades m und $m < 2n+1$ unter s die Zahl 1 und sonst die Zahl 0 zu verstehen.]   GHH[381](13d)

[In 2.524 1. and 2.524 2. we have $s = 1$ if m is odd and $m < 2n+1$ otherwise $s = 0$].   GHH[331](13d)

1. $\int \frac{dx}{\sin^{2m} x \cos^{2n} x} = \sum_{k=0}^{m+n-1} \binom{m+n-1}{k} \frac{\operatorname{tg}^{2k-2m+1} x}{2k-2m+1}$.   T(267)   2.525

2. $\int \frac{dx}{\sin^{2m+1} x \cos^{2n+1} x} = \sum_{k=0}^{m+n} \binom{m+n}{k} \frac{\operatorname{tg}^{2k-2m} x}{2k-2m} + \binom{m+n}{m} \ln \operatorname{tg} x$.

T(268), GHH[331](15f)

1. $\int \frac{dx}{\sin x} = \ln \operatorname{tg} \frac{x}{2}$.   2.526

2. $\int \frac{dx}{\sin^2 x} = -\operatorname{ctg} x$.

3. $\int \frac{dx}{\sin^3 x} = -\frac{1}{2} \frac{\cos x}{\sin^2 x} + \frac{1}{2} \ln \operatorname{tg} \frac{x}{2}$.

4. $\int \frac{dx}{\sin^4 x} = -\frac{\cos x}{3 \sin^3 x} - \frac{2}{3} \operatorname{ctg} x = -\frac{1}{3} \operatorname{ctg}^3 x - \operatorname{ctg} x$.

5. $\int \frac{dx}{\sin^5 x} = -\frac{\cos x}{4 \sin^4 x} - \frac{3}{8} \frac{\cos x}{\sin^2 x} + \frac{3}{8} \ln \operatorname{tg} \frac{x}{2}$.

6. $\int \frac{dx}{\sin^6 x} = -\frac{\cos x}{5 \sin^5 x} - \frac{4}{15} \operatorname{ctg}^3 x - \frac{4}{5} \operatorname{ctg} x$;

$$= -\frac{1}{5} \operatorname{ctg}^5 x - \frac{2}{3} \operatorname{ctg}^3 x - \operatorname{ctg} x.$$

7. $\int \frac{dx}{\sin^7 x} = -\frac{\cos x}{6 \sin^2 x} \left(\frac{1}{\sin^4 x} + \frac{5}{4 \sin^2 x} + \frac{15}{8} \right) + \frac{5}{16} \ln \operatorname{tg} \frac{x}{2}$.

8. $\int \frac{dx}{\sin^8 x} = -\left(\frac{1}{7} \operatorname{ctg}^7 x + \frac{3}{5} \operatorname{ctg}^5 x + \operatorname{ctg}^3 x + \operatorname{ctg} x \right)$.

9. $\int \frac{dx}{\cos x} = \ln \operatorname{tg} \left(\frac{\pi}{4} + \frac{x}{2} \right) = \ln \operatorname{ctg} \left(\frac{\pi}{4} - \frac{x}{2} \right) = \ln \sqrt{\frac{1+\sin x}{1-\sin x}}$.

10. $\int \frac{dx}{\cos^2 x} = \operatorname{tg} x$.

11. $\int \frac{dx}{\cos^3 x} = \frac{1}{2} \frac{\sin x}{\cos^2 x} + \frac{1}{2} \ln \operatorname{tg} \left(\frac{\pi}{4} + \frac{x}{2} \right)$.

12. $\int \frac{dx}{\cos^4 x} = \frac{\sin x}{3 \cos^3 x} + \frac{2}{3} \operatorname{tg} x = \frac{1}{3} \operatorname{tg}^3 x + \operatorname{tg} x$.

2.526

13. $\int \frac{dx}{\cos^5 x} = \frac{\sin x}{4\cos^4 x} + \frac{3}{8}\frac{\sin x}{\cos^2 x} + \frac{3}{8}\ln\operatorname{tg}\left(\frac{x}{2}+\frac{\pi}{4}\right).$

14. $\int \frac{dx}{\cos^6 x} = \frac{\sin x}{5\cos^5 x} + \frac{4}{15}\operatorname{tg}^3 x + \frac{4}{5}\operatorname{tg} x = \frac{1}{5}\operatorname{tg}^5 x + \frac{2}{3}\operatorname{tg}^3 x + \operatorname{tg} x.$

15. $\int \frac{dx}{\cos^7 x} = \frac{\sin x}{6\cos^6 x} + \frac{5\sin x}{24\cos^4 x} + \frac{5\sin x}{16\cos^2 x} + \frac{5}{16}\ln\operatorname{tg}\left(\frac{x}{2}+\frac{\pi}{4}\right).$

16. $\int \frac{dx}{\cos^8 x} = \frac{1}{7}\operatorname{tg}^7 x + \frac{3}{5}\operatorname{tg}^5 x + \operatorname{tg}^3 x + \operatorname{tg} x.$

17. $\int \frac{\sin x}{\cos x}dx = -\ln\cos x.$

18. $\int \frac{\sin^2 x}{\cos x}dx = -\sin x + \ln\operatorname{tg}\left(\frac{\pi}{4}+\frac{x}{2}\right).$

19. $\int \frac{\sin^3 x}{\cos x}dx = -\frac{\sin^2 x}{2} - \ln\cos x = \frac{1}{2}\cos^2 x - \ln\cos x.$

20. $\int \frac{\sin^4 x}{\cos x}dx = -\frac{1}{3}\sin^3 x - \sin x + \ln\operatorname{tg}\left(\frac{x}{2}+\frac{\pi}{4}\right).$

21. $\int \frac{\sin x\, dx}{\cos^2 x} = \frac{1}{\cos x}.$

22. $\int \frac{\sin^2 x\, dx}{\cos^2 x} = \operatorname{tg} x - x.$

23. $\int \frac{\sin^3 x\, dx}{\cos^2 x} = \cos x + \frac{1}{\cos x}.$

24. $\int \frac{\sin^4 x\, dx}{\cos^2 x} = \operatorname{tg} x + \frac{1}{2}\sin x\cos x - \frac{3}{2}x.$

25. $\int \frac{\sin x\, dx}{\cos^3 x} = \frac{1}{2\cos^2 x} = \frac{1}{2}\operatorname{tg}^2 x.$

26. $\int \frac{\sin^2 x\, dx}{\cos^3 x} = \frac{\sin x}{2\cos^2 x} - \frac{1}{2}\ln\operatorname{tg}\left(\frac{\pi}{4}+\frac{x}{2}\right).$

27. $\int \frac{\sin^3 x\, dx}{\cos^3 x} = \frac{1}{2\cos^2 x} + \ln\cos x.$

28. $\int \frac{\sin^4 x\, dx}{\cos^3 x} = \frac{1}{2}\frac{\sin x}{\cos^2 x} + \sin x - \frac{3}{2}\ln\operatorname{tg}\left(\frac{x}{2}+\frac{\pi}{4}\right).$

29. $\int \frac{\sin x\, dx}{\cos^4 x} = \frac{1}{3\cos^3 x}.$

30. $\int \frac{\sin^2 x\, dx}{\cos^4 x} = \frac{1}{3}\operatorname{tg}^3 x.$

31. $\int \frac{\sin^3 x\, dx}{\cos^4 x} = -\frac{1}{\cos x} + \frac{1}{3\cos^3 x}.$

32. $\int \frac{\sin^4 x\, dx}{\cos^4 x} = \frac{1}{3}\operatorname{tg}^3 x - \operatorname{tg} x + x.$

33. $\int \frac{\cos x\, dx}{\sin x} = \ln\sin x.$

34. $\int \frac{\cos^2 x\, dx}{\sin x} = \cos x + \ln \operatorname{tg} \frac{x}{2}.$ 2.526

35. $\int \frac{\cos^3 x\, dx}{\sin x} = \frac{\cos^2 x}{2} + \ln \sin x.$

36. $\int \frac{\cos^4 x\, dx}{\sin x} = \frac{1}{3} \cos^3 x + \cos x + \ln \operatorname{tg} \left(\frac{x}{2}\right).$

37. $\int \frac{\cos x}{\sin^2 x}\, dx = -\frac{1}{\sin x}.$

38. $\int \frac{\cos^2 x}{\sin^2 x}\, dx = -\operatorname{ctg} x - x.$

39. $\int \frac{\cos^3 x}{\sin^2 x}\, dx = -\sin x - \frac{1}{\sin x}.$

40. $\int \frac{\cos^4 x}{\sin^2 x}\, dx = -\operatorname{ctg} x - \frac{1}{2} \sin x \cos x - \frac{3}{2} x.$

41. $\int \frac{\cos x}{\sin^3 x}\, dx = -\frac{1}{2 \sin^2 x}.$

42. $\int \frac{\cos^2 x}{\sin^3 x}\, dx = -\frac{\cos x}{2 \sin^2 x} - \frac{1}{2} \ln \operatorname{tg} \frac{x}{2}.$

43. $\int \frac{\cos^3 x}{\sin^3 x}\, dx = -\frac{1}{2 \sin^2 x} - \ln \sin x.$

44. $\int \frac{\cos^4 x}{\sin^3 x}\, dx = -\frac{1}{2} \frac{\cos x}{\sin^2 x} - \cos x - \frac{3}{2} \ln \operatorname{tg} \frac{x}{2}.$

45. $\int \frac{\cos x}{\sin^4 x}\, dx = -\frac{1}{3 \sin^3 x}.$

46. $\int \frac{\cos^2 x}{\sin^4 x}\, dx = -\frac{1}{3} \operatorname{ctg}^3 x.$

47. $\int \frac{\cos^3 x}{\sin^4 x}\, dx = \frac{1}{\sin x} - \frac{1}{3 \sin^3 x}.$

48. $\int \frac{\cos^4 x}{\sin^4 x}\, dx = -\frac{1}{3} \operatorname{ctg}^3 x + \operatorname{ctg} x + x.$

49. $\int \frac{dx}{\sin x \cos x} = \ln \operatorname{tg} x.$

50. $\int \frac{dx}{\sin x \cos^2 x} = \frac{1}{\cos x} + \ln \operatorname{tg} \frac{x}{2}.$

51. $\int \frac{dx}{\sin x \cos^3 x} = \frac{1}{2 \cos^2 x} + \ln \operatorname{tg} x.$

52. $\int \frac{dx}{\sin x \cos^4 x} = \frac{1}{\cos x} + \frac{1}{3 \cos^3 x} + \ln \operatorname{tg} \frac{x}{2}.$

53. $\int \frac{dx}{\sin^2 x \cos x} = \ln \operatorname{tg} \left(\frac{\pi}{4} + \frac{x}{2}\right) - \operatorname{cosec} x.$

2.526

54. $\int \dfrac{dx}{\sin^2 x \cos^2 x} = -2\,\mathrm{ctg}\,2x.$

55. $\int \dfrac{dx}{\sin^2 x \cos^3 x} = \left(\dfrac{1}{2\cos^2 x} - \dfrac{3}{2}\right)\dfrac{1}{\sin x} + \dfrac{3}{2}\ln\left(\dfrac{\pi}{4} + \dfrac{x}{2}\right).$

56. $\int \dfrac{dx}{\sin^2 x \cos^4 x} = \dfrac{1}{3\sin x \cos^3 x} - \dfrac{8}{3}\,\mathrm{ctg}\,2x.$

57. $\int \dfrac{dx}{\sin^3 x \cos x} = -\dfrac{1}{2\sin^2 x} + \ln\mathrm{tg}\,x.$

58. $\int \dfrac{dx}{\sin^3 x \cos^2 x} = -\dfrac{1}{\cos x}\left(\dfrac{1}{2\sin^2 x} - \dfrac{3}{2}\right) + \dfrac{3}{2}\ln\mathrm{tg}\,\dfrac{x}{2}.$

59. $\int \dfrac{dx}{\sin^3 x \cos^3 x} = -\dfrac{2\cos 2x}{\sin^2 2x} + 2\ln\mathrm{tg}\,x.$

60. $\int \dfrac{dx}{\sin^3 x \cos^4 x} = \dfrac{2}{\cos x} + \dfrac{1}{3\cos^3 x} - \dfrac{\cos x}{2\sin^2 x} + \dfrac{5}{2}\ln\mathrm{tg}\,\dfrac{x}{2}.$

61. $\int \dfrac{dx}{\sin^4 x \cos x} = -\dfrac{1}{\sin x} - \dfrac{1}{3\sin^3 x} + \ln\mathrm{tg}\left(\dfrac{x}{2} + \dfrac{\pi}{4}\right).$

62. $\int \dfrac{dx}{\sin^4 x \cos^2 x} = -\dfrac{1}{3\cos x \sin^3 x} - \dfrac{8}{3}\,\mathrm{ctg}\,2x.$

63. $\int \dfrac{dx}{\sin^4 x \cos^3 x} = -\dfrac{2}{\sin x} - \dfrac{1}{3\sin^3 x} + \dfrac{\sin x}{2\cos^2 x} + \dfrac{5}{2}\ln\mathrm{tg}\left(\dfrac{x}{2} + \dfrac{\pi}{4}\right).$

64. $\int \dfrac{dx}{\sin^4 x \cos^4 x} = -8\,\mathrm{ctg}\,2x - \dfrac{8}{3}\,\mathrm{ctg}^3\,2x.$

2.527

1. $\int \mathrm{tg}^p x\,dx = \dfrac{\mathrm{tg}^{p-1} x}{p-1} - \int \mathrm{tg}^{p-2} x\,dx \quad [p \ne 1].$

2. $\int \mathrm{tg}^{2n+1} x\,dx = \sum\limits_{k=1}^{n}(-1)^{n+k}\binom{n}{k}\dfrac{1}{2k\cos^{2k} x} - (-1)^n \ln\cos x =$
$= \sum\limits_{k=1}^{n}\dfrac{(-1)^{k-1}\mathrm{tg}^{2n-2k+2} x}{2n-2k+2} - (-1)^n \ln\cos x.$

3. $\int \mathrm{tg}^{2n} x\,dx = \sum\limits_{k=1}^{n}(-1)^{k-1}\dfrac{\mathrm{tg}^{2n-2k+1} x}{2n-2k+1} + (-1)^n x.$ GHH[331](12)

4. $\int \mathrm{ctg}^p x\,dx = -\dfrac{\mathrm{ctg}^{p-1} x}{p-1} - \int \mathrm{ctg}^{p-2} x\,dx \quad [p \ne 1].$

5. $\int \mathrm{ctg}^{2n+1} x\,dx = \sum\limits_{k=1}^{n}(-1)^{n+k+1}\binom{n}{k}\dfrac{1}{2k\sin^{2k} x} + (-1)^n \ln\sin x =$
$= \sum\limits_{k=1}^{n}(-1)^k \dfrac{\mathrm{ctg}^{2n-2k+2} x}{2n-2k+2} + (-1)^n \ln\sin x.$

6. $\int \mathrm{ctg}^{2n} x\,dx = \sum\limits_{k=1}^{n}(-1)^k \dfrac{\mathrm{ctg}^{2n-2k+1} x}{2n-2k+1} + (-1)^n x.$ GHH[331](14)

Formeln speziellen Charakters für $p = 1, 2, 3, 4$ findet man unter 2.526 17., 2.526 33., 2.526 22., 2.526 38., 2.526 27., 2.526 43., 2.526 32., 2.526 48.

Formulae for special cases with $p = 1, 2, 3, 4$ are given in 2.526 17., 2.526 33., 2.526 22., 2.526 38., 2.526 27., 2.526 43., 2.526 32., and 2.526 48.

Sinus und Cosinus von Vielfachen des Bogens, von linearen und komplizierteren Funktionen des Arguments
Sines and cosines of multiple angles and of linear and more complicated functions of the argument

2.53–2.54

1. $\int \sin(ax+b)\,dx = -\dfrac{1}{a}\cos(ax+b).$ 2.531

2. $\int \cos(ax+b)\,dx = \dfrac{1}{a}\sin(ax+b).$

1. $\int \sin(ax+b)\sin(cx+d)\,dx = \dfrac{\sin[(a-c)x+b-d]}{2(a-c)} - \dfrac{\sin[(a+c)x+b+d]}{2(a+c)} \quad [a^2 \neq c^2].$ 2.532

2. $\int \sin(ax+b)\cos(cx+d)\,dx = -\dfrac{\cos[(a-c)x+b-d]}{2(a-c)} - \dfrac{\cos[(a+c)x+b+d]}{2(a+c)} \quad [a^2 \neq c^2].$

3. $\int \cos(ax+b)\cos(cx+d)\,dx = \dfrac{\sin[(a-c)x+b-d]}{2(a-c)} + \dfrac{\sin[(a+c)x+b+d]}{2(a+c)} \quad [a^2 \neq c^2].$

Im Fall $c = a$ gelten die Formeln | For $c = a$:

4. $\int \sin(ax+b)\sin(ax+d)\,dx = \dfrac{x}{2}\cos(b-d) - \dfrac{\sin(2ax+b+d)}{4a}.$

5. $\int \sin(ax+b)\cos(ax+d)\,dx = \dfrac{x}{2}\sin(b-d) - \dfrac{\cos(2ax+b+d)}{4a}.$

6. $\int \cos(ax+b)\cos(ax+d)\,dx = \dfrac{x}{2}\cos(b-d) + \dfrac{\sin(2ax+b+d)}{4a}.$ GHH[332](3)

1. $\int \sin ax \cos bx\,dx = -\dfrac{\cos(a+b)x}{2(a+b)} - \dfrac{\cos(a-b)x}{2(a-b)} \quad [a^2 \neq b^2].$ 2.533

2. $\int \sin ax \sin bx \sin cx\,dx = -\dfrac{1}{4}\left\{\dfrac{\cos(a-b+c)x}{a-b+c} + \dfrac{\cos(b+c-a)x}{b+c-a} + \dfrac{\cos(a+b-c)x}{a+b-c} - \dfrac{\cos(a+b+c)x}{a+b+c}\right\}.$ P(376)

3. $\int \sin ax \cos bx \cos cx\,dx = -\dfrac{1}{4}\left\{\dfrac{\cos(a+b+c)x}{a+b+c} - \dfrac{\cos(b+c-a)x}{b+c-a} + \dfrac{\cos(a+b-c)x}{a+b-c} + \dfrac{\cos(a+c-b)x}{a+c-b}\right\}.$ P(378)

2.533

4. $\int \cos ax \sin bx \sin cx \, dx = \frac{1}{4} \left\{ \frac{\sin(a+b-c)x}{a+b-c} + \frac{\sin(a+c-b)x}{a+c-b} \right.$
$\left. - \frac{\sin(a+b+c)x}{a+b+c} - \frac{\sin(b+c-a)x}{b+c-a} \right\}.$ P(379)

5. $\int \cos ax \cos bx \cos cx \, dx = \frac{1}{4} \left\{ \frac{\sin(a+b+c)x}{a+b+c} + \frac{\sin(b+c-a)x}{b+c-a} + \right.$
$\left. + \frac{\sin(a+c-b)x}{a+c-b} + \frac{\sin(a+b-c)x}{a+b-c} \right\}.$ P(377)

2.534

1. $\int \frac{\cos px + i \sin px}{\sin nx} \, dx = -2 \int \frac{z^{p+n-1}}{1-z^{2n}} \, dz$ P(374)

2. $\int \frac{\cos px + i \sin px}{\cos nx} \, dx = -2i \int \frac{z^{p+n-1}}{1+z^{2n}} \, dz$ $\quad [z = \cos x + i \sin x].$ P(373)

2.535

1. $\int \sin^p x \sin ax \, dx = \frac{1}{p+a} \left\{ -\sin^p x \cos ax + p \int \sin^{p-1} x \cos(a-1)x \, dx \right\}.$
GHH[332](5a)

2. $\int \sin^p x \sin(2n+1)x \, dx = (2n+1) \left\{ \int \sin^{p+1} x \, dx + \right.$
$\left. + \sum_{k=1}^{n} (-1)^k \frac{[(2n+1)^2 - 1^2][(2n+1)^2 - 3^2] \ldots [(2n+1)^2 - (2k-1)^2]}{(2k+1)!} \int \sin^{2k+p+1} x \, dx \right\};$
T(299)

$= \frac{\Gamma(p+1)}{\Gamma\left(\frac{p+3}{2}+n\right)} \left\{ \sum_{k=0}^{n-1} \left[\frac{(-1)^{k-1}\Gamma\left(\frac{p+1}{2}+n-2k\right)}{2^{2k+1}\Gamma(p-2k+1)} \sin^{p-2k} x \cos(2n-2k+1)x + \right. \right.$

$\left. + (-1)^k \frac{\Gamma\left(\frac{p-1}{2}+n-2k\right)}{2^{2k+2}\Gamma(p-2k)} \sin^{p-2k-1} x \sin(2n-2k)x \right] +$

$\left. + \frac{(-1)^n \Gamma\left(\frac{p+3}{2}-n\right)}{2^{2n}\Gamma(p-2n+1)} \int \sin^{p-2n+1} x \, dx \right\}.$ GHH[332](5c)

3. $\int \sin^p x \sin 2nx \, dx = 2n \left\{ \frac{\sin^{p+2} x}{p+2} + \right.$
$\left. + \sum_{k=1}^{n-1} (-1)^k \frac{(4n^2 - 2^2)(4n^2 - 4^2) \ldots [4n^2 - (2k)^2]}{(2k+1)!(2k+p+2)} \sin^{2k+p+2} x \right\};$ T(303)

$= \frac{\Gamma(p+1)}{\Gamma\left(\frac{p}{2}+n+1\right)} \left\{ \sum_{k=0}^{n-1} \frac{(-1)^{k-1}\Gamma\left(\frac{p}{2}+n-2k\right)}{2^{2k+1}\Gamma(p-2k+1)} \sin^{p-2k} x \cos(2n-2k)x - \right.$

$\left. - \frac{(-1)^k \Gamma\left(\frac{p}{2}+n-2k-1\right)}{2^{2k+2}\Gamma(p-2k)} \sin^{p-2k-1} x \sin(2n-2k-1)x \right\};$

$[p \neq -2, -4, \ldots, -2n].$ GHH[332](5c)

1. $$\int \sin^p x \cos ax \, dx = \frac{1}{p+a} \left\{ \sin^p x \sin ax - p \int \sin^{p-1} x \sin(a-1)x \, dx \right\}.$$ 2.536

GHH[332](6a)

2. $$\int \sin^p x \cos(2n+1)x \, dx = \frac{\sin^{p+1} x}{p+1} +$$

$$+ \sum_{k=1}^{n} (-1)^k \frac{[(2n+1)^2 - 1^2][(2n+1)^2 - 3^2] \ldots [(2n+1)^2 - (2k-1)^2]}{(2k)!(2k+p+1)} \sin^{2k+p+1} x;$$

T(301)

$$= \frac{\Gamma(p+1)}{\Gamma\left(\frac{p+3}{2}+n\right)} \left\{ \sum_{k=0}^{n-1} \left[\frac{(-1)^k \Gamma\left(\frac{p+1}{2}+n-2k\right)}{2^{2k+1}\Gamma(p-2k+1)} \sin^{p-2k} x \sin(2n-2k+1)x + \right. \right.$$

$$+ \frac{(-1)^k \Gamma\left(\frac{p-1}{2}+n-2k\right)}{2^{2k+2}\Gamma(p-2k)} \sin^{p-2k-1} x \cos(2n-2k)x \bigg] +$$

$$+ \frac{(-1)^n \Gamma\left(\frac{p+3}{2}-n\right)}{2^{2n}\Gamma(p-2n+1)} \int \sin^{p-2n} x \cos x \, dx \bigg\};$$

$[p \neq -3, -5, \ldots, -(2n+1)]$. GHH[332](6c)

3. $$\int \sin^p x \cos 2nx \, dx = \int \sin^p x \, dx +$$

$$+ \sum_{k=1}^{n} (-1)^k \frac{4n^2 \cdot (4n^2 - 2^2) \ldots [4n^2 - (2k-2)^2]}{(2k)!} \int \sin^{2k+p} x \, dx;$$ T(300)

$$= \frac{\Gamma(p+1)}{\Gamma\left(\frac{p}{2}+n+1\right)} \left\{ \sum_{k=0}^{n-1} \left[\frac{(-1)^k \Gamma\left(\frac{p}{2}+n-2k\right)}{2^{2k+1}\Gamma(p-2k+1)} \sin^{p-2k} x \sin(2n-2k)x + \right. \right.$$

$$+ \frac{(-1)^k \Gamma\left(\frac{p}{2}+n-2k-1\right)}{2^{2k+2}\Gamma(p-2k)} \sin^{p-2k-1} x \cos(2n-2k-1)x \bigg] +$$

$$+ \frac{(-1)^n \Gamma\left(\frac{p}{2}-n+1\right)}{2^{2n}\Gamma(p-2n+1)} \int \sin^{p-2n} x \, dx \bigg\}.$$ GHH[332](6c)

1. $$\int \cos^p x \sin ax \, dx = \frac{1}{p+a} \left\{ -\cos^p x \cos ax + p \int \cos^{p-1} x \sin(a-1)x \, dx \right\}.$$ 2.537

GHH[332](7a)

2. $$\int \cos^p x \sin(2n+1)x \, dx = (-1)^{n+1} \left\{ \frac{\cos^{p+1} x}{p+1} + \right.$$

$$+ \sum_{k=1}^{n} (-1)^k \frac{[(2n+1)^2 - 1^2][(2n+1)^2 - 3^2] \ldots [(2n+1)^2 - (2k-1)^2]}{(2k)!(2k+p+1)} \cos^{2k+p+1} x \bigg\};$$

T(295)

2.537

$$= \frac{\Gamma(p+1)}{\Gamma\left(\frac{p+3}{2}+n\right)} \left\{ -\sum_{k=0}^{n-1} \frac{\Gamma\left(\frac{p+1}{2}+n-k\right)}{2^{k+1}\Gamma(p-k+1)} \cos^{p-k}x \cos(2n-k+1)x \right.$$

$$\left. + \frac{\Gamma\left(\frac{p+3}{2}\right)}{2^n \Gamma(p-n+1)} \int \cos^{p-n} x \sin(n+1)x\, dx \right\};$$

$$[p \neq -3, -5, \ldots, -(2n+1)]. \qquad \text{GHH}[332](7\text{b})+$$

3. $\int \cos^p x \sin 2nx\, dx = (-1)^n \left\{ \frac{\cos^{p+2} x}{p+2} \right.$

$$\left. + \sum_{k=1}^{n-1} (-1)^k \frac{(4n^2-2^2)(4n^2-4^2)\ldots[4n^2-(2k)^2]}{(2k+1)!\,(2k+p+2)} \cos^{2k+p+2} x \right\}; \qquad \text{T(297)}$$

$$= \frac{\Gamma(p+1)}{\Gamma\left(\frac{p}{2}+n+1\right)} \left\{ -\sum_{k=0}^{n-1} \frac{\Gamma\left(\frac{p}{2}+n-k\right)}{2^{k+1}\Gamma(p-k+1)} \cos^{p-k}x \cos(2n-k)x \right.$$

$$\left. + \frac{\Gamma\left(\frac{p}{2}+1\right)}{2^n \Gamma(p-n+1)} \int \cos^{p-n} x \sin nx\, dx \right\};$$

$$[p \neq -2, -4, \ldots, -2n]. \qquad \text{GHH}[332](7\text{b})+$$

2.538

1. $\int \cos^p x \cos ax\, dx = \frac{1}{p+a} \left\{ \cos^p x \sin ax + p \int \cos^{p-1} x \cos(a-1)x\, dx \right\}.$

$$\text{GHH}[332](8\text{a})$$

2. $\int \cos^p x \cos(2n+1)x\, dx = (-1)^n (2n+1) \left\{ \int \cos^{p+1} x\, dx \right.$

$$\left. + \sum_{k=1}^{n} (-1)^k \frac{[(2n+1)^2-1^2][(2n+1)^2-3^2]\ldots[(2n+1)^2-(2k-1)^2]}{(2k+1)!} \int \cos^{2k+p+1} x\, dx \right\};$$

$$\text{T(293)}$$

$$= \frac{\Gamma(p+1)}{\Gamma\left(\frac{p+3}{2}+n\right)} \left\{ \sum_{k=0}^{n-1} \frac{\Gamma\left(\frac{p+1}{2}+n-k\right)}{2^{k+1}\Gamma(p-k+1)} \cos^{p-k} x \sin(2n-k+1)x \right.$$

$$\left. + \frac{\Gamma\left(\frac{p+3}{2}\right)}{2^n \Gamma(p-n+1)} \int \cos^{p-n} x \cos(n+1)x\, dx \right\}. \qquad \text{GHH}[332](8\text{b})+$$

3. $\int \cos^p x \cos 2nx\, dx = (-1)^n \left\{ \int \cos^p x\, dx \right.$

$$\left. + \sum_{k=1}^{n} (-1)^k \frac{4n^2 [4n^2-2^2]\ldots[4n^2-(2k-2)^2]}{(2k)!} \int \cos^{2k+p} x\, dx \right\}; \qquad \text{T(294)}$$

$$= \frac{\Gamma(p+1)}{\Gamma\left(\frac{p}{2}+n+1\right)} \left\{ \sum_{k=0}^{n-1} \frac{\Gamma\left(\frac{p}{2}+n-k\right)}{2^{k+1}\Gamma(p-k+1)} \cos^{p-k} x \sin(2n-k)x + \right.$$

$$\left. + \frac{\Gamma\left(\frac{p}{2}+1\right)}{2^n \Gamma(p-n+1)} \int \cos^{p-n} x \cos nx \, dx \right\}. \quad \text{GHH[332](8b)} +$$

2.538

1. $\int \frac{\sin(2n+1)x}{\sin x} dx = 2 \sum_{k=1}^{n} \frac{\sin 2kx}{2k} + x.$ 2.539

2. $\int \frac{\sin 2nx}{\sin x} dx = 2 \sum_{k=1}^{n} \frac{\sin(2k-1)x}{2k-1}.$ GHH[332](5e)

3. $\int \frac{\cos(2n+1)x}{\sin x} dx = 2 \sum_{k=1}^{n} \frac{\cos 2kx}{2k} + \ln \sin x.$

4. $\int \frac{\cos 2nx}{\sin x} dx = 2 \sum_{k=1}^{n} \frac{\cos(2k-1)x}{2k-1} + \ln \operatorname{tg} \frac{x}{2}.$ GHH[332](6e)

5. $\int \frac{\sin(2n+1)x}{\cos x} dx = 2 \sum_{k=1}^{n} (-1)^{n-k+1} \frac{\cos 2kx}{2k} + (-1)^{n+1} \ln \cos x.$

6. $\int \frac{\sin 2nx}{\cos x} dx = 2 \sum_{k=1}^{n} (-1)^{n-k+1} \frac{\cos(2k-1)x}{2k-1}.$ GHH[332](7d)

7. $\int \frac{\cos(2n+1)x}{\cos x} dx = 2 \sum_{k=1}^{n} (-1)^{n-k} \frac{\sin 2kx}{2k} + (-1)^n x.$

8. $\int \frac{\cos 2nx}{\cos x} dx = 2 \sum_{k=1}^{n} (-1)^{n-k} \frac{\sin(2k-1)x}{2k-1} + (-1)^n \ln \operatorname{tg}\left(\frac{\pi}{4}+\frac{x}{2}\right).$

GHH[332](8d)

1. $\int \sin(n+1)x \sin^{n-1} x \, dx = \frac{1}{n} \sin^n x \sin nx.$ BH[71](1)+ 2.541

2. $\int \sin(n+1)x \cos^{n-1} x \, dx = -\frac{1}{n} \cos^n x \cos nx.$ BH[71](2)+

3. $\int \cos(n+1)x \sin^{n-1} x \, dx = \frac{1}{n} \sin^n x \cos nx.$ BH[71](3)+

4. $\int \cos(n+1)x \cos^{n-1} x \, dx = \frac{1}{n} \cos^n x \sin nx.$ BH[71](4)+

5. $\int \sin\left[(n+1)\left(\frac{\pi}{2}-x\right)\right] \sin^{n-1} x \, dx = \frac{1}{n} \sin^n x \cos n\left(\frac{\pi}{2}-x\right).$ BH[71](5)+

6. $\int \cos\left[(n+1)\left(\frac{\pi}{2}-x\right)\right] \sin^{n-1} x \, dx = -\frac{1}{n} \sin^n x \sin n\left(\frac{\pi}{2}-x\right).$ BH[71](6)+

1. $\int \frac{\sin 2x}{\sin^n x} dx = -\frac{2}{(n-2)\sin^{n-2} x}.$ 2.542

2.542 Für $n=2$ gilt | For $n=2$:

2. $\int \frac{\sin 2x}{\sin^2 x} dx = 2 \ln \sin x.$

2.543 1. $\int \frac{\sin 2x \, dx}{\cos^n x} = \frac{2}{(n-2)\cos^{n-2} x}.$

Für $n=2$ gilt | For $n=2$:

2. $\int \frac{\sin 2x}{\cos^2 x} dx = -2 \ln \cos x.$

2.544 1. $\int \frac{\cos 2x \, dx}{\sin x} = 2 \cos x + \ln \operatorname{tg} \frac{x}{2}.$

2. $\int \frac{\cos 2x \, dx}{\sin^2 x} = -\operatorname{ctg} x - 2x.$

3. $\int \frac{\cos 2x \, dx}{\sin^3 x} = -\frac{\cos x}{2 \sin^2 x} - \frac{3}{2} \ln \operatorname{tg} \frac{x}{2}.$

4. $\int \frac{\cos 2x \, dx}{\cos x} = 2 \sin x - \ln \operatorname{tg} \left(\frac{\pi}{4} + \frac{x}{2}\right).$

5. $\int \frac{\cos 2x \, dx}{\cos^2 x} = 2x - \operatorname{tg} x.$

6. $\int \frac{\cos 2x \, dx}{\cos^3 x} = -\frac{\sin x}{2 \cos^2 x} + \frac{3}{2} \ln \operatorname{tg} \left(\frac{\pi}{4} + \frac{x}{2}\right).$

7. $\int \frac{\sin 3x \, dx}{\sin x} = x + \sin 2x.$

8. $\int \frac{\sin 3x}{\sin^2 x} dx = 3 \ln \operatorname{tg} \frac{x}{2} + 4 \cos x.$

9. $\int \frac{\sin 3x}{\sin^3 x} dx = -3 \operatorname{ctg} x - 4x.$

2.545 1. $\int \frac{\sin 3x}{\cos^n x} dx = \frac{4}{(n-3)\cos^{n-3} x} - \frac{1}{(n-1)\cos^{n-1} x}.$

Für $n=1$ und $n=3$ gilt | For $n=1$ and $n=3$:

2. $\int \frac{\sin 3x}{\cos x} dx = 2 \sin^2 x + \ln \cos x.$

3. $\int \frac{\sin 3x}{\cos^3 x} dx = -\frac{1}{2 \cos^2 x} - 4 \ln \cos x.$

2.546 1. $\int \frac{\cos 3x}{\sin^n x} dx = \frac{4}{(n-3)\sin^{n-3} x} - \frac{1}{(n-1)\sin^{n-1} x}.$

Für $n=1$ und $n=3$ gilt | For $n=1$ and $n=3$:

2. $\int \frac{\cos 3x}{\sin x} dx = -2 \sin^2 x + \ln \sin x.$

3. $\int \frac{\cos 3x}{\sin^3 x} dx = -\frac{1}{2 \sin^2 x} - 4 \ln \sin x.$

1. $$\int \frac{\sin nx}{\cos^p x}\,dx = 2\int \frac{\sin(n-1)x\,dx}{\cos^{p-1} x} - \int \frac{\sin(n-2)x\,dx}{\cos^p x}.$$ 2.547

2. $$\int \frac{\cos 3x}{\cos x}\,dx = \sin 2x - x.$$

3. $$\int \frac{\cos 3x}{\cos^2 x}\,dx = 4\sin x - 3\ln\operatorname{tg}\left(\frac{\pi}{4}+\frac{x}{2}\right).$$

4. $$\int \frac{\cos 3x}{\cos^3 x}\,dx = 4x - 3\operatorname{tg} x.$$

1. $$\int \frac{\sin^m x\,dx}{\sin(2n+1)x} = $$ 2.548

$$= \frac{1}{2n+1}\sum_{k=0}^{2n}(-1)^{n+k}\cos^m\left[\frac{2k+1}{2(2n+1)}\pi\right]\ln\frac{\sin\left[\dfrac{(k-n)\pi}{2(2n+1)}+\dfrac{x}{2}\right]}{\sin\left[\dfrac{k+n+1}{2(2n+1)}\pi-\dfrac{x}{2}\right]}$$

[m eine natürliche Zahl, $m \leq 2n$]. | [m positive integer, $m \leq 2n$].
T(378) | T(378)

2. $$\int \frac{\sin^{2m} x\,dx}{\sin 2nx} = \frac{(-1)^n}{2n}\left\{\ln\cos x + \sum_{k=1}^{n-1}(-1)^k\cos^{2m}\frac{k\pi}{2n}\ln\left(\cos^2 x - \sin^2\frac{k\pi}{2n}\right)\right\}$$

[m eine natürliche Zahl, $m \leq n$]. | [m positive integer, $m \leq n$].
T(379) | T(379)

3. $$\int \frac{\sin^{2m+1} x}{\sin 2nx}\,dx = \frac{(-1)^n}{2n}\left\{\ln\operatorname{tg}\left(\frac{\pi}{4}-\frac{x}{2}\right) + \right.$$
$$\left. + \sum_{k=1}^{n-1}(-1)^k\cos^{2m+1}\frac{k\pi}{2n}\ln\left[\operatorname{tg}\left(\frac{n+k}{4n}\pi-\frac{x}{2}\right)\operatorname{tg}\left(\frac{n-k}{4n}\pi-\frac{x}{2}\right)\right]\right\}$$

[m eine natürliche Zahl, $m < n$]. | [m positive integer, $m < n$].
T(380) | T(380)

4. $$\int \frac{\sin^{2m} x\,dx}{\cos(2n+1)x} = \frac{(-1)^{n+1}}{2n+1}\left\{\ln\operatorname{tg}\left(\frac{\pi}{4}-\frac{x}{2}\right)+\right.$$
$$\left.+\sum_{k=1}^{n}(-1)^k\cos^{2m}\frac{k\pi}{2n+1}\ln\left[\operatorname{tg}\left(\frac{2n+2k+1}{4(2n+1)}\pi-\frac{x}{2}\right)\operatorname{tg}\left(\frac{2n-2k+1}{4(2n+1)}\pi-\frac{x}{2}\right)\right]\right\}$$

[m eine natürliche Zahl, $m \leq n$]. | [m positive integer, $m \leq n$].
T(381) | T(381)

5. $$\int \frac{\sin^{2m+1} x\,dx}{\cos(2n+1)x} = \frac{(-1)^{n+1}}{2n+1}\left\{\ln\cos x + \right.$$
$$\left.+\sum_{k=1}^{n}(-1)^k\cos^{2m+1}\frac{k\pi}{2n+1}\ln\left(\cos^2 x - \sin^2\frac{k\pi}{2n+1}\right)\right\}$$

[m ist eine natürliche Zahl, $m \leq n$]. | [m positive integer, $m \leq n$].
T(382)+ | T(382)+

6. $$\int \frac{\sin^m x\,dx}{\cos 2nx} = \frac{1}{2n}\sum_{k=0}^{2n-1}(-1)^{n+k}\cos^m\left[\frac{2k+1}{4n}\pi\right]\ln\frac{\sin\left[\dfrac{2k-2n+1}{8n}\pi+\dfrac{x}{2}\right]}{\sin\left[\dfrac{2k+2n+1}{8n}\pi-\dfrac{x}{2}\right]}$$

2.548   [m ist eine natürliche Zahl, $m < 2n$]. | [m positive integer, $m < 2n$].
                    T(377)   T(377)

7. $$\int \frac{\cos^{2m+1} x \, dx}{\sin (2n+1) x} = \frac{1}{2n+1} \left\{ \ln \sin x + \sum_{k=1}^{n} (-1)^k \cos^{2m+1} \frac{k\pi}{2n+1} \ln \left(\sin^2 x - \sin^2 \frac{k\pi}{2n+1} \right) \right\}$$

[m ist eine natürliche Zahl, $m \leqslant n$]. | [m positive integer, $m \leqslant n$].
                  T(376)   T(376)

8. $$\int \frac{\cos^{2m} x \, dx}{\sin (2n+1) x} = \frac{1}{2n+1} \left\{ \ln \operatorname{tg} \frac{x}{2} + \sum_{k=1}^{n} (-1)^k \cos^{2m} \frac{k\pi}{2n+1} \ln \left[\operatorname{tg} \left(\frac{x}{2} + \frac{k\pi}{4n+2} \right) \operatorname{tg} \left(\frac{x}{2} - \frac{k\pi}{4n+2} \right) \right] \right\}$$

[m ist eine natürliche Zahl, $m \leqslant n$]. | [m positive integer, $m \leqslant n$].
                  T(375)   T(375)

9. $$\int \frac{\cos^{2m+1} x}{\sin 2nx} dx = \frac{1}{2n} \left\{ \ln \operatorname{tg} \frac{x}{2} + \sum_{k=1}^{n-1} (-1)^k \cos^{2m+1} \frac{k\pi}{2n} \ln \left[\operatorname{tg} \left(\frac{x}{2} + \frac{k\pi}{4n} \right) \operatorname{tg} \left(\frac{x}{2} - \frac{k\pi}{4n} \right) \right] \right\}$$

[m ist eine natürliche Zahl, $m < n$]. | [m positive integer, $m < n$].
                  T(374)   T(374)

10. $$\int \frac{\cos^{2m} x}{\sin 2nx} dx = \frac{1}{2n} \left\{ \ln \sin x + \sum_{k=1}^{n-1} (-1)^k \cos^{2m} \frac{k\pi}{2n} \ln \left(\sin^2 x - \sin^2 \frac{k\pi}{2n} \right) \right\}$$

[m ist eine natürliche Zahl, $m \leqslant n$]. | [m positive integer, $m \leqslant n$].
                  T(373)   T(373)

11. $$\int \frac{\cos^m x}{\cos nx} dx = \frac{1}{n} \sum_{k=0}^{n-1} (-1)^k \cos^m \frac{2k+1}{2n} \pi \ln \frac{\sin\left[\frac{2k+1}{4n}\pi + \frac{x}{2}\right]}{\sin\left[\frac{2k+1}{4n}\pi - \frac{x}{2}\right]}$$

[m ist eine natürliche Zahl, $m \leqslant n$]. | [m positive integer, $m \leqslant n$].
                  T(372)   T(372)

2.549

1. $$\int \sin x^2 \, dx = \sqrt{\frac{\pi}{2}} \, S(x).$$

2. $$\int \cos x^2 \, dx = \sqrt{\frac{\pi}{2}} \, C(x).$$

3. $$\int \sin(ax^2 + 2bx + c) \, dx = \sqrt{\frac{\pi}{2a}} \left\{ \cos \frac{ac-b^2}{a} S\left(\frac{ax+b}{\sqrt{a}}\right) + \sin \frac{ac-b^2}{a} C\left(\frac{ax+b}{\sqrt{a}}\right) \right\}.$$

4. $\int \cos(ax^2 + 2bx + c)\, dx = \sqrt{\dfrac{\pi}{2a}} \left\{ \cos \dfrac{ac - b^2}{a} C\left(\dfrac{ax + b}{\sqrt{a}}\right) - \sin \dfrac{ac - b^2}{a} S\left(\dfrac{ax + b}{\sqrt{a}}\right) \right\}.$ 2.549

5. $\int \sin \ln x\, dx = \dfrac{x}{2}(\sin \ln x - \cos \ln x).$ P(444)

6. $\int \cos \ln x\, dx = \dfrac{x}{2}(\sin \ln x + \cos \ln x).$ P(445)

Rationale Funktionen von Sinus und Cosinus
Rational functions of the sine and cosine
2.55–2.56

1. $\int \dfrac{A + B\sin x}{(a + b\sin x)^n}\, dx = \dfrac{1}{(n-1)(a^2 - b^2)} \left[\dfrac{(Ab - aB)\cos x}{(a + b\sin x)^{n-1}} + \int \dfrac{(Aa - Bb)(n-1) + (aB - bA)(n-2)\sin x}{(a + b\sin x)^{n-1}}\, dx \right].$ 2.551 T(358)+

Für $n = 1$ gilt | For $n = 1$:

2. $\int \dfrac{A + B\sin x}{a + b\sin x}\, dx = \dfrac{B}{b}x + \dfrac{Ab - aB}{b} \int \dfrac{dx}{a + b\sin x}$ (cf. 2.551 3.). T(342)

3. $\int \dfrac{dx}{a + b\sin x} = \dfrac{2}{\sqrt{a^2 - b^2}} \arctan \dfrac{a\,\mathrm{tg}\,\dfrac{x}{2} + b}{\sqrt{a^2 - b^2}}$ $[a^2 > b^2];$

$= \dfrac{1}{\sqrt{b^2 - a^2}} \ln \dfrac{a\,\mathrm{tg}\,\dfrac{x}{2} + b - \sqrt{b^2 - a^2}}{a\,\mathrm{tg}\,\dfrac{x}{2} + b + \sqrt{b^2 - a^2}}$ $[a^2 < b^2].$

1. $\int \dfrac{A + B\cos x}{(a + b\sin x)^n}\, dx = -\dfrac{B}{(n-1)b(a + b\sin x)^{n-1}} + A \int \dfrac{dx}{(a + b\sin x)^n}$ 2.552

(cf. 2.552 3.). T(361)

Für $n = 1$ gilt | For $n = 1$:

2. $\int \dfrac{A + B\cos x}{a + b\sin x}\, dx = \dfrac{B}{b}\ln(a + b\sin x) + A \int \dfrac{dx}{a + b\sin x}$

(cf. 2.551 3.). T(344)

3. $\int \dfrac{dx}{(a + b\sin x)^n} = \dfrac{1}{(n-1)(a^2 - b^2)} \left\{ \dfrac{b\cos x}{(a + b\sin x)^{n-1}} + \int \dfrac{(n-1)a - (n-2)b\sin x}{(a + b\sin x)^{n-1}}\, dx \right\}$ (cf. 2.551 1.). T(359)

1. $\int \dfrac{A + B\sin x}{(a + b\cos x)^n}\, dx = \dfrac{B}{(n-1)b(a + b\cos x)^{n-1}} + A \int \dfrac{dx}{(a + b\cos x)^n}$ 2.553

(cf. 2.554 3.). T(355)

Unbestimmte Integrale elem. Funktionen/Indefinite Integrals of Elem. Functions

2.553   Für $n = 1$ gilt | For $n = 1$:

2. $\int \dfrac{A + B \sin x}{a + b \cos x} dx = -\dfrac{B}{b} \ln (a + b \cos x) + A \int \dfrac{dx}{a + b \cos x}$

(cf. 2.553 3.).   T(343)

3. $\int \dfrac{dx}{a + b \cos x} = \dfrac{2}{\sqrt{a^2 - b^2}} \operatorname{arctg} \dfrac{\sqrt{a^2 - b^2} \operatorname{tg} \dfrac{x}{2}}{a + b}$ $[a^2 > b^2]$;

$= \dfrac{1}{\sqrt{b^2 - a^2}} \ln \dfrac{\sqrt{b^2 - a^2} \operatorname{tg} \dfrac{x}{2} + a + b}{\sqrt{b^2 - a^2} \operatorname{tg} \dfrac{x}{2} - a - b}$ $[a^2 < b^2]$.   F II 85, T(305)

2.554   1. $\int \dfrac{A + B \cos x}{(a + b \cos x)^n} dx = \dfrac{1}{(n-1)(a^2 - b^2)} \left[\dfrac{(aB - Ab) \sin x}{(a + b \cos x)^{n-1}} + \int \dfrac{(Aa - bB)(n-1) + (n-2)(aB - bA) \cos x}{a + b \cos x)^{n-1}} dx \right].$ T(353)

Für $n = 1$ gilt | For $n = 1$:

2. $\int \dfrac{A + B \cos x}{a + b \cos x} dx = \dfrac{B}{b} x + \dfrac{Ab - aB}{b} \int \dfrac{dx}{a + b \cos x}$ (cf. 2.553 3.).   T(341)

3. $\int \dfrac{dx}{(a + b \cos x)^n} = -\dfrac{1}{(n-1)(a^2 - b^2)} \left\{ \dfrac{b \sin x}{(a + b \cos x)^{n-1}} - \int \dfrac{(n-1)a - (n-2)b \cos x}{(a + b \cos x)^{n-1}} dx \right\}$ (cf. 2.554 1.).   T(354)

Bei der Integration in 2.551 3. und 2.553 3. darf nicht über Punkte integriert werden, in denen der Integrand unendlich wird, d.h. über die Punkte $x = \arcsin \left(-\dfrac{a}{b} \right)$ in 2.551 3. und die Punkte $x = \arccos \left(-\dfrac{a}{b} \right)$ in 2.553 3.

In 2.551 3. and 2.553 3. the function must not be integrated over the points where the integrand becomes infinite, that is, the point $x = \arcsin \left(-\dfrac{a}{b} \right)$ in 2.551 3. and the point $x = \arccos \left(-\dfrac{a}{b} \right)$ in 2.553 3.

2.555   Die Formeln 2.551 3. und 2.553 3. gelten nicht für $a^2 = b^2$. In diesen Fällen sind folgende Formeln zu benutzen:

Formulae 2.551 3. and 2.553 3. cannot be used if $a^2 = b^2$. In this case one has to use the following formulae:

1. $\int \dfrac{A + B \sin x}{(1 \pm \sin x)^n} dx = -\dfrac{1}{2^{n-1}} \left\{ 2B \sum_{k=0}^{n-2} \binom{n-2}{k} \dfrac{\operatorname{tg}^{2k+1}\left(\dfrac{\pi}{4} \mp \dfrac{x}{2} \right)}{2k+1} \pm \right.$

$\left. \pm (A \mp B) \sum_{k=0}^{n-1} \binom{n-1}{k} \dfrac{\operatorname{tg}^{2k+1}\left(\dfrac{\pi}{4} \mp \dfrac{x}{2} \right)}{2k+1} \right\}.$ T(361)+

2. $\int \dfrac{A + B \cos x}{(1 \pm \cos x)^n} dx = \dfrac{1}{2^{n-1}} \left\{ 2B \sum_{k=0}^{n-2} \binom{n-2}{k} \dfrac{\operatorname{tg}^{2k+1}\left[\dfrac{\pi}{4} \mp \left(\dfrac{\pi}{4} - \dfrac{x}{2} \right) \right]}{2k+1} \pm \right.$

$\left. \pm (A \mp B) \sum_{k=0}^{n-1} \binom{n-1}{k} \dfrac{\operatorname{tg}^{2k+1}\left[\dfrac{\pi}{4} \mp \left(\dfrac{\pi}{4} - \dfrac{x}{2} \right) \right]}{2k+1} \right\}.$ T(356)

Für $n=1$ gelten die Formeln | For $n=1$: 2.555

3. $\int \dfrac{A+B\sin x}{1\pm \sin x}\,dx = \pm Bx+(A\mp B)\,\text{tg}\left(\dfrac{\pi}{4}\mp\dfrac{x}{2}\right).$ T(250)

4. $\int \dfrac{A+B\cos x}{1\pm \cos x}\,dx = \pm Bx\pm(A\mp B)\,\text{tg}\left[\dfrac{\pi}{4}\mp\left(\dfrac{\pi}{4}-\dfrac{x}{2}\right)\right].$ T(248)

1. $\int \dfrac{(1-a^2)\,dx}{1-2a\cos x+a^2} = 2\,\text{arctg}\left(\dfrac{1+a}{1-a}\,\text{tg}\,\dfrac{x}{2}\right)\quad [0<a<1,\ |x|<\pi].$ F II 84 2.556

2. $\int \dfrac{(1-a\cos x)\,dx}{1-2a\cos x+a^2} = \dfrac{x}{2}+\text{arctg}\left(\dfrac{1+a}{1-a}\,\text{tg}\,\dfrac{x}{2}\right)\quad [0<a<1,\ |x|<\pi].$ F II 85

1. $\int \dfrac{dx}{(a\cos x+b\sin x)^n} = \dfrac{1}{\sqrt{(a^2+b^2)^n}}\int \dfrac{dx}{\sin^n\left(x+\text{arctg}\,\dfrac{a}{b}\right)}$ (cf. 2.515). MC 173+ 2.557

2. $\int \dfrac{\sin x\,dx}{a\sin x+b\cos x} = \dfrac{ax-b\ln\sin\left(x+\text{arctg}\,\dfrac{b}{a}\right)}{a^2+b^2}.$

3. $\int \dfrac{\cos x\,dx}{a\cos x+b\sin x} = \dfrac{ax+b\ln\sin\left(x+\text{arctg}\,\dfrac{a}{b}\right)}{a^2+b^2}.$ MC 174+

4. $\int \dfrac{dx}{a\cos x+b\sin x} = \dfrac{\ln\,\text{tg}\left[\dfrac{1}{2}\left(x+\text{arctg}\,\dfrac{a}{b}\right)\right]}{\sqrt{a^2+b^2}}.$

5. $\int \dfrac{dx}{(a\cos x+b\sin x)^2} = -\dfrac{\text{ctg}\left(x+\text{arctg}\,\dfrac{a}{b}\right)}{a^2+b^2} =$

$= \dfrac{1}{a^2+b^2}\cdot\dfrac{a\sin x-b\cos x}{a\cos x+b\sin x}.$ MC 174+

1. $\int \dfrac{A+B\cos x+C\sin x}{(a+b\cos x+c\sin x)^n}\,dx = \dfrac{(Bc-Cb)+(Ac-Ca)\cos x-(Ab-Ba)\sin x}{(n-1)(a^2-b^2-c^2)(a+b\cos x+c\sin x)^{n-1}}+$ 2.558

$+\dfrac{1}{(n-1)(a^2-b^2-c^2)}\int \dfrac{(n-1)(Aa-Bb-Cc)-(n-2)[(Ab-Ba)\cos x-(Ac-Ca)\sin x]}{(a+b\cos x+c\sin x)^{n-1}}\,dx$

$[n\neq 1,\ a^2\neq b^2+c^2];$

$= \dfrac{Cb-Bc+Ca\cos x-Ba\sin x}{(n-1)a(a+b\cos x+c\sin x)^n}+\left(\dfrac{A}{a}+\dfrac{n(Bb+Cc)}{(n-1)a^2}\right)(-c\cos x+b\sin x)\times$

$\times \dfrac{(n-1)!}{(2n-1)!!}\sum_{k=0}^{n-1}\dfrac{(2n-2k-3)!!}{(n-k-1)!\,a^k}\cdot\dfrac{1}{(a+b\cos x+c\sin x)^{n-k}}\quad [n\neq 1,\ a^2=b^2+c^2].$

Für $n=1$ gilt | For $n=1$:

2. $\int \dfrac{A+B\cos x+C\sin x}{a+b\cos x+c\sin x}\,dx = \dfrac{Bc-Cb}{b^2+c^2}\ln(a+b\cos x+c\sin x)+\dfrac{Bb+Cc}{b^2+c^2}x+$

$+\left(A-\dfrac{Bb+Cc}{b^2+c^2}a\right)\int \dfrac{dx}{a+b\cos x+c\sin x}$ (cf. 2.558 4.). GHH[331](18)

2.558   3. $\displaystyle\int\frac{dx}{(a+b\cos x+c\sin x)^n}=\int\frac{d(x-\alpha)}{[a+r\cos(x-\alpha)]^n}$

mit $b=r\cos\alpha$, $c=r\sin\alpha$ (vgl. 2.554 3.). | where $b=r\cos\alpha$, $c=r\sin\alpha$ (see 2.554 3.).

4. $\displaystyle\int\frac{dx}{a+b\cos x+c\sin x}=$

$\displaystyle=\frac{2}{\sqrt{a^2-b^2-c^2}}\,\mathrm{arctg}\,\frac{(a-b)\,\mathrm{tg}\,\frac{x}{2}+c}{\sqrt{a^2-b^2-c^2}}\quad [a^2>b^2+c^2];\qquad\text{T(253), F II 85–86}$

$\displaystyle=\frac{1}{\sqrt{b^2+c^2-a^2}}\,\ln\frac{(a-b)\,\mathrm{tg}\,\frac{x}{2}+c-\sqrt{b^2+c^2-a^2}}{(a-b)\,\mathrm{tg}\,\frac{x}{2}+c+\sqrt{b^2+c^2-a^2}}\quad [a^2<b^2+c^2];\text{T(253)}+$

$\displaystyle=\frac{1}{c}\ln\left(a+c\cdot\mathrm{tg}\,\frac{x}{2}\right)\quad [a=b];$

$\displaystyle=\frac{-2}{c+(a-b)\,\mathrm{tg}\,\frac{x}{2}}\quad [a^2=b^2+c^2].\qquad\text{T(253)}+$

2.559   1. $\displaystyle\int\frac{dx}{[a(1+\cos x)+c\sin x]^2}=\frac{1}{c^3}\left[\frac{c(a\sin x-c\cos x)}{a(1+\cos x)+c\sin x}-a\ln\left(a+c\,\mathrm{tg}\,\frac{x}{2}\right)\right].$

2. $\displaystyle\int\frac{A+B\cos x+C\sin x}{(a_1+b_1\cos x+c_1\sin x)(a_2+b_2\cos x+c_2\sin x)}\,dx=$

$\displaystyle=A_0\ln\frac{a_1+b_1\cos x+c_1\sin x}{a_2+b_2\cos x+c_2\sin x}+A_1\int\frac{dx}{a_1+b_1\cos x+c_1\sin x}+$

$\displaystyle\qquad\qquad\qquad\qquad\qquad\qquad\qquad +A_2\int\frac{dx}{a_2+b_2\cos x+c_2\sin x}$

mit | where

$$A_0=\frac{\begin{vmatrix}A&B&C\\a_1&b_1&c_1\\a_2&b_2&c_2\end{vmatrix}}{\begin{vmatrix}a_1&b_1\\a_2&b_2\end{vmatrix}^2-\begin{vmatrix}b_1&c_1\\b_2&c_2\end{vmatrix}^2+\begin{vmatrix}c_1&a_1\\c_2&a_2\end{vmatrix}^2},\qquad A_1=\frac{\begin{vmatrix}B&C\\b_1&c_1\end{vmatrix}a_1 + \begin{vmatrix}A&C\\a_1&c_1\end{vmatrix}b_1 + \begin{vmatrix}B&A\\b_1&a_1\end{vmatrix}c_1 }{\begin{vmatrix}a_1&b_1\\a_2&b_2\end{vmatrix}^2-\begin{vmatrix}b_1&c_1\\b_2&c_2\end{vmatrix}^2+\begin{vmatrix}c_1&a_1\\c_2&a_2\end{vmatrix}^2},$$

$$A_2=\frac{\begin{vmatrix}C&B\\c_2&b_2\end{vmatrix}a_1 + \begin{vmatrix}C&A\\c_2&a_2\end{vmatrix}b_1 + \begin{vmatrix}A&B\\a_2&b_2\end{vmatrix}c_1 }{\begin{vmatrix}a_1&b_1\\a_2&b_2\end{vmatrix}^2-\begin{vmatrix}b_1&c_1\\b_2&c_2\end{vmatrix}^2+\begin{vmatrix}c_1&a_1\\c_2&a_2\end{vmatrix}^2};$$

$\left[\begin{vmatrix}a_1&b_1\\a_2&b_2\end{vmatrix}^2+\begin{vmatrix}c_1&a_1\\c_2&a_2\end{vmatrix}^2\neq\begin{vmatrix}b_1&c_1\\b_2&c_2\end{vmatrix}^2\right]$ (cf. 2.5584.).   GHH[331](19)

3. $\int \dfrac{A\cos^2 x + 2B\sin x \cos x + C\sin^2 x}{a\cos^2 x + 2b\sin x \cos x + c\sin^2 x}\, dx =$ 2.559

$= \dfrac{1}{4b^2 + (a-c)^2}\{[4Bb + (A-C)(a-c)]x + [(A-C)b - B(a-c)] \times$

$\times \ln(a\cos^2 x + 2b\sin x \cos x + c\sin^2 x) +$

$+ [2(A+C)b^2 - 2Bb(a+c) + (aC - Ac)(a-c)]f(x)\}$

mit | where

$f(x) = \dfrac{1}{2\sqrt{b^2 - ac}} \ln \dfrac{c\,\mathrm{tg}\, x + b - \sqrt{b^2 - ac}}{c\,\mathrm{tg}\, x + b + \sqrt{b^2 - ac}} \quad [b^2 > ac];$

$= \dfrac{1}{\sqrt{ac - b^2}} \,\mathrm{arctg}\, \dfrac{c\,\mathrm{tg}\, x + b}{\sqrt{ac - b^2}} \quad [b^2 < ac];$

$= -\dfrac{1}{c\,\mathrm{tg}\, x + b} \quad [b^2 = ac].$ GHH [331](24)

1. $\int \dfrac{(A + B\sin x)\,dx}{\sin x(a + b\sin x)} = \dfrac{A}{a} \ln \mathrm{tg}\, \dfrac{x}{2} + \dfrac{Ba - Ab}{a}\int \dfrac{dx}{a + b\sin x}$ (cf. 2.551 3.). T(348) 2.561

2. $\int \dfrac{(A + B\sin x)\,dx}{\sin x(a + b\cos x)} = \dfrac{A}{a^2 - b^2}\left\{a \ln \mathrm{tg}\, \dfrac{x}{2} + b \ln \dfrac{a + b\cos x}{\sin x}\right\} +$

$+ B \int \dfrac{dx}{a + b\cos x}$ (cf. 2.553 3.). T(349)

Im Fall $a^2 = b^2 (= 1)$ gilt | For $a^2 = b^2 (= 1)$:

3. $\int \dfrac{(A + B\sin x)\,dx}{\sin x(1 + \cos x)} = \dfrac{A}{2}\left\{\ln \mathrm{tg}\, \dfrac{x}{2} + \dfrac{1}{1 + \cos x}\right\} + B\,\mathrm{tg}\, \dfrac{x}{2}.$

4. $\int \dfrac{(A + B\sin x)\,dx}{\sin x(1 - \cos x)} = \dfrac{A}{2}\left\{\ln \mathrm{tg}\, \dfrac{x}{2} - \dfrac{1}{1 - \cos x}\right\} - B\,\mathrm{ctg}\, \dfrac{x}{2}.$

5. $\int \dfrac{(A + B\sin x)\,dx}{\cos x(a + b\sin x)} = \dfrac{1}{a^2 - b^2}\left\{(Aa - Bb) \ln \mathrm{tg}\left(\dfrac{\pi}{4} + \dfrac{x}{2}\right) -\right.$

$\left. - (Ab - aB) \ln \dfrac{a + b\sin x}{\cos x}\right\}.$ T(346)

Im Fall $a^2 = b^2 (= 1)$ gilt | For $a^2 = b^2 (= 1)$:

6. $\int \dfrac{(A + B\sin x)\,dx}{\cos x(1 \pm \sin x)} = \dfrac{A \pm B}{2} \ln \mathrm{tg}\left(\dfrac{\pi}{4} + \dfrac{x}{2}\right) \mp \dfrac{A \mp B}{2(1 \pm \sin x)}.$

7. $\int \dfrac{(A + B\sin x)\,dx}{\cos x(a + b\cos x)} = \dfrac{A}{a} \ln \mathrm{tg}\left(\dfrac{\pi}{4} + \dfrac{x}{2}\right) + \dfrac{B}{a} \ln \dfrac{a + b\cos x}{\cos x} -$

$- \dfrac{Ab}{a}\int \dfrac{dx}{a + b\cos x}$ (cf. 2.553 3.). T(351)+

8. $\int \dfrac{(A + B\cos x)\,dx}{\sin x(a + b\sin x)} = \dfrac{A}{a} \ln \mathrm{tg}\, \dfrac{x}{2} - \dfrac{B}{a} \ln \dfrac{a + b\sin x}{\sin x} -$

$- \dfrac{Ab}{a}\int \dfrac{dx}{a + b\sin x}$ (cf. 2.551 3.). T(352)

2.561

9. $\int \dfrac{(A+B\cos x)\,dx}{\sin x(a+b\cos x)} = \dfrac{1}{a^2-b^2}\left\{(Aa-Bb)\ln \operatorname{tg}\dfrac{x}{2} + \right.$

$\left. + (Ab-Ba)\ln \dfrac{a+b\cos x}{\sin x}\right\}.$ T(345)

Im Fall $a^2 = b^2\,(=1)$ gilt | For $a^2 = b^2\,(=1)$:

10. $\int \dfrac{(A+B\cos x)\,dx}{\sin x\,(1\pm \cos x)} = \pm \dfrac{A\mp B}{2(1\pm \cos x)} + \dfrac{A\pm B}{2}\ln \operatorname{tg}\dfrac{x}{2}.$

11. $\int \dfrac{(A+B\cos x)\,dx}{\cos x(a+b\sin x)} = \dfrac{A}{a^2-b^2}\left\{a\ln \operatorname{tg}\left(\dfrac{\pi}{4}+\dfrac{x}{2}\right) - \right.$

$\left. - b\ln \dfrac{a+b\sin x}{\cos x}\right\} + B\int \dfrac{dx}{a+b\sin x}$ (cf. 2.551 3.).   T(350)

Im Fall $a^2 = b^2\,(=1)$ gilt | For $a^2 = b^2\,(=1)$:

12. $\int \dfrac{(A+B\sin x)\,dx}{\cos x\,(1\pm \sin x)} = \dfrac{A\pm B}{2}\ln \operatorname{tg}\left(\dfrac{\pi}{4}+\dfrac{x}{2}\right) \mp \dfrac{A\mp B}{2(1\pm \sin x)}.$

13. $\int \dfrac{(A+B\cos x)\,dx}{\cos x(a+b\cos x)} = \dfrac{A}{a}\ln \operatorname{tg}\left(\dfrac{\pi}{4}+\dfrac{x}{2}\right) + \dfrac{Ba-Ab}{a}\int \dfrac{dx}{a+b\cos x}$

(cf. 2.553 3.).   T(347)

2.562

1. $\int \dfrac{dx}{a+b\sin^2 x} = \dfrac{\operatorname{sign} a}{\sqrt{a(a+b)}}\operatorname{arctg}\left(\sqrt{\dfrac{a+b}{a}}\operatorname{tg} x\right)\quad \left[\dfrac{b}{a} > -1\right];$

$= \dfrac{\operatorname{sign} a}{\sqrt{-a(a+b)}}\operatorname{Arth}\left(\sqrt{-\dfrac{a+b}{a}}\operatorname{tg} x\right)$

$\left[\dfrac{b}{a} < -1,\ \sin^2 x < -\dfrac{a}{b}\right];$

$= \dfrac{\operatorname{sign} a}{\sqrt{-a(a+b)}}\operatorname{Arcth}\left(\sqrt{-\dfrac{a+b}{a}}\operatorname{tg} x\right)$

$\left[\dfrac{b}{a} < -1,\ \sin^2 x > -\dfrac{a}{b}\right].$ MC 155

2. $\int \dfrac{dx}{a+b\cos^2 x} = \dfrac{-\operatorname{sign} a}{\sqrt{a(a+b)}}\operatorname{arctg}\left(\sqrt{\dfrac{a+b}{a}}\operatorname{ctg} x\right)\quad \left[\dfrac{b}{a} > -1\right];$

$= \dfrac{-\operatorname{sign} a}{\sqrt{-a(a+b)}}\operatorname{Arth}\left(\sqrt{-\dfrac{a+b}{a}}\operatorname{ctg} x\right)$

$\left[\dfrac{b}{a} < -1,\ \cos^2 x < -\dfrac{a}{b}\right];$

$= \dfrac{-\operatorname{sign} a}{\sqrt{-a(a+b)}}\operatorname{Arcth}\left(\sqrt{-\dfrac{a+b}{a}}\operatorname{ctg} x\right)$

$\left[\dfrac{b}{a} < -1,\ \cos^2 x > -\dfrac{a}{b}\right].$ MC 162

3. $\int \dfrac{dx}{1+\sin^2 x} = \dfrac{1}{\sqrt{2}}\operatorname{arctg}(\sqrt{2}\operatorname{tg} x).$

4. $$\int \frac{dx}{1-\sin^2 x} = \operatorname{tg} x.$$ 2.562

5. $$\int \frac{dx}{1+\cos^2 x} = -\frac{1}{\sqrt{2}} \operatorname{arctg}(\sqrt{2}\operatorname{ctg} x).$$

6. $$\int \frac{dx}{1-\cos^2 x} = -\operatorname{ctg} x.$$

1. $$\int \frac{dx}{(a+b\sin^2 x)^2} = \frac{1}{2a(a+b)}\left[(2a+b)\int \frac{dx}{a+b\sin^2 x} + \frac{b\sin x\cos x}{a+b\sin^2 x}\right] \quad \text{(cf. 2.562 1.)}. \quad \text{MC 155}$$ 2.563

2. $$\int \frac{dx}{(a+b\cos^2 x)^2} = \frac{1}{2a(a+b)}\left[(2a+b)\int \frac{dx}{a+b\cos^2 x} - \frac{b\sin x\cos x}{a+b\cos^2 x}\right] \quad \text{(cf. 2.562 2.)}. \quad \text{M 163}$$

3. $$\int \frac{dx}{(a+b\sin^2 x)^3} = \frac{1}{8pa^3}\left[\left(3+\frac{2}{p^2}+\frac{3}{p^4}\right)\operatorname{arctg}(p\operatorname{tg} x) + \right.$$
$$\left. + \left(3+\frac{2}{p^2}-\frac{3}{p^4}\right)\frac{p\operatorname{tg} x}{1+p^2\operatorname{tg}^2 x} + \left(1-\frac{2}{p^2}-\frac{1}{p^2}\operatorname{tg}^2 x\right)\frac{2p\operatorname{tg} x}{(1+p^2\operatorname{tg}^2 x)^2}\right]$$
$$\left[p^2 = 1+\frac{b}{a} > 0\right];$$

$$= \frac{1}{8qa^3}\left[\left(3-\frac{2}{q^2}+\frac{3}{q^4}\right)\operatorname{Arth}(q\operatorname{tg} x) + \right.$$
$$\left. + \left(3-\frac{2}{q^2}-\frac{3}{q^4}\right)\frac{q\operatorname{tg} x}{1-q^2\operatorname{tg}^2 x} + \left(1+\frac{2}{q^2}+\frac{1}{q^2}\operatorname{tg}^2 x\right)\frac{2q\operatorname{tg} x}{(1-q^2\operatorname{tg}^2 x)^2}\right]$$

$\left[q^2 = -1-\frac{b}{a} > 0, \sin^2 x < -\frac{a}{b};\right.$ | $\left[q^2 = -1-\frac{b}{a} > 0, \sin^2 x < -\frac{a}{b};\right.$
im Fall $\sin^2 x > -\frac{a}{b}$ muß $\operatorname{Arth}(q\operatorname{tg} x)$ | if $\sin^2 x > -\frac{a}{b}$, then $\operatorname{Arth}(q\operatorname{tg} x)$
durch $\operatorname{Arcth}(q\operatorname{tg} x)$ ersetzt werden$\Big]$. | has to be replaced by $\operatorname{Arcth}(q\operatorname{tg} x)\Big]$.

MC 156 | MC 156

4. $$\int \frac{dx}{(a+b\cos^2 x)^3} = -\frac{1}{8pa^3}\left[\left(3+\frac{2}{p^2}+\frac{3}{p^4}\right)\operatorname{arctg}(p\operatorname{ctg} x) + \right.$$
$$\left. + \left(3+\frac{2}{p^2}-\frac{3}{p^4}\right)\frac{p\operatorname{ctg} x}{1+p^2\operatorname{ctg}^2 x} + \left(1-\frac{2}{p^2}-\frac{1}{p^2}\operatorname{ctg}^2 x\right)\frac{2p\operatorname{ctg} x}{(1+p^2\operatorname{ctg}^2 x)^2}\right]$$
$$\left[p^2 = 1+\frac{b}{a} > 0\right];$$

$$= -\frac{1}{8qa^3}\left[\left(3-\frac{2}{q^2}+\frac{3}{q^4}\right)\operatorname{Arth}(q\operatorname{ctg} x) + \right.$$
$$\left. + \left(3-\frac{2}{q^2}-\frac{3}{q^4}\right)\frac{q\operatorname{ctg} x}{1-q^2\operatorname{ctg}^2 x} + \left(1+\frac{2}{q^2}+\frac{1}{q^2}\operatorname{ctg}^2 x\right)\frac{2q\operatorname{ctg} x}{(1-q^2\operatorname{ctg}^2 x)^2}\right.$$

$$\left[q^2 = -1 - \frac{b}{a} > 0, \cos^2 x < -\frac{a}{b}; \right.$$
im Fall $\cos^2 x > -\frac{a}{b}$ muß $\text{Arth}(q \,\text{ctg}\, x)$
durch $\text{Arcth}(q \,\text{ctg}\, x)$ ersetzt werden $\Big]$.

$$\left[q^2 = -1 - \frac{b}{a} > 0, \cos^2 x < -\frac{a}{b}; \right.$$
if $\cos^2 x > -\frac{a}{b}$, then $\text{Arth}(q \,\text{ctg}\, x)$
has to be replaced by $\text{Arcth}(q \,\text{ctg}\, x)\Big]$.

MC 163+   MC 163+

2.564

1. $\displaystyle\int \frac{\text{tg}\, x\, dx}{1 + m^2 \,\text{tg}^2\, x} = \frac{\ln(\cos^2 x + m^2 \sin^2 x)}{2(m^2 - 1)}.$ La 210(10)

2. $\displaystyle\int \frac{\text{tg}\,\alpha - \text{tg}\, x}{\text{tg}\,\alpha + \text{tg}\, x}\, dx = \sin 2\alpha \ln \sin(x + \alpha) - x \cos 2\alpha.$ La 210(11)+

3. $\displaystyle\int \frac{\text{tg}\, x\, dx}{a + b\, \text{tg}\, x} = \frac{1}{a^2 + b^2}\{bx - a \ln(a \cos x + b \sin x)\}.$ P(335)

4. $\displaystyle\int \frac{dx}{a + b\, \text{tg}^2\, x} = \frac{1}{a - b}\left[x - \sqrt{\frac{b}{a}} \arctan\left(\sqrt{\frac{b}{a}}\, \text{tg}\, x\right)\right].$ P(334)

Funktionen von $\sqrt{a \pm b \sin x}$, $\sqrt{a \pm b \cos x}$
und darauf zurückführbare Funktionen

2.57

Functions of $\sqrt{a \pm b \sin x}$ **or** $\sqrt{a \pm b \cos x}$
and functions reducible to such roots

Abkürzungen:
Abbreviations:
$\alpha = \arcsin\sqrt{\dfrac{1 - \sin x}{2}},\quad \beta = \arcsin\sqrt{\dfrac{b(1 - \sin x)}{a + b}},$

$\gamma = \arcsin\sqrt{\dfrac{b(1 - \cos x)}{a + b}},\quad \delta = \arcsin\sqrt{\dfrac{(a + b)(1 - \cos x)}{2(a - b \cos x)}},\quad r = \sqrt{\dfrac{2b}{a + b}}.$

2.571

1. $\displaystyle\int \frac{dx}{\sqrt{a + b \sin x}} = \frac{-2}{\sqrt{a + b}}\, F(\alpha, r)\quad \left[a > b > 0,\ -\frac{\pi}{2} \leqslant x < \frac{\pi}{2}\right];$

$= -\sqrt{\dfrac{2}{b}}\, F\!\left(\beta, \dfrac{1}{r}\right)\quad \left[0 < |a| < b,\ -\arcsin\dfrac{a}{b} < x < \dfrac{\pi}{2}\right].$ BF(288.00, 288.50)

2. $\displaystyle\int \frac{\sin x\, dx}{\sqrt{a + b \sin x}} = \frac{2a}{b\sqrt{a + b}}\, F(\alpha, r) - \frac{2\sqrt{a + b}}{b}\, E(\alpha, r)$

$\left[a > b > 0,\ -\dfrac{\pi}{2} \leqslant x < \dfrac{\pi}{2}\right];$ BF(288.03)

$= \sqrt{\dfrac{2}{b}}\left\{F\!\left(\beta, \dfrac{1}{r}\right) - 2E\!\left(\beta, \dfrac{1}{r}\right)\right\}$

$\left[0 < |a| < b,\ -\arcsin\dfrac{a}{b} < x < \dfrac{\pi}{2}\right].$ BF(288.54)

3. $\displaystyle\int \frac{\sin^2 x\, dx}{\sqrt{a + b \sin x}} = \frac{4a\sqrt{a + b}}{3b^2}\, E(\alpha, r) - \frac{2(2a^2 + b^2)}{3b^2 \sqrt{a + b}}\, F(\alpha, r) -$

$- \dfrac{2}{3b} \cos x \sqrt{a + b \sin x}\quad \left[a > b > 0,\ -\dfrac{\pi}{2} \leqslant x < \dfrac{\pi}{2}\right];$

$$= \sqrt{\frac{2}{b}} \left\{ \frac{4a}{3b} E\left(\beta, \frac{1}{r}\right) - \frac{2a+b}{3b} F\left(\beta, \frac{1}{r}\right) \right\} - \frac{2}{3b} \cos x \sqrt{a+b \sin x} \quad 2.571$$

$$\left[0 < |a| < b, \; -\arcsin \frac{a}{b} < x < \frac{\pi}{2} \right]. \qquad \text{BF(288.03, 288.54)}$$

4. $\displaystyle\int \frac{dx}{\sqrt{a+b\cos x}} = \frac{2}{\sqrt{a+b}} F\left(\frac{x}{2}, r\right) \quad [a > b > 0, \; 0 \leqslant x \leqslant \pi];$ \qquad BF(289.00)

$$= \sqrt{\frac{2}{b}} F\left(\gamma, \frac{1}{r}\right)$$

$$\left[b \geqslant |a| > 0, \; 0 \leqslant x < \arccos\left(-\frac{a}{b}\right) \right]. \qquad \text{BF(290.00)}$$

5. $\displaystyle\int \frac{dx}{\sqrt{a-b\cos x}} = \frac{2}{\sqrt{a+b}} F(\delta, r) \quad [a > b > 0, \; 0 \leqslant x \leqslant \pi].$ \qquad BF(291.00)

6. $\displaystyle\int \frac{\cos x \, dx}{\sqrt{a+b\cos x}} = \frac{2}{b\sqrt{a+b}} \left\{ (a+b) E\left(\frac{x}{2}, r\right) - a F\left(\frac{x}{2}, r\right) \right\}.$

$$[a > b > 0, \; 0 \leqslant x \leqslant \pi]; \qquad \text{BF(289.03)}$$

$$= \sqrt{\frac{2}{b}} \left\{ 2E\left(\gamma, \frac{1}{r}\right) - F\left(\gamma, \frac{1}{r}\right) \right\}$$

$$\left[b > |a| > 0, \; 0 \leqslant x < \arccos\left(-\frac{a}{b}\right) \right]. \qquad \text{BF(290.04)}$$

7. $\displaystyle\int \frac{\cos x \, dx}{\sqrt{a-b\cos x}} = \frac{2}{b\sqrt{a+b}} \left\{ (b-a) \Pi(\delta, r^2, r) + a F(\delta, r) \right\}$

$$[a > b > 0, \; 0 \leqslant x \leqslant \pi]. \qquad \text{BF(291.03)}$$

8. $\displaystyle\int \frac{\cos^2 x \, dx}{\sqrt{a+b\cos x}} = \frac{2}{3b^2 \sqrt{a+b}} \left\{ (2a^2 + b^2) F\left(\frac{x}{2}, r\right) - \right.$

$$\left. - 2a(a+b) E\left(\frac{x}{2}, r\right) \right\} + \frac{2}{3b} \sin x \sqrt{a+b\cos x}$$

$$[a > b > 0, \; 0 \leqslant x \leqslant \pi]; \qquad \text{BF(289.03)}$$

$$= \frac{1}{3b} \sqrt{\frac{2}{b}} \left\{ (2a+b) F\left(\gamma, \frac{1}{r}\right) - 4a E\left(\gamma, \frac{1}{r}\right) \right\} +$$

$$+ \frac{2}{3b} \sin x \sqrt{a+b\cos x} \quad \left[b \geqslant |a| > 0, \; 0 \leqslant x < \arccos\left(-\frac{a}{b}\right) \right]. \qquad \text{BF(290.04)}$$

9. $\displaystyle\int \frac{\cos^2 x \, dx}{\sqrt{a-b\cos x}} = \frac{2}{3b^2 \sqrt{a+b}} \left\{ (2a^2 + b^2) F(\delta, r) - 2a(a+b) E(\delta, r) \right\} +$

$$+ \frac{2}{3b} \sin x \, \frac{a+b\cos x}{\sqrt{a-b\cos x}} \quad [a > b > 0, \; 0 \leqslant x < \pi]. \qquad \text{BF(291.04)+}$$

2.572 $\int \dfrac{\text{tg}^2 x\, dx}{\sqrt{a+b\sin x}} = \dfrac{1}{\sqrt{a+b}} F(\alpha, r) + \dfrac{a}{(a-b)\sqrt{a+b}} E(\alpha, r) -$

$\qquad - \dfrac{b - a\sin x}{(a^2 - b^2)\cos x} \sqrt{a + b\,\sin x}\ \ \left[0 < b < a,\ -\dfrac{\pi}{2} < x < \dfrac{\pi}{2}\right];$

$\qquad = \sqrt{\dfrac{2}{b}} \left\{ \dfrac{2a+b}{2(a+b)} F\!\left(\beta, \dfrac{1}{r}\right) + \dfrac{ab}{a^2 - b^2} E\!\left(\beta, \dfrac{1}{r}\right) \right\} -$

$\qquad - \dfrac{b - a\sin x}{(a^2 - b^2)\cos x}\sqrt{a+b\,\sin x}\ \ \left[0 < |a| < b,\ -\arcsin\dfrac{a}{b} < x < \dfrac{\pi}{2}\right].$

BF(288.08, 288.58)

2.573 1. $\int \dfrac{1-\sin x}{1+\sin x} \cdot \dfrac{dx}{\sqrt{a+b\sin x}} = \dfrac{2}{a-b}\left\{ \sqrt{a+b}\,E(\alpha, r) - \right.$

$\qquad \left. - \text{tg}\!\left(\dfrac{\pi}{4} - \dfrac{x}{2}\right) \sqrt{a + b\,\sin x} \right\}\ \ \left[0 < b < a,\ -\dfrac{\pi}{2} \leqslant x \leqslant \dfrac{\pi}{2}\right].$ BF(288.07)

2. $\int \dfrac{1 - \cos x}{1 + \cos x}\,\dfrac{dx}{\sqrt{a + b\cos x}} = \dfrac{2}{a - b}\,\text{tg}\,\dfrac{x}{2}\sqrt{a + b\cos x} -$

$\qquad - \dfrac{2\sqrt{a+b}}{a-b} E\!\left(\dfrac{x}{2}, r\right)\ \ [a > b > 0,\ 0 \leqslant x < \pi].$ BF(289.07)

2.574 1. $\int \dfrac{dx}{(2 - p^2 + p^2 \sin x)\sqrt{a + b\sin x}} = -\dfrac{1}{a+b}\,\Pi(\alpha, p^2, r)$

$\qquad \left[0 < b < a,\ -\dfrac{\pi}{2} \leqslant x < \dfrac{\pi}{2}\right].$ BF(288.02)

2. $\int \dfrac{dx}{(a + b - p^2 b + p^2 b\sin x)\sqrt{a + b\sin x}} = -\dfrac{1}{a+b}\sqrt{\dfrac{2}{b}}\,\Pi\!\left(\beta, p^2, \dfrac{1}{r}\right)$

$\qquad \left[0 < |a| < b,\ -\arcsin\dfrac{a}{b} < x < \dfrac{\pi}{2}\right].$ BF(288.52)

3. $\int \dfrac{dx}{(2 - p^2 + p^2 \cos x)\sqrt{a + b\cos x}} = \dfrac{1}{\sqrt{a+b}}\,\Pi\!\left(\dfrac{x}{2}, p^2, r\right)$

$\qquad [a > b > 0,\ 0 \leqslant x < \pi].$ BF(289.02)

4. $\int \dfrac{dx}{(a + b - p^2 b + p^2 b\cos x)\sqrt{a + b\cos x}} = \dfrac{\sqrt{2}}{(a+b)\sqrt{b}}\,\Pi\!\left(\gamma, p^2, \dfrac{1}{r}\right)$

$\qquad \left[b \geqslant |a| > 0,\ 0 \leqslant x < \arccos\!\left(-\dfrac{a}{b}\right)\right].$ BF(290.02)

2.575 1. $\int \dfrac{dx}{\sqrt{(a + b\sin x)^3}} = \dfrac{2b\cos x}{(a^2 - b^2)\sqrt{a + b\sin x}} - \dfrac{2}{(a-b)\sqrt{a+b}}\,E(\alpha, r)$

$\qquad \left[0 < b < a,\ -\dfrac{\pi}{2} \leqslant x \leqslant \dfrac{\pi}{2}\right];$ BF(288.05)

$\qquad = \sqrt{\dfrac{2}{b}} \left\{ \dfrac{2b}{b^2 - a^2} E\!\left(\beta, \dfrac{1}{r}\right) - \dfrac{1}{a+b} F\!\left(\beta, \dfrac{1}{r}\right) \right\} +$

$\qquad + \dfrac{2b}{b^2 - a^2} \cdot \dfrac{\cos x}{\sqrt{a + b\sin x}}\ \ \left[0 < |a| < b,\ -\arcsin\dfrac{a}{b} < x < \dfrac{\pi}{2}\right].$ BF(288.56)

2. $\int \dfrac{dx}{\sqrt{(a+b\sin x)^5}} = \dfrac{2}{3(a^2-b^2)^2 \sqrt{a+b}} \{(a^2-b^2) F(\alpha, r) -$

$\qquad - 4a(a+b) E(\alpha, r)\} + \dfrac{2b(5a^2-b^2+4ab\sin x)}{3(a^2-b^2)^2 \sqrt{(a+b\sin x)^3}} \cos x$

$\qquad \left[0 < b < a,\ -\dfrac{\pi}{2} \leqslant x < \dfrac{\pi}{2} \right];$ BF(288.05)

$\qquad = -\dfrac{1}{3(a^2-b^2)^2} \sqrt{\dfrac{2}{b}} \left\{ (3a-b)(a-b) F\left(\beta, \dfrac{1}{r}\right) + \right.$

$\qquad \left. + 8abE\left(\beta, \dfrac{1}{r}\right) \right\} + \dfrac{2b[a^2-b^2+4a(a+b\sin x)]}{3(a^2-b^2)^2 \sqrt{(a+b\sin x)^3}} \cos x$

$\qquad \left[0 < |a| < b,\ -\arcsin \dfrac{a}{b} < x < \dfrac{\pi}{2} \right].$ BF(288.56)

3. $\int \dfrac{dx}{\sqrt{(a+b\cos x)^3}} = \dfrac{2}{(a-b)\sqrt{a+b}} E\left(\dfrac{x}{2}, r\right) - \dfrac{2b}{a^2-b^2} \cdot \dfrac{\sin x}{\sqrt{a+b\cos x}}$

$\qquad [a > b > 0,\ 0 \leqslant x \leqslant \pi];$ BF(289.05)

$\qquad = \dfrac{1}{a^2-b^2} \cdot \sqrt{\dfrac{2}{b}} \left\{ (a-b) F\left(\gamma, \dfrac{1}{r}\right) + 2bE\left(\gamma, \dfrac{1}{r}\right) \right\} + \dfrac{2b}{b^2-a^2} \cdot \dfrac{\sin x}{\sqrt{a+b\cos x}}$

$\qquad \left[b \geqslant |a| > 0,\ 0 \leqslant x < \arccos\left(-\dfrac{a}{b}\right) \right].$ BF(290.06)

4. $\int \dfrac{dx}{\sqrt{(a-b\cos x)^3}} = \dfrac{2}{(a-b)\sqrt{a+b}} E(\delta, r)\quad [a>b>0,\ 0\leqslant x\leqslant \pi].$ BF(291.01)

5. $\int \dfrac{dx}{\sqrt{(a+b\cos x)^5}} = \dfrac{2\sqrt{a+b}}{3(a^2-b^2)^2} \left\{ 4aE\left(\dfrac{x}{2}, r\right) - (a-b) F\left(\dfrac{x}{2}, r\right) \right\} -$

$\qquad - \dfrac{2b}{3(a^2-b^2)^2} \cdot \dfrac{5a^2-b^2+4ab\cos x}{\sqrt{(a+b\cos x)^3}} \sin x \quad [a>b>0,\ 0\leqslant x\leqslant \pi];$ BF(289.05)

$\qquad = \dfrac{1}{3(a^2-b^2)^2} \sqrt{\dfrac{2}{b}} \left\{ (a-b)(3a-b) F\left(\gamma, \dfrac{1}{r}\right) + \right.$

$\qquad \left. + 8abE\left(\gamma, \dfrac{1}{r}\right) \right\} + \dfrac{2b(5a^2-b^2+4ab\cos x)\sin x}{3(a^2-b^2)^2 \sqrt{(a+b\cos x)^3}}$

$\qquad \left[b \geqslant |a| > 0,\ 0 \leqslant x < \arccos\left(-\dfrac{a}{b}\right) \right].$ BF(290.06)

1. $\int \sqrt{a+b\cos x}\, dx = 2\sqrt{a+b}\, E\left(\dfrac{x}{2}, r\right)\quad [a>b>0,\ 0\leqslant x\leqslant \pi];$ BF(289.01)   2.576

$\qquad = \sqrt{\dfrac{2}{b}} \left\{ (a-b) F\left(\gamma, \dfrac{1}{r}\right) + 2bE\left(\gamma, \dfrac{1}{r}\right) \right\}$

$\qquad \left[b \geqslant |a| > 0,\ 0 \leqslant x < \arccos\left(-\dfrac{a}{b}\right) \right].$ BF(290.03)

2. $\int \sqrt{a-b\cos x}\, dx = 2\sqrt{a+b}\, E(\delta, r) - \dfrac{2b\sin x}{\sqrt{a-b\cos x}}$

$\qquad [a > b > 0,\ 0 \leqslant x \leqslant \pi]$ BF(291.05)

2.577 $$\int \sqrt{\frac{a-b\cos x}{1+p\cos x}}\,dx = \frac{2(a-b)}{(1+p)\sqrt{a+b}}\,\Pi\left(\delta,\frac{2ap}{(a+b)(1+p)},r\right)$$
$$[a > b > 0,\ 0 \leqslant x \leqslant \pi,\ p \neq -1].\quad \text{BF(291.02)}$$

2.578 $$\int \frac{\operatorname{tg} x\,dx}{\sqrt{a+b\operatorname{tg}^2 x}} = \frac{1}{\sqrt{b-a}}\arccos\left(\frac{\sqrt{b-a}}{\sqrt{b}}\cos x\right) \quad [b > a,\ b > 0]. \quad \text{P(333)}$$

Integrale, die auf elliptische und pseudoelliptische Integrale führen
Integrals reducible to elliptic and pseudo-elliptic integrals

2.58–2.62

2.580 1. $$\int \frac{d\varphi}{\sqrt{a+b\cos\varphi+c\sin\varphi}} = 2\int \frac{d\psi}{\sqrt{a-p+2p\cos^2\psi}}$$
$$\left[\varphi = 2\psi + \alpha,\ \operatorname{tg}\alpha = \frac{c}{b},\ p = \sqrt{b^2+c^2}\right].$$

2. $$\int \frac{d\varphi}{\sqrt{a+b\cos\varphi+c\sin\varphi+d\cos^2\varphi+e\sin\varphi\cos\varphi+f\sin^2\varphi}} =$$
$$= 2\int \frac{dx}{\sqrt{A+Bx+Cx^2+Dx^3+Ex^4}}$$
$$\left[\operatorname{tg}\frac{\varphi}{2}=x,\ A=a+b+d,\ B=2c+2e,\ C=2a-2d+4f,\right.$$
$$\left. D=2c-2e,\ E=a-b+d\right].$$

Funktionen von $\sqrt{1-k^2\sin^2 x}$
Functions of $\sqrt{1-k^2\sin^2 x}$

Abkürzungen: $\Delta = \sqrt{1-k^2\sin^2 x};\ k' = \sqrt{1-k^2}$.
Abbreviations:

2.581 1. $$\int \sin^m x \cos^n x\,\Delta^r\,dx =$$
$$= \frac{1}{(m+n+r)k^2}\left\{\sin^{m-3} x \cos^{n+1} x \Delta^{r+2} + [m+n-2+(m+r-1)k^2]\,\times\right.$$
$$\left. \times \int \sin^{m-2} x \cos^n x \Delta^r\,dx - (m-3)\int \sin^{m-4} x \cos^n x \Delta^r\,dx\right\} =$$
$$= \frac{1}{(m+n+r)k^2}\left\{\sin^{m+1} x \cos^{n-3} x \Delta^{r+2} + [(n+r-1)k^2 - (m+n-2)k'^2]\,\times\right.$$
$$\left. \times \int \sin^m x \cos^{n-2} x \Delta^r\,dx + (n-3)k'^2 \int \sin^m x \cos^{n-4} x \Delta^r\,dx\right\}$$
$$[m+n+r \neq 0].$$

Für $r = -3$ und $r = -5$ gilt | For $r = -3$ and $r = -5$:

2. $$\int \frac{\sin^m x \cos^n x}{\Delta^3}\,dx = \frac{\sin^{m-1} x \cos^{n-1} x}{k^2 \Delta} -$$
$$- \frac{m-1}{k^2}\int \frac{\sin^{m-2} x \cos^n x}{\Delta}\,dx + \frac{n-1}{k^2}\int \frac{\sin^m x \cos^{n-2} x}{\Delta}\,dx.$$

3. $\int \dfrac{\sin^m x \cos^n x}{\Delta^5} dx = \dfrac{\sin^{m-1} x \cos^{n-1} x}{3k^2 \Delta^3} -$ 2.581

$$- \dfrac{m-1}{3k^2} \int \dfrac{\sin^{m-2} x \cos^n x}{\Delta^3} dx + \dfrac{n-1}{3k^2} \int \dfrac{\sin^m x \cos^{n-2} x}{\Delta^3} dx.$$

Im Fall $m=1$ oder $n=1$ gilt | For $m=1$ or $n=1$:

4. $\int \sin x \cos^n x \Delta^r dx = -\dfrac{\cos^{n-1} x \Delta^{r+2}}{(n+r+1) k^2} - \dfrac{(n-1) k'^2}{(n+r+1) k^2} \int \cos^{n-2} x \sin x \Delta^r dx.$

5. $\int \sin^m x \cos x \, \Delta^r dx = -\dfrac{\sin^{m-1} x \Delta^{r+2}}{(m+r+1) k^2} + \dfrac{m-1}{(m+r+1) k^2} \int \sin^{m-2} x \cos x \, \Delta^r dx.$

Im Fall $m=3$ oder $n=3$ gilt | For $m=3$ or $n=3$:

6. $\int \sin^3 x \cos^n x \Delta^r dx = \dfrac{(n+r+1) k^2 \cos^2 x - [(r+2) k^2 + n + 1]}{(n+r+1)(n+r+3) k^4} \cos^{n-1} x \Delta^{r+2} -$

$$- \dfrac{[(r+2) k^2 + n + 1](n-1) k'^2}{(n+r+1)(n+r+3) k^4} \int \cos^{n-2} x \sin x \, \Delta^r dx.$$

7. $\int \sin^m x \cos^3 x \Delta^r dx = \dfrac{(m+r+1) k^2 \sin^2 x - [(r+2) k^2 - (m+1) k'^2]}{(m+r+1)(m+r+3) k^4} \times$

$$\times \sin^{m-1} x \Delta^{r+2} + \dfrac{[(r+2) k^2 - (m+1) k'^2](m-1)}{(m+r+1)(m+r+3) k^4} \int \sin^{m-2} x \cos x \Delta^r dx.$$

1. $\int \Delta^n dx = \dfrac{n-1}{n} (2-k^2) \int \Delta^{n-2} dx - \dfrac{n-2}{n} (1-k^2) \int \Delta^{n-4} dx +$ 2.582

$$+ \dfrac{k^2}{n} \sin x \cos x \cdot \Delta^{n-2}. \quad \text{La 316(1)+}$$

2. $\int \dfrac{dx}{\Delta^{n+1}} = -\dfrac{k^2 \sin x \cos x}{(n-1) k'^2 \Delta^{n-1}} + \dfrac{n-2}{n-1} \dfrac{2-k^2}{k'^2} \int \dfrac{dx}{\Delta^{n-1}} - \dfrac{n-3}{n-1} \dfrac{1}{k'^2} \int \dfrac{dx}{\Delta^{n-3}}.$

La 317(8)+

3. $\int \dfrac{\sin^n x}{\Delta} dx = \dfrac{\sin^{n-3} x}{(n-1) k^2} \cos x \cdot \Delta + \dfrac{n-2}{n-1} \dfrac{1+k^2}{k^2} \int \dfrac{\sin^{n-2} x}{\Delta} dx -$

$$- \dfrac{n-3}{(n-1) k^2} \int \dfrac{\sin^{n-4} x}{\Delta} dx. \quad \text{La 316(1)+}$$

4. $\int \dfrac{\cos^n x}{\Delta} dx = \dfrac{\cos^{n-3} x}{(n-1) k^2} \sin x \cdot \Delta + \dfrac{n-2}{n-1} \dfrac{2k^2-1}{k^2} \int \dfrac{\cos^{n-2} x}{\Delta} dx +$

$$+ \dfrac{n-3}{n-1} \dfrac{k'^2}{k^2} \int \dfrac{\cos^{n-4} x}{\Delta} dx. \quad \text{La 316(2)+}$$

5. $\int \dfrac{\operatorname{tg}^n x}{\Delta} dx = \dfrac{\operatorname{tg}^{n-3} x}{(n-1) k'^2} \dfrac{\Delta}{\cos^2 x} - \dfrac{(n-2)(2-k^2)}{(n-1) k'^2} \int \dfrac{\operatorname{tg}^{n-2} x}{\Delta} dx -$

$$- \dfrac{n-3}{(n-1) k'^2} \int \dfrac{\operatorname{tg}^{n-4} x}{\Delta} dx. \quad \text{La 317(3)}$$

6. $\int \dfrac{\operatorname{ctg}^n x}{\Delta} dx = -\dfrac{\operatorname{ctg}^{n-1} x}{n-1} \dfrac{\Delta}{\cos^2 x} - \dfrac{n-2}{n-1} (2-k^2) \int \dfrac{\operatorname{ctg}^{n-2} x}{\Delta} dx -$

$$- \dfrac{n-3}{n-1} k'^2 \int \dfrac{\operatorname{ctg}^{n-4} x}{\Delta} dx. \quad \text{La 317(6)}$$

2.583

1. $$\int \Delta\, dx = E(x, k).$$

2. $$\int \Delta \sin x\, dx = -\frac{\Delta \cos x}{2} - \frac{k'^2}{2k} \ln(k \cos x + \Delta).$$

3. $$\int \Delta \cos x\, dx = \frac{\Delta \sin x}{2} + \frac{1}{2k} \arcsin(k \sin x).$$

4. $$\int \Delta \sin^2 x\, dx = -\frac{\Delta}{3} \sin x \cos x + \frac{k'^2}{3k^2} F(x, k) + \frac{2k^2 - 1}{3k^2} E(x, k).$$

5. $$\int \Delta \sin x \cos x\, dx = -\frac{\Delta^3}{3k^2}.$$

6. $$\int \Delta \cos^2 x\, dx = \frac{\Delta}{3} \sin x \cos x - \frac{k'^2}{3k^2} F(x, k) + \frac{k^2 + 1}{3k^2} E(x, k).$$

7. $$\int \Delta \sin^3 x\, dx = -\frac{2k^2 \sin^2 x + 3k^2 - 1}{8k^2} \Delta \cos x + \frac{3k^4 - 2k^2 - 1}{8k^3} \ln(k \cos x + \Delta).$$

8. $$\int \Delta \sin^2 x \cos x\, dx = \frac{2k^2 \sin^2 x - 1}{8k^2} \Delta \sin x + \frac{1}{8k^3} \arcsin(k \sin x).$$

9. $$\int \Delta \sin x \cos^2 x\, dx = -\frac{2k^2 \cos^2 x + k'^2}{8k^2} \Delta \cos x + \frac{k'^4}{8k^3} \ln(k \cos x + \Delta).$$

10. $$\int \Delta \cos^3 x\, dx = \frac{2k^2 \cos^2 x + 2k^2 + 1}{8k^2} \Delta \sin x + \frac{4k^2 - 1}{8k^3} \arcsin(k \sin x).$$

11. $$\int \Delta \sin^4 x\, dx = -\frac{3k^2 \sin^2 x + 4k^2 - 1}{15k^2} \Delta \sin x \cos x -$$
$$- \frac{2(2k^4 - k^2 - 1)}{15k^4} F(x, k) + \frac{8k^4 - 3k^2 - 2}{15k^4} E(x, k).$$

12. $$\int \Delta \sin^3 x \cos x\, dx = \frac{3k^4 \sin^4 x - k^2 \sin^2 x - 2}{15k^4} \Delta.$$

13. $$\int \Delta \sin^2 x \cos^2 x\, dx = -\frac{3k^2 \cos^2 x - 2k^2 + 1}{15k^2} \Delta \sin x \cos x -$$
$$- \frac{k'^2(1 + k'^2)}{15k^4} F(x, k) + \frac{2(k^4 - k^2 + 1)}{15k^4} E(x, k).$$

14. $$\int \Delta \sin x \cos^3 x\, dx = -\frac{3k^4 \sin^4 x - k^2(5k^2 + 1)\sin^2 x + 5k^2 - 2}{15k^4} \Delta.$$

15. $$\int \Delta \cos^4 x\, dx = \frac{3k^2 \cos^2 x + 3k^2 + 1}{15k^2} \Delta \sin x \cos x +$$
$$+ \frac{2k'^2(k'^2 - 2k^2)}{15k^4} F(x, k) + \frac{3k^4 + 7k^2 - 2}{15k^4} E(x, k).$$

16. $$\int \Delta \sin^5 x\, dx = \frac{-8k^4 \sin^4 x - 2k^2(5k^2 - 1)\sin^2 x - 15k^4 + 4k^2 + 3}{48k^4} \Delta \cos x +$$
$$+ \frac{5k^6 - 3k^4 - k^2 - 1}{16k^5} \ln(k \cos x + \Delta).$$

17. $\int \Delta \sin^4 x \cos x \, dx = \dfrac{8k^4 \sin^4 x - 2k^2 \sin^2 x - 3}{48k^4} \Delta \sin x +$

$\qquad + \dfrac{1}{16k^5} \arcsin(k \sin x).$

18. $\int \Delta \sin^3 x \cos^2 x \, dx = \dfrac{8k^4 \sin^4 x - 2k^2(k^2+1)\sin^2 x - 3k^4 + 2k^2 - 3}{48k^4} \Delta \cos x +$

$\qquad + \dfrac{k'^4(k^2+1)}{16k^5} \ln(k \cos x + \Delta).$

19. $\int \Delta \sin^2 x \cos^3 x \, dx = \dfrac{-8k^4 \sin^4 x + 2k^2(6k^2+1)\sin^2 x - 6k^2 + 3}{48k^4} \Delta \sin x +$

$\qquad + \dfrac{2k^2 - 1}{16k^5} \arcsin(k \sin x).$

20. $\int \Delta \sin x \cos^4 x \, dx = \dfrac{-8k^4 \sin^4 x + 2k^2(7k^2+1)\sin^2 x - 3k^4 - 8k^2 + 3}{48k^4} \Delta \cos x -$

$\qquad - \dfrac{k'^6}{16k^5} \ln(k \cos x + \Delta).$

21. $\int \Delta \cos^5 x \, dx = \dfrac{8k^4 \sin^4 x - 2k^2(12k^2+1)\sin^2 x + 24k^4 + 12k^2 - 3}{48k^4} \Delta \sin x +$

$\qquad + \dfrac{8k^4 - 4k^2 + 1}{16k^5} \arcsin(k \sin x).$

22. $\int \Delta^3 \, dx = \dfrac{2}{3}(1+k'^2) E(x,k) - \dfrac{k'^2}{3} F(x,k) + \dfrac{k^2}{3} \Delta \sin x \cos x.$

23. $\int \Delta^3 \sin x \, dx = \dfrac{2k^2 \sin^2 x + 3k^2 - 5}{8} \Delta \cos x - \dfrac{3k'^4}{8k} \ln(k \cos x + \Delta).$

24. $\int \Delta^3 \cos x \, dx = \dfrac{-2k^2 \sin^2 x + 5}{8} \Delta \sin x + \dfrac{3}{8k} \arcsin(k \sin x).$

25. $\int \Delta^3 \sin^2 x \, dx = \dfrac{3k^2 \sin^2 x + 4k^2 - 6}{15} \Delta \sin x \cos x + \dfrac{k'^2(3-4k^2)}{15k^2} F(x,k) -$

$\qquad - \dfrac{8k^4 - 13k^2 + 3}{15k^2} E(x,k).$

26. $\qquad \int \Delta^3 \sin x \cos x \, dx = -\dfrac{\Delta^5}{5k^2}.$

27. $\int \Delta^3 \cos^2 x \, dx = \dfrac{-3k^2 \sin^2 x + k^2 + 6}{15} \Delta \sin x \cos x - \dfrac{k'^2(k^2+3)}{15k^2} F(x,k) -$

$\qquad - \dfrac{2k^4 - 7k^2 - 3}{15k^2} E(x,k).$

28. $\int \Delta^3 \sin^3 x \, dx = \dfrac{8k^4 \sin^4 x + 2k^2(5k^2-7)\sin^2 x + 15k^4 - 22k^2 + 3}{48k^2} \Delta \cos x -$

$\qquad - \dfrac{5k^6 - 9k^4 + 3k^2 + 1}{16k^3} \ln(k \cos x + \Delta).$

2.583

2.583

29. $\int \Delta^3 \sin^2 x \cos x \, dx = \dfrac{-8k^4 \sin^4 x + 14k^2 \sin^2 x - 3}{48k^2} \Delta \sin x + \dfrac{1}{16k^3} \arcsin(k \sin x).$

30. $\int \Delta^3 \sin x \cos^2 x \, dx = \dfrac{-8k^4 \sin^4 x + 2k^2(k^2+7) \sin^2 x + 3k^4 - 8k^2 - 3}{48k^2} \times \Delta \cos x + \dfrac{k'^6}{16k^3} \ln(k \cos x + \Delta).$

31. $\int \Delta^3 \cos^3 x \, dx = \dfrac{8k^4 \sin^4 x - 2k^2(6k^2+7) \sin^2 x + 30k^2 + 3}{48k^2} \Delta \sin x + \dfrac{6k^2-1}{16k^3} \arcsin(k \sin x).$

32. $\int \dfrac{\Delta \, dx}{\sin x} = -\dfrac{1}{2} \ln \dfrac{\Delta + \cos x}{\Delta - \cos x} + k \ln(k \cos x + \Delta).$

33. $\int \dfrac{\Delta \, dx}{\cos x} = \dfrac{k'}{2} \ln \dfrac{\Delta + k' \sin x}{\Delta - k' \sin x} + k \arcsin(k \sin x).$

34. $\int \dfrac{\Delta \, dx}{\sin^2 x} = k'^2 F(x, k) - E(x, k) - \Delta \operatorname{ctg} x.$

35. $\int \dfrac{\Delta \, dx}{\sin x \cos x} = \dfrac{1}{2} \ln \dfrac{1-\Delta}{1+\Delta} + \dfrac{k'}{2} \ln \dfrac{\Delta + k'}{\Delta - k'}.$

36. $\int \dfrac{\Delta \, dx}{\cos^2 x} = F(x, k) - E(x, k) + \Delta \operatorname{tg} x.$

37. $\int \dfrac{\sin x}{\cos x} \Delta \, dx = \int \Delta \operatorname{tg} x \, dx = -\Delta + \dfrac{k'}{2} \ln \dfrac{\Delta + k'}{\Delta - k'}.$

38. $\int \dfrac{\cos x}{\sin x} \Delta \, dx = \int \Delta \operatorname{ctg} x \, dx = \Delta + \dfrac{1}{2} \ln \dfrac{1-\Delta}{1+\Delta}.$

39. $\int \dfrac{\Delta \, dx}{\sin^3 x} = -\dfrac{\Delta \cos x}{2 \sin^2 x} + \dfrac{k'^2}{4} \ln \dfrac{\Delta + \cos x}{\Delta - \cos x}.$

40. $\int \dfrac{\Delta \, dx}{\sin^2 x \cos x} = \dfrac{-\Delta}{\sin x} - \dfrac{1+k^2}{2k'} \ln \dfrac{\Delta - k' \sin x}{\Delta + k' \sin x}.$

41. $\int \dfrac{\Delta \, dx}{\sin x \cos^2 x} = \dfrac{\Delta}{\cos x} + \dfrac{1}{2} \ln \dfrac{\Delta + \cos x}{\Delta - \cos x}.$

42. $\int \dfrac{\Delta \, dx}{\cos^3 x} = \dfrac{\Delta \sin x}{2 \cos^2 x} + \dfrac{1}{4k'} \ln \dfrac{\Delta + k' \sin x}{\Delta - k' \sin x}.$

43. $\int \dfrac{\Delta \sin x \, dx}{\cos^2 x} = \dfrac{\Delta}{\cos x} - k \ln(k \cos x + \Delta).$

44. $\int \dfrac{\Delta \cos x \, dx}{\sin^2 x} = -\dfrac{\Delta}{\sin x} - k \arcsin(k \sin x).$

45. $\int \dfrac{\Delta \sin^2 x \, dx}{\cos x} = -\dfrac{\Delta \sin x}{2} + \dfrac{2k^2-1}{2k} \arcsin(k \sin x) + \dfrac{k'}{2} \ln \dfrac{\Delta + k' \sin x}{\Delta - k' \sin x}.$

46. $\int \dfrac{\Delta \cos^2 x \, dx}{\sin x} = \dfrac{\Delta \cos x}{2} + \dfrac{k^2+1}{2k} \ln(k \cos x + \Delta) + \dfrac{1}{2} \ln \dfrac{\Delta + \cos x}{\Delta - \cos x}.$

47. $\int \dfrac{\varDelta\, dx}{\sin^4 x} = \dfrac{1}{3}\{-\varDelta\, \text{ctg}^3 x + (k^2 - 3)\varDelta\, \text{ctg}\, x + 2k'^2 F(x, k) + (k^2 - 2)E(x, k)\}.$ 2.583

48. $\int \dfrac{\varDelta\, dx}{\sin^3 x \cos x} = -\dfrac{\varDelta}{2\sin^2 x} + \dfrac{k'}{2}\ln\dfrac{\varDelta + k'}{\varDelta - k'} + \dfrac{k^2 - 2}{4}\ln\dfrac{1 + \varDelta}{1 - \varDelta}.$

49. $\int \dfrac{\varDelta\, dx}{\sin^2 x \cos^2 x} = \left(\dfrac{1}{k'^2}\,\text{tg}\, x - \text{ctg}\, x\right)\varDelta + 2F(x, k) - \dfrac{1 + k'^2}{k'^2}E(x, k).$

50. $\int \dfrac{\varDelta\, dx}{\sin x \cos^3 x} = \dfrac{\varDelta}{2\cos^2 x} - \dfrac{1}{2}\ln\dfrac{1 + \varDelta}{1 - \varDelta} + \dfrac{2 - k^2}{4k'}\ln\dfrac{\varDelta + k'}{\varDelta - k'}.$

51. $\int \dfrac{\varDelta\, dx}{\cos^4 x} = \dfrac{1}{3k'^2}\{[k'^2\,\text{tg}^3 x - (2k^2 - 3)\,\text{tg}\, x]\varDelta + 2k'^2 F(x, k) +$
$\qquad\qquad\qquad\qquad\qquad + (k^2 - 2)E(x, k)\}.$

52. $\int \dfrac{\sin x}{\cos^3 x}\varDelta\, dx = \dfrac{\varDelta}{2\cos^2 x} + \dfrac{k^2}{4k'}\ln\dfrac{\varDelta + k'}{\varDelta - k'}.$

53. $\int \dfrac{\cos x}{\sin^3 x}\varDelta\, dx = -\dfrac{\varDelta}{2\sin^2 x} + \dfrac{k^2}{4}\ln\dfrac{1 + \varDelta}{1 - \varDelta}.$

54. $\int \dfrac{\sin^2 x}{\cos^2 x}\varDelta\, dx = \int \text{tg}^2 x\varDelta\, dx = \varDelta\,\text{tg}\, x + F(x, k) - 2E(x, k).$

55. $\int \dfrac{\cos^2 x}{\sin^2 x}\varDelta\, dx = \int \text{ctg}^2 x\varDelta\, dx = -\varDelta\,\text{ctg}\, x + k'^2 F(x, k) - 2E(x, k).$

56. $\int \dfrac{\sin^3 x}{\cos x}\varDelta\, dx = -\dfrac{k^2\sin^2 x + 3k^2 - 1}{3k^2}\varDelta + \dfrac{k'}{2}\ln\dfrac{\varDelta + k'}{\varDelta - k'}.$

57. $\int \dfrac{\cos^3 x}{\sin x}\varDelta\, dx = -\dfrac{k^2\sin^2 x - 3k^2 - 1}{3k^2}\varDelta + \dfrac{1}{2}\ln\dfrac{1 - \varDelta}{1 + \varDelta}.$

58. $\int \dfrac{\varDelta\, dx}{\sin^5 x} = \dfrac{(k^2 - 3)\sin^2 x + 2}{8\sin^4 x}\cos x\varDelta + \dfrac{k'^2(k^2 + 3)}{16}\ln\dfrac{\varDelta + \cos x}{\varDelta - \cos x}.$

59. $\int \dfrac{\varDelta\, dx}{\sin^4 x \cos x} = -\dfrac{(3 - k^2)\sin^2 x + 1}{3\sin^3 x}\varDelta - \dfrac{k'}{2}\ln\dfrac{\varDelta - k'\sin x}{\varDelta + k'\sin x}.$

60. $\int \dfrac{\varDelta\, dx}{\sin^3 x \cos^2 x} = \dfrac{3\sin^2 x - 1}{2\sin^2 x \cos x}\varDelta + \dfrac{k^2 - 3}{4}\ln\dfrac{\varDelta - \cos x}{\varDelta + \cos x}.$

61. $\int \dfrac{\varDelta\, dx}{\sin^2 x \cos^3 x} = \dfrac{3\sin^2 x - 2}{2\sin x \cos^2 x}\varDelta - \dfrac{2k^2 - 3}{4k'}\ln\dfrac{\varDelta + k'\sin x}{\varDelta - k'\sin x}.$

62. $\int \dfrac{\varDelta\, dx}{\sin x \cos^4 x} = \dfrac{(2k^2 - 3)\sin^2 x - 3k^2 + 4}{3k'^2 \cos^3 x}\varDelta + \dfrac{1}{2}\ln\dfrac{\varDelta + \cos x}{\varDelta - \cos x}.$

63. $\int \dfrac{\varDelta\, dx}{\cos^5 x} = \dfrac{(2k^2 - 3)\sin^2 x - 4k^2 + 5}{8k'^2 \cos^4 x}\sin x\varDelta - \dfrac{4k^2 - 3}{16k'^3}\ln\dfrac{\varDelta + k'\sin x}{\varDelta - k'\sin x}.$

64. $\int \dfrac{\sin x}{\cos^4 x}\varDelta\, dx = \dfrac{-(2k^2 + 1)k^2 \sin^2 x + 3k^4 - k^2 + 1}{3k'^2 \cos^3 x}\varDelta.$

65. $\int \dfrac{\cos x}{\sin^4 x}\varDelta\, dx = -\dfrac{\varDelta^3}{3\sin^3 x}.$

66. $\int \dfrac{\sin^2 x}{\cos^3 x}\varDelta\, dx = \dfrac{\sin x}{2\cos^2 x}\varDelta + \dfrac{2k^2 - 1}{4k'}\ln\dfrac{\varDelta + k'\sin x}{\varDelta - k'\sin x} - k\arcsin(k\sin x).$

67. $\int \dfrac{\cos^2 x}{\sin^3 x}\varDelta\, dx = -\dfrac{\cos x}{2\sin^2 x}\varDelta - \dfrac{k^2 + 1}{4}\ln\dfrac{\varDelta + \cos x}{\varDelta - \cos x} - k\ln(k\cos x + \varDelta).$

2.583

68. $\int \dfrac{\sin^3 x}{\cos^2 x} \Delta \, dx = -\dfrac{\sin^2 x - 3}{2 \cos x} \Delta - \dfrac{3k^2 - 1}{2k} \ln (k \cos x + \Delta)$.

69. $\int \dfrac{\cos^3 x}{\sin^2 x} \Delta \, dx = -\dfrac{\sin^2 x + 2}{2 \sin x} \Delta - \dfrac{2k^2 + 1}{2k} \arcsin (k \sin x)$.

70. $\int \dfrac{\sin^4 x}{\cos x} \Delta \, dx = -\dfrac{2k^2 \sin^2 x + 4k^2 - 1}{8k^2} \sin x \, \Delta +$
$\qquad + \dfrac{8k^4 - 4k^2 - 1}{8k^3} \arcsin (k \sin x) + \dfrac{k'}{2} \ln \dfrac{\Delta + k' \sin x}{\Delta - k' \sin x}$.

71. $\int \dfrac{\cos^4 x}{\sin x} \Delta \, dx = \dfrac{-2k^2 \sin^2 x + 5k^2 + 1}{8k^2} \cos x \, \Delta +$
$\qquad + \dfrac{1}{2} \ln \dfrac{\Delta + \cos x}{\Delta - \cos x} + \dfrac{3k^4 + 6k^2 - 1}{8k^3} \ln (k \cos x + \Delta)$.

2.584

1. $\int \dfrac{dx}{\Delta} = F(x, k)$.

2. $\int \dfrac{\sin x \, dx}{\Delta} = \dfrac{1}{2k} \ln \dfrac{\Delta - k \cos x}{\Delta + k \cos x} = -\dfrac{1}{k} \ln (k \cos x + \Delta)$.

3. $\int \dfrac{\cos x \, dx}{\Delta} = \dfrac{1}{k} \arcsin (k \sin x) = \dfrac{1}{k} \operatorname{arctg} \dfrac{k \sin x}{\Delta}$.

4. $\int \dfrac{\sin^2 x \, dx}{\Delta} = \dfrac{1}{k^2} F(x, k) - \dfrac{1}{k^2} E(x, k)$.

5. $\int \dfrac{\sin x \cos x \, dx}{\Delta} = -\dfrac{\Delta}{k^2}$.

6. $\int \dfrac{\cos^2 x \, dx}{\Delta} = \dfrac{1}{k^2} E(x, k) - \dfrac{k'^2}{k^2} F(x, k)$.

7. $\int \dfrac{\sin^3 x \, dx}{\Delta} = \dfrac{\cos x \, \Delta}{2k^2} - \dfrac{1 + k^2}{2k^3} \ln (k \cos x + \Delta)$.

8. $\int \dfrac{\sin^2 x \cos x \, dx}{\Delta} = -\dfrac{\sin x \, \Delta}{2k^2} + \dfrac{\arcsin (k \sin x)}{2k^3}$.

9. $\int \dfrac{\sin x \cos x \, dx}{\Delta} = -\dfrac{\cos x \, \Delta}{2k^2} + \dfrac{k'^2}{2k^3} \ln (k \cos x + \Delta)$.

10. $\int \dfrac{\cos^3 x \, dx}{\Delta} = \dfrac{\sin x \, \Delta}{2k^2} + \dfrac{2k^2 - 1}{2k^3} \arcsin (k \sin x)$.

11. $\int \dfrac{\sin^4 x \, dx}{\Delta} = \dfrac{\sin x \cos x \, \Delta}{3k^2} + \dfrac{2 + k^2}{3k^4} F(x, k) - \dfrac{2(1 + k^2)}{3k^4} E(x, k)$.

12. $\int \dfrac{\sin^3 x \cos x \, dx}{\Delta} = -\dfrac{1}{3k^4} (2 + k^2 \sin^2 x) \Delta$.

13. $\int \dfrac{\sin^2 x \cos^2 x \, dx}{\Delta} = -\dfrac{\sin x \cos x \, \Delta}{3k^2} + \dfrac{2 - k^2}{3k^4} E(x, k) + \dfrac{2k^2 - 2}{3k^4} F(x, k)$.

14. $\int \dfrac{\sin x \cos^3 x \, dx}{\Delta} = -\dfrac{1}{3k^4} (k^2 \cos^2 x - 2k'^2) \Delta$.

15. $\int \dfrac{\cos^4 x \, dx}{\Delta} = \dfrac{\sin x \cos x \, \Delta}{3k^2} + \dfrac{4k^2 - 2}{3k^4} E(x, k) + \dfrac{3k^4 - 5k^2 + 2}{3k^4} F(x, k)$.

16. $\int \frac{\sin^5 x \, dx}{\Delta} = \frac{2k^2 \sin^2 x + 3k^2 + 3}{8k^4} \cos x \Delta - \frac{3 + 2k^2 + 3k^4}{8k^5} \ln(k \cos x + \Delta).$ 2.584

17. $\int \frac{\sin^4 x \cos x \, dx}{\Delta} = -\frac{2k^2 \sin^2 x + 3}{8k^4} \sin x \Delta + \frac{3}{8k^5} \arcsin(k \sin x).$

18. $\int \frac{\sin^3 x \cos^2 x \, dx}{\Delta} = \frac{2k^2 \cos^2 x - k^2 - 3}{8k^4} \cos x \Delta - \frac{k^4 + 2k^2 - 3}{8k^5} \ln(k \cos x + \Delta).$

19. $\int \frac{\sin^2 x \cos^3 x \, dx}{\Delta} = -\frac{2k^2 \cos^2 x + 2k^2 - 3}{8k^4} \sin x \Delta + \frac{4k^2 - 3}{8k^5} \arcsin(k \sin x).$

20. $\int \frac{\sin x \cos^4 x \, dx}{\Delta} = \frac{3 - 5k^2 + 2k^2 \sin^2 x}{8k^4} \cos x \Delta -$
$$- \frac{3k^4 - 6k^2 + 3}{8k^5} \ln(k \cos x + \Delta).$$

21. $\int \frac{\cos^5 x \, dx}{\Delta} = \frac{2k^2 \cos^2 x + 6k^2 - 3}{8k^4} \sin x \Delta + \frac{8k^4 - 8k^2 + 3}{8k^5} \arcsin(k \sin x).$

22. $\int \frac{\sin^6 x \, dx}{\Delta} = \frac{3k^2 \sin^2 x + 4k^2 + 4}{15k^4} \sin x \cos x \Delta +$
$$+ \frac{4k^4 + 3k^2 + 8}{15k^6} F(x, k) - \frac{8k^4 + 7k^2 + 8}{15k^6} E(x, k).$$

23. $\int \frac{\sin^5 x \cos x \, dx}{\Delta} = -\frac{3k^4 \sin^4 x + 4k^2 \sin^2 x + 8}{15k^6} \Delta.$

24. $\int \frac{\sin^4 x \cos^2 x \, dx}{\Delta} = \frac{3k^2 \cos^2 x - 2k^2 - 4}{15k^4} \sin x \cos x \Delta +$
$$+ \frac{k^4 + 7k^2 - 8}{15k^6} F(x, k) - \frac{2k^4 + 3k^2 - 8}{15k^6} E(x, k).$$

25. $\int \frac{\sin^3 x \cos^3 x \, dx}{\Delta} = \frac{3k^4 \sin^4 x - (5k^4 - 4k^2) \sin^2 x - 10k^2 + 8}{15k^6} \Delta.$

26. $\int \frac{\sin^2 x \cos^4 x \, dx}{\Delta} = -\frac{3k^2 \cos^2 x + 3k^2 - 4}{15k^4} \sin x \cos x \Delta +$
$$+ \frac{9k^4 - 17k^2 + 8}{15k^6} F(x, k) - \frac{3k^4 - 13k^2 + 8}{15k^6} E(x, k).$$

27. $\int \frac{\sin x \cos^5 x \, dx}{\Delta} = \frac{-3k^4 \cos^4 x + 4k^2 k'^2 \cos^2 x - 8k^4 + 16k^2 - 8}{15k^6} \Delta.$

28. $\int \frac{\cos^6 x \, dx}{\Delta} = \frac{3k^2 \cos^2 x + 8k^2 - 4}{15k^4} \sin x \cos x \Delta +$
$$+ \frac{15k^6 - 34k^4 + 27k^2 - 8}{15k^6} F(x, k) + \frac{23k^4 - 23k^2 + 8}{15k^6} E(x, k).$$

29. $\int \frac{\sin^7 x \, dx}{\Delta} = \frac{8k^4 \sin^4 x + 10k^2(k^2 + 1)\sin^2 x + 15k^4 + 14k^2 + 15}{48k^6} \cos x \Delta -$
$$- \frac{(5k^4 - 2k^2 + 5)(k^2 + 1)}{16k^7} \ln(k \cos x + \Delta).$$

30. $\int \frac{\sin^6 x \cos x \, dx}{\Delta} = -\frac{8k^4 \sin^4 x + 10k^2 \sin^2 x + 15}{48k^6} \sin x \Delta + \frac{5}{16k^7} \arcsin(k \sin x).$

2.584

31. $\int \dfrac{\sin^5 x \cos^2 x\, dx}{\Delta} = \dfrac{-8k^4 \sin^4 x + 2k^2(k^2 - 5)\sin^2 x + 3k^4 + 4k^2 - 15}{48k^6} \cos x\, \Delta -$
$$- \dfrac{k^6 + k^4 + 3k^2 - 5}{16k^7} \ln(k \cos x + \Delta).$$

32. $\int \dfrac{\sin^4 x \cos^3 x\, dx}{\Delta} = \dfrac{8k^4 \sin^4 x - 2k^2(6k^2 - 5)\sin^2 x - 18k^2 + 15}{48k^6} \sin x\, \Delta +$
$$+ \dfrac{6k^2 - 5}{16k^7} \arcsin(k \sin x).$$

33. $\int \dfrac{\sin^3 x \cos^4 x\, dx}{\Delta} = \dfrac{8k^4 \sin^4 x - 2k^2(7k^2 - 5)\sin^2 x + 3k^4 - 22k^2 + 15}{48k^6} \cos x\, \Delta -$
$$- \dfrac{k^6 + 3k^4 - 9k^2 + 5}{16k^7} \ln(k \cos x + \Delta).$$

34. $\int \dfrac{\sin^2 x \cos^5 x\, dx}{\Delta} = \dfrac{-8k^4 \sin^4 x + 2k^2(12k^2 - 5)\sin^2 x - 24k^4 + 36k^2 - 15}{48k^6} \sin x\, \Delta +$
$$+ \dfrac{8k^4 - 12k^2 + 5}{16k^7} \arcsin(k \sin x).$$

35. $\int \dfrac{\sin x \cos^6 x\, dx}{\Delta} = \dfrac{-8k^4 \sin^4 x + 2k^2(13k^2 - 5)\sin^2 x - 33k^4 + 40k^2 - 15}{48k^6} \cos x\, \Delta +$
$$+ \dfrac{5k'^6}{16k^7} \ln(k \cos x + \Delta).$$

36. $\int \dfrac{\cos^7 x\, dx}{\Delta} = \dfrac{8k^4 \sin^4 x - 2k^2(18k^2 - 5)\sin^2 x + 72k^4 - 54k^2 + 15}{48k^6} \sin x\, \Delta +$
$$+ \dfrac{16k^6 - 24k^4 + 18k^2 - 5}{16k^7} \arcsin(k \sin x).$$

37. $\int \dfrac{dx}{\Delta^3} = \dfrac{1}{k'^2} E(x, k) - \dfrac{k^2}{k'^2} \dfrac{\sin x \cos x}{\Delta}.$

38. $\int \dfrac{\sin x\, dx}{\Delta^3} = - \dfrac{\cos x}{k'^2 \Delta}.$

39. $\int \dfrac{\cos x\, dx}{\Delta^3} = \dfrac{\sin x}{\Delta}.$

40. $\int \dfrac{\sin^2 x\, dx}{\Delta^3} = \dfrac{1}{k'^2 k^2} E(x, k) - \dfrac{1}{k^2} F(x, k) - \dfrac{1}{k'^2} \dfrac{\sin x \cos x}{\Delta}.$

41. $\int \dfrac{\sin x \cos x\, dx}{\Delta^3} = \dfrac{1}{k^2 \Delta}.$

42. $\int \dfrac{\cos^2 x\, dx}{\Delta^3} = \dfrac{1}{k^2} F(x, k) - \dfrac{1}{k^2} E(x, k) + \dfrac{\sin x \cos x}{\Delta}.$

43. $\int \dfrac{\sin^3 x\, dx}{\Delta^3} = - \dfrac{\cos x}{k^2 k'^2 \Delta} + \dfrac{1}{k^3} \ln(k \cos x + \Delta).$

44. $\int \dfrac{\sin^2 x \cos x\, dx}{\Delta^3} = \dfrac{\sin x}{k^2 \Delta} - \dfrac{1}{k^3} \arcsin(k \sin x).$

45. $\int \dfrac{\sin x \cos^2 x\, dx}{\Delta^3} = \dfrac{\cos x}{k^2 \Delta} - \dfrac{1}{k^3} \ln(k\cos x + \Delta).$ 2.584

46. $\int \dfrac{\cos^3 x\, dx}{\Delta^3} = -\dfrac{k'^2 \sin x}{k^2 \Delta} + \dfrac{1}{k^3} \arcsin(k \sin x).$

47. $\int \dfrac{\sin^4 x\, dx}{\Delta^3} = \dfrac{k'^2 + 1}{k'^2 k^4} E(x,k) - \dfrac{2}{k^4} F(x,k) - \dfrac{\sin x \cos x}{k^2 k'^2 \Delta}.$

48. $\int \dfrac{\sin^3 x \cos x\, dx}{\Delta^3} = \dfrac{2 - k^2 \sin^2 x}{k^4 \Delta}.$

49. $\int \dfrac{\sin^2 x \cos^2 x\, dx}{\Delta^3} = \dfrac{2-k^2}{k^4} F(x,k) - \dfrac{2}{k^4} E(x,k) + \dfrac{\sin x \cos x}{k^2 \Delta}.$

50. $\int \dfrac{\sin x \cos^3 x\, dx}{\Delta^3} = \dfrac{k^2 \sin^2 x + k^2 - 2}{k^4 \Delta}.$

51. $\int \dfrac{\cos^4 x\, dx}{\Delta^3} = \dfrac{k'^2 + 1}{k^4} E(x,k) - \dfrac{2k'^2}{k^4} F(x,k) - \dfrac{k'^2 \sin x \cos x}{k^2 \Delta}.$

52. $\int \dfrac{\sin^5 x\, dx}{\Delta^3} = \dfrac{k^2 k'^2 \sin^2 x + k^2 - 3}{2k^4 k'^2 \Delta} \cos x + \dfrac{k^2 + 3}{2k^5} \ln(k\cos x + \Delta).$

53. $\int \dfrac{\sin^4 x \cos x\, dx}{\Delta^3} = \dfrac{-k^2 \sin^2 x + 3}{2k^4 \Delta} \sin x - \dfrac{3}{2k^5} \arcsin(k \sin x).$

54. $\int \dfrac{\sin^3 x \cos^2 x\, dx}{\Delta^3} = \dfrac{-k^2 \sin^2 x + 3}{2k^4 \Delta} \cos x + \dfrac{k^2 - 3}{2k^5} \ln(k\cos x + \Delta).$

55. $\int \dfrac{\sin^2 x \cos^3 x\, dx}{\Delta^3} = \dfrac{k^2 \sin^2 x + 2k^2 - 3}{2k^4 \Delta} \sin x - \dfrac{2k^2 - 3}{2k^5} \arcsin(k \sin x).$

56. $\int \dfrac{\sin x \cos^4 x\, dx}{\Delta^3} = \dfrac{k^2 \sin^2 x + 2k^2 - 3}{2k^4 \Delta} \cos x + \dfrac{3k'^2}{2k^5} \ln(k\cos x + \Delta).$

57. $\int \dfrac{\cos^5 x\, dx}{\Delta^3} = \dfrac{-k^2 \sin^2 x + 2k^4 - 4k^2 + 3}{2k^4 \Delta} \sin x + \dfrac{4k^2 - 3}{2k^5} \arcsin(k \sin x).$

58. $\int \dfrac{dx}{\Delta^5} = \dfrac{-k^2 \sin x \cos x}{3k'^2 \Delta^3} - \dfrac{2k^2(k'^2 + 1)\sin x \cos x}{3k'^4 \Delta} - \dfrac{1}{3k'^2} F(x,k) +$
$\qquad\qquad\qquad + \dfrac{2(k'^2 + 1)}{3k'^4} E(x,k).$

59. $\int \dfrac{\sin x\, dx}{\Delta^5} = \dfrac{2k^2 \sin^2 x + k^2 - 3}{3k'^4 \Delta^3} \cos x.$

60. $\int \dfrac{\cos x\, dx}{\Delta^5} = \dfrac{-2k^2 \sin^2 x + 3}{3\Delta^3} \sin x.$

61. $\int \dfrac{\sin^2 x\, dx}{\Delta^5} = \dfrac{k^2 + 1}{3k'^4 k^2} E(x,k) - \dfrac{1}{3k'^2 k^2} F(x,k) +$
$\qquad\qquad\qquad + \dfrac{k^2(k^2 + 1)\sin^2 x - 2}{3k'^4 \Delta^3} \sin x \cos x.$

62. $\int \dfrac{\sin x \cos x\, dx}{\Delta^5} = \dfrac{1}{3k^2 \Delta^3}.$

2.584

63. $\int \frac{\cos^2 x \, dx}{\Delta^5} = \frac{1}{3k^2} F(x, k) + \frac{2k^2 - 1}{3k^2 k'^2} E(x, k) +$
$$+ \frac{k^2(2k^2 - 1) \sin^2 x - 3k^2 + 2}{3k' \Delta^3} \sin x \cos x.$$

64. $\int \frac{\sin^3 x}{\Delta^5} dx = \frac{(3k^2 - 1) \sin^2 x - 2}{3k'^4 \Delta^3} \cos x.$

65. $\int \frac{\sin^2 x \cos x}{\Delta^5} dx = \frac{\sin^3 x}{3\Delta^3}.$

66. $\int \frac{\sin x \cos^2 x}{\Delta^5} dx = -\frac{\cos^3 x}{3k'^2 \Delta^3}.$

67. $\int \frac{\cos^3 x \, dx}{\Delta^5} = \frac{-(2k^2 + 1) \sin^2 x + 3}{3\Delta^3} \sin x.$

68. $\int \frac{dx}{\Delta \sin x} = -\frac{1}{2} \ln \frac{\Delta + \cos x}{\Delta - \cos x}.$

69. $\int \frac{dx}{\Delta \cos x} = -\frac{1}{2k'} \ln \frac{\Delta - k' \sin x}{\Delta + k' \sin x}.$

70. $\int \frac{dx}{\Delta \sin^2 x} = \int \frac{1 + \operatorname{ctg}^2 x}{\Delta} dx = F(x, k) - E(x, k) - \Delta \operatorname{ctg} x.$

71. $\int \frac{dx}{\Delta \sin x \cos x} = \int (\operatorname{tg} x + \operatorname{ctg} x) \frac{dx}{\Delta} = \frac{1}{2} \ln \frac{1 - \Delta}{1 + \Delta} + \frac{1}{2k'} \ln \frac{\Delta + k'}{\Delta - k'}.$

72. $\int \frac{dx}{\Delta \cos^2 x} = \int (1 + \operatorname{tg}^2 x) \frac{dx}{\Delta} = F(x, k) - \frac{1}{k'^2} E(x, k) + \frac{1}{k'^2} \Delta \operatorname{tg} x.$

73. $\int \frac{\sin x}{\cos x} \frac{dx}{\Delta} = \int \operatorname{tg} x \frac{dx}{\Delta} = \frac{1}{2k'} \ln \frac{\Delta + k'}{\Delta - k'}.$

74. $\int \frac{\cos x}{\sin x} \frac{dx}{\Delta} = \int \operatorname{ctg} x \frac{dx}{\Delta} = \frac{1}{2} \ln \frac{1 - \Delta}{1 + \Delta}.$

75. $\int \frac{dx}{\Delta \sin^3 x} = -\frac{\Delta \cos x}{2 \sin^2 x} - \frac{1 + k^2}{4} \ln \frac{\Delta + \cos x}{\Delta - \cos x}.$

76. $\int \frac{dx}{\Delta \sin^2 x \cos x} = -\frac{\Delta}{\sin x} - \frac{1}{2k'} \ln \frac{\Delta - k' \sin x}{\Delta + k' \sin x}.$

77. $\int \frac{dx}{\Delta \sin x \cos^2 x} = \frac{\Delta}{k'^2 \cos x} + \frac{1}{2} \ln \frac{\Delta - \cos x}{\Delta + \cos x}.$

78. $\int \frac{dx}{\Delta \cos^3 x} = \frac{\Delta \sin x}{2k'^2 \cos^2 x} + \frac{2k^2 - 1}{4k'^3} \ln \frac{\Delta - k' \sin x}{\Delta + k' \sin x}.$

79. $\int \frac{\sin x}{\cos^2 x} \frac{dx}{\Delta} = \frac{\Delta}{k'^2 \cos x}.$

80. $\int \frac{\cos x}{\sin^2 x} \frac{dx}{\Delta} = -\frac{\Delta}{\sin x}.$

81. $\int \frac{\sin^2 x}{\cos x} \frac{dx}{\Delta} = \frac{1}{2k'} \ln \frac{\Delta + k' \sin x}{\Delta - k' \sin x} - \frac{1}{k} \arcsin (k \sin x).$

82. $\int \frac{\cos^2 x}{\sin x} \frac{dx}{\Delta} = \frac{1}{2} \ln \frac{\Delta + \cos x}{\Delta - \cos x} + \frac{1}{k} \ln (k \cos x + \Delta).$

83. $\int \frac{dx}{\Delta \sin^4 x} = \frac{1}{3}\{-\Delta \operatorname{ctg}^3 x - \Delta(2k^2+3)\operatorname{ctg} x + (k^2+2)F(x,k) -$
$$- 2(k^2+1)E(x,k)\}.$$
2.584

84. $\int \frac{dx}{\Delta \sin^3 x \cos x} = \int (\operatorname{tg} x + 2\operatorname{ctg} x + \operatorname{ctg}^3 x)\frac{dx}{\Delta} =$
$$= -\frac{\Delta}{2\sin^2 x} + \frac{1}{2k'}\ln\frac{\Delta+k'}{\Delta-k'} - \frac{k^2+2}{4}\ln\frac{1+\Delta}{1-\Delta}.$$

85. $\int \frac{dx}{\Delta \sin^2 x \cos^2 x} = \int (\operatorname{tg}^2 x + 2 + \operatorname{ctg}^2 x)\frac{dx}{\Delta} =$
$$= \left(\frac{\operatorname{tg} x}{k'^2} - \operatorname{ctg} x\right)\Delta + \frac{k^2-2}{k'^2}E(x,k) + 2F(x,k).$$

86. $\int \frac{dx}{\Delta \sin x \cos^3 x} = \int (\operatorname{ctg} x + 2\operatorname{tg} x + \operatorname{tg}^3 x)\frac{dx}{\Delta} =$
$$= \frac{\Delta}{2k'^2 \cos^2 x} - \frac{1}{2}\ln\frac{1+\Delta}{1-\Delta} + \frac{2-3k^2}{4k'^3}\ln\frac{\Delta+k'}{\Delta-k'}.$$

87. $\int \frac{dx}{\Delta \cos^4 x} = \frac{1}{3k'^2}\left\{\Delta \operatorname{tg}^3 x - \frac{5k^2-3}{k'^2}\Delta \operatorname{tg} x - (3k^2-2)F(x,k) + \right.$
$$\left. + \frac{2(2k^2-1)}{k'^2}E(x,k)\right\}.$$

88. $\int \frac{\sin x}{\cos^3 x}\frac{dx}{\Delta} = \int \operatorname{tg} x(1+\operatorname{tg}^2 x)\frac{dx}{\Delta} = \frac{\Delta}{2k'^2 \cos^2 x} - \frac{k^2}{4k'^3}\ln\frac{\Delta+k'}{\Delta-k'}.$

89. $\int \frac{\cos x}{\sin^3 x}\frac{dx}{\Delta} = -\frac{\Delta}{2\sin^2 x} - \frac{k^2}{4}\ln\frac{1+\Delta}{1-\Delta}.$

90. $\int \frac{\sin^2 x}{\cos^2 x}\frac{dx}{\Delta} = \int \frac{\operatorname{tg}^2 x}{\Delta}dx = \frac{\Delta}{k'^2}\operatorname{tg} x - \frac{1}{k'^2}E(x,k).$

91. $\int \frac{\cos^2 x}{\sin^2 x}\frac{dx}{\Delta} = \int \frac{\operatorname{ctg}^2 x}{\Delta}dx = -\Delta \operatorname{ctg} x - E(x,k).$

92. $\int \frac{\sin^3 x}{\cos x}\frac{dx}{\Delta} = \frac{\Delta}{k} + \frac{1}{2k'}\ln\frac{\Delta+k'}{\Delta-k'}.$

93. $\int \frac{\cos^3 x}{\sin x}\frac{dx}{\Delta} = \frac{\Delta}{k^2} - \frac{1}{2}\ln\frac{1+\Delta}{1-\Delta}.$

94. $\int \frac{dx}{\Delta \sin^5 x} = -\frac{[3(1+k^2)\sin^2 x + 2]}{8\sin^4 x}\Delta \cos x + \frac{3k^4+2k^2+3}{16}\ln\frac{\Delta+\cos x}{\Delta-\cos x}.$

95. $\int \frac{dx}{\Delta \sin^4 x \cos x} = -\frac{(3+2k^2)\sin^2 x + 1}{3\sin^3 x}\Delta - \frac{1}{2k'}\ln\frac{\Delta-k'\sin x}{\Delta+k'\sin x}.$

96. $\int \frac{dx}{\Delta \sin^3 x \cos^2 x} = \frac{(3-k^2)\sin^2 x - k'^2}{2k'^2 \sin^2 x \cos x}\Delta + \frac{k^2+3}{4}\ln\frac{\Delta-\cos x}{\Delta+\cos x}.$

97. $\int \frac{dx}{\Delta \sin^2 x \cos^3 x} = \frac{(3-2k^2)\sin^2 x - 2k'^2}{2k'^2 \sin x \cos^2 x}\Delta - \frac{4k^2-3}{4k'^3}\ln\frac{\Delta+k'\sin x}{\Delta-k'\sin x}.$

98. $\int \frac{dx}{\Delta \sin x \cos^4 x} = \frac{(5k^2-3)\sin^2 x - 6k^2+4}{3k'^4 \cos^3 x}\Delta - \frac{1}{2}\ln\frac{\Delta+\cos x}{\Delta-\cos x}.$

2.584

99. $\int \dfrac{dx}{\Delta \cos^5 x} = \dfrac{3(2k^2-1)\sin^2 x - 8k^2 + 5}{8k'^4 \cos^4 x} \Delta \sin x +$
$$+ \dfrac{8k^4 - 8k^2 + 3}{16k'^5} \ln \dfrac{\Delta + k' \sin x}{\Delta - k' \sin x}.$$

100. $\int \dfrac{\sin x}{\cos^4 x} \dfrac{dx}{\Delta} = -\dfrac{2k^2 \cos^2 x - k'^2}{3k'^4 \cos^3 x} \Delta.$

101. $\int \dfrac{\cos x}{\sin^4 x} \dfrac{dx}{\Delta} = -\dfrac{2k^2 \sin^2 x + 1}{3 \sin^3 x} \Delta.$

102. $\int \dfrac{\sin^2 x}{\cos^3 x} \dfrac{dx}{\Delta} = \dfrac{\Delta \sin x}{2k'^2 \cos^2 x} - \dfrac{1}{4k'^3} \ln \dfrac{\Delta + k' \sin x}{\Delta - k' \sin x}.$

103. $\int \dfrac{\cos^2 x}{\sin^3 x} \dfrac{dx}{\Delta} = -\dfrac{\Delta \cos x}{2 \sin^2 x} + \dfrac{k'^2}{4} \ln \dfrac{\Delta + \cos x}{\Delta - \cos x}.$

104. $\int \dfrac{\sin^3 x}{\cos^2 x} \dfrac{dx}{\Delta} = \dfrac{\Delta}{k'^2 \cos x} + \dfrac{1}{k} \ln(k \cos x + \Delta).$

105. $\int \dfrac{\cos^3 x}{\sin^2 x} \dfrac{dx}{\Delta} = \dfrac{-\Delta}{\sin x} - \dfrac{1}{k} \arcsin(k \sin x).$

106. $\int \dfrac{\sin^4 x}{\cos x} \dfrac{dx}{\Delta} = \dfrac{\Delta \sin x}{2k^2} + \dfrac{1}{2k'} \ln \dfrac{\Delta + k' \sin x}{\Delta - k' \sin x} - \dfrac{2k^2+1}{2k^3} \arcsin(k \sin x).$

107. $\int \dfrac{\cos^4 x}{\sin x} \dfrac{dx}{\Delta} = \dfrac{\Delta \cos x}{2k^2} + \dfrac{1}{2} \ln \dfrac{\Delta + \cos x}{\Delta - \cos x} + \dfrac{3k^2-1}{2k^3} \ln(k \cos x + \Delta).$

2.585

1. $\int \dfrac{(a + \sin x)^{p+3} dx}{\Delta} = \dfrac{1}{(p+2)k^2} \Big[(a + \sin x)^p \cos x \, \Delta +$
$$+ 2(2p+3)ak^2 \int \dfrac{(a + \sin x)^{p+2} dx}{\Delta} + (p+1)(1 + k^2 - 6a^2k^2) \int \dfrac{(a + \sin x)^{p+1} dx}{\Delta} -$$
$$- a(2p+1)(1 + k^2 - 2a^2k^2) \int \dfrac{(a + \sin x)^p dx}{\Delta} -$$
$$- p(1-a^2)(1-a^2k^2) \int \dfrac{(a + \sin x)^{p-1} dx}{\Delta} \Big]$$
$$\left[p \neq -2,\ a \neq \pm 1,\ a \neq \pm \dfrac{1}{k} \right].$$

Ist $p = n$ eine natürliche Ziahl, so läßt sich dieses Integral auf dee folgenden drei Integrale zurückführen: | If p is a positive integer n, this integral may be reduced to the following integrals:

2. $\int \dfrac{a + \sin x}{\Delta} dx = aF(x, k) + \dfrac{1}{2k} \ln \dfrac{\Delta - k \cos x}{\Delta + k \cos x}.$

3. $\int \dfrac{(a + \sin x)^2}{\Delta} dx = \dfrac{1 + k^2 a^2}{k^2} F(x, k) - \dfrac{1}{k^2} E(x, k) + \dfrac{a}{k} \ln \dfrac{\Delta - k \cos x}{\Delta + k \cos x}.$

4. $\int \dfrac{dx}{(a + \sin x) \Delta} = \dfrac{1}{a} \Pi\left(x, -\dfrac{1}{a^2}, k\right) - \int \dfrac{\sin x \, dx}{(a^2 - \sin^2 x) \Delta}$

mit | where

5. $\int \dfrac{\sin x \, dx}{(a^2 - \sin^2 x) \Delta} = \dfrac{-1}{2\sqrt{(1-a^2)(1-a^2k^2)}} \ln \dfrac{\sqrt{1-a^2}\,\Delta - \sqrt{1-k^2a^2}\cos x}{\sqrt{1-a^2}\,\Delta + \sqrt{1-k^2a^2}\cos x}.$

1. $\int \dfrac{dx}{(a+\sin x)^n \Delta} = \dfrac{1}{(n-1)(1-a^2)(1-a^2k^2)} \left[-\dfrac{\cos x\, \Delta}{(a+\sin x)^{n-1}} \right.$

$\qquad - (2n-3)(1+k^2-2a^2k^2)\, a \int \dfrac{dx}{(a+\sin x)^{n-1} \Delta}$

$\qquad - (n-2)(6a^2k^2 - k^2 - 1) \int \dfrac{dx}{(a+\sin x)^{n-2} \Delta}$

$\qquad \left. - (10-4n)\, ak^2 \int \dfrac{dx}{(a+\sin x)^{n-3}\Delta} - (n-3)k^2 \int \dfrac{dx}{(a+\sin x)^{n-4}\Delta} \right]$

$\qquad\qquad\qquad\qquad\qquad \left[n \neq 1,\ a \neq \pm 1,\ a \neq \pm \dfrac{1}{k} \right].$

2.586

Dieses Integral läßt sich auf folgende Integrale zurückführen: | This integral may be reduced to

2. $\int \dfrac{dx}{(a+\sin x)^2 \Delta} = \dfrac{1}{(1-a^2)(1-a^2k^2)} \left[-\dfrac{\cos x\, \Delta}{a+\sin x} \right.$

$\qquad - a(1+k^2 - 2a^2 k^2) \int \dfrac{dx}{(a+\sin x)\Delta} - 2ak^2 \int \dfrac{(a+\sin x)\, dx}{\Delta} +$

$\qquad \left. + k^2 \int \dfrac{(a+\sin x)^2\, dx}{\Delta} \right]$ (cf. 2.585 2., 3., 4.).

3. $\int \dfrac{dx}{(a+\sin x)^3 \Delta} = \dfrac{1}{2(1-a^2)(1-a^2k^2)} \left[-\dfrac{\cos x\, \Delta}{(a+\sin x)^2} \right.$

$\qquad \left. - 3a(1+k^2 - 2a^2k^2) \int \dfrac{dx}{(a+\sin x)^2 \Delta} - (6a^2k^2 - k^2 - 1)\int \dfrac{dx}{(a+\sin x)\Delta} + 2ak^2 F(x,k) \right]$

$\qquad\qquad\qquad\qquad\qquad$ (cf. 2.585 4., 2.586 2.).

Für $a = \pm 1$ gilt | For $a = \pm 1$:

4. $\int \dfrac{dx}{(1\pm \sin x)^n \Delta} = \dfrac{1}{(2n-1)k'^2} \left[\mp \dfrac{\cos x\, \Delta}{(1\pm \sin x)^n} + \right.$

$\qquad + (n-1)(1-5k^2) \int \dfrac{dx}{(1\pm \sin x)^{n-1}\Delta} + 2(2n-3)k^2 \int \dfrac{dx}{(1\pm \sin x)^{n-2}\Delta}$

$\qquad \left. - (n-2)k^2 \int \dfrac{dx}{(1\pm \sin x)^{n-3}\Delta} \right].$ GHH[241](6a)

Dieses Integral läßt sich auf folgende Integrale zurückführen: | This integral may be reduced to

5. $\int \dfrac{dx}{(1\pm \sin x)\Delta} = \dfrac{\mp \cos x\, \Delta}{k'^2(1\pm \sin x)} + F(x,k) - \dfrac{1}{k'^2} E(x,k).$ GHH[241](6c)

6. $\int \dfrac{dx}{(1\pm \sin x)^2 \Delta} = \dfrac{1}{3k'^4} \left[\mp \dfrac{k'^2 \cos x\, \Delta}{(1\pm \sin x)^2} \mp \dfrac{(1-5k^2)\cos x\, \Delta}{1\pm \sin x} + \right.$

$\qquad \left. + (1-3k^2)k'^2 F(x,k) - (1-5k^2) E(x,k) \right].$ GHH[241](6b)

2.586 Für $a = \pm \dfrac{1}{k}$ gilt | For $a = \pm \dfrac{1}{k}$:

7. $\displaystyle\int \frac{dx}{(1 \pm k \sin x)^n \Delta} = \frac{1}{(2n-1)k'^2}\left[\pm \frac{k \cos x\, \Delta}{(1 \pm k \sin x)^n} \right.$
$+ (n-1)(5-k^2)\displaystyle\int \frac{dx}{(1 \pm k \sin x)^{n-1}\Delta} - 2(2n-3)\int \frac{dx}{(1 \pm k \sin x)^{n-2}\Delta}$
$\left. + (n-2)\displaystyle\int \frac{dx}{(1 \pm k \sin x)^{n-3}\Delta}\right].$ GHH [241](7a)

Dieses Integral läßt sich auf folgende Integrale zurückführen: | This integral may be reduced to

8. $\displaystyle\int \frac{dx}{(1 \pm k \sin x)\Delta} = \pm \frac{k \cos x\, \Delta}{k'^2(1 \pm k \sin x)} + \frac{1}{k'^2} E(x,k).$ GHH [241](7b)

9. $\displaystyle\int \frac{dx}{(1 \pm k \sin x)^2 \Delta} = \frac{1}{3k'^4}\left[\pm \frac{kk'^2 \cos x\, \Delta}{(1 \pm k \sin x)^2} \pm \frac{k(5-k^2)\cos x\, \Delta}{1 \pm k \sin x} \right.$
$\left. - 2k'^2 F(x,k) + (5-k^2) E(x,k)\right].$ GHH [241](7c)

2.587

1. $\displaystyle\int \frac{(b + \cos x)^{p+3} dx}{\Delta} =$
$= \dfrac{1}{(p+2)k^2}\left[(b+\cos x)^p \sin x\, \Delta + 2(2p+3)bk^2 \displaystyle\int \frac{(b+\cos x)^{p+2} dx}{\Delta}\right.$
$- (p+1)(k'^2 - k^2 + 6b^2 k^2)\displaystyle\int \frac{(b+\cos x)^{p+1} dx}{\Delta}$
$+ (2p+1)b(k'^2 - k^2 + 2b^2 k^2)\displaystyle\int \frac{(b+\cos x)^p dx}{\Delta}$
$\left. + p(1-b^2)(k'^2 + k^2 b^2)\displaystyle\int \frac{(b+\cos x)^{p-1} dx}{\Delta}\right]$
$\left[p \neq -2,\ b \neq \pm 1,\ b \neq \dfrac{ik'}{k}\right].$

Ist p eine natürliche Zahl n, so kann dieses Integral auf die folgenden drei Integrale zurückgeführt werden: | If p is a positive integer n, this integral may be reduced to the following integrals:

2. $\displaystyle\int \frac{b + \cos x}{\Delta} dx = b F(x,k) + \frac{1}{k}\arcsin(k \sin x).$

3. $\displaystyle\int \frac{(b + \cos x)^2}{\Delta} dx = \frac{b^2 k^2 - k'^2}{k^2} F(x,k) + \frac{1}{k^2} E(x,k) + \frac{2b}{k}\arcsin(k \sin x).$

4. $\displaystyle\int \frac{dx}{(b + \cos x)\Delta} = \frac{b}{b^2 - 1}\Pi\left(x, \frac{b}{b^2 - 1}, k\right) + \int \frac{\cos x\, dx}{(1 - b^2 - \sin^2 x)\Delta}$

mit | where

5. $\displaystyle\int \frac{\cos x\, dx}{(1 - b^2 - \sin^2 x)\Delta} = \frac{1}{2\sqrt{(1-b^2)(k'^2 + k^2 b^2)}} \ln \frac{\sqrt{1-b^2}\, \Delta + k\sqrt{k'^2 + k^2 b^2}\sin x}{\sqrt{1-b^2}\, \Delta - k\sqrt{k'^2 + k^2 b^2}\sin x}.$

1. $\displaystyle\int\frac{dx}{(b+\cos x)^n \Delta} = \frac{1}{(n-1)(1-b^2)(k'^2+b^2k^2)} \left[\frac{\sin x\, \Delta}{(b+\cos x)^{n-1}} -\right.$ 2.588

$\displaystyle -(2n-3)(1-2k^2+2b^2k^2)\, b \int\frac{dx}{(b+\cos x)^{n-1} \Delta} -$

$\displaystyle -(n-2)(2k^2-1-6b^2k^2)\int\frac{dx}{(b+\cos x)^{n-2} \Delta} -$

$\displaystyle -(4n-10)\, bk^2 \int\frac{dx}{(b+\cos x)^{n-3} \Delta} + (n-3)k^2 \int\frac{dx}{(b+\cos x)^{n-4} \Delta}\left.\right]$

$\left[n\neq 1,\ b\neq \pm 1,\ b\neq \pm \dfrac{ik'}{k}\right].$

| Dieses Integral läßt sich auf folgende Integrale zurückführen: | This integral may be reduced to |

2. $\displaystyle\int\frac{dx}{(b+\cos x)^2 \Delta} =$

$\displaystyle = \frac{1}{(1-b^2)(k'^2+b^2k^2)} \left[\frac{\sin x\, \Delta}{b+\cos x} - (1-2k^2+2b^2k^2)\, b\int\frac{dx}{(b+\cos x)\Delta} +\right.$

$\displaystyle \left. + 2bk^2 \int\frac{b+\cos x}{\Delta}\, dx - k^2 \int\frac{(b+\cos x)^2}{\Delta}\, dx\right]$ (cf. 2.587 2., 3., 4.).

3. $\displaystyle\int\frac{dx}{(b+\cos x)^3 \Delta} = \frac{1}{2(1-b^2)(k'^2+b^2k^2)}\left[\frac{\sin x\, \Delta}{(b+\cos x)^2} -\right.$

$\displaystyle -3b(1-2k^2+2k^2b^2)\int\frac{dx}{(b+\cos x)^2 \Delta} -$

$\displaystyle -(2k^2-1-6b^2k^2)\int\frac{dx}{(b+\cos x)\Delta} - 2bk^2\, F(x,k)\left.\right]$ (cf. 2.588 2., 2.587 4.).

1. $\displaystyle\int\frac{(c+\operatorname{tg} x)^{p+3}\, dx}{\Delta} =$ 2.589

$\displaystyle = \frac{1}{(p+2)\, k'^2}\left[\frac{(c+\operatorname{tg} x)^p\, \Delta}{\cos^2 x} + 2(2n+3)\, ck'^2 \int\frac{(c+\operatorname{tg} x)^{p+2}\, dx}{\Delta} -\right.$

$\displaystyle -(p+1)(1+k'^2+6c^2k'^2)\int\frac{(c+\operatorname{tg} x)^{p+1}\, dx}{\Delta} +$

$\displaystyle + (2p+1)\, c(1+k'^2+2c^2k'^2)\int\frac{(c+\operatorname{tg} x)^p\, dx}{\Delta} -$

$\displaystyle - p(1+c^2)(1+k'^2c^2)\int\frac{(c+\operatorname{tg} x)^{p-1}\, dx}{\Delta}\left.\right]\quad [p\neq -2].$

| Ist p eine natürliche Zahl n, so kann dieses Integral auf die folgenden drei Integrale zurückgeführt werden: | If p is a positive integer n, this integral may be reduced to the following integrals: |

2. $\displaystyle\int\frac{c+\operatorname{tg} x}{\Delta}\, dx = cF(x,k) + \frac{1}{2k'}\ln\frac{\Delta+k'}{\Delta-k'}.$

3. $\displaystyle\int\frac{(c+\operatorname{tg} x)}{\Delta}\, dx = \frac{1}{k'^2}\operatorname{tg} x\, \Delta + c^2 F(x,k) - \frac{1}{k'^2} E(x,k) + \frac{c}{k'}\ln\frac{\Delta+k'}{\Delta-k'}.$

2.589 4. $\int \dfrac{dx}{(c+\operatorname{tg} x)\varDelta} = \dfrac{c}{1+c^2} F(x,k) + \dfrac{1}{c(1+c^2)} \varPi\left(x, -\dfrac{1+c^2}{c^2}, k\right) -$

$$- \int \dfrac{\sin x \cos x\, dx}{[c^2 - (1+c^2)\sin^2 x]\varDelta}$$

mit | where

5. $\int \dfrac{\sin x \cos x\, dx}{[c^2 - (1+c^2)\sin^2 x]\varDelta} = \dfrac{1}{2\sqrt{(1+c^2)(1+c^2 k'^2)}} \ln \dfrac{\sqrt{1+c^2 k'^2} + \sqrt{1+c^2}\,\varDelta}{\sqrt{1+c^2 k'^2} - \sqrt{1+c^2}\,\varDelta}.$

2.591 1. $\int \dfrac{dx}{(c+\operatorname{tg} x)^n \varDelta} = \dfrac{1}{(n-1)(1+c^2)(1+k'^2 c^2)}\Bigg[-\dfrac{\varDelta}{(c+\operatorname{tg} x)^{n-1}\cos^2 x} +$

$$+ (2n-3)c(1+k'^2 + 2c^2 k'^2)\int \dfrac{dx}{(c+\operatorname{tg} x)^{n-1}\varDelta} -$$

$$- (n-2)(1+k'^2 + 6c^2 k'^2)\int \dfrac{dx}{(c+\operatorname{tg} x)^{n-2}\varDelta} +$$

$$+ (4n-10)ck'^2 \int \dfrac{dx}{(c+\operatorname{tg} x)^{n-3}\varDelta} - (n-3)k'^2 \int \dfrac{dx}{(c+\operatorname{tg} x)^{n-4}\varDelta}\Bigg].$$

Dieses Integral läßt sich zurückführen auf | This integral may be reduced to

2. $\int \dfrac{dx}{(c+\operatorname{tg} x)^2 \varDelta} = \dfrac{1}{(1+c^2)(1+k'^2 c^2)}\Bigg[\dfrac{-\varDelta}{(c+\operatorname{tg} x)\cos^2 x} +$

$$+ c(1+k'^2 + 2c^2 k'^2)\int \dfrac{dx}{(c+\operatorname{tg} x)\varDelta} - 2ck'^2 \int \dfrac{c+\operatorname{tg} x}{\varDelta} dx + k'^2 \int \dfrac{(c+\operatorname{tg} x)^2}{\varDelta} dx\Bigg]$$

(cf. 2.589 2., 3., 4.).

3. $\int \dfrac{dx}{(c+\operatorname{tg} x)^3 \varDelta} = \dfrac{1}{2(1+c^2)(1+k'^2 c^2)}\Bigg[\dfrac{-\varDelta}{(c+\operatorname{tg} x)^2 \cos^2 x} +$

$$+ 3c(1+k'^2 + 2c^2 k'^2)\int \dfrac{dx}{(c+\operatorname{tg} x)^2 \varDelta} -$$

$$- (1+k'^2 + 6c^2 k'^2)\int \dfrac{dx}{(c+\operatorname{tg} x)\varDelta} + 2ck'^2 F(x,k)\Bigg]\quad \text{(cf. 2.591 2., 2.589 4.).}$$

2.592 1. $P_n = \int \dfrac{(a+\sin^2 x)^n}{\varDelta} dx.$

Die Rekursionsformel | The recurrence formula

$$P_{n+2} = \dfrac{1}{(2n+3)k^2}\{(a+\sin^2 x)^n \sin x \cos x\, \varDelta + (2n+2)(1+k^2 + 3ak^2)P_{n+1} -$$

$$- (2n+1)[1 + 2a(1+k^2) + 3a^2 k^2]P_n + 2na(1+a)(1+k^2 a)P_{n-1}\}$$

reduziert dieses Integral bei ganzzahligem n auf die Integrale | reduces this integral for n an integer to

2. $\quad P_1 \quad$ cf. 2.584 1., 2.584 4.

3. $\quad P_0 \quad$ cf. 2.584 1.

4. $P_{-1} = \int \dfrac{dx}{(a+\sin^2 x)\varDelta} = \dfrac{1}{a}\varPi\left(x, \dfrac{1}{a}, k\right).$

Im Fall $a = 0$ | If $a = 0$ 2.592

5. $\quad \int \dfrac{dx}{\sin^2 x\, \Delta} \quad$ cf. 2.584 70. $\hfill$ Z 37(124)+

6. $\quad T_n = \int \dfrac{dx}{(h + g \sin^2 x)^n\, \Delta}$

läßt sich ausrechnen mit Hilfe der Rekursionsformel | can be evaluated with the help of the recurrence formula

$$T_{n-3} = \dfrac{1}{(2n-5) k^2} \left\{ \dfrac{-g^2 \sin x \cos x\, \Delta}{(h + g \sin^2 x)^{n-1}} + 2(n-2)\, [g(1+k^2) + 3hk^2]\, T_{n-2} - \right.$$

$$\left. - (2n-3)[g^2 + 2hg(1+k^2) + 3h^2 k^2]\, T_{n-1} + 2(n-1)\, h(g+h)(g + hk^2)\, T_n \right\}.$$

1. $\quad Q_n = \int \dfrac{(b + \cos^2 x)^n}{\Delta}\, dx.$ 2.593

Die Rekursionsformel | The recurrence formula

$$Q_{n+2} = \dfrac{1}{(2n+3) k^2} \left\{ (b + \cos^2 x)^n \sin x \cos x\, \Delta - (2n+2)(1 - 2k^2 - 3bk^2)\, Q_{n+1} + \right.$$

$$\left. + (2n+1)[k'^2 + 2b(k'^2 - k^2) - 3b^2 k^2]\, Q_n - 2nb(1-b)(k'^2 - k^2 b)\, Q_{n-1} \right\}$$

führt dieses Integral bei ganzzahligem n auf folgende Integrale zurück: | reduces this integral for n an integer to

2. $\quad Q_1 \quad$ cf. 2.584 1., 2.584 6.

3. $\quad Q_0 \quad$ cf. 2.584 1.

4. $\quad Q_{-1} = \int \dfrac{dx}{(b + \cos^2 x)\, \Delta} = \dfrac{1}{b+1}\, \Pi\left(x, -\dfrac{1}{b+1}, k\right).$

Für $b = 0$ gilt | For $b = 0$:

5. $\quad \int \dfrac{dx}{\cos^2 x\, \Delta} \quad$ cf. 2.584 72. $\hfill$ Z 37(123)

1. $\quad R_n = \int \dfrac{(c + \operatorname{tg}^2 x)^n\, dx}{\Delta}.$ 2.594

Die Rekursionsformel | The recurrence formula

$$R_{n+2} = \dfrac{1}{(2n+3) k'^2} \left\{ \dfrac{(c + \operatorname{tg}^2 x)^n \operatorname{tg} x\, \Delta}{\cos^2 x} - (2n+2)(1 + k'^2 - 3ck'^2)\, R_{n+1} + \right.$$

$$\left. + (2n+1)[1 - 2c(1 + k'^2) + 3c^2 k'^2]\, R_n + 2nc(1-c)(1 - k'^2 c)\, R_{n-1} \right\}$$

führt dieses Integral bei ganzzahligem n in die Integrale über: | reduces this integral for n an integer to

2. $\quad R_1 \quad$ cf. 2.584 1., 2.584 90.

3. $\quad R_0 \quad$ cf. 2.584 1.

4. $\quad R_{-1} = \int \dfrac{dx}{(c + \operatorname{tg}^2 x)\, \Delta} = \dfrac{1}{c-1}\, F(x, k) + \dfrac{1}{c(1-c)}\, \Pi\left(x, \dfrac{1-c}{c}, k\right).$

Für $c = 0$ vgl. 2.582 5. | For $c = 0$ see 2.582 5.

2.595 Integrale der Gestalt: | Integrals of the form: $\int R(\sin x, \cos x, \sqrt{1 - p^2 \sin^2 x})\, dx$.

Im Fall $p^2 > 1$ | For $p^2 > 1$

Abkürzung: | Abbreviation: $\alpha = \arcsin(p \sin x)$.

Grundlegende Formeln | Basic formulae

1. $\int \dfrac{dx}{\sqrt{1 - p^2 \sin^2 x}} = \dfrac{1}{p} F\left(\alpha, \dfrac{1}{p}\right) \quad [p^2 > 1].$ BF(283.00)

2. $\int \sqrt{1 - p^2 \sin^2 x}\, dx = pE\left(\alpha, \dfrac{1}{p}\right) - \dfrac{p^2 - 1}{p} F\left(\alpha, \dfrac{1}{p}\right) \quad [p^2 > 1].$ BF(283.03)

3. $\int \dfrac{dx}{(1 - r^2 \sin^2 x)\sqrt{1 - p^2 \sin^2 x}} = \dfrac{1}{p} \Pi\left(\alpha, \dfrac{r^2}{p^2}, \dfrac{1}{p}\right) \quad [p^2 > 1].$ BF(283.02)

Zur Berechnung von Integralen der Gestalt | The integrals of the form

$$\int R(\sin x, \cos x, \sqrt{1 - p^2 \sin^2 x})\, dx$$

im Fall $p^2 > 1$ kann man die Formeln 2.583, 2.584 benutzen, wenn man darin zunächst folgende Änderungen durchführt: | can for $p^2 > 1$ be calculated with the help of 2.583 and 2.584 if the following changes have been carried out in the formulae:

1. k ersetzt man durch | is replaced by p;

2. k'^2 ersetzt man durch | is replaced by $1 - p^2$;

3. $F(x, k)$ ersetzt man durch | is replaced by $\dfrac{1}{p} F\left(\alpha, \dfrac{1}{p}\right)$;

4. $E(x, k)$ ersetzt man durch | is replaced by $pE\left(\alpha, \dfrac{1}{p}\right) - \dfrac{p^2 - 1}{p} F\left(\alpha, \dfrac{1}{p}\right).$

Beispielsweise ist (siehe 2.584 15.): | For instance (see 2.584 15.),

2.596

1. $\int \dfrac{\cos^4 x\, dx}{\sqrt{1 - p^2 \sin^2 x}} = \dfrac{\sin x \cos x \sqrt{1 - p^2 \sin^2 x}}{3p^2} + \dfrac{4p^2 - 2}{3p^4}\left[pE\left(\alpha; \dfrac{1}{p}\right) - \dfrac{p^2 - 1}{p} F\left(\alpha, \dfrac{1}{p}\right)\right] + \dfrac{2 - 5p^2 + 3p^4}{3p^4} \cdot \dfrac{1}{p} F\left(\alpha, \dfrac{1}{p}\right) =$

$= \dfrac{\sin x \cos x \sqrt{1 - p^2 \sin^2 x}}{3p^2} - \dfrac{p^2 - 1}{3p^3} F\left(\alpha, \dfrac{1}{p}\right) + \dfrac{4p^2 - 2}{3p^3} E\left(\alpha, \dfrac{1}{p}\right) \quad [p^2 > 1];$

(cf. 2.583 36.):

2. $\int \dfrac{\sqrt{1 - p^2 \sin^2 x}}{\cos^2 x}\, dx = \operatorname{tg} x \sqrt{1 - p^2 \sin^2 x} + \dfrac{1}{p} F\left(\alpha, \dfrac{1}{p}\right) - \left[pE\left(\alpha, \dfrac{1}{p}\right) - \dfrac{p^2 - 1}{p} F\left(\alpha, \dfrac{1}{p}\right)\right] =$

$= p\left[F\left(\alpha, \dfrac{1}{p}\right) - E\left(\alpha, \dfrac{1}{p}\right)\right] + \operatorname{tg} x \sqrt{1 - p^2 \sin^2 x} \quad [p^2 > 1];$

(cf. 2.584 37.):

3. $\displaystyle\int\frac{dx}{\sqrt{(1-p^2\sin^2 x)^3}} = \frac{-1}{p^2-1}\left[pE\left(\alpha,\frac{1}{p}\right) - \frac{p^2-1}{p}F\left(\alpha,\frac{1}{p}\right)\right] -$ 2.596

$\displaystyle - \frac{p^2}{1-p^2}\cdot\frac{\sin x \cos x}{\sqrt{1-p^2\sin^2 x}} = \frac{p^2}{p^2-1}\cdot\frac{\sin x \cos x}{\sqrt{1-p^2\sin^2 x}} +$

$\displaystyle + \frac{1}{p}F\left(\alpha,\frac{1}{p}\right) - \frac{p}{p^2-1}E\left(\alpha,\frac{1}{p}\right)\quad [p^2 > 1].$

Integrale der Gestalt: $\displaystyle\int R(\sin x, \cos x, \sqrt{1+p^2\sin^2 x})\,dx.$ 2.597
Integrals of the form:

Abkürzung:
Abbreviation: $\quad \alpha = \arcsin\dfrac{\sqrt{1+p^2}\sin x}{\sqrt{1+p^2\sin^2 x}}.$

Grundlegende Formeln | Basic formulae

1. $\displaystyle\int\frac{dx}{\sqrt{1+p^2\sin^2 x}} = \frac{1}{\sqrt{1+p^2}}F\left(\alpha,\frac{p}{\sqrt{1+p^2}}\right).$ BF(282.00)

2. $\displaystyle\int\sqrt{1+p^2\sin^2 x}\,dx = \sqrt{1+p^2}\,E\left(\alpha,\frac{p}{\sqrt{1+p^2}}\right) - p^2\frac{\sin x \cos x}{\sqrt{1+p^2\sin^2 x}}.$ BF(282.03)

3. $\displaystyle\int\frac{\sqrt{1+p^2\sin^2 x}\,dx}{1+(p^2-r^2p^2-r^2)\sin^2 x} = \frac{1}{\sqrt{1+p^2}}\Pi\left(\alpha,r^2,\frac{p}{\sqrt{1+p^2}}\right).$ BF(282.02)

4. $\displaystyle\int\frac{\sin x\,dx}{\sqrt{1+p^2\sin^2 x}} = -\frac{1}{p}\arcsin\left(\frac{p\cos x}{\sqrt{1+p^2}}\right).$

5. $\displaystyle\int\frac{\cos x\,dx}{\sqrt{1+p^2\sin^2 x}} = \frac{1}{p}\ln(p\sin x + \sqrt{1+p^2\sin^2 x}).$

6. $\displaystyle\int\frac{dx}{\sin x\sqrt{1+p^2\sin^2 x}} = \frac{1}{2}\ln\frac{\sqrt{1+p^2\sin^2 x} - \cos x}{\sqrt{1+p^2\sin^2 x} + \cos x}.$

7. $\displaystyle\int\frac{dx}{\cos x\sqrt{1+p^2\sin^2 x}} = \frac{1}{2\sqrt{1+p^2}}\ln\frac{\sqrt{1+p^2\sin^2 x} + \sqrt{1+p^2}\sin x}{\sqrt{1+p^2\sin^2 x} - \sqrt{1+p^2}\sin x}.$

8. $\displaystyle\int\frac{\operatorname{tg} x\,dx}{\sqrt{1+p^2\sin^2 x}} = \frac{1}{2\sqrt{1+p^2}}\ln\frac{\sqrt{1+p^2\sin^2 x} + \sqrt{1+p^2}}{\sqrt{1+p^2\sin^2 x} - \sqrt{1+p^2}}.$

9. $\displaystyle\int\frac{\operatorname{ctg} x\,dx}{\sqrt{1+p^2\sin^2 x}} = \frac{1}{2}\ln\frac{1-\sqrt{1+p^2\sin^2 x}}{1+\sqrt{1+p^2\sin^2 x}}.$

Zur Berechnung von Integralen der Gestalt $\int R(\sin x, \cos x, \sqrt{1+p^2\sin^2 x})\,dx$ kann man die Formeln 2.583, 2.584 benutzen, nachdem man in ihnen folgende Änderungen durchgeführt hat: | The integrals of the form $\int R(\sin x, \cos x,$ 2.598
$\sqrt{1+p^2\sin^2 x})\,dx$ can be evaluated with the help of 2.583 and 2.584 if the following changes have been carried out in the formulae:

1. k^2 ersetzt man durch / is replaced by $-p^2;$

2. k'^2 ersetzt man durch / is replaced by $1+p^2;$

3. $F(x,k)$ ersetzt man durch / is replaced by $\dfrac{1}{\sqrt{1+p^2}}F\left(\alpha,\dfrac{p}{\sqrt{1+p^2}}\right);$

2.598   4. $E(x, k)$ ersetzt man durch / is replaced by $\sqrt{1+p^2}\, E\left(\alpha, \dfrac{p}{\sqrt{1+p^2}}\right) - p^2 \dfrac{\sin x \cos x}{\sqrt{1+p^2 \sin^2 x}}$;

5. $\dfrac{1}{k}\ln(k \cos x + \Delta)$ ersetzt man durch / is replaced by $\dfrac{1}{p}\arcsin \dfrac{p \cos x}{\sqrt{1+p^2}}$;

6. $\dfrac{1}{k}\arcsin(k \sin x)$ ersetzt man durch / is replaced by $\dfrac{1}{p}\ln(p \sin x + \sqrt{1+p^2 \sin^2 x})$.

Beispielsweise ist (vgl. 2.584 90.): | For instance (cf. 2.584 90.),

1. $\displaystyle\int \dfrac{\operatorname{tg}^2 x\, dx}{\sqrt{1+p^2 \sin^2 x}} = \dfrac{1}{(1+p^2)}\left[\operatorname{tg} x\sqrt{1+p^2 \sin^2 x} -\right.$

$\left. -\sqrt{1+p^2}\, E\left(\alpha, \dfrac{p}{\sqrt{1+p^2}}\right) + p^2 \dfrac{\sin x \cos x}{\sqrt{1+p^2 \sin^2 x}}\right] =$

$= -\dfrac{1}{\sqrt{1+p^2}}\, E\left(\alpha, \dfrac{p}{\sqrt{1+p^2}}\right) + \dfrac{\operatorname{tg} x}{\sqrt{1+p^2 \sin^2 x}}$;

(cf. 2.584 37.):

2. $\displaystyle\int \dfrac{dx}{\sqrt{(1+p^2 \sin^2 x)^3}} = \dfrac{1}{\sqrt{1+p^2}}\, E\left(\alpha, \dfrac{p}{\sqrt{1+p^2}}\right).$

2.599 **Integrale der Gestalt:** / **Integrals of the form:** $\displaystyle\int R(\sin x, \cos x, \sqrt{a^2 \sin^2 x - 1})\, dx$ $[a^2 > 1]$.

Abkürzung: / Abbreviation: $\alpha = \arcsin \dfrac{a \cos x}{\sqrt{a^2 - 1}}$.

Grundlegende Formeln | **Basic formulae**

1. $\displaystyle\int \dfrac{dx}{\sqrt{a^2 \sin^2 x - 1}} = -\dfrac{1}{a} F\left(\alpha, \dfrac{\sqrt{a^2 - 1}}{a}\right)$ $[a^2 > 1]$.   BF(285.00)+

2. $\displaystyle\int \sqrt{a^2 \sin^2 x - 1}\, dx = \dfrac{1}{a} F\left(\alpha, \dfrac{\sqrt{a^2 - 1}}{a}\right) - a E\left(\alpha, \dfrac{\sqrt{a^2 - 1}}{a}\right)$ $[a^2 > 1]$.

   BF(285.06)+

3. $\displaystyle\int \dfrac{dx}{(1 - r^2 \sin^2 x)\sqrt{a^2 \sin^2 x - 1}} = \dfrac{1}{a(r^2 - 1)} \Pi\left(\alpha, \dfrac{r^2(a^2 - 1)}{a^2(r^2 - 1)}, \dfrac{\sqrt{a^2 - 1}}{a}\right)$

   $[a^2 > 1,\ r^2 > 1]$. BF(285.02)+

4. $\displaystyle\int \dfrac{\sin x\, dx}{\sqrt{a^2 \sin^2 x - 1}} = -\dfrac{\alpha}{a}$ $[a^2 > 1]$.

5. $\displaystyle\int \dfrac{\cos x\, dx}{\sqrt{a^2 \sin^2 x - 1}} = \dfrac{1}{a}\ln(a \sin x + \sqrt{a^2 \sin^2 x - 1})$ $[a^2 > 1]$.

6. $\displaystyle\int \dfrac{dx}{\sin x \sqrt{a^2 \sin^2 x - 1}} = -\arctan \dfrac{\cos x}{\sqrt{a^2 \sin^2 x - 1}}$ $[a^2 > 1]$.

7. $\displaystyle\int \dfrac{dx}{\cos x \sqrt{a^2 \sin^2 x - 1}} = \dfrac{1}{2\sqrt{a^2 - 1}}\ln \dfrac{\sqrt{a^2 - 1}\sin x + \sqrt{a^2 \sin^2 x - 1}}{\sqrt{a^2 - 1}\sin x - \sqrt{a^2 \sin^2 x - 1}}$

   $[a^2 > 1]$.

8. $\displaystyle\int \dfrac{\operatorname{tg} x\, dx}{\sqrt{a^2 \sin^2 x - 1}} = \dfrac{1}{2\sqrt{a^2 - 1}}\ln \dfrac{\sqrt{a^2 - 1} + \sqrt{a^2 \sin^2 x - 1}}{\sqrt{a^2 - 1} - \sqrt{a^2 \sin^2 x - 1}}$ $[a^2 > 1]$.

9. $\int \dfrac{\operatorname{ctg} x\, dx}{\sqrt{a^2\sin^2 x - 1}} = -\arcsin\dfrac{1}{a\sin x}\quad [a^2 > 1]$.   2.599

| Zur Berechnung von Integralen der Gestalt | The integrals of the form | 2.611 |

$$\int R(\sin x,\, \cos x,\, \sqrt{a^2\sin^2 x - 1})\, dx\quad (a^2 > 1)$$

kann man die Formeln 2.583, 2.584 benutzen. Dazu ist notwendig:
1. Auf den rechten Seiten dieser Formeln folgende Funktionen durch die mit ihnen übereinstimmenden Integrale zu ersetzen:

can be evaluated with the help of 2.583 and 2.584. In doing so it is necessary (1) to replace on the right-hand sides of these formulae the following functions by the integrals which are equal to them:

$F(x, k)$ durch by $\int \dfrac{dx}{\varDelta}$,

$E(x, k)$ durch by $\int \varDelta\, dx$,

$-\dfrac{1}{k}\ln(k\cos x + \varDelta)$ durch by $\int \dfrac{\sin x\, dx}{\varDelta}$,

$\dfrac{1}{k}\arcsin(k\sin x)$ durch by $\int \dfrac{\cos x\, dx}{\varDelta}$,

$\dfrac{1}{2}\ln\dfrac{\varDelta - \cos x}{\varDelta + \cos x}$ durch by $\int \dfrac{dx}{\varDelta \sin x}$,

$\dfrac{1}{2k'}\ln\dfrac{\varDelta + k'\sin x}{\varDelta - k'\sin x}$ durch by $\int \dfrac{dx}{\varDelta \cos x}$,

$\dfrac{1}{2k'}\ln\dfrac{\varDelta + k'}{\varDelta - k'}$ durch by $\int \dfrac{\operatorname{tg} x}{\varDelta}\, dx$,

$\dfrac{1}{2}\ln\dfrac{1 - \varDelta}{1 + \varDelta}$ durch by $\int \dfrac{\operatorname{ctg} x}{\varDelta}\, dx$.

2. Auf beiden Seiten der Gleichungen $\varDelta$ durch $i\sqrt{a^2\sin^2 x - 1}$, ferner k durch a und k'^2 durch $1 - a^2$ zu ersetzen.
3. Beide Seiten der erhaltenen Gleichungen mit i zu multiplizieren, so daß auf beiden Seiten der Gleichungen nur reellwertige Funktionen stehen $(a^2 > 1)$.
4. Anstelle der auf den rechten Seiten der Gleichungen stehenden Integrale die den Formeln 2.599 entnommenen Werte einzusetzen.
Beispiele:
1. Die Gleichung 2.584 4. schreiben wir in der Gestalt

(2) to replace $\varDelta$ on both sides of the equations by $i\sqrt{a^2\sin^2 x - 1}$, then k by a and k'^2 by $1 - a^2$;
(3) to multiply the resulting equations by i so that on both sides only real functions are to be found $(a^2 > 1)$;
(4) to replace the integrals on the right-hand sides of the equations by their values given by formulae 2.599.
Examples:
(1) Equation 2.584 4. is transformed as follows:

$$\int \dfrac{\sin^2 x}{i\sqrt{a^2\sin^2 x - 1}}\, dx = \dfrac{1}{a^2}\int \dfrac{dx}{i\sqrt{a^2\sin^2 x - 1}} - \dfrac{1}{a^2}\int i\sqrt{a^2\sin^2 x - 1}\, dx,$$

2.611   hieraus erhalten wir | which yields

$$\int \frac{\sin^2 x \, dx}{\sqrt{a^2 \sin^2 x - 1}} = \frac{1}{a^2} \left\{ \int \frac{dx}{\sqrt{a^2 \sin^2 x - 1}} + \int \sqrt{a^2 \sin^2 x - 1} \, dx \right\} =$$

$$= -\frac{1}{a} E\left(\alpha, \frac{\sqrt{a^2 - 1}}{a}\right) \quad [a^2 > 1].$$

2. Die Gleichung 2.584 58. schreiben wir in der Gestalt | (2) Equation 2.584 58. is transformed as follows:

$$\int \frac{dx}{i^5 \sqrt{(a^2 \sin^2 x - 1)^5}} = -\frac{2a^4(a^2 - 2) \sin^2 x - (3a^2 - 5) a^2}{3(1 - a^2)^2 \, i^3 \sqrt{(a^2 \sin^2 x - 1)^3}} \sin x \cos x -$$

$$- \frac{1}{3(1 - a^2)} \int \frac{dx}{i \sqrt{a^2 \sin^2 x - 1}} - \frac{2a^2 - 4}{3(1 - a^2)^2} \int i \sqrt{a^2 \sin^2 x - 1} \, dx,$$

hieraus erhalten wir | which yields

$$\int \frac{dx}{\sqrt{(a^2 \sin^2 x - 1)^5}} = \frac{2a^4(a^2 - 2) \sin^2 x - (3a^2 - 5) a^2}{3(1 - a^2)^2 \sqrt{(a^2 \sin^2 x - 1)^3}} \sin x \cos x + \frac{1}{3(1 - a^2)^2 a} \times$$

$$\times \left\{ (a^2 - 3) F\left(\alpha, \frac{\sqrt{a^2 - 1}}{a}\right) - 2a^2(a^2 - 2) E\left(\alpha, \frac{\sqrt{a^2 - 1}}{a}\right) \right\} \quad [a^2 > 1].$$

3. Die Gleichung 2.584 71. schreiben wir in der Gestalt | (3) Equation 2.584 71. is transformed as follows:

$$\int \frac{dx}{\sin x \cos x \, i \sqrt{a^2 \sin^2 x - 1}} = \int \frac{\ctg x \, dx}{i \sqrt{a^2 \sin^2 x - 1}} + \int \frac{\tg x \, dx}{i \sqrt{a^2 \sin^2 x - 1}},$$

hieraus erhalten wir | which yields

$$\int \frac{dx}{\sin x \cos x \sqrt{a^2 \sin^2 x - 1}} = \frac{1}{2 \sqrt{a^2 - 1}} \ln \frac{\sqrt{a^2 - 1} + \sqrt{a^2 \sin^2 x - 1}}{\sqrt{a^2 - 1} - \sqrt{a^2 \sin^2 x - 1}} -$$

$$- \arcsin \frac{1}{a \sin x} \quad [a^2 > 1].$$

2.612   Integrale der Gestalt: Integrals of the form: $\int R(\sin x, \cos x, \sqrt{1 - k^2 \cos^2 x}) \, dx.$

Um solche Integrale zu berechnen, macht man zunächst die Substitution $x = \frac{\pi}{2} - y$ und erhält | In integrals of this type the substitution $x = \frac{\pi}{2} - y$ leads to

$$\int R(\sin x, \cos x, \sqrt{1 - k^2 \cos^2 x}) \, dx = -\int R(\cos y, \sin y, \sqrt{1 - k^2 \sin^2 y}) \, dy.$$

Die erhaltenen Integrale findet man mittels der Formeln 2.583, 2.584. Sie führen (es wird vorausgesetzt, daß das Ausgangsintegral nur auf Integrale der ersten und zweiten LEGENDRESCHEN Normalform führt) nach der Ersetzung der Funktionen $F(x, k)$ und $E(x, k)$ | The integrals $\int R(\cos y, \sin y, \sqrt{1 - k^2 \sin^2 y}) \, dy$ can be evaluated with the help of. 2.583 and 2.584. If we suppose that the original integral leads only to integrals of the first and second LEGENDRE stand-

through the corresponding integrals, we arrive after replacing $F(x,k)$ and $E(x,k)$ by the corresponding integrals, at expressions of the form

$$-g(\cos y, \sin y) - A\int \frac{dy}{\sqrt{1-k^2\sin^2 y}} - B\int \sqrt{1-k^2\sin^2 y}\, dy.$$

Going back to variable x yields

$$\int R(\sin x, \cos x, \sqrt{1-k^2\cos^2 x})\, dx =$$

$$= -g(\sin x, \cos x) - A\int \frac{dx}{\sqrt{1-k^2\cos^2 x}} - B\int \sqrt{1-k^2\cos^2 x}\, dx.$$

The integrals in this expression can be calculated with the help of the formulae

1. $\displaystyle\int \frac{dx}{\sqrt{1-k^2\cos^2 x}} = F\left(\arcsin \frac{\sin x}{\sqrt{1-k^2\cos^2 x}},\, k\right).$

2. $\displaystyle\int \sqrt{1-k^2\cos^2 x}\, dx = E\left(\arcsin \frac{\sin x}{\sqrt{1-k^2\cos^2 x}},\, k\right) - \frac{k^2 \sin x \cos x}{\sqrt{1-k^2\cos^2 x}}.$

Integrals of the form: $\displaystyle\int R(\sin x, \cos x, \sqrt{1-p^2\cos^2 x})\, dx\ [p>1].$ 2.613

The integrals of this type may be evaluated as in 2.612 with the help of the formulae

1. $\displaystyle\int \frac{dx}{\sqrt{1-p^2\cos^2 x}} = -\frac{1}{p} F\left(\arcsin(p\cos x),\, \frac{1}{p}\right)\ [p>1].$

2. $\displaystyle\int \sqrt{1-p^2\cos^2 x}\, dx = \frac{p^2-1}{p} F\left(\arcsin(p\cos x),\, \frac{1}{p}\right) -$

$$- pE\left(\arcsin(p\cos x),\, \frac{1}{p}\right).$$

Integrals of the form: $\displaystyle\int R(\sin x, \cos x, \sqrt{1+p^2\cos^2 x})\, dx.$ 2.614

In integrals of this type the substitution $x = \frac{\pi}{2} - y$ leads to

$$\int R(\sin x, \cos x, \sqrt{1+p^2\cos^2 x})\, dx = -\int R(\cos y, \sin y, \sqrt{1+p^2\sin^2 y})\, dy.$$

In order to evaluate the integrals $\int R(\cos y, \sin y, \sqrt{1+p^2\sin^2 y})\, dy$ the

Veränderlichen x zurückgegangen ist, die Formeln | considerations of 2.598 and 2.612 are used and, after going back to variable x, the formulae

1. $$\int \frac{dx}{\sqrt{1+p^2\cos^2 x}} = \frac{1}{\sqrt{1+p^2}} F\left(x, \frac{p}{\sqrt{1+p^2}}\right).$$

2. $$\int \sqrt{1+p^2\cos^2 x}\, dx = \sqrt{1+p^2}\, E\left(x, \frac{p}{\sqrt{1+p^2}}\right).$$

2.615 Integrale der Gestalt:
Integrals of the form: $\int R(\sin x, \cos x, \sqrt{a^2\cos^2 x - 1})\, dx\ [a > 1]$.

Um solche Integrale zu berechnen, macht man die Substitution $x = \frac{\pi}{2} - y$ und erhält | In these integrals the substitution $x = \frac{\pi}{2} - y$ leads to

$$\int R(\sin x, \cos x, \sqrt{a^2\cos^2 x - 1})\, dx = -\int R(\cos y, \sin y, \sqrt{a^2\sin^2 y - 1})\, dy.$$

Zur Berechnung des rechten Integraltyps benutzt man das unter 2.611 Gesagte und, nachdem man wieder die Veränderliche x eingeführt hat, die Formeln | In order to evaluate the integrals $\int R(\cos y, \sin y, \sqrt{a^2\sin^2 y - 1})\, dy$ the considerations of 2.611 are used and, after going back to variable x, the formulae

1. $$\int \frac{dx}{\sqrt{a^2\cos^2 x - 1}} = \frac{1}{a} F\left(\arcsin \frac{a \sin x}{\sqrt{a^2-1}}, \frac{\sqrt{a^2-1}}{a}\right)\ [a > 1].$$

2. $$\int \sqrt{a^2\cos^2 x - 1}\, dx = a E\left(\arcsin \frac{a \sin x}{\sqrt{a^2-1}}, \frac{\sqrt{a^2-1}}{a}\right) -$$
$$- \frac{1}{a} F\left(\arcsin \frac{a \sin x}{\sqrt{a^2-1}}, \frac{\sqrt{a^2-1}}{a}\right)\ [a > 1].$$

2.616 Integrale der Gestalt:
Integrals of the form: $\int R(\sin x, \cos x, \sqrt{1-p^2\sin^2 x}, \sqrt{1-q^2\sin^2 x})\, dx$.

Abkürzung:
Abbreviation: $\alpha = \arcsin \frac{\sqrt{1-p^2}\sin x}{\sqrt{1-p^2\sin^2 x}}$.

1. $$\int \frac{dx}{\sqrt{(1-p^2\sin^2 x)(1-q^2\sin^2 x)}} = \frac{1}{\sqrt{1-p^2}} F\left(\alpha, \sqrt{\frac{q^2-p^2}{1-p^2}}\right)$$
$$\left[0 < p^2 < q^2 < 1,\ 0 < x \leqslant \frac{\pi}{2}\right].\quad \text{BF(284.00)}$$

2. $$\int \frac{\operatorname{tg}^2 x\, dx}{\sqrt{(1-p^2\sin^2 x)(1-q^2\sin^2 x)}} = \frac{\operatorname{tg} x \sqrt{1-q^2\sin^2 x}}{(1-q^2)\sqrt{1-p^2\sin^2 x}} -$$
$$- \frac{1}{(1-q^2)\sqrt{1-p^2}} E\left(\alpha, \sqrt{\frac{q^2-p^2}{1-p^2}}\right)\ \left[0 < p^2 < q^2 < 1,\ 0 < x \leqslant \frac{\pi}{2}\right].\quad \text{BF(284.07)}$$

3. $\int \dfrac{\operatorname{tg}^4 x\, dx}{\sqrt{(1-p^2 \sin^2 x)(1-q^2 \sin^2 x)}} = \dfrac{1}{3(1-q^2)^2(1-p^2)^{3/2}} \times$ 2.616

$\times \left[2(2-p^2-q^2) E\left(\alpha, \sqrt{\dfrac{q^2-p^2}{1-p^2}}\right) - (1-q^2) F\left(\alpha, \sqrt{\dfrac{q^2-p^2}{1-p^2}}\right) \right] +$

$+ \dfrac{2p^2+q^2-3+\sin^2 x(4-3p^2-2q^2+p^2q^2)}{3(1-p^2)(1-q^2)^2} \dfrac{\sin x}{\cos^3 x} \sqrt{\dfrac{1-q^2 \sin^2 x}{1-p^2 \sin^2 x}}$

$\left[0 < p^2 < q^2 < 1,\ 0 < x \leq \dfrac{\pi}{2} \right].$ BF(284.07)

4. $\int \dfrac{\sin^2 x\, dx}{\sqrt{(1-p^2 \sin^2 x)(1-q^2 \sin^2 x)^3}} = \dfrac{\sqrt{1-p^2}}{(1-q^2)(q^2-p^2)} E\left(\alpha, \sqrt{\dfrac{q^2-p^2}{1-p^2}}\right) -$

$- \dfrac{1}{(q^2-p^2)\sqrt{1-p^2}} F\left(\alpha, \sqrt{\dfrac{q^2-p^2}{1-p^2}}\right) -$

$- \dfrac{\sin x \cos x}{(1-q^2)\sqrt{(1-p^2 \sin^2 x)(1-q^2 \sin^2 x)}} \quad \left[0 < p^2 < q^2 < 1,\ 0 < x \leq \dfrac{\pi}{2} \right].$ BF(284.06)

5. $\int \dfrac{\cos^2 x\, dx}{\sqrt{(1-p^2 \sin^2 x)^3(1-q^2 \sin^2 x)}} =$

$= \dfrac{\sqrt{1-p^2}}{q^2-p^2} E\left(\alpha, \sqrt{\dfrac{q^2-p^2}{1-p^2}}\right) - \dfrac{1-q^2}{(q^2-p^2)\sqrt{1-p^2}} F\left(\alpha, \sqrt{\dfrac{q^2-p^2}{1-p^2}}\right)$

$\left[0 < p^2 < q^2 < 1,\ 0 < x \leq \dfrac{\pi}{2} \right].$ BF(284.05)

6. $\int \dfrac{\cos^4 x\, dx}{\sqrt{(1-p^2 \sin^2 x)^5(1-q^2 \sin^2 x)}} =$

$= \dfrac{(1-p^2)^{3/2}}{3(q^2-p^2)^2} \left[\dfrac{(2+p^2-3q^2)(1-q^2)}{(1-p^2)^2} F\left(\alpha, \sqrt{\dfrac{q^2-p^2}{1-p^2}}\right) + \right.$

$\left. + 2\, \dfrac{2q^2-p^2-1}{1-p^2} E\left(\alpha, \sqrt{\dfrac{q^2-p^2}{1-p^2}}\right) \right] + \dfrac{(1-p^2)\sin x \cos x \sqrt{1-q^2 \sin^2 x}}{3(q^2-p^2)\sqrt{(1-p^2 \sin^2 x)^3}}$

$\left[0 < p^2 < q^2 < 1,\ 0 < x \leq \dfrac{\pi}{2} \right].$ BF(284.05)

7. $\int \dfrac{dx}{1-p^2 \sin^2 x} \sqrt{\dfrac{1-q^2 \sin^2 x}{1-p^2 \sin^2 x}} = \dfrac{1}{\sqrt{1-p^2}} E\left(\alpha, \sqrt{\dfrac{q^2-p^2}{1-p^2}}\right)$

$\left[0 < p^2 < q^2 < 1,\ 0 < x \leq \dfrac{\pi}{2} \right].$ BF(284.01)

8. $\int \sqrt{\dfrac{1-p^2 \sin^2 x}{(1-q^2 \sin^2 x)^3}}\, dx = \dfrac{\sqrt{1-p^2}}{1-q^2} E\left(\alpha, \sqrt{\dfrac{q^2-p^2}{1-p^2}}\right) -$

$- \dfrac{q^2-p^2}{1-q^2} \dfrac{\sin x \cos x}{\sqrt{(1-p^2 \sin^2 x)(1-q^2 \sin^2 x)}} \quad \left[0 < p^2 < q^2 < 1,\ 0 < x \leq \dfrac{\pi}{2} \right].$

BF(284.04)

2.616   9. $\int \dfrac{dx}{1+(p^2r^2-p^2-r^2)\sin^2 x}\sqrt{\dfrac{1-p^2\sin^2 x}{1-q^2\sin^2 x}} =$

$$= \dfrac{1}{\sqrt{1-p^2}}\, \Pi\left(\alpha, r^2, \sqrt{\dfrac{q^2-p^2}{1-p^2}}\right)\quad \left[0<p^2<q^2<1,\ 0<x\leqslant \dfrac{\pi}{2}\right]. \qquad \text{BF(284.02)}$$

2.617   Abkürzungen: $\quad \alpha = \arcsin\sqrt{\dfrac{\sqrt{b^2+c^2}-b\sin x-c\cos x}{2\sqrt{b^2+c^2}}},\quad r=\sqrt{\dfrac{2\sqrt{b^2+c^2}}{a+\sqrt{b^2+c^2}}}.$
Abbreviations:

1. $\displaystyle \int \dfrac{dx}{\sqrt{a+b\sin x+c\cos x}} = -\dfrac{2}{\sqrt{a+\sqrt{b^2+c^2}}}\, F(\alpha, r)$

$$\left[0<\sqrt{b^2+c^2}<a,\ \arcsin\dfrac{b}{\sqrt{b^2+c^2}}-\pi\leqslant x < \arcsin\dfrac{b}{\sqrt{b^2+c^2}}\right]; \qquad \text{BF(294.00)}$$

$$=-\dfrac{\sqrt{2}}{\sqrt[4]{b^2+c^2}}\, F(\alpha, r)$$

$$\left[0<|a|<\sqrt{b^2+c^2},\ \arcsin\dfrac{b}{\sqrt{b^2+c^2}}-\arccos\left(-\dfrac{a}{\sqrt{b^2+c^2}}\right)\leqslant x < \right.$$
$$\left. < \arcsin\dfrac{b}{\sqrt{b^2+c^2}}\right]. \qquad \text{BF(293.00)}$$

2. $\displaystyle \int \dfrac{\sin x\, dx}{\sqrt{a+b\sin x+c\cos x}} = -\dfrac{\sqrt{2}\, b}{\sqrt[4]{(b^2+c^2)^3}}\,\{2E(\alpha, r)-F(\alpha, r)\}+$

$$+\dfrac{2c}{b^2+c^2}\sqrt{a+b\sin x+c\cos x}$$

$$\left[0<|a|<\sqrt{b^2+c^2},\ \arcsin\dfrac{b}{\sqrt{b^2+c^2}}-\arccos\left(-\dfrac{a}{\sqrt{b^2+c^2}}\right)\leqslant\right.$$
$$\left.\leqslant x < \arcsin\dfrac{b}{\sqrt{b^2+c^2}}\right]. \qquad \text{BF(293.05)}$$

3. $\displaystyle \int \dfrac{(b\cos x - c\sin x)\, dx}{\sqrt{a+b\sin x+c\cos x}} = 2\sqrt{a+b\sin x+c\cos x}.$

4. $\displaystyle \int \dfrac{\sqrt{b^2+c^2}+b\sin x+c\cos x}{\sqrt{a+b\sin x+c\cos x}}\, dx = -2\sqrt{a+\sqrt{b^2+c^2}}\, E(\alpha, r)+$

$$+\dfrac{2(a-\sqrt{b^2+c^2})}{\sqrt{a+\sqrt{b^2+c^2}}}\, F(\alpha, r)\quad \left[0<\sqrt{b^2+c^2}<a,\right.$$

$$\left.\arcsin\dfrac{b}{\sqrt{b^2+c^2}}-\pi\leqslant x < \arcsin\dfrac{b}{\sqrt{b^2+c^2}}\right]. \qquad \text{BF(294.04)}$$

$$=-2\sqrt{2}\sqrt[4]{b^2+c^2}\, E(\alpha, r)$$

$$\left[0<|a|<\sqrt{b^2+c^2},\ \arcsin\dfrac{b}{\sqrt{b^2+c^2}}-\arccos\left(-\dfrac{a}{\sqrt{b^2+c^2}}\right)\leqslant x <\right.$$
$$\left. < \arcsin\dfrac{b}{\sqrt{b^2+c^2}}\right]. \qquad \text{BF(293.01)}$$

5. $\int \sqrt{a + b \sin x + c \cos x}\, dx = -2\sqrt{a + \sqrt{b^2 + c^2}}\, E(\alpha, r)$ 2.617

$\left[0 < \sqrt{b^2 + c^2} < a,\ \arcsin \dfrac{b}{\sqrt{b^2 + c^2}} - \pi \leq x < \arcsin \dfrac{b}{\sqrt{b^2 + c^2}}\right]$; BF(294.01)

$= -2\sqrt{2}\sqrt[4]{b^2 + c^2}\, E(\alpha, r) +$

$+ \dfrac{\sqrt{2}(\sqrt{b^2 + c^2} - a)}{\sqrt[4]{b^2 + c^2}} F(\alpha, r)\ \left[0 < |a| < \sqrt{b^2 + c^2},\right.$

$\left.\arcsin \dfrac{b}{\sqrt{b^2 + c^2}} - \arccos\left(\dfrac{-a}{\sqrt{b^2 + c^2}}\right) \leq x < \arcsin \dfrac{b}{\sqrt{b^2 + c^2}}\right]$. BF(293.03)

Integrale der Gestalt:
Integrals of the form:
$$\int R(\sin ax, \cos ax, \sqrt{\cos 2ax})\, dx =$$
$$= \dfrac{1}{a} \int R(\sin t, \cos t, \sqrt{1 - 2 \sin^2 t})\, dt\quad (t = ax).$$ 2.618

Abkürzung:
Abbreviation: $\alpha = \arcsin (\sqrt{2} \sin ax)$.

Integrale $\int R(\sin ax, \cos ax, \sqrt{\cos 2ax})\, dx$ sind Spezialfälle der Integrale 2.595 (für $p = 2$). Wir führen folgende Formeln an: | The integrals $\int R(\sin ax, \cos ax, \sqrt{\cos 2ax})\, dx$ are special cases ($p = 2$) of the integrals 2.595. We note the following formulae:

1. $\int \dfrac{dx}{\sqrt{\cos 2ax}} = \dfrac{1}{a\sqrt{2}} F\left(\alpha, \dfrac{1}{\sqrt{2}}\right)\ \left[0 < ax \leq \dfrac{\pi}{4}\right]$.

2. $\int \dfrac{\cos^2 ax}{\sqrt{\cos 2ax}}\, dx = \dfrac{1}{a\sqrt{2}} E\left(\alpha, \dfrac{1}{\sqrt{2}}\right)\ \left[0 < ax \leq \dfrac{\pi}{4}\right]$.

3. $\int \dfrac{dx}{\cos^2 ax \sqrt{\cos 2ax}} = \dfrac{\sqrt{2}}{a} E\left(\alpha, \dfrac{1}{\sqrt{2}}\right) - \dfrac{\operatorname{tg} x}{a}\sqrt{\cos 2ax}\ \left[0 < ax \leq \dfrac{\pi}{4}\right]$.

4. $\int \dfrac{dx}{\cos^4 ax \sqrt{\cos 2ax}} = \dfrac{2\sqrt{2}}{a} E\left(\alpha, \dfrac{1}{\sqrt{2}}\right) -$
$- \dfrac{\sqrt{2}}{3a} F\left(\alpha, \dfrac{1}{\sqrt{2}}\right) - \dfrac{(6\cos^2 ax + 1) \sin ax}{3a \cos^3 ax}\sqrt{\cos 2ax}\ \left[0 < x \leq \dfrac{\pi}{4}\right]$.

5. $\int \dfrac{\operatorname{tg}^2 ax\, dx}{\sqrt{\cos 2ax}} = \dfrac{\sqrt{2}}{a} E\left(\alpha, \dfrac{1}{\sqrt{2}}\right) - \dfrac{1}{a\sqrt{2}} F\left(\alpha, \dfrac{1}{\sqrt{2}}\right) -$
$- \dfrac{1}{a} \operatorname{tg} ax \sqrt{\cos 2ax}\ \left[0 < x \leq \dfrac{\pi}{2}\right]$.

6. $\int \dfrac{\operatorname{tg}^4 ax\, dx}{\sqrt{\cos 2ax}} = \dfrac{1}{3a\sqrt{2}} F\left(\alpha, \dfrac{1}{\sqrt{2}}\right) - \dfrac{\sin ax}{3a \cos^3 ax}\sqrt{\cos 2ax}\ \left[0 < ax \leq \dfrac{\pi}{4}\right]$.

7. $\int \dfrac{dx}{(1 - 2r^2 \sin^2 ax)\sqrt{\cos 2ax}} = \dfrac{1}{a\sqrt{2}} \Pi\left(\alpha, r^2, \dfrac{1}{\sqrt{2}}\right)\ \left[0 < ax \leq \dfrac{\pi}{4}\right]$.

8. $\int \dfrac{dx}{\sqrt{\cos^3 2ax}} = \dfrac{1}{a\sqrt{2}} F\left(\alpha, \dfrac{1}{\sqrt{2}}\right) - \dfrac{\sqrt{2}}{a} E\left(\alpha, \dfrac{1}{\sqrt{2}}\right) + \dfrac{\sin 2ax}{a \sqrt{\cos 2ax}}$
$\left[0 < ax \leq \dfrac{\pi}{4}\right]$.

2.618

9. $\int \dfrac{\sin^2 ax\, dx}{\sqrt{\cos^3 2ax}} = \dfrac{\sin 2ax}{2a\sqrt{\cos 2ax}} - \dfrac{1}{a\sqrt{2}} E\left(\alpha, \dfrac{1}{\sqrt{2}}\right) \quad \left[0 < ax \leqslant \dfrac{\pi}{4}\right].$

10. $\int \dfrac{dx}{\sqrt{\cos^5 2ax}} = \dfrac{1}{3a\sqrt{2}} F\left(\alpha, \dfrac{1}{\sqrt{2}}\right) + \dfrac{\sin 2ax}{3a\sqrt{\cos^3 2ax}} \quad \left[0 < ax \leqslant \dfrac{\pi}{4}\right].$

11. $\int \sqrt{\cos 2ax}\, dx = \dfrac{\sqrt{2}}{a} E\left(\alpha, \dfrac{1}{\sqrt{2}}\right) - \dfrac{1}{a\sqrt{2}} F\left(\alpha, \dfrac{1}{\sqrt{2}}\right) \quad \left[0 < ax \leqslant \dfrac{\pi}{4}\right].$

12. $\int \dfrac{\sqrt{\cos 2ax}}{\cos^2 ax}\, dx = \dfrac{\sqrt{2}}{a}\left\{F\left(\alpha, \dfrac{1}{\sqrt{2}}\right) - E\left(\alpha, \dfrac{1}{\sqrt{2}}\right)\right\} +$

$+ \dfrac{1}{a}\operatorname{tg} ax \sqrt{\cos 2ax} \quad \left[0 < x \leqslant \dfrac{\pi}{4}\right].$

2.619 **Integrale der Gestalt:**
Integrals of the form:
$\begin{cases} \int R(\sin ax, \cos ax, \sqrt{-\cos 2ax})\, dx = \\ = \dfrac{1}{a} \int R(\sin x, \cos x, \sqrt{2\sin^2 x - 1})\, dx. \end{cases}$

Abkürzung:
Abbreviation: $\alpha = \arcsin(\sqrt{2} \cos ax).$

Integrale $\int R(\sin x, \cos x, \sqrt{2\sin^2 x - 1})\, dx$ sind Spezialfälle der Integrale 2.599, 2.611 (für $a = 2$). Wir geben folgende Formeln an: | The integrals $\int R(\sin x, \cos x, \sqrt{2\sin^2 x - 1})\, dx$ are special cases ($a = 2$) of the integrals 2.599 and 2.611. Note the following formulae:

1. $\int \dfrac{dx}{\sqrt{-\cos 2ax}} = -\dfrac{1}{a\sqrt{2}} F\left(\alpha, \dfrac{1}{\sqrt{2}}\right).$

2. $\int \dfrac{\cos^2 ax\, dx}{\sqrt{-\cos 2ax}} = \dfrac{1}{a\sqrt{2}}\left[E\left(\alpha, \dfrac{1}{\sqrt{2}}\right) - F\left(\alpha, \dfrac{1}{\sqrt{2}}\right)\right].$

3. $\int \dfrac{\cos^4 ax\, dx}{\sqrt{-\cos 2ax}} = \dfrac{1}{3a\sqrt{2}}\left[3F\left(\alpha, \dfrac{1}{\sqrt{2}}\right) - \dfrac{5}{2} E\left(\alpha, \dfrac{1}{\sqrt{2}}\right)\right] -$

$- \dfrac{1}{12a} \sin 2ax \sqrt{-\cos 2ax}.$

4. $\int \dfrac{dx}{\sin^2 ax \sqrt{-\cos 2ax}} = \dfrac{1}{a} \operatorname{ctg} ax \sqrt{-\cos 2ax} - \dfrac{\sqrt{2}}{a} E\left(\alpha, \dfrac{1}{\sqrt{2}}\right).$

5. $\int \dfrac{dx}{\sin^4 ax \sqrt{-\cos 2ax}} = \dfrac{2}{3a\sqrt{2}}\left[F\left(\alpha, \dfrac{1}{\sqrt{2}}\right) - 6E\left(\alpha, \dfrac{1}{\sqrt{2}}\right)\right] +$

$+ \dfrac{1}{3a} \dfrac{\cos ax}{\sin^3 ax} (6 \sin^2 ax + 1) \sqrt{-\cos 2ax}.$

6. $\int \dfrac{\operatorname{ctg}^2 ax\, dx}{\sqrt{-\cos 2ax}} = \dfrac{1}{a\sqrt{2}}\left[F\left(\alpha, \dfrac{1}{\sqrt{2}}\right) - 2E\left(\alpha, \dfrac{1}{\sqrt{2}}\right)\right] +$

$+ \dfrac{1}{a} \operatorname{ctg} ax \sqrt{-\cos 2ax}.$

7. $\int \dfrac{dx}{(1 - 2r^2 \cos^2 ax)\sqrt{-\cos 2ax}} = -\dfrac{1}{a\sqrt{2}} \Pi\left(\alpha, r^2, \dfrac{1}{\sqrt{2}}\right).$

Trigonometrische Funktionen/Trigonometric Functions

8. $\int \dfrac{dx}{\sqrt{-\cos^3 2ax}} = \dfrac{1}{a\sqrt{2}}\left[F\left(\alpha, \dfrac{1}{\sqrt{2}}\right) - 2E\left(\alpha, \dfrac{1}{\sqrt{2}}\right)\right] + \dfrac{\sin 2ax}{a\sqrt{-\cos 2ax}}.$ 2.619

9. $\int \dfrac{\cos^2 ax\, dx}{\sqrt{-\cos^3 2ax}} = \dfrac{\sin 2ax}{2a\sqrt{-\cos 2ax}} - \dfrac{1}{a\sqrt{2}} E\left(\alpha, \dfrac{1}{\sqrt{2}}\right).$

10. $\int \dfrac{dx}{\sqrt{-\cos^5 2ax}} = -\dfrac{1}{3a\sqrt{2}} F\left(\alpha, \dfrac{1}{\sqrt{2}}\right) - \dfrac{\sin 2ax}{3a\sqrt{-\cos^3 2ax}}.$

11. $\int \sqrt{-\cos 2ax}\, dx = \dfrac{1}{a\sqrt{2}}\left[F\left(\alpha, \dfrac{1}{\sqrt{2}}\right) - 2E\left(\alpha, \dfrac{1}{\sqrt{2}}\right)\right].$

Integrale der Gestalt:
Integrals of the form: $\int R(\sin ax, \cos ax, \sqrt{\sin 2ax})\, dx.$

Abkürzung:
Abbreviation: $\alpha = \arcsin\sqrt{\dfrac{2\sin ax}{1 + \sin ax + \cos ax}}.$

1. $\int \dfrac{dx}{\sqrt{\sin 2ax}} = \dfrac{\sqrt{2}}{a} F\left(\alpha, \dfrac{1}{\sqrt{2}}\right).$ BF(287.50) 2.621

2. $\int \dfrac{\sin ax\, dx}{\sqrt{\sin 2ax}} = \dfrac{\sqrt{2}}{a}\left\{\dfrac{1+i}{2}\Pi\left(\alpha, \dfrac{1+i}{2}, \dfrac{1}{\sqrt{2}}\right) + \right.$

$\left. + \dfrac{1-i}{2}\Pi\left(\alpha, \dfrac{1-i}{2}, \dfrac{1}{\sqrt{2}}\right) + F\left(\alpha, \dfrac{1}{\sqrt{2}}\right) - 2E\left(\alpha, \dfrac{1}{\sqrt{2}}\right)\right\}.$ BF(287.57)

3. $\int \dfrac{\sin ax\, dx}{(1 + \sin ax + \cos ax)\sqrt{\sin 2ax}} = \dfrac{\sqrt{2}}{a}\left[F\left(\alpha, \dfrac{1}{\sqrt{2}}\right) - E\left(\alpha, \dfrac{1}{\sqrt{2}}\right)\right].$ BF(287.54)

4. $\int \dfrac{\sin ax\, dx}{(1 - \sin ax + \cos ax)\sqrt{\sin 2ax}} = \dfrac{\sqrt{2}}{a}\left\{\sqrt{\operatorname{tg} ax} - E\left(\alpha, \dfrac{1}{\sqrt{2}}\right)\right\}\quad \left[ax \neq \dfrac{\pi}{2}\right].$ BF(287.55)

5. $\int \dfrac{(1 + \cos ax)\, dx}{(1 + \sin ax + \cos ax)\sqrt{\sin 2ax}} = \dfrac{\sqrt{2}}{a} E\left(\alpha, \dfrac{1}{\sqrt{2}}\right).$ BF(287.51)

6. $\int \dfrac{(1 + \cos ax)\, dx}{(1 - \sin ax + \cos ax)\sqrt{\sin 2ax}} =$

$= \dfrac{\sqrt{2}}{a}\left\{F\left(\alpha, \dfrac{1}{\sqrt{2}}\right) - E\left(\alpha, \dfrac{1}{\sqrt{2}}\right) + \sqrt{\operatorname{tg} ax}\right\}\quad \left[ax \neq \dfrac{\pi}{2}\right].$ BF(287.56)

7. $\int \dfrac{(1 - \sin ax + \cos ax)\, dx}{(1 + \sin ax + \cos ax)\sqrt{\sin 2ax}} = \dfrac{\sqrt{2}}{a}\left\{2E\left(\alpha, \dfrac{1}{\sqrt{2}}\right) - F\left(\alpha, \dfrac{1}{\sqrt{2}}\right)\right\}.$ BF(287.53)

8. $\int \dfrac{(1 + \sin ax + \cos ax)\, dx}{[1 + \cos ax + (1 - 2r^2)\sin ax]\sqrt{\sin 2ax}} = \dfrac{\sqrt{2}}{a} \Pi\left(\alpha, r^2, \dfrac{1}{\sqrt{2}}\right).$ BF(287.52)

Trigonometrische Funktionen und Potenzen
Trigonometric functions and powers

2.63–2.65

1. $\int x^r \sin^p x \cos^q x\, dx = \dfrac{1}{(p+q)^2}\left[(p+q) x^r \sin^{p+1} x \cos^{q-1} x + \right.$ 2.631

$+ rx^{r-1} \sin^p x \cos^q x - r(r-1)\int x^{r-2} \sin^p x \cos^q x\, dx -$

2.631

$$-rp\int x^{r-1}\sin^{p-1}x\cos^{q-1}x\,dx + (q-1)(p+q)\int x^r\sin^p x\cos^{q-2}x\,dx];$$

$$= \frac{1}{(p+q)^2}\Big[-(p+q)x^r\sin^{p-1}x\cos^{q+1}x +$$

$$+ rx^{r-1}\sin^p x\cos^q x - r(r-1)\int x^{r-2}\sin^p x\cos^q x\,dx +$$

$$+ rq\int x^{r-1}\sin^{p-1}x\cos^{q-1}x\,dx + (p-1)(p+q)\int x^r\sin^{p-2}x\cos^q x\,dx\Big].$$

GHH [333](1)

2. $\int x^m \sin^n x\,dx = \frac{x^{m-1}\sin^{n-1}x}{n^2}\{m\sin x - nx\cos x\} +$

$$+ \frac{n-1}{n}\int x^m \sin^{n-2} x\,dx - \frac{m(m-1)}{n^2}\int x^{m-2}\sin^n x\,dx.$$

3. $\int x^m \cos^n x\,dx = \frac{x^{m-1}\cos^{n-1}x}{n^2}\{m\cos x + nx\sin x\} +$

$$+ \frac{n-1}{n}\int x^m \cos^{n-2} x\,dx - \frac{m(m-1)}{n^2}\int x^{m-2}\cos^n x\,dx.$$

4. $\int x^n \sin^{2m} x\,dx = \binom{2m}{m}\frac{x^{n+1}}{2^{2m}(n+1)} +$

$$+ \frac{(-1)^m}{2^{2m-1}}\sum_{k=0}^{m-1}(-1)^k\binom{2m}{k}\int x^n \cos(2m-2k)x\,dx \quad (\text{cf. } 2.633\ 2.).$$

T 333

5. $\int x^n \sin^{2m+1} x\,dx = \frac{(-1)^m}{2^{2m}}\sum_{k=0}^{m}(-1)^k\binom{2m+1}{k}\int x^n \sin(2m-2k+1)x\,dx$

(cf. 2.633 1.). T 333

6. $\int x^n \cos^{2m} x\,dx = \binom{2m}{m}\frac{x^{n+1}}{2^{2m}(n+1)} +$

$$+ \frac{1}{2^{2m-1}}\sum_{k=0}^{m-1}\binom{2m}{k}\int x^n \cos(2m-2k)x\,dx \quad (\text{cf. } 2.633\ 2.).$$

T 333

7. $\int x^n \cos^{2m+1} x\,dx = \frac{1}{2^{2m}}\sum_{k=0}^{m}\binom{2m+1}{k}\int x^n \cos(2m-2k+1)x\,dx$

(cf. 2.633 2.). T 333

2.632

1. $\int x^{\mu-1}\sin\beta x\,dx = \frac{i}{2}(i\beta)^{-\mu}\gamma(\mu, i\beta x) - \frac{i}{2}(-i\beta)^{-\mu}\gamma(\mu, -i\beta x)$

[Re $\mu > -1, x > 0$]. Ba IV 317(2)

2. $\int x^{\mu-1}\sin ax\,dx = -\frac{1}{2a^\mu}\Big\{\exp\Big[\frac{\pi i}{2}(\mu-1)\Big]\Gamma(\mu, -iax) +$

$$+ \exp\Big[\frac{\pi i}{2}(1-\mu)\Big]\Gamma(\mu, iax)\Big\}\quad [\text{Re }\mu < 1, a > 0, x > 0].$$

Ba IV 317(3)

3. $\int x^{\mu-1}\cos\beta x\,dx = \frac{1}{2}\{(i\beta)^{-\mu}\gamma(\mu, i\beta x) + (-i\beta)^{-\mu}\gamma(\mu, -i\beta x)\}$

[Re $\mu > 0, x > 0$]. Ba IV 319(22)

4. $\int x^{\mu-1} \cos ax\, dx = -\frac{1}{2a^\mu}\left\{\exp\left(i\mu\frac{\pi}{2}\right)\Gamma(\mu,-iax) + \right.$

$\left. + \exp\left(-i\mu\frac{\pi}{2}\right)\Gamma(\mu, iax)\right\}$. Ba IV 319(23)

2.632

1. $\int x^n \sin ax\, dx = -\sum_{k=0}^{n} k!\binom{n}{k}\frac{x^{n-k}}{a^{k+1}}\cdot\cos\left(ax + \frac{1}{2}k\pi\right)$. T(487) 2.633

2. $\int x^n \cos ax\, dx = \sum_{k=0}^{n} k!\binom{n}{k}\frac{x^{n-k}}{a^{k+1}}\cdot\sin\left(ax + \frac{1}{2}k\pi\right)$. T(486)

3. $\int x^{2n}\sin x\, dx = (2n)!\left\{\sum_{k=0}^{n}(-1)^{k+1}\frac{x^{2n-2k}}{(2n-2k)!}\cos x + \right.$

$\left. + \sum_{k=0}^{n-1}(-1)^k\frac{x^{2n-2k-1}}{(2n-2k-1)!}\sin x\right\}$.

4. $\int x^{2n+1}\sin x\, dx = (2n+1)!\left\{\sum_{k=0}^{n}(-1)^{k+1}\frac{x^{2n-2k+1}}{(2n-2k+1)!}\cos x + \right.$

$\left. + \sum_{k=0}^{n}(-1)^k\frac{x^{2n-2k}}{(2n-2k)!}\sin x\right\}$.

5. $\int x^{2n}\cos x\, dx = (2n)!\left\{\sum_{k=0}^{n}(-1)^k\frac{x^{2n-2k}}{(2n-2k)!}\sin x + \right.$

$\left. + \sum_{k=0}^{n-1}(-1)^k\frac{x^{2n-2k-1}}{(2n-2k-1)!}\cos x\right\}$.

6. $\int x^{2n+1}\cos x\, dx = (2n+1)!\left\{\sum_{k=0}^{n}(-1)^k\frac{x^{2n-2k+1}}{(2n-2k+1)!}\sin x + \right.$

$\left. + \sum_{k=0}^{n}\frac{x^{2n-2k}}{(2n-2k)!}\cos x\right\}$.

1. $\int P_n(x)\sin mx\, dx =$

2.634

$= -\frac{\cos mx}{m}\sum_{k=0}^{E\left(\frac{n}{2}\right)}(-1)^k\frac{P_n^{(2k)}(x)}{m^{2k}} + \frac{\sin mx}{m}\sum_{k=1}^{E\left(\frac{n+1}{2}\right)}(-1)^{k-1}\frac{P_n^{(2k-1)}(x)}{m^{2k-1}}$.

2. $\int P_n(x)\cos mx\, dx =$

$= \frac{\sin mx}{m}\sum_{k=0}^{E\left(\frac{n}{2}\right)}(-1)^k\frac{P_n^{(2k)}(x)}{m^{2k}} + \frac{\cos mx}{m}\sum_{k=1}^{E\left(\frac{n+1}{2}\right)}(-1)^{k-1}\frac{P_n^{(2k-1)}(x)}{m^{2k-1}}$.

In den Formeln 2.634 ist $P_n(x)$ ein Polynom n-ten Grades, $P_n^{(k)}(x)$ seine k-te Ableitung nach x.

In the formulae 2.634 $P_n(x)$ is a polynomial of degree n and $P_n^{(k)}(x)$ is its kth derivative with respect to x.

Abkürzung:
Abbreviation: $z_1 = a + bx$.

1. $\int z_1 \sin kx\, dx = -\frac{1}{k}z_1\cos kx + \frac{b}{k^2}\sin kx$.

2.635

2.635

2. $$\int z_1 \cos kx \, dx = \frac{1}{k} z_1 \sin kx + \frac{b}{k^2} \cos kx.$$

3. $$\int z_1^2 \sin kx \, dx = \frac{1}{k}\left(\frac{2b^2}{k^2} - z_1^2\right) \cos kx + \frac{2bz_1}{k^2} \sin kx.$$

4. $$\int z_1^2 \cos kx \, dx = \frac{1}{k}\left(z_1^2 - \frac{2b^2}{k^2}\right) \sin kx + \frac{2bz_1}{k^2} \cos kx.$$

5. $$\int z_1^3 \sin kx \, dx = \frac{z_1}{k}\left(\frac{6}{k^2} - z_1^2\right) \cos kx + \frac{3b}{k^2}\left(z_1^2 - \frac{2b^2}{k^2}\right) \sin kx.$$

6. $$\int z_1^3 \cos kx \, dx = \frac{z_1}{k}\left(z_1^2 - \frac{6b^2}{k^2}\right) \sin kx + \frac{3b}{k^2}\left(z_1^2 - \frac{2b^2}{k^2}\right) \cos kx.$$

7. $$\int z_1^4 \sin kx \, dx = -\frac{1}{k}\left(z_1^4 - \frac{12b^2}{k^2} z_1^2 + \frac{24b^4}{k^4}\right) \cos kx +$$
$$+ \frac{4bz_1}{k^2}\left(z_1^2 - \frac{6b^2}{k^2}\right) \sin kx.$$

8. $$\int z_1^4 \cos kx \, dx = \frac{1}{k}\left(z_1^4 - \frac{12b^2}{k^2} z_1^2 + \frac{24b^4}{k^4}\right) \sin kx +$$
$$+ \frac{4bz_1}{k^2}\left(z_1^2 - \frac{6b^2}{k^2}\right) \cos kx.$$

9. $$\int z_1^5 \sin kx \, dx = \frac{5b}{k^2}\left(z_1^4 - \frac{12b^2}{k^2} z_1^2 + \frac{24b^4}{k^4}\right) \sin kx -$$
$$- \frac{z_1}{k}\left(z_1^4 - \frac{20b^2}{k^2} z_1^2 + \frac{120b^4}{k^4}\right) \cos kx.$$

10. $$\int z_1^5 \cos kx \, dx = \frac{5b}{k^2}\left(z_1^4 - \frac{12b^2}{k^2} z_1^2 + \frac{24b^4}{k^4}\right) \cos kx +$$
$$+ \frac{z_1}{k}\left(z_1^4 - \frac{20b^2}{k^2} z_1^2 + \frac{120b^4}{k^4}\right) \sin kx.$$

11. $$\int z_1^6 \sin kx \, dx = \frac{6bz_1}{k^2}\left(z_1^4 - \frac{20b^2}{k^2} z_1^2 + \frac{120b^4}{k^4}\right) \sin kx -$$
$$- \frac{1}{k}\left(z_1^6 - \frac{30b^2}{k^2} z_1^4 + \frac{360b^4}{k^4} z_1^2 - \frac{720b^6}{k^6}\right) \cos kx.$$

12. $$\int z_1^6 \cos kx \, dx = \frac{6bz_1}{k^2}\left(z_1^4 - \frac{20b^2}{k^2} z_1^2 + \frac{120b^4}{k^4}\right) \cos kx +$$
$$+ \frac{1}{k}\left(z_1^6 - \frac{30b^2}{k^2} z_1^4 + \frac{360b^4}{k^4} z_1^2 - \frac{720b^6}{k^6}\right) \sin kx.$$

2.636

1. $$\int x^n \sin^2 x \, dx = \frac{x^{n+1}}{2(n+1)} +$$
$$+ \frac{n!}{4}\left\{\sum_{k=0}^{E\left(\frac{n}{2}\right)} \frac{(-1)^{k+1} x^{n-2k}}{2^{2k}(n-2k)!} \sin 2x + \sum_{k=0}^{E\left(\frac{n-1}{2}\right)} \frac{(-1)^{k+1} x^{n-2k-1}}{2^{2k+1}(n-2k-1)!} \cos 2x\right\}.$$

GHH [333](2e)

2. $\int x^n \cos^2 x \, dx = \frac{x^{n+1}}{2(n+1)} -$ 2.636

$$- \frac{n!}{4} \left\{ \sum_{k=0}^{E\left(\frac{n}{2}\right)} \frac{(-1)^{k+1} x^{n-2k}}{2^{2k}(n-2k)!} \sin 2x + \sum_{k=0}^{E\left(\frac{n-1}{2}\right)} \frac{(-1)^{k+1} x^{n-2k-1}}{2^{2k+1}(n-2k-1)!} \cos 2x \right\}.$$

GHH [333](3e)

3. $\int x \sin^2 x \, dx = \frac{x^2}{4} - \frac{x}{4} \sin 2x - \frac{1}{8} \cos 2x.$

4. $\int x^2 \sin^2 x \, dx = \frac{x^3}{6} - \frac{x}{4} \cos 2x - \frac{1}{4}\left(x^2 - \frac{1}{2}\right) \sin 2x.$ MC 241

5. $\int x \cos^2 x \, dx = \frac{x^2}{4} + \frac{x}{4} \sin 2x + \frac{1}{8} \cos 2x.$

6. $\int x^2 \cos^2 x \, dx = \frac{x^3}{6} + \frac{x}{4} \cos 2x + \frac{1}{4}\left(x^2 - \frac{1}{2}\right) \sin 2x.$ MC 245

1. $\int x^n \sin^3 x \, dx = \frac{n!}{4} \left\{ \sum_{k=0}^{E\left(\frac{n}{2}\right)} \frac{(-1)^k x^{n-2k}}{(n-2k)!} \left(\frac{\cos 3x}{3^{2k+1}} - 3 \cos x \right) - \right.$ 2.637

$$\left. - \sum_{k=0}^{E\left(\frac{n-1}{2}\right)} (-1)^k \frac{x^{n-2k-1}}{(n-2k-1)!} \left(\frac{\sin 3x}{3^{2k+2}} - 3 \sin x \right) \right\}.$$

GHH [333](2f)

2. $\int x^n \cos^3 x \, dx = \frac{n!}{4} \left\{ \sum_{k=0}^{E\left(\frac{n}{2}\right)} \frac{(-1)^k x^{n-2k}}{(n-2k)!} \left(\frac{\sin 3x}{3^{2k+1}} + 3 \sin x \right) + \right.$

$$\left. + \sum_{k=0}^{E\left(\frac{n-1}{2}\right)} (-1)^k \frac{x^{n-2k-1}}{(n-2k-1)!} \left(\frac{\cos 3x}{3^{2k+2}} + 3 \cos x \right) \right\}.$$

GHH [333](3f)

3. $\int x \sin^3 x \, dx = \frac{3}{4} \sin x - \frac{1}{36} \sin 3x - \frac{3}{4} x \cos x + \frac{x}{12} \cos 3x.$

4. $\int x^2 \sin^3 x \, dx = -\left(\frac{3}{4} x^2 + \frac{3}{2}\right) \cos x + \left(\frac{x^2}{12} + \frac{1}{54}\right) \cos 3x +$

$$+ \frac{3}{2} x \sin x - \frac{x}{18} \sin 3x.$$ MC 241

5. $\int x \cos^3 x \, dx = \frac{3}{4} \cos x + \frac{1}{36} \cos 3x + \frac{3}{4} x \sin x + \frac{x}{12} \sin 3x.$

6. $\int x^2 \cos^3 x \, dx = \left(\frac{3}{4} x^2 - \frac{3}{2}\right) \sin x + \left(\frac{x^2}{12} - \frac{1}{54}\right) \sin 3x +$

$$+ \frac{3}{2} x \cos x + \frac{x}{18} \cos 3x.$$ MC 245, 246

2.638

1. $\int \dfrac{\sin^q x}{x^p}\,dx = -\dfrac{\sin^{q-1} x\,[(p-2)\sin x + qx\cos x]}{(p-1)(p-2)\,x^{p-1}}$
$-\dfrac{q^2}{(p-1)(p-2)}\int \dfrac{\sin^q x\,dx}{x^{p-2}} + \dfrac{q(q-1)}{(p-1)(p-2)}\int \dfrac{\sin^{q-2} x\,dx}{x^{p-2}}$
$[p \neq 1,\ p \neq 2]$.   T(496)

2. $\int \dfrac{\cos^q x}{x^p}\,dx = -\dfrac{\cos^{q-1} x\,[(p-2)\cos x - qx\sin x]}{(p-1)(p-2)\,x^{p-1}}$
$-\dfrac{q^2}{(p-1)(p-2)}\int \dfrac{\cos^q x\,dx}{x^{p-2}} + \dfrac{q(q-1)}{(p-1)(p-2)}\int \dfrac{\cos^{q-2} x\,dx}{x^{p-2}}$
$[p \neq 1,\ p \neq 2]$.   T(495)

3. $\int \dfrac{\sin x\,dx}{x^p} = -\dfrac{\sin x}{(p-1)\,x^{p-1}} + \dfrac{1}{p-1}\int \dfrac{\cos x\,dx}{x^{p-1}}$;

$\phantom{\int \dfrac{\sin x\,dx}{x^p}} = -\dfrac{\sin x}{(p-1)\,x^{p-1}} - \dfrac{\cos x}{(p-1)(p-2)\,x^{p-2}} - \dfrac{1}{(p-1)(p-2)}\int \dfrac{\sin x\,dx}{x^{p-2}}$
$(n > 2)$.   T(492)

4. $\int \dfrac{\cos x\,dx}{x^p} = -\dfrac{\cos x}{(p-1)\,x^{p-1}} - \dfrac{1}{p-1}\int \dfrac{\sin x\,dx}{x^{p-1}}$;

$\phantom{\int \dfrac{\cos x\,dx}{x^p}} = -\dfrac{\cos x}{(p-1)\,x^{p-1}} + \dfrac{\sin x}{(p-1)(p-2)\,x^{p-2}} - \dfrac{1}{(p-1)(p-2)}\int \dfrac{\cos x\,dx}{x^{p-2}}$
$(n > 2)$.   T(491)

2.639

1. $\int \dfrac{\sin x\,dx}{x^{2n}} = \dfrac{(-1)^{n+1}}{x(2n-1)!}\left\{\sum_{k=0}^{n-2} \dfrac{(-1)^k\,(2k+1)!}{x^{2k+1}}\cos x + \sum_{k=0}^{n-1} \dfrac{(-1)^{k+1}\,(2k)!}{x^{2k}}\sin x\right\} + \dfrac{(-1)^{n+1}}{(2n-1)!}\,\operatorname{ci}(x)$.   GHH[333](6b)+

2. $\int \dfrac{\sin x}{x^{2n+1}}\,dx = \dfrac{(-1)^{n+1}}{x(2n)!}\left\{\sum_{k=0}^{n-1} \dfrac{(-1)^{k+1}\,(2k)!}{x^{2k}}\cos x + \sum_{k=0}^{n-1} \dfrac{(-1)^{k+1}\,(2k+1)!}{x^{2k+1}}\sin x\right\} + \dfrac{(-1)^n}{(2n)!}\,\operatorname{si}(x)$.   GHH[333](6b)+

3. $\int \dfrac{\cos x}{x^{2n}}\,dx = \dfrac{(-1)^{n+1}}{x(2n-1)!}\left\{\sum_{k=0}^{n-1} \dfrac{(-1)^{k+1}\,(2k)!}{x^{2k}}\cos x - \sum_{k=0}^{n-2} \dfrac{(-1)^k\,(2k+1)!}{x^{2k+1}}\sin x\right\} + \dfrac{(-1)^n}{(2n-1)!}\,\operatorname{si}(x)$.   GHH[333](7b)

4. $\int \dfrac{\cos x\,dx}{x^{2n+1}} = \dfrac{(-1)^{n+1}}{x(2n)!}\left\{\sum_{k=0}^{n-1} \dfrac{(-1)^{k+1}\,(2k+1)!}{x^{2k+1}}\cos x - \sum_{k=0}^{n-1} \dfrac{(-1)^{k+1}\,(2k)!}{x^{2k}}\sin x\right\} + \dfrac{(-1)^n}{(2n)!}\,\operatorname{ci}(x)$.   GHH[333](7b)

2.641

1. $\int \dfrac{\sin kx}{a+bx}\,dx = \dfrac{1}{b}\left[\cos\dfrac{ka}{b}\,\operatorname{si}(u) - \sin\dfrac{ka}{b}\,\operatorname{ci}(u)\right]\left[u = \dfrac{k}{b}(a+bx)\right]$.

2. $\int \dfrac{\cos kx}{a+bx}\,dx = \dfrac{1}{b}\left[\cos\dfrac{ka}{b}\,\operatorname{ci}(u) + \sin\dfrac{ka}{b}\,\operatorname{si}(u)\right]\left[u = \dfrac{k}{b}(a+bx)\right]$.

3. $\int \frac{\sin kx}{(a+bx)^2}\,dx = -\frac{1}{b}\frac{\sin kx}{a+bx} + \frac{k}{b}\int \frac{\cos kx}{a+bx}\,dx$ (cf. 2.641 2.). 2.641

4. $\int \frac{\cos kx}{(a+bx)^2}\,dx = -\frac{1}{b}\frac{\cos kx}{a+bx} - \frac{k}{b}\int \frac{\sin kx}{a+bx}\,dx$ (cf. 2.641 1.).

5. $\int \frac{\sin kx}{(a+bx)^3}\,dx = -\frac{\sin kx}{2b(a+bx)^2} - \frac{k\cos kx}{2b^2(a+bx)} - \frac{k^2}{2b^2}\int \frac{\sin kx}{a+bx}\,dx$ (cf. 2.641 1).

6. $\int \frac{\cos kx}{(a+bx)^3}\,dx = -\frac{\cos kx}{2b(a+bx)^2} + \frac{k\sin kx}{2b^2(a+bx)} - \frac{k^2}{2b^2}\int \frac{\cos kx}{a+bx}\,dx$ (cf. 2.641 2.).

7. $\int \frac{\sin kx}{(a+bx)^4}\,dx = -\frac{\sin kx}{3b(a+bx)^3} - \frac{k\cos kx}{6b^2(a+bx)^2} +$
$+ \frac{k^2 \sin kx}{6b^3(a+bx)} - \frac{k^3}{6b^3}\int \frac{\cos kx}{a+bx}\,dx$ (cf. 2.641 2.).

8. $\int \frac{\cos kx}{(a+bx)^4}\,dx = -\frac{\cos kx}{3b(a+bx)^3} + \frac{k\sin kx}{6b^2(a+bx)^2} +$
$+ \frac{k^2 \cos kx}{6b^3(a+bx)} + \frac{k^3}{6b^3}\int \frac{\sin kx}{a+bx}\,dx$ (cf. 2.641 1.).

9. $\int \frac{\sin kx}{(a+bx)^5}\,dx = -\frac{\sin kx}{4b(a+bx)^4} - \frac{k\cos kx}{12b^2(a+bx)^3} +$
$+ \frac{k^2 \sin kx}{24b^3(a+bx)^2} + \frac{k^3 \cos kx}{24b^4(a+bx)} + \frac{k^4}{24b^4}\int \frac{\sin kx}{a+bx}\,dx$ (cf. 2.641 1.).

10. $\int \frac{\cos kx}{(a+bx)^5}\,dx = -\frac{\cos kx}{4b(a+bx)^4} + \frac{k\sin kx}{12b^2(a+bx)^3} +$
$+ \frac{k^2 \cos kx}{24b^3(a+bx)^2} - \frac{k^3 \sin kx}{24b^4(a+bx)} + \frac{k^4}{24b^4}\int \frac{\cos kx}{a+bx}\,dx$ (cf. 2.641 2.).

11. $\int \frac{\sin kx}{(a+bx)^6}\,dx = -\frac{\sin kx}{5b(a+bx)^5} - \frac{k\cos kx}{20b^2(a+bx)^4} +$
$+ \frac{k^2 \sin kx}{60b^3(a+bx)^3} + \frac{k^3 \cos kx}{120b^4(a+bx)^2} - \frac{k^4 \sin kx}{120b^5(a+bx)} + \frac{k^5}{120b^5}\int \frac{\cos kx}{a+bx}\,dx$ (cf. 2.641 2.).

12. $\int \frac{\cos kx}{(a+bx)^6}\,dx = -\frac{\cos kx}{5b(a+bx)^5} + \frac{k\sin kx}{20b^2(a+bx)^4} + \frac{k^2 \cos kx}{60b^3(a+bx)^3} -$
$- \frac{k^3 \sin kx}{120b^4(a+bx)^2} - \frac{k^4 \cos kx}{120b^5(a+bx)} - \frac{k^5}{120b^5}\int \frac{\sin kx}{a+bx}\,dx$ (cf. 2.641 1.).

1. $\int \frac{\sin^{2m} x}{x}\,dx = \binom{2m}{m}\frac{\ln x}{2^{2m}} + \frac{(-1)^m}{2^{2m-1}}\sum_{k=0}^{m-1}(-1)^k \binom{2m}{k}\operatorname{ci}[(2m-2k)x]$. 2.642

2. $\int \frac{\sin^{2m+1} x}{x}\,dx = \frac{(-1)^m}{2^{2m}}\sum_{k=0}^{m}(-1)^k \binom{2m+1}{k}\operatorname{si}[(2m-2k+1)x]$.

3. $\int \frac{\cos^{2m} x}{x}\,dx = \binom{2m}{m}\frac{\ln x}{2^{2m}} + \frac{1}{2^{2m-1}}\sum_{k=0}^{m-1}\binom{2m}{k}\operatorname{ci}[(2m-2k)x]$.

4. $\int \frac{\cos^{2m+1} x}{x}\,dx = \frac{1}{2^{2m}}\sum_{k=0}^{m}\binom{2m+1}{k}\operatorname{ci}[(2m-2k+1)x]$.

2.642

5. $\int \frac{\sin^{2m} x}{x^2} dx = -\binom{2m}{m} \frac{1}{2^{2m} x} +$
$+ \frac{(-1)^m}{2^{2m-1}} \sum_{k=0}^{m-1} (-1)^{k+1} \binom{2m}{k} \left\{ \frac{\cos(2m-2k)x}{x} + (2m-2k) \operatorname{si}[(2m-2k)x] \right\}.$

6. $\int \frac{\sin^{2m+1} x}{x^2} dx = \frac{(-1)^m}{2^{2m}} \sum_{k=0}^{m} (-1)^{k+1} \binom{2m+1}{k} \times$
$\times \left\{ \frac{\sin(2m-2k+1)x}{x} - (2m-2k+1) \operatorname{ci}[(2m-2k+1)x] \right\}.$

7. $\int \frac{\cos^{2m} x}{x^2} dx = -\binom{2m}{m} \frac{1}{2^{2m} x} -$
$- \frac{1}{2^{2m-1}} \sum_{k=0}^{m-1} \binom{2m}{k} \left\{ \frac{\cos(2m-2k)x}{x} + (2m-2k) \operatorname{si}[(2m-2k)x] \right\}.$

8. $\int \frac{\cos^{2m+1} x}{x^2} dx = -\frac{1}{2^{2m}} \sum_{k=0}^{m} \binom{2m+1}{k} \left\{ \frac{\cos(2m-2k+1)x}{x} + \right.$
$\left. + (2m-2k+1) \operatorname{si}[(2m-2k+1)x] \right\}.$

2.643

1. $\int \frac{x^p \, dx}{\sin^q x} = -\frac{x^{p-1}[p \sin x + (q-2) x \cos x]}{(q-1)(q-2) \sin^{q-1} x} +$
$+ \frac{q-2}{q-1} \int \frac{x^p \, dx}{\sin^{q-2} x} + \frac{p(p-1)}{(q-1)(q-2)} \int \frac{x^{p-2} \, dx}{\sin^{q-2} x}.$

2. $\int \frac{x^p \, dx}{\cos^q x} = -\frac{x^{p-1}[p \cos x - (q-2) x \sin x]}{(q-1)(q-2) \cos^{q-1} x} +$
$+ \frac{q-2}{q-1} \int \frac{x^p \, dx}{\cos^{q-2} x} + \frac{p(p-1)}{(q-1)(q-2)} \int \frac{x^{p-2} \, dx}{\cos^{q-2} x}.$

3. $\int \frac{x^n}{\sin x} dx = \frac{x^n}{n} + \sum_{k=1}^{\infty} (-1)^{k+1} \frac{2(2^{k-1}-1)}{(n+2k)(2k)!} B_{2k} x^{n+2k}$
$[|x| < \pi, \; n > 0]. \quad \text{GHH}[333](8b)$

4. $\int \frac{dx}{x^n \sin x} = -\frac{1}{nx^n} - [1 + (-1)^n] (-1)^{\frac{n}{2}} \frac{2^{n-1}-1}{n!} B_n \ln x -$
$- \sum_{\substack{k=1 \\ k \ne \frac{n}{2}}}^{\infty} (-1)^k \frac{2(2^{2n-1}-1)}{(2k-n) \cdot (2k)!} B_{2k} x^{2k-n} \quad [n > 1, \; |x| < \pi].$

$\text{GHH}[333](9b)$

5. $\int \frac{x^n \, dx}{\cos x} = \sum_{k=0}^{\infty} \frac{|E_{2k}| x^{n+2k+1}}{(n+2k+1)(2k)!} \quad \left[|x| < \frac{\pi}{2}, \; n > 0\right] \quad \text{GHH}[333](10b)$

6. $\int \frac{dx}{x^n \cos x} = \frac{1}{2} [1 - (-1)^n] \frac{|E_{n-1}|}{(n-1)!} \ln x + \sum_{\substack{k=0 \\ k \ne \frac{n-1}{2}}}^{\infty} \frac{|E_{2k}| x^{2k-n+1}}{(2k-n+1) \cdot (2k)!}$
$\left[|x| < \frac{\pi}{2}\right]. \quad \text{GHH}[333](11b)$

7. $\int \dfrac{x^n\, dx}{\sin^2 x} = -x^n \operatorname{ctg} x + \dfrac{n}{n-1} x^{n-1} +$ 2.643

$+ n \sum\limits_{k=1}^{\infty} (-1)^k \dfrac{2^{2k}\, x^{n+2k-1}}{(n+2k-1)(2k)!} B_{2k} \quad [\,|x| < \pi,\ n > 1\,]$. GHH [333](8c)

8. $\int \dfrac{dx}{x^n \sin^2 x} = -\dfrac{\operatorname{ctg} x}{x^n} + \dfrac{n}{(n+1)\, x^{n+1}} -$

$- [1 - (-1)^n](-1)^{\frac{n+1}{2}} \dfrac{2^n n}{(n+1)!} B_{n+1} \ln x - \dfrac{n}{x^{n+1}} \sum\limits_{\substack{k=1 \\ k \neq \frac{n+1}{2}}}^{\infty} \dfrac{(-1)^k (2x)^{2k}}{(2k-n-1)(2k)!} B_{2k}$

$[\,|x| < \pi\,]$. GHH [333](9c)

9. $\int \dfrac{x^n\, dx}{\cos^2 x} = x^n \operatorname{tg} x + n \sum\limits_{k=1}^{\infty} (-1)^k \dfrac{2^{2k}(2^{2k} - 1)\, x^{n+2k-1}}{(n+2k-1)\cdot (2k)!} B_{2k}$

$\left[\, n > 1,\ |x| < \dfrac{\pi}{2} \,\right]$. GHH [333](10c)

10. $\int \dfrac{dx}{x^n \cos^2 x} = \dfrac{\operatorname{tg} x}{x^n} - [1 - (-1)^n](-1)^{\frac{n+1}{2}} \dfrac{2^n n}{(n+1)!} (2^{n+1} - 1) B_{n+1} \ln x -$

$- \dfrac{n}{x^{n+1}} \sum\limits_{\substack{k=1 \\ k \neq \frac{n+1}{2}}}^{\infty} \dfrac{(-1)^k (2^{2k} - 1) (2x)^{2k}}{(2k-n-1)(2k)!} B_{2k} \quad \left[\, |x| < \dfrac{\pi}{2} \,\right]$. GHH [333](11c)

1. $\int \dfrac{x\, dx}{\sin^{2n} x} =$ 2.644

$= -\sum\limits_{k=0}^{n-1} \dfrac{(2n-2)(2n-4)\ldots(2n-2k+2)}{(2n-1)(2n-3)\ldots(2n-2k+3)} \dfrac{\sin x + (2n-2k)\, x \cos x}{(2n-2k+1)(2n-2k)\sin^{2n-2k+1} x} +$

$+ \dfrac{2^{n-1}(n-1)!}{(2n-1)!!} (\ln \sin x - x \operatorname{ctg} x)$.

2. $\int \dfrac{x\, dx}{\sin^{2n+1} x} =$

$= -\sum\limits_{k=0}^{n-1} \dfrac{(2n-1)(2n-3)\ldots(2n-2k+1)}{2n(2n-2)\ldots(2n-2k+2)} \dfrac{\sin x + (2n-2k-1)\, x \cos x}{(2n-2k)(2n-2k-1)\sin^{2n-2k} x} +$

$+ \dfrac{(2n-1)!!}{2^n n!} \int \dfrac{x\, dx}{\sin x}$ (cf. 2.644 5.).

3. $\int \dfrac{x\, dx}{\cos^{2n} x} =$

$= \sum\limits_{k=0}^{n-1} \dfrac{(2n-2)(2n-4)\ldots(2n-2k+2)}{(2n-1)(2n-3)\ldots(2n-2k+3)} \dfrac{(2n-2k)\, x \sin x - \cos x}{(2n-2k+1)(2n-2k)\cos^{2n-2k+1} x} +$

$+ \dfrac{2^{n-1}(n-1)!}{(2n-1)!!} (x \operatorname{tg} x + \ln \cos x)$.

4. $\int \dfrac{x\, dx}{\cos^{2n+1} x} =$

$= \sum\limits_{k=0}^{n-1} \dfrac{(2n-1)(2n-3)\ldots(2n-2k+1)}{2n(2n-2)\ldots(2n-2k+2)} \dfrac{(2n-2k+1)\, x \sin x - \cos x}{(2n-2k)(2n-2k-1)\cos^{2n-2k} x} +$

$+ \dfrac{(2n-1)!!}{2^n n!} \int \dfrac{x\, dx}{\cos x}$ (cf. 2.644 6.).

2.644

5. $\int \dfrac{x\,dx}{\sin x} = x + \sum\limits_{k=1}^{\infty} (-1)^{k+1} \dfrac{2(2^{2k-1}-1)}{(2k+1)!} B_{2k} x^{2k+1}.$

6. $\int \dfrac{x\,dx}{\cos x} = \sum\limits_{k=0}^{\infty} \dfrac{|E_{2k}|\,x^{2k+2}}{(2k+2)(2k)!}.$

7. $\int \dfrac{x\,dx}{\sin^2 x} = -x\,\operatorname{ctg} x + \ln \sin x.$

8. $\int \dfrac{x\,dx}{\cos^2 x} = x\,\operatorname{tg} x + \ln \cos x.$

9. $\int \dfrac{x\,dx}{\sin^3 x} = -\dfrac{\sin x + x \cos x}{2 \sin^2 x} + \dfrac{1}{2}\int \dfrac{x}{\sin x}\,dx \quad$ (cf. 2.644 5.).

10. $\int \dfrac{x\,dx}{\cos^3 x} = \dfrac{x \sin x - \cos x}{2 \cos^2 x} + \dfrac{1}{2}\int \dfrac{x\,dx}{\cos x} \quad$ (cf. 2.644 6.).

11. $\int \dfrac{x\,dx}{\sin^4 x} = -\dfrac{x \cos x}{3 \sin^3 x} - \dfrac{1}{6 \sin^2 x} - \dfrac{2}{3} x\,\operatorname{ctg} x + \dfrac{2}{3} \ln(\sin x).$

12. $\int \dfrac{x\,dx}{\cos^4 x} = \dfrac{x \sin x}{3 \cos^3 x} - \dfrac{1}{6 \cos^2 x} + \dfrac{2}{3} x\,\operatorname{tg} x - \dfrac{2}{3} \ln(\cos x).$

13. $\int \dfrac{x\,dx}{\sin^5 x} = -\dfrac{x \cos x}{4 \sin^4 x} - \dfrac{1}{12 \sin^3 x} - \dfrac{3x \cos x}{8 \sin^2 x}$
$\qquad - \dfrac{3}{8 \sin x} + \dfrac{3}{8}\int \dfrac{x\,dx}{\sin x} \quad$ (cf. 2.644 5.).

14. $\int \dfrac{x\,dx}{\cos^5 x} = \dfrac{x \sin x}{4 \cos^4 x} - \dfrac{1}{12 \cos^3 x} + \dfrac{3x \sin x}{8 \cos^2 x}$
$\qquad - \dfrac{3}{8 \cos x} + \dfrac{3}{8}\int \dfrac{x\,dx}{\cos x} \quad$ (cf. 2.644 6.).

2.645

1. $\int x^p \dfrac{\sin^{2m} x}{\cos^n x}\,dx = \sum\limits_{k=0}^{m} (-1)^k \binom{m}{k} \int \dfrac{x^p\,dx}{\cos^{n-2k} x} \quad$ (cf. 2.643 2.).

2. $\int x^p \dfrac{\sin^{2m+1} x}{\cos^n x}\,dx = \sum\limits_{k=0}^{m} (-1)^k \binom{m}{k} \int \dfrac{x^p \sin x}{\cos^{n-2k} x}\,dx \quad$ (cf. 2.645 3.).

3. $\int x^p \dfrac{\sin x\,dx}{\cos^n x} = \dfrac{x^p}{(n-1)\cos^{n-1} x} - \dfrac{p}{n-1}\int \dfrac{x^{p-1}}{\cos^{n-1} x}\,dx \quad [n>1]\,($cf. 2.643 2.).

GHH [333](12)

4. $\int x^p \dfrac{\cos^{2m} x}{\sin^n x}\,dx = \sum\limits_{k=0}^{m} (-1)^k \binom{m}{k} \int \dfrac{x^p\,dx}{\sin^{n-2k} x} \quad$ (cf. 2.643 1.).

5. $\int x^p \dfrac{\cos^{2m+1} x}{\sin^n x}\,dx = \sum\limits_{k=0}^{m} (-1)^k \binom{m}{k} \int \dfrac{x^p \cos x}{\sin^{n-2k} x}\,dx \quad$ (cf. 2.645 6.).

6. $\int x^p \dfrac{\cos x}{\sin^n x}\,dx = -\dfrac{x^p}{(n-1)\sin^{n-1} x} + \dfrac{p}{n-1}\int \dfrac{x^{p-1}\,dx}{\sin^{n-1} x} \quad [n>1] \quad$ (cf. 2.643 1.).

GHH [333](13)

7. $\int \dfrac{x \cos x}{\sin^2 x}\,dx = -\dfrac{x}{\sin x} + \ln \operatorname{tg} \dfrac{x}{2}.$

8. $\int \frac{x \sin x}{\cos^2 x} dx = \frac{x}{\cos x} - \ln \mathrm{tg}\left(\frac{x}{2} + \frac{\pi}{4}\right).$ 2.645

1. $\int x^p \mathrm{tg}\, x\, dx = \sum_{k=1}^{\infty} (-1)^{k+1} \frac{2^{2k}(2^{2k}-1)}{(p+2k)\cdot(2k)!} B_{2k}\, x^{p+2k} \quad \left[p \geqslant -1, |x| < \frac{\pi}{2}\right].$ 2.646

GHH[333](12d)

2. $\int x^p \mathrm{ctg}\, x\, dx = \sum_{k=0}^{\infty} (-1)^k \frac{2^{2k} B_{2k}}{(p+2k)(2k)!} x^{p+2k}$

$[p \geqslant 1,\ |x| < \pi].$ GHH[333](13d)

3. $\int x\, \mathrm{tg}^2 x\, dx = x\, \mathrm{tg}\, x + \ln \cos x - \frac{x^2}{2}.$

4. $\int x\, \mathrm{ctg}^2 x\, dx = -x\, \mathrm{ctg}\, x + \ln \sin x - \frac{x^2}{2}.$

1. $\int \frac{x^n \cos x\, dx}{(a + b \sin x)^m} = -\frac{x^n}{(m-1)b(a+b\sin x)^{m-1}} +$ 2.647

$+ \frac{n}{(m-1)b} \int \frac{x^{n-1}\, dx}{(a+b\sin x)^{m-1}} \quad [m \neq 1].$ MC 247

2. $\int \frac{x^n \sin x\, dx}{(a + b \cos x)^m} = \frac{x^n}{(m-1)b(a+b\cos x)^{m-1}} -$

$- \frac{n}{(m-1)b} \int \frac{x^{n-1}\, dx}{(a+b\cos x)^{m-1}} \quad [m \neq 1].$ MC 247

3. $\int \frac{x\, dx}{1+\sin x} = -x\, \mathrm{tg}\left(\frac{\pi}{4} - \frac{x}{2}\right) + 2 \ln \cos\left(\frac{\pi}{4} - \frac{x}{2}\right).$ P(329)

4. $\int \frac{x\, dx}{1-\sin x} = x\, \mathrm{ctg}\left(\frac{\pi}{4} - \frac{x}{2}\right) + 2 \ln \sin\left(\frac{\pi}{4} - \frac{x}{2}\right).$ P(330)

5. $\int \frac{x\, dx}{1+\cos x} = x\, \mathrm{tg}\frac{x}{2} + 2 \ln \cos\frac{x}{2}.$ P(331)

6. $\int \frac{x\, dx}{1-\cos x} = -x\, \mathrm{ctg}\frac{x}{2} + 2 \ln \sin\frac{x}{2}.$ P(332)

7. $\int \frac{x \cos x}{(1+\sin x)^2} dx = -\frac{x}{1+\sin x} + \mathrm{tg}\left(\frac{x}{2} - \frac{\pi}{4}\right).$

8. $\int \frac{x \cos x}{(1-\sin x)^2} dx = \frac{x}{1-\sin x} + \mathrm{tg}\left(\frac{x}{2} + \frac{\pi}{4}\right).$

9. $\int \frac{x \sin x}{(1+\cos x)^2} dx = \frac{x}{1+\cos x} - \mathrm{tg}\frac{x}{2}.$

10. $\int \frac{x \sin x}{(1-\cos x)^2} dx = -\frac{x}{1-\cos x} - \mathrm{ctg}\frac{x}{2}.$ MC 247+

1. $\int \frac{x + \sin x}{1 + \cos x} dx = x\, \mathrm{tg}\frac{x}{2}.$ 2.648

2. $\int \frac{x - \sin x}{1 - \cos x} dx = -x\, \mathrm{ctg}\frac{x}{2}.$ GHH[333](16)

2.649 $\int \dfrac{x^2\,dx}{[(ax-b)\sin x+(a+bx)\cos x]^2}=\dfrac{x\sin x+\cos x}{b[(ax-b)\sin x+(a+bx)\cos x]}$. GHH[333](17)

2.651 $\int \dfrac{dx}{[a+(ax+b)\operatorname{tg} x]^2}=\dfrac{\operatorname{tg} x}{a[a+(ax+b)\operatorname{tg} x]}$. GHH[333](18)

2.652 $\int \dfrac{x\,dx}{\cos(x+t)\cos(x-t)}=\operatorname{cosec} 2t\left\{x\ln\dfrac{\cos(x-t)}{\cos(x+t)}-L(x+t)+L(x-t)\right\}$

$$\left[t\neq n\pi;\ |x|<\left|\dfrac{\pi}{2}-|t_0|\right|\right],$$

wobei t_0 der Wert des Argumentes t ist, der sich mit Hilfe des Argumentes π im Intervall $\left(-\dfrac{\pi}{2},\dfrac{\pi}{2}\right)$ ergibt.

where t_0 is the value of $t\pm k\pi$, which lies in the interval $\left(-\dfrac{\pi}{2},\dfrac{\pi}{2}\right)$.

Lo III 288   Lo III 288

2.653 1. $\int \dfrac{\sin x}{\sqrt{x}}\,dx=\sqrt{2\pi}\,S(\sqrt{x})$ (cf. 2.528 1.).

2. $\int \dfrac{\cos x}{\sqrt{x}}\,dx=\sqrt{2\pi}\,C(\sqrt{x})$ (cf. 2.528 2.).

Abkürzungen: $\Delta=\sqrt{1-k^2\sin^2 x},\ k'=\sqrt{1-k^2}$.
Abbreviations:

2.654 1. $\int \dfrac{x\sin x\cos x}{\Delta}\,dx=-\dfrac{x\Delta}{k^2}+\dfrac{1}{k^2}E(x,k)$.

2. $\int \dfrac{x\sin^3 x\cos x}{\Delta}\,dx=\dfrac{k'^2}{9k^4}F(x,k)+\dfrac{2k^2+5}{9k^4}E(x,k)-$

$$-\dfrac{1}{9k^4}[3(3-\Delta^2)x+k^2\sin x\cos x]\Delta.$$

3. $\int \dfrac{x\sin x\cos^3 x}{\Delta}\,dx=-\dfrac{k'^2}{9k^4}F(x,k)+\dfrac{7k^2-5}{9k^4}E(x,k)-$

$$-\dfrac{1}{9k^4}[3(\Delta^2-3k'^2)x-k^2\sin x\cos x]\Delta.$$

4. $\int \dfrac{x\sin x\,dx}{\Delta^3}=-\dfrac{x\cos x}{k'^2\Delta}+\dfrac{1}{kk'^2}\arcsin(k\sin x)$.

5. $\int \dfrac{x\cos x\,dx}{\Delta^3}=\dfrac{x\sin x}{\Delta}+\dfrac{1}{k}\ln(k\cos x+\Delta)$.

6. $\int \dfrac{x\sin x\cos x\,dx}{\Delta^3}=\dfrac{x}{k^2\Delta}-\dfrac{1}{k^2}F(x,k)$.

7. $\int \dfrac{x\sin^3 x\cos x\,dx}{\Delta^3}=x\dfrac{2-k^2\sin^2 x}{k^4\Delta}-\dfrac{1}{k^4}[E(x,k)+F(x,k)]$.

8. $\int \dfrac{x\sin x\cos^3 x\,dx}{\Delta^3}=x\dfrac{k^2\sin^2 x+k^2-2}{k^4\Delta}+\dfrac{k'^2}{k^4}F(x,k)+\dfrac{1}{k^4}E(x,k)$.

Integrale, die $\sin x^2$ und $\cos x^2$ enthalten
Integrals containing $\sin x^2$ and $\cos x^2$

In derartigen Integralen macht man zweckmäßigerweise die Substitution $x^2 = u$.	In integrals of this type the substitution $x^2 = u$ is used with advantage.

1. $\int x^p \sin x^2 \, dx = -\dfrac{x^{p-1}}{2} \cos x^2 + \dfrac{p-1}{2} \int x^{p-2} \cos x^2 \, dx.$ 2.655

2. $\int x^p \cos x^2 \, dx = \dfrac{x^{p-1}}{2} \sin x^2 - \dfrac{p-1}{2} \int x^{p-2} \sin x^2 \, dx.$

3. $\int x^n \sin x^2 \, dx = (n-1)!! \left\{ \sum_{k=1}^{r} (-1)^k \left[\dfrac{x^{n-4k+3} \cos x^2}{2^{2k-1}(n-4k+3)!!} - \dfrac{x^{n-4k+1} \sin x^2}{2^{2k}(n-4k+1)!!} \right] + \dfrac{(-1)^r}{2^{2r}(n-4r-1)!!} \int x^{n-4r} \sin x^2 \, dx \right\}$

$\left[r = E\left(\dfrac{n}{4}\right) \right].$ GHH [336](4a)

4. $\int x^n \cos x^2 \, dx = (n-1)!! \left\{ \sum_{k=1}^{r} (-1)^{k-1} \left[\dfrac{x^{n-4k+3} \sin x^2}{2^{2k-1}(n-4k+3)!!} + \dfrac{x^{n-4k+1} \cos x^2}{2^{2k}(n-4k+1)!!} \right] + \dfrac{(-1)^r}{2^{2r}(n-4r-1)!!} \int x^{n-4r} \cos x^2 \, dx \right\}$

$\left[r = E\left(\dfrac{n}{4}\right) \right].$ GHH [336](5a)

5. $\int x \sin x^2 \, dx = -\dfrac{\cos x^2}{2}.$

6. $\int x \cos x^2 \, dx = \dfrac{\sin x^2}{2}.$

7. $\int x^2 \sin x^2 \, dx = -\dfrac{x}{2} \cos x^2 + \dfrac{1}{2} \sqrt{\dfrac{\pi}{2}} \, C(x).$

8. $\int x^2 \cos x^2 \, dx = \dfrac{x}{2} \sin x^2 - \dfrac{1}{2} \sqrt{\dfrac{\pi}{2}} \, S(x).$

9. $\int x^3 \sin x^2 \, dx = -\dfrac{x^2}{2} \cos x^2 + \dfrac{1}{2} \sin x^2.$

10. $\int x^3 \cos x^2 \, dx = \dfrac{x^2}{2} \sin x^2 + \dfrac{1}{2} \cos x^2.$

Trigonometrische Funktionen und Exponentialfunktion
Trigonometric functions and exponential functions 2.66

$\int e^{ax} \sin^p x \cos^q x \, dx =$ 2.661

$= \dfrac{1}{a^2 + (p+q)^2} \left\{ e^{ax} \sin^p x \cos^{q-1} x \left[a \cos x + (p+q) \sin x \right] - pa \int e^{ax} \sin^{p-1} x \cos^{q-1} x \, dx + (q-1)(p+q) \int e^{ax} \sin^p x \cos^{q-2} x \, dx \right\};$

T(523)

2.661
$$= \frac{1}{a^2+(p+q)^2}\left\{e^{ax}\sin^{p-1}x\cos^q x\,[a\sin x - (p+q)\cos x] + \right.$$
$$\left. + qa\int e^{ax}\sin^{p-1}x\cos^{q-1}x\,dx + (p-1)(p+q)\int e^{ax}\sin^{p-2}x\cos^q x\,dx\right\};$$
T(524)

$$= \frac{1}{a^2+(p+q)^2}\left\{e^{ax}\sin^{p-1}x\cos^{q-1}x\,[a\sin x\cos x + q\sin^2 x - p\cos^2 x] + \right.$$
$$\left. + q(q-1)\int e^{ax}\sin^p x\cos^{q-2}x\,dx + p(p-1)\int e^{ax}\sin^{p-2}x\cos^q x\,dx\right\};$$
T(525)

$$= \frac{1}{a^2+(p+q)^2}\left\{e^{ax}\sin^{p-1}x\cos^{q-1}x\,(a\sin x\cos x + q\sin^2 x - p\cos^2 x) + \right.$$
$$+ q(q-1)\int e^{ax}\sin^{p-2}x\cos^{q-2}x\,dx -$$
$$\left. - (q-p)(p+q-1)\int e^{ax}\sin^{p-2}x\cos^q x\,dx\right\}; \quad \text{T(526)}$$

$$= \frac{1}{a^2+(p+q)^2}\left\{e^{ax}\sin^{p-1}x\cos^{q-1}x\,(a\sin x\cos x + q\sin^2 x - p\cos^2 x) + \right.$$
$$+ p(p-1)\int e^{ax}\sin^{p-2}x\cos^{q-2}x\,dx +$$
$$\left. + (q-p)(p+q-1)\int e^{ax}\sin^p x\cos^{q-2}x\,dx\right\}. \quad \text{GHH [334](1a)}$$

For $p=m$ and $q=n$ positive even integers, the integrals $\int e^{ax}\sin^m x\cos^n x\,dx$ can be reduced with the help of these formulae to the integral $\int e^{ax}\,dx$; but if only m or n is even, to the integral $\int e^{ax}\cos^n x\,dx$ or $\int e^{ax}\sin^m x\,dx$ respectively.

2.662

1. $\int e^{ax}\sin^n bx\,dx = \frac{1}{a^2+n^2b^2}\left[(a\sin bx - nb\cos bx)e^{ax}\sin^{n-1}bx + \right.$
$$\left. + n(n-1)b^2\int e^{ax}\sin^{n-2}bx\,dx\right].$$

2. $\int e^{ax}\cos^n bx\,dx = \frac{1}{a^2+n^2b^2}\left[(a\cos bx + nb\sin bx)e^{ax}\cos^{n-1}bx + \right.$
$$\left. + n(n-1)b^2\int e^{ax}\cos^{n-2}bx\,dx\right].$$

3. $\int e^{ax}\sin^{2m}bx\,dx =$
$$= \sum_{k=0}^{m-1}\frac{(2m)!\,b^{2k}\,e^{ax}\sin^{2m-2k-1}bx}{(2m-2k)!\,[a^2+(2m)^2b^2][a^2+(2m-2)^2b^2]\ldots[a^2+(2m-2k)^2b^2]}\times$$

$$\times [a \sin bx - (2m - 2k) b \cos bx] + \frac{(2m)! \, b^{2m} e^{ax}}{[a^2 + (2m)^2 b^2][a^2 + (2m-2)^2 b^2] \ldots [a^2 + 4b^2] a} = \quad 2.662$$

$$= \binom{2m}{m} \frac{e^{ax}}{2^{2m} a} + \frac{e^{ax}}{2^{2m-1}} \sum_{k=1}^{m} (-1)^k \binom{2m}{m-k} \frac{1}{a^2 + 4b^2 k^2} (a \cos 2bkx + 2bk \sin 2bkx).$$

4. $\int e^{ax} \sin^{2m+1} bx \, dx =$

$$= \sum_{k=0}^{m} \frac{(2m+1)! \, b^{2k} e^{ax} \sin^{2m-2k} bx [a \sin bx - (2m-2k+1) b \cos bx]}{(2m-2k+1)! [a^2 + (2m+1)^2 b^2][a^2 + (2m-1)^2 b^2] \ldots [a^2 + (2m-2k+1)^2 b^2]} =$$

$$= \frac{e^{ax}}{2^{2m}} \sum_{k=0}^{m} \frac{(-1)^k}{a^2 + (2k+1)^2 b^2} \binom{2m+1}{m-k} [a \sin (2k+1) bx - (2k+1) b \cos (2k+1) bx].$$

5. $\int e^{ax} \cos^{2m} bx \, dx =$

$$= \sum_{k=0}^{m-1} \frac{(2m)! \, b^{2k} e^{ax} \cos^{2m-2k-1} bx [a \cos bx + (2m-2k) b \sin bx]}{(2m-2k)! [a^2 + (2m)^2 b^2][a^2 + (2m-2)^2 b^2] \ldots [a^2 + (2m-2k)^2 b^2]} +$$

$$+ \frac{(2m)! \, b^{2m} e^{ax}}{[a^2 + (2m)^2 b^2][a^2 + (2m-2)^2 b^2] \ldots [a^2 + 4b^2] a} =$$

$$= \binom{2m}{m} \frac{e^{ax}}{2^{2m} a} + \frac{e^{ax}}{2^{2m-1}} \sum_{k=1}^{m} \binom{2m}{m-k} \frac{1}{a^2 + 4b^2 k^2} [a \cos 2kbx + 2kb \sin 2kbx].$$

6. $\int e^{ax} \cos^{2m+1} bx \, dx =$

$$= \sum_{k=0}^{m} \frac{(2m+1)! \, b^{2k} e^{ax} \cos^{2m-2k} bx [a \cos bx + (2m-2k+1) b \sin bx]}{(2m-2k+1)! [a^2 + (2m+1)^2 b^2][a^2 + (2m-1)^2 b^2] \ldots [a^2 + (2m-2k+1)^2 b^2]} =$$

$$= \frac{e^{ax}}{2^{2m}} \sum_{k=0}^{m} \binom{2m+1}{m-k} \frac{1}{a^2 + (2k+1)^2 b^2} [a \cos (2k+1) bx + (2k+1) b \sin (2k+1) bx].$$

1. $\int e^{ax} \sin bx \, dx = \dfrac{e^{ax} (a \sin bx - b \cos bx)}{a^2 + b^2}.$ 2.663

2. $\int e^{ax} \sin^2 bx \, dx = \dfrac{e^{ax} \sin bx (a \sin bx - 2b \cos bx)}{4b^2 + a^2} + \dfrac{2b^2 e^{ax}}{(4b^2 + a^2) a} =$

$$= \frac{e^{ax}}{2a} - \frac{e^{ax}}{a^2 + 4b^2} \left(\frac{a}{2} \cos 2bx + b \sin 2bx \right).$$

3. $\int e^{ax} \cos bx \, dx = \dfrac{e^{ax} (a \cos bx + b \sin bx)}{a^2 + b^2}.$

4. $\int e^{ax} \cos^2 bx \, dx = \dfrac{e^{ax} \cos bx (a \cos bx + 2b \sin bx)}{4b^2 + a^2} + \dfrac{2b^2 e^{ax}}{(4b^2 + a^2) a} =$

$$= \frac{e^{ax}}{2a} + \frac{e^{ax}}{a^2 + 4b^2} \left(\frac{a}{2} \cos 2bx + b \sin 2bx \right).$$

1. $\int e^{ax} \sin bx \cos cx \, dx = \dfrac{e^{ax}}{2} \left[\dfrac{a \sin (b+c) x - (b+c) \cos (b+c) x}{a^2 + (b+c)^2} + \right.$ 2.664

$$\left. + \frac{a \sin (b-c) x - (b-c) \cos (b-c) x}{a^2 + (b-c)^2} \right].$$ GHH [334](6b)

2.664

2. $\int e^{ax} \sin^2 bx \cos cx \, dx = \dfrac{e^{ax}}{4} \Bigg[2\,\dfrac{a \cos cx + c \sin cx}{a^2 + c^2} -$

$- \dfrac{a \cos(2b+c)x + (2b+c)\sin(2b+c)x}{a^2 + (2b+c)^2} -$

$- \dfrac{a \cos(2b-c)x + (2b-c)\sin(2b-c)x}{a^2 + (2b-c)^2} \Bigg].$ GHH [334](6c)

3. $\int e^{ax} \sin bx \cos^2 cx \, dx = \dfrac{e^{ax}}{4} \Bigg[2\,\dfrac{a \sin bx - b \cos bx}{a^2 + b^2} +$

$+ \dfrac{a \sin(b+2c)x - (b+2c)\cos(b+2c)x}{a^2 + (b+2c)^2} +$

$+ \dfrac{a \sin(b-2c)x - (b-2c)\cos(b-2c)x}{a^2 + (b-2c)^2} \Bigg].$ GHH [334](6d)

2.665

1. $\int \dfrac{e^{ax} dx}{\sin^p bx} = -\dfrac{e^{ax}[a \sin bx + (p-2)b \cos bx]}{(p-1)(p-2)b^2 \sin^{p-1} bx} +$

$+ \dfrac{a^2 + (p-2)^2 b^2}{(p-1)(p-2)b^2} \int \dfrac{e^{ax} dx}{\sin^{p-2} bx}.$ T(530)+

2. $\int \dfrac{e^{ax} dx}{\cos^p bx} = -\dfrac{e^{ax}[a \cos bx - (p-2)b \sin bx]}{(p-1)(p-2)b^2 \cos^{p-1} bx} +$

$+ \dfrac{a^2 + (p-2)^2 b^2}{(p-1)(p-2)b^2} \int \dfrac{e^{ax} dx}{\cos^{p-2} bx}.$ T(529)+

Ist p eine natürliche Zahl, so gelangt man durch wiederholte Anwendung der Formeln 2.665 zu Integralen der Gestalt

For p a positive integer, the successive application of the formulae 2.665 leads to integrals of the form

$$\int \dfrac{e^{ax} dx}{\sin bx}, \quad \int \dfrac{e^{ax} dx}{\sin^2 bx}, \quad \int \dfrac{e^{ax} dx}{\cos bx}, \quad \int \dfrac{e^{ax} dx}{\cos^2 bx},$$

die sich nicht durch eine endliche Kombination elementarer Funktionen ausdrücken lassen.

which cannot be expressed in terms of any finite combination of elementary functions.

2.666

1. $\int e^{ax} \mathrm{tg}^p x \, dx = \dfrac{e^{ax}}{p-1} \mathrm{tg}^{p-1} x - \dfrac{a}{p-1} \int e^{ax} \mathrm{tg}^{p-1} x \, dx - \int e^{ax} \mathrm{tg}^{p-2} x \, dx.$

T(527)

2. $\int e^{ax} \mathrm{ctg}^p x \, dx =$

$= -\dfrac{e^{ax} \mathrm{ctg}^{p-1} x}{p-1} + \dfrac{a}{p-1} \int e^{ax} \mathrm{ctg}^{p-1} x \, dx - \int e^{ax} \mathrm{ctg}^{p-2} x \, dx.$ T(528)

3. $\int e^{ax} \mathrm{tg}\, x \, dx = \dfrac{e^{ax} \mathrm{tg}\, x}{a} - \dfrac{1}{a} \int \dfrac{e^{ax} dx}{\cos^2 x}$

(vgl. die Bemerkung zu 2.665). | (see remark under 2.665).

4. $\int e^{ax} \mathrm{tg}^2 x \, dx = \dfrac{e^{ax}}{a}(a\,\mathrm{tg}\, x - 1) - a \int e^{ax} \mathrm{tg}\, x \, dx$ (cf. 2.666 3.). T 355

5. $\int e^{ax} \mathrm{ctg}\, x \, dx = \dfrac{e^{ax} \mathrm{ctg}\, x}{a} + \dfrac{1}{a} \int \dfrac{e^{ax} dx}{\sin^2 x}$

(vgl. die Bemerkung zu 2.665). | (see remark under 2.665).

6. $\int e^{ax} \operatorname{ctg}^2 x \, dx = -\dfrac{e^{ax}}{a} (a \operatorname{ctg} x + 1) + a \int e^{ax} \operatorname{ctg} x \, dx$ (cf. 2.666 5.).

Integrale der Gestalt:
Integrals of the form: $\int R(x, e^{ax}, \sin bx, \cos cx) \, dx$.

Abkürzungen:
Abbreviations: $\sin t = -\dfrac{b}{\sqrt{a^2 + b^2}}$; $\cos t = \dfrac{a}{\sqrt{a^2 + b^2}}$.

1. $\int x^p e^{ax} \sin bx \, dx = \dfrac{x^p e^{ax}}{a^2 + b^2} (a \sin bx - b \cos bx) -$

$$- \dfrac{p}{a^2 + b^2} \int x^{p-1} e^{ax} (a \sin bx - b \cos bx) \, dx;$$

$$= \dfrac{x^p e^{ax}}{\sqrt{a^2 + b^2}} \sin (bx + t) - \dfrac{p}{\sqrt{a^2 + b^2}} \int x^{p-1} e^{ax} \sin (bx + t) \, dx.$$

2. $\int x^p e^{ax} \cos bx \, dx =$

$$= \dfrac{x^p e^{ax}}{a^2 + b^2} (a \cos bx + b \sin bx) - \dfrac{p}{a^2 + b^2} \int x^{p-1} e^{ax} (a \cos bx + b \sin bx) \, dx;$$

$$= \dfrac{x^p e^{ax}}{\sqrt{a^2 + b^2}} \cos (bx + t) - \dfrac{p}{\sqrt{a^2 + b^2}} \int x^{p-1} e^{ax} \cos (bx + t) \, dx.$$

3. $\int x^n e^{ax} \sin bx \, dx = e^{ax} \sum_{k=1}^{n+1} \dfrac{(-1)^{k+1} n! \, x^{n-k+1}}{(n-k+1)! \, (a^2 + b^2)^{k/2}} \sin (bx + kt)$.

4. $\int x^n e^{ax} \cos bx \, dx = e^{ax} \sum_{k=1}^{n+1} \dfrac{(-1)^{k+1} n! \, x^{n-k+1}}{(n-k+1)! \, (a^2 + b^2)^{k/2}} \cos (bx + kt)$.

5. $\int x e^{ax} \sin bx \, dx = \dfrac{e^{ax}}{a^2 + b^2} \left[\left(ax - \dfrac{a^2 - b^2}{a^2 + b^2} \right) \sin bx - \left(bx - \dfrac{2ab}{a^2 + b^2} \right) \cos bx \right]$.

6. $\int x e^{ax} \cos bx \, dx = \dfrac{e^{ax}}{a^2 + b^2} \left[\left(ax - \dfrac{a^2 - b^2}{a^2 + b^2} \right) \cos bx + \left(bx - \dfrac{2ab}{a^2 + b^2} \right) \sin bx \right]$.

7. $\int x^2 e^{ax} \sin bx \, dx =$

$$= \dfrac{e^{ax}}{a^2 + b^2} \left\{ \left[ax^2 - \dfrac{2(a^2 - b^2)}{a^2 + b^2} x + \dfrac{2a(a^2 - 3b^2)}{(a^2 + b^2)^2} \right] \sin bx - \left[bx^2 - \dfrac{4ab}{a^2 + b^2} x + \dfrac{2b(3a^2 - b^2)}{(a^2 + b^2)^2} \right] \cos bx \right\}.$$

8. $\int x^2 e^{ax} \cos bx \, dx =$

$$= \dfrac{e^{ax}}{a^2 + b^2} \left\{ \left[ax^2 - \dfrac{2(a^2 - b^2)}{a^2 + b^2} x + \dfrac{2a(a^2 - 3b^2)}{(a^2 + b^2)^2} \right] \cos bx + \left[bx^2 - \dfrac{4ab}{a^2 + b^2} x + \dfrac{2b(3a^2 - b^2)}{(a^2 + b^2)^2} \right] \sin bx \right\}.$$

GHH [335], MC 274–275

Trigonometrische Funktionen und Hyperbelfunktionen
Trigonometric and hyperbolic functions

2.67

2.671

1. $\int \operatorname{sh}(ax+b) \sin(cx+d)\, dx = \dfrac{a}{a^2+c^2} \operatorname{ch}(ax+b) \sin(cx+d) -$

$$- \dfrac{c}{a^2+c^2} \operatorname{sh}(ax+b) \cos(cx+d).$$

2. $\int \operatorname{sh}(ax+b) \cos(cx+d)\, dx = \dfrac{a}{a^2+c^2} \operatorname{ch}(ax+b) \cos(cx+d) +$

$$+ \dfrac{c}{a^2+c^2} \operatorname{sh}(ax+b) \sin(cx+d).$$

3. $\int \operatorname{ch}(ax+b) \sin(cx+d)\, dx = \dfrac{a}{a^2+c^2} \operatorname{sh}(ax+b) \sin(cx+d) -$

$$- \dfrac{c}{a^2+c^2} \operatorname{ch}(ax+b) \cos(cx+d).$$

4. $\int \operatorname{ch}(ax+b) \cos(cx+d)\, dx = \dfrac{a}{a^2+c^2} \operatorname{sh}(ax+b) \cos(cx+d) +$

$$+ \dfrac{c}{a^2+c^2} \operatorname{ch}(ax+b) \sin(cx+d).$$

GHH [354](1)

2.672

1. $\int \operatorname{sh} x \sin x\, dx = \dfrac{1}{2}(\operatorname{ch} x \sin x - \operatorname{sh} x \cos x).$

2. $\int \operatorname{sh} x \cos x\, dx = \dfrac{1}{2}(\operatorname{ch} x \cos x + \operatorname{sh} x \sin x).$

3. $\int \operatorname{ch} x \sin x\, dx = \dfrac{1}{2}(\operatorname{sh} x \sin x - \operatorname{ch} x \cos x).$

4. $\int \operatorname{ch} x \cos x\, dx = \dfrac{1}{2}(\operatorname{sh} x \cos x + \operatorname{ch} x \sin x).$

2.673

1. $\int \operatorname{sh}^{2m}(ax+b) \sin^{2n}(cx+d)\, dx = \dfrac{(-1)^m}{2^{2m+2n}} \binom{2m}{m}\binom{2n}{n} x +$

$$+ \dfrac{(-1)^{m+n}}{2^{2m+2n-1}} \binom{2m}{m} \sum_{k=0}^{n-1} \dfrac{(-1)^k}{(2n-2k)c} \binom{2n}{k} \sin[(2n-2k)(cx+d)] +$$

$$+ \dfrac{(-1)^n}{2^{2m+2n-2}} \sum_{j=0}^{m-1}\sum_{k=0}^{n-1} \dfrac{(-1)^{j+k}\binom{2m}{j}\binom{2n}{k}}{(2m-2j)^2 a^2 + (2n-2k)^2 c^2} \times$$

$$\times \{(2m-2j)\, a \operatorname{sh}[(2m-2j)(ax+b)] \cos[(2n-2k)(cx+d)] +$$

$$+ (2n-2k)\, c \operatorname{ch}[(2m-2j)(ax+b)] \sin[(2n-2k)(cx+d)]\}.$$

GHH [354](3a)

2. $\int \operatorname{sh}^{2m}(ax+b) \sin^{2n-1}(cx+d)\, dx =$

$$= \dfrac{(-1)^{m+n}}{2^{2m+2n-2}} \binom{2m}{m} \sum_{k=0}^{n-1} \dfrac{(-1)^k}{(2n-2k-1)c} \binom{2n-1}{k} \cos[(2n-2k-1)(cx+d)] +$$

$$+ \frac{(-1)^{n-1}}{2^{2m+2n-3}} \sum_{j=0}^{m-1} \sum_{k=0}^{n-1} \frac{(-1)^{j+k} \binom{2m}{j}\binom{2n-1}{k}}{(2m-2j)^2 a^2 + (2n-2k-1)^2 c^2} \times$$

$$\times \{(2m-2j) a \, \text{sh}\,[(2m-2j)(ax+b)] \sin[(2n-2k-1)(cx+d)] -$$

$$- (2n-2k-1) c \, \text{ch}\,[(2m-2j)(ax+b)] \cos[(2n-2k-1)(cx+d)]\}. \qquad \text{GHH [354](3b)}$$

3. $\int \text{sh}^{2m-1}(ax+b) \sin^{2n}(cx+d)\,dx =$

$$= \frac{\binom{2n}{n}}{2^{2m+2n-2}} \sum_{j=0}^{m-1} \frac{(-1)^j \binom{2m-1}{j}}{(2m-2j-1)a} \text{ch}\,[(2m-2j-1)(ax+b)] +$$

$$+ \frac{(-1)^n}{2^{2m+2n-3}} \sum_{j=0}^{m-1} \sum_{k=0}^{n-1} \frac{(-1)^{j+k} \binom{2m-1}{j}\binom{2n}{k}}{(2m-2j-1)^2 a^2 + (2n-2k)^2 c^2} \times$$

$$\times \{(2m-2j-1) a \, \text{ch}\,[(2m-2j-1)(ax+b)] \cos[(2n-2k)(cx+d)] +$$

$$+ (2n-2k) c \, \text{sh}\,[(2m-2j-1)(ax+b)] \sin[(2n-2k)(cx+d)]\}. \qquad \text{GHH [354](3c)}$$

4. $\int \text{sh}^{2m-1}(ax+b) \sin^{2n-1}(cx+d)\,dx =$

$$= \frac{(-1)^{n-1}}{2^{2m-2n-4}} \sum_{j=0}^{m-1} \sum_{k=0}^{n-1} \frac{(-1)^{j+k} \binom{2m-1}{j}\binom{2n-1}{k}}{(2m-2j-1)^2 a^2 + (2n-2k-1)^2 c^2} \times$$

$$\times \{(2m-2j-1) a \, \text{ch}\,[(2m-2j-1)(ax+b)] \sin[(2n-2k-1)(cx+d)] -$$

$$- (2n-2k-1) c \, \text{sh}\,[(2m-2j-1)(ax+b)] \cos[(2n-2k-1)(cx+d)]\}. \qquad \text{GHH [354](3d)}$$

5. $\int \text{sh}^{2m}(ax+b) \cos^{2n}(cx+d)\,dx = \frac{(-1)^m}{2^{2m+2n}} \binom{2m}{m}\binom{2n}{n} x +$

$$+ \frac{\binom{2n}{n}}{2^{2m+2n-1}} \sum_{j=0}^{m-1} \frac{(-1)^j \binom{2m}{j}}{(2m-2j)a} \text{sh}\,[(2m-2j)(ax+b)] +$$

$$+ \frac{(-1)^m \binom{2m}{m}}{2^{2m+2n-1}} \sum_{k=0}^{n-1} \frac{\binom{2n}{k}}{(2n-2k)c} \sin[(2n-2k)(cx+d)] +$$

$$+ \frac{1}{2^{2m+2n-2}} \sum_{j=0}^{m-1} \sum_{k=0}^{n-1} \frac{(-1)^j \binom{2m}{j}\binom{2n}{k}}{(2m-2j)^2 a^2 + (2n-2k)^2 c^2} \times$$

$$\times \{(2m-2j) a \, \text{sh}\,[(2m-2j)(ax+b)] \cos[(2n-2k)(cx+d)] +$$

$$+ (2n-2k) c \, \text{ch}\,[(2m-2j)(ax+b)] \sin[(2n-2k)(cx+d)]\}. \qquad \text{GHH [354](4a)}$$

2.673

6. $\int \operatorname{sh}^{2m}(ax+b)\cos^{2n-1}(cx+d)\,dx =$

$$= \frac{(-1)^m \binom{2m}{m}}{2^{2m+2n-2}} \sum_{k=0}^{n-1} \frac{\binom{2n-1}{k}}{(2n-2k-1)c} \sin[(2n-2k-1)(cx+d)] +$$

$$+ \frac{1}{2^{2m+2n-3}} \sum_{j=0}^{m-1} \sum_{k=0}^{n-1} \frac{(-1)^j \binom{2m}{j}\binom{2n-1}{k}}{(2m-2j)^2 a^2 + (2n-2k-1)^2 c^2} \times$$

$$\times \{(2m-2j)\,a\,\operatorname{sh}[(2m-2j)(ax+b)]\cos[(2n-2k-1)(cx+d)] +$$

$$+ (2n-2k-1)\,c\,\operatorname{ch}[(2m-2j)(ax+b)]\sin[(2n-2k-1)(cx+d)]\}.$$

GHH [354](4a)

7. $\int \operatorname{sh}^{2m-1}(ax+b)\cos^{2n}(cx+d)\,dx =$

$$= \frac{\binom{2n}{n}}{2^{2m+2n-2}} \sum_{j=0}^{m-1} \frac{(-1)^j \binom{2m-1}{j}}{(2m-2j-1)a} \operatorname{ch}[(2m-2j-1)(ax+b)] +$$

$$+ \frac{1}{2^{2m-2n-3}} \sum_{j=0}^{m-1} \sum_{k=0}^{n-1} \frac{(-1)^j \binom{2m-1}{j}\binom{2n}{k}}{(2m-2j-1)^2 a^2 + (2n-2k)^2 c^2} \times$$

$$\times \{(2m-2j-1)\,a\,\operatorname{ch}[(2m-2j-1)(ax+b)]\cos[(2n-2k)(cx+d)] +$$

$$+ (2n-2k)\,c\,\operatorname{sh}[(2m-2j-1)(ax+b)]\sin[(2n-2k)(cx+d)]\}.$$

GHH [354](4b)

8. $\int \operatorname{sh}^{2m-1}(ax+b)\cos^{2n-1}(cx+d)\,dx =$

$$= \frac{1}{2^{2m+2n-4}} \sum_{j=0}^{m-1} \sum_{k=0}^{n-1} \frac{(-1)^j \binom{2m-1}{j}\binom{2n-1}{k}}{(2m-2j-1)^2 a^2 + (2n-2k-1)^2 c^2} \times$$

$$\times \{(2m-2j-1)\,a\,\operatorname{ch}[(2m-2j-1)(ax+b)]\cos[(2n-2k-1)(cx+d)] +$$

$$+ (2n-2k-1)\,c\,\operatorname{sh}[(2m-2j-1)(ax+b)]\sin[(2n-2k-1)(cx+d)]\}.$$

GHH [354](4b)

9. $\int \operatorname{ch}^{2m}(ax+b)\sin^{2n}(cx+d)\,dx = \dfrac{\binom{2m}{m}\binom{2n}{n}}{2^{2m+2n}} x +$

$$+ \frac{(-1)^n \binom{2m}{m}}{2^{2m+2n-1}} \sum_{k=0}^{m-1} \frac{(-1)^k \binom{2n}{k}}{(2n-2k)c} \sin[(2n-2k)(cx+d)] +$$

$$+ \frac{\binom{2n}{n}}{2^{2m+2n-1}} \sum_{j=0}^{m-1} \frac{\binom{2m}{j}}{(2m-2j)a} \operatorname{sh}[(2m-2j)(ax+b)] +$$

$$+ \frac{(-1)^n}{2^{2m+2n-2}} \sum_{j=0}^{m-1} \sum_{k=0}^{n-1} \frac{(-1)^k \binom{2m}{j}\binom{2n}{k}}{(2m-2j)^2 a^2 + (2n-2k)^2 c^2} \times$$

$$\times \{(2m-2j)\, a\, \text{sh}\, [(2m-2j)(ax+b)] \cos[(2n-2k)(cx+d)] +$$
$$+ (2n-2k)\, c\, \text{ch}[(2m-2j)(ax+b)] \sin[(2n-2k)(cx+d)]\}. \qquad \text{GHH}[354](5a)$$

2.673

10. $\displaystyle\int \text{ch}^{2m-1}(ax+b)\sin^{2n}(cx+d)\,dx =$

$$= \frac{\binom{2n}{n}}{2^{2m+2n-2}} \sum_{j=0}^{m-1} \frac{\binom{2m-1}{j}}{(2m-2j-1)a}\, \text{sh}\,[(2m-2j-1)(ax+b)] +$$

$$+ \frac{(-1)^n}{2^{2m+2n-3}} \sum_{j=0}^{m-1}\sum_{k=0}^{n-1} \frac{(-1)^k \binom{2m-1}{j}\binom{2n}{k}}{(2m-2j-1)^2 a^2 + (2n-2k)^2 c^2} \times$$

$$\times \{(2m-2j-1)\,a\,\text{sh}\,[(2m-2j-1)(ax+b)]\cos[(2n-2k)(cx+d)] +$$
$$+ (2n-2k)\,c\,\text{ch}\,[(2m-2j-1)(ax+b)]\sin[(2n-2k)(cx+d)]\}. \qquad \text{GHH}[354](5a)$$

11. $\displaystyle\int \text{ch}^{2m}(ax+b)\sin^{2n-1}(cx+d)\,dx =$

$$= \frac{(-1)^{n-1}\binom{2m}{m}}{2^{2m+2n-2}} \sum_{k=0}^{n-1} \frac{(-1)^{k+1}\binom{2n-1}{k}}{(2n-2k-1)c}\cos[(2n-2k-1)(cx+d)] +$$

$$+ \frac{(-1)^{n-1}}{2^{2m+2n-3}} \sum_{j=0}^{m-1}\sum_{k=0}^{n-1} \frac{(-1)^k \binom{2m}{j}\binom{2n-1}{k}}{(2m-2j)^2 a^2 + (2n-2k-1)^2 c^2} \times$$

$$\times \{(2m-2j)\,a\,\text{sh}\,[(2m-2j)(ax+b)]\sin[(2n-2k-1)(cx+d)] -$$
$$- (2n-2k-1)\,c\,\text{ch}\,[(2m-2j)(ax+b)]\cos[(2n-2k-1)(cx+d)]\}.$$

GHH[354](5b)

12. $\displaystyle\int \text{ch}^{2m-1}(ax+b)\sin^{2n-1}(cx+d)\,dx =$

$$= \frac{(-1)^{n-1}}{2^{2m+2n-4}} \sum_{j=0}^{m-1}\sum_{k=0}^{n-1} \frac{(-1)^k \binom{2m-1}{j}\binom{2n-1}{k}}{(2m-2j-1)^2 a^2 + (2n-2k-1)^2 c^2} \times$$

$$\times \{(2m-2j-1)\,a\,\text{sh}\,[(2m-2j-1)(ax+b)]\sin[(2n-2k-1)(cx+d)] -$$
$$- (2n-2k-1)\,c\,\text{ch}\,[(2m-2j-1)(ax+b)]\cos[(2n-2k-1)(cx+d)]\}.$$

GHH[354](5b)

13. $\displaystyle\int \text{ch}^{2m}(ax+b)\cos^{2n}(cx+d)\,dx = \frac{\binom{2m}{m}\binom{2n}{n}}{2^{2m+2n}}x +$

$$+ \frac{\binom{2m}{m}}{2^{2m+2n-}} \sum_{k=0}^{n-1} \frac{\binom{2n}{k}}{(2n-2k)c}\sin[(2n-2k)(cx+d)] +$$

$$+ \frac{\binom{2n}{n}}{2^{2m+2n-1}} \sum_{j=0}^{m-1} \frac{\binom{2m}{j}}{(2m-2j)a}\text{sh}\,[(2m-2j)(ax+b)] +$$

2.673

$$+ \frac{1}{2^{2m+2n-2}} \sum_{j=0}^{m-1} \sum_{k=0}^{n-1} \frac{\binom{2m}{j}\binom{2n}{k}}{(2m-2j)^2 a^2 + (2n-2k)^2 c^2} \times$$

$$\times \{(2m-2j) a \operatorname{sh}[(2m-2j)(ax+b)] \cos[(2n-2k)(cx+d)] +$$

$$+ (2n-2k) c \operatorname{ch}[(2m-2j)(ax+b)] \sin[(2n-2k)(cx+d)]\}. \qquad \text{GHH[354](6)}$$

14. $\int \operatorname{ch}^{2m-1}(ax+b) \cos^{2n}(cx+d) \, dx =$

$$= \frac{\binom{2n}{n}}{2^{2m+2n-2}} \sum_{j=0}^{m-1} \frac{\binom{2m-1}{j}}{(2m-2j-1) a} \operatorname{sh}[(2m-2j-1)(ax+b)] +$$

$$+ \frac{1}{2^{2m+2n-3}} \sum_{j=0}^{m-1} \sum_{k=0}^{n-1} \frac{\binom{2m-1}{j}\binom{2n}{k}}{(2m-2j-1)^2 a^2 + (2n-2k)^2 c^2} \times$$

$$\times \{(2m-2j-1) a \operatorname{sh}[(2m-2j-1)(ax+b)] \cos[(2n-2k)(cx+d)] +$$

$$+ (2n-2k) c \operatorname{ch}[(2m-2j-1)(ax+b)] \sin[(2n-2k)(cx+d)]\}. \qquad \text{GHH[354](6)}$$

15. $\int \operatorname{ch}^{2m}(ax+b) \cos^{2n-1}(cx+d) \, dx =$

$$= \frac{\binom{2m}{m}}{2^{2m+2n-2}} \sum_{k=0}^{n-1} \frac{\binom{2n-1}{k}}{(2n-2k-1) c} \sin[(2n-2k-1)(cx+d)] +$$

$$+ \frac{1}{2^{2m+2n-3}} \sum_{j=0}^{m-1} \sum_{k=0}^{n-1} \frac{\binom{2m}{j}\binom{2n-1}{k}}{(2m-2j)^2 a^2 + (2n-2k-1)^2 c^2} \times$$

$$\times \{(2m-2j) a \operatorname{sh}[(2m-2j)(ax+b)] \cos[(2n-2k-1)(cx+d)] +$$

$$+ (2n-2k-1) c \operatorname{ch}[(2m-2j)(ax+b)] \sin[(2n-2k-1)(cx+d)]\}. \qquad \text{GHH[354](6)}$$

16. $\int \operatorname{ch}^{2m-1}(ax+b) \cos^{2n-1}(cx+d) \, dx =$

$$= \frac{1}{2^{2m+2n-4}} \sum_{j=0}^{m-1} \sum_{k=0}^{n-1} \frac{\binom{2m-1}{j}\binom{2n-1}{k}}{(2m-2j-1)^2 a^2 + (2n-2k-1)^2 c^2} \times$$

$$\times \{(2m-2j-1) a \operatorname{sh}[(2m-2j-1)(ax+b)] \cos[(2n-2k-1)(cx+d)] +$$

$$+ (2n-2k-1) c \operatorname{ch}[(2m-2j-1)(ax+b)] \sin[(2n-2k-1)(cx+d)]\}. \qquad \text{GHH[354](6)}$$

2.674

1. $\int e^{ax} \operatorname{sh} bx \sin cx \, dx = \dfrac{e^{(a+b)x}}{2[(a+b)^2 + c^2]} [(a+b) \sin cx - c \cos cx] -$

$$- \frac{e^{(a-b)x}}{2[(a-b)^2 + c^2]} [(a-b) \sin cx - c \cos cx].$$

2. $\int e^{ax} \operatorname{sh} bx \cos cx \, dx = \dfrac{e^{(a+b)x}}{2[(a+b)^2+c^2]}[(a+b)\cos cx + c\sin cx] - $ 2.674

$\qquad - \dfrac{e^{(a-b)x}}{2[(a-b)^2+c^2]}[(a-b)\cos cx + c\sin cx].$

3. $\int e^{ax} \operatorname{ch} bx \sin cx \, dx = \dfrac{e^{(a+b)x}}{2[(a+b)^2+c^2]}[(a+b)\sin cx - c\cos cx] +$

$\qquad + \dfrac{e^{(a-b)x}}{2[(a-b)^2+c^2]}[(a-b)\sin cx - c\cos cx].$

4. $\int e^{ax} \operatorname{ch} bx \cos cx \, dx = \dfrac{e^{(a+b)x}}{2[(a+b)^2+c^2]}[(a+b)\cos cx + c\sin cx] +$

$\qquad + \dfrac{e^{(a-b)x}}{2[(a-b)^2+c^2]}[(a-b)\cos cx + c\sin cx].$

MC 379

Der Logarithmus; Area-Funktionen
The Logarithmic and Inverse Hyperbolic Functions

2.7

Der Logarithmus
The logarithmic function

2.71

$\int \ln^m x \, dx = x \ln^m x - m \int \ln^{m-1} x \, dx =$ 2.711

$= \dfrac{x}{m+1} \sum\limits_{k=0}^{m} (-1)^k (m+1) m(m-1) \ldots (m-k+1) \ln^{m-k} x \quad (m>0).$ T(603)

Logarithmus und algebraische Funktionen
Logarithmic and algebraic functions

2.72–2.73

1. $\int x^n \ln^m x \, dx = \dfrac{x^{n+1} \ln^m x}{n+1} - \dfrac{m}{n+1} \int x^n \ln^{m-1} x \, dx \quad \text{(cf. 2.722)}.$ 2.721

Für $n = -1$ gilt | For $n = -1$:

2. $\int \dfrac{\ln^m x \, dx}{x} = \dfrac{\ln^{m+1} x}{m+1}.$

Für $n = -1$ und $m = -1$ gilt | For $n = -1$ and $m = 1$:

3. $\int \dfrac{dx}{x \ln x} = \ln(\ln x).$

$\int x^n \ln^m x \, dx = \dfrac{x^{n+1}}{m+1} \sum\limits_{k=0}^{m} (-1)^k (m+1) m(m-1) \ldots (m-k+1) \dfrac{\ln^{m-k} x}{(n+1)^{k+1}}.$ 2.722

T(604)

1. $\int x^n \ln x \, dx = x^{n+1} \left[\dfrac{\ln x}{n+1} - \dfrac{1}{(n+1)^2}\right].$ T 375   2.723

2.723 2. $$\int x^n \ln^2 x \, dx = x^{n+1} \left[\frac{\ln^2 x}{n+1} - \frac{2 \ln x}{(n+1)^2} + \frac{2}{(n+1)^3} \right].$$ T 375

3. $$\int x^n \ln^3 x \, dx = x^{n+1} \left[\frac{\ln^3 x}{n+1} - \frac{3 \ln^2 x}{(n+1)^2} + \frac{6 \ln x}{(n+1)^3} - \frac{6}{(n+1)^4} \right].$$

2.724 1. $$\int \frac{x^n \, dx}{(\ln x)^m} = -\frac{x^{n+1}}{(m-1)(\ln x)^{m-1}} + \frac{n+1}{m-1} \int \frac{x^n \, dx}{(\ln x)^{m-1}}.$$

Für $m = 1$ gilt | For $m = 1$:

2. $$\int \frac{x^n \, dx}{\ln x} = \mathrm{li}\,(x^{n+1}).$$

2.725 1. $$\int (a+bx)^m \ln x \, dx =$$
$$= \frac{1}{(m+1)b} \left[(a+bx)^{m+1} \ln x - \int \frac{(a+bx)^{m+1} \, dx}{x} \right].$$ T 374

2. $$\int (a+bx)^m \ln x \, dx = \frac{1}{(m+1)b} [(a+bx)^{m+1} - a^{m+1}] \ln x -$$
$$- \sum_{k=0}^{m} \frac{\binom{m}{k} a^{m-k} b^k x^{k+1}}{(k+1)^2}.$$

Für $m = -1$ vgl. 2.727 2. | For $m = -1$ see 2.727 2.

2.726 1. $$\int (a+bx) \ln x \, dx = \left[\frac{(a+bx)^2}{2b} - \frac{a^2}{2b} \right] \ln x - \left(ax + \frac{1}{4} bx^2 \right).$$

2. $$\int (a+bx)^2 \ln x \, dx = \frac{1}{3b} [(a+bx)^3 - a^3] \ln x - \left(a^2 x + \frac{abx^2}{2} + \frac{b^2 x^3}{9} \right).$$

3. $$\int (a+bx)^3 \ln x \, dx = \frac{1}{4b} [(a+bx)^4 - a^4] \ln x -$$
$$- \left(a^3 x + \frac{3}{4} a^2 b x^2 + \frac{1}{3} a b^2 x^3 + \frac{1}{16} b^3 x^4 \right).$$

2.727 1. $$\int \frac{\ln x \, dx}{(a+bx)^m} = \frac{1}{b(m-1)} \left[-\frac{\ln x}{(a+bx)^{m-1}} + \int \frac{dx}{x(a+bx)^{m-1}} \right].$$ T 376

Für $m = 1$ gilt | For $m = 1$:

2. $$\int \frac{\ln x \, dx}{a+bx} = \frac{1}{b} \ln x \ln(a+bx) - \frac{1}{b} \int \frac{\ln(a+bx)\, dx}{x} \quad (\text{cf. } 2.728 \ 2.).$$

3. $$\int \frac{\ln x \, dx}{(a+bx)^2} = -\frac{\ln x}{b(a+bx)} + \frac{1}{ab} \ln \frac{x}{a+bx}.$$

4. $$\int \frac{\ln x \, dx}{(a+bx)^3} = -\frac{\ln x}{2b(a+bx)^2} + \frac{1}{2ab(a+bx)} + \frac{1}{2a^2 b} \ln \frac{x}{a+bx}.$$

5. $$\int \frac{\ln x \, dx}{\sqrt{a+bx}} = \frac{2}{b} \left\{ (\ln x - 2) \sqrt{a+bx} + \sqrt{a} \ln \frac{\sqrt{a+bx} + \sqrt{a}}{\sqrt{a+bx} - \sqrt{a}} \right\} \quad [a > 0];$$
$$= \frac{2}{b} \left\{ (\ln x - 2) \sqrt{a+bx} + 2\sqrt{-a} \, \mathrm{arctg}\, \sqrt{\frac{a+bx}{-a}} \right\} \quad [a < 0].$$

1. $\displaystyle\int x^m \ln(a+bx)\,dx = \frac{1}{m+1}\left[x^{m+1}\ln(a+bx) - b\int\frac{x^{m+1}\,dx}{a+bx}\right].$ 2.728

2. $\displaystyle\int \frac{\ln(a+bx)}{x}\,dx$

läßt sich nicht durch eine endliche Kombination elementarer Funktionen ausdrücken; vgl. 1.511 und 0.312. | cannot be expressed in terms of any finite combination of elementary functions (see 1.511 and 0.312).

1. $\displaystyle\int x^m \ln(a+bx)\,dx = \frac{1}{m+1}\left[x^{m+1} - \frac{(-a)^{m+1}}{b^{m+1}}\right]\ln(a+bx) +$ 2.729

$$+ \frac{1}{m+1}\sum_{k=1}^{m+1}\frac{(-1)^k x^{m-k+2} a^{k-1}}{(m-k+2)b^{k-1}}.$$

2. $\displaystyle\int x \ln(a+bx)\,dx = \frac{1}{2}\left[x^2 - \frac{a^2}{b^2}\right]\ln(a+bx) - \frac{1}{2}\left[\frac{x^2}{2} - \frac{ax}{b}\right].$

3. $\displaystyle\int x^2 \ln(a+bx)\,dx = \frac{1}{3}\left[x^3 + \frac{a^3}{b^3}\right]\ln(a+bx) - \frac{1}{3}\left[\frac{x^3}{3} - \frac{ax^2}{2b} + \frac{a^2 x}{b}\right].$

4. $\displaystyle\int x^3 \ln(a+bx)\,dx = \frac{1}{4}\left[x^4 - \frac{a^4}{b^4}\right]\ln(a+bx) -$

$$- \frac{1}{4}\left[\frac{x^4}{4} - \frac{ax^3}{3b} + \frac{a^2 x^2}{2b^2} - \frac{a^3 x}{b^3}\right].$$

$\displaystyle\int x^{2n}\ln(x^2+a^2)\,dx = \frac{1}{2n+1}\left\{x^{2n+1}\ln(x^2+a^2) + (-1)^n 2a^{2n+1}\operatorname{arctg}\frac{x}{a} - \right.$ 2.731

$$\left. - 2\sum_{k=0}^{n}\frac{(-1)^{n-k}}{2k+1} a^{2n-2k} x^{2k+1}\right\}.$$

$\displaystyle\int x^{2n+1}\ln(x^2+a^2)\,dx = \frac{1}{2n+1}\left\{(x^{2n+2} + (-1)^n a^{2n+2})\ln(x^2+a^2) + \right.$ 2.732

$$\left. + \sum_{k=1}^{n+1}\frac{(-1)^{n-k}}{k} a^{2n-2k+2} x^{2k}\right\}.$$

1. $\displaystyle\int \ln(x^2+a^2)\,dx = x\ln(x^2+a^2) - 2x + 2a\operatorname{arctg}\frac{x}{a}.$ D(623)   2.733

2. $\displaystyle\int x\ln(x^2+a^2)\,dx = \frac{1}{2}\left[(x^2+a^2)\ln(x^2+a^2) - x^2\right].$ D(623.1)

3. $\displaystyle\int x^2\ln(x^2+a^2)\,dx = \frac{1}{3}\left[x^3\ln(x^2+a^2) - \frac{2}{3}x^3 + 2a^2 x - 2a^3\operatorname{arctg}\frac{x}{a}\right].$ D(623.2)

4. $\displaystyle\int x^3\ln(x^2+a^2)\,dx = \frac{1}{4}\left[(x^4 - a^4)\ln(x^2+a^2) - \frac{x^4}{2} + a^2 x^2\right].$ D(623.3)

5. $\displaystyle\int x^4\ln(x^2+a^2)\,dx = \frac{1}{5}\left[x^5\ln(x^2+a^2) - \frac{2}{5}x^5 + \frac{2}{3}a^2 x^3 - 2a^4 x + \right.$

$$\left. + 2a^5\operatorname{arctg}\frac{x}{a}\right].$$ D(623.4)

2.734 $\int x^{2n} \ln |x^2 - a^2| \, dx = \frac{1}{2n+1} \left\{ x^{2n+1} \ln |x^2 - a^2| + a^{2n+1} \ln \left| \frac{x+a}{x-a} \right| - \right.$

$$\left. - 2 \sum_{k=0}^{n} \frac{1}{2k+1} a^{2n-2k} x^{2k+1} \right\}.$$

2.735 $\int x^{2n+1} \ln |x^2 - a^2| \, dx = \frac{1}{2n+2} \left\{ (x^{2n+2} - a^{2n+2}) \ln |x^2 - a^2| - \right.$

$$\left. - \sum_{k=1}^{n+1} \frac{1}{k} a^{2n-2k+2} x^{2k} \right\}.$$

2.736 1. $\quad \int \ln |x^2 - a^2| \, dx = x \ln |x^2 - a^2| - 2x + a \ln \left| \frac{x+a}{x-a} \right|.$ \hfill D(624)

2. $\quad \int x \ln |x^2 - a^2| \, dx = \frac{1}{2} \{(x^2 - a^2) \ln |x^2 - a^2| - x^2\}.$ \hfill D(624.1)

3. $\quad \int x^2 \ln |x^2 - a^2| \, dx = \frac{1}{3} \left\{ x^3 \ln |x^2 - a^2| - \frac{2}{3} x^3 - 2a^2 x + a^3 \ln \left| \frac{x+a}{x-a} \right| \right\}.$ \hfill D(624.2)

4. $\quad \int x^3 \ln |x^2 - a^2| \, dx = \frac{1}{4} \left\{ (x^4 - a^4) \ln |x^2 - a^2| - \frac{x^4}{2} - a^2 x^2 \right\}.$ \hfill D(624.3)

5. $\quad \int x^4 \ln |x^2 - a^2| \, dx = \frac{1}{5} \left\{ x^5 \ln |x^2 - a^2| - \frac{2}{5} x^5 - \frac{2}{3} a^2 x^3 - 2a^4 x + \right.$

$$\left. + a^5 \ln \left| \frac{x+a}{x-a} \right| \right\}.$$ \hfill D(624.4)

Area-Funktionen
Inverse hyperbolic functions

2.74

2.741 1. $\quad \int \operatorname{Arsh} \frac{x}{a} \, dx = x \operatorname{Arsh} \frac{x}{a} - \sqrt{x^2 + a^2}.$ \hfill D(730)

2. $\quad \int \operatorname{Arch} \frac{x}{a} \, dx = x \operatorname{Arch} \frac{x}{a} - \sqrt{x^2 - a^2} \quad \left[\operatorname{Arch} \frac{x}{a} > 0 \right];$

$$= x \operatorname{Arch} \frac{x}{a} + \sqrt{x^2 - a^2} \quad \left[\operatorname{Arch} \frac{x}{a} < 0 \right].$$ \hfill D(732)

3. $\quad \int \operatorname{Arth} \frac{x}{a} \, dx = x \operatorname{Arth} \frac{x}{a} + \frac{a}{2} \ln(a^2 - x^2).$ \hfill D(734)

4. $\quad \int \operatorname{Arcth} \frac{x}{a} \, dx = x \operatorname{Arcth} \frac{x}{a} + \frac{a}{2} \ln(x^2 - a^2).$ \hfill D(736)

2.742 1. $\quad \int x \operatorname{Arsh} \frac{x}{a} \, dx = \left(\frac{x^2}{2} + \frac{a^2}{4} \right) \operatorname{Arsh} \frac{x}{a} - \frac{x}{4} \sqrt{x^2 + a^2}.$ \hfill D(730.1)

2. $\quad \int x \operatorname{Arch} \frac{x}{a} \, dx = \left(\frac{x^2}{2} - \frac{a^2}{4} \right) \operatorname{Arch} \frac{x}{a} - \frac{x}{4} \sqrt{x^2 - a^2} \quad \left[\operatorname{Arch} \frac{x}{a} > 0 \right];$

$$= \left(\frac{x^2}{2} - \frac{a^2}{4} \right) \operatorname{Arch} \frac{x}{a} + \frac{x}{4} \sqrt{x^2 - a^2} \quad \left[\operatorname{Arch} \frac{x}{a} < 0 \right].$$ \hfill D(732.1)

Zyklometrische Funktionen
Inverse Trigonometric Functions

2.8

Arkussinus und Arkuscosinus
Arcsine and arccosine

2.81

$$\int \left(\arcsin \frac{x}{a}\right)^n dx = x \sum_{k=0}^{E\left(\frac{n}{2}\right)} (-1)^k \binom{n}{2k} \cdot (2k)! \left(\arcsin \frac{x}{a}\right)^{n-2k} +$$

$$+ \sqrt{a^2-x^2} \sum_{k=1}^{E\left(\frac{n+1}{2}\right)} (-1)^{k-1} \binom{n}{2k-1} \cdot (2k-1)! \left(\arcsin \frac{x}{a}\right)^{n-2k+1}.$$

2.811

$$\int \left(\arccos \frac{x}{a}\right)^n dx = x \sum_{k=0}^{E\left(\frac{n}{2}\right)} (-1)^k \binom{n}{2k} \cdot (2k)! \left(\arccos \frac{x}{a}\right)^{n-2k} +$$

$$+ \sqrt{a^2-x^2} \sum_{k=1}^{E\left(\frac{n+1}{2}\right)} (-1)^k \binom{n}{2k-1} \cdot (2k-1)! \left(\arccos \frac{x}{a}\right)^{n-2k+1}.$$

2.812

1. $\int \arcsin \frac{x}{a} dx = x \arcsin \frac{x}{a} + \sqrt{a^2-x^2}.$

2. $\int \left(\arcsin \frac{x}{a}\right)^2 dx = x \left(\arcsin \frac{x}{a}\right)^2 + 2\sqrt{a^2-x^2} \arcsin \frac{x}{a} - 2x.$

3. $\int \left(\arcsin \frac{x}{a}\right)^3 dx = x \left(\arcsin \frac{x}{a}\right)^3 + 3\sqrt{a^2-x^2} \left(\arcsin \frac{x}{a}\right)^2 -$
$$- 6x \arcsin \frac{x}{a} - 6\sqrt{a^2-x^2}.$$

2.813

1. $\int \arccos \frac{x}{a} dx = x \arccos \frac{x}{a} - \sqrt{a^2-x^2}.$

2. $\int \left(\arccos \frac{x}{a}\right)^2 dx = x \left(\arccos \frac{x}{a}\right)^2 - 2\sqrt{a^2-x^2} \arccos \frac{x}{a} - 2x.$

3. $\int \left(\arccos \frac{x}{a}\right)^3 dx = x \left(\arccos \frac{x}{a}\right)^3 - 3\sqrt{a^2-x^2} \left(\arccos \frac{x}{a}\right)^2 -$
$$- 6x \arccos \frac{x}{a} + 6\sqrt{a^2-x^2}.$$

2.814

Arkussekans und Arkuscosekans, Arkustangens und Arkuscotangens
The arcsecant, the arccosecant, the arctangent and the arccotangent

2.82

1. $\int \operatorname{arccosec} \frac{x}{a} dx = \int \arcsin \frac{a}{x} dx =$

$$= x \arcsin \frac{a}{x} + a \ln (x + \sqrt{x^2-a^2}) \quad \left[0 < \arcsin \frac{a}{x} < \frac{\pi}{2}\right];$$

$$= x \arcsin \frac{a}{x} - a \ln (x + \sqrt{x^2-a^2}) \quad \left[-\frac{\pi}{2} < \arcsin \frac{a}{x} < 0\right]. \quad \text{D(534)}$$

2.821

2.821   2. $\int \operatorname{arcsec} \dfrac{x}{a}\, dx = \int \arccos \dfrac{a}{x}\, dx =$

$= x \arccos \dfrac{a}{x} - a \ln(x + \sqrt{x^2 - a^2}) \quad \left[0 < \arccos \dfrac{a}{x} < \dfrac{\pi}{2}\right];$

$= x \arccos \dfrac{a}{x} + a \ln(x + \sqrt{x^2 - a^2}) \quad \left[-\dfrac{\pi}{2} < \arccos \dfrac{a}{x} < 0\right].$ D(531)

2.822   1. $\int \operatorname{arctg} \dfrac{x}{a}\, dx = x \operatorname{arctg} \dfrac{x}{a} - \dfrac{a}{2} \ln(a^2 + x^2).$ D(525)

     2. $\int \operatorname{arcctg} \dfrac{x}{a}\, dx = x \operatorname{arcctg} \dfrac{x}{a} + \dfrac{a}{2} \ln(a^2 + x^2).$ D(528)

2.83   Arkussinus, Arkuscosinus und algebraische Funktionen
Arcsine or arccosine and algebraic functions

2.831 $\int x^n \arcsin \dfrac{x}{a}\, dx = \dfrac{x^{n+1}}{n+1} \arcsin \dfrac{x}{a} - \dfrac{1}{n+1} \int \dfrac{x^{n+1}\, dx}{\sqrt{a^2 - x^2}}$ (cf. 2.263 1., 2.264, 2.27).

2.832 $\int x^n \arccos \dfrac{x}{a}\, dx = \dfrac{x^{n+1}}{n+1} \arccos \dfrac{x}{a} + \dfrac{1}{n+1} \int \dfrac{x^{n+1}\, dx}{\sqrt{a^2 - x^2}}$ (cf. 2.263 1., 2.264, 2.27).

1. Für $n = -1$ lassen sich diese Integrale $\left(\text{d.h. } \int \dfrac{\arcsin x}{x}\, dx \text{ und } \int \dfrac{\arccos x}{x}\, dx\right)$ nicht mit Hilfe einer endlichen Kombination elementarer Funktionen ausdrücken. | 1. For $n = -1$ these integrals $\left(\text{that is, } \int \dfrac{\arcsin x}{x}\, dx \text{ and } \int \dfrac{\arccos x}{x}\, dx\right)$ cannot be expressed in terms of any finite combination of elementary functions.

     2. $\int \dfrac{\arccos x}{x}\, dx = -\dfrac{\pi}{2} \ln \dfrac{1}{x} - \int \dfrac{\arcsin x}{x}\, dx.$

2.833   1. $\int x \arcsin \dfrac{x}{a}\, dx = \left(\dfrac{x^2}{2} - \dfrac{a^2}{4}\right) \arcsin \dfrac{x}{a} + \dfrac{x}{4} \sqrt{a^2 - x^2}.$

     2. $\int x \arccos \dfrac{x}{a}\, dx = \left(\dfrac{x^2}{2} - \dfrac{a^2}{4}\right) \arccos \dfrac{x}{a} - \dfrac{x}{4} \sqrt{a^2 - x^2}.$

2.834   1. $\int \dfrac{1}{x^2} \arcsin \dfrac{x}{a}\, dx = -\dfrac{1}{x} \arcsin \dfrac{x}{a} - \dfrac{1}{a} \ln \dfrac{a + \sqrt{a^2 - x^2}}{x}.$

     2. $\int \dfrac{1}{x^2} \arccos \dfrac{x}{a}\, dx = -\dfrac{1}{x} \arccos \dfrac{x}{a} + \dfrac{1}{a} \ln \dfrac{a + \sqrt{a^2 - x^2}}{x}.$

2.835 $\int \dfrac{\arcsin x}{(a + bx)^2}\, dx = -\dfrac{\arcsin x}{b(a + bx)} - \dfrac{2}{b\sqrt{a^2 - b^2}} \operatorname{arctg} \sqrt{\dfrac{(a - b)(1 - x)}{(a + b)(1 + x)}}$ $[a^2 > b^2];$

$= -\dfrac{\arcsin x}{b(a + bx)} - \dfrac{1}{b\sqrt{b^2 - a^2}} \ln \dfrac{\sqrt{(a + b)(1 + x)} + \sqrt{(b - a)(1 - x)}}{\sqrt{(a + b)(1 + x)} - \sqrt{(b - a)(1 - x)}}$ $[a^2 < b^2].$

2.836 $\int \dfrac{x \arcsin x}{(1 + cx^2)^2}\, dx = \dfrac{\arcsin x}{2c(1 + cx^2)} + \dfrac{1}{2c\sqrt{c + 1}} \operatorname{arctg} \dfrac{\sqrt{c + 1}\, x}{\sqrt{1 - x^2}}$ $[c > -1];$

$= -\dfrac{\arcsin x}{2c(1 + cx^2)} + \dfrac{1}{4c\sqrt{-(c + 1)}} \ln \dfrac{\sqrt{1 - x^2} + x\sqrt{-(c + 1)}}{\sqrt{1 - x^2} - x\sqrt{-(c + 1)}}$ $[c < -1].$

Zyklometrische Funktionen/Inverse Trigonometric Functions

1. $\int \dfrac{x \arcsin x}{\sqrt{1-x^2}}\, dx = x - \sqrt{1-x^2}\, \arcsin x.$ 2.837

2. $\int \dfrac{x^2 \arcsin x}{\sqrt{1-x^2}}\, dx = \dfrac{x^2}{4} - \dfrac{x}{2}\sqrt{1-x^2}\, \arcsin x + \dfrac{1}{4}(\arcsin x)^2.$

3. $\int \dfrac{x^3 \arcsin x}{\sqrt{1-x^2}}\, dx = \dfrac{x^3}{9} + \dfrac{2x}{3} - \dfrac{1}{3}(x^2+2)\sqrt{1-x^2}\, \arcsin x.$

1. $\int \dfrac{\arcsin x}{\sqrt{(1-x^2)^3}}\, dx = \dfrac{x \arcsin x}{\sqrt{1-x^2}} + \dfrac{1}{2}\ln(1-x^2).$ 2.838

2. $\int \dfrac{x \arcsin x}{\sqrt{(1-x^2)^3}}\, dx = \dfrac{\arcsin x}{\sqrt{1-x^2}} + \dfrac{1}{2}\ln \dfrac{1-x}{1+x}.$

Arkussekans, Arkuscosekans und Potenzen von x
Arcsecant and arccosecant with powers of x 2.84

1. $\int x \arcsec \dfrac{x}{a}\, dx = \int x \arccos \dfrac{a}{x}\, dx =$ 2.841

$= \dfrac{1}{2}\left\{x^2 \arccos \dfrac{a}{x} - a\sqrt{x^2-a^2}\right\}\left[0 < \arccos \dfrac{a}{x} < \dfrac{\pi}{2}\right];$

$= \dfrac{1}{2}\left\{x^2 \arccos \dfrac{a}{x} + a\sqrt{x^2-a^2}\right\}\left[\dfrac{\pi}{2} < \arccos \dfrac{a}{x} < \pi\right].$ D(531.1)

2. $\int x^2 \arcsec \dfrac{x}{a}\, dx = \int x^2 \arccos \dfrac{a}{x}\, dx =$

$= \dfrac{1}{3}\left\{x^3 \arccos \dfrac{a}{x} - \dfrac{a}{2}x\sqrt{x^2-a^2} - \dfrac{a^3}{2}\ln(x+\sqrt{x^2-a^2})\right\}$

$\left[0 < \arccos \dfrac{a}{x} < \dfrac{\pi}{2}\right];$

$= \dfrac{1}{3}\left\{x^3 \arccos \dfrac{a}{x} + \dfrac{a}{2}x\sqrt{x^2-a^2} + \dfrac{a^3}{2}\ln(x+\sqrt{x^2-a^2})\right\}$

$\left[\dfrac{\pi}{2} < \arccos \dfrac{a}{x} < \pi\right].$ D(531.2)

3. $\int x \arccosec \dfrac{x}{a}\, dx = \int x \arcsin \dfrac{a}{x}\, dx =$

$= \dfrac{1}{2}\left\{x^2 \arcsin \dfrac{a}{x} + a\sqrt{x^2-a^2}\right\}\left[0 < \arcsin \dfrac{a}{x} < \dfrac{\pi}{2}\right];$

$= \dfrac{1}{2}\left\{x^2 \arcsin \dfrac{a}{x} - a\sqrt{x^2-a^2}\right\}\left[-\dfrac{\pi}{2} < \arcsin \dfrac{a}{x} < 0\right].$ D(534.1)

Arkustangens, Arkuscotangens und algebraische Funktionen
Arctangent and arccotangent with algebraic functions 2.85

$\int x^n \arctg \dfrac{x}{a}\, dx = \dfrac{x^{n+1}}{n+1}\arctg \dfrac{x}{a} - \dfrac{a}{n+1}\int \dfrac{x^{n+1}\, dx}{a^2+x^2}.$ 2.851

2.852 1. $\int x^n \operatorname{arcctg} \frac{x}{a} dx = \frac{x^{n+1}}{n+1} \operatorname{arcctg} \frac{x}{a} + \frac{a}{n+1} \int \frac{x^{n+1} dx}{a^2 + x^2}$.

Für $n=-1$ kann das Integral $\int \frac{\operatorname{arctg} x}{x} dx$ nicht mit Hilfe einer endlichen Kombination elementarer Funktionen ausgedrückt werden. | For $n=-1$ integral $\int \frac{\operatorname{arctg} x}{x} dx$ cannot be expressed in terms of any combination of elementary fuctions.

2. $\int \frac{\operatorname{arcctg} x}{x} dx = \frac{\pi}{2} \ln x - \int \frac{\operatorname{arctg} x}{x} dx$.

2.853 1. $\int x \operatorname{arctg} \frac{x}{a} dx = \frac{1}{2} (x^2 + a^2) \operatorname{arctg} \frac{x}{a} - \frac{ax}{2}$.

2. $\int x \operatorname{arcctg} \frac{x}{a} dx = \frac{1}{2} (x^2 + a^2) \operatorname{arcctg} \frac{x}{a} + \frac{ax}{2}$.

2.854 $\int \frac{1}{x^2} \operatorname{arctg} \frac{x}{a} dx = -\frac{1}{x} \operatorname{arctg} \frac{x}{a} - \frac{1}{2a} \ln \frac{a^2 + x^2}{x^2}$.

2.855 $\int \frac{\operatorname{arctg} x}{(\alpha + \beta x)^2} dx = \frac{1}{\alpha^2 + \beta} \left\{ \ln \frac{\alpha + \beta x}{\sqrt{1 + x^2}} - \frac{\beta - \alpha x}{\alpha + \beta x} \operatorname{arctg} x \right\}$.

2.856 1. $\int \frac{x \operatorname{arctg} x}{1 + x^2} dx = \frac{1}{2} \operatorname{arctg} x \ln (1 + x^2) - \frac{1}{2} \int \frac{\ln (1 + x^2) dx}{1 + x^2}$. T(689)

2. $\int \frac{x^2 \operatorname{arctg} x}{1 + x^2} dx = x \operatorname{arctg} x - \frac{1}{2} \ln (1 + x^2) - \frac{1}{2} (\operatorname{arctg} x)^2$. T(405)

3. $\int \frac{x^3 \operatorname{arctg} x}{1 + x^2} dx = -\frac{1}{2} x + \frac{1}{2} (1 + x^2) \operatorname{arctg} x - \int \frac{x \operatorname{arctg} x}{1 + x^2} dx$ (cf. 2.851 1.).

4. $\int \frac{x^4 \operatorname{arctg} x}{1 + x^2} dx = -\frac{1}{6} x^2 + \frac{2}{3} \ln (1 + x^2) +$
$+ \left(\frac{x^3}{3} - x \right) \operatorname{arctg} x + \frac{1}{2} (\operatorname{arctg} x)^2$.

2.857 $\int \frac{\operatorname{arctg} x \, dx}{(1 + x^2)^{n+1}} = \left[\sum_{k=1}^{n} \frac{(2n - 2k)!! (2n - 1)!!}{(2n)!! (2n - 2k + 1)!!} \frac{x}{(1 + x^2)^{n-k+1}} + \right.$
$\left. + \frac{1}{2} \frac{(2n - 1)!!}{(2n)!!} \operatorname{arctg} x \right] \operatorname{arctg} x +$
$+ \frac{1}{2} \sum_{k=1}^{n} \frac{(2n - 1)!! (2n - 2k)!!}{(2n)!! (2n - 2k + 1)!! (n - k + 1)} \frac{1}{(1 + x^2)^{n-k+1}}$.

2.858 $\int \frac{x \operatorname{arctg} x}{\sqrt{1 - x^2}} dx = -\sqrt{1 - x^2} \operatorname{arctg} x + \sqrt{2} \operatorname{arctg} \frac{x \sqrt{2}}{\sqrt{1 - x^2}} - \arcsin x$.

2.859 $\int \frac{\operatorname{arctg} x}{\sqrt{(a + bx^2)^3}} dx = \frac{x \operatorname{arctg} x}{a \sqrt{a + bx^2}} - \frac{1}{a \sqrt{b - a}} \operatorname{arctg} \sqrt{\frac{a + bx^2}{b - a}}$ $[a < b]$;

$= \frac{x \operatorname{arctg} x}{a \sqrt{a + bx^2}} - \frac{1}{2a \sqrt{a - b}} \ln \frac{\sqrt{a + bx^2} - \sqrt{a - b}}{\sqrt{a + bx^2} + \sqrt{a - b}}$ $[a > b]$.

Bestimmte Integrale elementarer Funktionen
Definite Integrals of Elementary Functions

Einführung*
Introduction*

3.0

Allgemeine Sätze
General theorems

3.01

Die Funktion $f(x)$ sei im größten der Intervalle (p, q), (p, r), (r, q) integrierbar.** Dann ist sie (unabhängig von der gegenseitigen Lage der Punkte p, q, r) in den beiden anderen Intervallen integrierbar, und es gilt

Let the function $f(x)$ be integrable** over the largest of the three intervals (p, q), (p, r) and (r, q). Then $f(x)$ is also integrable over the two other intervals (regardless of the relative position of the points p, q, r) and

3.011

$$\int_p^q f(x)\,dx = \int_p^r f(x)\,dx + \int_r^q f(x)\,dx.$$

F II 113

Der (erste) Mittelwertsatz. Folgende Voraussetzungen seien erfüllt:

1. Die Funktion $f(x)$ sei stetig und die Funktion $g(x)$ sei im Intervall (p, q) integrierbar;
2. es gelte die Ungleichung $m \leqslant f(x) \leqslant M$;

The first mean value theorem. Assume that

(1) $f(x)$ is continuous and $g(x)$ integrable over (p, q);

(2) $m \leqslant f(x) \leqslant M$, and

3.012

---

* Wir verzichten darauf, die Definition des bestimmten Integrals bzw. die Definition mehrfacher Integrale wiederzugeben, da wir sie als bekannt voraussetzen. Notfalls können sie jedem Lehrbuch entnommen werden. Lediglich werden einige Sätze allgemeinen Charakters angeführt, welche Abschätzungen liefern oder vorgegebenes Integral auf einfachere Integrale zurückführen.

** Eine Funktion $f(x)$ heißt *integrierbar* im Intervall (p, q), wenn $\int_p^q f(x)\,dx$ existiert.

Dabei verstehen wir die Existenz des Integrals im RIEMANNschen Sinne. Handelt es sich um die Existenz des Integrals im Sinne von STIELTJES, LEBESGUE usw., so sprechen wir von Integrierbarkeit im Sinne von STIELTJES, LEBESGUE usw.

* The definitions of the definite integral and of the multiple integral are supposed to be known, and they may be found in books on the subject. We will give only some general theorems, which lead to approximations or transformations of a given integral into simpler ones.

** A function $f(x)$ is said to be *integrable* over the interval (p, q) if the integral $\int_p^q f(x)\,dx$ exists. Usually the existence of the integral is meant in the sense of RIEMANN. If the integral exists in the sense of STIELTJES, LEBESGUE, etc., $f(x)$ is said to be integrable in the sense of STIELTJES, LEBESGUE, etc.

3. $g(x)$ does not change sign within (p, q). Then there exists at least one point ξ $(p \leqslant \xi \leqslant q)$ such that

$$\int_p^q f(x)\, g(x)\, dx = f(\xi) \int_p^q g(x)\, dx.$$ F II 119

3.013 The second mean value theorem. If $f(x)$ is a nonnegative and monotone nonincreasing function in (p, q) $[p < q]$ and if $g(x)$ is integrable, there exists at least one point ξ $(p \leqslant \xi \leqslant q)$ such that

1. $$\int_p^q f(x)\, g(x)\, dx = f(p) \int_p^\xi g(x)\, dx.$$

If instead of being nonincreasing $f(x)$ is monotone nondecreasing,

2. $$\int_p^q f(x)\, g(x)\, dx = f(q) \int_\xi^q g(x)\, dx \quad [p \leqslant \xi \leqslant q].$$

If the function $f(x)$ is monotonic and $g(x)$ integrable in the interval (p, q) $[p < q]$,

3. $$\int_p^q f(x)\, g(x)\, dx = f(p) \int_p^\xi g(x)\, dx + f(q) \int_\xi^q g(x)\, dx \quad [p \leqslant \xi \leqslant q],$$

or

4. $$\int_p^q f(x)\, g(x)\, dx = A \int_p^\xi g(x)\, dx + B \int_\xi^p g(x)\, dx \quad [p \leqslant \xi \leqslant q],$$

where A and B are any two numbers satisfying the conditions

$$A \geqslant f(p+0) \text{ and } B \leqslant f(q-0) \quad \begin{bmatrix} \text{für fallende } f \\ \text{for } f \text{ decreasing} \end{bmatrix},$$

$$A \leqslant f(p+0) \text{ and } B \geqslant f(q-0) \quad \begin{bmatrix} \text{für wachsende } f \\ \text{for } f \text{ increasing} \end{bmatrix}.$$

In particular,

5. $$\int_p^q f(x)\, g(x)\, dx = f(p+0) \int_p^\xi g(x)\, dx + f(q-0) \int_\xi^q g(x)\, dx.$$ F II 123

Variablentransformation in einem bestimmten Integral
Change of variable in a definite integral

3.02

$$\int_\alpha^\beta f(x)\,dx = \int_\varphi^\psi f[g(t)]\,g'(t)\,dt;\quad x = g(t).$$

3.020

Diese Formel gilt unter folgenden Bedingungen:
1. Die Funktion $f(x)$ ist in einem gewissen Intervall $A \leqslant x \leqslant B$, das die alten Grenzen α und β umfaßt, stetig.
2. Es gelten die Gleichungen $\alpha = g(\varphi)$, $\beta = g(\psi)$.
3. Die Funktion $g(t)$ und ihre Ableitung $g'(t)$ sind im Intervall $\varphi \leqslant t \leqslant \psi$ stetig.
4. Läuft t von φ bis ψ, so ändert sich $g(t)$ stets in ein und demselben Sinne von $g(\varphi) = \alpha$ bis $g(\psi) = \beta$*.

This formula holds under the following conditions:
(1) $f(x)$ is continuous in a certain interval $[A, B]$ containing the interval $[\alpha, \beta]$;
(2) $\alpha = g(\varphi,) \beta = g(\psi)$;
(3) the function $g(t)$ and its derivative $g'(t)$ are continuous in the interval $\varphi \leqslant t \leqslant \psi$;
(4) the function $g(t)$ is either uniformly increasing or uniformly decreasing in the interval $\varphi \leqslant t \leqslant \psi$ *.

Das Integral $\int_\alpha^\beta f(x)\,dx$ läßt sich in ein Integral mit den vorgegebenen Grenzen φ und ψ überführen, und zwar mit Hilfe der linearen Substitution

The integral $\int_\alpha^\beta f(x)\,dx$ can be transformed into an integral with prescribed limits φ and ψ; this is achieved by means of the linear substitution

3.021

$$x = \frac{\beta - \alpha}{\psi - \varphi}\,t + \frac{\alpha\psi - \beta\varphi}{\psi - \varphi}:$$

1. $$\int_\alpha^\beta f(x)\,dx = \frac{\beta - \alpha}{\psi - \varphi}\int_\varphi^\psi f\left(\frac{\beta - \alpha}{\psi - \varphi}\,t + \frac{\alpha\psi - \beta\varphi}{\psi - \varphi}\right)dt;$$

insbesondere kann $\varphi = 0$ und $\psi = 1$ sein

in particular, for $\varphi = 0$ and $\psi = 1$:

2. $$\int_\alpha^\beta f(x)\,dx = (\beta - \alpha)\int_0^1 f((\beta - \alpha)\,t + \alpha)\,dt.$$

Im Fall $\varphi = 0$, $\psi = \infty$ gilt

For $\varphi = 0$ and $\psi = \infty$:

3. $$\int_\alpha^\beta f(x)\,dx = (\beta - \alpha)\int_0^\infty f\left(\frac{\alpha + \beta t}{1 + t}\right)\frac{dt}{(1 + t)^2}.$$

---

* Ist die letzte Bedingung nicht erfüllt, so muss man das Intervall $\varphi \leqslant t \leqslant \psi$ in Teilintervalle zerlegen, in denen sie erfüllt ist, dann gilt

* If this latter condition is not fulfilled, the interval $[\varphi, \psi]$ has to be partitioned into subintervals in each of which the condition is satisfied; then

$$\int_\alpha^\beta f(x)\,dx = \int_\varphi^{\varphi_1} f[g(t)]\,g'(t)\,dt + \int_{\varphi_1}^{\varphi_2} f[g(t)]\,g'(t)\,dt + \ldots + \int_{\varphi_{n-1}}^\psi f[g(t)]\,g'(t)\,dt.$$

3.022 Weiterhin gelten die Identitäten | Furthermore,

1. $$\int_\alpha^\beta f(x)\,dx = \int_\alpha^\beta f(\alpha + \beta - x)\,dx.$$

2. $$\int_0^\beta f(x)\,dx = \int_0^\beta f(\beta - x)\,dx.$$

3. $$\int_{-\alpha}^\alpha f(x)\,dx = \int_{-\alpha}^\alpha f(-x)\,dx.$$

Allgemeine Formeln
General formulae

3.03

3.031 1. Die Funktion $f(x)$ sei im Intervall $(-p, p)$ integrierbar und genüge dort der Beziehung $f(-x) = f(x)$ (dann nennt man $f(x)$ eine *gerade* Funktion). Dann gilt | (1) Let $f(x)$ be *even*, that is, satisfy the condition $f(-x) = f(x)$, and integrable over the interval $(-p, p)$. Then

$$\int_{-p}^p f(x)\,dx = 2\int_0^p f(x)\,dx. \qquad \text{F II 141}$$

2. Die Funktion $f(x)$ sei im Intervall $(-p, p)$ integrierbar und genüge dort der Beziehung $f(-x) = -f(x)$ (dann nennt man $f(x)$ eine *ungerade* Funktion). Dann gilt | (2) Let $f(x)$ be *odd*, that is, satisfy the relation $f(-x) = -f(x)$, and integrable over the interval $(-p, p)$. Then

$$\int_{-p}^p f(x)\,dx = 0. \qquad \text{F II 141}$$

3.032 1.
$$\int_0^{\pi/2} f(\sin x)\,dx = \int_0^{\pi/2} f(\cos x)\,dx,$$

wobei $f(x)$ eine im Intervall $(0, 1)$ integrierbare Funktion ist. F II 141 | where $f(x)$ is integrable over the interval $(0, 1)$. F II 141

2. $$\int_0^{2\pi} f(p\cos x + q\sin x)\,dx = 2\int_0^\pi f(\sqrt{p^2 + q^2}\cos x)\,dx,$$

wobei $f(x)$ eine im Intervall $(-\sqrt{p^2+q^2}, \sqrt{p^2+q^2})$ integrierbare Funktion ist. F II 142 | where $f(x)$ is integrable over the interval $(-\sqrt{p^2+q^2}, \sqrt{p^2+q^2})$. F II 142

3. $$\int_0^{\pi/2} f(\sin 2x)\cos x\,dx = \int_0^{\pi/2} f(\cos^2 x)\cos x\,dx,$$

wobei $f(x)$ eine im Intervall $(0, 1)$ integrierbare Funktion ist. F II 142

where $f(x)$ is integrable over the interval $(0, 1)$. F II 142

3.033

1. Ist $f(x + \pi) = f(x)$ und $f(-x) = f(x)$, so ist

(1) If $f(x + \pi) = f(x)$ and $f(-x) = f(x)$,

$$\int_0^\infty f(x) \frac{\sin x}{x} dx = \int_0^{\pi/2} f(x) \, dx.$$

Lo V 277(3)

2. Ist $f(x + \pi) = -f(x)$ und $f(-x) = f(x)$, so ist

(2) If $f(x + \pi) = -f(x)$ and $f(-x) = f(x)$,

$$\int_0^\infty f(x) \frac{\sin x}{x} dx = \int_0^{\pi/2} f(x) \cos x \, dx.$$

Lo V 279(4)

In den Formeln 3.033 wird vorausgesetzt, daß die auf den linken Seiten der Formeln stehenden Integrale existieren.

It is assumed that the integrals on the left-hand side of formulae 3.033 exist.

$$\int_0^\infty \frac{f(px) - f(qx)}{x} dx = [f(0) - f(+\infty)] \ln \frac{q}{p},$$

3.034

wenn $f(x)$ für $x \geq 0$ stetig ist und $f(+\infty) = \lim_{x \to +\infty} f(x)$ existiert und endlich ist. F II 642–643

if $f(x)$ is continuous for $x \geq 0$ and if $\lim_{x \to +\infty} f(x) = f(+\infty)$ exists and is finite. F II 642–643

1. $$\int_0^\pi \frac{f(\alpha + e^{xi}) + f(\alpha + e^{-xi})}{1 + 2p \cos x + p^2} dx = \frac{2\pi}{1 - p^2} f(\alpha + p) \quad [|p| < 1].$$ La 230(16) 3.035

2. $$\int_0^\pi \frac{1 - p \cos x}{1 - 2p \cos x + p^2} \{f(\alpha + e^{xi}) + f(\alpha + e^{-xi})\} dx = \pi\{f(\alpha + p) + f(\alpha)\}$$

$$[|p| < 1]. \quad \text{B 169}$$

3. $$\int_0^\pi \frac{f(\alpha + e^{-xi}) - f(\alpha + e^{xi})}{1 - 2p \cos x + p^2} \sin x \, dx = \frac{\pi}{pi} \{f(\alpha + p) - f(\alpha)\} \quad [|p| < 1]. \quad \text{B 169}$$

In den Formeln 3.035 wird vorausgesetzt, die Funktion f sei auf der abgeschlossenen Einheitskreisscheibe um den Mittelpunkt α analytisch.

In formulae 3.035 it is assumed that the function $f(x)$ is analytic on the whole unit circle with the centre at α.

1. $$\int_0^\pi f\left(\frac{\sin^2 x}{1 + 2p \cos x + p^2}\right) dx = \int_0^\pi f(\sin^2 x) \, dx \quad [p^2 \geq 1];$$ 3.036

$$= \int_0^\pi f\left(\frac{\sin^2 x}{p^2}\right) dx \quad [p^2 < 1]. \quad \text{La 228(6)}$$

2. $$\int_0^\pi F^{(n)}(\cos x) \sin^{2n} x \, dx = (2n-1)!! \int_0^\pi F(\cos x) \cos nx \, dx. \quad \text{B 174}$$

3.037 Ist die Funktion f in einem Kreis vom Radius r analytisch und ist | If f is analytic within the circle of radius r and if

$$f[r(\cos x + i \sin x)] = f_1(r, x) + if_2(r, x),$$

so gelten die Formeln | then

1. $$\int_0^\infty \frac{f_1(r, x)}{p^2 + x^2} dx = \frac{\pi}{2p} f(re^{-p}).$$ La 230(19)

2. $$\int_0^\infty f_2(r, x) \frac{x\, dx}{p^2 + x^2} = \frac{\pi}{2} [f(re^{-p}) - f(0)].$$ La 230(20)

3. $$\int_0^\infty \frac{f_2(r, x)}{x} dx = \frac{\pi}{2} [f(r) - f(0)].$$ La 230(21)

4. $$\int_0^\infty \frac{f_2(r, x)}{x(p^2 + x^2)} dx = \frac{\pi}{2p^2} [f(r) - f(re^{-p})].$$ La 230(22)

3.038
$$\int_{-\infty}^\infty \frac{x\, dx}{\sqrt{1+x^2}} F(qx + p\sqrt{1+x^2}) = \int_{-\infty}^\infty F(p\, \text{ch}\, x + q\, \text{sh}\, x)\, \text{sh}\, x\, dx =$$

$$= 2q \int_0^\infty F'(\text{sign}\, p \cdot \sqrt{p^2 - q^2}\, \text{ch}\, x)\, \text{sh}^2 x\, dx$$

[dabei ist F eine Funktion, deren Ableitung im Intervall $(-\infty, \infty)$ stetig ist; alle vorkommenden Integrale müssen konvergieren]. | [F must have a continuous derivative in the whole interval $(-\infty, \infty)$ and all integrals must converge].

Lo III 281÷, Lo III 391÷ | Lo III 281÷, Lo III 391÷

Uneigentliche Integrale
Improper integrals

3.04

3.041 Die Funktion $f(x)$ sei im Intervall $(p, +\infty)$ definiert und in jedem endlichen Teilintervall (p, P) integrierbar; dann definiert man | Let the function $f(x)$ be defined in the interval $(p, +\infty)$, and integrable over any finite interval (p, P); then by definition

$$\int_p^{+\infty} f(x)\, dx = \lim_{P \to +\infty} \int_p^P f(x)\, dx,$$

wenn dieser Grenzwert existiert. Existiert dieser Grenzwert, so sagt man, das Integral $\int_p^{+\infty} f(x)\, dx$ *existiere* oder *konvergiere*. Anderenfalls sagt man, das Integral *divergiere*. | provided the limit exists. In this case the integral $\int_p^{+\infty} f(x)\, dx$ is said to *exist*, or to *converge;* otherwise the integral is said to *diverge.*

Let $f(x)$ be bounded and integrable over any interval $(p, q-\eta)$ ($0 < \eta < q-p$) yet unbounded in the interval $(q-\eta, q)$ to the left of q. The function f is then said to have a *singularity* at q. Then by definition

$$\int_p^q f(x)\, dx = \lim_{\eta \to 0} \int_p^{q-\eta} f(x)\, dx,$$

provided the limit exists. In this case the integral $\int_p^q f(x)\, dx$ is said to *exist*, or to *converge*.

3.042

If both the integral of $f(x)$ and the integral of $|f(x)|$ exist, the integral $\int f(x)\, dx$ is said to be *absolutely convergent*.

3.043

The integral $\int_p^{+\infty} f(x)\, dx$ converges absolutely if there is a certain number $\alpha > 1$ such that the limit

$$\lim_{x \to +\infty} \{x^\alpha |f(x)|\}$$

exists. But if $\alpha = 1$ and

$$\lim_{x \to +\infty} \{x |f(x)|\} = L > 0,$$

then the integral $\int_p^{+\infty} |f(x)|\, dx$ diverges.

3.044

The integral $\int_p^q f(x)\, dx$, where $f(x)$ has a singularity at the upper bound q, converges absolutely if there is a certain number $\alpha < 1$ such that the limit

$$\lim_{x \to q} [(q-x)^\alpha |f(x)|]$$

exists. But if

$$\lim_{x \to q} [(q-x) |f(x)|] = L > 0,$$

3.045

so divergiert $\int_p^q f(x)\,dx$.

3.046 Die Funktionen $f(x)$ und $g(x)$ seien im Intervall $(p, +\infty)$ definiert, und $f(x)$ sei in jedem endlichen Teilintervall (p, P) integrierbar. Ist das Integral

$$\int_p^P f(x)\,dx$$

eine beschränkte Funktion von P und $g(x)$ eine monotone Funktion mit $g(x) \to 0$ für $x \to +\infty$, so konvergiert

$$\int_p^{+\infty} f(x)\,g(x)\,dx.$$

then the integral $\int_p^q f(x)\,dx$ diverges.

Let the functions $f(x)$ and $g(x)$ be defined on the interval $(p, +\infty)$ and let $f(x)$ be integrable over every finite interval (p, P). If the integral

$$\int_p^P f(x)\,dx$$

considered as a function of P is bounded and if $g(x)$ is monotonic with $g(x) \to 0$ as $x \to +\infty$, then the integral converges.

F II 586

Hauptwerte uneigentlicher Integrale
Principle values of improper integrals

3.05

3.051 Die Funktion $f(x)$ besitze im Innern ihres Definitionsintervalls (p, q) einen einzigen singulären Punkt r und sei in jedem den Punkt r nicht enthaltenden Teilintervall integrierbar. Dann definiert man

Let the function $f(x)$ have only one singularity at the point r in the interval (p, q) and let $f(x)$ be integrable over any part of this interval that does not contain r. Then by definition

$$\int_p^q f(x)\,dx = \lim_{\substack{\eta \to 0 \\ \eta' \to 0}} \left\{ \int_p^{r-\eta} f(x)\,dx + \int_{r+\eta'}^q f(x)\,dx \right\},$$

wobei der Grenzwert u n a b h ä n - g i g von der Art der Grenzübergänge $\eta \to 0$ und $\eta' \to 0$ existieren muß. Wenn dieser Grenzwert nicht existiert, wohl aber der spezielle Grenzwert

which must exist i r r e s p e c t i v e of the order in which η and η' approach zero. If the limit does not exist while

$$\lim_{\eta \to 0} \left\{ \int_p^{r-\eta} f(x)\,dx + \int_{r+\eta}^q f(x)\,dx \right\}$$

so nennt man diesen den *Hauptwert des uneigentlichen Integrals* $\int_p^q f(x)\,dx$ und sagt: Das Integral $\int_p^q f(x)\,dx$ existiert im Sinne des Hauptwertes.

F II 613

does, the latter is called the *principal value of the improper integral* $\int_p^q f(x)\,dx$, and this integral is said to *exist in the sense of (having) a principle value*.

F II 613

Let $f(x)$ be continuous on the interval (p, q) and vanish only at one point r within this interval. In the neighbourhood of r let $f'(x)$ exist and let $f'(r) \neq 0$; finally, let the second derivative at r, that is $f''(r)$ exist. Then the integral

$$\int_p^q \frac{dx}{f(x)},$$

is divergent yet exists in the sense of having a principal value.   F II 614–615   **3.052**

A divergent integral of a positive-valued function cannot have a principle value.   F II 615   **3.053**

Let the function $f(x)$ be free from singularities in the interval $(-\infty, +\infty)$. Then by definition   **3.054**

$$\int_{-\infty}^{+\infty} f(x)\,dx = \lim_{\substack{P \to -\infty \\ Q \to +\infty}} \int_P^Q f(x)\,dx,$$

which must exist irrespective of the order in which P and Q approach $-\infty$ and $+\infty$, respectively. If this limit does not exist while the limit

$$\lim_{P \to +\infty} \int_{-P}^{+P} f(x)\,dx$$

does, the latter is called the *principal value of the improper integral*

$$\int_{-\infty}^{+\infty} f(x)\,dx.$$

   F II 616

For the improper integral of an even function the principal value exists only if this integral converges (in the usual sense).   F II 616–617   **3.055**

Potenzfunktionen und algebraische Funktionen
Power and Algebraic Functions

3.1–3.2

Rationale Funktionen
Rational functions

3.11

3.111
$$\int_{-\infty}^{\infty} \frac{p+qx}{r^2+2rx\cos\lambda+x^2}\, dx = \frac{\pi}{r\sin\lambda}(p - qr\cos\lambda) \qquad \text{BH [22](14)}$$

(Hauptwert) * | (principal value) *
(siehe auch 3.194 8. und 3.252 1. und 2.). | (see also 3.194 8., 3.252 1., 2).

3.112 Integrale der Gestalt: / Integrals of the form:
$$\int_{-\infty}^{\infty} \frac{g_n(x)\, dx}{h_n(x)\, h_n(-x)}$$

mit | where

$$g_n(x) = b_0 x^{2n-2} + b_1 x^{2n-4} + \ldots + b_{n-1},$$
$$h_n(x) = a_0 x^n + a_1 x^{n-1} + \ldots + a_n$$

[alle Nullstellen von $h_n(x)$ liegen in der oberen Halbebene]. | [Im $h_n > 0$ for all zeros of $h_n(x)$].

1.
$$\int_{-\infty}^{\infty} \frac{g_n(x)\, dx}{h_n(x)\, h_n(-x)} = \frac{\pi i}{a_0} \frac{M_n}{\Delta_n} \qquad \text{Jo 456}$$

mit | where

$$\Delta_n = \begin{vmatrix} a_1 & a_3 & a_5 & \ldots & 0 \\ a_0 & a_2 & a_4 & \ldots & 0 \\ 0 & a_1 & a_3 & \ldots & 0 \\ \vdots & \vdots & \vdots & & \vdots \\ 0 & 0 & 0 & \ldots & a_n \end{vmatrix},$$

$$M_n = \begin{vmatrix} b_0 & b_1 & b_2 & \ldots & b_{n-1} \\ a_0 & a_2 & a_4 & \ldots & 0 \\ 0 & a_1 & a_3 & \ldots & 0 \\ \vdots & \vdots & \vdots & & \vdots \\ 0 & 0 & 0 & \ldots & a_n \end{vmatrix}.$$

---

* In dieser Formelsammlung werden die Werte eigentlicher und uneigentlicher konvergierender Integrale angegeben, sowie die Hauptwerte divergenter Integrale (vgl. 3.05), sofern sie existieren. Hauptwerte werden im folgenden nicht besonders gekennzeichnet.

* Besides formulae for proper and convergent improper integrals the tables give the principal values of divergent integrals (see 3.05) if they exist. The latter will not be indicated as such.

2. $$\int_{-\infty}^{\infty} \frac{g_1(x)\,dx}{h_1(x)\,h_1(-x)} = \frac{\pi i b_0}{a_0 a_1}.$$ Jo 454 3.112

3. $$\int_{-\infty}^{\infty} \frac{g_2(x)\,dx}{h_2(x)\,h_2(-x)} = \pi i \frac{-b_0 + \dfrac{a_0 b_1}{a_2}}{a_0 a_1}.$$

4. $$\int_{-\infty}^{\infty} \frac{g_3(x)\,dx}{h_3(x)\,h_3(-x)} = \pi i \frac{-a_2 b_0 + a_0 b_1 - \dfrac{a_0 a_1 b_2}{a_3}}{a_0(a_0 a_3 - a_1 a_2)}.$$ Jo 454

5. $$\int_{-\infty}^{\infty} \frac{g_4(x)\,dx}{h_4(x)\,h_4(-x)} =$$

$$= \pi i \frac{b_0(-a_1 a_4 + a_2 a_3) - a_0 a_3 b_1 + a_0 a_1 b_2 + \dfrac{a_0 b_3}{a_4}(a_0 a_3 - a_1 a_2)}{a_0(a_0 a_3^2 + a_1^2 a_4 - a_1 a_2 a_3)}.$$ Jo 455

6. $$\int_{-\infty}^{\infty} \frac{g_5(x)\,dx}{h_5(x)\,h_5(-x)} = \pi i \frac{M_5}{a_0 \Delta_5}$$

mit | where

$$M_5 = b_0(-a_0 a_4 a_5 + a_1 a_4^2 + a_2^2 a_5 - a_2 a_3 a_4) + a_0 b_1(-a_2 a_5 + a_3 a_4) +$$
$$+ a_0 b_2(a_0 a_5 - a_1 a_4) + a_0 b_3(-a_0 a_3 + a_1 a_2) + \frac{a_0 b_4}{a_5}(-a_0 a_1 a_5 + a_0 a_3^2 + a_1^2 a_4 - a_1 a_2 a_3).$$

$$\Delta_5 = a_0^2 a_5^2 - 2 a_0 a_1 a_4 a_5 - a_0 a_2 a_3 a_5 + a_0 a_3^2 a_4 + a_1^2 a_4^2 + a_1 a_2^2 a_5 - a_1 a_2 a_3 a_4.$$ Jo 455

Produkte rationaler Funktionen mit Ausdrücken, die auf Quadratwurzeln aus linearen und quadratischen Polynomen führen
Products of rational functions with expressions reducible to square roots of first- and second-degree polynomials
3.12

1. $$\int_0^1 \frac{1}{1 - 2x\cos\lambda + x^2} \frac{dx}{\sqrt{x}} = 2\operatorname{cosec}\lambda \sum_{k=1}^{\infty} \frac{\sin k\lambda}{2k-1}.$$ BH[10](17) 3.121

2. $$\int_0^1 \frac{1}{q - px} \frac{dx}{\sqrt{x(1-x)}} = \frac{\pi}{\sqrt{q(q-p)}} \quad [0 < p < q].$$ BH[10](9)

3. $$\int_0^1 \frac{dx}{1 - 2rx + r^2} \sqrt{\frac{1 \mp x}{1 \pm x}} = \pm \frac{\pi}{4r} \mp \frac{1}{r}\frac{1 \mp r}{1 \pm r} \operatorname{arctg} \frac{1+r}{1-r}.$$ Li[14](5), Li[14](16)

Ausdrücke, die auf Quadratwurzeln aus Polynomen dritten und vierten Grades führen, und ihre Produkte mit rationalen Funktionen

3.13–3.17 Expressions reducible to square roots of third- and fourth-degree polynomials and their products with rational functions

In 3.131 – 3.137 werden folgende Abkürzungen verwendet: | In 3.131–3.137 the following notation is used:

$$\alpha = \arcsin\sqrt{\frac{a-c}{a-u}}, \quad \beta = \arcsin\sqrt{\frac{c-u}{b-u}},$$

$$\gamma = \arcsin\sqrt{\frac{u-c}{b-c}}, \quad \delta = \arcsin\sqrt{\frac{(a-c)(b-u)}{(b-c)(a-u)}},$$

$$\varkappa = \arcsin\sqrt{\frac{(a-c)(u-b)}{(a-b)(u-c)}}, \quad \lambda = \arcsin\sqrt{\frac{a-u}{a-b}},$$

$$\mu = \arcsin\sqrt{\frac{u-a}{u-b}}, \quad \nu = \arcsin\sqrt{\frac{a-c}{u-c}}, \quad p = \sqrt{\frac{a-b}{a-c}}, \quad q = \sqrt{\frac{b-c}{a-c}}.$$

3.131

1. $\displaystyle\int_{-\infty}^{u} \frac{dx}{\sqrt{(a-x)(b-x)(c-x)}} = \frac{2}{\sqrt{a-c}} F(\alpha, p) \quad [a > b > c \geqslant u]$.   BF(231.00)

2. $\displaystyle\int_{u}^{c} \frac{dx}{\sqrt{(a-x)(b-x)(c-x)}} = \frac{2}{\sqrt{a-c}} F(\beta, p) \quad [a > b > c > u]$.   BF(232.00)

3. $\displaystyle\int_{c}^{u} \frac{dx}{\sqrt{(a-x)(b-x)(x-c)}} = \frac{2}{\sqrt{a-c}} F(\gamma, q) \quad [a > b \geqslant u > c]$.   BF(233.00)

4. $\displaystyle\int_{u}^{b} \frac{dx}{\sqrt{(a-x)(b-x)(x-c)}} = \frac{2}{\sqrt{a-c}} F(\delta, q) \quad [a > b > u \geqslant c]$.   BF(234.00)

5. $\displaystyle\int_{b}^{u} \frac{dx}{\sqrt{(a-x)(x-b)(x-c)}} = \frac{2}{\sqrt{a-c}} F(\varkappa, p) \quad [a \geqslant u > b > c]$.   BF(235.00)

6. $\displaystyle\int_{u}^{a} \frac{dx}{\sqrt{(a-x)(x-b)(x-c)}} = \frac{2}{\sqrt{a-c}} F(\lambda, p) \quad [a > u \geqslant b > c]$.   BF(236.00)

7. $\displaystyle\int_{a}^{u} \frac{dx}{\sqrt{(x-a)(x-b)(x-c)}} = \frac{2}{\sqrt{a-c}} F(\mu, q) \quad [u > a > b > c]$.   BF(237.00)

8. $\displaystyle\int_{u}^{\infty} \frac{dx}{\sqrt{(x-a)(x-b)(x-c)}} = \frac{2}{\sqrt{a-c}} F(\nu, q) \quad [u \geqslant a > b > c]$.   BF(238.00)

1. $\displaystyle\int_u^c \frac{x\,dx}{\sqrt{(a-x)(b-x)(c-x)}} = \frac{2}{\sqrt{a-c}}[cF(\beta,p) +$

$+ (a-c)E(\beta,p)] - 2\sqrt{\dfrac{(a-u)(c-u)}{b-u}}$ $\quad [a>b>c>u]$.   BF(232.19)

3.132

2. $\displaystyle\int_c^u \frac{x\,dx}{\sqrt{(a-x)(b-x)(x-c)}} = \frac{2a}{\sqrt{a-c}}F(\gamma,q) - 2\sqrt{a-c}\,E(\gamma,q)$

$[a>b\geqslant u>c]$.   BF(233.17)

3. $\displaystyle\int_u^b \frac{x\,dx}{\sqrt{(a-x)(b-x)(x-c)}} = \frac{2}{\sqrt{a-c}}[(b-a)\Pi(\delta,q^2,q) + aF(\delta,q)]$

$[a>b>u\geqslant c]$.   BF(234.16)

4. $\displaystyle\int_b^u \frac{x\,dx}{\sqrt{(a-x)(x-b)(x-c)}} = \frac{2}{\sqrt{a-c}}[(b-c)\Pi(\varkappa,p^2,p) + cF(\varkappa,p)]$

$[a\geqslant u>b>c]$.   BF(235.16)

5. $\displaystyle\int_u^a \frac{x\,dx}{\sqrt{(a-x)(x-b)(x-c)}} = \frac{2c}{\sqrt{a-c}}F(\lambda,p) + 2\frac{a}{b}\sqrt{a-c}\,E(\lambda,p)$

$[a>u\geqslant b>c]$.   BF(236.16)

6. $\displaystyle\int_a^u \frac{x\,dx}{\sqrt{(x-a)(x-b)(x-c)}} = \frac{2}{b\sqrt{a-c}}[a(a-b)\Pi(\mu,1,q) + b^2 F(\mu,q)]$

$[u>a>b>c]$.   BF(237.16)

1. $\displaystyle\int_{-\infty}^u \frac{dx}{\sqrt{(a-x)^3(b-x)(c-x)}} = \frac{2}{(a-b)\sqrt{a-c}}[F(\alpha,p) - E(\alpha,p)]$

3.133

$[a>b>c\geqslant u]$.   BF(231.08)

2. $\displaystyle\int_u^c \frac{dx}{\sqrt{(a-x)^3(b-x)(c-x)}} = \frac{2}{(a-b)\sqrt{a-c}}[F(\beta,p) - E(\beta,p)] +$

$+ \dfrac{2}{a-c}\sqrt{\dfrac{c-u}{(a-u)(b-u)}}$ $\quad [a>b>c>u]$.   BF(232.13)

3. $\displaystyle\int_c^u \frac{dx}{\sqrt{(a-x)^3(b-x)(x-c)}} = \frac{2}{(a-b)\sqrt{a-c}}E(\gamma,q) -$

$- \dfrac{2}{(a-b)(a-c)}\sqrt{\dfrac{(b-u)(u-c)}{a-u}}$ $\quad [a>b\geqslant u>c]$.   BF(233.09)

3.133

4. $$\int_u^b \frac{dx}{\sqrt{(a-x)^3(b-x)(x-c)}} = \frac{2}{(a-b)\sqrt{a-c}} E(\delta, q)$$

    $[a > b > u \geqslant c]$. BF(234.05)

5. $$\int_b^u \frac{dx}{\sqrt{(a-x)^3(x-b)(x-c)}} = \frac{2}{(a-b)\sqrt{a-c}} [F(\varkappa, p) - E(\varkappa, p)] +$$

    $$+ \frac{2}{a-b} \sqrt{\frac{u-b}{(a-u)(u-c)}} \quad [a > u > b > c].$$ BF(235.04)

6. $$\int_u^\infty \frac{dx}{\sqrt{(x-a)^3(x-b)(x-c)}} = \frac{2}{(b-a)\sqrt{a-c}} E(v, q) +$$

    $$+ \frac{2}{a-b} \sqrt{\frac{u-b}{(u-a)(u-c)}} \quad [u > a > b > c].$$ BF(238.05)

7. $$\int_{-\infty}^u \frac{dx}{\sqrt{(a-x)(b-x)^3(c-x)}} = \frac{2\sqrt{a-c}}{(a-b)(b-c)} E(\alpha, p) -$$

    $$- \frac{2}{(a-b)\sqrt{a-c}} F(\alpha, p) - \frac{2}{b-c} \sqrt{\frac{c-u}{(a-u)(b-u)}} \quad [a > b > c \geqslant u].$$ BF(231.09)

8. $$\int_u^c \frac{dx}{\sqrt{(a-x)(b-x)^3(c-x)}} = \frac{2\sqrt{a-c}}{(a-b)(b-c)} E(\beta, p) -$$

    $$- \frac{2}{(a-b)\sqrt{a-c}} F(\beta, p) \quad [a > b > c > u].$$ BF(232.14)

9. $$\int_c^u \frac{dx}{\sqrt{(a-x)(b-x)^3(x-c)}} = \frac{2}{(b-c)\sqrt{a-c}} F(\gamma, q) -$$

    $$- \frac{2\sqrt{a-c}}{(a-b)(b-c)} E(\gamma, q) + \frac{2}{(a-b)(b-c)} \sqrt{\frac{(a-u)(u-c)}{b-u}}$$

    $[a > b > u > c]$. BF(233.10)

10. $$\int_u^a \frac{dx}{\sqrt{(a-x)(x-b)^3(x-c)}} = \frac{2}{(a-b)\sqrt{a-c}} F(\lambda, p) -$$

    $$- \frac{2\sqrt{a-c}}{(a-b)(b-c)} E(\lambda, p) + \frac{2}{(a-b)(b-c)} \sqrt{\frac{(a-u)(u-c)}{u-b}}$$

    $[a > u > b > c]$. BF(236.09)

11. $$\int_a^u \frac{dx}{\sqrt{(x-a)(x-b)^3(x-c)}} = \frac{2\sqrt{a-c}}{(a-b)(b-c)} E(\mu, q) -$$

    $$- \frac{2}{(b-c)\sqrt{a-c}} F(\mu, q) \quad [u > a > b > c].$$ BF(237.12)

12. $\displaystyle\int_u^\infty \frac{dx}{\sqrt{(x-a)(x-b)^3(x-c)}} = \frac{2\sqrt{a-c}}{(a-b)(b-c)} E(v,q) -$ 3.133

$\displaystyle - \frac{2}{(b-c)\sqrt{a-c}} F(v,q) - \frac{2}{a-b}\sqrt{\frac{u-a}{(u-b)(u-c)}}$

$[u \geq a > b > c]$. BF(238.04)

13. $\displaystyle\int_{-\infty}^u \frac{dx}{\sqrt{(a-x)(b-x)(c-x)^3}} = \frac{2}{(c-b)\sqrt{a-c}} E(\alpha,p) +$

$\displaystyle + \frac{2}{b-c}\sqrt{\frac{b-u}{(a-u)(c-u)}}$ $\quad [a > b > c > u]$. BF(231.10)

14. $\displaystyle\int_u^b \frac{dx}{\sqrt{(a-x)(b-x)(x-c)^3}} = \frac{2}{(b-c)\sqrt{a-c}}[F(\delta,q) -$

$\displaystyle - E(\delta,q)] + \frac{2}{b-c}\sqrt{\frac{b-u}{(a-u)(u-c)}}$ $\quad [a > b > u > c]$. BF(234.04)

15. $\displaystyle\int_b^u \frac{dx}{\sqrt{(a-x)(x-b)(x-c)^3}} = \frac{2}{(b-c)\sqrt{a-c}} E(\varkappa,p)$ $\quad [a \geq u > b > c]$. BF(235.01)

16. $\displaystyle\int_u^a \frac{dx}{\sqrt{(a-x)(x-b)(x-c)^3}} = \frac{2}{(b-c)\sqrt{a-c}} E(\lambda,p) -$

$\displaystyle - \frac{2}{(b-c)(a-c)}\sqrt{\frac{(a-u)(u-b)}{u-c}}$ $\quad [a > u \geq b > c]$. BF(236.10)

17. $\displaystyle\int_a^u \frac{dx}{\sqrt{(x-a)(x-b)(x-c)^3}} = \frac{2}{(b-c)\sqrt{a-c}}[F(\mu,q) - E(\mu,q)] +$

$\displaystyle + \frac{2}{a-c}\sqrt{\frac{u-a}{(u-b)(u-c)}}$

$[u > a > b > c]$. BF(237.13)

18. $\displaystyle\int_u^\infty \frac{dx}{\sqrt{(x-a)(x-b)(x-c)^3}} = \frac{2}{(b-c)\sqrt{a-c}}[F(v,q) - E(v,q)]$

$[u \geq a > b > c]$. BF(238.03)

1. $\displaystyle\int_{-\infty}^u \frac{dx}{\sqrt{(a-x)^5(b-x)(c-x)}} = \frac{2}{3(a-b)^2\sqrt{(a-c)^3}} \times$ 3.134

$\displaystyle \times [(3a-b-2c) F(\alpha,p) - 2(2a-b-c) E(\alpha,p)] +$

$\displaystyle + \frac{2}{3(a-c)(a-b)}\sqrt{\frac{(c-u)(b-u)}{(a-u)^3}}$

$[a > b > c \geq u]$. BF(231.08)

3.134

2. $\int_u^c \dfrac{dx}{\sqrt{(a-x)^5(b-x)(c-x)}} = \dfrac{2}{3(a-b)^2\sqrt{(a-c)^3}} \times$

$\times [(3a - b - 2c) F(\beta, p) - 2(2a - b - c) E(\beta, p)] +$

$+ \dfrac{2[4a^2 - 3ab - 2ac + bc - u(3a - 2b - c)]}{3(a-b)(a-c)^2} \sqrt{\dfrac{c-u}{(a-u)^3(b-u)}}$

$[a > b > c > u]$.   BF(232.13)

3. $\int_c^u \dfrac{dx}{\sqrt{(a-x)^5(b-x)(x-c)}} = \dfrac{2}{3(a-b)^2\sqrt{(a-c)^3}} \times$

$\times [2(2a - b - c) E(\gamma, q) - (a - b) F(\gamma, q)] -$

$- \dfrac{2[5a^2 - 3ab - 3ac + bc - 2u(2a - b - c)]}{3(a-b)^2(a-c)^2} \sqrt{\dfrac{(b-u)(u-c)}{(a-u)^3}}$

$[a > b \geqslant u > c]$.   BF(233.09)

4. $\int_u^b \dfrac{dx}{\sqrt{(a-x)^5(b-x)(x-c)}} = \dfrac{2}{3(a-b)^2\sqrt{(a-c)^3}} \times$

$\times [2(2a - b - c) E(\delta, q) - (a - b) F(\delta, q)] -$

$- \dfrac{2}{3(a-b)(a-c)} \sqrt{\dfrac{(b-u)(u-c)}{(a-u)^3}}$

$[a > b > u \geqslant c]$.   BF(234.05)

5. $\int_b^u \dfrac{dx}{\sqrt{(a-x)^5(x-b)(x-c)}} = \dfrac{2}{3(a-b)^2\sqrt{(a-c)^3}} \times$

$\times [(3a - b - 2c) F(\varkappa, p) - 2(2a - b - c) E(\varkappa, p)] +$

$+ \dfrac{2[4a^2 - 2ab - 3ac + bc - u(3a - b - 2c)]}{3(a-b)^2(a-c)} \sqrt{\dfrac{u-b}{(a-u)^3(u-c)}}$

$[a > u > b > c]$.   BF(235.04)

6. $\int_u^\infty \dfrac{dx}{\sqrt{(x-a)^5(x-b)(x-c)}} = \dfrac{2}{3(a-b)^2\sqrt{(a-c)^3}} \times$

$\times [2(2a - b - c) E(\nu, q) - (a - b) F(\nu, q)] +$

$+ \dfrac{2[4a^2 - 2ab - 3ac + bc + u(b + 2c - 3a)]}{3(a-b)^2(a-c)} \sqrt{\dfrac{u-b}{(u-a)^3(u-c)}}$

$[u > a > b > c]$.   BF(238.05)

7. $\displaystyle\int_{-\infty}^{u} \frac{dx}{\sqrt{(a-x)(b-x)^5(c-x)}} = \frac{2}{3(a-b)^2(b-c)^2\sqrt{a-c}} \times$ 3.134

$\times [2(a-c)(a+c-2b)E(\alpha, p) + (b-c)(3b-a-2c)F(\alpha, p)] -$

$- \dfrac{2[3ab - ac + 2bc - 4b^2 - u(2a - 3b + c)]}{3(a-b)(b-c)^2}\sqrt{\dfrac{c-u}{(a-u)(b-u)^3}}$

$[a > b > c \geqslant u].$ BF(231.09)

8. $\displaystyle\int_{u}^{c} \frac{dx}{\sqrt{(a-x)(b-x)^5(c-x)}} = \frac{2}{3(a-b)^2(b-c)^2\sqrt{a-c}} \times$

$\times [(b-c)(3b-a-2c)F(\beta, p) + 2(a-c)(a-2b+c)E(\beta, p)] +$

$+ \dfrac{2}{3(a-b)(b-c)}\sqrt{\dfrac{(a-u)(c-u)}{(b-u)^3}}$

$[a > b > c > u].$ BF(232.14)

9. $\displaystyle\int_{c}^{u} \frac{dx}{\sqrt{(a-x)(b-x)^5(x-c)}} = \frac{2}{3(a-b)^2(b-c)^2\sqrt{a-c}} \times$

$\times [(a-b)(2a-3b+c)F(\gamma, q) + 2(a-c)(2b-a-c)E(\gamma, q)] +$

$+ \dfrac{2[3ab + 3bc - ac - 5b^2 - 2u(a-2b+c)]}{3(a-b)^2(b-c)^2}\sqrt{\dfrac{(a-u)(u-c)}{(b-u)^3}}$

$[a > b > u > c].$ BF(233.10)

10. $\displaystyle\int_{u}^{a} \frac{dx}{\sqrt{(a-x)(x-b)^5(x-c)}} = \frac{2}{3(a-b)^2(b-c)^2\sqrt{a-c}} \times$

$\times [(b-c)(3b-2c-a)F(\lambda, p) + 2(a-c)(a+c-2b)E(\lambda, p)] +$

$+ \dfrac{2[3ab + 3bc - ac - 5b^2 + 2u(2b - a - c)]}{3(a-b)^2(b-c)^2}\sqrt{\dfrac{(a-u)(u-c)}{(u-b)^3}}$

$[a > u > b > c].$ BF(236.09)

11. $\displaystyle\int_{a}^{} \frac{dx}{\sqrt{(x-a)(x-b)^5(x-c)}} = \frac{2}{3(a-b)^2(b-c)^2\sqrt{a-c}} \times$

$\times [(a-b)(2a+c-3b)F(\mu, q) + 2(a-c)(2b-a-c)E(\mu, q)] +$

$+ \dfrac{2}{3(a-b)(b-c)}\sqrt{\dfrac{(u-a)(u-c)}{(u-b)^3}}$

$[u > a > b > c].$ BF(237.12)

12. $\displaystyle\int_{u}^{\infty} \frac{dx}{\sqrt{(x-a)(x-b)^5(x-c)}} = \frac{2}{3(a-b)^2(b-c)^2\sqrt{a-c}} \times$

$\times [(a-b)(2a+c-3b)F(\nu, q) + 2(a-c)(2b-c-a)E(\nu, q)] -$

$- \dfrac{2[3bc + 2ab - ac - 4b^2 + u(3b - a - 2c)]}{3(a-b)^2(b-c)}\sqrt{\dfrac{u-a}{(u-b)^3(u-c)}}$

$[u \geqslant a > b > c].$ BF(238.04)

3.134

13. $\displaystyle\int_{-\infty}^{u} \frac{dx}{\sqrt{(a-x)(b-x)(c-x)^5}} = \frac{2}{3(b-c)^2\sqrt{(a-c)^3}} \times$

$\times [2(a+b-2c)\,E(\alpha,p) - (b-c)\,F(\alpha,p)] +$

$+ \dfrac{2[ab - 3ac - 2bc + 4c^2 + u(2a+b-3c)]}{3(a-c)(b-c)^2} \sqrt{\dfrac{b-u}{(a-u)(c-u)^3}}$

$[a > b > c > u].$ BF(231.10)

14. $\displaystyle\int_{u}^{b} \frac{dx}{\sqrt{(a-x)(b-x)(x-c)^5}} = \frac{2}{3(b-c)^2\sqrt{(a-c)^3}} \times$

$\times [(2a+b-3c)\,F(\delta,q) - 2(a+b-2c)\,E(\delta,q)] +$

$+ \dfrac{2[ab - 3ac - 2bc + 4c^2 + u(2a+b-3c)]}{3(b-c)^2(a-c)} \sqrt{\dfrac{b-u}{(a-u)(u-c)^3}}$

$[a > b > u > c].$ BF(234.04)

15. $\displaystyle\int_{b}^{u} \frac{dx}{\sqrt{(a-x)(x-b)(x-c)^5}} = \frac{2}{3(b-c)^2\sqrt{(a-c)^3}} \times$

$\times [2(a+b-2c)\,E(\varkappa,p) - (b-c)\,F(\varkappa,p)] +$

$+ \dfrac{2}{3(a-c)(b-c)} \sqrt{\dfrac{(a-u)(u-b)}{(u-c)^3}} \quad [a \geqslant u > b > c].$ BF(235.20)

16. $\displaystyle\int_{u}^{a} \frac{dx}{\sqrt{(a-x)(x-b)(x-c)^5}} = \frac{2}{3(b-c)^2\sqrt{(a-c)^3}} \times$

$\times [2(a+b-2c)\,E(\lambda,p) - (b-c)\,F(\lambda,p)] -$

$- \dfrac{2[ab - 3ac - 3bc + 5c^2 + 2u(a+b-2c)]}{3(b-c)^2(a-c)^2} \sqrt{\dfrac{(a-u)(u-b)}{(u-c)^3}}$

$[a > u \geqslant b > c].$ BF(236.10)

17. $\displaystyle\int_{a}^{u} \frac{dx}{\sqrt{(x-a)(x-b)(x-c)^5}} = \frac{2}{3(b-c)^2\sqrt{(a-c)^3}} \times$

$\times [(2a+b-3c)\,F(\mu,q) - 2(a+b-2c)\,E(\mu,q)] +$

$+ \dfrac{2[4c^2 - ab - 2ac - bc + u(3a+2b-5c)]}{3(b-c)(a-c)^2} \sqrt{\dfrac{u-a}{(u-b)(u-c)^3}}$

$[u > a > b > c].$ BF(237.13)

18. $\displaystyle\int_{u}^{\infty} \frac{dx}{\sqrt{(x-a)(x-b)(x-c)^5}} = \frac{2}{3(b-c)^2\sqrt{(a-c)^3}} \times$

$\times [(2a+b-3c)\,F(v,q) - 2(a+b-2c)\,E(v,q)] +$

$+ \dfrac{2}{3(a-c)(b-c)} \sqrt{\dfrac{(u-a)(u-b)}{(u-c)^3}} \quad [u \geqslant a > b > c].$ BF(238.03)

1. $\int_{-\infty}^{u} \frac{dx}{\sqrt{(a-x)(b-x)^3(c-x)^3}} = \frac{2}{(a-b)(b-c)^2\sqrt{a-c}} \times$ 3.135

$$\times [(b-c)F(\alpha,p) - (a+b-2c)E(\alpha,p)] +$$

$$+ \frac{2(b+c-2u)}{(b-c)^2\sqrt{(a-u)(b-u)(c-u)}} \quad [a > b > c > u]. \quad \text{BF(231.13)}$$

2. $\int_{u}^{a} \frac{dx}{\sqrt{(a-x)(x-b)^3(x-c)^3}} = \frac{2}{(a-b)(b-c)^2\sqrt{a-c}} \times$

$$\times [(b-c)F(\lambda,p) - 2(2a-b-c)E(\lambda,p)] +$$

$$+ \frac{2(a-b-c+u)}{(a-b)(b-c)(a-c)}\sqrt{\frac{a-u}{(u-b)(u-c)}} \quad [a > u > b > c]. \quad \text{BF(236.15)}$$

3. $\int_{a}^{u} \frac{dx}{\sqrt{(x-a)(x-b)^3(x-c)^3}} = \frac{2}{(a-b)(b-c)^2\sqrt{a-c}} \times$

$$\times [(2a-b-c)E(\mu,q) - 2(a-b)F(\mu,q)] +$$

$$+ \frac{2}{(a-c)(b-c)}\sqrt{\frac{u-a}{(u-b)(u-c)}} \quad [u > a > b > c]. \quad \text{BF(237.14)}$$

4. $\int_{u}^{\infty} \frac{dx}{\sqrt{(x-a)(x-b)^3(x-c)^3}} = \frac{2}{(a-b)(b-c)^2\sqrt{a-c}} \times$

$$\times [(2a-b-c)E(v,q) - 2(a-b)F(v,q)] -$$

$$- \frac{2}{(a-b)(b-c)}\sqrt{\frac{u-a}{(u-b)(u-c)}} \quad [u \geqslant a > b > c]. \quad \text{BF(238.13)}$$

5. $\int_{-\infty}^{u} \frac{dx}{\sqrt{(a-x)^3(b-x)(c-x)^3}} = \frac{2}{(a-b)(b-c)\sqrt{(a-c)^3}} \times$

$$\times [(2b-a-c)E(\alpha,p) - (b-c)F(\alpha,p)] +$$

$$+ \frac{2}{(b-c)(a-c)}\sqrt{\frac{b-u}{(a-u)(c-u)}} \quad [a > b > c > u]. \quad \text{BH(231.12)}$$

6. $\int_{u}^{b} \frac{dx}{\sqrt{(a-x)^3(b-x)(x-c)^3}} = \frac{2}{(b-c)(a-b)\sqrt{(a-c)^3}} \times$

$$\times [(a-b)F(\delta,q) + (2b-a-c)E(\delta,q)] +$$

$$+ \frac{2}{(b-c)(a-c)}\sqrt{\frac{b-u}{(a-u)(u-c)}} \quad [a > b > u > c]. \quad \text{BF(234.03)}$$

3.135

7. $\int_b^u \dfrac{dx}{\sqrt{(a-x)^3(x-b)(x-c)^3}} = \dfrac{2}{(a-b)(b-c)\sqrt{(a-c)^3}} \times$

$\times [(b-c) F(\varkappa, p) - (2b-a-c) E(\varkappa, p)] +$

$+ \dfrac{2}{(a-b)(a-c)} \sqrt{\dfrac{u-b}{(a-u)(u-c)}} \quad [a > u > b > c].$ BF(235.15)

8. $\int_u^\infty \dfrac{dx}{\sqrt{(x-a)^3(x-b)(x-c)^3}} = \dfrac{2}{(a-b)(b-c)\sqrt{(a-c)^3}} \times$

$\times [(a+c-2b) E(v, q) - (a-b) F(v, q)] +$

$+ \dfrac{2}{(a-b)(a-c)} \sqrt{\dfrac{u-b}{(u-a)(u-c)}} \quad [u > a > b > c].$ BF(238.14)

9. $\int_{-\infty}^u \dfrac{dx}{\sqrt{(a-x)^3(b-x)^3(c-x)}} = \dfrac{2}{(b-c)(a-b)^2\sqrt{a-c}} \times$

$\times [(a+b-2c) E(\alpha, p) - 2(b-c) F(\alpha, p)] -$

$- \dfrac{2}{(a-b)(b-c)} \sqrt{\dfrac{c-u}{(a-u)(b-u)}} \quad [a > b > c \geqslant u].$ BF(231.11)

10. $\int_u^c \dfrac{dx}{\sqrt{(a-x)^3(b-x)^3(c-x)}} = \dfrac{2}{(a-b)^2(b-c)\sqrt{a-c}} \times$

$\times [(a+b-2c) E(\beta, p) - 2(b-c) F(\beta, p)] +$

$+ \dfrac{2}{(a-b)(a-c)} \sqrt{\dfrac{c-u}{(a-u)(b-u)}} \quad [a > b > c > u].$ BF(232.15)

11. $\int_c^u \dfrac{dx}{\sqrt{(a-x)^3(b-x)^3(x-c)}} = \dfrac{2}{(a-b)^2(b-c)\sqrt{a-c}} \times$

$\times [(a-b) F(\gamma, q) - (a+b-2c) E(\gamma, q)] +$

$+ \dfrac{2[a^2+b^2-ac-bc-u(a+b-2c)]}{(a-b)^2(b-c)(a-c)} \sqrt{\dfrac{u-c}{(a-u)(b-u)}}$

$[a > b > u > c].$ BF(233.11)

12. $\int_u^\infty \dfrac{dx}{\sqrt{(x-a)^3(x-b)^3(x-c)}} = \dfrac{2}{(a-b)^2(b-c)\sqrt{a-c}} \times$

$\times [(a-b) F(v, q) - (a+b-2c) E(v, q)] +$

$+ \dfrac{2u-a-b}{(a-b)^2\sqrt{(u-a)(u-b)(u-c)}} \quad [u > a > b > c].$ BF(238.15)

1. $\displaystyle\int_{-\infty}^{u} \frac{dx}{\sqrt{(a-x)^3(b-x)^3(c-x)^3}} = \frac{2}{(a-b)^2(b-c)^2\sqrt{(a-c)^3}} \times$ 3.136

$\times [(b-c)(a+b-2c)F(\alpha,p) - 2(c^2+a^2+b^2-ab-ac-bc)E(\alpha,p)] +$

$\displaystyle +\frac{2[c(a-c)+b(a-b)-u(2a-c-b)]}{(a-b)(a-c)(b-c)^2\sqrt{(a-u)(b-u)(c-u)}}$

$[a > b > c > u]$.   BF(231.14)

2. $\displaystyle\int_{u}^{\infty} \frac{dx}{\sqrt{(x-a)^3(x-b)^3(x-c)^3}} = \frac{2}{(a-b)^2(b-c)^2\sqrt{(a-c)^3}} \times$

$\times [(a-b)(2a-b-c)F(v,q) - 2(a^2+b^2+c^2-ab-ac-bc)E(v,q)] +$

$\displaystyle +\frac{2[u(a+b-2c)-a(a-c)-b(b-c)]}{(a-b)^2(a-c)(b-c)\sqrt{(u-a)(u-b)(u-c)}}$

$[u > a > b > c]$.   BF(238.16)

1. $\displaystyle\int_{-\infty}^{u} \frac{dx}{(r-x)\sqrt{(a-x)(b-x)(c-x)}} = \frac{2}{(a-r)\sqrt{a-c}} \times$ 3.137

$\times \left[\Pi\left(\alpha, \frac{a-r}{a-c}, p\right) - F(\alpha,p) \right] \quad [a > b > c \geqslant u]$.   BF(231.15)

2. $\displaystyle\int_{u}^{c} \frac{dx}{(r-x)\sqrt{(a-x)(b-x)(c-x)}} = \frac{2(c-b)}{(r-b)(r-c)\sqrt{a-c}} \times$

$\times \Pi\left(\beta, \frac{r-b}{r-c}, p\right) + \frac{2}{(r-b)\sqrt{a-c}} F(\beta,p)$

$[a > b > c > u, r \neq 0]$.   BF(232.17)

3. $\displaystyle\int_{c}^{u} \frac{dx}{(r-x)\sqrt{(a-x)(b-x)(x-c)}} = \frac{2}{(r-c)\sqrt{a-c}} \Pi\left(\gamma, \frac{b-c}{r-c}, q\right)$

$[a > b \geqslant u > c, r \neq c]$.   BF(233.02)

4. $\displaystyle\int_{u}^{b} \frac{dx}{(r-x)\sqrt{(a-x)(b-x)(x-c)}} = \frac{2}{(r-a)(r-b)\sqrt{a-c}} \times$

$\times \left[(b-a)\Pi\left(\delta, q^2\frac{r-a}{r-b}, q\right) + (r-b)F(\delta,q) \right]$

$[a > b > u \geqslant c, r \neq b]$.   BF(234.18)

5. $\displaystyle\int_{b}^{u} \frac{dx}{(x-r)\sqrt{(a-x)(x-b)(x-c)}} = \frac{2}{(c-r)(b-r)\sqrt{a-c}} \times$

$\times \left[(c-b)\Pi\left(\varkappa, p^2\frac{c-r}{b-r}, p\right) + (b-r)F(\varkappa,p) \right]$

$[a \geqslant u > b > c, r \neq b]$.   BF(235.17)

3.137

6. $\int_u^a \dfrac{dx}{(x-r)\sqrt{(a-x)(x-b)(x-c)}} = \dfrac{2}{(a-r)\sqrt{a-c}}\,\Pi\left(\lambda,\dfrac{a-b}{a-r},p\right)$

$[a > u \geqslant b > c,\ r \neq a]$. BF(236.02)

7. $\int_a^u \dfrac{dx}{(x-r)\sqrt{(x-a)(x-b)(x-c)}} = \dfrac{2}{(b-r)(a-r)\sqrt{a-c}} \times$

$\times \left[(b-a)\Pi\left(\mu,\dfrac{b-r}{a-b},q\right) + (a-p)F(\mu,q)\right]$

$[u > a > b > c,\ r \neq a]$. BF(237.17)

8. $\int_u^\infty \dfrac{dx}{(x-r)\sqrt{(x-a)(x-b)(x-c)}} = \dfrac{2}{(r-c)\sqrt{a-c}} \times$

$\times \left[\Pi\left(v,\dfrac{r-c}{a-c},q\right) - F(v,q)\right] \quad [u \geqslant a > b > c]$. BF(238.06)

3.138

1. $\int_0^u \dfrac{dx}{\sqrt{x(1-x)(1-k^2x)}} = 2F(\arcsin\sqrt{u},k) \quad [0<u<1]$. P(532), JE 125

2. $\int_u^1 \dfrac{dx}{\sqrt{x(1-x)(k'^2+k^2x)}} = 2F(\arccos\sqrt{u},k) \quad [0<u<1]$. P(533)

3. $\int_u^1 \dfrac{dx}{\sqrt{x(1-x)(x-k'^2)}} = 2F\left(\arcsin\dfrac{\sqrt{1-u}}{k},k\right) \quad [0<u<1]$. P(534)

4. $\int_0^u \dfrac{dx}{\sqrt{x(1+x)(1+k'^2x)}} = 2F(\arctan\sqrt{u},k) \quad [0<u<1]$. P(535)

5. $\int_0^u \dfrac{dx}{\sqrt{x[1+x^2+2(k'^2-k^2)x]}} = F(2\arctan\sqrt{u},k) \quad [0<u<1]$. JE 126

6. $\int_u^1 \dfrac{dx}{\sqrt{x[k'^2(1+x^2)+2(1+k^2)x]}} = F\left(\dfrac{\pi}{2}-2\arctan\sqrt{u},k\right) \quad [0<u<1]$. JE 126

7. $\int_\alpha^u \dfrac{dx}{\sqrt{(x-\alpha)[(x-m)^2+n^2]}} = \dfrac{1}{\sqrt{p}}F\left(2\arctan\sqrt{\dfrac{u-\alpha}{p}},\sqrt{\dfrac{p+m-\alpha}{2p}}\right) \quad [\alpha<u]$.

8. $\int_u^a \dfrac{dx}{\sqrt{(\alpha-x)[(x-m)^2+n^2]}} = \dfrac{1}{\sqrt{p}}F\left(2\operatorname{arcctg}\sqrt{\dfrac{\alpha-u}{p}},\sqrt{\dfrac{p-m+\alpha}{qp}}\right) \quad [u<\alpha]$.

mit $p = \sqrt{(m-\alpha)^2+n^2}$. | where $p = \sqrt{(m-\alpha)^2+n^2}$.

Abkürzungen:
Abbreviations:
$$\begin{cases} \alpha = \arccos\dfrac{1-\sqrt{3}-u}{1+\sqrt{3}-u}, & \beta = \arccos\dfrac{\sqrt{3}-1+u}{\sqrt{3}+1-u}, \\ \gamma = \arccos\dfrac{\sqrt{3}+1-u}{\sqrt{3}-1+u}, & \delta = \arccos\dfrac{u-1-\sqrt{3}}{u-1+\sqrt{3}}. \end{cases}$$
3.139

1. $\displaystyle\int_{-\infty}^{u}\frac{dx}{\sqrt{1-x^3}} = \frac{1}{\sqrt[4]{3}}F(\alpha, \sin 75°).$ Z 66(285)

2. $\displaystyle\int_{u}^{1}\frac{dx}{\sqrt{1-x^3}} = \frac{1}{\sqrt[4]{3}}F(\beta, \sin 75°).$ Z 65(284)

3. $\displaystyle\int_{1}^{u}\frac{dx}{\sqrt{x^3-1}} = \frac{1}{\sqrt[4]{3}}F(\gamma, \sin 15°).$ Z 65(283)

4. $\displaystyle\int_{u}^{\infty}\frac{dx}{\sqrt{x^3-1}} = \frac{1}{\sqrt[4]{3}}F(\delta, \sin 15°).$ Z 65(282)

5. $\displaystyle\int_{0}^{1}\frac{dx}{\sqrt{1-x^3}} = \frac{1}{2\pi\sqrt{3}\sqrt[3]{2}}\left\{\Gamma\left(\frac{1}{3}\right)\right\}^2.$ MO 9

6. $\displaystyle\int_{0}^{1}\frac{x\,dx}{\sqrt{1-x^3}} = \frac{1}{\pi}\frac{\sqrt{3}}{\sqrt[3]{4}}\left\{\Gamma\left(\frac{2}{3}\right)\right\}^2.$ MO 9

7. $\displaystyle\int_{u}^{1}\sqrt{1-x^3}\,dx = \frac{1}{5}\left\{\sqrt[4]{27}F(\beta, \sin 75°) - 2u\sqrt{1-u^3}\right\}.$ BF(244.01)

8. $\displaystyle\int_{u}^{1}\frac{x\,dx}{\sqrt{1-x^3}} = (3^{-\frac{1}{4}} - 3^{\frac{1}{4}})\,F(\beta, \sin 75°) +$

$\qquad\qquad + 2\sqrt[4]{3}\,E(\beta, \sin 75°) - \dfrac{2\sqrt{1-u^3}}{\sqrt{3}+1-u}.$ BF(244.05)

9. $\displaystyle\int_{u}^{1}\frac{x^m\,dx}{\sqrt{1-x^3}} = \frac{2u^{m-2}\sqrt{1-u^3}}{2m-1} + \frac{2(m-2)}{2m-1}\int_{u}^{1}\frac{x^{m-3}\,dx}{\sqrt{1-x^3}}.$ BF(244.07)

10. $\displaystyle\int_{1}^{u}\frac{x\,dx}{\sqrt{x^3-1}} = (3^{-\frac{1}{4}} + 3^{\frac{1}{4}})\,F(\gamma, \sin 15°) -$

$\qquad\qquad - 2\sqrt[4]{3}\,E(\gamma, \sin 15°) + \dfrac{2\sqrt{u^3-1}}{\sqrt{3}-1+u}.$ BF(240.05)

3.139

11. $\int_{-\infty}^{u} \dfrac{dx}{(1-x)\sqrt{1-x^3}} = \dfrac{1}{\sqrt[4]{27}}[F(\alpha, \sin 75°) - 2E(\alpha, \sin 75°)] +$

$+ \dfrac{2}{\sqrt{3}} \dfrac{\sqrt{1+u+u^2}}{(1+\sqrt{3}-u)\sqrt{1-u}} \quad [u \neq 1].$ BF(246.06)

12. $\int_{u}^{\infty} \dfrac{dx}{(x-1)\sqrt{x^3-1}} = \dfrac{1}{\sqrt[4]{27}}[F(\delta, \sin 15°) - 2E(\delta, \sin 15°)] +$

$+ \dfrac{2}{\sqrt{3}} \dfrac{\sqrt{1+u+u^2}}{(u-1+\sqrt{3})\sqrt{u-1}} \quad [u \neq 1].$ BF(242.03)

13. $\int_{-\infty}^{u} \dfrac{(1-x)\,dx}{(1+\sqrt{3}-x)^2 \sqrt{1-x^3}} =$

$= \dfrac{2-\sqrt{3}}{\sqrt[4]{27}}[F(\alpha, \sin 75°) - E(\alpha, \sin 75°)].$ BF(246.07)

14. $\int_{u}^{1} \dfrac{(1-x)\,dx}{(1+\sqrt{3}-x)^2 \sqrt{1-x^3}} = \dfrac{2-\sqrt{3}}{\sqrt[4]{27}}[F(\beta, \sin 75°) - E(\beta, \sin 75°)].$ BF(244.04)

15. $\int_{1}^{u} \dfrac{(x-1)\,dx}{(1+\sqrt{3}-x)^2 \sqrt{x^3-1}} = \dfrac{2(\sqrt{3}-2)}{\sqrt{3}} \dfrac{\sqrt{u^3-1}}{u^2-2u-2} - \dfrac{2-\sqrt{3}}{\sqrt[4]{27}} E(\gamma, \sin 15°).$

BF(240.08)

16. $\int_{u}^{\infty} \dfrac{(x-1)\,dx}{(1+\sqrt{3}-x)^2 \sqrt{x^3-1}} = \dfrac{2(2-\sqrt{3})}{\sqrt{3}} \dfrac{\sqrt{u^3-1}}{u^2-2u-2} - \dfrac{2-\sqrt{3}}{\sqrt[4]{27}} E(\delta, \sin 15°).$

BF(242.07)

17. $\int_{-\infty}^{u} \dfrac{(1-x)\,dx}{(1-\sqrt{3}-x)^2 \sqrt{1-x^3}} = \dfrac{2+\sqrt{3}}{\sqrt[4]{27}} \left[\dfrac{2\sqrt[4]{3}\sqrt{1-u^3}}{u^2-2u-2} - E(\alpha, \sin 75°) \right].$

BF(246.08)

18. $\int_{1}^{u} \dfrac{(x-1)\,dx}{(1-\sqrt{3}-x)^2 \sqrt{x^3-1}} = \dfrac{2+\sqrt{3}}{\sqrt[4]{27}}[F(\gamma, \sin 15°) - E(\gamma, \sin 15°)].$ BF(240.04)

19. $\int_{u}^{\infty} \dfrac{(x-1)\,dx}{(1-\sqrt{3}-x)^2 \sqrt{x^3-1}} = \dfrac{2+\sqrt{3}}{\sqrt[4]{27}}[F(\delta, \sin 15°) - E(\delta, \sin 15°)].$ BF(242.05)

20. $\int_{-\infty}^{u} \dfrac{(x^2+x+1)\,dx}{(1+\sqrt{3}-x)^2 \sqrt{1-x^3}} = \dfrac{1}{\sqrt[4]{3}} E(\alpha, \sin 75°).$ BF(246.01)

21. $\int_{u}^{1} \dfrac{(x^2+x+1)\,dx}{(x-1+\sqrt{3})^2 \sqrt{1-x^3}} = \dfrac{1}{\sqrt[4]{3}} E(\beta, \sin 75°).$ BF(244.02)

22. $\displaystyle\int_1^u \frac{(x^2+x+1)\,dx}{(\sqrt{3}+x-1)^2\sqrt{x^3-1}} = \frac{1}{\sqrt[4]{3}}\,E(\gamma,\sin 15°).$ BF(240.01)   3.139

23. $\displaystyle\int_u^\infty \frac{(x^2+x+1)\,dx}{(x-1+\sqrt{3})^2\sqrt{x^3-1}} = \frac{1}{\sqrt[4]{3}}\,E(\delta,\sin 15°).$ BF(242.01)

24. $\displaystyle\int_1^u \frac{(x-1)\,dx}{(x^2+x+1)\sqrt{x^3-1}} = \frac{4}{\sqrt[4]{27}}\,E(\gamma,\sin 15°) -$

$\qquad -\dfrac{2+\sqrt{3}}{\sqrt[4]{27}}\,F(\gamma,\sin 15°) - \dfrac{2-\sqrt{3}}{\sqrt{3}}\,\dfrac{2(u-1)(\sqrt{3}+1-u)}{(\sqrt{3}-1+u)\sqrt{u^3-1}}.$ BF(240.09)

25. $\displaystyle\int_{-\infty}^u \frac{(1+\sqrt{3}-x)^2\,dx}{[(1+\sqrt{3}-x)^2 - 4\sqrt{3}p^2(1-x)]\sqrt{1-x^3}} = \frac{1}{\sqrt[4]{3}}\,\Pi(\alpha,p^2,\sin 75°).$ BF(246.02)

26. $\displaystyle\int_u^1 \frac{(1+\sqrt{3}-x)^2\,dx}{[(1+\sqrt{3}-x)^2 - 4\sqrt{3}p^2(1-x)]\sqrt{1-x^3}} = \frac{1}{\sqrt[4]{3}}\,\Pi(\beta,p^2,\sin 75°).$ BF(244.03)

27. $\displaystyle\int_1^u \frac{(1-\sqrt{3}-x)^2\,dx}{[(1-\sqrt{3}-x)^2 - 4\sqrt{3}p^2(x-1)]\sqrt{x^3-1}} = \frac{1}{\sqrt[4]{3}}\,\Pi(\gamma,p^2,\sin 15°).$ BF(240.02)

28. $\displaystyle\int_u^\infty \frac{(1-\sqrt{3}-x)^2\,dx}{[(1-\sqrt{3}-x)^2 - 4\sqrt{3}p^2(x-1)]\sqrt{x^3-1}} = \frac{1}{\sqrt[4]{3}}\,\Pi(\delta,p^2,\sin 15°).$ BF(242.02)

In 3.141 und 3.142 werden folgende Abkürzungen verwendet: | In 3.141 and 3.142 the following notation is used:

$$\alpha = \arcsin\sqrt{\frac{a-c}{a-u}}, \quad \beta = \arcsin\sqrt{\frac{c-u}{b-u}},$$

$$\gamma = \arcsin\sqrt{\frac{u-c}{b-c}}, \quad \delta = \arcsin\sqrt{\frac{(a-c)(b-u)}{(b-c)(a-u)}}, \quad \varkappa = \arcsin\sqrt{\frac{(a-c)(u-b)}{(a-b)(u-c)}},$$

$$\lambda = \arcsin\sqrt{\frac{a-u}{a-b}}, \quad \mu = \arcsin\sqrt{\frac{u-a}{u-b}}, \quad \nu = \arcsin\sqrt{\frac{a-c}{u-c}}, \quad p = \sqrt{\frac{a-b}{a-c}},$$

$$q = \sqrt{\frac{b-c}{a-c}}.$$

1. $\displaystyle\int_u^c \sqrt{\frac{a-x}{(b-x)(c-x)}}\,dx = 2\sqrt{a-c}\,[F(\beta,p) - E(\beta,p)] +$ 3.141

$\qquad + 2\sqrt{\dfrac{(a-u)(c-u)}{b-u}}\quad [a>b>c>u].$ BF(232.06)

3.141

2. $\int_c^u \sqrt{\dfrac{a-x}{(b-x)(x-c)}}\, dx = 2\sqrt{a-c}\, E(\gamma, q) \quad [a > b \geq u > c].$ BF(233.01)

3. $\int_u^b \sqrt{\dfrac{a-x}{(b-x)(x-c)}}\, dx = 2\sqrt{a-c}\, E(\delta, q) - 2\sqrt{\dfrac{(b-u)(u-c)}{a-u}}$

$[a > b > u \geq c].$ BF(234.06)

4. $\int_b^u \sqrt{\dfrac{a-x}{(x-b)(x-c)}}\, dx = 2\sqrt{a-c}\,[F(\varkappa, p) - E(\varkappa, p)] +$

$+ 2\sqrt{\dfrac{(a-u)(u-b)}{u-c}} \quad [a \geq u > b > c].$ BF(235.07)

5. $\int_u^a \sqrt{\dfrac{a-x}{(x-b)(x-c)}}\, dx = 2\sqrt{a-c}\,[F(\lambda, p) - E(\lambda, p)]$

$[a > u \geq b > c].$ BF(236.04)

6. $\int_a^u \sqrt{\dfrac{x-a}{(x-b)(x-c)}}\, dx = -2\sqrt{a-c}\, E(\mu, q) + 2\sqrt{\dfrac{(u-a)(u-c)}{u-b}}$

$[u > a > b > c].$ BF(237.03)

7. $\int_u^c \sqrt{\dfrac{b-x}{(a-x)(c-x)}}\, dx = \dfrac{2(b-c)}{\sqrt{a-c}}\, F(\beta, p) - 2\sqrt{a-c}\, E(\beta, p) +$

$+ 2\sqrt{\dfrac{(a-u)(c-u)}{b-u}} \quad [a > b > c > u].$ BF(232.07)

8. $\int_c^u \sqrt{\dfrac{b-x}{(a-x)(x-c)}}\, dx = 2\sqrt{a-c}\, E(\gamma, q) - \dfrac{2(a-b)}{\sqrt{a-c}}\, F(\gamma, q)$

$[a > b \geq u > c].$ BF(233.04)

9. $\int_u^b \sqrt{\dfrac{b-x}{(a-x)(x-c)}}\, dx = 2\sqrt{a-c}\, E(\delta, q) - \dfrac{2(a-b)}{\sqrt{a-c}}\, F(\delta, q) -$

$- 2\sqrt{\dfrac{(b-u)(u-c)}{a-u}} \quad [a > b > u \geq c].$ BF(234.07)

10. $\int_b^u \sqrt{\dfrac{x-b}{(a-x)(x-c)}}\, dx = 2\sqrt{a-c}\, E(\varkappa, p) - \dfrac{2(b-c)}{\sqrt{a-c}}\, F(\varkappa, p) -$

$- 2\sqrt{\dfrac{(a-u)(u-b)}{u-c}} \quad [a \geq u > b > c].$ BF(235.06)

11. $\displaystyle\int_u^a \sqrt{\dfrac{x-b}{(a-x)(x-c)}}\,dx = 2\sqrt{a-c}\,E(\lambda, p) - \dfrac{2(b-c)}{\sqrt{a-c}}\,F(\lambda, p)$ 3.141

$[a > u \geq b > c]$. BF(236.03)

12. $\displaystyle\int_a^u \sqrt{\dfrac{x-b}{(x-a)(x-c)}}\,dx = \dfrac{2(a-b)}{\sqrt{a-c}}\,F(\mu, q) - 2\sqrt{a-c}\,E(\mu, q) +$

$+ 2\sqrt{\dfrac{(u-a)(u-c)}{u-b}}\quad [u > a > b > c]$. BF(237.04)

13. $\displaystyle\int_u^c \sqrt{\dfrac{c-x}{(a-x)(b-x)}}\,dx = -2\sqrt{a-c}\,E(\beta, p) +$

$+ 2\sqrt{\dfrac{(a-u)(c-u)}{b-u}}\quad [a > b > c > u]$. BF(232.08)

14. $\displaystyle\int_c^u \sqrt{\dfrac{x-c}{(a-x)(b-x)}}\,dx = 2\sqrt{a-c}\,[F(\gamma, q) - E(\gamma, q)]\quad [a > b \geq u > c]$.

BF(233.03)

15. $\displaystyle\int_u^b \sqrt{\dfrac{x-c}{(a-x)(b-x)}}\,dx = 2\sqrt{a-c}\,[F(\delta, q) - E(\delta, q)] +$

$+ 2\sqrt{\dfrac{(b-u)(u-c)}{a-u}}\quad [a > b > u \geq c]$. BF(234.08)

16. $\displaystyle\int_b^u \sqrt{\dfrac{x-c}{(a-x)(x-b)}}\,dx = 2\sqrt{a-c}\,E(\varkappa, p) - 2\sqrt{\dfrac{(a-u)(u-b)}{u-c}}$

$[a \geq u > b > c]$. BF(235.07)

17. $\displaystyle\int_u^a \sqrt{\dfrac{x-c}{(a-x)(x-b)}}\,dx = 2\sqrt{a-c}\,E(\lambda, p)\quad [a > u \geq b > c]$. BF(236.01)

18. $\displaystyle\int_a^u \sqrt{\dfrac{x-c}{(x-a)(x-b)}}\,dx = 2\sqrt{a-c}\,[F(\mu, q) - E(\mu, q)] +$

$+ 2\sqrt{\dfrac{(u-a)(u-c)}{u-b}}\quad [u > a > b > c]$. BF(237.05)

19. $\displaystyle\int_u^c \sqrt{\dfrac{(b-x)(c-x)}{a-x}}\,dx = \dfrac{2}{3}\sqrt{a-c}\,[(2a - b - c)\,E(\beta, p) -$

$- (b-c)\,F(\beta, p)] + \dfrac{2}{3}(2b - 2a + c - u)\sqrt{\dfrac{(a-u)(c-u)}{b-u}}$

$[a > b > c > u]$. BF(232.11)

3.141

20. $$\int_c^u \sqrt{\frac{(x-c)(b-x)}{a-x}}\,dx = \frac{2}{3}\sqrt{a-c}[(2a-b-c)E(\gamma,q) -$$
$$-2(a-b)F(\gamma,q)] - \frac{2}{3}\sqrt{(a-u)(b-u)(u-c)}$$
$$[a > b \geqslant u > c]. \qquad \text{BF(233.06)}$$

21. $$\int_u^b \sqrt{\frac{(x-c)(b-x)}{a-x}}\,dx = \frac{2}{3}\sqrt{a-c}[2(b-a)F(\delta,q) +$$
$$+ (2a-b-c)E(\delta,q)] + \frac{2}{3}(2c-b-u)\sqrt{\frac{(b-u)(u-c)}{a-u}}$$
$$[a > b > u \geqslant c]. \qquad \text{BF(234.11)}$$

22. $$\int_b^u \sqrt{\frac{(x-b)(x-c)}{a-x}}\,dx = \frac{2}{3}\sqrt{a-c}[(2a-b-c)E(\varkappa,p) -$$
$$-(b-c)F(\varkappa,p)] + \frac{2}{3}(b+2c-2a-u)\sqrt{\frac{(a-u)(u-b)}{u-c}}$$
$$[a \geqslant u > b > c]. \qquad \text{BF(235.10)}$$

23. $$\int_u^a \sqrt{\frac{(x-b)(x-c)}{a-x}}\,dx = \frac{2}{3}\sqrt{a-c}[(2a-b-c)E(\lambda,p) -$$
$$-(b-c)F(\lambda,p)] + \frac{2}{3}\sqrt{(a-u)(u-b)(u-c)} \quad [a > u \geqslant b > c].$$
$$\text{BF(236.07)}$$

24. $$\int_a^u \sqrt{\frac{(x-b)(x-c)}{x-a}}\,dx = \frac{2}{3}\sqrt{a-c}[2(a-b)F(\mu,q) +$$
$$+ (b+c-2a)E(\mu,q)] + \frac{2}{3}(u+2a-2b-c)\sqrt{\frac{(u-a)(u-b)}{u-c}}$$
$$[u > a > b > c]. \qquad \text{BF(237.08)}$$

25. $$\int_u^c \sqrt{\frac{(a-x)(c-x)}{b-x}}\,dx = \frac{2}{3}\sqrt{a-c}[(2b-a-c)E(\beta,p) -$$
$$-(b-c)F(\beta,p)] + \frac{2}{3}(a+c-b-u)\sqrt{\frac{(a-u)(c-u)}{b-u}}$$
$$[a > b > c > u]. \qquad \text{BF(232.10)}$$

26. $$\int_c^u \sqrt{\frac{(a-x)(x-c)}{b-x}}\,dx = \frac{2}{3}\sqrt{a-c}[(2b-a-c)E(\gamma,q) +$$
$$+ (a-b)F(\gamma,q)] - \frac{2}{3}\sqrt{(a-u)(b-u)(u-c)}$$
$$[a > b \geqslant u > c]. \qquad \text{BF(233.05)}$$

27. $\int_u^b \sqrt{\frac{(a-x)(x-c)}{b-x}}\,dx = \frac{2}{3}\sqrt{a-c}\,[(a-b)\,F(\delta, q) +$

$+ (2b - a - c)\,E(\delta, q)] + \frac{2}{3}(2a + c - 2b - u)\sqrt{\frac{(b-u)(u-c)}{a-u}}$

$[a > b > u \geqslant c]$. BF(234.10)

28. $\int_b^u \sqrt{\frac{(a-x)(x-c)}{x-b}}\,dx = \frac{2}{3}\sqrt{a-c}\,[(b-c)\,F(\varkappa, p) +$

$+ (a + c - 2b)\,E(\varkappa, p)] + \frac{2}{3}(2b - a - 2c + u)\sqrt{\frac{(a-u)(u-b)}{u-c}}$

$[a \geqslant u > b > c]$. BF(235.11)

29. $\int_u^a \sqrt{\frac{(a-x)(x-c)}{x-b}}\,dx = \frac{2}{3}\sqrt{a-c}\,[(a + c - 2b)E(\lambda, p) +$

$+ (b-c)\,F(\lambda, p)] - \frac{2}{3}\sqrt{(a-u)(u-b)(u-c)}$

$[a > u \geqslant b > c]$. BF(236.06)

30. $\int_a^u \sqrt{\frac{(x-a)(x-c)}{x-b}}\,dx = \frac{2}{3}\frac{\sqrt{(a-c)^3}}{b-c}\,[(a+c-2b)E(\mu, q) -$

$- (a-b)\,F(\mu, q)] + \frac{2}{3}\frac{a-c}{b-c}(u + b - a - c)\sqrt{\frac{(u-a)(u-c)}{u-b}}$

$[u > a > b > c]$. BF(237.06)

31. $\int_u^c \sqrt{\frac{(a-x)(b-x)}{c-x}}\,dx = \frac{2}{3}\sqrt{a-c}\,[2(b-c)F(\beta, p) +$

$+ (2c - a - b)E(\beta, p)] + \frac{2}{3}(a + 2b - 2c - u)\sqrt{\frac{(a-u)(c-u)}{b-u}}$

$[a > b > c > u]$. BF(232.09)

32. $\int_c^u \sqrt{\frac{(a-x)(b-x)}{x-c}}\,dx = \frac{2}{3}\sqrt{a-c}\,[(a + b - 2c)E(\gamma, q) -$

$-(a-b)F(\gamma, q)] + \frac{2}{3}\sqrt{(a-u)(b-u)(u-c)}$ $[a > b \geqslant u > c]$. BF(233.07)

33. $\int_u^b \sqrt{\frac{(a-x)(b-x)}{x-c}}\,dx = \frac{2}{3}\sqrt{a-c}\,[(a + b - 2c)\,E(\delta, q) -$

$-(a-b)F(\delta, q)] + \frac{2}{3}(2c - 2a - b + u)\sqrt{\frac{(b-u)(u-c)}{a-u}}$

$[a > b > u \geqslant c]$. BF(234.09)

3.141

34. $\int_b^u \sqrt{\dfrac{(a-x)(x-b)}{x-c}}\,dx = \dfrac{2}{3}\sqrt{a-c}\,[(a+b-2c)E(\varkappa, p)] -$

$\qquad - 2(b-c)F(\varkappa, p)] + \dfrac{2}{3}(u+c-a-b)\sqrt{\dfrac{(a-u)(u-b)}{u-c}}$

$\qquad\qquad\qquad [a \geqslant u > b > c].\qquad \text{BF(235.09)}$

35. $\int_u^a \sqrt{\dfrac{(a-x)(x-b)}{x-c}}\,dx = \dfrac{2}{3}\sqrt{a-c}\,[(a+b-2c)E(\lambda, p) -$

$\qquad - 2(b-c)F(\lambda, p)] - \dfrac{2}{3}\sqrt{(a-u)(u-b)(u-c)}$

$\qquad\qquad\qquad [a > u \geqslant b > c].\qquad \text{BF(236.05)}$

36. $\int_a^u \sqrt{\dfrac{(x-a)(x-b)}{x-c}}\,dx = \dfrac{2}{3}\sqrt{a-c}\,[(a+b-2c)E(\mu, q) -$

$\qquad -(a-b)F(\mu, q)] + \dfrac{2}{3}(u+2c-a-2b)\sqrt{\dfrac{(u-a)(u-c)}{u-b}}$

$\qquad\qquad\qquad [u > a > b > c].\qquad \text{BF(237.07)}$

3.142

1. $\int_{-\infty}^u \sqrt{\dfrac{a-x}{(b-x)(c-x)^3}}\,dx = \dfrac{2}{\sqrt{a-c}}F(\alpha, p) - \dfrac{2\sqrt{a-c}}{b-c}E(\alpha, p) +$

$\qquad + \dfrac{2(a-c)}{b-c}\sqrt{\dfrac{b-u}{(a-u)(c-u)}}\qquad [a > b > c > u].\qquad \text{BF(231.05)}$

2. $\int_u^b \sqrt{\dfrac{a-x}{(b-x)(x-c)^3}}\,dx = 2\dfrac{a-b}{(b-c)\sqrt{a-c}}F(\delta, q) -$

$\qquad - \dfrac{2\sqrt{a-c}}{b-c}E(\delta, q) + 2\dfrac{a-c}{b-c}\sqrt{\dfrac{b-u}{(a-u)(u-c)}}$

$\qquad\qquad\qquad [a > b > u > c].\qquad \text{BF(234.13)}$

3. $\int_b^u \sqrt{\dfrac{a-x}{(x-b)(x-c)^3}}\,dx = \dfrac{2\sqrt{a-c}}{b-c}E(\varkappa, p) - \dfrac{2}{\sqrt{a-c}}F(\varkappa, p)$

$\qquad\qquad\qquad [a \geqslant u > b > c].\qquad \text{BF(235.12)}$

4. $\int_u^a \sqrt{\dfrac{a-x}{(x-b)(x-c)^3}}\,dx = \dfrac{2\sqrt{a-c}}{b-c}E(\lambda, p) -$

$\qquad - \dfrac{2}{\sqrt{a-c}}F(\lambda, p) - \dfrac{2}{b-c}\sqrt{\dfrac{(a-u)(u-b)}{u-c}}$

$\qquad\qquad\qquad [a > u \geqslant b > c].\qquad \text{BF(236.12)}$

5. $\displaystyle\int_a^u \sqrt{\frac{x-a}{(x-b)(x-c)^3}}\,dx = \frac{2\sqrt{a-c}}{b-c}E(\mu,q) - \frac{2(a-b)}{(b-c)\sqrt{a-c}}F(\mu,q) -$ 3.142

$\displaystyle\qquad\qquad -2\sqrt{\frac{u-a}{(u-b)(u-c)}}\quad [u>a>b>c].$ BF(237.10)

6. $\displaystyle\int_u^\infty \sqrt{\frac{x-a}{(x-b)(x-c)^3}}\,dx = \frac{2\sqrt{a-c}}{b-c}E(\nu,q) -$

$\displaystyle\qquad\qquad -\frac{2(a-b)}{(b-c)\sqrt{a-c}}F(\nu,q)\quad [u\geqslant a>b>c].$ BF(238.09)

7. $\displaystyle\int_{-\infty}^u \sqrt{\frac{a-x}{(b-x)^3(c-x)}}\,dx = \frac{2\sqrt{a-c}}{b-c}E(\alpha,p) -$

$\displaystyle\qquad\qquad -2\frac{a-b}{b-c}\sqrt{\frac{c-u}{(a-u)(b-u)}}\quad [a>b>c\geqslant u].$ BF(231.03)

8. $\displaystyle\int_u^c \sqrt{\frac{a-x}{(b-x)^3(c-x)}}\,dx = \frac{2\sqrt{a-c}}{b-c}E(\beta,p)\quad [a>b>c>u].$ BF(232.01)

9. $\displaystyle\int_c^u \sqrt{\frac{a-x}{(b-x)^3(x-c)}}\,dx = \frac{2\sqrt{a-c}}{b-c}[F(\gamma,q)-E(\gamma,q)] +$

$\displaystyle\qquad\qquad +\frac{2}{b-c}\sqrt{\frac{(a-u)(u-c)}{b-u}}\quad [a>b>u>c].$ BF(233.15)

10. $\displaystyle\int_u^a \sqrt{\frac{a-x}{(x-b)^3(x-c)}}\,dx = \frac{2\sqrt{a-c}}{c-b}E(\lambda,p) +$

$\displaystyle\qquad\qquad +\frac{2}{b-c}\sqrt{\frac{(a-u)(u-c)}{u-b}}\quad [a>u>b>c].$ BF(236.11)

11. $\displaystyle\int_a^u \sqrt{\frac{x-a}{(x-b)^3(x-c)}}\,dx = \frac{2\sqrt{a-c}}{b-c}[F(\mu,q)-E(\mu,q)]\quad [u>a>b>c].$

BF(237.09)

12. $\displaystyle\int_u^\infty \sqrt{\frac{x-a}{(x-b)^3(x-c)}}\,dx = \frac{2\sqrt{a-c}}{b-c}[F(\nu,q)-E(\nu,q)] +$

$\displaystyle\qquad\qquad +2\sqrt{\frac{u-a}{(u-b)(u-c)}}\quad [u\geqslant a>b>c].$ BF(238.10)

13. $\displaystyle\int_{-\infty}^u \sqrt{\frac{b-x}{(a-x)^3(c-x)}}\,dx = \frac{2}{\sqrt{a-c}}E(\alpha,p)\quad [a>b>c\geqslant u].$ BF(231.01)

3.142

14. $\displaystyle\int_u^c \sqrt{\frac{b-x}{(a-x)^3(c-x)}}\,dx = \frac{2}{\sqrt{a-c}}E(\beta,\,p) -$

$\displaystyle -\frac{2(a-b)}{a-c}\sqrt{\frac{c-u}{(a-u)(b-u)}}\qquad [a>b>c>u].$ BF(232.05)

15. $\displaystyle\int_c^u \sqrt{\frac{b-x}{(a-x)^3(x-c)}}\,dx = \frac{2}{\sqrt{a-c}}[F(\gamma,\,q)-E(\gamma,\,q)] +$

$\displaystyle +\frac{2}{a-c}\sqrt{\frac{(b-u)(u-c)}{a-u}}\qquad [a>b\geqslant u>c].$ BF(233.13)

16. $\displaystyle\int_u^b \sqrt{\frac{b-x}{(a-x)^3(x-c)}}\,dx = \frac{2}{\sqrt{a-c}}[F(\delta,\,q)-E(\delta,\,q)]\qquad [a>b>u\geqslant c].$

BF(234.15)

17. $\displaystyle\int_b^u \sqrt{\frac{x-b}{(a-x)^3(x-c)}}\,dx = -\frac{2}{\sqrt{a-c}}E(\varkappa,\,p) +$

$\displaystyle +2\sqrt{\frac{u-b}{(a-u)(u-c)}}\qquad [a>u>b>c].$ BF(235.08)

18. $\displaystyle\int_u^\infty \sqrt{\frac{x-b}{(x-a)^3(x-c)}}\,dx = \frac{2}{\sqrt{a-c}}[F(\nu,\,q)-E(\nu,\,q)] +$

$\displaystyle +2\sqrt{\frac{u-b}{(u-a)(u-c)}}\qquad [u>a>b>c].$ BF(238.07)

19. $\displaystyle\int_{-\infty}^u \sqrt{\frac{b-x}{(a-x)(c-x)^3}}\,dx = \frac{2}{\sqrt{a-c}}[F(\alpha,\,p)-E(\alpha,\,p)] +$

$\displaystyle +2\sqrt{\frac{b-u}{(a-u)(c-u)}}\qquad [a>b>c>u].$ BF(231.04)

20. $\displaystyle\int_u^b \sqrt{\frac{b-x}{(a-x)(x-c)^3}}\,dx = -\frac{2}{\sqrt{a-c}}E(\delta,\,q) +$

$\displaystyle +2\sqrt{\frac{b-u}{(a-u)(u-c)}}\qquad [a>b>u>c].$ BF(234.14)

21. $\displaystyle\int_b^u \sqrt{\frac{x-b}{(a-x)(x-c)^3}}\,dx = \frac{2}{\sqrt{a-c}}[F(\varkappa,\,p)-E(\varkappa,\,p)]\qquad [a\geqslant u>b>c].$

BF(235.03)

22. $\displaystyle\int_u^a \sqrt{\frac{x-b}{(a-x)(x-c)^3}}\,dx = \frac{2}{\sqrt{a-c}}[F(\lambda,\,p)-E(\lambda,\,p)] +$

$\displaystyle +\frac{2}{a-c}\sqrt{\frac{(a-u)(u-b)}{u-c}}\qquad [a>u\geqslant b>c].$ BF(236.14)

23. $\int_a^u \sqrt{\frac{x-b}{(x-a)(x-c)^3}}\, dx = \frac{2}{\sqrt{a-c}} E(\mu,\ q) -$ 3.142

$$- 2\frac{b-c}{a-c}\sqrt{\frac{u-a}{(u-b)(u-c)}} \quad [u > a > b > c]. \quad \text{BF(237.11)}$$

24. $\int_u^\infty \sqrt{\frac{x-b}{(x-a)(x-c)^3}}\, dx = \frac{2}{\sqrt{a-c}} E(\nu,\ q) \quad [u \geq a > b > c]. \quad \text{BF(238.01)}$

25. $\int_{-\infty}^u \sqrt{\frac{c-x}{(a-x)^3(b-x)}}\, dx = \frac{2\sqrt{a-c}}{a-b} E(\alpha,\ p) - \frac{2(b-c)}{(a-b)\sqrt{a-c}} F(\alpha,\ p)$

$$[a > b > c \geq u]. \quad \text{BF(231.07)}$$

26. $\int_u^c \sqrt{\frac{c-x}{(a-x)^3(b-x)}}\, dx = \frac{2\sqrt{a-c}}{a-b} E(\beta,\ p) -$

$$- \frac{2(b-c)}{(a-b)\sqrt{a-c}} F(\beta,\ p) - 2\sqrt{\frac{c-u}{(a-u)(b-u)}} \quad [a > b > c > u]. \quad \text{BF(232.03)}$$

27. $\int_c^u \sqrt{\frac{x-c}{(a-x)^3(b-x)}}\, dx = \frac{2\sqrt{a-c}}{a-b} E(\gamma,\ q) -$

$$- \frac{2}{\sqrt{a-c}} F(\gamma,\ q) - \frac{2}{a-b}\sqrt{\frac{(b-u)(u-c)}{a-u}} \quad [a > b \geq u > c]. \quad \text{BF(233.14)}$$

28. $\int_u^b \sqrt{\frac{x-c}{(a-x)^3(b-x)}}\, dx = \frac{2\sqrt{a-c}}{a-b} E(\delta,\ q) - \frac{2}{\sqrt{a-c}} F(\delta,\ q)$

$$[a > b > u \geq c]. \quad \text{BF(234.20)}$$

29. $\int_b^u \sqrt{\frac{x-c}{(a-x)^3(x-b)}}\, dx = \frac{2(b-c)}{(a-b)\sqrt{a-c}} F(\varkappa,\ p) -$

$$- \frac{2\sqrt{a-c}}{a-b} E(\varkappa,\ p) + 2\frac{a-c}{a-b}\sqrt{\frac{u-b}{(a-u)(u-c)}} \quad [a > u > b > c]. \quad \text{BF(235.13)}$$

30. $\int_u^\infty \sqrt{\frac{x-c}{(x-a)^3(x-b)}}\, dx = \frac{2}{\sqrt{a-c}} F(\nu,\ q) - \frac{2\sqrt{a-c}}{a-b} E(\nu,\ q) +$

$$+ \frac{2(a-c)}{a-b}\sqrt{\frac{u-b}{(u-a)(u-c)}} \quad [u > a > b > c]. \quad \text{BF(238.08)}$$

31. $\int_{-\infty}^u \sqrt{\frac{c-x}{(a-x)(b-x)^3}}\, dx = \frac{2\sqrt{a-c}}{a-b}[F(\alpha,\ p) - E(\alpha,\ p)] +$

$$+ 2\sqrt{\frac{c-u}{(a-u)(b-u)}} \quad [a > b > c \geq u]. \quad \text{BF(231.06)}$$

3.142

32. $\int_u^c \sqrt{\dfrac{c-x}{(a-x)(b-x)^3}}\, dx = \dfrac{2\sqrt{a-c}}{a-b}\,[F(\beta, p) - E(\beta, p)]\quad [a > b > c > u].$ BF(232.04)

33. $\int_c^u \sqrt{\dfrac{x-c}{(a-x)(b-x)^3}}\, dx = -\dfrac{2\sqrt{a-c}}{a-b}\,E(\gamma, q) +$

$\qquad\qquad + \dfrac{2}{a-b}\sqrt{\dfrac{(a-u)(u-c)}{b-u}}\quad [a > b > u > c].$ BF(233.16)

34. $\int_u^a \sqrt{\dfrac{x-c}{(a-x)(x-b)^3}}\, dx = \dfrac{2\sqrt{a-c}}{a-b}\,[F(\lambda, p) - E(\lambda, p)] +$

$\qquad\qquad + \dfrac{2}{a-b}\sqrt{\dfrac{(a-u)(u-c)}{u-b}}\quad [a > u > b > c].$ BF(236.13)

35. $\int_a^u \sqrt{\dfrac{x-c}{(x-a)(x-b)^3}}\, dx = \dfrac{2\sqrt{a-c}}{a-b}\,E(\mu, q)\quad [u > a > b > c].$ BF(237.01)

36. $\int_u^\infty \sqrt{\dfrac{x-c}{(x-a)(x-b)^3}}\, dx = \dfrac{2\sqrt{a-c}}{a-b}\,E(\nu, q) -$

$\qquad\qquad - 2\,\dfrac{b-c}{a-b}\sqrt{\dfrac{u-a}{(u-b)(u-c)}}\quad [u \geqslant a > b > c].$ BF(238.11)

3.143

1. $\int_u^1 \dfrac{dx}{\sqrt{1+x^4}} = \dfrac{1}{2}\,F\left(\arccos\dfrac{u\sqrt{2}}{\sqrt{1+u^4}},\ \sin 80°7'15''\right)^{*}.$ Z 66(286)

2. $\int_u^\infty \dfrac{dx}{\sqrt{1+x^4}} = \dfrac{1}{2}\,F\left(\arccos\dfrac{u^2-1}{u^2+1},\ \dfrac{\sqrt{2}}{2}\right).$ Z 66(287)

3.144

Abkürzung: $\alpha = \arcsin\dfrac{1}{\sqrt{u^2 - u + 1}}.$
Abbreviation:

1. $\int_u^\infty \dfrac{dx}{\sqrt{x(x-1)(x^2-x+1)}} = F\left(\alpha, \dfrac{\sqrt{3}}{2}\right)\quad [u \geqslant 1].$ BF(261.50)

2. $\int_u^\infty \dfrac{dx}{\sqrt{x^3(x-1)^3(x^2-x+1)}} = \dfrac{2(2u-1)}{\sqrt{u(u-1)(u^2-u+1)}} - 4E\left(\alpha, \dfrac{\sqrt{3}}{2}\right)\quad [u > 1].$ BF(261.54)

3. $\int_u^\infty \dfrac{(2x-1)^2\, dx}{\sqrt{x^3(x-1)^3(x^2-x+1)}} = 4\left[F\left(\alpha, \dfrac{\sqrt{3}}{2}\right) - E\left(\alpha, \dfrac{\sqrt{3}}{2}\right) +\right.$

$\qquad\qquad \left. + \dfrac{2u-1}{2\sqrt{u(u-1)(u^2-u+1)}}\right]\quad [u > 1].$ BF(261.56)

4. $\int_u^\infty \dfrac{dx}{\sqrt{x(x-1)(x^2-x+1)^3}} = \dfrac{4}{3}\left[F\left(\alpha, \dfrac{\sqrt{3}}{2}\right) - E\left(\alpha, \dfrac{\sqrt{3}}{2}\right)\right]\quad [u \geqslant 1].$ BF(261.52)

---

* $\sin 80°7'15'' = 2\sqrt[4]{2}(\sqrt{2}-1) = 0{,}985\,171\ldots$

5. $\int_u^\infty \frac{(2x-1)^2 \, dx}{\sqrt{x(x-1)(x^2-x+1)^3}} = 4E\left(\alpha, \frac{\sqrt{3}}{2}\right)$ $\quad [u > 1]$. $\qquad$ BF(261.51) $\quad$ 3.144

6. $\int_u^\infty \sqrt{\frac{x(x-1)}{(x^2-x+1)^3}} \, dx = \frac{4}{3} E\left(\alpha, \frac{\sqrt{3}}{2}\right) - \frac{1}{3} F\left(\alpha, \frac{\sqrt{3}}{2}\right)$ $\quad [u > 1]$. $\qquad$ BF(261.53)

7. $\int_u^\infty \frac{dx}{(2x-1)^2} \sqrt{\frac{x(x-1)}{x^2-x+1}} = \frac{1}{3}\left[F\left(\alpha, \frac{\sqrt{3}}{2}\right) - E\left(\alpha, \frac{\sqrt{3}}{2}\right)\right] +$

$\qquad + \frac{1}{2(2u-1)} \sqrt{\frac{u(u-1)}{u^2-u+1}}$ $\quad [u > 1]$. $\qquad$ BF(261.57)

8. $\int_u^\infty \frac{dx}{(2x-1)^2} \sqrt{\frac{x^2-x+1}{x(x-1)}} = E\left(\alpha, \frac{\sqrt{3}}{2}\right) - \frac{3}{2(2u-1)} \sqrt{\frac{u(u-1)}{u^2-u+1}}$ $\quad [u > 1]$. $\qquad$ BF(261.58)

9. $\int_u^\infty \frac{dx}{(2x-1)^2 \sqrt{x(x-1)(x^2-x+1)}} = \frac{4}{3} E\left(\alpha, \frac{\sqrt{3}}{2}\right) - \frac{1}{3} F\left(\alpha, \frac{\sqrt{3}}{2}\right) -$

$\qquad - \frac{2}{2u-1} \sqrt{\frac{u(u-1)}{u^2-u+1}}$ $\quad [u > 1]$. $\qquad$ BF(261.55)

10. $\int_u^\infty \frac{dx}{\sqrt{x^5(x-1)^5(x^2-x+1)}} = \frac{40}{3} E\left(\alpha, \frac{\sqrt{3}}{2}\right) - \frac{4}{3} F\left(\alpha, \frac{\sqrt{3}}{2}\right) -$

$\qquad - \frac{2(2u-1)(9u^2-9u-1)}{3\sqrt{u^3(u-1)^3(u^2-u+1)}}$ $\quad [u > 1]$. $\qquad$ BF(261.55)

11. $\int_u^\infty \frac{dx}{\sqrt{x(x-1)(x^2-x+1)^5}} = \frac{44}{27} F\left(\alpha, \frac{\sqrt{3}}{2}\right) - \frac{56}{27} E\left(\alpha, \frac{\sqrt{3}}{2}\right) +$

$\qquad + \frac{2(2u-1)\sqrt{u(u-1)}}{9\sqrt{(u^2-u+1)^3}}$ $\quad [u > 1]$. $\qquad$ BF(261.52)

12. $\int_u^\infty \frac{dx}{(2x-1)^4 \sqrt{x(x-1)(x^2-x+1)}} = \frac{16}{27} E\left(\alpha, \frac{\sqrt{3}}{2}\right) -$

$\qquad - \frac{1}{27} F\left(\alpha, \frac{\sqrt{3}}{2}\right) - \frac{8(5u^2-5u+2)}{9(2u-1)^3} \sqrt{\frac{u(u-1)}{u^2-u+1}}$ $\quad [u > 1]$. $\qquad$ BF(261.55)

1. $\int_\alpha^u \frac{dx}{\sqrt{(x-\alpha)(x-\beta)[(x-m)^2+n^2]}} =$ $\qquad\qquad\qquad\qquad\qquad$ 3.145

$\qquad = \frac{1}{\sqrt{pq}} F\left(2 \arctan \sqrt{\frac{q(u-\alpha)}{p(u-\beta)}}, \frac{1}{2} \sqrt{\frac{(p+q)^2+(\alpha-\beta)^2}{pq}}\right)$ $\quad [\beta < \alpha < u]$.

3.145

2. $\int_{\beta}^{u} \dfrac{dx}{\sqrt{(\alpha - x)(x - \beta)[(x - m)^2 + n^2]}} =$

$= \dfrac{1}{\sqrt{pq}} F\left(2 \operatorname{arcctg} \sqrt{\dfrac{q(\alpha - u)}{p(u - \beta)}}, \dfrac{1}{2} \sqrt{\dfrac{-(p - q)^2 + (\alpha - \beta)^2}{pq}}\right)$ $[\beta < u < \alpha].$

3. $\int_{u}^{\beta} \dfrac{dx}{\sqrt{(x - \alpha)(x - \beta)[(x - m)^2 + n^2]}} =$

$= \dfrac{1}{\sqrt{pq}} F\left(2 \operatorname{arctg} \sqrt{\dfrac{q(\beta - u)}{p(\alpha - u)}}, \dfrac{1}{2} \sqrt{\dfrac{(p + q)^2 + (\alpha - \beta)^2}{pq}}\right)$ $[u < \beta < \alpha]$

mit | where

$(m - \alpha)^2 + n^2 = p^2, \ (m - \beta)^2 + n^2 = q^{2*}.$

4. Es sei | 4. Let

$(m_1 - m)^2 + (n_1 + n)^2 = p^2, \ (m_1 - m)^2 + (n_1 - n)^2 = p_1^2,$

$\operatorname{ctg} \alpha = \sqrt{\dfrac{(p + p_1)^2 - 4n^2}{4n^2 - (p - p_1)^2}};$

dann ist | then

$\int_{m - n \operatorname{tg} \alpha}^{u} \dfrac{dx}{\sqrt{[(x - m)^2 + n^2][(x - m_1)^2 + n_1^2]}} =$

$= \dfrac{2}{p + p_1} F\left(\alpha + \operatorname{arctg} \dfrac{u - m}{n}, \dfrac{2\sqrt{pp_1}}{p + p_1}\right)$ $[m - n \operatorname{tg} \alpha < u < m + n \operatorname{ctg} \alpha].$

3.146

1. $\int_{0}^{1} \dfrac{1}{1 + x^4} \dfrac{dx}{\sqrt{1 - x^4}} = \dfrac{\pi}{8} + \dfrac{1}{4} \sqrt{2} K\left(\dfrac{\sqrt{2}}{2}\right).$ BH[13](6)

2. $\int_{0}^{1} \dfrac{x^2}{1 + x^4} \dfrac{dx}{\sqrt{1 - x^4}} = \dfrac{\pi}{8}.$ BH[13](7)

3. $\int_{0}^{1} \dfrac{x^4}{1 + x^4} \dfrac{dx}{\sqrt{1 - x^4}} = -\dfrac{\pi}{8} + \dfrac{1}{4} \sqrt{2} K\left(\dfrac{\sqrt{2}}{2}\right).$ BH[13](8)

In 3.147—3.151 werden folgende Abkürzungen verwendet: | In 3.147—3.151 the following notation is used:

$\alpha = \arcsin \sqrt{\dfrac{(a - c)(d - u)}{(a - d)(c - u)}},$

---

* Im Fall $\alpha + \beta = 2m$ gelten die Formeln 3.145 nicht; dann kann man $x - m = z$ substituieren, was auf eine der Formeln 3.152 führt. | * For $\alpha + \beta = 2m$ formulae 3.145 do not hold. In this case the substitution $x - m = z$ may be used, which leads to one of the formulae 3.152.

$$\beta = \arcsin\sqrt{\frac{(a-c)(u-d)}{(c-d)(a-u)}}, \quad \gamma = \arcsin\sqrt{\frac{(b-d)(c-u)}{(c-d)(b-u)}},$$

$$\delta = \arcsin\sqrt{\frac{(b-d)(u-c)}{(b-c)(u-d)}}, \quad \varkappa = \arcsin\sqrt{\frac{(a-c)(b-u)}{(b-c)(a-u)}},$$

$$\lambda = \arcsin\sqrt{\frac{(a-c)(u-b)}{(a-b)(u-c)}}, \quad \mu = \arcsin\sqrt{\frac{(b-d)(a-u)}{(a-b)(u-d)}},$$

$$\nu = \arcsin\sqrt{\frac{(b-d)(u-a)}{(a-d)(u-b)}}, \quad q = \sqrt{\frac{(b-c)(a-d)}{(a-c)(b-d)}}, \quad r = \sqrt{\frac{(a-b)(c-d)}{(a-c)(b-d)}}.$$

1. $\displaystyle\int_u^d \frac{dx}{\sqrt{(a-x)(b-x)(c-x)(d-x)}} = \frac{2}{\sqrt{(a-c)(b-d)}} F(\alpha, q)$ 3.147

$[a > b > c > d > u]$. BF(251.00)

2. $\displaystyle\int_d^u \frac{dx}{\sqrt{(a-x)(b-x)(c-x)(x-d)}} = \frac{2}{\sqrt{(a-c)(b-d)}} F(\beta, r)$

$[a > b > c \geqslant u > d]$. BF(252.00)

3. $\displaystyle\int_u^c \frac{dx}{\sqrt{(a-x)(b-x)(c-x)(x-d)}} = \frac{2}{\sqrt{(a-c)(b-d)}} F(\gamma, r)$

$[a > b > c > u \geqslant d]$. BF(253.00)

4. $\displaystyle\int_c^u \frac{dx}{\sqrt{(a-x)(b-x)(x-c)(x-d)}} = \frac{2}{\sqrt{(a-c)(b-d)}} F(\delta, q)$

$[a > b \geqslant u > c > d]$. BF(254.00)

5. $\displaystyle\int_u^b \frac{dx}{\sqrt{(a-x)(b-x)(x-c)(x-d)}} = \frac{2}{\sqrt{(a-c)(b-d)}} F(\varkappa, q)$

$[a > b > u \geqslant c > d]$. BF(255.00)

6. $\displaystyle\int_b^u \frac{dx}{\sqrt{(a-x)(x-b)(x-c)(x-d)}} = \frac{2}{\sqrt{(a-c)(b-d)}} F(\lambda, r)$

$[a \geqslant u > b > c > d]$. BF(256.00)

7. $\displaystyle\int_u^a \frac{dx}{\sqrt{(a-x)(x-b)(x-c)(x-d)}} = \frac{2}{\sqrt{(a-c)(b-d)}} F(\mu, r)$

$[a > u \geqslant b > c > d]$. BF(257.00)

3.147 8. $$\int_a^u \frac{dx}{\sqrt{(x-a)(x-b)(x-c)(x-d)}} = \frac{2}{\sqrt{(a-c)(b-d)}} F(v, q)$$

$[u > a > b > c > d].$ BF(258.00)

3.148 1. $$\int_u^d \frac{x\, dx}{\sqrt{(a-x)(b-x)(c-x)(d-x)}} = \frac{2}{\sqrt{(a-c)(b-d)}} \left\{ c\Pi\left(\alpha, \frac{a-d}{a-c}, q\right) - \right.$$

$$\left. - (c-d) F(\alpha, q) \right\} \quad [a > b > c > d > u]. \quad \text{BF(251.03)}$$

2. $$\int_d^u \frac{x\, dx}{\sqrt{(a-x)(b-x)(c-x)(x-d)}} = \frac{2}{\sqrt{(a-c)(b-d)}} \left\{ (d-a)\Pi\left(\beta, \frac{d-c}{a-c}, r\right) + \right.$$

$$\left. + aF(\beta, r) \right\} \quad [a > b > c \geqslant u > d]. \quad \text{BF(252.11)}$$

3. $$\int_u^c \frac{x\, dx}{\sqrt{(a-x)(b-x)(c-x)(x-d)}} = \frac{2}{\sqrt{(a-c)(b-d)}} \left\{ (c-b)\Pi\left(\gamma, \frac{c-d}{b-d}, r\right) + \right.$$

$$\left. + bF(\gamma, r) \right\} \quad [a > b > c > u \geqslant d]. \quad \text{BF(253.11)}$$

4. $$\int_c^u \frac{x\, dx}{\sqrt{(a-x)(b-x)(x-c)(x-d)}} = \frac{2}{\sqrt{(a-c)(b-d)}} \left\{ (c-d)\Pi\left(\delta, \frac{b-c}{b-d}, q\right) + \right.$$

$$\left. + dF(\delta, q) \right\} \quad [a > b \geqslant u > c > d]. \quad \text{BF(254.10)}$$

5. $$\int_u^b \frac{x\, dx}{\sqrt{(a-x)(b-x)(x-c)(x-d)}} = \frac{2}{\sqrt{(a-c)(b-d)}} \left\{ (b-a)\Pi\left(\varkappa, \frac{b-c}{a-c}, q\right) + \right.$$

$$\left. + aF(\varkappa, q) \right\} \quad [a > b > u \geqslant c > d]. \quad \text{BF(255.17)}$$

6. $$\int_b^u \frac{x\, dx}{\sqrt{(a-x)(x-b)(x-c)(x-d)}} = \frac{2}{\sqrt{(a-c)(b-d)}} \left\{ (b-c)\Pi\left(\lambda, \frac{a-b}{a-c}, r\right) + \right.$$

$$\left. + cF(\lambda, r) \right\} \quad [a \geqslant u > b > c > d]. \quad \text{BF(256.11)}$$

7. $$\int_u^a \frac{x\, dx}{\sqrt{(a-x)(x-b)(x-c)(x-d)}} = \frac{2}{\sqrt{(a-c)(b-d)}} \left\{ (a-d)\Pi\left(\mu, \frac{b-a}{b-d}, r\right) + \right.$$

$$\left. + dF(\mu, r) \right\} \quad [a > u \geqslant b > c > d]. \quad \text{BF(257.11)}$$

8. $\int_a^u \dfrac{x\,dx}{\sqrt{(x-a)(x-b)(x-c)(x-d)}} = \dfrac{2}{\sqrt{(a-c)(b-d)}}\left\{(a-b)\Pi\left(v,\dfrac{a-d}{b-d},q\right)+\right.$ 3.148

$\left. + bF(v,q)\right\}\quad [u>a>b>c>d].$ BF(258.11)

1. $\int_u^d \dfrac{dx}{x\sqrt{(a-x)(b-x)(c-x)(d-x)}} =$ 3.149

$= \dfrac{2}{cd\sqrt{(a-c)(b-d)}}\left\{(c-d)\Pi\left(\alpha,\dfrac{c(a-d)}{d(a-c)},q\right)+dF(\alpha,q)\right\}$

$[a>b>c>d>u].$ BF(251.04)

2. $\int_d^u \dfrac{dx}{x\sqrt{(a-x)(b-x)(c-x)(x-d)}} =$

$= \dfrac{2}{ad\sqrt{(a-c)(b-d)}}\left\{(a-d)\Pi\left(\beta,\dfrac{a(d-c)}{d(a-c)},r\right)+dF(\beta,r)\right\}$

$[a>b>c\geqslant u>d].$ BF(252.12)

3. $\int_u^c \dfrac{dx}{x\sqrt{(a-x)(b-x)(c-x)(x-d)}} =$

$= \dfrac{2}{bc\sqrt{(a-c)(b-d)}}\left\{(b-c)\Pi\left(\gamma,\dfrac{b(c-d)}{c(b-d)},r\right)+cF(\gamma,r)\right\}$

$[a>b>c>u\geqslant d].$ BF(253.12)

4. $\int_c^u \dfrac{dx}{x\sqrt{(a-x)(b-x)(x-c)(x-d)}} =$

$= \dfrac{2}{cd\sqrt{(a-c)(b-d)}}\left\{(d-c)\Pi\left(\delta,\dfrac{d(b-c)}{c(b-d)},q\right)+cF(\delta,q)\right\}$

$[a>b\geqslant u>c>d].$ BF(254.11)

5. $\int_u^b \dfrac{dx}{x\sqrt{(a-x)(b-x)(x-c)(x-d)}} = \dfrac{2}{ab\sqrt{(a-c)(b-d)}}\times$

$\times\left\{(a-b)\Pi\left(\varkappa,\dfrac{a(b-c)}{b(a-c)},q\right)+bF(\varkappa,q)\right\}$

$[a>b>u\geqslant c>d].$ BF(255.18)

3.149

6. $\displaystyle\int_b^u \frac{dx}{x\sqrt{(a-x)(x-b)(x-c)(x-d)}} = \frac{2}{bc\sqrt{(a-c)(b-d)}} \times$

$$\times \left\{(c-b)\Pi\left(\lambda, \frac{c(a-b)}{b(a-c)}, r\right) + bF(\lambda, r)\right\}$$

$[a \geqslant u > b > c > d]$.   BF(256.12)

7. $\displaystyle\int_u^a \frac{dx}{x\sqrt{(a-x)(x-b)(x-c)(x-d)}} = \frac{2}{ad\sqrt{(a-c)(b-d)}} \times$

$$\times \left\{(d-a)\Pi\left(\mu, \frac{d(b-a)}{a(b-d)}, r\right) + aF(\mu, r)\right\}$$

$[a > u \geqslant b > c > d]$.   BF(257.12)

8. $\displaystyle\int_a^u \frac{dx}{x\sqrt{(x-a)(x-b)(x-c)(x-d)}} =$

$$= \frac{2}{ab\sqrt{(a-c)(b-d)}} \left\{(b-a)\Pi\left(\nu, \frac{b(a-d)}{a(b-d)}, q\right) + aF(\nu, q)\right\}$$

$[u > a > b > c > d]$.   BF(258.12)

3.151

1. $\displaystyle\int_u^d \frac{dx}{(p-x)\sqrt{(a-x)(b-x)(c-x)(d-x)}} = \frac{2}{(p-c)(p-d)\sqrt{(a-c)(b-d)}} \times$

$$\times \left[(d-c)\Pi\left(\alpha, \frac{(a-d)(p-c)}{(a-c)(p-d)}, q\right) + (p-d)F(\alpha, q)\right]$$

$[a > b > c > d > u, \, p \neq d]$.   BF(251.39)

2. $\displaystyle\int_d^u \frac{dx}{(p-x)\sqrt{(a-x)(b-x)(c-x)(x-d)}} = \frac{2}{(p-a)(p-d)\sqrt{(a-c)(b-d)}} \times$

$$\times \left[(d-a)\Pi\left(\beta, \frac{(d-c)(p-a)}{(a-c)(p-d)}, r\right) + (p-d)F(\beta, r)\right]$$

$[a > b > c \geqslant u > d, \, p \neq d]$.   BF(252.39)

3. $\displaystyle\int_u^c \frac{dx}{(p-x)\sqrt{(a-x)(b-x)(c-x)(x-d)}} = \frac{2}{(p-b)(p-c)\sqrt{(a-c)(b-d)}} \times$

$$\times \left[(c-b)\Pi\left(\gamma, \frac{(c-d)(p-b)}{(b-d)(p-c)}, r\right) + (p-c)F(\gamma, r)\right]$$

$[a > b > c > u \geqslant d, \, p \neq c]$.   BF(253.39)

4. $\displaystyle\int_c^u \frac{dx}{(p-x)\sqrt{(a-x)(b-x)(x-c)(x-d)}} = \frac{2}{(p-c)(p-d)\sqrt{(a-c)(b-d)}} \times$ 3.151

$$\times \left[(c-d)\Pi\left(\delta, \frac{(b-c)(p-d)}{(b-d)(p-c)}, q\right) + (p-c)\ F(\delta, q)\right]$$

$\qquad\qquad\qquad\qquad\qquad\qquad [a > b \geqslant u > c > d,\ p \neq c].$ BF(254.39)

5. $\displaystyle\int_u^b \frac{dx}{(p-x)\sqrt{(a-x)(b-x)(x-c)(x-d)}} = \frac{2}{(p-a)(p-b)\sqrt{(a-c)(b-d)}} \times$

$$\times \left[(b-a)\Pi\left(\varkappa, \frac{(b-c)(p-a)}{(a-c)(p-b)}, q\right) + (p-b)F(\varkappa, q)\right]$$

$\qquad\qquad\qquad\qquad\qquad\qquad [a > b > u \geqslant c > d,\ p \neq b].$ BF(255.38)

6. $\displaystyle\int_b^u \frac{dx}{(x-p)\sqrt{(a-x)(x-b)(x-c)(x-d)}} = \frac{2}{(b-p)(p-c)\sqrt{(a-c)(b-d)}} \times$

$$\times \left[(b-c)\Pi\left(\lambda, \frac{(a-b)(p-c)}{(a-c)(p-b)}, r\right) + (p-b)\ F(\lambda, r)\right]$$

$\qquad\qquad\qquad\qquad\qquad\qquad [a \geqslant u > b > c > d,\ p \neq b].$ BF(256.39)

7. $\displaystyle\int_u^a \frac{dx}{(p-x)\sqrt{(a-x)(x-b)(x-c)(x-d)}} = \frac{2}{(p-a)(p-d)\sqrt{(a-c)(b-d)}} \times$

$$\times \left[(a-d)\Pi\left(\mu, \frac{(b-a)(p-d)}{(b-d)(p-a)}, r\right) + (p-a)\ F(\mu, r)\right]$$

$\qquad\qquad\qquad\qquad\qquad\qquad [a > u \geqslant b > c > d,\ p \neq a].$ BF(257.39)

8. $\displaystyle\int_a^u \frac{dx}{(p-x)\sqrt{(x-a)(x-b)(x-c)(x-d)}} = \frac{2}{(p-a)(p-b)\sqrt{(a-c)(b-d)}} \times$

$$\times \left[(a-b)\Pi\left(\nu, \frac{(a-d)(p-b)}{(b-d)(p-a)}, q\right) + (p-a)F(\nu, q)\right]$$

$\qquad\qquad\qquad\qquad\qquad\qquad [u > a > b > c > d,\ p \neq a].$ BF(258.39)

In 3.152—3.163 werden folgende Abkürzungen verwendet: | In 3.152—3.163 the following notation is used:

$$\alpha = \operatorname{arctg} \frac{u}{b},\quad \beta = \operatorname{arcctg} \frac{u}{a},$$

$$\gamma = \arcsin \frac{u}{b}\sqrt{\frac{a^2+b^2}{a^2+u^2}},\quad \delta = \arccos \frac{u}{b},\quad \varepsilon = \arccos \frac{b}{u},\quad \xi = \arcsin \sqrt{\frac{a^2+b^2}{a^2+u^2}}$$

Bestimmte Integrale elem. Funktionen/Definite Integrals of Elem. Functions

$$\eta = \arcsin\frac{u}{b}, \quad \zeta = \arcsin\frac{a}{b}\sqrt{\frac{b^2-u^2}{a^2-u^2}}, \quad \varkappa = \arcsin\frac{a}{u}\sqrt{\frac{u^2-b^2}{a^2-b^2}},$$

$$\lambda = \arcsin\sqrt{\frac{a^2-u^2}{a^2-b^2}}, \quad \mu = \arcsin\sqrt{\frac{u^2-a^2}{u^2-b^2}}, \quad \nu = \arcsin\frac{a}{u}, \quad q = \frac{\sqrt{a^2-b^2}}{a},$$

$$r = \frac{b}{\sqrt{a^2+b^2}}, \quad s = \frac{a}{\sqrt{a^2+b^2}}, \quad t = \frac{b}{a}.$$

3.152

1. $\displaystyle\int_0^u \frac{dx}{\sqrt{(x^2+a^2)(x^2+b^2)}} = \frac{1}{a} F(\alpha, q) \quad [a > b > 0].$ Z 62(258), BF(221.00)

2. $\displaystyle\int_u^\infty \frac{dx}{\sqrt{(x^2+a^2)(x^2+b^2)}} = \frac{1}{a} F(\beta, q) \quad [a > b > 0].$ Z 63(259), BF(222.00)

3. $\displaystyle\int_0^u \frac{dx}{\sqrt{(x^2+a^2)(b^2-x^2)}} = \frac{1}{\sqrt{a^2+b^2}} F(\gamma, r) \quad [b \geqslant u > 0].$ Z 63(260)

4. $\displaystyle\int_u^b \frac{dx}{\sqrt{(x^2+a^2)(b^2-x^2)}} = \frac{1}{\sqrt{a^2+b^2}} F(\delta, r) \quad [b > u \geqslant 0].$ Z 63(261), BF(213.00)

5. $\displaystyle\int_b^u \frac{dx}{\sqrt{(x^2+a^2)(x^2-b^2)}} = \frac{1}{\sqrt{a^2+b^2}} F(\varepsilon, s) \quad [u > b > 0].$ Z 63(262), BF(211.00)

6. $\displaystyle\int_u^\infty \frac{dx}{\sqrt{(x^2+a^2)(x^2-b^2)}} = \frac{1}{\sqrt{a^2+b^2}} F(\xi, s) \quad [u > b > 0].$ Z 63(263), BF(212.00)

7. $\displaystyle\int_0^u \frac{dx}{\sqrt{(a^2-x^2)(b^2-x^2)}} = \frac{1}{a} F(\eta, t) \quad [a > b \geqslant u > 0].$ Z 63(264), BF(219.00)

8. $\displaystyle\int_u^b \frac{dx}{\sqrt{(a^2-x^2)(b^2-x^2)}} = \frac{1}{a} F(\zeta, t) \quad [a > b > u \geqslant 0].$ Z 63(265), BF(220.00)

9. $\displaystyle\int_b^u \frac{dx}{\sqrt{(a^2-x^2)(x^2-b^2)}} = \frac{1}{a} F(\varkappa, q) \quad [a \geqslant u > b > 0].$ Z 63(266), BF(217.00)

10. $\displaystyle\int_u^a \frac{dx}{\sqrt{(a^2-x^2)(x^2-b^2)}} = \frac{1}{a} F(\lambda, q) \quad [a > u \geqslant b > 0].$ Z 63(267), BF(218.00)

11. $\int_a^u \dfrac{dx}{\sqrt{(x^2-a^2)(x^2-b^2)}} = \dfrac{1}{a} F(\mu, t)$ $[u > a > b > 0]$.   Z 63(268), BF(216.00)   3.152

12. $\int_u^\infty \dfrac{dx}{\sqrt{(x^2-a^2)(x^2-b^2)}} = \dfrac{1}{a} F(\nu, t)$ $[u \geqslant a > b > 0]$.   Z 64(269), BF(215.00)

1. $\int_0^u \dfrac{x^2\, dx}{\sqrt{(x^2+a^2)(x^2+b^2)}} = u\sqrt{\dfrac{a^2+u^2}{b^2+u^2}} - aE(\alpha, q)$ $[u > 0, a > b]$.   BF(221.09)   3.153

2. $\int_0^u \dfrac{x^2\, dx}{\sqrt{(a^2+x^2)(b^2-x^2)}} = \sqrt{a^2+b^2}\, E(\gamma, r) - \dfrac{a^2}{\sqrt{a^2+b^2}} F(\gamma, r) - u\sqrt{\dfrac{b^2-u^2}{a^2+u^2}}$

$[b \geqslant u > 0]$.   BF(214.05)

3. $\int_u^b \dfrac{x^2\, dx}{\sqrt{(a^2+x^2)(b^2-x^2)}} = \sqrt{a^2+b^2}\, E(\delta, r) - \dfrac{a^2}{\sqrt{a^2+b^2}} F(\delta, r)$

$[b > u \geqslant 0]$.   BF(213.06)

4. $\int_b^u \dfrac{x^2\, dx}{\sqrt{(a^2+x^2)(x^2-b^2)}} = \dfrac{b^2}{\sqrt{a^2+b^2}} F(\varepsilon, s) - \sqrt{a^2+b^2}\, E(\varepsilon, s) +$

$+ \dfrac{1}{u} \sqrt{(u^2+a^2)(u^2-b^2)}$ $[u > b > 0]$.   BF(211.09)

5. $\int_0^u \dfrac{x^2\, dx}{\sqrt{(a^2-x^2)(b^2-x^2)}} = a\{F(\eta, t) - E(\eta, t)\}$ $[a > b \geqslant u > 0]$.   BF(219.05)

6. $\int_u^b \dfrac{x^2\, dx}{\sqrt{(a^2-x^2)(b^2-x^2)}} = a\{F(\zeta, t) - E(\zeta, t)\} + u\sqrt{\dfrac{b^2-u^2}{a^2-u^2}}$

$[a > b > u \geqslant 0]$.   BF(220.06)

7. $\int_b^u \dfrac{x^2\, dx}{\sqrt{(a^2-x^2)(x^2-b^2)}} = aE(\varkappa, q) - \dfrac{1}{u}\sqrt{(a^2-u^2)(u^2-b^2)}$

$[a \geqslant u > b > 0]$.   BF(217.05)

8. $\int_u^a \dfrac{x^2\, dx}{\sqrt{(a^2-x^2)(x^2-b^2)}} = aE(\lambda, q)$ $[a > u \geqslant b > 0]$.   BF(218.06)

3.153

9. $\int_a^u \dfrac{x^2\, dx}{\sqrt{(x^2 - a^2)(x^2 - b^2)}} = a\{F(v,\ t) - E(v,\ t)\} + u\sqrt{\dfrac{u^2 - a^2}{u^2 - b^2}}$

[$u > a > b > 0$]. BF(216.06)

10. $\int_0^1 \dfrac{x^2 dx}{\sqrt{(1+x^2)(1+k^2 x^2)}} = \dfrac{1}{k^2}\left\{\sqrt{\dfrac{1+k^2}{2}} - E\left(\dfrac{\pi}{4},\ \sqrt{1-k^2}\right)\right\}.$ BH[14](9)

3.154

1. $\int_0^u \dfrac{x^4\, dx}{\sqrt{(x^2 + a^2)(x^2 + b^2)}} = \dfrac{a}{3}\{2(a^2 + b^2)\,E(\alpha,\ q) - b^2 F(\alpha,\ q)\} +$

$+ \dfrac{u}{3}(u^2 - 2a^2 - b^2)\sqrt{\dfrac{a^2 + u^2}{b^2 + u^2}}$ \quad [$a > b,\ u > 0$]. BF(221.09)

2. $\int_0^u \dfrac{x^4\, dx}{\sqrt{(a^2 + x^2)(b^2 - x^2)}} = \dfrac{1}{3\sqrt{a^2 + b^2}}\{(2a^2 - b^2)a^2 F(\gamma,\ r) -$

$- 2(a^4 - b^4) E(\gamma,\ r)\} - \dfrac{u}{3}(2b^2 - a^2 + u^2)\sqrt{\dfrac{b^2 - u^2}{a^2 + u^2}}$

[$a \geqslant u > 0$]. BF(214.05)

3. $\int_u^b \dfrac{x^4\, dx}{\sqrt{(a^2 + x^2)(b^2 - x^2)}} = \dfrac{1}{3\sqrt{a^2 + b^2}}\{(2a^2 - b^2)a^2 F(\delta,\ r) -$

$- 2(a^4 - b^4) E(\delta,\ r)\} + \dfrac{u}{3}\sqrt{(a^2 + u^2)(b^2 - u^2)}$

[$b > u \geqslant 0$]. BF(213.06)

4. $\int_b^u \dfrac{x^4\, dx}{\sqrt{(a^2 + x^2)(x^2 - b^2)}} = \dfrac{1}{3\sqrt{a^2 + b^2}}\{(2b^2 - a^2)b^2 F(\varepsilon,\ s) +$

$+ 2(a^4 - b^4) E(\varepsilon,\ s)\} + \dfrac{2b^2 - 2a^2 + u^2}{3u}\sqrt{(u^2 + a^2)(u^2 - b^2)}$

[$u > b > 0$]. BF(211.09)

5. $\int_0^u \dfrac{x^4\, dx}{\sqrt{(a^2 - x^2)(b^2 - x^2)}} = \dfrac{a}{3}\{(2a^2 + b^2)\,F(\eta,\ t) - 2(a^2 + b^2)E(\eta,\ t)\} +$

$+ \dfrac{u}{3}\sqrt{(a^2 - u^2)(b^2 - u^2)}$ \quad [$a > b \geqslant u > 0$]. BF(219.05)

6. $\int_u^b \dfrac{x^4 \, dx}{\sqrt{(a^2 - x^2)(b^2 - x^2)}} = \dfrac{a}{3} \{(2a^2 + b^2)F(\zeta, t) - 2(a^2 + b^2)E(\zeta, t)\} +$ 3.154

$$+ \dfrac{u}{3}(u^2 + a^2 + 2b^2)\sqrt{\dfrac{b^2 - u^2}{a^2 - u^2}} \quad [a > b > u \geqslant 0].$$ BF(220.06)

7. $\int_b^u \dfrac{x^4 \, dx}{\sqrt{(a^2 - x^2)(x^2 - b^2)}} = \dfrac{a}{3}\{2(a^2 + b^2)E(\varkappa, q) - b^2 F(\varkappa, q)\} -$

$$- \dfrac{u^2 + 2a^2 + 2b^2}{3u}\sqrt{(a^2 - u^2)(u^2 - b^2)} \quad [a \geqslant u > b > 0].$$ BF(217.05)

8. $\int_u^a \dfrac{x^4 \, dx}{\sqrt{(a^2 - x^2)(x^2 - b^2)}} = \dfrac{a}{3}\{2(a^2 + b^2)E(\lambda, q) - b^2 F(\lambda, q)\} +$

$$+ \dfrac{u}{3}\sqrt{(a^2 - u^2)(u^2 - b^2)} \quad [a > u \geqslant b > 0].$$ BF(218.06)

9. $\int_a^u \dfrac{x^4 \, dx}{\sqrt{(x^2 - a^2)(x^2 - b^2)}} = \dfrac{a}{3}\{(2a^2 + b^2)F(\mu, t) - 2(a^2 + b^2)E(\mu, t)\} +$

$$+ \dfrac{u}{3}(u^2 + 2a^2 + b^2)\sqrt{\dfrac{u^2 - a^2}{u^2 - b^2}} \quad [u > a > b > 0].$$ BF(216.06)

1. $\int_u^a \sqrt{(a^2 - x^2)(x^2 - b^2)} \, dx = \dfrac{a}{3}\{(a^2 + b^2)E(\lambda, q) - 2b^2 F(\lambda, q)\} -$ 3.155

$$- \dfrac{u}{3}\sqrt{(a^2 - u^2)(u^2 - b^2)} \quad [a > u \geqslant b > 0].$$ BF(218.11)

2. $\int_a^u \sqrt{(x^2 - a^2)(x^2 - b^2)} \, dx = \dfrac{a}{3}\{(a^2 + b^2)E(\mu, t) - (a^2 - b^2)F(\mu, t)\} +$

$$+ \dfrac{u}{3}(u^2 - a^2 - 2b^2)\sqrt{\dfrac{u^2 - a^2}{u^2 - b^2}} \quad [u > a > b > 0].$$ BF(216.10)

3. $\int_0^u \sqrt{(x^2 + a^2)(x^2 + b^2)} \, dx = \dfrac{a}{3}\{2b^2 F(\alpha, q) - (a^2 + b^2)E(\alpha, q)\} +$

$$+ \dfrac{u}{3}(u^2 + a^2 + 2b^2)\sqrt{\dfrac{a^2 + u^2}{b^2 + u^2}} \quad [a > b, u > 0].$$ BF(221.08)

4. $\int_0^u \sqrt{(a^2 + x^2)(b^2 - x^2)} \, dx = \dfrac{1}{3}\sqrt{a^2 + b^2}\{a^2 F(\gamma, r) - (a^2 - b^2)E(\gamma, r)\} +$

$$+ \dfrac{u}{3}(u^2 + 2a^2 - b^2)\sqrt{\dfrac{b^2 - u^2}{a^2 + u^2}} \quad [a \geqslant u > 0].$$ BF(214.12)

3.155

5. $\int_u^b \sqrt{(a^2+x^2)(b^2-x^2)}\,dx = \frac{1}{3}\sqrt{a^2+b^2}\,\{a^2 F(\delta, r)+$

$+ 2(b^2 - a^2) E(\delta, r)\} + \frac{u}{3}\sqrt{(a^2+u^2)(b^2-u^2)}$ \quad $[b > u \geqslant 0]$. \hfill BF(213.13)

6. $\int_b^u \sqrt{(a^2+x^2)(x^2-b^2)}\,dx = \frac{1}{3}\sqrt{a^2+b^2}\,\{(b^2-a^2)E(\varepsilon, s) - b^2 F(\varepsilon, s)\} +$

$+ \frac{u^2+a^2-b^2}{3u}\sqrt{(a^2+u^2)(u^2-b^2)}$ \quad $[u > b > 0]$. \hfill BF(211.08)

7. $\int_0^u \sqrt{(a^2-x^2)(b^2-x^2)}\,dx = \frac{a}{3}\{(a^2+b^2)E(\eta, t) - (a^2-b^2)F(\eta, t)\} +$

$+ \frac{u}{3}\sqrt{(a^2-u^2)(b^2-u^2)}$ \quad $[a > b \geqslant u > 0]$. \hfill BF(219.11)

8. $\int_u^b \sqrt{(a^2-x^2)(b^2-x^2)}\,dx = \frac{a}{3}\{(a^2+b^2)E(\zeta, t) - (a^2-b^2)F(\zeta, t)\} +$

$+ \frac{u}{3}(u^2 - 2a^2 - b^2)\sqrt{\frac{b^2-u^2}{a^2-u^2}}$ \quad $[a > b > u \geqslant 0]$. \hfill BF(220.05)

9. $\int_b^u \sqrt{(a^2-x^2)(x^2-b^2)}\,dx = \frac{a}{3}\{(a^2+b^2)E(\varkappa, q) - 2b^2 F(\varkappa, q)\} +$

$+ \frac{u^2-a^2-b^2}{3u}\sqrt{(a^2-u^2)(u^2-b^2)}$ \quad $[a \geqslant u > b > 0]$. \hfill BF(217.09)

3.156

1. $\int_u^\infty \frac{dx}{x^2\sqrt{(x^2+a^2)(x^2+b^2)}} = \frac{1}{ub^2}\sqrt{\frac{b^2+u^2}{a^2+u^2}} - \frac{1}{ab^2} E(\alpha, q)$

$[a \geqslant b, u > 0]$. \hfill BF(222.04)

2. $\int_u^b \frac{dx}{x^2\sqrt{(x^2+a^2)(b^2-x^2)}} = \frac{1}{a^2b^2\sqrt{a^2+b^2}}\{a^2 F(\delta, r) - (a^2+b^2)E(\delta, r)\} +$

$+ \frac{1}{a^2b^2u}\sqrt{(a^2+u^2)(b^2-u^2)}$ \quad $[b > u > 0]$. \hfill BF(213.09)

3. $\int_b^u \frac{dx}{x^2\sqrt{(x^2+a^2)(x^2-b^2)}} = \frac{1}{a^2b^2\sqrt{a^2+b^2}}\{(a^2+b^2)E(\varepsilon, s) - b^2 F(\varepsilon, s)\}$

$[u > b > 0]$. \hfill BF(211.11)

4. $\int_u^\infty \dfrac{dx}{x^2\sqrt{(x^2+a^2)(x^2-b^2)}} = \dfrac{1}{a^2b^2\sqrt{a^2+b^2}}\{(a^2+b^2)E(\gamma,s) - b^2 F(\gamma,s)\} -$ 3.156

$$- \dfrac{1}{b^2 u}\sqrt{\dfrac{u^2-b^2}{a^2+u^2}} \quad [u \geqslant b > 0].$$ BF(212.06)

5. $\int_u^b \dfrac{dx}{x^2\sqrt{(a^2-x^2)(b^2-x^2)}} = \dfrac{1}{ab^2}\{F(\zeta,t) - E(\zeta,t)\} +$

$$+ \dfrac{1}{b^2 u}\sqrt{\dfrac{b^2-u^2}{a^2-u^2}} \quad [a > b > u > 0].$$ BF(220.09)

6. $\int_b^u \dfrac{dx}{x^2\sqrt{(a^2-x^2)(x^2-b^2)}} = \dfrac{1}{ab^2} E(\varkappa, q) \quad [a \geqslant u > b > 0].$ BF(217.01)

7. $\int_u^a \dfrac{dx}{x^2\sqrt{(a^2-x^2)(x^2-b^2)}} = \dfrac{1}{ab^2} E(\lambda, q) - \dfrac{1}{a^2 b^2 u}\sqrt{(a^2-u^2)(u^2-b^2)}$

$$[a > u \geqslant b > 0].$$ BF(218.12)

8. $\int_a^u \dfrac{dx}{x^2\sqrt{(x^2-a^2)(x^2-b^2)}} = \dfrac{1}{ab^2}\{F(\mu,t) - E(\mu,t)\} +$

$$+ \dfrac{1}{a^2 u}\sqrt{\dfrac{u^2-a^2}{u^2-b^2}} \quad [u > a > b > 0].$$ BF(216.09)

9. $\int_u^\infty \dfrac{dx}{x^2\sqrt{(x^2-a^2)(x^2-b^2)}} = \dfrac{1}{ab^2}\{F(\nu,t) - E(\nu,t)\} \quad [u \geqslant a > b > 0].$ BF(215.07)

1. $\int_0^u \dfrac{dx}{(p-x^2)\sqrt{(x^2+a^2)(x^2+b^2)}} =$ 3.157

$$= \dfrac{1}{a(p+b^2)}\left\{\dfrac{b^2}{p}\Pi\left(\alpha, \dfrac{p+b^2}{p}, q\right) + F(\alpha, q)\right\} \quad [p \neq 0].$$ BF(221.13)

2. $\int_u^\infty \dfrac{dx}{(p-x^2)\sqrt{(x^2+a^2)(x^2+b^2)}} =$

$$= -\dfrac{1}{a(a^2+p)}\left\{\Pi\left(\beta, \dfrac{a^2+p}{a^2}, q\right) - F(\beta, q)\right\}.$$ BF(222.11)

3.157

3. $\displaystyle\int_0^u \frac{dx}{(p-x^2)\sqrt{(a^2+x^2)(b^2-x^2)}} =$

$$= \frac{1}{p(p+a^2)\sqrt{a^2+b^2}} \left\{ a^2 \Pi\left(\gamma, \frac{b^2(p+a^2)}{p(a^2+b^2)}, r\right) + pF(\gamma, r) \right\}$$

$[b \geqslant u > 0, p \neq 0]$.   BF(214.13)+

4. $\displaystyle\int_u^b \frac{dx}{(p-x^2)\sqrt{(a^2+x^2)(b^2-x^2)}} =$

$$= \frac{1}{(p-b^2)\sqrt{a^2+b^2}} \Pi\left(\delta, \frac{b^2}{b^2-p}, r\right) \quad [b > u \geqslant 0, p \neq b^2].$$ BF(213.02)

5. $\displaystyle\int_b^u \frac{dx}{(p-x^2)\sqrt{(a^2+x^2)(x^2-b^2)}} =$

$$= \frac{1}{p(p-b^2)\sqrt{a^2+b^2}} \left\{ b^2 \Pi\left(\varepsilon, \frac{p}{p-b^2}, s\right) + (p-b^2)F(\varepsilon, s) \right\}$$

$[u > b > 0, p \neq b^2]$.   BF(211.14)

6. $\displaystyle\int_u^\infty \frac{dx}{(x^2-p)\sqrt{(a^2+x^2)(x^2-b^2)}} =$

$$= \frac{1}{(a^2+p)\sqrt{a^2+b^2}} = \left\{ \Pi\left(\xi, \frac{a^2+p}{a^2+b^2}, s\right) - F(\xi, s) \right\} \quad [u \geqslant b > 0].$$ BF(212.12)

7. $\displaystyle\int_0^u \frac{dx}{(p-x^2)\sqrt{(a^2-x^2)(b^2-x^2)}} = \frac{1}{ap} \Pi\left(\eta, \frac{b^2}{p}, t\right)$

$[a > b \geqslant u > 0; p \neq b]$.   BF(219.02)

8. $\displaystyle\int_u^b \frac{dx}{(p-x^2)\sqrt{(a^2-x^2)(b^2-x^2)}} =$

$$= \frac{1}{a(p-a^2)(p-b^2)} = \left\{ (b^2-a^2)\Pi\left(\zeta, \frac{b^2(p-a^2)}{a^2(p-b^2)}, t\right) + (p-b^2)F(\zeta, t) \right\}$$

$[a > b > u \geqslant 0; p \neq b^2]$.   BF(220.13)

9. $\displaystyle\int_b^u \frac{dx}{(p-x^2)\sqrt{(a^2-x^2)(x^2-b^2)}} =$

$$= \frac{1}{ap(p-b^2)} \left\{ b^2\Pi\left(\varkappa, \frac{p(a^2-b^2)}{a^2(p-b^2)}, q\right) + (p-b^2)F(\varkappa, q) \right\}$$

$[a \geqslant u > b > 0; p \neq b^2]$.   BF(217.12)

10. $\int_u^a \dfrac{dx}{(x^2-p)\sqrt{(a^2-x^2)(x^2-b^2)}} = \dfrac{1}{a(a^2-p)} \Pi\left(\lambda, \dfrac{a^2-b^2}{a^2-p}, q\right)$ 3.157

$[a > u \geq b > 0;\ p \neq a^2].$ BF(218.02)

11. $\int_a^u \dfrac{dx}{(p-x^2)\sqrt{(x^2-a^2)(x^2-b^2)}} =$

$= \dfrac{1}{a(p-a^2)(p-b^2)} \left\{(a^2-b^2)\Pi\left(\mu, \dfrac{p-b^2}{p-a^2}, t\right) + (p-a^2)F(\mu, t)\right\}$

$[u > a > b > 0;\ p \neq a^2,\ p \neq b^2].$ BF(216.12)

12. $\int_u^\infty \dfrac{dx}{(x^2-p)\sqrt{(x^2-a^2)(x^2-b^2)}} = \dfrac{1}{ap}\left\{\Pi\left(\nu, \dfrac{p}{a^2}, t\right) - F(\nu, t)\right\}$

$[u \geq a > b > 0;\ p \neq 0].$ BF(215.12)

1. $\int_0^u \dfrac{dx}{\sqrt{(x^2+a^2)(x^2+b^2)^3}} = \dfrac{1}{ab^2(a^2-b^2)}\{a^2 E(\alpha, q) - b^2 F(\alpha, q)\}$ 3.158

$[a > b;\ u > 0].$ BF(221.05)

2. $\int_u^\infty \dfrac{dx}{\sqrt{(x^2+a^2)(x^2+b^2)^3}} = \dfrac{1}{ab^2(a^2-b^2)}\{a^2 E(\beta, q) - b^2 F(\beta, q)\} -$

$- \dfrac{u}{b^2\sqrt{(a^2+u^2)(b^2+u^2)}}$ $[a > b,\ u \geq 0].$ BF(222.05)

3. $\int_0^u \dfrac{dx}{\sqrt{(x^2+a^2)^3(x^2+b^2)}} = \dfrac{1}{a(a^2-b^2)}\{F(\alpha, q) - E(\alpha, q)\} +$

$+ \dfrac{u}{a^2\sqrt{(u^2+a^2)(u^2+b^2)}}$ $[a > b,\ u > 0].$ BF(221.06)

4. $\int_u^\infty \dfrac{dx}{\sqrt{(a^2+x^2)^3(x^2+b^2)}} = \dfrac{1}{a(a^2-b^2)}\{F(\beta, q) - E(\beta, q)\}$

$[a > b,\ u \geq 0].$ BF(222.03)

5. $\int_0^u \dfrac{dx}{\sqrt{(a^2+x^2)^3(b^2-x^2)}} = \dfrac{1}{a^2\sqrt{a^2+b^2}} E(\gamma, r)$ $[b \geq u > 0].$ BF(214.01)+

6. $\int_u^b \dfrac{dx}{\sqrt{(a^2+x^2)^3(b^2-x^2)}} = \dfrac{1}{a^2\sqrt{a^2+b^2}} E(\delta, r) -$

$- \dfrac{u}{a^2(a^2+b^2)}\sqrt{\dfrac{b^2-u^2}{a^2+u^2}}$ $[b > u \geq 0].$ BF(213.08)

3.158

7. $\int_b^u \dfrac{dx}{\sqrt{(a^2+x^2)^3(x^2-b^2)}} = \dfrac{1}{a^2\sqrt{a^2+b^2}}\{F(\varepsilon,s)-E(\varepsilon,s)\}+$

$\qquad\qquad\qquad + \dfrac{1}{(a^2+b^2)u}\sqrt{\dfrac{u^2-b^2}{u^2+a^2}}\quad [u>b>0].\qquad\text{BF(211.05)}$

8. $\int_u^\infty \dfrac{dx}{\sqrt{(a^2+x^2)^3(x^2-b^2)}} = \dfrac{1}{a^2\sqrt{a^2+b^2}}\{F(\xi,s)-E(\xi,s)\}$

$\qquad\qquad\qquad\qquad\qquad\qquad [u\geqslant b>0].\qquad\text{BF(212.03)}$

9. $\int_0^u \dfrac{dx}{\sqrt{(a^2+x^2)(b^2-x^2)^3}} = \dfrac{1}{b^2\sqrt{a^2+b^2}}\{F(\gamma,r)-E(\gamma,r)\}+$

$\qquad\qquad\qquad + \dfrac{u}{b^2\sqrt{(a^2+u^2)(b^2-u^2)}}\quad [b>u>0].\qquad\text{BF(214.10)}$

10. $\int_u^\infty \dfrac{dx}{\sqrt{(a^2+x^2)(x^2-b^2)^3}} = \dfrac{u}{b^2\sqrt{(a^2+u^2)(u^2-b^2)}} -$

$\qquad\qquad\qquad - \dfrac{1}{b^2\sqrt{a^2+b^2}}E(\xi,s)\quad [u\geqslant b>0].\qquad\text{BF(212.04)}$

11. $\int_0^u \dfrac{dx}{\sqrt{(a^2-x^2)^3(b^2-x^2)}} = \dfrac{1}{a^2(a^2-b^2)}\left\{aE(\eta,t)-u\sqrt{\dfrac{b^2-u^2}{a^2-u^2}}\right\}$

$\qquad\qquad\qquad\qquad\qquad\qquad [a>b\geqslant u>0].\qquad\text{BF(219.07)}$

12. $\int_u^b \dfrac{dx}{\sqrt{(a^2-x^2)^3(b^2-x^2)}} = \dfrac{1}{a(a^2-b^2)}E(\zeta,t)\quad [a>b>u\geqslant 0].\qquad\text{BF(220.10)}$

13. $\int_b^u \dfrac{dx}{\sqrt{(a^2-x^2)^3(x^2-b^2)}} = \dfrac{1}{a(a^2-b^2)}\left\{F(\varkappa,q)-E(\varkappa,q)+\dfrac{a}{u}\sqrt{\dfrac{u^2-b^2}{a^2-u^2}}\right\}$

$\qquad\qquad\qquad\qquad\qquad\qquad [a>u>b>0].\qquad\text{BF(217.10)}$

14. $\int_u^\infty \dfrac{dx}{\sqrt{(x^2-a^2)^3(x^2-b^2)}} = \dfrac{1}{a(b^2-a^2)}\left\{E(v,t)-\dfrac{a}{u}\sqrt{\dfrac{u^2-b^2}{u^2-a^2}}\right\}$

$\qquad\qquad\qquad\qquad\qquad\qquad [u>a>b>0].\qquad\text{BF(215.04)}$

15. $\int_0^u \dfrac{dx}{\sqrt{(a^2-x^2)(b^2-x^2)^3}} = \dfrac{1}{ab^2}F(\eta,t) - \dfrac{b}{b^2(a^2-b^2)}\times$

$\qquad\qquad\times\left\{aE(\eta,t)-u\sqrt{\dfrac{a^2-u^2}{b^2-u^2}}\right\}\quad [a>b>u>0].\qquad\text{BF(219.06)}$

16. $\int_u^a \dfrac{dx}{\sqrt{(a^2-x^2)(x^2-b^2)^3}} = \dfrac{1}{ab^2(a^2-b^2)} \left\{ b^2 F(\lambda, q) - a^2 E(\lambda, q) + \right.$

$\left. + au \sqrt{\dfrac{a^2-u^2}{u^2-b^2}} \right\}$ $\quad [a > u > b > 0]$.   3.158   BF(218.04)

17. $\int_a^u \dfrac{dx}{\sqrt{(x^2-a^2)(x^2-b^2)^3}} = \dfrac{a}{b^2(a^2-b^2)} E(\mu, t) - \dfrac{1}{ab^2} F(\mu, t)$

$[u > a > b > 0]$.   BF(216.11)

18. $\int_u^\infty \dfrac{dx}{\sqrt{(x^2-a^2)(x^2-b^2)^3}} = \dfrac{1}{b^2(a^2-b^2)} \left\{ aE(v, t) - \dfrac{b^2}{u} \sqrt{\dfrac{u^2-a^2}{u^2-b^2}} \right\} -$

$- \dfrac{1}{ab^2} F(v, t)$ $\quad [u \geqslant a > b > 0]$.   BF(215.06)

1. $\int_0^u \dfrac{x^2\,dx}{\sqrt{(x^2+a^2)(x^2+b^2)^3}} = \dfrac{a}{a^2-b^2} \{F(\alpha, q) - E(\alpha, q)\}$ 3.159

$[a > b, u > 0]$.   BF(221.12)

2. $\int_u^\infty \dfrac{x^2\,dx}{\sqrt{(x^2+a^2)(x^2+b^2)^3}} = \dfrac{a}{a^2-b^2} \{F(\beta, q) - E(\beta, q)\} +$

$+ \dfrac{u}{\sqrt{(a^2+u^2)(b^2+u^2)}}$ $\quad [a > b, u \geqslant 0]$.   BF(220.10)

3. $\int_0^u \dfrac{x^2\,dx}{\sqrt{(x^2+a^2)^3(x^2+b^2)}} = \dfrac{1}{a(a^2-b^2)} \{a^2 E(\alpha, q) - b^2 F(\alpha, q)\} -$

$- \dfrac{u}{\sqrt{(a^2+u^2)(b^2+u^2)}}$ $\quad [a > b, u > 0]$.   BF(221.11)

4. $\int_u^\infty \dfrac{x^2\,dx}{\sqrt{(x^2+a^2)^3(x^2+b^2)}} = \dfrac{1}{a(a^2-b^2)} \{a^2 E(\beta, q) - b^2 F(\beta, q)\}$

$[a > b, u \geqslant 0]$.   BF(222.07)

5. $\int_0^u \dfrac{x^2\,dx}{\sqrt{(a^2+x^2)^3(b^2-x^2)}} = \dfrac{1}{\sqrt{a^2+b^2}} \{F(\gamma, r) - E(\gamma, r)\}$

$[b \geqslant u > 0]$.   BF(214.04)

6. $\int_u^b \dfrac{x^2\,dx}{\sqrt{(a^2+x^2)^3(b^2-x^2)}} = \dfrac{1}{\sqrt{a^2+b^2}} \{F(\delta, r) - E(\delta, r)\} +$

$+ \dfrac{u}{a^2+b^2} \sqrt{\dfrac{b^2-u^2}{a^2+u^2}}$ $\quad [b > u \geqslant 0]$.   BF(213.07)

3.159

7. $\int_b^u \dfrac{x^2\,dx}{\sqrt{(a^2+x^2)^3(x^2-b^2)}} = \dfrac{1}{\sqrt{a^2+b^2}} E(\varepsilon,\,s) -$

$\qquad\qquad\qquad - \dfrac{a^2}{u(a^2+b^2)}\sqrt{\dfrac{u^2-b^2}{u^2+a^2}}\quad [u>b>0].$ BF(211.13)

8. $\int_u^\infty \dfrac{x^2\,dx}{\sqrt{(a^2+x^2)^3(x^2-b^2)}} = \dfrac{1}{\sqrt{a^2+b^2}} E(\xi,\,s)\quad [u\geqslant b>0].$ BF(212.01)

9. $\int_0^u \dfrac{x^2\,dx}{\sqrt{(a^2+x^2)(b^2-x^2)^3}} = \dfrac{u}{\sqrt{(a^2+u^2)(b^2-u^2)}} - \dfrac{1}{\sqrt{a^2+b^2}} E(\gamma,r)$

$\qquad\qquad\qquad [b>u>0].$ BF(214.07)

10. $\int_u^\infty \dfrac{x^2\,dx}{\sqrt{(a^2+x^2)(x^2-b^2)^3}} = \dfrac{1}{\sqrt{a^2+b^2}}\{F(\xi,\,s) - E(\xi,\,s)\} +$

$\qquad\qquad\qquad + \dfrac{u}{\sqrt{(a^2+u^2)(u^2-b^2)}}\quad [u>b>0].$ BF(212.10)

11. $\int_0^u \dfrac{x^2\,dx}{\sqrt{(a^2-x^2)^3(b^2-x^2)}} = \dfrac{1}{a^2-b^2}\left\{aE(\eta,\,t) - u\sqrt{\dfrac{b^2-u^2}{a^2-u^2}}\right\} - \dfrac{1}{a} F(\eta,t)$

$\qquad\qquad\qquad [a>b\geqslant u>0].$ BF(219.04)

12. $\int_u^b \dfrac{x^2\,dx}{\sqrt{(a^2-x^2)^3(b^2-x^2)}} = \dfrac{a}{a^2-b^2} E(\zeta,\,t) - \dfrac{1}{a} F(\zeta,\,t)\quad [a>b>u\geqslant 0].$

BF(220.08)

13. $\int_b^u \dfrac{x^2\,dx}{\sqrt{(a^2-x^2)^3(x^2-b^2)}} = \dfrac{1}{a(a^2-b^2)}\left\{b^2 F(\varkappa,\,q) - a^2 E(\varkappa,\,q) + \dfrac{a^3}{u}\sqrt{\dfrac{u^2-b^2}{a^2-u^2}}\right\}$

$\qquad\qquad\qquad [a>u>b>0].$ BF(217.06)

14. $\int_u^\infty \dfrac{x^2\,dx}{\sqrt{(x^2-a^2)^3(x^2-b^2)}} = \dfrac{a}{a^2-b^2}\left\{\dfrac{a}{u}\sqrt{\dfrac{u^2-b^2}{u^2-a^2}} - E(v,\,t)\right\} + \dfrac{1}{a} F(v,\,t)$

$\qquad\qquad\qquad [u>a>b>0].$ BF(215.09)

15. $\int_0^u \dfrac{x^2\,dx}{\sqrt{(a^2-x^2)(b^2-x^2)^3}} = \dfrac{1}{a^2-b^2}\left\{u\sqrt{\dfrac{a^2-u^2}{b^2-u^2}} - aE(\eta,t)\right\}$

$\qquad\qquad\qquad [a>b>u>0].$ BF(219.12)

16. $\int_u^a \dfrac{x^2\,dx}{\sqrt{(a^2-x^2)(x^2-b^2)^3}} = \dfrac{1}{a^2-b^2}\left\{aF(\lambda,\,q) - aE(\lambda,\,q) + u\sqrt{\dfrac{a^2-u^2}{u^2-b^2}}\right\}$

$\qquad\qquad\qquad [a>u>b>0].$ BF(218.07)

17. $\int_a^u \dfrac{x^2\,dx}{\sqrt{(x^2-a^2)(x^2-b^2)^3}} = \dfrac{a}{a^2-b^2} E(\mu, t)$ $\quad [u > a > b > 0]$. $\qquad$ BF(216.01) $\qquad$ 3.159

18. $\int_u^\infty \dfrac{x^2\,dx}{\sqrt{(x^2-a^2)(x^2-b^2)^3}} = \dfrac{1}{a^2-b^2}\left\{ aE(v, t) - \dfrac{b^2}{u}\sqrt{\dfrac{u^2-a^2}{u^2-b^2}}\right\}$

$\qquad\qquad\qquad\qquad\qquad\qquad\qquad\qquad [u \geqslant a > b > 0]$. $\qquad$ BF(215.11)

1. $\int_u^\infty \dfrac{dx}{x^4\sqrt{(x^2+a^2)(x^2+b^2)}} = \dfrac{1}{3a^3 b^4}\left\{2(a^2+b^2) E(\beta, q) - b^2 F(\beta, q)\right\} +$ $\qquad$ 3.161

$\qquad\qquad\qquad\qquad + \dfrac{a^2 b^2 - u^2(2a^2+b^2)}{3a^2 b^4 u^3}$ $\quad [a > b,\, u > 0]$. $\qquad$ BF(222.04)

2. $\int_u^b \dfrac{dx}{x^4\sqrt{(x^2+a^2)(b^2-x^2)}} =$

$\qquad = \dfrac{1}{3a^4 b^4 \sqrt{a^2+b^2}}\left\{a^2(2a^2 - b^2)\, F(\delta, r) - 2(a^4 - b^4)\, E(\delta, r)\right\} +$

$\qquad\qquad + \dfrac{a^2 b^2 + 2u^2(a^2 - b^2)}{3a^4 b^4 u^3}\sqrt{(b^2-u^2)(a^2+u^2)}$ $\quad [b > u > 0]$. $\qquad$ BF(213.09)

3. $\int_b^u \dfrac{dx}{x^4\sqrt{(x^2+a^2)(x^2-b^2)}} = \dfrac{2b^2 - a^2}{3a^4 b^2 \sqrt{a^2+b^2}}\, F(\varepsilon, s) +$

$\qquad + \dfrac{2}{3}\dfrac{(a^2-b^2)\sqrt{a^2+b^2}}{a^4 b^4}\, E(\varepsilon, s) + \dfrac{1}{3a^2 b^2 u^3}\sqrt{(u^2+a^2)(u^2-b^2)}$

$\qquad\qquad\qquad\qquad\qquad\qquad\qquad\qquad [u > b > 0]$. $\qquad$ BF(211.11)

4. $\int_u^\infty \dfrac{dx}{x^4\sqrt{(x^2+a^2)(x^2-b^2)}} =$

$\qquad = \dfrac{1}{3a^4 b^4 \sqrt{a^2+b^2}}\left\{2(a^4 - b^4) E(\xi, s) + b^2(2b^2 - a^2)\, F(\xi, s)\right\} -$

$\qquad\qquad - \dfrac{a^2 b^2 + u^2(2a^2 - b^2)}{3a^2 b^4 u^3}\sqrt{\dfrac{u^2-b^2}{u^2+a^2}}$ $\quad [u \geqslant b > 0]$. $\qquad$ BF(212.06)

5. $\int_u^b \dfrac{dx}{x^4\sqrt{(a^2-x^2)(b^2-x^2)}} = \dfrac{1}{3a^3 b^4}\left\{(2a^2+b^2) F(\zeta, t) - 2(a^2+b^2)\, E(\zeta, t) +\right.$

$\qquad\qquad \left. + \dfrac{[(2a^2+b^2)u^2 + a^2 b^2]a}{u^3}\sqrt{\dfrac{b^2-u^2}{a^2-u^2}}\right\}$ $\quad [a > b > u > 0]$. $\qquad$ BF(220.09)

6. $\int_b^u \dfrac{dx}{x^4\sqrt{(a^2-x^2)(x^2-b^2)}} = \dfrac{1}{3a^3 b^4}\left\{2(a^2+b^2)\, E(\varkappa, q) - b^2 F(\varkappa, q)\right\} +$

$\qquad\qquad + \dfrac{1}{3a^2 b^2 u^3}\sqrt{(a^2-u^2)(u^2-b^2)}$ $\quad [a \geqslant u > b > 0]$. $\qquad$ BF(217.14)

3.161

7. $\int_u^a \dfrac{dx}{x^4\sqrt{(a^2-x^2)(x^2-b^2)}} = \dfrac{1}{3a^3b^4}\Big\{2(a^2+b^2)E(\lambda,q) - b^2F(\lambda,q) -$

$- \dfrac{2(a^2+b^2)u^2+a^2b^2}{au^3}\sqrt{(a^2-u^2)(u^2-b^2)}\Big\}$ $\quad [a>u\geqslant b>0]$. BF(218.12)

8. $\int_a^u \dfrac{dx}{x^4\sqrt{(x^2-a^2)(x^2-b^2)}} = \dfrac{1}{3a^3b^4}\Big\{(2a^2+b^2)F(\mu,t) -$

$-2(a^2+b^2)E(\mu,t) + \dfrac{[(a^2+2b^2)u^2+a^2b^2]b^2}{au^3}\sqrt{\dfrac{u^2-a^2}{u^2-b^2}}\Big\}$

$[u>a>b>0]$. BF(216.09)

9. $\int_u^\infty \dfrac{dx}{x^4\sqrt{(x^2-a^2)(x^2-b^2)}} = \dfrac{1}{3a^3b^4}\Big\{(2a^2+b^2)F(v,t) - 2(a^2+b^2)E(v,t) +$

$+ \dfrac{ab^2}{u^3}\sqrt{(u^2-a^2)(u^2-b^2)}\Big\}$ $[u\geqslant a>b>0]$. BF(215.07)

3.162

1. $\int_0^u \dfrac{dx}{\sqrt{(x^2+a^2)^5(x^2+b^2)}} =$

$= \dfrac{1}{3a^3(a^2-b^2)^2}\{(3a^2-b^2)F(\alpha,q) - 2(2a^2-b^2)E(\alpha,q)\} +$

$+ \dfrac{u[a^2(4a^2-3b^2)+u^2(3a^2-2b^2)]}{3a^4(a^2-b^2)\sqrt{(u^2+a^2)^3(u^2+b^2)}}$ $[a>b, u>0]$. BF(221.06)

2. $\int_u^\infty \dfrac{dx}{\sqrt{(x^2+a^2)^5(x^2+b^2)}} = \dfrac{1}{3a^3(a^2-b^2)^2}\{(3a^2-b^2)F(\beta,q) -$

$-2(2a^2-b^2)E(\beta,q)\} + \dfrac{u}{3a^2(a^2-b^2)}\sqrt{\dfrac{u^2+b^2}{(a^2+u^2)^3}}$ $[a>b, u\geqslant 0]$. BF(222.03)

3. $\int_0^u \dfrac{dx}{\sqrt{(x^2+a^2)(x^2+b^2)^5}} = \dfrac{3b^2-a^2}{3ab^2(a^2-b^2)^2}F(\alpha,q) + \dfrac{a(2a^2-4b^2)}{3b^4(a^2-b^2)^2}E(\alpha,q) +$

$+ \dfrac{u}{3b^2(a^2-b^2)}\sqrt{\dfrac{u^2+a^2}{(u^2+b^2)^3}}$ $[a>b, u>0]$. BF(221.05)

4. $\int_u^\infty \dfrac{dx}{\sqrt{(x^2+a^2)(x^2+b^2)^5}} =$

$= \dfrac{1}{3ab^4(a^2-b^2)^2}\{2a^2(a^2-2b^2)E(\beta,q) + b^2(3b^2-a^2)F(\beta,q)\} -$

$- \dfrac{u[b^2(3a^2-4b^2)+u^2(2a^2-3b^2)]}{3b^4(a^2-b^2)\sqrt{(u^2+a^2)(u^2+b^2)^3}}$ $[a>b, u\geqslant 0]$. BF(222.05)

5. $\int_0^u \dfrac{dx}{\sqrt{(a^2+x^2)^5(b^2-x^2)}} = \dfrac{1}{3a^4\sqrt{(a^2+b^2)^3}}\{2(b^2+2a^2)E(\gamma,r) - a^2F(\gamma,r)\} +$

$+ \dfrac{u}{3a^2(a^2+b^2)}\sqrt{\dfrac{b^2-u^2}{(a^2+u^2)^3}}$ $[b\geqslant u>0]$. BF(214.05)

6. $\displaystyle\int_u^b \frac{dx}{\sqrt{(a^2+x^2)^5(b^2-x^2)}} = \frac{1}{3a^4\sqrt{(a^2+b^2)^3}} \{(4a^2+2b^2)E(\delta,r) - a^2 F(\delta,r)\} -$ 3.162

$\displaystyle - \frac{u[a^2(5a^2+3b^2) + u^2(4a^2+2b^2)]}{3a^4(a^2+b^2)^2}\sqrt{\frac{b^2-u^2}{(a^2+u^2)^3}}\quad [b>u>0].$ BF(213.08)

7. $\displaystyle\int_b^u \frac{dx}{\sqrt{(a^2+x^3)^5(x^2-b^2)}} = \frac{1}{3a^4\sqrt{(a^2+b^2)^3}} \{(3a^2+2b^2)F(\varepsilon,s) -$

$\displaystyle -(4a^2+2b^2)E(\varepsilon,s)\} + \frac{(3a^2+b^2)u^2 + 2(2a^2+b^2)a^2}{3a^2(a^2+b^2)^2 u}\sqrt{\frac{u^2-b^2}{(u^2+a^2)^3}}$

$[u>b>0].$ BF(211.05)

8. $\displaystyle\int_u^\infty \frac{dx}{\sqrt{(a^2+x^2)^5(x^2-b^2)}} = \frac{1}{3a^4\sqrt{(a^2+b^2)^3}}\{(3a^2+2b^2)F(\xi,s) -$

$\displaystyle - (4a^2+2b^2)E(\xi,s)\} + \frac{u}{3a^2(a^2+b^2)}\sqrt{\frac{u^2-b^2}{(a^2+u^2)^3}}\quad [u>b>0].$ BF(212.03)

9. $\displaystyle\int_0^u \frac{dx}{\sqrt{(a^2+x^2)(b^2-x^2)^5}} = \frac{1}{3b^4\sqrt{(a^2+b^2)^3}}\{(2a^2+3b^2)F(\gamma,r) -$

$\displaystyle - (2a^2+4b^2)E(\gamma,r)\} + \frac{u[(3a^2+4b^2)b^2 - (2a^2+3b^2)u^2]}{3b^4(a^2+b^2)\sqrt{(a^2+u^2)(b^2-u^2)^3}}\quad [b>u>0].$ BF(214.10)

10. $\displaystyle\int_u^\infty \frac{dx}{\sqrt{(a^2+x^2)(x^2-b^2)^5}} = \frac{1}{3b^4\sqrt{(a^2+b^2)^3}}\{(2a^2+4b^2)E(\xi,s) - b^2 F(\xi,s)\} +$

$\displaystyle + \frac{u[(3a^2+4b^2)b^2 - (2a^2+3b^2)u^2]}{3b^4(a^2+b^2)\sqrt{(a^2+u^2)(u^2-b^2)^3}}\quad [u>b>0].$ BF(212.04)

11. $\displaystyle\int_0^u \frac{dx}{\sqrt{(a^2-x^2)(b^2-x^2)^5}} = \frac{2a^2-3b^2}{3ab^4(a^2-b^2)} F(\eta,t) +$

$\displaystyle + \frac{2a(2b^2-a^2)}{3b^4(a^2-b^2)^2} E(\eta,t) + \frac{u[(3a^2-5b^2)b^2 - 2(a^2-2b^2)u^2]}{3b^4(a^2-b^2)^2(b^2-u^2)}\sqrt{\frac{a^2-u^2}{b^2-u^2}}$

$[a>b>u>0].$ BF(219.06)

12. $\displaystyle\int_u^a \frac{dx}{\sqrt{(a^2-x^2)(x^2-b^2)^5}} = \frac{3b^2-a^2}{3ab^2(a^2-b^2)^2} F(\lambda,q) +$

$\displaystyle + \frac{2a(a^2-2b^2)}{3b^4(a^2-b^2)^2} E(\lambda,q) + \frac{u[2(2b^2-a^2)u^2 + (3a^2-5b^2)b^2]}{3b^4(a^2-b^2)^2(u^2-b^2)}\sqrt{\frac{a^2-u^2}{u^2-b^2}}$

$[a>u>b>0].$ BF(218.04)

3.162

13. $\int_a^u \dfrac{dx}{\sqrt{(x^2-a^2)(x^2-b^2)^5}} = \dfrac{2a^2-3b^2}{3ab^4(a^2-b^2)} F(\mu, t) +$

$+ \dfrac{2a(2b^2-a^2)}{3b^4(a^2-b^2)^2} E(\mu, t) + \dfrac{u}{3b^2(a^2-b^2)(u^2-b^2)} \sqrt{\dfrac{u^2-a^2}{u^2-b^2}}$

$[u > a > b > 0].$ BF(216.11)

14. $\int_u^\infty \dfrac{dx}{\sqrt{(x^2-a^2)(x^2-b^2)^5}} = \dfrac{(4b^2-2a^2)a}{3b^4(a^2-b^2)^2} E(v, t) +$

$+ \dfrac{2a^2-3b^2}{3ab^4(a^2-b^2)} F(v, t) - \dfrac{(3b^2-a^2)u^2-(4b^2-2a^2)b^2}{3b^2u(a^2-b^2)^2(u^2-b^2)} \sqrt{\dfrac{u^2-a^2}{u^2-b^2}}$

$[u \geqslant a > b > 0].$ BF(215.06)

15. $\int_0^u \dfrac{dx}{\sqrt{(a^2-x^2)^5(b^2-x^2)}} = \dfrac{1}{3a^3(a^2-b^2)^2} \Big\{(4a^2-2b^2)E(\eta, t) -$

$- (a^2-b^2)F(\eta, t) - \dfrac{u[(5a^2-3b^2)a^2-(4a^2-2b^2)u^2]}{a(a^2-u^2)} \sqrt{\dfrac{b^2-u^2}{a^2-u^2}}\Big\}$

$[a > b \geqslant u > 0].$ BF(219.07)

16. $\int_u^b \dfrac{dx}{\sqrt{(a^2-x^2)^5(b^2-x^2)}} = \dfrac{2(2a^2-b^2)}{3a^3(a^2-b^2)^2} E(\zeta, r) -$

$- \dfrac{1}{3a^3(a^2-b^2)} F(\zeta, t) + \dfrac{u}{3a^2(a^2-b^2)(a^2-u^2)} \sqrt{\dfrac{b^2-u^2}{a^2-u^2}}$

$[a > b > u \geqslant 0].$ BF(220.10)

17. $\int_b^u \dfrac{dx}{\sqrt{(a^2-x^2)^5(x^2-b^2)}} = \dfrac{1}{3a^3(a^2-b^2)^2} \Big\{(3a^2-b^2)F(\varkappa, q) -$

$- (4a^2-2b^2) E(\varkappa, q)\Big\} + \dfrac{2(2a^2-b^2)a^2 + (b^2-3a^2)u^2}{3a^2u(a^2-b^2)^2(a^2-u^2)} \sqrt{\dfrac{u^2-b^2}{a^2-u^2}}$

$[a > u > b > 0].$ BF(217.10)

18. $\int_u^\infty \dfrac{dx}{\sqrt{(x^2-a^2)^5(x^2-b^2)}} = \dfrac{1}{3a^3(a^2-b^2)^2} \{(4a^2-2b^2) E(v, t)-(a^2-b^2) F(v, t)\}+$

$+ \dfrac{(4a^2-2b^2)a^2+(b^2-3a^2)u^2}{3a^2u(a^2-b^2)^2(u^2-a^2)} \sqrt{\dfrac{u^2-b^2}{u^2-a^2}} \quad [u > a > b > 0].$ BF(215.04)

3.163

1. $\int_0^u \dfrac{dx}{\sqrt{(x^2+a^2)^3(x^2+b^2)^3}} = \dfrac{1}{ab^2(a^2-b^2)^2} \{(a^2+b^2)E(\alpha, q) - 2b^2 F(\alpha, q)\} -$

$- \dfrac{u}{a^2(a^2-b^2)\sqrt{(a^2+u^2)(b^2+u^2)}} \quad [a > b, u > 0].$ BF(221.07)

2. $\int_u^\infty \dfrac{dx}{\sqrt{(x^2+a^2)^3(x^2+b^2)^3}} = \dfrac{1}{ab^2(a^2-b^2)^2}\{(a^2+b^2)E(\beta,q) - 2b^2 F(\beta,q)\} -$

$\qquad - \dfrac{u}{b^2(a^2-b^2)\sqrt{(a^2+u^2)(b^2+u^2)}} \quad [a>b, u \geq 0].$ BF(222.12)   3.163

3. $\int_0^u \dfrac{dx}{\sqrt{(x^2+a^2)^3(b^2-x^2)^3}} = \dfrac{1}{a^2b^2\sqrt{(a^2+b^2)^3}}\{a^2 F(\gamma,r) - (a^2-b^2)E(\gamma,r)\} +$

$\qquad + \dfrac{u}{b^2(a^2+b^2)\sqrt{(a^2+u^2)(b^2-u^2)}} \quad [b>u>0].$ BF(214.15)

4. $\int_u^\infty \dfrac{dx}{\sqrt{(x^2+a^2)^3(x^2-b^2)^3}} = \dfrac{b^2-a^2}{a^2b^2\sqrt{(a^2+b^2)^3}} E(\xi,s) - \dfrac{1}{a^2\sqrt{(a^2+b^2)^3}} F(\xi,s) +$

$\qquad + \dfrac{u}{b^2(a^2+b^2)\sqrt{(u^2+a^2)(u^2-b^2)}} \quad [u>b>0].$ BF(212.05)

5. $\int_0^u \dfrac{dx}{\sqrt{(a^2-x^2)^3(b^2-x^2)^3}} = \dfrac{1}{ab^2(a^2-b^2)} F(\eta,t) - \dfrac{a^2+b^2}{ab^2(a^2-b^2)^2} E(\eta,t) +$

$\qquad + \dfrac{[a^4+b^4-(a^2+b^2)u^2]u}{a^2b^2(a^2-b^2)^2\sqrt{(a^2-u^2)(b^2-u^2)}} \quad [a>b>u>0].$ BF(279.08)

6. $\int_u^\infty \dfrac{dx}{\sqrt{(x^2-a^2)^3(x^2-b^2)^3}} = \dfrac{1}{ab^2(a^2-b^2)} F(v,t) - \dfrac{a^2+b^2}{ab^2(a^2-b^2)^2} E(v,t) +$

$\qquad + \dfrac{1}{u(a^2-b^2)\sqrt{(u^2-a^2)(u^2-b^2)}} \quad [u>a>b>0].$ BF(215.10)

Abkürzungen:
Abbreviations: $\quad \alpha = \arccos\dfrac{u^2-\rho\bar\rho}{u^2+\rho\bar\rho}, \quad r = \dfrac{1}{2}\sqrt{-\dfrac{(\rho-\bar\rho)^2}{\rho\bar\rho}}.$ 3.164

1. $\int_u^\infty \dfrac{dx}{\sqrt{(x^2+\rho^2)(x^2+\bar\rho^2)}} = \dfrac{1}{\sqrt{\rho\bar\rho}} F(\alpha,r).$ BF(225.00)

2. $\int_u^\infty \dfrac{x^2\, dx}{(x^2-\rho\bar\rho)^2\sqrt{(x^2+\rho^2)(x^2+\bar\rho^2)}} = \dfrac{2u\sqrt{(u^2+\rho^2)(u^2+\bar\rho^2)}}{(\rho+\bar\rho)^2(u^4-\rho^2\bar\rho^2)} -$

$\qquad - \dfrac{1}{(\rho+\bar\rho)^2\sqrt{\rho\bar\rho}} E(\alpha,r).$ BF(225.03)

3. $\int_u^\infty \dfrac{x^2\, dx}{(x^2+\rho\bar\rho)^2\sqrt{(x^2+\rho^2)(x^2+\bar\rho^2)}} = -\dfrac{1}{(\rho-\bar\rho)^2\sqrt{\rho\bar\rho}}[F(\alpha,r) - E(\alpha,r)].$

   BF(225.07)

4. $\int_u^\infty \dfrac{x^2\, dx}{\sqrt{(x^2+\rho^2)^3(x^2+\bar\rho^2)^3}} = -\dfrac{4\sqrt{\rho\bar\rho}}{(\rho^2-\bar\rho^2)^2} E(\alpha,r) + \dfrac{1}{(\ -\bar\rho)^2\sqrt{\rho\bar\rho}} F(\alpha,r) -$

$\qquad - \dfrac{2u(u^2-\rho\bar\rho)}{(\rho+\bar\rho)^2(u^2+\rho\bar\rho)\sqrt{(u^2+\rho^2)(u^2+\bar\rho^2)}}.$ BF(225.05)

3.164

5. $\int_u^\infty \frac{(x^2 - \rho\bar{\rho})^2 \, dx}{\sqrt{(x^2 + \rho^2)^3(x^2 + \bar{\rho}^2)^3}} = -\frac{4\sqrt{\rho\bar{\rho}}}{(\rho - \bar{\rho})^2}[F(\alpha, r) - E(\alpha, r)] +$

$+ \frac{2u(u^2 - \rho\bar{\rho})}{(u^2 + \rho\bar{\rho})\sqrt{(u^2 + \rho^2)(u^2 + \bar{\rho}^2)}}$.   BF(225.06)

6. $\int_u^\infty \frac{\sqrt{(x^2 + \rho^2)(x^2 + \bar{\rho}^2)}}{(x^2 + \rho\bar{\rho})^2} \, dx = \frac{1}{\sqrt{\rho\bar{\rho}}} E(\alpha, r).$ BF(225.01)

7. $\int_u^\infty \frac{(x^2 - \rho\bar{\rho})^2 \, dx}{(x^2 + \rho\bar{\rho})^2 \sqrt{(x^2 + \rho^2)(x^2 + \bar{\rho}^2)}} = -\frac{4\sqrt{\rho\bar{\rho}}}{(\rho - \bar{\rho})^2} E(\alpha, r) +$

$+ \frac{(\rho + \bar{\rho})^2}{(\rho - \bar{\rho})^2 \sqrt{\rho\bar{\rho}}} F(\alpha, r).$ BF(225.08)

8. $\int_u^\infty \frac{(x^2 + \rho\bar{\rho})^2 \, dx}{[(x^2 + \rho\bar{\rho})^2 - 4p^2\rho\bar{\rho}x^2]\sqrt{(x^2 + \rho^2)(x^2 + \bar{\rho}^2)}} = \frac{1}{\sqrt{\rho\bar{\rho}}} \Pi(\alpha, p^2, r).$ BF(225.02)

3.165   Abkürzungen: $\alpha = \arccos \frac{u^2 - a^2}{u^2 + a^2}, \; r = \frac{\sqrt{a^2 - b^2}}{a\sqrt{2}}.$
Abbreviations:

1. $\int_u^a \frac{dx}{\sqrt{x^4 + 2b^2x^2 + a^4}} = \frac{\sqrt{2}}{a\sqrt{2} + \sqrt{a^2 + b^2}} \times$

$\times F\left[\arctg\left(\frac{a\sqrt{2} + \sqrt{a^2 - b^2}}{\sqrt{a^2 + b^2}} \cdot \frac{a - u}{a + u} \right), \frac{2\sqrt{a\sqrt{2}(a^2 - b^2)}}{a\sqrt{2} + \sqrt{a^2 - b^2}} \right]$

$[a > b, \; a > u \geq 0].$ BF(264.00)

2. $\int_u^\infty \frac{dx}{\sqrt{x^4 + 2b^2x^2 + a^4}} = \frac{1}{2a} F(\alpha, r) \quad [a^2 > b^2 > -\infty, \; a^2 > 0, \; u \geq 0].$

BF(263.00, 266.00)

3. $\int_u^\infty \frac{dx}{x^2\sqrt{x^4 + 2b^2x^2 + a^4}} = \frac{1}{2a^3}[F(\alpha, r) - 2E(\alpha, r)] + \frac{\sqrt{u^4 + 2b^2u^2 + a^4}}{a^2u(u^2 + a^2)}$

$[a > b > 0, \; u > 0].$ BF(263.06)

4. $\int_u^\infty \frac{x^2 \, dx}{(x^2 + a^2)^2 \sqrt{x^4 + 2b^2x^2 + a^4}} = \frac{1}{4a(a^2 - b^2)}[F(\alpha, r) - E(\alpha, r)]$

$[a^2 > b^2 > -\infty, \; a^2 > 0, \; u \geq 0].$ BF(263.03, 266.05)

5. $\int_u^\infty \frac{x^2 \, dx}{(x^2 - a^2)^2 \sqrt{x^4 + 2b^2x^2 + a^4}} = \frac{u\sqrt{u^4 + 2b^2u^2 + a^4}}{2(a^2 + b^2)(u^4 - a^4)} - \frac{1}{4a(a^2 + b^2)} E(\alpha, r)$

$[a^2 > b^2 > -\infty, \; u^2 > a^2 > 0].$ BF(263.05, 266.02)

6. $\int_u^\infty \frac{x^2 \, dx}{\sqrt{(x^4 + 2b^2x^2 + a^4)^3}} = \frac{a}{2(a^4 - b^4)} E(\alpha, r) - \frac{1}{4a(a^2 - b^2)} F(\alpha, r) -$

$- \frac{u(u^2 - a^2)}{2(a^2 + b^2)(u^2 + a^2)\sqrt{u^4 + 2b^2u^2 + a^4}} \quad [a^2 > b^2 > -\infty, \; a^2 > 0, \; u \geq 0].$

BF(263.08, 266.03)

7. $\int_u^\infty \frac{(x^2-a^2)^2\,dx}{\sqrt{(x^4+2b^2x^2+a^4)^3}} = \frac{a}{a^2-b^2}[F(\alpha,r)-E(\alpha,r)] +$ 3.165

$+ \frac{u^2-a^2}{u^2+a^2}\frac{u}{\sqrt{u^4+2b^2u^2+a^4}}$ $\quad[|b^2|<a^2,\ u\geq 0]$. BF(266.08)

8. $\int_u^\infty \frac{(x^2+a^2)^2\,dx}{\sqrt{(x^2+2b^2x^2+a^4)^3}} = \frac{a}{a^2+b^2}E(\alpha,r) - \frac{a^2-b^2}{a^2+b^2}\cdot\frac{u^2-a^2}{u^2+a^2}\cdot\frac{u}{\sqrt{u^4+2b^2u^2+a^4}}$

$[|b^2|<a^2,\ u\geq 0]$. BF(266.06)+

9. $\int_u^\infty \frac{(x^2-a^2)^2\,dx}{(x^2+a^2)^2\sqrt{x^4+2b^2x^2+a^4}} = \frac{a}{a^2-b^2}E(\alpha,r) - \frac{a^2+b^2}{2a(a^2-b^2)}F(\alpha,r)$

$[a^2>b^2>-\infty,\ a^2>0,\ u\geq 0]$. BF(263.04, 266.07)

10. $\int_u^\infty \frac{\sqrt{x^4+2b^2x^2+a^4}}{(x^2+a^2)^2}\,dx = \frac{1}{2a}E(\alpha,r)$ $\quad[a^2>b^2>-\infty,\ a^2>0,\ u\geq 0]$.

BF(263.01, 266.01)

11. $\int_u^\infty \frac{\sqrt{x^4+2b^2x^2+a^4}}{(x^2-a^2)^2}\,dx = \frac{1}{2a}[F(\alpha,r)-E(\alpha,r)] +$

$+ \frac{u}{u^4-a^4}\sqrt{u^4+2b^2u^2+a^4}$ $\quad[a>b>0,\ u>a]$. BF(263.07)

12. $\int_u^\infty \frac{(x^2+a^2)^2\,dx}{[(x^2+a^2)^2-4a^2p^2x^2]\sqrt{x^4+2b^2x^2+a^4}} = \frac{1}{2a}\Pi(\alpha,p^2,r)$ $\quad[a>b>0,\ u\geq 0]$.

BF(263.02)

Abkürzungen:
Abbreviations: $\alpha = \arccos\frac{u^2-1}{u^2+1},\quad \beta = \operatorname{arctg}\left\{(1+\sqrt{2})\frac{1-u}{1+u}\right\}$, 3.166

$\gamma = \arccos u,\ \delta = \arccos\frac{1}{u},\ \varepsilon = \arccos\frac{1-u^2}{1+u^2},\ r = \frac{\sqrt{2}}{2},\ q = 2\sqrt{3\sqrt{2}-4} =$

$= 2\sqrt[4]{2}(\sqrt{2}-1) = \sin 80°7'15'' \approx 0{,}985\,171$.

1. $\int_u^\infty \frac{dx}{\sqrt{x^4+1}} = \frac{1}{2}F(\alpha,r)$ $\quad[u\geq 0]$. Z 66(287), BF(263.50)

2. $\int_u^\infty \frac{dx}{x^2\sqrt{x^4+1}} = \frac{1}{2}[F(\alpha,r)-2E(\alpha,r)] + \frac{\sqrt{u^4+1}}{u(u^2+1)}$ $\quad[u>0]$. BF(263.57)

3. $\int_u^\infty \frac{x^2\,dx}{(x^4+1)\sqrt{x^4+1}} = \frac{1}{2}E(\alpha,r) - \frac{1}{4}F(\alpha,r) - \frac{u(u^2-1)}{2(u^2+1)\sqrt{u^4+1}}$ $\quad[u\geq 0]$.

BF(263.59)

4. $\int_u^\infty \frac{x^2\,dx}{(x^2+1)^2\sqrt{x^4+1}} = \frac{1}{4}[F(\alpha,r)-E(\alpha,r)]$ $\quad[u\geq 0]$. BF(263.53)

5. $\int_u^\infty \frac{x^2\,dx}{(x^2-1)^2\sqrt{x^4+1}} = \frac{u\sqrt{u^4+1}}{2(u^4-1)} - \frac{1}{4}E(\alpha,r)$ $\quad[u>1]$. BF(263.55)

3.166

6. $\int_u^\infty \frac{\sqrt{x^4+1}}{(x^2-1)^2}\,dx = \frac{1}{2}[F(\alpha,r) - E(\alpha,r)] + \frac{u\sqrt{u^4+1}}{u^4-1}$ $[u>1]$.   BF(263.58)

7. $\int_u^\infty \frac{(x^2-1)^2\,dx}{(x^2+1)^2\sqrt{x^4+1}} = E(\alpha,r) - \frac{1}{2}F(\alpha,r)$ $[u \geqslant 0]$.   BF(263.54)

8. $\int_u^\infty \frac{\sqrt{x^4+1}\,dx}{(x^2+1)^2} = \frac{1}{2}E(\alpha,r)$ $[u \geqslant 0]$.   BF(263.51)

9. $\int_u^\infty \frac{(x^2+1)^2\,dx}{[(x^2+1)^2 - 4p^2x^2]\sqrt{x^4+1}} = \frac{1}{2}\Pi(\alpha, p^2, r)$ $[u \geqslant 0]$.   BF(263.52)

10. $\int_0^u \frac{dx}{\sqrt{x^4+1}} = \frac{1}{2}F(\varepsilon, r)$.   Z 66(288)

11. $\int_u^1 \frac{dx}{\sqrt{x^4+1}} = (2-\sqrt{2})F(\beta, q)$ $[0 \leqslant u < 1]$.   BF(264.50)

12. $\int_u^1 \frac{(x^2+x\sqrt{2}+1)\,dx}{(x^2-x\sqrt{2}+1)\sqrt{x^4+1}} = (2+\sqrt{2})E(\beta, q)$ $[0 \leqslant u < 1]$.   BF(264.51)

13. $\int_u^1 \frac{(1-x)^2\,dx}{(x^2-x\sqrt{2}+1)\sqrt{x^4+1}} = \frac{1}{\sqrt{2}}[F(\beta,q) - E(\beta,q)]$ $[0 \leqslant u < 1]$.   BF(264.55)

14. $\int_u^1 \frac{(1+x)^2\,dx}{(x^2-x\sqrt{2}+1)\sqrt{x^4+1}} = \frac{3\sqrt{2}+4}{2}E(\beta,q) - \frac{3\sqrt{2}-4}{2}F(\beta,q)$

$[0 \leqslant u < 1]$.   BF(264.56)

15. $\int_u^1 \frac{dx}{\sqrt{1-x^4}} = \frac{1}{\sqrt{2}}F(\gamma, r)$ $[u<1]$.   Z 66(290), BF(259.75)

16. $\int_0^1 \frac{dx}{\sqrt{1-x^4}} = \frac{1}{4\sqrt{2\pi}}\left\{\Gamma\left(\frac{1}{4}\right)\right\}^2$.

17. $\int_1^u \frac{dx}{\sqrt{x^4-1}} = \frac{1}{\sqrt{2}}F(\delta, r)$ $[u>1]$.   Z 66(289), BF(260.75)

18. $\int_u^1 \frac{x^2\,dx}{\sqrt{1-x^4}} = \sqrt{2}E(\gamma,r) - \frac{1}{\sqrt{2}}F(\gamma, r)$ $[u<1]$.   BF(259.76)

19. $\int_1^u \frac{x^2\,dx}{\sqrt{x^4-1}} = \frac{1}{\sqrt{2}}F(\delta,r) - \sqrt{2}E(\delta,r) + \frac{1}{u}\sqrt{u^4-1}$ $[u>1]$.   BF(260.77)

20. $\int_u^1 \frac{x^4\,dx}{\sqrt{1-x^4}} = \frac{1}{3\sqrt{2}}F(\gamma,r) + \frac{u}{3}\sqrt{1-u^4}$ $[u<1]$.   BF(259.76)

21. $\displaystyle\int_1^u \frac{x^4\,dx}{\sqrt{x^4-1}} = \frac{1}{3} F(\delta, r) + \frac{\sqrt{2}}{3} u\sqrt{u^4-1} \quad [u>1].$ BF(260.77) 3.166

22. $\displaystyle\int_0^u \frac{dx}{\sqrt{x(1+x^3)}} = \frac{1}{\sqrt[4]{3}} F\left(\arccos\frac{1+(1-\sqrt{3})u}{1+(1+\sqrt{3})u}, \frac{\sqrt{2+\sqrt{3}}}{2}\right) \quad [u>0].$ BF(260.50)

23. $\displaystyle\int_0^u \frac{dx}{\sqrt{x(1-x^3)}} = \frac{1}{\sqrt[4]{3}} F\left(\arccos\frac{1-(1+\sqrt{3})u}{1+(\sqrt{3}-1)u}, \frac{\sqrt{2-\sqrt{3}}}{2}\right) \quad [1\geq u>0].$ BF(259.50)

In 3.167 und 3.168 werden folgende Abkürzungen verwendet: | In 3.167 and 3.168 the following notation is used:

$$\alpha = \arcsin\sqrt{\frac{(a-c)(d-u)}{(a-d)(c-u)}},$$

$$\beta = \arcsin\sqrt{\frac{(a-c)(u-d)}{(c-d)(a-u)}}, \quad \gamma = \arcsin\sqrt{\frac{(b-d)(c-u)}{(c-d)(b-u)}},$$

$$\delta = \arcsin\sqrt{\frac{(b-d)(u-c)}{(b-c)(u-d)}}, \quad \varkappa = \arcsin\sqrt{\frac{(a-c)(b-u)}{(b-c)(a-u)}},$$

$$\lambda = \arcsin\sqrt{\frac{(a-c)(u-b)}{(a-b)(u-c)}}, \quad \mu = \arcsin\sqrt{\frac{(b-d)(a-u)}{(a-b)(u-d)}},$$

$$\nu = \arcsin\sqrt{\frac{(b-d)(u-a)}{(a-d)(u-b)}}, \quad q = \sqrt{\frac{(b-c)(a-d)}{(a-c)(b-d)}}, \quad r = \sqrt{\frac{(a-b)(c-d)}{(a-c)(b-d)}}.$$

1. $\displaystyle\int_u^d \sqrt{\frac{d-x}{(a-x)(b-x)(c-x)}}\,dx = \frac{2(c-d)}{\sqrt{(a-c)(b-d)}}\left\{\Pi\left(\alpha, \frac{a-d}{a-c}, q\right) - F(\alpha, q)\right\}$ 3.167

$[a>b>c>d\geq u].$ BF(251.05)

2. $\displaystyle\int_d^u \sqrt{\frac{x-d}{(a-x)(b-x)(c-x)}}\,dx = \frac{2(d-a)}{\sqrt{(a-c)(b-d)}}\left\{\Pi\left(\beta, \frac{d-c}{a-c}, r\right) - F(\beta, r)\right\}$

$[a>b>c\geq u>d].$ BF(252.14)

3. $\displaystyle\int_u^c \sqrt{\frac{x-d}{(a-x)(b-x)(c-x)}}\,dx =$

$= \dfrac{2}{\sqrt{(a-c)(b-d)}}\left\{(c-b)\Pi\left(\gamma, \frac{c-d}{b-d}, r\right) + (b-d)F(\gamma, r)\right\}$

$[a>b>c>u\geq d].$ BF(253.14)

4. $\displaystyle\int_c^u \sqrt{\frac{x-d}{(a-x)(b-x)(x-c)}}\,dx = \frac{2(c-d)}{\sqrt{(a-c)(b-d)}}\left(\delta, \frac{b-c}{b-d}, q\right).$

$[a>b\geq u>c>d].$ BF(254.02)

3.167

5. $\int_u^b \sqrt{\dfrac{x-d}{(a-x)(b-x)(x-c)}}\, dx =$

$= \dfrac{2}{\sqrt{(a-c)(b-d)}} \left\{(b-a)\Pi\left(\varkappa, \dfrac{b-c}{a-c}, q\right) + (a-d)F(\varkappa, q)\right\}$

$[a > b > u \geqslant c > d].$ BF(255.20)

6. $\int_b^u \sqrt{\dfrac{x-d}{(a-x)(x-b)(x-c)}}\, dx =$

$= \dfrac{2}{\sqrt{(a-c)(b-d)}} \left\{(b-c)\Pi\left(\lambda, \dfrac{a-b}{a-c}, r\right) + (c-d)F(\lambda, r)\right\}$

$[a \geqslant u > b > c > d].$ BF(256.13)

7. $\int_u^a \sqrt{\dfrac{x-d}{(a-x)(x-b)(x-c)}}\, dx = \dfrac{2(a-d)}{\sqrt{(a-c)(b-d)}} \Pi\left(\mu, \dfrac{b-a}{b-d}, r\right)$

$[a > u \geqslant b > c > d].$ BF(257.02)

8. $\int_a^u \sqrt{\dfrac{x-d}{(x-a)(x-b)(x-c)}}\, dx =$

$= \dfrac{2}{\sqrt{(a-c)(b-d)}} \left\{(a-b)\Pi\left(v, \dfrac{a-d}{b-d}, q\right) + (b-d)F(v, q)\right\}$

$[u > a > b > c > d].$ BF(258.14)

9. $\int_u^d \sqrt{\dfrac{c-x}{(a-x)(b-x)(d-x)}}\, dx = \dfrac{2(c-d)}{\sqrt{(a-c)(b-d)}} \Pi\left(\alpha, \dfrac{a-d}{a-c}, q\right)$

$[a > b > c > d > u].$ BF(251.02)

10. $\int_d^u \sqrt{\dfrac{c-x}{(a-x)(b-x)(x-d)}}\, dx = \dfrac{2}{\sqrt{(a-c)(b-d)}} \left[(a-d)\Pi\left(\beta, \dfrac{d-c}{a-c}, r\right) - \right.$

$\left. -(a-c)F(\beta, r)\right]$ $[a > b > c \geqslant u > d].$ BF(252.13)

11. $\int_u^c \sqrt{\dfrac{c-x}{(a-x)(b-x)(x-d)}}\, dx = \dfrac{2(b-c)}{\sqrt{(a-c)(b-d)}} \left[\Pi\left(\gamma, \dfrac{c-d}{b-d}, r\right) - F(\gamma, r)\right]$

$[a > b > c > u \geqslant d].$ BF(253.13)

12. $\int_c^u \sqrt{\dfrac{x-c}{(a-x)(b-x)(x-d)}}\, dx = \dfrac{2(c-d)}{\sqrt{(a-c)(b-d)}} \left[\Pi\left(\delta, \dfrac{b-c}{b-d}, q\right) - F(\delta, q)\right]$

$[a > b \geqslant u > c > d].$ BF(254.12)

13. $\int_u^b \sqrt{\dfrac{x-c}{(a-x)(b-x)(x-d)}}\, dx =$

$= \dfrac{2}{\sqrt{(a-c)(b-d)}} \left[(b-a)\Pi\left(\varkappa, \dfrac{b-c}{a-c}, q\right) + (a-c)F(\varkappa, q)\right]$

$[a > b > u \geqslant c > d].$ BF(259.19)

14. $\int_b^u \sqrt{\dfrac{x-c}{(a-x)(x-b)(x-d)}}\,dx = \dfrac{2(b-c)}{\sqrt{(a-c)(b-d)}}\,\Pi\!\left(\lambda,\dfrac{a-b}{a-c},r\right)$ 3.167

$[a \geqslant u > b > c > d]$. BF(256.02)

15. $\int_u^a \sqrt{\dfrac{x-c}{(a-x)(x-b)(x-d)}}\,dx = \dfrac{2}{\sqrt{(a-c)(b-d)}}\left[(a-d)\Pi\!\left(\mu,\dfrac{b-a}{b-d},r\right)+\right.$

$\left.+(d-c)F(\mu,r)\right]$ $[a > u \geqslant b > c > d]$. BF(257.13)

16. $\int_a^u \sqrt{\dfrac{x-c}{(x-a)(x-b)(x-d)}}\,dx =$

$=\dfrac{2}{\sqrt{(a-c)(b-d)}}\left[(a-b)\Pi\!\left(\nu,\dfrac{a-d}{b-d},q\right)+(b-c)F(\nu,q)\right]$

$[u > a > b > c > d]$. BF(258.13)

17. $\int_u^d \sqrt{\dfrac{b-x}{(a-x)(c-x)(d-x)}}\,dx =$

$=\dfrac{2}{\sqrt{(a-c)(b-d)}}\left[(c-d)\Pi\!\left(\alpha,\dfrac{a-d}{a-c},q\right)+(b-c)F(\alpha,q)\right]$

$[a > b > c > d > u]$. BF(251.07)

18. $\int_d^u \sqrt{\dfrac{b-x}{(a-x)(c-x)(x-d)}}\,dx =$

$=\dfrac{2}{\sqrt{(a-c)(b-d)}}\left[(a-d)\Pi\!\left(\beta,\dfrac{d-c}{a-c},r\right)-(a-b)F(\beta,r)\right]$

$[a > b > c \geqslant u > d]$. BF(252.15)

19. $\int_u^c \sqrt{\dfrac{b-x}{(a-x)(c-x)(x-d)}}\,dx = \dfrac{2(b-c)}{\sqrt{(a-c)(b-d)}}\,\Pi\!\left(\gamma,\dfrac{c-d}{b-d},r\right)$

$[a > b > c > u \geqslant d]$. BF(253.02)

20. $\int_c^u \sqrt{\dfrac{b-x}{(a-x)(x-c)(x-d)}}\,dx =$

$=\dfrac{2}{\sqrt{(a-c)(b-d)}}\left[(d-c)\Pi\!\left(\delta,\dfrac{b-c}{b-d},q\right)+(b-d)F(\delta,q)\right]$

$[a > b \geqslant u > c > d]$. BF(254.14)

21. $\int_u^b \sqrt{\dfrac{b-x}{(a-x)(x-c)(x-d)}}\,dx =$

$=\dfrac{2(a-b)}{\sqrt{(a-c)(b-d)}}\left[\Pi\!\left(\varkappa,\dfrac{b-c}{a-c},q\right)-F(\varkappa,q)\right]$ $[a > b > u \geqslant c > d]$. BF(255.21)

3.167

22. $$\int_b^u \sqrt{\frac{x-b}{(a-x)(x-c)(x-d)}}\,dx = \frac{2(b-c)}{\sqrt{(a-c)(b-d)}}\left[\Pi\left(\lambda, \frac{a-b}{a-c}, r\right) - F(\lambda, r)\right]$$

$$[a \geqslant u > b > c > d]. \qquad \text{BF(256.15)}$$

23. $$\int_u^a \sqrt{\frac{x-b}{(a-x)(x-c)(x-d)}}\,dx =$$

$$= \frac{2}{\sqrt{(a-c)(b-d)}}\left[(d-a)\,\Pi\left(\mu, \frac{b-a}{b-d}, r\right) - (b-d)F(\mu, r)\right]$$

$$[a > u \geqslant b > c > d]. \qquad \text{BF(257.15)}$$

24. $$\int_a^u \sqrt{\frac{x-b}{(x-a)(x-c)(x-d)}}\,dx = \frac{2(a-b)}{\sqrt{(a-c)(b-d)}}\,\Pi\left(\nu, \frac{a-d}{b-d}, q\right)$$

$$[u > a > b > c > d]. \qquad \text{BF(258.02)}$$

25. $$\int_u^d \sqrt{\frac{a-x}{(b-x)(c-x)(d-x)}}\,dx =$$

$$= \frac{2}{\sqrt{(a-c)(b-d)}}\left[(c-d)\,\Pi\left(\alpha, \frac{a-d}{a-c}, q\right) + (a-c)F(\alpha, q)\right]$$

$$[a > b > c > d > u]. \qquad \text{BF (251.06)}$$

26. $$\int_d^u \sqrt{\frac{a-x}{(b-x)(c-x)(x-d)}}\,dx = \frac{2(a-d)}{\sqrt{(a-c)(b-d)}}\,\Pi\left(\beta, \frac{d-c}{a-c}, r\right)$$

$$[a > b > c \geqslant u > d]. \qquad \text{BF(252.02)}$$

27. $$\int_u^c \sqrt{\frac{a-x}{(b-x)(c-x)(x-d)}}\,dx =$$

$$= \frac{2}{\sqrt{(a-c)(b-d)}}\left[(b-c)\,\Pi\left(\gamma, \frac{c-d}{b-d}, r\right) + (a-b)F(\gamma, r)\right]$$

$$[a > b > c > u \geqslant d]. \qquad \text{BF(253.15)}$$

28. $$\int_c^u \sqrt{\frac{a-x}{(b-x)(x-c)(x-d)}}\,dx =$$

$$= \frac{2}{\sqrt{(a-c)(b-d)}}\left[(d-c)\,\Pi\left(\delta, \frac{b-c}{b-d}, q\right) + (a-d)F(\delta, q)\right]$$

$$[a > b \geqslant u > c > d]. \qquad \text{BF(254.13)}$$

29. $$\int_u^b \sqrt{\frac{a-x}{(b-x)(x-c)(x-d)}}\,dx = \frac{2(a-b)}{\sqrt{(a-c)(b-d)}}\,\Pi\left(\varkappa, \frac{b-c}{a-c}, q\right)$$

$$[a > b > u \geqslant c > d]. \qquad \text{BF(255.02)}$$

30. $$\int_b^u \sqrt{\frac{a-x}{(x-b)(x-c)(x-d)}}\, dx =$$ 3.167

$$= \frac{2}{\sqrt{(a-c)(b-d)}} \left[(c-b)\, \Pi\!\left(\lambda, \frac{a-b}{a-c}, r\right) + (a-c)\, F(\lambda, r) \right]$$
$$[a \geq u > b > c > d]. \quad \text{BF(256.14)}$$

31. $$\int_u^a \sqrt{\frac{a-x}{(x-b)(x-c)(x-d)}}\, dx = \frac{2(d-a)}{\sqrt{(a-c)(b-d)}} \left[\Pi\!\left(\mu, \frac{b-a}{b-d}\right) - F(\mu, r) \right]$$
$$[a > u \geq b > c > d]. \quad \text{BF(257.14)}$$

32. $$\int_a^u \sqrt{\frac{x-a}{(x-b)(x-c)(x-d)}}\, dx = \frac{2(a-b)}{\sqrt{(a-c)(b-d)}} \left[\Pi\!\left(v, \frac{a-d}{b-d}, q\right) - F(v, q) \right]$$
$$[u > a > b > c > d]. \quad \text{BF(258.15)}$$

1. $$\int_u^c \sqrt{\frac{c-x}{(a-x)(b-x)(x-d)^3}}\, dx =$$ 3.168

$$= \frac{2}{d-a} \left[\sqrt{\frac{a-c}{b-d}}\, E(\gamma, r) - \sqrt{\frac{(a-u)(c-u)}{(b-u)(u-d)}} \right] \quad [a > b > c > u > d]. \quad \text{BF(253.06)}$$

2. $$\int_c^u \sqrt{\frac{x-c}{(a-x)(b-x)(x-d)^3}}\, dx = \frac{2}{a-d} \sqrt{\frac{a-c}{b-d}}\, [F(\delta, q) - E(\delta, q)]$$
$$[a > b \geq u > c > d]. \quad \text{BF(254.04)}$$

3. $$\int_u^b \sqrt{\frac{x-c}{(a-x)(b-x)(x-d)^3}}\, dx = \frac{2}{a-d} \sqrt{\frac{a-c}{b-d}}\, [F(\varkappa, q) - E(\varkappa, q)] +$$
$$+ \frac{2}{b-d} \sqrt{\frac{(b-u)(u-c)}{(a-u)(u-d)}} \quad [a > b > u \geq c > d]. \quad \text{BF(255.09)}$$

4. $$\int_b^u \sqrt{\frac{x-c}{(a-x)(x-b)(x-d)^3}}\, dx =$$

$$= \frac{2}{a-d} \left[\sqrt{\frac{a-c}{b-d}}\, E(\lambda, r) - \frac{c-d}{b-d} \sqrt{\frac{(a-u)(u-b)}{(u-c)(u-d)}} \right] \quad [a \geq u > b > c > d]. \quad \text{BF(256.06)}$$

5. $$\int_u^a \sqrt{\frac{x-c}{(a-x)(x-b)(x-d)^3}}\, dx = \frac{2}{a-d} \sqrt{\frac{a-c}{b-d}}\, E(\mu, r)$$
$$[a > u \geq b > c > d]. \quad \text{BF(257.01)}$$

6. $$\int_a^u \sqrt{\frac{x-c}{(x-a)(x-b)(x-d)^3}}\, dx =$$

$$= \frac{2}{a-d} \sqrt{\frac{a-c}{b-d}}\, [F(v, q) - E(v, q)] + \frac{2}{a-d} \sqrt{\frac{(u-a)(u-c)}{(u-b)(u-d)}}$$
$$[u > a > b > c > d]. \quad \text{BF(258.10)}$$

3.168

7. $$\int_u^c \sqrt{\frac{b-x}{(a-x)(c-x)(x-d)^3}}\,dx =$$
$$= \frac{2}{(a-d)(c-d)\sqrt{(a-c)(b-d)}}[(b-c)(a-d)F(\gamma,r) - (a-c)(b-d)E(\gamma,r)] +$$
$$+ \frac{2(b-d)}{(a-d)(c-d)}\sqrt{\frac{(a-u)(c-u)}{(b-u)(u-d)}} \quad [a > b > c > u > d]. \qquad \text{BF(253.03)}$$

8. $$\int_c^u \sqrt{\frac{b-x}{(a-x)(x-c)(x-d)^3}}\,dx = \frac{2}{(a-d)(c-d)\sqrt{(a-c)(b-d)}} \times$$
$$\times [(a-c)(b-d)E(\delta,q) - (a-b)(c-d)F(\delta,q)]$$
$$[a > b \geqslant u > c > d]. \qquad \text{BF(254.15)}$$

9. $$\int_u^b \sqrt{\frac{b-x}{(a-x)(x-c)(x-d)^3}}\,dx = \frac{2}{(a-d)(c-d)\sqrt{(a-c)(b-d)}} \times$$
$$\times [(a-c)(b-d)E(\varkappa,q) - (a-b)(c-d)F(\varkappa,q)] -$$
$$- \frac{2}{c-d}\sqrt{\frac{(b-u)(u-c)}{(a-u)(u-d)}} \quad [a > b > u \geqslant c > d]. \qquad \text{BF(255.06)}$$

10. $$\int_b^u \sqrt{\frac{x-b}{(a-x)(x-c)(x-d)^3}}\,dx = \frac{2}{(a-d)(c-d)\sqrt{(a-c)(b-d)}} \times$$
$$\times [(a-c)(b-d)E(\lambda,r) - (a-d)(b-c)F(\lambda,r)] -$$
$$- \frac{2}{a-d}\sqrt{\frac{(a-u)(u-b)}{(u-c)(u-d)}} \quad [a \geqslant u > b > c > d]. \qquad \text{BF(256.03)}$$

11. $$\int_u^a \sqrt{\frac{x-b}{(a-x)(x-c)(x-d)^3}}\,dx = 2\frac{\sqrt{(a-c)(b-d)}}{(a-d)(c-d)}E(\mu,r) -$$
$$- \frac{2(b-c)}{(c-d)\sqrt{(a-c)(b-d)}}F(\mu,r) \quad [a > u \geqslant b > c > d]. \qquad \text{BF(257.09)}$$

12. $$\int_a^u \sqrt{\frac{x-b}{(x-a)(x-c)(x-d)^3}}\,dx = \frac{2(b-d)}{(a-d)(c-d)}\sqrt{\frac{(u-a)(u-c)}{(u-b)(u-d)}} +$$
$$+ \frac{2(a-b)}{(a-d)\sqrt{(a-c)(b-d)}}F(\nu,q) + 2\frac{\sqrt{(a-c)(b-d)}}{(a-d)(c-d)}E(\nu,q)$$
$$[u > a > b > c > d]. \qquad \text{BF(258.09)}$$

13. $$\int_u^c \sqrt{\frac{a-x}{(b-x)(c-x)(x-d)^3}}\,dx = \frac{2}{c-d}\sqrt{\frac{a-c}{b-d}}[F(\gamma,r) - E(\gamma,r)] +$$
$$+ \frac{2}{c-d}\sqrt{\frac{(a-u)(c-u)}{(b-u)(u-d)}} \quad [a > b > c > u > d]. \qquad \text{BF(253.04)}$$

14. $$\int_c^u \sqrt{\frac{a-x}{(b-x)(x-c)(x-d)^3}}\,dx = \frac{2}{c-d}\sqrt{\frac{a-c}{b-d}}\,E(\delta, q) \qquad 3.168$$
$$[a > b \geq u > c > d]. \qquad \text{BF(254.01)}$$

15. $$\int_u^b \sqrt{\frac{a-x}{(b-x)(x-c)(x-d)^3}}\,dx = \frac{2}{c-d}\sqrt{\frac{a-c}{b-d}}\,E(\varkappa, q) -$$
$$-\frac{2(a-d)}{(b-d)(c-d)}\sqrt{\frac{(b-u)(u-c)}{(a-u)(u-d)}} \qquad [a > b > u \geq c > d]. \qquad \text{BF(255.08)}$$

16. $$\int_b^u \sqrt{\frac{a-x}{(x-b)(x-c)(x-d)^3}}\,dx =$$
$$= \frac{2}{c-d}\sqrt{\frac{a-c}{b-d}}\,[F(\lambda, r) - E(\lambda, r)] + \frac{2}{b-d}\sqrt{\frac{(a-u)(u-b)}{(u-c)(u-d)}}$$
$$[a \geq u > b > c > d]. \qquad \text{BF(256.05)}$$

17. $$\int_u^a \sqrt{\frac{a-x}{(x-b)(x-c)(x-d)^3}}\,dx = \frac{2}{c-d}\sqrt{\frac{a-c}{b-d}}\,[F(\mu, r) - E(\mu, r)]$$
$$[a > u \geq b > c > d]. \qquad \text{BF(257.06)}$$

18. $$\int_a^u \sqrt{\frac{x-a}{(x-b)(x-c)(x-d)^3}}\,dx = \frac{-2}{c-d}\sqrt{\frac{a-c}{b-d}}\,E(\nu, q) +$$
$$+\frac{2}{c-d}\sqrt{\frac{(u-a)(u-c)}{(u-b)(u-d)}} \qquad [u > a > b > c > d]. \qquad \text{BF(258.05)}$$

19. $$\int_u^d \sqrt{\frac{d-x}{(a-x)(b-x)(c-x)^3}}\,dx = \frac{2}{b-c}\sqrt{\frac{b-d}{a-c}}\,[F(\alpha, q) - E(\alpha, q)]$$
$$[a > b > c > d > u]. \qquad \text{BF(251.01)}$$

20. $$\int_d^u \sqrt{\frac{x-d}{(a-x)(b-x)(c-x)^3}}\,dx = \frac{-2}{b-c}\sqrt{\frac{b-d}{a-c}}\,E(\beta, r) +$$
$$+\frac{2}{b-c}\sqrt{\frac{(b-u)(u-d)}{(a-u)(c-u)}} \qquad [a > b > c \geq u > d]. \qquad \text{BF(252.06)}$$

21. $$\int_u^b \sqrt{\frac{x-d}{(a-x)(b-x)(x-c)^3}}\,dx =$$
$$= \frac{2}{b-c}\sqrt{\frac{b-d}{a-c}}\,[F(\varkappa, q) - E(\varkappa, q)] + \frac{2}{b-c}\sqrt{\frac{(b-u)(u-d)}{(a-u)(u-c)}}$$
$$[a > b > u > c > d]. \qquad \text{BF(255.05)}$$

3.168

22. $\int_b^u \sqrt{\dfrac{x-d}{(a-x)(x-b)(x-c)^3}}\,dx = \dfrac{2}{b-c}\sqrt{\dfrac{b-d}{a-c}}\,E(\lambda,r)$

$[a \geqslant u > b > c > d]$. BF(256.01)

23. $\int_u^a \sqrt{\dfrac{x-d}{(a-x)(x-b)(x-c)^3}}\,dx =$

$= \dfrac{2}{b-c}\sqrt{\dfrac{b-d}{a-c}}\,E(\mu,r) - \dfrac{2(c-d)}{(a-c)(b-c)}\sqrt{\dfrac{(a-u)(u-b)}{(u-c)(u-d)}}$

$[a > u \geqslant b > c > d]$. BF(257.06)

24. $\int_a^u \sqrt{\dfrac{x-d}{(x-a)(x-b)(x-c)^3}}\,dx =$

$= \dfrac{2}{b-c}\sqrt{\dfrac{b-d}{a-c}}\,[F(v,q) - E(v,q)] + \dfrac{2}{a-c}\sqrt{\dfrac{(u-a)(u-d)}{(u-b)(u-c)}}$

$[u > a > b > c > d]$. BF(258.06)

25. $\int_u^a \sqrt{\dfrac{b-x}{(a-x)(c-x)^3(d-x)}}\,dx = \dfrac{2}{c-d}\sqrt{\dfrac{b-d}{a-c}}\,E(\alpha,q)$

$[a > b > c > d > u]$. BF(251.01)

26. $\int_a^u \sqrt{\dfrac{b-x}{(a-x)(c-x)^3(x-d)}}\,dx =$

$= \dfrac{2}{c-d}\sqrt{\dfrac{b-d}{a-c}}\,[F(\beta,r) - E(\beta,r)] + \dfrac{2}{c-d}\sqrt{\dfrac{(b-u)(u-d)}{(a-u)(c-u)}}$

$[a > b > c > u > d]$. BF(252.03)

27. $\int_u^b \sqrt{\dfrac{b-x}{(a-x)(x-c)^3(x-d)}}\,dx =$

$= \dfrac{2}{d-c}\sqrt{\dfrac{b-d}{a-c}}\,E(\varkappa,q) + \dfrac{2}{c-d}\sqrt{\dfrac{(b-u)(u-d)}{(a-u)(u-c)}}$

$[a > b > u > c > d]$. BF(255.03)

28. $\int_b^u \sqrt{\dfrac{x-b}{(a-x)(x-c)^3(x-d)}}\,dx = \dfrac{2}{c-d}\sqrt{\dfrac{b-d}{a-c}}\,[F(\lambda,r) - E(\lambda,r)]$

$[a \geqslant u > b > c > d]$. BF(256.08)

29. $\int\limits_{u}^{a} \sqrt{\dfrac{x-b}{(a-x)(x-c)^3(x-d)}}\, dx =$ 3.168

$$= \dfrac{2}{c-d}\sqrt{\dfrac{b-d}{a-c}}[F(\mu,r) - E(\mu,r)] + \dfrac{2}{a-c}\sqrt{\dfrac{(a-u)(u-b)}{(u-c)(u-d)}}$$

$$[a > u \geqslant b > c > d]. \quad \text{BF(257.03)}$$

30. $\int\limits_{a}^{u} \sqrt{\dfrac{x-b}{(x-a)(x-c)^3(x-d)}}\, dx =$

$$= \dfrac{2}{c-d}\sqrt{\dfrac{b-d}{a-c}}E(v,q) - \dfrac{2(b-c)}{(a-c)(c-d)}\sqrt{\dfrac{(u-a)(u-d)}{(u-b)(u-c)}}$$

$$[u > a > b > c > d]. \quad \text{BF(258.03)}$$

31. $\int\limits_{u}^{d} \sqrt{\dfrac{a-x}{(b-x)(c-x)^3(d-x)}}\, dx =$

$$= \dfrac{2\sqrt{(a-c)(b-d)}}{(b-c)(c-d)}E(\alpha,q) - \dfrac{a-b}{b-c}\dfrac{2}{\sqrt{(a-c)(b-d)}}F(\alpha,q)$$

$$[a > b > c > d > u]. \quad \text{BF(251.08)}$$

32. $\int\limits_{d}^{u} \sqrt{\dfrac{a-x}{(b-x)(c-x)^3(x-d)}}\, dx = \dfrac{2(a-d)}{(c-d)\sqrt{(a-c)(b-d)}}F(\beta,r) -$

$$- 2\dfrac{\sqrt{(a-c)(b-d)}}{(b-c)(c-d)}E(\beta,r) + 2\dfrac{a-c}{(b-c)(c-d)}\sqrt{\dfrac{(b-u)(u-d)}{(a-u)(c-u)}}$$

$$[a > b > c > u > d]. \quad \text{BF(252.04)}$$

33. $\int\limits_{u}^{b} \sqrt{\dfrac{a-x}{(b-x)(x-c)^3(x-d)}}\, dx = \dfrac{2(a-b)}{(b-c)\sqrt{(a-c)(b-d)}}F(\varkappa,q) -$

$$- 2\dfrac{\sqrt{(a-c)(b-d)}}{(b-c)(c-d)}E(\varkappa,q) + \dfrac{2(a-c)}{(b-c)(c-d)}\sqrt{\dfrac{(b-u)(u-d)}{(a-u)(u-c)}}$$

$$[a > b > u > c > d]. \quad \text{BF(255.04)}$$

34. $\int\limits_{b}^{u} \sqrt{\dfrac{a-x}{(x-b)(x-c)^3(x-d)}}\, dx =$

$$= \dfrac{2\sqrt{(a-c)(b-d)}}{(b-c)(c-d)}E(\lambda,r) - \dfrac{2(a-d)}{(c-d)\sqrt{(a-c)(b-d)}}F(\lambda,r)$$

$$[a \geqslant u > b > c > d]. \quad \text{BF(256.09)}$$

35. $\int\limits_{u}^{a} \sqrt{\dfrac{a-x}{(x-b)(x-c)^3(x-d)}}\, dx = \dfrac{2\sqrt{(a-c)(b-d)}}{(b-c)(c-d)}E(\mu,r) -$

$$- \dfrac{2(a-d)}{(c-d)\sqrt{(a-c)(b-d)}}F(\mu,r) - \dfrac{2}{b-c}\sqrt{\dfrac{(a-u)(u-b)}{(u-c)(u-d)}}$$

$$[a > u \geqslant b > c > d]. \quad \text{BF(257.04)}$$

3.168

36. $$\int_a^u \sqrt{\frac{x-a}{(x-b)(x-c)^3(x-d)}}\,dx = \frac{2\sqrt{(a-c)(b-d)}}{(b-c)(c-d)} E(v,q) -$$
$$- \frac{2(a-b)}{(b-c)\sqrt{(a-c)(b-d)}} F(v,q) - \frac{2}{c-d}\sqrt{\frac{(u-a)(u-d)}{(u-b)(u-c)}}$$
$$[u > a > b > c > d]. \qquad \text{BF(258.04)}$$

37. $$\int_u^d \sqrt{\frac{d-x}{(a-x)(b-x)^3(c-x)}}\,dx = \frac{2\sqrt{(a-c)(b-d)}}{(a-b)(b-c)} E(\alpha,q) -$$
$$- \frac{2(c-d)}{(b-c)\sqrt{(a-c)(b-d)}} F(\alpha,q) - \frac{2}{a-b}\sqrt{\frac{(a-u)(d-u)}{(b-u)(c-u)}}$$
$$[a > b > c > d > u]. \qquad \text{BF(251.11)}$$

38. $$\int_d^u \sqrt{\frac{x-d}{(a-x)(b-x)^3(c-x)}}\,dx = \frac{2\sqrt{(a-c)(b-d)}}{(a-b)(b-c)} E(\beta,r) -$$
$$- \frac{2(a-d)}{(a-b)\sqrt{(a-c)(b-d)}} F(\beta,r) + \frac{2}{b-c}\sqrt{\frac{(c-u)(u-d)}{(a-u)(b-u)}}$$
$$[a > b > c \geqslant u > d]. \qquad \text{BF(252.07)}$$

39. $$\int_u^c \sqrt{\frac{x-d}{(a-x)(b-x)^3(c-x)}}\,dx = \frac{2\sqrt{(a-c)(b-d)}}{(a-b)(b-c)} E(\gamma,r) -$$
$$- \frac{2(a-d)}{(a-b)\sqrt{(a-c)(b-d)}} F(\gamma,r) \quad [a > b > c > u \geqslant d]. \qquad \text{BF(253.07)}$$

40. $$\int_c^u \sqrt{\frac{x-d}{(a-x)(b-x)^3(x-c)}}\,dx = \frac{2(c-d)}{(b-c)\sqrt{(a-c)(b-d)}} F(\delta,q) -$$
$$- \frac{2\sqrt{(a-c)(b-d)}}{(a-b)(b-c)} E(\delta,q) + \frac{2(b-d)}{(a-b)(b-c)}\sqrt{\frac{(a-u)(u-c)}{(b-u)(u-d)}}$$
$$[a > b > u > c > d]. \qquad \text{BF(254.05)}$$

41. $$\int_u^a \sqrt{\frac{x-d}{(a-x)(x-b)^3(x-c)}}\,dx = \frac{2(a-d)}{(a-b)\sqrt{(a-c)(b-d)}} F(\mu,r) -$$
$$- \frac{2\sqrt{(a-c)(b-d)}}{(a-b)(b-c)} E(\mu,r) + \frac{2(b-d)}{(a-b)(b-c)}\sqrt{\frac{(a-u)(u-c)}{(u-b)(u-d)}}$$
$$[a > u > b > c > d]. \qquad \text{BF(257.07)}$$

42. $$\int_a^u \sqrt{\frac{x-d}{(x-a)(x-b)^3(x-c)}}\,dx = \frac{2\sqrt{(a-c)(b-d)}}{(a-b)(b-c)} E(v,q) -$$
$$- \frac{2(c-d)}{(b-c)\sqrt{(a-c)(b-d)}} F(v,q) \quad [u > a > b > c > d]. \qquad \text{BF(258.07)}$$

43. $\displaystyle\int_u^d \sqrt{\frac{c-x}{(a-x)(b-x)^3(d-x)}}\,dx = \frac{2}{a-b}\sqrt{\frac{a-c}{b-d}}E(\alpha, q)-$ 3.168

$\qquad -\dfrac{2(b-c)}{(a-b)(b-d)}\sqrt{\dfrac{(a-u)(d-u)}{(b-u)(c-u)}}\quad [a>b>c>d>u].$ BF(251.14)

44. $\displaystyle\int_d^u \sqrt{\frac{c-x}{(a-x)(b-x)^3(x-d)}}\,dx = \frac{2}{a-b}\sqrt{\frac{a-c}{b-d}}[F(\beta, r)-E(\beta, r)]+$

$\qquad +\dfrac{2}{b-d}\sqrt{\dfrac{(c-u)(u-d)}{(a-u)(b-u)}}\quad [a>b>c\geqslant u>d].$ BF(252.10)

45. $\displaystyle\int_u^c \sqrt{\frac{c-x}{(a-x)(b-x)^3(x-d)}}\,dx = \frac{2}{a-b}\sqrt{\frac{a-c}{b-d}}[F(\gamma, r)-E(\gamma, r)]$

$\qquad\qquad\qquad [a>b>c>u\geqslant d].$ BF(254.08)

46. $\displaystyle\int_c^u \sqrt{\frac{x-c}{(a-x)(b-x)^3(x-d)}}\,dx = \frac{2}{b-a}\sqrt{\frac{a-c}{b-d}}E(\delta, q)+$

$\qquad +\dfrac{2}{a-b}\sqrt{\dfrac{(a-u)(u-c)}{(b-u)(u-d)}}\quad [a>b\geqslant u>c>d].$ BF(254.08)

47. $\displaystyle\int_u^a \sqrt{\frac{x-c}{(a-x)(x-b)^3(x-d)}}\,dx = \frac{2}{a-b}\sqrt{\frac{a-c}{b-d}}[F(\mu, r)-E(\mu, r)]+$

$\qquad +\dfrac{2}{a-b}\sqrt{\dfrac{(a-u)(u-c)}{(u-b)(u-d)}}\quad [a>u\geqslant b>c>d].$ BF(257.10)

48. $\displaystyle\int_a^u \sqrt{\frac{x-c}{(x-a)(x-b)^3(x-d)}}\,dx = \frac{2}{a-b}\sqrt{\frac{a-c}{b-d}}E(v, q)$

$\qquad\qquad\qquad [u>a>b>c>d].$ BF(258.01)

49. $\displaystyle\int_u^d \sqrt{\frac{a-x}{(b-x)^3(c-x)(d-x)}}\,dx =$

$\qquad = \dfrac{2}{b-c}\sqrt{\dfrac{a-c}{b-d}}[F(\alpha, q)-E(\alpha, q)] + \dfrac{2}{b-d}\sqrt{\dfrac{(a-u)(d-u)}{(b-u)(c-u)}}$

$\qquad\qquad\qquad [a>b>c>d>u].$ BF(251.12)

50. $\displaystyle\int_d^u \sqrt{\frac{a-x}{(b-x)^3(c-x)(x-d)}}\,dx = \frac{2}{b-c}\sqrt{\frac{a-c}{b-d}}E(\beta, r)-$

$\qquad -\dfrac{2(a-b)}{(b-c)(b-d)}\sqrt{\dfrac{(u-d)(c-u)}{(a-u)(b-u)}}\quad [a>b>c\geqslant u>d].$ BF(252.09)

3.168

51. $$\int_u^c \sqrt{\frac{a-x}{(b-x)^3(c-x)(x-d)}}\, dx = \frac{2}{b-c}\sqrt{\frac{a-c}{b-d}}\, E(\gamma, r)$$

$$[a > b > c > u \geqslant d].\quad \text{BF(253.01)}$$

52. $$\int_c^u \sqrt{\frac{a-x}{(b-x)^3(x-c)(x-d)}}\, dx =$$

$$= \frac{2}{b-c}\sqrt{\frac{a-c}{b-d}}\, [F(\delta, q) - E(\delta, q)] + \frac{2}{b-c}\sqrt{\frac{(a-u)(u-c)}{(b-u)(u-d)}}$$

$$[a > b > u > c > d].\quad \text{BF(254.06)}$$

53. $$\int_u^a \sqrt{\frac{a-x}{(x-b)^3(x-c)(x-d)}}\, dx = \frac{2}{c-b}\sqrt{\frac{a-c}{b-d}}\, E(\mu, r) +$$

$$+ \frac{2}{b-c}\sqrt{\frac{(a-u)(u-c)}{(u-b)(u-d)}}\quad [a > u > b > c > d].\quad \text{BF(257.08)}$$

54. $$\int_a^u \sqrt{\frac{x-a}{(x-b)^3(x-c)(x-d)}}\, dx = \frac{2}{b-c}\sqrt{\frac{a-c}{b-d}}\, [F(v, q) - E(v, q)]$$

$$[u > a > b > c > d].\quad \text{BF(258.08)}$$

55. $$\int_u^d \sqrt{\frac{d-x}{(a-x)^3(b-x)(c-x)}}\, dx = \frac{2}{b-a}\sqrt{\frac{b-d}{a-c}}\, E(\alpha, q) +$$

$$+ \frac{2}{a-b}\sqrt{\frac{(b-u)(d-u)}{(a-u)(c-u)}}\quad [a > b > c > d > u].\quad \text{BF(251.09)}$$

56. $$\int_d^u \sqrt{\frac{x-d}{(a-x)^3(b-x)(c-x)}}\, dx = \frac{2}{a-b}\sqrt{\frac{b-d}{a-c}}\, [F(\beta, q) - E(\beta, q)]$$

$$[a > b > c \geqslant u > d].\quad \text{BF(252.05)}$$

57. $$\int_u^c \sqrt{\frac{x-d}{(a-x)^3(b-x)(c-x)}}\, dx =$$

$$= \frac{2}{a-b}\sqrt{\frac{b-d}{a-c}}\, [F(\gamma, r) - E(\gamma, r)] + \frac{2}{a-c}\sqrt{\frac{(c-u)(u-d)}{(a-u)(b-u)}}$$

$$[a > b > c > u \geqslant d].\quad \text{BF(253.05)}$$

58. $$\int_c^u \sqrt{\frac{x-d}{(a-x)^3(b-x)(x-c)}}\, dx = \frac{2}{a-b}\sqrt{\frac{b-d}{a-c}}\, E(\delta, q) -$$

$$- \frac{2(a-d)}{(a-b)(a-c)}\sqrt{\frac{(b-u)(u-c)}{(a-u)(u-d)}}\quad [a > b \geqslant u > c > d].\quad \text{BF(254.03)}$$

59. $\int_u^b \sqrt{\dfrac{x-d}{(a-x)^3(b-x)(x-c)}}\,dx = \dfrac{2}{a-b}\sqrt{\dfrac{b-d}{a-c}}\,E(\varkappa,q)$ 3.168

$[a > b > u \geqslant c > d]$. BF(255.01)

60. $\int_b^u \sqrt{\dfrac{x-d}{(a-x)^3(x-b)(x-c)}}\,dx =$

$= \dfrac{2}{a-b}\sqrt{\dfrac{b-d}{a-c}}\,[F(\lambda,r) - E(\lambda,r)] + \dfrac{2}{a-b}\sqrt{\dfrac{(u-b)(u-d)}{(a-u)(u-c)}}$

$[a > u > b > c > d]$. BF(256.10)

61. $\int_u^d \sqrt{\dfrac{c-x}{[(a-x)^3(b-x)(d-x)}}\,dx = \dfrac{2(c-d)}{(a-d)\sqrt{(a-c)(b-d)}}\,F(\alpha,q) -$

$- \dfrac{2\sqrt{(a-c)(b-d)}}{(a-b)(a-d)}\,E(\alpha,q) + \dfrac{2(a-c)}{(a-b)(a-d)}\sqrt{\dfrac{(b-u)(d-u)}{(a-u)(c-u)}}$

$[a > b > c > d > u]$. BF(251.15)

62. $\int_d^u \sqrt{\dfrac{c-x}{(a-x)^3(b-x)(x-d)}}\,dx =$

$= \dfrac{2\sqrt{(a-c)(b-d)}}{(a-b)(a-d)}\,E(\beta,r) - \dfrac{2(b-c)}{(a-b)\sqrt{(a-c)(b-d)}}\,F(\beta,r)$

$[a > b > c \geqslant u > d]$. BF(252.08)

63. $\int_u^c \sqrt{\dfrac{c-x}{(a-x)^3(b-x)(x-d)}}\,dx = \dfrac{2\sqrt{(a-c)(b-d)}}{(a-b)(a-d)}\,E(\gamma,r) -$

$- \dfrac{2(b-c)}{(a-b)\sqrt{(a-c)(b-d)}}\,F(\gamma,r) - \dfrac{2}{a-d}\sqrt{\dfrac{(c-u)(u-d)}{(a-u)(b-u)}}$

$[a > b > c > u \geqslant d]$. BF(253.10)

64. $\int_c^u \sqrt{\dfrac{x-c}{(a-x)^3(b-x)(x-d)}}\,dx = \dfrac{2\sqrt{(a-c)(b-d)}}{(a-b)(a-d)}\,E(\delta,q) -$

$- \dfrac{2(c-d)}{(a-d)\sqrt{(a-c)(b-d)}}\,F(\delta,q) - \dfrac{2}{a-b}\sqrt{\dfrac{(b-u)(u-c)}{(a-u)(u-d)}}$

$[a > b \geqslant u > c > d]$. BF(254.09)

65. $\int_u^b \sqrt{\dfrac{x-c}{(a-x)^3(b-x)(x-d)}}\,dx =$

$= \dfrac{2\sqrt{(a-c)(b-d)}}{(a-b)(a-d)}\,E(\varkappa,q) - \dfrac{2(c-d)}{(a-d)\sqrt{(a-c)(b-d)}}\,F(\varkappa,q)$

$[a > b > u \geqslant c > d]$. BF(255.10)

3.168

66. $$\int\limits_{b}^{u}\sqrt{\frac{x-c}{(a-x)^3(x-b)(x-d)}}\,dx =$$

$$= \frac{2(b-c)}{(a-b)\sqrt{(a-c)(b-d)}}F(\lambda,r) - \frac{2\sqrt{(a-c)(b-d)}}{(a-b)(a-d)}E(\lambda,r) +$$

$$+ \frac{2(a-c)}{(a-b)(a-d)}\sqrt{\frac{(u-b)(u-d)}{(a-u)(a-c)}} \quad [a>u>b>c>d]. \qquad \text{BF(256.07)}$$

67. $$\int\limits_{u}^{d}\sqrt{\frac{b-x}{(a-x)^3(c-x)(d-x)}}\,dx =$$

$$= \frac{2}{a-d}\sqrt{\frac{b-d}{a-c}}[F(\alpha,q) - E(\alpha,q)] + \frac{2}{a-d}\sqrt{\frac{(b-u)(d-u)}{(a-u)(c-u)}}$$

$$[a>b>c>d>u]. \qquad \text{BF(251.13)}$$

68. $$\int\limits_{d}^{u}\sqrt{\frac{b-x}{(a-x)^3(c-x)(x-d)}}\,dx = \frac{2}{a-d}\sqrt{\frac{b-d}{a-c}}E(\beta,r)$$

$$[a>b>c\geqslant u>d]. \qquad \text{BF(252.01)}$$

69. $$\int\limits_{u}^{c}\sqrt{\frac{b-x}{(a-x)^3(c-x)(x-d)}}\,dx =$$

$$= \frac{2}{a-d}\sqrt{\frac{b-d}{a-c}}E(\gamma,r) - \frac{2(a-b)}{(a-c)(a-d)}\sqrt{\frac{(c-u)(u-d)}{(a-u)(b-u)}}$$

$$[a>b>c>u\geqslant d]. \qquad \text{BF(253.08)}$$

70. $$\int\limits_{c}^{u}\sqrt{\frac{b-x}{(a-x)^3(x-c)(x-d)}}\,dx =$$

$$= \frac{2}{a-d}\sqrt{\frac{b-d}{a-c}}[F(\delta,q) - E(\delta,q)] + \frac{2}{a-c}\sqrt{\frac{(b-u)(u-c)}{(a-u)(u-d)}}$$

$$[a>b\geqslant u>c>d]. \qquad \text{BF(254.07)}$$

71. $$\int\limits_{u}^{b}\sqrt{\frac{b-x}{(a-x)^3(x-c)(x-d)}}\,dx = \frac{2}{a-d}\sqrt{\frac{b-d}{a-c}}[F(\varkappa,q) - E(\varkappa,q)]$$

$$[a>b>u\geqslant c>d]. \qquad \text{BF(255.07)}$$

72. $$\int\limits_{b}^{u}\sqrt{\frac{x-b}{(a-x)^3(x-c)(x-d)}}\,dx =$$

$$= \frac{-2}{a-d}\sqrt{\frac{b-d}{a-c}}E(\lambda,r) + \frac{2}{a-d}\sqrt{\frac{(u-b)(u-d)}{(a-u)(u-c)}}$$

$$[a\geqslant u>b>c>d]. \qquad \text{BF(256.04)}$$

In 3.169—3.172 werden folgende Abkürzungen verwendet: | In 3.169—3.172 the following notation is used:

$$\alpha = \operatorname{arctg}\frac{u}{b}, \quad \beta = \operatorname{arctg}\frac{a}{u},$$

$$\gamma = \arcsin\frac{u}{b}\sqrt{\frac{a^2+b^2}{a^2+u^2}}, \quad \delta = \arccos\frac{u}{b}, \quad \varepsilon = \arccos\frac{b}{u}, \quad \zeta = \arcsin\sqrt{\frac{a^2+b^2}{a^2+u^2}},$$

$$\eta = \arcsin\frac{u}{b}, \quad \zeta = \arcsin\frac{a}{b}\sqrt{\frac{b^2-u^2}{a^2-u^2}}, \quad \varkappa = \arcsin\frac{a}{u}\sqrt{\frac{u^2-b^2}{a^2-b^2}},$$

$$\lambda = \arcsin\sqrt{\frac{a^2-u^2}{a^2-b^2}}, \quad \mu = \arcsin\sqrt{\frac{u^2-a^2}{u^2-b^2}}, \quad \nu = \arcsin\frac{a}{u}, \quad q = \frac{\sqrt{a^2-b^2}}{a},$$

$$r = \frac{b}{\sqrt{a^2+b^2}}, \quad s = \frac{a}{\sqrt{a^2+b^2}}, \quad t = \frac{b}{a}.$$

1. $\int_0^u \sqrt{\frac{x^2+a^2}{x^2+b^2}}\,dx = a\{F(\alpha,q) - E(\alpha,q)\} + u\sqrt{\frac{a^2+u^2}{b^2+u^2}}$ $[a>b,\ u>0]$.   3.169

BF(221.03)

2. $\int_0^u \sqrt{\frac{x^2+b^2}{x^2+a^2}}\,dx = \frac{b^2}{a}F(\beta,q) - aE(\beta,q) + u\sqrt{\frac{b^2+u^2}{a^2+u^2}}$ $[a>b,\ u>0]$.

BF(221.04)

3. $\int_0^u \sqrt{\frac{x^2+a^2}{b^2-x^2}}\,dx = \sqrt{a^2+b^2}\,E(\gamma,r) - u\sqrt{\frac{b^2-u^2}{a^2+u^2}}$ $[b \geq u > 0]$.   BF(214.11)

4. $\int_u \sqrt{\frac{a^2+x^2}{b^2-x^2}}\,dx = \sqrt{a^2+b^2}\,E(\delta,r)$ $[b > u \geq 0]$.   BF(213.01), Z 64(273)

5. $\int_b^u \sqrt{\frac{a^2+x^2}{x^2-b^2}}\,dx = \sqrt{a^2+b^2}\,\{F(\varepsilon,s) - E(\varepsilon,s)\} +$

$$+ \frac{1}{u}\sqrt{(u^2+a^2)(u^2-b^2)} \quad [u>b>0].$$ BF(211.03)

6. $\int_0^u \sqrt{\frac{b^2-x^2}{a^2+x^2}}\,dx = \sqrt{a^2+b^2}\,\{F(\gamma,r) - E(\gamma,r)\} + u\sqrt{\frac{b^2-u^2}{a^2+u^2}}$

$[b \geq u > 0]$.   BF(214.03)

7. $\int_u^b \sqrt{\frac{b^2-x^2}{a^2+x^2}}\,dx = \sqrt{a^2+b^2}\,\{F(\delta,r) - E(\delta,r)\}$ $[b>u \geq 0]$.   BF(213.03)

8. $\int_b^u \sqrt{\frac{x^2-b^2}{a^2+x^2}}\,dx = \frac{1}{u}\sqrt{(a^2+u^2)(u^2-b^2)} - \sqrt{a^2+b^2}\,E(\varepsilon,s)$

$[u>b>0]$.   BF(211.04)

3.169

9. $\int_0^u \sqrt{\dfrac{b^2 - x^2}{a^2 - x^2}}\, dx = aE(\eta, t) - \dfrac{a^2 - b^2}{a} F(\eta, t) \quad [a > b \geqslant u > 0].$ BF(219.03)

10. $\int_u^b \sqrt{\dfrac{b^2 - x^2}{a^2 - x^2}}\, dx = aE(\zeta, t) - \dfrac{a^2 - b^2}{a} F(\zeta, t) - u\sqrt{\dfrac{b^2 - u^2}{a^2 - u^2}}$

$[a > b > u \geqslant 0].$ BF(220.04)

11. $\int_b^u \sqrt{\dfrac{x^2 - b^2}{a^2 - x^2}}\, dx = aE(\varkappa, q) - \dfrac{b^2}{a} F(\varkappa, q) -$

$- \dfrac{1}{u}\sqrt{(a^2 - u^2)(u^2 - b^2)} \quad [a \geqslant u > b > 0].$ BF(217.04)

12. $\int_u^a \sqrt{\dfrac{x^2 - b^2}{a^2 - x^2}}\, dx = aE(\lambda, q) - \dfrac{b^2}{a} F(\lambda, q) \quad [a > u \geqslant b > 0].$ BF(218.03)

13. $\int_a^u \sqrt{\dfrac{x^2 - b^2}{x^2 - a^2}}\, dx = \dfrac{a^2 - b^2}{a} F(\mu, t) - aE(\mu, t) + u\sqrt{\dfrac{u^2 - a^2}{u^2 - b^2}}$

$[u > a > b > 0].$ BF(216.03)

14. $\int_0^u \sqrt{\dfrac{a^2 - x^2}{b^2 - x^2}}\, dx = aE(\eta, t) \quad [a > b \geqslant u > 0].$ Z 64(276), BF(219.01)

15. $\int_u^b \sqrt{\dfrac{a^2 - x^2}{b^2 - x^2}}\, dx = a\left\{E(\zeta, t) - \dfrac{u}{a}\sqrt{\dfrac{b^2 - u^2}{a^2 - u^2}}\right\} \quad [a > b > u \geqslant 0].$ BF(220.03)

16. $\int_b^u \sqrt{\dfrac{a^2 - x^2}{x^2 - b^2}}\, dx = a\{F(\varkappa, q) - E(\varkappa, q)\} +$

$+ \dfrac{1}{u}\sqrt{(a^2 - u^2)(u^2 - b^2)} \quad [a \geqslant u > b > 0].$ BF(217.03)

17. $\int_u^a \sqrt{\dfrac{a^2 - x^2}{x^2 - b^2}}\, dx = a\{F(\lambda, q) - E(\lambda, q)\} \quad [a > u \geqslant b > 0].$ BF(218.09)

18. $\int_a^u \sqrt{\dfrac{x^2 - a^2}{x - b^2}}\, dx = u\sqrt{\dfrac{u^2 - a^2}{u^2 - b^2}} - aE(\mu, t) \quad [u > a > b > 0].$ BF(216.04)

3.171

1. $\int_b^u \dfrac{dx}{x^2}\sqrt{\dfrac{a^2 + x^2}{x^2 - b^2}} = \dfrac{\sqrt{a^2 + b^2}}{b^2} E(\varepsilon, s) \quad [u > b > 0].$ BF(211.01), Z 64(274)

2. $\int_u^\infty \dfrac{dx}{x^2}\sqrt{\dfrac{a^2 + x^2}{x^2 - b^2}} = \dfrac{\sqrt{a^2 + b^2}}{b^2} E(\xi, s) - \dfrac{a^2}{b^2 u}\sqrt{\dfrac{u^2 - b^2}{a^2 + u^2}} \quad [u \geqslant b > 0].$ BF(212.09)

3. $\int_u^b \dfrac{dx}{x^2}\sqrt{\dfrac{a^2 - x^2}{b^2 - x^2}} = \dfrac{a^2 - b^2}{ab^2} F(\zeta, t) - \dfrac{a}{b^2} E(\zeta, t) + \dfrac{a^2}{b^2 u}\sqrt{\dfrac{b^2 - u^2}{a^2 - u^2}} \quad [a > b > u > 0].$

BF(220.12)

4. $\int_b^u \frac{dx}{x^2} \sqrt{\frac{a^2 - x^2}{x^2 - b^2}} = \frac{a}{b^2} E(\varkappa, q) - \frac{1}{a} F(\varkappa, q)$ $[a \geq u > b > 0]$. BF(217.11) 3.171

5. $\int_u^a \frac{dx}{x^2} \sqrt{\frac{a^2 - x^2}{x^2 - b^2}} = \frac{a}{b^2} E(\lambda, q) - \frac{1}{a} F(\lambda, q) - \frac{\sqrt{(a^2 - u^2)(u^2 - b^2)}}{b^2 u}$

$[a > u \geq b > 0]$. BF(218.10)

6. $\int_a^u \frac{dx}{x^2} \sqrt{\frac{x^2 - a^2}{x^2 - b^2}} = \frac{a}{b^2} E(\mu, t) - \frac{a^2 - b^2}{ab^2} F(\mu, t) - \frac{1}{u} \sqrt{\frac{u^2 - a^2}{u^2 - b^2}}$

$[u > a > b > 0]$. BF(216.08)

7. $\int_u^\infty \frac{dx}{x^2} \sqrt{\frac{x^2 + a^2}{x^2 + b^2}} = \frac{1}{a} F(\beta, q) - \frac{a}{b^2} E(\beta, q) + \frac{a^2}{b^2 u} \sqrt{\frac{b^2 + u^2}{a^2 + u^2}}$

$[a > b, u > 0]$. BF(222.08)

8. $\int_u^\infty \frac{dx}{x^2} \sqrt{\frac{x^2 + b^2}{x^2 + a^2}} = \frac{1}{a} \{F(\beta, q) - E(\beta, q)\} + \frac{1}{u} \sqrt{\frac{b^2 + u^2}{a^2 + u^2}}$

$[a > b, u > 0]$. BF(222.09)

9. $\int_u^b \frac{dx}{x^2} \sqrt{\frac{b^2 - x^2}{a^2 + x^2}} = \frac{\sqrt{(b^2 - u^2)(a^2 + u^2)}}{a^2 u} - \frac{\sqrt{a^2 + b^2}}{a^2} E(\delta, r)$ $[b > u > 0]$.

BF(213.10)

10. $\int_b^u \frac{dx}{x^2} \sqrt{\frac{x^2 - b^2}{a^2 + x^2}} = \frac{\sqrt{a^2 + b^2}}{a^2} \{F(\varepsilon, s) - E(\varepsilon, s)\}$ $[u > b > 0]$. BF(211.07)

11. $\int_u^\infty \frac{dx}{x^2} \sqrt{\frac{x^2 - b^2}{a^2 + x^2}} = \frac{\sqrt{a^2 + b^2}}{a^2} \{F(\xi, s) - E(\xi, s)\} + \frac{1}{u} \sqrt{\frac{u^2 - b^2}{a^2 + u^2}}$

$[u \geq b > 0]$. BF(212.11)

12. $\int_u^b \frac{dx}{x^2} \sqrt{\frac{a^2 + x^2}{b^2 - x^2}} = -\frac{a^2 + b^2}{b^2} \{F(\delta, r) - E(\delta, r)\} + \frac{\sqrt{(b^2 - u^2)(a^2 + u^2)}}{b^2 u}$

$[b > u > 0]$. BF(213.05)

13. $\int_u^\infty \frac{dx}{x^2} \sqrt{\frac{x^2 - a^2}{x^2 - b^2}} = \frac{a}{b^2} E(v, t) - \frac{a^2 - b^2}{ab^2} F(v, t)$ $[u \geq a > b > 0]$. BF(215.08)

14. $\int_u^b \frac{dx}{x^2} \sqrt{\frac{b^2 - x^2}{a^2 - x^2}} = \frac{1}{u} \sqrt{\frac{b^2 - u^2}{a^2 - u^2}} - \frac{1}{a} E(\zeta, t)$ $[a > b > u > 0]$. BF(220.11)

3.171

15. $\displaystyle\int_b^u \frac{dx}{x^2} \sqrt{\frac{x^2-b^2}{a^2-x^2}} = \frac{1}{a}\{F(\varkappa,q) - E(\varkappa,q)\}$ $[a \geqslant u > b > 0]$. BF(217.08)

16. $\displaystyle\int_u^a \frac{dx}{x^2} \sqrt{\frac{x^2-b^2}{a^2-x^2}} = \frac{1}{a}\{F(\lambda,q) - E(\lambda,q)\} + \frac{\sqrt{(a^2-u^2)(u^2-b^2)}}{a^2 u}$

$[a > u \geqslant b > 0]$. BF(218.08)

17. $\displaystyle\int_a^u \frac{dx}{x^2} \sqrt{\frac{x^2-b^2}{x^2-a^2}} = \frac{1}{a} E(\mu,t) - \frac{1}{u}\sqrt{\frac{u^2-a^2}{u^2-b^2}}$ $[u > a > b > 0]$. BF(216.07)

18. $\displaystyle\int_u^\infty \frac{dx}{x^2} \sqrt{\frac{x^2-b^2}{x^2-a^2}} = \frac{1}{a} E(\nu,t)$ $[u \geqslant a > b > 0]$. BF(215.01), Z 65(281)

3.172

1. $\displaystyle\int_0^u \sqrt{\frac{x^2+b^2}{(x^2+a^2)^3}}\, dx = \frac{1}{a} E(\alpha,q) - \frac{a^2-b^2}{a^2} \frac{u}{\sqrt{(a^2+u^2)(b^2+u^2)}}$

$[a > b,\ u > 0]$. BF(221.10)

2. $\displaystyle\int_u^\infty \sqrt{\frac{x^2+b^2}{(x^2+a^2)^3}}\, dx = \frac{1}{a} E(\beta,q)$ $[\ > b,\ u \geqslant 0]$. Z 64(271)

3. $\displaystyle\int_0^u \sqrt{\frac{x^2+a^2}{(x^2+b^2)^3}}\, dx = \frac{a}{b^2} E(\alpha,q)$ $[a > b,\ u > 0]$. Z 64(270)

4. $\displaystyle\int_u^\infty \sqrt{\frac{x^2+a^2}{(x^2+b^2)^3}}\, dx = \frac{a}{b^2} E(\beta,q) - \frac{a^2-b^2}{b^2} \frac{u}{\sqrt{(a^2+u^2)(b^2+u^2)}}$

$[a > b,\ u \geqslant 0]$. BF(222.06)

5. $\displaystyle\int_0^u \sqrt{\frac{b^2-x^2}{(a^2+x^2)^3}}\, dx = \frac{\sqrt{a^2+b^2}}{a^2} E(\gamma,r) - \frac{1}{\sqrt{a^2+b^2}} F(\gamma,r)$

$[b \geqslant u > 0]$. BF(214.08)

6. $\displaystyle\int_u^b \sqrt{\frac{b^2-x^2}{(a^2+x^2)^3}}\, dx = \frac{\sqrt{a^2+b^2}}{a^2} E(\delta,r) - \frac{1}{\sqrt{a^2+b^2}} F(\delta,r) -$

$- \frac{u}{a^2}\sqrt{\frac{b^2-u^2}{a^2+u^2}}$ $[b > u \geqslant 0]$. BF(213.04)

7. $\displaystyle\int_b^u \sqrt{\frac{x^2-b^2}{(a^2+x^2)^3}}\, dx = \frac{\sqrt{a^2+b^2}}{a^2} E(\varepsilon,s) - \frac{b^2}{a^2\sqrt{a^2+b^2}} F(\varepsilon,s) -$

$- \frac{1}{u}\sqrt{\frac{u^2-b^2}{u^2+a^2}}$ $[u > b > 0]$. BF(211.06)

8. $\displaystyle\int_u^\infty \sqrt{\frac{x^2-b^2}{(a^2+x^2)^3}}\,dx = \frac{\sqrt{a^2+b^2}}{a^2} E(\xi, s) - \frac{b^2}{a^2\sqrt{a^2+b^2}} F(\xi, s)$ 3.172

$\qquad\qquad\qquad\qquad\qquad\qquad\qquad [u \geqslant b > 0].$ BF(212.08)

9. $\displaystyle\int_0^u \sqrt{\frac{x^2+a^2}{(b^2-x^2)^3}}\,dx = \frac{a^2}{b^2\sqrt{a^2+b^2}} F(\gamma, r) - \frac{\sqrt{a^2+b^2}}{b^2} E(\gamma, r) +$

$\qquad\qquad\qquad\qquad + \dfrac{(a^2+b^2)u}{b^2\sqrt{(a^2+u^2)(b^2-u^2)}}\quad [b > u > 0].$ BF(214.09)

10. $\displaystyle\int_u^\infty \sqrt{\frac{x^2+a^2}{(x^2-b^2)^3}}\,dx = \frac{1}{\sqrt{a^2+b^2}} F(\xi, s) - \frac{\sqrt{a^2+b^2}}{b^2} E(\xi, s) +$

$\qquad\qquad\qquad\qquad + \dfrac{(a^2+b^2)u}{b^2\sqrt{(a^2+u^2)(u^2-b^2)}}\quad [u > b > 0].$ BF(212.07)

11. $\displaystyle\int_0^u \sqrt{\frac{b^2-x^2}{(a^2-x^2)^3}}\,dx = \frac{1}{a}\left\{F(\eta, t) - E(\eta, t) + \frac{u}{a}\sqrt{\frac{b^2-u^2}{a^2-u^2}}\right\}$

$\qquad\qquad\qquad\qquad\qquad\qquad\qquad [a > b \geqslant u > 0].$ BF(219.09)

12. $\displaystyle\int_u^b \sqrt{\frac{b^2-x^2}{(a^2-x^2)^3}}\,dx = \frac{1}{a}\{F(\zeta, t) - E(\zeta, t)\}\quad [a > b > u \geqslant 0].$ BF(220.07)

13. $\displaystyle\int_b^u \sqrt{\frac{x^2-b^2}{(a^2-x^2)^3}}\,dx = \frac{1}{u}\sqrt{\frac{u^2-b^2}{a^2-u^2}} - \frac{1}{a} E(\varkappa, q)$

$\qquad\qquad\qquad\qquad\qquad\qquad\qquad [a > u > b > 0].$ BF(217.07)

14. $\displaystyle\int_u^\infty \sqrt{\frac{x^2-b^2}{(x^2-a^2)^3}}\,dx = \frac{1}{a}[F(v, t) - E(v, t)] + \frac{1}{u}\sqrt{\frac{u^2-b^2}{u^2-a^2}}$

$\qquad\qquad\qquad\qquad\qquad\qquad\qquad [u > a > b > 0].$ BF(215.05)

15. $\displaystyle\int_0^u \sqrt{\frac{a^2-x^2}{(b^2-x^2)^3}}\,dx = \frac{a}{b^2}[F(\eta, t) - E(\eta, t)] + \frac{u}{b^2}\sqrt{\frac{a^2-u^2}{b^2-u^2}}$

$\qquad\qquad\qquad\qquad\qquad\qquad\qquad [a > b > u > 0].$ BF(219.10)

16. $\displaystyle\int_u^a \sqrt{\frac{a^2-x^2}{(x^2-b^2)^3}}\,dx = \frac{u}{b^2}\sqrt{\frac{a^2-u^2}{u^2-b^2}} - \frac{a}{b^2} E(\lambda, q)\quad [a > u > b > 0].$

$\qquad\qquad\qquad\qquad\qquad\qquad\qquad$ BF(218.05)

17. $\displaystyle\int_a^u \sqrt{\frac{x^2-a^2}{(x^2-b^2)^3}}\,dx = \frac{a}{b^2}[F(\mu, t) - E(\mu, t)]\quad [u > a > b > 0].$ BF(216.05)

18. $\displaystyle\int_u^\infty \sqrt{\frac{x^2-a^2}{(x^2-b^2)^3}}\,dx = \frac{a}{b^2}[F(v, t) - E(v, t)] + \frac{1}{u}\sqrt{\frac{u^2-a^2}{u^2-b^2}}$

$\qquad\qquad\qquad\qquad\qquad\qquad\qquad [u \geqslant a > b > 0].$ BF(215.03)

3.173

1. $\int_u^1 \dfrac{dx}{x^2}\sqrt{\dfrac{x^2+1}{1-x^2}} = \sqrt{2}\left[F\left(\arccos u, \dfrac{\sqrt{2}}{2}\right) - E\left(\arccos u, \dfrac{\sqrt{2}}{2}\right)\right] + \dfrac{\sqrt{1-u^4}}{u}$ $[u < 1]$.   BF(259.77)

2. $\int_1^u \dfrac{dx}{x^2}\sqrt{\dfrac{x^2+1}{x^2-1}} = \sqrt{2}\,E\left(\arccos \dfrac{1}{u}, \dfrac{\sqrt{2}}{2}\right)$ $[u > 1]$.   BF(260.76)

In 3.174 und 3.175 werden folgende Abkürzungen verwendet: | In 3.174 and 3.175 the following notation is used:

$$\alpha = \arccos \dfrac{1+(1-\sqrt{3})u}{1+(1+\sqrt{3})u},$$

$$\beta = \arccos \dfrac{1-(1+\sqrt{3})u}{1+(\sqrt{3}-1)u}, \quad p = \dfrac{\sqrt{2+\sqrt{3}}}{2}, \quad q = \dfrac{\sqrt{2-\sqrt{3}}}{2}.$$

3.174

1. $\int_0^u \dfrac{dx}{[1+(1+\sqrt{3})x]^2}\sqrt{\dfrac{1-x+x^2}{x(1+x)}} = \dfrac{1}{\sqrt[4]{3}}\,E(\alpha, p)$ $[u > 0]$.   BF(260.51)

2. $\int_0^u \dfrac{dx}{[1+(\sqrt{3}-1)x]^2}\sqrt{\dfrac{1+x+x^2}{x(1-x)}} = \dfrac{1}{\sqrt[4]{3}}\,E(\beta, q)$ $[1 \geqslant u > 0]$.   BF(259.51)

3. $\int_0^u \dfrac{dx}{1-x+x^2}\sqrt{\dfrac{x(1+x)}{1-x+x^2}} = \dfrac{1}{\sqrt[4]{27}}E(\alpha, p) - \dfrac{2-\sqrt{3}}{\sqrt[4]{27}}\,F(\alpha, p) - \dfrac{2(2+\sqrt{3})}{\sqrt{3}}\dfrac{1+(1-\sqrt{3})u}{1+(1+\sqrt{3})u}\sqrt{\dfrac{u(1+u)}{1-u+u^2}}$ $[u > 0]$.   BF(260.54)

4. $\int_0^u \dfrac{dx}{1+x+x^2}\sqrt{\dfrac{x(1-x)}{1+x+x^2}} = \dfrac{4}{\sqrt[4]{27}}\,E(\beta, q) - \dfrac{2+\sqrt{3}}{\sqrt[4]{27}}\,F(\beta, q) - \dfrac{2(2-\sqrt{3})}{\sqrt{3}}\dfrac{1-(1+\sqrt{3})u}{1+(\sqrt{3}-1)u}\sqrt{\dfrac{u(1-u)}{1+u+u^2}}$ $[1 \geqslant u > 0]$.   BF(259.55)

3.175

1. $\int_0^u \dfrac{dx}{1+x}\sqrt{\dfrac{x}{1+x^3}} = \dfrac{1}{\sqrt[4]{27}}\,[F(\alpha, p) - 2E(\alpha, p)] + \dfrac{2}{\sqrt{3}}\dfrac{\sqrt{u(1-u+u^2)}}{\sqrt{1+u}\,[1+(1+\sqrt{3})\,u]}$ $[u > 0]$.   BF(260.55)

2. $\int_0^u \dfrac{dx}{1-x}\sqrt{\dfrac{x}{1-x^3}} = \dfrac{1}{\sqrt[4]{27}}\,[F(\beta, q) - 2E(\beta, q)] + \dfrac{2}{\sqrt{3}}\dfrac{\sqrt{u(1+u+u^2)}}{\sqrt{1-u}\,[1+(\sqrt{3}-1)u]}$ $[0 < u < 1]$.   BF(259.52)

Ausdrücke, die auf vierte Wurzeln aus Polynomen zweiten Grades führen, und ihre Produkte mit rationalen Funktionen

Expressions reducible to fourth roots of second-degree polynomials and their products with rational functions

3.18

1. $\displaystyle\int_b^u \frac{dx}{\sqrt[4]{(a-x)(x-b)}} = \sqrt{a-b}\left\{2\left[E\left(\frac{1}{\sqrt{2}}\right) + E\left(\arccos\sqrt[4]{\frac{4(a-u)(u-b)}{(a-b)^2}}, \frac{1}{\sqrt{2}}\right)\right] - \left[K\left(\frac{1}{\sqrt{2}}\right) + F\left(\arccos\sqrt[4]{\frac{4(a-u)(u-b)}{(a-b)^2}}, \frac{1}{\sqrt{2}}\right)\right]\right\}$ $[a \geqslant u > b]$. BF(271.05)

3.181

2. $\displaystyle\int_a^u \frac{dx}{\sqrt[4]{(x-a)(x-b)}} = \sqrt{\frac{a-b}{2}}F\left[\left(\arccos\frac{a-b-2\sqrt{(u-a)(u-b)}}{a-b+2\sqrt{(u-a)(u-b)}}, \frac{1}{\sqrt{2}}\right) - 2E\left(\arccos\frac{a-b-2\sqrt{(u-a)(u-b)}}{a-b+2\sqrt{(u-a)(u-b)}}, \frac{1}{\sqrt{2}}\right)\right] + \frac{2(2u-a-b)\sqrt[4]{(u-a)(u-b)}}{a-b+2\sqrt{(u-a)(u-b)}}$ $[u > a > b]$. BF(272.05)

1. $\displaystyle\int_b^u \frac{dx}{\sqrt[4]{[(a-x)(x-b)]^3}} = \frac{2}{\sqrt{a-b}}\left[K\left(\frac{1}{\sqrt{2}}\right) + F\left(\arccos\sqrt[4]{\frac{4(a-u)(u-b)}{(a-b)^2}}, \frac{1}{\sqrt{2}}\right)\right]$ $[a \geqslant u > b]$. BF(271.01)

3.182

2. $\displaystyle\int_a^u \frac{dx}{\sqrt[4]{[(x-a)(x-b)]^3}} = \frac{\sqrt{2}}{\sqrt{a-b}}F\left(\arccos\frac{a-b-2\sqrt{(u-a)(u-b)}}{a-b+2\sqrt{(u-a)(u-b)}}, \frac{1}{\sqrt{2}}\right)$

$[u > a > b]$. BF(272.00)

In 3.183—3.186 werden folgende Abkürzungen verwendet:

In 3.183—3.186 the following notation is used:

$$\alpha = \arccos\frac{1}{\sqrt[4]{u^2+1}}, \quad \beta = \arccos\sqrt[4]{1-u^2}, \quad \gamma = \arccos\frac{1-\sqrt{u^2-1}}{1+\sqrt{u^2-1}}.$$

1. $\displaystyle\int_0^u \frac{dx}{\sqrt[4]{x^2+1}} = \sqrt{2}\left[F\left(\alpha, \frac{1}{\sqrt{2}}\right) - 2E\left(\alpha, \frac{1}{\sqrt{2}}\right)\right] + \frac{2u}{\sqrt[4]{u^2+1}}$

$[u > 0]$. BF(273.55)

3.183

2. $\displaystyle\int_0^u \frac{dx}{\sqrt[4]{1-x^2}} = \sqrt{2}\left[2E\left(\beta, \frac{1}{\sqrt{2}}\right) - F\left(\beta, \frac{1}{\sqrt{2}}\right)\right]$ $[0 < u \leqslant 1]$. BF(271.55)

3.183 3. $\int_1^u \dfrac{dx}{\sqrt[4]{x^2-1}} = F\left(\gamma, \dfrac{1}{\sqrt{2}}\right) - 2E\left(\gamma, \dfrac{1}{\sqrt{2}}\right) + \dfrac{2u\sqrt[4]{u^2-1}}{1+\sqrt{u^2-1}}$ $[u > 1]$.

BF(272.55)

3.184 1. $\int_0^u \dfrac{x^2\,dx}{\sqrt[4]{1-x^2}} = \dfrac{2\sqrt{2}}{5}\left[2E\left(\beta, \dfrac{1}{\sqrt{2}}\right) - F\left(\beta, \dfrac{1}{\sqrt{2}}\right)\right] - \dfrac{2u}{5}\sqrt[4]{(1-u^2)^3}$

$[0 < u \leqslant 1]$. BF(271.59)

2. $\int_1^u \dfrac{dx}{x^2\sqrt[4]{x^2-1}} = E\left(\gamma, \dfrac{1}{\sqrt{2}}\right) - \dfrac{1}{2}F\left(\gamma, \dfrac{1}{\sqrt{2}}\right) - \dfrac{1-\sqrt{u^2-1}}{1+\sqrt{u^2-1}} \cdot \dfrac{\sqrt{u^2-1}}{u}$

$[u > 1]$. BF(272.54)

3.185 1. $\int_0^u \dfrac{dx}{\sqrt[4]{(x^2+1)^3}} = \sqrt{2}\,F\left(\alpha, \dfrac{1}{\sqrt{2}}\right)$ $[u > 0]$. BF(273.50)

2. $\int_0^u \dfrac{dx}{\sqrt[4]{(1-x^2)^3}} = \sqrt{2}\,F\left(\beta, \dfrac{1}{\sqrt{2}}\right)$ $[0 < u \leqslant 1]$. BF(271.51)

3. $\int_1^u \dfrac{dx}{\sqrt[4]{(x^2-1)^3}} = F\left(\gamma, \dfrac{1}{\sqrt{2}}\right)$ $[u > 1]$. BF(272.50)

4. $\int_0^u \dfrac{x^2\,dx}{\sqrt[4]{(1-x^2)^3}} = \dfrac{2\sqrt{2}}{3}F\left(\beta, \dfrac{1}{\sqrt{2}}\right) - \dfrac{2}{3}u\sqrt[4]{1-u^2}$ $[0 < u \leqslant 1]$. BF(271.54)

5. $\int_0^u \dfrac{dx}{\sqrt[4]{(x^2+1)^5}} = 2\sqrt{2}\,E\left(\alpha, \dfrac{1}{\sqrt{2}}\right) - \sqrt{2}\,F\left(\alpha, \dfrac{1}{\sqrt{2}}\right)$ $[u > 0]$. BF(273.54)

6. $\int_0^u \dfrac{x^2\,dx}{\sqrt[4]{(x^2+1)^5}} = 2\sqrt{2}\left[F\left(\alpha, \dfrac{1}{\sqrt{2}}\right) - 2E\left(\alpha, \dfrac{1}{\sqrt{2}}\right)\right] + \dfrac{2u}{\sqrt[4]{u^2+1}}$

$[u > 0]$. BF(273.56)

7. $\int_0^u \dfrac{x^2\,dx}{\sqrt[4]{(x^2+1)^7}} = \dfrac{1}{3\sqrt{2}}F\left(\alpha, \dfrac{1}{\sqrt{2}}\right) - \dfrac{u}{6\sqrt[4]{(u^2+1)^3}}$ $[u > 0]$. BF(273.53)

3.186 1. $\int_0^u \dfrac{1+\sqrt{x^2+1}}{(x^2+1)\sqrt[4]{x^2+1}}\,dx = 2\sqrt{2}\,E\left(\alpha, \dfrac{1}{\sqrt{2}}\right)$ $[u > 0]$. BF(273.51)

2. $\int_0^u \dfrac{dx}{(1+\sqrt{1-x^2})\sqrt[4]{1-x^2}} = \sqrt{2}\left[F\left(\beta, \dfrac{1}{\sqrt{2}}\right) - E\left(\beta, \dfrac{1}{\sqrt{2}}\right)\right] +$

$+ \dfrac{u\sqrt[4]{1-u^2}}{1+\sqrt{1-u^2}}$ $[0 < u \leqslant 1]$. BF(271.58)

3. $\int_1^u \dfrac{dx}{(x^2 + 2\sqrt{x^2-1})\sqrt[4]{x^2-1}} = \dfrac{1}{2}\left[F\left(\gamma, \dfrac{1}{\sqrt{2}}\right) - E\left(\gamma, \dfrac{1}{\sqrt{2}}\right)\right]\quad [u>1].$ 3.186

BF(272.53)

4. $\int_0^u \dfrac{1-\sqrt{1-x^2}}{1+\sqrt{1-x^2}} \cdot \dfrac{dx}{\sqrt[4]{(1-x^2)^3}} = \sqrt{2}\left[2E\left(\beta, \dfrac{1}{\sqrt{2}}\right) - F\left(\beta, \dfrac{1}{\sqrt{2}}\right)\right] -$

$$- \dfrac{2u\sqrt[4]{1-u^2}}{1+\sqrt{1-u^2}}\quad [0<u\leqslant 1].$$

BF(271.57)

5. $\int_1^u \dfrac{x^2\,dx}{(x^2+2\sqrt{x^2-1})\sqrt[4]{(x^2-1)^3}} = E\left(\gamma, \dfrac{1}{\sqrt{2}}\right)\quad [u>1].$ BF(272.51)

Potenzen von x und von Binomen $\alpha + \beta x$
Powers of x and powers of binomials of the form $(\alpha + \beta x)$ 3.19–3.23

1. $\int_0^u x^{\nu-1}(u-x)^{\mu-1}\,dx = u^{\mu+\nu-1}\,\mathrm{B}(\mu, \nu)\quad [\operatorname{Re}\mu > 0,\ \operatorname{Re}\nu > 0].$ Ba V 185(7)   3.191

2. $\int_u^\infty x^{-\nu}(x-u)^{\mu-1}\,dx = u^{\mu-\nu}\,\mathrm{B}(\nu-\mu, \mu)\quad [\operatorname{Re}\nu > \operatorname{Re}\mu > 0].$ Ba V 201(6)

3. $\int_0^1 x^{\nu-1}(1-x)^{\mu-1}\,dx = \int_0^1 x^{\mu-1}(1-x)^{\nu-1}\,dx = \mathrm{B}(\mu, \nu)$

$$[\operatorname{Re}\mu > 0,\ \operatorname{Re}\nu > 0].$$ F II 778(1)

1. $\int_0^1 \dfrac{x^p\,dx}{(1-x)^p} = p\pi\,\operatorname{cosec} p\pi\quad [p^2 < 1].$ BH[3](4)   3.192

2. $\int_0^1 \dfrac{x^p\,dx}{(1-x)^{p+1}} = -\pi\,\operatorname{cosec} p\pi\quad [-1<p<0].$ BH[3](5)

3. $\int_0^1 \dfrac{(1-x)^p}{x^{p+1}}\,dx = -\pi\,\operatorname{cosec} p\pi\quad [-1<p<0].$ BH[4](6)

4. $\int_1^\infty (x-1)^{p-\frac{1}{2}}\,\dfrac{dx}{x} = \pi\,\operatorname{sec} p\pi\quad \left[-\dfrac{1}{2}<p<\dfrac{1}{2}\right].$ BH[23](7)

$\int_0^n x^{\nu-1}(n-x)^n\,dx = \dfrac{n!\,n^{\nu+n}}{\nu(\nu+1)(\nu+2)\ldots(\nu+n)}\quad [\operatorname{Re}\nu > 0].$ Ba I 2   3.193

3.194

1. $\int_0^u \dfrac{x^{\mu-1} dx}{(1+\beta x)^\nu} = \dfrac{u^\mu}{\mu} {}_2F_1(\nu, \mu; 1+\mu; -\beta u)$ $\quad [|\arg(1+\beta u)| < \pi,\ \operatorname{Re}\mu > 0].$

Ba IV 310(20)

2. $\int_u^\infty \dfrac{x^{\mu-1} dx}{(1+\beta x)^\nu} = \dfrac{u^{\mu-\nu}}{\beta^\nu(\nu-\mu)} {}_2F_1\left(\nu,\ \nu-\mu;\ \nu-\mu+1;\ -\dfrac{1}{\beta u}\right)$ $\quad [\operatorname{Re}\mu > \operatorname{Re}\nu].$

Ba IV 310(21)

3. $\int_0^\infty \dfrac{x^{\mu-1} dx}{(1+\beta x)^\nu} = \beta^{-\mu}\, \mathrm{B}(\mu, \nu-\mu)$ $\quad [|\arg\beta| < \pi,\ \operatorname{Re}\nu > \operatorname{Re}\mu > 0].$

F II 779+, Ba IV 310(19)

4. $\int_0^\infty \dfrac{x^{\mu-1} dx}{(1+\beta x)^{n+1}} = (-1)^n \dfrac{\pi}{\beta^\mu}\binom{\mu-1}{n}\operatorname{cosec}(\mu\pi)$

$[|\arg\beta| < \pi,\ 0 < \operatorname{Re}\nu < n+1].$ Ba IV 308(6)

5. $\int_0^u \dfrac{x^{\mu-1} dx}{1+\beta x} = \dfrac{u^\mu}{\mu} {}_2F_1(1, \mu;\ 1+\mu;\ -u\beta)$ $\quad [|\arg(1+u\beta)| < \pi,\ \operatorname{Re}\mu > 0].$

Ba IV 308(5)

6. $\int_0^\infty \dfrac{x^{\mu-1} dx}{(1+\beta x)^2} = \dfrac{(1-\mu)\pi}{\beta^\mu}\operatorname{cosec}\mu\pi$ $\quad [0 < \operatorname{Re}\mu < 2].$ BH[16](4)

7. $\int_0^\infty \dfrac{x^m dx}{(a+bx)^{n+\frac{1}{2}}} = 2^{m+1} m!\, \dfrac{(2n-2m-3)!!}{(2n-1)!!}\, \dfrac{a^{m-n+\frac{1}{2}}}{b^{m+1}}$

$\left[m < n - \dfrac{1}{2},\ a > 0,\ b > 0\right].$ BH[21](2)

8. $\int_0^1 \dfrac{x^{n-1} dx}{(1+x)^m} = 2^{-n} \sum_{k=0}^\infty \binom{m-n-1}{k}\dfrac{(-2)^{-k}}{n+k}.$ BH[3](1)

3.195 $\int_0^\infty \dfrac{(1+x)^{p-1}}{(x+a)^{p+1}} dx = \dfrac{1-a^{-p}}{p(a-1)}$ $\quad [a > 0].$ Li[19](6)

3.196

1. $\int_0^u (x+\beta)^\nu (u-x)^{\mu-1} dx = \dfrac{\beta^\nu u^\mu}{\mu} {}_2F_1\left(1,\ -\nu;\ 1+\mu;\ -\dfrac{u}{\beta}\right)$

$\left[\left|\arg\dfrac{u}{\beta}\right| < \pi\right].$ Ba V 185(8)

2. $\int_u^\infty (x+\beta)^{-\nu}(x-u)^{\mu-1} dx = (u+\beta)^{\mu-\nu}\, \mathrm{B}(\nu-\mu, \mu)$

$\left[\left|\arg\dfrac{u}{\beta}\right| < \pi,\ \operatorname{Re}\nu > \operatorname{Re}\mu > 0\right].$ Ba V 201(7)

3. $$\int_a^b (x-a)^{\mu-1}(b-x)^{\nu-1}\,dx = (b-a)^{\mu+\nu-1}\,B(\mu,\nu)$$
$$[b > a,\ \operatorname{Re}\mu > 0,\ \operatorname{Re}\nu > 0].\quad \text{Ba I 10(13)}$$

3.196

4. $$\int_1^\infty \frac{dx}{(a-bx)(x-1)^\nu} = -\frac{\pi}{b}\operatorname{cosec}\nu\pi \left(\frac{b}{b-a}\right)^\nu \quad [a<b,\ b>0,\ 0<\nu<1].$$
Li[23](5)

5. $$\int_{-\infty}^1 \frac{dx}{(a-bx)(1-x)^\nu} = \frac{\pi}{b}\operatorname{cosec}\nu\pi \left(\frac{b}{a-b}\right)^\nu \quad [a>b>0,\ 0<\nu<1].$$
Li[24](10)

1. $$\int_0^\infty x^{\nu-1}(\beta+x)^{-\mu}(x+\gamma)^{-\rho}\,dx = \beta^{-\mu}\gamma^{\nu-\rho}\,B(\nu,\mu-\nu+\rho) \times$$
$$\times {}_2F_1\left(\mu,\nu;\ \mu+\rho;\ 1-\frac{\gamma}{\beta}\right) \quad [|\arg\beta| < \pi,\ |\arg\gamma| < \pi,$$
$$\operatorname{Re}\nu > 0,\ \operatorname{Re}\mu > \operatorname{Re}(\nu-\rho)].\quad \text{Ba V 233(9)}$$

3.197

2. $$\int_u^\infty x^{-\lambda}(x+\beta)^\nu(x-u)^{\mu-1}\,dx = u^{\mu+\nu-\lambda}\,B(\lambda-\mu-\nu,\mu) \times$$
$$\times {}_2F_1\left(-\nu,\lambda-\mu-\nu;\ \lambda-\nu;\ -\frac{\beta}{u}\right)$$
$$\left[\left|\arg\frac{u}{\beta}\right| < \pi \ \text{oder or}\ \left|\frac{\beta}{u}\right| < 1,\ 0 < \operatorname{Re}\mu < \operatorname{Re}(\lambda-\nu)\right].\quad \text{Ba V 201(8)}$$

3. $$\int_0^1 x^{\lambda-1}(1-x)^{\mu-1}(1-\beta x)^{-\nu}\,dx = B(\lambda,\mu)\,{}_2F_1(\nu,\lambda;\lambda+\mu;\beta)$$
$$[\operatorname{Re}\lambda > 0,\ \operatorname{Re}\mu > 0,\ |\beta| < 1].\quad \text{WW 293}$$

4. $$\int_0^1 x^{\mu-1}(1-x)^{\nu-1}(1+ax)^{-\mu-\nu}\,dx = (1+a)^{-\mu}B(\mu,\nu)$$
$$[\operatorname{Re}\mu > 0,\ \operatorname{Re}\nu > 0,\ a > -1].\quad \text{BH[5](4), Ba I 10(11)}$$

5. $$\int_0^\infty x^{\lambda-1}(1+x)^\nu(1+\alpha x)^\mu\,dx = B(\lambda, -\mu-\nu-\lambda) \times$$
$$\times {}_2F_1(-\mu,\lambda;\ -\mu-\nu;\ 1-\alpha) \quad [|\arg\alpha| < \pi,\ -\operatorname{Re}(\mu+\nu) > \operatorname{Re}\lambda > 0].$$
Ba I 60(12), Ba IV 310(23)

6. $$\int_1^\infty x^{\lambda-\nu}(x-1)^{\nu-\mu-1}(\alpha x-1)^{-\lambda}\,dx = \alpha^{-\lambda}B(\mu,\nu-\mu)\,{}_2F_1(\nu,\mu;\lambda;\alpha^{-1})$$
$$[1+\operatorname{Re}\nu > \operatorname{Re}\lambda > \operatorname{Re}\mu,\ |\arg(\alpha-1)| < \pi].\quad \text{Ba I 115(6)}$$

3.197

7. $\int_0^\infty x^{\mu - \frac{1}{2}} (x+a)^{-\mu} (x+b)^{-\mu} \, dx = \sqrt{\pi} (\sqrt{a} + \sqrt{b})^{1-2\mu} \dfrac{\Gamma\left(\mu - \frac{1}{2}\right)}{\Gamma(\mu)}$

$[\operatorname{Re} \mu > 0]$. BH[19](5)

8. $\int_0^u x^{\nu-1} (x+a)^{\lambda} (u-x)^{\mu-1} \, dx = a^{\lambda} u^{\mu+\nu-1} B(\mu, \nu) \, _2F_1\left(-\lambda, \nu; \mu+\nu; -\dfrac{u}{a}\right)$

$\left[\left|\arg\left(\dfrac{u}{a}\right)\right| < \pi, \operatorname{Re} \mu > 0, \operatorname{Re} \nu > 0\right]$. Ba V 186(9)

9. $\int_0^\infty x^{\lambda-1} (1+x)^{-\mu+\nu} (x+\beta)^{-\nu} \, dx = B(\mu-\lambda, \lambda) \, _2F_1(\nu, \mu-\lambda; \mu; 1-\beta)$

$[\operatorname{Re} \mu > \operatorname{Re} \lambda > 0]$. Ba I 205

10. $\int_0^1 \dfrac{x^{q-1} \, dx}{(1-x)^q (1+px)^q} = \dfrac{\pi}{(1+p)^q} \operatorname{cosec} q\pi \quad [0 < q < 1, \, p > -1]$. BH[5](1)

11. $\int_0^1 \dfrac{x^{p - \frac{1}{2}} \, dx}{(1-x)^p (1+qx)^p} =$

$= \dfrac{2\Gamma\left(+\frac{1}{2}\right) \Gamma(1-p)}{\sqrt{\pi}} \cos^{2p}(\operatorname{arctg} \sqrt{q}) \dfrac{\sin[(2p-1) \operatorname{arctg}(\sqrt{q})]}{(2p-1) \sin[\operatorname{arctg}(\sqrt{q})]}$

$\left[-\dfrac{1}{2} < p < 1, \, q > 0\right]$. BH[11](1)

12. $\int_0^1 \dfrac{x^{p - \frac{1}{2}} \, dx}{(1-x)^p (1-qx)^p} = \dfrac{\Gamma\left(p + \frac{1}{2}\right) \Gamma(1-p)}{\sqrt{\pi}} \dfrac{(1-\sqrt{q})^{1-2p} - (1+\sqrt{q})^{1-2p}}{(2p-1) \sqrt{q}}$

$\left[-\dfrac{1}{2} < p < 1, \, 0 < q < 1\right]$. BH[11](2)

3.198 $\int_0^1 x^{\mu-1} (1-x)^{\nu-1} [ax + b(1-x) + c]^{-(\mu+\nu)} \, dx = (a+c)^{-\mu} (b+c)^{-\nu} B(\mu, \nu)$

$[a \geq 0, \, b \geq 0, \, c > 0, \, \operatorname{Re} \mu > 0, \, \operatorname{Re} \nu > 0]$. F II 793

3.199 $\int_a^b (x-a)^{\mu-1} (b-x)^{\nu-1} (x-c)^{-\mu-\nu} \, dx =$

$= (b-a)^{\mu+\nu-1} (b-c)^{-\mu} (a-c)^{-\nu} B(\mu, \nu)$

$[\operatorname{Re} \mu > 0, \, \operatorname{Re} \nu > 0, \, c < a < b]$. Ba I 10(14)

3.211 $\int_0^1 x^{\lambda-1} (1-x)^{\mu-1} (1-ux)^{-\rho} (1-vx)^{-\sigma} \, dx = B(\mu, \lambda) \, F_1(\lambda, \rho, \sigma, \lambda+\mu; u, v)$

$[\operatorname{Re} \lambda > 0, \, \operatorname{Re} \mu > 0]$. Ba I 231(5)

$$\int_0^\infty [(1+ax)^{-p} + (1+bx)^{-p}] x^{q-1}\, dx =$$ 3.212

$$= 2(ab)^{-\frac{q}{2}} B(q, p-q) \cos\left\{q \arccos\left[\frac{a+b}{2\sqrt{ab}}\right]\right\}$$

$$[p > q > 0].$$ BH[19](9)

$$\int_0^\infty [(1+ax)^{-p} - (1+bx)^{-p}] x^{q-1}\, dx =$$ 3.213

$$= -2i(ab)^{-\frac{q}{2}} B(q, p-q) \sin\left\{q \arccos\left[\frac{a+b}{2\sqrt{ab}}\right]\right\}$$

$$[p > q > 0].$$ BH[19](10)

$$\int_0^1 [(1+x)^{\mu-1}(1-x)^{\nu-1} + (1+x)^{\nu-1}(1-x)^{\mu-1}]\, dx = 2^{\mu+\nu-1} B(\mu, \nu)$$ 3.214

$$[\operatorname{Re}\mu > 0,\ \operatorname{Re}\nu > 0].$$ Li[1](15), Ba I 10(10)

$$\int_0^1 \{a^\mu x^{\mu-1}(1-ax)^{\nu-1} + (1-a)^\nu x^{\nu-1}[1-(1-a)x]^{\mu-1}\}\, dx = B(\mu, \nu)$$ 3.215

$$[\operatorname{Re}\mu > 0,\ \operatorname{Re}\nu > 0,\ |a| < 1].$$ BH[1](16)

1. $$\int_0^1 \frac{x^{\mu-1} + x^{\nu-1}}{(1+x)^{\mu+\nu}}\, dx = B(\mu, \nu) \quad [\operatorname{Re}\mu > 0,\ \operatorname{Re}\nu > 0].$$ F II 778 3.216

2. $$\int_1^\infty \frac{x^{\mu-1} + x^{\nu-1}}{(1+x)^{\mu+\nu}}\, dx = B(\mu, \nu) \quad [\operatorname{Re}\mu > 0,\ \operatorname{Re}\nu > 0].$$ F II 778

$$\int_0^\infty \left\{\frac{b^p x^{p-1}}{(1+bx)^p} - \frac{(1+bx)^{p-1}}{b^{p-1} x^p}\right\} dx = \pi \operatorname{ctg} p\pi \quad [0 < p < 1,\ b > 0].$$ BH[18](13) 3.217

$$\int_0^\infty \frac{x^{2p-1} - (a+x)^{2p-1}}{(a+x)^p x^p}\, dx = \pi \operatorname{ctg} p\pi \quad [p < 1] \quad \text{(cf. 3.217)}.$$ BH[18](7) 3.218

$$\int_0^\infty \left\{\frac{x^\nu}{(x+1)^{\nu+1}} - \frac{x^\mu}{(x+1)^{\mu+1}}\right\} dx = \psi(\mu+1) - \psi(\nu+1) \quad [\operatorname{Re}\mu > 1,\ \operatorname{Re}\nu > 1].$$ 3.219

BH[19](13)

1. $$\int_a^\infty \frac{(x-a)^{p-1}}{x-b}\, dx = \pi(a-b)^{p-1} \operatorname{cosec} p\pi \quad [a > b,\ 0 < p < 1].$$ Li[24](8) 3.221

2. $$\int_{-\infty}^a \frac{(a-x)^{p-1}}{x-b}\, dx = -\pi(b-a)^{p-1} \operatorname{cosec} p\pi \quad [a < b,\ 0 < p < 1].$$ Li[24](8)

3.222

1. $\int_0^1 \frac{x^{\mu-1} dx}{1+x} = \beta(\mu) \quad [\mathrm{Re}\,\mu > 0].$ WW 262

2. $\int_0^\infty \frac{x^{\mu-1} dx}{x+a} = \begin{cases} \pi \,\mathrm{cosec}\,(\mu\pi)\, a^{\mu-1} & \text{für} \\ & \text{for } a > 0, \\ -\pi \,\mathrm{ctg}\,(\mu\pi)\,(-a)^{\mu-1} & \text{für} \\ & \text{for } a < 0, \end{cases}$

F II 723—724, F II 743

BH[18](2), Ba V 249(28)

$[0 < \mathrm{Re}\,\mu < 1].$

3.223

1. $\int_0^\infty \frac{x^{\mu-1} dx}{(\beta+x)(\gamma+x)} = \frac{\pi}{\gamma-\beta}(\beta^{\mu-1} - \gamma^{\mu-1})\,\mathrm{cosec}\,(\mu\pi)$

$[|\arg\beta| < \pi,\ |\arg\gamma| < \pi,\ 0 < \mathrm{Re}\,\mu < 2].$ Ba IV 309(7)

2. $\int_0^\infty \frac{x^{\mu-1} dx}{(\beta+x)(\alpha-x)} = \frac{\pi}{\alpha+\beta}[\beta^{\mu-1}\,\mathrm{cosec}\,(\mu\pi) + \alpha^{\mu-1}\,\mathrm{ctg}(\mu\pi)]$

$[|\arg\beta| < \pi,\ \alpha > 0,\ 0 < \mathrm{Re}\,\mu < 2].$ Ba IV 309(8)

3. $\int_0^\infty \frac{x^{\mu-1} dx}{(a-x)(b-x)} = \pi \,\mathrm{ctg}\,(\mu\pi)\,\frac{a^{\mu-1} - b^{\mu-1}}{b-a}$

$[a > b > 0,\ 0 < \mathrm{Re}\,\mu < 2].$ Ba IV 309(9)

3.224

$\int_0^\infty \frac{(x+\beta)\,x^{\mu-1} dx}{(x+\gamma)(x+\delta)} = \pi \,\mathrm{cosec}\,(\mu\pi)\left\{\frac{\gamma-\beta}{\gamma-\delta}\gamma^{\mu-1} + \frac{\delta-\beta}{\delta-\gamma}\delta^{\mu-1}\right\}$

$[|\arg\gamma| < \pi,\ |\arg\delta| < \pi,\ 0 < \mathrm{Re}\,\mu < 1].$ Ba IV 309(10)

3.225

1. $\int_1^\infty \frac{(x-1)^{p-1}}{x^2}\,dx = (1-p)\,\pi\,\mathrm{cosec}\,p\pi \quad [-1 < p < 1].$ BH[23](8)

2. $\int_1^\infty \frac{(x-1)^{1-p}}{x^3}\,dx = \frac{1}{2}p(1-p)\,\pi\,\mathrm{cosec}\,p\pi \quad [0 < p < 1].$ BH[23](1)

3. $\int_0^\infty \frac{x^p\,dx}{(1+x)^3} = \frac{\pi}{2}p(1-p)\,\mathrm{cosec}\,p\pi \quad [-1 < p < 2].$ BH[16](5)

3.226

1. $\int_0^1 \frac{x^n\,dx}{\sqrt{1-x}} = 2\,\frac{(2n)!!}{(2n+1)!!}$ BH[8](1)

2. $\int_0^1 \frac{x^{n-\frac{1}{2}}\,dx}{\sqrt{1-x}} = \frac{(2n-1)!!}{(2n)!!}\pi.$ BH[8](2)

3.227

1. $\int_0^\infty \frac{x^{\nu-1}(\beta+x)^{1-\mu}}{\gamma+x}\,dx = \beta^{1-\mu}\gamma^{\nu-1}\mathrm{B}(\nu,\mu-\nu)\,{}_2F_1\left(\mu-1,\nu;\mu;1-\frac{\gamma}{\beta}\right)$

$[|\arg\beta| < \pi,\ |\arg\gamma| < \pi,\ 0 < \mathrm{Re}\,\nu < \mathrm{Re}\,\mu].$ Ba V 217(9)

2. $\displaystyle\int_0^\infty \frac{x^{-\rho}(\beta+x)^{-\sigma}}{\gamma+x}\,dx = \pi\gamma^{-\rho}(\beta-\gamma)^{-\sigma}\,\mathrm{cosec}\,(\rho\pi)\,I_{1-\gamma/\beta}(\sigma,\rho)$ 3.227

$$[|\arg\beta|<\pi,\ |\arg\gamma|<\pi,\ -\mathrm{Re}\,\sigma<\mathrm{Re}\,\rho<1].\quad \text{Ba V 217(10)}$$

1. $\displaystyle\int_a^b \frac{(x-a)^\nu(b-x)^{-\nu}}{x-c}\,dx = \pi\,\mathrm{cosec}\,(\nu\pi)\left[1-\left(\frac{a-c}{b-c}\right)^\nu\right]\ \begin{matrix}\text{für}\\\text{for}\end{matrix}\ c<a;$ 3.228

$\qquad\qquad = \pi\,\mathrm{cosec}\,(\nu\pi)\left[1-\cos(\nu\pi)\left(\frac{c-a}{b-c}\right)^\nu\right]\ \begin{matrix}\text{für}\\\text{for}\end{matrix}\ a<c<b;$

$\qquad\qquad = \pi\,\mathrm{cosec}\,(\nu\pi)\left[1-\left(\frac{c-a}{c-b}\right)^\nu\right]\ \begin{matrix}\text{für}\\\text{for}\end{matrix}\ c>b$

$$[|\mathrm{Re}\,\nu|<1].\quad \text{Ba V 250(31)}$$

2. $\displaystyle\int_a^b \frac{(x-a)^{\nu-1}(b-x)^{-\nu}}{x-c}\,dx = \frac{\pi\,\mathrm{cosec}\,(\nu\pi)}{b-c}\left|\frac{a-c}{b-c}\right|^{\nu-1}\ \begin{matrix}\text{für}\\\text{for}\end{matrix}\ c<a\ \begin{matrix}\text{oder}\\\text{or}\end{matrix}\ c>b;$

$\qquad\qquad = -\dfrac{\pi(c-a)^{\nu-1}}{(b-c)^\nu}\,\mathrm{ctg}\,(\nu\pi)\ \begin{matrix}\text{für}\\\text{for}\end{matrix}\ a<c<b$

$$[0<\mathrm{Re}\,\nu<1].\quad \text{Ba V 250(32)}$$

3. $\displaystyle\int_a^b \frac{(x-a)^{\nu-1}(b-x)^{\mu-1}}{x-c}\,dx = \frac{(b-a)^{\mu+\nu-1}}{b-c}\,\mathrm{B}(\mu,\nu)\,{}_2F_1\!\left(1,\mu;\mu+\nu;\frac{b-a}{b-c}\right)$

$$\begin{matrix}\text{für}\\\text{for}\end{matrix}\ c<a\ \begin{matrix}\text{oder}\\\text{or}\end{matrix}\ c>b;$$

$\qquad = (c-a)^{\nu-1}(b-c)^{\mu-1}\,\mathrm{ctg}\,\mu\pi - (b-a)^{\mu+\nu-2}\,\mathrm{B}(\mu-1,\nu)\,\times$

$\qquad\qquad \times\,{}_2F_1\!\left(2-\mu-\nu,1;2-\mu;\frac{b-c}{b-a}\right)\ \begin{matrix}\text{für}\\\text{for}\end{matrix}\ a<c<b$

$$[\mathrm{Re}\,\mu>0,\ \mathrm{Re}\,\nu>0].\quad \text{Ba V 250(33)}$$

4. $\displaystyle\int_0^1 \frac{(1-x)^{\nu-1}x^{-\nu}}{a-bx}\,dx = \frac{\pi(a-b)^{\nu-1}}{a^\nu}\,\mathrm{cosec}\,(\nu\pi)$

$$[0<\mathrm{Re}\,\nu<1,\ 0<b<a].\quad \text{BH [5](8)}$$

5. $\displaystyle\int_0^\infty \frac{x^{\nu-1}(x+a)^{1-\mu}}{x-c}\,dx = a^{1-\mu}(-c)^{\nu-1}\,\mathrm{B}(\mu-\nu,\nu)\,{}_2F_1\!\left(\mu-1,\nu;\mu;1+\frac{c}{a}\right)$

$$\begin{matrix}\text{für}\\\text{for}\end{matrix}\ c<0;$$

$\qquad = c^{\nu-1}(a+c)^{1-\mu}\,\mathrm{ctg}[(\mu-\nu)\pi]\,-$

$\qquad -\dfrac{a^{1-\mu+\nu}}{\pi(a+c)}\,\mathrm{B}(\mu-\nu-1,\nu)\,{}_2F_1\!\left(2-\mu,1;2-\mu+\nu;\dfrac{a}{a+c}\right)\ \begin{matrix}\text{für}\\\text{for}\end{matrix}\ c>0$

$$[a>0,\ 0<\mathrm{Re}\,\nu<\mathrm{Re}\,\mu].\quad \text{Ba V 251(34)}$$

3.229 $$\int_0^1 \frac{x^{\mu-1}\,dx}{(1-x)^\mu(1+ax)(1+bx)} = \frac{\pi \operatorname{cosec} \mu\pi}{a-b}\left[\frac{a}{(1+a)^\mu} - \frac{b}{(1+b)^\mu}\right]$$
$$[0 < \operatorname{Re} \mu < 1]. \qquad \text{BH[5](7)}$$

3.231
1. $$\int_0^1 \frac{x^{p-1} - x^{-p}}{1-x}\,dx = \pi \operatorname{ctg} p\pi \quad [p^2 < 1]. \qquad \text{BH[4](4)}$$

2. $$\int_0^1 \frac{x^{p-1} - x^{-p}}{1+x}\,dx = \pi \operatorname{cosec} p\pi \quad [p^2 < 1]. \qquad \text{BH[4](1)}$$

3. $$\int_0^1 \frac{x^p - x^{-p}}{x-1}\,dx = \frac{1}{p} - \pi \operatorname{ctg} p\pi \quad [p^2 < 1]. \qquad \text{BH[4](3)}$$

4. $$\int_0^1 \frac{x^p - x^{-p}}{1+x}\,dx = \frac{1}{p} - \pi \operatorname{cosec} p\pi \quad [p^2 < 1]. \qquad \text{BH[4](2)}$$

5. $$\int_0^1 \frac{x^{\mu-1} - x^{\nu-1}}{1-x}\,dx = \psi(\nu) - \psi(\mu) \quad [\operatorname{Re} \mu > 0, \ \operatorname{Re} \nu > 0]. \qquad \text{F II 819, BH[4](5)}$$

6. $$\int_0^\infty \frac{x^{p-1} - x^{q-1}}{1-x}\,dx = \pi(\operatorname{ctg} p\pi - \operatorname{ctg} q\pi) \quad [p > 0, q > 0]. \qquad \text{F II 724}$$

3.232 $$\int_0^\infty \frac{(c+ax)^{-\mu} - (c+bx)^{-\mu}}{x}\,dx = c^{-\mu} \ln\frac{b}{a} \quad [\operatorname{Re} \mu > -1; \ a > 0; \ b > 0; \ c > 0].$$
$$\text{BH[18](14)}$$

3.233 $$\int_0^\infty \left\{\frac{1}{1+x} - (1+x)^{-\nu}\right\}\frac{dx}{x} = \psi(\nu) + C \quad [\operatorname{Re} \nu > 0]. \qquad \text{Ba I 17, WW 260}$$

3.234
1. $$\int_0^1 \left(\frac{x^{q-1}}{1-ax} - \frac{x^{-q}}{a-x}\right)dx = \pi a^{-q} \operatorname{ctg} q\pi \quad [0 < q < 1, \ a > 0]. \qquad \text{BH[5](11)}$$

2. $$\int_0^1 \left(\frac{x^{q-1}}{1+ax} + \frac{x^{-q}}{a+x}\right)dx = \pi a^{-q} \operatorname{cosec} q\pi \quad [0 < q < 1, \ a > 0]. \qquad \text{BH[5](10)}$$

3.235 $$\int_0^\infty \frac{(1+x)^\mu - 1}{(1+x)^\nu}\,\frac{dx}{x} = \psi(\nu) - \psi(\nu-\mu) \quad [\operatorname{Re} \nu > \operatorname{Re} \mu > 0]. \qquad \text{BH[18](5)}$$

3.236 $$\int_0^1 \frac{x^{\frac{\mu}{2}}\,dx}{[(1-x)(1-a^2x)]^{\frac{\mu+1}{2}}} = \frac{(1-a)^{-\mu} - (1+a)^{-\mu}}{2a\mu\sqrt{\pi}}\,\Gamma\!\left(1+\frac{\mu}{2}\right)\Gamma\!\left(\frac{1-\mu}{2}\right)$$
$$[-2 < \mu < 1, \ |a| < 1]. \qquad \text{BH[12](32)}$$

$$\sum_{n=0}^{\infty}(-1)^{n+1}\int_{n}^{n+1}\frac{dx}{x+u}=\ln\frac{u\left[\Gamma\left(\frac{u}{2}\right)\right]^{2}}{2\left[\Gamma\left(\frac{u+1}{2}\right)\right]^{2}}\qquad[|\arg u|<\pi].$$

Ba V 216(1)   3.237

1. $$\int_{-\infty}^{\infty}\frac{|x|^{\nu-1}}{x-u}dx=-\pi\operatorname{ctg}\frac{\nu\pi}{2}|u|^{\nu-1}\operatorname{sign} u\qquad[0<\operatorname{Re}\nu<1].$$

Ba V 249(29)   3.238

2. $$\int_{-\infty}^{\infty}\frac{|x|^{\nu-1}}{x-u}\operatorname{sign} x\, dx=\pi\operatorname{tg}\frac{\nu\pi}{2}|u|^{\nu-1}\qquad[0<\operatorname{Re}\nu<1].$$

Ba V 249(30)

3. $$\int_{a}^{b}\frac{(b-x)^{\mu-1}(x-a)^{\nu-1}}{|x-u|^{\mu+\nu}}dx=\frac{(b-a)^{\mu+\nu-1}}{|a-u|^{\mu}|b-u|^{\nu}}\frac{\Gamma(\mu)\Gamma(\nu)}{\Gamma(\mu+\nu)}$$

$$\left[\operatorname{Re}\mu>0,\ \operatorname{Re}\nu>0,\ 0<u<a<b\ \text{oder}\atop\text{or}\ 0<a<b<u\right].$$

MO 7

Potenzen von x, von Binomen $\alpha+\beta x^p$ und von Polynomen in x
Powers of x, of binomials of the form $(\alpha+\beta x^p)$, and of polynomials in x

3.24–3.27

1. $$\int_{0}^{1}\frac{x^{\mu-1}dx}{1+x^p}=\frac{1}{p}\beta\left(\frac{\mu}{p}\right)\qquad[\operatorname{Re}\mu>0,\ p>0].$$

WW 262+, BH[2](13)   3.241

2. $$\int_{0}^{\infty}\frac{x^{\mu-1}dx}{1+x^{\nu}}=\frac{\pi}{\nu}\operatorname{cosec}\frac{\mu\pi}{\nu}=\frac{1}{\nu}\mathrm{B}\left(\frac{\mu}{\nu},\frac{\nu-\mu}{\nu}\right)\qquad[\operatorname{Re}\nu>\operatorname{Re}\mu>0].$$

Ba IV 309(15)+, BH[17](10)

3. $$\int_{0}^{\infty}\frac{x^{p-1}dx}{1-x^{q}}=\frac{\pi}{q}\operatorname{ctg}\frac{p\pi}{q}\qquad[p<q].$$

BH[17](11)

4. $$\int_{0}^{\infty}\frac{x^{\mu-1}dx}{(p+qx^{\nu})^{n+1}}=\frac{1}{\nu p^{n+1}}\left(\frac{p}{q}\right)^{\frac{\mu}{\nu}}\frac{\Gamma\left(\frac{\mu}{\nu}\right)\Gamma\left(1+n-\frac{\mu}{\nu}\right)}{\Gamma(1+n)}\qquad\left[0<\frac{\mu}{\nu}<n+1\right].$$

BH[17](22)+

5. $$\int_{0}^{\infty}\frac{x^{p-1}dx}{(1+x^{q})^{2}}=\frac{(p-q)\pi}{q^{2}}\operatorname{cosec}\frac{(p-q)\pi}{q}\qquad[p<2q].$$

BH[17](18)

$$\int_{-\infty}^{\infty}\frac{x^{2m}\,dx}{x^{4n}+2x^{2n}\cos t+1}=\frac{\pi}{n}\sin\left[\frac{(2n-2m-1)}{2n}t\right]\operatorname{cosec} t\operatorname{cosec}\frac{(2m+1)\pi}{2n}$$

3.242

$$[m<n,\ t^{2}<\pi^{2}].$$

F II 648–650

$$\int_{0}^{\infty}\frac{x^{\mu-1}dx}{(1+x^{2\nu})(1+x^{3\nu})}=-\frac{\pi}{8\nu}\frac{\operatorname{cosec}\left(\frac{\mu\pi}{3\nu}\right)}{1-4\cos^{2}\left(\frac{\mu\pi}{3\nu}\right)}\qquad[0<\operatorname{Re}\mu<5\operatorname{Re}\nu].$$

Ba IV 312(34)   3.243

3.244

1. $$\int_0^1 \frac{x^{p-1} + x^{q-p-1}}{1+x^q} dx = \frac{\pi}{q} \operatorname{cosec} \frac{p\pi}{q} \quad [q > p > 0].$$ BH[2](14)

2. $$\int_0^1 \frac{x^{p-1} - x^{q-p-1}}{1-x^q} dx = \frac{\pi}{q} \operatorname{ctg} \frac{p\pi}{q} \quad [q > p > 0].$$ BH[2](16)

3. $$\int_0^1 \frac{x^{\nu-1} - x^{\mu-1}}{1-x^\nu} dx = \frac{1}{\nu}\left[C + \psi\left(\frac{\mu}{\nu}\right)\right] \quad [\operatorname{Re} \mu > \operatorname{Re} \nu > 0].$$ BH[2](17)

4. $$\int_{-\infty}^{\infty} \frac{x^{2m} - x^{2n}}{1 - x^{2l}} dx = \frac{\pi}{l}\left[\operatorname{ctg}\left(\frac{2m+1}{2l}\pi\right) - \operatorname{ctg}\left(\frac{2n+1}{2l}\pi\right)\right] \quad [m < l, n < l].$$ F II 647–648

3.245 $$\int_0^\infty [x^{\nu-\mu} - x^\nu(1+x)^{-\mu}] dx = \frac{\nu}{\nu - \mu + 1} B(\nu, \mu - \nu) \quad [\operatorname{Re} \mu > \operatorname{Re} \nu > 0].$$ BH[16](13)

3.246 $$\int_0^\infty \frac{1-x^q}{1-x^r} x^{p-1} dx = \frac{\pi}{r} \sin\frac{q\pi}{r} \operatorname{cosec}\frac{p\pi}{r} \operatorname{cosec}\frac{(p+q)\pi}{r} \quad [p+q < r, p > 0].$$ Ba IV 311(33), BH[17](12)

Integrale der Gestalt $\int f(x^p \pm x^{-p}, x^q \pm \pm x^{-q}, \ldots)\frac{dx}{x}$ müssen durch die Substitution $x = e^t$ bzw. $x = e^{-t}$ transformiert werden; beispielsweise muß man anstelle von $\int_0^1 (x^{1+p} + x^{1-p})^{-1} dx$ das Integral $\int_0^\infty \operatorname{sech} px\, dx$ aufsuchen, anstelle von $\int_0^1 \frac{x^{n-m-1} + x^{n+m-1}}{1 + 2x^n \cos a + x^{2n}} dx$ das Integral $\int_0^\infty \operatorname{ch} mx\, (\operatorname{ch} nx - \cos a)^{-1} dx$ (vgl. 3.514 2.).

Integrals of the form $\int f(x^p \pm x^{-p}, x^q \pm x^{-q}, \ldots) \frac{dx}{x}$ should be transformed by means of the substitution $x = e^t$ or $x = e^{-t}$, respectively; for instance, instead of $\int_0^1 (x^{1+p} + x^{1-p})^{-1} dx$ one must look up $\int_0^\infty \operatorname{sech} px\, dx$, and instead of $\int_0^1 \frac{x^{n-m-1} + x^{n+m-1}}{1 + 2x^n \cos a + x^{2n}} dx$ the integral $\int_0^\infty \operatorname{ch} mx(\operatorname{ch} nx - \cos a)^{-1} dx$ (cf. 3.514 2.).

3.247

1. $$\int_0^1 \frac{x^{\alpha-1}(1-x)^{n-1}}{1-\xi x^b} dx = (n-1)! \sum_{k=0}^\infty \frac{\xi^k}{(\alpha+kb)(\alpha+kb+1)\ldots(\alpha+kb+k-1)}$$
$$[b > 0, |\xi| < 1].$$ A(6.704)

2. $$\int_0^\infty \frac{(1-x^p) x^{\nu-1}}{1-x^{np}} dx = \frac{\pi}{np} \sin\left(\frac{\pi}{n}\right) \operatorname{cosec}\frac{(p+\nu)\pi}{np} \operatorname{cosec}\frac{\pi\nu}{np}$$
$$[0 < \operatorname{Re} \nu < (n-1)p].$$ Ba IV 311(33)

3.248

1. $$\int_0^\infty \frac{x^{\mu-1}\,dx}{\sqrt{1+x^\nu}} = 2^{\frac{2\mu}{\nu}}\,B(\nu-2\mu,\mu) \quad [\nu > 2\mu].$$ BH[21](9)

2. $$\int_0^1 \frac{x^{2n+1}\,dx}{\sqrt{1-x^2}} = \frac{(2n)!!}{(2n+1)!!}.$$ BH[8](14)

3. $$\int_0^1 \frac{x^{2n}\,dx}{\sqrt{1-x^2}} = \frac{(2n-1)!!}{(2n)!!}\,\frac{\pi}{2}.$$ BH[8](13)

3.249

1. $$\int_0^\infty \frac{dx}{(x^2+a^2)^n} = \frac{(2n-3)!!}{2\cdot(2n-2)!!}\,\frac{\pi}{a^{2n-1}}.$$ F II 749

2. $$\int_0^a (a^2-x^2)^{n-\frac{1}{2}}\,dx = a^{2n}\frac{(2n-1)!!}{2(2n)!!}\,\pi.$$ F II 139

3. $$\int_{-1}^1 \frac{(1-x^2)^n\,dx}{(a-x)^{n+1}} = 2^{n+1}Q_n(a).$$ Ba II 181(31)

4. $$\int_0^1 \frac{x^\mu\,dx}{1+x^2} = \frac{1}{2}\beta\!\left(\frac{\mu+1}{2}\right) \quad [\operatorname{Re}\mu > -1].$$ BH[2](7)

5. $$\int_0^1 (1-x^2)^{\mu-1}\,dx = 2^{2\mu-2}B(\mu,\mu) = \frac{1}{2}B\!\left(\frac{1}{2},\mu\right) \quad [\operatorname{Re}\mu > 0].$$ F II 792–793

6. $$\int_0^1 (1-\sqrt{x})^{p-1}\,dx = \frac{2}{p(p+1)} \quad [p > 0].$$ BH[7](7)

7. $$\int_0^1 (1-x^\mu)^{-\frac{1}{\nu}}\,dx = \frac{1}{\mu}B\!\left(\frac{1}{\mu},1-\frac{1}{\nu}\right) \quad [\operatorname{Re}\mu > 0,\ |\nu| > 1].$$

3.251

1. $$\int_0^1 x^{\mu-1}(1-x^\lambda)^{\nu-1}\,dx = \frac{1}{\lambda}B\!\left(\frac{\mu}{\lambda},\nu\right) \quad [\operatorname{Re}\mu > 0,\ \operatorname{Re}\nu > 0,\ \lambda > 0].$$ F II 792–793

2. $$\int_0^\infty x^{\mu-1}(1+x^2)^{\nu-1}\,dx = \frac{1}{2}B\!\left(\frac{\mu}{2},1-\nu-\frac{\mu}{2}\right)$$

$$\left[\operatorname{Re}\mu > 0,\ \operatorname{Re}\!\left(\nu+\frac{1}{2}\mu\right) < 1\right].$$

3.251

3. $$\int_1^\infty x^{\mu-1}(x^p-1)^{\nu-1}\,dx = \frac{1}{p}\,\mathrm{B}\left(1-\nu-\frac{\mu}{p},\,\nu\right)$$

$[p>0,\ \mathrm{Re}\,\nu>0,\ \mathrm{Re}\,\mu < p - p\,\mathrm{Re}\,\nu].$ Ba IV 311(32)

4. $$\int_0^\infty \frac{x^{2m}\,dx}{(ax^2+c)^n} = \frac{(2m-1)!!\,(2n-2m-3)!!\,\pi}{2\cdot(2n-2)!!\,a^m c^{n-m-1}\sqrt{ac}} \quad [a>0,\ c>0,\ n>m+1].$$

GH[141](8a)

5. $$\int_0^\infty \frac{x^{2m+1}\,dx}{(ax^2+c)^n} = \frac{m!\,(n-m-2)!}{2(n-1)!\,a^{m+1}c^{n-m-1}} \quad [ac>0,\ n>m+1\geqslant 1].$$

GH[141](8b)

6. $$\int_0^\infty \frac{x^{\mu+1}}{(1+x^2)^2}\,dx = \frac{\mu\pi}{4\sin\frac{\mu\pi}{2}} \quad [-2<\mathrm{Re}\,\mu<2].$$

WW 119

7. $$\int_0^1 \frac{x^\mu\,dx}{(1+x^2)^2} = -\frac{1}{4} + \frac{\mu-1}{4}\beta\left(\frac{\mu-1}{2}\right) \quad [\mathrm{Re}\,\mu>1].$$

Li[3](11)

8. $$\int_0^1 x^{q+p-1}(1-x^q)^{-\frac{p}{q}}\,dx = \frac{p\pi}{q^2}\operatorname{cosec}\frac{p\pi}{q} \quad [q>p].$$

BH[9](22)

9. $$\int_0^1 x^{\frac{q}{p}-1}(1-x^q)^{-\frac{1}{p}}\,dx = \frac{\pi}{q}\operatorname{cosec}\frac{\pi}{p} \quad [p>1,\ q>0].$$

BH[9](23)+

10. $$\int_0^1 x^{p-1}(1-x^q)^{-\frac{p}{q}}\,dx = \frac{\pi}{q}\operatorname{cosec}\frac{p\pi}{q} \quad [q>p>0].$$

BH[9](20)

11. $$\int_0^\infty x^{\mu-1}(1+\beta x^p)^{-\nu}\,dx = \frac{1}{p}\beta^{-\frac{\mu}{p}}\mathrm{B}\left(\frac{\mu}{p},\,\nu-\frac{\mu}{p}\right)$$

$[|\arg\beta|<\pi,\ p>0,\ 0<\mathrm{Re}\,\mu<p\,\mathrm{Re}\,\nu].$ BH[17](20), Ba I 10(16)

3.252

1. $$\int_0^\infty \frac{dx}{(ax^2+2bx+c)^n} = \frac{(-1)^{n-1}}{(n-1)!}\frac{\partial^{n-1}}{\partial c^{n-1}}\left[\frac{1}{\sqrt{ac-b^2}}\operatorname{arctg}\frac{b}{\sqrt{ac-b^2}}\right]$$

$[a>0,\ ac>b^2].$ GH[131](4)

2. $$\int_{-\infty}^\infty \frac{dx}{(ax^2+2bx+c)^n} = \frac{(2n-3)!!\,\pi a^{n-1}}{(2n-2)!!\,(ac-b)^{n-\frac{1}{2}}} \quad [a>0,\ ac>b^2].$$

GH[131](5)

3. $$\int_0^\infty \frac{dx}{(ax^2+2bx+c)^{n+\frac{3}{2}}} = \frac{(-2)^n}{(2n+1)!!} \frac{\partial^n}{\partial c^n}\left\{\frac{1}{\sqrt{c}(\sqrt{ac}+b)}\right\}$$
$$[a \geq 0,\ c > 0,\ b > -\sqrt{ac}]. \qquad \text{GH [213](4)}$$

3.252

4. $$\int_0^\infty \frac{x\,dx}{(ax^2+2bx+c)^n} = \frac{(-1)^n}{(n-1)!}\frac{\partial^{n-2}}{\partial c^{n-2}}\left\{\frac{1}{2(ac-b^2)} - \frac{b}{2(ac-b^2)^{\frac{3}{2}}}\operatorname{arcctg}\frac{b}{\sqrt{ac-b^2}}\right\} \quad \text{für}\atop\text{for}\ ac > b^2;$$

$$= \frac{(-1)^n}{(n-1)!}\frac{\partial^{n-2}}{\partial c^{n-2}}\left\{\frac{1}{2(ac-b^2)} + \frac{b}{4(b^2-ac)^{\frac{3}{2}}}\ln\frac{b+\sqrt{b^2-ac}}{b-\sqrt{b^2-ac}}\right\}$$
$$\text{für}\atop\text{for}\ b^2 > ac > 0;$$

$$= \frac{a^{n-2}}{2(n-1)(2n-1)b^{2n-2}} \quad \text{für}\atop\text{for}\ ac = b^2 \quad [a > 0,\ b > 0,\ n \geq 2]. \qquad \text{GH [141](5)}$$

5. $$\int_{-\infty}^\infty \frac{x\,dx}{(ax^2+2bx+c)^n} = -\frac{(2n-3)!!\,\pi b a^{n-2}}{(2n-2)!!\,(ac-b^2)^{\frac{(2n-1)}{2}}} \quad [ac > b^2,\ a > 0,\ n \geq 2].$$

GH [141](6)

6. $$\int_{-\infty}^\infty \frac{x^m\,dx}{(ax^2+2bx+c)^n} = \frac{(-1)^m \pi a^{n-m-1} b^m}{(2n-2)!!\,(ac-b^2)^{n-\frac{1}{2}}} \times$$
$$\times \sum_{k=0}^{E\left(\frac{m}{2}\right)} \binom{m}{2k}(2k-1)!!(2n-2k-3)!!\left(\frac{ac-b^2}{b^2}\right)^k$$
$$[ac > b^2,\ 0 \leq m \leq 2n-2]. \qquad \text{GH [141](17)}$$

7. $$\int_0^\infty \frac{x^n\,dx}{(ax^2+2bx+c)^{n+\frac{3}{2}}} = \frac{n!}{(2n+1)!!\sqrt{c}\,(\sqrt{ac}+b)^{n+1}}$$
$$[a \geq 0,\ c > 0,\ b > -\sqrt{ac}]. \qquad \text{GH [213](5a)}$$

8. $$\int_0^\infty \frac{x^{n+1}\,dx}{(ax^2+2bx+c)^{n+\frac{3}{2}}} = \frac{n!}{(2n+1)!!\sqrt{a}\,(\sqrt{ac}+b)^{n+1}}$$
$$[a > 0,\ c \geq 0,\ b > -\sqrt{ac}]. \qquad \text{GH [213](5b)}$$

9. $$\int_0^\infty \frac{x^{n+\frac{1}{2}}\,dx}{(ax^2+2bx+c)^{n+1}} = \frac{(2n-1)!!\,\pi}{2^{2n+\frac{1}{2}}(b+\sqrt{ac})^{n+\frac{1}{2}}n!\sqrt{a}}$$
$$[a > 0,\ c > 0,\ b + \sqrt{ac} > 0]. \qquad \text{Li [21](19)}$$

3.252

10. $\int_0^\infty \dfrac{x^{\mu-1}\,dx}{(1+2x\cos t+x^2)^\nu} =$

$$= 2^{\nu-\frac{1}{2}} \sin^{\nu-\frac{1}{2}} t\, \Gamma\left(\nu+\frac{1}{2}\right) B(\mu,\, 2\nu-\mu)\, P_{\mu-\nu-\frac{1}{2}}^{\frac{1}{2}-\nu}(\cos t)$$

$[-\pi < t < \pi,\ 0 < \operatorname{Re}\mu < \operatorname{Re}2\nu].$ Ba IV 310(22)

11. $\int_0^\infty (1+2\beta x+x^2)^{\mu-\frac{1}{2}} x^{-\nu-1}\,dx = 2^{-\mu}(\beta^2-1)^{\frac{\mu}{2}} \Gamma(1-\mu) \times$

$$\times B(\nu-2\mu+1,\,-\nu)\, P_{\nu-\mu}^{\mu}(\beta)$$

$[\operatorname{Re}\nu < 0,\ \operatorname{Re}(2\mu-\nu) < 1,\ |\arg(\beta\pm 1)| < \pi];$ Ba I 160(33)

$$= -\pi\operatorname{cosec}\nu\pi\, C_\nu^{\frac{1}{2}-\mu}(\beta)$$

$\left[-2 < \operatorname{Re}\left(\dfrac{1}{2}-\mu\right) < \operatorname{Re}\nu < 0,\ |\arg(\beta\pm 1)| < \pi\right].$ Ba I 178(24)

12. $\int_0^\infty \dfrac{x^{\mu-1}\,dx}{x^2+2ax\cos t+a^2} = -\pi a^{\mu-2}\operatorname{cosec} t\,\operatorname{cosec}(\mu\pi) \sin[(\mu-1)t]$

$[a>0,\ |t|<\pi,\ 0<\operatorname{Re}\mu<2].$ F II 743, BH[20](3)

13. $\int_0^\infty \dfrac{x^{\mu-1}\,dx}{(x^2+2ax\cos t+a^2)^2} = \dfrac{\pi a^{\mu-4}}{2} \operatorname{cosec}\mu\pi\,\operatorname{cosec}^3 t \times$

$$\times \{(\mu-1)\sin t\cos[(\mu-2)t] - \sin[(\mu-1)t]\}$$

$[a>0,\ |t|<\pi,\ 0<\operatorname{Re}\mu<4].$ Li[20](8)+, Ba IV 309(13)

14. $\int_0^\infty \dfrac{x^{\mu-1}\,dx}{\sqrt{1+2x\cos t+x^2}} = \pi\operatorname{cosec}(\mu\pi)\, P_{\mu-1}(\cos t)$ $\quad[|t|<\pi,\ 0<\operatorname{Re}\mu<1].$

Ba IV 310(17)

3.253 $\int_{-1}^{1} \dfrac{(1+x)^{2\mu-1}(1-x)^{2\nu-1}}{(1+x^2)^{\mu+\nu}}\,dx = 2^{\mu+\nu-2} B(\mu,\nu)$ $\quad[\operatorname{Re}\mu>0,\ \operatorname{Re}\nu>0].$

F II 793-794

3.254

1. $\int_0^u x^{\lambda-1}(u-x)^{\mu-1}(x^2+\beta^2)^\nu\,dx = \beta^{2\nu}u^{\lambda+\mu-1}B(\lambda,\mu) \times$

$$\times {}_3F_2\left(-\nu,\,\dfrac{\lambda}{2},\,\dfrac{\lambda+1}{2};\,\dfrac{\lambda+\mu}{2},\,\dfrac{\lambda+\mu+1}{2};\,\dfrac{-u^2}{\beta^2}\right)$$

$\left[\operatorname{Re}\left(\dfrac{u}{\beta}\right)>0,\ \operatorname{Re}\lambda>0,\ \operatorname{Re}\mu>0\right].$ Ba V 186(10)

2. $\int_u^\infty x^{-\lambda}(x-u)^{\mu-1}(x^2+\beta^2)^\nu\,dx = u^{\mu-\lambda+2\nu}\,\dfrac{\Gamma(\mu)\,\Gamma(\lambda-\mu-2\nu)}{\Gamma(\lambda-\mu)} \times$ 3.254

$$\times\ {}_3F_2\left(-\nu,\ \frac{\lambda-\mu}{2}-\nu,\ \frac{1+\lambda-\mu}{2}-\nu;\ \frac{\lambda}{2}-\nu,\ \frac{1+\lambda}{2}-\nu;\ -\frac{\beta^2}{u^2}\right)$$

$\left[|u|>|\beta|\ \text{oder or}\ \operatorname{Re}\left(\dfrac{\beta}{u}\right)>0,\ 0<\operatorname{Re}\mu<\operatorname{Re}(\lambda-2\nu)\right].$ Ba V 202(9)

$\displaystyle\int_0^1 \frac{x^{\mu+\frac{1}{2}}(1-x)^{\mu-\frac{1}{2}}}{(c+2bx-ax^2)^{\mu+1}}\,dx = \frac{\sqrt{\pi}}{\{a+(\sqrt{c+2b-a}+\sqrt{c})^2\}^{\mu+\frac{1}{2}}\sqrt{c+2b-a}}\,\frac{\Gamma\!\left(\mu+\frac{1}{2}\right)}{\Gamma(\mu+1)}$ 3.255

$\left[a+(\sqrt{c+2b-a}+\sqrt{c})^2>0,\ c+2b-a>0,\ \operatorname{Re}\mu>-\dfrac{1}{2}\right].$ BH [14](2)

1. $\displaystyle\int_0^1 \frac{x^{p-1}+x^{q-1}}{(1-x^2)^{\frac{p+q}{2}}}\,dx = \frac{1}{2}\cos\!\left(\frac{q-p}{4}\pi\right)\sec\!\left(\frac{q+p}{4}\pi\right) B\!\left(\frac{p}{2},\frac{q}{2}\right)$ 3.256

$[p>0,\ q>0,\ p+q<2].$ BH[8](25)

2. $\displaystyle\int_0^1 \frac{x^{p-1}-x^{q-1}}{(1-x^2)^{\frac{p+q}{2}}}\,dx = \frac{1}{2}\sin\!\left(\frac{q-p}{4}\pi\right)\operatorname{cosec}\!\left(\frac{q+p}{4}\pi\right) B\!\left(\frac{p}{2},\frac{q}{2}\right)$

$[p>0,\ q>0,\ p+q<2].$ BH[8](26)

$\displaystyle\int_0^\infty \left[\left(ax+\frac{b}{x}\right)^2+c\right]^{-p-1} dx = \frac{2\sqrt{\pi}\,\Gamma\!\left(p+\frac{1}{2}\right)}{a\,c^{p+\frac{1}{2}}\,\Gamma(p+1)}.$ BH[20](4) 3.257

1. $\displaystyle\int_b^\infty (x-\sqrt{x^2-a^2})^n\,dx = \frac{a^2}{2(n-1)}(b-\sqrt{b^2-a^2})^{n-1} -$ 3.258

$\qquad -\dfrac{1}{2(n+1)}(b-\sqrt{b^2-a^2})^{n+1}\quad [0<a\leqslant b,\ n\geqslant 2].$ GH [215](5)

2. $\displaystyle\int_b^\infty (\sqrt{x^2+1}-x)^n\,dx = \frac{(\sqrt{b^2+1}-b)^{n-1}}{2(n-1)}+\frac{(\sqrt{b^2+1}-b)^{n+1}}{2(n+1)}$

$[n\geqslant 2].$ GH [214](7)

3. $\displaystyle\int_0^\infty (\sqrt{x^2+a^2}-x)^n\,dx = \frac{na^{n+1}}{n^2-1}\quad [n\geqslant 2].$ GH [214](6a)

3.258

4. $$\int_0^\infty \frac{dx}{(x+\sqrt{x^2+a^2})^n} = \frac{n}{a^{n-1}(n^2-1)} \quad [n \geqslant 2].$$ GH[214](5a)

5. $$\int_0^\infty x^m(\sqrt{x^2+a^2}-x)^n\,dx = \frac{n \cdot m!\, a^{m+n+1}}{(n-m-1)(n-m+1)\ldots(m+n+1)}$$
$$[a>0,\ 0 \leqslant m \leqslant n-2].$$ GH[214](6)

6. $$\int_0^\infty \frac{x^m\,dx}{(x+\sqrt{x^2+a^2})^n} = \frac{n \cdot m!}{(n-m-1)(n-m+1)\ldots(m+n+1)\,a^{n-m-1}}$$
$$[a>0,\ 0 \leqslant m \leqslant n-2].$$ GH[214](5)

7. $$\int_a^\infty (x-a)^m (x-\sqrt{x^2-a^2})^n\,dx = \frac{n \cdot (n-m-2)!\,(2m+1)!\, a^{m+n+1}}{2^m (n+m+1)!}$$
$$[a>0,\ n \geqslant m+2].$$ GH[215](6)

3.259

1. $$\int_0^1 x^{p-1}(1-x)^{n-1}(1+bx^m)^l\,dx = (n-1)!\sum_{k=0}^\infty \binom{l}{k}\frac{b^k \Gamma(p+km)}{\Gamma(p+n+km)}$$
$$[b^2 > 1].$$ BH[1](14)

2. $$\int_0^u x^{p-1}(u-x)^{\mu-1}(x^m+\beta^m)^\lambda\,dx = \beta^{m\lambda} u^{\mu+\nu+1} \mathbf{B}(\mu,\nu) \times$$
$$\times\,_{m+1}F_m\left(-\lambda,\frac{\nu}{m},\frac{\nu+1}{m},\ldots,\frac{\nu+m-1}{m};\frac{\mu+\nu}{m},\frac{\mu+\nu+1}{m},\ldots,\right.$$
$$\left.\ldots,\frac{\mu+\nu+m-1}{m};\frac{-u^m}{\beta^m}\right)$$
$$\left[\operatorname{Re}\mu > 0,\ \operatorname{Re}\nu > 0,\ \left|\arg\left(\frac{u}{\beta}\right)\right| < \frac{\pi}{m}\right].$$ Ba V 186(11)

3. $$\int_0^\infty x^{\lambda-1}(1+\alpha x^p)^{-\mu}(1+\beta x^p)^{-\nu}\,dx = \frac{1}{p}\alpha^{-\frac{\lambda}{p}}\mathbf{B}\left(\frac{\lambda}{p},\mu+\nu-\frac{\lambda}{p}\right)\times$$
$$\times\,_2F_1\left(\nu,\frac{\lambda}{p};\mu+\nu;1-\frac{\beta}{\alpha}\right)$$
$$[|\arg\alpha|<\pi,\ |\arg\beta|<\pi,\ p>0,\ 0<\operatorname{Re}\lambda<2\operatorname{Re}(\mu+\nu)].$$ Ba IV 312(35)

3.261

1. $$\int_0^1 \frac{(1-x\cos t)x^{\mu-1}\,dx}{1-2x\cos t+x^2} = \sum_{k=0}^\infty \frac{\cos kt}{\mu+k}$$
$$[\operatorname{Re}\mu > 0,\ t \neq 2n\pi].$$ BH[6](9)

2. $\int_0^1 \frac{(x^\nu + x^{-\nu})\,dx}{1 + 2x\cos t + x^2} = \frac{\pi \sin \nu t}{\sin t \sin \nu\pi}$ $\quad [\nu^2 < 1,\ t \neq (2n+1)\pi]$. BH[6](8) 3.261

3. $\int_0^1 \frac{(x^{1+p} + x^{1-p})\,dx}{(1 + 2x\cos t + x^2)^2} = \frac{\pi(p \sin t \cos pt - \cos t \sin pt)}{2 \sin^3 t \sin p\pi}$

$\quad [p^2 < 1,\ t \neq (2n+1)\pi]$. BH[6](18)

4. $\int_0^1 \frac{x^{\mu-1}}{1 + 2ax\cos t + a^2 x^2} \cdot \frac{dx}{(1-x)^\mu} =$

$= \frac{\pi \operatorname{cosec} t \operatorname{cosec} \mu\pi}{(1 + 2a\cos t + a^2)^{\frac{\mu}{2}}} \sin\left(t - \mu \operatorname{arctg} \frac{a \sin t}{1 + a \cos t}\right)$

$\quad [a > 0,\ 0 < \operatorname{Re} \mu < 1]$. BH[6](21)

$\int_0^\infty \frac{x^{-p}\,dx}{1 + x^3} = \frac{\pi}{3} \operatorname{cosec} \frac{(1-p)\pi}{3}$ $\quad [-2 < p < 1]$. Li[18](3) 3.262

$\int_0^\infty \frac{x^\nu\,dx}{(x+\gamma)(x^2+\beta^2)} = \frac{\pi}{2(\beta^2+\gamma^2)}\left[\gamma\beta^{\nu-1}\sec\frac{\nu\pi}{2} + \beta^\nu \operatorname{cosec}\frac{\nu\pi}{2} - 2\gamma^\nu \operatorname{cosec}(\nu\pi)\right]$ 3.263

$\quad [\operatorname{Re}\beta > 0,\ |\arg\gamma| < \pi,\ -1 < \operatorname{Re}\nu < 2]$. Ba V 216(7)

1. $\int_0^\infty \frac{x^{p-1}\,dx}{(a^2+x^2)(b^2-x^2)} = \frac{\pi}{2} \cdot \frac{a^{p-2} + b^{p-2}\cos\frac{p\pi}{2}}{a^2+b^2} \operatorname{cosec}\frac{p\pi}{2}$ $\quad [p < 4]$. BH[19](14) 3.264

2. $\int_0^\infty \frac{x^{\mu-1}\,dx}{(\beta+x^2)(\gamma+x^2)} = \frac{\pi}{2} \cdot \frac{\gamma^{\frac{\mu}{2}-1} - \beta^{\frac{\mu}{2}-1}}{\beta - \gamma} \operatorname{cosec}\frac{\mu\pi}{2}$

$\quad [|\arg\beta| < \pi,\ |\arg\gamma| < \pi,\ 0 < \operatorname{Re}\mu < 4]$. Ba IV 309(14)

$\int_0^1 \frac{1-x^{\mu-1}}{1-x}\,dx = \psi(\mu) + C$ $\quad [\operatorname{Re}\mu > 0]$; F II 802, WW 260, Ba I 16(13) 3.265

$= \psi(1-\mu) + C - \pi \operatorname{ctg}(\mu\pi)$ $\quad [\operatorname{Re}\mu > 0]$. Ba I 16(15)+

$\int_0^\infty \frac{(x^\nu - a^\nu)\,dx}{(x-a)(\beta+x)} = \frac{\pi}{a+\beta}\left\{\beta^\nu \operatorname{cosec}(\nu\pi) - a^\nu \operatorname{ctg}(\nu\pi) - \frac{a^\nu}{\pi}\ln\frac{\beta}{a}\right\}$ 3.266

$\quad [|\arg\beta| < \pi,\ |\operatorname{Re}\nu| < 1]$. Ba V 216(8)

1. $\int_0^1 \frac{x^{3n}\,dx}{\sqrt[3]{1-x^3}} = \frac{2\pi}{3\sqrt{3}} \cdot \frac{\Gamma\left(n+\frac{1}{3}\right)}{\Gamma\left(\frac{1}{3}\right)\Gamma(n+1)}$. BH[9](6) 3.267

3.267 2. $$\int_0^1 \frac{x^{3n-1} dx}{\sqrt[3]{1-x^3}} = \frac{(n-1)!\, \Gamma\left(\frac{2}{3}\right)}{3\Gamma\left(n+\frac{2}{3}\right)}.$$ BH[9](7)

3.268 1. $$\int_0^1 \left(\frac{1}{1-x} - \frac{px^{p-1}}{1-x^p}\right) dx = \ln p.$$ BH[5](14)

2. $$\int_0^1 \frac{1-x^\mu}{1-x} x^{\nu-1} dx = \psi(\mu+\nu) - \psi(\nu) \quad [\text{Re } \nu > 0,\ \text{Re } \mu > 0].$$ BH[2](3)

3. $$\int_0^1 \left[\frac{n}{1-x} - \frac{x^{\mu-1}}{1-\sqrt[n]{x}}\right] dx = nC + \sum_{k=1}^n \psi\left(\mu + \frac{n-k}{n}\right)$$
$$[\text{Re } \mu > 0].$$ BH[13](10)

3.269 1. $$\int_0^1 \frac{x^p - x^{-p}}{1-x^2} x\, dx = \frac{\pi}{2} \operatorname{ctg} \frac{p\pi}{2} - \frac{1}{p} \quad [p^2 < 1].$$ BH[4](12)

2. $$\int_0^\infty \frac{x^p - x^{-p}}{1+x^2} x\, dx = \frac{1}{p} - \frac{\pi}{2} \operatorname{cosec} \frac{p\pi}{2} \quad [p^2 < 1].$$ BH[4](8)

3. $$\int_0^1 \frac{x^\mu - x^\nu}{1-x^2}\, dx = \frac{1}{2}\psi\left(\frac{\nu+1}{2}\right) - \frac{1}{2}\psi\left(\frac{\mu+1}{2}\right)$$
$$[\text{Re } \mu > -1,\ \text{Re } \nu > -1].$$ BH[2](9)

3.271 1. $$\int_0^\infty \frac{x^p - x^q}{x-1} \frac{dx}{x+a} = \frac{\pi}{1+a}\left(\frac{a^p - \cos p\pi}{\sin p\pi} - \frac{a^q - \cos q\pi}{\sin q\pi}\right)$$
$$[p^2 < 1,\ q^2 < 1,\ a > 0].$$ BH[19](2)

2. $$\int_0^\infty \frac{x^p - a^p}{x-a} \frac{x^p - 1}{x-1}\, dx = \frac{\pi}{a-1}\left\{\frac{a^{2p}-1}{\sin(2p\pi)} - \frac{1}{\pi} a^p \ln a\right\}$$
$$\left[p^2 < \frac{1}{4}\right].$$ BH[19](3)

3. $$\int_0^\infty \frac{x^p - a^p}{x-a} \frac{x^{-p} - 1}{x-1}\, dx = \frac{\pi}{a-1}\left\{2(a^p - 1)\operatorname{ctg} p\pi - \frac{1}{\pi}(a^p+1)\ln a\right\}$$
$$[p^2 < 1].$$ BH[18](9)

4. $$\int_0^\infty \frac{x^p - a^p}{x-a} \frac{1-x^{-p}}{1-x} x^q\, dx = \frac{\pi}{a-1}\left\{\frac{a^{p+q}-1}{\sin[(p+q)\pi]} + \right.$$
$$\left. + \frac{a^p - a^q}{\sin[(q-p)\pi]}\right\} \frac{\sin p\pi}{\sin q\pi} \quad [(p+q)^2 < 1,\ (p-q)^2 < 1].$$ BH[19](4)

5. $\int_0^\infty \left(\dfrac{x^p - x^{-p}}{1-x}\right)^2 dx = 2(1 - 2p\pi \operatorname{ctg} 2p\pi) \quad \left[p^2 < \dfrac{1}{4}\right].$ BH[16](3) 3.271

1. $\int_0^1 \dfrac{x^{n-1} + x^{n-\frac{1}{2}} - 2x^{2n-1}}{1-x} dx = 2 \ln 2.$ BH[8](8) 3.272

2. $\int_0^1 \dfrac{x^{n-1} + x^{n-\frac{2}{3}} + x^{n-\frac{1}{3}} - 3x^{3n-1}}{1-x} dx = 3 \ln 3.$ BH[8](9)

1. $\int_0^1 \dfrac{\sin t - a^n x^n \sin[(n+1)t] + a^{n+1} x^{n+1} \sin nt}{1 - 2ax \cos t + a^2 x^2} (1-x)^{p-1} dx =$ 3.273

$= \Gamma(p) \sum_{k=1}^n \dfrac{(k-1)! \, a^{k-1} \sin kt}{\Gamma(p+k)} \quad [p > 0].$ BH[6](13)

2. $\int_0^1 \dfrac{\cos t - ax - a^n x^n \cos[(n+1)t] + a^{n+1} x^{n+1} \cos nt}{1 - 2ax \cos t + a^2 x^2} (1-x)^{p-1} dx =$

$= \Gamma(p) \sum_{k=1}^n \dfrac{(k-1)! \, a^{k-1} \cos kt}{\Gamma(p+k)} \quad [p > 0].$ BH[6](14)

3. $\int_0^1 x \dfrac{\sin t - x^n \sin[(n+1)t] + x^{n+1} \sin nt}{1 - 2x \cos t + x^2} dx = \sum_{k=1}^n \dfrac{\sin kt}{k+1}.$ BH[6](12)

4. $\int_0^1 \dfrac{1 - x\cos t - x^{n+1} \cos[(n+1)t] + x^{n+2} \cos nt}{1 - 2x \cos t + x^2} dx = \sum_{k=0}^n \dfrac{\cos kt}{k+1}.$ BH[6](11)

1. $\int_0^\infty \dfrac{x^{\mu-1}(1-x)}{1-x^n} dx = \dfrac{\pi}{n} \sin \dfrac{\pi}{n} \operatorname{cosec} \dfrac{\mu\pi}{n} \operatorname{cosec} \dfrac{(\mu+1)\pi}{n} \quad [0 < \operatorname{Re} \mu < n-1].$ 3.274 BH[20](13)

2. $\int_0^1 \dfrac{1 - x^n}{(1+x)^{n+1}} \dfrac{dx}{1-x} = \dfrac{1}{2^{n+1}} \sum_{k=1}^n \dfrac{2^k}{k}.$ BH[5](3)

3. $\int_0^\infty \dfrac{x^q - 1}{x^p - x^{-p}} \dfrac{dx}{x} = \dfrac{\pi}{2p} \dfrac{q\pi}{2p} \quad [p > q].$ BH[18](6)

1. $\int_0^1 \left\{\dfrac{x^{n-1}}{1 - x^{\frac{1}{p}}} - \dfrac{px^{np-1}}{1-x}\right\} dx = p \ln p \quad [p > 0].$ BH[13](9) 3.275

2. $\int_0^1 \left\{\dfrac{nx^{n-1}}{1-x^n} - \dfrac{x^{mn-1}}{1-x}\right\} dx = C + \dfrac{1}{n} \sum_{k=1}^n \psi\left(m + \dfrac{n-k}{n}\right).$ BH[5](13)

3. $\int_0^1 \left(\dfrac{x^{p-1}}{1-x} - \dfrac{qx^{pq-1}}{1-x^q}\right) dx = \ln q \quad [q > 0].$ BH[5](12)

3.275

4.
$$\int_0^\infty \left\{\frac{1}{1+x^{2n}} - \frac{1}{1+x^{2m}}\right\} \frac{dx}{x} = 0.$$
BH[18](17)

3.276

1.
$$\int_0^\infty \frac{\left[\left(ax+\frac{b}{x}\right)^2+c\right]^{-p-1} dx}{x^2} = \frac{\sqrt{\pi}}{2bc^{p+\frac{1}{2}}} \frac{\Gamma\left(p+\frac{1}{2}\right)}{\Gamma(p+1)} \quad \left[p > -\frac{1}{2}\right].$$
BH[20](19)

2.
$$\int_0^\infty \left(a+\frac{b}{x^2}\right)\left[\left(ax+\frac{b}{x}\right)^2+c\right]^{-p-1} dx = \frac{\sqrt{\pi}}{c^{p+\frac{1}{2}}} \frac{\Gamma\left(p+\frac{1}{2}\right)}{\Gamma(p+1)}$$
$$\left[p > -\frac{1}{2}\right].$$
BH[20](5)

3.277

1.
$$\int_0^\infty \frac{x^{\mu-1}[\sqrt{1+x^2}+\beta]^\nu}{\sqrt{1+x^2}} dx =$$
$$= 2^{\frac{\mu}{2}-1}(\beta^2-1)^{\frac{\nu}{2}+\frac{\mu}{4}} \Gamma\left(\frac{\mu}{2}\right)\Gamma(1-\mu-\nu) P_{\frac{\mu}{2}-1}^{\frac{\nu+\mu}{2}}(\beta)$$
$$[\operatorname{Re}\beta > -1, 0 < \operatorname{Re}\mu < 1 - \operatorname{Re}\nu].$$
Ba IV 310(25)

2.
$$\int_0^\infty \frac{x^{\mu-1}[\sqrt{\beta^2+x^2}+x]^\nu}{\sqrt{\beta^2+x^2}} dx = \frac{\beta^{\mu+\nu-1}}{2^\mu} B\left(\mu, \frac{1-\mu-\nu}{2}\right)$$
$$[\operatorname{Re}\beta > 0, 0 < \operatorname{Re}\mu < 1 - \operatorname{Re}\nu].$$
Ba IV 311(28)

3.
$$\int_0^\infty \frac{x^{\mu-1}[\cos t \pm i \sin t \sqrt{1+x^2}]^\nu}{\sqrt{1+x^2}} dx = 2^{\frac{\mu-1}{2}} \sin^{\frac{1-\mu}{2}} t \frac{\Gamma\left(\frac{\mu}{2}\right)\Gamma(1-\mu-\nu)}{\Gamma(-\nu)} \times$$
$$\times \left[\pi^{-\frac{1}{2}} Q_{-\frac{\mu+1}{2}-\nu}^{\frac{\mu-1}{2}}(\cos t) \mp \frac{i}{2} \pi^{\frac{1}{2}} P_{-\frac{\mu+1}{2}-\nu}^{\frac{\mu-1}{2}}(\cos t)\right] \quad [\operatorname{Re}\mu > 0].$$
Ba IV 311(27)

4.
$$\int_0^\infty \frac{x^{\mu-1}[\sqrt{(\beta^2-1)(x^2+1)}+\beta]^\nu}{\sqrt{x^2+1}} dx = \frac{2^{\frac{\mu-1}{2}}}{\sqrt{\pi}} \times$$
$$\times e^{-\frac{1}{2}i\pi(\mu-1)} \frac{\Gamma\left(\frac{\mu}{2}\right)\Gamma(1-\mu-\nu)}{\Gamma(-\nu)} (\beta^2-1)^{\frac{1-\mu}{4}} Q_{-\nu-\frac{\mu+1}{2}}^{\frac{\mu-1}{2}}(\beta)$$
$$[\operatorname{Re}\beta > 1, \operatorname{Re}\nu < 0, \operatorname{Re}\mu < 1 - \operatorname{Re}\nu].$$
Ba IV 311(26)

5.
$$\int_u^\infty \frac{(x-u)^{\mu-1}(\sqrt{x+1}-\sqrt{x-1})^{2\nu}}{\sqrt{x^2-1}} dx = \frac{2^{\nu+\frac{1}{2}}}{\sqrt{\pi}} e^{\left(\mu-\frac{1}{2}\right)\pi i}(u^2-1)^{\frac{2\mu-1}{4}} Q_{\nu-\frac{1}{2}}^{\frac{1}{2}-\mu}(u)$$
$$[|\arg(u-1)| < \pi, 0 < \operatorname{Re}\mu < 1 + \operatorname{Re}\nu].$$
Ba V 202(10)

6.
$$\int_1^\infty \frac{x^{\mu-1}[(x-\sqrt{x^2-1})^\nu+(x-\sqrt{x^2-1})^{-\nu}]}{\sqrt{x^2-1}} dx =$$
$$= 2^{-\mu} B\left(\frac{1-\mu+\nu}{2}, \frac{1-\mu-\nu}{2}\right) \quad [\operatorname{Re}\mu < 1 + \operatorname{Re}\nu].$$
Ba IV 311(29)

7. $$\int_0^u \frac{(u-x)^{\mu-1}[(\sqrt{x+2}+\sqrt{x})^{2\nu}+(\sqrt{x+2}-\sqrt{x})^{2\nu}]}{\sqrt{x(x+2)}}\,dx =$$
$$= 2^{\frac{2\mu+1}{2}}\sqrt{\pi[u(u+2)]}^{\mu-\frac{1}{2}} P_{\nu-\frac{1}{2}}^{\frac{1}{2}-\mu}(u+1)$$
$$[|\arg u| < \pi,\ \operatorname{Re}\mu > 0]. \qquad \text{Ba V 186(12)}$$

3.277

$$\int_0^\infty \left(\frac{x^p}{1+x^{2p}}\right)^q \cdot \frac{dx}{1-x^2} = 0.$$

3.278

Die Exponentialfunktion
The Exponential Function

3.3–3.4

Die Exponentialfunktion
The exponential function

3.31

$$\int_0^\infty e^{-px}\,dx = \frac{1}{p} \quad [\operatorname{Re} p > 0].$$

3.310

1. $$\int_0^\infty \frac{dx}{1+e^{px}} = \frac{\ln 2}{p}. \qquad \text{Lo III 284+}$$

3.311

2. $$\int_0^\infty \frac{e^{-\mu x}}{1+e^{-x}}\,dx = \beta(\mu) \quad [\operatorname{Re}\mu > 0]. \qquad \text{Ba I 20(3), Ba IV 144(7)}$$

3. $$\int_{-\infty}^\infty \frac{e^{-px}}{1+e^{-qx}}\,dx = \frac{\pi}{q}\operatorname{cosec}\frac{p\pi}{q} \quad \left[\begin{array}{c} q > p > 0 \text{ oder} \\ \text{or } 0 > p > q \end{array}\right] \quad (\text{cf. 3.241 2.}).$$
BH[28](7)

4. $$\int_0^\infty \frac{e^{-qx}\,dx}{1-ae^{-px}} = \sum_{k=0}^\infty \frac{a^k}{q+kp} \quad [0 < a < 1]. \qquad \text{BH[27](7)}$$

5. $$\int_0^\infty \frac{1-e^{\nu x}}{e^x - 1}\,dx = \psi(\nu) + C + \pi\operatorname{ctg}(\pi\nu) \quad [\operatorname{Re}\nu < 1] \quad (\text{cf. 3.266}). \qquad \text{Ba I 16(16)}$$

6. $$\int_0^\infty \frac{e^{-x}-e^{-\nu x}}{1-e^{-x}}\,dx = \psi(\nu) + C \quad [\operatorname{Re}\nu > 0]. \qquad \text{WW 260, Ba I 16(14)}$$

7. $$\int_0^\infty \frac{e^{-\mu x}-e^{-\nu x}}{1-e^{-x}}\,dx = \psi(\nu) - \psi(\mu) \quad [\operatorname{Re}\mu > 0, \operatorname{Re}\nu > 0] \quad (\text{cf. 3.231 5.}).$$
BH[27](8)

8. $$\int_{-\infty}^\infty \frac{e^{-\mu x}\,dx}{b-e^{-x}} = \pi b^{\mu-1}\operatorname{ctg}(\mu\pi) \quad [b > 0,\ 0 < \operatorname{Re}\mu < 1]. \qquad \text{Ba IV 120(14)+}$$

3.311

9. $\displaystyle\int_{-\infty}^{\infty}\frac{e^{-\mu x}\,dx}{b+e^{-x}} = \pi b^{\mu-1}\operatorname{cosec}(\mu\pi)$ $\quad[|\arg b|<\pi,\ 0<\operatorname{Re}\mu<1].$ Ba IV 120(15)+

10. $\displaystyle\int_{0}^{\infty}\frac{e^{-px}-e^{-qx}}{1-e^{-(p+q)x}}\,dx = \frac{\pi}{p+q}\operatorname{ctg}\frac{p\pi}{p+q}$ $\quad[p>0,\ q>0].$ GH[311](16c)

11. $\displaystyle\int_{0}^{\infty}\frac{e^{px}-e^{qx}}{e^{rx}-e^{sx}}\,dx = \frac{1}{r-s}\left[\psi\left(\frac{r-q}{r-s}\right)-\psi\left(\frac{r-p}{r-s}\right)\right]$ $\quad[r>s,\ r>p,\ r>q].$ GH[311](16)

12. $\displaystyle\int_{0}^{\infty}\frac{a^{x}-b^{x}}{c^{x}-d^{x}}\,dx = \frac{1}{\ln\frac{c}{d}}\left\{\psi\left(\frac{\ln\frac{c}{b}}{\ln\frac{c}{d}}\right)-\psi\left(\frac{\ln\frac{c}{a}}{\ln\frac{c}{d}}\right)\right\}$ $\quad[c>a>0,\ b>0,\ d>0].$ GH[311](16a)

3.312

1. $\displaystyle\int_{0}^{\infty}(1-e^{-\frac{x}{\beta}})^{\nu-1}e^{-\mu x}\,dx = \beta B(\beta\mu,\nu)$ $\quad[\operatorname{Re}\beta>0,\ \operatorname{Re}\nu>0,\ \operatorname{Re}\mu>0].$ Li[26](13), Ba I 11(24)

2. $\displaystyle\int_{0}^{\infty}(1-e^{-x})^{-1}(1-e^{-\alpha x})(1-e^{-\beta x})e^{-px}\,dx =$

$= \psi(p+\alpha)+\psi(p+\beta)-\psi(p+\alpha+\beta)-\psi(p)$

$[\operatorname{Re}p>0,\ \operatorname{Re}p>-\operatorname{Re}\alpha,\ \operatorname{Re}p>-\operatorname{Re}\beta,\ \operatorname{Re}p>-\operatorname{Re}(\alpha+\beta)].$ Ba IV 145(15)

3. $\displaystyle\int_{0}^{\infty}(1-e^{-x})^{\nu-1}(1-\beta e^{-x})^{-\rho}e^{-\mu x}\,dx = B(\mu,\nu)\,{}_{2}F_{1}(\rho,\mu;\mu+\nu;\beta)$

$[\operatorname{Re}\mu>0,\ \operatorname{Re}\nu>0,\ |\arg(1-\beta)|<\pi].$ Ba I 116(15)

3.313 $\displaystyle\int_{-\infty}^{\infty}\frac{e^{-\mu x}\,dx}{(1-e^{-x})^{n}} = \pi\operatorname{cosec}\mu\pi\prod_{k=1}^{n-1}\frac{k-\mu}{(n-1)!}$ $\quad\left[0<\operatorname{Re}\mu<n,\ n\ \begin{array}{l}\text{ungeradzahlig}\\ \text{odd}\end{array}\right].$ Ba IV 120(20)

3.314 $\displaystyle\int_{-\infty}^{\infty}\frac{e^{-\mu x}\,dx}{(e^{\beta/\gamma}+e^{-x/\gamma})^{\nu}} = \gamma\exp\left[\beta\left(\mu-\frac{\nu}{\gamma}\right)\right]B(\gamma\mu,\nu-\gamma\mu)$

$\left[\operatorname{Re}\left(\frac{\nu}{\gamma}\right)>\operatorname{Re}\mu>0,\ |\operatorname{Im}\beta|<\pi\operatorname{Re}\gamma\right].$ Ba IV 120(21)

3.315

1. $\displaystyle\int_{-\infty}^{\infty}\frac{e^{-\mu x}\,dx}{(e^{\beta}+e^{-x})^{\nu}(e^{\gamma}+e^{-x})^{\rho}} = \exp[\gamma(\mu-\rho)-\beta\nu]\times$

$\times B(\mu,\nu+\rho-\mu)\,{}_{2}F_{1}(\nu,\mu;\nu+\rho;1-e^{\gamma-\beta})$

$[|\operatorname{Im}\beta|<\pi,\ |\operatorname{Im}\gamma|<\pi,\ 0<\operatorname{Re}\mu<\operatorname{Re}(\nu+\rho)].$ Ba IV 121(22)

2. $\displaystyle\int_{-\infty}^{\infty}\frac{e^{-\mu x}\,dx}{(\beta+e^{-x})(\gamma+e^{-x})} = \frac{\pi(\beta^{\mu-1}-\gamma^{\mu-1})}{\gamma-\beta}\operatorname{cosec}(\mu\pi)$

$[|\arg\beta|<\pi,\ |\arg\gamma|<\pi,\ \beta\neq\gamma,\ 0<\operatorname{Re}\mu<2].$ Ba IV 120(18)

$$\int_{-\infty}^{\infty} \frac{(1+e^{-x})^{\nu}-1}{(1+e^{-x})^{\mu}} dx = \psi(\mu) - \psi(\mu-\nu) \quad [\operatorname{Re}\mu > \operatorname{Re}\nu > 0] \quad (\text{cf. } 3.235).$$

BH[28](8) 3.316

1. $$\int_{-\infty}^{\infty} \left\{\frac{1}{1+e^{-x}} - \frac{1}{(1+e^{-x})^{\mu}}\right\} dx = C + \psi(\mu) \quad [\operatorname{Re}\mu > 0] \quad (\text{cf. } 3.233\,2.).$$
3.317

BH[28](10)

2. $$\int_{-\infty}^{\infty} \left\{\frac{1}{(1+e^{-x})^{\nu}} - \frac{1}{(1+e^{-x})^{\mu}}\right\} dx = \psi(\mu) - \psi(\nu)$$

$$[\operatorname{Re}\mu > 0, \operatorname{Re}\nu > 0] \quad (\text{cf. } 3.219).$$

BH[28](11)

1. $$\int_{0}^{\infty} \frac{[\beta+\sqrt{1-e^{-x}}]^{-\nu} + [\beta-\sqrt{1-e^{-x}}]^{-\nu}}{\sqrt{1-e^{-x}}} e^{-\mu x} dx =$$
3.318

$$= \frac{2^{\mu+1} e^{(\mu-\nu)\pi i} (\beta^2 - 1)^{(\mu-\nu)/2} \Gamma(\mu) Q_{\mu-1}^{\nu-\mu}(\beta)}{\Gamma(\nu)} \quad [\operatorname{Re}\mu > 0].$$

Ba IV 145(18)

2. $$\int_{u}^{\infty} \frac{1}{\sqrt{1-e^{-2x}}} \{e^{-u}\sqrt{1-e^{-2x}} - e^{-x}\sqrt{1-e^{-2u}}\}^{\nu} e^{-\mu x} dx =$$

$$= \frac{2^{-\frac{1}{2}(\mu+\nu)} \sqrt{\pi} e^{-\frac{u}{2}(\mu+\nu)} \Gamma(\mu)\Gamma(\nu+1) P_{-\frac{1}{2}(\mu-\nu)}^{-\frac{1}{2}(\mu+\nu)}(\sqrt{1-e^{-2u}})}{\Gamma[(\mu+\nu+1)/2]}$$

$$[u > 0, \operatorname{Re}\mu > 0, \operatorname{Re}\nu > -1].$$

Ba IV 145(19)

Die Exponentialfunktion von komplizierteren Argumenten
Exponentials of compound arguments

3.32–3.34

1. $$\int_{0}^{u} e^{-x^2} dx = \sum_{k=0}^{\infty} \frac{(-1)^k u^{2k+1}}{k!(2k+1)};$$
3.321

$$= e^{-u^2} \sum_{k=0}^{\infty} \frac{2^k u^{2k+1}}{(2k+1)!!}.$$

A(6.700)

2. $$\int_{0}^{u} e^{-q^2 x^2} dx = \frac{\sqrt{\pi}}{2q} \Phi(qu) \quad [q > 0].$$

3. $$\int_{0}^{\infty} e^{-q^2 x^2} dx = \frac{\sqrt{\pi}}{2q} \quad [q > 0].$$

F II 634–635

1. $$\int_{u}^{\infty} \exp\left(-\frac{x^2}{4\beta} - \gamma x\right) dx = \sqrt{\pi\beta}\, e^{\beta\gamma^2} \left[1 - \Phi\left(\gamma\sqrt{\beta} + \frac{u}{2\sqrt{\beta}}\right)\right]$$
3.322

$$[\operatorname{Re}\beta > 0, u > 0].$$

Ba IV 146(21)

3.322 2. $\int_0^\infty \exp\left(-\dfrac{x^2}{4\beta} - \gamma x\right) dx = \sqrt{\pi\beta}\, \exp(\beta\gamma^2)\, [1 - \Phi(\gamma\sqrt{\beta})]$ [Re $\beta > 0$].

NI 27(1)+

3.323 1. $\int_1^\infty \exp(-qx - x^2)\, dx = \dfrac{e^{-q-1}}{q+2} \sum_{k=0}^\infty (-1)^k\, 2^k\, \dfrac{(2k-1)!!}{(q+2)^{2k}}$.

BH[29](4)

2. $\int_{-\infty}^\infty \exp(-p^2 x^2 \pm qx)\, dx = \exp\left(\dfrac{q^2}{4p^2}\right) \dfrac{\sqrt{\pi}}{p}$ [$p > 0$].

BH[28](1)

3.324 1. $\int_0^\infty \exp\left(-\dfrac{\beta}{4x} - \gamma x\right) dx = \sqrt{\dfrac{\beta}{\gamma}}\, K_1(\sqrt{\beta\gamma})$ [Re $\beta \geq 0$, Re $\gamma > 0$]. Ba IV 146(25)

2. $\int_{-\infty}^\infty \exp\left[-\left(x - \dfrac{b}{x}\right)^{2n}\right] dx = \dfrac{1}{n}\, \Gamma\left(\dfrac{1}{2n}\right)$.

BH[28](6)

3.325 $\int_0^\infty \exp\left(-ax^2 - \dfrac{b}{x^2}\right) dx = \dfrac{1}{2}\sqrt{\dfrac{\pi}{a}}\, \exp(-2\sqrt{ab})$ [$a > 0$, $b > 0$]. F II 652–653

3.326 $\int_0^\infty \exp(-x^\mu)\, dx = \dfrac{1}{\mu}\, \Gamma\left(\dfrac{1}{\mu}\right)$ [Re $\mu > 0$].

BH[26](4)

Die Exponentialfunktion im Exponenten
Exponentials of exponential functions

3.327 $\int_0^\infty \exp(-ae^{nx})\, dx = -\dfrac{1}{n}\, \mathrm{Ei}(-a)$.

Li[26](5)

3.328 $\int_{-\infty}^\infty \exp(-e^x)\, e^{\mu x}\, dx = \Gamma(\mu)$ [Re $\mu > 0$].

NG 145(14)

3.329 $\int_0^\infty \left[\dfrac{a\exp(-ce^{ax})}{1 - e^{-ax}} - \dfrac{b\exp(-ce^{bx})}{1 - e^{-bx}}\right] dx = e^{-c}\, \ln\dfrac{b}{a}$ [$a>0, b>0, c>0$].

BH[27](12)

3.331 1. $\int_0^\infty \exp(-\beta e^{-x} - \mu x)\, dx = \beta^{-\mu}\, \gamma(\mu, \beta)$ [Re $\mu > 0$].

Ba IV 147(36)

2. $\int_0^\infty \exp(-\beta e^x - \mu x)\, dx = \beta^\mu\, \Gamma(-\mu, \beta)$ [Re $\beta > 0$].

Ba IV 147(37)

3. $\int_0^\infty (1 - e^{-x})^{\nu-1}\, \exp(\beta e^{-x} - \mu x)\, dx = \mathrm{B}(\mu, \nu)\, \beta^{-\frac{\mu+\nu}{2}}\, e^{\frac{\beta}{2}}\, M_{\frac{\nu-\mu}{2},\, \frac{\nu+\mu-1}{2}}(\beta)$

[Re $\mu > 0$, Re $\nu > 0$].   Ba IV 147(38)

4. $\int_0^\infty (1-e^{-x})^{\nu-1} \exp(-\beta e^x - \mu x)\, dx = \Gamma(\nu) \beta^{\frac{\mu-1}{2}} e^{-\frac{\beta}{2}} W_{\frac{1-\mu-2\nu}{2}, \frac{-\mu}{2}}(\beta)$ 3.331

$[\operatorname{Re}\beta > 0,\ \operatorname{Re}\nu > 0]$. Ba IV 147(39)

$\int_0^\infty (1-e^{-x})^{\nu-1}(1-\lambda e^{-x})^{-\rho} \exp(\beta e^{-x} - \mu x)\, dx = B(\mu,\nu)\, \Phi_1(\mu,\rho,\nu;\lambda,\beta)$ 3.332

$[\operatorname{Re}\mu > 0,\ \operatorname{Re}\nu > 0,\ |\arg(1-\lambda)| < \pi]$. Ba IV 147(40)

1. $\int_{-\infty}^\infty \frac{e^{-\mu x}\, dx}{\exp(e^{-x}) - 1} = \Gamma(\mu)\, \zeta(\mu)$ $[\operatorname{Re}\mu < 1]$. Ba IV 121(24) 3.333

2. $\int_{-\infty}^\infty \frac{e^{-\mu x}\, dx}{\exp(e^{-x}) + 1} = (1 - 2^{1-\mu})\, \Gamma(\mu)\, \zeta(\mu)$ $[\operatorname{Re}\mu > 0]$. Ba IV 121(25)

$\int_0^\infty (e^x - 1)^{\nu-1} \exp\left[-\frac{\beta}{e^x - 1} - \mu x\right] dx =$ 3.334

$= \Gamma(\mu - \nu + 1)\, e^{\frac{\beta}{2}} \beta^{\frac{\nu-1}{2}}\, W_{\frac{\nu-2\mu-1}{2}, \frac{\nu}{2}}(\beta)$ $[\operatorname{Re}\beta > 0,\ \operatorname{Re}\mu > \operatorname{Re}\nu - 1]$.

Ba IV 147(41)

Hyperbelfunktionen im Exponenten
Exponentials of hyperbolic functions

$\int_0^\infty (e^{\nu x} + e^{-\nu x} \cos\nu\pi) \exp(-\beta\operatorname{sh} x)\, dx = -\pi[\mathbf{E}_\nu(\beta) + N_\nu(\beta)]$ $[\operatorname{Re}\beta > 0]$. 3.335

Ba II 35(34)

1. $\int_0^\infty \exp(-\nu x - \beta\operatorname{sh} x)\, dx = \pi\operatorname{cosec}\nu\pi\, [\mathbf{J}_\nu(\beta) - J_\nu(\beta)]$ 3.336

$\left[|\arg\beta| < \frac{\pi}{2}\ \text{und/and}\ |\arg\beta| = \frac{\pi}{2}\ \text{für/for}\ \operatorname{Re}\nu > 0;\ \nu\ \text{nicht ganzzahlig/noninteger}\right]$.

W 312(2)

2. $\int_0^\infty \exp(nx - \beta\operatorname{sh} x)\, dx = \frac{1}{2}[S_n(\beta) - \pi\mathbf{E}_n(\beta) - \pi N_n(\beta)]$

$[\operatorname{Re}\beta > 0;\ n = 0, 1, 2 \ldots]$. W 313(6)

3. $\int_0^\infty \exp(-nx - \beta\operatorname{sh} x)\, dx = \frac{1}{2}(-1)^{n+1}[S_n(\beta) + \pi\mathbf{E}_n(\beta) + \pi N_n(\beta)]$

$[\operatorname{Re}\beta > 0;\ n = 0, 1, 2 \ldots]$. Ba II 84(47)

1. $\int_{-\infty}^\infty \exp(-\alpha x - \beta\operatorname{ch} x)\, dx = 2K_\alpha(\beta)$ $\left[|\arg\beta| < \frac{\pi}{2}\right]$. W 182(7) 3.337

3.337

2. $\int_{-\infty}^{\infty} \exp(-\nu x + i\beta \operatorname{ch} x)\, dx = i\pi e^{\frac{i\nu\pi}{2}} H^1_\nu(\beta) \quad [0 < \arg z < \pi].$ Ba II 21(27)

3. $\int_{-\infty}^{\infty} \exp(-\nu x - i\beta \operatorname{ch} x)\, dx = -i\pi e^{-\frac{i\nu\pi}{2}} H^2_\nu(\beta) \quad [-\pi < \arg z < 0].$ Ba II 21(30)

Trigonometrische Funktionen und Logarithmen im Exponenten
Exponentials of trigonometric functions and logarithms

3.338

1. $\int_0^\pi \{\exp i[(\nu-1)x - \beta \sin x] - \exp i[(\nu+1)x - \beta \sin x]\}\, dx =$
$= 2\pi[\mathbf{J}'_\nu(\beta) + i\mathbf{E}'_\nu(\beta)] \quad [\operatorname{Re}\beta > 0].$ Ba II 36

2. $\int_0^\pi \exp[\pm i(\nu x - \beta \sin x)]\, dx = \pi[\mathbf{J}_\nu(\beta) \pm i\mathbf{E}_\nu(\beta)] \quad [\operatorname{Re}\beta > 0].$ Ba II 35(32)

3. $\int_0^\infty \exp[-\gamma(x - \beta \sin x)]\, dx = \frac{1}{\gamma} + 2\sum_{k=1}^\infty \frac{\gamma J_k(k\beta)}{\gamma^2 + k^2} \quad [\operatorname{Re}\gamma > 0].$ W 563(4)

3.339 $\int_0^\pi \exp(2\cos x)\, dx = \pi I_0(2).$ BH[277](2)+

3.341 $\int_0^{\pi/2} \exp(-p \operatorname{tg} x)\, dx = \operatorname{ci}(p)\sin p - \operatorname{si}(p)\cos(p) \quad [p > 0].$ BH[271](2)+

3.342 $\int_0^1 \exp(-px \ln x)\, dx = \int_0^1 x^{-px}\, dx = \sum_{k=1}^\infty \frac{p^{k-1}}{k^k}.$ BH[29](1)

Die Exponentialfunktion und rationale Funktionen
Exponentials and rational functions

3.35

3.351

1. $\int_0^u x^n e^{-\mu x}\, dx = \frac{n!}{\mu^{n+1}} - e^{-u\mu}\sum_{k=0}^n \frac{n!}{k!}\frac{u^k}{\mu^{n-k+1}} \quad [u>0,\ \operatorname{Re}\mu > 0].$ Ba IV 134(5)

2. $\int_u^\infty x^n e^{-\mu x}\, dx = e^{-u\mu}\sum_{k=0}^n \frac{n!}{k!}\frac{u^k}{\mu^{n-k+1}} \quad [u>0,\ \operatorname{Re}\mu > 0].$ Ba IV 133(4)

3. $\int_0^\infty x^n e^{-\mu x}\, dx = n!\,\mu^{-n-1} \quad [\operatorname{Re}\mu > 0].$ Ba IV 133(3)

4. $\int_u^\infty \dfrac{e^{-px}\,dx}{x^{n+1}} = (-1)^{n+1}\dfrac{p^n \operatorname{Ei}(-pu)}{n!} + \dfrac{e^{-pu}}{u^n}\sum_{k=0}^{n-1}\dfrac{(-1)^k p^k u^k}{n(n-1)\dots(n-k)}$ $[p>0]$.   3.351

NI 21(3)

5. $\int_1^\infty \dfrac{e^{-\mu x}\,dx}{x} = -\operatorname{Ei}(-\mu)$ $[\operatorname{Re}\mu > 0]$.   BH[104](10)

6. $\int_{-\infty}^u \dfrac{e^x}{x}\,dx = \operatorname{li}(e^u) = \operatorname{Ei}(u)$ $[u<0]$.

1. $\int_0^u \dfrac{e^{-\mu x}\,dx}{x+\beta} = e^{\mu\beta}[\operatorname{Ei}(-\mu u - \mu\beta) - \operatorname{Ei}(-\mu\beta)]$ $[|\arg\beta| < \pi]$.   Ba V 217(12)   3.352

2. $\int_u^\infty \dfrac{e^{-\mu x}\,dx}{x+\beta} = -e^{\beta\mu}\operatorname{Ei}(-\mu u - \mu\beta)$ $[u\geqslant 0,\ |\arg(u+\beta)|<\pi,\ \operatorname{Re}\mu>0]$.

Ba IV 134(6), JE 80

3. $\int_u^v \dfrac{e^{-\mu x}\,dx}{x+\alpha} = e^{\alpha\mu}\{\operatorname{Ei}[-(\alpha+v)\mu] - \operatorname{Ei}[-(\alpha+u)\mu]\}$

$\left[-\alpha < u,\ \text{bzw.}\atop\text{or}\ -\alpha > v,\ \operatorname{Re}\mu>0\right]$.   Ba IV 134(7)

4. $\int_0^\infty \dfrac{e^{-\mu x}\,dx}{x+\beta} = -e^{\beta\mu}\operatorname{Ei}(-\mu\beta)$ $[|\arg\beta|<\pi,\ \operatorname{Re}\mu>0]$.   Ba V 217(11)

5. $\int_u^\infty \dfrac{e^{-px}\,dx}{a-x} = e^{-pa}\operatorname{Ei}(pa-pu)$

[$p>0$; $a<u$; im Fall $a>u$ muß $\underline{\operatorname{Ei}(pa-pu)}$ in dieser Formel durch $\overline{\operatorname{Ei}(pa-pu)}$ ersetzt werden]. Ba V 251(37) | [$p>0$ and $a<u$; if $a>u$, then in this formula $\underline{\operatorname{Ei}(pa-pu)}$ must be replaced by $\overline{\operatorname{Ei}}(pa-pu)$]. Ba V 251(37)

6. $\int_0^\infty \dfrac{e^{-\mu x}\,dx}{a-x} = e^{-\mu a}\operatorname{Ei}(a\mu)$ $[a>0,\ \operatorname{Re}\mu>0]$.   BH[91](4)

7. $\int_{-\infty}^\infty \dfrac{e^{ipx}\,dx}{x-a} = i\pi e^{iap}$ $[p>0]$.   Ba V 251(38)

1. $\int_u^\infty \dfrac{e^{-\mu x}\,dx}{(x+\beta)^n} = e^{-u\mu}\sum_{k=1}^{n-1}\dfrac{(k-1)!(-\mu)^{n-k-1}}{(n-1)!(u+\beta)^k} - \dfrac{(-\mu)^{n-1}}{(n-1)!}e^{\beta\mu}\operatorname{Ei}[-(u+\beta)\mu]$ 3.353

$[n\geqslant 2,\ |\arg(u+\beta)|<\pi,\ \operatorname{Re}\mu>0]$.   Ba IV 134(10)

3.353

2. $\int_0^\infty \dfrac{e^{-\mu x}\,dx}{(x+\beta)^n} = \dfrac{1}{(n-1)!} \sum_{k=1}^{n-1} (k-1)!(-\mu)^{n-k-1}\beta^{-k} - \dfrac{(-\mu)^{n-1}}{(n-1)!} e^{\beta\mu} \text{Ei}(-\beta\mu)$

$[n > 2,\ |\arg \beta| < \pi,\ \text{Re}\,\mu > 0]$.   Ba IV 134(9), BH[92](2)

3. $\int_0^\infty \dfrac{e^{-px}\,dx}{(a+x)^2} = pe^{ap}\,\text{Ei}(-ap) + \dfrac{1}{a}$ $[p>0,\ a>0]$.   Li[281](28), Li[281](29)

4. $\int_0^1 \dfrac{xe^x}{(1+x)^2}\,dx = \dfrac{e}{2} - 1$.   BH[80](6)

5. $\int_0^\infty \dfrac{x^n e^{-\mu x}}{x+\beta}\,dx = (-1)^{n-1}\beta^n e^{\beta\mu}\,\text{Ei}(-\beta\mu) + \sum_{k=1}^{n} (k-1)!(-\beta)^{n-k}\mu^{-k}$

$[|\arg \beta| < \pi,\ \text{Re}\,\mu > 0]$.   BH[91](3)+, Ba IV 135(11)

3.354

1. $\int_0^\infty \dfrac{e^{-\mu x}\,dx}{\beta^2 + x^2} = \dfrac{1}{\beta}[\text{ci}(\beta\mu)\sin\beta\mu - \text{si}(\beta\mu)\cos\beta\mu]$ $[\text{Re}\,\beta > 0,\ \text{Re}\,\mu > 0]$.   BH[91](7)

2. $\int_0^\infty \dfrac{xe^{-\mu x}\,dx}{\beta^2 + x^2} = -\text{ci}(\beta\mu)\cos\beta\mu - \text{si}(\beta\mu)\sin\beta\mu$ $[\text{Re}\,\beta > 0,\ \text{Re}\,\mu > 0]$.   BH[91](8)

3. $\int_0^\infty \dfrac{e^{-\mu x}\,dx}{\beta^2 - x^2} = \dfrac{1}{2\beta}[e^{-\beta\mu}\text{Ei}(\beta\mu) - e^{\beta\mu}\text{Ei}(-\beta\mu)]$ BH[91](14)

[$|\arg(\pm\beta)| < \pi$, $\text{Re}\,\mu > 0$; im Fall $\beta > 0$ muß in dieser Formel Ei$(\beta\mu)$ durch $\overline{\text{Ei}}(\beta\mu)$ ersetzt werden]. | [$|\arg(\pm\beta)| < \pi$ and $\text{Re}\,\mu > 0$; if $\beta > 0$, then in this formula Ei$(\beta\mu)$ must be replaced by $\overline{\text{Ei}}(\beta\mu)$].

4. $\int_0^\infty \dfrac{xe^{-\mu x}\,dx}{\beta^2 - x^2} = \dfrac{1}{2}[e^{-\beta\mu}\text{Ei}(\beta\mu) + e^{\beta\mu}\text{Ei}(-\beta\mu)]$ BH[91](15)

[$|\arg(\pm\beta)| < \pi$, $\text{Re}\,\mu > 0$; im Fall $\beta > 0$ muß in dieser Formel Ei$(\beta\mu)$ durch $\overline{\text{Ei}}(\beta\mu)$ ersetzt werden]. | [$|\arg(\pm\beta)| < \pi$ and $\text{Re}\,\mu > 0$; if $\beta > 0$, then in this formula Ei$(\beta\mu)$ must be replaced by $\overline{\text{Ei}}(\beta\mu)$].

5. $\int_{-\infty}^\infty \dfrac{e^{-ipx}\,dx}{a^2 + x^2} = \dfrac{\pi}{a} e^{-|ap|}$ $[a > 0]$.   Ba IV 118(1)+

3.355

1. $\int_0^\infty \dfrac{e^{-\mu x}\,dx}{(\beta^2 + x^2)^2} = \dfrac{1}{2\beta^3}\{\text{ci}(\beta\mu)\sin\beta\mu - \text{si}(\beta\mu)\cos\beta\mu -$

$- \beta\mu[\text{ci}(\beta\mu)\cos\beta\mu + \text{si}(\beta\mu)\sin\beta\mu]\}$.   Li[92](6)

2. $\int_0^\infty \dfrac{xe^{-\mu x}\,dx}{(\beta^2 + x^2)^2} = \dfrac{1}{2\beta^2}\{1 - \beta\mu[\text{ci}(\beta\mu)\sin\beta\mu - \text{si}(\beta\mu)\cos\beta\mu]\}$

$[\text{Re}\,\beta > 0,\ \text{Re}\,\mu > 0]$.   BH[92](7)

3.356

1. $\int_0^\infty \dfrac{x^{2n+1} e^{-px}}{a^2 + x^2}\, dx = (-1)^{n-1} a^{2n} [\operatorname{ci}(ap) \cos ap + \operatorname{si}(ap) \sin ap] +$

$\qquad + \dfrac{1}{p^{2n}} \sum_{k=1}^{n} (2n - 2k + 1)! (-a^2 p^2)^{k-1} \quad [p > 0].$ BH[91](12)

2. $\int_0^\infty \dfrac{x^{2n} e^{-px}}{a^2 + x^2}\, dx = (-1)^n a^{2n-1} [\operatorname{ci}(ap) \sin ap - \operatorname{si}(ap) \cos ap] +$

$\qquad + \dfrac{1}{p^{2n-1}} \sum_{k=1}^{n} (2n - 2k)! (-a^2 p^2)^{k-1} \quad [p > 0].$ BH[91](11)

3. $\int_0^\infty \dfrac{x^{2n+1} e^{-px}}{a^2 - x^2}\, dx = \dfrac{1}{2} a^{2n} [e^{ap} \operatorname{Ei}(-ap) + e^{-ap} \operatorname{Ei}(ap)] -$

$\qquad - \dfrac{1}{p^{2n}} \sum_{k=1}^{n} (2n - 2k + 1)! (a^2 p^2)^{k-1} \quad [p > 0].$ BH[91](17)

4. $\int_0^\infty \dfrac{x^{2n} e^{-px}}{a^2 - x^2}\, dx = \dfrac{1}{2} a^{2n-1} [e^{-ap} \operatorname{Ei}(ap) - e^{ap} \operatorname{Ei}(-ap)] -$

$\qquad - \dfrac{1}{p^{2n-1}} \sum_{k=1}^{n} (2n - 2k)! (a^2 p^2)^{k-1} \quad [p > 0].$ BH[91](16)

3.357

1. $\int_0^\infty \dfrac{e^{-\mu x}\, dx}{a^3 + a^2 x + a x^2 + x^3} = \dfrac{1}{2a^2} \{\operatorname{ci}(a\mu) (\sin a\mu + \cos a\mu) +$

$\qquad + \operatorname{si}(a\mu) (\sin a\mu - \cos a\mu) - e^{a\mu} \operatorname{Ei}(-a\mu)\} \quad [\operatorname{Re} \mu > 0,\ a > 0].$ BH[92](18)

2. $\int_0^\infty \dfrac{x e^{-\mu x}\, dx}{a^3 + a^2 x + a x^2 + x^3} = \dfrac{1}{2a} \{\operatorname{ci}(a\mu) (\sin a\mu - \cos a\mu) -$

$\qquad - \operatorname{si}(a\mu) (\sin a\mu + \cos a\mu) + e^{a\mu} \operatorname{Ei}(-a\mu)\} \quad [\operatorname{Re} \mu > 0,\ a > 0].$ BH[92](19)

3. $\int_0^\infty \dfrac{x^2 e^{-\mu x}\, dx}{a^3 + a^2 x + a x^2 + x^3} = \dfrac{1}{2} \{-\operatorname{ci}(a\mu) (\sin a\mu + \cos a\mu) -$

$\qquad - \operatorname{si}(a\mu) (\sin a\mu - \cos a\mu) - e^{a\mu} \operatorname{Ei}(-a\mu)\} \quad [\operatorname{Re} \mu > 0,\ a > 0].$ BH[92](20)

4. $\int_0^\infty \dfrac{e^{-\mu x}\, dx}{a^3 - a^2 x + a x^2 - x^3} = \dfrac{1}{2a^2} \{\operatorname{ci}(a\mu) (\sin a\mu - \cos a\mu) -$

$\qquad - \operatorname{si}(a\mu) (\sin a\mu + \cos a\mu) + e^{-a\mu} \operatorname{Ei}(a\mu)\} \quad [\operatorname{Re} \mu > 0,\ a > 0].$ BH[92](21)

5. $\int_0^\infty \dfrac{x e^{-\mu x}\, dx}{a^3 - a^2 x + a x^2 - x^3} = \dfrac{1}{2a} \{-\operatorname{ci}(a\mu) (\sin a\mu + \cos a\mu) -$

$\qquad - \operatorname{si}(a\mu) (\sin a\mu - \cos a\mu) + e^{-a\mu} \operatorname{Ei}(a\mu)\} \quad [\operatorname{Re} \mu > 0,\ a > 0].$ BH[92](22)

3.357

6. $$\int_0^\infty \frac{x^2 e^{-\mu x}\,dx}{a^3 - a^2 x + ax^2 - x^3} = \frac{1}{2}\{\operatorname{ci}(a\mu)(\cos a\mu - \sin a\mu) +$$

$$+ \operatorname{si}(a\mu)(\cos a\mu + \sin a\mu) + e^{-a\mu}\operatorname{Ei}(a\mu)\} \quad [\operatorname{Re} \mu > 0,\ a > 0].$$
BH[92](23)

3.358

1. $$\int_0^\infty \frac{e^{-px}}{a^4 - x^4}\,dx = \frac{1}{4a^3}\{e^{-ap}\operatorname{Ei}(ap) - e^{ap}\operatorname{Ei}(-ap) +$$

$$+ 2\operatorname{ci}(ap)\sin ap - 2\operatorname{si}(ap)\cos ap\} \quad [p > 0,\ a > 0].$$
BH[91](18)

2. $$\int_0^\infty \frac{xe^{-px}\,dx}{a^4 - x^4} = \frac{1}{4a^2}\{e^{ap}\operatorname{Ei}(-ap) + e^{-ap}\operatorname{Ei}(ap) -$$

$$- 2\operatorname{ci}(ap)\cos ap - 2\operatorname{si}(ap)\sin ap\} \quad [p > 0,\ a > 0].$$
BH[91](19)

3. $$\int_0^\infty \frac{x^2 e^{-px}\,dx}{a^4 - x^4} = \frac{1}{4a}\{e^{-ap}\operatorname{Ei}(ap) - e^{ap}\operatorname{Ei}(-ap) -$$

$$- 2\operatorname{ci}(ap)\sin ap + 2\operatorname{si}(ap)\cos ap\} \quad [p > 0,\ a > 0].$$
BH[91](20)

4. $$\int_0^\infty \frac{x^3 e^{-px}\,dx}{a^4 - x^4} = \frac{1}{4}\{e^{ap}\operatorname{Ei}(-ap) + e^{-ap}\operatorname{Ei}(ap) +$$

$$+ 2\operatorname{ci}(ap)\cos ap + 2\operatorname{si}(ap)\sin ap\} \quad [p > 0,\ a > 0].$$
BH[91](21)

5. $$\int_0^\infty \frac{x^{4n}e^{-px}}{a^4 - x^4}\,dx = \frac{1}{4}a^{4n-3}[e^{-ap}\operatorname{Ei}(ap) - e^{ap}\operatorname{Ei}(-ap) +$$

$$+ 2\operatorname{ci}(ap)\sin ap - 2\operatorname{si}(ap)\cos ap] - \frac{1}{p^{4n-3}}\sum_{k=1}^n (4n-4k)!\,(a^4 p^4)^{k-1}$$

$$[p > 0,\ a > 0].$$
BH[91](22)

6. $$\int_0^\infty \frac{x^{4n+1} e^{-px}}{a^4 - x^4}\,dx = \frac{1}{4}a^{4n-2}[e^{ap}\operatorname{Ei}(-ap) + e^{-ap}\operatorname{Ei}(ap) -$$

$$- 2\operatorname{ci}(ap)\cos ap - 2\operatorname{si}(ap)\sin ap] - \frac{1}{p^{4n-2}}\sum_{k=1}^n (4n-4k+1)!\,(a^4 p^4)^{k-1}$$

$$[p > 0,\ a > 0].$$
BH[91](23)

7. $$\int_0^\infty \frac{x^{4n+2} e^{-px}}{a^4 - x^4}\,dx = \frac{1}{4}a^{4n-1}[e^{-ap}\operatorname{Ei}(ap) - e^{ap}\operatorname{Ei}(-ap) -$$

$$- 2\operatorname{ci}(ap)\sin ap + 2\operatorname{si}(ap)\cos ap] - \frac{1}{p^{4n-1}}\sum_{k=1}^n (4n-4k+2)!\,(a^4 p^4)^{k-1}$$

$$[p > 0,\ a > 0].$$
BH[91](24)

8. $\int_0^\infty \frac{x^{4n+3} e^{-px}}{a^4 - x^4} dx = \frac{1}{4} a^{4n} [e^{ap} \text{Ei}(-ap) + e^{-ap} \text{Ei}(ap) +$ 3.358

$+ 2 \text{ci}(ap) \cos ap + 2 \text{si}(ap) \sin ap] - \frac{1}{p^{4n}} \sum_{k=1}^{n} (4n - 4k + 3)! \, (a^4 p^4)^{k-1}$

$[p > 0, \, a > 0]$. BH [91](25)

$\int_{-\infty}^{\infty} \frac{(i-x)^n}{(i+x)^n} \frac{e^{-ipx}}{1+x^2} dx = (-1)^{n-1} 2\pi p e^{-p} L_{n-1}^1(2p)$ für/for $p > 0$; 3.359

$= 0$ für/for $p < 0$. Ba IV 118(2)

Die Exponentialfunktion und algebraische Funktionen
Exponentials and algebraic functions

3.36—3.37

1. $\int_0^u \frac{e^{-qx}}{\sqrt{x}} dx = \sqrt{\frac{\pi}{q}} \Phi(\sqrt{qu})$. 3.361

2. $\int_0^\infty \frac{e^{-qx}}{\sqrt{x}} dx = \sqrt{\frac{\pi}{q}} \quad [q > 0]$. BH [98](10)

3. $\int_{-1}^\infty \frac{e^{-qx}}{\sqrt{1+x}} dx = e^q \sqrt{\frac{\pi}{q}} \quad [q > 0]$. BH[104](16)

1. $\int_1^\infty \frac{e^{-\mu x} dx}{\sqrt{x-1}} = \sqrt{\frac{\pi}{\mu}} e^{-\mu} \quad [\text{Re } \mu > 0]$. BH [104](11)+ 3.362

2. $\int_0^\infty \frac{e^{-\mu x} dx}{\sqrt{x+\beta}} = \sqrt{\frac{\pi}{\mu}} e^{\beta\mu} [1 - \Phi(\sqrt{\beta\mu})] \quad [\text{Re } \mu > 0, \, |\arg \beta| < \pi]$.

Ba IV 135(18)

1. $\int_u^\infty \frac{\sqrt{x-u}}{x} e^{-\mu x} dx = \sqrt{\frac{\pi}{\mu}} e^{-u\mu} - \pi \sqrt{u} [1 - \Phi(\sqrt{u\mu})]$ 3.363

$[u > 0, \, \text{Re } \mu > 0]$. Ba IV 136(23)

2. $\int_u^\infty \frac{e^{-\mu x} dx}{x \sqrt{x-u}} = \frac{\pi}{\sqrt{u}} [1 - \Phi(\sqrt{u\mu})] \quad [u > 0, \, \text{Re } \mu \geq 0]$. Ba IV 136(26)

1. $\int_0^2 \frac{e^{-px} dx}{\sqrt{x(2-x)}} = \pi e^{-p} I_0(p) \quad [p > 0]$. GH [312](7a) 3.364

2. $\int_{-1}^1 \frac{e^{2x} dx}{\sqrt{1-x^2}} = \pi I_0(2)$. BH [277](2)+

3.364 3. $\int_0^\infty \dfrac{e^{-px}\,dx}{\sqrt{x(x+a)}} = e^{\frac{ap}{2}} K_0\left(\dfrac{ap}{2}\right)$ $[a > 0,\ p > 0]$. GH [312](8a)

3.365 1. $\int_0^u \dfrac{xe^{-\mu x}\,dx}{\sqrt{u^2-x^2}} = \dfrac{\pi u}{2}[\mathbf{L}_1(\mu u) - I_1(\mu u)] + u$ $[u > 0,\ \operatorname{Re}\mu > 0]$. Ba IV 136(28)

2. $\int_u^\infty \dfrac{xe^{-\mu x}\,dx}{\sqrt{x^2-u^2}} = uK_1(u\mu)$ $[u > 0,\ \operatorname{Re}\mu > 0]$. Ba IV 136(29)

3.366 1. $\int_0^{2u} \dfrac{(u-x)e^{-\mu x}\,dx}{\sqrt{2ux-x^2}} = \pi u e^{-u\mu} I_1(u\mu)$ $[\operatorname{Re}\mu > 0]$. Ba IV 136(31)

2. $\int_0^\infty \dfrac{(x+\beta)e^{-\mu x}\,dx}{\sqrt{x^2+2\beta x}} = \beta e^{\beta\mu} K_1(\beta\mu)$ $[\operatorname{Re}\mu > 0,\ |\arg\beta| < \pi]$. Ba IV 136(30)

3. $\int_0^\infty \dfrac{xe^{-\mu x}\,dx}{\sqrt{x^2+\beta^2}} = \dfrac{\beta\pi}{2}[\mathbf{H}_1(\beta\mu) - N_1(\beta\mu)] - \beta$ $\left[|\arg\beta| < \dfrac{\pi}{2},\ \operatorname{Re}\mu > 0\right]$.

Ba IV 136(27)

3.367 $\int_0^\infty \dfrac{e^{-\mu x}\,dx}{(1+\cos t + x)\sqrt{x^2+2x}} = \dfrac{\exp\left(2\mu\cos^2\dfrac{t}{2}\right)}{\sin t} \times$

$\times \left(t - \sin t \int_0^\mu K_0(v)\,e^{-v\cos t}\,dv\right)$ $[\operatorname{Re}\mu > 0]$. Ba IV 136(33)

3.368 $\int_0^\infty \dfrac{e^{-\mu x}\,dx}{x + \sqrt{x^2+\beta^2}} = \dfrac{\pi}{2\beta\mu}[\mathbf{H}_1(\beta\mu) - N_1(\beta\mu)] - \dfrac{1}{\beta^2\mu^2}$

$\left[|\arg\beta| < \dfrac{\pi}{2},\ \operatorname{Re}\mu > 0\right]$. Ba IV 136(32)

3.369 $\int_0^\infty \dfrac{e^{-\mu x}\,dx}{\sqrt{(x+a)^3}} = \dfrac{2}{\sqrt{a}} - 2\sqrt{\pi\mu}\,e^{a\mu}(1 - \Phi(\sqrt{a\mu}))$ $[|\arg a| < \pi,\ \operatorname{Re}\mu > 0]$.

Ba IV 135(20)

3.371 $\int_0^\infty x^{n-\frac{1}{2}} e^{-\mu x}\,dx = \sqrt{\pi}\cdot\dfrac{1}{2}\cdot\dfrac{3}{2}\cdots\dfrac{2n-1}{2}\,\mu^{-n-\frac{1}{2}}$ $[\operatorname{Re}\mu > 0]$. Ba IV 135(17)

3.372 $\int_0^\infty x^{n-\frac{1}{2}}(2+x)^{n-\frac{1}{2}} e^{-px}\,dx = \dfrac{(2n-1)!!}{p^n} e^p K_n(p)$ $[p > 0,\ n = 0, 1, 2, \ldots]$.

GH [312](8)

3.373 $\int_0^\infty [(x + \sqrt{x^2+\beta^2})^n + (x - \sqrt{x^2+\beta^2})^n] e^{-\mu x}\,dx = 2\beta^{n+1} O_n(\beta\mu)$ $[\operatorname{Re}\mu > 0]$.

W 278(1)

1. $\int_0^\infty \frac{(x+\sqrt{1+x^2})^n}{\sqrt{1+x^2}} e^{-\mu x} dx = \frac{1}{2}[S_n(\mu) - \pi \mathbf{E}_n(\mu) - \pi N_n(\mu)] \quad [\operatorname{Re}\mu > 0].$ 3.374

Ba IV 137(35)

2. $\int_0^\infty \frac{(x-\sqrt{1+x^2})^n}{\sqrt{1+x^2}} e^{-\mu x} dx = -\frac{1}{2}[S_n(\mu) + \pi \mathbf{E}_n(\mu) + \pi N_n(\mu)] \quad [\operatorname{Re}\mu > 0].$

Ba IV 137(36)

Die Exponentialfunktion und Potenzen mit beliebigen Exponenten
Exponentials and arbitrary powers

3.38–3.39

1. $\int_0^u x^{\nu-1} e^{-\mu x} dx = \mu^{-\nu} \gamma(\nu, \mu u) \quad [\operatorname{Re}\nu > 0].$ Ba I 266(22), Ba II 133(1) 3.381

2. $\int_0^u x^{p-1} e^{-x} dx = \sum_{k=0}^\infty (-1)^k \frac{u^{p+k}}{k!(p+k)};$

$= e^{-u} \sum_{k=0}^\infty \frac{u^{p+k}}{p(p+1)\dots(p+k)}.$ A(6.705)

3. $\int_u^\infty x^{\nu-1} e^{-\mu x} dx = \mu^{-\nu} \Gamma(\nu, \mu u) \quad [u > 0, \ \operatorname{Re}\mu > 0].$ Ba I 256(21), Ba II 133(2)

4. $\int_0^\infty x^{\nu-1} e^{-\mu x} dx = \frac{1}{\mu^\nu} \Gamma(\nu) \quad [\operatorname{Re}\mu > 0, \ \operatorname{Re}\nu > 0].$ F II 784

5. $\int_0^\infty x^{\nu-1} e^{-(p+iq)x} dx = \Gamma(\nu)(p^2+q^2)^{-\frac{\nu}{2}} \exp\left(-i\nu \operatorname{arctg}\frac{q}{p}\right)$

$\left[p > 0, \ \operatorname{Re}\nu > 0 \ \begin{array}{c}\text{bzw.}\\ \text{or}\end{array} \ p = 0, \ 0 < \operatorname{Re}\nu < 1\right].$ Ba I 12(32)

6. $\int_u^\infty \frac{e^{-x}}{x^\nu} dx = u^{-\frac{\nu}{2}} e^{-\frac{u}{2}} W_{-\frac{\nu}{2},\frac{(1-\nu)}{2}}(u) \quad [u > 0].$ WW 352

1. $\int_0^u (u-x)^\nu e^{-\mu x} dx = \mu^{-\nu-1} e^{-u\mu} \gamma(\nu+1, -u\mu) \quad [\operatorname{Re}\nu > -1, \ u > 0].$ 3.382

Ba IV 137(6)

2. $\int_u^\infty (x-u)^\nu e^{-\mu x} dx = \mu^{-\nu-1} e^{-u\mu} \Gamma(\nu+1) \quad [u > 0, \ \operatorname{Re}\nu > -1, \ \operatorname{Re}\mu > 0].$

Ba IV 137(5), Ba V 202(11)

3.382

3. $\int_0^\infty (1+x)^{-\nu} e^{-\mu x}\, dx = \mu^{\frac{\nu}{2}-1} e^{\frac{\mu}{2}} W_{-\frac{\nu}{2},\,\frac{(1-\nu)}{2}}(\mu)$ [Re $\mu > 0$]. WW 352

4. $\int_0^\infty (x+\beta)^\nu e^{-\mu x}\, dx = \mu^{-\nu-1} e^{\beta\mu} \Gamma(\nu+1, \beta\mu)$ [|arg $\beta| < \pi$, Re $\mu > 0$].

Ba IV 137(4), Ba V 233(10)

5. $\int_0^u (a+x)^{\mu-1} e^{-x}\, dx = e^a [\gamma(\mu, a+u) - \gamma(\mu, a)]$ [Re $\mu > 0$]. Ba II 139

6. $\int_{-\infty}^\infty (\beta + ix)^{-\nu} e^{-ipx}\, dx = 0 \quad \begin{matrix}\text{für}\\ \text{or}\end{matrix}\, p > 0;$

$$= \frac{2\pi(-p)^{\nu-1} e^{\beta p}}{\Gamma(\nu)} \begin{matrix}\text{für}\\ \text{for}\end{matrix}\, p < 0.$$

[Re $\nu > 0$, Re $\beta > 0$]. Ba IV 118(4)

7. $\int_{-\infty}^\infty (\beta - ix)^{-\nu} e^{-ipx}\, dx = \frac{2\pi p^{\nu-1} e^{-\beta p}}{\Gamma(\nu)} \begin{matrix}\text{für}\\ \text{for}\end{matrix}\, p > 0;$

$$= 0 \quad \begin{matrix}\text{für}\\ \text{for}\end{matrix}\, p < 0$$

[Re $\nu > 0$, Re $\beta > 0$]. Ba IV 118(3)

3.383

1. $\int_0^u x^{\nu-1}(u-x)^{\mu-1} e^{\beta x}\, dx = B(\mu, \nu)\, u^{\mu+\nu-1}\, {}_1F_1(\nu; \mu+\nu; \beta u)$

[Re $\mu > 0$, Re $\nu > 0$]. Ba V 187(14)

2. $\int_0^u x^{\mu-1}(u-x)^{\mu-1} e^{\beta x}\, dx = \sqrt{\pi}\left(\frac{u}{\beta}\right)^{\mu-\frac{1}{2}} \exp\left(\frac{\beta u}{2}\right) \Gamma(\mu)\, I_{\mu-\frac{1}{2}}\left(\frac{\beta u}{2}\right)$

[Re $\mu > 0$]. Ba V 187(13)

3. $\int_u^\infty x^{\mu-1}(x-u)^{\mu-1} e^{-\beta x}\, dx = \frac{1}{\sqrt{\pi}}\left(\frac{u}{\beta}\right)^{\mu-\frac{1}{2}} \Gamma(\mu) \exp\left(-\frac{\beta u}{2}\right) K_{\mu-\frac{1}{2}}\left(\frac{\beta u}{2}\right)$

[Re $\mu > 0$, Re $\beta u > 0$]. Ba V 202(12)

4. $\int_u^\infty x^{\nu-1}(x-u)^{\mu-1} e^{-\beta x}\, dx = \beta^{-\frac{\mu+\nu}{2}} u^{\frac{\mu+\nu-2}{2}} \Gamma(\mu) \exp\left(-\frac{\beta u}{2}\right) W_{\frac{\nu-\mu}{2},\,\frac{1-\mu-\nu}{2}}(\beta u)$

[Re $\mu > 0$, Re $\beta u > 0$]. Ba V 202(13)

5. $\int_0^\infty x^{\nu-1}(x+\beta)^{-\nu+\frac{1}{2}} e^{-\mu x}\, dx = 2^{\nu-\frac{1}{2}} \Gamma(\nu)\, \mu^{-\frac{1}{2}} e^{\frac{\beta\mu}{2}} D_{1-2\nu}(\sqrt{2\beta\mu})$

[|arg $\beta| < \pi$, Re $\nu > 0$, Re $\mu \geqslant 0$]. Ba IV 139(20), Ba II 119(2)+

6. $$\int_0^\infty x^{\nu-1}(x+\beta)^{-\nu-\frac{1}{2}} e^{-\mu x}\, dx = 2^\nu \Gamma(\nu)\, \beta^{-\frac{1}{2}} e^{\frac{\beta\mu}{2}} D_{-2\nu}(\sqrt{2\beta\mu})$$
$$[|\arg\beta|<\pi,\ \operatorname{Re}\nu>0,\ \operatorname{Re}\mu\geqslant 0].$$
3.383

Ba IV 139(21), Ba II 119(1)+

7. $$\int_0^\infty x^{\nu-1}(x+\beta)^{-\rho} e^{-\mu x}\, dx = \beta^{\frac{\nu-\rho-1}{2}} \mu^{\frac{\rho-\nu-1}{2}} e^{\frac{\beta\mu}{2}} \Gamma(\nu) W_{\frac{1-\nu-\rho}{2},\,\frac{\nu-\rho}{2}}(\beta\mu)$$
$$[|\arg\beta|<\pi,\ \operatorname{Re}\mu>0,\ \operatorname{Re}\nu>0].$$

WW 340+, Ba V 234(12), Ba I 255(2)+

8. $$\int_u^\infty \frac{(x-u)^\nu e^{-\mu x}}{x}\, dx = u^\nu \Gamma(\nu+1)\, \Gamma(-\nu, u\mu)$$
$$[u>0,\ \operatorname{Re}\nu>-1,\ \operatorname{Re}\mu>0].$$

Ba IV 138(8)

9. $$\int_0^\infty \frac{x^{\nu-1} e^{-\mu}}{x+\beta}\, dx = \beta^{\nu-1} e^{\beta\mu} \Gamma(\nu)\, \Gamma(1-\nu, \beta\mu)$$
$$[|\arg\beta|<\pi,\ \operatorname{Re}\mu>0,\ \operatorname{Re}\nu>0].$$

Ba II 137(3)

1. $$\int_{-1}^1 (1-x)^{\nu-1}(1+x)^{\mu-1} e^{-ipx}\, dx = 2^{\mu+\nu-1} B(\mu,\nu)\, e^{ip}\, {}_1F_1(\mu;\nu+\mu;-2ip)$$
$$[\operatorname{Re}\nu>0,\ \operatorname{Re}\mu>0].$$
3.384

Ba IV 119(13)

2. $$\int_u^v (x-u)^{2\mu-1}(v-x)^{2\nu-1} e^{-px}\, dx = B(2\mu, 2\nu)(v-u)^{\mu+\nu-1} \times$$
$$\times p^{-\mu-\nu} \exp\left(-p\frac{u+v}{2}\right) M_{\mu-\nu,\,\mu+\nu-\frac{1}{2}}(vp-up)$$
$$[v>u>0,\ \operatorname{Re}\mu>0,\ \operatorname{Re}\nu>0].$$

Ba IV 139(23)

3. $$\int_u^\infty (x+\beta)^{2\nu-1}(x-u)^{2\rho-1} e^{-\mu x}\, dx = \frac{(u+\beta)^{\nu+\rho-1}}{\mu^{\nu+\rho}} \exp\left[\frac{(\beta-u)\mu}{2}\right]\times$$
$$\times \Gamma(2\rho)\, W_{\nu-\rho,\,\nu+\rho-\frac{1}{2}}(u\mu+\beta\mu)\quad [u>0,\ |\arg(\beta+u)|<\pi,\ \operatorname{Re}\mu>0,\ \operatorname{Re}\rho>0].$$

Ba IV 139(22)

4. $$\int_u^\infty (x+\beta)^\nu (x-u)^{-\nu} e^{-\mu x}\, dx = \frac{1}{\mu}\, \nu\pi \operatorname{cosec}(\nu\pi)\, e^{-\frac{(\beta+u)\mu}{2}} k_{2\nu}\left[\frac{(\beta+u)\mu}{2}\right]$$
$$[u>0,\ |\arg(u+\beta)|<\pi,\ \operatorname{Re}\mu>0,\ \operatorname{Re}\nu<1].$$

Ba IV 139(17)

5. $$\int_u^\infty (x-u)^{\nu-1}(x+u)^{-\nu+\frac{1}{2}} e^{-\mu x}\, dx = \frac{1}{\sqrt{\mu}}\, 2^{\nu-\frac{1}{2}}\, \Gamma(\nu)\, D_{1-2\nu}(2\sqrt{u\mu})$$
$$[u>0,\ \operatorname{Re}\mu>0,\ \operatorname{Re}\nu>0].$$

Ba IV 139(18)

3.384

6. $$\int_u^\infty (x-u)^{\nu-1}(x+u)^{-\nu-\frac{1}{2}} e^{-\mu x}\, dx = \frac{1}{\sqrt{u}} 2^{\nu-\frac{1}{2}} \Gamma(\nu) D_{-2\nu}(2\sqrt{u\mu})$$

$$[u > 0,\ \operatorname{Re}\mu \geq 0,\ \operatorname{Re}\nu > 0].$$ Ba IV 139(19)

7. $$\int_{-\infty}^\infty (\beta - ix)^{-\mu}(\gamma - ix)^{-\nu} e^{-ipx}\, dx =$$

$$= \frac{2\pi e^{-\beta p} p^{\mu+\nu-1}}{\Gamma(\mu+\nu)}\, {}_1F_1(\nu;\mu+\nu;(\beta-\gamma)p) \quad \begin{matrix}\text{für}\\\text{for}\end{matrix}\ p > 0;$$

$$= 0 \quad \begin{matrix}\text{für}\\\text{for}\end{matrix}\ p < 0$$

$$[\operatorname{Re}\beta > 0,\ \operatorname{Re}\gamma > 0,\ \operatorname{Re}(\mu+1) > \nu].$$ Ba IV 119(10)

8. $$\int_{-\infty}^\infty (\beta + ix)^{-\mu}(\gamma + ix)^{-\nu} e^{-ixp}\, dx = 0 \quad \begin{matrix}\text{für}\\\text{for}\end{matrix}\ p > 0;$$

$$= \frac{2\pi e^{\beta p}(-p)^{\mu+\nu-1}}{\Gamma(\mu+\nu)}\, {}_1F_1[\mu;\mu+\nu;(\beta-\gamma)p] \quad \begin{matrix}\text{für}\\\text{for}\end{matrix}\ p < 0$$

$$[\operatorname{Re}\beta > 0,\ \operatorname{Re}\gamma > 0,\ \operatorname{Re}(\mu+\nu) > 1].$$ Ba IV 119(11)

9. $$\int_{-\infty}^\infty (\beta + ix)^{-2\mu}(\gamma - ix)^{-2\nu} e^{-ipx}\, dx =$$

$$= -2\pi(\beta+\gamma)^{-\mu-\nu}\frac{p^{\mu+\nu-1}}{\Gamma(2\nu)}\exp\left(\frac{\gamma-\beta}{2}p\right) \times$$

$$\times W_{\nu-\mu,\,\frac{1}{2}-\nu-\mu}(\beta p + \gamma p) \quad \begin{matrix}\text{für}\\\text{for}\end{matrix}\ p > 0;$$

$$= 2\pi(\beta+\gamma)^{-\mu-\nu}\frac{(-p)^{\mu+\nu-1}}{\Gamma(2\mu)}\exp\left(\frac{\beta-\gamma}{2}p\right) \times$$

$$\times W_{\mu-\nu,\,\frac{1}{2}-\nu-\mu}(-\beta p - \gamma p) \quad \begin{matrix}\text{für}\\\text{for}\end{matrix}\ p < 0$$

$$\left[\operatorname{Re}\beta > 0,\ \operatorname{Re}\gamma > 0,\ \operatorname{Re}(\mu+\nu) > \frac{1}{2}\right].$$ Ba IV 119(12)

3.385

$$\int_0^1 x^{\nu-1}(1-x)^{\lambda-1}(1-\beta x)^{-\rho} e^{-\mu x}\, dx = B(\nu,\lambda)\,\Phi_1(\nu,\rho,\lambda+\nu;\beta,-\mu)$$

$$[\operatorname{Re}\lambda > 0,\ \operatorname{Re}\nu > 0,\ |\arg(1-\beta)| < \pi].$$ Ba IV 139(24)

3.386

1. $$\int_{-\infty}^\infty \frac{(ix)^{\nu_0}\prod_{k=1}^n (\beta_k + ix)^{\nu_k} e^{-ipx}\, dx}{\beta_0 - ix} = 2\pi e^{-\beta_0 p}\beta_0^{\nu_0}\prod_{k=1}^n (\beta_0 + \beta_k)^{\nu_k}$$

$$\left[\operatorname{Re}\nu_0 > -1,\ \operatorname{Re}\beta_k > 0,\ \sum_{k=0}^n \operatorname{Re}\nu_k < 1,\ \arg ix = \frac{\pi}{2}\operatorname{sign} x,\ p > 0\right].$$ Ba IV 118(8)

2.
$$\int_{-\infty}^{\infty} \frac{(ix)^{\nu_0} \prod_{k=1}^{n}(\beta_k + ix)^{\nu_k} e^{-ipx} dx}{\beta_0 + ix} = 0$$
3.386

$\left[\operatorname{Re} \nu_0 > -1, \operatorname{Re} \beta_k > 0, \sum_{k=0}^{n} \operatorname{Re} \nu_k < 1, \arg ix = \frac{\pi}{2} \operatorname{sign} x, p > 0\right].$ Ba IV 119(9)

1.
$$\int_{-1}^{1} (1-x^2)^{\nu-1} e^{-\mu x} dx = \sqrt{\pi} \left(\frac{2}{\mu}\right)^{\nu-\frac{1}{2}} \Gamma(\nu) I_{\nu-\frac{1}{2}}(\mu)$$
3.387

$\left[\operatorname{Re} \nu > 0, |\arg \mu| < \frac{\pi}{2}\right].$ W 172(2)+

2.
$$\int_{-1}^{1} (1-x^2)^{\nu-1} e^{i\mu x} dx = \sqrt{\pi} \left(\frac{2}{\mu}\right)^{\nu-\frac{1}{2}} \Gamma(\nu) J_{\nu-\frac{1}{2}}(\mu) \quad [\operatorname{Re} \nu > 0].$$

W 25(3)+, W 48(4)+

3.
$$\int_{1}^{\infty} (x^2-1)^{\nu-1} e^{-\mu x} dx = \frac{1}{\sqrt{\pi}} \left(\frac{2}{\mu}\right)^{\nu-\frac{1}{2}} \Gamma(\nu) K_{\nu-\frac{1}{2}}(\mu)$$

$\left[|\arg \mu| < \frac{\pi}{2}, \operatorname{Re} \nu > 0\right].$ W 172(4)+

4.
$$\int_{1}^{\infty} (x^2-1)^{\nu-1} e^{i\mu x} dx = i \frac{\sqrt{\pi}}{2} \left(\frac{2}{\mu}\right)^{\nu-\frac{1}{2}} \Gamma(\nu) H^{(1)}_{\frac{1}{2}-\nu}(\mu)$$

$[\operatorname{Im} \mu > 0, \operatorname{Re} \nu > 0];$ Ba II 83(28)+

$$= -i \frac{\sqrt{\pi}}{2} \left(-\frac{2}{\mu}\right)^{\nu-\frac{1}{2}} \Gamma(\nu) H^{(2)}_{\frac{1}{2}-\nu}(-\mu)$$

$[\operatorname{Im} \mu < 0, \operatorname{Re} \nu > 0].$ Ba II 83(29)+

5.
$$\int_{0}^{u} (u^2-x^2)^{\nu-1} e^{\mu x} dx = \frac{\sqrt{\pi}}{2} \left(\frac{2u}{\mu}\right)^{\nu-\frac{1}{2}} \Gamma(\nu) [I_{\nu-\frac{1}{2}}(u\mu) + \mathbf{L}_{\nu-\frac{1}{2}}(u\mu)]$$

$[u > 0, \operatorname{Re} \nu > 0].$ Ba V 188(20)+

6.
$$\int_{u}^{\infty} (x^2-u^2)^{\nu-1} e^{-\mu x} dx = \frac{1}{\sqrt{\pi}} \left(\frac{2u}{\mu}\right)^{\nu-\frac{1}{2}} \Gamma(\nu) K_{\nu-\frac{1}{2}}(u\mu)$$

$[u > 0, \operatorname{Re} \mu > 0, \operatorname{Re} \nu > 0].$ Ba V 203(17)+

7.
$$\int_{0}^{\infty} (x^2+u^2)^{\nu-1} e^{-\mu x} dx = \frac{\sqrt{\pi}}{2} \left(\frac{2u}{\mu}\right)^{\nu-\frac{1}{2}} \Gamma(\nu) [\mathbf{H}_{\nu-\frac{1}{2}}(u\mu) - N_{\nu-\frac{1}{2}}(u\mu)]$$

$[|\arg u| < \pi, \operatorname{Re} \mu > 0].$ Ba IV 138(10)

1.
$$\int_{0}^{2u} (2ux-x^2)^{\nu-1} e^{-\mu x} dx = \sqrt{\pi} \left(\frac{2u}{\mu}\right)^{\nu-\frac{1}{2}} e^{-u\mu} \Gamma(\nu) I_{\nu-\frac{1}{2}}(u\mu)$$
3.388

$[u > 0, \operatorname{Re} \nu > 0].$ Ba IV 138(14)

3.388

2.
$$\int_0^\infty (2\beta x + x^2)^{\nu-1} e^{-\mu x}\, dx = \frac{1}{\sqrt{\pi}} \left(\frac{2\beta}{\mu}\right)^{\nu-\frac{1}{2}} e^{\beta\mu} \Gamma(\nu) K_{\nu-\frac{1}{2}}(\beta\mu)$$

$$[|\arg \beta| < \pi, \ \operatorname{Re} \nu > 0, \ \operatorname{Re} \mu > 0].$$
Ba IV 138(13)

3.
$$\int_0^\infty (x^2 + ix)^{\nu-1} e^{-\mu x}\, dx = -\frac{i\sqrt{\pi}\, e^{\frac{i\mu}{2}}}{2\mu^{\nu-\frac{1}{2}}} \Gamma(\nu) H^{(2)}_{\nu-\frac{1}{2}}\left(\frac{\mu}{2}\right)$$

$$[\operatorname{Re} \mu > 0, \ \operatorname{Re} \nu > 0].$$
Ba IV 138(15)

4.
$$\int_0^\infty (x^2 - ix)^{\nu-1} e^{-\mu x}\, dx = \frac{i\sqrt{\pi}\, e^{-\frac{i\mu}{2}}}{2\mu^{\nu-\frac{1}{2}}} \Gamma(\nu) H^{(1)}_{\nu-\frac{1}{2}}\left(\frac{\mu}{2}\right)$$

$$[\operatorname{Re} \mu > 0, \ \operatorname{Re} \nu > 0].$$
Ba IV 138(16)

3.389

1.
$$\int_0^u x^{2\nu-1}(u^2 - x^2)^{\rho-1} e^{\mu x}\, dx =$$

$$= \frac{1}{2} B(\nu,\rho)\, u^{2\nu+2\rho-2}\, {}_1F_2\left(\nu;\, \frac{1}{2},\, \nu+\rho;\, \frac{\mu^2 u^2}{4}\right) +$$

$$+ \frac{\mu}{2} B\left(\nu+\frac{1}{2},\, \rho\right) u^{2\nu+2\rho-1}\, {}_1F_2\left(\nu+\frac{1}{2};\, \frac{3}{2},\, \nu+\rho+\frac{1}{2};\, \frac{\mu^2 u^2}{4}\right)$$

$$[\operatorname{Re} \rho > 0, \ \operatorname{Re} \nu > 0].$$
Ba V 188(21)

2.
$$\int_0^\infty x^{2\nu-1}(u^2 + x^2)^{\rho-1} e^{-\mu x}\, dx = \frac{u^{2\nu+2\rho-2}}{2\sqrt{\pi}\, \Gamma(1-\rho)} G^{31}_{13}\left(\frac{\mu^2 u^2}{4}\, \middle|\, \begin{matrix} 1-\nu \\ 1-\rho-\nu,\, 0,\, \frac{1}{2} \end{matrix}\right)$$

$$\left[|\arg u| < \frac{\pi}{2},\ \operatorname{Re} \mu > 0,\ \operatorname{Re} \nu > 0\right].$$
Ba V 234(15)+

3.
$$\int_0^u x(u^2 - x^2)^{\nu-1} e^{\mu x}\, dx = \frac{u^{2\nu}}{2\nu} + \frac{\sqrt{\pi}}{2} \left(\frac{\mu}{2}\right)^{\frac{1}{2}-\nu} u^{\nu+\frac{1}{2}} \Gamma(\nu) \times$$

$$\times [I_{\nu+\frac{1}{2}}(\mu u) + \mathbf{L}_{\nu+\frac{1}{2}}(\mu u)] \quad [\operatorname{Re} \nu > 0].$$
Ba V 188(19)+

4.
$$\int_0^\infty x(x^2 - u^2)^{\nu-1} e^{-\mu x}\, dx = 2^{\nu-\frac{1}{2}} (\sqrt{\pi})^{-1} \mu^{\frac{1}{2}-\nu} u^{\nu+\frac{1}{2}} \Gamma(\nu) K_{\nu+\frac{1}{2}}(u\mu)$$

$$[\operatorname{Re}(u\mu) > 0].$$
Ba V 203(16)+

5.
$$\int_{-\infty}^\infty \frac{(ix)^{-\nu} e^{-ipx}\, dx}{\beta^2 + x^2} = \pi \beta^{-\nu-1} e^{-|p|\beta}$$

$$\left[|\nu| < 1,\ \operatorname{Re} \beta > 0,\ \arg ix = \frac{\pi}{2} \operatorname{sign} x\right].$$
Ba IV 118(5)

6. $\int_0^\infty \frac{x^\nu e^{-\mu x}}{\beta^2 + x^2} dx = \frac{1}{2} \Gamma(\nu) \beta^{\nu-1} \left[\exp\left(i\mu\beta + i\frac{(\nu-1)\pi}{2} \right) \times \right.$ 3.389

$\left. \times \Gamma(1-\nu, i\beta\mu) + \exp\left(-i\beta\mu - i\frac{(\nu-1)\pi}{2} \right) \Gamma(1-\nu, -i\beta\mu) \right]$
[Re $\beta > 0$, Re $\mu > 0$, Re $\nu > -1$].   Ba V 218(22)

7. $\int_0^\infty \frac{x^{\nu-1} e^{-\mu x} dx}{1 + x^2} = \pi \operatorname{cosec}(\nu\pi) V_\nu(2\mu, 0)$ [Re $\mu > 0$, Re $\nu > 0$].   Ba IV 138(9)

8. $\int_{-\infty}^\infty \frac{(\beta + ix)^{-\nu} e^{-ipx}}{\gamma^2 + x^2} dx = \frac{\pi}{\gamma} (\beta + \gamma)^{-\nu} e^{-p\gamma}$

[Re $\nu > -1$, $p > 0$, Re $\beta > 0$, Re $\gamma > 0$].   Ba IV 118(6)

9. $\int_{-\infty}^\infty \frac{(\beta - ix)^{-\nu} e^{-ipx}}{\gamma^2 + x^2} dx = \frac{\pi}{\gamma} (\beta - \gamma)^{-\nu} e^{\gamma p}$

[$p > 0$, Re $\beta > 0$, Re $\gamma > 0$, $\beta \neq \gamma$, Re $\nu > -1$].   Ba IV 118(7)

$\int_0^\infty [(\sqrt{x + 2\beta} + \sqrt{x})^{2\nu} - (\sqrt{x + 2\beta} - \sqrt{x})^{2\nu}] e^{-\mu x} dx = 2^{\nu+1} \frac{\nu}{\mu} \beta^\nu e^{\beta\mu} K_\nu(\beta\mu)$ 3.391

[$|\arg \beta| < \pi$, Re $\mu > 0$].   Ba IV 140(30)

1. $\int_0^\infty (x + \sqrt{1 + x^2})^\nu e^{-\mu x} dx = \frac{1}{\mu} S_{1,\nu}(\mu) + \frac{\nu}{\mu} S_{0,\nu}(\mu)$ [Re $\mu > 0$].   3.392

Ba IV 140(25)

2. $\int_0^\infty (\sqrt{1 + x^2} - x)^\nu e^{-\mu x} dx = \frac{1}{\mu} S_{1,\nu}(\mu) - \frac{\nu}{\mu} S_{0,\nu}(\mu)$ [Re $\mu > 0$].

Ba IV 140(26)

3. $\int_0^\infty \frac{(x + \sqrt{1 + x^2})^\nu}{\sqrt{1 + x^2}} e^{-\mu x} dx = \pi \operatorname{cosec} \nu\pi [\mathbf{J}_{-\nu}(\mu) - J_{-\nu}(\mu)]$

[Re $\mu > 0$].   Ba IV 140(27), Ba II 35(33)

4. $\int_0^\infty \frac{(\sqrt{1 + x^2} - x)^\nu}{\sqrt{1 + x^2}} e^{-\mu x} dx = S_{0,\nu}(\mu) - \nu S_{-1,\nu}(\mu)$ [Re $\mu > 0$].

Ba IV 140(28)

$\int_0^\infty \frac{(x + \sqrt{x^2 + 4\beta^2})^{2\nu}}{\sqrt{x^3 + 4\beta^2 x}} e^{-\mu x} dx =$ 3.393

$= \frac{\sqrt{\mu\pi^3}}{2^{2\nu + \frac{3}{2}} \beta^{2\nu}} [J_{\nu + \frac{1}{4}}(\beta\mu) N_{\nu - \frac{1}{4}}(\beta\mu) - J_{\nu - \frac{1}{4}}(\beta\mu) N_{\nu + \frac{1}{4}}(\beta\mu)]$

[Re $\beta > 0$, Re $\mu > 0$].   Ba IV 140(33)

3.394 $$\int_0^\infty \frac{(1+\sqrt{1+x^2})^{\nu+\frac{1}{2}}}{x^{\nu+1}\sqrt{1+x^2}} e^{-\mu x}\, dx = \sqrt{2}\,\Gamma(-\nu) D_\nu(\sqrt{2i\mu})\, D_\nu(\sqrt{-2i\mu})$$

$[\operatorname{Re}\mu \geqslant 0,\ \operatorname{Re}\nu > 0]$. Ba IV 140(32)

3.395

1. $$\int_1^\infty \frac{(\sqrt{x^2-1}+x)^\nu + (\sqrt{x^2-1}+x)^{-\nu}}{\sqrt{x^2-1}} e^{-\mu x}\, dx = 2K_\nu(\mu) \quad [\operatorname{Re}\mu > 0].$$

Ba IV 140(29)

2. $$\int_1^\infty \frac{(x+\sqrt{x^2-1})^{2\nu} + (x-\sqrt{x^2-1})^{2\nu}}{\sqrt{x(x^2-1)}} e^{-\mu x}\, dx =$$

$$= \sqrt{\frac{2\mu}{\pi}}\, K_{\nu+\frac{1}{4}}\!\left(\frac{\mu}{2}\right) K_{\nu-\frac{1}{4}}\!\left(\frac{\mu}{2}\right) \quad [\operatorname{Re}\mu > 0].$$ Ba IV 140(34)

3. $$\int_0^\infty \frac{(x+\sqrt{x^2+1})^\nu + \cos\nu\pi\,(x+\sqrt{x^2+1})^{-\nu}}{\sqrt{x^2+1}} e^{-\mu x}\, dx =$$

$$= -\pi[\mathbf{E}_\nu(\mu) + N_\nu(\mu)] \quad [\operatorname{Re}\mu > 0].$$ Ba II 35(34)

Rationale Funktionen von Potenzen und der Exponentialfunktion
3.41–3.44 **Rational functions of powers and exponentials**

3.411

1. $$\int_0^\infty \frac{x^{\nu-1}\, dx}{e^{\mu x}-1} = \frac{1}{\mu^\nu}\,\Gamma(\nu)\,\zeta(\nu) \quad [\operatorname{Re}\mu > 0,\ \operatorname{Re}\nu > 1].$$ F II 798+

2. $$\int_0^\infty \frac{x^{2n-1}\, dx}{e^{px}-1} = (-1)^{n-1}\left(\frac{2\pi}{p}\right)^{2n}\frac{B_{2n}}{4n}.$$ F II 726–727

3. $$\int_0^\infty \frac{x^{\nu-1}\, dx}{e^{\mu x}+1} = \frac{1}{\mu^\nu}(1-2^{1-\nu})\Gamma(\nu)\zeta(\nu) \quad [\operatorname{Re}\mu > 0,\ \operatorname{Re}\nu > 0].$$

F II 798+, WW 267

4. $$\int_0^\infty \frac{x^{2n-1}\, dx}{e^{px}+1} = (1-2^{1-2n})\left(\frac{2\pi}{p}\right)^{2n}\frac{|B_{2n}|}{4n}.$$ BH[83](2), Ba I 39(25)

5. $$\int_0^{\ln 2} \frac{x\, dx}{1-e^{-x}} = \frac{\pi^2}{12}.$$ BH[104](5)

6. $$\int_0^\infty \frac{x^{\nu-1} e^{-\mu x}}{1-\beta e^{-x}}\, dx = \Gamma(\nu)\sum_{n=0}^\infty (\mu+n)^{-\nu}\beta^n$$ Ba I 27(3)

[$\operatorname{Re}\mu > 0$ und entweder $|\beta|\leqslant 1$, $\beta\neq 1$, $\operatorname{Re}\nu > 0$ oder $\beta = 1$, $\operatorname{Re}\nu > 1$]. | [$\operatorname{Re}\mu > 0$, and either $|\beta|\leqslant 1$, $\beta\neq 1$, $\operatorname{Re}\nu > 0$ or $\beta = 1$, $\operatorname{Re}\nu > 1$].

7. $$\int_0^\infty \frac{x^{\nu-1} e^{-\mu x}}{1-e^{-\beta x}}\, dx = \frac{1}{\beta^\nu}\,\Gamma(\nu)\,\zeta\!\left(\nu,\frac{\mu}{\beta}\right) \quad [\operatorname{Re}\mu > 0,\ \operatorname{Re}\nu > 1].$$ Ba IV 144(10)

Die Exponentialfunktion/The Exponential Function

8. $\int_0^\infty \dfrac{x^{n-1}e^{-px}}{1+e^x}\,dx = (n-1)!\sum_{k=1}^{\infty}\dfrac{(-1)^{k-1}}{(p+k)^n}$ $[p>-1;\ n=1,2,\ldots]$.   BH[83](9) 3.411

9. $\int_0^\infty \dfrac{xe^{-x}\,dx}{e^x-1} = \dfrac{\pi^2}{6} - 1$ (cf. 4.231 2.).   BH[82](1)

10. $\int_0^\infty \dfrac{xe^{-2x}\,dx}{e^{-x}+1} = 1 - \dfrac{\pi^2}{12}$ (cf. 4.251 6.).   BH[82](2)

11. $\int_0^\infty \dfrac{xe^{-3x}}{e^{-x}+1}\,dx = \dfrac{\pi^2}{12} - \dfrac{3}{4}$ (cf. 4.251 5.).   BH[82](3)

12. $\int_0^\infty \dfrac{xe^{-2nx}}{1+e^x}\,dx = -\dfrac{\pi^2}{12} + \sum_{k=1}^{2n-1}\dfrac{(-1)^{k-1}}{k^2}$ (cf. 4.251 6.).   BH[82](5)

13. $\int_0^\infty \dfrac{xe^{-(2n-1)x}}{1+e^x}\,dx = \dfrac{\pi^2}{12} + \sum_{k=1}^{2n}\dfrac{(-1)^k}{k^2}$ (cf. 4.251 5.).   BH[82](4)

14. $\int_0^\infty \dfrac{x^2 e^{-nx}}{1-e^{-x}}\,dx = 2\sum_{k=n}^{\infty}\dfrac{1}{k^3}$ (cf. 4.261 12.).   BH[82](9)

15. $\int_0^\infty \dfrac{x^2 e^{-nx}}{1+e^{-x}}\,dx = 2\sum_{k=n}^{\infty}\dfrac{(-1)^{n+k}}{k^3}$ (cf. 4.261 11.).   Li[82](10)

16. $\int_{-\infty}^\infty \dfrac{x^2 e^{-\mu x}}{1+e^{-x}}\,dx = \pi^3 \cos^3 \mu\pi\,(2-\sin^2 \mu\pi)$ $[0<\operatorname{Re}\mu<1]$.   Ba IV 120(17)+

17. $\int_0^\infty \dfrac{x^3 e^{-nx}}{1-e^{-x}}\,dx = \dfrac{\pi^4}{15} - 6\sum_{k=1}^{n-1}\dfrac{1}{k^4}$ (cf. 4.262 5.).   BH[82](12)

18. $\int_0^\infty \dfrac{x^3 e^{-nx}}{1+e^{-x}}\,dx = 6\sum_{k=n}^{\infty}\dfrac{(-1)^{n+k}}{k^4}$ (cf. 4.262 4.).   Li[82](13)

19. $\int_0^\infty e^{-px}(e^{-x}-1)^n\,\dfrac{dx}{x} = -\sum_{k=0}^{n}(-1)^k \binom{n}{k}\ln(p+n-k)$.   Li[89](10)

20. $\int_0^\infty e^{-px}(e^{-x}-1)^n\,\dfrac{dx}{x^2} = \sum_{k=0}^{n}(-1)^k\binom{n}{k}(p+n-k)\ln(p+n-k)$.

    Li[89](15)

21. $\int_0^\infty x^{n-1}\,\dfrac{1-e^{-mx}}{1-e^x}\,dx = (n-1)!\sum_{k=1}^{m}\dfrac{1}{k^n}$ (cf. 4.272 11.).   Li[83](8)

3.411

22. $\int_0^\infty \dfrac{x^{p-1}}{e^{rx}-q}\,dx = \dfrac{1}{qr^p}\,\Gamma(p)\sum_{k=1}^\infty \dfrac{q^k}{k^p}$ $[p>0]$.   BH[83](5)

23. $\int_{-\infty}^\infty \dfrac{xe^{\mu x}\,dx}{\beta+e^x} = \pi\beta^{\mu-1}\operatorname{cosec}(\mu\pi)\,[\ln\beta - \pi\operatorname{ctg}(\mu\pi)]$

$[|\arg\beta|<\pi,\ 0<\operatorname{Re}\mu<1]$.   BH[101](5), Ba IV 120(16)+

24. $\int_{-\infty}^\infty \dfrac{xe^{\mu x}}{e^{\nu x}-1}\,dx = \left(\dfrac{\pi}{\nu}\operatorname{cosec}\dfrac{\mu\pi}{\nu}\right)^2$ $[\operatorname{Re}\nu>\operatorname{Re}\mu>0]$ (cf. 4.254 2.).

Li[101](3)

25. $\int_0^\infty x\,\dfrac{1+e^{-x}}{e^x-1}\,dx = \dfrac{\pi^2}{3}-1$ (cf. 4.231 3.).   BH[82](6)

26. $\int_0^\infty x\,\dfrac{1-e^{-x}}{1+e^{-3x}}\,e^{-x}\,dx = \dfrac{2\pi^2}{27}$.   Li[82](7)+

27. $\int_0^\infty \dfrac{1-e^{-\mu x}}{1+e^x}\,\dfrac{dx}{x} = \ln\left[\dfrac{\Gamma\!\left(\dfrac{\mu}{2}+1\right)}{\Gamma\!\left(\dfrac{\mu+1}{2}\right)}\sqrt{\pi}\,\right]$ $[\operatorname{Re}\mu>-1]$.   BH[93](4)

28. $\int_0^\infty \dfrac{e^{-\nu x}-e^{-\mu x}}{e^{-x}+1}\,\dfrac{dx}{x} = \ln\dfrac{\Gamma\!\left(\dfrac{\nu}{2}\right)\Gamma\!\left(\dfrac{\mu+1}{2}\right)}{\Gamma\!\left(\dfrac{\mu}{2}\right)\Gamma\!\left(\dfrac{\nu+1}{2}\right)}$ $[\operatorname{Re}\mu>0,\ \operatorname{Re}\nu>0]$.   BH[93](6)

29. $\int_{-\infty}^\infty \dfrac{e^{px}-e^{qx}}{1+e^{rx}}\,\dfrac{dx}{x} = \ln\left[\operatorname{tg}\dfrac{p\pi}{2r}\operatorname{ctg}\dfrac{q\pi}{2r}\right]$

$[|r|>|p|,\ |r|>|q|,\ rp>0,\ rq>0]$
(cf. 4.267 18.).   BH[103](3)

30. $\int_{-\infty}^\infty \dfrac{e^{px}-e^{qx}}{1-e^{rx}}\,\dfrac{dx}{x} = \ln\left[\sin\dfrac{p\pi}{r}\operatorname{cosec}\dfrac{q\pi}{r}\right]$

$[|r|>|p|,\ |r|>|q|,\ rp>0,\ rq>0]$
(cf. 4.267 19.).   BH[103](4)

31. $\int_0^\infty \dfrac{e^{-qx}+e^{(q-p)x}}{1-e^{-px}}\,x\,dx = \left(\dfrac{\pi}{p}\operatorname{cosec}\dfrac{q\pi}{p}\right)^2$ $[0<q<p]$.   BH[82](8)

32. $\int_0^\infty \dfrac{e^{-px}-e^{(p-q)x}}{e^{-qx}+1}\,\dfrac{dx}{x} = \ln\operatorname{ctg}\dfrac{p\pi}{2q}$ $[0<p<q]$.   BH[93](7)

3.412

$\int_0^\infty \left\{\dfrac{a+be^{-px}}{ce^{px}+g+he^{-px}} - \dfrac{a+be^{-qx}}{ce^{qx}+g+he^{-qx}}\right\}\dfrac{dx}{x} = \dfrac{a+b}{c+g+h}\ln\dfrac{p}{q}$

$[p>0,\ q>0]$.   BH[96](7)

1. $\int_0^\infty \dfrac{(1-e^{-\beta x})(1-e^{-\gamma x})e^{-\mu x}}{1-e^{-x}}\,\dfrac{dx}{x} = \ln\dfrac{\Gamma(\mu)\Gamma(\beta+\gamma+\mu)}{\Gamma(\mu+\beta)\Gamma(\mu+\gamma)}$ 3.413

[Re $\mu > 0$, Re $\mu > -$ Re β, Re $\mu > -$Re γ, Re $\mu > -$ Re $(\beta+\gamma)$]

(cf. 4.267 25.). BH[93](13)

2. $\int_0^\infty \dfrac{\{1-e^{(q-p)x}\}^2}{e^{qx}-e^{(q-2p)x}}\,\dfrac{dx}{x} = \ln\operatorname{cosec}\dfrac{q\pi}{2p}$ $[0 < q < p]$. BH[95](6)

3. $\int_0^\infty \dfrac{e^{-px}-e^{-qx}}{1+e^{-x}}\,\dfrac{1+e^{-(2n+1)x}}{x}\,dx =$

$= \ln\left\{\dfrac{q(q+2)(q+4)\dots(q+2n)}{p(p+2)(p+4)\dots(p+2n)}\,\dfrac{(p+1)(p+3)\dots(p+2n-1)}{(q+1)(q+3)\dots(q+2n-1)}\right\}$

[Re $p > -2n$, Re $q > -2n$] (cf. 4.267 14.). BH[93](11)

$\int_0^\infty \dfrac{(1-e^{-\beta x})(1-e^{-\gamma x})(1-e^{-\delta x})e^{-\mu x}}{1-e^{-x}}\,\dfrac{dx}{x} =$ 3.414

$= \ln\dfrac{\Gamma(\mu)\,\Gamma(\mu+\beta+\gamma)\,\Gamma(\mu+\beta+\delta)\,\Gamma(\mu+\gamma+\delta)}{\Gamma(\mu+\beta)\,\Gamma(\mu+\gamma)\,\Gamma(\mu+\delta)\,\Gamma(\mu+\beta+\gamma+\delta)}$

[2 Re $\mu > |\operatorname{Re}\beta| + |\operatorname{Re}\gamma| + |\operatorname{Re}\delta|$] (cf. 4.267 31.). BH[93](14), Ba IV 145(17)

1. $\int_0^\infty \dfrac{x\,dx}{(x^2+\beta^2)(e^{\mu x}-1)} = \dfrac{1}{2}\left[\ln\left(\dfrac{\beta\mu}{2\pi}\right) - \dfrac{\pi}{\beta\mu} - \psi\left(\dfrac{\beta\mu}{2\pi}\right)\right]$ 3.415

[Re $\beta > 0$, Re $\mu > 0$]. BH[97](20), Ba I 18(27)

2. $\int_0^\infty \dfrac{x\,dx}{(x^2+\beta^2)^2(e^{2\pi x}-1)} = -\dfrac{1}{8\beta^3} - \dfrac{1}{4\beta^2} + \dfrac{1}{4\beta}\psi'(\beta);$

$= \dfrac{1}{4\beta^4}\sum_{k=0}^\infty \dfrac{|B_{2k+2}|}{\beta^{2k}}$ [Re $\beta > 0$]. BH[97](22), Ba I 22(12)

1. $\int_0^\infty \dfrac{(1+ix)^{2n}-(1-ix)^{2n}}{i}\,\dfrac{dx}{e^{2\pi x}-1} = \dfrac{1}{2}\,\dfrac{2n-1}{2n+1}$. BH[88](4) 3.416

2. $\int_0^\infty \dfrac{(1+ix)^{2n}-(1-ix)^{2n}}{i}\,\dfrac{dx}{e^{\pi x}+1} = \dfrac{1}{2n+1}$. BH[87](1)

3. $\int_0^\infty \dfrac{(1+ix)^{2n-1}-(1-ix)^{2n-1}}{i}\,\dfrac{dx}{e^{\pi x}+1} = \dfrac{1}{2n}[1-2^{2n}B_{2n}]$. BH[87](2)

1. $\int_{-\infty}^\infty \dfrac{x\,dx}{a^2 e^x + b^2 e^{-x}} = \dfrac{\pi}{2ab}\ln\dfrac{b}{a}$ $[ab > 0]$ (cf. 4.231 6.). BH[101](1) 3.417

2. $\int_{-\infty}^\infty \dfrac{x\,dx}{a^2 e^x - b^2 e^{-x}} = \dfrac{\pi^2}{4ab}$ (cf. 4.231 8.). Li[101](2)

3.418

1. $\int_0^\infty \dfrac{x\,dx}{e^x + e^{-x} - 1} = 1{,}171\,953\,619\,3\ldots$ Li[88](1)

2. $\int_0^\infty \dfrac{xe^{-x}\,dx}{e^x + e^{-x} - 1} = 0{,}311\,821\,131\,9\ldots$ Li[88](2)

3. $\int_0^{\ln 2} \dfrac{x\,dx}{e^x + 2e^{-x} - 2} = \dfrac{\pi}{8} \ln 2.$ BH[104](7)

3.419

1. $\int_{-\infty}^{\infty} \dfrac{x\,dx}{(\beta + e^x)(1 + e^{-x})} = \dfrac{(\ln \beta)^2}{2(\beta - 1)}$ $[|\arg \beta| < \pi]$ (cf. 4.232 2.).   BH[101](16)

2. $\int_{-\infty}^{\infty} \dfrac{x\,dx}{(\beta + e^x)(1 - e^{-x})} = \dfrac{\pi^2 + (\ln \beta)^2}{2(\beta + 1)}$ $[|\arg \beta| < \pi]$ (cf. 4.232 3.).   BH[101](17)

3. $\int_{-\infty}^{\infty} \dfrac{x^2\,dx}{(\beta + e^x)(1 - e^{-x})} = \dfrac{[\pi^2 + (\ln \beta)^2]\ln \beta}{3(\beta + 1)}$ $[|\arg \beta| < \pi]$ (cf. 4.261 4.).   BH[102](6)

4. $\int_{-\infty}^{\infty} \dfrac{x^3\,dx}{(\beta + e^x)(1 - e^{-x})} = \dfrac{\pi^2 + (\ln \beta)^2}{4(\beta + 1)}$ $[|\arg \beta| < \pi]$ (cf. 4.262 3.).   BH[102](9)

5. $\int_{-\infty}^{\infty} \dfrac{x^4\,dx}{(\beta + e^x)(1 - e^{-x})} = \dfrac{[\pi^2 + (\ln \beta)^2]^2}{15(\beta + 1)}[7\pi^2 + 3(\ln \beta)^2]\ln \beta$ (cf. 4.263 1.).   BH[102](10)

6. $\int_{-\infty}^{\infty} \dfrac{x^5\,dx}{(\beta + e^x)(1 - e^{-x})} = \dfrac{[\pi^2 + (\ln \beta)^2]^2}{6(\beta + 1)}[3\pi^2 + (\ln \beta)^2]^2$ (cf. 4.264 3.).   BH[102](7)

7. $\int_{-\infty}^{\infty} \dfrac{(x - \ln \beta)x\,dx}{(\beta - e^x)(1 - e^{-x})} = \dfrac{-[4\pi^2 + (\ln \beta)^2]\ln \beta}{6(\beta - 1)}$ $[|\arg \beta| < \pi]$ (cf. 4.257 4.).   BH[102](7)

3.421

1. $\int_0^\infty (e^{-\nu x} - 1)^n (e^{-\rho x} - 1)^m e^{-\mu x} \dfrac{dx}{x^2} = \sum_{k=0}^n (-1)^k \binom{n}{k} \sum_{l=0}^m (-1)^l \binom{m}{l} \times$
$\times \{(m-l)\rho + (n-k)\nu + \mu\} \ln\,[(m-l)\rho + (n-k)\nu + \mu]$
$[\operatorname{Re} \nu > 0,\ \operatorname{Re} \mu > 0,\ \operatorname{Re} \rho > 0].$ BH[89](17)

2. $\int_0^\infty (1 - e^{-\nu x})^n (1 - e^{-\rho x}) e^{-x} \dfrac{dx}{x^3} = \dfrac{1}{2} \sum_{k=0}^n (-1)^k \binom{n}{k} (\rho + k\nu + 1)^2 \times$
$\times \ln(\rho + k\nu + 1) + \dfrac{1}{2} \sum_{k=1}^n (-1)^{k-1} \binom{n}{k} (k\nu + 1)^2 \ln(k\nu + 1)$
$[n \geqslant 2,\ \operatorname{Re} \nu > 0,\ \operatorname{Re} \rho > 0].$ BH[89](31)

3. $\int_{-\infty}^{\infty} \dfrac{xe^{-\mu x}\,dx}{(\beta + e^{-x})(\gamma + e^{-x})} = \dfrac{\pi(\beta^{\mu-1}\ln \beta - \gamma^{\mu-1}\ln \gamma)}{(\beta - \gamma)\sin \mu\pi} + \dfrac{\pi^2(\beta^{\mu-1} - \gamma^{\mu-1})\cos \mu\pi}{(\gamma - \beta)\sin^2 \mu\pi}$
$[|\arg \beta| < \pi,\ |\arg \gamma| < \pi,\ \beta \neq \gamma,\ 0 < \operatorname{Re} \mu < 2].$ Ba IV 120(19)

4. $\int_0^\infty (e^{-px} - e^{-qx})(e^{-rx} - e^{-sx}) e^{-x} \dfrac{dx}{x} = \ln \dfrac{(p+s+1)(q+r+1)}{(p+r+1)(q+s+1)}$ 3.421

$[p+s > -1, \; p+r > -1, \; q > p]$ (cf. 4.267 24.). BH[89](11)

5. $\int_0^\infty (1 - e^{-px})(1 - e^{-qx})(1 - e^{-rx}) e^{-x} \dfrac{dx}{x^2} = (p+q+1)\ln(p+q+1) +$

$+ (p+r+1)\ln(p+r+1) + (q+r+1)\ln(q+r+1) - (p+1)\ln(p+1) -$

$- (q+1)\ln(q+1) - (r+1)\ln(r+1) - (p+q+r)\ln(p+q+r)$

$[p > 0, \; q > 0, \; r > 0]$ (cf. 4.268 3.). BH[89](14)

$\int_{-\infty}^\infty \dfrac{x(x-a)\, e^{\mu x}\, dx}{(\beta - e^x)(1 - e^{-x})} = \dfrac{-\pi^2}{e^a - 1} \operatorname{cosec}^2 \mu\pi \left[(e^{a\mu} + 1)\ln\mu - 2\pi \operatorname{ctg} \mu\pi(z^{a\mu} - 1)\right]$ 3.422

$[a > 0, \; |\arg \beta| < \pi, \; |\operatorname{Re} \mu| < 1]$ (cf. 4.257 5.). BH[102](8)+

1. $\int_0^\infty \dfrac{x^{\nu - 1}}{(e^x - 1)^2}\, dx = \Gamma(\nu)[\zeta(\nu - 1) - \zeta(\nu)]$ $[\operatorname{Re} \nu > 2]$. Ba IV 313(10) 3.423

2. $\int_0^\infty \dfrac{x^{\nu - 1} e^{-\mu x}}{(e^x - 1)^2}\, dx = \Gamma(\nu)[\zeta(\nu - 1, \mu + 1) - (\mu + 1)\zeta(\nu, \mu + 1)]$

$[\operatorname{Re} \mu > -2, \; \operatorname{Re} \nu > 2]$. Ba IV 313(11)

3. $\int_0^\infty \dfrac{x^q e^{-px}\, dx}{(1 - ae^{-px})^2} = \dfrac{\Gamma(q+1)}{ap^{q+1}} \sum_{k=1}^\infty \dfrac{a^k}{k^q}$ $[a < 1, \; q > -1, \; p > 0]$. BH[85](13)

4. $\int_0^\infty \dfrac{x^{\nu-1} e^{-\mu x}}{(1 - \beta e^{-x})^2}\, dx = \Gamma(\nu)[{}_1F_1(\beta;\, \nu - 1;\, \mu - 1) - (\mu - 1)\, {}_1F_1(\beta;\, \nu;\, \mu - 1)]$

$[\operatorname{Re} \nu > 0, \; \operatorname{Re} \mu > 0, \; |\arg(1 - \beta)| < \pi]$. Ba IV 313(12)

5. $\int_{-\infty}^\infty \dfrac{xe^x\, dx}{(\beta + e^x)^2} = \dfrac{1}{\beta} \ln \beta$ $[|\arg \beta| < \pi]$ (cf. 4.231 3.). BH[101](10)

1. $\int_0^\infty \dfrac{(1 + a)e^x - a}{(1 - e^x)^2}\, e^{-ax} x^n\, dx = n!\, \zeta(n, a)$. BH[85](15) 3.424

2. $\int_0^\infty \dfrac{(1 + a)e^x + a}{(1 + e^x)^2}\, e^{-ax} x^n\, dx = n! \sum_{k=1}^\infty \dfrac{(-1)^k}{(a+k)^n}$. BH[85](14)

3. $\int_{-\infty}^\infty \dfrac{a^2 e^x + b^2 e^{-x}}{(a^2 e^x - b^2 e^{-x})^2}\, x^2\, dx = \dfrac{\pi^2}{2ab}$ $[ab > 0]$. BH[102](3)+

3.424

4. $$\int_{-\infty}^{\infty} \frac{a^2 e^x - b^2 e^{-x}}{(a^2 e^x + b^2 e^{-x})^2} x^2 \, dx = \frac{\pi}{ab} \ln \frac{b}{a} \quad [ab > 0].$$ BH[102](1)

5. $$\int_{0}^{\infty} \frac{e^x - e^{-x} + 2}{(e^x - 1)^2} x^2 \, dx = \frac{2}{3} \pi^2 - 2.$$ BH[85](7)

3.425

1. $$\int_{-\infty}^{\infty} \frac{x e^x \, dx}{(a^2 + b^2 e^{2x})^n} = \frac{\sqrt{\pi}\, \Gamma\left(n - \frac{1}{2}\right)}{4 a^{2n-1} b \Gamma(n)} \left[2 \ln \frac{a}{2b} - C - \psi\left(n - \frac{1}{2}\right) \right]$$

$[ab > 0, n > 0]$ (cf. 4.231 5.). BH[101](13), Li[101](13)

2. $$\int_{-\infty}^{\infty} \frac{(a^2 e^x - e^{-x}) x^2 \, dx}{(a^2 e^x + e^{-x})^{p+1}} = -\frac{1}{a^{p+1}} B\left(\frac{p}{2}, \frac{p}{2}\right) \ln a \quad [a > 0, p > 0].$$ BH[102](5)

3.426

1. $$\int_{-\infty}^{\infty} \frac{(e^x - a e^{-x}) x^2 \, dx}{(a + e^x)^2 (1 + e^{-x})^2} = \frac{(\ln a)^2}{a - 1}.$$ BH[102](12)

2. $$\int_{-\infty}^{\infty} \frac{(e^x - a e^{-x}) x^2 \, dx}{(a + e^x)^2 (1 - e^{-x})^2} = \frac{\pi^2 + (\ln a)^2}{a + 1}.$$ BH[102](13)

3.427

1. $$\int_{0}^{\infty} \left(\frac{e^{-x}}{x} + \frac{e^{-\mu x}}{e^{-x} - 1} \right) dx = \psi(\mu) \quad [\operatorname{Re} \mu > 0] \quad (\text{cf. } 4.281 \, 4.).$$ WW 247

2. $$\int_{0}^{\infty} \left(\frac{1}{1 - e^{-x}} - \frac{1}{x} \right) e^{-x} \, dx = C \quad (\text{cf. } 4.281 \, 1.).$$ BH[94](1)

3. $$\int_{0}^{\infty} \left(\frac{1}{2} - \frac{1}{1 + e^{-x}} \right) \frac{e^{-2x}}{x} \, dx = \frac{1}{2} \ln \frac{\pi}{4}.$$ BH[94](5)

4. $$\int_{0}^{\infty} \left(\frac{1}{2} - \frac{1}{x} + \frac{1}{e^x - 1} \right) \frac{e^{-\mu x}}{x} \, dx = \ln \Gamma(\mu) - \left(\mu - \frac{1}{2} \right) \ln \mu + \mu - \frac{1}{2} \ln(2\pi)$$

$[\operatorname{Re} \mu > 0].$ WW 249

5. $$\int_{0}^{\infty} \left(\frac{1}{2} e^{-2x} - \frac{1}{e^x + 1} \right) \frac{dx}{x} = -\frac{1}{2} \ln \pi.$$ BH[94](6)

6. $$\int_{0}^{\infty} \left(\frac{e^{\mu x} - 1}{1 - e^{-x}} - \mu \right) \frac{e^{-x}}{x} \, dx = -\ln \Gamma(\mu) - \ln \sin(\pi \mu) + \ln \pi \quad [\operatorname{Re} \mu < 1].$$ Ba I 21(6)

7. $$\int_{0}^{\infty} \left(\frac{e^{-\nu x}}{1 - e^{-x}} - \frac{e^{-\mu x}}{x} \right) dx = \ln \mu - \psi(\nu) \quad (\text{cf. } 4.281 \, 5.).$$ BH[94](3)

8. $\displaystyle\int_0^\infty \left(\frac{n}{x} - \frac{e^{-\mu x}}{1-e^{-x/n}}\right) e^{-x}\, dx = n\psi(n\mu + n) - n\ln n \quad [\operatorname{Re}\mu > 0].$ BH [94](4) 3.427

9. $\displaystyle\int_0^\infty \left(\mu - \frac{1-e^{-\mu x}}{1-e^{-x}}\right)\frac{e^{-x}}{x}\, dx = \ln\Gamma(\mu+1) \quad [\operatorname{Re}\mu > -1].$ WW 249

10. $\displaystyle\int_0^\infty \left(ve^{-x} - \frac{e^{-\mu x} - e^{-(\mu+v)x}}{e^x - 1}\right)\frac{dx}{x} = \ln\frac{\Gamma(\mu+v+1)}{\Gamma(\mu+1)} \quad [\operatorname{Re}\mu > -1,\ \operatorname{Re} v > 0]$

(cf. 4.267 33.). BH [94](8)

11. $\displaystyle\int_0^\infty [(1-e^x)^{-1} + x^{-1} - 1]e^{-xz}\, dx = \psi(z) - \ln z \quad [\operatorname{Re} z > 0].$ Ba I 18(24)

1. $\displaystyle\int_0^\infty \left(ve^{-\mu x} - \frac{1}{\mu}e^{-x} - \frac{1}{\mu}\cdot\frac{e^{-x} - e^{-\mu v x}}{1-e^{-x}}\right)\frac{dx}{x} = \frac{1}{\mu}\ln\Gamma(\mu v) - v\ln\mu$ 3.428

$[\operatorname{Re}\mu > 0,\ \operatorname{Re} v > 0].$ BH [94](18)

2. $\displaystyle\int_0^\infty \left(\frac{n-1}{2} + \frac{n-1}{1-e^{-x}} + \frac{e^{(1-\mu)x}}{1-e^{x/n}} + \frac{e^{-n\mu x}}{1-e^{-x}}\right) e^{-x}\frac{dx}{x} =$

$\displaystyle = \frac{n-1}{2}\ln 2\pi - \left(n\mu + \frac{1}{2}\right)\ln n \quad [\operatorname{Re}\mu > 0].$ BH [94](14)

3. $\displaystyle\int_0^\infty \left(n\mu - \frac{n-1}{2} - \frac{n}{1-e^{-x}} - \frac{e^{(1-\mu)x}}{1-e^{\frac{x}{n}}}\right)\frac{e^{-x}}{x}\, dx = \sum_{k=0}^{n-1}\ln\Gamma\left(\mu - \frac{k}{n} + 1\right)$

$[\operatorname{Re}\mu > 0].$ BH [94](13)

4. $\displaystyle\int_0^\infty \left(\frac{e^{-vx}}{1-e^x} - \frac{e^{-\mu v x}}{1-e^{\mu x}} - \frac{e^x}{1-e^x} + \frac{e^{\mu x}}{1-e^{\mu x}}\right)\frac{dx}{x} = v\ln\mu$

$[\operatorname{Re}\mu > 0,\ \operatorname{Re} v > 0].$ Li [94](15)

5. $\displaystyle\int_0^\infty \left[\frac{1}{e^x-1} - \frac{\mu e^{-\mu x}}{1-e^{-\mu x}} + \left(a\mu - \frac{\mu+1}{2}\right)e^{-\mu x} + (1-a\mu)e^{-x}\right]\frac{dx}{x} =$

$\displaystyle = \frac{\mu-1}{2}\ln(2\pi) + \left(\frac{1}{2} - a\mu\right)\ln\mu \quad [\operatorname{Re}\mu > 0].$ BH [94](16)

6. $\displaystyle\int_0^\infty \left[\frac{e^{-vx}}{1-e^{-x}} - \frac{e^{-\mu v x}}{1-e^{-\mu x}} - \frac{(\mu-1)e^{-\mu x}}{1-e^{-\mu x}} - \frac{\mu-1}{2}e^{-\mu x}\right]\frac{dx}{x} =$

$\displaystyle = \frac{\mu-1}{2}\ln(2\pi) + \left(\frac{1}{2} - \mu v\right)\ln\mu$

$[\operatorname{Re}\mu > 0,\ \operatorname{Re} v > 0]$ (cf. 4.267 37.). BH [94](17)

3.428 7. $\int_0^\infty \left[1 - e^{-x} - \frac{(1-e^{-\nu x})(1-e^{-\mu x})}{1-e^{-x}}\right] \frac{dx}{x} = \ln B(\mu, \nu)$

[Re $\mu > 0$, Re $\nu > 0$] (cf. 4.267 35.). BH[94](12)

3.429 $\int_0^\infty [e^{-x} - (1+x)^{-\mu}] \frac{dx}{x} = \psi(\mu)$ [Re $\mu > 0$]. Ba I 17(20), NG 184(7)

3.431 1. $\int_0^\infty \left(e^{-\mu x} - 1 + \mu x - \frac{1}{2}\mu^2 x^2\right) x^{\nu-1} dx = \frac{-1}{\nu(\nu+1)(\nu+2)\mu^\nu} \Gamma(\nu+3)$

[Re $\mu > 0$, $-2 > $ Re $\nu > -3$]. Li[90](5)

2. $\int_0^\infty \left[x^{-1} - \frac{1}{2}x^{-2}(x+2)(1-e^{-x})\right] e^{-px} dx = -1 + \left(p + \frac{1}{2}\right)\ln\left(1 + \frac{1}{p}\right)$

[Re $p > 0$]. Ba IV 144(6)

3.432 1. $\int_0^\infty x^{\nu-1} e^{-mx}(e^{-x} - 1)^n dx = \Gamma(\nu) \sum_{k=0}^{n} (-1)^k \binom{n}{k} \frac{1}{(n+m-k)^\nu}$

[Re $\nu > 0$]. Li[90](10)

2. $\int_0^\infty [x^{\nu-1} e^{-x} - e^{-\mu x}(1-e^{-x})^{\nu-1}] dx = \Gamma(\nu) - \frac{\Gamma(\mu)}{\Gamma(\mu+\nu)}$

[Re $\mu > 0$, Re $\nu > 0$]. Li[81](14)

3.433 $\int_0^\infty x^{p-1} \left[e^{-x} + \sum_{k=1}^{n} (-1)^k \frac{x^{k-1}}{(k-1)!}\right] dx = \Gamma(p)$ $[-n < p < -n+1]$. F II 811

3.434 1. $\int_0^\infty \frac{e^{-\nu x} - e^{-\mu x}}{x^{\rho+1}} dx = \frac{\mu^\rho - \nu^\rho}{\rho} \Gamma(1-\rho)$ [Re $\mu > 0$, Re $\nu > 0$, Re $\rho < 1$].

BH[90](6)

2. $\int_0^\infty \frac{e^{-\mu x} - e^{-\nu x}}{x} dx = \ln \frac{\nu}{\mu}$ [Re $\mu > 0$, Re $\nu > 0$]. F II 643

3.435 1. $\int_0^\infty \{(x+1)e^{-x} - e^{-\frac{x}{2}}\} \frac{dx}{x} = 1 - \ln 2$. Li[89](19)

2. $\int_0^\infty \frac{1-e^{-\mu x}}{x(x+\beta)} dx = \frac{1}{\beta}[\ln(\beta\mu) - e^{\beta\mu}\text{Ei}(-\beta\mu)]$ [$|\arg \beta| < \pi$, Re $\mu > 0$].

Ba V 217(18)

3. $\int_0^\infty \left(\frac{1}{1+x} - e^{-x}\right) \frac{dx}{x} = C$. F II 805–807

4. $\int_0^\infty \left(e^{-\mu x} - \frac{1}{1+ax}\right)\frac{dx}{x} = \ln\frac{a}{\mu} - C$ $\quad [a > 0, \; \operatorname{Re}\mu > 0]$. $\quad$ BH[92](10) $\quad$ 3.435

$\int_0^\infty \left\{\frac{e^{-npx} - e^{-nqx}}{n} - \frac{e^{-mpx} - e^{-mqx}}{m}\right\}\frac{dx}{x^2} = (q-p)\ln\frac{m}{n}$ $\quad [p > 0, q > 0]$. $\quad$ BH[89](28) $\quad$ 3.436

$\int_0^\infty \left\{pe^{-x} - \frac{1 - e^{-px}}{x}\right\}\frac{dx}{x} = p\ln p - p$ $\quad [p > 0]$. $\quad$ BH[89](24) $\quad$ 3.437

1. $\int_0^\infty \left\{\left(\frac{1}{2} + \frac{1}{x}\right)e^{-x} - \frac{1}{x}e^{-\frac{x}{2}}\right\}\frac{dx}{x} = \frac{\ln 2 - 1}{2}$. $\quad$ BH[89](19) $\quad$ 3.438

2. $\int_0^\infty \left\{\frac{p^2}{6}e^{-x} - \frac{p^2}{2x} - \frac{p}{x^2} - \frac{1 - e^{-px}}{x^3}\right\}\frac{dx}{x} = \frac{p^2}{6}\ln p - \frac{11}{36}p^3$ $\quad [p > 0]$.
$\quad$ BH[89](33)

3. $\int_0^\infty \left(e^{-x} - e^{-2x} - \frac{1}{x}e^{-2x}\right)\frac{dx}{x} = 1 - \ln 2$. $\quad$ BH[89](25)

4. $\int_0^\infty \left\{\left(p - \frac{1}{2}\right)e^{-x} + \frac{x+2}{2x}(e^{-px} - e^{-\frac{x}{2}})\right\}\frac{dx}{x} = \left(p - \frac{1}{2}\right)(\ln p - 1)$
$\quad [p > 0]$. $\quad$ BH[89](22)

$\int_0^\infty \left\{(p-q)e^{-rx} + \frac{1}{mx}(e^{-mpx} - e^{-mqx})\right\}\frac{dx}{x} =$ $\quad$ 3.439

$= p\ln p - q\ln q - (p-q)\left(1 + \ln\frac{r}{m}\right)$ $\quad [p > 0, q > 0, r > 0]$. $\quad$ Li[89](26), Li[89](27)

$\int_0^\infty \{(p-r)e^{-qx} + (r-q)e^{-px} + (q-p)e^{-rx}\}\frac{dx}{x^2} = (r-q)p\ln p +$ $\quad$ 3.441

$+ (p-r)q\ln q + (q-p)r\ln r$
$\quad [p > 0, q > 0, r > 0]$ $\quad$ (cf. 4.268 6.). $\quad$ BH[89](18)

1. $\int_0^\infty \left\{1 - \frac{x+2}{2x}(1 - e^{-x})\right\}e^{-qx}\frac{dx}{x} = -1 + \left(q + \frac{1}{2}\right)\ln\frac{q+1}{q}$ $\quad [q > 0]$. $\quad$ 3.442
$\quad$ BH[89](23)

2. $\int_0^\infty \left(\frac{e^{-x} - 1}{x} + \frac{1}{1+x}\right)\frac{dx}{x} = C - 1$. $\quad$ BH[92](16)

3. $\int_0^\infty \left(e^{-px} - \frac{1}{1+a^2x^2}\right)\frac{dx}{x} = -C + \ln\frac{a}{p}$ $\quad [p > 0]$. $\quad$ BH[92](11)

3.443

1. $\int_0^\infty \left\{ \frac{e^{-x}p^2}{2} - \frac{p}{x} + \frac{1-e^{-px}}{x^2} \right\} \frac{dx}{x} = \frac{p^2}{2} \ln p - \frac{3}{4} p^2 \quad [p > 0]$. BH[89](32)

2. $\int_0^\infty \frac{(1-e^{-px})^n e^{-qx}}{x^3} dx = \frac{1}{2} \sum_{k=2}^n (-1)^{k-1} \binom{n}{k} (q+kp)^2 \ln (q+kp)$

$[n > 2, \; q > 0, \; pn + q > 0]$ (cf. 4.268 4.). BH[89](30)

3. $\int_0^\infty (1-e^{-px})^2 e^{-qx} \frac{dx}{x^2} = (2p+q) \ln (2p+q) - 2(p+q) \ln (p+q) + q \ln q$

$[q > 0, \; 2p > -q]$ (cf. 4.268 2.). BH[89](13)

Algebraische Funktionen von der Exponentialfunktion und Potenzen
3.45   Algebraic functions of exponentials and powers

3.451

1. $\int_0^\infty x e^{-x} \sqrt{1-e^{-x}} \; dx = \frac{4}{3} \left(\frac{4}{3} - \ln 2 \right)$. BH[99](1)

2. $\int_0^\infty x e^{-x} \sqrt{1-e^{-2x}} \; dx = \frac{\pi}{4} \left(\frac{1}{2} + \ln 2 \right)$ (cf. 4.241 9.). BH[99](2)

3.452

1. $\int_0^\infty \frac{x \, dx}{\sqrt{e^x - 1}} = 2\pi \ln 2$. F II 644, 652+, BH[99](4)

2. $\int_0^\infty \frac{x^2 \, dx}{\sqrt{e^x - 1}} = 4\pi \left\{ (\ln 2)^2 + \frac{\pi^2}{12} \right\}$. BH[99](5)

3. $\int_0^\infty \frac{x e^{-x} dx}{\sqrt{e^x - 1}} = \frac{\pi}{?} [2 \ln 2 - 1]$. BH[99](6)

4. $\int_0^\infty \frac{x e^{-x} dx}{\sqrt{e^{2x} - 1}} = 1 - \ln 2$. BH[99](8)

5. $\int_0^\infty \frac{x e^{-2x} dx}{\sqrt{e^x - 1}} = \frac{3}{4} \pi \left(\ln 2 - \frac{7}{12} \right)$. BH[99](7)

3.453

1. $\int_0^\infty \frac{x e^x}{a^2 e^x - (a^2 - b^2)} \frac{dx}{\sqrt{e^x - 1}} = \frac{2\pi}{ab} \ln \left(1 + \frac{b}{a} \right) \quad [ab > 0]$ (cf. 4.298 18.). BH[99](16)

2. $\int_0^\infty \frac{x e^x \, dx}{[a^2 e^x - (a^2 + b^2)] \sqrt{e^x - 1}} = \frac{2\pi}{ab} \arctg \frac{b}{a} \quad [ab > 0]$ (cf. 4.298 19.). BH[99](17)

1. $$\int_0^\infty \frac{xe^{-2nx}dx}{\sqrt{e^{2x}+1}} = \frac{(2n-1)!!}{(2n)!!}\frac{\pi}{2}\left\{\ln 2 + \sum_{k=1}^{2n}\frac{(-1)^k}{k}\right\}.$$ Li[99](10)   3.454

2. $$\int_0^\infty \frac{xe^{-(2n-1)x}dx}{\sqrt{e^{2x}-1}} = -\frac{(2n-2)!!}{(2n-1)!!}\left\{\ln 2 + \sum_{k=1}^{2n-1}\frac{(-1)^k}{k}\right\}.$$ Li[99](9)

1. $$\int_0^\infty \frac{x^2 e^x\, dx}{\sqrt{(e^x-1)^3}} = 8\pi \ln 2.$$ BH[99](11)   3.455

2. $$\int_0^\infty \frac{x^3 e^x\, dx}{\sqrt{(e^x-1)^3}} = 24\pi\left[(\ln 2)^2 + \frac{\pi^2}{12}\right].$$ BH[99](12)

1. $$\int_0^\infty \frac{x\, dx}{\sqrt[3]{e^{3x}-1}} = \frac{\pi}{3\sqrt{3}}\left[\ln 3 + \frac{\pi}{3\sqrt{3}}\right].$$ BH[99](13)   3.456

2. $$\int_0^\infty \frac{x\, dx}{\sqrt[3]{(e^{3x}-1)^2}} = \frac{\pi}{3\sqrt{3}}\left[\ln 3 - \frac{\pi}{3\sqrt{3}}\right] \quad \text{(cf. 4.244 3.).}$$ BH[99](14)

1. $$\int_0^\infty xe^{-x}(1-e^{-2x})^{n-\frac{1}{2}}\, dx = \frac{(2n-1)!!}{4\cdot(2n)!!}\pi[C + \psi(n+1) + 2\ln 2]$$ 3.457

(cf. 4.241 5.).   BH[99](3)

2. $$\int_{-\infty}^\infty \frac{xe^x\, dx}{(a+e^x)^{n+\frac{3}{2}}} = \frac{2}{(2n+1)a^{n+\frac{1}{2}}}[\ln(4a) - 3C - 2\psi(2n) - \psi(n)].$$ BH[101](12)

3. $$\int_{-\infty}^\infty \frac{x\, dx}{(a^2 e^x + e^{-x})^\mu} = \frac{-1}{2a^\mu} B\left(\frac{\mu}{2},\frac{\mu}{2}\right)\ln a \quad [a>0,\ \mathrm{Re}\,\mu > 0].$$ BH[101](14)

1. $$\int_0^{\ln 2} xe^x(e^x-1)^{p-1}dx = \frac{1}{p}\left[\ln 2 + \sum_{k=0}^\infty \frac{(-1)^{k-1}}{p+k+1}\right].$$ BH[104](4)   3.458

2. $$\int_{-\infty}^\infty \frac{xe^x\, dx}{(a+e^x)^{v+1}} = \frac{1}{va^v}[\ln a - C - \psi(v)] \quad [a>0];$$

$$= \frac{1}{va^v}\left[\ln a - \sum_{k=1}^{v-1}\frac{1}{k}\right]\left[\begin{array}{l}v \text{ ganzzahlig}\\ \text{an integer.}\end{array}\right].$$ BH[101](11)

Exponentialfunktionen mit komplizierteren Argumenten und Potenzen
Exponentials of compound arguments and powers   3.46–3.48

1. $$\int_u^\infty \frac{e^{-p^2 x^2}}{x^{2n}} dx = \frac{(-1)^n 2^{n-1} p^{2n-1}\sqrt{\pi}}{(2n-1)!!}[1 - \Phi(pu)] +$$ 3.461

$$+ \frac{e^{-p^2 u^2}}{2u^{2n-1}}\sum_{k=0}^{n-1}\frac{(-1)^k 2^{k+1}(pu)^{2k}}{(2n-1)(2n-3)\ldots(2n-2k-1)} \quad [p>0].$$ NI 21(4)

3.461

2. $$\int_0^\infty x^{2n} e^{-px^2} dx = \frac{(2n-1)!!}{2(2p)^n} \sqrt{\frac{\pi}{p}} \quad [p > 0].$$
F II 749

3. $$\int_0^\infty x^{2n+1} e^{-px^2} dx = \frac{n!}{2p^{n+1}} \quad [p > 0].$$
BH[81](7)

4. $$\int_{-\infty}^\infty (x+ai)^{2n} e^{-x^2} dx = \frac{(2n-1)!!}{2^n} \sqrt{\pi} \sum_{k=0}^{n} (-1)^k \frac{(2a)^{2k} n!}{(2k)!(n-k)!}.$$
BH[100](12)

5. $$\int_u^\infty e^{-\mu x^2} \frac{dx}{x^2} = \frac{1}{u} e^{-\mu u^2} - \sqrt{\mu \pi} \left[1 - \Phi(u\sqrt{\mu})\right] \quad \left[|\arg \mu| < \frac{\pi}{4},\ u > 0\right].$$
Ba IV 135(19)+

3.462

1. $$\int_0^\infty x^{\nu-1} e^{-\beta x^2 - \gamma x} dx = (2\beta)^{-\frac{\nu}{2}} \Gamma(\nu) \exp\left(\frac{\gamma^2}{8\beta}\right) D_{-\nu}\left(\frac{\gamma}{\sqrt{2\beta}}\right)$$

$[\operatorname{Re} \beta > 0, \ \operatorname{Re} \nu > 0].$
Ba II 119(3)+, Ba IV 313(13)

2. $$\int_{-\infty}^\infty x^n e^{-px^2 + 2qx} dx = \frac{1}{2^{n-1} p} \sqrt{\frac{\pi}{p}} \frac{d^{n-1}}{dq^{n-1}} (q e^{\frac{q^2}{p}}) \quad [p > 0];$$
BH[100](8)

$$= n!\, e^{\frac{q^2}{p}} \sqrt{\frac{\pi}{p}} \left(\frac{q}{p}\right)^n \sum_{k=0}^{E\left(\frac{n}{2}\right)} \frac{1}{(n-2k)!(k)!} \left(\frac{p}{4q^2}\right)^k \quad [p > 0].$$
Li[100](8)

3. $$\int_{-\infty}^\infty (ix)^\nu e^{-\beta^2 x^2 - iqx} dx = 2^{-\frac{\nu}{2}} \sqrt{\pi} \beta^{-\nu-1} \exp\left(-\frac{q^2}{8\beta^2}\right) D_\nu\left(\frac{q}{\beta\sqrt{2}}\right)$$

$\left[\operatorname{Re} \beta > 0,\ \operatorname{Re} \nu > -1,\ \arg ix = \frac{\pi}{2} \operatorname{sign} x\right].$
Ba IV 121(23)

4. $$\int_{-\infty}^\infty x^n \exp[-(x-\beta)^2] dx = (2i)^{-n} \sqrt{\pi}\, H_n(i\beta).$$
Ba II 195(31)

5. $$\int_0^\infty x e^{-\mu x^2 - 2\nu x} dx = \frac{1}{2\mu} - \frac{\nu}{2\mu} \sqrt{\frac{\pi}{\mu}} e^{\frac{\nu^2}{\mu}} \left[1 - \Phi\left(\frac{\nu}{\sqrt{\mu}}\right)\right]$$

$\left[|\arg \nu| < \frac{\pi}{2},\ \operatorname{Re} \mu > 0\right].$
Ba IV 146(31)+

6. $$\int_{-\infty}^\infty x e^{-px^2 + 2qx} dx = \frac{q}{p} \sqrt{\frac{\pi}{p}} \exp\left(\frac{q^2}{p}\right) \quad [\operatorname{Re} p > 0].$$
BH[100](7)

Die Exponentialfunktion/The Exponential Function

7. $\int_0^\infty x^2 e^{-\mu x^2 - 2\nu x}\,dx = -\frac{\nu}{2\mu^2} + \sqrt{\frac{\pi}{\mu^5}}\,\frac{2\nu^2+\mu}{4}\,e^{\frac{\nu^2}{\mu}}\left[1-\Phi\left(\frac{\nu}{\sqrt{\mu}}\right)\right]$ 3.462

$\left[|\arg \nu|<\frac{\pi}{2},\ \operatorname{Re}\mu>0\right].$ Ba IV 146(32)

8. $\int_{-\infty}^\infty x^2 e^{-\mu x^2 + 2\nu x}\,dx = \frac{1}{2\mu}\sqrt{\frac{\pi}{\mu}}\left(1+2\frac{\nu^2}{\mu}\right)e^{\frac{\nu^2}{\mu}}$

$[|\arg \nu|<\pi,\ \operatorname{Re}\mu>0].$ BH[100](8)+

$\int_0^\infty (e^{-x^2}-e^{-x})\frac{dx}{x} = \frac{1}{2}C.$ BH[89](5) 3.463

$\int_0^\infty (e^{-\mu x^2}-e^{-\nu x^2})\frac{dx}{x^2} = \sqrt{\pi}(\sqrt{\nu}-\sqrt{\mu})\quad [\operatorname{Re}\mu>0,\ \operatorname{Re}\nu>0].$ F II 653 3.464

$\int_0^\infty (1+2\beta x^2)e^{-\mu x^2}\,dx = \frac{\mu+\beta}{2}\sqrt{\frac{\pi}{\mu^3}}\quad [\operatorname{Re}\mu>0].$ Ba IV 136(24)+ 3.465

1. $\int_0^\infty \frac{e^{-\mu^2 x^2}}{x^2+\beta^2}\,dx = [1-\Phi(\beta\mu)]\frac{\pi}{2\beta}\,e^{\beta^2\mu^2}\quad \left[\operatorname{Re}\beta>0,\ |\arg\mu|<\frac{\pi}{4}\right].$ NI 19(13) 3.466

2. $\int_0^\infty \frac{x^2 e^{-\mu^2 x^2}}{x^2+\beta^2}\,dx = \frac{\sqrt{\pi}}{2\mu} - \frac{\pi\beta}{2}e^{\mu^2\beta^2}[1-\Phi(\beta\mu)]\quad \left[\operatorname{Re}\beta>0,\ |\arg\mu|<\frac{\pi}{4}\right].$

Ba V 217(16)

3. $\int_0^1 \frac{e^{x^2}-1}{x^2}\,dx = \sum_{k=1}^\infty \frac{1}{k!(2k-1)}.$ F II 692

$\int_0^\infty \left(e^{-x^2}-\frac{1}{1+x^2}\right)\frac{dx}{x} = -\frac{1}{2}C.$ BH[92](12) 3.467

1. $\int_{u\sqrt{2}}^\infty \frac{e^{-x^2}}{\sqrt{x^2-u^2}}\frac{dx}{x} = \frac{\pi}{4u}[1-\Phi(u)]^2\quad [u>0.$ NI 33(17) 3.468

2. $\int_0^\infty \frac{xe^{-\mu x^2}\,dx}{\sqrt{a^2+x^2}} = \frac{1}{2}\sqrt{\frac{\pi}{\mu}}\,e^{a^2\mu}[1-\Phi(a\sqrt{\mu})]\quad [\operatorname{Re}\mu>0,\ a>0].$ NI 19(11)

3.469

1. $$\int_0^\infty (e^{-x^4} - e^{-x}) \frac{dx}{x} = \frac{3}{4} C.$$ BH[89](7)

2. $$\int_0^\infty (e^{-x^4} - e^{-x^2}) \frac{dx}{x} = \frac{1}{4} C.$$ BH[89](6)

3.471

1. $$\int_0^u \exp\left(-\frac{\beta}{x}\right) \frac{dx}{x^2} = \frac{1}{\beta} \exp\left(-\frac{\beta}{u}\right).$$ Ba V 188(22)

2. $$\int_0^u x^{\nu-1}(u-x)^{\mu-1} e^{-\frac{\beta}{x}} dx = \beta^{\frac{\nu-1}{2}} u^{\frac{2\mu+\nu-1}{2}} \exp\left(-\frac{\beta}{2u}\right) \times$$

$$\times \Gamma(\mu) W_{\frac{1-2\mu-\nu}{2}, \frac{\nu}{2}}\left(\frac{\beta}{u}\right) \quad [\operatorname{Re}\mu > 0,\ \operatorname{Re}\beta > 0,\ u > 0].$$ Ba V 187(18)

3. $$\int_0^u x^{-\mu-1}(u-x)^{\mu-1} e^{-\frac{\beta}{x}} dx = \beta^{-\mu} u^{\mu-1} \Gamma(\mu) \exp\left(-\frac{\beta}{u}\right)$$

$$[\operatorname{Re}\mu > 0,\ u > 0].$$ Ba V 187(16)

4. $$\int_0^u x^{-2\mu}(u-x)^{\mu-1} e^{-\frac{\beta}{x}} dx = \frac{1}{\sqrt{\pi u}} \beta^{\frac{1}{2}-\mu} e^{-\frac{\beta}{2u}} \Gamma(\mu) K_{\mu-\frac{1}{2}}\left(\frac{\beta}{2u}\right)$$

$$[u > 0,\ \operatorname{Re}\beta > 0,\ \operatorname{Re}\mu > 0].$$ Ba V 187(17)

5. $$\int_u^\infty x^{\nu-1}(x-u)^{\mu-1} e^{\frac{\beta}{x}} dx =$$

$$= B(1-\mu-\nu,\ \mu)\, u^{\mu+\nu-1}\, {}_1F_1\left(1-\mu-\nu;\ 1-\nu;\ \frac{\beta}{u}\right)$$

$$[0 < \operatorname{Re}\mu < \operatorname{Re}(1-\nu),\ u > 0].$$ Ba V 203(15)

6. $$\int_u^\infty x^{-2\mu}(x-u)^{\mu-1} e^{\frac{\beta}{x}} dx = \sqrt{\frac{\pi}{u}}\, \beta^{\frac{1}{2}-\mu}\, \Gamma(\mu) \exp\left(\frac{\beta}{2u}\right) I_{\mu-\frac{1}{2}}\left(\frac{\beta}{2u}\right)$$

$$[\operatorname{Re}\mu > 0,\ u > 0].$$ Ba V 202(14)

7. $$\int_0^\infty x^{\nu-1}(x+\gamma)^{\mu-1} e^{-\frac{\beta}{x}} dx =$$

$$= \beta^{\frac{\nu-1}{2}} \gamma^{\frac{\nu-1}{2}+\mu}\, \Gamma(1-\mu-\nu) e^{\frac{\beta}{2\gamma}}\, W_{\frac{\nu-1}{2}+\mu,\ -\frac{\nu}{2}}\left(\frac{\beta}{\gamma}\right)$$

$$[|\arg\gamma| < \pi,\ \operatorname{Re}(1-\mu) > \operatorname{Re}\nu > 0].$$ Ba V 234(13)+

3.471

8. $\int_0^u x^{-2\mu}(u^2-x^2)^{\mu-1} e^{-\frac{\beta}{x}} dx =$

$= \frac{1}{\sqrt{\pi}} \left(\frac{2}{\beta}\right)^{\mu-\frac{1}{2}} u^{\mu-\frac{3}{2}} \Gamma(\mu) K_{\mu-\frac{1}{2}}\left(\frac{\beta}{u}\right)$ [Re $\beta > 0$, $u > 0$, Re $\mu > 0$].

Ba V 188(23)+

9. $\int_0^\infty x^{\nu-1} e^{-\frac{\beta}{x}-\gamma x} dx = 2\left(\frac{\beta}{\gamma}\right)^{\frac{\nu}{2}} K_\nu(2\sqrt{\beta\gamma})$ [Re $\beta > 0$, Re $\gamma > 0$].

Ba II 82(23)+, Ba IV 146(29)

10. $\int_0^\infty x^{\nu-1} \exp\left[\frac{i\mu}{2}\left(x-\frac{\beta^2}{x}\right)\right] dx = 2\beta^\nu e^{\frac{i\nu\pi}{2}} K_{-\nu}(\beta\mu)$

[Im $\mu > 0$, Im $(\beta^2\mu) < 0$].   Ba II 82(24)

11. $\int_0^\infty x^{\nu-1} \exp\left[\frac{i\mu}{2}\left(x+\frac{\beta^2}{x}\right)\right] dx = i\pi\beta^\nu e^{-\frac{i\nu\pi}{2}} H_{-\nu}^{(1)}(\beta\mu)$

[Im $\mu > 0$, Im $(\beta^2\mu) > 0$].   Ba II 21(33)

12. $\int_0^\infty x^{\nu-1} \exp\left(-x-\frac{\mu^2}{4x}\right) dx = 2\left(\frac{\mu}{2}\right)^\nu K_{-\nu}(\mu)$

$\left[|\arg \mu| < \frac{\pi}{2},\ \text{Re}\,\mu^2 > 0\right]$.   W 183(15)

13. $\int_0^\infty \frac{x^{\nu-1} e^{-\frac{\beta}{x}}}{x+\gamma} dx = \gamma^{\nu-1} e^{\frac{\beta}{\gamma}} \Gamma(1-\nu) \Gamma\left(\nu, \frac{\beta}{\gamma}\right)$

[$|\arg \gamma| < \pi$, Re $\beta > 0$, Re $\nu < 1$].   Ba V 218(19)

14. $\int_0^1 \frac{\exp\left(1-\frac{1}{x}\right) - x^\nu}{x(1-x)} dx = \psi(\nu)$ [Re $\nu > 0$].   BH[80](7)

3.472

1. $\int_0^\infty \left(\exp\left(-\frac{a}{x^2}\right) - 1\right) e^{-\mu x^2} dx = \frac{1}{2}\sqrt{\frac{\pi}{\mu}} [\exp(-2\sqrt{a\mu}) - 1]$

[Re $\mu > 0$, Re $a > 0$].   Ba IV 146(30)

2. $\int_0^\infty x^2 \exp\left(-\frac{a}{x^2} - \mu x^2\right) dx = \frac{1}{4}\sqrt{\frac{\pi}{\mu^3}} (1+2\sqrt{a\mu}) \exp(-2\sqrt{a\mu})$

[Re $\mu > 0$, Re $a > 0$].   Ba IV 146(26)

3. $\int_0^\infty \exp\left(-\frac{a}{x^2} - \mu x^2\right) \frac{dx}{x^2} = \frac{1}{2}\sqrt{\frac{\pi}{a}} \exp(-2\sqrt{a\mu})$ [Re $\mu > 0$, $a > 0$].

Ba IV 146(28)+

3.472

4. $\int_0^\infty \exp\left[-\frac{1}{2a}\left(x^2 + \frac{1}{x^2}\right)\right] \frac{dx}{x^4} = \sqrt{\frac{a\pi}{2}}(1+a)e^{-\frac{1}{a}}$ $[a > 0]$.   BH [98](14)

3.473

$\int_0^\infty \exp(-x^n) x^{\left(m+\frac{1}{2}\right)n - 1} dx = \frac{(2m-1)!!}{2^m n}\sqrt{\pi}$.   BH [98](6)

3.474

1. $\int_0^1 \left\{\frac{n \exp(1 - x^{-n})}{1 - x^n} - \frac{x^{np}}{1-x}\right\} \frac{dx}{x} = \frac{1}{n}\sum_{k=1}^n \psi\left(p + \frac{k-1}{n}\right)$ $[p > 0]$.

BH [80](8)

2. $\int_0^1 \left\{\frac{n \exp(1 - x^{-n})}{1 - x^n} - \frac{\exp\left(1 - \frac{1}{x}\right)}{1 - x}\right\} \frac{dx}{x} = -\ln n$.   BH [80](9)

3.475

1. $\int_0^\infty \left\{\exp(-x^{2^n}) - \frac{1}{1 + x^{2^{n+1}}}\right\} \frac{dx}{x} = -\frac{1}{2^n} C$.   BH [92](14)

2. $\int_0^\infty \left\{\exp(-x^{2^n}) - \frac{1}{1 + x^2}\right\} \frac{dx}{x} = -2^{-n} C$.   BH [92](13)

3. $\int_0^\infty \{\exp(-x^{2^n}) - e^{-x}\} \frac{dx}{x} = (1 - 2^{-n}) C$.   BH [89](8)

3.476

1. $\int_0^\infty [\exp(-\nu x^p) - \exp(-\mu x^p)] \frac{dx}{x} = \frac{1}{p} \ln \frac{\mu}{\nu}$ $[\operatorname{Re}\mu > 0, \ \operatorname{Re}\nu > 0]$.

BH [89](3)

2. $\int_0^\infty [\exp(-x^p) - \exp(-x^q)] \frac{dx}{x} = \frac{p-q}{pq} C$ $[p > 0, q > 0]$.   BH [89](9)

3.477

1. $\int_{-\infty}^\infty \frac{\exp(-a|x|)}{x - u} dx = \frac{\operatorname{sign} u}{\pi}[\exp(a|u|) \operatorname{Ei}(-a|u|) -$

$- \exp(-a|u|) \overline{\operatorname{Ei}}(a|u|)]$ $[a > 0]$.   Ba V 251(35)

2. $\int_{-\infty}^\infty \frac{\operatorname{sign} x \exp(-a|x|)}{x - u} dx = -[\exp(a|u|) \operatorname{Ei}(-a|u|) -$

$- \exp(-a|u|) \overline{\operatorname{Ei}}(a|u|)]$ $[a > 0]$.   Ba V 251(36)

1. $\int_0^\infty x^{\nu-1} \exp(-\mu x^p) \, dx = \frac{1}{p} \mu^{-\frac{\nu}{p}} \Gamma\left(\frac{\nu}{p}\right)$ [Re $\mu > 0$, Re $\nu > 0$, $p > 0$].   3.478

BH[81](8)+, Ba IV 313(15, 16)

2. $\int_0^\infty x^{\nu-1} [1 - \exp(-\mu x^p)] \, dx = -\frac{1}{|p|} \mu^{-\frac{\nu}{p}} \Gamma\left(\frac{\nu}{p}\right)$ Ba IV 313(18, 19)

[Re $\mu > 0$ und $-p <$ Re $\nu < 0$ für $p>0$, aber $0 <$ Re $\nu < -p$ für $p < 0$]. | [Re $\mu > 0$ and $-p <$ Re $\nu < 0$ for $p > 0$; $0 <$ Re $\nu < -p$ for $p < 0$].

3. $\int_0^u x^{\nu-1}(u-x)^{\mu-1} \exp(\beta x^n) dx = B(\mu, \nu) u^{\mu+\nu-1} \times$

$\times {}_nF_n\left(\frac{\nu}{n}, \frac{\nu+1}{n}, \ldots, \frac{\nu+n-1}{n}; \frac{\mu+\nu}{n}, \frac{\mu+\nu+1}{n}, \ldots, \frac{\mu+\nu+n-1}{n}; \beta u^n\right)$

[Re $\mu > 0$, Re $\nu > 0$, $n = 2, 3, \ldots$].   Ba V 187(15)

4. $\int_0^\infty x^{\nu-1} \exp(-\beta x^p - \gamma x^{-p}) dx = \frac{2}{p} \left(\frac{\gamma}{\beta}\right)^{\frac{\nu}{2p}} K_{\frac{\nu}{p}}(2\sqrt{\beta\gamma})$

[Re $\beta > 0$, Re $\gamma > 0$].   Ba IV 313(17)

1. $\int_0^\infty \frac{x^{\nu-1} \exp(-\beta\sqrt{1+x})}{\sqrt{1+x}} dx = \frac{2}{\sqrt{\pi}} \left(\frac{\beta}{2}\right)^{\frac{1}{2}-\nu} \Gamma(\nu) K_{\frac{1}{2}-\nu}(\beta)$ 3.479

[Re $\beta > 0$, Re $\nu > 0$].   Ba IV 313(14)

2. $\int_0^\infty \frac{x^{\nu-1} \exp(i\mu\sqrt{1+x^2})}{\sqrt{1+x^2}} dx = i\frac{\sqrt{\pi}}{2} \left(\frac{\mu}{2}\right)^{\frac{1-\nu}{2}} \Gamma\left(\frac{\nu}{2}\right) H_{\frac{1-\nu}{2}}^{(1)}(\mu)$

[Im $\mu > 0$, Re $\nu > 0$].   Ba II 83(30)

1. $\int_{-\infty}^\infty xe^x \exp(-\mu e^x) \, dx = -\frac{1}{\mu}(C + \ln \mu)$ [Re $\mu > 0$].   BH[100](13)   3.481

2. $\int_{-\infty}^\infty xe^x \exp(-\mu e^{2x}) dx = -\frac{1}{4}[C + \ln(4\mu)]\sqrt{\frac{\pi}{\mu}}$ [Re $\mu > 0$].   BH[100](14)

1. $\int_0^\infty \exp(nx - \beta \operatorname{sh} x) \, dx = \frac{1}{2}[S_n(\beta) - \pi E_n(\beta) + \pi N_n(\beta)]$ [Re $\beta > 0$].   3.482

Ba IV 168(11)

2. $\int_0^\infty \exp(-nx - \beta \operatorname{sh} x) dx = (-1)^{n+1} \frac{1}{2}[S_n(\beta) + \pi E_n(\beta) + \pi N_n(\beta)]$

[Re $\beta > 0$].   Ba IV 168(12)

3.482

3. $\int_0^\infty \exp(-vx - \beta \operatorname{sh} x)\, dx = \dfrac{\pi}{\sin v\pi}[\mathbf{J}_v(\beta) - J_v(\beta)] \quad [\operatorname{Re}\beta > 0].$

Ba IV 168(13)

3.483

$\int_{-\infty}^\infty \dfrac{\exp(v \operatorname{Arsh} x - iax)}{\sqrt{1+x^2}}\, dx = \begin{cases} -2\exp\left(-\dfrac{iv\pi}{2}\right) K_v(a) & \text{falls if } a > 0, \\ -2\exp\left(\dfrac{iv\pi}{2}\right) K_v(a) & \text{falls if } a < 0 \end{cases}$

$[|\operatorname{Re} v| < 1].$ Ba IV 122(32)

3.484

$\int_0^\infty \left[\left(1 + \dfrac{a}{qx}\right)^{qx} - \left(1 + \dfrac{a}{px}\right)^{px}\right] \dfrac{dx}{x} = (e^a - 1) \ln \dfrac{q}{p} \quad [p > 0,\ q > 0].$

BH=89](34)

3.485

$\int_0^{\pi/2} \exp(-\operatorname{tg}^2 x)\, dx = \dfrac{\pi e}{2}[1 - \Phi(1)].$

3.486

$\int_0^1 x^{-x}\, dx = \int_0^1 e^{-x \ln x}\, dx = \sum_{k=1}^\infty k^{-k}.$

F II 477

Hyperbelfunktionen
Hyperbolic functions

3.5

Hyperbelfunktionen
Hyperbolic Functions

3.51

3.511

1. $\int_0^\infty \dfrac{dx}{\operatorname{ch} ax} = \dfrac{\pi}{2a} \quad [a > 0].$

2. $\int_0^\infty \dfrac{\operatorname{sh} ax}{\operatorname{sh} bx}\, dx = \dfrac{\pi}{2b} \operatorname{tg} \dfrac{a\pi}{2b} \quad [b > |a|].$ BH[27](10)+

3. $\int_0^\infty \dfrac{\operatorname{sh} ax}{\operatorname{ch} bx}\, dx = \dfrac{\pi}{2b} \sec \dfrac{a\pi}{2b} - \dfrac{1}{b}\beta\left(\dfrac{a+b}{2b}\right) \quad [b > |a|].$ GH[351](3b)

4. $\int_0^\infty \dfrac{\operatorname{ch} ax}{\operatorname{ch} bx}\, dx = \dfrac{\pi}{2b} \sec \dfrac{a\pi}{2b} \quad [b > |a|].$ BH[4](14)+

5. $\int_0^\infty \dfrac{\operatorname{sh} ax \operatorname{ch} bx}{\operatorname{sh} cx}\, dx = \dfrac{\pi}{2c} \dfrac{\sin \dfrac{a\pi}{c}}{\cos \dfrac{a\pi}{c} + \cos \dfrac{b\pi}{c}} \quad [c > |a| + |b|].$ BH[27](11)

6. $$\int_0^\infty \frac{\operatorname{ch} ax \operatorname{ch} bx}{\operatorname{ch} cx} dx = \frac{\pi}{c} \frac{\cos \dfrac{a\pi}{2c} \cos \dfrac{b\pi}{2c}}{\cos \dfrac{a\pi}{c} + \cos \dfrac{b\pi}{c}} \quad [c > |a| + |b|]. \qquad \text{BH[27](5)+}$$

7. $$\int_0^\infty \frac{\operatorname{sh} ax \operatorname{sh} bx}{\operatorname{ch} cx} dx = \frac{\pi}{c} \frac{\sin \dfrac{a\pi}{2c} \sin \dfrac{b\pi}{2c}}{\cos \dfrac{a\pi}{c} + \cos \dfrac{b\pi}{c}} \quad [c > |a| + |b|]. \qquad \text{BH[27](6)+}$$

8. $$\int_0^\infty \frac{dx}{\operatorname{ch} x^2} = \sqrt{\pi} \sum_{k=0}^\infty \frac{(-1)^k}{\sqrt{2k+1}}. \qquad \text{BH[98](25)}$$

9. $$\int_{-\infty}^\infty \frac{\operatorname{sh}^2 ax}{\operatorname{sh}^2 x} dx = 1 - a\pi \operatorname{ctg} a\pi \quad [a^2 < 1]. \qquad \text{BH[16](3)+}$$

10. $$\int_0^\infty \frac{\operatorname{sh} ax \operatorname{sh} bx}{\operatorname{ch}^2 bx} dx = \frac{a\pi}{2b^2} \sec \frac{a\pi}{2b} \quad [b > |a|]. \qquad \text{BH[27](16)+}$$

3.512

1. $$\int_0^\infty \frac{\operatorname{ch} 2\beta x}{\operatorname{ch}^{2\nu} ax} dx = \frac{4^{\nu-1}}{a} \operatorname{B}\left(\nu + \frac{\beta}{a}, \nu - \frac{\beta}{a}\right) \quad [\operatorname{Re} \nu > |\operatorname{Re} \beta|, a > 0]. \qquad \text{Li[27](17)+, Ba I 11(26)}$$

2. $$\int_0^\infty \frac{\operatorname{sh}^\mu x}{\operatorname{ch}^\nu x} dx = \frac{1}{2} \operatorname{B}\left(\frac{\mu+1}{2}, \frac{\nu-\mu}{2}\right) \quad [\operatorname{Re} \mu > -1, \operatorname{Re}(\mu+\nu) > 0]. \qquad \text{Ba I 11(23)}$$

3.513

1. $$\int_0^\infty \frac{dx}{a + b \operatorname{sh} x} = \frac{1}{\sqrt{a^2 + b^2}} \ln \frac{a + b + \sqrt{a^2 + b^2}}{a + b - \sqrt{a^2 + b^2}} \quad [ab \neq 0]. \qquad \text{GH[351](8)}$$

2. $$\int_0^\infty \frac{dx}{a + b \operatorname{ch} x} = \frac{2}{\sqrt{b^2 - a^2}} \operatorname{arctg} \frac{\sqrt{b^2 - a^2}}{a + b} \quad [b^2 > a^2];$$
$$= \frac{1}{\sqrt{a^2 - b^2}} \ln \frac{a + b + \sqrt{a^2 - b^2}}{a + b - \sqrt{a^2 - b^2}} \quad [b^2 < a^2]. \qquad \text{GH[351](7)}$$

3. $$\int_0^\infty \frac{dx}{a \operatorname{sh} x + b \operatorname{ch} x} = \frac{2}{\sqrt{b^2 - a^2}} \operatorname{arctg} \frac{\sqrt{b^2 - a^2}}{a + b} \quad [b^2 > a^2];$$
$$= \frac{1}{\sqrt{a^2 - b^2}} \ln \frac{a + b + \sqrt{a^2 - b^2}}{a + b - \sqrt{a^2 - b^2}} \quad [a^2 > b^2]. \qquad \text{GH[351](9)}$$

3.513 4. $\int_0^\infty \dfrac{dx}{a+b\,\text{ch}\,x+c\,\text{sh}\,x} = \dfrac{2}{\sqrt{b^2-a^2-c^2}}\left[\text{arctg}\,\dfrac{\sqrt{b^2-a^2-c^2}}{a+b+c} + \varepsilon\pi\right]$

[$b^2 > a^2 + c^2$; $\varepsilon = 0$ für $(b-a)(a+b+c) > 0$, aber $|\varepsilon| = 1$ für $(b-a)(a+b+c) < 0$; überdies ist $\varepsilon = 1$ im Fall $a < b+c$ und $\varepsilon = -1$ im Fall $a > b+c$]; | [$b^2 > a^2 + c^2$; $\varepsilon = 0$ for $(b-a)(a+b+c) > 0$ but $|\varepsilon| = 1$ for $(b-a)(a+b+c) < 0$; moreover, $\varepsilon = 1$ if $a < b+c$ and $\varepsilon = -1$ if $a > b+c$];

$$= \dfrac{1}{\sqrt{a^2-b^2+c^2}} \ln \dfrac{a+b+c+\sqrt{a^2-b^2+c^2}}{a+b+c-\sqrt{a^2-b^2+c^2}}$$

$$[b^2 < a^2 + c^2,\ a^2 \ne b^2];$$

$$= \dfrac{1}{c} \ln \dfrac{a+c}{a} \quad [a = b \ne 0,\ c \ne 0];$$

$$= \dfrac{2(a-b)}{c(a-b-c)} \quad [b^2 = a^2 + c^2,\ c(a-b-c) < 0]. \qquad \text{GH[351](6)}$$

3.514 1. $\int_0^\infty \dfrac{dx}{\text{ch}\,ax + \cos t} = \dfrac{t}{a}\,\text{cosec}\,t \quad [0 < t < \pi].$ \hfill BH[27](22)+

2. $\int_0^\infty \dfrac{\text{ch}\,ax - \cos t_1}{\text{ch}\,bx - \cos t_2}\,dx = \dfrac{\pi}{b}\,\dfrac{\sin\dfrac{a(\pi - t_2)}{b}}{\sin t_2 \sin\dfrac{a}{b}\pi} - \dfrac{\pi - t_2}{b \sin t_2}\cos t_1 \quad [b > |a|,\ 0 < t < \pi].$ \hfill BH[6](20)+

3. $\int_0^\infty \dfrac{\text{ch}\,ax\,dx}{(\text{ch}\,x + \cos t)^2} = \dfrac{\pi(-\cos t \sin at + a \sin t \cos at)}{\sin^3 t \sin a\pi} \quad [a^2 < 1,\ 0 < t < \pi].$ \hfill BH[6](18)+

4. $\int_0^\infty \dfrac{\text{sh}\,ax\,\text{sh}\,bx}{(\text{ch}\,ax + \cos t)^2}\,dx = \dfrac{b\pi}{a^2}\,\text{cosec}\,t\,\text{cosec}\,\dfrac{b\pi}{a}\sin\dfrac{bt}{a} \quad [a > |b|,\ 0 < t < \pi].$ \hfill BH[27](27)+

3.515 $\int_{-\infty}^\infty \left(1 - \dfrac{\sqrt{2}\,\text{ch}\,x}{\sqrt{\text{ch}\,2x}}\right)dx = -\ln 2.$ \hfill BH[21](12)+

3.516 1. $\int_0^\infty \dfrac{dx}{(z + \sqrt{z^2 - 1}\,\text{ch}\,x)^\mu} = \dfrac{1}{2}\int_{-\infty}^\infty \dfrac{dx}{(z + \sqrt{z^2 - 1}\,\text{ch}\,x)^\mu} = Q_{\mu-1}(z) \quad [\text{Re}\,\mu > -1].$

Bei passender Wahl des eindeutigen Zweiges des Integranden gilt diese Formel für alle Werte von z in der von -1 bis $+1$ aufgeschnittenen z-Ebene, sofern nur $\mu < 0$ ist. Ist aber $\mu > 0$, so gilt diese Formel nicht mehr in Punkten, in denen der Nenner Null wird. \hfill CH 436, WW 319 | For $\mu < 0$ the formula holds for all values of z in a plane cut from -1 to $+1$ provided a single-valued branch of the integrand is chosen; for $\mu > 0$ the formula does not hold any more at points where the denominator vanishes. \hfill CH 436, WW 319

2. $$\int_0^\infty \frac{dx}{(\beta + \sqrt{\beta^2-1}\,\mathrm{ch}\,x)^{n+1}} = Q_n(\beta).$$ Ba II 181(32) 3.516

3. $$\int_0^\infty \frac{\mathrm{ch}\,\gamma x\,dx}{(\beta + \sqrt{\beta^2-1}\,\mathrm{ch}\,x)^{\nu+1}} = \frac{e^{-i\gamma\pi}\,\Gamma(\nu-\gamma+1)Q_\nu^\gamma(\beta)}{\Gamma(\nu+1)}$$

[Re $(\nu \pm \gamma) > -1$, $\nu \ne -1, -2, -3, \ldots$]. Ba I 157(12)

4. $$\int_0^\infty \frac{\mathrm{sh}^{2\mu}x\,dx}{(\beta + \sqrt{\beta^2-1}\,\mathrm{ch}\,x)^{\nu+1}} = \frac{2^\mu e^{-i\mu\pi}\Gamma(\nu-2\mu+1)\,\Gamma\left(\mu+\frac{1}{2}\right)}{\sqrt{\pi}\,(\beta^2-1)^{\frac{\mu}{2}}\,\Gamma(\nu+1)}\,Q_{\nu-\mu}^\mu(\beta)$$

[Re$(\nu - 2\mu + 1) > 0$, Re$(\nu + 1) > 0$]. Ba I 155(2)

1. $$\int_0^\infty \frac{\mathrm{ch}\left(\gamma+\frac{1}{2}\right)x\,dx}{(\beta + \mathrm{ch}\,x)^{\nu+\frac{1}{2}}} = \sqrt{\frac{\pi}{2}}\,(\beta^2-1)^{-\frac{\nu}{2}}\,\frac{\Gamma(\nu+\gamma+1)\,\Gamma(\nu-\gamma)\,P_\gamma^{-\nu}(\beta)}{\Gamma\left(\nu+\frac{1}{2}\right)}$$ 3.517

[Re $(\nu - \gamma) > 0$, Re $(\nu + \gamma + 1) > 0$]. Ba I 156(11)

2. $$\int_0^a \frac{\mathrm{ch}\left(\gamma+\frac{1}{2}\right)x\,dx}{(\mathrm{ch}\,a - \mathrm{ch}\,x)^{\nu+\frac{1}{2}}} = \sqrt{\frac{\pi}{2}}\,\frac{\Gamma\left(\frac{1}{2}-\nu\right)}{\mathrm{sh}^\nu a}\,P_\gamma^\nu(\mathrm{ch}\,a)$$

$\left[\mathrm{Re}\,\nu < \frac{1}{2},\, a > 0\right]$. Ba I 156(8)

1. $$\int_0^\infty \frac{\mathrm{sh}^{2\mu}x\,dx}{(\mathrm{ch}\,a + \mathrm{sh}\,a\,\mathrm{ch}\,x)^{\nu+1}} = \frac{2^\mu e^{-i\mu\pi}}{\sqrt{\pi}\,\mathrm{sh}^\mu a}\,\frac{\Gamma(\nu-2\mu+1)\,\Gamma\left(\mu+\frac{1}{2}\right)}{\Gamma(\nu+1)}\,Q_{\nu-\mu}^\mu(\mathrm{ch}\,a)$$ 3.518

[Re$(\nu+1) > 0$, Re $(\nu - 2\mu + 1) > 0$, $a > 0$]. Ba I 155(3)+

2. $$\int_0^\infty \frac{\mathrm{sh}^{2\mu+1}x\,dx}{(\beta + \mathrm{ch}\,x)^{\nu+1}} = 2^\mu (\beta^2-1)^{\frac{\mu-\nu}{2}}\,\Gamma(\nu-2\mu)\,\Gamma(\mu+1)\,P_\mu^{\mu-\nu}(\beta)$$

[Re $(-\mu-\nu) >$ Re $\mu > -1$, β liegt nicht auf dem Strahl $(-1, +\infty)$ der reellen Achse]. Ba I 155(1)

[Re $(-\mu-\nu) <$ Re $\mu > -1$; β does not lie on the half-axis $(-1, +\infty)$ of the real axis]. Ba I 155(1)

3. $$\int_0^\infty \frac{\mathrm{sh}^{2\mu-1}x\,\mathrm{ch}\,x\,dx}{(1+a\,\mathrm{sh}^2 x)^\nu} = \frac{1}{2}a^{-\mu}B(\mu, \nu-\mu)$$ [Re $\nu >$ Re$\mu > 0$, $a > 0$]. Ba I 11(22)

3.518 4. $$\int_0^\infty \frac{\operatorname{sh}^{\mu-1} x (\operatorname{ch} x + 1)^{\nu-1} dx}{(\beta + \operatorname{ch} x)^\rho} = \frac{2^{-\rho+\frac{\mu}{2}} {}_2F_1\left(\rho, \rho - \frac{\mu}{2}; 1 + \rho - \frac{\mu}{4} - \frac{\nu}{2}; \frac{1-\beta}{2}\right)}{B\left(\rho - \frac{\mu}{2}, 1 + \frac{\mu}{4} - \frac{\nu}{2}\right)}$$

$$\left[\operatorname{Re} 2\rho > \operatorname{Re} \mu; \operatorname{Re}\left(1 + \frac{\mu}{4}\right) > \operatorname{Re}\left(\frac{\nu}{2}\right)\right].$$
Ba I 115(11)

5. $$\int_0^\infty \frac{\operatorname{sh}^{\mu-1} x (\operatorname{ch} x - 1)^{\nu-1} dx}{(\beta + \operatorname{ch} x)^\rho} =$$

$$= \frac{2^{-(2-\mu-\nu+\rho)} {}_2F_1\left(\rho, 2 - \mu - \nu + \rho; 1 + \rho - \frac{\mu}{2}, \frac{1-\beta}{2}\right)}{B\left(2 - \mu - \nu + \rho, -1 + \nu + \frac{\mu}{2}\right)}$$

$[\operatorname{Re}(1 + \rho) > \operatorname{Re}(\mu + \nu), \operatorname{Re}(4\rho + 2\nu + \mu) > 0].$ Ba I 115(10)

6. $$\int_0^\infty \frac{\operatorname{sh}^{\mu-1} x \operatorname{ch}^{\nu-1} x}{(\operatorname{ch}^2 x - \beta)^\rho} dx = \frac{{}_2F_1\left(\rho, 1 + \rho - \frac{\mu+\nu}{2}; 1 + \rho - \frac{\nu}{2}; \beta\right)}{2B\left(\frac{\mu}{2}, 1 + \rho - \frac{\mu+\nu}{2}\right)}$$

$[2\operatorname{Re}(1 + \rho) > \operatorname{Re} \nu, 2\operatorname{Re}(1+\rho) > \operatorname{Re}(\mu+\nu)].$ Ba I 115(9)

3.519 $$\int_0^{\pi/2} \frac{\operatorname{sh}[(r-p)\operatorname{tg} x]}{\operatorname{sh}(r \operatorname{tg} x)} dx = \pi \sum_{k=1}^\infty \frac{1}{k\pi + r} \sin \frac{pk\pi}{r} \quad [p^2 < r^2].$$
BH[274](13)

Hyperbelfunktionen und algebraische Funktionen
Hyperbolic functions and algebraic functions

3.52–3.53

3.521 1. $$\int_0^\infty \frac{x\,dx}{\operatorname{sh} ax} = \frac{\pi^2}{4a^2} \quad [a > 0].$$
GH[352](2b)

2. $$\int_0^\infty \frac{x\,dx}{\operatorname{ch} x} = 2G = \pi \ln 2 - 4L\left(\frac{\pi}{4}\right) = 1{,}831\,931\,188\ldots$$
Lo III 225(103a), BH[84](1)+

3. $$\int_1^\infty \frac{dx}{x \operatorname{sh} ax} = -2 \sum_{k=0}^\infty \operatorname{Ei}[-(2k+1)a].$$
Li[104](14)

4. $$\int_1^\infty \frac{dx}{x \operatorname{ch} ax} = 2 \sum_{k=0}^\infty (-1)^{k+1} \operatorname{Ei}[-(2k+1)a].$$
Li[104](13)

1. $$\int_0^\infty \frac{x\,dx}{(b^2+x^2)\,\text{sh}\,ax} = \frac{\pi}{2ab} + \pi \sum_{k=1}^\infty \frac{(-1)^k}{ab+k\pi} \quad [a>0,\ b>0].$$ 3.522

2. $$\int_0^\infty \frac{x\,dx}{(b^2+x^2)\,\text{sh}\,\pi x} = \frac{1}{2b} - \beta(b+1) \quad [b>0].$$ BH[97](16), GH[352](8)

3. $$\int_0^\infty \frac{dx}{(b^2+x^2)\,\text{ch}\,ax} = \frac{2\pi}{b} \sum_{k=1}^\infty \frac{(-1)^{k-1}}{2ab+(2k-1)\pi} \quad [a>0,\ b>0].$$ BH[97](5)

4. $$\int_0^\infty \frac{dx}{(b^2+x^2)\,\text{ch}\,\pi x} = \frac{1}{b}\beta\left(b+\frac{1}{2}\right) \quad [b>0].$$ BH[97](4)

5. $$\int_0^\infty \frac{x\,dx}{(1+x^2)\,\text{sh}\,\pi x} = \ln 2 - \frac{1}{2}.$$ BH[97](7)

6. $$\int_0^\infty \frac{dx}{(1+x^2)\,\text{ch}\,\pi x} = 2 - \frac{\pi}{2}.$$ BH[97](1)

7. $$\int_0^\infty \frac{x\,dx}{(1+x^2)\,\text{sh}\,\frac{\pi x}{2}} = \frac{\pi}{2} - 1.$$ BH[97](8)

8. $$\int_0^\infty \frac{dx}{(1+x^2)\,\text{ch}\,\frac{\pi x}{2}} = \ln 2.$$ BH[97](2)

9. $$\int_0^\infty \frac{x\,dx}{(1+x^2)\,\text{sh}\,\frac{\pi x}{4}} = \frac{1}{\sqrt{2}}[\pi + 2\ln(\sqrt{2}+1)] - 2.$$ BH[97](9)

10. $$\int_0^\infty \frac{dx}{(1+x^2)\,\text{ch}\,\frac{\pi x}{4}} = \frac{1}{\sqrt{2}}[\pi - 2\ln(\sqrt{2}+1)].$$ BH[97](3)

1. $$\int_0^\infty \frac{x^{\beta-1}}{\text{sh}\,ax}\,dx = \frac{2^\beta - 1}{2^{\beta-1} a^\beta}\,\Gamma(\beta)\zeta(\beta) \quad [\text{Re}\,\beta > 1,\ a>0].$$ WW 267+ 3.523

2. $$\int_0^\infty \frac{x^{2n-1}}{\text{sh}\,ax}\,dx = \frac{2^{2n}-1}{2n}\left(\frac{\pi}{a}\right)^{2n}|B_{2n}| \quad [a>0].$$ WW 126+, GH[352](2a)

3. $$\int_0^\infty \frac{x^{\beta-1}}{\text{ch}\,ax}\,dx = \frac{2}{(2a)^\beta}\,\Gamma(\beta)\,{}_1F_1\left(-1;\beta;\frac{1}{2}\right) = \frac{2}{(2a)^\beta}\,\Gamma(\beta)\sum_{k=0}^\infty (-1)^k \left(\frac{2}{2k+1}\right)^\beta$$

$$[\text{Re}\,\beta > 0,\ a>0].$$ Ba I 35, Ba IV 322(1)

3.523

4. $\displaystyle\int_0^\infty \frac{x^{2n}}{\operatorname{ch} ax}\,dx = \left(\frac{\pi}{2a}\right)^{2n+1} |E_{2n}| \quad [a > 0].$ BH[84](12)+, GH[352](1a)

5. $\displaystyle\int_0^\infty \frac{x^2\,dx}{\operatorname{ch} x} = \frac{\pi^3}{8}$ (cf. 4. 261 6.). BH[84](3)

6. $\displaystyle\int_0^\infty \frac{x^3\,dx}{\operatorname{sh} x} = \frac{\pi^4}{8}$ (cf. 4.262 1., 2.). BH[84](5)

7. $\displaystyle\int_0^\infty \frac{x^4\,dx}{\operatorname{ch} x} = \frac{5}{32}\pi^5.$ BH[84](7)

8. $\displaystyle\int_0^\infty \frac{x^5}{\operatorname{sh} x}\,dx = \frac{\pi^6}{4}.$ BH[84](8)

9. $\displaystyle\int_0^\infty \frac{x^6}{\operatorname{ch} x}\,dx = \frac{61}{128}\pi^7.$ BH[84](9)

10. $\displaystyle\int_0^\infty \frac{x^7}{\operatorname{sh} x}\,dx = \frac{17}{16}\pi^8.$ BH[84](10)

11. $\displaystyle\int_0^\infty \frac{\sqrt{x}\,dx}{\operatorname{ch} x} = \sqrt{\pi}\sum_{k=0}^\infty (-1)^k \frac{1}{\sqrt{(2k+1)^3}}.$ BH[98](7)+

12. $\displaystyle\int_0^\infty \frac{dx}{\sqrt{x}\,\operatorname{ch} x} = 2\sqrt{\pi}\sum_{k=0}^\infty \frac{(-1)^k}{\sqrt{2k+1}}.$ BH[98](25)+

3.524

1. $\displaystyle\int_0^\infty x^{\mu-1}\frac{\operatorname{sh}\beta x}{\operatorname{sh}\gamma x}\,dx = \frac{\Gamma(\mu)}{(2\gamma)^\mu}\left\{\zeta\left[\mu,\frac{1}{2}\left(1-\frac{\beta}{\gamma}\right)\right] - \zeta\left[\mu,\frac{1}{2}\left(1+\frac{\beta}{\gamma}\right)\right]\right\}$

[Re γ > |Re β|, Re μ > -1]. Ba IV 323(10)

2. $\displaystyle\int_0^\infty x^{2m}\frac{\operatorname{sh} ax}{\operatorname{sh} bx}\,dx = \frac{\pi}{2b}\frac{d^{2m}}{da^{2m}}\operatorname{tg}\frac{a\pi}{2b} \quad [b > |a|].$ BH[112](20)+

3. $\displaystyle\int_0^\infty \frac{\operatorname{sh} ax}{\operatorname{sh} bx}\frac{dx}{x^p} = \Gamma(1-p)\sum_{k=0}^\infty \left\{\frac{1}{[b(2k+1)-a]^{1-p}} - \frac{1}{[b(2k+1)+a]^{1-p}}\right\}$

$[b > |a|,\ p < 1].$ BH[131](2)+

4. $\displaystyle\int_0^\infty x^{2m+1}\frac{\operatorname{sh} ax}{\operatorname{ch} bx}\,dx = \frac{\pi}{2b}\frac{d^{2m+1}}{da^{2m+1}}\sec\frac{a\pi}{2b} \quad [b > |a|].$ BH[112](18)+

5. $\int_0^\infty x^{\mu-1} \frac{\operatorname{ch}\beta x}{\operatorname{sh}\gamma x} dx = \frac{\Gamma(\mu)}{(2\gamma)^\mu} \left\{ \zeta\left[\mu, \frac{1}{2}\left(1-\frac{\beta}{\gamma}\right)\right] + \zeta\left[\mu, \frac{1}{2}\left(1+\frac{\beta}{\gamma}\right)\right] \right\}$ 3.524

[Re γ > | Re β|, Re μ > 1].   Ba IV 323(12)

6. $\int_0^\infty x^{2m} \frac{\operatorname{ch} ax}{\operatorname{ch} bx} dx = \frac{\pi}{2b} \frac{d^{2m}}{da^{2m}} \sec\frac{a\pi}{2b}$ $[b > |a|]$.   BH[112](17)

7. $\int_0^\infty \frac{\operatorname{ch} ax}{\operatorname{ch} bx} \cdot \frac{dx}{x^p} = \Gamma(1-p) \sum_{k=0}^\infty (-1)^k \left\{ \frac{1}{[b(2k+1)-a]^{1-p}} + \frac{1}{[b(2k+1)+a]^{1-p}} \right\}$

$[b > |a|, p < 1]$.   BH[131](1)+

8. $\int_0^\infty x^{2m+1} \frac{\operatorname{ch} ax}{\operatorname{sh} bx} dx = \frac{\pi}{2b} \frac{d^{2m+1}}{da^{2m+1}} \operatorname{tg}\frac{a\pi}{2b}$ $[b > |a|]$.   BH[112](19)+

9. $\int_0^\infty x^{2m-1} \operatorname{cth} ax \, dx = \frac{2^{2m-1}-1}{m} \left(\frac{\pi}{2a}\right)^{2m} |B_{2m}|$ $[a > 0]$.   BH[83](11)

10. $\int_0^\infty x^2 \frac{\operatorname{sh} ax}{\operatorname{sh} bx} dx = \frac{\pi^3}{4b^3} \sin\frac{a\pi}{2b} \sec^3\frac{a\pi}{2b}$ $[b > |a|]$.   BH[84](18)

11. $\int_0^\infty x^4 \frac{\operatorname{sh} ax}{\operatorname{sh} bx} dx = 8\left(\frac{\pi}{2b}\sec\frac{a\pi}{2b}\right)^5 \cdot \sin\frac{a\pi}{2b} \cdot \left(2 + \sin^2\frac{a\pi}{2b}\right)$ $[b > |a|]$.   BH[82](17)+

12. $\int_0^\infty x^6 \frac{\operatorname{sh} ax}{\operatorname{sh} bx} dx = 16\left(\frac{\pi}{2b}\sec\frac{a\pi}{2b}\right)^7 \sin\frac{a\pi}{2b}\left(45 - 30\cos^2\frac{a\pi}{2b} + 2\cos^4\frac{a\pi}{2b}\right)$

$[b > |a|]$.   BH[82](21)+

13. $\int_0^\infty x \frac{\operatorname{sh} ax}{\operatorname{ch} bx} dx = \frac{\pi^2}{4b^2} \sin\frac{a\pi}{2b} \sec^2\frac{a\pi}{2b}$ $[b > |a|]$.   BH[84](15)+

14. $\int_0^\infty x^3 \frac{\operatorname{sh} ax}{\operatorname{ch} bx} dx = \left(\frac{\pi}{2b}\sec\frac{a\pi}{2b}\right)^4 \sin\frac{a\pi}{2b} \cdot \left(6 - \cos^2\frac{a\pi}{2b}\right)$ $[b > |a|]$.   BH[82](14)+

15. $\int_0^\infty x^5 \frac{\operatorname{sh} ax}{\operatorname{ch} bx} dx = \left(\frac{\pi}{2b}\sec\frac{a\pi}{2b}\right)^6 \sin\frac{a\pi}{2b}\left(120 - 60\cos^2\frac{a\pi}{2b} + \cos^4\frac{a\pi}{2b}\right)$

$[b > |a|]$.   BH[82](18)+

16. $\int_0^\infty x^7 \frac{\operatorname{sh} ax}{\operatorname{ch} bx} dx = \left(\frac{\pi}{2b}\sec\frac{a\pi}{2b}\right)^8 \sin\frac{a\pi}{2b} \times$

$\times \left(5040 - 4200\cos^2\frac{a\pi}{2b} + 546\cos^4\frac{a\pi}{2b} - \cos^6\frac{a\pi}{2b}\right)$ $[b > |a|]$.   BH[82](22)+

3.524

17. $$\int_0^\infty x \frac{\operatorname{ch} ax}{\operatorname{sh} bx}\, dx = \left(\frac{\pi}{2b} \sec \frac{a\pi}{2b}\right)^2 \quad [b > |a|].$$ BH[84](16)+

18. $$\int_0^\infty x^3 \frac{\operatorname{ch} ax}{\operatorname{sh} bx}\, dx = 2\left(\frac{\pi}{2b} \sec \frac{a\pi}{2b}\right)^4 \left(1 + 2\sin^2 \frac{a\pi}{2b}\right) \quad [b > |a|].$$ BH[82](15)+

19. $$\int_0^\infty x^5 \frac{\operatorname{ch} ax}{\operatorname{sh} bx}\, dx = 8\left(\frac{\pi}{2b} \sec \frac{a\pi}{2b}\right)^6 \left(15 - 15\cos^2 \frac{a\pi}{2b} + 2\cos^4 \frac{a\pi}{2b}\right)$$

$$[b > |a|].$$ BH[82](19)+

20. $$\int_0^\infty x^7 \frac{\operatorname{ch} ax}{\operatorname{sh} bx}\, dx = 16\left(\frac{\pi}{2b} \sec \frac{a\pi}{2b}\right)^8 \times$$

$$\times \left(315 - 420\cos^2 \frac{a\pi}{2b} + 126\cos^4 \frac{a\pi}{2b} - 4\cos^6 \frac{a\pi}{2b}\right) \quad [b > |a|].$$ BH[82](23)+

21. $$\int_0^\infty x^2 \frac{\operatorname{ch} ax}{\operatorname{ch} bx}\, dx = \frac{\pi^3}{8b^3}\left(2\sec^3 \frac{a\pi}{2b} - \sec \frac{a\pi}{2b}\right) \quad [b > |a|].$$ BH[84](17)+

22. $$\int_0^\infty x^4 \frac{\operatorname{ch} ax}{\operatorname{ch} bx}\, dx = \left(\frac{\pi}{2b} \sec \frac{a\pi}{2b}\right)^5 \left(24 - 20\cos^2 \frac{a\pi}{2b} + \cos^4 \frac{a\pi}{2b}\right) \quad [b > |a|].$$ BH[82](16)+

23. $$\int_0^\infty x^6 \frac{\operatorname{ch} ax}{\operatorname{ch} bx}\, dx = \left(\frac{\pi}{2b} \sec \frac{a\pi}{2b}\right)^7 \left(720 - 840\cos^2 \frac{a\pi}{2b} + \right.$$

$$\left. + 182\cos^4 \frac{a\pi}{2b} - \cos^6 \frac{a\pi}{2b}\right) \quad [b > |a|].$$ BH[82](20)+

24. $$\int_0^\infty \frac{\operatorname{sh} ax}{\operatorname{ch} bx} \cdot \frac{dx}{x} = \ln \operatorname{tg}\left(\frac{a\pi}{4b} + \frac{\pi}{4}\right) \quad [b > |a|].$$ BH[95](3)+

3.525

1. $$\int_0^\infty \frac{\operatorname{sh} ax}{\operatorname{sh} \pi x} \cdot \frac{dx}{1+x^2} = -\frac{a}{2}\cos a + \frac{1}{2}\sin a \ln[2(1+\cos a)] \quad [\pi \geqslant |a|].$$

BH[97](10)+

2. $$\int_0^\infty \frac{\operatorname{sh} ax}{\operatorname{sh} \frac{\pi}{2} x} \cdot \frac{dx}{1+x^2} = \frac{\pi}{2}\sin a + \frac{1}{2}\cos a \ln \frac{1-\sin a}{1+\sin a} \quad [\pi \geqslant 2|a|].$$ BH[97](11)+

3. $$\int_0^\infty \frac{\operatorname{ch} ax}{\operatorname{sh} \pi x} \cdot \frac{x\, dx}{1+x^2} = \frac{1}{2}(a\sin a - 1) + \frac{1}{2}\cos a \ln[2(1+\cos a)]$$

$$[\pi > |a|].$$ BH[97](12)+

4. $\int_0^\infty \dfrac{\operatorname{ch} ax}{\operatorname{sh} \dfrac{\pi}{2} x} \cdot \dfrac{x\, dx}{1+x^2} = \dfrac{\pi}{2} \cos a - 1 + \dfrac{1}{2} \sin a \ln \dfrac{1+\sin a}{1-\sin a} \quad \left[\dfrac{\pi}{2} > |a|\right].$ 3.525

BH[97](13)+

5. $\int_0^\infty \dfrac{\operatorname{sh} ax}{\operatorname{ch} \pi x} \cdot \dfrac{x\, dx}{1+x^2} = -2 \sin \dfrac{a}{2} + \dfrac{\pi}{2} \sin a - \cos a \ln \operatorname{tg} \dfrac{a+\pi}{4} \quad [\pi > |a|].$

GH[352](12)

6. $\int_0^\infty \dfrac{\operatorname{ch} ax}{\operatorname{ch} \pi x} \cdot \dfrac{dx}{1+x^2} = 2\cos \dfrac{a}{2} - \dfrac{\pi}{2} \cos a - \sin a \ln \operatorname{tg} \dfrac{a+\pi}{4} \quad [\pi > |a|].$

GH[352](11)

7. $\int_0^\infty \dfrac{\operatorname{sh} ax}{\operatorname{sh} bx} \cdot \dfrac{dx}{c^2+x^2} = \dfrac{\pi}{c} \sum_{k=1}^\infty \dfrac{\sin \dfrac{k(b-a)}{b} \pi}{bc + k\pi} \quad [b \geq |a|].$ BH[97](18)

8. $\int_0^\infty \dfrac{\operatorname{ch} ax}{\operatorname{sh} bx} \cdot \dfrac{x\, dx}{c^2+x^2} = \dfrac{\pi}{2bc} + \pi \sum_{k=1}^\infty \dfrac{\cos \dfrac{k(b-a)}{b} \pi}{bc+k\pi} \quad [b > |a|].$ BH[97](19)

1. $\int_0^\infty \dfrac{\operatorname{sh} ax \operatorname{ch} bx}{\operatorname{ch} cx} \cdot \dfrac{dx}{x} = \dfrac{1}{2} \ln \left\{ \operatorname{tg} \dfrac{(a+b+c)\pi}{4c} \operatorname{ctg} \dfrac{(b+c-a)\pi}{4c} \right\}$ 3.526

$[c > |a| + |b|].$ BH[93](10)+

2. $\int_0^\infty \dfrac{\operatorname{sh}^2 ax}{\operatorname{sh} bx} \cdot \dfrac{dx}{x} = \dfrac{1}{2} \ln \sec \dfrac{a}{b} \pi \quad [b > |2a|].$ BH[95](5)+

3. $\int_0^\infty \dfrac{x^{\mu-1}}{\operatorname{sh} \beta x \operatorname{ch} \gamma x} dx = \dfrac{\Gamma(\mu)}{(2\gamma)^\mu} \left\{ {}_1F_1\left[-1; \mu; \dfrac{1}{2}\left(1+\dfrac{\beta}{\gamma}\right)\right] + \right.$

$\left. + {}_1F_1\left[-1; \mu; \dfrac{1}{2}\left(1-\dfrac{\beta}{\gamma}\right)\right] \right\} \quad [\operatorname{Re} \gamma > |\operatorname{Re} \beta|, \operatorname{Re} \mu > 0],$

Ba IV 323(11)

1. $\int_0^\infty \dfrac{x^{\mu-1}}{\operatorname{sh}^2 ax} dx = \dfrac{4}{(2a)^\mu} \Gamma(\mu) \zeta(\mu-1) \quad [\operatorname{Re} a > 0,\ \operatorname{Re} \mu > 2].$ BH[86](7)+   3.527

2. $\int_0^\infty \dfrac{x^{2m}}{\operatorname{sh}^2 ax} dx = \dfrac{\pi^{2m}}{a^{2m+1}} |B_{2m}| \quad [a > 0].$ BH[86](5)+

3. $\int_0^\infty \dfrac{x^{\mu-1}}{\operatorname{ch}^2 ax} dx = \dfrac{4}{(2a)^\mu} (1-2^{2-\mu}) \Gamma(\mu) \zeta(\mu-1) \quad [\operatorname{Re} a > 0,\ \operatorname{Re} \mu > 0].$ BH[86](6)+

4. $\int_0^\infty \dfrac{x\, dx}{\operatorname{ch}^2 ax} = \dfrac{\ln 2}{a^2}.$ Lo III 39 6

3.527

5. $\int_0^\infty \dfrac{x^{2m}}{\operatorname{ch}^2 ax} dx = \dfrac{(2^{2m}-2)\pi^{2m}}{(2a)^{2m} a} |B_{2m}|$ $[a>0]$.   BH[86](2)+

6. $\int_0^\infty x^{\mu-1} \dfrac{\operatorname{sh} ax}{\operatorname{ch}^2 ax} dx = \dfrac{2\Gamma(\mu)}{a^\mu} \sum_{k=0}^\infty \dfrac{(-1)^k}{(2k+1)^{\mu-1}}$ $[\operatorname{Re}\mu>0,\ a>0]$.   BH[86](15)+

7. $\int_0^\infty \dfrac{x \operatorname{sh} ax}{\operatorname{ch}^2 ax} dx = \dfrac{\pi}{2a^2}$.   BH[86](8)+

8. $\int_0^\infty x^{2m+1} \dfrac{\operatorname{sh} ax}{\operatorname{ch}^2 ax} dx = \dfrac{2m+1}{a} \left(\dfrac{\pi}{2a}\right)^{2m+1} |E_{2m}|$ $[a>0]$.   BH[86](12)+

9. $\int_0^\infty x^{2m+1} \dfrac{\operatorname{ch} ax}{\operatorname{sh}^2 ax} dx = \dfrac{2^{2m+1}-1}{a^2 (2a)^{2m}} (2m+1)!\, \zeta(2m+1)$.   BH[86](13)+

10. $\int_0^\infty x^{2m} \dfrac{\operatorname{ch} ax}{\operatorname{sh}^2 ax} dx = \dfrac{2^{2m}-1}{a} \left(\dfrac{\pi}{a}\right)^{2m} |B_{2m}|$ $[a>0]$.   BH[86](14)+

11. $\int_0^\infty \dfrac{x \operatorname{sh} ax}{\operatorname{ch}^{2\mu+1} ax} dx = \dfrac{\sqrt{\pi}}{4\mu a^2} \dfrac{\Gamma(\mu)}{\Gamma\left(\mu+\dfrac{1}{2}\right)}$ $[\mu>0]$.   Li[86](9)

12. $\int_{-\infty}^\infty \dfrac{x^2\, dx}{\operatorname{sh}^2 x} = \dfrac{\pi^2}{3}$.   BH[102](2)+

13. $\int_0^\infty x^2 \dfrac{\operatorname{ch} ax}{\operatorname{sh}^2 ax} dx = \dfrac{\pi^2}{2a^3}$.   BH[86](11)+

14. $\int_0^\infty x^2 \dfrac{\operatorname{sh} ax}{\operatorname{ch}^2 ax} dx = \dfrac{\ln 2}{2a}$.   BH[86](10)+

15. $\int_0^\infty \dfrac{\operatorname{th} \dfrac{x}{2}}{\operatorname{ch} x} \dfrac{dx}{x} = \ln 2$.   BH[93](17)+

3.528

1. $\int_0^\infty \dfrac{(1+xi)^{2n-1} - (1-xi)^{2n-1}}{\operatorname{sh} \dfrac{\pi x}{2}} dx = 2$.   BH[86](8)

2. $\int_0^\infty \dfrac{(1+xi)^{2n} - (1-xi)^{2n}}{i \operatorname{sh} \dfrac{\pi x}{2}} dx = (-1)^{n+1} 2|E_{2n}| + 2$.   BH[87](7)

1. $$\int_0^\infty \left(\frac{1}{\operatorname{sh} x} - \frac{1}{x}\right) \frac{dx}{x} = -\ln 2.$$ BH[94](10)+   3.529

2. $$\int_0^\infty \frac{\operatorname{ch} ax - 1}{\operatorname{sh} bx} \cdot \frac{dx}{x} = -\ln \cos \frac{a\pi}{2b} \quad [b > |a|].$$ GH[352](66)

3. $$\int_0^\infty \left(\frac{a}{\operatorname{sh} ax} - \frac{b}{\operatorname{sh} bx}\right) \frac{dx}{x} = (b - a) \ln 2.$$ BH[94](11)+

1. $$\int_0^\infty \frac{x\, dx}{2\operatorname{ch} x - 1} = 1{,}171\,953\,619\,3\ldots$$ Li[88[(1)   3.531

2. $$\int_0^\infty \frac{x\, dx}{\operatorname{ch} 2x + \cos 2t} = \frac{t \ln 2 - L(t)}{\sin t \cos t}.$$ Lo III 402

3. $$\int_0^\infty \frac{x^2\, dx}{\operatorname{ch} x + \cos t} = \frac{t}{3} \frac{\pi^2 - t^2}{\sin t} \quad [0 < t < \pi].$$ BH[88](3)+

4. $$\int_0^\infty \frac{x^4\, dx}{\operatorname{ch} x + \cos t} = \frac{t}{15} \frac{(\pi^2 - t^2)(7\pi^2 - 3t^2)}{\sin t} \quad [0 < t < \pi].$$ BH[88](4)+

5. $$\int_0^\infty \frac{x^{2m}\, dx}{\operatorname{ch} x - \cos 2a\pi} = 2 \cdot (2m)!\, \operatorname{cosec} 2a\pi \sum_{k=1}^\infty \frac{\sin 2ka\pi}{k^{2m+1}}.$$ BH[88](5)+

6. $$\int_0^\infty \frac{x^{\mu-1}\, dx}{\operatorname{ch} x - \cos t} = \frac{i\Gamma(\mu)}{\sin t} [e^{-it}\,{}_1F_1(e^{-it};\mu;1) - e^{-it}\,{}_1F_1(e^{it};\mu;1)]$$

$$[\operatorname{Re} \mu > 0,\ 0 < t < 2\pi].$$ Ba IV 323(5)

7. $$\int_0^\infty \frac{x^\mu\, dx}{\operatorname{ch} x + \cos t} = \frac{2\Gamma(\mu+1)}{\sin t} \sum_{k=1}^\infty (-1)^{k-1} \frac{\sin kt}{k^{\mu+1}} \quad [\mu > -1].$$ BH[96](14)+

8. $$\int_0^u \frac{x\, dx}{\operatorname{ch} 2x - \cos 2t} = \frac{1}{2}\operatorname{cosec} 2t\, [L(\theta + t) - L(\theta - t) - 2L(t)]$$

$$[\theta = \operatorname{arctg}(\operatorname{th} u\, \operatorname{ctg} t),\ t \neq n\pi].$$ Lo III 402

1. $$\int_0^\infty \frac{x^n\, dx}{a\operatorname{ch} x + b\operatorname{sh} x} = \frac{(2n)!}{a + b} \sum_{k=0}^\infty \frac{1}{(2k+1)^{n+1}} \left(\frac{b-a}{b+a}\right)^k$$ 3.532

$$[a > 0,\ b > 0,\ n > -1].$$ GH[352](5)

3.532

2. $\displaystyle\int_0^u \frac{x\,\text{ch}\,x\,dx}{\text{ch}\,2x - \cos 2t} = \frac{1}{2}\cosec t \left\{ L\left(\frac{\theta+t}{2}\right) - L\left(\frac{\theta-t}{2}\right) + \right.$

$$+ L\left(\pi - \frac{\Psi+t}{2}\right) + L\left(\frac{\Psi-t}{2}\right) - 2L\left(\frac{t}{2}\right) - 2L\left(\frac{\pi-t}{2}\right)\right\}$$

$\left[\text{tg}\,\dfrac{\theta}{2} = \text{th}\,\dfrac{u}{2}\,\text{ctg}\,\dfrac{t}{2}, \; \text{tg}\,\dfrac{\Psi}{2} = \text{cth}\,\dfrac{u}{2}\,\text{ctg}\,\dfrac{t}{2}; \; t \ne n\pi\right]$ Lo III 288+

3.533

1. $\displaystyle\int_0^\infty \frac{x\,\text{ch}\,x\,dx}{\text{ch}\,2x - \cos 2t} = \cosec t \left[\frac{\pi}{2}\ln 2 - L\left(\frac{t}{2}\right) - L\left(\frac{\pi-t}{2}\right)\right]$ $[t \ne m\pi]$.

Lo III 403

2. $\displaystyle\int_0^\infty x\,\frac{\text{sh}\,ax\,dx}{(\text{ch}\,ax - \cos t)^2} = \frac{t}{a^2}\cosec t$ $[0 < t < \pi]$ (cf. 3.514 1.). BH[88](11)+

3. $\displaystyle\int_0^\infty x^3\,\frac{\text{sh}\,x\,dx}{(\text{ch}\,x + \cos t)^2} = \frac{t(\pi^2 - t^2)}{\sin t}$ $[0 < t < \pi]$ (cf. 3.531 3.).

BH[88](13)

4. $\displaystyle\int_0^\infty x^{2m+1}\,\frac{\text{sh}\,x\,dx}{(\text{ch}\,x - \cos 2a\pi)^2} = 2(2m+1)!\cosec 2a\pi \sum_{k=1}^\infty \frac{\cos 2ka\pi}{k^{2m+1}}$ $[0 < a < \pi]$.

BH[88](14)

3.534

1. $\displaystyle\int_0^1 \sqrt{1-x^2}\,\text{ch}\,ax\,dx = \frac{\pi}{2a}I_1(a)$. W 79(9)

2. $\displaystyle\int_0^1 \frac{\text{ch}\,ax}{\sqrt{1-x^2}}\,dx = \frac{\pi}{2}I_0(a)$. W 79(9)

3.535 $\displaystyle\int_0^1 \frac{x}{\sqrt{\text{ch}\,2a - \text{ch}\,2ax}}\cdot\frac{dx}{\text{sh}\,ax} = \frac{\pi}{2\sqrt{2a^2}}\cdot\frac{\arcsin(\text{th}\,a)}{\text{sh}\,a}$. BH[80](11)

3.536

1. $\displaystyle\int_0^\infty \frac{x^2}{\text{ch}\,x^2}\,dx = \frac{\sqrt{\pi}}{2}\sum_{k=0}^\infty \frac{(-1)^k}{\sqrt{(2k+1)^3}}$. BH[98](7)

2. $\displaystyle\int_0^\infty \frac{x^2\,\text{th}\,x^2\,dx}{\text{ch}\,x^2} = \frac{\sqrt{\pi}}{2}\sum_{k=0}^\infty \frac{(-1)^k}{\sqrt{2k+1}}$. BH[98](8)

3. $\displaystyle\int_0^\infty \text{sh}(\nu\,\text{Arsh}\,x)\frac{x^{\mu-1}}{\sqrt{1+x^2}}\,dx = \frac{\sin\frac{\mu\pi}{2}\sin\frac{\nu\pi}{2}}{2^\mu\pi}\Gamma(\mu)\Gamma\left(\frac{1-\mu-\nu}{2}\right) \times$

$\times \Gamma\left(\dfrac{1-\mu+\nu}{2}\right)$ $[-1 < \text{Re}\,\mu < 1 - |\text{Re}\,\nu|]$. Ba IV 324(14)

4. $\int_0^\infty \mathrm{ch}(v\,\mathrm{Arch}\,x)\,\dfrac{x^{\mu-1}}{\sqrt{1+x^2}}\,dx = \dfrac{\cos\dfrac{\mu\pi}{2}\cos\dfrac{v\pi}{2}}{2^\mu \pi}\,\Gamma(\mu)\Gamma\!\left(\dfrac{1-\mu-v}{2}\right)\times$ 3.536

$\times \Gamma\!\left(\dfrac{1-\mu+v}{2}\right)\quad [0 < \mathrm{Re}\,\mu < 1 - |\mathrm{Re}\,v|].$ Ba IV 324(15)

Hyperbelfunktionen und die Exponentialfunktion
Hyperbolic functions and exponentials 3.54

1. $\int_0^\infty e^{-\mu x}\,\mathrm{sh}^v\,\beta x\,dx = \dfrac{1}{2^{v+1}\beta}\,\mathrm{B}\!\left(\dfrac{\mu}{2\beta} - \dfrac{v}{2},\, v+1\right)$ 3.541

$[\mathrm{Re}\,\beta > 0,\ \mathrm{Re}\,v > -1,\ \mathrm{Re}\,\mu > \mathrm{Re}\,\beta v].$ Ba I 11(25), Ba IV 163(5)

2. $\int_0^\infty e^{-\mu x}\,\dfrac{\mathrm{sh}\,\beta x}{\mathrm{sh}\,bx}\,dx = \dfrac{1}{2b}\left[\psi\!\left(\dfrac{1}{2} + \dfrac{\mu+\beta}{2b}\right) - \psi\!\left(\dfrac{1}{2} + \dfrac{\mu-\beta}{2b}\right)\right]$

$[\mathrm{Re}\,(\mu + b\pm\beta) > 0].$ Ba I 16

3. $\int_{-\infty}^\infty e^{-\mu x}\,\dfrac{\mathrm{sh}\,\mu x}{\mathrm{sh}\,\beta x}\,dx = \dfrac{\pi}{2\beta}\,\mathrm{tg}\,\dfrac{\mu\pi}{\beta}\quad [\mathrm{Re}\,\beta > 2|\mathrm{Re}\,\mu|].$ BH[18](6)

4. $\int_0^\infty e^{-x}\,\dfrac{\mathrm{sh}\,ax}{\mathrm{sh}\,x}\,dx = \dfrac{1}{a} - \dfrac{\pi}{2}\,\mathrm{ctg}\,\dfrac{a\pi}{2}\quad [0 < a < 2].$ BH[4](3)

5. $\int_0^\infty \dfrac{e^{-px}\,dx}{(\mathrm{ch}\,px)^{2q+1}} = \dfrac{2^{2q-2}}{p}\,\mathrm{B}(q,q) - \dfrac{1}{2qp}.$ Li[27](19)

6. $\int_0^\infty e^{-\mu x}\,\dfrac{dx}{\mathrm{ch}\,x} = \beta\!\left(\dfrac{\mu+1}{2}\right)\quad [\mathrm{Re}\,\mu > 0].$ Ba IV 163(7)

7. $\int_0^\infty e^{-\mu x}\,\mathrm{th}\,x\,dx = \beta\!\left(\dfrac{\mu}{2}\right) - \dfrac{1}{\mu}\quad [\mathrm{Re}\,\mu > 0].$ Ba IV 163(9)

8. $\int_0^\infty \dfrac{e^{-\mu x}}{\mathrm{ch}^2 x}\,dx = \mu\beta\!\left(\dfrac{\mu}{2}\right) - 1\quad [\mathrm{Re}\,\mu > 0].$ Ba IV 163(8)

9. $\int_0^\infty e^{-\mu x}\,\dfrac{\mathrm{sh}\,\mu x}{\mathrm{ch}^2 \mu x}\,dx = \dfrac{1}{\mu}(1 - \ln 2)\quad [\mathrm{Re}\,\mu > 0].$ Li[27](15)

10. $\int_0^\infty e^{-qx}\,\dfrac{\mathrm{sh}\,px}{\mathrm{sh}\,qx}\,dx = \dfrac{1}{p} - \dfrac{\pi}{2q}\,\mathrm{ctg}\,\dfrac{p\pi}{2q}\quad [0 < p < 2q].$ BH[27](9)+

3.542

1. $$\int_0^\infty e^{-\mu x}(\operatorname{ch}\beta x - 1)^\nu\,dx = \frac{1}{2^\nu \beta}\,\mathrm{B}\left(\frac{\mu}{\beta} - \nu,\,2\nu + 1\right)$$

    $$\left[\operatorname{Re}\beta > 0,\ \operatorname{Re}\nu > -\frac{1}{2},\ \operatorname{Re}\mu > \operatorname{Re}\beta\nu\right].\quad \text{Ba IV 163(6)}$$

2. $$\int_0^\infty e^{-\mu x}(\operatorname{ch} x - \operatorname{ch} u)^{\nu-1}\,dx = -i\sqrt{\frac{2}{\pi}}\,e^{i\pi\nu}\,\Gamma(\nu)\,\operatorname{sh}^{\nu-\frac{1}{2}} u\,Q_{\mu-\frac{1}{2}}^{\frac{1}{2}-\nu}(\operatorname{ch} u)$$

    $$[\operatorname{Re}\nu > 0,\ \operatorname{Re}\mu > \operatorname{Re}\nu - 1].\quad \text{Ba I 155(4), Ba IV 164(23)}$$

3.543

1. $$\int_{-\infty}^\infty \frac{e^{-ibx}\,dx}{\operatorname{sh} x + \operatorname{sh} t} = -\frac{i\pi e^{itb}}{\operatorname{sh}\pi b \operatorname{ch} t}(\operatorname{ch}\pi b - e^{-2itb})\quad [t > 0].\quad \text{Ba IV 121(30)}$$

2. $$\int_0^\infty \frac{e^{-\mu x}}{\operatorname{ch} x - \cos t}\,dx = 2\operatorname{cosec} t \sum_{k=1}^\infty \frac{\sin kt}{\mu + k}\quad [\operatorname{Re}\mu > -1,\ t \neq 2n\pi].$$

    BH[6](10)+

3. $$\int_0^\infty \frac{1 - e^{-x}\cos t}{\operatorname{ch} x - \cos t}\,e^{-(\mu-1)x}\,dx = 2\sum_{k=0}^\infty \frac{\cos kt}{\mu + k}\quad [\operatorname{Re}\mu > 0,\ t \neq 2n\pi].\quad \text{BH[6](9)+}$$

4. $$\int_0^\infty \frac{e^{px} + \cos t}{(\operatorname{ch} px + \cos t)^2}\,dx = \frac{1}{p}\left(t\operatorname{cosec} t + \frac{1}{1 + \cos t}\right)\quad [p > 0].\quad \text{BH[27](26)+}$$

3.544

$$\int_u^\infty \frac{\exp\left[-\left(n + \frac{1}{2}\right)x\right]}{\sqrt{2(\operatorname{ch} x - \operatorname{ch} u)}}\,dx = Q_n(\operatorname{ch} u).\quad \text{Ba II 181(33)}$$

3.545

1. $$\int_0^\infty \frac{\operatorname{sh} ax}{e^{px} + 1}\,dx = \frac{\pi}{2p}\operatorname{cosec}\frac{a\pi}{p} - \frac{1}{2a}\quad [p > a,\ p > 0].\quad \text{BH[27](3)}$$

2. $$\int_0^\infty \frac{\operatorname{sh} ax}{e^{px} - 1}\,dx = \frac{1}{2a} - \frac{\pi}{2p}\operatorname{ctg}\frac{a\pi}{p}\quad [p > a,\ p > 0].\quad \text{BH[27](9)}$$

3.546

1. $$\int_0^\infty e^{-\beta x^2}\operatorname{sh} ax\,dx = \frac{1}{2}\frac{\sqrt{\pi}}{\sqrt{\beta}}\exp\frac{a^2}{4\beta}\,\Phi\left(\frac{a}{2\sqrt{\beta}}\right)\quad [\operatorname{Re}\beta > 0].\quad \text{Ba IV 166(38)+}$$

2. $$\int_0^\infty e^{-\beta x^2}\operatorname{ch} ax\,dx = \frac{1}{2}\sqrt{\frac{\pi}{\beta}}\exp\frac{a^2}{4\beta}\quad [\operatorname{Re}\beta > 0].\quad \text{F II 725–726}$$

3. $\int_0^\infty e^{-\beta x^2} \operatorname{sh}^2 ax\, dx = \frac{1}{4}\sqrt{\frac{\pi}{\beta}} \left(\exp\frac{a^2}{\beta} - 1\right)$ [Re $\beta > 0$]. Ba IV 166(40) 3.546

4. $\int_0^\infty e^{-\beta x^2} \operatorname{ch}^2 ax\, dx = \frac{1}{4}\sqrt{\frac{\pi}{\beta}} \left(\exp\frac{a^2}{\beta} + 1\right)$ [Re $\beta > 0$]. Ba IV 166(41)

1. $\int_0^\infty \exp(-\beta \operatorname{sh} x) \operatorname{sh} \gamma x\, dx = \frac{\pi}{2} \operatorname{ctg} \frac{\gamma\pi}{2} [\mathbf{J}_\gamma(\beta) - J_\gamma(\beta)] -$ 3.547

$- \frac{\pi}{2} [\mathbf{E}_\gamma(\beta) + N_\gamma(\beta)] = \gamma S_{-1,\gamma}(\beta)$ [Re $\beta > 0$]. W 312(5), Ba IV 168(14)+

2. $\int_0^\infty \exp(-\beta \operatorname{ch} x) \operatorname{sh} \gamma x \operatorname{sh} x\, dx = \frac{\gamma}{\beta} K_\gamma(\beta)$. Ba IV 168(9)+

3. $\int_0^\infty \exp(-\beta \operatorname{sh} x) \operatorname{ch} \gamma x\, dx = \frac{\pi}{2} \operatorname{tg} \frac{\pi\gamma}{2} [\mathbf{J}_\gamma(\beta) - J_\gamma(\beta)] -$

$- \frac{\pi}{2} [\mathbf{E}_\gamma(\beta) + N_\gamma(\beta)] = S_{0,\gamma}(\beta)$ [Re $\beta > 0$, γ nicht ganzzahlig / noninteger].

Ba IV 168(16)+, W 312(4), Ba II 84(50)

4. $\int_0^\infty \exp(-\beta \operatorname{ch} x) \operatorname{ch} \gamma x\, dx = K_\gamma(\beta)$ [Re $\beta > 0$]. Ba IV 168(16)+, W 181(5)

5. $\int_0^\infty \exp(-\beta \operatorname{sh} x) \operatorname{sh} \gamma x \operatorname{ch} x\, dx = \frac{\gamma}{\beta} S_{0,\gamma}(\beta)$ [Re $\beta > 0$].

Ba IV 168(7), Ba II 85(51)

6. $\int_0^\infty \exp(-\beta \operatorname{sh} x) \operatorname{sh}[(2n+1)x] \operatorname{ch} x\, dx = O_{2n+1}(\beta)$ [Re $\beta > 0$].

Ba IV 167(5)

7. $\int_0^\infty \exp(-\beta \operatorname{sh} x) \operatorname{ch} \gamma x \operatorname{ch} x\, dx = \frac{1}{\beta} S_{1,\gamma}(\beta)$ [Re $\beta > 0$].

Ba IV 168(8), Ba II 85(52)

8. $\int_0^\infty \exp(-\beta \operatorname{sh} x) \operatorname{ch} 2nx \operatorname{ch} x\, dx = O_{2n}(\beta)$ [Re $\beta > 0$]. Ba IV 168(6)

3.547

9. $$\int_0^\infty \exp(-\beta \operatorname{ch} x) \operatorname{sh}^{2\nu} x \, dx = \frac{1}{\sqrt{\pi}} \left(\frac{2}{\beta}\right)^\nu \Gamma\left(\nu + \frac{1}{2}\right) K_\nu(\beta)$$

$$\left[\operatorname{Re} \beta > 0, \quad \operatorname{Re} \nu > -\frac{1}{2}\right].$$ Ba II 82(20)

10. $$\int_0^\infty \exp[-2(\beta \operatorname{cth} x + \mu x)] \operatorname{sh}^{2\nu} x \, dx = \frac{1}{4} \beta^{\nu - \frac{1}{2}} \Gamma(\mu - \nu) \times$$

$$\times [W_{-\mu + \frac{1}{2}, \nu}(4\beta) - (\mu - \nu) W_{-\mu - \frac{1}{2}, \nu}(4\beta)] \quad [\operatorname{Re}\beta > 0, \quad \operatorname{Re}\mu > \operatorname{Re}\nu].$$

Ba IV 165(31)

11. $$\int_0^\infty \exp\left(-\frac{\beta^2}{2} \operatorname{sh} x\right) \operatorname{sh}^{\nu-1} x \operatorname{ch}^\nu x \, dx =$$

$$= -\pi D_\nu(\beta e^{\frac{i\pi}{4}}) D_\nu(\beta e^{-\frac{i\pi}{4}}) \quad \left[\operatorname{Re} \nu > 0, \, |\arg \beta| \leq \frac{\pi}{4}\right].$$ Ba II 120 (10)

12. $$\int_0^\infty \frac{\exp(2\nu x - 2\beta \operatorname{sh} x)}{\sqrt{\operatorname{sh} x}} \, dx = \frac{1}{2} \sqrt{\pi^3 \beta} \, [J_{\nu + \frac{1}{4}}(\beta) J_{\nu - \frac{1}{4}}(\beta) +$$

$$+ N_{\nu + \frac{1}{4}}(\beta) N_{\nu - \frac{1}{4}}(\beta)] \quad [\operatorname{Re}\beta > 0].$$ Ba IV 169(20)

13. $$\int_0^\infty \frac{\exp(-2\nu x - 2\beta \operatorname{sh} x)}{\sqrt{\operatorname{sh} x}} \, dx = \frac{1}{2} \sqrt{\pi^3 \beta} \, [J_{\nu + \frac{1}{4}}(\beta) N_{\nu - \frac{1}{4}}(\beta) -$$

$$- J_{\nu - \frac{1}{4}}(\beta) N_{\nu + \frac{1}{4}}(\beta)] \quad [\operatorname{Re} \beta > 0].$$ Ba IV 169(21)

14. $$\int_0^\infty \frac{\exp(-2\beta \operatorname{sh} x) \operatorname{sh} 2\nu x}{\sqrt{\operatorname{sh} x}} \, dx = \frac{1}{4i} \sqrt{\frac{\pi^3 \beta}{2}} [e^{\nu\pi i} H^{(1)}_{\frac{1}{2} + \nu}(\beta) H^{(2)}_{\frac{1}{2} - \nu}(\beta) -$$

$$- e^{-\nu\pi i} H^{(1)}_{\frac{1}{2} - \nu}(\beta) H^{(2)}_{\frac{1}{2} + \nu}(\beta) \quad [\operatorname{Re} \beta > 0].$$ Ba IV 170(24)

15. $$\int_0^\infty \frac{\exp(-2\beta \operatorname{sh} x) \operatorname{ch} 2\nu x}{\sqrt{\operatorname{sh} x}} \, dx = \frac{1}{4} \sqrt{\frac{\pi^3 \beta}{2}} [e^{\nu\pi i} H^{(1)}_{\frac{1}{2} + \nu}(\beta) H^{(2)}_{\frac{1}{2} - \nu}(\beta) +$$

$$+ e^{-\nu\pi i} H^{(1)}_{\frac{1}{2} - \nu}(\beta) H^{(2)}_{\frac{1}{2} + \nu}(\beta)] \quad [\operatorname{Re}\beta > 0].$$ Ba IV 170(25)

16. $$\int_0^\infty \frac{\exp(-2\beta \operatorname{ch} x) \operatorname{ch} 2\nu x}{\sqrt{\operatorname{ch} x}} \, dx = \sqrt{\frac{\beta}{\pi}} K_{\nu + \frac{1}{4}}(\beta) K_{\nu - \frac{1}{4}}(\beta) \quad [\operatorname{Re}\beta > 0].$$

Ba IV 170(26)

17. $\int_0^\infty \dfrac{\exp[-2\beta(\operatorname{ch} x - 1)] \operatorname{ch} 2\nu x}{\sqrt{\operatorname{ch} x}} dx = \sqrt{\dfrac{\beta}{\pi}} \cdot e^{2\beta} K_{\nu+\frac{1}{2}}(\beta) K_{\nu-\frac{1}{2}}(\beta) \quad [\operatorname{Re} \beta > 0].$ 3.547

Ba IV 170(27)

18. $\int_0^\infty \dfrac{\cos\left[\left(\nu+\frac{1}{4}\right)\pi\right] \exp(-2\nu x - 2\beta \operatorname{sh} x) + \sin\left[\left(\nu+\frac{1}{4}\right)\pi\right] \exp(2\nu x - 2\beta \operatorname{sh} x)}{\sqrt{\operatorname{sh} x}} dx =$

$= \dfrac{1}{2}\sqrt{\pi^3 \beta}\, [J_{\frac{1}{4}+\nu}(\beta) J_{\frac{1}{4}-\nu}(\beta) + N_{\frac{1}{4}+\nu}(\beta) N_{\frac{1}{4}-\nu}(\beta)] \quad [\operatorname{Re}\beta > 0].$

Ba IV 169(22)

19. $\int_0^\infty \dfrac{\sin\left[\left(\nu+\frac{1}{4}\right)\pi\right] \exp(-2\nu x - 2\beta \operatorname{sh} x) - \cos\left[\left(\nu+\frac{1}{4}\right)\pi\right] \exp(2\nu x - 2\beta \operatorname{sh} x)}{\sqrt{\operatorname{sh} x}} dx =$

$= \dfrac{1}{2}\sqrt{\pi^3 \beta}\, [J_{\frac{1}{4}+\nu}(\beta) N_{\frac{1}{4}-\nu}(\beta) - J_{\frac{1}{4}-\nu}(\beta) N_{\frac{1}{4}+\nu}(\beta)]$

$[\operatorname{Re}\beta > 0].$ Ba IV 169(23)

20. $\int_0^\infty \dfrac{\exp[-\beta(\operatorname{ch} x - 1)] \operatorname{ch} \nu x \operatorname{sh} x}{\sqrt{\operatorname{ch} x (\operatorname{ch} x - 1)}} dx = e^\beta K_\nu(\beta) \quad [\operatorname{Re}\beta > 0].$ Ba IV 169(19)

1. $\int_0^\infty e^{-\mu x^4} \operatorname{sh} ax^2\, dx = \dfrac{\pi}{4}\sqrt{\dfrac{a}{2\mu}} \exp\left(\dfrac{a^2}{8\mu}\right) I_{\frac{1}{4}}\left(\dfrac{a^2}{8\mu}\right) \quad [\operatorname{Re}\mu > 0].$ Ba IV 166(42)   3.548

2. $\int_0^\infty e^{-\mu x^4} \operatorname{ch} ax^2\, dx = \dfrac{\pi}{4}\sqrt{\dfrac{a}{2\mu}} \exp\left(\dfrac{a^2}{8\mu}\right) I_{-\frac{1}{4}}\left(\dfrac{a^2}{8\mu}\right) \quad [\operatorname{Re}\mu > 0].$ Ba IV 166(43)

1. $\int_0^\infty e^{-\beta x} \operatorname{sh}[(2n+1)\operatorname{Arsh} x]\, dx = O_{2n+1}(\beta) \quad [\operatorname{Re}\beta > 0] \quad (\text{cf. } 3.547\ 6.).$ 3.549

Ba IV 167(5)

2. $\int_0^\infty e^{-\beta x} \operatorname{ch}(2n \operatorname{Arsh} x)\, dx = O_{2n}(\beta) \quad [\operatorname{Re}\beta > 0] \quad (\text{cf. } 3.547\ 8.).$

Ba IV 168(6)

3. $\int_0^\infty e^{-\beta x} \operatorname{sh}(\nu \operatorname{Arsh} x)\, dx = \dfrac{\nu}{\beta} S_{0,\nu}(\beta) \quad [\operatorname{Re}\beta > 0] \quad (\text{cf. } 3.547\ 5.).$

Ba IV 168(7)

4. $\int_0^\infty e^{-\beta x} \operatorname{ch}(\nu \operatorname{Arsh} x)\, dx = \dfrac{1}{\beta} S_{1,\nu}(\beta) \quad [\operatorname{Re}\beta > 0] \quad (\text{cf. } 3.547\ 7.).$

Ba IV 168(8)

Eine Reihe anderer Integrale, in denen Hyperbelfunktionen und die Exponentialfunktion vorkommen, die von Arsh x oder Arch x abhängen, findet man in dieser Sammlung, wenn man vorher die Substitutionen $x = \text{sh } t$ bzw. $x = \text{ch } t$ durchgeführt hat.

Some other integrals with hyperbolic functions and exponential functions depending on Arsh x or Arch x although not included in these tables in the original form are to be found after the substitution $x = \text{sh } t$ or $x = \text{ch } t$.

Hyperbel-, Exponential- und Potenzfunktionen
Hyperbolic functions, exponentials and powers

3.55–3.56

3.551

1. $$\int_0^\infty x^{\mu-1} e^{-\beta x} \text{sh } \gamma x \, dx = \frac{1}{2} \Gamma(\mu)[(\beta-\gamma)^{-\mu} - (\beta+\gamma)^{-\mu}]$$
    [Re $\mu > -1$, Re$\beta > |$Re $\gamma|$].   Ba IV 164(18)

2. $$\int_0^\infty x^{\mu-1} e^{-\beta x} \text{ch } \gamma x \, dx = \frac{1}{2} \Gamma(\mu)[(\beta-\gamma)^{-\mu} + (\beta+\gamma)^{-\mu}]$$
    [Re $\mu > 0$, Re$\beta > |$Re $\gamma|$].   Ba 164(19)

3. $$\int_0^\infty x^{\mu-1} e^{-\beta x} \text{cth } x \, dx = \Gamma(\mu) \left[2^{1-\mu} \zeta\left(\mu, \frac{\beta}{2}\right) - \beta^{-\mu} \right]$$
    [Re $\mu > 1$,   Re $\beta > 0$].   Ba 164(21)

4. $$\int_0^\infty x^n e^{-(p+mq)x} \text{sh}^m qx \, dx = 2^{-m} n! \sum_{k=0}^{m} \binom{m}{k} \frac{(-1)^k}{(p+2kq)^{n+1}}$$
    [$p > 0$, $q > 0$, $m < p + qm$].   Li[81](4)

5. $$\int_0^1 \frac{e^{-\beta x}}{x} \cdot \text{sh } \gamma x \, dx = \frac{1}{2}\left[\ln\frac{\beta+\gamma}{\beta-\gamma} + \text{Ei}(\gamma-\beta) - \text{Ei}(-\gamma-\beta)\right].$$
    BH[80](4)

6. $$\int_0^\infty \frac{e^{-\beta x}}{x} \text{sh } \gamma x \, dx = \frac{1}{2} \ln \frac{\beta+\gamma}{\beta-\gamma} \quad [\text{Re } \beta > |\text{Re } \gamma|].$$
    Ba IV 163(12)

7. $$\int_1^\infty \frac{e^{-\beta x}}{x} \text{ch } \gamma x \, dx = \frac{1}{2}[-\text{Ei}(\gamma-\beta) - \text{Ei}(-\gamma-\beta)] \quad [\text{Re } \beta > |\text{Re } \gamma|].$$
    Ba IV 164(15)

8. $$\int_0^\infty x e^{-x} \text{cth } x \, dx = \frac{\pi^2}{3} - 1.$$
    BH[82](6)

9. $$\int_0^\infty e^{-\beta x} \text{th } x \frac{dx}{x} = \ln \frac{\beta}{4} + 2 \ln \frac{\Gamma\left(\frac{\beta}{4}\right)}{\Gamma\left(\frac{\beta}{4} + \frac{1}{2}\right)} \quad [\text{Re } \beta > 0].$$
    Ba IV 164(16)

1. $\int_0^\infty \dfrac{x^{\mu-1} e^{-\beta x}}{\operatorname{sh} x}\, dx = 2^{1-\mu}\, \Gamma(\mu)\, \zeta\!\left[\mu, \dfrac{1}{2}(\beta+1)\right]$ [Re $\mu > 1$, Re $\beta > -1$]. 3.552

Ba IV 164(20)

2. $\int_0^\infty \dfrac{x^{2m-1} e^{-ax}}{\operatorname{sh} ax}\, dx = \dfrac{1}{2m}\, |B_{2m}|\!\left(\dfrac{\pi}{a}\right)^{2m}.$

Ba I 38(24)+

3. $\int_0^\infty \dfrac{x^{\mu-1} e^{-x}}{\operatorname{ch} x}\, dx = 2^{1-\mu}\,(1 - 2^{1-\mu})\,\Gamma(\mu)\,\zeta(\mu)$ [Re $\mu > 0$].

Ba I 32(5)

4. $\int_0^\infty \dfrac{x^{2m-1} e^{-ax}}{\operatorname{ch} ax}\, dx = \dfrac{1 - 2^{1-2m}}{2m}\, |B_{2m}|\!\left(\dfrac{\pi}{a}\right)^{2m}.$

Ba I 39(25)+

5. $\int_0^\infty \dfrac{x^2 e^{-2nx}}{\operatorname{sh} x}\, dx = 4 \sum_{k=n}^\infty \dfrac{1}{(2k+1)^3}$ (cf. 4.261 13.).

BH [84](4)

6. $\int_0^\infty \dfrac{x^3 e^{-2nx}}{\operatorname{sh} x}\, dx = \dfrac{\pi^4}{8} - 12 \sum_{k=1}^n \dfrac{1}{(2k-1)^4}$ (cf. 4.262 6.).

BH [84](6)

1. $\int_0^\infty \dfrac{\operatorname{sh}^2 ax}{\operatorname{sh} x} \cdot \dfrac{e^{-x}\, dx}{x} = \dfrac{1}{2} \ln(a\pi \operatorname{cosec} a\pi)$ [$a < 1$]. 3.553

BH [95](7)

2. $\int_0^\infty \dfrac{\operatorname{sh}^2 \dfrac{x}{2}}{\operatorname{ch} x} \cdot \dfrac{e^{-x}\, dx}{x} = \dfrac{1}{2} \ln \dfrac{4}{\pi}$ (cf. 4.267 2.).

BH [95](4)

1. $\int_0^\infty e^{-\beta x}(1 - \operatorname{sech} x) \dfrac{dx}{x} = 2 \ln \dfrac{\Gamma\!\left(\dfrac{\beta+3}{4}\right)}{\Gamma\!\left(\dfrac{\beta+1}{4}\right)} - \ln \dfrac{\beta}{4}$ [Re $\beta > 0$]. 3.554

Ba IV 164(7)

2. $\int_0^\infty e^{-\beta x}\!\left(\dfrac{1}{x} - \operatorname{cosech} x\right) dx = \psi\!\left(\dfrac{\beta+1}{2}\right) - \ln \dfrac{\beta}{2}$ [Re $\beta > 0$].

Ba IV 163(10)

3. $\int_0^\infty \left[\dfrac{\operatorname{sh}\!\left(\dfrac{1}{2} - \beta\right) x}{\operatorname{sh} \dfrac{x}{2}} - (1 - 2\beta)\, e^{-x}\right] \dfrac{dx}{x} = 2 \ln \Gamma(\beta) - \ln \pi + \ln(\sin \pi \beta)$

[$0 < \operatorname{Re} \beta < 1$]. Ba I 21(7)

4. $\int_0^\infty e^{-\beta x}\!\left(\dfrac{1}{x} - \operatorname{cth} x\right) dx = \psi\!\left(\dfrac{\beta}{2}\right) - \ln \dfrac{\beta}{2} + \dfrac{1}{\beta}$ [Re $\beta > 0$].

Ba IV 163(11)

3.554

5. $\int_0^\infty \left\{ -\frac{\operatorname{sh} qx}{\operatorname{sh} \frac{x}{2}} + 2qe^{-x} \right\} \frac{dx}{x} = 2 \ln \Gamma\left(q + \frac{1}{2}\right) + \ln \cos \pi q - \ln \pi$

$\left[q^2 < \frac{1}{2} \right]$. WW 250

6. $\int_0^\infty x^{\mu-1} e^{-\beta x} (\operatorname{cth} x - 1)\, dx = 2^{1-\mu} \Gamma(\mu) \zeta\left(\mu, \frac{\beta}{2} + 1\right)$

$[\operatorname{Re} \beta > 0,\ \operatorname{Re} \mu > 1]$. Ba IV 164(22)

3.555

1. $\int_0^\infty \frac{\operatorname{sh}^2 ax}{1 - e^{px}} \cdot \frac{dx}{x} = \frac{1}{4} \ln\left(\frac{p}{2a\pi} \sin \frac{2a\pi}{p}\right)$ $[2a < p]$ (cf. 3.545 2.). BH[93](15)

2. $\int_0^\infty \frac{\operatorname{sh}^2 ax}{e^x + 1} \cdot \frac{dx}{x} = -\frac{1}{4} \ln(a\pi \operatorname{ctg} a\pi)$ $\left[a < \frac{1}{2} \right]$ (cf. 3.545 1.). BH[93](9)

3.556

1. $\int_{-\infty}^\infty x \frac{1 - e^{px}}{\operatorname{sh} x}\, dx = -\frac{\pi^2}{2} \operatorname{tg}^2 \frac{p\pi}{2}$ $[p < 1]$. (cf. 4.255 3.). BH[101](4)

2. $\int_0^\infty \frac{1 - e^{-px}}{\operatorname{sh} x} \cdot \frac{1 - e^{-(p+1)x}}{x}\, dx = 2p \ln 2$ $[p > -1]$. BH[95](8)

3.557

1. $\int_0^\infty \frac{e^{-px} - e^{-qx}}{\operatorname{ch} x - \cos \frac{m}{n} \pi} \cdot \frac{dx}{x} =$

$= 2 \operatorname{cosec} \frac{m}{n} \pi \sum_{k=1}^{n-1} (-1)^{k-1} \sin\left(\frac{km}{n} \pi\right) \ln \frac{\Gamma\left(\frac{n+q+k}{2n}\right) \Gamma\left(\frac{p+k}{2n}\right)}{\Gamma\left(\frac{n+p+k}{2n}\right) \Gamma\left(\frac{q+k}{2n}\right)}$

$\left[m+n \quad \begin{matrix} \text{ungeradzahlig} \\ \text{odd} \end{matrix} \right];$

$= 2 \operatorname{cosec} \frac{m}{n} \pi \sum_{k=1}^{\frac{n-1}{2}} (-1)^{k-1} \sin\left(\frac{km}{n} \pi\right) \ln \frac{\Gamma\left(\frac{n+q-k}{n}\right) \Gamma\left(\frac{p+k}{n}\right)}{\Gamma\left(\frac{n+p-k}{n}\right) \Gamma\left(\frac{q+k}{n}\right)}$

$\left[m+n \quad \begin{matrix} \text{geradzahlig} \\ \text{even} \end{matrix} \right];$

$[p > -1,\ q > -1]$. BH[96](1)

2. $\int_0^\infty \frac{(1 - e^{-x})^2}{\operatorname{ch} x + \cos \frac{m}{n} \pi} \cdot \frac{dx}{x} =$

$= 2 \operatorname{cosec} \frac{m}{n} \pi \sum_{k=1}^{n-1} (-1)^{k-1} \sin\left(\frac{km}{n} \pi\right) \times$

$$\times \ln \frac{\left[\Gamma\left(\frac{n+k+1}{2n}\right)\right]^2 \Gamma\left(\frac{k+2}{2n}\right) \Gamma\left(\frac{k}{2n}\right)}{\left[\Gamma\left(\frac{k+1}{2n}\right)\right]^2 \Gamma\left(\frac{n+k}{2n}\right) \Gamma\left(\frac{n+k+2}{2n}\right)} \quad \left[m+n \quad \begin{matrix}\text{ungeradzahlig}\\ \text{odd}\end{matrix}\right]; \quad 3.557$$

$$= 2 \operatorname{cosec} \frac{m}{n} \pi \sum_{k=1}^{\frac{n-1}{2}} (-1)^{k-1} \sin\left(\frac{km}{n}\pi\right) \times$$

$$\times \ln \frac{\left[\Gamma\left(\frac{n-k+1}{n}\right)\right]^2 \Gamma\left(\frac{k+2}{n}\right) \Gamma\left(\frac{k}{n}\right)}{\left[\Gamma\left(\frac{k+1}{n}\right)\right]^2 \Gamma\left(\frac{n-k}{n}\right) \Gamma\left(\frac{n-k+2}{n}\right)} \quad \left[m+n \quad \begin{matrix}\text{geradzahlig}\\ \text{even}\end{matrix}\right].$$

BH [96](2)

3. $\displaystyle\int_0^\infty \left[e^{-x} \operatorname{tg} \frac{m}{2n}\pi - \frac{e^{-px} \sin\frac{m}{n}\pi}{\operatorname{ch} x + \cos\frac{m}{n}\pi} \right] \cdot \frac{dx}{x} =$

$$= \operatorname{tg}\left(\frac{m}{2n}\pi\right) \ln(2n) + 2 \sum_{k=1}^{n-1} (-1)^{k-1} \sin\left(\frac{km}{n}\pi\right) \ln \frac{\Gamma\left(\frac{p+n+k}{2n}\right)}{\Gamma\left(\frac{p+k}{2n}\right)}$$

$$\left[m+n \quad \begin{matrix}\text{ungeradzahlig}\\ \text{odd}\end{matrix}\right];$$

$$= \operatorname{tg}\left(\frac{m}{2n}\pi\right) \ln n + 2 \sum_{k=1}^{\frac{n-1}{2}} (-1)^{k-1} \sin\left(\frac{km}{n}\pi\right) \ln \frac{\Gamma\left(\frac{p+n-k}{n}\right)}{\Gamma\left(\frac{p+k}{n}\right)}$$

$$\left[m+n \quad \begin{matrix}\text{geradzahlig}\\ \text{even}\end{matrix}\right]. \qquad \text{BH [96](3)}$$

4. $\displaystyle\int_0^\infty \frac{1+e^{-x}}{\operatorname{ch} x + \cos a} \cdot \frac{dx}{x^{1-p}} = 2 \sec\frac{a}{2} \Gamma(p) \sum_{k=1}^\infty (-1)^{k-1} \frac{\cos\left(k-\frac{1}{2}\right)a}{k^p} \quad [p>0].$

Li [96](5)

5. $\displaystyle\int_0^\infty \frac{x^q e^{-\frac{x}{2}} \operatorname{ch}\frac{x}{2}}{\operatorname{ch} x + \cos\lambda} dx = \frac{\Gamma(q+1)}{\cos\frac{\lambda}{2}} \sum_{k=1}^\infty (-1)^{k-1} \frac{\cos\left(k-\frac{1}{2}\right)\lambda}{k^{q+1}} \quad [q>-1].$

Li 96(5)+

6. $\displaystyle\int_0^\infty x \frac{e^{-x}-\cos a}{\operatorname{ch} x - \cos a} dx = a\pi - \frac{a^2}{2} - \frac{\pi^2}{3}.$ BH[88](8)

7. $\displaystyle\int_0^\infty x^{2m+1} \frac{e^{-x}-\cos a\pi}{\operatorname{ch} x - \cos a\pi} dx = 2\cdot(2m+1)! \sum_{k=1}^\infty \frac{\cos ka\pi}{k^{2m+2}}.$ BH[88](6)

3.558

1. $$\int_0^\infty x \frac{1-e^{-nx}}{\operatorname{sh}^2 \frac{x}{2}} dx = \frac{2n\pi^2}{3} - 4 \sum_{k=1}^{n-1} \frac{n-k}{k^2}.$$ BH[85](3)

2. $$\int_0^\infty x \frac{1-(-1)^n e^{-nx}}{\operatorname{ch}^2 \frac{x}{2}} dx = \frac{n\pi^2}{3} + 4 \sum_{k=1}^{n-1} (-1)^k \frac{n-k}{k^2}.$$ Li[85](1)

3. $$\int_0^\infty x^2 \frac{1-e^{-nx}}{\operatorname{sh}^2 \frac{x}{2}} dx = 8n\zeta(3) - 8 \sum_{k=1}^{n-1} \frac{n-k}{k^3}.$$ BH[85](5)

4. $$\int_0^\infty x^2 e^x \frac{1-e^{-2nx}}{\operatorname{sh}^2 x} dx = 8n \sum_{k=1}^\infty \frac{1}{(2k-1)^3} - 8 \sum_{k=1}^{n-1} \frac{n-k}{(2k-1)^3}.$$ Li[85](6)

5. $$\int_0^\infty x^2 \frac{1+(-1)^n e^{-nx}}{\operatorname{ch}^2 \frac{x}{2}} dx = 6n\zeta(3) - 8 \sum_{k=1}^{n-1} \frac{n-k}{k^3}.$$ Li[85](4)

6. $$\int_0^\infty x^3 \frac{1-e^{-nx}}{\operatorname{sh}^2 \frac{x}{2}} dx = \frac{4}{15} n\pi^4 - 24 \sum_{k=1}^{n-1} \frac{n-k}{k^4}.$$ BH[85](9)

7. $$\int_0^\infty x^3 \frac{1+(-1)^n e^{-nx}}{\operatorname{ch}^2 \frac{x}{2}} dx = \frac{7}{30} n\pi^4 + 24 \sum_{k=1}^{n-1} (-1)^k \frac{n-k}{k^4}.$$ BH[85](8)

3.559

$$\int_0^\infty e^{-x} \left[a - \frac{1}{2} + \frac{(1-e^{-x})(1-ax) - xe^{-x}}{4 \operatorname{sh}^2 \frac{x}{2}} e^{(2-a)x} \right] \frac{dx}{x} =$$

$$= a - \frac{1}{2} + \ln \Gamma(a) - \frac{1}{2} \ln(2\pi) \quad [a > 0].$$ BH[96](6)

3.561

$$\int_0^\infty \frac{e^{-2x} \operatorname{th} \frac{x}{2}}{x \operatorname{ch} x} dx = 2 \ln \frac{\pi}{2\sqrt{2}}.$$ BH[93](18)

3.562

1. $$\int_0^\infty x^{2\mu-1} e^{-\beta x^2} \operatorname{sh} \gamma x \, dx = \frac{1}{2} \Gamma(2\mu)(2\beta)^{-\mu} \exp\left(\frac{\gamma^2}{8\beta}\right) \times$$

$$\times \left[D_{-2\mu}\left(-\frac{\gamma}{\sqrt{2\beta}}\right) - D_{-2\mu}\left(\frac{\gamma}{\sqrt{2\beta}}\right) \right] \quad \left[\operatorname{Re} \mu > -\frac{1}{2}, \operatorname{Re} \beta > 0\right].$$ Ba IV 166(44)

2. $$\int_0^\infty x^{2\mu-1} e^{-\beta x^2} \operatorname{ch} \gamma x \, dx = \frac{1}{2} \Gamma(2\mu)(2\beta)^{-\mu} \exp\left(\frac{\gamma^2}{8\beta}\right) \times$$

$$\times \left[D_{-2\mu}\left(-\frac{\gamma}{\sqrt{2\beta}}\right) + D_{-2\mu}\left(\frac{\gamma}{\sqrt{2\beta}}\right) \right] \quad [\operatorname{Re} \mu > 0, \operatorname{Re} \beta > 0].$$ Ba IV 166(45)

3. $$\int_0^\infty x e^{-\beta x^2} \operatorname{sh} \gamma x \, dx = \frac{\gamma}{4\beta} \sqrt{\frac{\pi}{\beta}} \exp \frac{\gamma^2}{4\beta} \quad [\operatorname{Re} \beta > 0].$$

BH[81](12)+, Ba IV 165(34)

4. $\int_0^\infty x e^{-\beta x^2} \operatorname{ch} \gamma x \, dx = \frac{\gamma}{4\beta} \sqrt{\frac{\pi}{\beta}} \exp \frac{\gamma^2}{4\beta} \Phi\left(\frac{\gamma}{2\sqrt{\beta}}\right) + \frac{1}{2\beta}$ $[\operatorname{Re} \beta > 0]$.

Ba IV 166(35)

5. $\int_0^\infty x^2 e^{-\beta x^2} \operatorname{sh} \gamma x \, dx = \frac{\sqrt{\pi}(2\beta + \gamma^2)}{8\beta^2 \sqrt{\beta}} \exp\left(\frac{\gamma^2}{4\beta}\right) \Phi\left(\frac{\gamma}{2\sqrt{\beta}}\right) + \frac{\gamma}{4\beta^2}$ $[\operatorname{Re} \beta > 0]$.

Ba IV 166(36)

6. $\int_0^\infty x^2 e^{-\beta x^2} \operatorname{ch} \gamma x \, dx = \frac{\sqrt{\pi}(2\beta + \gamma^2)}{8\beta^2 \sqrt{\beta}} \exp\left(\frac{\gamma^2}{4\beta}\right)$ $[\operatorname{Re} \beta > 0]$. Ba IV 166(37)

Trigonometrische Funktionen
Trigonometric Functions

3.6–4.1

Rationale Funktionen von Sinus und Cosinus und trigonometrische Funktionen von Vielfachen des Bogens
Rational functions of sines and cosines and trigonometric functions of multiple angles

3.61

1. $\int_0^{2\pi} (1 - \cos x)^n \sin nx \, dx = 0.$ BH[68](10) 3.611

2. $\int_0^{2\pi} (1 - \cos x)^n \cos nx \, dx = (-1)^n \frac{\pi}{2^{n-1}}.$ BH[68](11)

3. $\int_0^\pi (\cos t + i \sin t \cos x)^n \, dx = \int_0^\pi (\cos t + i \sin t \cos x)^{-n-1} \, dx = \pi P_n(\cos t).$

Ba I 158(23)+

1. $\int_0^\pi \frac{\sin nx \cos mx}{\sin x} \, dx = 0$ für/for $n \leqslant m$; 3.612

    $= \pi$ für/for $n > m$, wenn/if $m+n$ ungeradzahlig ist/odd;

    $= 0$ für/for $n > m$, wenn/if $m+n$ geradzahlig ist/even.

Li[64](3)

2. $\int_0^\pi \frac{\sin nx}{\sin x} \, dx = 0,$ wenn/if n geradzahlig ist/even;

    $= \pi,$ wenn/if n ungeradzahlig ist/odd. BH[64](1, 2)

3.612 3. $\displaystyle\int_0^{\frac{\pi}{2}} \frac{\sin(2n-1)x}{\sin x}\,dx = \frac{\pi}{2}.$ F II 129

4. $\displaystyle\int_0^{\frac{\pi}{2}} \frac{\sin 2nx}{\sin x}\,dx = 2\left(1 - \frac{1}{3} + \frac{1}{5} - \cdots + \frac{(-1)^{n-1}}{2n-1}\right).$ GH[332](21b)

5. $\displaystyle\int_0^{\pi} \frac{\sin 2nx}{\cos x}\,dx = 2\int_0^{\frac{\pi}{2}} \frac{\sin 2nx}{\cos x}\,dx = (-1)^{n-1} 4\left(1 - \frac{1}{3} + \frac{1}{5} - \cdots + \frac{(-1)^{n-1}}{2n-1}\right).$ GH[332](22a)

6. $\displaystyle\int_0^{\pi} \frac{\cos(2n+1)x}{\cos x}\,dx = 2\int_0^{\frac{\pi}{2}} \frac{\cos(2n+1)x}{\cos x}\,dx = (-1)^n \pi.$ GH[332](22b)

7. $\displaystyle\int_0^{\frac{\pi}{2}} \frac{\sin 2nx \cos x}{\sin x}\,dx = \frac{\pi}{2}.$ Li[45](17)

3.613 1. $\displaystyle\int_0^{\pi} \frac{\cos nx\,dx}{1 + a\cos x} = \frac{\pi}{\sqrt{1-a^2}}\left(\frac{\sqrt{1-a^2}-1}{a}\right)^n \quad [a^2 < 1].$ BH[64](12)

2. $\displaystyle\int_0^{\pi} \frac{\cos nx\,dx}{1 - 2a\cos x + a^2} = \frac{\pi a^n}{1-a^2} \quad [a^2 < 1];$

$\displaystyle\qquad\qquad\qquad\qquad = \frac{\pi}{(a^2-1)a^n} \quad [a^2 > 1].$ BH[65](3)

3. $\displaystyle\int_0^{\pi} \frac{\sin nx \sin x\,dx}{1 - 2a\cos x + a^2} = \begin{cases} \dfrac{\pi}{2} a^{n-1} & [n \geq 1] \\ 0 & [n = 0] \end{cases} \quad [a^2 < 1];$

$\displaystyle\qquad\qquad\qquad\qquad = \begin{cases} \dfrac{\pi}{2a^{n+1}} & [n \geq 1] \\ 0 & [n = 0] \end{cases} \quad [a^2 > 1].$

BH[65](4), GH[332](34a)

4. $\displaystyle\int_0^{\pi} \frac{\cos nx \cos x\,dx}{1 - 2a\cos x + a^2} = \begin{cases} \dfrac{\pi}{2} \cdot \dfrac{1+a^2}{1-a^2} a^{n-1} & [n \geq 1] \\ \dfrac{\pi a}{1-a^2} & [n = 0] \end{cases} \quad [a^2 < 1];$

$\displaystyle\qquad\qquad\qquad\qquad = \begin{cases} \dfrac{\pi}{2a^{n+1}} \cdot \dfrac{a^2+1}{a^2-1} & [n \geq 1] \\ \dfrac{\pi}{a(a^2-1)} & [n = 0] \end{cases} \quad [a^2 > 1].$

BH[65](5), GH[332](34b)

5. $\int_0^\pi \dfrac{\cos(2n-1)x\,dx}{1-2a\cos 2x+a^2} = \int_0^\pi \dfrac{\cos 2nx \cos x\,dx}{1-2a\cos 2x+a^2} = 0 \quad [a^2 \neq 1]$.   BH[65](9, 10)   3.613

6. $\int_0^\pi \dfrac{\cos(2n-1)x \cos 2x\,dx}{1-2a\cos 2x+a^2} = 0 \quad [a^2 \neq 1]$.   BH[65](12)

7. $\int_0^\pi \dfrac{\sin 2nx \sin x\,dx}{1-2a\cos 2x+a^2} = \int_0^\pi \dfrac{\sin(2n-1)x \sin 2x\,dx}{1-2a\cos 2x+a^2} = 0 \quad [a^2 \neq 1]$.   BH[65](6, 7)

8. $\int_0^\pi \dfrac{\sin(2n-1)x \sin x\,dx}{1-2a\cos 2x+a^2} = \dfrac{\pi}{2}\cdot\dfrac{a^{n-1}}{1+a} \quad [a^2 < 1]$;

$\qquad\qquad\qquad\qquad\qquad = \dfrac{\pi}{2}\cdot\dfrac{1}{(1+a)a^n} \quad [a^2 > 1]$.   BH[65](8)

9. $\int_0^\pi \dfrac{\cos(2n-1)x \cos x\,dx}{1-2a\cos 2x+a^2} = \dfrac{\pi}{2}\cdot\dfrac{a^{n-1}}{1-a} \quad [a^2 < 1]$;

$\qquad\qquad\qquad\qquad\qquad = \dfrac{\pi}{2}\cdot\dfrac{1}{(a-1)a^n} \quad [a^2 > 1]$.   BH[65](11)

10. $\int_0^\pi \dfrac{\sin nx - a\sin(n-1)x}{1-2a\cos x+a^2}\sin mx\,dx = 0 \quad \text{für/for } m < n$;

$\qquad\qquad\qquad\qquad\qquad = \dfrac{\pi}{2}a^{m-n} \quad \text{für/for } m \geq n;\ [a^2 < 1]$.   Li[65](13)

11. $\int_0^\pi \dfrac{\cos nx - a\cos(n-1)x}{1-2a\cos x+a^2}\cos mx\,dx = \dfrac{\pi}{2}(a^{m-n}-1) \quad [a^2 < 1]$.   BH[65](14)

12. $\int_0^\pi \dfrac{\sin nx - a\sin[(n+1)x]}{1-2a\cos x+a^2}\,dx = 0 \quad [a^2 < 1]$.   BH[68](13)

13. $\int_0^\pi \dfrac{\cos nx - a\cos[(n+1)x]}{1-2a\cos x+a^2}\,dx = 2\pi a^n \quad [a^2 < 1]$.   BH[68](14)

$\int_0^\pi \dfrac{\sin x}{a^2-2ab\cos x+b^2}\cdot\dfrac{\sin px\,dx}{1-2a^p\cos px+a^{2p}} = \dfrac{\pi b^{p-1}}{2a^{p+1}(1-b^p)}$ 3.614

$\qquad\qquad\qquad [0 < a < 1,\ 0 < a < b,\ p > 0]$.   BH[66](9)

1. $\int_0^{\pi/2} \dfrac{\cos 2nx\,dx}{1-a^2\sin^2 x} = \dfrac{(-1)^n \pi}{2\sqrt{1-a^2}}\left(\dfrac{1-\sqrt{1-a^2}}{a}\right)^{2n} \quad [a^2 < 1]$.   BH[47](27)   3.615

2. $\int_0^\pi \dfrac{\cos x \sin 2nx\,dx}{1+(a+b\sin x)^2} = -\dfrac{\pi}{b}\sin\left\{2n\,\text{arctg}\sqrt{\dfrac{s}{2}}\right\}\text{tg}^{2n}\left(\dfrac{1}{2}\arccos\sqrt{\dfrac{s}{2a^2}}\right)$.

3.615

3. $\int_0^\pi \dfrac{\cos x \cos (2n+1) x \, dx}{1 + (a + b \sin x)^2} =$

$= \dfrac{\pi}{b} \cos\left\{(2n+1) \operatorname{arctg} \sqrt{\dfrac{s}{2}}\right\} \operatorname{tg}^{2n+1}\left(\dfrac{1}{2} \arccos \sqrt{\dfrac{s}{2a^2}}\right),$

wobei
where $\quad s = -(1 + b^2 - a^2) + \sqrt{(1 + b^2 - a^2)^2 + 4a^2}.$ BH[65](21, 22)

3.616

1. $\int_0^\pi (1 - 2a \cos x + a^2)^n \, dx = \pi \sum_{k=0}^n \binom{n}{k}^2 a^{2k}.$ BH[63](1)

2. $\int_0^\pi \dfrac{dx}{(1 - 2a \cos x + a^2)^n} = \dfrac{1}{2} \int_0^{2\pi} \dfrac{dx}{(1 - 2a \cos x + a^2)^n} =$

$= \dfrac{\pi}{(1-a^2)^n} \sum_{k=0}^{n-1} \dfrac{(n+k-1)!}{(k!)^2 (n-k-1)!} \cdot \left(\dfrac{a^2}{1-a^2}\right)^k \quad [a^2 < 1];$

$= \dfrac{\pi}{(a^2-1)^n} \sum_{k=0}^{n-1} \dfrac{(n+k-1)!}{(k!)^2 (n-k-1)!} \cdot \dfrac{1}{(a^2-1)^k} \quad [a^2 > 1].$ GH[331](63)

3. $\int_0^\pi (1 - 2a \cos x + a^2)^n \cos nx \, dx = (-1)^n \pi a^n.$ BH[63](2)

4. $\int_0^\pi (1 - 2a \cos x + a^2)^n \cos mx \, dx = \dfrac{1}{2} \int_0^{2\pi} (1 - 2a \cos x + a^2)^n \cos mx \, dx =$

$= 0 \quad [n < m];$

$= \pi(-a)^m (1 + a^2)^{n-m} \sum_{k=0}^{E\left(\frac{n-m}{2}\right)} \binom{n}{k} \binom{n-k}{m+k} \left(\dfrac{a}{1+a^2}\right)^{2k} \quad [n \geqslant m].$
GH[332](35a)

5. $\int_0^{2\pi} \dfrac{\sin nx \, dx}{(1 - 2a \cos 2x + a^2)^m} = 0.$ GH[332](32a)

6. $\int_0^\pi \dfrac{\sin x \, dx}{(1 - 2a \cos 2x + a^2)^m} = \dfrac{1}{2(m-1)a} \left[\dfrac{1}{(1-a)^{2m-2}} - \dfrac{1}{(1+a)^{2m-2}}\right]$

$[a \neq 0, \pm 1].$ GH[332](32c)

7. $\int_0^\pi \dfrac{\cos nx \, dx}{(1 - 2a \cos x + a^2)^m} = \dfrac{1}{2} \int_0^{2\pi} \dfrac{\cos nx \, dx}{(1 - 2a \cos x + a^2)^m} =$

$= \dfrac{a^{2m+n-2}\pi}{(1-a^2)^{2m-1}} \sum_{k=0}^{m-1} \binom{m+n-1}{k} \binom{2m-k-2}{m-1} \left(\dfrac{1-a^2}{a^2}\right)^k \quad [a^2 < 1];$

$= \dfrac{\pi}{a^n (a^2-1)^{2m-1}} \sum_{k=0}^{m-1} \binom{m+n-1}{k} \binom{2m-k-2}{m-1} (a^2-1)^k \quad [a^2 > 1].$
GH[332](31)

8. $\int_0^{\frac{\pi}{2}} \dfrac{\cos 2nx \, dx}{(a^2 \cos^2 x + b^2 \sin^2 x)^{n+1}} = \binom{2n}{n} \dfrac{(b^2 - a^2)^n}{(2ab)^{2n+1}} \pi \quad [a > 0, \, b > 0].$ GH[332](30b)

Potenzen trigonometrischer Funktionen
Powers of trigonometric functions   3.62

1. $\int_0^{\pi/2} \sin^{\mu-1} x \, dx = \int_0^{\pi/2} \cos^{\mu-1} x \, dx = 2^{\mu-2} B\left(\frac{\mu}{2}, \frac{\mu}{2}\right).$ F II 794–795   3.621

2. $\int_0^{\pi/2} \sin^{\frac{3}{2}} x \, dx = \int_0^{\pi/2} \cos^{\frac{3}{2}} x \, dx = \frac{1}{6\sqrt{2\pi}} \Gamma\left(\frac{1}{4}\right).$

3. $\int_0^{\pi/2} \sin^{2m} x \, dx = \int_0^{\pi/2} \cos^{2m} x \, dx = \frac{(2m-1)!!}{(2m)!!} \frac{\pi}{2}.$ F II 137

4. $\int_0^{\pi/2} \sin^{2m+1} x \, dx = \int_0^{\pi/2} \cos^{2m+1} x \, dx = \frac{(2m)!!}{(2m+1)!!}.$ F II 137

5. $\int_0^{\pi/2} \sin^{\mu-1} x \cos^{\nu-1} x \, dx = \frac{1}{2} B\left(\frac{\mu}{2}, \frac{\nu}{2}\right) \quad [\operatorname{Re} \mu > 0, \operatorname{Re} \nu > 0].$

Lo V 113(50), Lo V 122, F II 794

1. $\int_0^{\pi/2} \operatorname{tg}^{\pm\mu} x \, dx = \frac{\pi}{2} \sec \frac{\mu\pi}{2} \quad [|\operatorname{Re} \mu| < 1].$ BH[42](1)   3.622

2. $\int_0^{\pi/4} \operatorname{tg}^{\mu} x \, dx = \frac{1}{2} \beta\left(\frac{\mu+1}{2}\right) \quad [\operatorname{Re} \mu > -1].$ BH[34](1)

3. $\int_0^{\pi/4} \operatorname{tg}^{2n} x \, dx = (-1)^n \frac{\pi}{4} + \sum_{k=0}^{n-1} \frac{(-1)^k}{2n-2k-1}.$ BH[34](2)

4. $\int_0^{\pi/4} \operatorname{tg}^{2n+1} x \, dx = (-1)^{n+1} \frac{\ln 2}{2} + \sum_{k=0}^{n-1} \frac{(-1)^k}{2n-2k}.$ BH[34](3)

1. $\int_0^{\pi/2} \operatorname{tg}^{\mu-1} x \cos^{2\nu-2} x \, dx = \int_0^{\pi/2} \operatorname{ctg}^{\mu-1} x \sin^{2\nu-2} x \, dx =$ 3.623

$= \frac{1}{2} B\left(\frac{\mu}{2}, \nu - \frac{\mu}{2}\right) \quad [0 < \operatorname{Re} \mu < 2 \operatorname{Re} \nu].$ BH[42](6), BH[45](22)

2. $\int_0^{\pi/4} \operatorname{tg}^{\mu} x \sin^2 x \, dx = \frac{1+\mu}{4} \beta\left(\frac{\mu+1}{2}\right) \quad [\operatorname{Re} \mu > -1].$ BH[34](4)

3.623   3. $\int_0^{\pi/4} \operatorname{tg}^\mu x \cos^2 x \, dx = \dfrac{1-\mu}{4} \mathrm{B}\!\left(\dfrac{\mu+1}{2}\right)$ [Re $\mu > -1$].   BH[34](5)

3.624   1. $\int_0^{\pi/4} \dfrac{\sin^p x}{\cos^{p+2} x} \, dx = \dfrac{1}{p+1}$ [$p > -1$].   GH[331](34b)

2. $\int_0^{\pi/2} \dfrac{\sin^{\mu-\frac{1}{2}} x}{\cos^{2\mu-1} x} \, dx = \int_0^{\pi/2} \dfrac{\cos^{\mu-\frac{1}{2}} x}{\sin^{2\mu-1} x} \, dx = \dfrac{\Gamma\!\left(\dfrac{\mu}{2}+\dfrac{1}{4}\right) \Gamma(1-\mu)}{\Gamma\!\left(\dfrac{5}{4}-\dfrac{\mu}{2}\right)}$

$\left[-\dfrac{1}{2} < \operatorname{Re} \mu < 1\right]$.   Li[55](12)

3. $\int_0^{\pi/4} \dfrac{\cos^{n-\frac{1}{2}} 2x}{\cos^{2n+1} x} \, dx = \dfrac{(2n-1)!!}{2 \cdot (2n)!!} \pi$.   BH[38](3)

4. $\int_0^{\pi/4} \dfrac{\cos^\mu 2x}{\cos^{2(\mu+1)} x} \, dx = 2^{2\mu} \mathrm{B}(\mu+1, \mu+1)$ [Re $\mu > -1$].   BH[35](1)

5. $\int_0^{\pi/4} \dfrac{\sin^{2\mu-2} x}{\cos^\mu 2x} \, dx = 2^{1-2\mu} \mathrm{B}(2\mu-1, 1-\mu) = \dfrac{\Gamma\!\left(\mu-\dfrac{1}{2}\right) \Gamma(1-\mu)}{2\sqrt{\pi}}$

$\left[\dfrac{1}{2} < \operatorname{Re} \mu < 1\right]$.   BH[35](4)

6. $\int_0^{\pi/2} \left(\dfrac{\sin ax}{\sin x}\right)^2 dx = \dfrac{\pi |a|}{2} - \dfrac{\sin \pi |a|}{2} [2|a| \mathrm{B}(|a|) - 1]$.   F II 129+

3.625   1. $\int_0^{\pi/4} \dfrac{\sin^{2n-1} x \cos^p 2x}{\cos^{2p+2n+1} x} \, dx = \dfrac{(n-1)!}{2} \cdot \dfrac{\Gamma(p+1)}{\Gamma(p+n+1)} =$

$= \dfrac{(n-1)!}{2(p+n)(p+n-1)\ldots(p+1)} = \dfrac{1}{2} \mathrm{B}(n, p+1)$ [$p > -1$]   (cf. 3.251 1.).   BH[35](2)

2. $\int_0^{\pi/4} \dfrac{\sin^{2n} x \cos^p 2x}{\cos^{2p+2n+2} x} \, dx = \dfrac{1}{2} \mathrm{B}\!\left(n+\dfrac{1}{2}, p+1\right)$ [$p > -1$]   (cf. 3.251 1.).   BH[35](3)

3. $\int_0^{\pi/4} \dfrac{\sin^{2n-1} x \cos^{m-\frac{1}{2}} 2x}{\cos^{2n+2m} x} \, dx = \dfrac{(2n-2)!! (2m-1)!!}{(2n+2m-1)!!}$.   BH[38](6)

4. $$\int_0^{\pi/4} \frac{\sin^{2n} x \cos^{m-\frac{1}{2}} 2x}{\cos^{2n+2m+1} x} dx = \frac{(2n-1)!!\,(2m-1)!!}{(2n+2m)!!} \cdot \frac{\pi}{2};$$ BH[38](7)   3.625

1. $$\int_0^{\pi/4} \frac{\sin^{2n-1} x}{\cos^{2n+2} x} \sqrt{\cos 2x}\, dx = \frac{(2n-2)!!}{(2n+1)!!} \quad \text{(cf. 3.251 1.).}$$ BH[38](4)   3.626

2. $$\int_0^{\pi/4} \frac{\sin^{2n} x}{\cos^{2n+3} x} \sqrt{\cos 2x}\, dx = \frac{(2n-1)!!}{(2n+2)!!} \cdot \frac{\pi}{2} \quad \text{(cf. 3.251 1.).}$$ BH[38](5)

$$\int_0^{\pi/2} \frac{\operatorname{tg}^\mu x}{\cos^\mu x} dx = \int_0^{\pi/2} \frac{\operatorname{ctg}^\mu x}{\sin^\mu x} dx = \frac{\Gamma(\mu)\Gamma\left(\frac{1}{2}-\mu\right)}{2^\mu \sqrt{\pi}} \sin \frac{\mu\pi}{2}$$
$$\left[-1 < \operatorname{Re} \mu < \frac{1}{2}\right].$$ BH[55](12)+   3.627

$$\int_0^{\pi/2} \sec^{2p+1} x \frac{d \sin^{2p} x}{dx} dx = \frac{1}{\sqrt{\pi}} \Gamma(p+1) \Gamma\left(\frac{1}{2}-p\right) \quad \left[\frac{1}{2} > p > 0\right].$$ W 630   3.628

Potenzen trigonometrischer Funktionen und trigonometrische Funktionen linearer Argumentsfunktionen
Powers of trigonometric functions and trigonometric functions of linear functions
3.63

1. $$\int_0^\pi \sin^{\nu-1} x \sin ax\, dx = \frac{\pi \sin \frac{a\pi}{2}}{2^{\nu-1} \nu B\left(\frac{\nu+a+1}{2}, \frac{\nu-a+1}{2}\right)} \quad [\operatorname{Re} \nu > 0].$$ 3.631

Lo V 121(67)+, W 309+

2. $$\int_0^{\pi/2} \sin^{\nu-2} x \sin \nu x\, dx = \frac{-1}{\nu-1} \cos \frac{\nu\pi}{2} \quad [\operatorname{Re} \nu > 1].$$ GH[332](16d), F II 135

3. $$\int_0^\pi \sin^\nu x \sin \nu x\, dx = 2^{-\nu} \pi \sin \frac{\nu\pi}{2} \quad [\operatorname{Re} \nu > -1].$$ Lo V 121(69)

4. $$\int_0^\pi \sin^n x \sin 2mx\, dx = 0.$$ GH[332](11a)

5. $$\int_0^\pi \sin^{2n} x \sin(2m+1)x\, dx = 2\int_0^{\pi/2} \sin^{2n} x \sin(2m+1)x\, dx =$$
$$= \frac{(-1)^m\, 2^{n+1}\, n!\, (2n-1)!!}{(2n-2m-1)!!\,(2m+2n+1)!!} \quad [m \leq n]^*;$$
$$= \frac{(-1)^n\, 2^{n+1}\, n!\, (2m-2n-1)!!\,(2n-1)!!}{(2m+2n+1)!!} \quad [m \geq n]^*.$$ GH[332](11b)

* Im Fall $m = n$ setzt man $(2n - 2m - 1)!! = 1$. | * For $m = n$ one must put $(2n - 2m - 1)!! = 1$.

3.631

6. $\int_0^\pi \sin^{2n+1} x \sin(2m+1)x\, dx = 2\int_0^{\frac{\pi}{2}} \sin^{2n+1} x \sin(2m+1)x\, dx =$

$$= \frac{(-1)^m \pi}{2^{2n+1}} \binom{2n+1}{n-m} \quad [n \geq m];$$

$$= 0 \quad [n < m]. \qquad \text{BH[40](12), GH[332](11c)}$$

7. $\int_0^\pi \sin^n x \cos(2m+1)x\, dx = 0.$ \qquad GH[332](12a)

8. $\int_0^\pi \sin^{\nu-1} x \cos ax\, dx = \dfrac{\pi \cos\frac{a\pi}{2}}{2^{\nu-1} \nu B\left(\frac{\nu+a+1}{2}, \frac{\nu-a+1}{2}\right)}$

$[\operatorname{Re} \nu > 0].$ \qquad Lo V 121(68)+, W 309+

9. $\int_0^{\frac{\pi}{2}} \cos^{\nu-1} x \cos ax\, dx = \dfrac{\pi}{2^\nu \nu B\left(\frac{\nu+a+1}{2}, \frac{\nu-a+1}{2}\right)} \quad [\operatorname{Re} \nu > 0].$

GH[332](9c)

10. $\int_0^{\frac{\pi}{2}} \sin^{\nu-2} x \cos \nu x\, dx = \dfrac{1}{\nu-1} \sin\frac{\nu\pi}{2} \quad [\operatorname{Re} \nu > 1].$ \qquad GH[332](16b), F II 135

11. $\int_0^\pi \sin^\nu x \cos \nu x\, dx = \dfrac{\pi}{2^\nu} \cos\frac{\nu\pi}{2} \quad [\operatorname{Re} \nu > -1].$ \qquad Lo V 121(70)+

12. $\int_0^\pi \sin^{2n} x \cos 2mx\, dx = 2\int_0^{\frac{\pi}{2}} \sin^{2n} x \cos 2mx\, dx =$

$$= \frac{(-1)^m}{2^{2n}} \binom{2n}{n-m} \quad [n \geq m];$$

$$= 0 \quad [n < m]. \qquad \text{BH[40](16), GH[332](12b)}$$

13. $\int_0^\pi \sin^{2n+1} x \cos 2mx\, dx = 2\int_0^{\frac{\pi}{2}} \sin^{2n+1} x \cos 2mx\, dx =$

$$= \frac{(-1)^m 2^{2n+1} n!\, (2n+1)!!}{(2n-2m+1)!!\, (2m+2n+1)!!} \quad [n \geq m-1];$$

$$= \frac{(-1)^{n+1} 2^{2n+1} n!\, (2m-2n-1)!!\, (2n+1)!!}{(2m+2n+1)!!} \quad [n < m-1]. \qquad \text{GH[332](12c)}$$

14. $\int_0^{\frac{\pi}{2}} \cos^{\nu-2} x \sin \nu x\, dx = \dfrac{1}{\nu-1} \quad [\operatorname{Re} \nu > 1].$ \qquad GH[332](16c), F II 135

15. $\displaystyle\int_0^{\pi} \cos^m x \sin nx\, dx = [1 - (-1)^{m+n}]\int_0^{\frac{\pi}{2}} \cos^m x \sin nx\, dx =$ 3.631

$= [1 - (-1)^{m+n}]\left\{\displaystyle\sum_{k=0}^{r-1} \frac{m!}{(m-k)!}\frac{(m+n-2k-2)!!}{(m+n)!!} + s\frac{m!\,(n-m-2)!!}{(m+n)!!}\right\}$

$\left[r = \begin{cases} m & [m \leqslant n], \\ n & [m \geqslant n], \end{cases}\; s = \begin{cases} 2 & [n-m = 4l+2 > 0], \\ 1 & [n-m = 2l+1 > 0], \\ 0 & [n-m = 4l\;\text{oder}\;n-m<0] \end{cases}\right].$

GH [332](13a)

16. $\displaystyle\int_0^{\frac{\pi}{2}} \cos^n x \sin nx\, dx = \frac{1}{2^{n+1}} \sum_{k=1}^n \frac{2^k}{k}.$ F II 136

17. $\displaystyle\int_0^{\pi} \cos^n x \cos mx\, dx = [1 + (-1)^{m+n}]\int_0^{\frac{\pi}{2}} \cos^n x \cos mx\, dx =$

$= [1 + (-1)^{m+n}]\begin{cases} s\dfrac{n!}{(m-n)(m-n+2)\dots(m+n)} & [n < m]; \\[2mm] \dfrac{\pi}{2^{n+1}}\dbinom{n}{k} & \left[m \leqslant n\;\text{und}\atop\quad\text{and}\; n-m=2k\right]; \\[2mm] \dfrac{n!}{(2k+1)!!\,(2m+2k+1)!!} & \left[m < n\;\text{und}\atop\quad\text{and}\; n-m = 2k+1\right]; \end{cases}$

$\text{wobei}\atop\text{where}\; s = \begin{cases} 0 & [m-n = 2k], \\ 1 & [m-n = 4k+1], \\ -1 & [m-n = 4k-1]. \end{cases}$ GH [332](14a)

18. $\displaystyle\int_0^{\pi} \cos^m x \cos ax\, dx = \frac{(-1)^m \sin a\pi}{2^m(m+a)}\; _2F_1\left(-m, \frac{a+m}{2}; 1-\frac{a+m}{2}; -1\right)$

$[a \neq 0, \pm 1, \pm 2, \dots].$ W 313

19. $\displaystyle\int_0^{\frac{\pi}{2}} \cos^{\nu-2} x \cos \nu x\, dx = 0\quad [\text{Re}\,\nu > 1].$ GH [332](16a), F II 135

20. $\displaystyle\int_0^{\frac{\pi}{2}} \cos^n x \cos nx\, dx = \frac{\pi}{2^{n+1}}.$ Lo V 122(78), F II 135–136

1. $\displaystyle\int_0^{\pi} \sin^{p-1} x \cos\left[a\left(\frac{\pi}{2}-x\right)\right]dx = 2^{p-1}\frac{\Gamma\!\left(\frac{p-a}{2}\right)\Gamma\!\left(\frac{p+a}{2}\right)}{\Gamma(p-a)\,\Gamma(p+a)}\Gamma(p)$ 3.632

$[p^2 < a^2].$ BH [62](11)

3.632

2. $\int_{-\pi/2}^{\pi/2} \cos^{\nu-1} x \sin\left[a\left(x+\frac{\pi}{2}\right)\right] dx = \dfrac{\pi \sin\frac{a\pi}{2}}{2^{\nu-1} \nu B\left(\frac{\nu+a+1}{2}, \frac{\nu-a+1}{2}\right)}$

[Re $\nu > 0$].   W 309+

3. $\int_0^{\pi/2} \cos^p x \sin[(p+2n)x] dx = (-1)^{n-1} \sum_{k=0}^{n-1} \dfrac{(-1)^k 2^k}{p+k+1} \binom{n-1}{k}$.   Li[41](12)

4. $\int_{-\pi}^{\pi} \cos^{n-1} x \cos[m(x-a)] dx = [1-(-1)^{n+m}] = \int_{-\pi/2}^{\pi/2} \cos^{n-1} x \cos[m(x-a)] dx =$

$= \dfrac{[1-(-1)^{n+m}]\pi \cos ma}{2^{n-1} nB\left(\frac{n+m+1}{2}, \frac{n-m+1}{2}\right)}$ [$n \geqslant m$].

Lo V 123(80), Lo V 139(94a)

5. $\int_0^{\pi/2} \cos^{p+q-2} x \cos[(p-q)x] dx = \dfrac{\pi}{2^{p+q-1}(p+q-1) B(p,q)}$

[$p+q > 1$].   WW 263

3.633

1. $\int_0^{\pi/2} \cos^{p-1} x \sin ax \sin x \, dx = \dfrac{a\pi}{2^{p+1} p(p+1) B\left(\frac{p+a}{2}+1, \frac{p-a}{2}+1\right)}$.

Lo V 150(110)

2. $\int_0^{\pi/2} \cos^n x \sin nx \sin 2mx \, dx = \int_0^{\pi/2} \cos^n x \cos nx \cos 2mx \, dx = \dfrac{\pi}{2^{n+2}} \binom{n}{m}$.

BH[42](19, 20)

3. $\int_0^{\pi/2} \cos^{n-1} x \cos[(n+1)x] \cos 2mx \, dx = \dfrac{\pi}{2^{n+1}} \binom{n-1}{m-1}$ [$n > m-1$].

BH[42](21)

4. $\int_0^{\pi/2} \cos^{p+q} x \cos px \cos qx \, dx = \dfrac{\pi}{2^{p+q+2}} \left[1 + \dfrac{1}{(p+q+1) B(p+1, q+1)}\right]$

[$p+q > -1$].   GH[332](10c)

5. $\int_0^{\pi/2} \cos^{p+q} x \sin px \sin qx \, dx = \dfrac{\pi}{2^{p+q+2}} \sum_{k=1}^{\infty} \binom{p}{k}\binom{q}{k}$ [$p+q > -1$].   BH[42](16)

3.634

1. $\int_0^{\pi/2} \sin^{\mu-1} x \cos^{\nu-1} x \sin(\mu+\nu) x \, dx = \sin\frac{\mu\pi}{2} B(\mu, \nu)$

[Re $\mu > 0$, Re $\nu > 0$].   BH[42](23), F II 818+

3.634

2. $\int_0^{\frac{\pi}{2}} \sin^{\mu-1} x \cos^{\nu-1} x \cos(\mu + \nu) x \, dx = \cos\frac{\mu\pi}{2} B(\mu, \nu)$

$[\operatorname{Re}\mu > 0, \operatorname{Re}\nu > 0]$. BH[42](24), F II 818+

3. $\int_0^{\frac{\pi}{2}} \cos^{p+n-1} x \sin px \cos[(n+1)x] \sin x \, dx = \frac{\pi}{2^{p+n+1}} \frac{\Gamma(p+n)}{n!\,\Gamma(p)}$ $[p > -n]$.

BH[42](15)

3.635

1. $\int_0^{\frac{\pi}{4}} \cos^{\mu-1} 2x \operatorname{tg} x \, dx = \frac{1}{4}\left[\psi\left(\frac{\mu+1}{2}\right) - \psi\left(\frac{\mu}{2}\right)\right]$ $[\operatorname{Re}\mu > 0]$. BH[34](7)

2. $\int_0^{\frac{\pi}{2}} \cos^{p+2n} x \sin px \operatorname{tg} x \, dx = \frac{\pi}{2^{p+2n+1}\Gamma(p)} \sum_{k=0}^{\infty} \binom{n}{k} \frac{\Gamma(p+n-k)}{(n-k)!}$ $[p > -2n]$.

BH[42](22)

3. $\int_0^{\frac{\pi}{2}} \cos^{n-1} x \sin[(n+1)x] \operatorname{ctg} x \, dx = \frac{\pi}{2}$. BH[45](18)

3.636

1. $\int_0^{\frac{\pi}{2}} \operatorname{tg}^{\pm\mu} x \sin 2x \, dx = \frac{\mu\pi}{2} \operatorname{cosec} \frac{\mu\pi}{2}$ $[0 < \operatorname{Re}\mu < 2]$. BH[45](20)+

2. $\int_0^{\frac{\pi}{2}} \operatorname{tg}^{\pm\mu} x \cos 2x \, dx = \mp \frac{\mu\pi}{2} \sec \frac{\mu\pi}{2}$ $[|\operatorname{Re}\mu| < 1]$. BH[45](21)

3. $\int_0^{\frac{\pi}{2}} \frac{\operatorname{tg}^{2\mu} x}{\cos x} dx = \int_0^{\frac{\pi}{2}} \frac{\operatorname{ctg}^{2\mu} x}{\sin x} dx = \frac{\Gamma\left(\mu + \frac{1}{2}\right)\Gamma(-\mu)}{2\sqrt{\pi}}$

$\left[-\frac{1}{2} < \operatorname{Re}\mu < 1\right]$ (cf. 3.251 1.). BH[45](13, 14)

3.637

1. $\int_0^{\frac{\pi}{2}} \operatorname{tg}^p x \sin^{q-2} x \sin qx \, dx = -\cos\frac{(p+q)\pi}{2} B(p+q-1, 1-p)$

$[p+q > 1 > p]$. GH[332](15d)

2. $\int_0^{\frac{\pi}{2}} \operatorname{tg}^p x \sin^{q-2} x \cos qx \, dx = \sin\frac{(p+q)\pi}{2} B(p+q-1, 1-p)$

$[p+q > 1 > p]$. GH[332](15b)

3. $\int_0^{\frac{\pi}{2}} \operatorname{ctg}^p x \cos^{q-2} x \sin qx \, dx = \cos\frac{p\pi}{2} B(p+q-1, 1-p)$

$[p+q > 1 > p]$. GH[332](15c)

3.637 4. $\int_0^{\frac{\pi}{2}} \operatorname{ctg}^p x \cos^{q-2} x \cos qx \, dx = \sin\frac{p\pi}{2} B(p+q-1, 1-p)$

$[p+q > 1 > p]$. GH[332](15a)

3.638 1. $\int_0^{\frac{\pi}{4}} \frac{\sin^{2\mu} x \, dx}{\cos^{\mu+\frac{1}{2}} 2x \cos x} = \frac{\pi}{2} \sec \mu\pi \quad \left[|\operatorname{Re} \mu| < \frac{1}{2}\right]$ (cf. 3.192 2.). BH[38](8)

2. $\int_0^{\frac{\pi}{4}} \frac{\sin^{\mu-\frac{1}{2}} 2x \, dx}{\cos^\mu 2x \cos x} = \frac{2}{2\mu - 1} \cdot \frac{\Gamma\left(\mu + \frac{1}{2}\right) \Gamma(1-\mu)}{\sqrt{\pi}} \sin\left(\frac{2\mu - 1}{4}\pi\right)$

$\left[-\frac{1}{2} < \operatorname{Re} \mu < 1\right]$. BH[38](17)

3. $\int_0^{\frac{\pi}{2}} \frac{\cos^{p-1} x \sin px}{\sin x} \, dx = \frac{\pi}{2} \quad [p > 0]$. GH[332](17), BH[45](5)

Potenzen trigonometrischer Funktionen und rationale Funktionen trigonometrischer Funktionen

3.64–3.65 **Powers of trigonometric functions and rational functions of trigonometric functions**

3.641 1. $\int_0^{\frac{\pi}{2}} \frac{\sin^{p-1} x \cos^{-p} x}{a \cos x + b \sin x} \, dx = \int_0^{\frac{\pi}{2}} \frac{\sin^{-p} x \cos^{p-1} x}{a \sin x + b \cos x} \, dx =$

$= \frac{\pi \operatorname{cosec} p\pi}{a^{1-p} b^p} \quad [ab > 0, \; 0 < p < 1]$. GH[331](62)

2. $\int_0^{\frac{\pi}{2}} \frac{\sin^{1-p} x \cos^p x}{(\sin x + \cos x)^3} \, dx = \int_0^{\frac{\pi}{2}} \frac{\sin^p x \cos^{1-p} x}{(\sin x + \cos x)^3} \, dx =$

$= \frac{(1-p)p}{2} \pi \operatorname{cosec} p\pi \quad [-1 < p < 2]$. BH[48](5)

3.642 1. $\int_0^{\frac{\pi}{2}} \frac{\sin^{2\mu-1} x \cos^{2\nu-1} x \, dx}{(a^2 \sin^2 x + b^2 \cos^2 x)^{\mu+\nu}} = \frac{1}{2 a^{2\mu} b^{2\nu}} B(\mu, \nu)$

$[\operatorname{Re} \mu > 0, \; \operatorname{Re} \nu > 0]$. BH[48](28)

2. $\int_0^{\frac{\pi}{2}} \frac{\sin^{n-1} x \cos^{n-1} x \, dx}{(a^2 \cos^2 x + b^2 \sin^2 x)^n} = \frac{B\left(\frac{n}{2}, \frac{n}{2}\right)}{2 (ab)^n} \quad [ab > 0]$. GH[331](59a)

3. $\displaystyle\int_0^{\pi/2} \frac{\sin^{2n} x \, dx}{(a^2\cos^2 x + b^2\sin^2 x)^{n+1}} = \frac{1}{2}\int_0^{\pi} \frac{\sin^{2n} x \, dx}{(a^2\cos^2 x + b^2\sin^2 x)^{n+1}} =$ 3.642

$\displaystyle = \int_0^{\pi/2} \frac{\cos^{2n} x \, dx}{(a^2\sin^2 x + b^2\cos^2 x)^{n+1}} = \frac{1}{2}\int_0^{\pi} \frac{\cos^{2n} x \, dx}{(a^2\sin^2 x + b^2\cos^2 x)^{n+1}} = \frac{(2n-1)!!\,\pi}{2^{n+1}\,n!\,ab^{2n+1}}$

$\qquad\qquad\qquad\qquad\qquad\qquad [ab > 0].$ GH[331](58)

4. $\displaystyle\int_0^{\pi/2} \frac{\cos^{p+2n} x \cos px \, dx}{(a^2\cos^2 x + b^2\sin^2 x)^{n+1}} = \pi\sum_{k=0}^{n}\binom{2n-k}{n}\binom{p+k-1}{k}\frac{b^{p-1}}{(2a)^{2n-k+1}(a+b)^{p+k}}$

$\qquad\qquad\qquad [a > 0,\ b > 0,\ p > -2n - 1].$ GH[332](30)

1. $\displaystyle\int_0^{\pi/2} \frac{\cos^p x \cos px \, dx}{1 - 2a\cos 2x + a^2} = \frac{\pi}{2^{p+1}}\cdot\frac{(1+a)^{p-1}}{1-a}\quad [a^2 < 1,\ p > -1].$ GH[332](33c)   3.643

2. $\displaystyle\int_0^{\pi/2} \frac{\sin^{2n} x \cos^{\mu} x \cos \beta x}{(1 - 2a\cos 2x + a^2)^m}\, dx =$

$\displaystyle = \frac{(-1)^n \pi (1-a)^{2n-2m+1}}{2^{2m-\beta-1}(1+a)^{2m+\beta+1}}\sum_{k=0}^{m-1}\sum_{l=0}^{m-k-1}\binom{\beta}{k}\binom{2n}{l}\binom{2m-k-l-2}{m-1}(-2)^l(a-1)^k$

$\qquad\qquad [a^2 < 1,\ \beta = 2m - 2n - \mu - 2,\ \mu > -1].$ GH[332](33)

1. $\displaystyle\int_0^{\pi} \frac{\sin^m x}{p + q\cos x}\,dx =$ 3.644*

$\displaystyle = 2^{m-2}\frac{p}{q^2}\sum_{\nu=1}^{k}\left(\frac{p^2-q^2}{-4q^2}\right)^{\nu-1} B\left(\frac{m+1-2\nu}{2}, \frac{m+1-2\nu}{2}\right) + \left(\frac{p^2-q^2}{-q^2}\right)^{k} A;$

$\displaystyle A = \frac{\pi p}{q^2}\left(1 - \sqrt{1 - \frac{q^2}{p^2}}\right)\qquad [m = 2k + 2];$

$\displaystyle A = \frac{1}{q}\ln\frac{p+q}{p-q}\qquad [m = 2k + 1]$

$\qquad\qquad\qquad\qquad\qquad [k \geqslant 1,\ q \neq 0,\ p^2 - q^2 \geqslant 0].$

2. $\displaystyle\int_0^{\pi} \frac{\sin^m x}{1 + \cos x}\,dx = 2^{m-1} B\left(\frac{m-1}{2}, \frac{m+1}{2}\right)\quad [m \geqslant 2].$

---

* Das Integral 3.644 ist angegeben in: „Ob Integrale $\int_0^{\pi}\frac{\sin^m x}{p+q\cos x}dx$" von K. W. BRODOWITSKIJ, DAN 120, Nr. 6 (1958).

* The integrals in 3.644 were taken from K. V. Brodovitskii, "On the integral $\int_0^{\pi}\frac{\sin^m x}{p+q\cos x}dx$", Dokl. Akad. Nauk SSSR 120, No. 6 (1958).

3.644 3. $$\int_0^\pi \frac{\sin^m x}{1-\cos x}\,dx = 2^{m-1} B\left(\frac{m-1}{2}, \frac{m+1}{2}\right) \quad [m \geq 2].$$

4. $$\int_0^\pi \frac{\sin^2 x}{p+q\cos x}\,dx = \frac{p\pi}{q^2}\left(1 - \sqrt{1-\frac{q^2}{p^2}}\right).$$

5. $$\int_0^\pi \frac{\sin^3 x}{p+q\cos x}\,dx = 2\frac{p}{q^2} + \frac{1}{q}\left(1 - \frac{p^2}{q^2}\right)\ln\frac{p+q}{p-q}.$$

3.645 $$\int_0^\pi \frac{\cos^n x\,dx}{(a+b\cos x)^{n+1}} = \frac{\pi}{2^n(a+b)^n\sqrt{a^2-b^2}} \times$$
$$\times \sum_{k=0}^n (-1)^k \frac{(2n-2k-1)!!(2k-1)!!}{(n-k)!\,k!}\left(\frac{a+b}{a-b}\right)^k \quad [a^2 > b^2]. \qquad \text{Li [64](16)}$$

3.646 1. $$\int_0^{\pi/2} \frac{\cos^n x \sin nx \sin 2x}{1-2a\cos 2x + a^2}\,dx = \frac{\pi}{4a}\left[\left(\frac{1+a}{2}\right)^n - \frac{1}{2^n}\right] \quad [a^2 < 1]. \qquad \text{BH [50](6)}$$

2. $$\int_0^{\pi/2} \frac{1-a\cos 2nx}{1-2a\cos 2nx + a^2}\cos^m x \cos mx\,dx =$$
$$= \frac{\pi}{2^{m+2}}\sum_{k=1}^\infty \binom{m}{kn} a^k + \frac{\pi}{2^{m+1}} \quad [a^2 < 1]. \qquad \text{Li [50](7)}$$

3.647 $$\int_0^{\pi/2} \frac{\cos^p x \cos px\,dx}{a^2\sin^2 x + b^2\cos^2 x} = \frac{\pi}{2b}\cdot\frac{a^{p-1}}{(a+b)^p} \quad [p > -1,\ a > 0,\ b > 0]. \qquad \text{BH [47](20)}$$

3.648 1. $$\int_0^{\pi/4} \frac{\operatorname{tg}^l x\,dx}{1+\cos\frac{m}{n}\pi\sin 2x} = \frac{1}{2n}\operatorname{cosec}\frac{m}{n}\pi\sum_{k=0}^{n-1}(-1)^{k-1}\sin\frac{km}{n}\pi \times$$
$$\times \left[\psi\left(\frac{n+l+k}{2n}\right) - \psi\left(\frac{l+k}{2n}\right)\right]\left[m+n\ \begin{array}{l}\text{ungeradzahlig}\\\text{odd}\end{array}\right];$$

$$= \frac{1}{n}\operatorname{cosec}\frac{m}{n}\pi\sum_{k=0}^{\frac{n-1}{2}}(-1)^{k-1}\sin\frac{km}{n}\pi \times$$
$$\times \left[\psi\left(\frac{n+l-k}{n}\right) - \psi\left(\frac{l+k}{n}\right)\right]\left[m+n\ \begin{array}{l}\text{geradzahlig}\\\text{even}\end{array}\right]$$

$$\left[l\ \begin{array}{l}\text{eine natürliche Zahl}\\\text{a positive integer}\end{array}\right]. \qquad \text{BH [36](5)}$$

2. $\int_0^{\pi/2} \frac{\operatorname{tg}^{\pm\mu} x \, dx}{1 + \cos t \sin 2x} = \pi \operatorname{cosec} t \sin \mu t \operatorname{cosec}(\mu\pi)$ $\quad [|\operatorname{Re} \mu| < 1, \; t^2 < \pi^2]$. $\qquad$ 3.648

BH [47](4)

1. $\int_0^{\pi/2} \frac{\operatorname{tg}^{\pm\mu} x \sin 2x \, dx}{1 \mp 2a \cos 2x + a^2} = \frac{\pi}{4a} \operatorname{cosec} \frac{\mu\pi}{2} \left[1 - \left(\frac{1-a}{1+a}\right)^\mu\right]$ $\quad [a^2 < 1];$ $\qquad$ 3.649

$\qquad = \frac{\pi}{4a} \operatorname{cosec} \frac{\mu\pi}{2} \left[1 + \left(\frac{a-1}{a+1}\right)^\mu\right]$ $\quad [a^2 > 1]$

$\qquad [-2 < \operatorname{Re} \mu < 1]$. $\quad$ BH [50](3)

2. $\int_0^{\pi/2} \frac{\operatorname{tg}^{\pm\mu} x (1 \mp a \cos 2x)}{1 \mp 2a \cos 2x + a^2} \, dx = \frac{\pi}{4} \sec \frac{\mu\pi}{2} \left[1 + \left(\frac{1-a}{1+a}\right)^\mu\right]$ $\quad [a^2 < 1];$

$\qquad = \frac{\pi}{4} \sec \frac{\mu\pi}{2} \left[1 - \left(\frac{a-1}{a+1}\right)^\mu\right]$ $\quad [a^2 > 1]$

$\qquad [|\operatorname{Re} \mu| < 1]$. $\quad$ BH [50](4)

1. $\int_0^{\pi/4} \frac{\operatorname{tg}^\mu x \, dx}{1 + \sin x \cos x} = \frac{1}{3} \left[\psi\left(\frac{\mu+2}{3}\right) - \psi\left(\frac{\mu+1}{3}\right)\right]$ $\quad [\operatorname{Re} \mu > -1]$. $\quad$ BH [36](3) $\quad$ 3.651

2. $\int_0^{\pi/4} \frac{\operatorname{tg}^\mu x \, dx}{1 - \sin x \cos x} = \frac{1}{3} \left[\beta\left(\frac{\mu+2}{3}\right) + \beta\left(\frac{\mu+1}{3}\right)\right]$ $\quad [\operatorname{Re} \mu > -1]$. $\quad$ BH [36](4)+

1. $\int_0^{\pi/2} \frac{\operatorname{tg}^\mu x \, dx}{(\sin x + \cos x) \sin x} = \int_0^{\pi/2} \frac{\operatorname{ctg}^\mu x \, dx}{(\sin x + \cos x) \cos x} = \pi \operatorname{cosec} \mu\pi$ $\quad [0 < \operatorname{Re} \mu < 1]$. $\quad$ 3.652

BH [49](1)

2. $\int_0^{\pi/2} \frac{\operatorname{tg}^\mu x \, dx}{(\sin x - \cos x) \sin x} = \int_0^{\pi/2} \frac{\operatorname{ctg}^\mu x \, dx}{(\cos x - \sin x) \cos x} = -\pi \operatorname{ctg} \mu\pi$ $\quad [0 < \operatorname{Re} \mu < 1]$.

BH [49](2)

3. $\int_0^{\pi/2} \frac{\operatorname{ctg}^{\mu + \frac{1}{2}} x \, dx}{(\sin x + \cos x) \cos x} = \int_0^{\pi/2} \frac{\operatorname{tg}^{\mu - \frac{1}{2}} x \, dx}{(\sin x + \cos x) \cos x} = \pi \sec \mu\pi$ $\quad \left[|\operatorname{Re} \mu| < \frac{1}{2}\right]$.

BH [61](1), BH [61](2)

1. $\int_0^{\pi/2} \frac{\operatorname{tg}^{1-2\mu} x \, dx}{a^2 \cos^2 x + b^2 \sin^2 x} = \int_0^{\pi/2} \frac{\operatorname{ctg}^{1-2\mu} x \, dx}{a^2 \sin^2 x + b^2 \cos^2 x} = \frac{\pi}{2 a^{2\mu} b^{2-2\mu} \sin \mu\pi}$ $\qquad$ 3.653

$[0 < \operatorname{Re} \mu < 1]$. $\quad$ GH [331](59b)

2. $\int_0^{\pi/2} \frac{\operatorname{tg}^\mu x \, dx}{1 - a \sin^2 x} = \int_0^{\pi/2} \frac{\operatorname{ctg}^\mu x \, dx}{1 - a \cos^2 x} = \frac{\pi \sec \frac{\mu\pi}{2}}{2\sqrt{(1-a)^{\mu+1}}}$ $\quad [|\operatorname{Re} \mu| < 1, \; a < 1]$.

BH [49](6)

3.653

3. $$\int_0^{\frac{\pi}{2}} \frac{\operatorname{tg}^{\pm\mu} x \, dx}{1 - \cos^2 t \sin^2 2x} = \frac{\pi}{2} \operatorname{cosec} t \sec \frac{\mu\pi}{2} \cos\left[\left(\frac{\pi}{2} - t\right)\mu\right]$$

[|Re μ| < 1, $t^2 < \pi^2$]. BH[49](7), BH[47](21)

4. $$\int_0^{\frac{\pi}{2}} \frac{\operatorname{tg}^{\pm\mu} x \sin 2x}{1 - \cos^2 t \sin^2 2x} \, dx = \pi \operatorname{cosec} 2t \operatorname{cosec} \frac{\mu\pi}{2} \sin\left[\left(\frac{\pi}{2} - t\right)\mu\right]$$

[|Re μ| < 1, $t^2 < \pi^2$]. BH[47](22)+

5. $$\int_0^{\frac{\pi}{2}} \frac{\operatorname{tg}^\mu x \sin^2 x \, dx}{1 - \cos^2 t \sin^2 2x} = \int_0^{\frac{\pi}{2}} \frac{\operatorname{ctg}^\mu x \cos^2 x \, dx}{1 - \cos^2 t \sin^2 2x} =$$

$$= \frac{\pi}{2} \operatorname{cosec} 2t \sec \frac{\mu\pi}{2} \cos\left[\frac{\mu\pi}{2} - (\mu+1)t\right] \quad [|\operatorname{Re} \mu| < 1, \, t^2 < \pi^2].$$

BH[47](23)+, BH[49](10)

6. $$\int_0^{\frac{\pi}{2}} \frac{\operatorname{tg}^\mu x \cos^2 x \, dx}{1 - \cos^2 t \sin^2 2x} = \int_0^{\frac{\pi}{2}} \frac{\operatorname{ctg}^\mu x \sin^2 x \, dx}{1 - \cos^2 t \sin^2 2x} =$$

$$= \frac{\pi}{2} \operatorname{cosec} 2t \sec \frac{\mu\pi}{2} \cos\left[\frac{\mu\pi}{2} - (\mu-1)t\right] \quad [|\operatorname{Re} \mu| < 1, \, t^2 < \pi^2].$$

BH[47](24)+, BH[49](9)

3.654

1. $$\int_0^{\frac{\pi}{2}} \frac{\operatorname{tg}^{\mu+1} x \cos^2 x \, dx}{(1 + \cos t \sin 2x)^2} = \int_0^{\frac{\pi}{2}} \frac{\operatorname{ctg}^{\mu+1} x \sin^2 x \, dx}{(1 + \cos t \sin 2x)^2} = \frac{\pi(\mu \sin t \cos \mu t - \cos t \sin \mu t)}{2 \sin \mu\pi \sin^3 t}$$

[|Re μ| < 1, $t^2 < \pi^2$]. BH[48](3), BH[49](22)

2. $$\int_0^{\frac{\pi}{2}} \frac{\operatorname{tg}^{\pm\mu} x \, dx}{(\sin x + \cos x)^2} = \frac{\mu\pi}{\sin \mu\pi} \quad [0 < \operatorname{Re} \mu < 1].$$ BH[56](9)+

3. $$\int_0^{\frac{\pi}{2}} \frac{\operatorname{tg}^{\pm(\mu-1)} x \, dx}{\cos^2 x - \sin^2 x} = \pm \frac{\pi}{2} \operatorname{ctg} \frac{\mu\pi}{2} \quad [0 < \operatorname{Re} \mu < 2.$$ BH[45](27, 29)

3.655

$$\int_0^{\frac{\pi}{2}} \frac{\operatorname{tg}^{2\mu-1} x \, dx}{1 - 2a(\cos t_1 \sin^2 x + \cos t_2 \cos^2 x) + a^2} =$$

$$= \int_0^{\frac{\pi}{2}} \frac{\operatorname{ctg}^{2\mu-1} x \, dx}{1 - 2a(\cos t_1 \cos^2 x + \cos t_2 \sin^2 x) + a^2} =$$

$$= \frac{\pi \operatorname{cosec} \mu\pi}{(1 - 2a \cos t_2 + a^2)^\mu (1 - 2a \cos t_1 + a^2)^{1-\mu}}$$

[0 < Re μ < 1, $t_1^2 < \pi^2$, $t^2 < \pi^2$]. BH[50](18)

1. $\int_0^{\frac{\pi}{4}} \frac{\text{tg}^\mu x \, dx}{1 - \sin^2 x \cos^2 x} = \frac{1}{12} \left\{ -\psi\left(\frac{\mu+1}{6}\right) - \psi\left(\frac{\mu+2}{6}\right) + \right.$ 3.656

$$\left. + \psi\left(\frac{\mu+4}{6}\right) + \psi\left(\frac{\mu+5}{6}\right) + 2\psi\left(\frac{\mu+2}{3}\right) - 2\psi\left(\frac{\mu+1}{3}\right) \right\}$$

[Re $\mu > -1$]   (cf. 3.651 1., 2.).   Li[36](10)

2. $\int_0^{\frac{\pi}{2}} \frac{\text{tg}^{\mu-1} x \cos^2 x \, dx}{1 - \sin^2 x \cos^2 x} = \int_0^{\frac{\pi}{2}} \frac{\text{ctg}^{\mu-1} x \sin^2 x \, dx}{1 - \sin^2 x \cos^2 x} =$

$$= \frac{\pi}{4\sqrt{3}} \text{cosec} \frac{\mu\pi}{6} \text{cosec}\left(\frac{2+\mu}{6}\pi\right) \quad [0 < \text{Re}\,\mu < 4].$$ Li[47](26)

Potenzen linearer Funktionen von trigonometrischen Funktionen
Powers of linear functions of trigonometric functions   3.66

1. $\int_0^{2\pi} (a \sin x + b \cos x)^{2n+1} dx = 0.$ BH[68](9)   3.661

2. $\int_0^{2\pi} (a \sin x + b \cos x)^{2n} dx = \frac{(2n-1)!!}{(2n)!!} \cdot 2\pi(a^2 + b^2)^n.$ BH[68](8)

3. $\int_0^\pi (a + b \cos x)^n \, dx = \frac{1}{2} \int_0^{2\pi} (a + b \cos x)^n \, dx = \pi(a^2 - b^2)^{\frac{n}{2}} P_n\left(\frac{a}{\sqrt{a^2 - b^2}}\right) =$

$$= \frac{\pi}{2^n} \sum_{k=0}^{E\left(\frac{n}{2}\right)} \frac{(-1)^k (2n-2k)!}{k!(n-k)!(n-2k)!} a^{n-2k}(a^2 - b^2)^k \quad [a^2 > b^2].$$ GH[332](37a)

4. $\int_0^\pi \frac{dx}{(a + b \cos x)^{n+1}} = \frac{1}{2} \int_0^{2\pi} \frac{dx}{(a + b \cos x)^{n+1}} = \frac{\pi}{(a^2 - b^2)^{\frac{n+1}{2}}} P_n\left(\frac{a}{\sqrt{a^2 - b^2}}\right) =$

$$= \frac{\pi}{2^n(a+b)^n \sqrt{a^2 - b^2}} \sum_{k=0}^n \frac{(2n-2k-1)!!(2k-1)!!}{(n-k)!k!} \cdot \left(\frac{a+b}{a-b}\right)^k$$

[$a > |b|$].   GH[332](38), Li[64](14)

1. $\int_0^{\frac{\pi}{2}} (\sec x - 1)^\mu \sin x \, dx = \int_0^{\frac{\pi}{2}} (\text{cosec } x - 1)^\mu \cos x \, dx =$ 3.662

$$= \mu\pi \text{ cosec } \mu\pi \quad [|\text{Re}\,\mu| < 1].$$ BH[55](13)

2. $\int_0^{\frac{\pi}{2}} (\text{cosec } x - 1)^\mu \sin 2x \, dx = (1 - \mu) \mu\pi \text{ cosec } \mu\pi \quad [-1 < \text{Re}\,\mu < 2].$

BH[48](7)

3.662 3. $\int_0^{\pi/2} (\sec x - 1)^\mu \operatorname{tg} x \, dx = \int_0^{\pi/2} (\operatorname{cosec} x - 1)^\mu \operatorname{ctg} x \, dx = -\pi \operatorname{cosec} \mu\pi$

$[-1 < \operatorname{Re} \mu < 0]$ (cf. 3.192 2.).   BH [46](4, 6)

4. $\int_0^{\pi/4} (\operatorname{ctg} x - 1)^\mu \frac{dx}{\sin 2x} = -\frac{\pi}{2} \operatorname{cosec} \mu\pi \quad [-1 < \operatorname{Re} \mu < 0]$.

BH [38](22)+

5. $\int_0^{\pi/4} (\operatorname{ctg} x - 1)^\mu \frac{dx}{\cos^2 x} = \mu\pi \operatorname{cosec} \mu\pi \quad [|\operatorname{Re} \mu| < 1]$.   BH [38](11)+

3.663 1. $\int_0^u (\cos x - \cos u)^{v-\frac{1}{2}} \cos ax \, dx = \sqrt{\frac{\pi}{2}} \sin^v u \, \Gamma\left(v + \frac{1}{2}\right) P_{a-\frac{1}{2}}^{-v}(\cos u)$

$\left[\operatorname{Re} v > -\frac{1}{2}; \, a > 0, \, 0 < u < \pi\right]$.   Ba I 159(27), Ba IV 22(28)

2. $\int_0^u (\cos x - \cos u)^{v-1} \cos [(v + \beta) x] \, dx =$

$$= \frac{\sqrt{\pi}\, \Gamma(\beta + 1)\, \Gamma(v)\, \Gamma(2v) \sin^{2v-1} u}{2^v \Gamma(\beta + 2v)\, \Gamma\left(v + \frac{1}{2}\right)} C_\beta^v(\cos u)$$

$[\operatorname{Re} v > 0, \, \operatorname{Re} \beta > -1, \, 0 < u < \pi]$.   Ba I 178(23)

3.664 1. $\int_0^\pi (z + \sqrt{z^2 - 1} \cos x)^q \, dx = \pi P_q(z)$

$\left[\operatorname{Re} z > 0, \, \arg(z + \sqrt{z^2 - 1} \cos x) = \arg z \begin{array}{c} \text{für} \\ \text{for} \end{array} x = \frac{\pi}{2}\right]$.   Sm III₂ 415

2. $\int_0^\pi \frac{dx}{(z + \sqrt{z^2 - 1} \cos x)^q} = \pi P_{q-1}(z)$

$\left[\operatorname{Re} z > 0, \, \arg(z + \sqrt{z^2 - 1} \cos x) = \arg z \begin{array}{c} \text{für} \\ \text{for} \end{array} x = \frac{\pi}{2}\right]$.   WW 314

3. $\int_0^\pi (z + \sqrt{z^2 - 1} \cos x)^q \cos nx \, dx = \frac{\pi}{(q+1)(q+2)\ldots(q+n)} P_q^n(z)$

$\left[\operatorname{Re} z > 0, \, \arg(z + \sqrt{z^2 - 1} \cos x) = \arg z \begin{array}{c} \text{für} \\ \text{for} \end{array} x = \frac{\pi}{2}\right.$,

z liegt außerhalb des Intervalls (−1, 1) der reellen Achse]. WW 326, Sm III₂ 416(15) | z lies outside the interval (−1, 1) on the real axis]. WW 326, Sm III₂ 416(15)

4. $$\int_0^\pi (z+\sqrt{z^2-1}\cos x)^\mu \sin^{2\nu-1} x\, dx = \frac{2^{2\nu-1}\Gamma(\mu+1)[\Gamma(\nu)]^2}{\Gamma(2\nu+\mu)} C_\mu^\nu(z) =$$ 3.664

$$= \frac{\sqrt{\pi}\,\Gamma(\nu)\Gamma(2\nu)\Gamma(\mu+1)}{\Gamma(2\nu+\mu)\Gamma\left(\nu+\frac{1}{2}\right)} C_\mu^\nu(z) = 2^\nu \sqrt{\frac{\pi}{2}} (z^2-1)^{\frac{1}{4}-\frac{\nu}{2}} \Gamma(\nu) P_{\mu+\nu-\frac{1}{2}}^{\frac{1}{2}-\nu}(z)$$

[Re $\nu > 0$]. Ba I 155(6)+, Ba I 178(22)

5. $$\int_0^{2\pi} [\beta+\sqrt{\beta^2-1}\cos(a-x)]^\nu (\gamma+\sqrt{\gamma^2-1}\cos x)^{\nu-1} dx =$$

$$= 2\pi P_\nu[\beta\gamma - \sqrt{\beta^2-1}\sqrt{\gamma^2-1}\cos a]\quad [\text{Re }\beta > 0,\ \text{Re }\gamma > 0].$$ Ba I 157(18)

1. $$\int_0^\pi \frac{\sin^{\mu-1} x\, dx}{(a+b\cos x)^\mu} = \frac{2^{\mu-1}}{\sqrt{(a^2-b^2)^\mu}} B\left(\frac{\mu}{2}, \frac{\mu}{2}\right)\quad [\text{Re }\mu > 0,\ 0 < b < a].$$ 3.665

F II 796+

2. $$\int_0^\pi \frac{\sin^{2\mu-1} x\, dx}{(1+2a\cos x + a^2)^\nu} = B\left(\mu, \frac{1}{2}\right) F\left(\nu, \nu-\mu+\frac{1}{2}; \mu+\frac{1}{2}; a^2\right)$$

[Re $\mu > 0$, $|a| < 1$]. Ba I 81(9)

1. $$\int_0^\pi (\beta+\cos x)^{-\nu-\frac{1}{2}} \sin^{2\nu} x\, dx = \frac{2^{\nu+\frac{1}{2}} e^{-i\mu\pi} (\beta^2-1)^{\frac{\mu}{2}} \Gamma\left(\nu+\frac{1}{2}\right) Q_{\nu-\frac{1}{2}}^\mu(\beta)}{\Gamma\left(\nu+\mu+\frac{1}{2}\right)}$$ 3.666

$$\left[\text{Re}\left(\nu+\mu+\frac{1}{2}\right) > 0,\ \text{Re }\nu > -\frac{1}{2}\right].$$ Ba I 155(5)+

2. $$\int_0^{\pi/2} (\operatorname{ch}\beta + \operatorname{sh}\beta\cos x)^{\mu+\nu} \sin^{-2\nu} x\, dx =$$

$$= \frac{\sqrt{\pi}}{2^\nu} \operatorname{sh}^\nu(\beta)\, \Gamma\left(\frac{1}{2}-\nu\right) P_\mu^\nu(\operatorname{ch}\beta)\quad \left[\text{Re }\nu < \frac{1}{2}\right].$$ Ba I 156(7)

3. $$\int_0^\pi (\cos t + i\sin t\cos x)^\mu \sin^{2\nu-1} x\, dx =$$

$$= 2^{\nu-\frac{1}{2}} \sqrt{\pi} \sin^{\frac{1}{2}-\nu} t\, \Gamma(\nu) P_{\mu+\nu-\frac{1}{2}}^{\frac{1}{2}-\nu}(\cos t)\quad [\text{Re }\nu > 0,\ t^2 < \pi^2].$$ Ba I 158(23)

4. $$\int_0^{2\pi} [\cos t + i\sin t\cos(a-x)]^\nu \cos mx\, dx =$$

$$= \frac{i^{3m} 2\pi \Gamma(\nu+1)}{\Gamma(\nu+m+1)} \cos ma\, P_\nu^m(\cos t)\quad \left[0 < t < \frac{\pi}{2}\right].$$ Ba I 159(25)

5. $$\int_0^{2\pi} [\cos t + i\sin t\cos(a-x)]^\nu \sin mx\, dx =$$

$$= \frac{i^{3m} 2\pi \Gamma(\nu+1)}{\Gamma(\nu+m+1)} \sin ma\, P_\nu^m(\cos t)\quad \left[0 < t < \frac{\pi}{2}\right].$$ Ba I 159(26)

3.667

1. $$\int_0^{\pi/4} \frac{\sin^{\mu-1} 2x \, dx}{(\cos x + \sin x)^{2\mu}} = \frac{\sqrt{\pi}}{2^{\mu+1}} \frac{\Gamma(\mu)}{\Gamma\left(\mu + \frac{1}{2}\right)} \quad [\operatorname{Re} \mu > 0].$$ BH[37](1)

2. $$\int_0^{\pi/4} \frac{\sin^\mu x \, dx}{(\cos x - \sin x)^{\mu+1} \cos x} = -\pi \operatorname{cosec} \mu\pi$$

$$[-1 < \operatorname{Re} \mu < 0]. \quad (\text{cf. } 3.192 \, 2.).$$ BH[37](16)

3. $$\int_0^{\pi/4} \frac{(\cos x - \sin x)^\mu}{\sin^\mu x \sin 2x} dx = -\frac{\pi}{2} \operatorname{cosec} \mu\pi \quad [-1 < \operatorname{Re} \mu < 0].$$ BH[35](27)

4. $$\int_0^{\pi/4} \frac{\sin^\mu x \, dx}{(\cos x - \sin x)^\mu \sin 2x} = \frac{\pi}{2} \operatorname{cosec} \mu\pi \quad [0 < \operatorname{Re} \mu < 1].$$ Li[37](20)+

5. $$\int_0^{\pi/4} \frac{\sin^\mu x \, dx}{(\cos x - \sin x)^\mu \cos^2 x} = \mu\pi \operatorname{cosec} \mu\pi \quad [|\operatorname{Re} \mu| < 1].$$ BH[37](17)

6. $$\int_0^{\pi/4} \frac{\sin^\mu x \, dx}{(\cos x - \sin x)^{\mu-1} \cos^3 x} = \frac{1-\mu}{2} \mu\pi \operatorname{cosec} \mu\pi \quad [|\operatorname{Re} \mu| < 1].$$

BH[35](24), BH[37](18)

7. $$\int_0^{\pi/2} \frac{\sin^{\mu-1} x \cos^{\nu-1} x}{(\sin x + \cos x)^{\mu+\nu}} dx = B(\mu, \nu) \quad [\operatorname{Re} \mu > 0, \operatorname{Re} \nu > 0].$$ BH[48](8)

3.668

1. $$\int_{-\pi/4}^{\pi/4} \left(\frac{\cos x + \sin x}{\cos x - \sin x}\right)^{\cos 2t} dx = \frac{\pi}{2 \sin(\pi \cos^2 t)}.$$ F II 794

2. $$\int_u^v \frac{(\cos u - \cos x)^{\mu-1}}{(\cos x - \cos v)^\mu} \cdot \frac{\sin x \, dx}{1 - 2a \cos x + a^2} = \frac{(1 - 2a \cos u + a^2)^{\mu-1}}{(1 - 2a \cos v + a^2)^\mu} \cdot \frac{\pi}{\sin \mu\pi}$$

$$[0 < \operatorname{Re} \mu < 1, \, a^2 < 1].$$ BH[73](28)

3.669

$$\int_0^{\pi/2} \frac{\sin^{p-1} x \cos^{q-p-1} x \, dx}{(a \cos x + b \sin x)^q} = \int_0^{\pi/2} \frac{\sin^{q-p-1} x \cos^{p-1} x}{(a \sin x + b \cos x)^q} dx = \frac{B(p, q-p)}{a^{q-p} b^p}$$

$$[q > p > 0, \, ab > 0].$$ GH[331](90)

Quadratwurzeln aus Ausdrücken, die trigonometrische Funktionen enthalten
Square roots of expressions containing trigonometric functions
3.67

1. $\int_0^{\frac{\pi}{2}} \sin^\alpha x \cos^\beta x \sqrt{1-k^2 \sin^2 x}\, dx =$
3.671

$$= \frac{1}{2} B\left(\frac{\alpha+1}{2}, \frac{\beta+1}{2}\right) F\left(\frac{\alpha+1}{2}, -\frac{1}{2}; \frac{\alpha+\beta+2}{2}; k^2\right)$$
$[\alpha > -1, \beta > -1, |k| < 1]$.   GH[331](93)

2. $\int_0^{\frac{\pi}{2}} \frac{\sin^\alpha x \cos^\beta x}{\sqrt{1-k^2 \sin^2 x}}\, dx = \frac{1}{2} B\left(\frac{\alpha+1}{2}, \frac{\beta+1}{2}\right) F\left(\frac{\alpha+1}{2}, \frac{1}{2}; \frac{\alpha+\beta+2}{2}; k^2\right)$

$[\alpha > -1, \beta > -1, |k| < 1]$.   GH[331](92)

3. $\int_0^\pi \frac{\sin^{2n} x \, dx}{\sqrt{1-k^2 \sin^2 x}} = \frac{\pi}{2^n} \sum_{j=0}^\infty \frac{(2j-1)!!(2n+2j-1)!!}{2^{2j} j!(n+j)!} k^{2j} \quad [k^2 < 1];$

$$= \frac{(2n-1)!!\pi}{2^n \sqrt{1-k^2}} \sum_{j=0}^\infty \frac{[(2j-1)!!]^2}{2^{2j} j!(n+j)!} \left(\frac{k^2}{k^2-1}\right)^j \quad \left[k^2 < \frac{1}{2}\right].$$ Li[67](2)

1. $\int_0^{\frac{\pi}{4}} \frac{\sin^n x}{\cos^{n+1} x} \cdot \frac{dx}{\sqrt{\cos x(\cos x - \sin x)}} = 2 \cdot \frac{(2n)!!}{(2n+1)!!}.$ BH[39](5)   3.672

2. $\int_0^{\frac{\pi}{4}} \frac{\sin^n x}{\cos^{n+1} x} \cdot \frac{dx}{\sqrt{\sin x(\cos x - \sin x)}} = \frac{(2n-1)!!}{(2n)!!} \pi.$ BH[39](6)

$\int_u^{\frac{\pi}{2}} \frac{dx}{\sqrt{\sin x - \sin u}} = \sqrt{2}\, K\left(\sin \frac{\pi - 2u}{4}\right).$ BH[74](11)   3.673

1. $\int_0^\pi \frac{dx}{\sqrt{1 \pm 2p \cos x + p^2}} = 2K(p) \quad [p^2 < 1]$.   BH[67](5)   3.674

2. $\int_0^\pi \frac{\sin x \, dx}{\sqrt{1-2p \cos x + p^2}} = 2 \quad [p^2 \leq 1];$

$$= \frac{2}{p} \quad [p^2 \geq 1].$$ BH[67](6)

3. $\int_0^\pi \frac{\cos x \, dx}{\sqrt{1-2p \cos x + p^2}} = \frac{2}{p}[K(p) - E(p)] \quad [p^2 < 1]$.   BH[67](7)

1. $\int_u^\pi \frac{\sin\left(n+\frac{1}{2}\right)x \, dx}{\sqrt{2(\cos u - \cos x)}} = \frac{\pi}{2} P_n(\cos u).$ WW 315   3.675

3.675

2. $\int_0^u \dfrac{\cos\left(n + \dfrac{1}{2}\right) x\, dx}{\sqrt{2(\cos x - \cos u)}} = \dfrac{\pi}{2} P_n(\cos u).$ F II 693, WW 315

3.676

1. $\int_0^{\pi/2} \dfrac{\sin x\, dx}{\sqrt{1 + p^2 \sin^2 x}} = \dfrac{1}{p} \operatorname{arctg} p.$ BH [60](5)

2. $\int_0^{\pi/2} \operatorname{tg}^2 x \sqrt{1 - p^2 \sin^2 x}\, dx = \infty.$ BH [53](8)

3. $\int_0^{\pi/2} \dfrac{dx}{\sqrt{p^2 \cos^2 x + q^2 \sin^2 x}} = \dfrac{1}{p} K\left(\dfrac{\sqrt{p^2 - q^2}}{p}\right) \quad [0 < q < p].$ F II 144–146

3.677

1. $\int_0^{\pi/2} \dfrac{\sin^2 x\, dx}{\sqrt{1 + \sin^2 x}} = \sqrt{2}\, E\left(\dfrac{\sqrt{2}}{2}\right) - \dfrac{1}{\sqrt{2}} K\left(\dfrac{\sqrt{2}}{2}\right).$ BH [60](2)

2. $\int_0^{\pi/2} \dfrac{\cos^2 x\, dx}{\sqrt{1 + \sin^2 x}} = \sqrt{2}\left[K\left(\dfrac{\sqrt{2}}{2}\right) - E\left(\dfrac{\sqrt{2}}{2}\right)\right].$ BH [60](3)

3.678

1. $\int_0^{\pi/4} (\sec^2 2x - 1) \dfrac{dx}{\operatorname{tg} x} = \ln 2.$ BH [38](23)

2. $\int_0^{\pi/4} \dfrac{\operatorname{tg}^2 x\, dx}{\sqrt{1 - k^2 \sin^2 2x}} = \sqrt{1 - k^2} - E(k) + \dfrac{1}{2} K(k).$ BH [39](2)

3. $\int_0^u \sqrt{\dfrac{\cos 2x - \cos 2u}{\cos 2x + 1}}\, dx = \dfrac{\pi}{2}(1 - \cos u) \quad \left[u^2 < \dfrac{\pi^2}{4}\right].$ Li [74](6)

4. $\int_0^{\pi/4} \dfrac{(\cos x - \sin x)^{n - 1/2}}{\cos^{n+1} x} \sqrt{\operatorname{cosec} x}\, dx = \dfrac{(2n-1)!!}{(2n)!!} \pi.$ BH [38](24)

5. $\int_0^{\pi/4} \dfrac{(\cos x - \sin x)^{n - 1/2}}{\cos^{n+1} x} \operatorname{tg}^m x \sqrt{\operatorname{cosec} x}\, dx = \dfrac{(2n-1)!!\,(2m-1)!!}{(2n+2m)!!} \pi.$ BH [38](25)

3.679

1. $\int_0^{\pi/2} \dfrac{\cos^2 x}{1 - \cos^2 \beta \cos^2 x} \cdot \dfrac{dx}{\sqrt{1 - k^2 \sin^2 x}} =$

    $= \dfrac{1}{\sin \beta \cos \beta \sqrt{1 - k'^2 \sin^2 \beta}} \left\{\dfrac{\pi}{2} - KE(\beta, k') - EF(\beta, k') + KF(\beta, k')\right\}.$

    MO 138

2.
$$\int_0^{\frac{\pi}{2}} \frac{\sin^2 x}{1-(1-k'^2\sin^2\beta)\sin^2 x} \cdot \frac{dx}{\sqrt{1-k^2\sin^2 x}} =$$
$$= \frac{1}{k'^2 \sin\beta\cos\beta\sqrt{1-k'^2\sin^2\beta}} \left\{\frac{\pi}{2} - \mathbf{K}E(\beta, k') - \mathbf{E}F(\beta, k') + \mathbf{K}F(\beta, k')\right\}.$$
MO 138

3.679

3.
$$\int_0^{\frac{\pi}{2}} \frac{\sin^2 x}{1-k^2\sin^2\beta\sin^2 x} \cdot \frac{dx}{\sqrt{1-k^2\sin^2 x}} = \frac{\mathbf{K}E(\beta, k) - \mathbf{E}F(\beta, k)}{k^2 \sin\beta\cos\beta\sqrt{1-k^2\sin^2\beta}}.$$
MO 138

Funktionen von Potenzen trigonometrischer Funktionen
Functions of powers of trigonometric functions

3.68

1.
$$\int_0^{\frac{\pi}{2}} \frac{\sin^{2\mu-1} x \cos^{2\nu-1} x \, dx}{(1-k^2\sin^2 x)^\rho} = \frac{1}{2} B(\mu, \nu) F(\rho, \mu; \mu+\nu; k^2)$$
$[\operatorname{Re}\mu > 0, \operatorname{Re}\nu > 0].$ Ba I 115(7)

3.681

2.
$$\int_0^{\frac{\pi}{2}} \frac{\sin^{2\mu-1} x \cos^{2\nu-1} x \, dx}{(1-k^2\sin^2 x)^{\mu+\nu}} = \frac{B(\mu, \nu)}{2(1-k^2)^\mu} \quad [\operatorname{Re}\mu > 0, \operatorname{Re}\nu > 0].$$
Ba I 10(20)

3.
$$\int_0^{\frac{\pi}{2}} \frac{\sin^\mu x \, dx}{\cos^{\mu-3} x (1-k^2\sin^2 x)^{\frac{\mu}{2}-1}} =$$
$$= \frac{\Gamma\left(\frac{\mu+1}{2}\right)\Gamma\left(2-\frac{\mu}{2}\right)}{k^3\sqrt{\pi}(\mu-1)(\mu-3)(\mu-5)} \left\{\frac{1+(\mu-3)k+k^2}{(1+k)^{\mu-3}} - \frac{1-(\mu-3)k+k^2}{(1-k)^{\mu-3}}\right\}$$
$[-1 < \operatorname{Re}\mu < 4].$ BH [54](10)

4.
$$\int_0^{\frac{\pi}{2}} \frac{\sin^{\mu+1} x \, dx}{\cos^\mu x (1-k^2\sin^2 x)^{\frac{\mu+1}{2}}} = \frac{(1-k)^{-\mu} - (1+k)^{-\mu}}{4k\mu\sqrt{\pi}} \Gamma\left(1+\frac{\mu}{2}\right)\Gamma\left(\frac{1-\mu}{2}\right)$$
$[-2 < \operatorname{Re}\mu < 1].$ BH [61](5)

$$\int_0^{\frac{\pi}{2}} \frac{\sin^\mu x \cos^\nu x}{(a-b\cos^2 x)^\rho} dx =$$
$$= \frac{1}{2a^\rho} B\left(\frac{\mu+1}{2}, \frac{\nu+1}{2}\right) F\left(\frac{\nu+1}{2}, \rho; \frac{\mu+\nu}{2}+1; \frac{b}{a}\right)$$
$[\operatorname{Re}\mu > -1, \operatorname{Re}\nu > -1, a > |b| \geqslant 0].$ GH [331](64)

3.682

1.
$$\int_0^{\frac{\pi}{4}} (\sin^n 2x - 1)\operatorname{tg}\left(\frac{\pi}{4}+x\right) dx = \int_0^{\frac{\pi}{4}} (\cos^n 2x - 1)\operatorname{ctg} x \, dx =$$
$$= -\frac{1}{2}\sum_{k=1}^n \frac{1}{k} = -\frac{1}{2}[C+\psi(n+1)].$$
BH [34](8), BH [35](11)

3.683

3.683

2. $\int_0^{\pi/4} (\sin^\mu 2x - 1) \operatorname{cosec}^\mu 2x \, \operatorname{tg}\left(\frac{\pi}{4} + x\right) dx =$

$= \int_0^{\pi/4} (\cos^\mu 2x - 1) \sec^\mu 2x \operatorname{ctg} x \, dx = \frac{1}{2} [C + \psi(1 - \mu)]$

[Re $\mu < 1$]. BH[35](20)

3. $\int_0^{\pi/4} (\sin^{2\mu} 2x - 1) \operatorname{cosec}^\mu 2x \, \operatorname{tg}\left(\frac{\pi}{4} + x\right) dx =$

$= \int_0^{\pi/4} (\cos^{2\mu} 2x - 1) \sec^\mu 2x \operatorname{ctg} x \, dx = -\frac{1}{2\mu} + \frac{\pi}{2} \operatorname{ctg} \mu\pi.$ BH[35](21)

4. $\int_0^{\pi/4} (1 - \sec^\mu 2x) \operatorname{ctg} x \, dx = \int_0^{\pi/4} (1 - \operatorname{cosec}^\mu 2x) \operatorname{tg}\left(\frac{\pi}{4} + x\right) dx =$

$= \frac{1}{2}[C + \psi(1 - \mu)]$ [Re $\mu < 1$]. BH[35](13)

3.684

$\int_0^{\pi/4} \frac{(\operatorname{ctg}^\mu x - 1) dx}{(\cos x - \sin x) \sin x} = \int_0^{\pi/2} \frac{(\operatorname{tg}^\mu x - 1) \, dx}{(\sin x - \cos x) \cos x} = -C - \psi(1 - \mu)$

[Re $\mu < 1$]. BH[37](9)

3.685

1. $\int_0^{\pi/4} (\sin^{\mu-1} 2x - \sin^{\nu-1} 2x) \operatorname{tg}\left(\frac{\pi}{4} + x\right) dx =$

$= \int_0^{\pi/4} (\cos^{\mu-1} 2x - \cos^{\nu-1} 2x) \operatorname{ctg} x \, dx = \frac{1}{2} [\psi(\nu) - \psi(\mu)]$

[Re $\mu > 0$, Re $\nu > 0$]. BH[34](9), BH[35](12)

2. $\int_0^{\pi/2} (\sin^{\mu-1} x - \sin^{\nu-1} x) \frac{dx}{\cos x} = \int_0^{\pi/2} (\cos^{\mu-1} x - \cos^{\nu-1} x) \frac{dx}{\sin x} =$

$= \frac{1}{2}\left[\psi\left(\frac{\nu}{2}\right) - \psi\left(\frac{\mu}{2}\right)\right]$ [Re $\mu > 0$, Re $\nu > 0$]. BH[46](2)

3. $\int_0^{\pi/2} (\sin^\mu x - \operatorname{cosec}^\mu x) \frac{dx}{\cos x} = \int_0^{\pi/2} (\cos^\mu x - \sec^\mu x) \frac{dx}{\sin x} = -\frac{\pi}{2} \operatorname{tg} \frac{\mu\pi}{2}$

[|Re $\mu| < 1$]. BH[46](1, 3)

4. $$\int_0^{\frac{\pi}{4}} (\sin^\mu 2x - \operatorname{cosec}^\mu 2x) \operatorname{ctg}\left(\frac{\pi}{4} + x\right) dx =$$ 3.685

$$= \int_0^{\frac{\pi}{4}} (\cos^\mu 2x - \sec^\mu 2x) \operatorname{tg} x \, dx = \frac{1}{2\mu} - \frac{\pi}{2} \operatorname{cosec} \mu\pi$$

$$[|\operatorname{Re} \mu| < 1]. \qquad \text{BH[35](19, 22)}$$

5. $$\int_0^{\frac{\pi}{4}} (\sin^\mu 2x - \operatorname{cosec}^\mu 2x) \operatorname{tg}\left(\frac{\pi}{4} + x\right) dx =$$

$$= \int_0^{\frac{\pi}{4}} (\cos^\mu 2x - \sec^\mu 2x) \operatorname{ctg} x \, dx = -\frac{1}{2\mu} + \frac{\pi}{2} \operatorname{ctg} \mu\pi$$

$$[|\operatorname{Re} \mu| < 1]. \qquad \text{BH[35](14)}$$

6. $$\int_0^{\frac{\pi}{4}} (\sin^{\mu-1} 2x + \operatorname{cosec}^\mu 2x) \operatorname{ctg}\left(\frac{\pi}{4} + x\right) dx =$$

$$= \int_0^{\frac{\pi}{4}} (\cos^{\mu-1} 2x + \sec^\mu 2x) \operatorname{tg} x \, dx = \frac{\pi}{2} \operatorname{cosec} \mu\pi$$

$$[0 < \operatorname{Re} \mu < 1]. \qquad \text{BH[35](18, 8)}$$

7. $$\int_0^{\frac{\pi}{4}} (\sin^{\mu-1} 2x - \operatorname{cosec}^\mu 2x) \operatorname{tg}\left(\frac{\pi}{4} + x\right) dx =$$

$$= \int_0^{\frac{\pi}{4}} (\cos^{\mu-1} 2x - \sec^\mu 2x) \operatorname{ctg} x \, dx = \frac{\pi}{2} \operatorname{ctg} \mu\pi$$

$$[0 < \operatorname{Re} \mu < 1]. \qquad \text{BH[35](7), Li[34](10)}$$

$$\int_0^{\frac{\pi}{2}} \frac{\operatorname{tg} x \, dx}{\cos^\mu x + \sec^\mu x} = \int_0^{\frac{\pi}{2}} \frac{\operatorname{ctg} x \, dx}{\sin^\mu x + \operatorname{cosec}^\mu x} = \frac{\pi}{4\mu} \cdot \qquad \text{BH[47](28), BH[49](14)} \qquad 3.686$$

1. $$\int_0^{\frac{\pi}{2}} \frac{\sin^{\mu-1} x + \sin^{\nu-1} x}{\cos^{\mu+\nu-1} x} dx = \int_0^{\frac{\pi}{2}} \frac{\cos^{\mu-1} x + \cos^{\nu-1} x}{\sin^{\mu+\nu-1} x} dx =$$ 3.687

$$= \frac{\cos\left(\frac{\nu - \mu}{4} \pi\right)}{2 \cos\left(\frac{\nu + \mu}{4} \pi\right)} B\left(\frac{\mu}{2}, \frac{\nu}{2}\right) \quad [\operatorname{Re} \mu > 0, \operatorname{Re} \nu > 0]. \qquad \text{BH[46](7)}$$

3.687 2. $$\int_0^{\pi/2} \frac{\sin^{\mu-1} x - \sin^{\nu-1} x}{\cos^{\mu+\nu-1} x} dx = \int_0^{\pi/2} \frac{\cos^{\mu-1} x - \cos^{\nu-1} x}{\sin^{\mu+\nu-1} x} dx =$$

$$= \frac{\sin\left(\frac{\nu-\mu}{4}\pi\right)}{2\sin\left(\frac{\nu+\mu}{4}\pi\right)} B\left(\frac{\mu}{2}, \frac{\nu}{2}\right) \quad [\operatorname{Re}\mu > 0,\ \operatorname{Re}\nu > 0]. \qquad \text{BH[46](8)}$$

3. $$\int_0^{\pi/2} \frac{\sin^\mu x + \sin^\nu x}{\sin^{\mu+\nu} x + 1}\operatorname{ctg} x\, dx = \int_0^{\pi/2} \frac{\cos^\mu x + \cos^\nu x}{\cos^{\mu+\nu} x + 1}\operatorname{tg} x\, dx = \frac{\pi}{\mu+\nu}\sec\left(\frac{\mu-\nu}{\mu+\nu}\cdot\frac{\pi}{2}\right)$$

$$[\operatorname{Re}\mu > 0,\ \operatorname{Re}\nu > 0]. \qquad \text{BH[49](15)+, BH[47](29)}$$

4. $$\int_0^{\pi/2} \frac{\sin^\mu x - \sin^\nu x}{\sin^{\mu+\nu} x - 1}\operatorname{ctg} x\, dx = \int_0^{\pi/2} \frac{\cos^\mu x - \cos^\nu x}{\cos^{\mu+\nu} x - 1}\operatorname{tg} x\, dx = \frac{\pi}{\mu+\nu}\operatorname{tg}\left(\frac{\mu-\nu}{\mu+\nu}\cdot\frac{\pi}{2}\right)$$

$$[\operatorname{Re}\mu > 0,\ \operatorname{Re}\nu > 0]. \qquad \text{BH[49](16)+, BH[47](30)}$$

5. $$\int_0^{\pi/2} \frac{\cos^\mu x + \sec^\mu x}{\cos^\nu x + \sec^\nu x}\operatorname{tg} x\, dx = \frac{\pi}{2\nu}\sec\left(\frac{\mu}{\nu}\cdot\frac{\pi}{2}\right) \quad [|\operatorname{Re}\nu| > |\operatorname{Re}\mu|]. \qquad \text{BH[49](12)}$$

6. $$\int_0^{\pi/2} \frac{\cos^\mu x - \sec^\mu x}{\cos^\nu x - \sec^\nu x}\operatorname{tg} x\, dx = \frac{\pi}{2\nu}\operatorname{tg}\left(\frac{\mu}{\nu}\cdot\frac{\pi}{2}\right) \quad [|\operatorname{Re}\nu| > |\operatorname{Re}\mu|]. \qquad \text{BH[49](13)}$$

3.688 1. $$\int_0^{\pi/4} \frac{\operatorname{tg}^\nu x - \operatorname{tg}^\mu x}{\cos x - \sin x}\cdot\frac{dx}{\sin x} = \psi(\mu) - \psi(\nu) \quad [\operatorname{Re}\mu > 0,\ \operatorname{Re}\nu > 0]. \qquad \text{BH[37](10)}$$

2. $$\int_0^{\pi/4} \frac{\operatorname{tg}^\mu x - \operatorname{tg}^{1-\mu} x}{\cos x - \sin x}\cdot\frac{dx}{\sin x} = \pi \operatorname{ctg}\mu\pi \quad [0 < \operatorname{Re}\mu < 1]. \qquad \text{BH[37](11)}$$

3. $$\int_0^{\pi/4} (\operatorname{tg}^\mu x + \operatorname{ctg}^\mu x)\, dx = \frac{\pi}{2}\sec\frac{\mu\pi}{2} \quad [|\operatorname{Re}\mu| < 1]. \qquad \text{BH[35](9)}$$

4. $$\int_0^{\pi/4} (\operatorname{tg}^\mu x - \operatorname{ctg}^\mu x)\operatorname{tg} x\, dx = \frac{1}{\mu} - \frac{\pi}{2}\operatorname{cosec}\frac{\mu\pi}{2} \quad [0 < \operatorname{Re}\mu < 2]. \qquad \text{BH[35](15)}$$

5. $$\int_0^{\pi/4} \frac{\operatorname{tg}^{\mu-1} x - \operatorname{ctg}^{\mu-1} x}{\cos 2x}\, dx = \frac{\pi}{2}\operatorname{ctg}\frac{\mu\pi}{2} \quad [|\operatorname{Re}\mu| < 2]. \qquad \text{BH[35](10)}$$

6. $\int_0^{\pi/4} \dfrac{\operatorname{tg}^\mu x - \operatorname{ctg}^\mu x}{\cos 2x} \operatorname{tg} x\, dx = -\dfrac{1}{\mu} + \dfrac{\pi}{2} \operatorname{ctg} \dfrac{\mu\pi}{2}$ $[-2 < \operatorname{Re} \mu < 0]$.   BH[35](23)   3.688

7. $\int_0^{\pi/4} \dfrac{\operatorname{tg}^\mu x + \operatorname{ctg}^\mu x}{1 + \cos t \sin 2x}\, dx = \pi \operatorname{cosec} t \operatorname{cosec} \mu\pi \sin \mu t$ $[t \neq n\pi,\ |\operatorname{Re} \mu| < 1]$.   BH[36](6)

8. $\int_0^{\pi/4} \dfrac{\operatorname{tg}^{\mu-1} x + \operatorname{ctg}^\mu x}{(\sin x + \cos x)\cos x}\, dx = \pi \operatorname{cosec} \mu\pi$ $[0 < \operatorname{Re} \mu < 1]$.   BH[37](3)

9. $\int_0^{\pi/4} \dfrac{\operatorname{tg}^\mu x - \operatorname{ctg}^\mu x}{(\sin x + \cos x)\cos x}\, dx = -\pi \operatorname{cosec} \mu\pi + \dfrac{1}{\mu}$ $[0 < \operatorname{Re} \mu < 1]$.   BH[37](4)

10. $\int_0^{\pi/4} \dfrac{\operatorname{tg}^\nu x - \operatorname{ctg}^\mu x}{(\cos x - \sin x)\cos x}\, dx = \psi(1-\mu) - \psi(1+\nu)$

$[\operatorname{Re} \mu < 1,\ \operatorname{Re} \nu > -1]$.   BH[37](5)

11. $\int_0^{\pi/4} \dfrac{\operatorname{tg}^{\mu-1} x - \operatorname{ctg}^\mu x}{(\cos x - \sin x)\cos x}\, dx = \pi \operatorname{ctg} \mu\pi$ $[0 < \operatorname{Re} \mu < 1]$.   BH[37](7)

12. $\int_0^{\pi/4} \dfrac{\operatorname{tg}^\mu x - \operatorname{ctg}^\mu x}{(\cos x - \sin x)\cos x}\, dx = \pi \operatorname{ctg} \mu\pi - \dfrac{1}{\mu}$ $[0 < \operatorname{Re} \mu < 1]$.   BH[37](8)

13. $\int_0^{\pi/4} \dfrac{1}{\operatorname{tg}^\mu x + \operatorname{ctg}^\mu x} \cdot \dfrac{dx}{\sin 2x} = \dfrac{\pi}{8\mu}$ $[\operatorname{Re} \mu \neq 0]$.   BH[37](12)

14. $\int_0^{\pi/2} \dfrac{1}{(\operatorname{tg}^\mu x + \operatorname{ctg}^\mu x)^\nu} \cdot \dfrac{dx}{\operatorname{tg} x} = \int_0^{\pi/2} \dfrac{1}{(\operatorname{tg}^\mu x + \operatorname{ctg}^\mu x)^\nu} \cdot \dfrac{dx}{\sin 2x} =$

$= \dfrac{\sqrt{\pi}}{2^{2\nu+1}\mu} \dfrac{\Gamma(\nu)}{\Gamma\left(\nu + \dfrac{1}{2}\right)}$ $[\nu > 0]$.   BH[49](25), BH[49](26)

15. $\int_0^{\pi/4} (\operatorname{tg}^\mu x - \operatorname{ctg}^\mu x)(\operatorname{tg}^\nu x - \operatorname{ctg}^\nu x)\, dx = \dfrac{2\pi \sin \dfrac{\mu\pi}{2} \sin \dfrac{\nu\pi}{2}}{\cos \mu\pi + \cos \nu\pi}$

$[|\operatorname{Re} \mu| < 1,\ |\operatorname{Re} \nu| < 1]$.   BH[35](17)

3.688

16. $\int\limits_0^{\pi/4} (\operatorname{tg}^\mu x + \operatorname{ctg}^\mu x)(\operatorname{tg}^\nu x + \operatorname{ctg}^\nu x)\, dx = \dfrac{2\pi \cos\dfrac{\mu\pi}{2}\cos\dfrac{\nu\pi}{2}}{\cos\mu\pi + \cos\nu\pi}$

$[|\operatorname{Re}\mu| < 1,\ |\operatorname{Re}\nu| < 1]$. BH[35](16)

17. $\int\limits_0^{\pi/4} \dfrac{(\operatorname{tg}^\mu x - \operatorname{ctg}^\mu x)(\operatorname{tg}^\nu x + \operatorname{ctg}^\nu x)}{\cos 2x}\, dx = -\pi\dfrac{\sin\mu\pi}{\cos\mu\pi + \cos\nu\pi}$

$[|\operatorname{Re}\mu| < 1,\ |\operatorname{Re}\nu| < 1]$. BH[35](25)

18. $\int\limits_0^{\pi/4} \dfrac{\operatorname{tg}^\nu x - \operatorname{ctg}^\nu x}{\operatorname{tg}^\mu x - \operatorname{ctg}^\mu x}\cdot\dfrac{dx}{\sin 2x} = \dfrac{\pi}{4\mu}\operatorname{tg}\dfrac{\nu\pi}{2\mu}$ $[0 < \operatorname{Re}\nu < 1]$. BH[37](14)

19. $\int\limits_0^{\pi/4} \dfrac{\operatorname{tg}^\nu x + \operatorname{ctg}^\nu x}{\operatorname{tg}^\mu x + \operatorname{ctg}^\mu x}\cdot\dfrac{dx}{\sin 2x} = \dfrac{\pi}{4\mu}\sec\dfrac{\nu\pi}{2\mu}$ $[0 < \operatorname{Re}\nu < 1]$. BH[37](13)

20. $\int\limits_0^{\pi/2} \dfrac{(1+\operatorname{tg} x)^\nu - 1}{(1+\operatorname{tg} x)^{\mu+\nu}}\dfrac{dx}{\sin x \cos x} = \psi(\mu+\nu) - \psi(\mu)$ $[\mu > 0,\ \nu > 0]$. BH[49](29)

3.689

1. $\int\limits_0^{\pi/2} \dfrac{(\sin^\mu x + \operatorname{cosec}^\mu x)\operatorname{ctg} x\, dx}{\sin^\nu x - 2\cos t + \operatorname{cosec}^\nu x} = \dfrac{\pi}{\nu}\operatorname{cosec} t\,\operatorname{cosec}\dfrac{\mu\pi}{\nu}\sin\dfrac{\mu t}{\nu}$ $[\mu < \nu]$. Li[50](14)

2. $\int\limits_0^{\pi/2} \dfrac{\sin^\mu x - 2\cos t_1 + \operatorname{cosec}^\mu x}{\sin^\nu x + 2\cos t_2 + \operatorname{cosec}^\nu x}\cdot\operatorname{ctg} x\cdot dx =$

$= \dfrac{\pi}{\nu}\operatorname{cosec} t_2\,\operatorname{cosec}\dfrac{\mu\pi}{\nu}\sin\dfrac{\mu t_2}{\nu} - \dfrac{t_2}{\nu}\operatorname{cosec} t_2\cos t_1$

$[\nu > \mu > 0$ oder $\nu < \mu < 0$ oder $\mu > 0$, $\nu < 0$ und $\mu+\nu < 0$ oder $\mu < 0,\ \nu > 0$ und $\mu+\nu > 0]$. BH[50](15)

$[\nu > \mu > 0$ or $\nu < \mu < 0$ or $\mu > 0$, $\nu < 0$ and $\mu+\nu < 0$ or $\mu < 0,\ \nu > 0$ and $\mu+\nu > 0]$. BH[50](15)

Trigonometrische Funktionen mit komplizierteren Argumenten
Trigonometric functions of compound arguments

3.69–3.71

3.691

1. $\int\limits_0^\infty \sin(ax^2)\, dx = \int\limits_0^\infty \cos ax^2\, dx = \dfrac{1}{2}\sqrt{\dfrac{\pi}{2a}}$ $[a > 0]$. F II 747, 748+, Ba IV 64(7)+

2. $\int\limits_0^1 \sin(ax^2)\, dx = \sqrt{\dfrac{\pi}{2a}}\, S(\sqrt{a})$ $[a > 0]$.

3. $$\int_0^1 \cos(ax^2)\,dx = \sqrt{\frac{\pi}{2a}}\,C(\sqrt{a}) \quad [a>0].$$ Ba IV 8(5)+   3.691

4. $$\int_0^\infty \sin(ax^2)\sin 2bx\,dx = \sqrt{\frac{\pi}{2a}}\left\{\cos\frac{b^2}{a}C\left(\frac{b}{\sqrt{a}}\right)+\sin\frac{b^2}{a}S\left(\frac{b}{\sqrt{a}}\right)\right\}$$
$$[a>0,\ b>0]. \quad \text{Ba IV 82(1)+}$$

5. $$\int_0^\infty \sin(ax^2)\cos 2bx\,dx = \frac{1}{2}\sqrt{\frac{\pi}{2a}}\left\{\cos\frac{b^2}{a}-\sin\frac{b^2}{a}\right\} = \frac{1}{2}\sqrt{\frac{\pi}{a}}\cos\left(\frac{b^2}{a}+\frac{\pi}{4}\right)$$
$$[a>0,\ b>0]. \quad \text{Ba IV 23(1), BH[70](13), GH[334](5a)}$$

6. $$\int_0^\infty \cos ax^2 \sin 2bx\,dx = \sqrt{\frac{\pi}{2a}}\left\{\sin\frac{b^2}{a}C\left(\frac{b}{\sqrt{a}}\right)-\cos\frac{b^2}{a}S\left(\frac{b}{\sqrt{a}}\right)\right\}$$
$$[a>0,\ b>0]. \quad \text{Ba IV 83(3)+}$$

7. $$\int_0^\infty \cos ax^2 \cos 2bx\,dx = \frac{1}{2}\sqrt{\frac{\pi}{2a}}\left\{\cos\frac{b^2}{a}+\sin\frac{b^2}{a}\right\} \quad [a>0,\ b>0].$$
GH[334](5a), BH[70](14), Ba IV 24(7)

8. $$\int_0^\infty \sin(a^2x^2)\sin 2bx\sin 2cx\,dx = \frac{\sqrt{\pi}}{2a}\sin\frac{2bc}{a^2}\cos\left(\frac{b^2+c^2}{a^2}-\frac{\pi}{4}\right)$$
$$[a>0,\ b>0,\ c>0]. \quad \text{Ba IV 84(15)}$$

9. $$\int_0^\infty \sin(a^2x^2)\cos 2bx\cos 2cx\,dx = \frac{\sqrt{\pi}}{2a}\cos\frac{2bc}{a^2}\cos\left(\frac{b^2+c^2}{a^2}+\frac{\pi}{4}\right)$$
$$[a>0,\ b>0,\ c>0]. \quad \text{Ba IV 84(21)}$$

10. $$\int_0^\infty \cos(a^2x^2)\sin 2bx\sin 2cx\,dx = \frac{\sqrt{\pi}}{2a}\sin\frac{2bc}{a^2}\sin\left(\frac{b^2+c^2}{a^2}-\frac{\pi}{4}\right)$$
$$[a>0,\ b>0,\ c>0]. \quad \text{Ba IV 25(19)}$$

11. $$\int_0^\infty \sin(ax^2)\cos(bx^2)\,dx = \frac{1}{4}\sqrt{\frac{\pi}{2}}\left(\frac{1}{\sqrt{a+b}}+\frac{1}{\sqrt{a-b}}\right) \quad [a>b>0];$$
$$= \frac{1}{4}\sqrt{\frac{\pi}{2}}\left(\frac{1}{\sqrt{b+a}}-\frac{1}{\sqrt{b-a}}\right) \quad [b>a>0].$$
BH[177](21)

12. $$\int_0^\infty (\sin^2 ax^2 - \sin^2 bx^2)\,dx = \frac{1}{8}\left(\sqrt{\frac{\pi}{b}}-\sqrt{\frac{\pi}{a}}\right) \quad [a>0,\ b>0]. \quad \text{BH[178](1)}$$

13. $$\int_0^\infty (\cos^2 ax^2 - \sin^2 bx^2)\,dx = \frac{1}{8}\left(\sqrt{\frac{\pi}{b}}+\sqrt{\frac{\pi}{a}}\right) \quad [a>0,\ b>0]. \quad \text{BH[178](3)}$$

14. $$\int_0^\infty (\cos^2 ax^2 - \cos^2 bx^2)\,dx = \frac{1}{8}\left(\sqrt{\frac{\pi}{a}}-\sqrt{\frac{\pi}{b}}\right) \quad [a>0,\ b>0]. \quad \text{BH[178](5)}$$

3.691

15. $\int_0^\infty (\sin^4 ax^2 - \sin^4 bx^2)\, dx = \frac{1}{64}(8 - \sqrt{2})\left(\sqrt{\frac{\pi}{b}} - \sqrt{\frac{\pi}{a}}\right)$

$[a > 0,\ b > 0]$. BH[178](2)

16. $\int_0^\infty (\cos^4 ax^2 - \sin^4 bx^2)\, dx = \frac{1}{8}\left(\sqrt{\frac{\pi}{a}} + \sqrt{\frac{\pi}{b}}\right) + \frac{1}{32}\left(\sqrt{\frac{\pi}{2a}} - \sqrt{\frac{\pi}{2b}}\right)$

$[a > 0,\ b > 0]$. BH[178](4)

17. $\int_0^\infty (\cos^4 ax^2 - \cos^4 bx^2)\, dx = \frac{1}{64}(8 + \sqrt{2})\left(\sqrt{\frac{\pi}{a}} - \sqrt{\frac{\pi}{b}}\right)$

$[a > 0,\ b > 0]$. BH[178](6)

18. $\int_0^\infty \sin^{2n} ax^2\, dx = \int_0^\infty \cos^{2n} ax^2\, dx = \infty$.

BH[177](5, 6)

19. $\int_0^\infty \sin^{2n+1}(ax^2)\, dx = \frac{1}{2^{2n+1}} \sum_{k=0}^{n} (-1)^{n+k} \binom{2n+1}{k} \sqrt{\frac{\pi}{2(2n-2k+1)a}}$

$[a > 0]$. BH[70](9)

20. $\int_0^\infty \cos^{2n+1}(ax^2)\, dx = \frac{1}{2^{2n+1}} \sum_{k=0}^{n} \binom{2n+1}{k} \sqrt{\frac{\pi}{2(2n-2k+1)a}}$

$[a > 0]$. BH[177](7)÷, BH[70](10)

3.692

1. $\int_0^\infty [\sin(a - x^2) + \cos(a - x^2)]\, dx = \sqrt{\frac{\pi}{2}} \sin a$. GH[333](30c), BH[178](7)+

2. $\int_0^\infty \cos\left(\frac{x^2}{2} - \frac{\pi}{8}\right) \cos ax\, dx = \sqrt{\frac{\pi}{2}} \cos\left(\frac{a^2}{2} - \frac{\pi}{8}\right)$ $[a > 0]$. Ba IV 24(8)

3. $\int_0^\infty \sin[a(1 - x^2)] \cos bx\, dx = -\frac{1}{2}\sqrt{\frac{\pi}{a}} \cos\left(a + \frac{b^2}{4a} + \frac{\pi}{4}\right)$ $[a > 0]$. Ba IV 23(2)

4. $\int_0^\infty \cos[a(1 - x^2)] \cos bx\, dx = \frac{1}{2}\sqrt{\frac{\pi}{a}} \sin\left(a + \frac{b^2}{4a} + \frac{\pi}{4}\right)$ $[a > 0]$. Ba IV 24(10)

5. $\int_0^\infty \sin\left(ax^2 + \frac{b^2}{a}\right) \cos 2bx\, dx = \int_0^\infty \cos\left(ax^2 + \frac{b^2}{a}\right) \cos 2bx\, dx = \frac{1}{2}\sqrt{\frac{\pi}{2a}}$

$[a > 0]$. BH[70](19, 20)

In 3.693—3.694 ist | In 3.693—3.694 we put

$$S_2(x) = \frac{1}{\sqrt{2\pi}} \int_0^x \frac{\sin t}{\sqrt{t}}\, dt,\ C_2(x) = \frac{1}{\sqrt{2\pi}} \int_0^x \frac{\cos t}{\sqrt{t}}\, dt.$$

1. $\int_0^\infty \sin(ax^2 + 2bx)\, dx = \sqrt{\dfrac{\pi}{2a}} \left\{ \cos\dfrac{b^2}{a}\left(\dfrac{1}{2} - S_2\left(\dfrac{b^2}{a}\right)\right) - \sin\dfrac{b^2}{a}\left(\dfrac{1}{2} - C_2\left(\dfrac{b^2}{a}\right)\right) \right\}$ 3.693

$[a > 0]$. BH [70](3)

2. $\int_0^\infty \cos(ax^2 + 2bx)\, dx = \sqrt{\dfrac{\pi}{2a}} \left\{ \cos\dfrac{b^2}{a}\left(\dfrac{1}{2} - C_2\left(\dfrac{b^2}{a}\right)\right) + \sin\dfrac{b^2}{a}\left(\dfrac{1}{2} - S_2\left(\dfrac{b^2}{a}\right)\right) \right\}$

$[a > 0]$. BH [70](4)

1. $\int_0^\infty \sin(ax^2 + 2bx + c)\, dx = \sqrt{\dfrac{\pi}{2a}} \cos\dfrac{b^2}{a}\left\{\left(\dfrac{1}{2} - C_2\left(\dfrac{b^2}{a}\right)\right) \sin c +$ 3.694

$+ \left(\dfrac{1}{2} - S_2\left(\dfrac{b^2}{a}\right)\right) \cos c \right\} + \sqrt{\dfrac{\pi}{2a}} \sin\dfrac{b^2}{a}\left\{\left(\dfrac{1}{2} - S_2\left(\dfrac{b^2}{a}\right)\right) \sin c -\right.$

$\left. - \left(\dfrac{1}{2} - C_2\left(\dfrac{b^2}{a}\right)\right) \cos c \right\}$ $[a > 0]$. GH [334](4a)

2. $\int_0^\infty \cos(ax^2 + 2bx + c)\, dx = \sqrt{\dfrac{\pi}{2a}} \cos\dfrac{b^2}{a}\left\{\left(\dfrac{1}{2} - C_2\left(\dfrac{b^2}{a}\right)\right) \cos c -\right.$

$\left. - \left(\dfrac{1}{2} - S_2\left(\dfrac{b^2}{a}\right)\right) \sin c \right\} + \sqrt{\dfrac{\pi}{2a}} \sin\dfrac{b^2}{a}\left\{\left(\dfrac{1}{2} - S_2\left(\dfrac{b^2}{a}\right)\right) \cos c +\right.$

$\left. + \left(\dfrac{1}{2} - C_2\left(\dfrac{b^2}{a}\right)\right) \sin c \right\}$ $[a > 0]$. GH [334](4b)

1. $\int_0^\infty \sin(a^3 x^3) \sin(bx)\, dx = \dfrac{\pi}{6a}\sqrt{\dfrac{b}{3a}} \left\{ J_{\frac{1}{3}}\left(\dfrac{2b}{3a}\sqrt{\dfrac{b}{3a}}\right) +\right.$ 3.695

$\left. + J_{-\frac{1}{3}}\left(\dfrac{2b}{3a}\sqrt{\dfrac{b}{3a}}\right) - \dfrac{\sqrt{3}}{\pi} K_{\frac{1}{3}}\left(\dfrac{2b}{3a}\sqrt{\dfrac{b}{3a}}\right) \right\}$ $[a > 0,\, b > 0]$. Ba IV 83(5)

2. $\int_0^\infty \cos(a^3 x^3) \cos(bx)\, dx = \dfrac{\pi}{6a}\sqrt{\dfrac{b}{3a}} \left\{ J_{\frac{1}{3}}\left(\dfrac{2b}{3a}\sqrt{\dfrac{b}{3a}}\right) +\right.$

$\left. + J_{-\frac{1}{3}}\left(\dfrac{2b}{3a}\sqrt{\dfrac{b}{3a}}\right) + \dfrac{\sqrt{3}}{\pi} K_{\frac{1}{3}}\left(\dfrac{2b}{3a}\sqrt{\dfrac{b}{3a}}\right) \right\}$ $[a > 0,\, b > 0]$. Ba IV 24(11)

1. $\int_0^\infty \sin(ax^4) \sin(bx^2)\, dx = -\dfrac{\pi}{4}\sqrt{\dfrac{b}{2a}} \sin\left(\dfrac{b^2}{8a} - \dfrac{3}{8}\pi\right) J_{\frac{1}{4}}\left(\dfrac{b^2}{8a}\right)$ 3.696

$[a > 0,\, b > 0]$. Ba IV 83(2)

2. $\int_0^\infty \sin(ax^4) \cos(bx^2)\, dx = -\dfrac{\pi}{4}\sqrt{\dfrac{b}{2a}} \sin\left(\dfrac{b^2}{8a} - \dfrac{\pi}{8}\right) J_{-\frac{1}{4}}\left(\dfrac{b^2}{8a}\right)$

$[a > 0,\, b > 0]$. Ba IV 84(19)

3. $\int_0^\infty \cos(ax^4) \sin(bx^2)\, dx = \dfrac{\pi}{4}\sqrt{\dfrac{b}{2a}} \cos\left(\dfrac{b^2}{8a} - \dfrac{3}{8}\pi\right) J_{\frac{1}{4}}\left(\dfrac{b^2}{8a}\right)$

$[a > 0,\, b > 0]$. Ba IV 83(4), Ba IV 25(24)

3.696 4. $$\int_0^\infty \cos(ax^4)\cos(bx^2)\,dx = \frac{\pi}{4}\sqrt{\frac{b}{2a}}\cos\left(\frac{b^2}{8a}-\frac{\pi}{8}\right)J_{-\frac{1}{4}}\left(\frac{b^2}{8a}\right)$$
$[a > 0, b > 0]$. Ba IV 25(25)

3.697 $$\int_0^\infty \sin\left(\frac{a^2}{x}\right)\sin(bx)\,dx = \frac{a\pi}{2\sqrt{b}}\,J_1(2a\sqrt{b}) \quad [a>0, b>0].$$ Ba IV 83(6)

3.698 1. $$\int_0^\infty \sin\left(\frac{a^2}{x^2}\right)\sin(b^2x^2)\,dx = \frac{1}{4b}\sqrt{\frac{\pi}{2}}\,[\sin 2ab - \cos 2ab + e^{-2ab}]$$
$[a > 0, b > 0]$. Ba IV 83(9)

2. $$\int_0^\infty \sin\left(\frac{a^2}{x^2}\right)\cos(b^2x^2)\,dx = \frac{1}{4b}\sqrt{\frac{\pi}{2}}\,[\sin 2ab + \cos 2ab + e^{-2ab}]$$
$[a > 0, b > 0]$. Ba IV 24(13)

3. $$\int_0^\infty \cos\left(\frac{a^2}{x^2}\right)\sin(b^2x^2)\,dx = \frac{1}{4b}\sqrt{\frac{\pi}{2}}\,[\sin 2ab + \cos 2ab + e^{-2ab}]$$
$[a > 0, b > 0]$. Ba IV 84(12)

4. $$\int_0^\infty \cos\left(\frac{a^2}{x^2}\right)\cos(b^2x^2)\,dx = \frac{1}{4b}\sqrt{\frac{\pi}{2}}\,[\cos 2ab - \sin 2ab + e^{-2ab}]$$
$[a > 0, b > 0]$. Ba IV 24(14)

3.699 1. $$\int_0^\infty \sin\left(a^2x^2 + \frac{b^2}{x^2}\right)dx = \frac{\sqrt{2\pi}}{4a}(\cos 2ab + \sin 2ab) \quad [a > 0, b > 0].$$ BH[70](27)

2. $$\int_0^\infty \cos\left(a^2x^2 + \frac{b^2}{x^2}\right)dx = \frac{\sqrt{2\pi}}{4a}(\cos 2ab - \sin 2ab) \quad [a > 0, b > 0].$$ BH[70](28)

3. $$\int_0^\infty \sin\left(a^2x^2 - 2ab + \frac{b^2}{x^2}\right)dx = \int_0^\infty \cos\left(a^2x^2 - 2ab + \frac{b^2}{x^2}\right)dx = \frac{\sqrt{2\pi}}{4a}$$
$[a > 0, b > 0]$. BH[179](11, 12)+, Ba IV 83(6)+

4. $$\int_0^\infty \sin\left(a^2x^2 - \frac{b^2}{x^2}\right)dx = \frac{\sqrt{2\pi}}{4a}\,e^{-2ab} \quad [a > 0, b > 0].$$ GH[334](9b)+

5. $$\int_0^\infty \cos\left(a^2x^2 - \frac{b^2}{x^2}\right)dx = \frac{\sqrt{2\pi}}{4a}\,e^{-2ab} \quad [a > 0, b > 0].$$ GH[334](9b)+

3.711 $$\int_0^u \sin(a\sqrt{u^2-x^2})\cos bx\,dx = \frac{\pi au}{2\sqrt{a^2+b^2}}\,J_1(u\sqrt{a^2+b^2})$$
$[a > 0, b > 0, u > 0]$. Ba IV 27(37)

3.712 1. $$\int_0^\infty \sin(ax^p)\,dx = \frac{\Gamma\left(\frac{1}{p}\right)\sin\frac{\pi}{2p}}{pa^{\frac{1}{p}}} \quad [a > 0, p > 1].$$ Ba I 13(40)

2. $\int_0^\infty \cos(ax^p)\, dx = \dfrac{\Gamma\left(\dfrac{1}{p}\right)\cos\dfrac{\pi}{2p}}{pa^{\frac{1}{p}}}$ $[a > 0,\ p > 1]$.   Ba I 13(39)   3.712

1. $\int_0^\infty \sin(ax^p + bx^q)\, dx = \dfrac{1}{p}\sum_{k=0}^\infty \dfrac{(-b)^k}{k!}\, a^{-\frac{kq+1}{p}}\, \Gamma\left(\dfrac{kq+1}{p}\right) \times$ 3.713

$\times \sin\left[\dfrac{k(q-p)+1}{2p}\pi\right]$ $[a > 0,\ b > 0,\ p > 0,\ q > 0]$.   BH[70](7)

2. $\int_0^\infty \cos(ax^p + bx^q)\, dx = \dfrac{1}{p}\sum_{k=0}^\infty \dfrac{(-b)^k}{k!}\, a^{-\frac{kq+1}{p}}\, \Gamma\left(\dfrac{kq+1}{p}\right) \times$

$\times \cos\left[\dfrac{k(q-p)+1}{2p}\pi\right]$ $[a > 0,\ b > 0,\ p > 0,\ q > 0]$.   BH[70](8)

1. $\int_0^\infty \cos(z\, \text{sh}\, x)\, dx = K_0(z)$ $[\text{Re}\, z > 0]$.   W 183(14)   3.714

2. $\int_0^\infty \sin(z\, \text{ch}\, x)\, dx = \dfrac{\pi}{2} J_0(z)$ $[\text{Re}\, z > 0]$.   MO 36

3. $\int_0^\infty \cos(z\, \text{ch}\, x)\, dx = -\dfrac{\pi}{2} N_0(z)$ $[\text{Re}\, z > 0]$.   MO 37

4. $\int_0^\infty \cos(z\, \text{sh}\, x)\, \text{ch}\, \mu x\, dx = \cos\dfrac{\mu\pi}{2} K_\mu(z)$ $[\text{Re}\, z > 0,\ |\text{Re}\, \mu| < 1]$.   W 183(13)

5. $\int_0^\pi \cos(z\, \text{ch}\, x)\, \sin^{2\mu} x\, dx = \sqrt{\pi}\left(\dfrac{2}{z}\right)^\mu \Gamma\left(\mu + \dfrac{1}{2}\right) I_\mu(z)$

$\left[\text{Re}\, z > 0,\ \text{Re}\, \mu > -\dfrac{1}{2}\right]$.   WW 384

1. $\int_0^\pi \sin(z \sin x)\, \sin ax\, dx = \sin a\pi\, s_{0,a}(z) =$ 3.715

$= \sin a\pi \sum_{k=1}^\infty \dfrac{(-1)^{k-1} z^{2k-1}}{(1^2 - a^2)(3^2 - a^2)\ldots[(2k-1)^2 - a^2]}$ $[a > 0]$.   W 310(13)

2. $\int_0^\pi \sin(z \sin x)\, \sin nx\, dx = \dfrac{1}{2}\int_{-\pi}^\pi \sin(z \sin x)\, \sin nx\, dx =$

$= [1 - (-1)^n]\int_0^{\frac{\pi}{2}} \sin(z \sin x)\, \sin nx\, dx =$

$= [1 - (-1)^n]\dfrac{\pi}{2} J_n(z)$ $[n = 0, \pm 1, \pm 2, \ldots]$.   W 20(6), GH[334](53a)

3.715

3. $\int_0^{\pi/2} \sin(z \sin x) \sin 2x\, dx = \dfrac{2}{z^2}(\sin z - z \cos z)$. Li[43](14)

4. $\int_0^{\pi} \sin(z \sin x) \cos ax\, dx = (1 + \cos a\pi)\, s_{0,a}(z) =$

$= (1 + \cos a\pi) \sum_{k=1}^{\infty} \dfrac{(-1)^{k-1} z^{2k-1}}{(1^2 - a^2)(3^2 - a^2) \ldots [(2k-1)^2 - a^2]}$ $[a > 0]$. W 310(14)

5. $\int_0^{\pi} \sin(z \sin x) \cos[(2n + 1) x]\, dx = 0$. GH[334](53b)

6. $\int_0^{\pi} \cos(z \sin x) \sin ax\, dx = -a(1 - \cos a\pi)\, s_{-1,a}(z) =$

$= -a(1 - \cos a\pi) \left\{ -\dfrac{1}{a^2} + \sum_{k=1}^{\infty} \dfrac{(-1)^{k-1} z^{2k}}{a^2(2^2 - a^2)(4^2 - a^2) \ldots [(2k)^2 - a^2]} \right\}$

$[a > 0]$. W 310(12)

7. $\int_0^{\pi} \cos(z \sin x) \sin 2nx\, dx = 0$. GH[334](54a)

8. $\int_0^{\pi} \cos(z \sin x) \cos ax\, dx = -a \sin a\pi\, s_{-1,a}(z) =$

$= -a \sin a\pi \left\{ -\dfrac{1}{a^2} + \sum_{k=1}^{\infty} \dfrac{(-1)^{k-1} z^{2k}}{a^2(2^2 - a^2)(4^2 - a^2) \ldots [(2k)^2 - a^2]} \right\}$ $[a > 0]$. W 310(11)

9. $\int_0^{\pi} \cos(z \sin x) \cos nx\, dx = \dfrac{1}{2} \int_{-\pi}^{\pi} \cos(z \sin x) \cos nx\, dx =$

$= [1 + (-1)^n] \int_0^{\pi/2} \cos(z \sin x) \cos nx\, dx = [1 + (-1)^n] \dfrac{\pi}{2} J_n(z)$. GH[334](54b)

10. $\int_0^{\pi/2} \cos(z \sin x) \cos^{2n} x\, dx = \dfrac{\pi}{2} \dfrac{(2n-1)!!}{z^n} J_n(z)$ $\left[\operatorname{Re} z > -\dfrac{1}{2} \right]$.

F II 480, W 26–27+

11. $\int_0^{\pi/2} \sin(z \cos x) \sin 2x\, dx = \dfrac{2}{z^2}(\sin z - z \cos z)$. Li[43](15)

12. $\int_0^{\pi/2} \sin(z\cos x) \cos ax \, dx = \cos\dfrac{a\pi}{2} s_{0,a}(z) =$ 3.715

$= \dfrac{\pi}{4} \operatorname{cosec}\dfrac{a\pi}{2} [\mathbf{J}_\nu(z) - \mathbf{J}_{-\nu}(z)] = -\dfrac{\pi}{4} \sec\dfrac{a\pi}{4} [\mathbf{E}_\nu(z) + \mathbf{E}_{-\nu}(z)] =$

$= \cos\dfrac{a\pi}{2} \sum_{k=1}^{\infty} \dfrac{(-1)^{k-1} z^{2k-1}}{(1^2 - a^2)(3^2 - a^2)\cdots[(2k-1)^2 - a^2]} \quad [a > 0].$ W 310–311

13. $\int_0^{\pi} \sin(z\cos x) \cos nx \, dx = \dfrac{1}{2}\int_{-\pi}^{\pi} \sin(z\cos x) \cos nx \, dx = \pi \sin\dfrac{n\pi}{2} J_n(z).$

GH [334](55b)

14. $\int_0^{\pi/2} \sin(z\cos x) \cos[(2n+1)x]\, dx = (-1)^n \dfrac{\pi}{2} J_{2n+1}(z).$ W 21(8)

15. $\int_0^{\pi/2} \sin(a\cos x) \operatorname{tg} x \, dx = \operatorname{si}(a) + \dfrac{\pi}{2} \quad [a > 0].$ BH [43](17)

16. $\int_0^{\pi/2} \sin(z\cos x) \sin^{2\nu} x \, dx = \dfrac{\sqrt{\pi}}{2}\left(\dfrac{2}{z}\right)^\nu \Gamma\left(\nu+\dfrac{1}{2}\right) \mathbf{H}_\nu(z) \left[\operatorname{Re}\nu > -\dfrac{1}{2}\right].$ W 328(1)

17. $\int_0^{\pi/2} \cos(z\cos x)\cos ax \, dx = -a\sin\dfrac{a\pi}{2} s_{-1,a}(z) =$

$= \dfrac{\pi}{4}\sec\dfrac{a\pi}{2}[\mathbf{J}_\nu(z)+\mathbf{J}_{-\nu}(z)] = \dfrac{\pi}{4}\operatorname{cosec}\dfrac{a\pi}{2}[\mathbf{E}_\nu(z)-\mathbf{E}_{-\nu}(z)] =$

$= -a\sin\dfrac{a\pi}{2}\left\{-\dfrac{1}{a^2}+\sum_{k=1}^{\infty}\dfrac{(-1)^{k-1}z^{2k}}{a^2(2^2-a^2)(4^2-a^2)\cdots[(2k)^2-a^2]}\right\} \quad [a > 0].$ W 310

18. $\int_0^{\pi} \cos(z\cos x) \cos nx \, dx = \dfrac{1}{2}\int_{-\pi}^{\pi} \cos(z\cos x)\cos nx \, dx = \pi \cos\dfrac{n\pi}{2} J_n(z).$

GH [334](56b)

19. $\int_0^{\pi/2} \cos(z\cos x) \cos 2nx \, dx = (-1)^n \cdot \dfrac{\pi}{2} J_{2n}(z).$ W 21(9)

20. $\int_0^{\pi/2} \cos(z\cos x) \sin^{2\nu} x \, dx = \dfrac{\sqrt{\pi}}{2}\left(\dfrac{2}{z}\right)^\nu \Gamma\left(\nu+\dfrac{1}{2}\right) J_\nu(z) \quad \left[\operatorname{Re}\nu > -\dfrac{1}{2}\right].$

W 24, WW 366

21. $\int_0^{\pi} \cos(z\cos x) \sin^{2\mu} x \, dx = \sqrt{\pi}\left(\dfrac{2}{z}\right)^\mu \Gamma\left(\mu+\dfrac{1}{2}\right) J_\mu(z) \quad \left[\operatorname{Re}\mu > -\dfrac{1}{2}\right].$

WW 366

1. $\int_0^{\pi/2} \sin(a \operatorname{tg} x) \, dx = \dfrac{1}{2}[e^{-a}\overline{\operatorname{Ei}}(a) - e^a \operatorname{Ei}(-a)] \quad (\text{cf. 3.723 1.}).$ BH [43](1) 3.716

3.716

2. $$\int_0^{\pi/2} \cos(a \operatorname{tg} x)\, dx = \frac{\pi}{2} e^{-a}.$$ BH[43](2)

3. $$\int_0^{\pi/2} \sin(a \operatorname{tg} x) \sin 2x\, dx = \frac{a\pi}{2} e^{-a}.$$ BH[43](7)

4. $$\int_0^{\pi/2} \cos(a \operatorname{tg} x) \sin^2 x\, dx = \frac{1-a}{4} \pi e^{-a}.$$ BH[43](8)

5. $$\int_0^{\pi/2} \cos(a \operatorname{tg} x) \cos^2 x\, dx = \frac{1+a}{4} \pi e^{-a}.$$ BH[43](9)

6. $$\int_0^{\pi/2} \sin(a \operatorname{tg} x) \operatorname{tg} x\, dx = \frac{\pi}{2} e^{-a}.$$ BH[43](5)

7. $$\int_0^{\pi/2} \cos(a \operatorname{tg} x) \operatorname{tg} x\, dx = -\frac{1}{2}[e^{-a}\overline{\operatorname{Ei}}(a) + e^{a}\operatorname{Ei}(-a)] \quad \text{(cf. 3.723 5.).}$$ BH[43](6)

8. $$\int_0^{\pi/2} \sin(a \operatorname{tg} x) \sin^2 x \operatorname{tg} x\, dx = \frac{2-a}{4} \pi e^{-a}.$$ BH[43](11)

9. $$\int_0^{\pi/2} \sin^2(a \operatorname{tg} x)\, dx = \frac{\pi}{4}(1 - e^{-2a}) \quad \text{(cf. 3.742 1.).}$$ BH[43](3)

10. $$\int_0^{\pi/2} \cos^2(a \operatorname{tg} x)\, dx = \frac{\pi}{4}(1 + e^{-2a}) \quad \text{(cf. 3.742 3.).}$$ BH[43](4)

11. $$\int_0^{\pi/2} \sin^2(a \operatorname{tg} x) \operatorname{ctg}^2 x\, dx = \frac{\pi}{4}(e^{-2a} + 2a - 1).$$ BH[43](19)

12. $$\int_0^{\pi/2} [1 - \sec^2 x \cos(\operatorname{tg} x)] \frac{dx}{\operatorname{tg} x} = C.$$ BH[51](14)

13. $$\int_0^{\pi/2} \sin(a \operatorname{ctg} x) \sin 2x\, dx = \frac{a\pi}{2} e^{-a} \quad \text{(cf. 3.716 3.).}$$

Die Formeln 3.716 bleiben gültig, wenn man darin die im Argument des Sinus bzw. Cosinus vorkommende

Formulae 3.716 remain valid if in the argument of the sine or the cosine tg x is replaced by ctg x, provided that in the

Funktion tg x durch ctg x ersetzt, vorausgesetzt, man hat in den übrigen Faktoren ersetzt: sin x durch cos x, cos x durch sin x, also tg x durch ctg x, ctg x durch tg x, sec x durch cosec x und cosec x durch sec x. Analog gilt

other factors sin x is replaced by cos x, cos x by sin x, and also tg x by ctg x, ctg x by tg x, sec x by cosec x, cosec x by sec x. Similarly

$$\int_0^{\pi/2} \sin(a \operatorname{cosec} x) \sin(a \operatorname{ctg} x) \frac{dx}{\cos x} =$$

3.717

$$= \int_0^{\pi/2} \sin(a \sec x) \sin(a \operatorname{tg} x) \frac{dx}{\sin x} = \frac{\pi}{2} \sin a \quad [a > 0].$$

BH [52](11, 12)

1. $\displaystyle\int_0^{\pi/2} \sin\left(\frac{\pi}{2} p - a \operatorname{tg} x\right) \operatorname{tg}^{p-1} x \, dx =$

3.718

$$= \int_0^{\pi/2} \cos\left(\frac{\pi}{2} p - a \operatorname{tg} x\right) \operatorname{tg}^p x \, dx = \frac{\pi}{2} e^{-a} \quad [p^2 < 1].$$

BH [44](5, 6)

2. $\displaystyle\int_0^{\pi/2} \sin(a \operatorname{tg} x - vx) \sin^{v-2} x \, dx = 0 \quad [\operatorname{Re} v > 0].$

NG 157(15)

3. $\displaystyle\int_0^{\pi/2} \sin(n \operatorname{tg} x + vx) \frac{\cos^{v-1} x}{\sin x} \, dx = \frac{\pi}{2} \quad [\operatorname{Re} v > 0].$

BH [51](15)

4. $\displaystyle\int_0^{\pi/2} \cos(a \operatorname{tg} x - vx) \cos^{v-2} x \, dx = \frac{\pi e^{-a} a^{v-1}}{\Gamma(v)} \quad [\operatorname{Re} v > 1].$

Lo V 153(112), NG 157(14)

5. $\displaystyle\int_0^{\pi/2} \cos(a \operatorname{tg} x + vx) \cos^v x \, dx = 2^{-v-1} \pi e^{-a} \quad [\operatorname{Re} v > -1].$

BH [44](4)

6. $\displaystyle\int_0^{\pi/2} \cos(a \operatorname{tg} x - \gamma x) \cos^v x \, dx =$

$$= \frac{\pi a^{\frac{v}{2}}}{2^{\frac{v}{2}+1}} \cdot \frac{W_{\frac{\gamma}{2}, -\frac{v+1}{2}}(2a)}{\Gamma\left(1 + \frac{\gamma + v}{2}\right)} \left[a > 0, \; \operatorname{Re} v > -1, \; \frac{v+\gamma}{2} \neq -1, -2, \ldots\right].$$

Ba I 274(13)

7. $\displaystyle\int_0^{\pi/2} \frac{\sin nx - \sin(nx - a \operatorname{tg} x)}{\sin x} \cos^{n-1} x \, dx = \pi.$

Lo V 153(114)

3.719

1. $\int_0^\pi \sin(vx - z \sin x)\, dx = \pi \mathbf{E}_v(z).$ W 308(2)

2. $\int_0^\pi \cos(nx - z \sin x)\, dx = \pi J_n(z).$ WW 362

3. $\int_0^\pi \cos(vx - z \sin x)\, dx = \pi \mathbf{J}_v(z).$ W 308(1)

3.72–3.74

Trigonometrische und rationale Funktionen
Trigonometric and rational functions

3.721

1. $\int_0^\infty \dfrac{\sin(ax)}{x}\, dx = \dfrac{\pi}{2} \operatorname{sign} a.$ F II 654

2. $\int_1^\infty \dfrac{\sin(ax)}{x}\, dx = -\operatorname{si}(a).$ BH[203](1)

3. $\int_1^\infty \dfrac{\cos(ax)}{x}\, dx = -\operatorname{ci}(a).$ BH[203](5)

3.722

1. $\int_0^\infty \dfrac{\sin(ax)}{x+\beta}\, dx = \operatorname{ci}(a\beta) \sin(a\beta) - \cos(a\beta) \operatorname{si}(a\beta)$ $[|\arg \beta| < \pi,\ a > 0].$
    BH[160](1), FII 655+

2. $\int_{-\infty}^\infty \dfrac{\sin(ax)}{x+\beta}\, dx = \pi \cos(a\beta)$ $[|\arg \beta| < \pi,\ a > 0].$ BH[202](1)

3. $\int_0^\infty \dfrac{\cos(ax)}{x+\beta}\, dx = -\sin(a\beta) \operatorname{si}(a\beta) - \cos(a\beta) \operatorname{ci}(a\beta)$ $[|\arg \beta| < \pi,\ a > 0].$
    Ba IV 8(7), BH[160](2)

4. $\int_{-\infty}^\infty \dfrac{\cos(ax)}{x+\beta}\, dx = \pi \sin(a\beta)$ $[|\arg \beta| < \pi,\ a > 0].$ BH[202](4)

5. $\int_0^\infty \dfrac{\sin(ax)}{\beta - x}\, dx = \sin(\beta a) \operatorname{ci}(\beta a) - \cos(\beta a) [\operatorname{si}(\beta a) + \pi]$ $[a > 0].$
    F II 655, BH[161](1)

6. $\int_{-\infty}^\infty \dfrac{\sin(ax)}{\beta - x}\, dx = -\pi \cos(a\beta)$ $[a > 0].$ BH[202](3)

7. $\int_0^\infty \dfrac{\cos(ax)}{\beta - x}\, dx = \cos(a\beta) \operatorname{ci}(a\beta) + \sin(a\beta) [\operatorname{si}(a\beta) + \pi]$ $[a > 0].$
    Ba IV 8(8), BH[161](2)+

8. $\int_{-\infty}^\infty \dfrac{\cos(ax)}{\beta - x}\, dx = \pi \sin(a\beta)$ $[a > 0].$ BH[202](6)

1. $\int_0^\infty \frac{\sin(ax)}{\beta^2 + x^2} dx = \frac{1}{2\beta} [e^{-a\beta}\overline{\text{Ei}}(a\beta) - e^{a\beta}\text{Ei}(-a\beta)]$ $[a > 0, \text{ Re } \beta > 0]$.   3.723

Ba IV 65(14), BH[160](3)

2. $\int_0^\infty \frac{\cos(ax)}{\beta^2 + x^2} dx = \frac{\pi}{2\beta} e^{-a\beta}$ $[a > 0, \text{ Re } \beta > 0]$.

F II 746, 756–757, Ba V 8(11), WW 116

3. $\int_0^\infty \frac{x \sin(ax)}{\beta^2 + x^2} dx = \frac{\pi}{2} e^{-a\beta}$ $[a > 0, \text{ Re } \beta > 0]$.

F II 746, 756–757, Ba IV 65(15), WW 116

4. $\int_{-\infty}^\infty \frac{x \sin(ax)}{\beta^2 + x^2} dx = \pi e^{-a\beta}$ $[a > 0, \text{ Re } \beta > 0]$.   BH[202](10)

5. $\int_0^\infty \frac{x \cos(ax)}{\beta^2 + x^2} dx = -\frac{1}{2} [e^{-a\beta}\overline{\text{Ei}}(a\beta) + e^{a\beta}\text{Ei}(-a\beta)]$ $[a > 0, \text{ Re } \beta > 0]$.

BH[160](6)

6. $\int_{-\infty}^\infty \frac{\sin[a(b-x)]}{c^2 + x^2} dx = \frac{\pi}{c} e^{-ac} \sin(ab)$ $[a > 0, b > 0, c > 0]$.   Li[202](9)

7. $\int_{-\infty}^\infty \frac{\cos[a(b-x)]}{c^2 + x^2} dx = \frac{\pi}{c} e^{-ac} \cos(ab)$ $[a > 0, b > 0, c > 0]$.   Li[202](11)+

8. $\int_0^\infty \frac{\sin(ax)}{\beta^2 - x^2} dx = \frac{1}{\beta} \left[\sin(a\beta)\text{ci}(a\beta) - \cos(a\beta)\left(\text{si}(a\beta) + \frac{\pi}{2}\right) \right]$

$[|\arg \beta| < \pi, a > 0]$.   BH[161](3)

9. $\int_0^\infty \frac{\cos(ax)}{b^2 - x^2} dx = \frac{\pi}{2b} \sin(ab)$ $[a > 0, b > 0]$.   BH[161](5), Ba IV 9(15)

10. $\int_0^\infty \frac{x \sin(ax)}{b^2 - x^2} dx = -\frac{\pi}{2} \cos(ab)$ $[a > 0]$.   F II 655, Ba V 252(45)

11. $\int_0^\infty \frac{x \cos(ax)}{\beta^2 - x^2} dx = \cos(a\beta)\,\text{ci}(a\beta) + \sin(a\beta)\left[\text{si}(a\beta) + \frac{\pi}{2}\right]$

$[|\arg \beta| < \pi, a > 0]$.   BH[161](6)

12. $\int_{-\infty}^\infty \frac{\sin(ax)}{x(x-b)} dx = \pi \frac{\cos(ab) - 1}{b}$ $[a > 0, b > 0]$.   Ba V 252(44)

1. $\int_{-\infty}^\infty \frac{b + cx}{p + 2qx + x^2} \sin(ax)\, dx = \left(\frac{cq - b}{\sqrt{p - q^2}} \sin(aq) + c \cos(aq) \right) \pi e^{-a\sqrt{p - q^2}}$ 3.724

$[a > 0, p > q^2]$.   BH[202](12)

2. $\int_{-\infty}^\infty \frac{b + cx}{p + 2qx + x^2} \cos(ax)\, dx = \left(\frac{b - cq}{\sqrt{p - q^2}} \cos(aq) + c \sin(aq) \right) \pi e^{-a\sqrt{p - q^2}}$

$[a > 0, p > q^2]$.   BH[202](13)

3.724 3. $\int_{-\infty}^{\infty} \dfrac{\cos[(b-1)t] - x\cos(bt)}{1 - 2x\cos t + x^2} \cos(ax)\,dx = \pi e^{-a\sin t} \sin(bt + a\cos t)$

$[a > 0,\ t^2 < \pi^2]$.   BH[202](14)

3.725 1. $\int_0^{\infty} \dfrac{\sin(ax)\,dx}{x(\beta^2 + x^2)} = \dfrac{\pi}{2\beta^2}(1 - e^{-a\beta})$ $[\operatorname{Re}\beta > 0,\ a > 0]$.   BH[172](1)

2. $\int_0^{\infty} \dfrac{\sin(ax)\,dx}{x(b^2 - x^2)} = \dfrac{\pi}{2b^2}(1 - \cos(ab))$ $[a > 0]$.   BH[172](4)

3. $\int_0^{\infty} \dfrac{\sin(ax)\cos(bx)}{x(x^2 + \beta^2)}\,dx = \dfrac{\pi}{2\beta^2} e^{-\beta b}\operatorname{sh}(a\beta)$ $[0 < a < b]$;

$= -\dfrac{\pi}{2\beta^2} e^{-a\beta}\operatorname{ch}(b\beta) + \dfrac{\pi}{2\beta}$ $[a > b > 0]$.   Ba IV 19(4)

3.726 1. $\int_0^{\infty} \dfrac{x\sin(ax)\,dx}{b^3 \pm b^2 x + bx^2 \pm x^3} = \pm \dfrac{1}{4b}\left[e^{-ab}\overline{\operatorname{Ei}}(ab) - e^{ab}\operatorname{Ei}(-ab) - \right.$

$\left. - 2\operatorname{ci}(ab)\sin(ab) + 2\cos(ab)\left(\operatorname{si}(ab) + \dfrac{\pi}{2}\right) \right] + \dfrac{\pi e^{-ab} - \pi\cos(ab)}{4b}$

[$a > 0,\ b > 0$; für das untere Vorzeichen ist der Hauptwert des Integrals angegeben].

[$a > 0,\ b > 0$; in the case of the lower sign the principal value is meant].

Ba IV 65(21)+, BH[176](10, 13)

2. $\int_0^{\infty} \dfrac{x^2 \sin(ax)\,dx}{b^3 \pm b^2 x + bx^2 \pm x^3} = \dfrac{1}{4}\left[e^{ab}\operatorname{Ei}(-ab) - e^{-ab}\overline{\operatorname{Ei}}(ab) + \right.$

$\left. + 2\operatorname{ci}(ab)\sin(ab) - 2\cos(ab)\left(\operatorname{si}(ab) + \dfrac{\pi}{2}\right) \right] \pm \pi(e^{-ab} + \cos(ab))$

[$a > 0,\ b > 0$; für das untere Vorzeichen ist der Hauptwert des Integrals angegeben].

[$a > 0,\ b > 0$; in the case of the lower sign the principal value is meant].

Ba IV 66(22), BH[176](11, 14)

3.727 1. $\int_0^{\infty} \dfrac{\cos(ax)\,dx}{b^4 + x^4} = \dfrac{\pi\sqrt{2}}{4b^3}\exp\left(-\dfrac{ab}{\sqrt{2}}\right)\left(\cos\dfrac{ab}{\sqrt{2}} + \sin\dfrac{ab}{\sqrt{2}}\right)$ $[a > 0,\ b > 0]$.

BH[160](25)+, Ba IV 9(19)

2. $\int_0^{\infty} \dfrac{\sin(ax)\,dx}{b^4 - x^4} = \dfrac{1}{4b^3}\left[2\sin(ab)\operatorname{ci}(ab) - 2\cos(ab)\left(\operatorname{si}(ab) + \dfrac{\pi}{2}\right) + \right.$

$\left. + e^{-ab}\overline{\operatorname{Ei}}(ab) - e^{ab}\operatorname{Ei}(-ab) \right]$ $[a > 0,\ b > 0]$ (cf. 3.723 1., 3.723 8.).

BH[161](12)

3. $$\int_0^\infty \frac{\cos(ax)\,dx}{b^4 - x^4} = \frac{\pi}{4b^3}[e^{-ab} + \sin(ab)] \quad [a > 0, b > 0]$$

(cf. 3.723 2., 3.723 9.). BH[161](16)

4. $$\int_0^\infty \frac{x \sin(ax)\,dx}{b^4 + x^4} = \frac{\pi}{2b^2} \exp\left(-\frac{ab}{\sqrt{2}}\right) \sin\frac{ab}{\sqrt{2}} \quad [a > 0, b > 0].$$ BH[160](23)+

5. $$\int_0^\infty \frac{x \sin(ax)}{b^4 - x^4}\,dx = \frac{\pi}{4b^2}[e^{-ab} - \cos(ab)] \quad [a > 0, b > 0].$$

(cf. 3.723 3., 3.723 10.). BH[161](13)

6. $$\int_0^\infty \frac{x \cos(ax)\,dx}{b^4 - x^4} = \frac{1}{4b^2}\left[2\cos(ab)\,\text{ci}(ab) + 2\sin(ab)\left(\text{si}(ab) + \frac{\pi}{2}\right) - \right.$$
$$\left. - e^{-ab}\overline{\text{Ei}}(ab) - e^{ab}\text{Ei}(-ab)\right] \quad [a > 0, b > 0]. \quad \text{(cf. 3.723 5., 3.723 11.).}$$ BH[161](17)

7. $$\int_0^\infty \frac{x^2 \cos(ax)\,dx}{b^4 + x^4} = \frac{\pi\sqrt{2}}{4b} \exp\left(-\frac{ab}{\sqrt{2}}\right)\left(\cos\frac{ab}{\sqrt{2}} - \sin\frac{ab}{\sqrt{2}}\right)$$

$$[a > 0, b > 0].$$ BH[160](26)+

8. $$\int_0^\infty \frac{x^2 \sin(ax)\,dx}{b^4 - x^4} = \frac{1}{4b}\left[2\sin(ab)\,\text{ci}(ab) - \right.$$
$$\left. - 2\cos(ab)\left(\text{si}(ab) + \frac{\pi}{2}\right) - e^{-ab}\overline{\text{Ei}}(ab) + e^{ab}\text{Ei}(-ab)\right]$$
$$[a > 0, b > 0] \quad \text{(cf. 3.723 1., 3.723 8.).}$$ BH[161](14)

9. $$\int_0^\infty \frac{x^2 \cos(ax)\,dx}{b^4 - x^4} = \frac{\pi}{4b}(\sin(ab) - e^{-ab}) \quad [a > 0, b > 0]$$

(cf. 3.723 2., 3.723 9.). BH[161](18)

10. $$\int_0^\infty \frac{x^3 \sin(ax)}{b^4 + x^4}\,dx = \frac{\pi}{2}\exp\left(-\frac{ab}{\sqrt{2}}\right)\cos\frac{ab}{\sqrt{2}} \quad [a > 0, b > 0].$$ BH[160](24)

11. $$\int_0^\infty \frac{x^3 \sin(ax)}{b^4 - x^4}\,dx = \frac{-\pi}{4}[e^{-ab} + \cos(ab)] \quad [a > 0, b > 0]$$

(cf. 3.723 4., 3.723 10.). BH[161](15)

12. $$\int_0^\infty \frac{x^3 \cos(ax)\,dx}{b^4 - x^4} = \frac{1}{4}\left[2\cos(ab)\,\text{ci}(ab) + 2\sin(ab)\left(\text{si}(ab) + \frac{\pi}{2}\right) + \right.$$
$$\left. + e^{-ab}\overline{\text{Ei}}(ab) + e^{ab}\text{Ei}(-ab)\right] \quad [a > 0, b > 0] \quad \text{(cf. 3.723 5., 3.723 11.).}$$ BH[161](19)

3.728

1. $\int_0^\infty \dfrac{\cos(ax)\,dx}{(\beta^2+x^2)(\gamma^2+x^2)} = \dfrac{\pi(\beta e^{-a\gamma} - \gamma e^{-a\beta})}{2\beta\gamma(\beta^2-\gamma^2)}$ $\quad [a>0,\ \operatorname{Re}\beta>0,\ \operatorname{Re}\gamma>0]$. BH[175](1)

2. $\int_0^\infty \dfrac{x\sin(ax)\,dx}{(\beta^2+x^2)(\gamma^2+x^2)} = \dfrac{\pi(e^{-a\beta} - e^{-a\gamma})}{2(\gamma^2-\beta^2)}$ $\quad [a>0,\ \operatorname{Re}\beta>0,\ \operatorname{Re}\gamma>0]$. BH[174](1)

3. $\int_0^\infty \dfrac{x^2\cos(ax)\,dx}{(\beta^2+x^2)(\gamma^2+x^2)} = \dfrac{\pi(\beta e^{-a\beta} - \gamma e^{-a\gamma})}{2(\beta^2-\gamma^2)}$ $\quad [a>0,\ \operatorname{Re}\beta>0,\ \operatorname{Re}\gamma>0]$. BH[175](2)

4. $\int_0^\infty \dfrac{x^3\sin(ax)\,dx}{(\beta^2+x^2)(\gamma^2+x^2)} = \dfrac{\pi(\beta^2 e^{-a\beta} - \gamma^2 e^{-a\gamma})}{2(\beta^2-\gamma^2)}$ $\quad [a>0,\ \operatorname{Re}\beta>0,\ \operatorname{Re}\gamma>0]$. BH[174](2)

5. $\int_0^\infty \dfrac{\cos(ax)\,dx}{(b^2-x^2)(c^2-x^2)} = \dfrac{\pi(b\sin(ac) - c\sin(ab))}{2bc(b^2-c^2)}$ $\quad [a>0,\ b>0,\ c>0]$. BH[175](3)

6. $\int_0^\infty \dfrac{x\sin(ax)\,dx}{(b^2-x^2)(c^2-x^2)} = \dfrac{\pi(\cos(ab) - \cos(ac))}{2(b^2-c^2)}$ $\quad [a>0]$. BH[174](3)

7. $\int_0^\infty \dfrac{x^2\cos(ax)\,dx}{(b^2-x^2)(c^2-x^2)} = \dfrac{\pi(c\sin(ac) - b\sin(ab))}{2(b^2-c^2)}$ $\quad [a>0,\ b>0,\ c>0]$. BH[175](4)

8. $\int_0^\infty \dfrac{x^3\sin(ax)\,dx}{(b^2-x^2)(c^2-x^2)} = \dfrac{\pi(b^2\cos(ab) - c^2\cos(ac))}{2(b^2-c^2)}$ $\quad [a>0,\ b>0,\ c>0]$. BH[174](4)

3.729

1. $\int_0^\infty \dfrac{\cos(ax)\,dx}{(b^2+x^2)^2} = \dfrac{\pi}{4b^3}(1+ab)e^{-ab}$ $\quad [a>0,\ b>0]$. BH[170](7)

2. $\int_0^\infty \dfrac{x\sin(ax)\,dx}{(b^2+x^2)^2} = \dfrac{\pi}{4b}ae^{-ab}$ $\quad [a>0,\ b>0]$. BH[170](3)

3. $\int_0^\infty \cos(px)\,\dfrac{1-x^2}{(1+x^2)^2}\,dx = \dfrac{\pi p}{2}e^{-p}$. BH[43](10)+

4. $\int_0^\infty \dfrac{x^3\sin(ax)\,dx}{(b^2+x^2)^2} = \dfrac{\pi}{4}(2-ab)e^{-ab}$ $\quad [a>0,\ b>0]$. BH[170](4)

3.731 Abkürzungen:/Abbreviations: $2A^2 = \sqrt{b^4+c^2} + b^2,\ 2B^2 = \sqrt{b^4+c^2} - b^2$.

1. $\int_0^\infty \dfrac{\cos(ax)\,dx}{(x^2+b^2)^2+c^2} = \dfrac{\pi}{2c}\cdot\dfrac{e^{-aA}(B\cos(aB) + A\sin(aB))}{\sqrt{b^4+c^2}}$

$\quad [a>0,\ b>0,\ c>0]$. BH[176](3)

2. $\int_0^\infty \frac{x \sin(ax)\, dx}{(x^2+b^2)^2 + c^2} = \frac{\pi}{2c} e^{-aA} \sin(aB)$ $\quad [a>0,\ b>0,\ c>0]$.   BH[176](1)   3.731

3. $\int_0^\infty \frac{(x^2+b^2)\cos(ax)\, dx}{(x^2+b^2)^2 + c^2} = \frac{\pi}{2} \frac{e^{-aA}(A\cos(aB) - B\sin(aB))}{\sqrt{b^4+c^2}}$

$[a>0,\ b>0,\ c>0]$.   BH[176](4)

4. $\int_0^\infty \frac{x(x^2+b^2)\sin(ax)\, dx}{(x^2+b^2)^2 + c^2} = \frac{\pi}{2} e^{-aA} \cos(aB)$ $\quad [a>0,\ b>0,\ c>0]$.

BH[176](2)

1. $\int_0^\infty \left[\frac{1}{\beta^2 + (\gamma-x)^2} - \frac{1}{\beta^2 + (\gamma+x)^2} \right] \sin(ax)\, dx = \frac{\pi}{\beta} e^{-a\beta} \sin(a\gamma)$ 3.732

[$a>0$, Re $\beta > 0$, $\gamma + i\beta$ ist keine reelle Zahl]. | [$a>0$, Re $\beta > 0$, $\gamma + i\beta$ is not a real number].

Ba IV 65(16) | Ba IV 65(16)

2. $\int_0^\infty \left[\frac{1}{\beta^2 + (\gamma-x)^2} + \frac{1}{\beta^2 + (\gamma+x)^2} \right] \cos(ax)\, dx = \frac{\pi}{\beta} e^{-a\beta} \cos(a\gamma)$

$[a>0,\ |\text{Im } \gamma| < \text{Re } \beta]$.   Ba IV 8(13)

3. $\int_0^\infty \left[\frac{\gamma+x}{\beta^2 + (\gamma+x)^2} - \frac{\gamma-x}{\beta^2 + (\gamma-x)^2} \right] \sin(ax)\, dx = \pi e^{-a\beta} \cos(a\gamma)$

[$a>0$, Re $\beta > 0$, $\gamma + i\beta$ ist keine reelle Zahl]. | [$a>0$, Re $\beta > 0$, $\gamma + i\beta$ is not a real number].

Li[175](17) | Li[175](17)

4. $\int_0^\infty \left[\frac{\gamma+x}{\beta^2 + (\gamma+x)^2} + \frac{\gamma-x}{\beta^2 + (\gamma-x)^2} \right] \cos(ax)\, dx = \pi e^{-a\beta} \sin(a\gamma)$

$[a>0,\ |\text{Im } a| < \text{Re } \beta]$.   Li[176](21)

1. $\int_0^\infty \frac{\cos(ax)\, dx}{x^4 + 2b^2 x^2 \cos 2t + b^4} = \frac{\pi}{2b^3} \exp(-ab\cos t) \frac{\sin(t + ab \sin t)}{\sin 2t}$ 3.733

$\left[a > 0,\ b > 0,\ |t| < \frac{\pi}{2} \right]$.   BH[176](7)

2. $\int_0^\infty \frac{x \sin(ax)\, dx}{x^4 + 2b^2 x^2 \cos 2t + b^4} = \frac{\pi}{2b^2} \exp(-ab\cos t) \frac{\sin(ab \sin t)}{\sin 2t}$

$\left[a > 0,\ b > 0,\ |t| < \frac{\pi}{2} \right]$.   BH[176](5), Ba IV 66(23)

3.733

3. $$\int_0^\infty \frac{x^2 \cos(ax)\, dx}{x^4 + 2b^2 x^2 \cos 2t + b^4} = \frac{\pi}{2b} \exp(-ab \cos t) \frac{\sin(t - ab \sin t)}{\sin 2t}$$

$$\left[a > 0,\ b > 0,\ |t| < \frac{\pi}{2}\right].$$ BH[176](8)

4. $$\int_0^\infty \frac{x^3 \sin(ax)\, dx}{x^4 + 2b^2 x^2 \cos 2t + b^4} = \frac{\pi}{2} \exp(-ab \cos t) \frac{\sin(2t - ab \sin t)}{\sin 2t}$$

$$\left[a > 0,\ b > 0,\ |t| < \frac{\pi}{2}\right].$$ BH[176](6)

5. $$\int_0^\infty \frac{\sin(ax)\, dx}{x(x^4 + 2b^2 x^2 \cos 2t + b^4)} = \frac{\pi}{2b^4} \left[1 - \exp(-ab \cos t) \frac{\sin(2t + ab \sin t)}{\sin 2t}\right]$$

$$\left[a > 0,\ b > 0, |t| < \frac{\pi}{2}\right].$$ BH[176](22)

3.734

1. $$\int_0^\infty \frac{\sin(ax)\, dx}{x(b^4 + x^4)} = \frac{\pi}{2b^4}\left[1 - \exp\left(-\frac{ab}{\sqrt{2}}\right)\cos\frac{ab}{\sqrt{2}}\right]\quad [a > 0,\ b > 0].$$

BH[172](7)

2. $$\int_0^\infty \frac{\sin(ax)\, dx}{x(b^4 - x^4)} = \frac{\pi}{4b^4}\left[2 - e^{-ab} - \cos(ab)\right]\quad [a > 0,\ b > 0].$$ BH[172](10)

3.735 $$\int_0^\infty \frac{\sin(ax)\, dx}{x(b^2 + x^2)^2} = \frac{\pi}{2b^4}\left[1 - \frac{1}{2} e^{-ab}(2 + ab)\right]\quad [a > 0,\ b > 0].$$

WW 116, BH[172](22)

3.736

1. $$\int_0^\infty \frac{\cos(ax)\, dx}{(b^2 + x^2)(b^4 - x^4)} = \frac{\pi}{8b^5}\left[\sin(ab) + (2 + ab)\, e^{-ab}\right]\ [a > 0,\ b > 0]$$

(cf. 3.723 2., 9., 3.729 1.). BH[176](5)

2. $$\int_0^\infty \frac{x \sin(ax)\, dx}{(b^2 + x^2)(b^4 - x^4)} = \frac{\pi}{8b^4}\left[(1 + ab)\, e^{-ab} - \cos(ab)\right]\ [a > 0,\ b > 0]$$

(cf. 3.723 3., 10., 3.729 2.). BH[174](5)

3. $$\int_0^\infty \frac{x^2 \cos(ax)\, dx}{(b^2 + x^2)(b^4 - x^4)} = \frac{\pi}{8b^3}\left[\sin(ab) - ab e^{-ab}\right]\ [a > 0,\ b > 0]$$

(cf. 3.723 2., 9., 3.729 1.). BH[175](6)

4. $$\int_0^\infty \frac{x^3 \sin(ax)\, dx}{(b^2 + x^2)(b^4 - x^4)} = \frac{\pi}{8b^2}\left[(1 - ab)\, e^{-ab} - \cos(ab)\right]\ [a > 0,\ b > 0]$$

(cf. 3.723 3., 10., 3.729 2.). BH[174](6)

5. $\int_0^\infty \dfrac{x^4 \cos(ax)\, dx}{(b^2+x^2)(b^4-x^4)} = \dfrac{\pi}{8b}\left[\sin(ab) + (ab-2)\,e^{-ab}\right]$ $\quad [a>0,\ b>0]$ 3.736

(cf. 3.723 2., 9., 3.729 1.).   BH[175](7)

6. $\int_0^\infty \dfrac{x^5 \sin(ax)\, dx}{(b^2+x^2)(b^4-x^4)} = \dfrac{\pi}{8}\left[(ab-3)\,e^{-ab} - \cos(ab)\right]$ $\quad [a>0,\ b>0]$

(cf. 3.723 3., 10., 3.729 2.).   BH[174](7)

1. $\int_0^\infty \dfrac{\cos(ax)\, dx}{(b^2+x^2)^n} = \dfrac{\pi e^{-ab}}{(2b)^{2n-1}(n-1)!} \sum_{k=0}^{n-1} \dfrac{(2n-k-2)!\,(2ab)^k}{k!\,(n-k-1)!};$ 3.737

$$= \dfrac{(-1)^{n-1}\pi}{b^{2n-1}(n-1)!}\left[\dfrac{d^{n-1}}{dp^{n-1}}\left(\dfrac{e^{-ab\sqrt{p}}}{\sqrt{p}}\right)\right]_{p=1};$$

$$= \dfrac{(-1)^{n-1}\pi}{2b^{2n-1}(n-1)!}\left[\dfrac{d^{n-1}}{dp^{n-1}}\dfrac{e^{-abp}}{(1+p)^n}\right]_{p=1} \quad [a>0,\ b>0].$$

GH[333](67b), W 189, W 172–173

2. $\int_0^\infty \dfrac{x \sin(ax)\, dx}{(x^2+\beta^2)^{n+1}} = \dfrac{\pi a e^{-a\beta}}{2^{2n} n!\,\beta^{2n-1}} \sum_{k=0}^{n-1} \dfrac{(2n-k-2)!\,(2a\beta)^k}{k!\,(n-k-1)!}$ $\quad [a>0,\ \operatorname{Re}\beta>0]$.

GH[333](66c)

3. $\int_0^\infty \dfrac{\sin(ax)\, dx}{x(\beta^2+x^2)^{n+1}} = \dfrac{\pi}{2\beta^{2n+2}}\left[1 - \dfrac{e^{-a\beta}}{2^n n!}\,F_n(a\beta)\right]$

$[a>0,\ \operatorname{Re}\beta>0,\ F_0(z)=1,\ F_1(z)=z+2,\ldots,\ F_n(z) =$

$= (z+2n)\,F_{n-1}(z) - z F'_{n-1}(z)]$.   GH[333](66e)

4. $\int_0^\infty \dfrac{x \sin(ax)\, dx}{(b^2+x^2)^3} = \dfrac{\pi a}{16 b^3}(1+ab)\,e^{-ab}$ $\quad [a>0,\ b>0]$.

BH[170](5), Ba IV 67(35)+

5. $\int_0^\infty \dfrac{x \sin(ax)\, dx}{(b^2+x^2)^4} = \dfrac{\pi a}{96 b^5}(3 + 3ab + a^2 b^2)\,e^{-ab}$ $\quad [a>0,\ b>0]$.

BH[170](6), Ba IV 67(35)+

1. $\int_0^\infty \dfrac{x^{m-1}\sin(ax)}{x^{2n}+\beta^{2n}}\,dx = 0$ $\quad \left[m\ \text{ungeradzahlig odd}\right];$ 3.738

$$= -\dfrac{\pi \beta^{m-2n}}{2n}\sum_{k=1}^{n}\exp\left[-a\beta \sin\dfrac{(2k-1)\pi}{2n}\right] \times$$

$$\times\left\{\cos\dfrac{(2k-1)m\pi}{n} + a\beta \cos\dfrac{(2k-1)\pi}{2n}\right\} \quad \left[m\ \text{geradzahlig even}\right];$$

$\left[a>0,\ |\arg\beta|<\dfrac{\pi}{2n},\ 0\leqslant m<2n\right]$.   Ba IV 67(38)

3.738 2. $\int_0^\infty \dfrac{x^{m-1}\cos(ax)}{x^{2n}+\beta^{2n}}\,dx = 0 \quad \left[m \text{ geradzahlig even}\right];$

$$= \dfrac{\pi\beta^{m-2n}}{2n}\sum_{k=1}^{n}\exp\left[-a\beta\sin\dfrac{(2k-1)\pi}{2n}\right]\times$$

$$\times\left\{\sin\dfrac{(2k-1)m\pi}{2n} + a\beta\cos\dfrac{(2k-1)\pi}{2n}\right\}\quad\left[m\text{ ungeradzahlig odd}\right];$$

$\left[a>0,\ |\arg\beta|<\dfrac{\pi}{2n},\ 0<m<2n+1\right].$ BH[161](20)+, Ba IV 10(29)

3.739 1. $\int_0^\infty \dfrac{\sin(ax)\,dx}{x(x^2+2^2)(x^2+4^2)\cdots(x^2+4n^2)} =$

$$= \dfrac{\pi(-1)^n}{(2n)!\,2^{2n+1}}\left[2\sum_{k=0}^{n-1}(-1)^k\binom{2n}{k}e^{2(k-n)a} + (-1)^n\binom{2n}{n}\right].\quad \text{Li}[174](8)$$

2. $\int_0^\infty \dfrac{\cos(ax)\,dx}{(x^2+1^2)(x^2+3^2)\cdots[x^2+(2n+1)^2]} =$

$$= \dfrac{(-1)^n}{(2n+1)!}\dfrac{\pi}{2^{2n+1}}\sum_{k=0}^{n}(-1)^k\binom{2n+1}{k}e^{(2k-2n-1)a}.\quad \text{BH}[175](8)$$

3. $\int_0^\infty \dfrac{x\sin(ax)\,dx}{(x^2+1^2)(x^2+3^2)\cdots[x^2+(2n+1)^2]} =$

$$= \dfrac{\pi(-1)^n}{(2n+1)!\,2^{2n+1}}\sum_{k=0}^{n}(-1)^k\binom{2n+1}{k}(2n-2k+1)\,e^{(2k-2n-1)a}.\quad \text{Li}[174](9)$$

3.741 1. $\int_0^\infty \dfrac{\sin(ax)\sin(bx)}{x}\,dx = \dfrac{1}{4}\ln\left(\dfrac{a+b}{a-b}\right)^2 \quad [a>0,\ b>0,\ a\neq b].\quad \text{F II 657}$

2. $\int_0^\infty \dfrac{\sin(ax)\cos(bx)}{x}\,dx = \dfrac{\pi}{2}\quad [a>b\geqslant 0];$

$\qquad\qquad = \dfrac{\pi}{4}\quad [a=b>0];$

$\qquad\qquad = 0 \quad [b>a\geqslant 0].\quad \text{F II 654}$

3. $\int_0^\infty \dfrac{\sin(ax)\sin(bx)}{x^2}\,dx = \dfrac{a\pi}{2}\quad [0<a\leqslant b];$

$\qquad\qquad = \dfrac{b\pi}{2}\quad [0<b\leqslant a].\quad \text{BH}[157](1)$

3.742 1. $\int_0^\infty \dfrac{\sin(ax)\sin(bx)}{\beta^2+x^2}\,dx = \dfrac{\pi}{4\beta}\left(e^{-|a-b|\beta} - e^{-(a+b)\beta}\right)$

$[a>0,\ b>0,\ \operatorname{Re}\beta>0].\quad \text{BH}[162](1)+,\ \text{GH}[333](71a)$

2. $\int_0^\infty \dfrac{\sin(ax)\cos(bx)}{\beta^2+x^2}\,dx = \dfrac{1}{4\beta}e^{-a\beta}\{e^{b\beta}\mathrm{Ei}[\beta(a-b)] +$ 3.742

$+ e^{-b\beta}\mathrm{Ei}[\beta(a+b)]\} - \dfrac{1}{4\beta}e^{a\beta}\{e^{b\beta}\mathrm{Ei}[-\beta(a+b)] +$

$+ e^{-b\beta}\mathrm{Ei}[\beta(b-a)]\}$. BH[162](3)

3. $\int_0^\infty \dfrac{\cos(ax)\cos(bx)}{\beta^2+x^2}\,dx = \dfrac{\pi}{4\beta}[e^{-|a-b|\beta} + e^{-(a+b)\beta}]$

$[a>0,\ b>0,\ \mathrm{Re}\,\beta>0]$. BH[163](1)+, GH[333](71c)

4. $\int_0^\infty \dfrac{x\cos(ax)\cos(bx)}{\beta^2+x^2}\,dx = -\dfrac{1}{4}e^{a\beta}\{e^{b\beta}\mathrm{Ei}[-\beta(a+b)] + e^{-b\beta}\mathrm{Ei}[\beta(b-a)]\} -$

$- \dfrac{1}{4}e^{-a\beta}\{e^{b\beta}\mathrm{Ei}[\beta(a-b)] + e^{-b\beta}\mathrm{Ei}[\beta(a+b)]\}$ $[a\ne b]$;

$= \infty$ $[a=b]$. BH[163](2)

5. $\int_0^\infty \dfrac{x\sin(ax)\cos(bx)}{x^2+\beta^2}\,dx = \dfrac{\pi}{2}e^{-a\beta}\mathrm{ch}(b\beta)$ $[0<b<a]$;

$= \dfrac{\pi}{4}e^{-2a\beta}$ $[0<b=a]$;

$= -\dfrac{\pi}{2}e^{-b\beta}\mathrm{sh}(a\beta)$ $[0<a<b]$. BH[162](4)

6. $\int_0^\infty \dfrac{\sin(ax)\sin(bx)}{p^2-x^2}\,dx = -\dfrac{\pi}{2p}\cos(ap)\sin(bp)$ $[a>b>0]$;

$= -\dfrac{\pi}{4p}\sin(2ap)$ $[a=b>0]$;

$= -\dfrac{\pi}{2p}\sin(ap)\cos(bp)$ $[b>a>0]$. BH[166](1)

7. $\int_0^\infty \dfrac{\sin(ax)\cos(bx)}{p^2-x^2}x\,dx = -\dfrac{\pi}{2}\cos(ap)\cos(bp)$ $[a>b>0]$;

$= -\dfrac{\pi}{4}\cos(2ap)$ $[a=b>0]$;

$= \dfrac{\pi}{2}\sin(ap)\sin(bp)$ $[b>a>0]$. BH[166](2)

8. $\int_0^\infty \dfrac{\cos(ax)\cos(bx)}{p^2-x^2}\,dx = \dfrac{\pi}{2p}\sin(ap)\cos(bp)$ $[a>b>0]$;

$= \dfrac{\pi}{4p}\sin(2ap)$ $[a=b>0]$;

$= \dfrac{\pi}{2p}\cos(ap)\sin(bp)$ $[b>a>0]$. BH[166](3)

3.743

1. $\int_0^\infty \frac{\sin(ax)}{\sin(bx)} \cdot \frac{dx}{x^2+\beta^2} = \frac{\pi}{2\beta} \cdot \frac{\sh(a\beta)}{\sh(b\beta)}$ $\quad [0 < a < b,\ \text{Re}\,\beta > 0]$. $\hfill$ Ba IV 80(21)

2. $\int_0^\infty \frac{\sin(ax)}{\cos(bx)} \cdot \frac{x\,dx}{x^2+\beta^2} = -\frac{\pi}{2} \cdot \frac{\sh(a\beta)}{\ch(b\beta)}$ $\quad [0 < a < b,\ \text{Re}\,\beta > 0]$. $\hfill$ Ba IV 81(30)

3. $\int_0^\infty \frac{\cos(ax)}{\sin(bx)} \cdot \frac{x\,dx}{x^2+\beta^2} = \frac{\pi}{2} \cdot \frac{\ch(a\beta)}{\sh(b\beta)}$ $\quad [0 < a < b,\ \text{Re}\,\beta > 0]$. $\hfill$ Ba IV 23(37)

4. $\int_0^\infty \frac{\cos(ax)}{\cos(bx)} \cdot \frac{dx}{x^2+\beta^2} = \frac{\pi}{2\beta} \cdot \frac{\ch(a\beta)}{\ch(b\beta)}$ $\quad [0 < a < b,\ \text{Re}\,\beta > 0]$. $\hfill$ Ba IV 23(36)

5. $\int_0^\infty \frac{\sin(2ax)}{\sin x} \cdot \frac{dx}{b^2-x^2} = \frac{\pi}{b} \cdot \frac{\sin^2(ab)}{\sin b}$ $\quad [0 < a < 1,\ b > 0]$. $\hfill$ BH[191](18)

3.744

1. $\int_0^\infty \frac{\sin(ax)}{\cos(bx)} \cdot \frac{dx}{x(x^2+\beta^2)} = \frac{\pi}{2\beta^2} \cdot \frac{\sh(a\beta)}{\ch(b\beta)}$ $\quad [0 < a < b,\ \text{Re}\,\beta > 0]$. $\hfill$ Ba IV 82(32)

2. $\int_0^\infty \frac{\sin(ax)}{\cos(bx)} \cdot \frac{dx}{x(c^2-x^2)} = 0$ $\quad [0 < a < b,\ c > 0]$. $\hfill$ Ba IV 82(31)

3.745

1. $\int_0^\infty \frac{dx}{x^{n+1}} \prod_{k=0}^n \sin(a_k x) = \frac{\pi}{2} \prod_{k=1}^n a_k$ $\quad \left[a_0 > \sum_{k=1}^n a_k,\ a_k > 0\right]$. $\hfill$ F II 654

2. $\int_0^\infty \frac{\sin(ax)}{x^{n+1}}\,dx \prod_{k=1}^n \sin(a_k x) \prod_{j=1}^m \cos(b_j x) = \frac{\pi}{2} \prod_{k=1}^n a_k$

$\left[a > \sum_{k=1}^n |a_k| + \sum_{j=1}^m |b_j|\right]$. $\hfill$ WW 122

3.746

1. $\int_0^{\pi/2} \frac{x^m}{\sin x}\,dx = \left(\frac{\pi}{2}\right)^m \left[\frac{1}{m} + \sum_{k=1}^\infty \frac{2^{2k-1}-1}{4^{2k-1}(m+2k)}\,\zeta(2k)\right]$. $\hfill$ Li[206](2)

2. $\int_0^{\pi/2} \frac{x\,dx}{\sin x} = \int_0^{\pi/2} \frac{\left(\frac{\pi}{2}-x\right)dx}{\cos x} = 2G.$ $\hfill$ BH[204](18), BH[206](1), GH[333](32)

3. $\int_0^\infty \frac{x\,dx}{(x^2+b^2)\sin(ax)} = \frac{\pi}{2\,\sh(ab)}$ $\quad [b > 0]$. $\hfill$ GH[333](79c)

4. $\int_0^\pi x\,\tg x\,dx = -\pi \ln 2.$ $\hfill$ BH[218](4)

5. $$\int_0^{\frac{\pi}{2}} x \operatorname{tg} x \, dx = \infty.$$ BH[205](2) 3.746

6. $$\int_0^{\frac{\pi}{4}} x \operatorname{tg} x \, dx = -\frac{\pi}{8} \ln 2 + \frac{1}{2} G = 0{,}185\,784\,535\,8\ldots$$ BH[204](1)

7. $$\int_0^{\frac{\pi}{2}} x \operatorname{ctg} x \, dx = \frac{\pi}{2} \ln 2.$$ F II 633

8. $$\int_0^{\frac{\pi}{4}} x \operatorname{ctg} x \, dx = \frac{\pi}{8} \ln 2 + \frac{1}{2} G = 0{,}730\,181\,058\,4\ldots$$ BH[204](2)

9. $$\int_0^{\frac{\pi}{2}} \left(\frac{\pi}{2} - x\right) \operatorname{tg} x \, dx = \frac{1}{2} \int_0^{\pi} \left(\frac{\pi}{2} - x\right) \operatorname{tg} x \, dx = \frac{\pi}{2} \ln 2.$$ GH[333],(33b), BH[218](12)

10. $$\int_0^{\infty} \operatorname{tg} ax \, \frac{dx}{x} = \frac{\pi}{2} \quad [a > 0].$$ Lo V 279(5)

11. $$\int_0^{\frac{\pi}{2}} \frac{x \operatorname{ctg} x}{\cos 2x} \, dx = \frac{\pi}{4} \ln 2.$$ BH[206](12)

1. $$\int_0^{\frac{\pi}{4}} x^m \operatorname{tg} x \, dx = \frac{1}{2} \left(\frac{\pi}{4}\right)^m \sum_{k=1}^{\infty} \frac{(4^k - 1)\,\zeta(2k)}{4^{2k-1}(m + 2k)}.$$ Li[204](5) 3.747

2. $$\int_0^{\frac{\pi}{2}} x^p \operatorname{ctg} x \, dx = \left(\frac{\pi}{2}\right)^p \left\{ \frac{1}{p} - 2 \sum_{k=1}^{\infty} \frac{1}{4^k(p + 2k)} \zeta(2k) \right\}.$$ Li[205](7)

3. $$\int_0^{\frac{\pi}{4}} x^m \operatorname{ctg} x \, dx = \frac{1}{2} \left(\frac{\pi}{4}\right)^m \left[\frac{2}{m} - \sum_{k=1}^{\infty} \frac{\zeta(2k)}{4^{2k-1}(m + 2k)} \right].$$ Li[204](6)

1. $$\int_0^{\infty} \frac{x \operatorname{tg}(ax) \, dx}{x^2 + b^2} = \frac{\pi}{e^{2ab} + 1} \quad [a > 0,\, b > 0].$$ GH[333](79a) 3.748

2. $$\int_0^{\infty} \frac{x \operatorname{ctg}(ax) \, dx}{x^2 + b^2} = \frac{\pi}{e^{2ab} - 1} \quad [a > 0,\, b > 0].$$ GH[333](79b)

3. $$\int_0^{\infty} \frac{x \operatorname{tg}(ax) \, dx}{b^2 - x^2} = \int_0^{\infty} \frac{x \operatorname{ctg}(ax) \, dx}{b^2 - x^2} = \int_0^{\infty} \frac{x \operatorname{cosec}(ax) \, dx}{b^2 - x^2} = \infty.$$ BH[161](7, 8, 9)

Trigonometrische und algebraische Funktionen
Trigonometric and algebraic functions

3.751

1. $\int_0^\infty \dfrac{\sin(ax)\,dx}{\sqrt{x+\beta}} = \sqrt{\dfrac{\pi}{2a}}\,[\cos(a\beta) - \sin(a\beta) + 2C(\sqrt{a\beta})\sin(a\beta) -$

$- 2S(\sqrt{a\beta})\cos(a\beta)]$ $[a > 0,\ |\arg\beta| < \pi]$.   Ba IV 65(12)+

2. $\int_0^\infty \dfrac{\cos(ax)\,dx}{\sqrt{x+\beta}} = \sqrt{\dfrac{\pi}{2a}}\,[\cos a\beta + \sin(a\beta) - 2C(\sqrt{a\beta})\cos(a\beta) -$

$- 2S(\sqrt{a\beta})\sin(a\beta)]$ $[a > 0,\ |\arg\beta| < \pi]$.   Ba IV 8(9)+

3. $\int_u^\infty \dfrac{\sin(ax)}{\sqrt{x-u}}\,dx = \sqrt{\dfrac{\pi}{2a}}\,[\sin(au) + \cos(au)]$ $[a > 0,\ u > 0]$.   Ba IV 65(13)

4. $\int_u^\infty \dfrac{\cos(ax)}{\sqrt{x-u}}\,dx = \sqrt{\dfrac{\pi}{2a}}\,[\cos(au) - \sin(au)]$ $[a > 0,\ u > 0]$.   Ba IV 8(10)

3.752

1. $\int_0^1 \sin(ax)\sqrt{1-x^2}\,dx = \sum_{k=0}^\infty \dfrac{(-1)^k a^{2k+1}}{(2k+1)!!(2k+3)!!}$ $[a > 0]$.   BH[149](6)

2. $\int_0^1 \cos(ax)\sqrt{1-x^2}\,dx = \dfrac{\pi}{2a}\,J_1(a)$.   Ku 65(6)+

3.753

1. $\int_0^1 \dfrac{\sin(ax)\,dx}{\sqrt{1-x^2}} = \sum_{k=0}^\infty \dfrac{(-1)^k a^{2k+1}}{[(2k+1)!!]^2}$ $[a > 0]$.   BH[149](9)

2. $\int_0^1 \dfrac{\cos(ax)\,dx}{\sqrt{1-x^2}} = \dfrac{\pi}{2}\,J_0(a)$.   W 20(7)+

3. $\int_1^\infty \dfrac{\sin(ax)\,dx}{\sqrt{x^2-1}} = \dfrac{\pi}{2}\,J_0(a)$ $[a > 0]$.   W 180(14)

4. $\int_1^\infty \dfrac{\cos(ax)}{\sqrt{x^2-1}}\,dx = -\dfrac{\pi}{2}\,N_0(a)$.   W 180(15)

5. $\int_0^1 \dfrac{x\sin(ax)}{\sqrt{1-x^2}}\,dx = \dfrac{\pi}{2}\,J_1(a)$ $[a > 0]$.   W 20(6)

3.754

1. $\int_0^\infty \dfrac{\sin(ax)\,dx}{\sqrt{\beta^2+x^2}} = \dfrac{\pi}{2}\,[I_0(a\beta) - \mathbf{L}_0(a\beta)]$ $[a > 0,\ \operatorname{Re}\beta > 0]$.   Ba IV 66(26)

2. $\int_0^\infty \dfrac{\cos(ax)\,dx}{\sqrt{\beta^2+x^2}} = K_0(a\beta)$ $[a > 0,\ \operatorname{Re}\beta > 0]$.   W 172(1), GH[333](78a)

3.754

3. $\int_0^\infty \frac{x \sin(ax)}{\sqrt{(\beta^2 + x^2)^3}} dx = aK_0(a\beta) \quad [a > 0, \text{Re } \beta > 0].$ Ba IV 66(27)

3.755

1. $\int_0^\infty \frac{\sqrt{\sqrt{x^2 + \beta^2} - \beta} \sin(ax) \, dx}{\sqrt{x^2 + \beta^2}} = \sqrt{\frac{\pi}{2a}} e^{-a\beta} \quad [a > 0].$ Ba IV 66(31)

2. $\int_0^\infty \frac{\sqrt{\sqrt{x^2 + \beta^2} + \beta} \cos(ax) \, dx}{\sqrt{x^2 + \beta^2}} = \sqrt{\frac{\pi}{2a}} e^{-a\beta} \quad [a > 0, \text{Re } \beta > 0].$ Ba IV 10(25)

3.756

1. $\int_0^\infty \frac{\sin(ax)}{x^{\frac{n}{2}-1}} \prod_{k=2}^n \sin(a_k x) \, dx = 0 \quad \left[a_k > 0, \, a > \sum_{k=2}^n a_k \right].$ Ba IV 80(22)

2. $\int_0^\infty x^{\frac{n}{2}-1} \cos(ax) \prod_{k=1}^n \cos(a_k x) \, dx = 0 \quad \left[a_k > 0, \, a > \sum_{k=1}^n a_k \right].$ Ba IV 22(26)

3.757

1. $\int_0^\infty \frac{\sin(ax)}{\sqrt{x}} dx = \sqrt{\frac{\pi}{2a}}.$ BH[177](1)

2. $\int_0^\infty \frac{\cos(ax)}{\sqrt{x}} dx = \sqrt{\frac{\pi}{2a}}.$ BH[177](2)

Produkte von trigonometrischen Funktionen mit Potenzen
Trigonometric functions and powers

3.76–3.77

3.761

1. $\int_0^1 x^{\mu-1} \sin(ax) \, dx = \frac{-i}{2\mu} [{}_1F_1(\mu; \mu+1; ia) - {}_1F_1(\mu; \mu+1; -ia)]$

$[a > 0, \text{Re } \mu > -1].$ Ba IV 68(2)+

2. $\int_u^\infty x^{\mu-1} \sin x \, dx = \frac{i}{2} [e^{-\frac{\pi}{2}i\mu} \Gamma(\mu, iu) - e^{\frac{\pi}{2}i\mu} \Gamma(\mu, -iu)]$

$[\text{Re } \mu < 1].$ Ba II 149(2)

3. $\int_1^\infty \frac{\sin(ax)}{x^{2n}} dx = \frac{a^{2n-1}}{(2n-1)!} \left[\sum_{k=1}^{2n-1} \frac{(2n-k-1)!}{a^{2n-k}} \sin\left(a + (k-1)\frac{\pi}{2}\right) + (-1)^n \text{ci}(a) \right]$

$[a > 0].$ Li[203](15)

4. $\int_0^\infty x^{\mu-1} \sin(ax) \, dx = \frac{\Gamma(\mu)}{a^\mu} \sin\frac{\mu\pi}{2} = \frac{\pi \sec\frac{\mu\pi}{2}}{2a^\mu \Gamma(1-\mu)}$

$[a > 0, \, 0 < |\text{Re } \mu| < 1].$ F II 814–815+, BH[150](1)

5. $\int_0^\pi x^m \sin(nx) \, dx = \frac{(-1)^{n+1}}{n^{m+1}} \sum_{k=0}^{E\left(\frac{m}{2}\right)} (-1)^k \frac{m!}{(m-2k)!} (n\pi)^{m-2k} -$

$- (-1)^{E\left(\frac{m}{2}\right)} \frac{m! \left[m - 2E\left(\frac{m}{2}\right) - 1 \right]}{n^{m+1}}.$ GH[333](6)

3.761

6. $\int_0^1 x^{\mu-1} \cos(ax)\, dx = \frac{1}{2\mu} \left[{}_1F_1(\mu; \mu+1; ia) + {}_1F_1(\mu, \mu+1; -ia) \right]$

$\qquad [a > 0, \operatorname{Re}\mu > 0].$ Ba IV 11(2)

7. $\int_u^\infty x^{\mu-1} \cos x\, dx = \frac{1}{2} \left[e^{-\frac{\pi}{2}i\mu} \Gamma(\mu, iu) + e^{\frac{\pi}{2}i\mu} \Gamma(\mu, -iu) \right]$

$\qquad [\operatorname{Re}\mu < 1].$ Ba II 149(1)

8. $\int_1^\infty \frac{\cos(ax)}{x^{2n+1}}\, dx = \frac{a^{2n}}{(2n)!} \left[\sum_{k=1}^{2n} \frac{(2n-k)!}{a^{2n-k+1}} \cos\left(a + (k-1)\frac{\pi}{2}\right) + (-1)^{n+1} \operatorname{ci}(a) \right]$

$\qquad [a > 0].$ Li[203](16)

9. $\int_0^\infty x^{\mu-1} \cos(ax)\, dx = \frac{\Gamma(\mu)}{a^\mu} \cos\frac{\mu\pi}{2} = \frac{\pi \operatorname{cosec}\frac{\mu\pi}{2}}{2a^\mu \Gamma(1-\mu)} \quad [a > 0,\, 0 < \operatorname{Re}\mu < 1].$

F II 817+, BH[150](2)

10. $\int_0^\pi x^m \cos(nx)\, dx = \frac{(-1)^n}{n^{m+1}} \sum_{k=0}^{E\left(\frac{m-1}{2}\right)} (-1)^k \frac{m!}{(m-2k-1)!} (n\pi)^{m-2k-1} +$

$\qquad + (-1)^{E\left(\frac{m+1}{2}\right)} \frac{2E\left(\frac{m+1}{2}\right) - m}{n^{m+1}} \cdot m!.$ GH[333](7)

11. $\int_0^{\pi/2} x^m \cos x\, dx = \sum_{k=0}^{E\left[\frac{m}{2}\right]} (-1)^k \frac{m!}{(m-2k)!} \left(\frac{\pi}{2}\right)^{m-2k} +$

$\qquad + (-1)^{E\left(\frac{m}{2}\right)} \left[2E\left(\frac{m}{2}\right) - m \right] m!.$ GH[333](9c)

12. $\int_0^{2n\pi} x^m \cos kx\, dx = -\sum_{j=0}^{m-1} \frac{j!}{k^{j+1}} \binom{m}{j} (2n\pi)^{m-j} \cos\frac{j+1}{2}\pi.$ BH[226](2)

3.762

1. $\int_0^\infty x^{\mu-1} \sin(ax) \sin(bx)\, dx = \frac{1}{2} \cos\frac{\mu\pi}{2} \Gamma(\mu) \left[|b-a|^{-\mu} - (b+a)^{-\mu} \right]$

$\qquad [a > 0,\, b > 0,\, a \neq b,\, -2 < \operatorname{Re}\mu < 1].$ BH[159](7), Ba IV 321(40)

(für $\mu = 0$ siehe 3.741 1., für $\mu = -1$ siehe 3.741 3.). $\quad$ (for $\mu = 0$ see 3.741 1., and for $\mu = -1$ see 3.741 3.).

2. $\int_0^\infty x^{\mu-1} \sin(ax) \cos(bx)\, dx = \frac{1}{2} \sin\frac{\mu\pi}{2} \Gamma(\mu) \left[(a+b)^{-\mu} + \right.$

$\left. + |a-b|^{-\mu} \operatorname{sign}(a-b) \right] \quad [a > 0,\, b > 0,\, |\operatorname{Re}\mu| < 1]$ BH[159](8)+, Ba IV 321(41)

(für $\mu = 0$ siehe 3.741 2.). $\quad$ (for $\mu = 0$ see 3.741 2.).

3. $\int_0^\infty x^{\mu-1} \cos(ax) \cos(bx)\, dx = \frac{1}{2} \cos\frac{\mu\pi}{2} \Gamma(\mu) \left[(a+b)^{-\mu} + |a-b|^{-\mu} \right]$

$\qquad [a > 0,\, b > 0,\, 0 < \operatorname{Re}\mu < 1].$ Ba IV 20(17)

1. $\int_0^\infty \dfrac{\sin(ax)\sin(bx)\sin(cx)}{x^\nu}\,dx = \dfrac{1}{4}\cos\dfrac{\nu\pi}{2}\,\Gamma(1-\nu)\,[(c+a-b)^{\nu-1} -$ 3.763

$- (c+a+b)^{\nu-1} - |c-a+b|^{\nu-1}\,\text{sign}(a-b-c) +$

$+ |c-a-b|^{\nu-1}\,\text{sign}(a+b-c)]\quad [c>0,\ 0<\text{Re}\,\nu<4,$

$\nu \ne 1, 2, 3,\ a \geqslant b > 0].$ GH [333](26a)+, Ba IV 79(13)

2. $\int_0^\infty \dfrac{\sin(ax)\sin(bx)\sin(cx)}{x}\,dx = 0 \quad [c<a-b,\ c>a+b];$

$= \dfrac{\pi}{8} \quad [c=a-b,\ c=a+b];$

$= \dfrac{\pi}{4} \quad [a-b<c<a+b]$

$[a \geqslant b > 0,\ c>0].$ F II 654

3. $\int_0^\infty \dfrac{\sin(ax)\sin(bx)\sin(cx)}{x^2}\,dx = \dfrac{1}{4}(c+a+b)\ln(c+a+b) -$

$- \dfrac{1}{4}(c+a-b)\ln(c+a-b) - \dfrac{1}{4}|c-a-b|\ln|c-a-b|\times$

$\times \text{sign}(a+b-c) + \dfrac{1}{4}|c-a+b|\ln|c-a+b|\,\text{sign}(a-b-c)$

$[a \geqslant b > 0,\ c>0].$ BH [157](8)+, Ba IV 79(11)

4. $\int_0^\infty \dfrac{\sin(ax)\sin(bx)\sin(cx)}{x^3}\,dx = \dfrac{\pi bc}{2} \quad [0<c<a-b,\quad c>a+b];$

$= \dfrac{\pi bc}{2} - \dfrac{\pi(a-b-c)^2}{8} \quad [a-b<c<a+b];$

$[a \geqslant b > 0,\ c>0].$ BH [157](20), Ba IV 79(12)

1. $\int_0^\infty x^p \sin(ax+b)\,dx = \dfrac{1}{a^{p+1}}\Gamma(1+p)\cos\left(b+\dfrac{p\pi}{2}\right) \quad [a>0,\ -1<p<0].$ 3.764

GH [333](30a)

2. $\int_0^\infty x^p \cos(ax+b)\,dx = -\dfrac{1}{a^{p+1}}\Gamma(1+p)\sin\left(b+\dfrac{\pi p}{2}\right)$

$[a>0,\ -1<p<0].$ GH [333](30b)

1. $\int_0^\infty \dfrac{\sin(ax)\,dx}{x^\nu(x+\beta)} = \dfrac{i}{2\beta^\nu}\Gamma(1-\nu)[e^{-ia\beta}\Gamma(\nu,-ia\beta) - e^{ia\beta}\Gamma(\nu,ia\beta)]$ 3.765

$[a>0,\ -1<\text{Re}\,\nu<2,\ |\arg\beta|<\pi].$ Ba IV 219(34)

2. $\int_0^\infty \dfrac{\cos(ax)\,dx}{x^\nu(x+\beta)} = \dfrac{\Gamma(1-\nu)}{2\beta^\nu}[e^{ia\beta}\Gamma(\nu,ia\beta) + e^{-ia\beta}\Gamma(\nu,-ia\beta)]$

$[a>0,\ |\text{Re}\,\nu|<1,\ |\arg\beta|<\pi].$ Ba V 221(52)

3.766

1. $$\int_0^\infty \frac{x^{\mu-1}\sin(ax)}{1+x^2}\,dx = \frac{\pi}{2}\sec\frac{\mu\pi}{2}\operatorname{sh} a +$$
$$+ \frac{1}{2}\sin\frac{\mu\pi}{2}\Gamma(\mu)\{\exp[-a+i\pi(1-\mu)]\gamma(1-\mu,-a) - e^a\gamma(1-\mu,a)\}$$
$$[a>0,\ -1<\operatorname{Re}\mu<3].\qquad \text{Ba IV 317(4)}$$

2. $$\int_0^\infty \frac{x^{\mu-1}\cos(ax)}{1+x^2}\,dx = \frac{\pi}{2}\operatorname{cosec}\frac{\mu\pi}{2}\operatorname{ch} a +$$
$$+ \frac{1}{2}\cos\frac{\mu\pi}{2}\Gamma(\mu)\{\exp[-a+i\pi(1-\mu)]\gamma(1-\mu,-a) - e^a\gamma(1-\mu,a)\}$$
$$[a>0,\ 0<\operatorname{Re}\mu<3].\qquad \text{Ba IV 319(24)}$$

3. $$\int_0^\infty \frac{x^{2\mu+1}\sin(ax)\,dx}{x^2+b^2} = -\frac{\pi}{2}b^{2\mu}\sec(\mu\pi)\operatorname{sh}(ab) -$$
$$- \frac{\sin(\mu\pi)}{a^{2\mu}}\Gamma(2\mu)[{}_1F_1(1;1-2\mu;ab) + {}_1F_1(1;1-2\mu;-ab)]$$
$$\left[a>0,\ -\frac{3}{2}<\operatorname{Re}\mu<\frac{1}{2}\right].\qquad \text{Ba V 220(39)}$$

4. $$\int_0^\infty \frac{x^{2\mu+1}\cos(ax)\,dx}{x^2+b^2} = -\frac{\pi}{2}b^{2\mu}\operatorname{cosec}(\mu\pi)\operatorname{ch}(ab) -$$
$$- \frac{\cos(\mu\pi)}{2a^{2\mu}}\Gamma(2\mu)[{}_1F_1(1;1-2\mu;ab) + {}_1F_1(1;1-2\mu;-ab)]$$
$$\left[a>0,\ -1<\operatorname{Re}\mu<\frac{1}{2}\right].\qquad \text{Ba V 221(56)}$$

3.767

1. $$\int_0^\infty \frac{x^{\beta-1}\sin\left(ax-\frac{\beta\pi}{2}\right)}{\gamma^2+x^2}\,dx = -\frac{\pi}{2}\gamma^{\beta-2}e^{-a\gamma}$$
$$[a>0,\ \operatorname{Re}\gamma>0,\ 0<\operatorname{Re}\beta<2].\qquad \text{BH[160](20)}$$

2. $$\int_0^\infty \frac{x^\beta \cos\left(ax-\frac{\beta\pi}{2}\right)}{\gamma^2+x^2}\,dx = \frac{\pi}{2}\gamma^{\beta-1}e^{-a\gamma}\qquad [a>0,\ \operatorname{Re}\gamma>0,\ |\operatorname{Re}\beta|<1].$$
$$\text{BH[160](21)}$$

3. $$\int_0^\infty \frac{x^{\beta-1}\sin\left(ax-\frac{\beta\pi}{2}\right)}{x^2-b^2}\,dx = \frac{\pi}{2}b^{\beta-2}\cos\left(ab-\frac{\pi\beta}{2}\right)$$
$$[a>0,\ b>0,\ 0<\operatorname{Re}\beta<2].\qquad \text{BH[161](11)}$$

4. $$\int_0^\infty \frac{x^\beta \cos\left(ax-\frac{\beta\pi}{2}\right)}{x^2-b^2}\,dx = -\frac{\pi}{2}b^{\beta-1}\sin\left(ab-\frac{\beta\pi}{2}\right)$$
$$[a>0,\ b>0,\ |\beta|<1].\qquad \text{GH[333](82)}$$

3.768

1. $$\int_u^\infty (x-u)^{\mu-1}\sin(ax)\,dx = \frac{\Gamma(\mu)}{a^\mu}\sin\left(au+\frac{\mu\pi}{2}\right)\qquad [a>0,\ 0<\operatorname{Re}\mu<1].$$
$$\text{Ba V 203(19)}$$

2. $\int_u^\infty (x-u)^{\mu-1} \cos(ax)\, dx = \frac{\Gamma(\mu)}{a^\mu} \cos\left(au + \frac{\mu\pi}{2}\right)$ $\quad [a > 0, 0 < \text{Re}\,\mu < 1]$. 3.768

Ba IV 204(24)

3. $\int_0^1 (1-x)^\nu \sin(ax)\, dx = \frac{1}{a} - \frac{\Gamma(\nu+1)}{a^{\nu+1}} C_\nu(a)$ $\quad [a > 0, \text{Re}\,\nu > -1]$.

Ba IV 68(3)

4. $\int_0^1 (1-x)^\nu \cos(ax)\, dx = \frac{i}{2} a^{-\nu-1} \left\{ \exp\left[\frac{i}{2}(\nu\pi - 2a)\right] \gamma(\nu+1, -ia) - \right.$

$\left. - \exp\left[-\frac{i}{2}(\nu\pi - 2a)\right] \gamma(\nu+1, ia) \right\}$ $\quad [a > 0, \text{Re}\,\nu > -1]$. Ba IV 11(3)+

5. $\int_0^u x^{\nu-1}(u-x)^{\mu-1} \sin(ax)\, dx =$

$$= \frac{u^{\mu+\nu-1}}{2i} B(\mu, \nu) \left[{}_1F_1(\nu; \mu+\nu; iau) - {}_1F_1(\nu; \mu+\nu; -iau)\right]$$

$[a > 0, \text{Re}\,\mu > 0, \text{Re}\,\nu > -1]$. Ba V 189(26)

6. $\int_0^u x^{\nu-1}(u-x)^{\mu-1} \cos(ax)\, dx =$

$$= \frac{u^{\mu+\nu-1}}{2} B(\mu, \nu) \left[{}_1F_1(\nu; \mu+\nu; iau) + {}_1F_1(\nu; \mu+\nu; -iau)\right]$$

$[a > 0, \text{Re}\,\mu > 0, \text{Re}\,\nu > 0]$. Ba V 189(32)

7. $\int_0^u x^{\mu-1}(u-x)^{\mu-1} \sin(ax)\, dx = \sqrt{\pi}\left(\frac{u}{a}\right)^{\mu-\frac{1}{2}} \sin\frac{au}{2}\, \Gamma(\mu)\, J_{\mu-\frac{1}{2}}\left(\frac{au}{2}\right)$

$[\text{Re}\,\mu > 0]$. Ba V 189(25)

8. $\int_u^\infty x^{\mu-1}(x-u)^{\mu-1} \sin(ax)\, dx =$

$$= \frac{\sqrt{\pi}}{2}\left(\frac{u}{a}\right)^{\mu-\frac{1}{2}} \Gamma(\mu)\left[\cos\frac{au}{2} J_{\frac{1}{2}-\mu}\left(\frac{au}{2}\right) - \sin\frac{au}{2} N_{\frac{1}{2}-\mu}\left(\frac{au}{2}\right)\right]$$

$\left[a > 0, 0 < \text{Re}\,\mu < \frac{1}{2}\right]$. Ba V 203(20)

9. $\int_0^u x^{\mu-1}(u-x)^{\mu-1} \cos(ax)\, dx = \sqrt{\pi}\left(\frac{u}{a}\right)^{\mu-\frac{1}{2}} \cos\frac{au}{2}\, \Gamma(\mu)\, J_{\mu-\frac{1}{2}}\left(\frac{au}{2}\right)$

$[\text{Re}\,\mu > 0]$. Ba V 189(31)

10. $\int_u^\infty x^{\mu-1}(x-u)^{\mu-1} \cos(ax)\, dx =$

$$= -\frac{\sqrt{\pi}}{2}\left(\frac{u}{a}\right)^{\mu-\frac{1}{2}} \Gamma(\mu)\left[\sin\frac{au}{2} J_{\frac{1}{2}-\mu}\left(\frac{au}{2}\right) - \cos\frac{au}{2} N_{\frac{1}{2}-\mu}\left(\frac{au}{2}\right)\right]$$

$\left[a > 0, 0 < \text{Re}\,\mu < \frac{1}{2}\right]$. Ba V 204(25)

3.768

11. $\int_0^1 x^{\nu-1}(1-x)^{\mu-1} \sin(ax)\,dx = -\frac{i}{2} B(\mu,\nu) [{}_1F_1(\nu;\nu+\mu;ia) -$

$- {}_1F_1(\nu;\nu+\mu;-ia)]$ $[a>0,\ \operatorname{Re}\mu>0,\ \operatorname{Re}\nu>0]$.

Ba IV 68(5)+, Ba IV 317(5)

12. $\int_0^1 x^{\nu-1}(1-x)^{\mu-1} \cos(ax)\,dx = \frac{1}{2} B(\mu,\nu) [{}_1F_1(\nu;\nu+\mu;ia) +$

$+ {}_1F_1(\nu;\nu+\mu;-ia)]$ $[a>0,\ \operatorname{Re}\mu>0,\ \operatorname{Re}\nu>0]$.   Ba IV 11(5)

13. $\int_0^1 x^\mu (1-x)^\mu \sin(2ax)\,dx = \dfrac{\sqrt{\pi}}{(2a)^{\mu+\frac{1}{2}}} \Gamma(\mu+1) \sin a\, J_{\mu+\frac{1}{2}}(a)$

$[a>0,\ \operatorname{Re}\mu>-1]$.   Ba IV 68(4)

14. $\int_0^1 x^\mu (1-x)^\mu \cos(2ax)\,dx = \dfrac{\sqrt{\pi}}{(2a)^{\mu+\frac{1}{2}}} \Gamma(\mu+1) \cos a\, J_{\mu+\frac{1}{2}}(a)$

$[a>0,\ \operatorname{Re}\mu>-1]$.   Ba IV 11(4)

3.769

1. $\int_0^\infty [(\beta+ix)^{-\nu} - (\beta-ix)^{-\nu}] \sin(ax)\,dx = \dfrac{\pi i a^{\nu-1} e^{-a\beta}}{\Gamma(\nu)}$.

$[a>0,\ \operatorname{Re}\beta>0,\ \operatorname{Re}\nu>0]$.   Ba IV 70(15)

2. $\int_0^\infty [(\beta+ix)^{-\nu} + (\beta-ix)^{-\nu}] \cos(ax)\,dx = \dfrac{\pi a^{\nu-1} e^{-a\beta}}{\Gamma(\nu)}$

$[a>0,\ \operatorname{Re}\beta>0,\ \operatorname{Re}\nu>0]$.   Ba IV 13(19)

3. $\int_0^\infty x[(\beta+ix)^{-\nu} + (\beta-ix)^{-\nu}] \sin(ax)\,dx = \dfrac{\pi a^{\nu-2}(1-a\beta)}{\Gamma(\nu)} e^{-a\beta}$

$[a>0,\ \operatorname{Re}\beta>0,\ \operatorname{Re}\nu>0]$.   Ba IV 70(16)

4. $\int_0^\infty x^{2n}[(\beta-ix)^{-\nu} - (\beta+ix)^{-\nu}] \sin(ax)\,dx =$

$= \dfrac{(-1)^{n+1} i}{\Gamma(\nu)} (2n)!\, \pi a^{\nu-2n-1} e^{-a\beta} L_{2n}^{\nu-2n-1}(a\beta)$

$[a>0,\ \operatorname{Re}\beta>0,\ 0\leqslant 2n\leqslant \operatorname{Re}\nu]$.   Ba IV 70(17)

5. $\int_0^\infty x^{2n}[(\beta+ix)^{-\nu} + (\beta-ix)^{-\nu}] \cos(ax)\,dx =$

$= \dfrac{(-1)^n}{\Gamma(\nu)} (2n)!\, \pi a^{\nu-2n-1} e^{-a\beta} L_{2n}^{\nu-2n-1}(a\beta)$

$[a>0,\ \operatorname{Re}\beta>0,\ 0\leqslant 2n<\operatorname{Re}\nu]$.   Ba IV 13(20)

6. $\int_0^\infty x^{2n+1} [(\beta + ix)^{-\nu} + (\beta - ix)^{-\nu}] \sin(ax) \, dx =$ 3.769

$$= \frac{(-1)^{n+1}}{\Gamma(\nu)} (2n+1)! \, \pi a^{\nu-2n-2} e^{-a\beta} L_{2n+1}^{\nu-2n-2}(a\beta)$$

$[a > 0, \text{Re } \beta > 0, -1 \leqslant 2n + 1 < \text{Re } \nu]$. Ba IV 70(18)

7. $\int_0^\infty x^{2n+1} [(\beta + ix)^{-\nu} - (\beta - ix)^{-\nu}] \cos(ax) \, dx =$

$$= \frac{(-1)^{n+1} i}{\Gamma(\nu)} (2n+1)! \, \pi a^{\nu-2n-2} e^{-a\beta} L_{2n+1}^{\nu-2n-2}(a\beta)$$

$[a > 0, \text{Re } \beta > 0, 0 \leqslant 2n < \text{Re } \nu - 1]$. Ba IV 13(21)

1. $\int_0^\infty (\beta^2 + x^2)^{\nu - \frac{1}{2}} \sin(ax) \, dx = \frac{\sqrt{\pi}}{2} \left(\frac{2\beta}{a}\right)^\nu \Gamma\left(\nu + \frac{1}{2}\right) [I_{-\nu}(a\beta) - \mathbf{L}_\nu(a\beta)]$ 3.771

$\left[a > 0, \text{ Re } \beta > 0, \text{ Re } \nu < \frac{1}{2}, \nu \neq -\frac{1}{2}, -\frac{3}{2}, -\frac{5}{2}, \ldots\right]$.
Ba II 38+, Ba IV 68(6)

2. $\int_0^\infty (\beta^2 + x^2)^{\nu - \frac{1}{2}} \cos(ax) \, dx = \frac{1}{\sqrt{\pi}} \left(\frac{2\beta}{a}\right)^\nu \cos(\pi\nu) \, \Gamma\left(\nu + \frac{1}{2}\right) K_{-\nu}(a\beta)$

$\left[a > 0, \text{ Re } \beta > 0, \text{ Re } \nu < \frac{1}{2}\right]$. W 172(1)+, GH[333](78)+

3. $\int_0^u x^{2\nu - 1} (u^2 - x^2)^{\mu - 1} \sin(ax) \, dx =$

$$= \frac{a}{2} u^{2\mu + 2\nu - 1} \mathrm{B}\left(\mu, \nu + \frac{1}{2}\right) {}_1F_2\left(\nu + \frac{1}{2}; \frac{3}{2}, \mu + \nu + \frac{1}{2}; -\frac{a^2 u^2}{4}\right)$$

$\left[\text{Re } \mu > 0, \text{ Re } \nu > -\frac{1}{2}\right]$. Ba V 189(29)

4. $\int_0^u x^{2\nu - 1} (u^2 - x^2)^{\mu - 1} \cos(ax) \, dx = \frac{1}{2} u^{2\mu + 2\nu - 2} \mathrm{B}(\mu, \nu) \times$

$\times {}_1F_2\left(\nu; \frac{1}{2}, \mu + \nu; -\frac{a^2 u^2}{4}\right)$ $[\text{Re } \mu > 0, \text{ Re}\nu > 0]$. Ba V 190(35)

5. $\int_0^\infty x(x^2 + \beta^2)^{\nu - \frac{1}{2}} \sin(ax) \, dx = \frac{1}{\sqrt{\pi}} \beta \left(\frac{2\beta}{a}\right)^\nu \cos \nu\pi \, \Gamma\left(\nu + \frac{1}{2}\right) K_{\nu+1}(a\beta)$

$[a > 0, \text{Re } \beta > 0, \text{Re } \nu > -2]$. Ba IV 69(11)

6. $\int_0^u (u^2 - x^2)^{\nu - \frac{1}{2}} \sin(ax) \, dx = \frac{\sqrt{\pi}}{2} \left(\frac{2u}{a}\right)^\nu \Gamma\left(\nu + \frac{1}{2}\right) \mathbf{H}_\nu(au)$

$\left[a > 0, \, u > 0, \, \text{Re } \nu > -\frac{1}{2}\right]$. Ba IV 69(7), W 330(1)+

3.771

7. $$\int_u^\infty (x^2-u^2)^{\nu-\frac{1}{2}} \sin(ax)\,dx = \frac{\sqrt{\pi}}{2}\left(\frac{2u}{a}\right)^\nu \Gamma\left(\nu+\frac{1}{2}\right) J_{-\nu}(au)$$

$$\left[a>0,\ u>0,\ |\operatorname{Re}\nu|<\frac{1}{2}\right].\qquad \text{Ba II 81(12)+, Ba IV 69(8), W 170(3)+}$$

8. $$\int_0^u (u^2-x^2)^{\nu-\frac{1}{2}} \cos(ax)\,dx = \frac{\sqrt{\pi}}{2}\left(\frac{2u}{a}\right)^\nu \Gamma\left(\nu+\frac{1}{2}\right) J_\nu(au)$$

$$\left[a>0,\ u>0,\ \operatorname{Re}\nu>-\frac{1}{2}\right].\qquad \text{Ba IV 11(8)}$$

9. $$\int_u^\infty (x^2-u^2)^{\nu-\frac{1}{2}} \cos(ax)\,dx = -\frac{\sqrt{\pi}}{2}\left(\frac{2u}{a}\right)^\nu \Gamma\left(\nu+\frac{1}{2}\right) N_{-\nu}(au)$$

$$\left[a>0,\ u>0,\ |\operatorname{Re}\nu|<\frac{1}{2}\right].\qquad \text{W 170(4)+, Ba II 82(13)+, Ba IV 11(9)}$$

10. $$\int_0^u x(u^2-x^2)^{\nu-\frac{1}{2}} \sin(ax)\,dx = \frac{\sqrt{\pi}}{2} u\left(\frac{2u}{a}\right)^\nu \Gamma\left(\nu+\frac{1}{2}\right) J_{\nu+1}(au)$$

$$\left[a>0,\ u>0,\ \operatorname{Re}\nu>-\frac{1}{2}\right].\qquad \text{Ba IV 69(9)}$$

11. $$\int_u^\infty x(x^2-u^2)^{\nu-\frac{1}{2}} \sin(ax)\,dx = \frac{\sqrt{\pi}}{2} u\left(\frac{2u}{a}\right)^\nu \Gamma\left(\nu+\frac{1}{2}\right) N_{-\nu-1}(au)$$

$$\left[a>0,\ u>0,\ -\frac{1}{2}<\operatorname{Re}\nu<0\right].\qquad \text{Ba IV 69(10)}$$

12. $$\int_0^u x(u^2-x^2)^{\nu-\frac{1}{2}} \cos(ax)\,dx = -\frac{u^{\nu+1}}{a^\nu} s_{\nu,\nu+1}(au) =$$

$$= \frac{1}{2}\left(\nu+\frac{1}{2}\right) u^{2\nu+1} - \frac{\sqrt{\pi}}{2} u\left(\frac{2u}{a}\right)^\nu \Gamma\left(\nu+\frac{1}{2}\right) \mathbf{H}_{\nu+1}(au)$$

$$\left[a>0,\ u>0,\ \operatorname{Re}\nu>-\frac{1}{2}\right].\qquad \text{Ba IV 12(10)}$$

13. $$\int_u^\infty x(x^2-u^2)^{\nu-\frac{1}{2}} \cos(ax)\,dx = \frac{\sqrt{\pi}\,u}{2}\left(\frac{2u}{a}\right)^\nu \Gamma\left(\nu+\frac{1}{2}\right) J_{-\nu-1}(au)$$

$$\left[a>0,\ u>0,\ 0<\operatorname{Re}\nu<\frac{1}{2}\right].\qquad \text{Ba IV 12(11)}$$

3.772

1. $$\int_0^\infty (x^2+2\beta x)^{\nu-\frac{1}{2}} \sin(ax)\,dx =$$

$$= \frac{\sqrt{\pi}}{2}\left(\frac{2\beta}{a}\right)^\nu \Gamma\left(\nu+\frac{1}{2}\right)[J_{-\nu}(a\beta)\cos(a\beta)+N_{-\nu}(a\beta)\sin(a\beta)]$$

$$\left[a>0,\ |\arg\beta|<\pi,\ \frac{1}{2}>\operatorname{Re}\nu>-\frac{3}{2}\right].\qquad \text{Ba IV 69(12)}$$

2. $\int_0^\infty (x^2 + 2\beta x)^{\nu - \frac{1}{2}} \cos(ax)\, dx =$

3.772

$$= -\frac{\sqrt{\pi}}{2}\left(\frac{2\beta}{a}\right)^\nu \Gamma\left(\nu + \frac{1}{2}\right) [N_{-\nu}(a\beta)\cos(a\beta) - J_{-\nu}(a\beta)\sin(a\beta)]$$

$$\left[a > 0,\ |\operatorname{Re}\nu| < \frac{1}{2}\right]. \quad \text{Ba IV 12(13)}$$

3. $\int_0^{2u} (2ux - x^2)^{\nu - \frac{1}{2}} \sin(ax)\, dx = \sqrt{\pi}\left(\frac{2u}{a}\right)^\nu \Gamma\left(\nu + \frac{1}{2}\right) \sin(au)\, J_\nu(au)$

$$\left[a > 0,\ u > 0,\ \operatorname{Re}\nu > -\frac{1}{2}\right]. \quad \text{Ba IV 69(13)+}$$

4. $\int_{2u}^\infty (x^2 - 2ux)^{\nu - \frac{1}{2}} \sin(ax)\, dx =$

$$= \frac{\sqrt{\pi}}{2}\left(\frac{2u}{a}\right)^\nu \Gamma\left(\nu + \frac{1}{2}\right) [J_{-\nu}(au)\cos(au) - N_{-\nu}(au)\sin(au)]$$

$$\left[a > 0,\ u > 0,\ |\operatorname{Re}\nu| < \frac{1}{2}\right]. \quad \text{Ba IV 70(14)}$$

5. $\int_0^{2u} (2ux - x^2)^{\nu - \frac{1}{2}} \cos(ax)\, dx = \sqrt{\pi}\left(\frac{2u}{a}\right)^\nu \Gamma\left(\nu + \frac{1}{2}\right) J_\nu(au) \cos(au)$

$$\left[a > 0,\ u > 0,\ \operatorname{Re}\nu > -\frac{1}{2}\right]. \quad \text{Ba IV 12(14)}$$

6. $\int_{2u}^\infty (x^2 - 2ux)^{\nu - \frac{1}{2}} \cos(ax)\, dx =$

$$= -\frac{\sqrt{\pi}}{2}\left(\frac{2u}{a}\right)^\nu \Gamma\left(\nu + \frac{1}{2}\right) [J_{-\nu}(au)\sin(au) + N_{-\nu}(au)\cos(au)]$$

$$\left[a > 0,\ u > 0,\ |\operatorname{Re}\nu| < \frac{1}{2}\right]. \quad \text{Ba IV 12(12)}$$

1. $\int_0^\infty \frac{x^{2\nu}}{(x^2 + \beta^2)^{\mu+1}} \sin(ax)\, dx =$

3.773

$$= \frac{1}{2}\beta^{2\nu - 2\mu}\, a\, \mathrm{B}(1 + \nu,\ \mu - \nu)\, {}_1F_2\left(\nu + 1;\ \nu + 1 - \mu,\ \frac{3}{2};\ \frac{\beta^2 a^2}{4}\right) +$$

$$+ \frac{\sqrt{\pi}\, a^{2\mu - 2\nu + 1}}{4^{\mu - \nu + 1}} \cdot \frac{\Gamma(\nu - \mu)}{\Gamma\left(\mu - \nu + \frac{3}{2}\right)}\, {}_1F_2\left(\mu + 1;\ \mu - \nu + \frac{3}{2},\ \mu - \nu + 1;\ \frac{\beta^2 a^2}{4}\right) =$$

$$= \frac{\sqrt{\pi}}{2\Gamma(\mu + 1)} \beta^{2\nu - 2\mu - 1} G_{13}^{21}\left(\frac{a^2\beta^2}{4} \middle| \begin{matrix} -\nu + \frac{1}{2} \\ \mu - \nu + \frac{1}{2},\ \frac{1}{2},\ 0 \end{matrix}\right) \quad \text{Ba IV 71(28)+, Ba V 234(17)}$$

$$[a > 0,\ \operatorname{Re}\beta > 0,\ -1 < \operatorname{Re}\nu < \operatorname{Re}\mu + 1].$$

3.773

2. $$\int_0^\infty \frac{x^{2m+1} \sin(ax)}{(z+x^2)^{n+1}}\,dx = \frac{(-1)^{n+m}}{n!} \cdot \frac{\pi}{2} \frac{d^n}{dz^n}\left(z^m e^{-a\sqrt{z}}\right)$$

$$[a > 0,\ 0 \leq m \leq n,\ |\arg z| < \pi].\quad \text{Ba IV 68(39)}$$

3. $$\int_0^\infty \frac{x^{2m+1} \sin(ax)\,dx}{(\beta^2 + x^2)^{n+\frac{1}{2}}} = \frac{(-1)^{m+1}\sqrt{\pi}}{2^n \beta^n \Gamma\left(n+\frac{1}{2}\right)} \cdot \frac{d^{2m+1}}{da^{2m+1}}[a^n K_n(a\beta)]$$

$$[a > 0,\ \operatorname{Re}\beta > 0,\ -1 \leq m \leq n].\quad \text{Ba IV 67(37)}$$

4. $$\int_0^\infty \frac{x^{2\nu} \cos(ax)\,dx}{(x^2 + \beta^2)^{\mu+1}} =$$

$$= \frac{1}{2}\beta^{2\nu-2\mu-1}\, \mathbf{B}\left(\nu + \frac{1}{2},\ \mu - \nu + \frac{1}{2}\right){}_1F_2\left(\nu + \frac{1}{2};\ \nu - \mu + \frac{1}{2},\ \frac{1}{2};\ \frac{\beta^2 a^2}{4}\right) +$$

$$+ \frac{\sqrt{\pi}\,a^{2\mu-2\nu+1}}{4^{\mu-\nu+1}} \frac{\Gamma\left(\nu - \mu - \frac{1}{2}\right)}{\Gamma(\mu - \nu + 1)}\, {}_1F_2\left(\mu + 1;\ \mu - \nu + 1,\ \mu - \nu + \frac{3}{2};\ \frac{\beta^2 a^2}{4}\right) =$$

$$= \frac{\sqrt{\pi}}{2\Gamma(\mu+1)} \beta^{2\nu-2\mu-1} G^{21}_{13}\left(\frac{a^2\beta^2}{4}\,\Big|\, \begin{matrix}-\nu + \frac{1}{2}\\ \mu-\nu+\frac{1}{2},\ 0,\ \frac{1}{2}\end{matrix}\right)$$

$$\left[a > 0,\ \operatorname{Re}\beta > 0,\ -\frac{1}{2} < \operatorname{Re}\nu < \operatorname{Re}\mu + 1\right].\quad \text{Ba IV 14(29)+, Ba V 235(19)}$$

5. $$\int_0^\infty \frac{x^{2m}\cos(ax)\,dx}{(z+x^2)^{n+1}} = (-1)^{m+n}\frac{\pi}{2\cdot n!}\cdot \frac{d^n}{dz^n}\left(z^{m-\frac{1}{2}} e^{-a\sqrt{z}}\right)$$

$$[a > 0,\ n+1 > m \geq 0,\ |\arg z| < \pi].\quad \text{Ba IV 10(28)}$$

6. $$\int_0^\infty \frac{x^{2m}\cos(ax)\,dx}{(\beta^2 + x^2)^{n+\frac{1}{2}}} = \frac{(-1)^m \sqrt{\pi}}{2^n \beta^n \Gamma\left(n+\frac{1}{2}\right)} \cdot \frac{d^{2m}}{da^{2m}}\{a^n K_n(a\beta)\}$$

$$\left[a > 0,\ \operatorname{Re}\beta > 0,\ 0 \leq m < n + \frac{1}{2}\right].\quad \text{Ba IV 14(28)}$$

3.774

1. $$\int_0^\infty \frac{\sin(ax)\,dx}{\sqrt{x^2+b^2}\,(x+\sqrt{x^2+b^2})^\nu} = \frac{\pi}{b^\nu \sin(\nu\pi)}\left[\sin\frac{\nu\pi}{2} I_\nu(ab) + \frac{i}{2} \mathbf{J}_\nu(iab) - \right.$$

$$\left. - \frac{i}{2} \mathbf{J}_\nu(-iab)\right]\quad [a > 0,\ b > 0,\ \operatorname{Re}\nu > -1].\quad \text{Ba IV 70(19)}$$

2. $$\int_0^\infty \frac{\cos(ax)\,dx}{\sqrt{x^2+b^2}\,(x+\sqrt{x^2+b^2})^\nu} = \frac{\pi}{b^\nu \sin\nu\pi}\left[\frac{1}{2}\mathbf{J}_\nu(iab) + \frac{1}{2}\mathbf{J}_\nu(-iab) - \right.$$

$$\left. - \cos\frac{\nu\pi}{2} I_\nu(ab)\right]\quad [a > 0, b > 0,\ \operatorname{Re}\nu > -1].\quad \text{Ba IV 12(15)}$$

3. $\int_0^\infty \dfrac{(x+\sqrt{x^2+\beta^2})^\nu}{\sqrt{x(x^2+\beta^2)}} \sin(ax)\,dx = \sqrt{\dfrac{a\pi}{2}}\,\beta^\nu I_{\frac{1}{4}-\frac{\nu}{2}}\left(\dfrac{a\beta}{2}\right) K_{\frac{1}{4}+\frac{\nu}{2}}\left(\dfrac{a\beta}{2}\right)$ 3.774

$\left[a>0,\ \operatorname{Re}\beta>0,\ \operatorname{Re}\nu<\dfrac{3}{2}\right].$ Ba IV 71(23)

4. $\int_0^\infty \dfrac{(\sqrt{x^2+\beta^2}-x)^\nu}{\sqrt{x(x^2+\beta^2)}} \cos(ax)\,dx = \sqrt{\dfrac{a\pi}{2}}\,\beta^\nu I_{-\frac{1}{4}+\frac{\nu}{2}}\left(\dfrac{a\beta}{2}\right) K_{-\frac{1}{4}-\frac{\nu}{2}}\left(\dfrac{a\beta}{2}\right)$

$\left[a>0,\ \operatorname{Re}\beta>0,\ \operatorname{Re}\nu>-\dfrac{3}{2}\right].$ Ba IV 12(17)

5. $\int_0^\infty \dfrac{(\beta+\sqrt{x^2+\beta^2})^\nu}{x^{\nu+\frac{1}{2}}\sqrt{x^2+\beta^2}} \sin(ax)\,dx = \dfrac{1}{\beta}\sqrt{\dfrac{2}{a}}\,\Gamma\left(\dfrac{3}{4}-\dfrac{\nu}{2}\right) W_{\frac{\nu}{2},\,\frac{1}{4}}(a\beta) M_{-\frac{\nu}{2},\,\frac{1}{4}}(a\beta)$

$\left[a>0,\ \operatorname{Re}\beta>0,\ \operatorname{Re}\nu<\dfrac{3}{2}\right].$ Ba IV 71(27)

6. $\int_0^\infty \dfrac{(\beta+\sqrt{x^2+\beta^2})^\nu}{x^{\nu+\frac{1}{2}}\sqrt{\beta^2+x^2}} \cos(ax)\,dx = \dfrac{1}{\beta\sqrt{2a}}\,\Gamma\left(\dfrac{1}{4}-\dfrac{\nu}{2}\right) W_{\frac{\nu}{2},\,-\frac{1}{4}}(a\beta) M_{-\frac{\nu}{2},\,-\frac{1}{4}}(a\beta)$

$\left[a>0,\ \operatorname{Re}\beta>0,\ \operatorname{Re}\nu<\dfrac{1}{2}\right].$ Ba IV 12(18)

1. $\int_0^\infty \dfrac{(\sqrt{x^2+\beta^2}+x)^\nu - (\sqrt{x^2+\beta^2}-x)^\nu}{\sqrt{x^2+\beta^2}} \sin(ax)\,dx = 2\beta^\nu \sin\dfrac{\nu\pi}{2} K_\nu(a\beta)$ 3.775

$[a>0,\ \operatorname{Re}\beta>0,\ |\operatorname{Re}\nu|<1].$ Ba IV 70(20)

2. $\int_0^\infty \dfrac{(\sqrt{x^2+\beta^2}+x)^\nu + (\sqrt{x^2+\beta^2}-x)^\nu}{\sqrt{x^2+\beta^2}} \cos(ax)\,dx = 2\beta^\nu \cos\dfrac{\nu\pi}{2} K_\nu(a\beta)$

$[a>0,\ \operatorname{Re}\beta>0,\ |\operatorname{Re}\nu|<1].$ Ba IV 13(22)

3. $\int_u^\infty \dfrac{(x+\sqrt{x^2-u^2})^\nu + (x-\sqrt{x^2-u^2})^\nu}{\sqrt{x^2-u^2}} \sin(ax)\,dx =$

$= \pi u^\nu \left[J_\nu(au)\cos\dfrac{\nu\pi}{2} - N_\nu(au)\sin\dfrac{\nu\pi}{2}\right]$

$[a>0,\ u>0,\ |\operatorname{Re}\nu|<1].$ Ba IV 70(22)

4. $\int_u^\infty \dfrac{(x+\sqrt{x^2-u^2})^\nu + (x-\sqrt{x^2-u^2})^\nu}{\sqrt{x^2-u^2}} \cos(ax)\,dx =$

$= -\pi u^\nu \left[N_\nu(au)\cos\dfrac{\nu\pi}{2} + J_\nu(au)\sin\dfrac{\nu\pi}{2}\right]$

$[a>0,\ u>0,\ |\operatorname{Re}\nu|<1].$ Ba IV 13(25)

3.775

5. $$\int_0^u \frac{(x+i\sqrt{u^2-x^2})^\nu + (x-i\sqrt{u^2-x^2})^\nu}{\sqrt{u^2-x^2}} \sin(ax)\, dx =$$

$$= \frac{\pi}{2} u^\nu \operatorname{cosec} \frac{\nu\pi}{2} [\mathbf{J}_\nu(au) - \mathbf{J}_{-\nu}(au)] \quad [a>0,\ u>0].$$ Ba IV 70(21)

6. $$\int_0^u \frac{(x+i\sqrt{u^2-x^2})^\nu + (x-i\sqrt{u^2-x^2})^\nu}{\sqrt{u^2-x^2}} \cos(ax)\, dx =$$

$$= \frac{\pi}{2} u^\nu \sec \frac{\nu\pi}{2} [\mathbf{J}_\nu(au) + \mathbf{J}_{-\nu}(au)] \quad [a>0,\ u>0,\ |\operatorname{Re}\nu|<1].$$ Ba IV 13(24)

7. $$\int_u^\infty \frac{(x+\sqrt{x^2-u^2})^\nu + (x^2-\sqrt{x^2-u^2})^\nu}{\sqrt{x(x^2-u^2)}} \sin(ax)\, dx =$$

$$= -\sqrt{\left(\frac{\pi}{2}\right)^3} a u^\nu \left[J_{\frac{1}{4}+\frac{\nu}{2}}\!\left(\frac{au}{2}\right) N_{\frac{1}{4}-\frac{\nu}{2}}\!\left(\frac{au}{2}\right) + \right.$$

$$\left. + J_{\frac{1}{4}-\frac{\nu}{2}}\!\left(\frac{au}{2}\right) N_{\frac{1}{4}+\frac{\nu}{2}}\!\left(\frac{au}{2}\right) \right] \quad \left[a>0,\ u>0,\ \operatorname{Re}\nu<\frac{3}{2}\right].$$ Ba IV 71(25)

8. $$\int_u^\infty \frac{(x+\sqrt{x^2-u^2})^\nu + (x-\sqrt{x^2-u^2})^\nu}{\sqrt{x(x^2-u^2)}} \cos(ax)\, dx =$$

$$= -\sqrt{\left(\frac{\pi}{2}\right)^3} a u^\nu \left[J_{-\frac{1}{4}+\frac{\nu}{2}}\!\left(\frac{au}{2}\right) N_{-\frac{1}{4}-\frac{\nu}{2}}\!\left(\frac{au}{2}\right) + \right.$$

$$\left. + J_{-\frac{1}{4}-\frac{\nu}{2}}\!\left(\frac{au}{2}\right) N_{-\frac{1}{4}+\frac{\nu}{2}}\!\left(\frac{au}{2}\right) \right] \quad \left[a>0,\ u>0,\ \operatorname{Re}\nu<\frac{3}{2}\right].$$ Ba IV 13(26)

9. $$\int_0^\infty \frac{(x+\beta+\sqrt{x^2+2\beta x})^\nu + (x+\beta-\sqrt{x^2+2\beta x})^\nu}{\sqrt{x^2+2\beta x}} \sin(ax)\, dx =$$

$$= \pi\beta^\nu \left[N_\nu(\beta a) \sin\!\left(\beta a - \frac{\nu\pi}{2}\right) + J_\nu(\beta a) \cos\!\left(\beta a - \frac{\nu\pi}{2}\right) \right]$$

$$[a>0,\ |\arg\beta|<\pi,\ |\operatorname{Re}\nu|<1].$$ Ba IV 71(26)

10. $$\int_0^\infty \frac{(x+\beta+\sqrt{x^2+2\beta x})^\nu + (x+\beta-\sqrt{x^2+2\beta x})^\nu}{\sqrt{x^2+2\beta x}} \cos(ax)\, dx =$$

$$= \pi\beta^\nu \left[J_\nu(\beta a) \sin\!\left(\beta a - \frac{\nu\pi}{2}\right) - N_\nu(\beta a) \cos\!\left(\beta a - \frac{\nu\pi}{2}\right) \right]$$

$$[a>0,\ |\arg\beta|<\pi,\ |\operatorname{Re}\nu|<1].$$ Ba IV 13(23)

11. $\int_0^{2u} \frac{(\sqrt{2u+x} + i\sqrt{2u-x})^{4v} + (\sqrt{2u+x} - i\sqrt{2u-x})^{4v}}{\sqrt{4u^2 x - x^3}} \cos(ax) \, dx =$ 3.775

$$= (4u)^{2v} \pi^{\frac{3}{2}} \sqrt{\frac{a}{2}} J_{v-\frac{1}{4}}(au) J_{-v-\frac{1}{4}}(au) \quad [a>0, \ u>0].$$ Ba IV 14(27)

1. $\int_0^\infty \frac{a^2(b+x)^2 + p(p+1)}{(b+x)^{p+2}} \sin(ax) \, dx = \frac{a}{b^p} \quad [a>0, \ b>0, \ p>0].$ BH[170](1) 3.776

2. $\int_0^\infty \frac{a^2(b+x)^2 + p(p+1)}{(b+x)^{p+2}} \cos(ax) \, dx = \frac{p}{b^{p+1}} \quad [a>0, b>0, p>0].$ BH[170](2)

Rationale Funktionen von x und von trigonometrischen Funktionen
Rational functions of x and of trigonometric functions 3.78–3.81

1. $\int_0^\infty \left(\frac{\sin x}{x} - \frac{1}{1+x} \right) \frac{dx}{x} = 1 - C$ (cf. 3.784 4., 3.781. 2.). BH[173](7) 3.781

2. $\int_0^\infty \left(\cos x - \frac{1}{1+x} \right) \frac{dx}{x} = -C.$ BH[173](8)

1. $\int_0^u \frac{1-\cos x}{x} \, dx - \int_u^\infty \frac{\cos x}{x} \, dx = C + \ln u \quad [u>0].$ GH[333](31) 3.782

2. $\int_0^\infty \frac{1-\cos ax}{x^2} \, dx = \frac{a\pi}{2}.$ BH[158](1)

3. $\int_{-\infty}^\infty \frac{1-\cos ax}{x(x-b)} \, dx = \frac{\sin ab}{b} \quad [a>0]$ Ba V 253(48)

1. $\int_0^\infty \left[\frac{\cos x - 1}{x^2} + \frac{1}{2(1+x)} \right] \frac{dx}{x} = \frac{1}{2} C - \frac{3}{4}.$ BH[173](19) 3.783

2. $\int_0^\infty \left(\cos x - \frac{1}{1+x^2} \right) \frac{dx}{x} = -C.$ Ba I 17, BH[273](21)

1. $\int_0^\infty \frac{\cos ax - \cos bx}{x} \, dx = \ln \frac{b}{a} \quad [ab \neq 0].$ F II 644, GH[333](20) 3.784

3.784

2. $\int_0^\infty \dfrac{a \sin bx - b \sin ax}{x^2} dx = ab \ln \dfrac{a}{b}$ $\quad [a > 0,\ b > 0].$ \hfill F II 657

3. $\int_0^\infty \dfrac{\cos ax - \cos bx}{x^2} dx = \dfrac{(b-a)\pi}{2}$ $\quad [a \geq 0,\ b \geq 0].$ \hfill BH[158](2), F II 653

4. $\int_0^\infty \dfrac{\sin x - x \cos x}{x^2} dx = 1.$ \hfill BH[158](3)

5. $\int_0^\infty \dfrac{\cos ax - \cos bx}{x(x+\beta)} dx = \dfrac{1}{\beta}\left[\text{ci}(a\beta)\cos a\beta + \text{si}(a\beta)\sin a\beta - \right.$

   $\left. - \text{ci}(b\beta)\cos b\beta - \text{si}(b\beta)\sin b\beta + \ln \dfrac{b}{a}\right]$

   $[a > 0,\ b > 0,\ |\arg \beta| < \pi].$ \hfill Ba V 221(49)

6. $\int_0^\infty \dfrac{\cos ax + x \sin ax}{1 + x^2} dx = \pi e^{-a}$ $\quad [a > 0].$ \hfill GH[333](73)

7. $\int_0^\infty \dfrac{\sin ax - ax \cos ax}{x^3} dx = \dfrac{\pi}{4} a^2 \operatorname{sign} a.$ \hfill Li[158](5)

8. $\int_0^\infty \dfrac{\cos ax - \cos bx}{x^2(x^2 + \beta^2)} dx = \dfrac{\pi[(b-a)\beta + e^{-b\beta} - e^{-a\beta}]}{2\beta^3}$

   $[a > 0,\ b > 0,\ |\arg \beta| < \pi].$ \hfill BH[173](20)+, Ba V 222(59)

3.785

$\int_0^\infty \dfrac{1}{x} \sum_{k=1}^n a_k \cos b_k x\, dx = -\sum_{k=1}^n a_k \ln b_k$ $\quad \left[b_k > 0,\ \sum_{k=1}^n a_k = 0\right].$ \hfill F II 658

3.786

1. $\int_0^\infty \dfrac{(1 - \cos ax) \sin bx}{x^2} dx = \dfrac{b}{2} \ln \dfrac{b^2 - a^2}{b^2} + \dfrac{a}{2} \ln \dfrac{a+b}{a-b}$ $\quad [a > 0,\ b > 0].$

\hfill Ba IV 81(29)

2. $\int_0^\infty \dfrac{(1 - \cos ax) \cos bx}{x} dx = \ln \dfrac{\sqrt{|a^2 - b^2|}}{b}$ $\quad [a > 0,\ b > 0,\ a \neq b].$ \hfill F II 657

3. $\int_0^\infty \dfrac{(1 - \cos ax) \cos bx}{x^2} dx = \dfrac{\pi}{2}(a - b)$ $\quad [0 < b \leq a];$

   $= 0$ $\quad [0 < a \leq b].$ \hfill Ba IV 20(16)

3.787

1. $\int_0^\infty \dfrac{(\cos a - \cos nax) \sin mx}{x} dx = \dfrac{\pi}{2}(\cos a - 1)$ $\quad [m > na > 0];$

   $= \dfrac{\pi}{2} \cos a$ $\quad [na > m].$ \hfill BH[155](7)

2. $\int_0^\infty \dfrac{\sin^2 ax - \sin^2 bx}{x}\,dx = \dfrac{1}{2}\ln\dfrac{a}{b}$ $[ab \neq 0]$. GH[333](20b) 3.787

3. $\int_0^\infty \dfrac{x^3 - \sin^3 x}{x^5}\,dx = \dfrac{13}{32}\pi$. BH[158](6)

4. $\int_0^\infty \dfrac{(3 - 4\sin^2 ax)\sin^2 ax}{x}\,dx = \dfrac{1}{2}\ln 2$. BH[155](6)

$\int_0^{\pi/2} \left(\dfrac{1}{x} - \operatorname{ctg} x\right) dx = \ln\dfrac{\pi}{2}$. GH[333](61a) 3.788

$\int_0^{\pi/2} \dfrac{4x^2 \cos x + (\pi - x)x}{\sin x}\,dx = \pi^2 \ln 2$. Li[206](10) 3.789

1. $\int_0^{\pi/2} \dfrac{x\,dx}{1 + \sin x} = \ln 2$. GH[333](55a) 3.791

2. $\int_0^\pi \dfrac{x \cos x}{1 + \sin x}\,dx = \pi \ln 2 - 4G$. GH[333](55c)

3. $\int_0^{\pi/2} \dfrac{x \cos x}{1 + \sin x}\,dx = \pi \ln 2 - 2G$. GH[333](55b)

4. $\int_0^\pi \dfrac{\left(\dfrac{\pi}{2} - x\right)\cos x}{1 - \sin x}\,dx = 2\int_0^{\pi/2} \dfrac{\left(\dfrac{\pi}{2} - x\right)\cos x}{1 - \sin x}\,dx =$

$= \pi \ln 2 + 4G = 5{,}841\,448\,4669\ldots$ BH[207](3), GH[333](56c)

5. $\int_0^{\pi/2} \dfrac{x^2\,dx}{1 - \cos x} = -\dfrac{\pi}{4} + \pi \ln 2 + 4G = 3{,}374\,047\,366\,7\ldots$ BH[207](3)

6. $\int_0^\pi \dfrac{x^2\,dx}{1 - \cos x} = 4\pi \ln 2$. BH[219](1)

7. $\int_0^{\pi/2} \dfrac{x^{p+1}\,dx}{1 - \cos x} = -\left(\dfrac{\pi}{2}\right)^{p+1} + \left(\dfrac{\pi}{2}\right)^p (p+1) \left\{\dfrac{2}{p} - \sum_{k=1}^\infty \dfrac{1}{4^{2k-1}(p+2k)} \zeta(2k)\right\}$

$[p > 0]$. Li[207](4)

3.791

8. $$\int_0^{\frac{\pi}{2}} \frac{x\,dx}{1+\cos x} = \frac{\pi}{2} - \ln 2.$$ GH[333](55a)

9. $$\int_0^{\frac{\pi}{2}} \frac{x \sin x\,dx}{1-\cos x} = \frac{\pi}{2} \ln 2 + 2G.$$ GH[333](56a)

10. $$\int_0^{\pi} \frac{x \sin x\,dx}{1-\cos x} = 2\pi \ln 2.$$ GH[333](56b)

11. $$\int_0^{\pi} \frac{x-\sin x}{1-\cos x}\,dx = \frac{\pi}{2} + \int_0^{\frac{\pi}{2}} \frac{x-\sin x}{1-\cos x}\,dx = 2.$$ GH[333](57a)

12. $$\int_0^{\frac{\pi}{2}} \frac{x \sin x}{1+\cos x}\,dx = -\frac{\pi}{2}\ln 2 + 2G.$$ GH[333](55b)

3.792

1. $$\int_{-\pi}^{\pi} \frac{dx}{1-2a\cos x + a^2} = \frac{2\pi}{1-a^2} \quad [a^2 < 1].$$ F II 480

2. $$\int_0^{\frac{\pi}{2}} \frac{x \cos x\,dx}{1+2a\sin x + a^2} = \frac{\pi}{2a}\ln(1+a) - \sum_{k=0}^{\infty}(-1)^k \frac{a^{2k}}{(2k+1)^2} \quad [a^2 < 1].$$ Li[241](2)

3. $$\int_0^{\pi} \frac{x \sin x\,dx}{1-2a\cos x + a^2} = \frac{\pi}{a}\ln(1+a) \quad [a^2 < 1];$$

$$= \frac{\pi}{a}\ln\left(1+\frac{1}{a}\right) \quad [a^2 > 1].$$ BH[221](2)

4. $$\int_0^{2\pi} \frac{x \sin x\,dx}{1-2a\cos x + a^2} = \frac{2\pi}{a}\ln(1-a) \quad [a^2 < 1];$$

$$= \frac{2\pi}{a}\ln\left(1-\frac{1}{a}\right) \quad [a^2 > 1].$$ BH[223](4)

5. $$\int_0^{2\pi} \frac{x \sin nx\,dx}{1-2a\cos x + a^2} = \frac{2\pi}{1-a^2}\left[(a^{-n} - a^n)\ln(1-a) + \sum_{k=1}^{n-1}\frac{a^{-k}-a^k}{n-k}\right]$$

$$[a^2 < 1].$$ BH[223](5)

6. $$\int_0^{\infty} \frac{\sin x}{1-2a\cos x + a^2} \cdot \frac{dx}{x} = \frac{\pi}{4a}\left[\left|\frac{1+a}{1-a}\right| - 1\right].$$ GH[333](62b)

7. $\int_0^\infty \dfrac{\sin bx}{1-2a\cos x + a^2} \cdot \dfrac{dx}{x} = \dfrac{\pi}{2} \dfrac{1+a-2a^{E(b)+1}}{(1-a^2)(1-a)}$ $[b \neq 0, 1, 2, \ldots]$; 3.792

$\phantom{\int_0^\infty \dfrac{\sin bx}{1-2a\cos x + a^2} \cdot \dfrac{dx}{x}} = \dfrac{\pi}{2} \dfrac{1+a-a^b-a^{b+1}}{(1-a^2)(1-a)}$ $[b = 0, 1, 2, \ldots]$
$[0 < a < 1]$. Ba IV 81(26)

8. $\int_0^\infty \dfrac{\sin x \cos bx}{1-2a\cos x + a^2} \cdot \dfrac{dx}{x} = \dfrac{\pi}{2(1-a)} a^{E(b)}$ $[b \neq 0, 1, 2, \ldots]$;

$ = \dfrac{\pi}{2(1-a)} a^b + \dfrac{\pi}{4} a^{b-1}$ $[b = 0, 1, 2, \ldots$
$[0 < a < 1,\ b > 0]$. Ba IV 19(5)

9. $\int_0^\infty \dfrac{(1-a\cos x)\sin bx}{1-2a\cos x + a^2} \cdot \dfrac{dx}{x} = \dfrac{\pi}{2} \cdot \dfrac{1-a^{E(b)+1}}{1-a}$ $[b \neq 1, 2, 3, \ldots]$;

$ = \dfrac{\pi}{2} \cdot \dfrac{1-a^b}{1-a} + \dfrac{\pi a^b}{4}$ $[b = 1, 2, 3, \ldots]$
$[0 < a < 1]$. Ba IV 82(33)

10. $\int_0^\infty \dfrac{1}{1-2a\cos bx + a^2} \dfrac{dx}{\beta^2 + x^2} = \dfrac{\pi}{2\beta(1-a^2)} \dfrac{1+ae^{-b\beta}}{1-ae^{-b\beta}}$ $[a^2 < 1]$. BH[192](1)

11. $\int_0^\infty \dfrac{1}{1-2a\cos bx + a^2} \dfrac{dx}{\beta^2 - x^2} = \dfrac{a\pi}{\beta(1-a^2)} \dfrac{\sin b\beta}{1-2a\cos b\beta + a^2}$ $[a^2 < 1]$. BH[193](1)

12. $\int_0^\infty \dfrac{\sin bcx}{1-2a\cos bx + a^2} \dfrac{x\,dx}{\beta^2 + x^2} = \dfrac{\pi}{2} \dfrac{e^{-\beta bc} - a^c}{(1-ae^{-b\beta})(1-ae^{b\beta})}$ $[a^2 < 1]$. BH[192](8)

13. $\int_0^\infty \dfrac{\sin bx}{1-2a\cos bx + a^2} \dfrac{x\,dx}{\beta^2 + x^2} = \dfrac{\pi}{2} \dfrac{1}{e^{b\beta} - a}$ $[a^2 < 1]$;

$ = \dfrac{\pi}{2a} \dfrac{1}{ae^{b\beta} - 1}$ $[a^2 > 1]$. BH[192](2)

14. $\int_0^\infty \dfrac{\sin bcx}{1-2a\cos bx + a^2} \dfrac{x\,dx}{\beta^2 - x^2} = \dfrac{\pi}{2} \dfrac{a^c - \cos \beta bc}{1-2a\cos \beta b + a^2}$ $[a^2 < 1]$. BH[193](5)

15. $\int_0^\infty \dfrac{\cos bcx}{1-2a\cos bx + a^2} \dfrac{dx}{\beta^2 - x^2} = \dfrac{\pi}{2\beta(1-a^2)} \dfrac{(1-a^2)\sin \beta bc + 2a^{c+1}\sin \beta b}{1-2a\cos \beta b + a^2}$
$[a^2 < 1]$. BH[193](9)

16. $\int_0^\infty \dfrac{1-a\cos bx}{1-2a\cos bx + a^2} \dfrac{dx}{1+x^2} = \dfrac{\pi}{2} \dfrac{e^b}{e^b - a}$. F II 724–725

3.792

17. $$\int_0^\infty \frac{\cos bx}{1 - 2a \cos x + a^2} \cdot \frac{dx}{x^2 + \beta^2} = \frac{\pi (e^{\beta - \beta b} + a e^{\beta b})}{2\beta(1 - a^2)(e^\beta - a)}$$

$$[0 \leqslant b < 1, \ |a| < 1, \ \operatorname{Re} \beta > 0].$$ Ba IV 21(21)

18. $$\int_0^\infty \frac{\sin bx \sin x}{1 - 2a \cos x + a^2} \cdot \frac{dx}{x^2 + \beta^2} = \frac{\pi}{2\beta} \frac{\operatorname{sh} b\beta}{e^\beta - a} \quad [0 \leqslant b < 1];$$

$$= \frac{\pi}{4\beta(ae^\beta - 1)} [a^m e^{\beta(m+1-b)} - e^{(1-b)\beta}] -$$

$$- \frac{\pi}{4\beta(ae^{-\beta} - 1)} [a^m e^{-(m+1-b)\beta} - e^{-(1-b)\beta}] \quad [m \leqslant b \leqslant m+1]$$

$$[0 < a < 1, \ \operatorname{Re} \beta > 0].$$ Ba IV 81(27)

19. $$\int_0^\infty \frac{(\cos x - a) \cos bx}{1 - 2a \cos x + a^2} \cdot \frac{dx}{x^2 + \beta^2} = \frac{\pi \operatorname{ch} \beta b}{2\beta(e^\beta - a)}$$

$$[0 \leqslant b < 1, \ |a| < 1, \ \operatorname{Re} \beta > 0].$$ Ba IV 21(23)

20. $$\int_0^\infty \frac{\sin x}{(1 - 2a \cos 2x + a^2)^{n+1}} \frac{dx}{x} = \int_0^\infty \frac{\operatorname{tg} x}{(1 - 2a \cos 2x + a^2)^{n+1}} \frac{dx}{x} =$$

$$= \int_0^\infty \frac{\operatorname{tg} x}{(1 - 2a \cos 4x + a^2)^{n+1}} \frac{dx}{x} = \frac{\pi}{2(1 - a^2)^{2n+1}} \sum_{k=0}^n \binom{n}{k}^2 a^{2k}.$$

BH[187](14, 15, 16)

3.793

1. $$\int_0^{2\pi} \frac{\sin nx - a \sin[(n+1)x]}{1 - 2a \cos x + a^2} x \, dx = 2\pi a^n \left[\ln(1 - a) + \sum_{k=1}^n \frac{1}{ka^k}\right] \quad [|a| < 1].$$

BH[223](9)

2. $$\int_0^{2\pi} \frac{\cos nx - a \cos[(n+1)x]}{1 - 2a \cos x + a^2} x \, dx = 2\pi^2 a^n \quad [a^2 < 1].$$ BH[223](13)

3.794

1. $$\int_0^\pi \frac{x \, dx}{a \pm \cos x} = \frac{\pi^2}{2\sqrt{a^2 - 1}} \pm \frac{4}{\sqrt{a^2 - 1}} \sum_{k=1}^\infty \frac{(a - \sqrt{a^2 - 1})^{2k+1}}{(2k+1)^2} \quad [a > 1].$$

Li[219](2)

2. $$\int_0^{2\pi} \frac{x \sin nx}{1 \pm a \cos x} dx = \frac{2\pi}{\sqrt{1 - a^2}} \left[(\mp 1)^n \frac{(1 + \sqrt{1 - a^2})^n - (1 - \sqrt{1 - a^2})^n}{a^n} \times \right.$$

$$\left. \times \ln \frac{2\sqrt{1 \pm a}}{\sqrt{1 + a} + \sqrt{1 - a}} + \sum_{k=1}^{n-1} \frac{(\mp 1)^k}{n - k} \frac{(1 + \sqrt{1 - a^2})^k - (1 - \sqrt{1 - a^2})^k}{a^k}\right]$$

$$[a^2 < 1].$$ BH[223](2)

3. $\int_0^{2\pi} \frac{x \cos nx}{1 \pm a \cos x} dx = \frac{2\pi^2}{\sqrt{1-a^2}} \left(\frac{1 - \sqrt{1-a^2}}{\pm a} \right)^n$ $[a^2 < 1]$. BH[223](3) 3.794

4. $\int_0^{\pi} \frac{x \sin x \, dx}{a + b \cos x} = \frac{\pi}{b} \ln \frac{a + \sqrt{a^2 - b^2}}{2(a - b)}$ $[a > |b| > 0]$. GH[333](53a)

5. $\int_0^{2\pi} \frac{x \sin x \, dx}{a + b \cos x} = \frac{2\pi}{b} \ln \frac{a + \sqrt{a^2 - b^2}}{2(a + b)}$ $[a > |b| > 0]$. GH[333](53b)

6. $\int_0^{\infty} \frac{\sin x}{a \pm b \cos 2x} \cdot \frac{dx}{x} = \frac{\pi}{2\sqrt{a^2 - b^2}}$ $[a^2 > b^2]$;

$= 0$ $[a^2 < b^2]$. BH[181](1)

$\int_{-\infty}^{\infty} \frac{(b^2 + c^2 + x^2) x \sin ax - (b^2 - c^2 - x^2) c \, \text{sh} \, ac}{[x^2 + (b-c)^2][x^2 + (b+c)^2](\cos ax + \text{ch} \, ac)} dx = \pi$ $[c > b > 0]$; 3.795

$= \frac{2\pi}{e^{ab} + 1}$ $[b > c > 0]$;

$[a > 0]$. BH[202](18)

1. $\int_0^{\pi/2} \frac{\cos x \pm \sin x}{\cos x \mp \sin x} x \, dx = \mp \frac{\pi}{4} \ln 2 - G$. BH[207](8, 9) 3.796

2. $\int_0^{\pi/4} \frac{\cos x - \sin x}{\cos x + \sin x} x \, dx = \frac{\pi}{4} \ln 2 - \frac{1}{2} G$. BH[204](23)

1. $\int_0^{\pi/4} \left(\frac{\pi}{4} - x \, \text{tg} \, x \right) \text{tg} \, x \, dx = \frac{1}{2} \ln 2 + \frac{\pi^2}{32} - \frac{\pi}{4} + \frac{\pi}{8} \ln 2$. BH[204](8) 3.797

2. $\int_0^{\pi/4} \frac{\left(\frac{\pi}{4} - x \right) \text{tg} \, x \, dx}{\cos 2x} = -\frac{\pi}{8} \ln 2 + \frac{1}{2} G$. BH[204](19)

3. $\int_0^{\pi/4} \frac{\frac{\pi}{4} - x \, \text{tg} \, x}{\cos 2x} dx = \frac{\pi}{8} \ln 2 + \frac{1}{2} G$. BH[204](20)

1. $\int_0^{\infty} \frac{\text{tg} \, x}{a + b \cos 2x} \cdot \frac{dx}{x} = \frac{\pi}{2\sqrt{a^2 - b^2}}$ $[a^2 > b^2]$; 3.798

$= 0$ $[a^2 < b^2]$; $[a > 0]$. BH[181](2)

3.798 2. $\int_0^\infty \dfrac{\operatorname{tg} x}{a + b \cos 4x} \cdot \dfrac{dx}{x} = \dfrac{\pi}{2\sqrt{a^2 - b^2}}$ $[a^2 > b^2]$;

$\qquad\qquad\qquad\qquad\qquad\qquad = 0$ $[a^2 < b^2]$; $[a > 0]$.   BH[181](3)

3.799 1. $\int_0^{\pi/2} \dfrac{x\,dx}{(\sin x + a \cos x)^2} = \dfrac{a}{1+a^2}\dfrac{\pi}{2} - \dfrac{\ln a}{1+a^2}$ $[a > 0]$.   BH[208](5)

2. $\int_0^{\pi/4} \dfrac{x\,dx}{(\cos x + a \sin x)^2} = \dfrac{1}{1+a^2}\ln\dfrac{1+a}{\sqrt{2}} + \dfrac{\pi}{4}\cdot\dfrac{1-a}{(1+a)(1+a^2)}$ $[a > 0]$.   BH[204](24)

3. $\int_0^\pi \dfrac{a\cos x + b}{(a + b \cos x)^2} x^2\,dx = \dfrac{2\pi}{b}\ln\dfrac{2(a-b)}{a + \sqrt{a^2 - b^2}}$ $[a > |b| > 0]$.   GH[333](58a)

3.811 1. $\int_0^\pi \dfrac{\sin x}{1 - \cos t_1 \cos x}\cdot\dfrac{x\,dx}{1 - \cos t_2 \cos x} = \pi\operatorname{cosec}\dfrac{t_1 + t_2}{2}\operatorname{cosec}\dfrac{t_1 - t_2}{2}\ln\dfrac{1 + \operatorname{tg}\dfrac{t_1}{2}}{1 + \operatorname{tg}\dfrac{t_2}{2}}$

$\qquad\qquad\qquad\qquad\qquad\qquad\qquad$ (cf. 3.794 4.).   BH[222](5)

2. $\int_0^{\pi/2} \dfrac{x\,dx}{(\cos x \pm \sin x)\sin x} = \dfrac{\pi}{4}\ln 2 \pm G$.   BH[208](16, 17)

3. $\int_0^{\pi/4} \dfrac{x\,dx}{(\cos x + \sin x)\sin x} = -\dfrac{\pi}{8}\ln 2 + G$.   BH[204](29)

4. $\int_0^{\pi/4} \dfrac{x\,dx}{(\cos x + \sin x)\cos x} = \dfrac{\pi}{8}\ln 2$.   BH[204](28)

5. $\int_0^{\pi/4} \dfrac{\sin x}{\sin x + \cos x}\dfrac{x\,dx}{\cos^2 x} = -\dfrac{\pi}{8}\ln 2 + \dfrac{\pi}{4} - \dfrac{1}{2}\ln 2$.   BH[204](30)

3.812 1. $\int_0^\pi \dfrac{x \sin x\,dx}{a + b \cos^2 x} = \dfrac{\pi}{\sqrt{ab}}\operatorname{arctg}\sqrt{\dfrac{b}{a}}$ $[a > 0, b > 0]$;

$\qquad\qquad\qquad\qquad\qquad\quad = \dfrac{\pi}{2\sqrt{-ab}}\ln\dfrac{\sqrt{a} + \sqrt{-b}}{\sqrt{a} - \sqrt{-b}}$ $[a > -b > 0]$.   GH[333](60a)

2. $\int_0^{\pi/2} \dfrac{x \sin 2x\,dx}{1 + a\cos^2 x} = \dfrac{\pi}{a}\ln\dfrac{1 + \sqrt{1+a}}{2}$ $[a > -1, a \neq 0]$.   BH[207](10)

3. $\int_0^{\pi/2} \dfrac{x \sin 2x\,dx}{1 + a\sin^2 x} = \dfrac{\pi}{a}\ln\dfrac{2(1 + a - \sqrt{1+a})}{a}$ $[a > -1, a \neq 0]$.   BH[207](2)

4. $\int_0^\pi \frac{x\,dx}{a^2 - \cos^2 x} = \frac{\pi^2}{2a\sqrt{a^2-1}}$ $[a^2 > 1]$;   3.812

$\qquad\qquad\qquad = 0$ $[a^2 < 1]$.   BH[219](10)

5. $\int_0^\pi \frac{x \sin x\,dx}{a^2 - \cos^2 x} = \frac{\pi}{2a} \ln \frac{1+a}{1-a}$ $[a \neq 1]$.   BH[219](13)

6. $\int_0^\pi \frac{x \sin 2x\,dx}{a^2 - \cos^2 x} = \pi \ln\{4(1 - a^2)\}$ $[a^2 < 1]$;

$\qquad\qquad\qquad = 2\pi \ln[2(1 - a^2 + a\sqrt{a^2-1})]$ $[a^2 > 1]$.   BH[219](19)

7. $\int_0^{\pi/2} \frac{x \sin x\,dx}{\cos^2 t - \sin^2 x} = -2 \csc t \sum_{k=0}^{\infty} \frac{\sin(2k+1)t}{(2k+1)^2}$.   BH[207](1)

8. $\int_0^\pi \frac{x \sin x\,dx}{1 - \cos^2 t \sin^2 x} = \pi(\pi - 2t) \csc 2t$.   BH[219](12)

9. $\int_0^\pi \frac{x \cos x\,dx}{\cos^2 t - \cos^2 x} = 4 \csc t \sum_{k=0}^{\infty} \frac{\sin(2k+1)t}{(2k+1)^2}$.   BH[219](17)

10. $\int_0^\pi \frac{x \sin x\,dx}{\operatorname{tg}^2 t + \cos^2 x} = \frac{\pi}{2}(\pi - 2t) \operatorname{ctg} t$.   BH[219](14)

11. $\int_0^\infty \frac{x(a\cos x + b)\sin x\,dx}{\operatorname{ctg}^2 t + \cos^2 x} = 2a\pi \ln \cos \frac{t}{2} + \pi bt \operatorname{tg} t$.   BH[219](18)

1. $\int_0^\pi \frac{x\,dx}{a^2 \cos^2 x + b^2 \sin^2 x} = \frac{1}{4} \int_0^{2\pi} \frac{x\,dx}{a^2 \cos^2 x + b^2 \sin^2 x} = \frac{\pi^2}{2ab}$ $[a>0, b>0]$.   3.813

GH[333](36)

2. $\int_0^\infty \frac{1}{\beta^2 \sin^2 ax + \gamma^2 \cos^2 ax} \cdot \frac{dx}{x^2 + \delta^2} = \frac{\pi \operatorname{sh}(2a\delta)}{4\delta(\beta^2 \operatorname{sh}^2(a\delta) - \gamma^2 \operatorname{ch}^2(a\delta))} \left[\frac{\beta}{\gamma} - \frac{\gamma}{\beta} - \frac{2}{\operatorname{sh}(2a\delta)}\right]$

$\left[\left|\arg \frac{\beta}{\gamma}\right| < \pi,\ \operatorname{Re} \delta > 0,\ a > 0\right].$ GH[333](81), Ba V 222(63)

3. $\int_0^\infty \frac{\sin x\,dx}{x(a^2 \sin^2 x + b^2 \cos^2 x)} = \frac{\pi}{2ab}$ $[ab > 0]$.   BH[181](8)

4. $\int_0^\infty \frac{\sin^2 x\,dx}{x(a^2 \cos^2 x + b^2 \sin^2 x)} = \frac{\pi}{2b(a+b)}$ $[a > 0, b > 0]$.   BH[181](11)

5. $\int_0^{\pi/2} \frac{x \sin 2x\,dx}{a^2 \cos^2 x + b^2 \sin^2 x} = \frac{\pi}{a^2 - b^2} \ln \frac{a+b}{2b}$ $[a > 0, b > 0, a \neq b]$.   GH[333](52a)

3.813

6. $\int_0^\pi \dfrac{x \sin 2x \, dx}{a^2 \cos^2 x + b^2 \sin^2 x} = \dfrac{2\pi}{a^2 - b^2} \ln \dfrac{a+b}{2a} \quad [a > 0, \, b > 0, \, a \ne b]$.

GH[333](52b)

7. $\int_0^\infty \dfrac{\sin 2x}{a^2 \cos^2 x + b^2 \sin^2 x} \cdot \dfrac{dx}{x} = \dfrac{\pi}{a(a+b)} \quad [a > 0, \, b > 0]$.

BH[182](3)

8. $\int_0^\infty \dfrac{\sin 2ax}{\beta^2 \sin^2 ax + \gamma^2 \cos^2 ax} \cdot \dfrac{x \, dx}{x^2 + \delta^2} = \dfrac{\pi}{2(\beta^2 \operatorname{sh}^2(a\delta) - \gamma^2 \operatorname{ch}^2(a\delta))} \left[\dfrac{\beta - \gamma}{\beta + \gamma} - e^{-2a\delta} \right]$

$\left[a > 0, \, \left| \arg \dfrac{\beta}{\gamma} \right| < \pi, \, \operatorname{Re} \delta > 0 \right]$.

Ba V 222(64), GH[333](80)

9. $\int_0^\infty \dfrac{(1 - \cos x) \sin x}{a^2 \cos^2 x + b^2 \sin^2 x} \cdot \dfrac{dx}{x} = \dfrac{\pi}{2b(a+b)} \quad [a > 0, \, b > 0]$.

BH [182](7)+

10. $\int_0^\infty \dfrac{\sin x \cos^2 x}{a^2 \cos^2 x + b^2 \sin^2 x} \cdot \dfrac{dx}{x} = \dfrac{\pi}{2a(a+b)} \quad [a > 0, \, b > 0]$.

BH[182](4)

11. $\int_0^\infty \dfrac{\sin^3 x}{a^2 \cos^2 x + b^2 \sin^2 x} \cdot \dfrac{dx}{x} = \dfrac{\pi}{2b} \cdot \dfrac{1}{a+b} \quad [a > 0, \, b > 0]$.

BH[182](1)

3.814

1. $\int_0^{\pi/2} \dfrac{(1 - x \operatorname{ctg} x) \, dx}{\sin^2 x} = \dfrac{\pi}{4}$.

BH[206](9)

2. $\int_0^{\pi/4} \dfrac{x \operatorname{tg} x \, dx}{(\sin x + \cos x) \cos x} = -\dfrac{\pi}{8} \ln 2 + \dfrac{\pi}{4} - \dfrac{1}{2} \ln 2$.

BH[204](30)

3. $\int_0^\infty \dfrac{\operatorname{tg} x}{a^2 \cos^2 x + b^2 \sin^2 x} \cdot \dfrac{dx}{x} = \dfrac{\pi}{2ab} \quad [a > 0, \, b > 0]$.

BH[181](9)

4. $\int_0^{\pi/2} \dfrac{x \operatorname{ctg} x \, dx}{a^2 \cos^2 x + b^2 \sin^2 x} = \dfrac{\pi}{2a^2} \ln \dfrac{a+b}{b} \quad [a > 0, \, b > 0]$.

Li[208](20)

5. $\int_0^{\pi/2} \dfrac{\left(\dfrac{\pi}{2} - x\right) \operatorname{tg} x \, dx}{a^2 \cos^2 x + b^2 \sin^2 x} = \dfrac{1}{2} \int_0^\pi \dfrac{\left(\dfrac{\pi}{2} - x\right) \operatorname{tg} x \, dx}{a^2 \cos^2 x + b^2 \sin^2 x} =$

$= \dfrac{\pi}{2b^2} \ln \dfrac{a+b}{a} \quad [a > 0, \, b > 0]$.

GH[333](59)

6. $\int_0^\infty \dfrac{\sin^2 x \operatorname{tg} x}{a^2 \cos^2 x + b^2 \sin^2 x} \cdot \dfrac{dx}{x} = \dfrac{\pi}{2b(a+b)} \quad [a > 0, \, b > 0]$.

BH[182](6)

7. $\int_0^\infty \dfrac{\operatorname{tg} x}{a^2 \cos^2 2x + b^2 \sin^2 2x} \cdot \dfrac{dx}{x} = \dfrac{\pi}{2ab} \quad [a > 0, \, b > 0]$.

BH[181](10)+

8. $\int_0^\infty \dfrac{\sin^2 2x \, \text{tg } x}{a^2 \cos^2 2x + b^2 \sin^2 2x} \cdot \dfrac{dx}{x} = \dfrac{\pi}{2b} \cdot \dfrac{1}{a+b}$ $[a > 0, \ b > 0]$.   BH[182](2)+   3.814

9. $\int_0^\infty \dfrac{\cos^2 2x \, \text{tg } x}{a^2 \cos^2 2x + b^2 \sin^2 2x} \cdot \dfrac{dx}{x} = \dfrac{\pi}{2a} \cdot \dfrac{1}{a+b}$ $[a > 0, \ b > 0]$.   BH[182](5)+

10. $\int_0^\infty \dfrac{\sin^2 x \cos x}{a^2 \cos^2 2x + b^2 \sin^2 2x} \cdot \dfrac{dx}{x \cos 4x} = -\dfrac{\pi}{8b} \cdot \dfrac{a}{a^2 + b^2}$ $[a > 0, \ b > 0]$.

   BH[186](12)+

11. $\int_0^\infty \dfrac{\sin x}{a^2 \cos^2 x + b^2 \sin^2 x} \cdot \dfrac{dx}{x \cos 2x} = \dfrac{\pi}{2ab} \cdot \dfrac{b^2 - a^2}{b^2 + a^2}$ $[a > 0, \ b > 0]$.   BH[186](4)+

12. $\int_0^\infty \dfrac{\sin x \cos x}{a^2 \cos^2 x + b^2 \sin^2 x} \cdot \dfrac{dx}{x \cos 2x} = \dfrac{\pi}{2a} \cdot \dfrac{b}{a^2 + b^2}$ $[a > 0, \ b > 0]$.   BH[186](7)+

13. $\int_0^\infty \dfrac{\sin x \cos^2 x}{a^2 \cos^2 x + b^2 \sin^2 x} \cdot \dfrac{dx}{x \cos 2x} = \dfrac{\pi}{2ab} \cdot \dfrac{b^2}{a^2 + b^2}$ $[a > 0, \ b > 0]$.   BH[186](8)+

14. $\int_0^\infty \dfrac{\sin^3 x}{a^2 \cos^2 x + b^2 \sin^2 x} \cdot \dfrac{dx}{x \cos 2x} = -\dfrac{\pi}{2b} \cdot \dfrac{a}{a^2 + b^2}$ $[a > 0, \ b > 0]$.   BH[186](10)

15. $\int_0^\infty \dfrac{1 - \cos x}{a^2 \cos^2 x + b^2 \sin^2 x} \cdot \dfrac{dx}{x \sin x} = \dfrac{\pi}{2ab}$ $[a > 0, \ b > 0]$.   BH[186](3)+

1. $\int_0^{\pi/2} \dfrac{x \sin 2x \, dx}{(1 + a \sin^2 x)(1 + b \sin^2 x)} = \dfrac{\pi}{a - b} \ln \left\{ \dfrac{1 + \sqrt{1+b}}{1 + \sqrt{1+a}} \cdot \dfrac{\sqrt{1+a}}{\sqrt{1+b}} \right\}$ 3.815

    $[a > 0, \ b > 0]$ (cf. 3.812 3.).   BH[208](22)

2. $\int_0^{\pi/2} \dfrac{x \sin 2x \, dx}{(1 + a \sin^2 x)(1 + b \cos^2 x)} = \dfrac{\pi}{a + ab + b} \ln \dfrac{(1 + \sqrt{1+b})\sqrt{1+a}}{1 + \sqrt{1+a}}$

    $[a > 0, \ b > 0]$ (cf. 3.812 2., 3.).   BH[208](24)

3. $\int_0^{\pi/2} \dfrac{x \sin 2x \, dx}{(1 + a \cos^2 x)(1 + b \cos^2 x)} = \dfrac{\pi}{a - b} \ln \dfrac{1 + \sqrt{1+a}}{1 + \sqrt{1+b}}$

    $[a > 0, \ b > 0]$ (cf. 3.812 2.).   BH[208](23)

4. $\int_0^{\pi/2} \dfrac{x \sin 2x \, dx}{(1 - \sin^2 t_1 \cos^2 x)(1 - \sin^2 t_2 \cos^2 x)} = \dfrac{2\pi}{\cos^2 t_1 - \cos^2 t_2} \ln \dfrac{\cos \frac{t_1}{2}}{\cos \frac{t_2}{2}}$

    $[-\pi < t_1 < \pi, \ -\pi < t_2 < \pi]$.   BH[208](21)

3.816

1. $\int_0^\pi \dfrac{x^2 \sin 2x}{(a^2 - \cos^2 x)^2}\, dx = \pi^2 \dfrac{\sqrt{a^2 - 1} - a}{a(a^2 - 1)}$ $\quad [a > 1]$. $\qquad$ Li[220](9)

2. $\int_0^\pi \dfrac{(a^2 - 1 - \sin^2 x) \cos x}{(a^2 - \cos^2 x)^2}\, x^2\, dx = \dfrac{\pi}{a} \ln \dfrac{1 - a}{1 + a}$ $\quad [a > 0,\ a \neq 1]$.

(cf. 3.812 5.). $\qquad$ BH[220](12)

3. $\int_0^\pi \dfrac{a \cos 2x - \sin^2 x}{(a + \sin^2 x)^2}\, x^2\, dx = -2\pi \ln [2(-a + \sqrt{a(a+1)}\,)]$ $\quad [a > 0]$.

$\qquad$ Li[220](10)

4. $\int_0^\pi \dfrac{a \cos 2x + \sin^2 x}{(a - \sin^2 x)^2}\, x^2\, dx = \pi \ln (4a)$ $\quad [a > 1]$ (cf. 3.812 6.). $\qquad$ Li[220](11)

5. $\int_0^{\pi/2} \dfrac{(\cos^2 t + \sin^2 x) \cos x}{(\cos^2 t - \sin^2 x)^2} \cdot x^2\, dx = -\dfrac{\pi^2}{4 \sin^2 t} + \dfrac{4}{\sin t} \sum_{k=0}^{\infty} \dfrac{\sin [(2k+1)t]}{(2k+1)^2}$

(cf. 3.812 7.). $\qquad$ BH[208](14)

3.817

1. $\int_0^\infty \dfrac{\sin x}{(a^2 \cos^2 x + b^2 \sin^2 x)^2} \cdot \dfrac{dx}{x} = \dfrac{\pi}{4} \cdot \dfrac{a^2 + b^2}{a^3 b^3}$ $\quad [ab > 0]$. $\qquad$ BH[181](12)

2. $\int_0^\infty \dfrac{\sin x \cos x}{(a^2 \cos^2 x + b^2 \sin^2 x)^2} \cdot \dfrac{dx}{x} = \dfrac{\pi}{4 a^3 b}$ $\quad [ab > 0]$. $\qquad$ BH[182](8)

3. $\int_0^\infty \dfrac{\sin^3 x}{(a^2 \cos^2 x + b^2 \sin^2 x)^2} \cdot \dfrac{dx}{x} = \dfrac{\pi}{4 a b^3}$ $\quad [ab > 0]$. $\qquad$ BH[181](15)

4. $\int_0^\infty \dfrac{\sin x \cos^2 x}{(a^2 \cos^2 x + b^2 \sin^2 x)^2} \cdot \dfrac{dx}{x} = \dfrac{\pi}{4 a^3 b}$ $\quad [ab > 0]$. $\qquad$ BH[182](9)

5. $\int_0^\infty \dfrac{\operatorname{tg} x}{(a^2 \cos^2 x + b^2 \sin^2 x)^2} \cdot \dfrac{dx}{x} = \dfrac{\pi}{4} \cdot \dfrac{a^2 + b^2}{a^3 b^3}$ $\quad [ab > 0]$. $\qquad$ BH[181](13)

6. $\int_0^\infty \dfrac{\operatorname{tg} x}{(a^2 \cos^2 2x + b^2 \sin^2 2x)^2} \cdot \dfrac{dx}{x} = \dfrac{\pi}{4} \dfrac{a^2 + b^2}{a^3 b^3}$ $\quad [ab > 0]$. $\qquad$ BH[181](14)

7. $\int_0^\infty \dfrac{\sin^2 x\, \operatorname{tg} x}{(a^2 \cos^2 x + b^2 \sin^2 x)^2} \cdot \dfrac{dx}{x} = \dfrac{\pi}{4 a b^3}$ $\quad [ab > 0]$. $\qquad$ BH[182](11)

8. $\int_0^\infty \dfrac{\operatorname{tg} x \cos^2 2x}{(a^2 \cos^2 2x + b^2 \sin^2 2x)^2} \cdot \dfrac{dx}{x} = \dfrac{\pi}{4 a^3 b}$ $\quad [ab > 0]$. $\qquad$ BH[182](10)

1. $\int_0^\infty \frac{\sin x}{(a^2 \cos^2 x + b^2 \sin^2 x)^3} \cdot \frac{dx}{x} = \frac{\pi}{16} \cdot \frac{3a^4 + 2a^2b^2 + 3b^4}{a^5 b^5}$ $\quad [ab > 0].$ BH[181](16)   3.818

2. $\int_0^\infty \frac{\sin x \cos x}{(a^2 \cos^2 x + b^2 \sin^2 x)^3} \cdot \frac{dx}{x} = \frac{\pi}{16} \cdot \frac{a^2 + 3b^2}{a^5 b^3}$ $\quad [ab > 0].$ BH[182](13)

3. $\int_0^\infty \frac{\sin x \cos^2 x}{(a^2 \cos^2 x + b^2 \sin^2 x)^3} \cdot \frac{dx}{x} = \frac{\pi}{16} \cdot \frac{a^2 + 3b^2}{a^5 b^3}$ $\quad [ab > 0].$ BH[182](14)

4. $\int_0^\infty \frac{\sin^3 x}{(a^2 \cos^2 x + b^2 \sin^2 x)^3} \cdot \frac{dx}{x} = \frac{\pi}{16} \cdot \frac{3a^2 + b^2}{a^3 b^5}$ $\quad [ab > 0].$ Li[181](19)

5. $\int_0^\infty \frac{\sin^3 x \cos x}{(a^2 \cos^2 2x + b^2 \sin^2 2x)^3} \cdot \frac{dx}{x} = \frac{\pi}{64} \cdot \frac{3a^2 + b^2}{a^3 b^5}$ $\quad [ab > 0].$ BH[182](17)

6. $\int_0^\infty \frac{\operatorname{tg} x}{(a^2 \cos^2 x + b^2 \sin^2 x)^3} \cdot \frac{dx}{x} = \frac{\pi}{16} \cdot \frac{3a^4 + 2a^2b^2 + 3b^4}{a^5 b^5}$ $\quad [ab > 0].$ BH[181](17)

7. $\int_0^\infty \frac{\sin^2 x \operatorname{tg} x}{(a^2 \cos^2 x + b^2 \sin^2 x)^3} \cdot \frac{dx}{x} = \frac{\pi}{16} \cdot \frac{3a^2 + b^2}{a^3 b^5}$ $\quad [ab > 0].$ BH[182](16)

8. $\int_0^\infty \frac{\operatorname{tg} x}{(a^2 \cos^2 2x + b^2 \sin^2 2x)^3} \cdot \frac{dx}{x} = \frac{\pi}{16} \cdot \frac{3a^4 + 2a^2b^2 + 3b^4}{a^5 b^5}$ $\quad [ab > 0].$ BH[181](18)

9. $\int_0^\infty \frac{\operatorname{tg} x \cos^2 2x}{(a^2 \cos^2 2x + b^2 \sin^2 2x)^3} \cdot \frac{dx}{x} = \frac{\pi}{16} \cdot \frac{a^2 + 3b^2}{a^5 b^3}$ $\quad [ab > 0].$ BH[182](15)

1. $\int_0^\infty \frac{\sin x}{(a^2 \cos^2 x + b^2 \sin^2 x)^4} \cdot \frac{dx}{x} = \frac{\pi}{32} \cdot \frac{5a^6 + 3a^4 b^2 + 3a^2 b^4 + 5b^6}{a^7 b^7}$ $\quad [ab > 0].$ BH[181](20)   3.819

2. $\int_0^\infty \frac{\sin x \cos x}{(a^2 \cos^2 x + b^2 \sin^2 x)^4} \cdot \frac{dx}{x} = \frac{\pi}{32} \cdot \frac{a^4 + 2a^2 b^2 + 5b^4}{a^7 b^5}$ $\quad [ab > 0].$ BH[182](18)

3. $\int_0^\infty \frac{\sin x \cos^2 x}{(a^2 \cos^2 x + b^2 \sin^2 x)^4} \cdot \frac{dx}{x} = \frac{\pi}{32} \cdot \frac{a^4 + 2a^2 b^2 + 5b^4}{a^7 b^5}$ $\quad [ab > 0].$ BH[182](19)

4. $\int_0^\infty \frac{\sin^3 x}{(a^2 \cos^2 x + b^2 \sin^2 x)^4} \cdot \frac{dx}{x} = \frac{\pi}{32} \cdot \frac{5a^4 + a^2 b^2 + b^4}{a^5 b^7}$ $\quad [ab > 0].$ BH[181](23)

5. $\int_0^\infty \frac{\sin^3 x \cos x}{(a^2 \cos^2 x + b^2 \sin^2 x)^4} \cdot \frac{dx}{x} = \frac{\pi}{32} \cdot \frac{a^2 + b^2}{a^5 b^5}$ $\quad [ab > 0].$ BH[182](26)

6. $\int_0^\infty \frac{\sin x \cos^3 x}{(a^2 \cos^2 x + b^2 \sin^2 x)^4} \cdot \frac{dx}{x} = \frac{\pi}{32} \cdot \frac{a^2 + 5b^2}{a^7 b^3}$ $\quad [ab > 0].$ BH[182](23)

7. $\int_0^\infty \frac{\sin^3 x \cos^2 x}{(a^2 \cos^2 x + b^2 \sin^2 x)^4} \cdot \frac{dx}{x} = \frac{\pi}{32} \cdot \frac{a^2 + b^2}{a^5 b^5}$ $\quad [ab > 0].$ BH[182](27)

3.819

8. $\int_0^\infty \frac{\sin x \cos^4 x}{(a^2 \cos^2 x + b^2 \sin^2 x)^4} \cdot \frac{dx}{x} = \frac{\pi}{32} \cdot \frac{a^2 + 5b^2}{a^7 b^3} \quad [ab > 0].$ BH[182](24)

9. $\int_0^\infty \frac{\sin^5 x}{(a^2 \cos^2 x + b^2 \sin^2 x)^4} \cdot \frac{dx}{x} = \frac{\pi}{32} \cdot \frac{5a^2 + b^2}{a^3 b^7} \quad [ab > 0].$ BH[181](24)

10. $\int_0^\infty \frac{\sin^3 x \cos x}{(a^2 \cos^2 2x + b^2 \sin^2 2x)^4} \cdot \frac{dx}{x} = \frac{\pi}{128} \cdot \frac{5a^4 + 2a^2 b^2 + b^4}{a^5 b^7} \quad [ab > 0].$ BH[182](22)

11. $\int_0^\infty \frac{\sin^5 x \cos^3 x}{(a^2 \cos^2 2x + b^2 \sin^2 2x)^4} \cdot \frac{dx}{x} = \frac{\pi}{512} \cdot \frac{5a^2 + b^2}{a^3 b^7} \quad [ab > 0].$ BH[182](30)

12. $\int_0^\infty \frac{\sin^2 x \, \text{tg} \, x}{(a^2 \cos^2 x + b^2 \sin^2 x)^4} \cdot \frac{dx}{x} = \frac{\pi}{32} \cdot \frac{5a^4 + 2a^2 b^2 + b^4}{a^5 b^7} \quad [ab > 0].$ BH[182](21)

13. $\int_0^\infty \frac{\sin^4 x \, \text{tg} \, x}{(a^2 \cos^2 x + b^2 \sin^2 x)^4} \cdot \frac{dx}{x} = \frac{\pi}{32} \cdot \frac{5a^2 + b^2}{a^3 b^7} \quad [ab > 0].$ BH[182](29)

14. $\int_0^\infty \frac{\cos^2 2x \, \text{tg} \, x}{(a^2 \cos^2 2x + b^2 \sin^2 2x)^4} \cdot \frac{dx}{x} = \frac{\pi}{32} \cdot \frac{a^4 + 2a^2 b^2 + 5b^4}{a^7 b^5} \quad [ab > 0].$ BH[182](29)

15. $\int_0^\infty \frac{\sin^2 4x \, \text{tg} \, x}{(a^2 \cos^2 2x + b^2 \sin^2 2x)^4} \cdot \frac{dx}{x} = \frac{\pi}{8} \cdot \frac{a^2 + b^2}{a^5 b^5} \quad [ab > 0].$ BH[182](28)

16. $\int_0^\infty \frac{\cos^4 2x \, \text{tg} \, x}{(a^2 \cos^2 2x + b^2 \sin^2 2x)^4} \cdot \frac{dx}{x} = \frac{\pi}{32} \cdot \frac{a^2 + 5b^2}{a^7 b^3} \quad [ab > 0].$ BH[182](25)

3.82–3.83

Potenzen trigonometrischer Funktionen und Potenzen
Powers of trigonometric functions combined with other powers

3.821

1. $\int_0^\pi x \sin^p x \, dx = \frac{\pi^2}{2^{p+1}} \cdot \frac{\Gamma(p+1)}{\left[\Gamma\left(\frac{p}{2}+1\right)\right]^2} \quad [p > -1]$ BH[218](7), Lo V 121(71)

2. $\int_0^{r\pi} x \sin^n x \, dx = \frac{\pi^2}{2} \cdot \frac{(2m-1)!!}{(2m)!!} r^2 \quad [n = 2m];$

$= (-1)^{r+1} \pi \frac{(2m)!!}{(2m+1)!!} r \quad [n = 2m+1],$

$\left[\begin{array}{l} r \text{ eine natürliche Zahl} \\ \text{a positive integer} \end{array}\right]$ GH[333](8c)

3. $\int_0^{\frac{\pi}{2}} x \cos^n x \, dx = -\sum_{k=0}^{m-1} \frac{(n-2k+1)(n-2k+3) \ldots (n-1)}{(n-2k)(n-2k+2) \ldots n} \cdot \frac{1}{n-2k} +$

$+ \begin{cases} \frac{\pi}{2} \cdot \frac{(2m-2)!!}{(2m-1)!!} & [n = 2m-1]; \\ \frac{\pi^2}{8} \cdot \frac{(2m-1)!!}{(2m)!!} & [n = 2m]. \end{cases}$ GH[333](9b)

4. $$\int_0^\pi x \cos^{2m} x \, dx = \frac{\pi^2}{2} \cdot \frac{(2m-1)!!}{(2m)!!}.$$ BH[218](10)   3.821

5. $$\int_{r\pi}^{s\pi} x \cos^{2m} x \, dx = \frac{\pi^2}{2}(s^2 - r^2) \cdot \frac{(2m-1)!!}{(2m)!!}.$$ BH[226](3)

6. $$\int_0^\infty \frac{\sin^p x}{x} dx = \frac{\sqrt{\pi}}{2} \cdot \frac{\Gamma\left(\frac{p}{2}\right)}{\Gamma\left(\frac{p+1}{2}\right)} = 2^{p-2} B\left(\frac{p}{2}, \frac{p}{2}\right); \text{ Lo V 278, F II 814}$$

[p ist ein rationaler Bruch mit ungeradzahligem Zähler und ungeradzahligem Nenner].

[p is any rational fraction with odd numerator and denominator].

7. $$\int_0^\infty \frac{\sin^{2n+1} x}{x} dx = \frac{(2n-1)!!}{(2n)!!} \cdot \frac{\pi}{2}.$$ BH[151](4)

8. $$\int_0^\infty \frac{\sin^{2n} x}{x} dx = \infty.$$ BH[151](3)

9. $$\int_0^\infty \frac{\sin^2 ax}{x^2} dx = \frac{a\pi}{2} \quad [a > 0].$$ Lo V 307, 312, F II 641

10. $$\int_0^\infty \frac{\sin^{2m} ax}{x^2} dx = \frac{(2m-3)!!}{(2m-2)!!} \cdot \frac{a\pi}{2} \quad [a > 0].$$ GH[333](14b)

11. $$\int_0^\infty \frac{\sin^{2m+1} ax}{x^3} dx = \frac{(2m-3)!!}{(2m)!!}(2m+1)\frac{a^2\pi}{4} \quad [a > 0].$$ GH[333](14d)

12. $$\int_0^\infty \frac{\sin^p x}{x^m} dx = \frac{p}{m-1} \int_0^\infty \frac{\sin^{p-1} x}{x^{m-1}} \cos x \, dx \quad [p > m-1 > 0];$$

$$= \frac{p(p-1)}{(m-1)(m-2)} \int_0^\infty \frac{\sin^{p-2} x}{x^{m-2}} dx - \frac{p^2}{(m-1)(m-2)} \int_0^\infty \frac{\sin^p x}{x^{m-2}} dx$$
$$[p > m-1 > 1]. \quad \text{GH[333](17)}$$

13. $$\int_0^\infty \frac{\sin^{2n} px}{\sqrt{x}} dx = \infty.$$ BH[177](5)

14. $$\int_0^\infty \sin^{2n+1} px \frac{dx}{\sqrt{x}} = \frac{1}{2^{2n}} \sqrt{\frac{\pi}{2p}} \sum_{k=0}^n (-1)^k \binom{2n+1}{n+k+1} \frac{1}{\sqrt{2k+1}}.$$ BH[177](7)

1. $$\int_0^{\pi/2} x^p \cos^m x \, dx = -\frac{p(p-1)}{m^2} \int_0^{\pi/2} x^{p-2} \cos^m x \, dx + \frac{m-1}{m} \int_0^{\pi/2} x^p \cos^{m-2} x \, dx$$ 3.822
$$[m > 1, p > 1]. \quad \text{GH[333](9a)}$$

2. $$\int_0^\infty x^{-\frac{1}{2}} \cos^{2n+1}(px) \, dx = \frac{1}{2^{2n}} \sqrt{\frac{\pi}{2p}} \sum_{k=0}^n \binom{2n+1}{n+k+1} \frac{1}{\sqrt{2k+1}}.$$ BH[177](8)

3.823 $\int_0^\infty x^{\mu-1} \sin^2 ax \, dx = -\dfrac{\Gamma(\mu) \cos \dfrac{\mu\pi}{2}}{2^{\mu+1} a^\mu}$ $[a > 0, -2 < \text{Re}\,\mu < 0]$.

Ba IV 319(15), GH[333](19c)+

3.824 1. $\int_0^\infty \dfrac{\sin^2 ax}{x^2 + \beta^2} dx = \dfrac{\pi}{4\beta}(1 - e^{-2a\beta})$ $[a > 0,\ \text{Re}\,\beta > 0]$. BH[160](10)

2. $\int_0^\infty \dfrac{\cos^2 ax}{x^2 + \beta^2} dx = \dfrac{\pi}{4\beta}(1 + e^{-2a\beta})$ $[a > 0,\ \text{Re}\,\beta > 0]$. BH[160](11)

3. $\int_0^\infty \sin^{2m} x \, \dfrac{dx}{a^2 + x^2} = \dfrac{(-1)^m}{2^{2m+1}} \cdot \dfrac{\pi}{a} \Big\{ 2^{2m} \text{sh}^{2m} a -$

$- 2 \sum_{k=0}^{m} (-1)^k \binom{2m}{k} \text{sh}[2(m-k)a] \Big\}$ $[a > 0]$. BH[160](12)

4. $\int_0^\infty \sin^{2m+1} x \, \dfrac{dx}{a^2 + x^2} =$

$= \dfrac{(-1)^{m-1}}{2^{2m+2} a} \Big\{ e^{(2m+1)a} \sum_{k=0}^{2m+1} (-1)^k \binom{2m+1}{k} e^{-2ka} \, \text{Ei}\,[(2k - 2m - 1)a] +$

$+ e^{-(2m+1)a} \sum_{k=1}^{2m+1} (-1)^{k-1} \binom{2m+1}{k} e^{2ka} \, \text{Ei}\,[(2m + 1 - 2k)a] \Big\}$

$[a > 0]$. BH[160](14)

5. $\int_0^\infty \sin^{2m+1} x \, \dfrac{x \, dx}{a^2 + x^2} = \dfrac{(-1)^{m-1}}{2^{2m+2}} e^{-(2m+1)a} \Big\{ (1 - e^{2(2m+1)a})(1 - e^{-2a})^{2m+1} -$

$- 2 \sum_{k=0}^{m} (-1)^k \binom{2m+1}{k} e^{2ka} \Big\}$ $[a > 0]$. BH[160](15)

6. $\int_0^\infty \cos^{2m} x \, \dfrac{dx}{a^2 + x^2} = \dfrac{\pi}{2^{2m+1} a} \binom{2m}{m} + \dfrac{\pi}{2^{2m}} \sum_{k=1}^{m} \binom{2m}{m+k} e^{-2ka}$ $[a > 0]$.

BH[160](16)

7. $\int_0^\infty \cos^{2m+1} x \, \dfrac{dx}{a^2 + x^2} = \dfrac{\pi}{2^{2m+1} a} \sum_{k=1}^{m} \binom{2m+1}{m+k+1} e^{-(2k+1)a}$ $[a > 0]$ BH[160](17)

8. $\int_0^\infty \cos^{2m+1} x \, \dfrac{x \, dx}{a^2 + x^2} = -\dfrac{e^{-(2m+1)a}}{2^{2m+2}} \sum_{k=0}^{2m+1} \binom{2m+1}{k} e^{2ka} \text{Ei}\,[(2m - 2k + 1)a] -$

$- \dfrac{e^{(2m+1)a}}{2^{2m+2}} \sum_{k=0}^{2m+1} \binom{2m+1}{k} e^{-2ka} \text{Ei}\,[(2k - 2m - 1)a]$. BH[160](18)

9. $\int_0^\infty \dfrac{\cos^2 ax}{b^2 - x^2} dx = \dfrac{\pi}{4b} \sin 2ab$ $[a > 0,\ b > 0]$. BH[161](10)

10. $\int_0^\infty \dfrac{\sin^2 ax \cos^2 bx}{\beta^2 + x^2} dx = \dfrac{\pi}{8\beta}\left[1 - \dfrac{1}{2}e^{-2(a+b)\beta} + e^{-2b\beta} - \dfrac{1}{2}e^{2(b-a)\beta} - e^{-2a\beta}\right]$ 3.824

$[a > b];$

$= \dfrac{\pi}{16\beta}[1 - e^{-4a\beta}] \quad [a = b];$

$= \dfrac{\pi}{8\beta}\left[1 - \dfrac{1}{2}e^{-2(a+b)\beta} + e^{-2b\beta} - \dfrac{1}{2}e^{2(a-b)\beta} - e^{-2a\beta}\right] \quad [a < b];$

$[a > 0, \ b > 0]$ (cf. 3.824 1., 3.). BH[162](6)

11. $\int_0^\infty \dfrac{x \sin 2ax \cos^2 bx}{\beta^2 + x^2} dx = \dfrac{\pi}{8}[2e^{-2a\beta} + e^{-2(a+b)\beta} + e^{2(b-a)\beta}] \quad [a > b];$

$= \dfrac{\pi}{8}[e^{-4a\beta} + 2e^{-2a\beta}] \quad [a = b];$

$= \dfrac{\pi}{8}[2e^{-2a\beta} + e^{-2(a+b)\beta} - e^{2(a-b)\beta}] \quad [a < b].$ Li[162](5)

1. $\int_0^\infty \dfrac{\sin^2 ax \, dx}{(b^2 + x^2)(c^2 + x^2)} = \dfrac{\pi(b - c + ce^{-2ab} - be^{-2ac})}{4bc(b^2 - c^2)} \quad [a > 0, b > 0, c > 0].$ 3.825

BH[174](15)

2. $\int_0^\infty \dfrac{\cos^2 ax \, dx}{(b^2 + x^2)(c^2 + x^2)} = \dfrac{\pi(b - c + be^{-2ac} - ce^{-2ab})}{4bc(b^2 - c^2)} \quad [a > 0, b > 0, c > 0].$

BH[175](14)

3. $\int_0^\infty \dfrac{\sin^2 ax \, dx}{(b^2 - x^2)(c^2 - x^2)} = \dfrac{\pi(c \sin 2ab - b \sin 2ac)}{4bc(b^2 - c^2)} \quad [a > 0, b > 0, c > 0].$ Li[174](16)

4. $\int_0^\infty \dfrac{\cos^2 ax \, dx}{(b^2 - x^2)(c^2 - x^2)} = \dfrac{\pi(b \sin 2ac - c \sin 2ab)}{4bc(b^2 - c^2)} \quad [a > 0, b > 0, c > 0].$

Li[175](15)

1. $\int_0^\infty \dfrac{\sin^2 ax \, dx}{x^2(b^2 + x^2)} = \dfrac{\pi}{4b^2}\left[2a - \dfrac{1}{b}(1 - e^{-2ab})\right] \quad [a > 0, b > 0].$ BH[172](13) 3.826

2. $\int_0^\infty \dfrac{\sin^2 ax \, dx}{x^2(b^2 - x^2)} = \dfrac{\pi}{4b^2}\left(2a - \dfrac{1}{b}\sin 2ab\right) \quad [a > 0, b > 0].$ BH[172](14)

1. $\int_0^\infty \dfrac{\sin^3 ax}{x^\nu} dx = \dfrac{3 - 3^{\nu-1}}{4} a^{\nu-1} \cos\dfrac{\nu\pi}{2} \Gamma(1 - \nu) \quad [a > 0, \ 0 < \operatorname{Re}\nu < 2].$ 3.827

GH[333](19f)

2. $\int_0^\infty \dfrac{\sin^3 ax}{x} dx = \dfrac{\pi}{4}\operatorname{sign} a.$ Lo V 277

3.827

3. $\int_0^\infty \frac{\sin^3 ax}{x^2} dx = \frac{3}{4} a \ln 3.$ BH[156](2)

4. $\int_0^\infty \frac{\sin^3 ax}{x^3} dx = \frac{3}{8} a^2 \pi \operatorname{sign} a.$ BH[156](7)+, Lo V 312

5. $\int_0^\infty \frac{\sin^4 ax}{x^2} dx = \frac{a\pi}{4}$ $[a > 0].$ BH[156](3)

6. $\int_0^\infty \frac{\sin^4 ax}{x^3} dx = a^2 \ln 2.$ BH[156](8)

7. $\int_0^\infty \frac{\sin^4 ax}{x^4} dx = \frac{a^3 \pi}{3}$ $[a > 0].$ BH[156](11), Lo V 312

8. $\int_0^\infty \frac{\sin^5 ax}{x^2} dx = \frac{5}{16} a(3\ln 3 - \ln 5).$ BH[156](4)

9. $\int_0^\infty \frac{\sin^5 ax}{x^3} dx = \frac{5}{32} a^2 \pi$ $[a > 0].$ BH[156](9)

10. $\int_0^\infty \frac{\sin^5 ax}{x^4} dx = \frac{5}{96} a^3 (25\ln 5 - 27\ln 3)$ $[a > 0].$ BH[156](12)

11. $\int_0^\infty \frac{\sin^5 ax}{x^5} dx = \frac{115}{384} a^4 \pi$ $[a > 0].$ BH[156](13), Lo V 312

12. $\int_0^\infty \frac{\sin^6 ax}{x^2} dx = \frac{3}{16} a\pi$ $[a > 0].$ BH[156](5)

13. $\int_0^\infty \frac{\sin^6 ax}{x^3} dx = \frac{3}{16} a^2 (8\ln 2 - 3\ln 3).$ BH[156](10)

14. $\int_0^\infty \frac{\sin^6 ax}{x^5} dx = \frac{1}{16} a^4 (27\ln 3 - 32\ln 2).$ BH[156](14)

15. $\int_0^\infty \frac{\sin^6 ax}{x^6} dx = \frac{11}{40} a^5 \pi$ $[a > 0].$ Lo V 312

1. $\int_0^\infty \dfrac{\sin px \sin qx}{x}\,dx = \ln\sqrt{\dfrac{p+q}{|p-q|}} \quad [p \ne q].$ \hfill F II 657 \quad 3.828

2. $\int_0^\infty \sin qx \sin px \dfrac{dx}{x^2} = \dfrac{1}{2} p\pi \quad [p \leqslant q];$

$\qquad\qquad\qquad = \dfrac{1}{2} q\pi \quad [p \geqslant q].$ \hfill BH[157](1)

3. $\int_0^\infty \dfrac{\sin^2 ax \sin bx}{x}\,dx = \dfrac{\pi}{4} \quad [0 < b < 2a];$

$\qquad\qquad\qquad = \dfrac{\pi}{8} \quad [b = 2a];$

$\qquad\qquad\qquad = 0 \quad [b > 2a].$ \hfill BH[151](10)

4. $\int_0^\infty \dfrac{\sin^2 ax \cos bx}{x}\,dx = \dfrac{1}{4} \ln \dfrac{4a^2 - b^2}{b^2}.$ \hfill BH[151](12)

5. $\int_0^\infty \dfrac{\sin^2 ax \cos 2bx}{x^2}\,dx = \dfrac{\pi}{2}(a - b) \quad [b < a];$

$\qquad\qquad\qquad = 0 \quad [b > a].$ \hfill F III 528+, BH[157](5)+

6. $\int_0^\infty \dfrac{\sin 2ax \cos^2 bx}{x}\,dx = \dfrac{\pi}{2} \quad [a > b];$

$\qquad\qquad\qquad = \dfrac{3}{8}\pi \quad [a = b];$

$\qquad\qquad\qquad = \dfrac{\pi}{4} \quad [a < b].$ \hfill BH[151](9)

7. $\int_0^\infty \dfrac{\sin^2 ax \sin bx \sin cx}{x^2}\,dx = \dfrac{\pi}{16}(|b - 2a - c| - |2a - b - c| + 2c)$

$\qquad\qquad\qquad [a > 0,\ b > 0,\ c > 0].$ \hfill BH[157](9)+, Ba IV 79(15)

8. $\int_0^\infty \dfrac{\sin^2 ax \sin bx \sin cx}{x}\,dx = \dfrac{1}{4} \ln \dfrac{b+c}{b-c} +$

$+ \dfrac{1}{8} \ln \dfrac{(2a - b + c)(2a + b - c)}{(2a + b + c)(2a - b - c)} \quad [a > 0,\ b > 0,\ c > 0,\ b \ne c].$ \hfill Li[152](2)

9. $\int_0^\infty \dfrac{\sin^2 ax \sin^2 xb}{x^2}\,dx = \dfrac{\pi}{4} a \quad [0 \leqslant a \leqslant b];$

$\qquad\qquad\qquad = \dfrac{\pi}{4} b \quad [0 \leqslant b \leqslant a].$ \hfill BH[157](3)

3.828

10. $\int_0^\infty \frac{\sin^2 ax \sin^2 bx}{x^4} dx = \frac{1}{6} a^2 \pi (3b - a) \quad [0 \leqslant a \leqslant b];$

$= \frac{1}{6} b^2 \pi (3a - b) \quad [0 \leqslant b \leqslant a].$ BH[157](27)

11. $\int_0^\infty \frac{\sin^2 ax \cos^2 bx}{x^2} dx = \frac{2a - b}{4} \pi \quad [a \geqslant b > 0];$

$= \frac{a\pi}{4} \quad [0 < a \leqslant b].$ BH[157](6)

12. $\int_0^\infty \frac{\sin^3 ax \sin 3bx}{x^4} dx = \frac{a^3 \pi}{2} \quad [b > a];$

$= \frac{\pi}{16} [8a^3 - 9(a - b)^3] \quad [a \leqslant 3b \leqslant 3a];$ BH[157](28)

$= \frac{9b\pi}{8} (a^2 - b^2) \quad [3b \leqslant a].$ Li[157](28)

13. $\int_0^\infty \frac{\sin^3 ax \cos bx}{x} dx = 0 \quad [b > 3a];$

$= -\frac{\pi}{16} \quad [b = 3a];$

$= -\frac{\pi}{8} \quad [3a > b > a];$

$= \frac{\pi}{16} \quad [b = a];$

$= \frac{\pi}{4} \quad [a > b] \ [a > 0, \ b > 0].$ BH [151](15)

14. $\int_0^\infty \frac{\sin^3 ax \cos 3bx}{x^2} dx = \frac{3}{8} \{(a + b) \ln [3(a + b)] + (b - a) \ln [3(b - a)] -$

$- \frac{1}{3} (a + 3b) \ln (a + 3b) - \frac{1}{3} (3b - a) \ln (3b - a) \}$

$[a > 0, b > 0].$ BH[157](7)+, Ba IV 19(9)

15. $\int_0^\infty \frac{\sin^3 ax \cos bx}{x^3} dx = \frac{\pi}{8} (3a^2 - b^2) \quad [b < a];$

$= \frac{\pi b^2}{4} \quad [a = b];$

$= \frac{\pi}{16} (3a - b^2) \quad [a < b < 3a];$

$= 0 \quad [3a < b] \quad [a > 0, b > 0].$ BH[157](19), Ba IV 19(10)

16. $\int_0^\infty \dfrac{\sin^3 ax \sin bx}{x^4} dx = \dfrac{b\pi}{24}(9a^2 - b^2) \quad [0 < b \leqslant a];$ 3.828

$\qquad = \dfrac{\pi}{48}[24a^3 - (3a - b)^3] \quad [0 < a \leqslant b \leqslant 3a];$

$\qquad = \dfrac{\pi a^3}{2} \quad [0 < 3a \leqslant b].$ Ba IV 79(16)

17. $\int_0^\infty \dfrac{\sin^3 ax \sin^2 bx}{x} dx = \dfrac{\pi}{8} \quad [2b > 3a];$

$\qquad = \dfrac{5\pi}{32} \quad [2b = 3a];$

$\qquad = \dfrac{3\pi}{16} \quad [3a > 2b > a];$

$\qquad = \dfrac{3\pi}{32} \quad [2b = a];$

$\qquad = 0 \quad [a > 2b] \quad [a > 0, b > 0].$ BH[151](14)

18. $\int_0^\infty \dfrac{\sin^2 ax \cos^3 bx}{x} dx = \dfrac{1}{16} \ln \dfrac{(2a+b)^3(b-2a)^3(2a+3b)(3b-2a)}{9b^8}$

$\qquad \left[b > 2a > 0 \;\; \substack{\text{oder}\\ \text{or}} \;\; 2a > 3b > 0\right];$

$\qquad = \dfrac{1}{16} \ln \dfrac{(2a+b)^3(2a-b)^3(2a+3b)(3b-2a)}{9b^8}$

$\qquad [3b > 2a > b].$ BH[151](13)

19. $\int_0^\infty \dfrac{\sin^2 ax \sin^2 bx \sin 2cx}{x} dx =$

$= \dfrac{\pi}{16}[1 + \text{sign}(c-a+b) + \text{sign}(c+a-b) - 2\,\text{sign}(c-a) - 2\,\text{sign}(c-b)]$

$\qquad [a > 0, b > 0, c > 0].$ Ba IV 80(17)

20. $\int_0^\infty \dfrac{\sin^2 ax \sin^2 bx \sin 2cx \, dx}{x^2} = \dfrac{a-b-c}{16} \ln 4(a-b-c)^2 -$

$\quad - \dfrac{a+b+c}{16} \ln 4(a+b+c)^2 + \dfrac{a+b-c}{16} \ln 4(a+b-c)^2 -$

$\quad - \dfrac{a-b+c}{16} \ln 4(a-b+c)^2 + \dfrac{a+c}{8} \ln 4(a+c)^2 - \dfrac{a-c}{8} \ln 4(a-c)^2 +$

$\quad + \dfrac{b+c}{8} \ln 4(b+c)^2 - \dfrac{b-c}{8} \ln 4(b-c)^2 - \dfrac{1}{2} c \ln 2c$

$\qquad [a > 0, b > 0, c > 0].$ BH[157](10)

3.828 21.
$$\int_0^\infty \frac{\sin^2 ax \, \sin^3 bx}{x^3} \, dx = \frac{3b^2 \pi}{16} \quad [2a > 3b];$$

$$= \frac{a^2 \pi}{12} \quad [2a = 3b];$$

$$= \frac{6b^2 - (3b - 2a)^2}{32} \pi \quad [3b > 2a > b];$$

$$= \frac{a^2 \pi}{4} \quad [b > 2a]; \quad [a > 0, b > 0].$$

BH[157](18)

3.829 1.
$$\int_0^\infty \frac{x^n - \sin^n x}{x^{n+2}} \, dx = \frac{\pi}{2^n (n+1)!} \sum_{k=0}^{E\left(\frac{n-1}{2}\right)} (-1)^k \binom{n}{k} (n - 2k)^{n+1}.$$

GH[333](63)

2.
$$\int_0^\infty (1 - \cos^{2m-1} x) \frac{dx}{x^2} = \int_0^\infty (1 - \cos^{2m} x) \frac{dx}{x^2} = \frac{m\pi}{2^{2m}} \binom{2m}{m}.$$

BH[158](7), BH[158](8)

3.831 1.
$$\int_0^\infty \frac{\sin^{2n} ax - \sin^{2n} bx}{x} \, dx = \frac{(2n-1)!!}{(2n)!!} \ln \frac{b}{a} \quad [a > 0, b > 0].$$
F II 660

2.
$$\int_0^\infty \frac{\cos^{2n} ax - \cos^{2n} bx}{x} \, dx = \left[1 - \frac{(2n-1)!!}{(2n)!!} \right] \ln \frac{b}{a} \quad [a > 0, b > 0].$$
F II 661

3.
$$\int_0^\infty \frac{\cos^{2m+1} ax - \cos^{2m+1} bx}{x} \, dx = \ln \frac{b}{a} \quad [a > 0, b > 0].$$
F II 661

4.
$$\int_0^\infty \frac{\cos^m ax \cos max - \cos^m bx \cos mbx}{x} \, dx = \left(1 - \frac{1}{2^m} \right) \ln \frac{b}{a} \quad [ab > 0].$$

Li[155](8)

3.832 1.
$$\int_0^{\frac{\pi}{2}} x \cos^{p-1} x \sin ax \, dx = \frac{\pi}{2^{p+1}} \Gamma(p) \frac{\Psi\left(\frac{p+a+1}{2}\right) - \Psi\left(\frac{p-a+1}{2}\right)}{\Gamma\left(\frac{p+a+1}{2}\right) \Gamma\left(\frac{p-a+1}{2}\right)}$$
$$[p > 0, \, -(p+1) < a < p+1].$$
BH[205](6)

2.
$$\int_0^\infty \sin^{2m+1} x \sin 2mx \frac{dx}{a^2 + x^2} = \frac{(-1)^m \pi}{2^{2m+1} a} [(1 - e^{-2a})^{2m} - 1] \, \text{sh} \, a \quad [a > 0].$$

BH[162](17)

3.
$$\int_0^\infty \sin^{2m-1} x \sin [(2m-1)x] \frac{dx}{a^2 + x^2} = \frac{(-1)^{m+1} \pi}{2^{2m} a} (1 - e^{-2a})^{2m-1} \quad [a > 0].$$

BH[162](11)

4. $\int_0^\infty \sin^{2m-1} x \sin[(2m+1)x] \frac{dx}{a^2+x^2} = \frac{(-1)^{m-1}\pi}{2^{2m}a} e^{-2a}(1-e^{-2a})^{2m-1}$ $[a>0]$. 3.832

BH[162](12)

5. $\int_0^\infty \sin^{2m+1} x \sin[3(2m+1)x] \frac{dx}{a^2+x^2} = \frac{(-1)^m \pi}{2a} e^{-3(2m+1)a} \text{sh}^{2m+1} a$

$[a>0]$. BH[162](18)

6. $\int_0^\infty \sin^{2m} x \sin[(2m-1)x)] \frac{x\,dx}{a^2+x^2} = \frac{(-1)^m \pi}{2^{2m+1}} e^a[(1-e^{-2a})^{2m} - 1]$ $[a>0]$.

BH[162](13)

7. $\int_0^\infty \sin^{2m} x \sin(2mx) \frac{x\,dx}{a^2+x^2} = \frac{(-1)^m \pi}{2^{2m+1}}[(1-e^{-2a})^{2m} - 1]$ $[a>0]$. BH[162](14)

8. $\int_0^\infty \sin^{2m} x \sin[(2m+2)x] \frac{x\,dx}{a^2+x^2} = \frac{(-1)^m \pi}{2^{2m+1}} e^{-2a}(1-e^{-2a})^{2m}$ $[a>0]$.

BH[162](15)

9. $\int_0^\infty \sin^{2m} x \sin 4mx \frac{x\,dx}{a^2+x^2} = \frac{(-1)^m \pi}{2} e^{-4ma} \text{sh}^{2m} a$ $[a>0]$. BH[162](16)

10. $\int_0^\infty \sin^{2m} x \cos x \frac{dx}{x^2} = \frac{(2m-3)!!}{(2m)!!} \cdot \frac{\pi}{2}$. GH[333](15a)

11. $\int_0^\infty \sin^{2m} x \cos[(2m-1)x] \frac{dx}{a^2+x^2} = \frac{(-1)^m \pi}{2^{2m}a}[(1-e^{-2a})^{2m-1} - 1] \text{sh } a$

$[a>0]$. BH[162](25)

12. $\int_0^\infty \sin^{2m} x \cos(2mx) \frac{dx}{a^2+x^2} = \frac{(-1)^m \pi}{2^{2m+1}a}(1-e^{-2a})^{2m}$ $[a>0]$. BH[162](26)

13. $\int_0^\infty \sin^{2m} x \cos[(2m+2)x] \frac{dx}{a^2+x^2} = \frac{(-1)^m \pi}{2^{2m+1}a} e^{-2a}(1-e^{-2a})^{2m}$ $[a>0]$.

BH[162](27)

14. $\int_0^\infty \sin^{2m} x \cos 4mx \frac{dx}{a^2+x^2} = \frac{(-1)^m \pi}{2a} e^{-4ma} \text{sh}^{2m} a$ $[a>0]$. BH[162](28)

15. $\int_0^\infty \sin^{2m+1} x \cos x \frac{dx}{x} = \frac{(2m-1)!!}{(2m+2)!!} \cdot \frac{\pi}{2}$. GH[333](15)

3.832

16. $\int_0^\infty \sin^{2m+1} x \cos x \, \frac{dx}{x^3} = \frac{(2m-3)!!}{(2m)!!} \cdot \frac{\pi}{2}$. GH[333](15b)

17. $\int_0^\infty \sin^{2m-1} x \cos[(2m-1)x] \frac{x \, dx}{a^2+x^2} = \frac{(-1)^m \pi}{2^{2m}} [(1-e^{-2a})^{2m-1} - 1]$

$[a > 0]$. BH[162](23)

18. $\int_0^\infty \sin^{2m+1} x \cos 2mx \, \frac{x \, dx}{a^2+x^2} = \frac{(-1)^{m-1} \pi}{2^{2m+2}} e^{-a}[(1-e^{-2a})^{2m+1} - 1]$

$[a > 0]$. BH[162](29)

19. $\int_0^\infty \sin^{2m-1} x \cos[(2m+1)x] \frac{x \, dx}{a^2+x^2} = \frac{(-1)^m \pi}{2^{2m}} e^{-2a} (1 - e^{-2a})^{2m-1}$

$[a > 0]$. BH[162](24)

20. $\int_0^\infty \sin^{2m+1} x \cos[2(2m+1)x] \frac{x \, dx}{a^2+x^2} = \frac{(-1)^{m-1} \pi}{2} e^{-2(2m+1)a} \sh^{2m+1} a$

$[a > 0]$. BH[162](30)

21. $\int_0^\infty \cos^m x \sin mx \, \frac{dx}{a^2+x^2} = \frac{1}{2^{m+1}a} \sum_{k=1}^m \binom{m}{k} [e^{-2ka} \text{Ei}(2ka) - e^{2ka} \text{Ei}(-2ka)]$

$[a > 0]$. BH[163](8)

22. $\int_0^\infty \cos^n sx \sin nsx \, \frac{x \, dx}{a^2+x^2} = \frac{\pi}{2^{n+1}} [(1 + e^{-2as})^n - 1]$. BH[163](9)

23. $\int_0^\infty \cos^n sx \sin nsx \, \frac{x \, dx}{a^2-x^2} = \frac{\pi}{2} (2^{-n} - \cos^n as \cos nas)$. BH[166](10)

24. $\int_0^\infty \cos^{m-1} x \sin[(m+1)x] \frac{x \, dx}{a^2+x^2} = \frac{\pi}{2^m} e^{-2a} (1 + e^{-2a})^{m-1} \quad [a > 0]$.

BH[163](6)

25. $\int_0^\infty \cos^m x \sin[(m+1)x] \frac{x \, dx}{a^2+x^2} = \frac{\pi}{2^{m+1}} e^{-a} (1 + e^{-2a})^m \quad [a > 0]$.

BH[163](10)

26. $\int_0^\infty \cos^m x \sin[(m-1)x] \frac{x \, dx}{a^2+x^2} = \frac{\pi}{2^{m+1}} e^a (1 + e^{-2a})^m \quad [a > 0]$. BH[163](7)

27. $\int_0^\infty \cos^m x \sin(3mx) \frac{x \, dx}{a^2+x^2} = \frac{\pi}{2} e^{-3a} \ch^m a \quad [a > 0]$. BH[163](11)

28. $\int_0^\infty \cos^n sx \cos nsx \frac{dx}{a^2+x^2} = \frac{\pi}{2^{n+1}a}(1+e^{-2as})^n.$ BH[163](16) 3.832

29. $\int_0^\infty \cos^n sx \cos nsx \frac{dx}{a^2-x^2} = \frac{\pi}{2a}\cos^n as \sin nas.$ BH[166](11)

30. $\int_0^\infty \cos^{m-1}x \cos[(m+1)x]\frac{dx}{a^2+x^2} = \frac{\pi}{2^m a}e^{-2a}(1+e^{-2a})^{m-1} \quad [a>0].$

BH[163](14)

31. $\int_0^\infty \cos^m x \cos[(m-1)x]\frac{dx}{a^2+x^2} = \frac{\pi}{2^{m+1}a}e^a(1+e^{-2a})^m \quad [a>0].$ BH[163](15)

32. $\int_0^\infty \cos^m x \cos[(m+1)x]\frac{dx}{a^2+x^2} = \frac{\pi}{2^{m+1}a}e^{-a}(1+e^{-2a})^m \quad [a>0].$

BH[163](17)

33. $\int_0^\infty \sin^p x \cos x \frac{dx}{x^q} = \frac{p}{q-1}\int_0^\infty \frac{\sin^{p-1}x}{x^{q-1}}dx - \frac{p+1}{q-1}\int_0^\infty \frac{\sin^{p+1}x}{x^{q-1}}dx \quad [p>q-1>0];$

$$= \frac{p(p-1)}{(q-1)(q-2)}\int_0^\infty \sin^{p-2}x \cos x \frac{dx}{x^{q-2}} -$$

$$- \frac{(p+1)^2}{(q-1)(q-2)}\int_0^\infty \sin^p x \cos x \frac{dx}{x^{q-2}} \quad [p>q-1>1].$$ GH[333](18)

34. $\int_0^\infty \cos^{2m}x \cos 2nx \sin x \frac{dx}{x} = \int_0^\infty \cos^{2m-1}x \cos 2nx \sin x \frac{dx}{x} = \frac{\pi}{2^{2m+1}}\binom{2m}{m+n}.$

BH[152](5, 6)

35. $\int_0^\infty \cos^p ax \sin bx \cos x \frac{dx}{x} = \frac{\pi}{2} \quad [b>ap, p>-1].$ BH[153](12)

36. $\int_0^\infty \cos^p ax \sin pax \cos x \frac{dx}{x} = \frac{\pi}{2^{p+1}}(2^p - 1) \quad [p>-1].$ BH[153](2)

37. $\int_0^\infty \frac{dx}{x^2}\prod_{k=1}^n \cos^{p_k} a_k x \cdot \sin bx \sin x = \frac{\pi}{2} \quad \left[b>\sum_{k=1}^n a_k p_k; \ a_k>0, \ p_k>0\right].$

BH[157](15)

1. $\int_0^\infty \sin^{2m+1}x \cos^{2n}x \frac{dx}{x} = \int_0^\infty \sin^{2m+1}x \cos^{2n-1}x \frac{dx}{x} = \frac{(2m-1)!!(2n-1)!!}{2^{m+n+1}(m+n)!}\pi.$ 3.833

BH[151](24), BH[151](25)

$$= \frac{1}{2}B\left(m+\frac{1}{2}, n+\frac{1}{2}\right).$$ GH[333](24)

3.833

2. $\int_0^\infty \sin^{2m+1} 2x \cos^{2n-1} 2x \cos^2 x \frac{dx}{x} = \frac{\pi}{2} \cdot \frac{(2m-1)!!\,(2n-1)!!}{(2m+2n)!!}.$ Li[152](4)

3.834

1. $\int_0^\infty \frac{\sin^{2m+1} x}{1 - 2a\cos x + a^2} \cdot \frac{dx}{x} = \frac{(-1)^m \pi (1+a)^{4m}}{2^{2m+2} a^{2m+1}} \left\{ \left|\frac{1-a}{1+a}\right|^{2m-1} - \sum_{k=1}^{2m} (-1)^k \binom{m - \frac{1}{2}}{k} \left(\frac{4a}{(1+a)^2}\right)^k \right\}$ $[|a| \neq 1].$ GH[333](62a)

2. $\int_0^\infty \frac{\sin^{2m+1} x \cos^n x}{(1 - 2a\cos x + a^2)^p} \cdot \frac{dx}{x} =$

$= \frac{n!\,\pi}{2^{n+1}(2m+n+1)!\,(1+a)^{2p}} \sum_{k=0}^n \frac{(-1)^k (2m+2n-2k+1)!!\,(2m+2k-1)!!}{k!\,(n-k)!} \times$

$\times F\left(m+n-k+\frac{3}{2},\, p;\, 2m+n+2;\, \frac{4a}{(1+a)^2}\right)$ $[a \neq \pm 1].$ GH[333](62)

3.835

1. $\int_0^\infty \frac{\cos^{2m} x \cos 2mx \sin x}{a^2 \cos^2 x + b^2 \sin^2 x} \cdot \frac{dx}{x} = \frac{\pi}{2} \frac{b^{2m-1}}{a(a+b)^{2m}}$ $[ab > 0].$ BH[182](31)+

2. $\int_0^\infty \frac{\cos^{2m-1} x \cos 2mx \sin x}{a^2 \cos^2 x + b^2 \sin^2 x} \cdot \frac{dx}{x} = \frac{\pi}{2a} \frac{b^{2m-1}}{(a+b)^{2m}}$ $[ab > 0].$ BH[182](32)+

3.836

1. $\int_0^\infty \left(\frac{\sin x}{x}\right)^n \frac{\sin mx}{x} dx = \frac{\pi}{2}$ $[m \geqslant n].$ Li[159](12)

2. $\int_0^\infty \left(\frac{\sin x}{x}\right)^n \cos mx\, dx = \frac{n\pi}{2^n} \sum_{0 < k < \frac{m+n}{2}} \frac{(-1)^k (n+m-2k)^{n-1}}{k!\,(n-k)!}$ $[0 < m < n];$

$= 0$ $[m \geqslant n]\,[n \geqslant 2].$ BH[159](14), Ba IV 20(11)

3. $\int_0^\infty \left(\frac{\sin x}{x}\right)^{n-1} \sin nx \cos x \frac{dx}{x} = \frac{\pi}{2}.$ BH[159](20)

4. $\int_0^\infty \left(\frac{\sin x}{x}\right)^n \frac{\sin(anx)}{x} dx =$

$= \frac{\pi}{2} \left[1 - \frac{1}{2^{n-1} n!} \sum_{0 < k < \frac{n}{2}(1-a)} (-1)^k \binom{n}{k} (n - an - 2k)^n \right]$ $[0 < a \leqslant 1].$ Lo V 341(15)

3.837

1. $\int_0^{\pi/2} \frac{x^2\, dx}{\sin^2 x} = \pi \ln 2.$ BH[206](9)

2. $\int_0^{\pi/4} \frac{x^2\,dx}{\sin^2 x} = -\frac{\pi^2}{16} + \frac{\pi}{4}\ln 2 + G = 0{,}843\,511\,841\,7\ldots$ BH[204](10) 3.837

3. $\int_0^{\pi/4} \frac{x^2\,dx}{\cos^2 x} = \frac{\pi^2}{16} + \frac{\pi}{4}\ln 2 - G.$ GH[333](35a)

4. $\int_0^{\pi/4} \frac{x^{p+1}}{\sin^2 x}\,dx = -\left(\frac{\pi}{4}\right)^{p+1} + (p+1)\left(\frac{\pi}{4}\right)^p \left\{\frac{1}{p} - \frac{1}{2}\sum_{k=1}^{\infty} \frac{1}{4^{2k-1}(p+2k)}\zeta(2k)\right\}$
$[p > 0].$ Li[204](14)

5. $\int_0^{\pi/2} \frac{x^2 \cos x}{\sin^2 x}\,dx = -\frac{\pi^2}{4} + 4G = 1{,}196\,461\,276\,4\ldots$ BH[206](7)

6. $\int_0^{\pi/2} \frac{x^3 \cos x}{\sin^3 x}\,dx = -\frac{\pi^3}{16} + \frac{3}{2}\pi \ln 2.$ BH[206](8)

7. $\int_0^{\infty} \frac{\cos 2nx}{\cos x}\sin^{2n} x\,\frac{dx}{x^m} = 0 \quad \left[n > \frac{m-1}{2},\, m > 0\right].$ BH[180](16)

8. $\int_0^{\infty} \frac{\cos 2nx}{\cos x}\sin^{2n+1} x\,\frac{dx}{x^m} = 0 \quad \left[n > \frac{m-2}{2},\, m > 0\right].$ BH[180](17)

9. $\int_0^1 \frac{x\,dx}{\cos ax \cos[a(1-x)]} = \frac{1}{a}\operatorname{cosec} a \cdot \ln \sec a \quad \left[a < \frac{\pi}{2}\right].$ BH[149](20)

1. $\int_0^{\pi/2} \frac{x\cos^{p-1} x}{\sin^{p+1} x}\,dx = \frac{\pi}{2p}\sec \frac{\pi p}{2} \quad [p < 1].$ BH[206](13)+ 3.838

2. $\int_0^{\pi/4} \frac{x\sin^{p-1} x}{\cos^{p+1} x}\,dx = \frac{\pi}{4p} - \frac{1}{2p}\beta\left(\frac{p+1}{2}\right) \quad [p > -1].$ Li[204](15)

3. $\int_0^{\pi/4} \frac{x\sin^{2m-1} x}{\cos^{2m+1} x}\,dx = \frac{\pi}{8m}(1 - \cos m\pi) + \frac{1}{2m}\sum_{k=0}^{m-1} \frac{(-1)^{k-1}}{2m - 2k - 1}.$ BH[204](17)

4. $\int_0^{\pi/4} \frac{x\sin^{2m} x}{\cos^{2m+2} x}\,dx = \frac{1}{2(2m+1)}\left[\frac{\pi}{2} + (-1)^{m-1}\ln 2 + \sum_{k=0}^{m-1}\frac{(-1)^{k-1}}{m-k}\right].$
BH[204](16)

3.839

1. $$\int_0^{\frac{\pi}{4}} x \, \text{tg}^2 \, x \, dx = \frac{\pi}{4} - \frac{\pi^2}{32} - \frac{1}{2} \ln 2.$$ BH[204](3)

2. $$\int_0^{\frac{\pi}{4}} x \, \text{tg}^3 \, x \, dx = \frac{\pi}{4} - \frac{1}{2} + \frac{\pi}{8} \ln 2 - \frac{1}{2} G.$$ BH[204](7)

3. $$\int_0^{\frac{\pi}{4}} \frac{x^2 \, \text{tg} \, x}{\cos^2 x} \, dx = \frac{1}{2} \ln 2 - \frac{\pi}{4} + \frac{\pi^2}{16} \quad \text{(cf. 3.839 1.)}.$$ BH[204](13)

4. $$\int_0^{\frac{\pi}{4}} \frac{x^2 \, \text{tg}^2 x}{\cos^2 x} \, dx = \frac{1}{3} \left(1 - \frac{\pi}{4} \ln 2 - \frac{\pi}{2} + \frac{\pi^2}{16} + G \right) \quad \text{(cf. 3.839 2.)}.$$ BH[204](12)

5. $$\int_0^{\frac{\pi}{2}} x \cos^p x \, \text{tg} \, x \, dx = \frac{\pi}{2^{p+1} p} \cdot \frac{\Gamma(p+1)}{\left[\Gamma\left(\frac{p}{2}+1\right)\right]^2} \quad [p > -1].$$ BH[205](3)

6. $$\int_0^{\frac{\pi}{2}} x \sin^p x \, \text{ctg} \, x \, dx = \frac{\pi}{2p} - \frac{2^{p-1}}{p} B\left(\frac{p+1}{2}, \frac{p+1}{2}\right) \quad [p > -1].$$ BH[206](11)

7. $$\int_0^\infty \sin^{2n} x \, \text{tg} \, x \, \frac{dx}{x} = \frac{\pi}{2} \cdot \frac{(2n-1)!!}{(2n)!!}.$$ GH[333](16)

8. $$\int_0^\infty \cos^s rx \, \text{tg} \, qx \, \frac{dx}{x} = \frac{\pi}{2} \quad [s > -1].$$ BH[151](26)

9. $$\int_0^\infty \frac{\cos[(2n-1)x]}{\cos x} \cdot \left(\frac{\sin x}{x}\right)^{2n} dx = (-1)^{n-1} \frac{2^{2n}-1}{(2n)!} \cdot 2^{2n-1} \pi |B_{2n}|.$$ BH[180](15)

10. $$\int_0^\infty \text{tg}^r \, px \, \frac{dx}{q^2+x^2} = \frac{\pi}{2q} \sec \frac{r\pi}{2} \, \text{th}^r \, pq \quad [r^2 < 1].$$ BH[160](19)

3.84 **Integrale, die $\sqrt{1-k^2\sin^2 x}$, $\sqrt{1-k^2\cos^2 x}$ und ähnliche Ausdrücke enthalten**
Integrals containing $\sqrt{1-k^2 \sin^2 x}$, $\sqrt{1-k^2 \cos^2 x}$, and similar expressions

3.841

1. $$\int_0^\infty \sin x \sqrt{1-k^2 \sin^2 x} \, \frac{dx}{x} = E(k).$$ BH[154](8)

2. $\int_0^\infty \sin x \sqrt{1 - k^2 \cos^2 x} \, \frac{dx}{x} = E(k).$ BH[154](20)   3.841

3. $\int_0^\infty \operatorname{tg} x \sqrt{1 - k^2 \sin^2 x} \, \frac{dx}{x} = E(k).$ BH[154](9)

4. $\int_0^\infty \operatorname{tg} x \sqrt{1 - k^2 \cos^2 x} \, \frac{dx}{x} = E(k).$ BH[154](21)

1. $\int_0^\infty \frac{\sin x}{\sqrt{1 + \sin^2 x}} \frac{dx}{x} = \int_0^\infty \frac{\operatorname{tg} x}{\sqrt{1 + \sin^2 x}} \cdot \frac{dx}{x} =$ 3.842

$$= \int_0^\infty \frac{\sin x}{\sqrt{1 + \cos^2 x}} \frac{dx}{x} = \int_0^\infty \frac{\operatorname{tg} x}{\sqrt{1 + \cos^2 x}} \frac{dx}{x} = \frac{1}{\sqrt{2}} K\left(\frac{1}{\sqrt{2}}\right).$$

BH[183] (4), BH[183](5), BH[183](9), BH[183](10)

2. $\int_u^{\pi/2} \frac{x \cos x \, dx}{\sqrt{\sin^2 x - \sin^2 u}} = \frac{\pi}{2} \ln(1 + \cos u).$ BH[226](4)

3. $\int_0^\infty \frac{\sin x}{\sqrt{1 - k^2 \sin^2 x}} \frac{dx}{x} = \int_0^\infty \frac{\operatorname{tg} x}{\sqrt{1 - k^2 \sin^2 x}} \frac{dx}{x} =$

$$= \int_0^\infty \frac{\sin x}{\sqrt{1 - k^2 \cos^2 x}} \frac{dx}{x} = \int_0^\infty \frac{\operatorname{tg} x}{\sqrt{1 - k^2 \cos^2 x}} \frac{dx}{x} = K(k).$$

BH[183](12), BH[183](13), BH[183](21), BH[183](22)

4. $\int_0^{\pi/2} \frac{x \sin x \cos x}{\sqrt{1 - k^2 \sin^2 x}} \, dx = \frac{1}{2k^2} [-\pi k' + 2E(k)].$ BH[211](1)

5. $\int_0^{\pi/2} \frac{x \sin x \cos x}{\sqrt{1 - k^2 \cos^2 x}} \, dx = \frac{1}{2k^2} [\pi - 2E(k)].$ BH[214](1)

6. $\int_0^\alpha \frac{x \sin x \, dx}{\cos^2 x \sqrt{\sin^2 \alpha - \sin^2 x}} = \frac{\pi \sin^2 \frac{\alpha}{2}}{\cos^2 \alpha}.$ Lo III 284

7. $\int_0^\beta \frac{x \sin x \, dx}{(1 - \sin^2 \alpha \sin^2 x)\sqrt{\sin^2 \beta - \sin^2 x}} = \frac{\pi \ln \dfrac{\cos \alpha + \sqrt{1 - \sin^2 \alpha \sin^2 \beta}}{2 \cos \beta \cos^2 \dfrac{\alpha}{2}}}{2 \cos \alpha \sqrt{1 - \sin^2 \alpha \sin^2 \beta}}.$ Lo III 284

1. $\int_0^\infty \operatorname{tg} x \sqrt{1 - k^2 \sin^2 2x} \, \frac{dx}{x} = E(k).$ BH[154](10)   3.843

3.843

2. $$\int_0^\infty \operatorname{tg} x \sqrt{1 - k^2 \cos^2 2x}\, \frac{dx}{x} = E(k).$$ BH[154](22)

3. $$\int_0^\infty \frac{\operatorname{tg} x}{\sqrt{1 + \sin^2 2x}}\, \frac{dx}{x} = \int_0^\infty \frac{\operatorname{tg} x}{\sqrt{1 + \cos^2 2x}}\, \frac{dx}{x} = \frac{1}{\sqrt{2}} K\left(\frac{1}{\sqrt{2}}\right).$$ BH[183](6), BH[183](11)

4. $$\int_0^\infty \frac{\operatorname{tg} x}{\sqrt{1 - k^2 \sin^2 2x}}\, \frac{dx}{x} = \int_0^\infty \frac{\operatorname{tg} x}{\sqrt{1 - k^2 \cos^2 2x}}\, \frac{dx}{x} = K(k).$$ BH[183](14), BH[183](23)

3.844

1. $$\int_0^\infty \frac{\sin x \cos x}{\sqrt{1 - k^2 \cos^2 x}}\, \frac{dx}{x} = \frac{1}{k^2}[K(k) - E(k)].$$ BH[185](20)

2. $$\int_0^\infty \frac{\sin x \cos^2 x}{\sqrt{1 - k^2 \cos^2 x}} \cdot \frac{dx}{x} = \frac{1}{k^2}[K(k) - E(k)].$$ BH[185](21)

3. $$\int_0^\infty \frac{\sin x \cos^3 x}{\sqrt{1 - k^2 \cos^2 x}} \cdot \frac{dx}{x} = \frac{1}{3k^4}[(2 + k^2) K(k) - 2(1 + k^2) E(k)].$$ BH[185](22)

4. $$\int_0^\infty \frac{\sin x \cos^4 x}{\sqrt{1 - k^2 \cos^2 x}} \cdot \frac{dx}{x} = \frac{1}{3k^4}[(2 + k^2) K(k) - 2(1 + k^2) E(k)].$$ BH[185](23)

5. $$\int_0^\infty \frac{\sin^3 x \cos x}{\sqrt{1 - k^2 \cos^2 x}} \cdot \frac{dx}{x} = \frac{1}{3k^4}[(1 + k'^2) E(k) - 2k'^2 K(k)].$$ BH[185](24)

6. $$\int_0^\infty \frac{\sin^3 x \cos^2 x}{\sqrt{1 - k^2 \cos^2 x}} \cdot \frac{dx}{x} = \frac{1}{3k^4}[(1 + k'^2) E(k) - 2k'^2 K(k)].$$ BH[185](25)

7. $$\int_0^\infty \frac{\sin^2 x\, \operatorname{tg} x}{\sqrt{1 - k^2 \cos^2 x}} \cdot \frac{dx}{x} = \frac{1}{k^2}[E(k) - k'^2 K(k)].$$ BH[184](16)

8. $$\int_0^\infty \frac{\sin^4 x\, \operatorname{tg} x}{\sqrt{1 - k^2 \cos^2 x}} \cdot \frac{dx}{x} = \frac{1}{3k^4}[(2 + 3k^2)k'^2 K(k) - 2(k'^2 - k^2) E(k)].$$ BH[184](18)

3.845

1. $$\int_0^\infty \frac{\sin x \cos x}{\sqrt{1 + \cos^2 x}} \cdot \frac{dx}{x} = \sqrt{2}\left[E\left(\frac{\sqrt{2}}{2}\right) - \frac{1}{2} K\left(\frac{\sqrt{2}}{2}\right)\right].$$ BH[185](6)

2. $$\int_0^\infty \frac{\sin x \cos^2 x}{\sqrt{1 + \cos^2 x}} \cdot \frac{dx}{x} = \sqrt{2}\left[E\left(\frac{\sqrt{2}}{2}\right) - \frac{1}{2} K\left(\frac{\sqrt{2}}{2}\right)\right].$$ BH[185](7)

3. $\int_0^\infty \frac{\sin^2 x \,\mathrm{tg}\, x}{\sqrt{1+\cos^2 x}} \cdot \frac{dx}{x} = \sqrt{2}\left[K\left(\frac{\sqrt{2}}{2}\right) - E\left(\frac{\sqrt{2}}{2}\right)\right].$ BH[184](8) 3.845

1. $\int_0^\infty \frac{\sin x \cos x}{\sqrt{1-k^2\sin^2 x}} \cdot \frac{dx}{x} = \frac{1}{k^2}[E(k) - k'^2 K(k)].$ BH[185](9) 3.846

2. $\int_0^\infty \frac{\sin x \cos^2 x}{\sqrt{1-k^2\sin^2 x}} \cdot \frac{dx}{x} = \frac{1}{k^2}[E(k) - k'^2 K(k)].$ BH[185](10)

3. $\int_0^\infty \frac{\sin x \cos^3 x}{\sqrt{1-k^2\sin^2 x}} \cdot \frac{dx}{x} = \frac{1}{3k^4}[(2-3k^2)k'^2 K(k) - 2(k'^2-k^2) E(k)].$ BH[185](11)

4. $\int_0^\infty \frac{\sin x \cos^4 x}{\sqrt{1-k^2\sin^2 x}} \cdot \frac{dx}{x} = \frac{1}{3k^4}\cdot[(2-3k^2)k'^2 K(k) - 2(k'^2-k^2) E(k)].$ BH[185](12)

5. $\int_0^\infty \frac{\sin^3 x \cos x}{\sqrt{1-k^2\sin^2 x}} \cdot \frac{dx}{x} = \frac{1}{3k^4}[(1+k'^2)E(k) - 2k'^2 K(k)].$ BH[185](13)

6. $\int_0^\infty \frac{\sin^3 x \cos^2 x}{\sqrt{1-k^2\sin^2 x}} \cdot \frac{dx}{x} = \frac{1}{3k^4}[(1+k'^2)E(k) - 2k'^2 K(k)].$ BH[185](14)

7. $\int_0^\infty \frac{\sin^2 x \,\mathrm{tg}\, x}{\sqrt{1-k^2\sin^2 x}} \cdot \frac{dx}{x} = \frac{1}{k^2}[K(k) - E(k)].$ BH[184](9)

8. $\int_0^\infty \frac{\sin^4 x \,\mathrm{tg}\, x}{\sqrt{1-k^2\sin^2 x}} \cdot \frac{dx}{x} = \frac{1}{3k^4}[(2+k^2) K(k) - 2(1+k^2) E(k)].$ BH[184](11)

$\int_0^\infty \frac{\sin x \cos x}{\sqrt{1+\sin^2 x}} \cdot \frac{dx}{x} = \int_0^\infty \frac{\sin x \cos^2 x}{\sqrt{1+\sin^2 x}} \cdot \frac{dx}{x} = \sqrt{2}\left[K\left(\frac{\sqrt{2}}{2}\right) - E\left(\frac{\sqrt{2}}{2}\right)\right].$ 3.847

BH[185](3, 4)

1. $\int_0^\infty \frac{\sin^3 x \cos x}{\sqrt{1-k^2\sin^2 2x}} \cdot \frac{dx}{x} = \frac{1}{4k^2}[K(k) - E(k)].$ BH[185](15) 3.848

2. $\int_0^\infty \frac{\cos^2 2x \,\mathrm{tg}\, x}{\sqrt{1-k^2\sin^2 2x}} \cdot \frac{dx}{x} = \frac{1}{k^2}[E(k) - k'^2 K(k)].$ BH[184](12)

3. $\int_0^\infty \frac{\cos^4 2x \,\mathrm{tg}\, x}{\sqrt{1-k^2\sin^2 2x}} \cdot \frac{dx}{x} = \frac{1}{3k^4}[(2-3k^2) k'^2 K(k) - 2(k'^2-k^2) E(k)].$

BH[184](13)

3.848

4. $\int_0^\infty \dfrac{\sin^2 4x \, \mathrm{tg}\, x}{\sqrt{1 - k^2 \sin^2 2x}} \cdot \dfrac{dx}{x} = \dfrac{4}{3k^4}[(1 + k'^2)\, E(k) - 2k'^2 K(k)].$ BH[184](17)

5. $\int_0^\infty \dfrac{\sin^3 x \cos x}{\sqrt{1 - k^2 \cos^2 2x}} \cdot \dfrac{dx}{x} = \dfrac{1}{4k^2}[E(k) - k'^2 K(k)].$ BH[185](26)

6. $\int_0^\infty \dfrac{\cos^2 2x \, \mathrm{tg}\, x}{\sqrt{1 - k^2 \cos^2 2x}} \cdot \dfrac{dx}{x} = \dfrac{1}{k^2}[K(k) - E(k)].$ BH[184](19)

7. $\int_0^\infty \dfrac{\cos^4 2x \, \mathrm{tg}\, x}{\sqrt{1 - k^2 \cos^2 2x}} \cdot \dfrac{dx}{x} = \dfrac{1}{3k^4}[(2 + k^2) K(k) - 2(1 + k^2) E(k)].$ BH[184](20)

3.849

1. $\int_0^\infty \dfrac{\sin^3 x \cos x}{\sqrt{1 + \cos^2 2x}} \cdot \dfrac{dx}{x} = \dfrac{1}{2\sqrt{2}}\left[K\left(\dfrac{\sqrt{2}}{2}\right) - E\left(\dfrac{\sqrt{2}}{2}\right)\right].$ BH[185](8)

2. $\int_0^\infty \dfrac{\sin^3 x \cos x}{\sqrt{1 + \sin^2 2x}} \cdot \dfrac{dx}{x} = \dfrac{\sqrt{2}}{8}\left[2E\left(\dfrac{\sqrt{2}}{2}\right) - K\left(\dfrac{\sqrt{2}}{2}\right)\right].$ BH[185](5)

3. $\int_0^\infty \dfrac{\cos^2 2x \, \mathrm{tg}\, x}{\sqrt{1 + \sin^2 2x}} \cdot \dfrac{dx}{x} = \sqrt{2}\left[K\left(\dfrac{\sqrt{2}}{2}\right) - E\left(\dfrac{\sqrt{2}}{2}\right)\right].$ BH[184](7)

3.85–3.88

Trigonometrische Funktionen mit komplizierteren Argumenten und Potenzen
Trigonometric functions of compound arguments combined with powers

3.851

1. $\int_0^\infty x \sin(ax^2) \sin(2bx) dx = \dfrac{b}{2a}\sqrt{\dfrac{\pi}{2a}}\left(\cos\dfrac{b^2}{a} + \sin\dfrac{b^2}{a}\right)$

$[a > 0,\ b > 0].$ BH[150](4)

2. $\int_0^\infty x \sin(ax^2) \cos(2bx) dx =$

$= \dfrac{1}{2a} - \dfrac{b}{a}\sqrt{\dfrac{\pi}{2a}}\left[\sin\dfrac{b^2}{a} C\left(\dfrac{b}{\sqrt{a}}\right) - \cos\dfrac{b^2}{a} S\left(\dfrac{b}{\sqrt{a}}\right)\right].$ BH[150](5)+

3. $\int_0^\infty x \cos(ax^2) \sin(2bx) dx = \dfrac{b}{2a}\sqrt{\dfrac{\pi}{2a}}\left(\sin\dfrac{b^2}{a} - \cos\dfrac{b^2}{a}\right)$

$[a > 0, b > 0]$ (cf. 3.691 7.).   BH[150](7)

4. $\int_0^\infty x \cos(ax^2) \cos(2bx) dx =$

$= \dfrac{b}{a}\sqrt{\dfrac{\pi}{2a}}\left[\cos\dfrac{b^2}{a} C\left(\dfrac{b}{\sqrt{a}}\right) + \sin\dfrac{b^2}{a} S\left(\dfrac{b}{\sqrt{a}}\right)\right].$ BH[150](6)+

Trigonometrische Funktionen/Trigonometric Functions

5. $\int_0^\infty \sin(ax^2) \cos(bx) \frac{dx}{x^2} = \frac{b\pi}{2} \left\{ S\left(\frac{b}{2\sqrt{a}}\right) - C\left(\frac{b}{2\sqrt{a}}\right) + \right.$

$\left. + \sqrt{a\pi} \sin\left(\frac{b^2}{4a} + \frac{\pi}{4}\right) \right\}$ $[a > 0, \ b > 0]$ (cf. 3.691 7.).   Ba IV 23(3)+   3.851

1. $\int_0^\infty \frac{\sin(ax^2)}{x^2} dx = \sqrt{\frac{a\pi}{2}}.$ BH[177](10)+   3.852

2. $\int_0^\infty \sin(ax^2) \cos(bx^2) \frac{dx}{x^2} = \frac{1}{2}\sqrt{\frac{\pi}{2}}(\sqrt{a+b}+\sqrt{a-b});$ $[a > b > 0]$

$= \frac{1}{2}\sqrt{\pi a}$ $[b = a \geq 0];$

$= \frac{1}{2}\sqrt{\frac{\pi}{2}}(\sqrt{a+b} - \sqrt{b-a})$ $[b > a > 0]$
(cf. 3.852 1.).   BH[177](23)

3. $\int_0^\infty \frac{\sin^2(a^2x^2)}{x^4} dx = \frac{2\sqrt{\pi}}{3} a^3$ $[a \geq 0]$.   GH[333](19e)

4. $\int_0^\infty \frac{\sin^3(a^2x^2)}{x^2} dx = \frac{3-\sqrt{3}}{8} \sqrt{\pi a}$ $[a \geq 0]$.   GH[333](19g)

5. $\int_0^\infty (\sin x^2 - x^2 \cos x^2) \frac{dx}{x^4} = -\frac{1}{3}\sqrt{\frac{\pi}{2}}.$ BH[178](8)

6. $\int_0^\infty \left\{ \cos x^2 - \frac{1}{1+x^2} \right\} \frac{dx}{x} = -\frac{1}{2} C.$ BH[173](22)

1. $\int_0^\infty \frac{\sin(ax^2)}{\beta^2 + x^2} dx = \frac{\pi}{2\beta} \left[\sqrt{2} \sin\left(a\beta^2 + \frac{\pi}{4}\right) C(\sqrt{a}\,\beta) - \right.$ 3.853

$\left. - \sqrt{2} \cos\left(a\beta^2 + \frac{\pi}{4}\right) S(\sqrt{a}\,\beta) - \sin(a\beta^2) \right]$ $[a > 0, \ \mathrm{Re}\,\beta > 0]$.   Ba V 219(33)+

2. $\int_0^\infty \frac{\cos(ax^2)}{\beta^2 + x^2} dx = \frac{\pi}{2\beta} \left[\cos(a\beta^2) - \sqrt{2}\cos\left(a\beta^2 + \frac{\pi}{4}\right) C(\sqrt{a}\,\beta) - \right.$

$\left. - \sqrt{2} \sin\left(a\beta^2 + \frac{\pi}{4}\right) S(\sqrt{a}\,\beta) \right]$ $[a > 0, \mathrm{Re}\,\beta > 0]$.   Ba V 221(51)+

3. $\int_0^\infty \frac{x^2 \sin(ax^2)}{\beta^2 + x^2} dx = \frac{\beta\pi}{2} \left[\sin(a\beta^2) - \sqrt{2}\sin\left(a\beta^2 + \frac{\pi}{4}\right) C(\sqrt{a}\,\beta) + \right.$

$\left. + \sqrt{2}\cos\left(a\beta^2 + \frac{\pi}{4}\right) S(\sqrt{a}\,\beta) \right] - \frac{1}{2}\sqrt{\frac{\pi}{2a}}$

$[a > 0, \ \mathrm{Re}\,\beta > 0]$.   Ba V 219(32)+

3.853

4. $\int_0^\infty \dfrac{x^2 \cos(ax^2)}{\beta^2 + x^2}\, dx = \dfrac{1}{2}\sqrt{\dfrac{\pi}{2a}} - \dfrac{\beta\pi}{2}\Big[\cos(a\beta^2) -$

$- \sqrt{2}\cos\left(a\beta^2 + \dfrac{\pi}{4}\right) C(\sqrt{a}\,\beta) - \sqrt{2}\sin\left(a\beta^2 + \dfrac{\pi}{4}\right) S(\sqrt{a}\,\beta)\Big]$

$[a > 0,\ \operatorname{Re}\beta > 0].$ Ba V 221(50)+

3.854

1. $\int_0^\infty (\cos(ax)^2 - \sin(ax^2))\, \dfrac{dx}{x^4 + b^4} = \dfrac{\pi e^{-ab^2}}{2b^3\sqrt{2}}\quad [a>0,\ b>0].$

Li]178](11)÷, BH[168](25)

2. $\int_0^\infty (\cos(ax^2) + \sin(ax^2))\, \dfrac{x^2\, dx}{x^4 + b^4} = \dfrac{\pi e^{-ab^2}}{2b\sqrt{2}}\quad [a>0, b>0].$ Li[178](12)

3. $\int_0^\infty (\cos(ax^2) + \sin(ax^2))\, \dfrac{x^2\, dx}{(x^4 + b^4)^2} = \dfrac{\pi e^{-ab^2}}{4\sqrt{2}\,b^3}\left(a + \dfrac{1}{2b^2}\right)\quad [a>0,\ b>0].$

Li[178](14)

4. $\int_0^\infty (\cos(ax^2) - \sin(ax^2))\, \dfrac{x^4\, dx}{(x^4 + b^4)^2} = \dfrac{\pi e^{-ab^2}}{4\sqrt{2}\,b}\left(\dfrac{1}{2b^2} - a\right)\quad [a>0,\ b>0].$

BH[178](15)

3.855

1. $\int_0^\infty \dfrac{\sin(ax^2)}{\sqrt{\beta^2 + x^4}}\, dx = \dfrac{1}{2}\sqrt{\dfrac{a\pi}{2}}\, I_{\frac{1}{4}}\!\left(\dfrac{a\beta}{2}\right) K_{\frac{1}{4}}\!\left(\dfrac{a\beta}{2}\right)\quad [a > 0,\ \operatorname{Re}\beta > 0].$

Ba IV 66(28)

2. $\int_0^\infty \dfrac{\cos(ax^2)}{\sqrt{\beta^2 + x^4}}\, dx = \dfrac{1}{2}\sqrt{\dfrac{a\pi}{2}}\, I_{-\frac{1}{4}}\!\left(\dfrac{a\beta}{2}\right) K_{\frac{1}{4}}\!\left(\dfrac{a\beta}{2}\right)\quad [a>0,\ \operatorname{Re}\beta>0].$ Ba IV 9(22)

3. $\int_0^u \dfrac{\sin(a^2 x^2)}{\sqrt{u^4 - x^4}}\, dx = \dfrac{a}{4}\sqrt{\dfrac{\pi^3}{2}}\left[J_{\frac{1}{4}}\!\left(\dfrac{a^2 u^2}{2}\right)\right]^2\quad [a > 0].$ Ba IV 66(29)

4. $\int_u^\infty \dfrac{\sin(a^2 x^2)}{\sqrt{x^4 - u^4}}\, dx = -\dfrac{a}{4}\sqrt{\dfrac{\pi^3}{2}}\, J_{\frac{1}{4}}\!\left(\dfrac{a^2 u^2}{2}\right) N_{\frac{1}{4}}\!\left(\dfrac{a^2 u^2}{2}\right)\quad [a > 0].$ Ba IV 66(30)

5. $\int_0^u \dfrac{\cos(a^2 x^2)}{\sqrt{u^4 - x^4}}\, dx = \dfrac{a}{4}\sqrt{\dfrac{\pi^3}{2}}\left[J_{-\frac{1}{4}}\!\left(\dfrac{a^2 u^2}{2}\right)\right]^2.$ Ba IV 9(23)

6. $\int_u^\infty \dfrac{\cos(a^2 x^2)}{\sqrt{x^4 - u^4}}\, dx = -\dfrac{a}{4}\sqrt{\dfrac{\pi^3}{2}}\, J_{-\frac{1}{4}}\!\left(\dfrac{a^2 u^2}{2}\right) N_{-\frac{1}{4}}\!\left(\dfrac{a^2 u^2}{2}\right).$ Ba IV 10(24)

3.856

1. $\int_0^\infty \dfrac{(\sqrt{\beta^4 + x^4} + x^2)^\nu}{\sqrt{\beta^4 + x^4}}\, \sin(a^2 x^2)\, dx =$

$= \dfrac{a}{2}\sqrt{\dfrac{\pi}{2}}\, \beta^{2\nu}\, I_{\frac{1}{4} - \frac{\nu}{2}}\!\left(\dfrac{a^2\beta^2}{2}\right) K_{\frac{1}{4} + \frac{\nu}{2}}\!\left(\dfrac{a^2\beta^2}{2}\right)$

$\left[\operatorname{Re}\nu < \dfrac{3}{2},\ |\arg\beta| < \dfrac{\pi}{4}\right].$ Ba IV 71(23)

3.856

2. $\int_0^\infty \dfrac{(\sqrt{\beta^4+x^4}+x^2)^\nu}{\sqrt{\beta^4+x^4}} \cos(a^2 x^2)\, dx =$

$= \dfrac{a}{2}\sqrt{\dfrac{\pi}{2}}\, \beta^{2\nu}\, I_{-\frac{1}{4}-\frac{\nu}{2}}\!\left(\dfrac{a^2\beta^2}{2}\right) K_{-\frac{1}{4}+\frac{\nu}{2}}\!\left(\dfrac{a^2\beta^2}{2}\right)$

$\left[\operatorname{Re}\nu < \dfrac{3}{2},\ |\arg\beta|<\dfrac{\pi}{4}\right].$ Ba IV 12(16)

3. $\int_0^\infty \dfrac{(\sqrt{\beta^4+x^4}-x^2)^\nu}{\sqrt{\beta^4+x^4}} \cos(a^2 x^2)\, dx =$

$= \dfrac{a}{2}\sqrt{\dfrac{\pi}{2}}\, \beta^{2\nu}\, I_{-\frac{1}{4}+\frac{\nu}{2}}\!\left(\dfrac{a^2\beta^2}{2}\right) K_{-\frac{1}{4}-\frac{\nu}{2}}\!\left(\dfrac{a^2\beta^2}{2}\right)$

$\left[\operatorname{Re}\nu > -\dfrac{3}{2},\ |\arg\beta|<\dfrac{\pi}{4}\right].$ Ba IV 12(17)

4. $\int_0^\infty \dfrac{\sin(a^2 x^2)\, dx}{\sqrt{\beta^4+x^4}\,\sqrt{x^2+\sqrt{\beta^4+x^4}}} = \dfrac{\operatorname{sh}\dfrac{a^2\beta^2}{2}}{\sqrt{2}\,\beta^2} K_0\!\left(\dfrac{a^2\beta^2}{2}\right)\ \left[|\arg\beta|<\dfrac{\pi}{4}\right].$ Ba IV 66(32)

5. $\int_0^\infty \dfrac{\cos(a^2 x^2)\, dx}{\sqrt{\beta^4+x^4}\,\sqrt{(x^2+\sqrt{\beta^4+x^4})^3}} = \dfrac{\operatorname{sh}\dfrac{a^2\beta^2}{2}}{2\sqrt{2}\,\beta^4} K_1\!\left(\dfrac{a^2\beta^2}{2}\right)\ \left[|\arg\beta|<\dfrac{\pi}{4}\right].$ Ba IV 10(27)

6. $\int_0^\infty \dfrac{\sqrt{\sqrt{\beta^4+x^4}+x^2}}{\sqrt{\beta^4+x^4}} \sin(a^2 x^2)\, dx = \dfrac{\pi}{2\sqrt{2}}\, e^{-\frac{a^2\beta^2}{2}} I_0\!\left(\dfrac{a^2\beta^2}{2}\right)$

$\left[|\arg\beta|<\dfrac{\pi}{4}\right].$ Ba IV 67(33)

3.857

1. $\int_0^\infty \dfrac{x^2}{R_1 R_2}\sqrt{\dfrac{R_2-R_1}{R_2+R_1}}\sin(ax^2)\, dx = \dfrac{1}{2\sqrt{b}} K_0(ac)\sin ab$

$[R_1=\sqrt{c^2+(b-x^2)^2},\quad R_2=\sqrt{c^2+(b+x^2)^2},\quad a>0,\ c>0].$ Ba IV 67(34)

2. $\int_0^\infty \dfrac{x^2}{R_1 R_2}\sqrt{\dfrac{R_2+R_1}{R_2-R_1}}\cos(ax^2)\, dx = \dfrac{1}{2\sqrt{b}} K_0(ac)\cos ab$

$[R_1=\sqrt{c^2+(b-x^2)^2},\quad R_2=\sqrt{c^2+(b+x^2)^2},\quad a>0,\ c>0].$ Ba IV 10(26)

3.858

1. $\int_u^\infty \dfrac{(x^2+\sqrt{x^4-u^4})^\nu+(x^2-\sqrt{x^4-u^4})^\nu}{\sqrt{x^4-u^4}}\sin(a^2 x^2)\, dx =$

$= -\dfrac{a}{4}\sqrt{\dfrac{\pi^3}{2}}\, u^{2\nu}\left[J_{\frac{1}{4}+\frac{\nu}{2}}\!\left(\dfrac{a^2 u^2}{2}\right) N_{\frac{1}{4}-\frac{\nu}{2}}\!\left(\dfrac{a^2 u^2}{2}\right)+\right.$

$\left.+ J_{\frac{1}{4}-\frac{\nu}{2}}\!\left(\dfrac{a^2 u^2}{2}\right) N_{\frac{1}{4}+\frac{\nu}{2}}\!\left(\dfrac{a^2 u^2}{2}\right)\right]\ \left[\operatorname{Re}\nu<\dfrac{3}{2}\right].$ Ba IV 71(25)

2. $\int_u^\infty \dfrac{(x^2+\sqrt{x^4-u^4})^\nu+(x^2-\sqrt{x^4-u^4})^\nu}{\sqrt{x^4-u^4}}\cos(a^2 x^2)\, dx =$

$= -\dfrac{a}{4}\sqrt{\dfrac{\pi^3}{2}}\, u^{2\nu}\left[J_{-\frac{1}{4}+\frac{\nu}{2}}\!\left(\dfrac{a^2 u^2}{2}\right) N_{-\frac{1}{4}-\frac{\nu}{2}}\!\left(\dfrac{a^2 u^2}{2}\right)+\right.$

$\left.+ J_{-\frac{1}{4}-\frac{\nu}{2}}\!\left(\dfrac{a^2 u^2}{2}\right) N_{-\frac{1}{4}+\frac{\nu}{2}}\!\left(\dfrac{a^2 u^2}{2}\right)\right]\ \left[\operatorname{Re}\nu<\dfrac{3}{2}\right].$ Ba IV 13(26)

3.859
$$\int_0^\infty \left[\cos(x^{2^n}) - \frac{1}{1+x^{2^{n+1}}}\right] \frac{dx}{x} = -\frac{1}{2^n} C.$$
BH [173](24)

3.861   1.
$$\int_0^\infty \sin^{2n+1}(ax^2) \frac{dx}{x^{2m}} = \pm \frac{\sqrt{\pi}\, a^{m-\frac{1}{2}}}{2^{2n-m+\frac{1}{2}}(2m-1)!!} \times$$
$$\times \sum_{k=1}^{n+1} (-1)^{k-1} \binom{2n+1}{n+k} (2k-1)^{m-\frac{1}{2}}$$

[das Pluszeichen ist im Fall $m \equiv 0$ (mod 4) oder $m \equiv 1$ (mod 4), das Minuszeichen im Fall $m \equiv 2$ (mod 4) oder $m \equiv 3$ (mod 4) zu nehmen].
BH[177](19)+

[the plus sign has to be chosen if $m \equiv 0$ (mod 4) or $m \equiv 1$ (mod 4), and the minus sign if $m \equiv 2$ (mod 4) or $m \equiv 3$ (mod 4)].
BH[177](19)+

2.
$$\int_0^\infty \sin^{2n}(ax^2) \frac{dx}{x^{2m}} = \pm \frac{\sqrt{\pi}\, a^{m-\frac{1}{2}}}{2^{3n-2m+1}(2m-1)!!} \sum_{k=1}^n (-1)^k \binom{2n}{n+k} k^{m-\frac{1}{2}}$$

[das Pluszeichen ist im Fall $m \equiv 0$ (mod 4) oder $m \equiv 3$ (mod 4), das Minuszeichen im Fall $m \equiv 2$ (mod 4) oder $m \equiv 1$ (mod 4) zu nehmen].
BH[177](18)+, Li[177](18)

[the plus sign has to be chosen if $m \equiv 0$ (mod 4) or $m \equiv 3$ (mod 4), and the minus sign if $m \equiv 2$ (mod 4) or $m \equiv 1$ (mod 4)].
BH[177](18)+, Li[177](18)

3.862
$$\int_0^\infty [\cos(ax^2 \sqrt{n}) + \sin(ax^2 \sqrt{n})] \left(\frac{\sin x^2}{x^2}\right)^n dx =$$
$$= \frac{\sqrt{\pi}}{(2n-1)!!\sqrt{2}} \sum_{k=0}^n (-1)^k \binom{n}{k}(n-2k+a\sqrt{n})^{n-\frac{1}{2}}$$
$[a > \sqrt{n} > 0]$.
BH[178](9)

3.863   1.
$$\int_0^\infty x^2 \cos(ax^4) \sin(2bx^2) dx = -\frac{\pi}{8}\sqrt{\frac{b^3}{a^3}} \left[\sin\left(\frac{b^2}{2a} - \frac{\pi}{8}\right) J_{-\frac{1}{4}}\left(\frac{b^2}{2a}\right) + \right.$$
$$\left. + \cos\left(\frac{b^2}{2a} - \frac{\pi}{8}\right) J_{\frac{3}{4}}\left(\frac{b^2}{2a}\right)\right] \quad [a > 0, b > 0].$$
Ba IV 25(22)

2.
$$\int_0^\infty x^2 \cos(ax^4) \cos(2bx^2) dx =$$
$$= -\frac{\pi}{8}\sqrt{\frac{b^3}{a^3}} \left[\sin\left(\frac{b^2}{2a} + \frac{\pi}{8}\right) J_{-\frac{3}{4}}\left(\frac{b^2}{2a}\right) + \right.$$
$$\left. + \cos\left(\frac{b^2}{2a} + \frac{\pi}{8}\right) J_{-\frac{1}{4}}\left(\frac{b^2}{2a}\right)\right] \quad [a > 0, b > 0].$$
Ba IV 25(23)

3.864   1.
$$\int_0^\infty \sin\frac{b}{x} \sin ax \frac{dx}{x} = \frac{\pi}{2} N_0(2\sqrt{ab}) + K_0(2\sqrt{ab}) \quad [a > 0, b > 0].$$
W 184(3)+

2. $\int_0^\infty \cos\dfrac{b}{x} \cos ax \,\dfrac{dx}{x} = -\dfrac{\pi}{2} N_0(2\sqrt{ab}) + K_0(2\sqrt{ab})$ $\quad [a>0,\ b>0]$. $\qquad$ 3.864

$\hfill$ W 184(4)+, Ba IV 24(12)

1. $\int_0^u \dfrac{(u^2-x^2)^{\mu-1}}{x^{2\mu}} \sin\dfrac{a}{x} dx = \dfrac{\sqrt{\pi}}{2}\left(\dfrac{2}{a}\right)^{\mu-\frac{1}{2}} u^{\mu-\frac{3}{2}} \Gamma(\mu) J_{\frac{1}{2}-\mu}\left(\dfrac{a}{u}\right)$ $\qquad$ 3.865

$\hfill [a>0,\ u>0,\ 0<\mathrm{Re}\,\mu<1]. \quad$ Ba V 189(30)

2. $\int_u^\infty \dfrac{(x-u)^{\mu-1}}{x^{2\mu}} \sin\dfrac{a}{x} dx = \sqrt{\dfrac{\pi}{u}} a^{\frac{1}{2}-\mu} \Gamma(\mu) \sin\dfrac{a}{2u} J_{\mu-\frac{1}{2}}\left(\dfrac{a}{2u}\right)$

$\hfill [a>0,\ u>0,\ \mathrm{Re}\,\mu>0]. \quad$ Ba V 203(21)

3. $\int_0^u \dfrac{(u^2-x^2)^{\mu-1}}{x^{2\mu}} \cos\dfrac{a}{x} dx = -\dfrac{\sqrt{\pi}}{2}\left(\dfrac{2}{a}\right)^{\mu-\frac{1}{2}} \Gamma(\mu) u^{\mu-\frac{3}{2}} N_{\frac{1}{2}-\mu}\left(\dfrac{a}{u}\right)$

$\hfill [a>0,\ u>0,\ 0<\mathrm{Re}\,\mu<1]. \quad$ Ba V 190(36)

4. $\int_u^\infty \dfrac{(x-u)^{\mu-1}}{x^{2\mu}} \cos\dfrac{a}{x} dx = \sqrt{\dfrac{\pi}{u}} a^{\frac{1}{2}-\mu} \Gamma(\mu) \cos\dfrac{a}{2u} J_{\mu-\frac{1}{2}}\left(\dfrac{a}{2u}\right)$

$\hfill [a>0,\ u>0,\ \mathrm{Re}\,\mu>0]. \quad$ Ba V 204(26)

1. $\int_0^\infty x^{\mu-1} \sin\dfrac{b^2}{x} \sin(a^2 x)\, dx = \dfrac{\pi}{4}\left(\dfrac{b}{a}\right)^\mu \mathrm{cosec}\,\dfrac{\mu\pi}{2} \times$ $\qquad$ 3.866

$\hfill \times [J_\mu(2ab) - J_{-\mu}(2ab) + I_{-\mu}(2ab) - I_\mu(2ab)]$

$\hfill [a>0,\ b>0,\ |\mathrm{Re}\,\mu|<1]. \quad$ W 184(1)+, Ba IV 322(42)

2. $\int_0^\infty x^{\mu-1} \sin\dfrac{b^2}{x} \cos(a^2 x)\, dx = \dfrac{\pi}{4}\left(\dfrac{b}{a}\right)^\mu \sec\dfrac{\mu\pi}{2} \times$

$\hfill \times [J_\mu(2ab) + J_{-\mu}(2ab) + I_\mu(2ab) - I_{-\mu}(2ab)]$

$\hfill [a>0,\ b>0,\ |\mathrm{Re}\,\mu|<1]. \quad$ Ba IV 322(43)

3. $\int_0^\infty x^{\mu-1} \cos\dfrac{b^2}{x} \cos(a^2 x)\, dx = \dfrac{\pi}{4}\left(\dfrac{b}{a}\right)^\mu \mathrm{cosec}\,\dfrac{\mu\pi}{2} \times$

$\hfill \times [J_{-\mu}(2ab) - J_\mu(2ab) + I_{-\mu}(2ab) - I_\mu(2ab)]$

$\hfill [a>0,\ b>0,\ |\mathrm{Re}\,\mu|<1]. \quad$ W 184(2)+, Ba IV 322(44)

1. $\int_0^1 \dfrac{\cos ax - \cos\dfrac{a}{x}}{1-x^2} dx = \dfrac{1}{2}\int_0^\infty \dfrac{\cos ax - \cos\dfrac{a}{x}}{1-x^2} dx = \dfrac{\pi}{2}\sin a \quad [a>0]$. $\qquad$ 3.867

$\hfill$ GH[334](7a)

3.867 2. $\int_0^1 \dfrac{\cos ax + \cos \dfrac{a}{x}}{1+x^2}\,dx = \dfrac{1}{2}\int_0^\infty \dfrac{\cos ax + \cos \dfrac{a}{x}}{1+x^2}\,dx = \dfrac{\pi}{2}e^{-a}$ $[a>0]$.

GH[334](7b)

3.868 1. $\int_0^\infty \sin\left(a^2 x + \dfrac{b^2}{x}\right)\dfrac{dx}{x} = \pi J_0(2ab)$ $[a>0,\ b>0]$. GH[334](11a), W 180(16)

2. $\int_0^\infty \cos\left(a^2 x + \dfrac{b^2}{x}\right)\dfrac{dx}{x} = -\pi N_0(2ab)$ $[a>0,\ b>0]$. GH[334](11a)

3. $\int_0^\infty \sin\left(a^2 x - \dfrac{b^2}{x}\right)\dfrac{dx}{x} = 0$ $[a>0,\ b>0]$. GH[334](11b)

4. $\int_0^\infty \cos\left(a^2 x - \dfrac{b^2}{x}\right)\dfrac{dx}{x} = 2K_0(2ab)$ $[a>0,\ b>0]$. GH[334](11b)

3.869 1. $\int_0^\infty \sin\left(ax - \dfrac{b}{x}\right)\dfrac{x\,dx}{\beta^2+x^2} = \dfrac{\pi}{2}\exp\left(-a\beta - \dfrac{b}{\beta}\right)$ $[a>0,\ b>0,\ \operatorname{Re}\beta>0]$.

Ba V 220(42)

2. $\int_0^\infty \cos\left(ax - \dfrac{b}{x}\right)\dfrac{dx}{\beta^2+x^2} = \dfrac{\pi}{2\beta}\exp\left(-a\beta - \dfrac{b}{\beta}\right)$ $[a>0, b>0, \operatorname{Re}\beta>0]$.

Ba V 222(58)

3.871 1. $\int_0^\infty x^{\mu-1}\sin\left[a\left(x+\dfrac{b^2}{x}\right)\right]dx = \pi b^\mu \left[J_\mu(2ab)\cos\dfrac{\mu\pi}{2} - N_\mu(2ab)\sin\dfrac{\mu\pi}{2}\right]$

$[a>0,\ b>0,\ \operatorname{Re}\mu<1]$. Ba IV 319(17)

2. $\int_0^\infty x^{\mu-1}\cos\left[a\left(x+\dfrac{b^2}{x}\right)\right]dx = -\pi b^\mu \left[J_\mu(2ab)\sin\dfrac{\mu\pi}{2} + N_\mu(2ab)\cos\dfrac{\mu\pi}{2}\right]$

$[a>0,\ b>0,\ |\operatorname{Re}\mu|<1]$. Ba IV 321(35)

3. $\int_0^\infty x^{\mu-1}\sin\left[a\left(x-\dfrac{b^2}{x}\right)\right]dx = 2b^\mu K_\mu(2ab)\sin\dfrac{\mu\pi}{2}$

$[a>0,\ b>0,\ |\operatorname{Re}\mu|<1]$. Ba IV 319(16)

4. $\int_0^\infty x^{\mu-1}\cos\left[a\left(x-\dfrac{b^2}{x}\right)\right]dx = 2b^\mu K_\mu(2ab)\cos\dfrac{\mu\pi}{2}$

$[a>0,\ b>0,\ |\operatorname{Re}\mu|<1]$. Ba IV 321(36)

1. $\int_0^1 \sin\left[a\left(x+\frac{1}{x}\right)\right] \sin\left[a\left(x-\frac{1}{x}\right)\right] \frac{dx}{1-x^2} =$ 　　3.872

$= \frac{1}{2} \int_0^\infty \sin\left[a\left(x+\frac{1}{x}\right)\right] \sin\left[a\left(x-\frac{1}{x}\right)\right] \frac{dx}{1-x^2} = -\frac{\pi}{4} \sin 2a$

$[a \geq 0]$.　BH[149](15), GH[334](8a)

2. $\int_0^1 \cos\left[a\left(x+\frac{1}{x}\right)\right] \cos\left[a\left(x-\frac{1}{x}\right)\right] \frac{dx}{1+x^2} =$

$= \frac{1}{2} \int_0^\infty \cos\left[a\left(x+\frac{1}{x}\right)\right] \cos\left[a\left(x-\frac{1}{x}\right)\right] \frac{dx}{1+x^2} = \frac{\pi}{4} e^{-2a}$

$[a \geq 0]$.　GH[334](8b)

1. $\int_0^\infty \sin\frac{a^2}{x^2} \cos b^2 x^2 \frac{dx}{x^2} = \frac{\sqrt{\pi}}{4\sqrt{2a}} [\sin(2ab) + \cos(2ab) + e^{-2ab}]$ 　　3.873

$[a > 0, b > 0]$.　Ba IV 24(15)

2. $\int_0^\infty \cos\frac{a^2}{x^2} \cos b^2 x^2 \frac{dx}{x^2} = \frac{\sqrt{\pi}}{4\sqrt{2a}} [\cos(2ab) - \sin(2ab) + e^{-2ab}]$

$[a > 0, b > 0]$.　Ba IV 24(16)

1. $\int_0^\infty \sin\left(a^2 x^2 + \frac{b^2}{x^2}\right) \frac{dx}{x^2} = \frac{\sqrt{\pi}}{2b} \sin\left(2ab + \frac{\pi}{4}\right)$ $[a > 0, b > 0]$. 　　3.874

BH[179](6)+, GH[334](10a)

2. $\int_0^\infty \cos\left(a^2 x^2 + \frac{b^2}{x^2}\right) \frac{dx}{x^2} = \frac{\sqrt{\pi}}{2b} \cos\left(2ab + \frac{\pi}{4}\right)$ $[a > 0, b > 0]$.

BH[179](8)+, GH[334](10a)

3. $\int_0^\infty \sin\left(a^2 x^2 - \frac{b^2}{x^2}\right) \frac{dx}{x^2} = -\frac{\sqrt{\pi}}{2\sqrt{2}b} e^{-2ab}$ $[a \geq 0, b > 0]$.　GH[334](10b)

4. $\int_0^\infty \cos\left(a^2 x^2 - \frac{b^2}{x^2}\right) \frac{dx}{x^2} = \frac{\sqrt{\pi}}{2\sqrt{2}b} e^{-2ab}$ $[a \geq 0, b > 0]$.　GH[334](10b)

5. $\int_0^\infty \sin\left(ax - \frac{b}{x}\right)^2 \frac{dx}{x^2} = \frac{\sqrt{2\pi}}{4b}$ $[a > 0, b > 0]$.　BH[179](13)+

6. $\int_0^\infty \cos\left(ax - \frac{b}{x}\right)^2 \frac{dx}{x^2} = \frac{\sqrt{2\pi}}{4b}$ $[a > 0, b > 0]$.　BH[179](14)+

3.875

1. $\int_u^\infty \dfrac{x \sin(p\sqrt{x^2-u^2})}{x^2+a^2} \cos bx\, dx = \dfrac{\pi}{2} \exp(-p\sqrt{a^2+u^2})\,\text{ch}\,ab$

$[0 < b < p]$.   Ba IV 27(39)

2. $\int_u^\infty \dfrac{x \sin(p\sqrt{x^2-u^2})}{a^2+x^2-u^2} \cos bx\, dx = \dfrac{\pi}{2} e^{-ap} \cos(b\sqrt{u^2-a^2})$

$[0 < b < p,\ a > 0]$.   Ba IV 27(38)

3. $\int_0^\infty \dfrac{\sin(p\sqrt{a^2+x^2})}{(a^2+x^2)^2} \cos bx\, dx = \dfrac{\pi p}{2a} e^{-ab}$ $[b > a > 0]$.   Ba IV 26(29)

3.876

1. $\int_0^\infty \dfrac{\sin(p\sqrt{x^2+a^2})}{\sqrt{x^2+a^2}} \cos bx\, dx = \dfrac{\pi}{2} J_0(a\sqrt{p^2-b^2})$ $[0 < b < p]$;

$= 0$ $[b > p > 0]$;

$[a > 0]$.   Ba IV 26(30)

2. $\int_0^\infty \dfrac{\cos(p\sqrt{x^2+a^2})}{\sqrt{x^2+a^2}} \cos bx\, dx = -\dfrac{\pi}{2} N_0(a\sqrt{b^2-p^2})$ $[0 < p < b]$;

$= K_0(a\sqrt{p^2-b^2})$ $[0 < b < p]$;

$[a > 0]$.   Ba IV 26(34)

3. $\int_0^\infty \dfrac{\cos(p\sqrt{x^2+a^2})}{x^2+c^2} \cos bx\, dx = \dfrac{\pi}{2c} e^{-bc} \cos(p\sqrt{a^2-c^2})$

$[c > 0,\ b > p]$.   Ba IV 26(33)

4. $\int_0^\infty \dfrac{\sin(p\sqrt{x^2+a^2})}{(x^2+c^2)\sqrt{x^2+a^2}} \cos bx\, dx = \dfrac{\pi}{2c} \dfrac{e^{-bc} \sin(p\sqrt{a^2-c^2})}{\sqrt{a^2-c^2}}$ $[c \neq a]$;

$= \dfrac{\pi}{2} e^{-ba} \dfrac{p}{a}$ $[c = a]$;

$[b > p,\ c > 0]$.   Ba IV 26(31)+

5. $\int_0^\infty \dfrac{\cos(p\sqrt{x^2+a^2})}{x^2+a^2} \cos bx\, dx = \dfrac{\pi}{2a} e^{-ab}$ $[b > a > 0]$.   Ba IV 27(35)+

6. $\int_0^\infty \dfrac{x \cos(p\sqrt{x^2+a^2})}{x^2+a^2} \sin bx\, dx = \dfrac{\pi}{2} e^{-ab}$ $[a > 0,\ b > p > 0]$.   Ba IV 85(29)+

7. $\int_0^u \dfrac{\cos(p\sqrt{u^2-x^2})}{\sqrt{u^2-x^2}} \cos bx\, dx = \dfrac{\pi}{2} J_0(u\sqrt{b^2+p^2})$.   Ba IV 28(42)

8. $\int_u^\infty \frac{\cos(p\sqrt{x^2-u^2})}{\sqrt{x^2-u^2}} \cos bx\, dx = K_0(u\sqrt{p^2-b^2})$ $\qquad [0 < b < |p|];$ $\qquad$ 3.876

$\qquad\qquad\qquad = -\frac{\pi}{2} N_0(u\sqrt{b^2-p^2})\quad [b > |p|].$

Ba IV 28(43)

1. $\int_0^u \frac{\sin(p\sqrt{u^2-x^2})}{\sqrt[4]{(u^2-x^2)^3}} \cos bx\, dx =$ $\qquad$ 3.877

$\qquad = \sqrt{\frac{\pi^3 p}{8}} J_{\frac{1}{4}}\left[\frac{u}{2}(\sqrt{b^2+p^2}-b)\right] J_{\frac{1}{4}}\left[\frac{u}{2}(\sqrt{b^2+p^2}+b)\right]$

$\qquad\qquad\qquad\qquad [b > 0,\ p > 0].$ $\qquad$ Ba IV 27(40)

2. $\int_u^\infty \frac{\sin(p\sqrt{x^2-u^2})}{\sqrt[4]{(x^2-u^2)^3}} \cos bx\, dx =$

$\qquad = -\sqrt{\frac{\pi^3 p}{8}} J_{\frac{1}{4}}\left[\frac{u}{2}(b-\sqrt{b^2-p^2})\right] N_{\frac{1}{4}}\left[\frac{u}{2}(b+\sqrt{b^2-p^2})\right]$

$\qquad\qquad\qquad\qquad [b > p > 0].$ $\qquad$ Ba IV 27(41)

3. $\int_0^u \frac{\cos(p\sqrt{u^2-x^2})}{\sqrt[4]{(u^2-x^2)^3}} \cos bx\, dx =$

$\qquad = \sqrt{\frac{\pi^3 p}{8}} J_{-\frac{1}{4}}\left[\frac{u}{2}(\sqrt{p^2+b^2}-b)\right] J_{-\frac{1}{4}}\left[\frac{u}{2}(\sqrt{p^2+b^2}+b)\right]$

$\qquad\qquad\qquad\qquad [u > 0,\ p > 0].$ $\qquad$ Ba IV 28(44)

4. $\int_u^\infty \frac{\cos(p\sqrt{x^2-u^2})}{\sqrt[4]{(x^2-u^2)^3}} \cos bx\, dx =$

$\qquad = -\sqrt{\frac{\pi^3 p}{8}} J_{-\frac{1}{4}}\left[\frac{u}{2}(b-\sqrt{b^2-p^2})\right] N_{\frac{1}{4}}\left[\frac{u}{2}(b+\sqrt{b^2-p^2})\right]$

$\qquad\qquad\qquad\qquad [b > p > 0].$ $\qquad$ Ba IV 28(45)

1. $\int_0^\infty \frac{\sin(p\sqrt{x^4+a^4})}{\sqrt{x^4+a^4}} \cos bx^2\, dx =$ $\qquad$ 3.878

$\qquad = \frac{1}{2}\sqrt{\left(\frac{\pi}{2}\right)^3 b}\, J_{-\frac{1}{4}}\left[\frac{a^2}{2}(p-\sqrt{p^2-b^2})\right] J_{\frac{1}{4}}\left[\frac{a^2}{2}(p+\sqrt{p^2-b^2})\right]$

$\qquad\qquad\qquad\qquad [p > b > 0].$ $\qquad$ Ba IV 26(32)

2. $\int_0^\infty \frac{\cos(p\sqrt{x^4+a^4})}{\sqrt{x^4+a^4}} \cos bx^2\, dx =$

$\qquad = -\frac{1}{2}\sqrt{\left(\frac{\pi}{2}\right)^3 b}\, J_{-\frac{1}{4}}\left[\frac{a^2}{2}(p-\sqrt{p^2-b^2})\right] N_{\frac{1}{4}}\left[\frac{a^2}{2}(p+\sqrt{p^2-b^2})\right]$

$\qquad\qquad\qquad\qquad [a > 0,\ p > b > 0].$ $\qquad$ Ba IV 27(36)

3.878

3. $\int_0^u \dfrac{\cos(p\sqrt{u^4-x^4})}{\sqrt{u^4-x^4}} \cos bx^2\, dx =$

$$= \dfrac{1}{2}\sqrt{\left(\dfrac{\pi}{2}\right)^3 b}\, J_{-\frac{1}{4}}\left[\dfrac{u^2}{2}(\sqrt{p^2+b^2}-p)\right] J_{-\frac{1}{4}}\left[\dfrac{u^2}{2}(\sqrt{p^2+b^2}+p)\right]$$

$[p > 0,\ b > 0]$.   Ba IV 28(46)

3.879

$$\int_0^\infty \sin ax^p\, \dfrac{dx}{x} = \dfrac{\pi}{2p} \quad [a>0,\ p>0].$$ GH[334](6)

3.881

1. $\displaystyle\int_0^{\pi/2} x\sin(a\,\text{tg}\,x)\, dx = \dfrac{\pi}{4} e^{-a}[C+\ln 2a - e^{2a}\text{Ei}(-2a)] \quad [a>0].$ BH[205](9)

2. $\displaystyle\int_0^\infty \sin(a\,\text{tg}\,x)\, \dfrac{dx}{x} = \dfrac{\pi}{2}(1-e^{-a}) \quad [a>0].$ BH[151](6)

3. $\displaystyle\int_0^\infty \sin(a\,\text{tg}\,x)\cos x\, \dfrac{dx}{x} = \dfrac{\pi}{2}(1-e^{-a}) \quad [a>0].$ BH[151](19)

4. $\displaystyle\int_0^\infty \cos(a\,\text{tg}\,x)\sin x\, \dfrac{dx}{x} = \dfrac{\pi}{2} e^{-a} \quad [a>0].$ BH[151](20)

5. $\displaystyle\int_0^\infty \sin(a\,\text{tg}\,x)\sin 2x\, \dfrac{dx}{x} = \dfrac{1+a}{2}\pi e^{-a} \quad [a>0].$ BH[152](11)

6. $\displaystyle\int_0^\infty \cos(a\,\text{tg}\,x)\sin^3 x\, \dfrac{dx}{x} = \dfrac{1-a}{4}\pi e^{-a} \quad a>0].$ BH[151](23)

7. $\displaystyle\int_0^\infty \sin(a\,\text{tg}\,x)\,\text{tg}\,\dfrac{x}{2}\cos^2 x\, \dfrac{dx}{x} = \dfrac{1+a}{4}\pi e^{-a} \quad [a>0].$ BH[152](13)

8. $\displaystyle\int_0^{\pi/2} \cos(a\,\text{tg}\,x)\, \dfrac{x\,dx}{\sin 2x} = -\dfrac{\pi}{4}\text{Ei}(-a) \quad [a>0].$ BH[206](15)

9. $\displaystyle\int_0^{\pi/2} \sin(a\,\text{ctg}\,x)\, \dfrac{x\,dx}{\sin^2 x} = \dfrac{1-e^{-a}}{2a}\pi \quad [a>0].$ Li[206](14)

10. $\displaystyle\int_0^{\pi/2} x\cos(a\,\text{tg}\,x)\,\text{tg}\,x\, dx = -\dfrac{\pi}{4}e^{-a}[C+\ln 2a + e^{2a}\text{Ei}(-2a)] \quad [a>0].$

   BH[205](10)

11. $\int_0^\infty \cos(a\,\operatorname{tg} x)\,\operatorname{tg} x\,\dfrac{dx}{x} = \dfrac{\pi}{2}\,e^{-a}\quad [a>0].$ BH[151](21) 3.881

12. $\int_0^\infty \cos(a\,\operatorname{tg} x)\sin^2 x\,\operatorname{tg} x\,\dfrac{dx}{x} = \dfrac{1-a}{16}\,\pi e^{-a}\quad [a>0].$ BH[152](15)

13. $\int_0^\infty \sin(a\,\operatorname{tg} x)\,\operatorname{tg}^2 x\,\dfrac{dx}{x} = \dfrac{\pi}{2}\,e^{-a}\quad [a>0].$ BH[152](9)

14. $\int_0^\infty \cos(a\,\operatorname{tg} 2x)\,\operatorname{tg} x\,\dfrac{dx}{x} = \dfrac{\pi}{2}\,e^{-a}\quad [a>0].$ BH[151](22)

15. $\int_0^\infty \sin(a\,\operatorname{tg} 2x)\cos^2 2x\,\operatorname{tg} x\,\dfrac{dx}{x} = \dfrac{1+a}{4}\,\pi e^{-a}\quad [a>0].$ BH[152](13)

16. $\int_0^\infty \sin(a\,\operatorname{tg} 2x)\,\operatorname{tg} x\,\operatorname{tg} 2x\,\dfrac{dx}{x} = \dfrac{\pi}{2}\,e^{-a}\quad [a>0].$ BH[152](10)

17. $\int_0^\infty \sin(a\,\operatorname{tg} 2x)\,\operatorname{tg} x\,\operatorname{ctg} 2x\,\dfrac{dx}{x} = \dfrac{\pi}{2}(1-e^{-a})\quad [a>0].$ BH[180](6)

1. $\int_0^\infty \sin(a\,\operatorname{tg}^2 x)\,\dfrac{x\,dx}{b^2+x^2} = \dfrac{\pi}{2}\,[\exp(-a\,\operatorname{th} b) - e^{-a}]\quad [a>0,\,b>0].$ 3.882
BH[160](22)

2. $\int_0^\infty \cos(a\,\operatorname{tg}^2 x)\cos x\,\dfrac{dx}{b^2+x^2} = \dfrac{\pi}{2b}\,[\operatorname{ch} b\,\exp(-a\,\operatorname{th} b) - e^{-a}\operatorname{sh} b]$

$[a>0,\,b>0].$ BH[163](3)

3. $\int_0^\infty \cos(a\,\operatorname{tg}^2 x)\operatorname{cosec} 2x\,\dfrac{x\,dx}{b^2+x^2} = \dfrac{\pi}{2\operatorname{sh} 2b}\,\exp(-a\,\operatorname{th} b)\quad [a>0,\,b>0].$

BH[191](10)

4. $\int_0^\infty \cos(a\,\operatorname{tg}^2 x)\,\operatorname{tg} x\,\dfrac{x\,dx}{b^2+x^2} = \dfrac{\pi}{2\operatorname{ch} b}\,[e^{-a}\operatorname{ch} b - \exp(-a\,\operatorname{th} b)\,\operatorname{sh} b]$

$[a>0,\,b>0].$ BH[163](4)

3.882

5. $\int_0^\infty \cos(a\,\text{tg}^2 x)\,\text{ctg}\,x\,\dfrac{x\,dx}{b^2+x^2} = \dfrac{\pi}{2}[\text{cth}\,b\,\exp(-a\,\text{th}\,b) - e^{-a}]$

$[a>0,\ b>0].$ BH[163](5)

6. $\int_0^\infty \cos(a\,\text{tg}^2 x)\,\text{ctg}\,2x\,\dfrac{x\,dx}{b^2+x^2} = \dfrac{\pi}{2}[\text{cth}\,2b\,\exp(-a\,\text{th}\,b) - e^{-a}]$

$[a>0,\ b>0].$ BH[191](11)

3.883

1. $\int_0^1 \cos(a\ln x)\dfrac{dx}{(1+x)^2} = \dfrac{a\pi}{2\,\text{sh}\,a\pi}.$ BH[404](4)

2. $\int_0^1 x^{\mu-1}\sin(\beta\ln x)\,dx = -\dfrac{\beta}{\beta^2+\mu^2}\quad[\text{Re}\,\mu > |\,\text{Im}\,\beta|].$ Ba IV 319(19)

3. $\int_0^1 x^{\mu-1}\cos(\beta\ln x)\,dx = \dfrac{\mu}{\beta^2+\mu^2}\quad[\text{Re}\,\mu > |\,\text{Im}\,\beta|].$ Ba IV 321(38)

3.884

$\int_{-\infty}^\infty \dfrac{\sin a\sqrt{|x|}}{x-b}\,\text{sign}\,x\,dx = \cos a\sqrt{|b|} + \exp(-a\sqrt{|b|})$

$[a>0].$ Ba V 253(46)

Trigonometrische Funktionen und die Exponentialfunktion
Trigonometric functions and exponentials

3.89–3.91

3.891

1. $\int_0^{2\pi} e^{imx}\sin nx\,dx = 0\ [m\neq n;\ m=n=0];$

$= \pi i\ [m=n\neq 0].$

2. $\int_0^{2\pi} e^{imx}\cos nx\,dx = 0\ [m\neq n];$

$= \pi\ [m=n\neq 0];$

$= 2\pi\ [m=n=0].$

3.892

1. $\int_0^\pi e^{i\beta x}\sin^{\nu-1}x\,dx = \dfrac{\pi e^{i\beta\frac{\pi}{2}}}{2^{\nu-1}\nu B\left(\dfrac{\nu+\beta+1}{2},\dfrac{\nu-\beta+1}{2}\right)}$

$[\text{Re}\,\nu > -1].$ NG 158, Ba I 12(29)

2. $\int_{-\frac{\pi}{2}}^{\frac{\pi}{2}} e^{i\beta x}\cos^{\nu-1}x\,dx = \dfrac{\pi}{2^{\nu-1}\nu B\left(\dfrac{\nu+\beta+1}{2},\dfrac{\nu-\beta+1}{2}\right)}$

$[\text{Re}\,\nu > -1].$ GH[335](19)

3.
$$\int_0^{\frac{\pi}{2}} e^{i2\beta x} \sin^{2\mu} x \cos^{2\nu} x \, dx =$$
3.892

$$= \frac{1}{2^{2\mu+2\nu+1}} \left\{ \exp\left[i\pi\left(\beta - \nu - \frac{1}{2}\right)\right] B(\beta - \mu - \nu, 2\nu + 1) \times \right.$$

$$\times F(-2\mu, \beta - \mu - \nu; 1 + \beta - \mu + \nu; -1) + \exp\left[i\pi\left(\mu + \frac{1}{2}\right)\right] \times$$

$$\left. \times B(\beta - \mu - \nu, 2\mu + 1) F(-2\nu, \beta - \mu - \nu; 1 + \beta + \mu - \nu; -1) \right\}$$

$$\left[\operatorname{Re}\mu > -\frac{1}{2}, \ \operatorname{Re}\nu > -\frac{1}{2}\right]. \qquad \text{Ba I 80(6)}$$

4.
$$\int_0^\pi e^{i2\beta x} \sin^{2\mu} x \cos^{2\nu} x \, dx =$$

$$= \frac{\pi \exp[i\pi(\beta - \nu)] F(-2\nu, \beta - \mu - \nu; 1 + \beta + \mu - \nu; -1)}{4^{\mu+\nu}(2\mu+1) B(1 - \beta + \mu + \nu, 1 + \beta + \mu - \nu)}. \qquad \text{Ba I 80(8)}$$

5.
$$\int_0^{\frac{\pi}{2}} e^{i(\mu+\nu)x} \sin^{\mu-1} x \cos^{\nu-1} x \, dx = e^{i\mu\frac{\pi}{2}} B(\mu, \nu) =$$

$$= \frac{1}{2^{\mu+\nu-1}} e^{i\mu\frac{\pi}{2}} \left\{ \frac{1}{\mu} F(1-\nu, 1; \mu+1; -1) + \frac{1}{\nu} F(1-\mu, 1; \nu+1; -1) \right\}$$

$$[\operatorname{Re}\mu > 0, \ \operatorname{Re}\nu > 0]. \qquad \text{Ba I 80(7)}$$

1.
$$\int_0^\infty e^{-px} \sin(qx + \lambda) \, dx = \frac{1}{p^2 + q^2}(q \cos \lambda + p \sin \lambda) \quad [p > 0]. \qquad \text{BH[261](3)}$$
3.893

2.
$$\int_0^\infty e^{-px} \cos(qx + \lambda) \, dx = \frac{1}{p^2 + q^2}(p \cos \lambda - q \sin \lambda) \quad [p > 0]. \qquad \text{BH[261](4)}$$

3.
$$\int_0^\infty e^{-x \cos t} \cos(t - x \sin t) \, dx = 1. \qquad \text{BH[261](7)}$$

4.
$$\int_0^\infty \frac{e^{-\beta x} \sin ax}{\sin bx} \, dx = \frac{1}{2bi} \left[\psi\left(\frac{a+b}{2b} - i\frac{\beta}{2b}\right) - \psi\left(\frac{b-a}{2b} - i\frac{\beta}{2b}\right) \right]$$

$$[\operatorname{Re}\beta > 0, \ b \neq 0]. \qquad \text{GH[335](15)}$$

5.
$$\int_0^\infty \frac{e^{-2px} \sin[(2n+1)x]}{\sin x} \, dx = \frac{1}{2p} + \sum_{k=1}^n \frac{p}{p^2 + k^2} \quad [p > 0]. \qquad \text{BH[267](15)}$$

6.
$$\int_0^\infty \frac{e^{-px} \sin 2nx}{\sin x} \, dx = 2p \sum_{k=0}^{n-1} \frac{1}{p^2 + (2k+1)^2} \quad [p > 0]. \qquad \text{GH[335](15c)}$$

3.893

7. $\int_0^\infty e^{-px} \cos[(2n+1)x] \, \text{tg}\, x \, dx = \frac{2n+1}{p^2+(2n+1)^2} +$

$$+ (-1)^n 2 \sum_{k=0}^{n-1} \frac{(-1)^k (2k+1)}{p^2+(2k+1)^2} \quad [p>0]. \qquad \text{Li[267](16)}$$

3.894

$$\int_{-\pi}^\pi [\beta + \sqrt{\beta^2-1} \cos x]^\nu e^{inx} \, dx = \frac{2\pi \Gamma(\nu+1) P_\nu^m(\beta)}{\Gamma(\nu+m+1)} \quad [\text{Re}\, \beta > 0]. \qquad \text{Ba I 157(15)}$$

3.895

1. $\int_0^\infty e^{-\beta x} \sin^{2m} x \, dx = \frac{(2m)!}{\beta(\beta^2+2^2)(\beta^2+4^2)\cdots[\beta^2+(2m)^2]}; \quad [\text{Re}\, \beta > 0].$

F II 625, W 563+

2. $\int_0^\pi e^{-px} \sin^{2m} x \, dx = \frac{(2m)!(1-e^{-p\pi})}{p(p^2+2^2)(p^2+4^2)\cdots[p^2+(2m)^2]} \quad [p \neq 0]. \qquad \text{GH[335](4a)}$

3. $\int_0^{\pi/2} e^{-px} \sin^{2m} x \, dx = \frac{(2m)!}{p(p^2+2^2)(p^2+4^2)\cdots[p^2+(2m)^2]} \times$

$$\times \left\{ 1 - e^{-\frac{p\pi}{2}} \left[1 + \frac{p^2}{2!} + \frac{p^2(p^2+2^2)}{4!} + \cdots + \frac{p^2(p^2+2^2)\cdots[p^2+(2m-2)^2]}{(2m)!} \right] \right\}$$

$[p \neq 0]$.

BH[270](4)

4. $\int_0^\infty e^{-\beta x} \sin^{2m+1} x \, dx = \frac{(2m+1)!}{(\beta^2+1^2)(\beta^2+3^2)\cdots[\beta^2+(2m+1)^2]}$

$[\text{Re}\, \beta > 0]$.

F II 625, W 564+

5. $\int_0^\pi e^{-px} \sin^{2m+1} x \, dx = \frac{(2m+1)!(1+e^{-p\pi})}{(p^2+1^2)(p^2+3^2)\cdots[p^2+(2m+1)^2]} \quad [p \neq 0]. \qquad \text{GH[335](4b)}$

6. $\int_0^{\pi/2} e^{-px} \sin^{2m+1} x \, dx = \frac{(2m+1)!}{(p^2+1^2)(p^2+3^2)\cdots[p^2+(2m+1)^2]} \times$

$$\times \left\{ 1 - p e^{\frac{p\pi}{2}} \left[1 + \frac{p^2+1^2}{3!} + \cdots + \frac{(p^2+1^2)(p^2+3^2)\cdots[p^2+(2m-1)^2]}{(2m+1)!} \right] \right\}$$

$[p \neq 0]$.

BH[270](5)

7. $\int_0^\infty e^{-px} \cos^{2m} x \, dx = \frac{(2m)!}{p(p^2+2^2)\cdots[p^2+(2m)^2]} \times$

$$\times \left\{ 1 + \frac{p^2}{2!} + \frac{p^2(p^2+2^2)}{4!} + \cdots + \frac{p^2(p^2+2^2)\cdots[p^2+(2m-2)^2]}{(2m)!} \right\}$$

$[p > 0]$.

BH[262](3)

8. $\int_0^{\frac{\pi}{2}} e^{-px} \cos^{2m} x \, dx = \frac{(2m)!}{p(p^2 + 2^2) \ldots [p^2 + (2m)^2]} \times$ 3.895

$\times \left\{ -e^{-p\frac{\pi}{2}} + 1 + \frac{p^2}{2!} + \frac{p^2(p^2 + 2^2)}{4!} + \ldots + \frac{p^2(p^2 + 2^2) \ldots [p^2 + (2m-2)^2]}{(2m)!} \right\}$

$[p \neq 0].$ BH[270](6)

9. $\int_0^\infty e^{-px} \cos^{2m+1} x \, dx = \frac{(2m+1)! \, p}{(p^2 + 1^2)(p^2 + 3^2) \ldots [p^2 + (2m+1)^2]} \times$

$\times \left\{ 1 + \frac{p^2 + 1^2}{3!} + \frac{(p^2 + 1^2)(p^2 + 3^2)}{5!} + \ldots + \frac{(p^2 + 1^2)(p^2 + 3^2) \ldots [p^2 + (2m-1)^2]}{(2m+1)!} \right\}$

$[p > 0].$ BH[262](4)

10. $\int_0^{\frac{\pi}{2}} e^{-px} \cos^{2m+1} x \, dx = \frac{(2m+1)!}{(p^2 + 1^2)(p^2 + 3^2) \ldots [p^2 + (2m+1)^2]} \times$

$\times \left\{ e^{-p\frac{\pi}{2}} + p \left[1 + \frac{p^2 + 1^2}{3!} + \ldots + \frac{(p^2 + 1)(p^2 + 3^2) \ldots [p^2 + (2m-1)^2]}{(2m+1)!} \right] \right\}$

$[p \neq 0].$ BH[270](7)

11. $\int_0^\infty e^{-\beta x} \sin^{2n} x \sin ax \, dx =$

$= -\frac{1}{(-4)^{n+1}(2n+1)} \left\{ \frac{1}{\binom{\frac{a}{2} + i\frac{\beta}{2} + n}{2n+1}} + \frac{1}{\binom{\frac{a}{2} - i\frac{\beta}{2} + n}{2n+1}} \right\}$

$[\text{Re}\,\beta > 0, a > 0].$ Ba IV 80(19)

12. $\int_0^\infty e^{-\beta x} \sin^{2n-1} x \sin ax \, dx =$

$= \frac{-i}{(-4)^{n+1} n} \left\{ \frac{1}{\binom{\frac{a}{2} - i\frac{\beta}{2} + n - \frac{1}{2}}{2n}} - \frac{1}{\binom{\frac{a}{2} + i\frac{\beta}{2} + n - \frac{1}{2}}{2n}} \right\}$

$[\text{Re}\,\beta > 0, \; a > 0].$ Ba IV 80(20)+

13. $\int_0^\infty e^{-\beta x} \sin^{2n} x \cos ax \, dx =$

$= \frac{(-1)^n i}{(2n+1) 2^{2n+2}} \left\{ \frac{1}{\binom{\frac{a}{2} + i\frac{\beta}{2} + n}{2n+1}} - \frac{1}{\binom{\frac{a}{2} - i\frac{\beta}{2} + n}{2n+1}} \right\}$

$[\text{Re}\,\beta > 0, \; a > 0].$ Ba IV 20(12)+

3.895 14. $\int_0^\infty e^{-\beta x} \sin^{2n-1} x \cos ax \, dx =$

$$= \frac{(-1)^n}{2^{2n+2} n} \left\{ \frac{1}{\binom{\frac{a}{2} - i\frac{\beta}{2} + n - \frac{1}{2}}{2n}} + \frac{1}{\binom{\frac{a}{2} + i\frac{\beta}{2} + n - \frac{1}{2}}{2n}} \right\}$$

[Re $\beta > 0$, $a > 0$]. Ba IV 20(13)+

3.896 1. $\int_{-\infty}^\infty e^{-q^2 x^2} \sin [p(x+\lambda)] \, dx = \frac{\sqrt{\pi}}{q} e^{-\frac{p^2}{4q^2}} \sin p\lambda.$ BH[269](2)

2. $\int_{-\infty}^\infty e^{-q^2 x^2} \cos [p(x+\lambda)] \, dx = \frac{\sqrt{\pi}}{q} e^{-\frac{p^2}{4q^2}} \cos p\lambda.$ BH[269](3)

3. $\int_0^\infty e^{-ax^2} \sin bx \, dx = \frac{b}{2a} \exp\left(-\frac{b^2}{4a}\right) {}_1F_1\left(\frac{1}{2}; \frac{3}{2}; \frac{b^2}{4a}\right) =$

$$= \frac{b}{2a} {}_1F_1\left(1; \frac{3}{2}; -\frac{b^2}{4a}\right);$$ Ba IV 73(18)

$$= \frac{b}{2a} \sum_{k=1}^\infty \frac{1}{(2k-1)!!} \left(-\frac{b^2}{2a}\right)^{k-1} \quad [a > 0].$$ F II 725–726

4. $\int_0^\infty e^{-\beta x^2} \cos bx \, dx = \frac{1}{2} \sqrt{\frac{\pi}{\beta}} \exp\left(-\frac{b^2}{4\beta}\right)$ [Re $\beta > 0$]. BH[263](2)

3.897 1. $\int_0^\infty e^{-\beta x^2 - \gamma x} \sin bx \, dx = -\frac{i}{4} \sqrt{\frac{\pi}{\beta}} \left\{ \exp\frac{(\gamma - ib)^2}{4\beta} \left[1 - \Phi\left(\frac{\gamma - ib}{2\sqrt{\beta}}\right)\right] - \right.$

$$\left. - \exp\frac{(\gamma + ib)^2}{4\beta} \left[1 - \Phi\left(\frac{\gamma + ib}{2\sqrt{\beta}}\right)\right] \right\} \quad [\text{Re } \beta > 0].$$ Ba IV 74(27)

2. $\int_0^\infty e^{-\beta x^2 - \gamma x} \cos bx \, dx = \frac{1}{4} \sqrt{\frac{\pi}{\beta}} \left\{ \exp\frac{(\gamma - ib)^2}{4\beta} \left[1 - \Phi\left(\frac{\gamma - ib}{2\sqrt{\beta}}\right)\right] + \right.$

$$\left. + \exp\frac{(\gamma + ib)^2}{4\beta} \left[1 - \Phi\left(\frac{\gamma + ib}{2\sqrt{\beta}}\right)\right] \right\}$$

[Re $\beta > 0$]. Ba IV 15(16)

3.898 1. $\int_0^\infty e^{-\beta x^2} \sin ax \sin bx \, dx = \frac{1}{4} \sqrt{\frac{\pi}{\beta}} \left\{ e^{-\frac{(a-b)^2}{4\beta}} - e^{-\frac{(a+b)^2}{4\beta}} \right\}$ [Re $\beta > 0$, $ab > 0$].

BH[263](4)

Trigonometrische Funktionen/Trigonometric Functions

2. $\int_0^\infty e^{-\beta x^2} \cos ax \cos bx \, dx = \frac{1}{4}\sqrt{\frac{\pi}{\beta}} \left\{ e^{-\frac{(a-b)^2}{4\beta}} + e^{-\frac{(a+b)^2}{4\beta}} \right\}$ $[\operatorname{Re}\beta > 0]$.   BH[263](5)   3.898

3. $\int_0^\infty e^{-px^2} \sin^2 ax \, dx = \frac{1}{4}\sqrt{\frac{\pi}{p}} \left(1 - e^{-\frac{a^2}{p}} \right)$ $[p > 0]$.   BH[263](6)

1. $\int_0^\infty \frac{e^{-p^2x^2} \sin[(2n+1)x]}{\sin x} \, dx = \frac{\sqrt{\pi}}{p}\left[\frac{1}{2} + \sum_{k=1}^n e^{-\left(\frac{k}{p}\right)^2} \right]$ $[p > 0]$.   BH[267](17)   3.899

2. $\int_0^\infty \frac{e^{-p^2x^2} \sin[(4n+1)x]}{\cos x} \, dx = \frac{\sqrt{\pi}}{p}\left[\frac{1}{2} + \sum_{k=1}^{2n} (-1)^k e^{\left(\frac{k}{p}\right)^2} \right]$ $[p > 0]$.   BH[267](18)

3. $\int_0^\infty \frac{e^{-px^2} \, dx}{1 - 2a \cos x + a^2} = \frac{\sqrt{\frac{\pi}{p}}}{1 - a^2}\left\{ \frac{1}{2} + \sum_{k=1}^\infty a^k \exp\left(-\frac{k^2}{4p} \right) \right\}$ $[a^2 < 1, \, p > 0]$;

       BH[266](1)

$= \frac{\sqrt{\frac{\pi}{p}}}{a^2 - 1}\left\{ \frac{1}{2} + \sum_{k=1}^\infty a^{-k} \exp\left(-\frac{k^2}{4p} \right) \right\}$ $[a^2 > 1, \, p > 0]$.

       Li[266](1)

1. $\int_0^\infty \frac{\sin ax}{e^{\beta x} + 1} \, dx = \frac{1}{2a} - \frac{\pi}{2\beta \operatorname{sh}\frac{a\pi}{\beta}}$ $[a > 0, \, \operatorname{Re}\beta > 0]$.   BH[264](1)   3.911

2. $\int_0^\infty \frac{\sin ax}{e^{\beta x} - 1} \, dx = \frac{\pi}{2\beta} \operatorname{cth}\left(\frac{\pi a}{\beta} \right) - \frac{1}{2a}$ $[a > 0, \, \operatorname{Re}\beta > 0]$.   BH[264](2), WW 122+

3. $\int_0^\infty \frac{\sin ax}{e^x - 1} e^{\frac{x}{2}} \, dx = -\frac{1}{2}\operatorname{th}(a\pi)$ $[a > 0]$.   Ba IV 73(13)

4. $\int_0^\infty \frac{\sin ax}{1 - e^{-x}} e^{-nx} \, dx = \frac{\pi}{2} - \frac{1}{2a} + \frac{\pi}{e^{2\pi a} - 1} - \sum_{k=1}^{n-1} \frac{a}{a^2 + k^2}$ $[a > 0]$.   BH[264](8)

5. $\int_0^\infty \frac{\sin ax}{e^{\beta x} - e^{\gamma x}} \, dx = \frac{1}{2i(\beta - \gamma)}\left[\psi\left(\frac{\beta + ia}{\beta - \gamma} \right) - \psi\left(\frac{\beta - ia}{\beta - \gamma} \right) \right]$

       $[\operatorname{Re}\beta > 0, \, \operatorname{Re}\gamma > 0]$.   GH[335](8)

6. $\int_0^\infty \frac{\sin ax \, dx}{e^{\beta x}(e^{-x} - 1)} = \frac{i}{2}[\psi(\beta + ia) - \psi(\beta - ia)]$ $[\operatorname{Re}\beta > -1]$.   Ba IV 73(15)

1. $\int_0^\infty e^{-\beta x}(1 - e^{-\gamma x})^{\nu - 1} \sin ax \, dx = -\frac{i}{2\gamma}\left[B\left(\nu, \frac{\beta - ia}{\gamma} \right) - B\left(\nu, \frac{\beta + ia}{\gamma} \right) \right]$ 3.912

       $[\operatorname{Re}\beta > 0, \, \operatorname{Re}\gamma > 0, \, \operatorname{Re}\nu > 0, \, a > 0]$.   Ba IV 73(17)

2. $\int_0^\infty e^{-\beta x}(1 - e^{-\gamma x})^{\nu - 1} \cos ax \, dx = \frac{1}{2\gamma}\left[B\left(\nu, \frac{\beta - ia}{\gamma} \right) + B\left(\nu, \frac{\beta + ia}{\gamma} \right) \right]$

       $[\operatorname{Re}\beta > 0, \, \operatorname{Re}\gamma > 0, \, \operatorname{Re}\nu > 0, \, a > 0]$.   Ba IV 15(10)

3.913

1.
$$\int_{-\frac{\pi}{2}}^{\frac{\pi}{2}} e^{i\beta x} \cos^\nu x (\beta^2 e^{ix} + \nu^2 e^{-ix})^\mu \, dx =$$

$$= \frac{\pi\, _2F_1\left(-\mu, \frac{\beta}{2} - \frac{\nu}{2} - \frac{\mu}{2}; 1 + \frac{\beta}{2} + \frac{\nu}{2} - \frac{\mu}{2}; \frac{\beta^2}{\nu^2}\right)}{2^\nu (\nu + 1)\, \mathrm{B}\left(1 + \frac{\beta}{2} + \frac{\nu}{2} - \frac{\mu}{2},\, 1 - \frac{\beta}{2} + \frac{\nu}{2} + \frac{\mu}{2}\right)}$$

$[\mathrm{Re}\,\nu > -1,\ |\nu| > |\beta|].$ Ba I 81(11)+

2.
$$\int_{-\frac{\pi}{2}}^{\frac{\pi}{2}} e^{-iux} \cos^\mu x (a^2 e^{ix} + b^2 e^{-ix})^\nu \, dx =$$

$$= \frac{\pi b^{2\nu}\, _2F_1\left(-\nu, \frac{u + \mu + \nu}{2}; 1 + \frac{\mu - \nu - u}{2}; \frac{a^2}{b^2}\right)}{2^\mu (\mu + 1)\, \mathrm{B}\left(1 - \frac{u + \nu - \mu}{2},\, 1 + \frac{u + \mu + \nu}{2}\right)} \quad \text{für } a^2 < b^2;$$

$$= \frac{\pi a^{2\nu}\, _2F_1\left(-\nu, \frac{\mu + \nu - u}{2}; 1 + \frac{\mu - \nu + u}{2}; \frac{b^2}{a^2}\right)}{2^\mu (\mu + 1)\, \mathrm{B}\left(1 + \frac{u + \mu - \nu}{2},\, 1 + \frac{\mu + \nu - u}{2}\right)} \quad \text{für } b^2 < a^2$$

$[\mathrm{Re}\,\mu > -1].$ Ba IV 122(31)+

3.914
$$\int_0^\infty e^{-\beta\sqrt{\gamma^2 + x^2}} \cos bx \, dx = \frac{\beta\gamma}{\sqrt{\beta^2 + b^2}} K_1(\gamma \sqrt{\beta^2 + b^2})$$

$[\mathrm{Re}\,\beta > 0,\ \mathrm{Re}\,\gamma > 0].$ Ba IV 16(26)

3.915

1.
$$\int_0^\pi e^{a\cos x} \sin x \, dx = \frac{2}{a} \operatorname{sh} a.$$ GH[337](15c)

2.
$$\int_0^\pi e^{i\beta\cos x} \cos nx \, dx = i^n \pi J_n(\beta).$$ Ba II 81(2)

3.
$$\int_{-\frac{\pi}{2}}^{\frac{\pi}{2}} e^{i\beta\cos x} \cos^{2\nu} x \, dx = \sqrt{\pi} \left(\frac{2}{\beta}\right)^\nu \Gamma\left(\nu + \frac{1}{2}\right) J_\nu(\beta)$$

$\left[\mathrm{Re}\,\nu > -\frac{1}{2}\right].$ Ba II 81(6)

4.
$$\int_0^\pi e^{\pm\beta\cos x} \sin^{2\nu} x \, dx = \sqrt{\pi} \left(\frac{2}{\beta}\right)^\nu \Gamma\left(\nu + \frac{1}{2}\right) I_\nu(\beta)$$

$\left[\mathrm{Re}\,\nu > -\frac{1}{2}\right].$ GH[337](15b)

5.
$$\int_0^\pi e^{i\beta\cos x} \sin^{2\nu} x \, dx = \sqrt{\pi} \left(\frac{2}{\beta}\right)^\nu \Gamma\left(\nu + \frac{1}{2}\right) J_\nu(\beta) \quad \left[\mathrm{Re}\,\nu > -\frac{1}{2}\right].$$

W 25(2), W 48(6)

3.916

1. $\int_0^{\frac{\pi}{2}} e^{-p^2 \operatorname{tg} x} \frac{\sin\frac{x}{2}\sqrt{\cos x}}{\sin 2x} dx = \left[C(p) - \frac{1}{2}\right]^2 + \left[S(p) - \frac{1}{2}\right]^2.$ NI 33(18)+

2. $\int_0^{\frac{\pi}{2}} \frac{\exp(-p \operatorname{tg} x) dx}{\sin 2x + a \cos 2x + a} = -\frac{1}{2} e^{ap} \operatorname{Ei}(-ap) \quad [p > 0] \quad (\text{cf. } 3.5524., 6.).$ BH[273](11)

3. $\int_0^{\frac{\pi}{2}} \frac{\exp(-p \operatorname{ctg} x) dx}{\sin 2x + a \cos 2x - a} = -\frac{1}{2} e^{-ap} \operatorname{Ei}(ap) \quad [p > 0] \quad (\text{cf. } 3.552\ 4., 6.).$ BH[273](12)

4. $\int_0^{\frac{\pi}{2}} \frac{\exp(-p \operatorname{tg} x) \sin 2x\, dx}{(1-a^2) - 2a^2 \cos 2x - (1+a^2)\cos^2 2x} = -\frac{1}{4}[e^{-ap}\operatorname{Ei}(ap) + e^{ap}\operatorname{Ei}(-ap)]$

   $[p > 0].$ BH[273](13)

5. $\int_0^{\frac{\pi}{2}} \frac{\exp(-p \operatorname{ctg} x) \sin 2x\, dx}{(1-a^2) + 2a^2 \cos 2x - (1+a^2)\cos^2 2x} = -\frac{1}{4}[e^{-ap}\operatorname{Ei}(ap) + e^{ap}\operatorname{Ei}(-ap)]$

   $[p > 0].$ BH[273](14)

3.917

1. $\int_0^{\frac{\pi}{2}} e^{-2\beta \operatorname{ctg} x} \cos^{\nu-\frac{1}{2}} x \sin^{-(\nu+1)} x \sin\left[\beta - \left(\nu - \frac{1}{2}\right)x\right] dx =$

   $= \frac{\sqrt{\pi}}{2 \cdot (2\beta)^\nu} \Gamma\left(\nu + \frac{1}{2}\right) J_\nu(\beta) \quad \left[\operatorname{Re}\nu > -\frac{1}{2}\right].$ W 169(7)

2. $\int_0^{\frac{\pi}{2}} e^{-2\beta \operatorname{ctg} x} \cos^{\nu-\frac{1}{2}} x \sin^{-(\nu+1)} x \cos\left[\beta - \left(\nu - \frac{1}{2}\right)x\right] dx =$

   $= \frac{\sqrt{\pi}}{2 \cdot (2\beta)^\nu} \Gamma\left(\nu + \frac{1}{2}\right) N_\nu(\beta) \quad \left[\operatorname{Re}\nu > -\frac{1}{2}\right].$ W 169(8)

3.918

1. $\int_0^{\frac{\pi}{2}} \frac{\cos^\mu x}{\sin^{2\mu+2} x} e^{i\gamma(\beta-\mu x) - 2\beta \operatorname{ctg} x} dx = \frac{i\gamma}{2}\sqrt{\frac{\pi}{2\beta}} (2\beta)^{-\mu} \Gamma(\mu+1) H^{(\varepsilon)}_{\mu+\frac{1}{2}}(\beta)$

   $[\varepsilon = 1, 2;\ \nu = (-1)^{\varepsilon+1};\ \operatorname{Re}\beta > 0,\ \operatorname{Re}\mu > -1].$ GH[337](16)

2. $\int_0^{\frac{\pi}{2}} \frac{\cos^\mu x \sin(\beta - \mu x)}{\sin^{2\mu+2} x} e^{-2\beta \operatorname{ctg} x} dx = \frac{1}{2}\sqrt{\frac{\pi}{2\beta}} (2\beta)^{-\mu} \Gamma(\mu+1) J_{\mu+\frac{1}{2}}(\beta)$

   $[\operatorname{Re}\beta > 0,\ \operatorname{Re}\mu > -1].$ WW 369

3. $\int_0^{\frac{\pi}{2}} \frac{\cos^\mu x \cos(\beta - \mu x)}{\sin^{2\mu+2} x} e^{-2\beta \operatorname{ctg} x} dx = -\frac{1}{2}\sqrt{\frac{\pi}{2\beta}} (2\beta)^{-\mu} \Gamma(\mu+1) N_{\mu+\frac{1}{2}}(\beta)$

   $[\operatorname{Re}\beta > 0,\ \operatorname{Re}\mu > -1].$ GH[337](17b)

3.919

1. $\int_0^{\pi/2} \frac{\sin 2nx}{\sin^{2n+2} x} \cdot \frac{dx}{\exp(2\pi \operatorname{ctg} x) - 1} = (-1)^{n-1} \frac{2n-1}{4(2n+1)}.$ BH[275](6), Li[275](6)

2. $\int_0^{\pi/2} \frac{\sin 2nx}{\sin^{2n+2} x} \cdot \frac{dx}{\exp(\pi \operatorname{ctg} x) - 1} = (-1)^{n-1} \frac{n}{2n+1}.$ BH[275](7), Li[275](7)

3.92 Trigonometrische Funktionen mit komplizierteren Argumenten und die Exponentialfunktion
Trigonometric functions of compound arguments combined with exponentials

3.921 $\int_0^\infty e^{-\beta x^2} \cos ax^2 \, dx = \frac{1}{2} \sqrt{\frac{\pi}{\sqrt{\beta^2 + a^2}}} \cos\left(\frac{1}{2} \operatorname{arctg} \frac{a}{\beta}\right)$ [Re β > |Im a|].

3.922

1. $\int_0^\infty e^{-\beta x^2} \sin ax^2 \, dx = \frac{1}{2} \int_{-\infty}^\infty e^{-\beta x^2} \sin ax^2 \, dx = \sqrt{\frac{\pi}{8}} \sqrt{\frac{\sqrt{\beta^2 + a^2} - \beta}{\beta^2 + a^2}} =$

$= \frac{\sqrt{\pi}}{2 \sqrt[4]{\beta^2 + a^2}} \sin\left(\frac{1}{2} \operatorname{arctg} \frac{a}{\beta}\right)$ [Re β > 0, a > 0]. F II 754–756, BH[263](8)

2. $\int_0^\infty e^{-\beta x^2} \cos ax^2 \, dx = \frac{1}{2} \int_{-\infty}^\infty e^{-\beta x^2} \cos ax^2 \, dx =$

$= \sqrt{\frac{\pi}{8}} \sqrt{\frac{\sqrt{\beta^2 + a^2} + \beta}{\beta^2 + a^2}} = \frac{\sqrt{\pi}}{2 \sqrt[4]{\beta^2 + a^2}} \cos\left(\frac{1}{2} \operatorname{arctg} \frac{a}{\beta}\right)$

[Re β > 0, a > 0]. F II 754–756, BH[263](9)

3. $\int_0^\infty e^{-\beta x^2} \sin ax^2 \cos bx \, dx = -\frac{1}{2} \sqrt{\frac{\pi}{\beta^2 + a^2}} e^{-A\beta} (B \sin Aa - C \cos Aa) =$

$= \frac{\sqrt{\pi}}{2 \sqrt[4]{\beta^2 + a^2}} \exp\left(-\frac{\beta b^2}{4(\beta^2 + a^2)}\right) \sin\left\{\frac{1}{2} \operatorname{arctg} \frac{a}{\beta} - \frac{ab^2}{4(\beta^2 + a^2)}\right\}.$

Li[263](10), GH[337](5)

4. $\int_0^\infty e^{-\beta x^2} \cos ax^2 \cos bx \, dx = \frac{1}{2} \sqrt{\frac{\pi}{\beta^2 + a^2}} e^{-A\beta} (B \cos Aa + C \sin Aa) =$

$= \frac{\sqrt{\pi}}{2 \sqrt[4]{\beta^2 + a^2}} \exp\left(-\frac{\beta b^2}{4(\beta^2 + a^2)}\right) \cos\left\{\frac{1}{2} \operatorname{arctg} \frac{a}{\beta} - \frac{ab^2}{4(\beta^2 + a^2)}\right\}.$

Li[263](11), GH[337](5)

[In den Formeln 3.922 3., 3.922 4. ist | [In the formulae 3.922 3. and 3.922 4.

$a > 0, b > 0, \operatorname{Re} \beta > 0, A = \frac{b^2}{4(a^2 + \beta^2)},$

$B = \sqrt{\frac{1}{2} (\sqrt{\beta^2 + a^2} + \beta)}, \quad C = \sqrt{\frac{1}{2} (\sqrt{\beta^2 + a^2} - \beta)}.$

Für komplexes a ist | If a is a complex number, $\operatorname{Re} \beta > |\operatorname{Im} a|$.]

3.923

1. $\int_{-\infty}^{\infty} \exp[-(ax^2 + 2bx + c)] \sin(px^2 + 2qx + r) \, dx =$

$$= \frac{\sqrt{\pi}}{\sqrt[4]{a^2+p^2}} \exp \frac{a(b^2 - ac) - (aq^2 - 2bpq + cp^2)}{a^2 + p^2} \times$$

$$\times \sin\left\{\frac{1}{2} \operatorname{arctg} \frac{p}{a} - \frac{p(q^2 - pr) - (b^2p - 2abq + a^2r)}{a^2 + p^2}\right\} \quad [a > 0].$$

GH[337](3), BH[269](6)

2. $\int_{-\infty}^{\infty} \exp[-(ax^2 + 2bx + c)] \cos(px^2 + 2qx + r) \, dx =$

$$= \frac{\sqrt{\pi}}{\sqrt[4]{a^2+p^2}} \exp \frac{a(b^2 - ac) - (aq^2 - 2bpq + cp^2)}{a^2 + p^2} \times$$

$$\times \cos\left\{\frac{1}{2} \operatorname{arctg} \frac{p}{a} - \frac{p(q^2 - pr) - (b^2p - 2abq + a^2r)}{a^2 + p^2}\right\} \quad [a > 0].$$

GH[337](3), BH[269](7)

3.924

1. $\int_0^{\infty} e^{-\beta x^4} \sin bx^2 \, dx = \frac{\pi}{4} \sqrt{\frac{b}{2\beta}} \exp\left(-\frac{b^2}{8\beta}\right) I_{\frac{1}{4}}\left(\frac{b^2}{8\beta}\right) \quad [\operatorname{Re} \beta > 0, b > 0].$

Ba IV 73(22)

2. $\int_0^{\infty} e^{-\beta x^4} \cos bx^2 \, dx = \frac{\pi}{4} \sqrt{\frac{b}{2\beta}} \exp\left(-\frac{b^2}{8\beta}\right) I_{-\frac{1}{4}}\left(\frac{b^2}{8\beta}\right) \quad [\operatorname{Re} \beta > 0, b > 0].$

Ba IV 15(12)

3.925

1. $\int_0^{\infty} e^{-\frac{p^2}{x^2}} \sin 2a^2 x^2 \, dx = \frac{1}{2} \int_{-\infty}^{\infty} e^{-\frac{p^2}{x^2}} \sin 2a^2 x^2 \, dx =$

$$= \frac{\sqrt{\pi}}{4a} e^{-2ap} (\cos 2ap + \sin 2ap) \quad [a > 0, b > 0]. \qquad \text{BH[268](12)}$$

2. $\int_0^{\infty} e^{-\frac{p^2}{x^2}} \cos 2a^2 x^2 \, dx = \frac{1}{2} \int_{-\infty}^{\infty} e^{-\frac{p^2}{x^2}} \cos 2a^2 x^2 \, dx =$

$$= \frac{\sqrt{\pi}}{4a} e^{-2ap}(\cos 2ap - \sin 2ap) \quad [a > 0, b > 0]. \qquad \text{BH[268](13)}$$

3.926

1. $\int_0^{\infty} e^{-\left(\beta x^2 + \frac{\gamma}{x^2}\right)} \sin ax^2 \, dx = \frac{1}{2} \sqrt{\frac{\pi}{a^2 + \beta^2}} e^{-2u\sqrt{\gamma}} \times$

$$\times [v \cos(2v\sqrt{\gamma}) + u \sin(2v\sqrt{\gamma})] \quad [\operatorname{Re} \beta > 0, \operatorname{Re} \gamma > 0]. \qquad \text{BH[268](14)}$$

2. $\int_0^{\infty} e^{-\left(\beta x^2 + \frac{\gamma}{x^2}\right)} \cos ax^2 \, dx = \frac{1}{2} \sqrt{\frac{\pi}{a^2 + \beta^2}} e^{-2u\sqrt{\gamma}} \times$

$$\times [u \cos(2v\sqrt{\gamma}) - v \sin(2v\sqrt{\gamma})] \quad [\operatorname{Re} \beta > 0, \operatorname{Re} \gamma > 0]. \qquad \text{BH[268](15)}$$

[In den Formeln 3.926 1., 3.926 2. ist | [In 3.926 1. and 3.926 2. put

$$u = \sqrt{\frac{\sqrt{a^2+\beta^2}+\beta}{2}}, \quad v = \sqrt{\frac{\sqrt{a^2+\beta^2}-\beta}{2}}.]$$

3.927 $\int_0^\infty e^{-\frac{p}{x}} \sin^2 \frac{a}{x} \, dx = a \, \text{arctg} \, \frac{2a}{p} + \frac{p}{4} \ln \frac{p^2}{p^2+4a^2}$ $[a > 0, \, p > 0]$. LI[268](4)

3.928 1. $\int_0^\infty \exp\left[-\left(p^2 x^2 + \frac{q^2}{x^2}\right)\right] \sin\left(a^2 x^2 + \frac{b^2}{x^2}\right) dx =$

$$= \frac{\sqrt{\pi}}{2r} e^{-2rs \cos(A+B)} \sin\{A + 2rs \sin(A+B)\}.$$ BH[268](22)

2. $\int_0^\infty \exp\left[-\left(p^2 x^2 + \frac{q^2}{x^2}\right)\right] \cos\left(a^2 x^2 + \frac{b^2}{x^2}\right) dx =$

$$= \frac{\sqrt{\pi}}{2r} e^{-2rs \cos(A+B)} \cos\{A + 2rs \sin(A+B)\}.$$ BH[268](23)

[In den Formeln 3.928 1., 3.928 2. | [In 3.928 1. and 3.928 2. $a^2+p^2 > 0$
ist $a^2+p^2 > 0$ und | and

$$r = \sqrt[4]{a^4+p^4}, \, s = \sqrt[4]{b^4+q^4}, \, A = \frac{1}{2} \text{arctg} \, \frac{a^2}{p^2}, \, B = \frac{1}{2} \text{arctg} \, \frac{b^2}{q^2}.]$$

3.929 $\int_0^\infty [e^{-x} \cos(p\sqrt{x}) + p e^{-x^2} \sin px] \, dx = 1$. LI[268](3)

Trigonometrische Funktionen und Exponentialfunktionen von trigonometrischen Funktionen

3.93

Trigonometric and exponential functions of trigonometric functions

3.931 1. $\int_0^{\pi/2} e^{-p \cos x} \sin(p \sin x) \, dx = \text{Ei}(-p) - \text{ci}(p)$. NI 13(27)

2. $\int_0^\pi e^{-p \cos x} \sin(p \sin x) \, dx = -\int_{-\pi}^0 e^{-p \cos x} \sin(p \sin x) \, dx = -2 \, \text{shi}(p)$.

GH[337](11b)

3. $\int_0^{\pi/2} e^{-p \cos x} \cos(p \sin x) \, dx = -\text{si}(p)$. NI 13(26)

4. $\int_0^\pi e^{-p \cos x} \cos(p \sin x) \, dx = \frac{1}{2} \int_0^{2\pi} e^{-p \cos x} \cos(p \sin x) \, dx = \pi$. GH[337](11a)

3.932 1. $\int_0^\pi e^{p \cos x} \sin(p \sin x) \sin mx \, dx =$

$$= \frac{1}{2} \int_0^{2\pi} e^{p \cos x} \sin(p \sin x) \sin mx \, dx = \frac{\pi}{2} \cdot \frac{p^m}{m!}.$$ BH[277](7), GH[337](13a)

2. $\int_0^\pi e^{p\cos x}\cos(p\sin x)\cos mx\,dx =$ 3.932

$$= \frac{1}{2}\int_0^{2\pi} e^{p\cos x}\cos(p\sin x)\cos mx\,dx = \frac{\pi}{2}\cdot\frac{p^m}{m!}.$$ BH[277](8), GH[337](13b)

$$\int_0^\pi e^{p\cos x}\sin(p\sin x)\cosec x\,dx = \pi\,\text{sh}\,p.$$ BH[278](1) 3.933

1. $\int_0^\pi e^{p\cos x}\sin(p\sin x)\,\text{tg}\,\frac{x}{2}\,dx = \pi(1-e^p).$ BH[271](8) 3.934

2. $\int_0^\pi e^{p\cos x}\sin(p\sin x)\,\text{ctg}\,\frac{x}{2}\,dx = \pi(e^p-1).$ BH[272](5)

$$\int_0^\pi e^{p\cos x}\cos(p\sin x)\frac{\sin 2nx}{\sin x}\,dx = \pi\sum_{k=0}^{n-1}\frac{p^{2k+1}}{(2k+1)!}\quad[p>0].$$ Li[278](3) 3.935

1. $\int_0^{2\pi} e^{p\cos x}\cos(p\sin x - mx)\,dx = 2\int_0^\pi e^{p\cos x}\cos(p\sin x - mx)\,dx = \frac{2\pi p^m}{m!}.$ 3.936

BH[277](9), GH[337](14a)

2. $\int_0^{2\pi} e^{p\sin x}\sin(p\cos x + mx)\,dx = \frac{2\pi p^m}{m!}\sin\frac{m\pi}{2}\quad[p>0].$ GH[337](14b)

3. $\int_0^{2\pi} e^{p\sin x}\cos(p\cos x + mx)\,dx = \frac{2\pi p^m}{m!}\cos\frac{m\pi}{2}\quad[p>0].$ GH[337](14b)

4. $\int_0^{2\pi} e^{\cos x}\sin(mx - \sin x)\,dx = 0.$ WW 113

5. $\int_0^\pi e^{\beta\cos x}\cos(ax + \beta\sin x)\,dx = \beta^{-a}\sin(a\pi)\,\gamma(a,\beta).$ Ba II 137(2)

1. $\int_0^{2\pi}\exp(p\cos x + q\sin x)\sin(a\cos x + b\sin x - mx)\,dx=$ 3.937

$$= i\pi[(b-p)^2 + (a+q)^2]^{-\frac{m}{2}}\{(A+iB)^{\frac{m}{2}}I_m(\sqrt{C-iD}) -$$

$$- (A-iB)^{\frac{m}{2}}I_m(\sqrt{C+iD})\}.$$ GH[337](9b)

3.937

2. $\int_0^{2\pi} \exp(p\cos x + q\sin x)\cos(a\cos x + b\sin x - mx)\,dx =$

$$= \pi[(b-p)^2 + (a+q)^2]^{-\frac{m}{2}}\{(A+iB)^{\frac{m}{2}} I_m(\sqrt{C-iD}) +$$
$$+ (A-iB)^{\frac{m}{2}} I_m(\sqrt{C+iD})\}.$$

[In den Formeln 3.937 1., 3.937 2. ist | [In 3.937 1. and 3.937 2.
$(b-p)^2 + (a+q)^2 > 0$, $m = 0, 1, 2, \ldots$, $A = p^2 - q^2 + a^2 - b^2$,
$B = 2(pq + ab)$, $C = p^2 + q^2 - a^2 - b^2$, $D = -2(ap + bq)$.] GH[337](9a)

3. $\int_0^{2\pi} \exp(p\cos x + q\sin x)\sin(q\cos x - p\sin x + mx)\,dx =$

$$= \frac{2\pi}{m!}(p^2 + q^2)^{\frac{m}{2}} \sin\left(m \arctg \frac{q}{p}\right).$$ GH[337](12)

4. $\int_0^{2\pi} \exp(p\cos x + q\sin x)\cos(q\cos x - p\sin x + mx)\,dx =$

$$= \frac{2\pi}{m!}(p^2 + q^2)^{\frac{m}{2}} \cos\left(m \arctg \frac{q}{p}\right).$$ GH[337](12)

3.938

1. $\int_0^{\pi} e^{r(\cos px + \cos qx)} \sin(r\sin px)\sin(r\sin qx)\,dx =$

$$= \frac{\pi}{2} \sum_{k=1}^{\infty} \frac{1}{\Gamma(pk+1)\Gamma(qk+1)} r^{(p+q)k}.$$ BH[277](14)

2. $\int_0^{\pi} e^{r(\cos px + \cos qx)} \cos(r\sin px)\cos(r\sin qx)\,dx =$

$$= \frac{\pi}{2}\left(2 + \sum_{k=1}^{\infty} \frac{r^{(p+q)k}}{\Gamma(pk+1)\Gamma(qk+1)}\right).$$ BH[277](15)

3.939

1. $\int_0^{\pi} e^{q\cos x} \frac{\sin rx}{1 - 2p^r \cos rx + p^{2r}} \sin(q\sin x)\,dx = \frac{\pi}{2pr} \sum_{k=1}^{\infty} \frac{(pq)^{kr}}{\Gamma(kr+1)}$ $[p^2 < 1]$.

BH[278](15)

2. $\int_0^{\pi} e^{q\cos x} \frac{1 - p^r \cos rx}{1 - 2p^r \cos rx + p^{2r}} \cos(q\sin x)\,dx = \frac{\pi}{2}\left[2 + \sum_{k=1}^{\infty} \frac{(pq)^{kr}}{\Gamma(kr+1)}\right]$

$[p^2 < 1]$. BH[278](16)

3. $\int_0^{\pi/2} \frac{e^{p\cos 2x} \cos(p\sin 2x)\,dx}{\cos^2 x + q^2 \sin^2 x} = \frac{\pi}{2q} \exp\left(p\frac{q-1}{q+1}\right)$. BH[273](8)

Trigonometrische Funktionen, Exponentialfunktionen und Potenzen
Trigonometric functions, exponentials, and powers 3.94–3.97

3.941

1. $\int_0^\infty e^{-px} \sin qx \, \dfrac{dx}{x} = \operatorname{arctg} \dfrac{q}{p} \quad [p > 0].$ BH[365](1)

2. $\int_0^\infty e^{-px} \cos qx \, \dfrac{dx}{x} = \infty.$ BH[365](2)

3.942

1. $\int_0^\infty e^{-px} \cos px \, \dfrac{x \, dx}{b^4 + x^4} = \dfrac{\pi}{4b^2} \exp(-bp\sqrt{2}) \quad [p > 0, b > 0].$ BH[386](6)+

2. $\int_0^\infty e^{-px} \cos px \, \dfrac{x \, dx}{b^4 - x^4} = \dfrac{\pi}{4b^2} e^{-bp} \sin bp \quad [p > 0, b > 0].$ BH[386](7)+

3.943

$\int_0^\infty e^{-\beta x} (1 - \cos ax) \, \dfrac{dx}{x} = \dfrac{1}{2} \ln \dfrac{a^2 + \beta^2}{\beta^2} \quad [\operatorname{Re} \beta > 0].$ BH[367](6)

3.944

1. $\int_0^u x^{\mu-1} e^{-\beta x} \sin \delta x \, dx = \dfrac{i}{2} (\beta + i\delta)^{-\mu} \gamma[\mu, (\beta + i\delta) u] -$

$\quad - \dfrac{i}{2} (\beta - i\delta)^{-\mu} \gamma[\mu, (\beta - i\delta) u] \quad [\operatorname{Re}\mu > -1].$ Ba IV 318(8)

2. $\int_u^\infty x^{\mu-1} e^{-\beta x} \sin \delta x \, dx = \dfrac{i}{2} (\beta + i\delta)^{-\mu} \Gamma[\mu, (\beta + i\delta) u] -$

$\quad - \dfrac{i}{2} (\beta - i\delta)^{-\mu} \Gamma[\mu, (\beta - i\delta) u] \quad [\operatorname{Re} \beta > |\operatorname{Im} \delta|].$ Ba IV 318(9)

3. $\int_0^u x^{\mu-1} e^{-\beta x} \cos \delta x \, dx = \dfrac{1}{2} (\beta + i\delta)^{-\mu} \gamma[\mu, (\beta + i\delta)u] +$

$\quad + \dfrac{1}{2} (\beta - i\delta)^{-\mu} \gamma[\mu, (\beta - i\delta)u] \quad [\operatorname{Re}\mu > 0].$ Ba IV 320(28)

4. $\int_u^\infty x^{\mu-1} e^{-\beta x} \cos \delta x \, dx = \dfrac{1}{2} (\beta + i\delta)^{-\mu} \Gamma[\mu, (\beta + i\delta) u] +$

$\quad + \dfrac{1}{2} (\beta - i\delta)^{-\mu} \Gamma[\mu, (\beta - i\delta) u] \quad [\operatorname{Re} \beta > |\operatorname{Im} \delta|].$ Ba IV 320(29)

5. $\int_0^\infty x^{\mu-1} e^{-\beta x} \sin \delta x \, dx = \dfrac{\Gamma(\mu)}{(\beta^2 + \delta^2)^{\frac{\mu}{2}}} \sin\left(\mu \operatorname{arctg} \dfrac{\delta}{\beta}\right)$

$\quad [\operatorname{Re} \mu > -1, \ \operatorname{Re} \beta > |\operatorname{Im} \delta|].$ F II 817, BH[361](9)

3.944

6. $$\int_0^\infty x^{\mu-1} e^{-\beta x} \cos \delta x \, dx = \frac{\Gamma(\mu)}{(\delta^2 + \beta^2)^{\frac{\mu}{2}}} \cos\left(\mu \arctg \frac{\delta}{\beta}\right)$$
   $$[\operatorname{Re} \mu > 0, \ \operatorname{Re} \beta > |\operatorname{Im} \delta|].$$ F II 817, BH[361](10)

7. $$\int_0^\infty x^{\mu-1} \exp(-ax \cos t) \sin(ax \sin t) \, dx = \Gamma(\mu) \, a^{-\mu} \sin(\mu t)$$
   $$\left[\operatorname{Re} \mu > -1, \ a > 0, \ |t| < \frac{\pi}{2}\right].$$ Ba I 13(36)

8. $$\int_0^\infty x^{\mu-1} \exp(-ax \cos t) \cos(ax \sin t) \, dx = \Gamma(\mu) \, a^{-\mu} \cos(\mu t)$$
   $$\left[\operatorname{Re} \mu > -1, \ a > 0, \ |t| < \frac{\pi}{2}\right].$$ Ba I 13(35)

9. $$\int_0^\infty x^{p-1} e^{-qx} \sin(qx \operatorname{tg} t) \, dx = \frac{1}{q^p} \Gamma(p) \cos^p t \sin pt \quad \left[|t| < \frac{\pi}{2}, \ q > 0\right].$$
   Lo V 288(16)

10. $$\int_0^\infty x^{p-1} e^{-qx} \cos(qx \operatorname{tg} t) \, dx = \frac{1}{q^p} \Gamma(p) \cos^p(t) \cos pt \quad \left[|t| < \frac{\pi}{2}, \ q > 0\right].$$
    Lo V 288(15)

11. $$\int_0^\infty x^n e^{-\beta x} \sin bx \, dx = n! \left(\frac{\beta}{\beta^2 + b^2}\right)^{n+1} \sum_{0 \leq 2k \leq n} (-1)^k \binom{n+1}{2k+1} \left(\frac{b}{\beta}\right)^{2k+1} =$$
    $$= (-1)^n \frac{\partial^n}{\partial \beta^n} \left(\frac{b}{b^2 + \beta^2}\right) \quad [\operatorname{Re} \beta > 0, \ b > 0].$$ GH[336](3), Ba IV 72(3)

12. $$\int_0^\infty x^n e^{-\beta x} \cos bx \, dx = n! \left(\frac{\beta}{\beta^2 + b^2}\right)^{n+1} \sum_{0 \leq 2k \leq n+1} (-1)^k \binom{n+1}{2k} \left(\frac{b}{\beta}\right)^{2k} =$$
    $$= (-1)^n \frac{\partial^n}{\partial \beta^n} \left(\frac{\beta}{b^2 + \beta^2}\right) \quad [\operatorname{Re} \beta > 0, \ b > 0].$$ GH[336](4), Ba IV 14(5)

13. $$\int_0^\infty x^{n-\frac{1}{2}} e^{-\beta x} \sin bx \, dx = (-1)^n \sqrt{\frac{\pi}{2}} \frac{d^n}{d\beta^n} \frac{\sqrt{\sqrt{\beta^2 + b^2} - \beta}}{\sqrt{\beta^2 + b^2}}$$
    $$[\operatorname{Re} \beta > 0, \ b > 0].$$ Ba IV 72(6)

14. $$\int_0^\infty x^{n-\frac{1}{2}} e^{-\beta x} \cos bx \, dx = (-1)^n \sqrt{\frac{\pi}{2}} \frac{d^n}{d\beta^n} \frac{\sqrt{\sqrt{\beta^2 + b^2} + \beta}}{\sqrt{\beta^2 + b^2}}$$
    $$[\operatorname{Re} \beta > 0, \ b > 0].$$ Ba IV 15(6)

3.945

1. $$\int_0^\infty (e^{-\beta x} \sin ax - e^{-\gamma x} \sin bx) \frac{dx}{x^r} =$$
   $$= \Gamma(1-r) \left\{ (b^2 + \gamma^2)^{\frac{r-1}{2}} \sin\left[(r-1) \arctg \frac{b}{\gamma}\right] - \right.$$
   $$\left. - (a^2 + \beta^2)^{\frac{r-1}{2}} \sin\left[(r-1) \arctg \frac{a}{\beta}\right] \right\} \quad [\operatorname{Re} \beta > 0, \operatorname{Re} \gamma > 0, r < 2, r \neq 1].$$ BH[371](6)

2. $\int_0^\infty (e^{-\beta x} \cos ax - e^{-\gamma x} \cos bx) \frac{dx}{x^r} =$ 3.945

$$= \Gamma(1-r)\left\{(a^2+\beta^2)^{\frac{r-1}{2}} \cos\left[(r-1)\arctg \frac{a}{\beta}\right] - (b^2+\gamma^2)^{\frac{r-1}{2}} \times \right.$$
$$\left. \times \cos\left[(r-1)\arctg \frac{b}{\gamma}\right]\right\}$$

[Re $\beta > 0$, Re $\gamma > 0$, $r < 2$, $r \neq 1$].   BH[371](7)

3. $\int_0^\infty (ae^{-\beta x} \sin bx - be^{-\gamma x} \sin ax) \frac{dx}{x^2} =$

$$= ab\left[\frac{1}{2} \ln \frac{a^2+\gamma^2}{b^2+\beta^2} + \frac{\gamma}{a} \arcctg \frac{\gamma}{a} - \frac{\beta}{b} \arcctg \frac{\beta}{b}\right]$$

[Re $\beta > 0$, Re $\gamma > 0$].   BH[368](22)

1. $\int_0^\infty e^{-px} \sin^{2m+1} ax \frac{dx}{x} = \frac{(-1)^m}{2^{2m}} \sum_{k=0}^m (-1)^k \binom{2m+1}{k} \arctg \frac{(2m-2k+1)a}{p}$ 3.946

[$p > 0$].   GH[336](9a)

2. $\int_0^\infty e^{-px} \sin^{2m} ax \frac{dx}{x} = \frac{(-1)^{m+1}}{2^{2m}} \sum_{k=0}^{m-1} (-1)^k \binom{2m}{k} \ln[p^2 + (2m-2k)^2 a^2] -$

$$- \frac{1}{2^{2m}} \binom{2m}{m} \ln p \quad [p > 0].$$ GH[336](9b)

1. $\int_0^\infty e^{-\beta x} \sin \gamma x \sin ax \frac{dx}{x} = \frac{1}{4} \ln \frac{\beta^2 + (a+\gamma)^2}{\beta^2 + (a-\gamma)^2}$ 3.947

[Re $\beta > |\text{Im } \gamma|$, $a > 0$].   BH[365](5)

2. $\int_0^\infty e^{-px} \sin ax \sin bx \frac{dx}{x^2} = \frac{a}{2} \arctg \frac{2pb}{p^2+a^2-b^2} + \frac{b}{2} \arctg \frac{2pa}{p^2+b^2-a^2} +$

$$+ \frac{p}{4} \ln \frac{p^2+(a-b)^2}{p^2+(a+b)^2} \quad [p > 0].$$ BH[368](1) F II 749–750

3. $\int_0^\infty e^{-px} \sin ax \cos bx \frac{dx}{x} = \frac{1}{2} \arctg \frac{2pa}{p^2-a^2+b^2} + s\frac{\pi}{2}$

$\left[a \geq 0, p > 0, s = 0 \begin{array}{c} \text{für} \\ \text{for} \end{array} p^2-a^2+b^2 \geq 0 \begin{array}{c} \text{und} \\ \text{and} \end{array} s = 1 \begin{array}{c} \text{für} \\ \text{for} \end{array} p^2-a^2+b^2 < 0\right]$.

GH 336(10b)

1. $\int_0^\infty e^{-\beta x} (\sin ax - \sin bx) \frac{dx}{x} = \arctg \frac{(a-b)\beta}{ab+\beta^2}$ 3.948

[Re $\beta > 0$]   (cf. 3.951 2.).   BH[367](7)

2. $\int_0^\infty e^{-\beta x} (\cos ax - \cos bx) \frac{dx}{x} = \frac{1}{2} \ln \frac{b^2+\beta^2}{a^2+\beta^2}$

[Re $\beta > 0$]   (cf. 3.951 3.).   BH[367](8), F II 754+

3. $\int_0^\infty e^{-\beta x} (\cos ax - \cos bx) \frac{dx}{x^2} = \frac{\beta}{2} \ln \frac{a^2+\beta^2}{b^2+\beta^2} +$

$$+ b \arctg \frac{b}{\beta} - a \arctg \frac{a}{\beta} \quad [\text{Re } p > 0].$$ BH[368](20)

3.948

4. $\int_0^\infty e^{-px}(\sin^2 ax - \sin^2 bx)\frac{dx}{x^2} = a\,\mathrm{arctg}\,\frac{2a}{p} -$
$- b\,\mathrm{arctg}\,\frac{2b}{p} - \frac{p}{4}\ln\frac{p^2+4a^2}{p^2+4b^2}$ \quad $[p>0]$. \hfill BH[368](25)

5. $\int_0^\infty e^{-px}(\cos^2 ax - \cos^2 bx)\frac{dx}{x^2} = -a\,\mathrm{arctg}\,\frac{2a}{p} +$
$+ b\,\mathrm{arctg}\,\frac{2b}{p} + \frac{p}{4}\ln\frac{p^2+4a^2}{p^2+4b^2}$ \quad $[p>0]$. \hfill BH[368](26)

3.949

1. $\int_0^\infty e^{-px}\sin ax \sin bx \sin cx\,\frac{dx}{x} = -\frac{1}{4}\mathrm{arctg}\,\frac{a+b+c}{p} +$
$+ \frac{1}{4}\mathrm{arctg}\,\frac{a+b-c}{p} + \frac{1}{4}\mathrm{arctg}\,\frac{a-b+c}{p} + \frac{1}{4}\mathrm{arctg}\,\frac{-a+b+c}{p}$ \quad $[p>0]$. \hfill BH[365](11)

2. $\int_0^\infty e^{-px}\sin^2 ax \sin bx\,\frac{dx}{x} = \frac{1}{2}\mathrm{arctg}\,\frac{b}{p} - \frac{1}{4}\mathrm{arctg}\,\frac{2pb}{p^2+a^2-b^2}$
$[p>0]$. \hfill BH[365](8)

3. $\int_0^\infty e^{-px}\sin^2 ax \cos bx\,\frac{dx}{x} = \frac{1}{8}\ln\frac{[p^2+(2a+b)^2][p^2+(2a-b)^2]}{(p^2+b^2)^2}$
$[p>0]$. \hfill BH[365](9)

4. $\int_0^\infty e^{-px}\sin ax \cos^2 bx\,\frac{dx}{x} = \frac{1}{2}\mathrm{arctg}\,\frac{a}{p} + \frac{1}{4}\mathrm{arctg}\,\frac{2pa}{p^2+b^2-a^2}$
$[p>0]$. \hfill BH[365](10)

5. $\int_0^\infty e^{-px}\sin^2 ax \sin bx \sin cx\,\frac{dx}{x} = \frac{1}{8}\ln\frac{p^2+(b+c)^2}{p^2+(b-c)^2} +$
$+ \frac{1}{16}\ln\frac{[p^2+(2a-b+c)^2][p^2+(2a+b-c)^2]}{[p^2+(2a+b+c)^2][p^2+(2a-b-c)^2]}$ \quad $[p>0]$. \hfill BH[365](15)

3.951

1. $\int_0^\infty (1-e^{-x})\cos x\,\frac{dx}{x} = \ln\sqrt{2}$. \hfill F II 750

2. $\int_0^\infty \frac{e^{-\gamma x} - e^{-\beta x}}{x}\sin bx\,dx = \mathrm{arctg}\,\frac{(\beta-\gamma)b}{b^2+\beta\gamma}$ \quad $[\mathrm{Re}\,\beta>0,\ \mathrm{Re}\,\gamma\geq 0]$. \hfill BH[367](3)

3. $\int_0^\infty \frac{e^{-\gamma x} - e^{-\beta x}}{x}\cos bx\,dx = \frac{1}{2}\ln\frac{b^2+\beta^2}{b^2+\gamma^2}$ \quad $[\mathrm{Re}\,\beta>0,\ \mathrm{Re}\,\gamma\geq 0]$. \hfill BH[367](4)

4. $\int_0^\infty \frac{e^{-\gamma x} - e^{-\beta x}}{x^2}\sin bx\,dx = \frac{b}{2}\ln\frac{b^2+\beta^2}{b^2+\gamma^2} + \beta\,\mathrm{arctg}\,\frac{b}{\beta} - \gamma\,\mathrm{arctg}\,\frac{b}{\gamma}$
$[\mathrm{Re}\,\beta>0,\ \mathrm{Re}\,\gamma>0]$. \hfill BH[368](21)+

5. $\int_0^\infty \frac{x}{e^{\beta x}-1}\cos bx\,dx = \frac{1}{2b^2} - \frac{\pi^2}{2\beta^2}\mathrm{cosech}^2\frac{b\pi}{\beta}$ \quad $[\mathrm{Re}\,\beta>0]$. \hfill Ba IV 15(8)

6. $\int_0^\infty \left(\frac{1}{e^x - 1} - \frac{1}{x}\right) \cos bx \, dx = \ln b - \frac{1}{2}[\psi(ib) + \psi(-ib)]$ $[b > 0]$.   Ba IV 15(9)   3.951

7. $\int_0^\infty \frac{1 - \cos ax}{e^{2\pi x} - 1} \cdot \frac{dx}{x} = \frac{a}{4} + \frac{1}{2} \ln \frac{1 - e^{-a}}{a}$ $[a > 0]$.   BH[387](10)

8. $\int_0^\infty (e^{-\beta x} - e^{-\gamma x} \cos ax) \frac{dx}{x} = \frac{1}{2} \ln \frac{a^2 + \gamma^2}{\beta^2}$ $[\operatorname{Re} \beta > 0, \operatorname{Re} \gamma > 0]$.   BH[367](10)

9. $\int_0^\infty \frac{\cos px - e^{-px}}{b^4 + x^4} \frac{dx}{x} = \frac{\pi}{2b^4} \exp\left(-\frac{1}{2} bp \sqrt{2}\right) \sin\left(\frac{1}{2} bp \sqrt{2}\right)$ $[p > 0]$.   BH[390](6)

10. $\int_0^\infty \left(\frac{1}{e^x - 1} - \frac{\cos x}{x}\right) dx = C$.   NI 65(8)

11. $\int_0^\infty \left(ae^{-px} - \frac{e^{-qx}}{x} \sin ax\right) \frac{dx}{x} = \frac{a}{2} \ln \frac{a^2 + q^2}{p^2} + q \operatorname{arctg} \frac{a}{q} - a$

     $[p > 0, q > 0]$.   BH[368](24)

12. $\int_0^\infty \frac{x^{2m} \sin bx}{e^x - 1} dx = (-1)^m \frac{\partial^{2m}}{\partial b^{2m}} \left[\frac{\pi}{2} \operatorname{cth} b\pi - \frac{1}{2b}\right]$ $[b > 0]$.   GH[336](15a)

13. $\int_0^\infty \frac{x^{2m+1} \cos bx}{e^x - 1} dx = (-1)^m \frac{\partial^{2m+1}}{\partial b^{2m+1}} \left[\frac{\pi}{2} \operatorname{cth} b\pi - \frac{1}{2b}\right]$ $[b > 0]$.   GH[336](15b)

14. $\int_0^\infty \frac{x^{2m} \sin bx \, dx}{e^{(2n+1)cx} - e^{(2n-1)cx}} = (-1)^m \frac{\partial^{2m}}{\partial b^{2m}} \left[\frac{\pi}{4c} \operatorname{th} \frac{b\pi}{2c} - \sum_{k=1}^n \frac{b}{b^2 + (2k-1)^2 c^2}\right]^*$

     $[b > 0]$.   GH[336](14a)

15. $\int_0^\infty \frac{x^{2m+1} \cos bx \, dx}{e^{(2n+1)cx} - e^{(2n-1)cx}} = (-1)^m \frac{\partial^{2m+1}}{\partial b^{2m+1}} \left[\frac{\pi}{4c} \operatorname{th} \frac{b\pi}{2c} - \sum_{k=1}^n \frac{b}{b^2 + (2k-1)^2 c^2}\right]^*$

     $[b > 0]$.   GH[336](14b)

16. $\int_0^\infty \frac{x^{2m} \sin bx \, dx}{e^{2ncx} - e^{(2n-2)cx}} = (-1)^m \frac{\partial^{2m}}{\partial b^{2m}} \left[\frac{\pi}{4c} \operatorname{cth} \frac{b\pi}{2c} - \frac{1}{2b} - \sum_{k=1}^{n-1} \frac{b}{b^2 + (2k)^2 c^2}\right]^{**}$

     $[b > 0, c > 0]$.   GH[336](14c)

17. $\int_0^\infty \frac{x^{2m+1} \cos bx \, dx}{e^{2ncx} - e^{(2n-2)cx}} = (-1)^m \frac{\partial^{2m+1}}{\partial b^{2m+1}} \left[\frac{\pi}{4c} \operatorname{cth} \frac{b\pi}{2c} - \frac{1}{2b} - \sum_{k=1}^{n-1} \frac{b}{b^2 + (2k)^2 c^2}\right]^{**}$

     $[b > 0, c > 0]$.   GH[336](14d)

18. $\int_0^\infty \frac{\cos ax - \cos bx}{e^{(2m+1)px} - e^{(2m-1)px}} \frac{dx}{x} = \frac{1}{2} \ln \frac{\operatorname{ch} \frac{b\pi}{2p}}{\operatorname{ch} \frac{a\pi}{2p}} -$

     $- \frac{1}{2} \sum_{k=1}^m \ln \frac{b^2 + (2k-1)^2 p^2}{a^2 + (2k-1)^2 p^2}{}^{***}$ $[p > 0]$.   GH[336](16a)

---

* Für $n = 0$ entfällt die Summe.   * For $n = 0$ the sum vanishes.
** Für $n = 1$ entfällt die Summe.   ** For $n = 1$ the sum vanishes.
*** Für $m = 0$ entfällt die Summe.   *** For $m = 0$ the sum vanishes.

3.951

19. $\int_0^\infty \dfrac{\cos ax - \cos bx}{e^{2mpx} - e^{(2m-2)px}} \dfrac{dx}{x} = \dfrac{1}{2} \ln \dfrac{a \operatorname{sh} \dfrac{b\pi}{2p}}{b \operatorname{sh} \dfrac{a\pi}{2p}} - \dfrac{1}{2} \sum_{k=1}^{m-1} \ln \dfrac{b^2 + 4k^2 p^2}{a^2 + 4k^2 p^2}$ *

[$p > 0$].   GH[336](16b)

20. $\int_0^\infty \dfrac{\sin x \sin bx}{1 - e^x} \cdot \dfrac{dx}{x} = \dfrac{1}{4} \ln \dfrac{(b+1) \operatorname{sh}[(b-1)\pi]}{(b-1) \operatorname{sh}[(b+1)\pi]}$ [$b^2 \neq 1$].   Lo V 305

21. $\int_0^\infty \dfrac{\sin^2 ax}{1 - e^x} \cdot \dfrac{dx}{x} = \dfrac{1}{4} \ln \dfrac{2a\pi}{\operatorname{sh} 2a\pi}$.   Lo V 306, BH[387](5)

3.952

1. $\int_0^\infty x e^{-p^2 x^2} \sin ax \, dx = \dfrac{a\sqrt{\pi}}{4p^3} \exp\left(-\dfrac{a^2}{4p^2}\right)$.   BH[362](1)

2. $\int_0^\infty x e^{-p^2 x^2} \cos ax \, dx = \dfrac{1}{2p^2} - \dfrac{a}{4p^3} \sum_{k=0}^\infty \dfrac{(-1)^k k!}{(2k+1)!} \left(\dfrac{a}{p}\right)^{2k+1}$ [$a > 0$].   BH[362](2)

3. $\int_0^\infty x^2 e^{-p^2 x^2} \sin ax \, dx = \dfrac{a}{4p^4} + \dfrac{2p^2 - a^2}{8p^5} \sum_{k=0}^\infty \dfrac{(-1)^k k!}{(2k+1)!} \left(\dfrac{a}{p}\right)^{2k+1}$ [$a > 0$].

BH[362](4)

4. $\int_0^\infty x^2 e^{-p^2 x^2} \cos ax \, dx = \sqrt{\pi} \, \dfrac{2p^2 - a^2}{8p^5} \exp\left(-\dfrac{a^2}{4p^2}\right)$.   BH[362](5)

5. $\int_0^\infty x^3 e^{-p^2 x^2} \sin ax \, dx = \sqrt{\pi} \, \dfrac{6ap^2 - a^3}{16p^7} \exp\left(-\dfrac{a^2}{4p^2}\right)$.   BH[362](6)

6. $\int_0^\infty e^{-p^2 x^2} \sin ax \, \dfrac{dx}{x} = \dfrac{a\sqrt{\pi}}{2p} \sum_{k=0}^\infty \dfrac{(-1)^k}{k!(2k+1)} \left(\dfrac{a}{2p}\right)^{2k}$.   BH[365](21)

7. $\int_0^\infty x^{\mu-1} e^{-\beta x^2} \sin \gamma x \, dx = \dfrac{\gamma e^{-\frac{\gamma^2}{4\beta}}}{2\beta^{\frac{\mu+1}{2}}} \Gamma\left(\dfrac{1+\mu}{2}\right) {}_1F_1\left(1 - \dfrac{\mu}{2}; \dfrac{3}{2}; \dfrac{\gamma^2}{4\beta}\right)$

[Re $\beta > 0$, Re $\mu > -1$].   Ba IV 318(10)

8. $\int_0^\infty x^{\mu-1} e^{-\beta x^2} \cos ax \, dx = \dfrac{\Gamma\left(\dfrac{\mu}{2}\right)}{2\beta^{\frac{\mu}{2}}} {}_1F_1\left(\dfrac{\mu}{2}; \dfrac{1}{2}; -\dfrac{a^2}{4\beta}\right)$

[Re $\beta > 0$, Re $\mu > 0$, $a > 0$].   Ba IV 15(14)

---

* Für $m = 1$ entfällt die Summe. | * For $m = 1$ the sum vanishes.

9. $\int_0^\infty x^{2n} e^{-\beta^2 x^2} \cos ax\, dx = (-1)^n \frac{\sqrt{\pi}}{2^{n+1}\beta^{2n+1}} \exp\left(-\frac{a^2}{8\beta^2}\right) D_{2n}\left(\frac{a}{\beta\sqrt{2}}\right) =$ 3.952

$= (-1)^n \frac{\sqrt{\pi}}{(2\beta)^{2n+1}} \exp\left(-\frac{a^2}{4\beta^2}\right) H_{2n}\left(\frac{a}{2\beta}\right)$

$\left[|\arg \beta| < \frac{\pi}{4},\ a > 0\right].$ WW 353, Ba IV 15(13)

10. $\int_0^\infty x^{2n+1} e^{-\beta^2 x^2} \sin ax\, dx = (-1)^n \frac{\sqrt{\pi}}{2^{n+\frac{3}{2}}\beta^{2n+2}} \exp\left(-\frac{a^2}{8\beta^2}\right) D_{2n+1}\left(\frac{a}{\beta\sqrt{2}}\right) =$

$= (-1)^n \frac{\sqrt{\pi}}{(2\beta)^{2n+2}} \exp\left(-\frac{a^2}{4\beta^2}\right) H_{2n+1}\left(\frac{a}{2\beta}\right)$

$\left[|\arg \beta| < \frac{\pi}{4},\ a > 0\right].$ WW 353+, Ba IV 74(23)

1. $\int_0^\infty x^{\mu-1} e^{-\gamma x - \beta x^2} \sin ax\, dx = -\frac{i}{2(2\beta)^{\frac{\mu}{2}}} \exp\frac{\gamma^2 - a^2}{8\beta} \times$ 3.953

$\times \Gamma(\mu)\left\{\exp\left(-\frac{ia\gamma}{4\beta}\right) D_{-\mu}\left(\frac{\gamma - ia}{\sqrt{2\beta}}\right) - \exp\frac{ia\gamma}{4\beta} D_{-\mu}\left(\frac{\gamma + ia}{\sqrt{2\beta}}\right)\right\}$

$[\operatorname{Re}\mu > -1,\ \operatorname{Re}\beta > 0,\ a > 0].$ Ba IV 318(11)

2. $\int_0^\infty x^{\mu-1} e^{-\gamma x - \beta x^2} \cos ax\, dx = \frac{1}{2(2\beta)^{\frac{\mu}{2}}} \exp\frac{\gamma^2 - a^2}{8\beta} \times$

$\times \Gamma(\mu)\left\{\exp\left(-\frac{ia\gamma}{4\beta}\right) D_{-\mu}\left(\frac{\gamma - ia}{\sqrt{2\beta}}\right) + \exp\frac{ia\gamma}{4\beta} D_{-\mu}\left(\frac{\gamma + ia}{\sqrt{2\beta}}\right)\right\}$

$[\operatorname{Re}\mu > 0,\ \operatorname{Re}\beta > 0,\ a > 0].$ Ba IV 16(18)

3. $\int_0^\infty x e^{-\gamma x - \beta x^2} \sin ax\, dx =$

$= \frac{i\sqrt{\pi}}{8\sqrt{\beta^3}}\left\{(\gamma - ia)\exp\left[\frac{(\gamma - ia)^2}{4\beta}\right]\left[1 - \Phi\left(\frac{\gamma - ia}{2\sqrt{\beta}}\right)\right] -\right.$

$\left. - (\gamma + ia)\exp\left[\frac{(\gamma + ia)^2}{4\beta}\right]\left[1 - \Phi\left(\frac{\gamma + ia}{2\sqrt{\beta}}\right)\right]\right\}\quad [\operatorname{Re}\beta > 0,\ a > 0].$ Ba IV 74(28)

4. $\int_0^\infty x e^{-\gamma x - \beta x^2} \cos ax\, dx =$

$= -\frac{\sqrt{\pi}}{8\sqrt{\beta^3}}\left\{(\gamma - ia)\exp\frac{(\gamma - ia)^2}{4\beta}\left[1 - \Phi\left(\frac{\gamma - ia}{2\sqrt{\beta}}\right)\right] +\right.$

$\left. + (\gamma + ia)\exp\frac{(\gamma + ia)^2}{4\beta}\left[1 - \Phi\left(\frac{\gamma + ia}{2\sqrt{\beta}}\right)\right]\right\} + \frac{1}{2\beta}$

$[\operatorname{Re}\beta > 0,\ a > 0].$ Ba IV 16(17)

3.954 1. $\int_0^\infty e^{-\beta x^2} \sin ax \, \frac{x \, dx}{\gamma^2 + x^2} =$

$$= -\frac{\pi}{4} e^{\beta \gamma^2} \left[2 \operatorname{sh} a\gamma - e^{a\gamma} \Phi\left(\frac{a}{2\sqrt{\beta}} + \gamma\sqrt{\beta}\right) - e^{-a\gamma} \Phi\left(\frac{a}{2\sqrt{\beta}} - \gamma\sqrt{\beta}\right) \right]$$

[Re $\beta > 0$, Re $\gamma > 0$, $a > 0$].   Ba IV 74(26)+

2. $\int_0^\infty e^{-\beta x^2} \cos ax \, \frac{dx}{\gamma^2 + x^2} = \frac{\pi e^{\beta \gamma^2}}{4\gamma} \left[2 \operatorname{ch} a\gamma - e^{a\gamma} \Phi\left(\frac{a}{2\sqrt{\beta}} + \gamma\sqrt{\beta}\right) + \right.$

$$\left. + e^{-a\gamma} \Phi\left(\frac{a}{2\sqrt{\beta}} - \gamma\sqrt{\beta}\right) \right] \quad [\text{Re } \beta > 0, \text{Re } \gamma > 0, a > 0].$$ Ba IV 15(15)

3.955 $\int_0^\infty x^\nu e^{-\frac{x^2}{2}} \cos\left(\beta x - \nu \frac{\pi}{2}\right) dx = \sqrt{\frac{\pi}{2}} e^{-\frac{\beta^2}{4}} D_\nu(\beta)$ [Re $\nu > -1$].   Ba II 120(4)

3.956 $\int_0^\infty e^{-x^2} (2x \cos x - \sin x) \sin x \, \frac{dx}{x^2} = \sqrt{\pi} \, \frac{e-1}{2e}$.   BH[369](19)

3.957 1. $\int_0^\infty x^{\mu-1} \exp\left(\frac{-\beta^2}{4x}\right) \sin ax \, dx = \frac{i}{2^\mu} \beta^\mu a^{-\frac{\mu}{2}} \times$

$$\times \left[\exp\left(-\frac{i}{4}\mu\pi\right) K_\mu\left(\beta e^{\frac{\pi i}{4}} \sqrt{a}\right) - \exp\left(\frac{i}{4}\mu\pi\right) K_\mu\left(\beta e^{-\frac{\pi i}{4}} \sqrt{a}\right) \right]$$

[Re $\beta > 0$, Re $\mu < 1$, $a > 0$].   Ba IV 318(12)

2. $\int_0^\infty x^{\mu-1} \exp\left(\frac{-\beta^2}{4x}\right) \cos ax \, dx =$

$$= \frac{1}{2^\mu} \beta^\mu a^{-\frac{\mu}{2}} \left[\exp\left(-\frac{i}{4}\mu\pi\right) K_\mu\left(\beta e^{\frac{\pi i}{4}} \sqrt{a}\right) + \exp\left(\frac{i}{4}\mu\pi\right) K_\mu\left(\beta e^{-\frac{\pi i}{4}} \sqrt{a}\right) \right]$$

[Re $\beta > 0$, Re $\mu < 1$, $a > 0$].   Ba IV 320(32)+

3.958 1. $\int_{-\infty}^\infty x^n e^{-(ax^2+bx+c)} \sin(px+q) \, dx =$

$$= -\left(\frac{-1}{2a}\right)^n \sqrt{\frac{\pi}{a}} \exp\left(\frac{b^2-p^2}{4a} - c\right) \sum_{k=0}^{E\left(\frac{n}{2}\right)} \frac{n!}{(n-2k)!k!} a^k \times$$

$$\times \sum_{j=0}^{n-2k} \binom{n-2k}{j} b^{n-2k-j} p^j \sin\left(\frac{pb}{2a} - q + \frac{\pi}{2}j\right)$$

[$a > 0$].   GH[337](1b)

2. $\int_{-\infty}^{\infty} x^n e^{-(ax^2+bx+c)} \cos(px+q)\,dx =$ 3.958

$$= \left(\frac{-1}{2a}\right)^n \sqrt{\frac{\pi}{a}} \exp\left(\frac{b^2-p^2}{4a}-c\right) \sum_{k=0}^{E\left(\frac{n}{2}\right)} \frac{n!}{(n-2k)!k!} a^k \times$$

$$\times \sum_{j=0}^{n-2k} \binom{n-2k}{j} b^{n-2k-j} p^j \cos\left(\frac{pb}{2a}-q+\frac{\pi}{2}j\right)$$

$[a > 0].$ GH[337](1a)

$\int_0^{\infty} x e^{-p^2 x^2} \operatorname{tg} ax\,dx = \frac{a\sqrt{\pi}}{p^3} \sum_{k=1}^{\infty} (-1)^k k \exp\left(-\frac{a^2 k^2}{p^2}\right)$ 3.959

$[p > 0].$ BH[362](15)

1. $\int_0^{\infty} \exp(-\beta\sqrt{\gamma^2+x^2}) \sin ax \frac{x\,dx}{\sqrt{\gamma^2+x^2}} = \frac{a\gamma}{\sqrt{a^2+\beta^2}} K_1(\gamma\sqrt{a^2+\beta^2})$ 3.961

$[\operatorname{Re}\beta > 0,\ \operatorname{Re}\gamma > 0,\ a > 0].$ Ba IV 75(36)

2. $\int_0^{\infty} \exp[-\beta\sqrt{\gamma^2+x^2}] \cos ax \frac{dx}{\sqrt{\gamma^2+x^2}} = K_0(\gamma\sqrt{a^2+\beta^2})$

$[\operatorname{Re}\beta > 0,\ \operatorname{Re}\gamma > 0,\ a > 0].$ Ba IV 17(27)

1. $\int_0^{\infty} \frac{\sqrt{\sqrt{\gamma^2+x^2}-\gamma}\,\exp(-\beta\sqrt{\gamma^2+x^2})}{\sqrt{\gamma^2+x^2}} \sin ax\,dx =$ 3.962

$$= \sqrt{\frac{\pi}{2}} \frac{a\exp(-\gamma\sqrt{a^2+\beta^2})}{\sqrt{\beta^2+a^2}\sqrt{\beta+\sqrt{a^2+\beta^2}}}$$

$[\operatorname{Re}\beta > 0,\ \operatorname{Re}\gamma > 0,\ a > 0].$ Ba IV 75(38)

2. $\int_0^{\infty} \frac{x\exp(-\beta\sqrt{\gamma^2+x^2})}{\sqrt{\gamma^2+x^2}\sqrt{\sqrt{\gamma^2+x^2}-\gamma}} \cos ax\,dx =$

$$= \sqrt{\frac{\pi}{2}} \frac{\sqrt{\beta+\sqrt{a^2+\beta^2}}}{\sqrt{a^2+\beta^2}} \exp[-\gamma\sqrt{a^2+\beta^2}]$$

$[\operatorname{Re}\beta > 0,\ \operatorname{Re}\gamma > 0,\ a > 0].$ Ba IV 17(29)

1. $\int_0^{\infty} e^{-\operatorname{tg}^2 x} \frac{\sin x}{\cos^2 x} \frac{dx}{x} = \frac{\sqrt{\pi}}{2}.$ BH[391](1)   3.963

2. $\int_0^{\frac{\pi}{2}} e^{-p\,\operatorname{tg} x} \frac{x\,dx}{\cos^2 x} = \frac{1}{p}\left[\operatorname{ci}(p)\sin p - \cos p\,\operatorname{si}(p)\right].$

$[p > 0]$ (cf. 3.339).   BH[396](3)

3.963

3. $\int_0^{\frac{\pi}{2}} x e^{-\operatorname{tg}^2 x} \sin 4x \, \frac{dx}{\cos^2 x} = -\frac{3}{2}\sqrt{\pi}.$ BH[396](5)

4. $\int_0^{\frac{\pi}{2}} x e^{-\operatorname{tg}^2 x} \sin^2 2x \, \frac{dx}{\cos^2 x} = 2\sqrt{\pi}.$ BH[396](6)

3.964

1. $\int_0^{\frac{\pi}{2}} x e^{-p \operatorname{tg} x} \frac{p \sin x - \cos x}{\cos^3 x} \, dx = -\sin p \operatorname{si}(p) - \operatorname{ci}(p) \cos p \quad [p > 0].$ Li[396](4)

2. $\int_0^{\frac{\pi}{2}} x e^{-p \operatorname{tg}^2 x} \frac{p - \cos^2 x}{\cos^4 x \operatorname{ctg} x} \, dx = \frac{1}{4} \sqrt{\frac{\pi}{p}} \quad [p > 0].$ BH[396](7)

3. $\int_0^{\frac{\pi}{2}} x e^{-p \operatorname{tg}^2 x} \frac{p - 2\cos^2 x}{\cos^6 x \operatorname{ctg} x} \, dx = \frac{1+2p}{8} \sqrt{\frac{\pi}{p}} \quad [p > 0].$ BH[396](8)

3.965

1. $\int_0^\infty x e^{-\beta x} \sin ax^2 \sin \beta x \, dx = \frac{\beta}{4} \sqrt{\frac{\pi}{2a^3}} e^{-\frac{\beta^2}{2a}} \quad \left[|\arg \beta| < \frac{\pi}{4}, \, a > 0\right].$

   Ba IV 84(17)

2. $\int_0^\infty x e^{-\beta x} \cos ax^2 \cos \beta x \, dx = \frac{\beta}{4} \sqrt{\frac{\pi}{2a^3}} e^{-\frac{\beta^2}{2a}} \quad [a > 0, \, \operatorname{Re}\beta > |\operatorname{Im}\beta|].$

   Ba IV 26(27)

3.966

1. $\int_0^\infty x e^{-px} \cos(2x^2 + px) \, dx = 0 \quad [p > 0].$ BH[361](16)

2. $\int_0^\infty x e^{-px} \cos(2x^2 - px) \, dx = \frac{p\sqrt{\pi}}{8} \exp\left(-\frac{1}{4}p^2\right) \quad [p > 0].$ BH[361](17)

3. $\int_0^\infty x^2 e^{-px} [\sin(2x^2 + px) + \cos(2x^2 + px)] \, dx = 0 \quad [p > 0].$ BH[361](18)

4. $\int_0^\infty x^2 e^{-px} [\sin(2x^2 - px) - \cos(2x^2 - px)] \, dx = \frac{\sqrt{\pi}}{16}(2-p^2) \exp\left(-\frac{1}{4}p^2\right).$

   BH[361](19)

1. $$\int_0^\infty e^{-\frac{\beta^2}{x^2}} \sin a^2x^2 \frac{dx}{x^2} = \frac{\sqrt{\pi}}{2\beta} e^{-\sqrt{2}\,a\beta} \sin(\sqrt{2}\,a\beta) \quad [\text{Re } \beta > 0,\ a > 0].$$ 3.967

Ba IV 75(30)+, BH[369](3)+

2. $$\int_0^\infty e^{-\frac{\beta^2}{x^2}} \cos a^2x^2 \frac{dx}{x^2} = \frac{\sqrt{\pi}}{2\beta} e^{-\sqrt{2}\,a\beta} \cos(\sqrt{2}\,a\beta) \quad [\text{Re } \beta > 0,\ a > 0].$$

BH[369](4), Ba IV 16(20)

3. $$\int_0^\infty x^2 e^{-\beta x^2} \cos ax^2\, dx = \frac{\sqrt{\pi}}{4\sqrt[4]{(a^2+\beta^2)^3}} \cos\left(\frac{3}{2} \arctan \frac{a}{\beta}\right)$$

$[\text{Re } \beta > 0].$ BaIV 14(3)+

1. $$\int_0^\infty e^{-\beta x^2} \sin ax^4\, dx = -\frac{\pi}{8}\sqrt{\frac{\beta}{a}}\left[J_{\frac{1}{4}}\!\left(\frac{\beta^2}{8a}\right)\cos\left(\frac{\beta^2}{8a}+\frac{\pi}{8}\right)+\right.$$ 3.968

$$\left.+N_{\frac{1}{4}}\!\left(\frac{\beta^2}{8a}\right)\sin\left(\frac{\beta^2}{8a}+\frac{\pi}{8}\right)\right]$$

$[\text{Re } \beta > 0,\ a > 0].$ Ba IV 75(34)

2. $$\int_0^\infty e^{-\beta x^2} \cos ax^4\, dx = \frac{\pi}{8}\sqrt{\frac{\beta}{a}}\left[J_{\frac{1}{4}}\!\left(\frac{\beta^2}{8a}\right)\sin\left(\frac{\beta^2}{8a}+\frac{\pi}{8}\right)-\right.$$

$$\left.-N_{\frac{1}{4}}\!\left(\frac{\beta^2}{8a}\right)\cos\left(\frac{\beta^2}{8a}+\frac{\pi}{8}\right)\right]$$

$[\text{Re } \beta > 0,\ a > 0].$ Ba IV 16(24)

1. $$\int_0^\infty e^{-p^2x^4+q^2x^2}[2px \cos(2pqx^3) + q\sin(2pqx^3)]\, dx = \frac{\sqrt{\pi}}{2}.$$ BH[363](7) 3.969

2. $$\int_0^\infty e^{-p^2x^4+q^2x^2}[2px \sin(2pqx^3) - q\cos(2pqx^3)]\, dx = 0.$$ BH[363](8)

1. $$\int_0^\infty \exp\left(-px^2 - \frac{q}{x^2}\right) \sin\left(ax^2 + \frac{b}{x^2}\right) \frac{dx}{x^2} =$$ 3.971

$$= \frac{1}{2}\int_{-\infty}^\infty \exp\left(-px^2 - \frac{q}{x^2}\right) \sin\left(ax^2 + \frac{b}{x^2}\right) \frac{dx}{x^2} =$$

$$= \frac{\sqrt{\pi}}{2s} \exp[-2rs \cos(A+B)] \sin[A + 2rs \sin(A+B)].$$ BH[369](16, 17)

3.971

2. $\int_0^\infty \exp\left(-px^2 - \frac{q}{x^2}\right) \cos\left(ax^2 + \frac{b}{x^2}\right) \frac{dx}{x^2} =$

$= \frac{1}{2} \int_{-\infty}^\infty \exp\left(-px^2 - \frac{q}{x^2}\right) \cos\left(ax^2 + \frac{b}{x^2}\right) \frac{dx}{x^2} =$

$= \frac{\sqrt{\pi}}{2s} \exp\left[-2rs \cos(A+B)\right] \cos\left[A + 2rs \sin(A+B)\right].$

[Ind en Formeln 3.971 1., 3.971 2. ist | [In 3.971 1. and 3.971 2.

$p \geq 0, \ q \geq 0, \ r = \sqrt[4]{a^2 + p^2}, \ s = \sqrt[4]{b^2 + q^2},$

$A = \operatorname{arctg} \frac{a}{p}, \quad B = \operatorname{arctg} \frac{b}{q}.$]

BH[369](15, 18)

3.972

1. $\int_0^\infty \exp\left[-\beta\sqrt{\gamma^4 + x^4}\right] \sin ax^2 \frac{dx}{\sqrt{\gamma^4 + x^4}} =$

$= \sqrt{\frac{a\pi}{8}} I_{\frac{1}{4}}\left[\frac{\gamma^2}{2}(\sqrt{\beta^2 + a^2} - \beta)\right] K_{\frac{1}{4}}\left[\frac{\gamma^2}{4}(\sqrt{\beta^2 + a^2} + \beta)\right]$

$\left[\operatorname{Re} \beta > 0, \ |\arg \gamma| < \frac{\pi}{4}, \ a > 0\right].$

Ba IV 75(37)

2. $\int_0^\infty \exp\left[-\beta\sqrt{\gamma^4 + x^4}\right] \cos ax^2 \frac{dx}{\sqrt{\gamma^4 + x^4}} =$

$= \sqrt{\frac{a\pi}{8}} I_{-\frac{1}{4}}\left[\frac{\gamma^2}{2}(\sqrt{\beta^2 + a^2} - \beta)\right] K_{\frac{1}{4}}\left[\frac{\gamma^2}{4}(\sqrt{\beta^2 + a^2} + \beta)\right]$

$\left[\operatorname{Re} \beta > 0, \ |\arg \gamma| < \frac{\pi}{4}, \ a > 0\right].$

Ba IV 17(28)

3.973

1. $\int_0^\infty \exp(p \cos ax) \sin(p \sin ax) \frac{dx}{x} = \frac{\pi}{2}(e^p - 1) \quad [p > 0, \ a > 0].$

WW 122, F II 731–733

2. $\int_0^\infty \exp(p \cos ax) \sin(p \sin ax + bx) \frac{x \, dx}{c^2 + x^2} = \frac{\pi}{2} \exp(-cb + pe^{-ac})$

$[a > 0, \ b > 0, \ c > 0, \ p > 0].$

BH[372](3)

3. $\int_0^\infty \exp(p \cos ax) \cos(p \sin ax + bx) \frac{dx}{c^2 + x^2} = \frac{\pi}{2c} \exp(-cb + pe^{-ac})$

$[a > 0, \ b > 0, \ c > 0, \ p > 0].$

BH[372](4)

4. $\int_0^\infty \exp(p \cos x) \sin(p \sin x + nx) \frac{dx}{x} = \frac{\pi}{2} e^p \quad [p > 0].$

BH[366](2)

5. $\int_0^\infty \exp(p\cos x)\sin(p\sin x)\cos nx \frac{dx}{x} = \frac{p^n}{n!}\cdot\frac{\pi}{4} + \frac{\pi}{2}\sum_{k=n+1}^\infty \frac{p^k}{k!}$ 3.973

$[p > 0]$.   Li[366](3)

6. $\int_0^\infty \exp(p\cos x)\cos(p\sin x)\sin nx \frac{dx}{x} = \frac{\pi}{2}\sum_{k=0}^{n-1}\frac{p^k}{k!} + \frac{p^n}{n!}\frac{\pi}{4}$

$[p > 0]$.   Li[366](4)

1. $\int_0^\infty \exp(p\cos ax)\sin(p\sin ax)\operatorname{cosec} ax \frac{dx}{b^2+x^2} = \frac{\pi[e^p - \exp(pe^{-ab})]}{2b\,\operatorname{sh} ab}$ 3.974

$[a > 0,\ b > 0,\ p > 0]$.   BH[391](4)

2. $\int_0^\infty [1 - \exp(p\cos ax)\cos(p\sin ax)]\operatorname{cosec} ax \frac{x\,dx}{b^2+x^2} = \frac{\pi[e^p - \exp(pe^{-ab})]}{2\,\operatorname{sh} ab}$

$[a > 0,\ b > 0,\ p > 0]$.   BH[391](5)

3. $\int_0^\infty \exp(p\cos ax)\sin(p\sin ax + ax)\operatorname{cosec} ax \frac{dx}{b^2+x^2} =$

$= \frac{\pi[e^p - \exp(pe^{-ab} - ab)]}{2b\,\operatorname{sh} ab}$ $[a > 0,\ b > 0,\ p > 0]$.   BH[391](6)

4. $\int_0^\infty \exp(p\cos ax)\cos(p\sin ax + ax)\operatorname{cosec} ax \frac{x\,dx}{b^2+x^2} =$

$= \frac{\pi[e^p - \exp(pe^{-ab} - ab)]}{2\,\operatorname{sh} ab}$ $[a > 0,\ b > 0,\ p > 0]$.   BH[391](7)

5. $\int_0^\infty \exp(p\cos ax)\sin(p\sin ax)\frac{x\,dx}{b^2-x^2} =$

$= \frac{\pi}{2}[1 - \exp(p\cos ab)\cos(p\sin ab)]$ $[p > 0,\ a > 0]$.   BH[378](1)

6. $\int_0^\infty \exp(p\cos ax)\cos(p\sin ax)\frac{dx}{b^2-x^2} =$

$= \frac{\pi}{2b}\exp(p\cos ab)\sin(p\sin ab)$ $[a > 0,\ b > 0,\ p > 0]$.   BH[378](2)

7. $\int_0^\infty \exp(p\cos ax)\sin(p\sin ax)\operatorname{tg} ax \frac{dx}{b^2+x^2} =$

$= \frac{\pi}{2b}\cdot\operatorname{th} ab\,[\exp(pe^{-ab}) - e^p]$ $[a > 0,\ b > 0,\ p > 0]$.   BH[372](14)

3.974

8. $\int_0^\infty \exp(p\cos ax)\sin(p\sin ax)\operatorname{ctg} ax \dfrac{dx}{b^2+x^2} =$

$= \dfrac{\pi}{2b}\operatorname{cth} ab\,[e^p - \exp(pe^{-ab})]\quad [a>0, b>0, p>0].$ BH[372](15)

9. $\int_0^\infty \exp(p\cos ax)\sin(p\sin ax)\operatorname{cosec} ax \dfrac{dx}{b^2-x^2} =$

$= \dfrac{\pi}{2b}\operatorname{cosec} ab\,[e^p - \exp(p\cos ab)\cos(p\sin ab)]$

$[a>0, b>0, p>0].$ BH[391](12)

10. $\int_0^\infty [1 - \exp(p\cos ax)\cos(p\sin ax)]\operatorname{cosec} ax \dfrac{x\,dx}{b^2-x^2} =$

$= -\dfrac{\pi}{2}\exp(p\cos ab)\sin(p\sin ab)\operatorname{cosec} ab$

$[a>0, b>0, p>0].$ BH[391](13)

3.975

1. $\int_0^\infty \dfrac{\sin\left(\beta\arctan\dfrac{x}{\gamma}\right)}{(\gamma^2+x^2)^{\frac{\beta}{2}}}\cdot\dfrac{dx}{e^{2\pi x}-1} = \dfrac{1}{2}\zeta(\beta,\gamma) - \dfrac{1}{4\gamma^\beta} - \dfrac{\gamma^{1-\beta}}{2(\beta-1)}$

$[\operatorname{Re}\beta > 1,\ \operatorname{Re}\gamma > 0].$ WW 270, Ba I 26(7)

2. $\int_0^\infty \dfrac{\sin(\beta\arctan x)}{(1+x^2)^{\frac{\beta}{2}}}\cdot\dfrac{dx}{e^{2\pi x}+1} = \dfrac{1}{2(\beta-1)} - \dfrac{\zeta(\beta)}{2^\beta}\quad [\operatorname{Re}\beta > 1].$ Ba I 33(13)

3.976 $\int_0^\infty (1+x^2)^{\beta-\frac{1}{2}} e^{-px^2}\cos[2px + (2\beta-1)\arctan x]\,dx = \dfrac{e^{-p}}{2p^\beta}\sin\pi\beta\,\Gamma(\beta)$

$[\operatorname{Re}\beta > 0,\ p > 0].$ WW 246

Trigonometrische und hyperbolische Funktionen
Trigonometric and hyperbolic functions

3.98–3.99

3.981

1. $\int_0^\infty \dfrac{\sin ax}{\operatorname{sh}\beta x}\,dx = \dfrac{\pi}{2\beta}\operatorname{th}\dfrac{a\pi}{2\beta}\quad [\operatorname{Re}\beta > 0, a > 0].$ BH[264](16)

2. $\int_0^\infty \dfrac{\sin ax}{\operatorname{ch}\beta x}\,dx = -\dfrac{\pi}{2\beta}\operatorname{th}\dfrac{a\pi}{2\beta} - \dfrac{i}{2\beta}\left[\psi\left(\dfrac{\beta+ai}{4\beta}\right) - \psi\left(\dfrac{\beta-ai}{4\beta}\right)\right]$

$[\operatorname{Re}\beta > 0, a > 0].$ GH[335](12), Ba IV 88(1)

3. $\int_0^\infty \dfrac{\cos ax}{\operatorname{ch}\beta x}\,dx = \dfrac{\pi}{2\beta}\operatorname{sech}\dfrac{a\pi}{2\beta}\quad [\operatorname{Re}\beta > 0,\ a > 0].$ BH[264](14)

4. $\int_0^\infty \sin ax \frac{\operatorname{sh}\beta x}{\operatorname{sh}\gamma x} dx = \frac{\pi}{2\gamma} \frac{\operatorname{sh}\frac{a\pi}{\gamma}}{\operatorname{ch}\frac{a\pi}{\gamma} + \cos\frac{\beta\pi}{\gamma}} +$
    $$+ \frac{i}{2\gamma}\left[\psi\left(\frac{\beta+\gamma+ia}{2\gamma}\right) - \psi\left(\frac{\beta+\gamma-ia}{2\gamma}\right)\right]$$
    $[|\operatorname{Re}\beta| < \operatorname{Re}\gamma,\ a > 0].$ Ba IV 88(5)   3.981

5. $\int_0^\infty \cos ax \frac{\operatorname{sh}\beta x}{\operatorname{sh}\gamma x} dx = \frac{\pi}{2\gamma} \frac{\sin\frac{\pi\beta}{\gamma}}{\operatorname{ch}\frac{a\pi}{\gamma} + \cos\frac{\beta\pi}{\gamma}}$ $[|\operatorname{Re}\beta| < \operatorname{Re}\gamma,\ a > 0].$ BH[265](7)

6. $\int_0^\infty \sin ax \frac{\operatorname{sh}\beta x}{\operatorname{ch}\gamma x} dx = \frac{\pi}{\gamma} \frac{\sin\frac{\beta\pi}{2\gamma}\operatorname{sh}\frac{a\pi}{2\gamma}}{\operatorname{ch}\frac{a\pi}{\gamma} + \cos\frac{\beta\pi}{\gamma}}$ $[|\operatorname{Re}\beta| < \operatorname{Re}\gamma,\ a > 0].$ BH[265](2)

7. $\int_0^\infty \cos ax \cdot \frac{\operatorname{sh}\beta x}{\operatorname{ch}\gamma x} dx = \frac{1}{4\gamma}\left\{\psi\left(\frac{3\gamma-\beta+ia}{4\gamma}\right) + \right.$
    $$\left. + \psi\left(\frac{3\gamma-\beta-ia}{4\gamma}\right) - \psi\left(\frac{3\gamma+\beta-ia}{4\gamma}\right) - \psi\left(\frac{3\gamma+\beta+ia}{4\gamma}\right) + \frac{2\pi\sin\frac{\pi\beta}{\gamma}}{\cos\frac{\pi\beta}{\gamma}+\operatorname{ch}\frac{\pi a}{\gamma}}\right\}$$
    $[|\operatorname{Re}\beta| < \operatorname{Re}\gamma,\ a > 0].$ Ba IV 31(13)

8. $\int_0^\infty \sin ax \frac{\operatorname{ch}\beta x}{\operatorname{sh}\gamma x} dx = \frac{\pi}{2\gamma} \cdot \frac{\operatorname{sh}\frac{\pi a}{\gamma}}{\operatorname{ch}\frac{\pi a}{\gamma} + \cos\frac{\pi\beta}{\gamma}}$ $[|\operatorname{Re}\beta| < \operatorname{Re}\gamma,\ a > 0].$

    BH[265](4)

9. $\int_0^\infty \sin ax \cdot \frac{\operatorname{ch}\beta x}{\operatorname{ch}\gamma x} dx = \frac{i}{4\gamma}\left[\psi\left(\frac{3\gamma+\beta+ai}{4\gamma}\right) - \right.$
    $$\left. - \psi\left(\frac{3\gamma+\beta-ai}{4\gamma}\right) + \psi\left(\frac{3\gamma-\beta+ia}{4\gamma}\right) - \psi\left(\frac{3\gamma-\beta-ai}{4\gamma}\right) - \frac{2\pi i\operatorname{sh}\frac{\pi a}{\gamma}}{\operatorname{ch}\frac{a\pi}{\gamma}+\cos\frac{\beta\pi}{\gamma}}\right]$$
    $[|\operatorname{Re}\beta| < \operatorname{Re}\gamma,\ a > 0].$ Ba IV 88(6)

10. $\int_0^\infty \cos ax \frac{\operatorname{ch}\beta x}{\operatorname{ch}\gamma x} dx = \frac{\pi}{\gamma} \frac{\cos\frac{\beta\pi}{2\gamma}\operatorname{ch}\frac{a\pi}{2\gamma}}{\operatorname{ch}\frac{a\pi}{\gamma} + \cos\frac{\beta\pi}{\gamma}}$ $[|\operatorname{Re}\beta| < \operatorname{Re}\gamma,\ a > 0].$ BH[265](6)

11. $\int_0^{\frac{\pi}{2}} \cos^{2m}x \operatorname{ch}\beta x\, dx = \frac{(2m)!\operatorname{sh}\frac{\pi\beta}{2}}{\beta(\beta^2+2^2)\dots[\beta^2+(2m)^2]}$ $[\operatorname{Re}\beta > 0].$ W 563+

3.981

12. $$\int_0^{\frac{\pi}{2}} \cos^{2m-1}x \, \operatorname{ch} \beta x \, dx = \frac{(2m-1)! \, \operatorname{ch} \frac{\pi\beta}{2}}{(\beta^2 + 1^2)(\beta^2 + 3^2) \cdots [\beta^2 + (2m+1)^2]}$$
[Re $\beta > 0$]. W 564+

3.982

1. $$\int_0^\infty \frac{\cos ax}{\operatorname{ch}^2 \beta x} dx = \frac{a\pi}{2\beta^2 \operatorname{sh} \frac{a\pi}{2\beta}}$$ [Re $\beta > 0$, $a > 0$]. BH[264](16)

2. $$\int_0^\infty \sin ax \, \frac{\operatorname{sh} \beta x}{\operatorname{ch}^2 \gamma x} dx = -\frac{\pi \left(a \sin \frac{\beta\pi}{2\gamma} \operatorname{ch} \frac{a\pi}{2\gamma} - \beta \cos \frac{\beta\pi}{2\gamma} \operatorname{sh} \frac{a\pi}{2\gamma} \right)}{\gamma^2 \left(\operatorname{ch} \frac{a\pi}{\gamma} - \cos \frac{\beta\pi}{\gamma} \right)}$$

[|Re β| < 2 Re γ, $a > 0$]. Ba IV 88(9)

3.983

1. $$\int_0^\infty \frac{\cos ax \, dx}{b \operatorname{ch} \beta x + c} = \frac{\pi \sin \left(\frac{a}{\beta} \operatorname{arch} \frac{c}{b} \right)}{\beta \sqrt{c^2 - b^2} \operatorname{sh} \frac{a\pi}{\beta}}$$ [$c > b > 0$];

$$= \frac{\pi \operatorname{sh} \left(\frac{a}{\beta} \arccos \frac{c}{b} \right)}{\beta \sqrt{b^2 - c^2} \operatorname{sh} \frac{a\pi}{\beta}}$$ [$b > |c| > 0$];

[Re $\beta > 0$, $a > 0$]. GH[335](13a)

2. $$\int_0^\infty \frac{\cos ax \, dx}{\operatorname{ch} \beta x + \cos \gamma} = \frac{\pi}{\beta} \frac{\operatorname{sh} \frac{a\gamma}{\beta}}{\sin \gamma \operatorname{sh} \frac{a\pi}{\beta}}$$ [π Re β < Im $\overline{\beta}\gamma$, $a > 0$]. BH[267](3)

3. $$\int_0^\infty \frac{\cos ax \, dx}{\operatorname{ch} x - \operatorname{ch} b} = -\pi \operatorname{ch} a\pi \, \frac{\sin ab}{\operatorname{sh} b}$$ [$a > 0$, $b > 0$]. BH[267](4), Ba IV 30(8)

4. $$\int_0^\infty \frac{\cos ax \, dx}{1 + 2 \operatorname{ch} \left(\sqrt{\frac{2}{3}} \pi x \right)} = \frac{\sqrt{\frac{\pi}{2}}}{1 + 2 \operatorname{ch} \left(\sqrt{\frac{2}{3}} \pi a \right)}$$ [$a > 0$]. Ba IV 30(9)

5. $$\int_0^\infty \frac{\sin ax \, \operatorname{sh} \beta x}{\operatorname{ch} \gamma x + \cos \delta} dx =$$

$$= \frac{\pi \left\{ \sin \left[\frac{\beta}{\gamma} (\pi - \delta) \right] \operatorname{sh} \left[\frac{a}{\gamma} (\pi + \delta) \right] - \sin \left[\frac{\beta}{\gamma} (\pi + \delta) \right] \operatorname{sh} \left[\frac{a}{\gamma} (\pi - \delta) \right] \right\}}{\gamma \sin \delta \left(\operatorname{ch} \frac{2\pi a}{\gamma} - \cos \frac{2\pi\beta}{\gamma} \right)}$$

[π Re γ > |Re $\overline{\gamma}\delta$|, |Re β| < Re γ, $a > 0$]. BH[267](2)

6. $\int_0^\infty \dfrac{\cos ax \operatorname{ch} \beta x}{\operatorname{ch} \gamma x + \cos b}\, dx =$

$$= \dfrac{\pi \left\{ \cos\left[\dfrac{\beta}{\gamma}(\pi - b)\right] \operatorname{ch}\left[\dfrac{a}{\gamma}(\pi + b)\right] - \cos\left[\dfrac{\beta}{\gamma}(\pi + b)\right] \operatorname{ch}\left[\dfrac{a}{\gamma}(\pi - b)\right] \right\}}{\gamma \sin b \left(\operatorname{ch}\dfrac{2\pi a}{\gamma} - \cos\dfrac{2\pi \beta}{\gamma} \right)}$$

$[|\operatorname{Re} \beta| < \operatorname{Re} \gamma,\; 0 < b < \pi,\; a > 0]$.   BH[267](6)

3.983

7. $\int_0^\infty \dfrac{\cos ax\, dx}{(\beta + \sqrt{\beta^2 - 1}\operatorname{ch} x)^{\nu+1}} = \Gamma(\nu + 1 - ai)\, e^{a\pi} \dfrac{Q_\nu^{ai}(\beta)}{\Gamma(\nu + 1)}$

$[\operatorname{Re} \nu > -1,\; |\arg(\beta \pm 1)| < \pi,\; a > 0]$.   Ba IV 30(10)

1. $\int_0^\infty \dfrac{\sin ax \operatorname{sh} x}{\operatorname{ch} x + \cos b}\, dx = \pi \dfrac{\operatorname{ch} ab}{\operatorname{ch} a\pi}$ $[b \leqslant \pi,\; a > 0]$.   BH[267](1)

3.984

2. $\int_0^\infty \dfrac{\cos ax \operatorname{ch} x}{\operatorname{ch} x + \cos b}\, dx = -\pi \operatorname{ctg} b\, \dfrac{\operatorname{sh} ab}{\operatorname{sh} a\pi}$ $[b \leqslant \pi]$.   BH[267](5)

3. $\int_0^\infty \dfrac{\sin ax \operatorname{sh}\dfrac{x}{2}}{\operatorname{ch} x + \cos \beta}\, dx = \dfrac{\operatorname{sh} a\beta}{2 \sin \dfrac{\beta}{2}\, \operatorname{ch} a\pi}$ $[\operatorname{Re} \beta < \pi,\; a > 0]$.   Ba IV 89(10)

4. $\int_0^\infty \dfrac{\cos ax \operatorname{ch}\dfrac{\beta}{2}x}{\operatorname{ch}\beta x + \operatorname{ch}\gamma}\, dx = \dfrac{\pi \cos\dfrac{a\gamma}{\beta}}{2\beta \operatorname{ch}\dfrac{\gamma}{2}\operatorname{ch}\dfrac{a\pi}{\beta}}$ $[\pi \operatorname{Re}\beta > |\operatorname{Im}(\bar\beta \gamma)|]$.   Ba IV 31(16)

5. $\int_0^\infty \dfrac{\sin ax \operatorname{sh} \beta x}{\operatorname{ch} 2\beta x + \cos 2ax}\, dx = \dfrac{a\pi}{4(a^2 + \beta^2)}$ $[a > 0,\; \operatorname{Re}\beta > 0]$.   BH[267](7)

6. $\int_0^\infty \dfrac{\cos ax \operatorname{ch} \beta x}{\operatorname{ch} 2\beta x + \cos 2ax}\, dx = \dfrac{\beta\pi}{4(a^2 + \beta^2)}$ $[\operatorname{Re}\beta > 0,\; a > 0]$.   BH[267](8)

7. $\int_0^\infty \dfrac{\operatorname{sh}^{2\mu-1} x \operatorname{ch}^{2\rho - 2\nu + 1} x}{(\operatorname{ch}^2 x - \beta \operatorname{sh}^2 x)^\rho}\, dx = \dfrac{1}{2} B(\mu, \nu - \mu)\, _2F_1(\rho, \mu;\, \nu;\, \beta)$

$[\operatorname{Re} \nu > \operatorname{Re} \mu > 0]$.   Ba I 115(12)

1. $\int_0^\infty \dfrac{\cos ax\, dx}{\operatorname{ch}^\nu \beta x} = \dfrac{2^{\nu-2}}{\beta \Gamma(\nu)} \Gamma\left(\dfrac{\nu}{2} + \dfrac{ai}{2\beta}\right) \Gamma\left(\dfrac{\nu}{2} - \dfrac{ai}{2\beta}\right)$

3.985

$[\operatorname{Re}\beta > 0,\; \operatorname{Re}\nu > 0,\; a > 0]$.   Ba IV 30(5)

2. $\int_0^\infty \dfrac{\cos ax\, dx}{\operatorname{ch}^{2n} \beta x} = \dfrac{4^{n-1} \pi a}{2(2n-1)!\, \beta^2 \operatorname{sh}\dfrac{a\pi}{2\beta}} \prod_{k=1}^{n-1}\left(\dfrac{a^2}{4\beta^2} + k^2\right);$

$= \dfrac{\pi a(a^2 + 2^2\beta^2)(a^2 + 4^2\beta^2)\cdots[a^2 + (2n-2)^2\beta^2]}{2(2n-1)!\, \beta^{2n} \operatorname{sh}\dfrac{a\pi}{2\beta}}$

$[n \geqslant 2,\; a > 0]$.   Ba IV 30(3)

3.985

3. $\int_0^\infty \dfrac{\cos ax\, dx}{\operatorname{ch}^{2n+1}\beta x} = \dfrac{\pi \cdot 2^{2n-1}}{(2n)!\,\beta\,\operatorname{ch}\dfrac{a\pi}{2\beta}} \prod_{k=1}^n \left[\dfrac{a^2}{4\beta^2} + \left(\dfrac{2k-1}{2}\right)^2\right]$;

$\qquad\qquad = \dfrac{\pi(a^2+\beta^2)(a^2+3^2\beta^2)\ldots[a^2+(2n-1)^2\beta^2]}{2(2n)!\,\beta^{2n+1}\operatorname{ch}\dfrac{a\pi}{2\beta}}$ $[a>0]$. Ba IV 30(4)

3.986

1. $\int_0^\infty \dfrac{\sin\beta x \sin\gamma x}{\operatorname{ch}\delta x}\, dx = \dfrac{\pi}{\delta} \cdot \dfrac{\operatorname{sh}\dfrac{\beta\pi}{2\delta}\operatorname{sh}\dfrac{\gamma\pi}{2\delta}}{\operatorname{ch}\dfrac{\beta}{\delta}\pi + \operatorname{ch}\dfrac{\gamma}{\delta}\pi}$ $[|\operatorname{Im}(\beta+\gamma)| < \operatorname{Re}\delta]$. BH[264](19)

2. $\int_0^\infty \dfrac{\sin\alpha x \cos\beta x}{\operatorname{sh}\gamma x}\, dx = \dfrac{\pi\operatorname{sh}\dfrac{\pi\alpha}{\gamma}}{2\gamma\left(\operatorname{ch}\dfrac{\alpha\pi}{\gamma} + \operatorname{ch}\dfrac{\beta\pi}{\gamma}\right)}$ $[|\operatorname{Im}(\alpha+\beta)| < \operatorname{Re}\gamma]$. Li[264](20)

3. $\int_0^\infty \dfrac{\cos\beta x \cos\gamma x}{\operatorname{ch}\delta x}\, dx = \dfrac{\pi}{\delta} \cdot \dfrac{\operatorname{ch}\dfrac{\beta\pi}{2\delta}\operatorname{ch}\dfrac{\gamma\pi}{2\delta}}{\operatorname{ch}\dfrac{\beta\pi}{\delta} + \operatorname{ch}\dfrac{\gamma\pi}{\delta}}$ $[|\operatorname{Im}(\beta+\gamma)| < \operatorname{Re}\delta]$. BH[264](21)

4. $\int_0^\infty \dfrac{\sin^2\beta x}{\operatorname{sh}^2\pi x}\, dx = \dfrac{\beta}{\pi(e^{2\beta}-1)} + \dfrac{\beta-1}{2\pi}$ $[|\operatorname{Im}\beta| < \pi]$. Ba I 44(3)

3.987

1. $\int_0^\infty \sin ax\,(1 - \operatorname{th}\beta x)\, dx = \dfrac{1}{a} - \dfrac{\pi}{2\beta\operatorname{sh}\dfrac{a\pi}{2\beta}}$ $[\operatorname{Re}\beta > 0]$. Ba IV 88(4)+

2. $\int_0^\infty \sin ax\,(\operatorname{cth}\beta x - 1)\, dx = \dfrac{\pi}{2\beta}\operatorname{cth}\dfrac{a\pi}{2\beta} - \dfrac{1}{a}$ $[\operatorname{Re}\beta > 0]$. Ba IV 88(3)

3.988

1. $\int_0^{\pi/2} \dfrac{\cos ax\,\operatorname{sh}(2b\cos x)}{\sqrt{\cos x}}\, dx = \dfrac{\pi}{2}\sqrt{\pi b}\, I_{\frac{a}{2}+\frac{1}{4}}(b)\, I_{-\frac{a}{2}+\frac{1}{4}}(b)$ $[a>0]$.

Ba IV 37(66)

2. $\int_0^{\pi/2} \dfrac{\cos ax\,\operatorname{ch}(2b\cos x)}{\sqrt{\cos x}}\, dx = \dfrac{\pi}{2}\sqrt{\pi b}\, I_{\frac{a}{2}-\frac{1}{4}}(b)\, I_{-\frac{a}{2}-\frac{1}{4}}(b)$ $[a>0]$.

Ba IV 37(67)

3.989

1. $\int_0^\infty \dfrac{\sin\dfrac{a^2 x^2}{\pi}\sin bx}{\operatorname{sh} ax}\, dx = \dfrac{\pi}{2a}\sin\dfrac{\pi b^2}{4a^2}\operatorname{cosech}\dfrac{\pi b}{2a}$ $[a>0,\,b>0]$. Ba IV 93(44)

2. $\displaystyle\int_0^\infty \frac{\cos\dfrac{a^2x^2}{\pi}\sin bx}{\operatorname{sh} ax}\,dx = \frac{\pi}{2a}\,\frac{\operatorname{ch}\dfrac{\pi b}{a} - \cos\dfrac{\pi b^2}{4a^2}}{\operatorname{sh}\dfrac{\pi b}{2a}}$ $\quad [a>0,\ b>0]$. $\qquad$ Ba IV 93(45) $\quad$ 3.989

3. $\displaystyle\int_0^\infty \frac{\sin\dfrac{x^2}{\pi}\cos ax}{\operatorname{ch} x}\,dx = \frac{\pi}{2}\,\frac{\cos\dfrac{a^2\pi}{4} - \dfrac{1}{\sqrt{2}}}{\operatorname{ch}\dfrac{a\pi}{2}}$ $\quad [a>0]$. $\qquad$ Ba IV 36(54)

4. $\displaystyle\int_0^\infty \frac{\cos\dfrac{x^2}{\pi}\cos ax}{\operatorname{ch} x}\,dx = \frac{\pi}{2}\,\frac{\sin\dfrac{a^2\pi}{4} + \dfrac{1}{\sqrt{2}}}{\operatorname{ch}\dfrac{a\pi}{2}}$ $\quad [a>0]$. $\qquad$ Ba IV 36(55)

5. $\displaystyle\int_0^\infty \frac{\sin(\pi ax^2)\cos bx}{\operatorname{ch}\pi x}\,dx = -\sum_{k=0}^\infty \exp\left[-\left(k+\tfrac{1}{2}\right)b\right]\sin\left[\left(k+\tfrac{1}{2}\right)^2\pi a\right] +$

$\displaystyle\qquad\qquad + \frac{1}{\sqrt{a}}\sum_{k=0}^\infty \exp\left[-\frac{b\left(k+\tfrac{1}{2}\right)}{a}\right]\sin\left[\frac{\pi}{4} - \frac{b^2}{4\pi a} + \frac{\left(k+\tfrac{1}{2}\right)^2\pi}{a}\right]$

$[a>0,\ b>0]$. $\qquad$ Ba IV 36(56)

6. $\displaystyle\int_0^\infty \frac{\cos(\pi ax^2)\cos bx}{\operatorname{ch}\pi x}\,dx =$

$\displaystyle = \sum_{k=0}^\infty (-1)^k \exp\left[-\left(k+\tfrac{1}{2}\right)b\right]\cos\left[\left(k+\tfrac{1}{2}\right)^2\pi a\right] +$

$\displaystyle\qquad\qquad + \frac{1}{\sqrt{a}}\sum_{k=0}^\infty \exp\left[-\frac{b\left(k+\tfrac{1}{2}\right)}{a}\right]\cos\left[\frac{\pi}{4} - \frac{b^2}{4\pi a} + \frac{\left(k+\tfrac{1}{2}\right)^2\pi}{a}\right]$

$[a>0,\ b>0]$. $\qquad$ Ba IV 36(57)

1. $\displaystyle\int_0^\infty \sin\pi x^2 \sin ax \operatorname{cth}\pi x\,dx = \frac{1}{2}\operatorname{th}\frac{a}{2}\sin\left(\frac{\pi}{4} + \frac{a^2}{4\pi}\right)$ $\quad [a>0]$. $\qquad$ Ba IV 93(42) $\quad$ 3.991

2. $\displaystyle\int_0^\infty \cos\pi x^2 \sin ax \operatorname{cth}\pi x\,dx = \frac{1}{2}\operatorname{th}\frac{a}{2}\left[1 - \cos\left(\frac{\pi}{4} + \frac{a^2}{4\pi}\right)\right]$ $\quad [a>0]$.

Ba IV 93(43)

1. $\displaystyle\int_0^\infty \frac{\sin\pi x^2 \cos ax}{1 + 2\operatorname{ch}\left(\dfrac{2}{\sqrt{3}}\pi x\right)}\,dx = -\sqrt{3} + \frac{\cos\left(\dfrac{\pi}{12} - \dfrac{a^2}{4\pi}\right)}{4\operatorname{ch}\dfrac{a}{\sqrt{3}} - 2}$ $\quad [a>0]$. $\qquad$ Ba IV 37(60) $\quad$ 3.992

3.992   2. $\displaystyle\int_0^\infty \frac{\cos \pi x^2 \cos ax}{1 + 2\,\text{ch}\left(\frac{2}{\sqrt{3}}\pi x\right)}\, dx = 1 - \frac{\sin\left(\frac{\pi}{12} - \frac{a^2}{4\pi}\right)}{4\,\text{ch}\frac{a}{\sqrt{3}} - 2}$ $[a > 0]$.   Ba IV 37(61)

3.993 $\displaystyle\int_0^\infty \frac{\sin x^2 + \cos x^2}{\text{ch}(\sqrt{\pi}\, x)} \cos ax\, dx = \frac{\sqrt{\pi}}{2} \cdot \frac{\sin a^2 + \cos a^2}{\text{ch}(\sqrt{\pi}\, a)}$ $[a > 0]$.   Ba IV 37(58)

3.994   1. $\displaystyle\int_0^\infty \frac{\sin(2a\,\text{ch}\,x)\cos bx}{\sqrt{\text{ch}\,x}}\, dx = -\frac{\pi}{4}\sqrt{a\pi}\,[J_{\frac{1}{4}+\frac{ib}{2}}(a) N_{\frac{1}{4}-\frac{ib}{2}}(a) +$

$+ J_{\frac{1}{4}-\frac{ib}{2}}(a) N_{\frac{1}{4}+\frac{ib}{2}}(a)]$ $[a > 0,\ b > 0]$.   Ba IV 37(62)

2. $\displaystyle\int_0^\infty \frac{\cos(2a\,\text{ch}\,x)\cos bx}{\sqrt{\text{ch}\,x}}\, dx = -\frac{\pi}{4}\sqrt{a\pi}\,[J_{-\frac{1}{4}+\frac{ib}{2}}(a) N_{-\frac{1}{4}-\frac{ib}{2}}(a) +$

$+ J_{-\frac{1}{4}-\frac{ib}{2}}(a) N_{-\frac{1}{4}+\frac{ib}{2}}(a)]$ $[a > 0,\ b > 0]$.   Ba IV 37(63)

3. $\displaystyle\int_0^\infty \frac{\sin(2a\,\text{sh}\,x)\sin bx}{\sqrt{\text{sh}\,x}}\, dx = -\frac{i}{2}\sqrt{\pi a}\,[I_{\frac{1}{4}-\frac{ib}{2}}(a) K_{-\frac{1}{4}+\frac{ib}{2}}(a) -$

$- I_{\frac{1}{4}+\frac{ib}{2}}(a) K_{\frac{1}{4}-\frac{ib}{2}}(a)]$ $[a > 0,\ b > 0]$.   Ba IV 93(47)

4. $\displaystyle\int_0^\infty \frac{\cos(2a\,\text{sh}\,x)\sin bx}{\sqrt{\text{sh}\,x}}\, dx =$

$= -\frac{i}{2}\sqrt{\pi a}\,[I_{-\frac{1}{4}-\frac{ib}{2}}(a) K_{-\frac{1}{4}+\frac{ib}{2}}(a) - I_{-\frac{1}{4}+\frac{ib}{2}}(a) K_{-\frac{1}{4}-\frac{ib}{2}}(a)]$

$[a > 0,\ b > 0]$.   Ba IV 93(48)

5. $\displaystyle\int_0^\infty \frac{\sin(2a\,\text{sh}\,x)\cos bx}{\sqrt{\text{sh}\,x}}\, dx = \frac{\sqrt{\pi a}}{2}\,[I_{\frac{1}{4}-\frac{ib}{2}}(a) K_{\frac{1}{4}+\frac{ib}{2}}(a) +$

$+ I_{\frac{1}{4}+\frac{ib}{2}}(a) K_{\frac{1}{4}-\frac{ib}{2}}(a)]$ $[a > 0,\ b > 0]$.   Ba IV 37(64)

6. $\displaystyle\int_0^\infty \frac{\cos(2a\,\text{sh}\,x)\cos bx}{\sqrt{\text{sh}\,x}}\, dx = \frac{\sqrt{\pi a}}{2}\,[I_{-\frac{1}{4}-\frac{ib}{2}}(a) K_{-\frac{1}{4}+\frac{ib}{2}}(a) +$

$+ I_{-\frac{1}{4}+\frac{ib}{2}}(a) K_{-\frac{1}{4}-\frac{ib}{2}}(a)]$ $[a > 0,\ b > 0]$.   Ba IV 37(65)

7. $\displaystyle\int_0^\infty \sin(a\,\text{ch}\,x)\sin(a\,\text{sh}\,x)\frac{dx}{\text{sh}\,x} = \frac{\pi}{2}\sin a$ $[a > 0]$.   BH[264](22)

3.995

1. $\int_0^{\pi/2} \dfrac{\sin(2a\cos^2 x)\,\operatorname{ch}(a\sin 2x)}{b^2\cos^2 x + c^2\sin^2 x}\,dx = \dfrac{\pi}{2bc}\sin\dfrac{2ac}{b+c}$ $[b>0,\ c>0]$. BH[273](9)

2. $\int_0^{\pi/2} \dfrac{\cos(2a\cos^2 x)\,\operatorname{ch}(a\sin 2x)}{b^2\cos^2 x + c^2\sin^2 x}\,dx = \dfrac{\pi}{2bc}\cos\dfrac{2ac}{b+c}$ $[b>0,\ c>0]$. BH[273](10)

3.996

1. $\int_0^\infty \sin(a\operatorname{sh} x)\operatorname{sh}\beta x\,dx = \sin\dfrac{\beta\pi}{2}K_\beta(a)$ $[|\operatorname{Re}\beta|<1,\ a>0]$. Ba II 82(26)

2. $\int_0^\infty \cos(a\operatorname{sh} x)\operatorname{ch}\beta x\,dx = \cos\dfrac{\beta\pi}{2}K_\beta(a)$ $[|\operatorname{Re}\beta|<1,\ a>0]$. W 183(13)

3. $\int_0^{\pi/2} \cos(a\sin x)\operatorname{ch}(\beta\cos x)\,dx = \dfrac{\pi}{2}J_0(\sqrt{a^2-\beta^2})$. MO 40

4. $\int_0^\infty \sin\!\left(a\operatorname{ch} x - \tfrac{1}{2}\beta\pi\right)\operatorname{ch}\beta x\,dx = \dfrac{\pi}{2}J_\beta(a)$ $[|\operatorname{Re}\beta|<1,\ a>0]$. W 180(12)

5. $\int_0^\infty \cos\!\left(a\operatorname{ch} x - \tfrac{1}{2}\beta\pi\right)\operatorname{ch}\beta x\,dx = -\dfrac{\pi}{2}N_\beta(a)$ $[|\operatorname{Re}\beta|<1,\ a>0]$. W 180(13)

3.997

1. $\int_0^{\pi/2} \sin^\nu x\,\operatorname{sh}(\beta\cos x)\,dx = \dfrac{\sqrt{\pi}}{2}\left(\dfrac{2}{\beta}\right)^{\nu/2}\Gamma\!\left(\dfrac{\nu+1}{2}\right)\mathbf{L}_{\nu/2}(\beta)$ $[\operatorname{Re}\nu>-1]$. Ba II 38(53)

2. $\int_0^\pi \sin^\nu x\,\operatorname{ch}(\beta\cos x)\,dx = \sqrt{\pi}\left(\dfrac{2}{\beta}\right)^{\nu/2}\Gamma\!\left(\dfrac{\nu+1}{2}\right)I_{\nu/2}(\beta)$ $[\operatorname{Re}\nu>-1]$. WW 373+

3. $\int_0^{\pi/2} \dfrac{dx}{\operatorname{ch}(\operatorname{tg} x)\cos x\sqrt{\sin 2x}} = \sqrt{2\pi}\sum_{k=0}^\infty \dfrac{(-1)^k}{\sqrt{2k+1}}$. BH[276](13)

4. $\int_0^{\pi/2} \dfrac{\operatorname{tg}^q x}{\operatorname{ch}(\operatorname{tg} x)+\cos\lambda}\dfrac{dx}{\sin 2x} = \dfrac{\Gamma(q)}{\sin\lambda}\sum_{k=1}^\infty (-1)^{k-1}\dfrac{\sin k\lambda}{k^q}$ $[q>0]$. BH[275](20)

Trigonometrische Funktionen, Hyperbelfunktionen und Potenzen
Trigonometric and hyperbolic functions and powers

4.11–4.12

4.111

1. $$\int_0^\infty \frac{\sin ax}{\operatorname{sh} \beta x} \cdot x^{2m}\, dx = (-1)^m \frac{\pi}{2\beta} \cdot \frac{\partial^{2m}}{\partial a^{2m}}\left(\operatorname{th} \frac{a\pi}{2\beta}\right)$$

[Re $\beta > 0$] (cf. 3.981 1.). GH[336](17a)

2. $$\int_0^\infty \frac{\cos ax}{\operatorname{sh} \beta x} \cdot x^{2m+1}\, dx = (-1)^m \frac{\pi}{2\beta} \frac{\partial^{2m+1}}{\partial a^{2m+1}}\left(\operatorname{th} \frac{a\pi}{2\beta}\right)$$

[Re $\beta > 0$] (cf. 3.981 1.). GH[336](17b)

3. $$\int_0^\infty \frac{\sin ax}{\operatorname{ch} \beta x} \cdot x^{2m+1}\, dx = (-1)^{m+1} \frac{\pi}{2\beta} \cdot \frac{\partial^{2m+1}}{\partial a^{2m+1}}\left(\frac{1}{\operatorname{ch} \frac{a\pi}{2\beta}}\right)$$

[Re $\beta > 0$] (cf. 3.981 3.). GH[336](18b)

4. $$\int_0^\infty \frac{\cos ax}{\operatorname{ch} \beta x} \cdot x^{2m}\, dx = (-1)^m \frac{\pi}{2\beta} \cdot \frac{\partial^{2m}}{\partial a^{2m}}\left(\frac{1}{\operatorname{ch} \frac{a\pi}{2\beta}}\right)$$

[Re $\beta > 0$] (cf. 3.981 3.). GH[336](18a)

5. $$\int_0^\infty x \frac{\sin 2ax}{\operatorname{ch} \beta x}\, dx = \frac{\pi^2}{4\beta^2} \cdot \frac{\operatorname{sh}\frac{a\pi}{\beta}}{\operatorname{ch}^2 \frac{a\pi}{\beta}} \qquad [\text{Re } \beta>0,\; a>0].$$ BH[364](6)+

6. $$\int_0^\infty x \frac{\cos 2ax}{\operatorname{sh} \beta x}\, dx = \frac{\pi^2}{4\beta^2} \cdot \frac{1}{\operatorname{ch}^2 \frac{a\pi}{\beta}} \qquad [\text{Re } \beta > 0,\; a > 0].$$ BH[364](1)+

7. $$\int_0^\infty \frac{\sin ax}{\operatorname{ch} \beta x} \cdot \frac{dx}{x} = 2 \arctan\left(\exp \frac{\pi a}{2\beta}\right) - \frac{\pi}{2} \qquad [\text{Re } \beta > 0,\; a > 0].$$

BH[387](1), Ba IV 89(13), Li[298](17)

4.112

1. $$\int_0^\infty (x^2 + \beta^2) \frac{\cos ax}{\operatorname{ch}\frac{\pi x}{2\beta}}\, dx = \frac{2\beta^3}{\operatorname{ch}^3 a\beta} \qquad [\text{Re } \beta > 0,\; a > 0].$$ Ba IV 32(19)

2. $$\int_0^\infty x(x^2 + 4\beta^2) \frac{\cos ax}{\operatorname{sh}\frac{\pi x}{2\beta}}\, dx = \frac{6\beta^4}{\operatorname{ch}^4 a\beta} \qquad [\text{Re } \beta > 0,\; a > 0].$$ Ba IV 32(20)

1. $\displaystyle\int_0^\infty \frac{\sin ax}{\operatorname{sh}\pi x}\cdot\frac{dx}{x^2+\beta^2} = -\frac{1}{2\beta^2} - \frac{\pi e^{-a\beta}}{\beta\sin\pi\beta} +$ 4.113

$\displaystyle + \frac{1}{2\beta^2}\left[{}_2F_1(1,-\beta;1-\beta;-e^{-a}) + {}_2F_1(1,\beta;1+\beta;-e^{-a})\right] =$

$\displaystyle = \frac{1}{2\beta^2} - \frac{\pi e^{-a\beta}}{2\beta\sin\pi\beta} - \sum_{k=1}^\infty \frac{(-1)^k e^{-ak}}{k^2-\beta^2}$

$[\operatorname{Re}\beta > 0, \quad \beta \neq 0, 1, 2, \ldots, a > 0]$. Ba IV 90(18)

2. $\displaystyle\int_0^\infty \frac{\sin ax}{\operatorname{sh}\pi x}\cdot\frac{dx}{x^2+m^2} = \frac{(-1)^m a e^{-ma}}{2m} +$

$\displaystyle + \frac{1}{2m}\sum_{k=1}^{m-1}\frac{(-1)^k e^{-ka}}{m-k} + \frac{(-1)^m e^{-ma}}{2m}\ln(1+e^{-a}) +$

$\displaystyle + \frac{1}{2m!}\frac{d^{m-1}}{dz^{m-1}}\left[\frac{(1+z)^{m-1}}{z}\ln(1+z)\right]_{z=e^{-a}} \quad [a>0]$. Ba IV 89(17)

3. $\displaystyle\int_0^\infty \frac{\sin ax}{\operatorname{sh}\pi x}\cdot\frac{dx}{1+x^2} = \frac{1}{2}\int_{-\infty}^\infty \frac{\sin ax}{\operatorname{sh}\pi x}\cdot\frac{dx}{1+x^2} = -\frac{a}{2}\operatorname{ch} a + \operatorname{sh} a \ln\left(2\operatorname{ch}\frac{a}{2}\right)$.

GH[336](21b)

4. $\displaystyle\int_0^\infty \frac{\sin ax}{\operatorname{sh}\frac{\pi}{2}x}\cdot\frac{dx}{1+x^2} = \frac{1}{2}\int_{-\infty}^\infty \frac{\sin ax}{\operatorname{sh}\frac{\pi}{2}x}\cdot\frac{dx}{1+x^2} = \frac{\pi}{2}\operatorname{sh} a - \operatorname{ch} a\operatorname{arctg}(\operatorname{sh} a)$.

GH[336](21a)

5. $\displaystyle\int_0^\infty \frac{\sin ax}{\operatorname{sh}\frac{\pi}{4}x}\cdot\frac{dx}{1+x^2} = -\frac{\pi}{\sqrt{2}}e^{-a} + \frac{\operatorname{sh} a}{\sqrt{2}}\ln\frac{2\operatorname{ch} a+\sqrt{2}}{2\operatorname{ch} a-\sqrt{2}} +$

$\displaystyle + \sqrt{2}\operatorname{ch} a\operatorname{arctg}\frac{\sqrt{2}}{2\operatorname{sh} a} \quad [a>0]$. Li[389](1)

6. $\displaystyle\int_0^\infty \frac{\sin ax}{\operatorname{ch}\frac{\pi}{4}x}\cdot\frac{x\,dx}{1+x^2} = \frac{\pi}{\sqrt{2}}e^{-a} + \frac{\operatorname{sh} a}{\sqrt{2}}\ln\frac{2\operatorname{ch} a+\sqrt{2}}{2\operatorname{ch} a-\sqrt{2}} -$

$\displaystyle - \sqrt{2}\operatorname{ch} a\operatorname{arctg}\left(\frac{1}{\sqrt{2}\operatorname{sh} a}\right) \quad [a>0]$. BH[388](1)

7. $\displaystyle\int_0^\infty \frac{\cos ax}{\operatorname{sh}\pi x}\cdot\frac{x\,dx}{1+x^2} = -\frac{1}{2} + \frac{a}{2}e^{-a} + \operatorname{ch} a\ln(1+e^{-a}) \quad [a>0]$.

BH[389](14), Ba IV 32(24)

4.113 8. $$\int_0^\infty \frac{\cos ax}{\sh \frac{\pi}{2} x} \cdot \frac{x\,dx}{1+x^2} = 2\sh a\,\arctg(e^{-a}) + \frac{\pi}{2} e^{-a} - 1 \quad [a > 0].$$ BH[389](11)

9. $$\int_0^\infty \frac{\cos ax}{\ch \pi x} \cdot \frac{dx}{x^2 + \beta^2} = \sum_{k=0}^\infty (-1)^k \frac{\left(k+\frac{1}{2}\right)^2 e^{-a\beta} - \beta e^{-\left(k+\frac{1}{2}\right)a}}{\beta\left[\left(k+\frac{1}{2}\right)^2 - \beta^2\right]}$$

$$[\Re \beta > 0,\ a > 0].$$ Ba IV 32(26)

10. $$\int_0^\infty \frac{\cos ax}{\ch \pi x} \cdot \frac{dx}{\left(m+\frac{1}{2}\right)^2 + x^2} = \frac{(-1)^m e^{-\left(m+\frac{1}{2}\right)a}}{2m+1}[a + \ln(1+e^{-a})] +$$

$$+ \frac{e^{-\frac{a}{2}}}{2m+1} \sum_{k=0}^{m-1} \frac{(-1)^k e^{-ak}}{k-m} + \frac{e^{-\frac{a}{2}}}{(2m+1)(m+1)} \cdot {}_2F_1(1, m+1;\ m+2;\ -e^{-a})$$

$$[a > 0].$$ Ba IV 32(25)

11. $$\int_0^\infty \frac{\cos ax}{\ch \pi x} \cdot \frac{dx}{1+x^2} = 2\ch\frac{a}{2} - [e^a \arctg(e^{-\frac{a}{2}}) + e^{-a} \arctg(e^{\frac{a}{2}})] \quad [a > 0].$$

Ba IV 32(21)

12. $$\int_0^\infty \frac{\cos ax}{\ch \frac{\pi}{2} x} \cdot \frac{dx}{1+x^2} = ae^{-a} + \ch a \ln(1+e^{-2a}) \quad [a > 0].$$ BH[388](6)

13. $$\int_0^\infty \frac{\cos ax}{\ch \frac{\pi}{4} x} \cdot \frac{dx}{1+x^2} = \frac{\pi}{\sqrt{2}} e^{-a} + \frac{2\sh a}{\sqrt{2}} \arctg\left(\frac{1}{\sqrt{2}\sh a}\right) -$$

$$- \frac{\ch a}{\sqrt{2}} \ln \frac{2\ch a + \sqrt{2}}{2\ch a - \sqrt{2}} \quad [a > 0].$$ BH[388](5)

4.114 1. $$\int_0^\infty \frac{\sin ax}{x} \cdot \frac{\sh \beta x}{\sh \gamma x}\, dx = \arctg\left(\tg\frac{\beta\pi}{2\gamma}\,\th\frac{a\pi}{2\gamma}\right) \quad [|\Re \beta| < \Re \gamma,\ a > 0].$$

BH[387](6)+

2. $$\int_0^\infty \frac{\cos ax}{x} \cdot \frac{\sh \beta x}{\ch \gamma x}\, dx = \frac{1}{2} \ln \frac{\ch\frac{a\pi}{2\gamma} + \sin\frac{\beta\pi}{2\gamma}}{\ch\frac{a\pi}{2\gamma} - \sin\frac{\beta\pi}{2\gamma}} \quad [|\Re \beta| < \Re \gamma].$$ Ba IV 33(34)

4.115 1. $$\int_0^\infty \frac{x \sin ax}{x^2 + b^2} \cdot \frac{\sh \beta x}{\sh \pi x}\, dx = \frac{\pi}{2} \frac{e^{-ab} \sin b\beta}{\sin b\pi} + \sum_{k=1}^\infty (-1)^k \frac{k e^{-ak} \sin k\beta}{k^2 - b^2}$$

$$[0 < \Re \beta < \pi,\ a > 0,\ b > 0].$$ BH[389](23)

2. $\int_0^\infty \dfrac{x \sin ax}{x^2+1} \cdot \dfrac{\operatorname{sh} \beta x}{\operatorname{sh} \pi x} \, dx = \dfrac{1}{2} e^{-a}(a \sin \beta - \beta \cos \beta) -$ 4.115

$\qquad - \dfrac{1}{2} \operatorname{sh} a \sin \beta \ln [1 + 2 e^{-a} \cos \beta + e^{-2a}] + \operatorname{ch} a \cos \beta \operatorname{arctg} \dfrac{\sin \beta}{e^a + \cos \beta}$

$\qquad\qquad [|\operatorname{Re} \beta| < \pi, \, a > 0]$. Li[389](10)

3. $\int_0^\infty \dfrac{x \sin ax}{x^2+1} \cdot \dfrac{\operatorname{sh} \beta x}{\operatorname{sh} \dfrac{\pi}{2} x} \, dx = \dfrac{\pi}{2} e^{-a} \sin \beta +$

$\qquad + \dfrac{1}{2} \cos \beta \operatorname{sh} a \ln \dfrac{\operatorname{ch} a + \sin \beta}{\operatorname{ch} a - \sin \beta} - \sin \beta \operatorname{ch} a \operatorname{arctg}\left(\dfrac{\cos \beta}{\operatorname{sh} a}\right)$

$\qquad\qquad \left[|\operatorname{Re} \beta| < \dfrac{\pi}{2}, \, a > 0\right]$. BH[389](8)

4. $\int_0^\infty \dfrac{\cos ax}{x^2+b^2} \cdot \dfrac{\operatorname{sh} \beta x}{\operatorname{sh} \pi x} \, dx = \dfrac{\pi}{2b} \cdot \dfrac{e^{-ab} \sin b\beta}{\sin b\pi} + \sum_{k=1}^\infty (-1)^k \dfrac{e^{-ak} \sin k\beta}{k^2 - b^2}$

$\qquad\qquad [0 < \operatorname{Re} \beta < \pi, \, a > 0, \, b > 0]$. BH[389](22)

5. $\int_0^\infty \dfrac{\cos ax}{x^2+1} \cdot \dfrac{\operatorname{sh} \beta x}{\operatorname{sh} \pi x} \, dx = \dfrac{1}{2} e^{-a}(a \sin \beta - \beta \cos \beta) +$

$\qquad + \dfrac{1}{2} \operatorname{ch} a \sin \beta \ln (1 + 2 e^{-a} \cos \beta + e^{-2a}) - \operatorname{sh} a \cos \beta \operatorname{arctg} \dfrac{\sin \beta}{e^a + \cos \beta}$

$\qquad\qquad [|\operatorname{Re} \beta| < \pi, \, a > 0, \, b > 0]$. BH[389](20)+

6. $\int_0^\infty \dfrac{\cos ax}{x^2+1} \cdot \dfrac{\operatorname{sh} \beta x}{\operatorname{sh} \dfrac{\pi}{2} x} \, dx = \dfrac{\pi}{2} e^{-a} \sin \beta -$

$\qquad - \dfrac{1}{2} \operatorname{ch} a \cos \beta \ln \dfrac{\operatorname{ch} a + \sin \beta}{\operatorname{ch} a - \sin \beta} + \operatorname{sh} a \sin \beta \operatorname{arctg} \dfrac{\cos \beta}{\operatorname{sh} a}$

$\qquad\qquad \left[|\operatorname{Re} \beta| < \dfrac{\pi}{2}, \, a > 0, \, b > 0\right]$. BH[389](18)

7. $\int_0^\infty \dfrac{\sin ax}{x^2+\dfrac{1}{4}} \cdot \dfrac{\operatorname{sh} \beta x}{\operatorname{ch} \pi x} \, dx = e^{-\frac{a}{2}}\left(a \sin \dfrac{\beta}{2} - \beta \cos \dfrac{\beta}{2}\right) -$

$\qquad - \operatorname{sh} \dfrac{a}{2} \sin \dfrac{\beta}{2} \ln (1 + 2 e^{-a} \cos \beta + e^{-2a}) + \operatorname{ch} \dfrac{a}{2} \cos \dfrac{\beta}{2} \operatorname{arctg} \dfrac{\sin \beta}{1 + e^{-a} \cos \beta}$

$\qquad\qquad [|\operatorname{Re} \beta| < \pi, \, a > 0]$. Ba IV 91(26)

8. $\int_0^\infty \dfrac{\sin ax}{x^2+\beta^2} \cdot \dfrac{\operatorname{ch} \gamma x}{\operatorname{sh} \pi x} \, dx = \dfrac{1}{2\beta^2} - \dfrac{\pi}{2\beta} \cdot \dfrac{e^{-a\beta} \cos \beta \gamma}{\sin \beta \pi} + \sum_{k=1}^\infty (-1)^{k-1} \dfrac{e^{-ak} \cos k\gamma}{k^2 - \beta^2}$

$\qquad\qquad [0 \leqslant \operatorname{Re} \beta, \, |\operatorname{Re} \gamma| < \pi, \, a > 0]$. BH[389](21)

4.115

9. $\int_0^\infty \frac{\sin ax}{x^2+1} \cdot \frac{\operatorname{ch}\beta x}{\operatorname{sh}\pi x}\, dx = -\frac{1}{2}e^{-a}(a\cos\beta + \beta\sin\beta) +$

$$+ \frac{1}{2}\operatorname{sh} a \cos\beta \ln(1 + 2e^{-a}\cos\beta + e^{-2a}) +$$

$$+ \operatorname{ch} a \sin\beta \operatorname{arctg}\frac{\sin\beta}{e^a + \cos\beta} \quad [|\operatorname{Re}\beta| < \pi,\, a > 0].$$

Ba IV 91(25), Li[389](9)

10. $\int_0^\infty \frac{\sin ax}{x^2+1} \cdot \frac{\operatorname{ch}\beta x}{\operatorname{sh}\frac{\pi}{2}x}\, dx = -\frac{\pi}{2}e^{-a}\cos\beta +$

$$+ \frac{1}{2}\operatorname{sh} a \sin\beta \ln\frac{\operatorname{ch} a + \sin\beta}{\operatorname{ch} a - \sin\beta} + \operatorname{ch} a \cos\beta \operatorname{arctg}\frac{\cos\beta}{\operatorname{sh} a}$$

$$\left[|\operatorname{Re}\beta| < \frac{\pi}{2},\, a > 0\right]. \qquad \text{BH[389](7)}$$

11. $\int_0^\infty \frac{x\cos ax}{x^2+b^2} \cdot \frac{\operatorname{ch}\beta x}{\operatorname{sh}\pi x}\, dx = \frac{\pi}{2} \cdot \frac{e^{-ab}\cos b\beta}{\sin b\pi} + \sum_{k=1}^\infty (-1)^k \frac{ke^{-ak}\cos k\beta}{k^2 - b^2}$

$$[|\operatorname{Re}\beta| < \pi,\, a > 0]. \qquad \text{BH[389](24)}$$

12. $\int_0^\infty \frac{x\cos ax}{x^2+1} \cdot \frac{\operatorname{ch}\beta x}{\operatorname{sh}\pi x}\, dx = \frac{1}{2}e^{-a}(a\cos\beta + \beta\sin\beta) -$

$$- \frac{1}{2} + \frac{1}{2}\operatorname{ch} a \cos\beta \ln[1 + 2e^{-a}\cos\beta + e^{-2a}] + \operatorname{sh} a \sin\beta \operatorname{arctg}\frac{\sin\beta}{e^a + \cos\beta}$$

$$[|\operatorname{Re}\beta| < \pi,\, a > 0]. \qquad \text{BH[389](19)}$$

13. $\int_0^\infty \frac{x\cos ax}{x^2+1} \cdot \frac{\operatorname{ch}\beta x}{\operatorname{sh}\frac{\pi}{2}x}\, dx = -1 + \frac{\pi}{2}e^{-a}\cos\beta +$

$$+ \frac{1}{2}\operatorname{ch} a \sin\beta \ln\frac{\operatorname{ch} a + \sin\beta}{\operatorname{ch} a - \sin\beta} + \operatorname{sh} a \cos\beta \operatorname{arctg}\frac{\cos\beta}{\operatorname{sh} a}$$

$$\left[|\operatorname{Re}\beta| < \frac{\pi}{2},\, a > 0\right]. \qquad \text{BH[389](17)}$$

14. $\int_0^\infty \frac{\cos ax}{x^2+1} \cdot \frac{\operatorname{ch}\beta x}{\operatorname{ch}\frac{\pi}{2}x}\, dx = ae^{-a}\cos\beta + \beta e^{-a}\sin\beta +$

$$+ \operatorname{sh} a \sin\beta \operatorname{arctg}\frac{e^{-2a}\sin 2\beta}{1 + e^{-2a}\cos 2\beta} + \frac{1}{2}\operatorname{ch} a \cos\beta \ln(1 + 2e^{-2a}\cos 2\beta + e^{-4a})$$

$$\left[|\operatorname{Re}\beta| < \frac{\pi}{2},\, a > 0\right]. \qquad \text{Ba IV 34(37)}$$

4.116

1. $\int_0^\infty x\cos 2ax\,\operatorname{th} x\, dx = -\frac{\pi^2}{4} \cdot \frac{\operatorname{ch} a\pi}{\operatorname{sh}^2 a\pi} \quad [a > 0]. \qquad \text{BH[364](2)}$

2. $\int_0^\infty \cos ax \, \text{th} \, \beta x \, \frac{dx}{x} = \ln \text{cth} \frac{a\pi}{4\beta}$ [Re $\beta > 0$, $a > 0$]. BH[387](8) 4.116

3. $\int_0^\infty \cos ax \, \text{cth} \, \beta x \, \frac{dx}{x} = -\ln\left(2 \, \text{sh} \frac{a\pi}{2\beta}\right)$ [Re $\beta > 0$, $a > 0$]. BH[387](9)

1. $\int_0^\infty \frac{\sin ax}{1+x^2} \, \text{th} \frac{\pi x}{2} \, dx = a \, \text{ch} \, a - \text{sh} \, a \ln(2 \, \text{sh} \, a)$ [$a > 0$]. BH[388](3) 4.117

2. $\int_0^\infty \frac{\sin ax}{1+x^2} \, \text{th} \frac{\pi x}{4} \, dx = -\frac{\pi}{2} e^a + \text{sh} \, a \ln \text{cth} \frac{a}{2} + 2 \, \text{ch} \, a \, \text{arctg}(e^a)$.

BH[388](4)

3. $\int_0^\infty \frac{\sin ax}{1+x^2} \, \text{cth} \, \pi x \, dx = \frac{a}{2} e^{-a} - \text{sh} \, a \ln(1-e^{-a})$ [$a > 0$]. BH[389](5)

4. $\int_0^\infty \frac{\sin ax}{1+x^2} \, \text{cth} \frac{\pi}{2} x \, dx = \text{sh} \, a \ln \text{cth} \frac{a}{2}$ [$a > 0$]. BH[389](6)

5. $\int_0^\infty \frac{x \cos ax}{1+x^2} \, \text{th} \frac{\pi}{2} x \, dx = -ae^{-a} + \text{sh} \, a \ln(1-e^{-2a})$ [$a > 0$]. BH[388](7)

6. $\int_0^\infty \frac{x \cos ax}{1+x^2} \, \text{th} \frac{\pi}{4} x \, dx = -\frac{\pi}{2} e^a + \text{ch} \, a \ln \text{cth} \frac{a}{2} + 2 \, \text{sh} \, a \, \text{arctg}(e^a)$ [$a > 0$].

BH[388](8)

7. $\int_0^\infty \frac{x \cos ax}{1+x^2} \, \text{cth} \, \pi x \, dx = -\frac{a}{2} e^{-a} - \frac{1}{2} - \text{ch} \, a \ln(1-e^{-a})$.

BH[389](15)+, Ba IV 33(31)+

8. $\int_0^\infty \frac{x \cos ax}{1+x^2} \, \text{cth} \frac{\pi}{2} x \, dx = -1 + \text{ch} \, a \ln \text{cth} \frac{a}{2}$ [$a > 0$]. BH[389](12)

9. $\int_0^\infty \frac{x \cos ax}{1+x^2} \, \text{cth} \frac{\pi}{4} x \, dx = -2 + \frac{\pi}{2} e^{-a} + \text{ch} \, a \ln \text{cth} \frac{a}{2} + 2 \, \text{sh} \, a \, \text{arctg}(e^{-a})$

[$a > 0$]. BH[389](13)

$\int_0^\infty \frac{x \sin ax}{\text{ch}^2 x} \, dx = -\frac{d}{da}\left(\frac{\pi a}{2 \, \text{sh} \frac{\pi a}{2}}\right)$ [$a > 0$]. Ba IV 89(14) 4.118

4.119
$$\int_0^\infty \frac{1-\cos px}{\operatorname{sh} qx} \cdot \frac{dx}{x} = \ln\left(\operatorname{ch} \frac{p\pi}{2q}\right).$$
BH[387](2)+

4.121 1.
$$\int_0^\infty \frac{\sin ax - \sin bx}{\operatorname{ch} \beta x} \cdot \frac{dx}{x} = 2\operatorname{arctg} \frac{\exp\frac{a\pi}{2\beta} - \exp\frac{b\pi}{2\beta}}{1 + \exp\frac{(a+b)\pi}{2\beta}} \quad [\operatorname{Re}\beta > 0].$$
GH[336](19b)

2.
$$\int_0^\infty \frac{\cos ax - \cos bx}{\operatorname{sh} \beta x} \cdot \frac{dx}{x} = \ln \frac{\operatorname{ch}\frac{b\pi}{2\beta}}{\operatorname{ch}\frac{a\pi}{2\beta}} \quad [\operatorname{Re}\beta > 0].$$
GH[336](19a)

4.122 1.
$$\int_0^\infty \frac{\cos \beta x \sin \gamma x}{\operatorname{ch} \delta x} \cdot \frac{dx}{x} = \operatorname{arctg} \frac{\operatorname{sh}\frac{\gamma\pi}{2\delta}}{\operatorname{ch}\frac{\beta\pi}{2\delta}} \quad [\operatorname{Re}\delta > |\operatorname{Im}(\beta+\gamma)|].$$
Ba IV 93(46)+

2.
$$\int_0^\infty \sin^2 ax \, \frac{\operatorname{ch} \beta x}{\operatorname{sh} x} \cdot \frac{dx}{x} = \frac{1}{4} \ln \frac{\operatorname{ch} 2a\pi + \cos\beta\pi}{1+\cos\beta\pi} \quad [|\operatorname{Re}\beta| < 1].$$
BH[387](7)

4.123 1.
$$\int_0^\infty \frac{\sin x}{\operatorname{ch} ax + \cos x} \cdot \frac{x\,dx}{x^2 - \pi^2} = \operatorname{arctg}\frac{1}{a} - \frac{1}{a}.$$
BH[390](1)

2.
$$\int_0^\infty \frac{\sin x}{\operatorname{ch} ax - \cos x} \cdot \frac{x\,dx}{x^2 - \pi^2} = \frac{a}{1+a^2} - \operatorname{arctg}\frac{1}{a}.$$
BH[390](2)

3.
$$\int_0^\infty \frac{\sin 2x}{\operatorname{ch} 2ax - \cos 2x} \cdot \frac{x\,dx}{x^2 - \pi^2} = \frac{1}{2a} \cdot \frac{1+2a^2}{1+a^2} - \operatorname{arctg}\frac{1}{a}.$$
BH[390](4)

4.
$$\int_0^\infty \frac{\operatorname{ch} ax \sin x}{\operatorname{ch} 2ax - \cos 2x} \cdot \frac{x\,dx}{x^2 - \pi^2} = \frac{-1}{2a(1+a^2)}.$$
Li[390](3)

5.
$$\int_0^\infty \frac{\cos ax}{\operatorname{ch}\pi x + \cos\pi\beta} \cdot \frac{dx}{x^2+\gamma^2} = \frac{\pi e^{-a\gamma}}{2\gamma(\cos\gamma\pi+\cos\beta\pi)} +$$
$$+ \frac{1}{\operatorname{sh}\beta\pi} \sum_{k=0}^\infty \left\{ \frac{\exp[-(2k+1-\beta)a]}{\gamma^2-(2k+1-\beta)^2} - \frac{\exp[-(2k+1+\beta)a]}{\gamma^2-(2k+1+\beta)^2} \right\}$$
$$[0 < \operatorname{Re}\beta < 1,\ \operatorname{Re}\gamma > 0,\ a > 0].$$
Ba IV 33(27)

6.
$$\int_0^\infty \frac{\sin ax \operatorname{sh} bx}{\cos 2ax + \operatorname{ch} 2bx} x^{p-1} dx = \frac{\Gamma(p)}{(a^2+b^2)^{\frac{p}{2}}} \sin\left(p \operatorname{arctg}\frac{a}{b}\right) \sum_{k=0}^\infty \frac{(-1)^k}{(2k+1)^p}$$
$$[p > 0].$$
BH[364](8)

7. $\int_0^\infty \sin ax^2 \dfrac{\sin\frac{\pi x}{2} \operatorname{sh}\frac{\pi x}{2}}{\cos \pi x + \operatorname{ch} \pi x} \cdot x\, dx = \dfrac{1}{4}\left[\dfrac{\partial \theta_1(z,q)}{\partial z}\right]_{z=0,\ q=e^{-2a}}$ $[a>0]$. Ba IV 93(49) 4.123

1. $\int_0^1 \dfrac{\cos px\, \operatorname{ch}(q\sqrt{1-x^2})}{\sqrt{1-x^2}}\, dx = \dfrac{\pi}{2} J_0(\sqrt{p^2-q^2}).$ MO (40) 4.124

2. $\int_u^\infty \cos ax\, \operatorname{ch}\sqrt{\beta(u^2-x^2)} \cdot \dfrac{dx}{\sqrt{u^2-x^2}} = \dfrac{\pi}{2} J_0\left(\dfrac{u}{\sqrt{a^2-\beta^2}}\right).$ Ba IV 34(38)

1. $\int_0^\infty \operatorname{sh}(a\sin x)\cos(a\cos x)\sin x \sin 2nx \dfrac{dx}{x} =$ 4.125

$= \dfrac{(-1)^{n-1} a^{2n-1}}{(2n-1)!}\dfrac{\pi}{8}\left[1 + \dfrac{a^2}{2n(2n+1)}\right].$ Li[367](14)

2. $\int_0^\infty \operatorname{ch}(a\sin x)\cos(a\cos x)\sin x \cos(2n-1)x \dfrac{dx}{x} =$

$= \dfrac{(-1)^{n-1} a^{2(n-1)}}{[2(n-1)]!}\dfrac{\pi}{8}\left[1 - \dfrac{a^2}{2n(2n-1)}\right].$ Li[367](15)

3. $\int_0^\infty \operatorname{sh}(a\sin x)\cos(a\cos x)\cos x \cos 2nx \dfrac{dx}{x} =$

$= \dfrac{\pi}{2}\sum_{k=n+1}^\infty \dfrac{(-1)^k a^{2k+1}}{(2k+1)!} + \dfrac{(-1)^n a^{2n+1}}{(2n+1)!}\dfrac{3\pi}{8} + \dfrac{(-1)^{n-1} a^{2n-1}}{(2n-1)!}\dfrac{\pi}{8}.$ Li[367](21)

1. $\int_0^\infty \sin(a\cos bx)\operatorname{sh}(a\sin bx)\dfrac{x\,dx}{c^2-x^2} = \dfrac{\pi}{2}[\cos(a\cos bc)\operatorname{ch}(a\sin bc) - 1]$ 4.126

$[b>0].$ BH[381](2)

2. $\int_0^\infty \sin(a\cos bx)\operatorname{ch}(a\sin bx)\dfrac{dx}{c^2-x^2} = \dfrac{\pi}{2c}\cos(a\cos bc)\operatorname{sh}(a\sin bc)$

$[b>0,\ c>0].$ BH[381](1)

3. $\int_0^\infty \cos(a\cos bx)\operatorname{sh}(a\sin bx)\dfrac{x\,dx}{c^2-x^2} = \dfrac{\pi}{2}[a\cos bc - \sin(a\cos bc)\operatorname{ch}(a\sin bc)]$

$[b>0].$ BH[381](4)

4. $\int_0^\infty \cos(a\cos bx)\operatorname{ch}(a\sin bx)\dfrac{dx}{c^2-x^2} = -\dfrac{\pi}{2c}\sin(a\cos bc)\operatorname{sh}(a\sin bc)$

$[b>0].$ BH[381](3)

Trigonometrische Funktionen, Hyperbelfunktionen und die Exponentialfunktion
Trigonometric and hyperbolic functions and exponentials

4.131

1. $\int_0^\infty \sin ax \, \text{sh}^\nu \gamma x \cdot e^{-\beta x} \, dx =$

$$= -\frac{i\Gamma(\nu+1)}{2^{\nu+2}\gamma} \left\{ \frac{\Gamma\left(\frac{\beta-\nu\gamma-ai}{2\gamma}\right)}{\Gamma\left(\frac{\beta+\nu\gamma-ai}{2\gamma}+1\right)} - \frac{\Gamma\left(\frac{\beta-\nu\gamma+ai}{2\gamma}\right)}{\Gamma\left(\frac{\beta+\gamma\nu+ai}{2\gamma}+1\right)} \right\}$$

[Re $\nu > -2$, Re $\gamma > 0$, | Re $(\gamma\nu)$| < Re β].   Ba IV 91(30)+

2. $\int_0^\infty \cos ax \, \text{sh}^\nu \gamma x \cdot e^{-\beta x} \, dx =$

$$= \frac{\Gamma(\nu+1)}{2^{\nu+2}\gamma} \left\{ \frac{\Gamma\left(\frac{\beta-\nu\gamma-ai}{2\gamma}\right)}{\Gamma\left(\frac{\beta+\gamma\nu-ai}{2\gamma}+1\right)} + \frac{\Gamma\left(\frac{\beta-\nu\gamma+ai}{2\gamma}\right)}{\Gamma\left(\frac{\beta+\nu\gamma+ai}{2\gamma}+1\right)} \right\}$$

[Re $\nu > -1$, Re $\gamma > 0$, | Re $(\gamma\nu)$| < Re β].   Ba IV 34(40)+

3. $\int_0^\infty e^{-\beta x} \frac{\sin ax}{\text{sh} \, \gamma x} \, dx = \sum_{k=1}^\infty \frac{2a}{a^2 + [\beta + (2k-1)\gamma]^2};$ BH[264](9)+

$$= \frac{1}{2\gamma i} \left[\psi\left(\frac{\beta+\gamma+ia}{2\gamma}\right) - \psi\left(\frac{\beta+\gamma-ia}{2\gamma}\right) \right]$$

[Re $\beta >$ | Re γ|].   Ba IV 91(28)

4. $\int_0^\infty e^{-x} \frac{\sin ax}{\text{sh} \, x} \, dx = \frac{\pi}{2} \text{cth} \frac{a\pi}{2} - \frac{1}{a}.$ Ba IV 91(29)

4.132

1. $\int_0^\infty \frac{\sin ax \, \text{sh} \, \beta x}{e^{\gamma x} - 1} \, dx = -\frac{a}{2(a^2+\beta^2)} + \frac{\pi}{2\gamma} \cdot \frac{\text{sh} \frac{2\pi a}{\gamma}}{\text{ch} \frac{2\pi a}{\gamma} - \cos \frac{2\pi\beta}{\gamma}} +$

$$+ \frac{i}{2\gamma} \left[\psi\left(\frac{\beta}{\gamma}+i\frac{a}{\gamma}+1\right) - \psi\left(\frac{\beta}{\gamma}-i\frac{a}{\gamma}+1\right) \right]$$

[Re $\gamma >$ | Re β|, $a > 0$].   Ba IV 92(23)

2. $\int_0^\infty \frac{\sin ax \, \text{ch} \, \beta x}{e^{\gamma x} - 1} \, dx = -\frac{a}{2(a^2+\beta^2)} + \frac{\pi}{2\gamma} \cdot \frac{\text{sh} \frac{2\pi a}{\gamma}}{\text{ch} \frac{2\pi a}{\gamma} - \cos \frac{2\pi\beta}{\gamma}}$ [Re $\gamma >$ |Re β|].

BH[265](5)+, Ba IV 92(34)

3. $\int_0^\infty \dfrac{\sin ax\, \operatorname{ch}\beta x}{e^{\gamma x}+1}\,dx = \dfrac{a}{2(a^2+\beta^2)} - \dfrac{\pi}{\gamma}\cdot\dfrac{\operatorname{sh}\dfrac{a\pi}{\gamma}\cos\dfrac{\beta\pi}{\gamma}}{\operatorname{ch}\dfrac{2a\pi}{\gamma}-\cos\dfrac{2\beta\pi}{\gamma}}$ [Re $\gamma >$ | Re β|]. 4.132

Ba IV 92(35)

4. $\int_0^\infty \dfrac{\cos ax\, \operatorname{sh}\beta x}{e^{\gamma x}-1}\,dx = \dfrac{\beta}{2(a^2+\beta^2)} - \dfrac{\pi}{2\gamma}\cdot\dfrac{\sin\dfrac{2\pi\beta}{\gamma}}{\operatorname{ch}\dfrac{2a\pi}{\gamma}-\cos\dfrac{2\beta\pi}{\gamma}}$ [Re $\gamma >$ | Re β|].

Li[265](8)

5. $\int_0^\infty \dfrac{\cos ax\, \operatorname{sh}\beta x}{e^{\gamma x}+1}\,dx = -\dfrac{\beta}{2(a^2+\beta^2)} + \dfrac{\pi}{\gamma}\cdot\dfrac{\sin\dfrac{\pi\beta}{\gamma}\operatorname{ch}\dfrac{\pi a}{\gamma}}{\operatorname{ch}\dfrac{2a\pi}{\gamma}-\cos\dfrac{2\beta\pi}{\gamma}}$ [Re $\gamma >$|Re β|].

Ba IV 34(39)

1. $\int_0^\infty \sin ax\, \operatorname{sh}\beta x\, \exp\left(-\dfrac{x^2}{4\gamma}\right)dx =$ 4.133

$= \sqrt{\pi\gamma}\, \exp\gamma(\beta^2-a^2)\sin(2a\beta\gamma)$ [Re $\gamma > 0$].

Ba IV 92(37)

2. $\int_0^\infty \cos ax\, \operatorname{ch}\beta x\, \exp\left(-\dfrac{x^2}{4\gamma}\right)dx =$

$= \sqrt{\pi\gamma}\, \exp\gamma(\beta^2-a^2)\cos(2a\beta\gamma)$ [Re $\gamma > 0$].

Ba IV 35(41)

1. $\int_0^\infty e^{-\beta x^2}(\operatorname{ch}x+\cos x)dx = \sqrt{\dfrac{\pi}{\beta}}\operatorname{ch}\dfrac{1}{4\beta}$ [Re $\beta > 0$]. MH 24 4.134

2. $\int_0^\infty e^{-\beta x^2}(\operatorname{ch}x-\cos x)\,dx = \sqrt{\dfrac{\pi}{\beta}}\operatorname{sh}\dfrac{1}{4\beta}$ [Re $\beta > 0$]. MH 24

1. $\int_0^\infty \sin ax^2\, \operatorname{ch}2\gamma x\cdot e^{-\beta x^2}\,dx = \dfrac{1}{2}\sqrt[4]{\dfrac{\pi^2}{a^2+\beta^2}}\exp\left(-\dfrac{\beta\gamma^2}{a^2+\beta^2}\right)\times$ 4.135

$\times\sin\left(\dfrac{a\gamma^2}{a^2+\beta^2}+\dfrac{1}{2}\operatorname{arctg}\dfrac{a}{\beta}\right)$ [Re $\beta > 0$]. Li[268](7)

2. $\int_0^\infty \cos ax^2\, \operatorname{ch}2\gamma x\cdot e^{-\beta x^2}\,dx = \dfrac{1}{2}\sqrt[4]{\dfrac{\pi^2}{a^2+\beta^2}}\exp\left(-\dfrac{\beta\gamma^2}{a^2+\beta^2}\right)\times$

$\times\cos\left(\dfrac{a\gamma^2}{a^2+\beta^2}+\dfrac{1}{2}\operatorname{arctg}\dfrac{a}{\beta}\right)$ [Re $\beta > 0$]. Li[268](8)

1. $\int_0^\infty (\operatorname{sh}x^2+\sin x^2)e^{-\beta x^4}\,dx = \dfrac{\sqrt{2}\,\pi}{4\sqrt{\beta}} I_{\frac{1}{4}}\!\left(\dfrac{1}{8\beta}\right)\operatorname{ch}\dfrac{1}{8\beta}$ [Re $\beta > 0$]. MH 24 4.136

4.136

2. $\int_0^\infty (\operatorname{sh} x^2 - \sin x^2) e^{-\beta x^4} dx = \dfrac{\sqrt{2}\,\pi}{4\sqrt{\beta}} I_{\frac{1}{4}}\left(\dfrac{1}{8\beta}\right) \operatorname{sh} \dfrac{1}{8\beta}$ [Re $\beta > 0$]. MH 24

3. $\int_0^\infty (\operatorname{ch} x^2 + \cos x^2) e^{-\beta x^4} dx = \dfrac{\sqrt{2}\,\pi}{4\sqrt{\beta}} I_{-\frac{1}{4}}\left(\dfrac{1}{8\beta}\right) \operatorname{ch} \dfrac{1}{8\beta}$ [Re $\beta > 0$]. MH 24

4. $\int_0^\infty (\operatorname{ch} x^2 - \cos x^2) e^{-\beta x^4} dx = \dfrac{\sqrt{2}\,\pi}{4\sqrt{\beta}} I_{-\frac{1}{4}}\left(\dfrac{1}{8\beta}\right) \operatorname{sh} \dfrac{1}{8\beta}$ [Re $\beta > 0$]. MH 24

4.137

1. $\int_0^\infty \sin 2x^2 \operatorname{sh} 2x^2 e^{-\beta x^4} dx = \dfrac{\pi}{\sqrt[4]{128\beta^2}} J_{-\frac{1}{4}}\left(\dfrac{1}{\beta}\right) \cos\left(\dfrac{1}{\beta} + \dfrac{\pi}{4}\right)$

[Re $\beta > 0$]. MHP 32

2. $\int_0^\infty \sin 2x^2 \operatorname{ch} 2x^2 e^{-\beta x^4} dx = \dfrac{\pi}{\sqrt[4]{128\beta^2}} J_{\frac{1}{4}}\left(\dfrac{1}{\beta}\right) \cos\left(\dfrac{1}{\beta} - \dfrac{\pi}{4}\right)$

[Re $\beta > 0$]. MHP 32

3. $\int_0^\infty \cos 2x^2 \operatorname{sh} 2x^2 e^{-\beta x^4} dx = \dfrac{-\pi}{\sqrt[4]{128\beta^2}} J_{\frac{1}{4}}\left(\dfrac{1}{\beta}\right) \sin\left(\dfrac{1}{\beta} - \dfrac{\pi}{4}\right)$

[Re $\beta > 0$]. MHP 32

4. $\int_0^\infty \cos 2x^2 \operatorname{ch} 2x^2 e^{-\beta x^4} dx = \dfrac{\pi}{\sqrt[4]{128\beta^2}} J_{-\frac{1}{4}}\left(\dfrac{1}{\beta}\right) \sin\left(\dfrac{1}{\beta} + \dfrac{\pi}{4}\right)$

[Re $\beta > 0$]. MHP 32

4.138

1. $\int_0^\infty (\sin 2x^2 \operatorname{ch} 2x^2 + \cos 2x^2 \operatorname{sh} 2x^2) e^{-\beta x^4} dx = \dfrac{\pi}{\sqrt[4]{32\beta^2}} J_{\frac{1}{4}}\left(\dfrac{1}{\beta}\right) \cos\left(\dfrac{1}{\beta}\right)$

[Re $\beta > 0$]. MHP 32

2. $\int_0^\infty (\sin 2x^2 \operatorname{ch} 2x^2 - \cos 2x^2 \operatorname{sh} 2x^2) e^{-\beta x^4} dx = \dfrac{\pi}{\sqrt[4]{32\beta^2}} J_{\frac{1}{4}}\left(\dfrac{1}{\beta}\right) \sin\left(\dfrac{1}{\beta}\right)$

[Re $\beta > 0$]. MHP 32

3. $\int_0^\infty (\cos 2x^2 \operatorname{ch} 2x^2 + \sin 2x^2 \operatorname{sh} 2x^2) e^{-\beta x^4} dx = \dfrac{\pi}{\sqrt[4]{32\beta^2}} J_{-\frac{1}{4}}\left(\dfrac{1}{\beta}\right) \cos\left(\dfrac{1}{\beta}\right)$

[Re $\beta > 0$]. MHP 32

4. $\int_0^\infty (\cos 2x^2 \operatorname{ch} 2x^2 - \sin 2x^2 \operatorname{sh} 2x^2) e^{-\beta x^4} dx = \dfrac{\pi}{\sqrt[4]{32\beta^2}} J_{-\frac{1}{4}}\left(\dfrac{1}{\beta}\right) \sin\left(\dfrac{1}{\beta}\right)$

[Re $\beta > 0$]. MHP 32

Trigonometrische Funktionen, Hyperbelfunktionen, die Exponentialfunktion und Potenzen
Trigonometric and hyperbolic functions, exponentials, and powers

4.14

1. $\int_0^\infty x e^{-\beta x^2} \operatorname{ch} x \sin x \, dx = \frac{1}{4}\sqrt{\frac{\pi}{\beta^3}} \left(\cos \frac{1}{2\beta} + \sin \frac{1}{2\beta} \right)$ [Re $\beta > 0$]. MHP 32   4.141

2. $\int_0^\infty x e^{-\beta x^2} \operatorname{sh} x \cos x \, dx = \frac{1}{4}\sqrt{\frac{\pi}{\beta^3}} \left(\cos \frac{1}{2\beta} - \sin \frac{1}{2\beta} \right)$ [Re $\beta > 0$]. MHP 32

3. $\int_0^\infty x^2 e^{-\beta x^2} \operatorname{ch} x \cos x \, dx = \frac{1}{4}\sqrt{\frac{\pi}{\beta^3}} \left(\cos \frac{1}{2\beta} - \frac{1}{\beta} \sin \frac{1}{2\beta} \right)$ [Re $\beta > 0$].

   MHP 32

4. $\int_0^\infty x^2 e^{-\beta x^2} \operatorname{sh} x \sin x \, dx = \frac{1}{4}\sqrt{\frac{\pi}{\beta^3}} \left(\sin \frac{1}{2\beta} + \frac{1}{\beta} \cos \frac{1}{2\beta} \right)$ [Re $\beta > 0$].

   MHP 32

1. $\int_0^\infty x e^{-\beta x^2} (\operatorname{sh} x + \sin x) \, dx = \frac{1}{2}\sqrt{\frac{\pi}{\beta^3}} \operatorname{ch} \frac{1}{4\beta}$ [Re $\beta > 0$]. MH 24   4.142

2. $\int_0^\infty x e^{-\beta x^2} (\operatorname{sh} x - \sin x) \, dx = \frac{1}{2}\sqrt{\frac{\pi}{\beta^3}} \operatorname{sh} \frac{1}{4\beta}$ [Re $\beta > 0$]. MH 24

3. $\int_0^\infty x^2 e^{-\beta x^2} (\operatorname{ch} x + \cos x) \, dx = \frac{1}{2}\sqrt{\frac{\pi}{\beta^3}} \left(\operatorname{ch} \frac{1}{4\beta} + \frac{1}{2\beta} \operatorname{sh} \frac{1}{4\beta} \right)$ [Re $\beta > 0$].

   MH 24

4. $\int_0^\infty x^2 e^{-\beta x^2} (\operatorname{ch} x - \cos x) \, dx = \frac{1}{2}\sqrt{\frac{\pi}{\beta^3}} \left(\operatorname{sh} \frac{1}{4\beta} + \frac{1}{2\beta} \operatorname{ch} \frac{1}{4\beta} \right)$ [Re $\beta > 0$].

   MH 24

1. $\int_0^\infty x e^{-\beta x^2} (\operatorname{ch} x \sin x + \operatorname{sh} x \cos x) \, dx = \frac{1}{2\beta}\sqrt{\frac{\pi}{\beta}} \cos \frac{1}{2\beta}$ [Re $\beta > 0$]. MHP 32   4.143

2. $\int_0^\infty x e^{-\beta x^2} (\operatorname{ch} x \sin x - \operatorname{sh} x \cos x) \, dx = \frac{1}{2\beta}\sqrt{\frac{\pi}{\beta}} \sin \frac{1}{2\beta}$ [Re $\beta > 0$]. MHP 32

$\int_0^\infty e^{-x^2} \operatorname{sh} x^2 \cos ax \frac{dx}{x^2} = \sqrt{\frac{\pi}{2}} e^{-\frac{a^2}{8}} - \frac{\pi a}{4} \left[1 - \Phi\left(\frac{a}{\sqrt{8}} \right) \right]$ 4.144

[$a > 0$].   Ba IV 35(44)

1. $\int_0^\infty x e^{-\beta x^2} \operatorname{ch}(2ax \sin t) \sin(2ax \cos t) \, dx =$ 4.145

$= \frac{a}{2}\sqrt{\frac{\pi}{\beta^3}} \exp\left(-\frac{a^2}{\beta} \cos 2t \right) \cos\left(t - \frac{a^2}{\beta} \sin 2t \right)$

[Re $\beta > 0$].   BH[363](5)

4.145

2. $\int_0^\infty x e^{-\beta x^2} \operatorname{sh}(2ax \sin t) \cos(2ax \cos t) \, dx =$

$$= \frac{a}{2} \sqrt{\frac{\pi}{\beta^3}} \exp\left(-\frac{a^2}{\beta} \cos 2t\right) \sin\left(t - \frac{a^2}{\beta} \sin 2t\right)$$

[Re $\beta > 0$]. BH[363](6)

Der Logarithmus
The Logarithmic Function

4.2–4.4

Der Logarithmus
The logarithmic function

4.21

4.211

1. $\int_e^\infty \dfrac{dx}{\ln \dfrac{1}{x}} = -\infty.$ BH[33](9)

2. $\int_0^u \dfrac{dx}{\ln x} = \operatorname{li} u.$ F III 533, F II 615

4.212

1. $\int_0^1 \dfrac{dx}{a + \ln x} = e^{-a}\, \overline{\operatorname{Ei}}(a).$ BH[31](4)

2. $\int_0^1 \dfrac{dx}{a - \ln x} = -e^a \operatorname{Ei}(-a).$ BH[31](5)

3. $\int_0^1 \dfrac{dx}{(a - \ln x)^2} = \dfrac{1}{a} + e^a \operatorname{Ei}(-a)$ [$a > 0$]. BH[31](16)

4. $\int_0^1 \dfrac{\ln x \, dx}{(a - \ln x)^2} = 1 + (1+a) e^a \operatorname{Ei}(-a)$ [$a > 0$]. BH[31](17)

5. $\int_1^e \dfrac{\ln x \, dx}{(1 + \ln x)^2} = \dfrac{e}{2} - 1.$ BH[33](10)

6. $\int_0^1 \dfrac{dx}{(a + \ln x)^n} = \dfrac{1}{(n-1)!} e^{-a}\, \overline{\operatorname{Ei}}(a) - \dfrac{1}{(n-1)!} \sum_{k=1}^{n-1} (n-k-1)!\, a^{k-n}$

[$a > 0$, n ungeradzahlig / odd]. BH[31](22)

7. $\int_0^1 \dfrac{dx}{(a - \ln x)^n} = \dfrac{(-1)^n}{(n-1)!} e^a \operatorname{Ei}(-a) + \dfrac{(-1)^{n-1}}{(n-1)!} \sum_{k=1}^{n-1} (n-k-1)!\, (-a)^{k-n}$

[$a > 0$]. BH[31](23)

| In Integralen der Gestalt | In integrals of the form |

$$\int \frac{(\ln x)^m}{[a^n + (\ln x)^n]^l} dx$$

| substituiert man zweckmäßigerweise $x = e^{-t}$. | the substitution $x = e^{-t}$ is used with advantage. |

1. $\int_0^1 \frac{dx}{a^2 + (\ln x)^2} = \frac{1}{a} [\operatorname{ci}(a) \sin a - \operatorname{si}(a) \cos a] \quad [a > 0].$ BH[31](16) 4.213

2. $\int_0^1 \frac{dx}{a^2 - (\ln x)^2} = \frac{1}{2a} [e^{-a} \overline{\operatorname{Ei}}(a) - e^a \operatorname{Ei}(-a)] \quad [a > 0]$

(cf. 4.212 1., 2.). BH[31](8)

3. $\int_0^1 \frac{\ln x \, dx}{a^2 + (\ln x)^2} = \operatorname{ci}(a) \cos(a) + \operatorname{si}(a) \sin a \quad [a > 0].$ BH[31](7)

4. $\int_0^1 \frac{\ln x \, dx}{a^2 - (\ln x)^2} = -\frac{1}{2} [e^{-a} \overline{\operatorname{Ei}}(a) + e^a \operatorname{Ei}(-a)] \quad [a > 0]$

(cf. 4.212 1., 2.). BH[31](9)

5. $\int_0^1 \frac{dx}{[a^2 + (\ln x)^2]^2} = \frac{1}{2a^3} [\operatorname{ci}(a) \sin a - \operatorname{si}(a) \cos a] -$

$- \frac{1}{2a^2} [\operatorname{ci}(a) \cos a + \operatorname{si}(a) \sin a] \quad [a > 0].$ Li[31](18)

6. $\int_0^1 \frac{\ln x \, dx}{[a^2 + (\ln x)^2]^2} = \frac{1}{2a} [\operatorname{ci}(a) \sin a - \operatorname{si}(a) \cos a] - \frac{1}{2a^2} \quad [a > 0].$ BH[31](19)

1. $\int_0^1 \frac{dx}{a^4 - (\ln x)^4} = -\frac{1}{4a^3} [e^a \operatorname{Ei}(-a) - e^{-a} \overline{\operatorname{Ei}}(a) -$ 4.214

$- 2 \operatorname{ci}(a) \sin a + 2 \operatorname{si}(a) \cos a] \quad [a > 0].$ BH[31](10)

2. $\int_0^1 \frac{\ln x \, dx}{a^4 - (\ln x)^4} = -\frac{1}{4a^2} [e^a \operatorname{Ei}(-a) + e^{-a} \overline{\operatorname{Ei}}(a) -$

$- 2 \operatorname{ci}(a) \cos a - 2 \operatorname{si}(a) \sin a] \quad [a > 0].$ BH[31](11)

3. $\int_0^1 \frac{(\ln x)^2 \, dx}{a^4 - (\ln x)^4} = -\frac{1}{4a} [e^a \operatorname{Ei}(-a) - e^{-a} \overline{\operatorname{Ei}}(a) +$

$+ 2 \operatorname{ci}(a) \sin a - 2 \operatorname{si}(a) \cos a] \quad [a > 0].$ BH[31](12)

4.214 4. $\int_0^1 \dfrac{(\ln x)^3 \, dx}{a^4 - (\ln x)^4} = -\dfrac{1}{4}\left[e^a \operatorname{Ei}(-a) + e^{-a} \overline{\operatorname{Ei}}(a) + \right.$

$\left. + 2\operatorname{ci}(a)\cos a + 2\operatorname{si}(a)\sin a\right] \quad [a > 0].$ BH[31](13)

4.215 1. $\int_0^1 \left(\ln \dfrac{1}{x}\right)^{\mu-1} dx = \Gamma(\mu) \quad [\operatorname{Re}\mu > 0].$ F II 780

2. $\int_0^1 \dfrac{dx}{\left(\ln \dfrac{1}{x}\right)^\mu} = \dfrac{\pi}{\Gamma(\mu)} \operatorname{cosec} \mu\pi \quad [\operatorname{Re}\mu < 1].$ BH[31](1)

3. $\int_0^1 \sqrt{\ln \dfrac{1}{x}}\, dx = \dfrac{\sqrt{\pi}}{2}.$ BH[32](1)

4. $\int_0^1 \dfrac{dx}{\sqrt{\ln \dfrac{1}{x}}} = \sqrt{\pi}.$ BH[32](3)

4.216 $\int_0^{1/e} \dfrac{dx}{\sqrt{(\ln x)^2 - 1}} = K_0(1).$ GH[321](2)

Logarithmen von zusammengesetzten Argumenten
Logarithms of compound arguments

4.22

4.221 1. $\int_0^1 \ln x \ln(1-x)\, dx = 2 - \dfrac{\pi^2}{6}.$ BH[30](7)

2. $\int_0^1 \ln x \ln(1+x)\, dx = 2 - \dfrac{\pi^2}{12} - 2\ln 2.$ BH[30](8)

3. $\int_0^1 \ln \dfrac{1-ax}{1-a} \dfrac{dx}{\ln x} = -\sum_{k=1}^\infty a^k \dfrac{\ln(1+k)}{k} \quad [a < 1].$ BH[31](3)

4.222 1. $\int_0^\infty \ln \dfrac{a^2 + x^2}{b^2 + x^2}\, dx = (a-b)\pi \quad [a > 0, b > 0].$ GH[322](20)

2. $\int_0^\infty \ln x \ln \dfrac{a^2 + x^2}{b^2 + x^2}\, dx = \pi(b-a) + \pi \ln \dfrac{a^a}{b^b} \quad [a > 0, b > 0].$ BH[33](1)

3. $\int_0^\infty \ln x \ln\left(1 + \dfrac{b^2}{x^2}\right) dx = \pi b(\ln b - 1) \quad [b > 0].$ BH[33](2)

4. $\int_0^\infty \ln(1+a^2x^2) \ln\left(1+\frac{b^2}{x^2}\right) dx = 2\pi\left[\frac{1+ab}{a} \ln(1+ab) - b\right]$

$[a > 0, \ b > 0]$.   BH[33](3)   4.222

5. $\int_0^\infty \ln(a^2+x^2) \ln\left(1+\frac{b^2}{x^2}\right) dx = 2\pi[(a+b)\ln(a+b) - a\ln a - b]$

$[a > 0, \ b > 0]$.   BH[33](4)

6. $\int_0^\infty \ln\left(1+\frac{a^2}{x^2}\right) \ln\left(1+\frac{b^2}{x^2}\right) dx = 2\pi[(a+b)\ln(a+b) - a\ln a - b\ln b]$

$[a > 0, \ b > 0]$.   BH[33](5)

7. $\int_0^\infty \ln\left(a^2+\frac{1}{x^2}\right) \ln\left(1+\frac{b^2}{x^2}\right) dx = 2\pi\left[\frac{1+ab}{a} \ln(1+ab) - b\ln b\right]$

$[a > 0, \ b > 0]$.   BH[33](7)

1. $\int_0^\infty \ln(1+e^{-x}) dx = \frac{\pi^2}{12}.$ BH[256](10)   4.223

2. $\int_0^\infty \ln(1-e^{-x}) dx = -\frac{\pi^2}{6}.$ BH[256](11)

3. $\int_0^\infty \ln(1+2e^{-x}\cos t + e^{-2x}) dx = \frac{\pi^2}{6} - \frac{t^2}{2}$ $[|t| < \pi]$.   BH[256](18)

1. $\int_0^u \ln \sin x \, dx = L\left(\frac{\pi}{2}-u\right) - L\left(\frac{\pi}{2}\right).$ Lo III 186(15)   4.224

2. $\int_0^{\pi/4} \ln \sin x \, dx = -\frac{\pi}{4}\ln 2 - \frac{1}{2}G.$ BH[285](1)

3. $\int_0^{\pi/2} \ln \sin x \, dx = \frac{1}{2}\int_0^{\pi} \ln \sin x \, dx = -\frac{\pi}{2}\ln 2.$ F II 633, 652

4. $\int_0^u \ln \cos x \, dx = -L(u).$ Lo III 184(10)

5. $\int_0^{\pi/4} \ln \cos x \, dx = -\frac{\pi}{4}\ln 2 + \frac{1}{2}G.$ BH[286](1)

6. $\int_0^{\pi/2} \ln \cos x \, dx = -\frac{\pi}{2}\ln 2.$ BH[306](1)

4.224

7.
$$\int_0^{\pi/2} (\ln \sin x)^2 \, dx = \frac{\pi}{2}\left[(\ln 2)^2 + \frac{\pi^2}{12}\right].$$
BH[305](19)

8.
$$\int_0^{\pi/2} (\ln \cos x)^2 \, dx = \frac{\pi}{2}\left[(\ln 2)^2 + \frac{\pi^2}{12}\right].$$
BH[306](14)

9.
$$\int_0^{\pi} \ln(a + b \cos x) \, dx = \pi \ln \frac{a + \sqrt{a^2 - b^2}}{2} \quad [a \geqslant |b| > 0].$$
GH[322](15)

10.
$$\int_0^{\pi} \ln(1 \pm \sin x) \, dx = -\pi \ln 2 \pm 4G.$$
GH[322](16a)

11.
$$\int_0^{\pi/2} \ln(1 + a \sin x)^2 \, dx = \int_0^{\pi/2} \ln(1 + a \cos x)^2 \, dx =$$
$$= \pi \ln\left(\frac{a}{2}\right) + 4G + 4S(b)$$

wobei / where $b = \dfrac{1-a}{1+a}$ und / and

$$S(b) = \sum_{k=1}^{\infty} \frac{b^k}{k} \sum_{n=1}^{k} \frac{(-1)^{n+1}}{2n-1} \quad [a > 0].$$
BH[308](5,6,7,8)

12.
$$\int_0^{\pi} \ln(1 + a \cos x)^2 \, dx = 2\pi \ln \frac{1 + \sqrt{1-a^2}}{2} \quad [a^2 \leqslant 1].$$
BH[330](1)

13.
$$\int_0^{\pi/2} \ln(1 + 2a \sin x + a^2) \, dx = \sum_{k=0}^{\infty} \frac{2^{2k}(k!)^2}{(2k+1)\cdot(2k+1)!!}\left(\frac{2a}{1+a^2}\right)^{2k+1}$$
$$[a^2 \leqslant 1].$$
BH[308](24)

14.
$$\int_0^{n\pi} \ln(1 - 2a \cos x + a^2) \, dx = 0 \quad [a^2 < 1];$$
$$= n\pi \ln a^2 \quad [a^2 > 1].$$
F II 127, 144, 697

4.225

1.
$$\int_0^{\pi/4} \ln(\cos x - \sin x) \, dx = -\frac{\pi}{8} \ln 2 - \frac{1}{2} G.$$
GH[322](9b)

2.
$$\int_0^{\pi/4} \ln(\cos x + \sin x) \, dx = \frac{1}{2} \int_0^{\pi/2} \ln(\cos x + \sin x) \, dx = -\frac{\pi}{8} \ln 2 + \frac{1}{2} G.$$
GH[322](9a)

3. $\int_0^{2\pi} \ln(1 + a\sin x + b\cos x)\,dx = 2\pi \ln \dfrac{1+\sqrt{1-a^2-b^2}}{2}$ $\quad [a^2+b^2 < 1]$. $\qquad$ 4.225

BH[332](2)

4. $\int_0^{2\pi} \ln(1 + a^2 + b^2 + 2a\sin x + 2b\cos x)\,dx =$

$$= 0 \qquad [a^2 + b^2 \leqslant 1];$$
$$= 2\pi \ln(a^2 + b^2) \quad [a^2 + b^2 \geqslant 1].$$

BH[332](3)

1. $\int_0^{\pi/2} \ln(a^2 - \sin^2 x)^2\,dx = -2\pi \ln 2 \quad [a^2 \leqslant 1];$ $\qquad$ 4.226

$$= 2\pi \ln \dfrac{a + \sqrt{a^2 - 1}}{2} = 2\pi(\operatorname{Arch} a - \ln 2) \quad [a > 1].$$

F II 652, 696

2. $\int_0^{\pi/2} \ln(1 + a\sin^2 x)\,dx = \dfrac{1}{2}\int_0^{\pi} \ln(1 + a\sin^2 x)\,dx =$

$= \int_0^{\pi/2} \ln(1 + a\cos^2 x)\,dx = \dfrac{1}{2}\int_0^{\pi} \ln(1 + a\cos^2 x)\,dx = \pi \ln \dfrac{1 + \sqrt{1+a}}{2}$

$[a \geqslant -1].$ $\qquad$ BH[308](15), GH[322](12)

3. $\int_0^u \ln(1 - \sin^2 \alpha \sin^2 x)\,dx = (\pi - 2\theta)\ln \operatorname{ctg}\dfrac{\alpha}{2} +$

$+ 2u \ln\left(\dfrac{1}{2}\sin \alpha\right) - \dfrac{\pi}{2}\ln 2 + L(\theta + u) - L(\theta - u) + L\left(\dfrac{\pi}{2} - 2u\right)$

$\left[\operatorname{ctg}\theta = \cos \alpha \operatorname{tg} u;\, -\pi \leqslant \alpha \leqslant \pi,\, -\dfrac{\pi}{2} \leqslant u \leqslant \dfrac{\pi}{2}\right].$ $\qquad$ Lo III 287

4. $\int_0^{\pi/2} \ln[1 - \cos^2 x(\sin^2 \alpha - \sin^2 \beta \sin^2 x)]\,dx =$

$= \pi \ln\left[\dfrac{1}{2}\left(\cos^2\dfrac{\alpha}{2} + \sqrt{\cos^4\dfrac{\alpha}{2} + \sin^2\dfrac{\beta}{2}\cos^2\dfrac{\beta}{2}}\right)\right]$

$[\alpha > \beta > 0].$ $\qquad$ Lo III 283

5. $\int_0^u \ln\left(1 - \dfrac{\sin^2 x}{\sin^2 \alpha}\right)dx = -u \ln \sin^2 \alpha - L\left(\dfrac{\pi}{2} - \alpha + u\right) + L\left(\dfrac{\pi}{2} - \alpha - u\right)$

$\left[-\dfrac{\pi}{2} \leqslant u \leqslant \dfrac{\pi}{2},\, |\sin u| \leqslant |\sin \alpha|\right].$ $\qquad$ Lo III 287

4.226

6. $$\int_0^{\frac{\pi}{2}} \ln(a^2\cos^2 x + b^2\sin^2 x)\,dx = \frac{1}{2}\int_0^{\pi} \ln(a^2\cos^2 x + b^2\sin^2 x)\,dx =$$
$$= \pi \ln \frac{a+b}{2} \quad [a>0, b>0].$$ GH[322] (13)

7. $$\int_0^{\frac{\pi}{2}} \ln \frac{1+\sin t\cos^2 x}{1-\sin t\cos^2 x}\,dx = \pi \ln \frac{1+\sin\frac{t}{2}}{\cos\frac{t}{2}} = \pi \ln \operatorname{ctg}\frac{\pi-t}{4} \quad \left[|t| < \frac{\pi}{2}\right].$$

Lo III 283

4.227

1. $$\int_0^u \ln \operatorname{tg} x\,dx = L(u) + L\left(\frac{\pi}{2} - u\right) - L\left(\frac{\pi}{2}\right).$$ Lo III 186(16)

2. $$\int_0^{\frac{\pi}{4}} \ln \operatorname{tg} x\,dx = -\int_{\frac{\pi}{4}}^{\frac{\pi}{2}} \ln \operatorname{tg} x\,dx = -G.$$ BH[286](11)

3. $$\int_0^{\frac{\pi}{2}} \ln(a\operatorname{tg} x)\,dx = \frac{\pi}{2}\ln a \quad [a>0].$$ BH[307](2)

4. $$\int_0^{\frac{\pi}{4}} (\ln \operatorname{tg} x)^n\,dx = n!(-1)^n \sum_{k=0}^{\infty} \frac{(-1)^k}{(2k+1)^{n+1}}.$$ BH[286](21)

5. $$\int_0^{\frac{\pi}{2}} (\ln \operatorname{tg} x)^{2n}\,dx = 2(2n)!\sum_{k=0}^{\infty} \frac{(-1)^k}{(2k+1)^{2n+1}}.$$ BH[307](15)

6. $$\int_0^{\frac{\pi}{2}} (\ln \operatorname{tg} x)^{2n+1}\,dx = 0.$$ BH[307](14)

7. $$\int_0^{\frac{\pi}{4}} (\ln \operatorname{tg} x)^2\,dx = \frac{\pi^3}{16}.$$ BH[286](16)

8. $$\int_0^{\frac{\pi}{4}} (\ln \operatorname{tg} x)^4\,dx = \frac{5}{64}\pi^5.$$ BH[286](19)

9. $$\int_0^{\frac{\pi}{4}} \ln(1+\operatorname{tg} x)\,dx = \frac{\pi}{8}\ln 2.$$ BH[287](1)

10. $$\int_0^{\frac{\pi}{2}} \ln(1+\operatorname{tg} x)\,dx = \frac{\pi}{4}\ln 2 + G.$$ BH[308](9) 4.227

11. $$\int_0^{\frac{\pi}{4}} \ln(1-\operatorname{tg} x)\,dx = \frac{\pi}{8}\ln 2 - G.$$ BH[287](2)

12. $$\int_0^{\frac{\pi}{2}} \ln(1-\operatorname{tg} x)^2\,dx = \frac{\pi}{2}\ln 2 - 2G.$$ BH[308](10)

13. $$\int_0^{\frac{\pi}{4}} \ln(1+\operatorname{ctg} x)\,dx = \frac{\pi}{8}\ln 2 + G.$$ BH[287](3)

14. $$\int_0^{\frac{\pi}{4}} \ln(\operatorname{ctg} x - 1)\,dx = \frac{\pi}{8}\ln 2.$$ BH[287](4)

15. $$\int_0^{\frac{\pi}{4}} \ln(\operatorname{tg} x + \operatorname{ctg} x)\,dx = \frac{1}{2}\int_0^{\frac{\pi}{2}} \ln(\operatorname{tg} x + \operatorname{ctg} x)\,dx = \frac{\pi}{2}\ln 2.$$

BH[287](5), BH[308](11)

16. $$\int_0^{\frac{\pi}{4}} \ln(\operatorname{ctg} x - \operatorname{tg} x)^2\,dx = \frac{1}{2}\int_0^{\frac{\pi}{2}} \ln(\operatorname{ctg} x - \operatorname{tg} x)^2\,dx = \frac{\pi}{2}\ln 2.$$

BH[287](6), BH[308](12)

17. $$\int_0^{\frac{\pi}{2}} \ln(a^2 + b^2 \operatorname{tg}^2 x)\,dx = \frac{1}{2}\int_0^{\pi} \ln(a^2 + b^2 \operatorname{tg}^2 x)\,dx = \pi \ln(a+b)$$

$$[a > 0, \quad b > 0].$$ GH[322](17)

1. $$\int_0^{\frac{\pi}{2}} \ln\left(\sin t \sin x + \sqrt{1 - \cos^2 t \sin^2 x}\right) dx =$$ 4.228

$$= \frac{\pi}{2}\ln 2 - 2L\left(\frac{t}{2}\right) - 2L\left(\frac{\pi - t}{2}\right).$$ Lo III 290

2. $$\int_0^u \ln\left(\cos x + \sqrt{\cos^2 x - \cos^2 t}\right) dx = -\left(\frac{\pi}{2} - t - \varphi\right)\ln \cos t +$$

$$+ \frac{1}{2}L(u+\varphi) - \frac{1}{2}L(u-\varphi) - L(\varphi)$$

$$\left[\cos\varphi = \frac{\sin u}{\sin t},\ 0 \leqslant u \leqslant t \leqslant \frac{\pi}{2}\right].$$ Lo III 290

4.228

3. $\int_0^t \ln(\cos x + \sqrt{\cos^2 x - \cos^2 t})\, dx = -\left(\frac{\pi}{2} - t\right) \ln \cos t.$ Lo III 285

4. $\int_0^u \ln \frac{\sin u + \sin t \cos x \sqrt{\sin^2 u - \sin^2 x}}{\sin u - \sin t \cos x \sqrt{\sin^2 u - \sin^2 x}}\, dx =$

$= \pi \ln \left[\operatorname{tg} \frac{t}{2} \sin u + \sqrt{\operatorname{tg}^2 \frac{t}{2} \sin^2 u + 1}\right]$ $[t > 0,\ u > 0].$ Lo III 283

5. $\int_0^{\pi/4} \sqrt{\ln \operatorname{ctg} x}\, dx = \frac{\sqrt{\pi}}{2} \sum_{k=0}^{\infty} \frac{(-1)^k}{\sqrt{(2k+1)^3}}.$ BH[297](9)

6. $\int_0^{\pi/4} \frac{dx}{\sqrt{\ln \operatorname{ctg} x}} = \sqrt{\pi} \sum_{k=0}^{\infty} \frac{(-1)^k}{\sqrt{2k+1}}.$ BH[304](24)

7. $\int_0^{\pi/4} \ln(\sqrt{\operatorname{tg} x} + \sqrt{\operatorname{ctg} x})\, dx = \frac{1}{2} \int_0^{\pi/2} \ln(\sqrt{\operatorname{tg} x} + \sqrt{\operatorname{ctg} x})\, dx =$

$= \frac{\pi}{8} \ln 2 + \frac{1}{2} G.$ BH[287](7), BH[308](22)

8. $\int_0^{\pi/4} \ln(\sqrt{\operatorname{ctg} x} - \sqrt{\operatorname{tg} x})^2\, dx = \frac{1}{2} \int_0^{\pi/2} \ln(\sqrt{\operatorname{ctg} x} - \sqrt{\operatorname{tg} x})^2\, dx =$

$= \frac{\pi}{4} \ln 2 - G.$ BH[287](8), BH[308](23)

4.229

1. $\int_0^1 \ln\left(\ln \frac{1}{x}\right) dx = -C.$ F II 813

2. $\int_0^1 \frac{dx}{\ln\left(\ln \frac{1}{x}\right)} = 0.$ BH[31](2)

3. $\int_0^1 \ln\left(\ln \frac{1}{x}\right) \frac{dx}{\sqrt{\ln \frac{1}{x}}} = -(C + 2\ln 2)\sqrt{\pi}.$ BH[32](4)

4. $\int_0^1 \ln\left(\ln \frac{1}{x}\right) \left(\ln \frac{1}{x}\right)^{\mu-1} dx = \psi(\mu)\Gamma(\mu)$ $[\operatorname{Re} \mu > 0].$ BH[30](10)

In Integralen mit $\ln\left(\ln\frac{1}{x}\right)$ im Integranden substituiert man zweckmäßigerweise $\ln\frac{1}{x} = u$, also $x = e^{-u}$.

In integrals with $\ln\left(\ln\frac{1}{x}\right)$ in the integrand the substitution $\ln\frac{1}{x} = u$, that is, $x = e^{-u}$, is used with advantage.

5. $\int_0^1 \ln(a + \ln x)\, dx = \ln a - e^{-a}\,\overline{\mathrm{Ei}}(a) \quad [a > 0].$ BH[30](5) 4.229

6. $\int_0^1 \ln(a - \ln x)\, dx = \ln a - e^{a}\,\mathrm{Ei}(-a) \quad [a > 0].$ BH[30](6)

7. $\int_{\pi/4}^{\pi/2} \ln\ln\mathrm{tg}\, x\, dx = \frac{\pi}{2} \ln\left\{\frac{\Gamma\left(\frac{3}{4}\right)}{\Gamma\left(\frac{1}{4}\right)} \sqrt{2\pi}\right\}.$ BH[308](28)

Logarithmus und rationale Funktionen
Logarithms and rational functions
4.23

1. $\int_0^1 \frac{\ln x}{1+x}\, dx = -\frac{\pi^2}{12}.$ F II 478+ 4.231

2. $\int_0^1 \frac{\ln x}{1-x}\, dx = -\frac{\pi^2}{6}.$ F II 720, 721

3. $\int_0^1 \frac{x \ln x}{1-x}\, dx = 1 - \frac{\pi^2}{6}.$ BH[108](7)

4. $\int_0^1 \frac{1+x}{1-x} \ln x\, dx = 1 - \frac{\pi^2}{3}.$ BH[108](9)

5. $\int_0^\infty \frac{\ln x\, dx}{(x+a)^2} = \frac{\ln a}{a} \quad [0 < a < 1].$ BH[139](1)

6. $\int_0^1 \frac{\ln x}{(1+x)^2}\, dx = -\ln 2.$ BH[111](1)

7. $\int_0^\infty \ln x \frac{dx}{(a^2 + b^2 x^2)^n} = \frac{\Gamma\left(n - \frac{1}{2}\right)\sqrt{\pi}}{4\cdot(n-1)!\, a^{2n-1} b}\left[2\ln\frac{a}{2b} - C - \psi\left(n - \frac{1}{2}\right)\right]$

$[a > 0,\ b > 0].$ Li[139](3)

4.231

8. $$\int_0^\infty \frac{\ln x\, dx}{a^2 + b^2 x^2} = \frac{\pi}{2ab} \ln \frac{a}{b} \quad [ab > 0].$$ BH[135](6)

9. $$\int_0^\infty \frac{\ln px}{q^2 + x^2} dx = \frac{\pi}{2q} \ln pq \quad [p > 0, q > 0].$$ BH[135](4)

10. $$\int_0^\infty \frac{\ln x\, dx}{a^2 - b^2 x^2} = -\frac{\pi^2}{4ab} \quad [ab > 0].$$ Li[135](6)

11. $$\int_0^a \frac{\ln x\, dx}{x^2 + a^2} = \frac{\pi \ln a}{4a} - \frac{G}{a} \quad [a > 0].$$ GH[324](7b)

12. $$\int_0^1 \frac{\ln x}{1 + x^2} dx = -\int_1^\infty \frac{\ln x}{1 + x^2} dx = -G.$$ F II 476, F II 611–624

13. $$\int_0^1 \frac{\ln x\, dx}{1 - x^2} = -\frac{\pi^2}{8}.$$ BH[108](11)

14. $$\int_0^1 \frac{x \ln x}{1 + x^2} dx = -\frac{\pi^2}{48}.$$ GH[324](7b)

15. $$\int_0^1 \frac{x \ln x}{1 - x^2} dx = -\frac{\pi^2}{24}.$$

16. $$\int_0^1 \ln x\, \frac{1 - x^{2n+2}}{(1 - x^2)^2} dx = -\frac{(n+1)\pi^2}{8} + \sum_{k=1}^n \frac{n-k+1}{(2k-1)^2}.$$ BH[111](5)

17. $$\int_0^1 \ln x\, \frac{1 + (-1)^n x^{n+1}}{(1 + x)^2} dx = -\frac{(n+1)\pi^2}{12} - \sum_{k=1}^n (-1)^k \frac{n-k+1}{k^2}.$$ BH[111](2)

18. $$\int_0^1 \ln x\, \frac{1 - x^{n+1}}{(1 - x)^2} dx = -\frac{(n+1)\pi^2}{6} + \sum_{k=1}^n \frac{n-k+1}{k^2}.$$ BH[111](3)

4.232

1. $$\int_u^v \frac{\ln x\, dx}{(x+u)(x+v)} = \frac{\ln uv}{2(v-u)} \ln \frac{(u+v)^2}{4uv}.$$ BH[145](32)

2. $$\int_0^\infty \frac{\ln x\, dx}{(x+\beta)(x+\gamma)} = \frac{(\ln \beta)^2 - (\ln \gamma)^2}{2(\beta - \gamma)} \quad [|\arg \beta| < \pi,\ |\arg \gamma| < \pi].$$ Ba V 218(24)

3. $$\int_0^\infty \frac{\ln x}{x+a} \frac{dx}{x-1} = \frac{\pi^2 + (\ln a)^2}{2(a+1)} \quad [a > 0].$$ BH[140](10)

4.233

1. $$\int_0^1 \frac{\ln x \, dx}{1+x+x^2} = -0{,}781\,302\,412\,9\ldots$$ Li[113](1)

2. $$\int_0^1 \frac{\ln x \, dx}{1-x+x^2} = -1{,}171\,953\,619\,34\ldots$$ Li[113](3)

3. $$\int_0^1 \frac{x \ln x \, dx}{1+x+x^2} = -0{,}157\,660\,149\,17\ldots$$ Li[113](2)

4. $$\int_0^1 \frac{x \ln x \, dx}{1-x+x^2} = -0{,}311\,821\,131\,9\ldots$$ Li[113](4)

5. $$\int_0^\infty \frac{\ln x \, dx}{x^2 + 2xa\cos t + a^2} = \frac{t \ln a}{a \sin t} \quad [a>0,\ 0<t<\pi].$$ GH[324](13c)

4.234

1. $$\int_1^\infty \frac{\ln x \, dx}{(1+x^2)^2} = \ln 2.$$ BH[144](18)+

2. $$\int_0^1 \frac{x \ln x \, dx}{(1+x^2)^2} = -\frac{1}{4} \ln 2.$$ BH[111](4)

3. $$\int_0^\infty \frac{1+x^2}{(1-x^2)^2} \ln x \, dx = 0.$$ BH[142](2)+

4. $$\int_0^\infty \frac{1-x^2}{(1+x^2)^2} \ln x \, dx = -\frac{\pi}{2}.$$ BH[142](1)+

5. $$\int_0^1 \frac{x^2 \ln x \, dx}{(1-x^2)(1+x^4)} = -\frac{\pi^2}{16(2+\sqrt{2})}.$$ BH[112](21)

6. $$\int_0^\infty \frac{\ln x \, dx}{(a^2+b^2 x^2)(1+x^2)} = \frac{b\pi}{2a(b^2-a^2)} \ln \frac{a}{b} \quad [ab>0].$$ BH[317](16)+

7. $$\int_0^\infty \frac{\ln x}{x^2+a^2} \cdot \frac{dx}{1+b^2 x^2} = \frac{\pi}{2(1-a^2 b^2)}\left(\frac{1}{a} \ln a + b \ln b\right)$$
$$[a>0,\ b>0].$$ Li[140](12)

8. $$\int_0^\infty \frac{x^2 \ln x \, dx}{(a^2+b^2 x^2)(1+x^2)} = \frac{a\pi}{2b(b^2-a^2)} \ln \frac{b}{a} \quad [ab>0].$$ Li[140](12), BH[317](15)+

4.235

1. $\int_0^\infty \ln x \frac{(1-x) x^{n-2}}{1-x^{2n}} dx = -\frac{\pi^2}{4n^2} \operatorname{tg}^2 \frac{\pi}{2n}$ $[n > 1]$. BH[135](10)

2. $\int_0^\infty \ln x \frac{(1-x^2) x^{m-1}}{1-x^{2n}} dx = -\frac{\pi^2 \sin \frac{m+1}{n} \pi \sin \frac{\pi}{n}}{4n^2 \sin^2 \frac{m\pi}{2n} \sin^2 \left(\frac{m+2}{2n} \pi\right)} \cdot$ Li[135](12)

3. $\int_0^\infty \ln x \frac{(1-x^2) x^{n-2}}{1-x^{2n}} dx = -\frac{\pi^2}{4n^2} \operatorname{tg}^2 \frac{\pi}{n}$ $[n > 2]$. BH[135](11)

4. $\int_0^1 \ln x \frac{x^{m-1} + x^{n-m-1}}{1-x^n} dx = -\frac{\pi^2}{n^2 \sin^2 \left(\frac{m}{n} \pi\right)}$ $[n > m]$. BH[108](15)

4.236

1. $\int_0^1 \left\{ \frac{1+(p-1)\ln x}{1-x} + \frac{x \ln x}{(1-x)^2} \right\} x^{p-1} dx = -1 + \psi'(p)$ $[p > 0]$.

BH[111](6)+, GH[326](13)

2. $\int_0^1 \left[\frac{1}{1-x} + \frac{x \ln x}{(1-x)^2} \right] dx = \frac{\pi^2}{6} - 1.$ GH[326](13a)

Logarithmus und algebraische Funktionen
4.24 **Logarithms and algebraic functions**

4.241

1. $\int_0^1 \frac{x^{2n} \ln x}{\sqrt{1-x^2}} dx = \frac{(2n-1)!!}{(2n)!!} \cdot \frac{\pi}{2} \left(\sum_{k=1}^{2n} \frac{(-1)^{k-1}}{k} - \ln 2 \right).$ BH[118](5)+

2. $\int_0^1 \frac{x^{2n+1} \ln x}{\sqrt{1-x^2}} dx = \frac{(2n)!!}{(2n+1)!!} \left(\ln 2 + \sum_{k=1}^{2n+1} \frac{(-1)^k}{k} \right).$ BH[118](5)+

3. $\int_0^1 x^{2n} \sqrt{1-x^2} \ln x \, dx = \frac{(2n-1)!!}{(2n+2)!!} \cdot \frac{\pi}{2} \left(\sum_{k=1}^{2n} \frac{(-1)^{k-1}}{k} - \frac{1}{2n+2} - \ln 2 \right).$

Li[117](4), GH[324](53a)

4. $\int_0^1 x^{2n+1} \sqrt{1-x^2} \ln x \, dx = \frac{(2n)!!}{(2n+3)!!} \left(\ln 2 + \sum_{k=1}^{2n+1} \frac{(-1)^k}{k} - \frac{1}{2n+3} \right).$

BH[117](5), GH[324](53b)

5. $\int_0^1 \ln x \cdot \sqrt{(1-x^2)^{2n-1}} \, dx = -\frac{(2n-1)!!}{4 \cdot (2n)!!} \pi [\psi(n+1) + C + \ln 4].$ BH[117](3)

6. $\displaystyle\int_0^{\sqrt{1/2}} \frac{\ln x\,dx}{\sqrt{1-x^2}} = -\frac{\pi}{4}\ln 2 - \frac{1}{2}G.$ BH[145](1) 4.241

7. $\displaystyle\int_0^1 \frac{\ln x\,dx}{\sqrt{1-x^2}} = -\frac{\pi}{2}\ln 2.$ F II 610, 652

8. $\displaystyle\int_1^\infty \frac{\ln x\,dx}{x^2\sqrt{x^2-1}} = 1 - \ln 2.$ BH[144](17)

9. $\displaystyle\int_0^1 \sqrt{1-x^2}\,\ln x\,dx = -\frac{\pi}{8} - \frac{\pi}{4}\ln 2.$ BH[117](1), GH[324](53c)

10. $\displaystyle\int_0^1 x\sqrt{1-x^2}\,\ln x\,dx = \frac{1}{3}\ln 2 - \frac{4}{9}.$ BH[117](2)

11. $\displaystyle\int_0^1 \frac{\ln x\,dx}{\sqrt{x(1-x^2)}} = -\frac{\sqrt{2\pi}}{8}\left[\Gamma\left(\frac{1}{4}\right)\right]^2.$ GH[324](54a)

1. $\displaystyle\int_0^\infty \frac{\ln x\,dx}{\sqrt{(a^2+x^2)(x^2+b^2)}} = \frac{1}{2a}K\!\left(\frac{\sqrt{a^2-b^2}}{a}\right)\ln ab \quad [a>b>0].$ BF(800.04) 4.242

2. $\displaystyle\int_0^b \frac{\ln x\,dx}{\sqrt{(a^2+x^2)(b^2-x^2)}} = \frac{1}{2\sqrt{a^2+b^2}}\left[K\!\left(\frac{b}{\sqrt{a^2+b^2}}\right)\ln ab \right.$

$\displaystyle\left. -\frac{\pi}{2}K\!\left(\frac{a}{\sqrt{a^2+b^2}}\right)\right] \quad [a>0,\ b>0].$ BF(800.02)

3. $\displaystyle\int_b^\infty \frac{\ln x\,dx}{\sqrt{(x^2+a^2)(x^2-b^2)}} = \frac{1}{2\sqrt{a^2+b^2}}\left[K\!\left(\frac{a}{\sqrt{a^2+b^2}}\right)\ln ab \right.$

$\displaystyle\left. +\frac{\pi}{2}K\!\left(\frac{b}{\sqrt{a^2+b^2}}\right)\right] \quad [a>0,\ b>0].$ BF(800.06)

4. $\displaystyle\int_0^b \frac{\ln x\,dx}{\sqrt{(a^2-x^2)(b^2-x^2)}} = \frac{1}{2a}\left[K\!\left(\frac{b}{a}\right)\ln ab - \frac{\pi}{2}K\!\left(\frac{\sqrt{a^2-b^2}}{a}\right)\right]$

$[a>b>0].$ BF(800.01)

5. $\displaystyle\int_b^a \frac{\ln x\,dx}{\sqrt{(a^2-x^2)(x^2-b^2)}} = \frac{1}{2a}K\!\left(\frac{\sqrt{a^2-b^2}}{a}\right)\ln ab.$ BF(800.03)

4.242   6. $$\int_a^\infty \frac{\ln x \, dx}{\sqrt{(x^2-a^2)(x^2-b^2)}} = \frac{1}{2a}\left[K\left(\frac{b}{a}\right)\ln ab + \frac{\pi}{2}K\left(\frac{\sqrt{a^2-b^2}}{a}\right)\right]$$

$$[a > b > 0].$$ BF(800.05)

4.243 $$\int_0^1 \frac{x \ln x}{\sqrt{1-x^4}} \, dx = -\frac{\pi}{8}\ln 2.$$ GH[324](56b)

4.244   1. $$\int_0^1 \frac{\ln x \, dx}{\sqrt[3]{x(1-x^2)^2}} = -\frac{1}{8}\left[\Gamma\left(\frac{1}{3}\right)\right]^3.$$ GH[324](54b)

2. $$\int_0^1 \frac{\ln x \, dx}{\sqrt[3]{1-x^3}} = -\frac{\pi}{3\sqrt{3}}\left(\ln 3 + \frac{\pi}{3\sqrt{3}}\right).$$ BH[118](7)

3. $$\int_0^1 \frac{x \ln x \, dx}{\sqrt[3]{(1-x^3)^2}} = \frac{\pi}{3\sqrt{3}}\left(\frac{\pi}{3\sqrt{3}} - \ln 3\right).$$ BH[118](8)

4.245   1. $$\int_0^1 \frac{x^{4n+1} \ln x}{\sqrt{1-x^4}} \, dx = \frac{(2n-1)!!}{(2n)!!} \cdot \frac{\pi}{8}\left(\sum_{k=1}^{2n} \frac{(-1)^{k-1}}{k} - \ln 2\right).$$ GH[324](56a)

2. $$\int_0^1 \frac{x^{4n+3} \ln x}{\sqrt{1-x^4}} \, dx = \frac{(2n)!!}{4 \cdot (2n+1)!!}\left(\ln 2 + \sum_{k=1}^{2n+1} \frac{(-1)^k}{k}\right).$$ GH[324](56c)

4.246 $$\int_0^1 (1-x^2)^{n-\frac{1}{2}} \ln x \, dx = -\frac{(2n-1)!!}{(2n)!!} \cdot \frac{\pi}{4}\left[2\ln 2 + \sum_{k=1}^n \frac{1}{k}\right].$$ GH[324](55)

4.247   1. $$\int_0^1 \frac{\ln x}{\sqrt[n]{1-x^{2n}}} \, dx = -\frac{\pi B\left(\frac{1}{2n}, \frac{1}{2n}\right)}{8n^2 \sin\frac{\pi}{2n}} \quad [n > 1].$$ GH[324](54c)+

2. $$\int_0^1 \frac{\ln x \, dx}{\sqrt[n]{x^{n-1}(1-x^2)}} = -\frac{\pi B\left(\frac{1}{2n}, \frac{1}{2n}\right)}{8 \sin\frac{\pi}{2n}}.$$ GH[324](54)

4.25 Logarithmus und Potenzen
Logarithms and powers

4.251   1. $$\int_0^\infty \frac{x^{\mu-1} \ln x}{\beta + x} \, dx = \frac{\pi \beta^{\mu-1}}{\sin \mu\pi}(\ln \beta - \pi \operatorname{ctg} \mu\pi)$$

$$[|\arg \beta| < \pi, \ 0 < \operatorname{Re} \mu < 1].$$ BH[135](1)

2. $$\int_0^\infty \frac{x^{\mu-1} \ln x}{a - x} dx = \pi a^{\mu-1} \left(\text{ctg } \mu\pi \ln a - \frac{\pi}{\sin^2 \mu\pi} \right)$$ 4.251

$$[a > 0, \ 0 < \text{Re } \mu < 1].$$ Ba IV 314(5)

3. $$\int_0^1 \frac{x^{\mu-1} \ln x}{x + 1} dx = \frac{1}{2} \beta'(\mu) \quad [\text{Re } \mu > 0].$$ GH[324](6), Ba IV 314(3)

4. $$\int_0^1 \frac{x^{\mu-1} \ln x}{1 - x} dx = -\psi'(\mu) = -\zeta(2, \mu) \quad [\text{Re } \mu > 0].$$ BH[108](8)

5. $$\int_0^1 \ln x \frac{x^{2n} dx}{1 + x} = -\frac{\pi^2}{12} + \sum_{k=1}^{2n} \frac{(-1)^{k-1}}{k^2}.$$ BH[108](4)

6. $$\int_0^1 \ln x \frac{x^{2n-1} dx}{1 + x} = \frac{\pi^2}{12} + \sum_{k=1}^{2n-1} \frac{(-1)^k}{k^2}.$$ BH[108](5)

1. $$\int_0^\infty \frac{x^{\mu-1} \ln x}{(x + \beta)(x + \gamma)} dx = \frac{\pi}{(\gamma - \beta) \sin \mu\pi} [\beta^{\mu-1} \ln \beta - \gamma^{\mu-1} \ln \gamma -$$ 4.252

$$- \pi \text{ ctg } \mu\pi (\beta^{\mu-1} - \gamma^{\mu-1})]$$

$$[|\arg \beta| < \pi, \ |\arg \gamma| < \pi, \ 0 < \text{Re } \mu < 2, \ \mu \neq 1].$$ BH[140](9)+, Ba IV314(6)

2. $$\int_0^\infty \frac{x^{\mu-1} \ln x \, dx}{(x + \beta)(x - 1)} = \frac{\pi}{(\beta + 1) \sin^2 \mu\pi} [\pi - \beta^{\mu-1} (\sin \mu\pi \ln \beta - \pi \cos \mu\pi)]$$

$$[|\arg \beta| < \pi, \ 0 < \text{Re } \mu < 2, \ \mu \neq 1].$$ BH[140](11)

3. $$\int_0^\infty \frac{x^{p-1} \ln x}{1 - x^2} dx = -\frac{\pi^2}{4} \text{cosec}^2 \frac{p\pi}{2} \quad [0 < p < 2]$$

(vgl. auch 4.254 2.). (see also 4.254 2.).

4. $$\int_0^\infty \frac{x^{\mu-1} \ln x}{(x + a)^2} dx = \frac{(1 - \mu) a^{\mu-2} \pi}{\sin \mu\pi} \left(\ln a - \pi \text{ ctg } \mu\pi + \frac{1}{\mu - 1} \right)$$

$$[a > 0, \ 0 < \text{Re } \mu < 2 \ (\mu \neq 1)].$$ GH[324](13b)

1. $$\int_0^1 x^{\mu-1} (1 - x^r)^{\nu-1} \ln x \, dx = \frac{1}{r^2} B\left(\frac{\mu}{r}, \nu\right) \left[\psi\left(\frac{\mu}{r}\right) - \psi\left(\frac{\mu}{r} + \nu\right) \right]$$ 4.253

$$[\text{Re } \mu > 0, \ \text{Re } \nu > 0, \ r > 0].$$ GH[324](3b)+,BH[107](5)+

4.253

2. $\displaystyle\int_0^1 \frac{x^{p-1}}{(1-x)^{p+1}} \ln x \, dx = -\frac{\pi}{p} \operatorname{cosec} p\pi \quad [0 < p < 1].$ BH[319](10)+

3. $\displaystyle\int_u^\infty \frac{(x-u)^{\mu-1} \ln x \, dx}{x^\lambda} = u^{\mu-\lambda} \, \mathrm{B}(\lambda - \mu, \mu) \, [\ln u + \psi(\lambda) - \psi(\lambda - \mu)]$

$[0 < \operatorname{Re} \mu < \operatorname{Re} \lambda].$ Ba V 203(18

4. $\displaystyle\int_0^1 \ln x \left(\frac{x}{a^2 + x^2}\right)^p \frac{dx}{x} = \frac{\ln a}{2a^p} \, \mathrm{B}\left(\frac{p}{2}, \frac{p}{2}\right)$

$[a > 0, \, p > 0].$ BH[140](6)

5. $\displaystyle\int_1^\infty (x-1)^{p-1} \ln x \, dx = \frac{\pi}{p} \operatorname{cosec} \pi p \quad [-1 < p < 0].$ BH[289](12)+

6. $\displaystyle\int_0^\infty \ln x \, \frac{dx}{(a+x)^{\mu+1}} = \frac{1}{\mu a^\mu} (\ln a - \mathbf{C} - \psi(\mu))$

[Re $\mu > 0$, $a \neq 0$, $\mu - a$ ist keine natürliche Zahl].   NI 68(7) | [Re $\mu > 0$, $a \neq 0$, $\mu - a$ is not a positive integer].   NI 68(7)

7. $\displaystyle\int_0^\infty \ln x \, \frac{dx}{(a+x)^{n+\frac{1}{2}}} = \frac{2}{(2n-1)a^{n-\frac{1}{2}}} \left(\ln a + 2\ln 2 - \sum_{k=1}^{n-2} \frac{1}{k} - 2\sum_{k=n-1}^{2n-3} \frac{1}{k}\right)$

$[a > 0].$ BH[142](5)

4.254

1. $\displaystyle\int_0^1 \frac{x^{p-1} \ln x}{1 - x^q} \, dx = -\frac{1}{q^2} \psi'\left(\frac{p}{q}\right) \quad [p > 0, q > 0].$ GH[324](5)

2. $\displaystyle\int_0^\infty \frac{x^{p-1} \ln x}{1 - x^q} \, dx = -\frac{\pi^2}{q^2 \sin^2 \frac{p\pi}{q}} \quad [0 < p < q].$ BH[135](8)

3. $\displaystyle\int_0^\infty \frac{\ln x}{x^q - 1} \frac{dx}{x^p} = \frac{\pi^2}{q^2 \sin^2 \frac{p-1}{q}\pi} \quad [p < 1, \, p + q > 1].$ BH[140](2)

4. $\displaystyle\int_0^1 \frac{x^{p-1} \ln x}{1 + x^q} \, dx = \frac{1}{2q^2} \beta\left(\frac{p}{q}\right) \quad [p > 0, q > 0].$ GH[324](7)

5. $\displaystyle\int_0^\infty \frac{x^{p-1} \ln x}{1 + x^q} \, dx = -\frac{\pi^2}{q^2} \frac{\cos \frac{p\pi}{q}}{\sin^2 \frac{p\pi}{q}} \quad [0 < p < q].$ BH[135](7)

6. $$\int_0^1 \frac{x^{q-1} \ln x}{1 - x^{2q}} dx = -\frac{\pi^2}{8q^2} \quad [q > 0].$$ BH[108](12) 4.254

1. $$\int_0^1 \ln x \, \frac{(1 - x^2) x^{p-2}}{1 + x^{2p}} dx = -\left(\frac{\pi}{2p}\right)^2 \frac{\sin \frac{\pi}{2p}}{\cos^3 \frac{\pi}{2p}} \quad [p > 1].$$ BH[108](13) 4.255

2. $$\int_0^1 \ln x \, \frac{(1 + x^2) x^{p-2}}{1 - x^{2p}} dx = -\left(\frac{\pi}{2p}\right)^2 \sec^2 \frac{\pi}{2p} \quad [p > 1].$$ BH[108](14)

3. $$\int_0^\infty \ln x \, \frac{1 - x^p}{1 - x^2} dx = \frac{\pi^2}{4} \operatorname{tg}^2 \frac{p\pi}{2} \quad [p < 1].$$ BH[140](3)

$$\int_0^1 \ln \frac{1}{x} \, \frac{x^{\mu-1} dx}{\sqrt[n]{(1 - x^n)^{n-m}}} = \frac{1}{n^2} B\left(\frac{\mu}{n}, \frac{m}{n}\right) \left[\psi\left(\frac{\mu + m}{n}\right) - \psi\left(\frac{\mu}{n}\right)\right]$$ 4.256

$$[\operatorname{Re} \mu > 0].$$ Li[118](12)

1. $$\int_0^\infty \frac{x^\nu \ln \frac{x}{\beta} \, dx}{(x + \beta)(x + \gamma)} = \frac{\pi \left[\gamma^\nu \ln \frac{\gamma}{\beta} + \pi (\beta^\nu - \gamma^\nu) \operatorname{ctg} \nu\pi\right]}{\sin \nu\pi (\gamma - \beta)}$$ 4.257

$$[|\arg \beta| < \pi, \, |\arg \gamma| < \pi, \, |\operatorname{Re} \nu| < 1].$$ Ba V 219(30)

2. $$\int_0^\infty \ln \frac{x}{q} \left(\frac{x^p}{q^{2p} + x^{2p}}\right) \frac{dx}{x} = 0 \quad [q > 0].$$ BH[140](4)+

3. $$\int_0^\infty \ln \frac{x}{q} \left(\frac{x^p}{q^{2p} + x^{2p}}\right)^r \frac{dx}{q^2 + x^2} = 0 \quad [q > 0].$$ BH[140](4)+

4. $$\int_0^\infty \ln x \ln \frac{x}{a} \, \frac{dx}{(x - 1)(x - a)} = \frac{[4\pi^2 + (\ln a)^2] \ln a}{6(a - 1)} \quad [a > 0],$$

$$[a = 1 \text{ cf. } 4.261 \, 5.].$$ BH[141](5)

5. $$\int_0^\infty \ln x \ln \frac{x}{a} \, \frac{x^p \, dx}{(x - 1)(x - a)} = \frac{\pi^2 [(a^p + 1) \ln a - 2\pi (a^p - 1) \operatorname{ctg} p\pi]}{(a - 1) \sin^2 p\pi}$$

$$[p^2 < 1, \, a > 0].$$ BH[141](6)

Potenzen des Logarithmus und Potenzen
Powers of the logarithm and other powers

4.26—4.27

1. $$\int_0^1 (\ln x)^2 \frac{dx}{1 + 2x \cos t + x^2} = \frac{t(\pi^2 - t^2)}{6 \sin t}.$$ BH[113](7) 4.261

4.261

2. $$\int_0^1 \frac{(\ln x)^2\, dx}{x^2 - x + 1} = \frac{1}{2}\int_0^\infty \frac{(\ln x)^2\, dx}{x^2 - x + 1} = \frac{10\pi^3}{81\sqrt{3}}.$$ GH[324](16c)

3. $$\int_0^1 \frac{(\ln x)^2\, dx}{x^2 + x + 1} = \frac{1}{2}\int_0^\infty \frac{(\ln x)^2\, dx}{x^2 + x + 1} = \frac{8\pi^3}{81\sqrt{3}}.$$ GH[324](16b)

4. $$\int_0^\infty (\ln x)^2 \frac{dx}{(x-1)(x+a)} = \frac{[\pi^2 + (\ln a)^2]\ln a}{3(1+a)}.$$ BH[141](1)

5. $$\int_0^\infty (\ln x)^2 \frac{dx}{(1-x)^2} = \frac{2}{3}\pi^2.$$ BH[139](4)

6. $$\int_0^1 (\ln x)^2 \frac{dx}{1+x^2} = \frac{\pi^3}{16}.$$ BH[109](3)

7. $$\int_0^1 (\ln x)^2 \frac{1+x^2}{1+x^4}\, dx = \frac{1}{2}\int_0^\infty (\ln x)^2 \frac{1+x^2}{1+x^4}\, dx = \frac{3\sqrt{2}}{64}\pi^3.$$ BH[109](5), BH[135](13)

8. $$\int_0^1 (\ln x)^2 \frac{1-x}{1-x^6}\, dx = \frac{\sqrt{3}}{27}\pi^3.$$ BH[109](6)

9. $$\int_0^1 (\ln x)^2 \frac{dx}{\sqrt{1-x^2}} = \frac{\pi}{2}\left[(\ln 2)^2 + \frac{\pi^2}{12}\right].$$ BH[118](13)

10. $$\int_0^\infty (\ln x)^2 \frac{x^{\mu-1}}{1+x}\, dx = \frac{\pi^3(2 - \sin^2\mu\pi)}{\sin^3 \mu\pi} \quad [0 < \operatorname{Re}\mu < 1].$$ Ba IV 315(10)

11. $$\int_0^1 (\ln x)^2 \frac{x^n\, dx}{1+x} = 2\sum_{k=n}^\infty \frac{(-1)^{n+k}}{(k+1)^3}.$$ BH[109](1)

12. $$\int_0^1 (\ln x)^2 \frac{x^n\, dx}{1-x} = 2\sum_{k=n}^\infty \frac{1}{(k+1)^3}.$$ BH[109](2)

13. $$\int_0^1 (\ln x)^2 \frac{x^{2n}\, dx}{1-x^2} = 2\sum_{k=n}^\infty \frac{1}{(2k+1)^3}.$$ BH[109](4)

14. $$\int_0^\infty (\ln x)^2 \frac{x^{p-1}\, dx}{x^2 + 2x\cos t + 1} = \frac{\pi \sin(1-p)t}{\sin t \sin p\pi} \times$$
$$\times \{\pi^2 - t^2 + 2\pi \operatorname{ctg} p\pi \,[\pi \operatorname{ctg} p\pi + t \operatorname{ctg}(1-p)t]\}$$
$$[0 < t < \pi,\ 0 < p < 2\ (p \neq 1)].$$ GH[324](17)

15. $\int_0^1 (\ln x)^2 \dfrac{x^{2n}\,dx}{\sqrt{1-x^2}} = \dfrac{(2n-1)!!}{2\cdot(2n)!!}\,\pi\left\{\dfrac{\pi^2}{12} + \sum_{k=1}^{2n}\dfrac{(-1)^k}{k^2} + \left[\sum_{k=1}^{2n}\dfrac{(-1)^k}{k} + \ln 2\right]^2\right\}.$ 4.261

GH[324](60a)

16. $\int_0^1 (\ln x)^2 \dfrac{x^{2n+1}\,dx}{\sqrt{1-x^2}} = \dfrac{(2n)!!}{(2n+1)!!}\left\{-\dfrac{\pi^2}{12} - \sum_{k=1}^{2n+1}\dfrac{(-1)^k}{k^2} + \right.$

$\left. + \left[\sum_{k=1}^{2n+1}\dfrac{(-1)^k}{k} + \ln 2\right]^2\right\}.$

GH[324](60b)

17. $\int_0^1 (\ln x)^2 x^{\mu-1}(1-x)^{\nu-1}\,dx = B(\mu,\nu)\left\{[\psi(\mu) - \psi(\nu+\mu)]^2 + \right.$

$\left. + \psi'(\mu) - \psi'(\mu+\nu)\right\}$ [Re $\mu > 0$, Re $\nu > 0$]. Ba IV 315(11)

18. $\int_0^1 (\ln x)^2 \dfrac{1-x^{n+1}}{(1-x)^2}\,dx = 2(n+1)\zeta(3) - 2\sum_{k=1}^n \dfrac{n-k+1}{k^3}.$ Li[111](8)

19. $\int_0^1 (\ln x)^2 \dfrac{1+(-1)^n x^{n+1}}{(1+x)^2}\,dx = \dfrac{3}{2}(n+1)\zeta(3) - 2\sum_{k=1}^n (-1)^{k-1}\dfrac{n-k+1}{k^3}.$ Li[111](7)

20. $\int_0^1 (\ln x)^2 \dfrac{1-x^{2n+2}}{(1-x)^2}\,dx = 2(n+1)\sum_{k=0}^\infty \dfrac{1}{(2k+1)^3} - 2\sum_{k=1}^n \dfrac{n-k+1}{(2k-1)^3}.$ Li[111](9)

21. $\int_0^1 (\ln x)^2 x^{p-1}(1-x^r)^{q-1}\,dx = \dfrac{1}{r^3} B\left(\dfrac{p}{r}, q\right)\left\{\psi'\left(\dfrac{p}{r}\right) - \right.$

$\left. - \psi'\left(\dfrac{p}{r}+q\right) + \left[\psi\left(\dfrac{p}{r}\right) - \psi\left(\dfrac{p}{r}+q\right)\right]^2\right\}$
[$p > 0$, $q > 0$, $r > 0$]. GH[324](8a)

1. $\int_0^1 (\ln x)^3 \dfrac{dx}{1+x} = -\dfrac{7}{120}\pi^4.$ BH[109](9) 4.262

2. $\int_0^1 (\ln x)^3 \dfrac{dx}{1-x} = -\dfrac{\pi^4}{15}.$ BH[109](11)

3. $\int_0^\infty (\ln x)^3 \dfrac{dx}{(x+a)(x-1)} = \dfrac{[\pi^2 + (\ln a)^2]^2}{4(a+1)}$ [$a > 0$]. BH[141](2)

4. $\int_0^1 (\ln x)^3 \dfrac{x^n\,dx}{1+x} = (-1)^{n+1}\left[\dfrac{7\pi^4}{120} - 6\sum_{k=0}^{n-1} \dfrac{(-1)^k}{(k+1)^4}\right].$ BH[109](10)

4.262

5. $$\int_0^1 (\ln x)^3 \frac{x^n\,dx}{1-x} = -\frac{\pi^4}{15} + 6\sum_{k=0}^{n-1} \frac{1}{(k+1)^4}.$$ BH[109](12)

6. $$\int_0^1 (\ln x)^3 \frac{x^{2n}\,dx}{1-x^2} = -\frac{\pi^4}{16} + 6\sum_{k=0}^{n-1} \frac{1}{(2k+1)^4}.$$ BH[109](14)

7. $$\int_0^1 (\ln x)^3 \frac{1-x^{n+1}}{(1-x)^2}\,dx = -\frac{(n+1)\pi^4}{15} + 6\sum_{k=1}^{n} \frac{n-k+1}{k^4}.$$ BH[111](11)

8. $$\int_0^1 (\ln x)^3 \frac{1+(-1)^n x^{n+1}}{(1+x)^2}\,dx = -\frac{7(n+1)\pi^4}{120} + 6\sum_{k=1}^{n} (-1)^{k-1}\frac{n-k+1}{k^4}.$$ BH[111](10)

9. $$\int_0^1 (\ln x)^3 \frac{1-x^{2n+2}}{(1-x^2)^2}\,dx = -\frac{(n+1)\pi^4}{16} + 6\sum_{k=1}^{n} \frac{n-k+1}{(2k-1)^4}.$$ BH[111](12)

4.263

1. $$\int_0^\infty (\ln x)^4 \frac{dx}{(x-1)(x+a)} = \frac{\ln a[\pi^2 + (\ln a)^2]^2 [7\pi^2 + 3(\ln a)^2]}{15(1+a)} \quad [a > 0].$$ BH[141](3)

2. $$\int_0^1 (\ln x)^4 \frac{dx}{1+x^2} = \frac{5\pi^5}{64}.$$ BH[109](17)

3. $$\int_0^1 (\ln x)^4 \frac{dx}{1+2x\cos t + x^2} = \frac{t(\pi^2 - t^2)(7\pi^2 - 3t^2)}{30\sin t} \quad [|t| < \pi].$$ BH[113](8)

4.264

1. $$\int_0^1 (\ln x)^5 \frac{dx}{1+x} = -\frac{31\pi^6}{252}.$$ BH[109](20)

2. $$\int_0^1 (\ln x)^5 \frac{dx}{1-x} = -\frac{8\pi^6}{63}.$$ BH[109](21)

3. $$\int_0^\infty (\ln x)^5 \frac{dx}{(x-1)(x+a)} = \frac{[\pi^2 + (\ln a)^2]^2 [3\pi^2 + (\ln a)^2]}{6(1+a)} \quad [a > 0].$$ BH[141](4)

4.265

$$\int_0^1 (\ln x)^6 \frac{dx}{1+x^2} = \frac{61\pi^7}{256}.$$ BH[109](25)

4.266

1. $$\int_0^1 (\ln x)^7 \frac{dx}{1+x} = -\frac{127\pi^8}{240}.$$ BH[109](28)

2. $$\int_0^1 (\ln x)^7 \frac{dx}{1-x} = -\frac{8\pi^8}{15}.$$ BH[109](29)   4.266

1. $$\int_0^1 \frac{1-x}{1+x} \frac{dx}{\ln x} = \ln \frac{2}{\pi}.$$ BH[127](3)   4.267

2. $$\int_0^1 \frac{(1-x)^2}{1+x^2} \frac{dx}{\ln x} = \ln \frac{\pi}{4}.$$ BH[128](2)

3. $$\int_0^1 \frac{(1-x)^2}{1+2x\cos\frac{mx}{n}+x^2} \cdot \frac{dx}{\ln x} = \frac{1}{\sin\frac{m\pi}{n}} \sum_{k=1}^{n-1} (-1)^k \sin\frac{km\pi}{n} \times$$

$$\times \ln \frac{\left\{\Gamma\left(\frac{n+k+1}{2n}\right)\right\}^2 \Gamma\left(\frac{k+2}{2n}\right)\Gamma\left(\frac{k}{2n}\right)}{\left\{\Gamma\left(\frac{k+1}{2n}\right)\right\}^2 \Gamma\left(\frac{n+k}{2n}\right)\Gamma\left(\frac{n+k+2}{2n}\right)} \quad \left[m+n\ \begin{matrix}\text{ungeradzahlig}\\ \text{odd}\end{matrix}\right];$$

$$= \frac{1}{\sin\frac{m\pi}{n}} \sum_{k=1}^{\frac{1}{2}(n-1)} (-1)^k \sin\frac{km\pi}{n} \times$$

$$\times \ln \frac{\left\{\Gamma\left(\frac{n-k+1}{n}\right)\right\}^2 \Gamma\left(\frac{k+2}{n}\right)\Gamma\left(\frac{k}{n}\right)}{\left\{\Gamma\left(\frac{k+1}{n}\right)\right\}^2 \Gamma\left(\frac{n-k}{n}\right)\Gamma\left(\frac{n-k+2}{n}\right)} \quad \left[m+n\ \begin{matrix}\text{geradzahlig}\\ \text{even}\end{matrix}\right];$$

$$[m<n].$$ BH[130](3)

4. $$\int_0^1 \frac{1-x}{1+x} \cdot \frac{1}{1+x^2} \cdot \frac{dx}{\ln x} = -\frac{\ln 2}{2}.$$ BH[130](16)

5. $$\int_0^1 \frac{1-x}{1+x} \cdot \frac{x^2}{1+x^2} \cdot \frac{dx}{\ln x} = \ln\frac{2\sqrt{2}}{\pi}.$$ BH[130](17)

6. $$\int_0^1 (1-x)^p \frac{dx}{\ln x} = \sum_{k=1}^\infty (-1)^k \binom{p}{k} \ln(1+k) \quad [p\geqslant 1].$$ BH[123](2)

7. $$\int_0^1 \left(\frac{1-x^p}{1-x} - p\right) \frac{dx}{\ln x} = \ln \Gamma(p+1) \quad [p>-1].$$ GH[326](10)

8. $$\int_0^1 \frac{x^{p-1}-x^{q-1}}{\ln x} dx = \ln \frac{p}{q} \quad [p>0,\ q>0].$$ F II 657

4.267

9. $\int_0^1 \frac{x^{p-1} - x^{q-1}}{\ln x} \cdot \frac{dx}{1+x} = \ln \frac{\Gamma\left(\frac{q}{2}\right) \Gamma\left(\frac{p+1}{2}\right)}{\Gamma\left(\frac{p}{2}\right) \Gamma\left(\frac{q+1}{2}\right)}$ $[p > 0,\ q > 0]$. F II 820

10. $\int_0^1 \frac{x^{p-1} - x^{-p}}{(1+x) \ln x} dx = \frac{1}{2} \int_0^\infty \frac{x^{p-1} - x^{-p}}{(1+x) \ln x} dx = \ln \left(\text{tg}\, \frac{p\pi}{2} \right)$ $[0 < p < 1]$. F II 820

11. $\int_0^1 (x^p - x^q) x^{r-1} \frac{dx}{\ln x} = \ln \frac{p+r}{r+q}$ $[r > 0,\ p > 0,\ q > 0]$. Li[123](5)

12. $\int_0^1 \frac{x^p - x^q}{(1-ax)^n} \frac{dx}{x \ln x} = \ln \frac{p}{q} + \sum_{k=1}^\infty \binom{n+k-1}{k} a^k \ln \frac{p+k}{q+k}$

$[p > 0,\ q > 0,\ a^2 < 1]$. BH [130](15)

13. $\int_0^1 (x^p - 1)(x^q - 1) \frac{dx}{\ln x} = \ln \frac{p+q+1}{(p+1)(q+1)}$ $[p > -1,\ q > -1,\ p+q > -1]$.

GH [324](19b)

14. $\int_0^1 \frac{x^p - x^q}{1+x} \cdot \frac{1 + x^{2n+1}}{x \ln x} dx = \ln \frac{\Gamma\left(\frac{p}{2}+n+1\right) \Gamma\left(\frac{q+1}{2}+n\right) \Gamma\left(\frac{p+1}{2}\right) \Gamma\left(\frac{q}{2}\right)}{\Gamma\left(\frac{q}{2}+n+1\right) \Gamma\left(\frac{p+1}{2}+n\right) \Gamma\left(\frac{q+1}{2}\right) \Gamma\left(\frac{p}{2}\right)}$

$[p > 0,\ q > 0]$. BH [127](7)

15. $\int_0^1 \frac{x^p - x^q}{1-x} \cdot \frac{1 - x^r}{\ln x} dx = \ln \frac{\Gamma(q+1) \Gamma(p+r+1)}{\Gamma(p+1) \Gamma(q+r+1)}$

$[p > -1,\ q > -1,\ p+r > -1,\ q+r > -1]$. GH [324](23)

16. $\int_0^1 \frac{x^{p-1} - x^{q-1}}{(1+x^r) \ln x} dx = \ln \frac{\Gamma\left(\frac{p+r}{2r}\right) \Gamma\left(\frac{q}{2r}\right)}{\Gamma\left(\frac{q+r}{2r}\right) \Gamma\left(\frac{p}{2r}\right)}$ $[p > 0,\ q > 0,\ r > 0]$. GH[324](21)

17. $\int_0^1 \frac{1 - x^{2p-2q}}{1 + x^{2p}} \frac{x^{q-1} dx}{\ln x} = \ln \text{tg}\, \frac{q\pi}{4p}$ $[0 < q < p]$ (cf. 3.524 27.). BH [128](6)

18. $\int_0^\infty \frac{x^{p-1} - x^{q-1}}{(1+x^r) \ln x} dx = \ln \left(\text{tg}\, \frac{p\pi}{2r} \text{ctg}\, \frac{q\pi}{2r} \right)$ $[0 < p < r,\ 0 < q < r]$.

GH[324](22), BH[143](2)

19. $\int_0^\infty \frac{x^{p-1} - x^{q-1}}{(1-x^r) \ln x} dx = \ln \left(\frac{\sin \frac{p\pi}{r}}{\sin \frac{q\pi}{r}} \right)$ $[0 < p < r,\ 0 < q < r]$. BH [143](4)

20. $\int_0^1 \frac{x^{p-1} - x^{q-1}}{1 - x^{2n}} \cdot \frac{1 - x^2}{\ln x} dx = \ln \frac{\Gamma\left(\frac{p+2}{2n}\right) \Gamma\left(\frac{q}{2n}\right)}{\Gamma\left(\frac{q+2}{2n}\right) \Gamma\left(\frac{p}{2n}\right)}$ $[p > 0,\ q > 0]$. BH[128](11)

21. $$\int_0^1 \frac{x^{p-1} - x^{q-1}}{1 + x^{2(2n+1)}} \cdot \frac{1+x^2}{\ln x} dx =$$

$$= \ln \frac{\Gamma\left(\dfrac{p+4n+4}{4(2n+1)}\right) \Gamma\left(\dfrac{q+2}{4(2n+1)}\right) \Gamma\left(\dfrac{p+4n+2}{4(2n+1)}\right) \Gamma\left(\dfrac{q}{4(2n+1)}\right)}{\Gamma\left(\dfrac{q+4n+4}{4(2n+1)}\right) \Gamma\left(\dfrac{p+2}{4(2n+1)}\right) \Gamma\left(\dfrac{q+4n+2}{4(2n+1)}\right) \Gamma\left(\dfrac{p}{4(2n+1)}\right)}$$

$[p > 0, \ q > 0]$.   BH [128](7)

22. $$\int_0^\infty \frac{x^{p-1} - x^{q-1}}{1 + x^{2(2n+1)}} \cdot \frac{1+x^2}{\ln x} dx =$$

$$= \ln \left\{ \operatorname{tg} \frac{p\pi}{4(2n+1)} \cdot \operatorname{tg} \frac{(p+2)\pi}{4(2n+1)} \cdot \operatorname{ctg} \frac{q\pi}{4(2n+1)} \cdot \operatorname{ctg} \frac{(q+2)\pi}{4(2n+1)} \right\}$$

$[0 < p < 4n, \ 0 < q < 4n]$.   BH [143](5)

23. $$\int_0^\infty \frac{x^{p-1} - x^{q-1}}{1 - x^{2n}} \cdot \frac{1-x^2}{\ln x} dx = \ln \frac{\sin \dfrac{p\pi}{2n} \cdot \sin \dfrac{(q+2)\pi}{2n}}{\sin \dfrac{q\pi}{2n} \cdot \sin \dfrac{(p+2)\pi}{2n}}$$

$[0 < p < 2n, \ 0 < q < 2n]$.   BH [143](6)

24. $$\int_0^1 (1 - x^p)(1 - x^q) \frac{x^{r-1} dx}{\ln x} = \ln \frac{(p+q+r)r}{(p+r)(q+r)}$$

$[p > 0, \ q > 0, \ r > 0]$.   BH [123](8)

25. $$\int_0^1 (1 - x^p)(1 - x^q) \frac{x^{r-1} dx}{(1-x)\ln x} = \ln \frac{\Gamma(p+r)\Gamma(q+r)}{\Gamma(p+q+r)\Gamma(r)}$$

$[r > 0, \ r + p > 0, \ r + q > 0, \ r + p + q > 0]$.   F II 820+

26. $$\int_0^1 (1 - x^p)(1 - x^q)(1 - x^r) \frac{dx}{\ln x} = \ln \frac{(p+q+1)(q+r+1)(r+p+1)}{(p+q+r+1)(p+1)(q+1)(r+1)}$$

$[p > -1, \ q > -1, \ r > -1, \ p+q > -1, \ p+r > -1, \ q+r > -1,$
$p+q+r > -1]$.   GH[324](19c)

27. $$\int_0^1 (1 - x^p)(1 - x^q)(1 - x^r) \frac{dx}{(1-x)\ln x} =$$

$$= \ln \frac{\Gamma(p+1)\Gamma(q+1)\Gamma(r+1)\Gamma(p+q+r+1)}{\Gamma(p+q+1)\Gamma(p+r+1)\Gamma(q+r+1)}$$

$[p > -1, \ q > -1, \ r > -1, \ p+q > -1, \ p+r > -1, \ q+r > -1,$
$p+q+r > -1]$.   F II 820

4.267

28. $$\int_0^1 (1-x^p)(1-x^q)(1-x^r)\frac{x^{s-1}\,dx}{\ln x} = \ln\frac{(p+q+s)(p+r+s)(q+r+s)\,s}{(p+s)(q+s)(r+s)(p+q+r+s)}$$
$$[p>0,\, q>0,\, r>0,\, s>0].\qquad \text{BH[123](10)}$$

29. $$\int_0^1 (1-x^p)(1-x^q)\frac{x^{s-1}\,dx}{(1-x^r)\ln x} = \ln\frac{\Gamma\!\left(\frac{p+s}{r}\right)\Gamma\!\left(\frac{q+s}{r}\right)}{\Gamma\!\left(\frac{s}{r}\right)\Gamma\!\left(\frac{p+q+s}{r}\right)}$$
$$[p>0,\, q>0,\, r>0,\, s>0].\qquad \text{GH[324](23a)}$$

30. $$\int_0^\infty (1-x^p)(1-x^q)\frac{x^{s-1}\,dx}{(1-x^{p+q+2s})\ln s} = 2\int_0^1 (1-x^p)(1-x^q)\frac{x^{s-1}\,dx}{(1-x^{p+q+2s})\ln x} =$$
$$= 2\ln\left\{\sin\frac{s\pi}{p+q+2s}\operatorname{cosec}\frac{(p+s)\pi}{p+q+2s}\right\}\,[s>0,\, s+p>0,\, s+p+q>0].\qquad \text{GH[324](23b)+}$$

31. $$\int_0^1 (1-x^p)(1-x^q)(1-x^r)\frac{x^{s-1}\,dx}{(1-x)\ln x} =$$
$$= \ln\frac{\Gamma(p+s)\,\Gamma(q+s)\,\Gamma(r+s)\,\Gamma(p+q+r+s)}{\Gamma(p+q+s)\,\Gamma(p+r+s)\,\Gamma(q+r+s)\,\Gamma(s)}$$
$$[p>0,\, q>0\ \ r>0,\, s>0]^*.\qquad \text{BH[127](11)}$$

32. $$\int_0^1 (1-x^p)(1-x^q)(1-x^r)\frac{x^{s-1}\,dx}{(1-x^t)\ln x} =$$
$$= \ln\frac{\Gamma\!\left(\frac{p+s}{t}\right)\Gamma\!\left(\frac{q+s}{t}\right)\Gamma\!\left(\frac{r+s}{t}\right)\Gamma\!\left(\frac{p+q+r+s}{t}\right)}{\Gamma\!\left(\frac{p+q+s}{t}\right)\Gamma\!\left(\frac{q+r+s}{t}\right)\Gamma\!\left(\frac{p+r+s}{t}\right)\Gamma\!\left(\frac{s}{t}\right)}$$
$$[p>0,\, q>0,\, r>0,\, s>0,\, t>0]^*.\qquad \text{GH[324](23b)}$$

33. $$\int_0^1 \left\{\frac{x^p - x^{p+q}}{1-x} - q\right\}\frac{dx}{\ln x} = \ln\frac{\Gamma(p+q+1)}{\Gamma(p+1)}\quad [p>-1,\, p+q>-1].\qquad \text{BH[127](19)}$$

34. $$\int_0^1 \left\{\frac{x^\mu - x}{x-1} - x(\mu-1)\right\}\frac{dx}{x\ln x} = \ln\Gamma(\mu)\quad [\operatorname{Re}\mu>0].\qquad \text{WW 261, BH[127](18)}$$

35. $$\int_0^1 \left\{1 - x - \frac{(1-x^p)(1-x^q)}{1-x}\right\}\frac{dx}{x\ln x} = -\ln\{B(p,q)\}\quad [p>0,\, q>0].$$
$$\text{BH[130](18)}$$

---

* Diese Einschränkungen können etwas abgeschwächt werden, wenn man z. B. in 4.267 31. und 32. voraussetzt:
$s>0,\ p+s>0,\ q+s>0,\ r+s>0,$
$p+q+s>0,\qquad p+r+s>0,$
$q+r+s>0,\qquad p+q+r+s>0.$

* These restrictions may be weakened if, for instance in 4.267 31. and 32. one writes $s>0,\ p+s>0,\ q+s>0,\ r+s>0,$ $p+q+s>0, p+r+s>0, q+r+s>0,$ $p+q+r+s>0.$

36. $\int_0^1 \left\{ \frac{x^{p-1}}{1-x} - \frac{x^{pq-1}}{1-x^q} - \frac{1}{x(1-x)} + \frac{1}{x(1-x^q)} \right\} \frac{dx}{\ln x} = q \ln p \quad [p > 0].$ 4.267

BH[130](20)

37. $\int_0^1 \left\{ \frac{x^{q-1}}{1-x} - \frac{x^{pq-1}}{1-x^p} - \frac{p-1}{1-x^p} x^{p-1} - \frac{p-1}{2} x^{p-1} \right\} \frac{dx}{\ln x} =$

$= \frac{1-p}{2} \ln (2\pi) + \left(pq - \frac{1}{2} \right) \ln p \quad [p > 0, q > 0].$ BH[130](22)

38. $\int_0^1 \frac{(1-x^p)(1-x^q) - (1-x)^2}{x(1-x) \ln x} dx = \ln B(p, q) \quad [p > 0, q > 0].$ GH[324](24)

39. $\int_0^1 (x^p - 1)^n \frac{dx}{\ln x} = \sum_{k=0}^n \binom{n}{n-k} (-1)^{n-k} \ln (pk + 1) \quad \left[p > -\frac{1}{n} \right].$

GH[324](19d) BH[123](12)+

40. $\int_0^1 \frac{(1-x^p)^n}{1-x} \frac{dx}{\ln x} = \sum_{k=0}^n (-1)^{k-1} \ln \Gamma[(n-k) p + 1] \quad \left[p > -\frac{1}{n} \right].$ BH[127](12)

41. $\int_0^1 (x^p - 1)^n x^{q-1} \frac{dx}{\ln x} = \sum_{k=0}^n (-1)^k \binom{n}{k} \ln [q + (n-k) p]$

$\left[p > -\frac{1}{n}, q > 0 \right].$ BH[123](12)

42. $\int_0^1 (1-x^p)^n x^{q-1} \frac{dx}{(1-x) \ln x} = \sum_{k=0}^n (-1)^{k-1} \ln \Gamma[(n-k) p + q]$

$\left[p > -\frac{1}{n}, q > 0 \right].$ BH[127](13)

43. $\int_0^1 (x^p - 1)^n (x^q - 1)^m \frac{x^{r-1} dx}{\ln x} =$

$= \sum_{j=0}^n (-1)^j \binom{n}{j} \sum_{k=0}^m (-1)^k \binom{m}{k} [r + (m-k) q + (n-j) p]$

$\left[p > -\frac{1}{n}, q > -\frac{1}{m}, r > 0 \right].$ BH[123](16)

1. $\int_0^1 \frac{(x^p - x^q)(1 - x^r)}{(\ln x)^2} dx = (p+1) \ln (p+1) - (q+1) \ln (q+1) -$ 4.268

$- (p+r+1) \ln (p+r+1) + (q+r+1) \ln (q+r+1)$

$[p > -1, q > -1, p+r > -1, q+r > -1].$ GH[324](26)

4.268

2. $$\int_0^1 (x^p - x^q)^2 \frac{dx}{(\ln x)^2} = (2p+1) \ln (2p+1) + (2q+1) \ln (2q+1) -$$

$$- 2(p+q+1) \ln (p+q+1) \quad \left[p > -\frac{1}{2}, q > -\frac{1}{2}\right].$$
GH[324](26a)

3. $$\int_0^1 (1 - x^p)(1 - x^q)(1 - x^r) \frac{dx}{(\ln x)^2} = (p+q+1) \ln (p+q+1) +$$

$$+ (q+r+1) \ln (q+r+1) + (p+r+1) \ln (p+r+1) - (p+1) \ln (p+1) -$$
$$- (q+1) \ln (q+1) - (r+1) \ln (r+1) - (p+q+r) \ln (p+q+r)$$
$$[p > -1, q > -1, r > -1, p+q > -1, p+r > -1, q+r > -1,$$
$$p+q+r > 0].$$
BH[124](4)

4. $$\int_0^1 (1 - x^p)^n x^{q-1} \frac{dx}{(\ln x)^2} = \frac{1}{2} \sum_{k=0}^n (-1)^k \binom{n}{k} (pk+q)^2 \ln (pk+q)$$

$$\left[q > 0, p > -\frac{q}{n}\right].$$
BH[124](14)

5. $$\int_0^1 (1 - x^p)^n (1 - x^q)^m x^{r-1} \frac{dx}{(\ln x)^2} = \sum_{k=0}^n (-1)^j \binom{n}{j} \sum_{k=0}^m (-1)^k \binom{m}{k} \times$$

$$\times [(m-k)q + (n-j)p + r] \ln [(m-k)q + (n-j)p + r]$$
$$[r > 0, mq + r > 0, np + r > 0, mq + np + r > 0].$$
BH[124](8)

6. $$\int_0^1 [(q-r) x^{p-1} + (r-p) x^{q-1} + (p-q) x^{r-1}] \frac{dx}{(\ln x)^2} =$$

$$= (q-r) p \ln p + (r-p) q \ln q + (p-q) r \ln r$$
$$[p > 0, q > 0, r > 0].$$
BH[124](9)

7. $$\int_0^1 \left[\frac{x^{p-1}}{(p-q)(p-r)(p-s)} + \frac{x^{q-1}}{(q-p)(q-r)(q-s)} + \frac{x^{r-1}}{(r-p)(r-q)(r-s)} + \right.$$

$$\left. + \frac{x^{s-1}}{(s-p)(s-q)(s-r)} \right] \frac{dx}{(\ln x)^2} = \frac{1}{2} \left[\frac{p^2 \ln p}{(p-q)(p-r)(p-s)} + \right.$$

$$\left. + \frac{q^2 \ln q}{(q-p)(q-r)(q-s)} + \frac{r^2 \ln r}{(r-p)(r-q)(r-s)} + \frac{s^2 \ln s}{(s-p)(s-q)(s-r)} \right]$$
$$[p > 0, q > 0, r > 0, s > 0].$$
BH[124](16)

4.269

1. $$\int_0^1 \sqrt{\ln \frac{1}{x}} \cdot \frac{dx}{1+x^2} = \frac{\sqrt{\pi}}{2} \sum_{k=0}^\infty \frac{(-1)^k}{\sqrt{(2k+1)^3}}.$$
BH[115](33)

2. $$\int_0^1 \frac{dx}{\sqrt{\ln\frac{1}{x}} \cdot (1+x)^2} = \sqrt{\pi} \sum_{k=0}^{\infty} \frac{(-1)^k}{\sqrt{2k+1}}.$$ BH[133](2) 4.269

3. $$\int_0^1 \sqrt{\ln\frac{1}{x}} \cdot x^{p-1}\, dx = \frac{1}{2}\sqrt{\frac{\pi}{p^3}} \quad [p>0].$$ GH[324](1c)

4. $$\int_0^1 \frac{x^{p-1}}{\sqrt{\ln\frac{1}{x}}}\, dx = \sqrt{\frac{\pi}{p}} \quad [p>0].$$ BH[133](1)

5. $$\int_0^1 \frac{\sin t - x^n \sin[(n+1)t] + x^{n+1}\sin nt}{1 - 2x\cos t + x^2} \cdot \frac{dx}{\sqrt{\ln\frac{1}{x}}} = \sqrt{\pi} \sum_{k=1}^{n} \frac{\sin kt}{\sqrt{k}}$$

$[|t| < \pi].$ BH[133](5)

6. $$\int_0^1 \frac{\cos t - x - x^{n-1}\cos nt + x^n \cos[(n-1)t]}{1 - 2x\cos t + x^2} \cdot \frac{dx}{\sqrt{\ln\frac{1}{x}}} = \sqrt{\pi} \sum_{k=1}^{n-1} \frac{\cos kt}{\sqrt{k}}$$

$[|t| < \pi].$ BH[133](6)

7. $$\int_u^v \frac{dx}{x \cdot \sqrt{\ln\frac{x}{u} \ln\frac{v}{x}}} = \pi \quad [uv > 0].$$ BH[145](37)

1. $$\int_0^1 (\ln x)^{2n} \frac{dx}{1+x} = \frac{2^{2n}-1}{2^{2n}} \cdot (2n)!\, \zeta(2n+1).$$ BH[110](1) 4.271

2. $$\int_0^1 (\ln x)^{2n-1} \frac{dx}{1+x} = \frac{1 - 2^{2n-1}}{2n} \pi^{2n} |B_{2n}|.$$ BH[110](2)

3. $$\int_0^1 (\ln x)^{2n-1} \frac{dx}{1-x} = -\frac{1}{n} 2^{2n-2} \pi^{2n} |B_{2n}|.$$ BH[110](5), GH[324](9a)

4. $$\int_0^1 (\ln x)^{p-1} \frac{dx}{1-x} = e^{i(p-1)\pi} \Gamma(p)\zeta(p) \quad [p>1].$$ GH[324](9b)

5. $$\int_0^1 (\ln x)^n \frac{dx}{1+x^2} = (-1)^n n! \sum_{k=0}^{\infty} \frac{(-1)^k}{(2k+1)^{n+1}}.$$ BH[110](11)

6. $$\int_0^1 (\ln x)^{2n} \frac{dx}{1+x^2} = \frac{1}{2} \int_0^{\infty} (\ln x)^{2n} \frac{dx}{1+x^2} = \frac{\pi^{2n+1}}{2^{2n+2}} |E_{2n}|.$$ GH[324](10)+

4.271

7. $\int_0^\infty \frac{(\ln x)^{2n+1}}{1+bx+x^2} dx = 0$ $\quad [|b|<2]$. BH[135](2)

8. $\int_0^1 (\ln x)^{2n} \frac{dx}{1-x^2} = \frac{2^{2n+1}-1}{2^{2n+1}} \cdot (2n)! \, \zeta(2n+1)$. BH[110](12)

9. $\int_0^\infty (\ln x)^{2n} \frac{dx}{1-x^2} = 0$. BH[312](7)+

10. $\int_0^1 (\ln x)^{2n-1} \frac{dx}{1-x^2} = \frac{1}{2} \int_0^\infty (\ln x)^{2n-1} \frac{dx}{1-x^2} = \frac{1-2^{2n}}{4n} \pi^{2n} |B_{2n}|$. BH[290](17)+, BH[312](6)+

11. $\int_0^1 (\ln x)^{2n-1} \frac{x\,dx}{1-x^2} = -\frac{1}{4n} \pi^{2n} |B_{2n}|$. BH[290](19)+

12. $\int_0^1 (\ln x)^{2n} \frac{1+x^2}{(1-x^2)^2} dx = \frac{2^{2n}-1}{2} \pi^{2n} |B_{2n}|$. BH[296](17)+

13. $\int_0^1 (\ln x)^{2n+1} \frac{(\cos 2a\pi - x)\,dx}{1 - 2x\cos 2a\pi + x^2} = -(2n+1)! \sum_{k=1}^\infty \frac{\cos 2ak\pi}{k^{2n+2}}$

$\left[a \begin{array}{c} \text{nicht ganzzahlig} \\ \text{noninteger} \end{array} \right]$ Li[113](10)

14. $\int_0^1 (\ln x)^n \frac{x^{\nu-1}\,dx}{a^2 + 2ax\cos t + x^2} = -\pi \cos t \frac{d^n}{d\nu^n} \left[a^{\nu-2} \frac{\sin(\nu-1)t}{\sin \nu\pi} \right]$

$[a>0,\, 0<\mathrm{Re}\,\nu<2,\, |t|<\pi]$. Ba IV 315(12)

15. $\int_0^1 (\ln x)^n \frac{x^{p-1}}{1-x^q} dx = -\frac{1}{q^{n+1}} \psi^{(n)}\left(\frac{p}{q}\right)$ $\quad [p>0,\, q>0]$. GH[324](9)

16. $\int_0^1 (\ln x)^n \frac{x^{p-1}}{1+x^q} dx = \frac{1}{2^n q^{n+1}} \beta\left(\frac{p}{q}\right)$ $\quad [p>0,\, q>0]$. GH[324](10)

4.272

1. $\int_0^1 \frac{\left[\ln\left(\frac{1}{x}\right)\right]^{q-1} dx}{1 + 2x\cos t + x^2} = \operatorname{cosec} t\, \Gamma(q) \sum_{k=1}^\infty (-1)^{k-1} \frac{\sin kt}{k^q}$

$[|t|<\pi,\, q<1]$. Li[130](1)

2. $\int_0^1 \left(\ln\frac{1}{x}\right)^{q-1} \frac{(1+x)\,dx}{1+2x\cos t + x^2} = \sec\frac{t}{2}\cdot\Gamma(q)\sum_{k=1}^\infty (-1)^{k-1}\frac{\cos\left[\left(k-\frac{1}{2}\right)t\right]}{k^q}$, 4.272

$$\left[|t|<\pi,\ q<\frac{1}{2}\right].$$ Li[130](5)

3. $\int_0^\infty \left[\ln\left(\frac{1}{x}\right)\right]^\mu \frac{x^{\nu-1}\,dx}{1-2ax\cos t + x^2 a^2} = \frac{\Gamma(\mu+1)}{a\sin t}\sum_{k=0}^\infty \frac{a^k\sin kt}{(\nu+k-1)^{\mu+1}}$

$[a>0,\ \operatorname{Re}\mu>0,\ 0<\operatorname{Re}\nu<2,\ |t|<\pi].$ BH[140](14)+

4. $\int_0^1 \left(\ln\frac{1}{x}\right)^{r-1} \frac{\cos\lambda - px}{1+p^2x^2 - 2px\cos\lambda} x^{q-1}\,dx = \Gamma(r)\sum_{k=1}^\infty \frac{p^{k-1}\cos k\lambda}{(q+k-1)^r}$

$[r>0,\ q>0].$ BH[113](11)

5. $\int_1^\infty (\ln x)^p \frac{dx}{x^2} = \Gamma(1+p)\quad [p>-1].$ BH[144](1)

6. $\int_0^1 \left(\ln\frac{1}{x}\right)^{\mu-1} x^{\nu-1}\,dx = \frac{1}{\nu^\mu}\Gamma(\mu)\quad [\operatorname{Re}\mu>0,\ \operatorname{Re}\nu>0].$ BH[107](3)

7. $\int_0^1 \left(\ln\frac{1}{x}\right)^{n-\frac{1}{2}} x^{\nu-1}\,dx = \frac{(2n-1)!!}{(2\nu)^n}\sqrt{\frac{\pi}{\nu}}\quad [\operatorname{Re}\nu>0].$ BH[107](2)

8. $\int_0^1 \left(\ln\frac{1}{x}\right)^{n-1} \frac{x^{\nu-1}}{1+x}\,dx = (n-1)!\sum_{k=0}^\infty \frac{(-1)^k}{(\nu+k)^n}\quad [\operatorname{Re}\nu>0].$ BH[110](4)

9. $\int_0^1 \left(\ln\frac{1}{x}\right)^{n-1} \frac{x^{\nu-1}}{1-x}\,dx = (n-1)!\,\zeta(n,\nu)\quad [\operatorname{Re}\nu>0].$ BH[110](7)

10. $\int_0^1 \left(\ln\frac{1}{x}\right)^{\mu-1} (x-1)^n \left(a+\frac{nx}{x-1}\right) x^{a-1}\,dx =$

$= \Gamma(\mu)\sum_{k=0}^n \frac{(-1)^k n(n-1)\ldots(n-k+1)}{(a+n-k)^{\mu-1}\,k!}\quad [\operatorname{Re}\mu>0].$ Li[110](10)

11. $\int_0^1 \left(\ln\frac{1}{x}\right)^{n-1} \frac{1-x^m}{1-x}\,dx = (n-1)!\sum_{k=1}^m \frac{1}{k^n}.$ Li[110](9)

12. $\int_0^1 \left(\ln\frac{1}{x}\right)^{\mu-1} \frac{x^{\nu-1}\,dx}{1-x^2} = \Gamma(\mu)\sum_{k=0}^\infty \frac{1}{(\nu+2k)^\mu} = \frac{1}{2^\mu}\Gamma(\mu)\zeta\left(\mu,\frac{\nu}{2}\right)$

$[\operatorname{Re}\mu>0,\ \operatorname{Re}\nu>0].$ BH[110](13)

4.272

13. $\int_0^1 \dfrac{x^q - x^{-q}}{1-x^2}\left(\ln\dfrac{1}{x}\right)^p dx = \Gamma(p+1)\sum_{k=1}^{\infty}\left\{\dfrac{1}{(2k+q-1)^{p+1}} - \dfrac{1}{(2k-q-1)^{p+1}}\right\}$ $\quad [p > -1,\ q^2 < 1]$. Li[326](12)+

14. $\int_0^1 \left(\ln\dfrac{1}{x}\right)^{r-1}\dfrac{x^{p-1}dx}{(1+x^q)^s} = \Gamma(r)\sum_{k=0}^{\infty}\binom{-s}{k}\dfrac{1}{(p+kq)^r}$

$\quad [p>0,\ q>0,\ r>0,\ 0<s<r+2]$. GH[324](11)

15. $\int_0^1 \left(\ln\dfrac{1}{x}\right)^n (1+x^q)^m x^{p-1} dx = n!\sum_{k=0}^{m}\binom{m}{k}\dfrac{1}{(p+kq)^{n+1}}$

$\quad [p>0,\ q>0]$. BH[107](6)

16. $\int_0^1 \left(\ln\dfrac{1}{x}\right)^n (1-x^q)^m x^{p-1} dx = n!\sum_{k=0}^{m}\binom{m}{k}\dfrac{(-1)^k}{(p+kq)^{n+1}}$

$\quad [p>0,\ q>0]$. BH[107](7)

17. $\int_0^1 \left(\ln\dfrac{1}{x}\right)^{p-1}\dfrac{x^{q-1}dx}{1-ax^q} = \dfrac{1}{aq^p}\Gamma(p)\sum_{k=1}^{\infty}\dfrac{a^k}{k^p}$ $\quad [p>0,\ q>0,\ a<1]$. Li[110](8)

18. $\int_0^1 \left(\ln\dfrac{1}{x}\right)^{2-\frac{1}{n}}(x^{p-1} - x^{q-1})\,dx = \dfrac{n}{n-1}\cdot\Gamma\left(\dfrac{1}{n}\right)\left(q^{1-\frac{1}{n}} - p^{1-\frac{1}{n}}\right)$

$\quad [q>p>0]$. BH[133](4)

19. $\int_0^1 \left(\ln\dfrac{1}{x}\right)^{2n-1}\dfrac{x^p - x^{-p}}{1-x^q}x^{q-1}dx = \dfrac{1}{p^{2n}}\sum_{k=n}^{\infty}\left(\dfrac{2p\pi}{q}\right)^k \dfrac{|B_{2k}|}{2k\cdot(2k-2n)!}$

$\quad \left[p < \dfrac{q}{2}\right]$. Li[110](16)

4.273

$\int_u^v \left(\ln\dfrac{x}{u}\right)^{p-1}\left(\ln\dfrac{v}{x}\right)^{q-1}\dfrac{dx}{x} = B(p,q)\left(\ln\dfrac{v}{u}\right)^{p+q-1}$

$\quad [p>0,\ q>0,\ uv>0]$. BH[145](36)

4.274

$\int_0^{1/e} \dfrac{\sqrt[q]{x}\,dx}{x\sqrt[q]{-(1+\ln x)}} = \dfrac{\sqrt[q]{q\pi}}{\sqrt[q]{e}}$ $\quad [q>0]$. BH[145](4)

4.275

1. $\int_0^1 \left[\left(\ln\dfrac{1}{x}\right)^{q-1} - x^{p-1}(1-x)^{q-1}\right] dx = \dfrac{\Gamma(q)}{\Gamma(p+q)}[\Gamma(p+q) - \Gamma(p)]$

$\quad [p>0,\ q>0]$. Brf[107](8)

2. $\quad\int_0^1 \left[x - \left(\dfrac{1}{1-\ln x}\right)^q\right] \dfrac{dx}{x \ln x} = -\psi(q) \quad [q > 0].$ BH[126](5)   4.275

Rationale Funktionen von ln x und Potenzen
Rational functions of ln x and powers   4.28

1. $\quad\int_0^1 \left[\dfrac{1}{\ln x} + \dfrac{1}{1-x}\right] dx = \mathbf{C}.$ BH[127](15)   4.281

2. $\quad\int_1^\infty \dfrac{dx}{x^2(\ln p - \ln x)} = \dfrac{1}{p}\, \mathrm{li}\,(p).$ La 281(30)

3. $\quad\int_0^1 \dfrac{x^{p-1}\, dx}{q \pm \ln x} = \pm\, e^{\mp pq}\, \mathrm{Ei}\,(\pm pq) \quad [p > 0, q > 0].$ Li[144](11, 12)

4. $\quad\int_0^1 \left[\dfrac{1}{\ln x} + \dfrac{x^{\mu-1}}{1-x}\right] dx = -\psi(\mu) \quad [\mathrm{Re}\,\mu > 0].$ WW 248

5. $\quad\int_0^1 \left[\dfrac{x^{p-1}}{\ln x} + \dfrac{x^{q-1}}{1-x}\right] dx = \ln p - \psi(q) \quad [p > 0, q > 0].$ BH[127](17)

6. $\quad\int_0^1 \left[\dfrac{1}{1-x^2} + \dfrac{1}{2x \ln x}\right] \dfrac{dx}{\ln x} = \dfrac{\ln 2}{2}.$ Li[130](19)

7. $\int_0^1 \left[q - \dfrac{1}{2} + \dfrac{(1-x)(1+q\ln x) + x \ln x}{(1-x)^2}\, x^{q-1}\right] \dfrac{dx}{\ln x} =$

$\quad = \dfrac{1}{2} - q - \ln \Gamma(q) + \dfrac{\ln 2\pi}{2} \quad [q > 0].$ BH[128](15)

1. $\quad\int_0^1 \dfrac{\ln x}{4\pi^2 + (\ln x)^2} \cdot \dfrac{dx}{1-x} = \dfrac{1}{4} - \dfrac{1}{2}\mathbf{C}.$ BH[129](1)   4.282

2. $\quad\int_0^1 \dfrac{1}{a^2 + (\ln x)^2} \cdot \dfrac{dx}{1+x^2} = \dfrac{1}{2a}\beta\left(\dfrac{2a+\pi}{4\pi}\right) \quad \left[a > -\dfrac{\pi}{2}\right].$ BH[129](9)

3. $\quad\int_0^1 \dfrac{1}{\pi^2 + (\ln x)^2} \cdot \dfrac{dx}{1+x^2} = \dfrac{4-\pi}{4\pi}.$ BH[129](6)

4. $\quad\int_0^1 \dfrac{\ln x}{\pi^2 + (\ln x)^2} \cdot \dfrac{dx}{1-x^2} = \dfrac{1}{2}\left(\dfrac{1}{2} - \ln 2\right).$ BH[129](10)

4.282

5. $\int_0^1 \frac{\ln x}{a^2 + (\ln x)^2} \cdot \frac{x\,dx}{1-x^2} = \frac{1}{2}\left[\frac{\pi}{2a} + \ln\frac{\pi}{a} + \psi\left(\frac{a}{\pi}\right)\right]$ $[a > 0]$.   BH[129](14)

6. $\int_0^1 \frac{\ln x}{\pi^2 + (\ln x)^2} \cdot \frac{x\,dx}{1-x^2} = \frac{1}{2}\left(\frac{1}{2} - C\right)$.   BH[129](13)

7. $\int_0^1 \frac{1}{\pi^2 + 4(\ln x)^2} \cdot \frac{dx}{1+x^2} = \frac{\ln 2}{4\pi}$.   BH[129](7)

8. $\int_0^1 \frac{\ln x}{\pi^2 + 4(\ln x)^2} \cdot \frac{dx}{1-x^2} = \frac{2-\pi}{16}$.   BH[129](11)

9. $\int_0^1 \frac{1}{\pi^2 + 16(\ln x)^2} \cdot \frac{dx}{1+x^2} = \frac{1}{8\pi\sqrt{2}}\left[\pi + 2\ln(\sqrt{2}-1)\right]$.   BH[129](8)

10. $\int_0^1 \frac{\ln x}{\pi^2 + 16(\ln x)^2} \cdot \frac{dx}{1-x^2} = -\frac{\pi}{32\sqrt{2}} + \frac{1}{16} + \frac{1}{16\sqrt{2}}\ln(\sqrt{2}-1)$.   BH[129](12)

11. $\int_0^1 \frac{\ln x}{[a^2 + (\ln x)^2]^2} \cdot \frac{dx}{1-x} = -\frac{\pi^2}{a^4}\sum_{k=1}^{\infty} |B_{2k}| \left(\frac{2\pi}{a}\right)^{2k-2}$.   BH[129](4)

12. $\int_0^1 \frac{\ln x}{[a^2 + (\ln x)^2]^2} \cdot \frac{x\,dx}{1-x^2} = -\frac{\pi^2}{4a^4}\sum_{k=1}^{\infty} |B_{2k}| \left(\frac{\pi}{a}\right)^{2k-2}$.   BH[129](16)

13. $\int_0^1 \frac{x^p - x^{-p}}{x^2 - 1} \cdot \frac{dx}{q^2 + \ln^2 x} = \frac{2\pi}{q}\sum_{k=1}^{\infty} (-1)^{k-1} \frac{\sin k p\pi}{2q + k\pi}$ $[p^2 < 1]$.   BH[132](13)+

4.283

1. $\int_0^1 \left(\frac{x-1}{\ln x} - x\right) \frac{dx}{\ln x} = \ln 2 - 1$.   BH[132](17)+

2. $\int_0^1 \left(\frac{1}{\ln x} + \frac{1}{1-x} - \frac{1}{2}\right) \frac{dx}{\ln x} = \frac{\ln 2\pi}{2} - 1$.   BH[127](20)

3. $\int_0^1 \left(\frac{1}{\ln x} + \frac{x}{1-x} + \frac{x}{2}\right) \frac{dx}{x\ln x} = \frac{\ln 2\pi}{2}$.   BH[127](23)

4. $\int_0^1 \left[\frac{1}{(\ln x)^2} - \frac{x}{(1-x)^2}\right] dx = C - \frac{1}{2}$.   GH[326](8a)

5. $\int_0^1 \left(\frac{1}{1-x^2} + \frac{1}{2\ln x} - \frac{1}{2}\right) \frac{dx}{\ln x} = \frac{\ln 2 - 1}{2}.$ BH[128](14)   4.283

6. $\int_0^1 \left(\frac{1}{\ln x} + \frac{1}{2} \cdot \frac{1+x}{1-x} - \ln x\right) \frac{dx}{\ln x} = \frac{\ln 2\pi}{2}.$ BH[127](22)

7. $\int_0^1 \left[\frac{1}{1-\ln x} - x\right] \frac{dx}{x \ln x} = -\mathbf{C}.$ GH[326](11a)

8. $\int_0^1 \left[\frac{x^q - 1}{x(\ln x)^2} - \frac{q}{\ln x}\right] dx = q \ln q - q \quad [q > 0].$ BH[126](2)

9. $\int_0^1 \left[x + \frac{1}{a\ln x - 1}\right] \frac{dx}{x \ln x} = \ln \frac{a}{q} + C \quad [a > 0, \, q > 0].$ BH[126](8)

10. $\int_0^1 \left[\frac{1}{\ln x} + \frac{1+x}{2(1-x)}\right] \frac{x^{p-1}}{\ln x} dx = -\ln \Gamma(p) + \left(p - \frac{1}{2}\right) \ln p - p + \frac{\ln 2\pi}{2}$

$[p > 0].$ GH[326](9)

11. $\int_0^1 \left[p - 1 - \frac{1}{1-x} + \left(\frac{1}{2} - \frac{1}{\ln x}\right) x^{p-1}\right] \frac{dx}{\ln x} = \left(\frac{1}{2} - p\right) \ln p + p - \frac{\ln 2\pi}{2}$

$[p > 0].$ BH[127](25)

12. $\int_0^1 \left[-\frac{1}{(\ln x)^2} + \frac{(p-2)x^p - (p-1)x^{p-1}}{(1-x)^2}\right] dx = -\psi(p) + p - \frac{3}{2} \quad [p > 0].$

GH[326](8)

13. $\int_0^1 \left[\left(p - \frac{1}{2}\right) x^3 + \frac{1}{2}\left(1 - \frac{1}{\ln x}\right)(x^{2p-1} - 1)\right] \frac{dx}{\ln x} = \left(\frac{1}{2} - p\right)(\ln p - 1)$

$[p > 0].$ BH[132](23)+

14. $\int_0^1 \left[\left(q - \frac{1}{2}\right) \frac{x^{p-1} - x^{r-1}}{\ln x} + \frac{px^{pq-1}}{1 - x^p} - \frac{rx^{rq-1}}{1 - x^r}\right] \frac{dx}{\ln x} =$

$= (p - r) \left[\frac{1}{2} - q - \ln \Gamma(q) + \frac{\ln 2\pi}{2}\right] \quad [q > 0].$ BH[132](13)

1. $\int_0^1 \left[\frac{x^q - 1}{x(\ln x)^3} - \frac{q}{x(\ln x)^2} - \frac{q^2}{2 \ln x}\right] dx = \frac{q^2}{2} \ln q - \frac{3}{4} q^2 \quad [q > 0].$ BH[126](3)   4.284

4.284

2. $\int_0^1 \left[\dfrac{x^q-1}{x(\ln x)^4} - \dfrac{q}{x(\ln x)^3} - \dfrac{q^2}{2x(\ln x)^2} - \dfrac{q^3}{6\ln x}\right]dx = \dfrac{q^3}{6}\ln q - \dfrac{11}{36}q^3$

$[q>0]$.   BH[126](4)

4.285

$\int_0^1 \dfrac{x^{p-1}\,dx}{(q+\ln x)^n} = \dfrac{p^{n-1}}{(n-1)!}\,e^{-pq}\,\text{Ei}(pq) - \dfrac{1}{(n-1)!\,q^{n-1}}\sum_{k=1}^{n-1}(n-k-1)!\,(pq)^{k-1}$

$[p>0,\ q<0]$.   BH[125](21)

In Integralen der Gestalt | In integrals of the form

$$\int \dfrac{x^a(\ln x)^n\,dx}{[b \pm (\ln x)^m]^l}$$

substituiert man $x = e^t$ bzw. $x = e^{-t}$, und sucht dann die sich ergebenden Integrale in 3.351–3.356 auf. | the substitutions $x = e^t$ and $x = e^{-t}$, respectively, lead to integrals included under 3.351–3.356.

4.29–4.32

Der Logarithmus von zusammengesetzten Argumenten und Potenzen
Logarithmic functions of compound arguments and powers

4.291

1. $\int_0^1 \dfrac{\ln(1+x)}{x}\,dx = \dfrac{\pi^2}{12}$.   F II 478

2. $\int_0^1 \dfrac{\ln(1-x)}{x}\,dx = -\dfrac{\pi^2}{6}$.   F II 720, 721

3. $\int_0^{\frac{1}{2}} \dfrac{\ln(1-x)}{x}\,dx = \dfrac{1}{2}(\ln 2)^2 - \dfrac{\pi^2}{12}$.   BH[145](2)

4. $\int_0^1 \ln\left(1-\dfrac{x}{2}\right)\dfrac{dx}{x} = \dfrac{1}{2}(\ln 2)^2 - \dfrac{\pi^2}{12}$.   BH[114](18)

5. $\int_0^1 \dfrac{\ln\dfrac{1+x}{2}}{1-x}\,dx = \dfrac{1}{2}(\ln 2)^2 - \dfrac{\pi^2}{12}$.   BH[115](1)

6. $\int_0^1 \dfrac{\ln(1+x)}{1+x}\,dx = \dfrac{1}{2}(\ln 2)^2$.   BH[114](14)+

7. $\int_0^1 \dfrac{\ln(1+x)}{1+x^2}\,dx = \dfrac{\pi}{8}\ln 2$.   F II 139, 140

8. $\int_0^\infty \dfrac{\ln(1+x)}{1+x^2}\,dx = \dfrac{\pi}{4}\ln 2 + \mathbf{G}$.   BH[136](1)

9. $\int_0^1 \dfrac{\ln(1-x)}{1+x^2}\,dx = \dfrac{\pi}{8}\ln 2 - G.$ BH[114](17) 4.291

10. $\int_1^\infty \dfrac{\ln(x-1)}{1+x^2}\,dx = \dfrac{\pi}{8}\ln 2.$ BH[144](4)

11. $\int_0^1 \dfrac{\ln(1+x)}{x(1+x)}\,dx = \dfrac{\pi^2}{12} - \dfrac{1}{2}(\ln 2)^2.$ BH[144](4)

12. $\int_0^\infty \dfrac{\ln(1+x)}{x(1+x)}\,dx = \dfrac{\pi^2}{6}.$ BH[141](9)+

13. $\int_0^1 \dfrac{\ln(1+x)}{(ax+b)^2}\,dx = \dfrac{1}{a(a-b)}\ln\dfrac{a+b}{b} + \dfrac{2\ln 2}{b^2-a^2}$ $[a \neq b,\ ab > 0];$

$\qquad = \dfrac{1}{2a^2}(1-\ln 2)\quad [a=b].$ Li[114](5)+

14. $\int_0^\infty \dfrac{\ln(1+x)}{(ax+b)^2}\,dx = \dfrac{\ln\dfrac{a}{b}}{a(a-b)}\quad [ab > 0].$ BH[139](5)

15. $\int_0^1 \ln(a+x)\dfrac{dx}{a+x^2} = \dfrac{1}{2\sqrt{a}}\operatorname{arcctg}\sqrt{a}\,\ln[(1+a)a]\quad [a > 0].$ BH[114](20)

16. $\int_0^\infty \ln(a+x)\dfrac{dx}{(b+x)^2} = \dfrac{a\ln a - b\ln b}{b(a-b)}\quad [a > 0,\ b > 0,\ a \neq b].$ Li[139](6)

17. $\int_0^a \dfrac{\ln(1+ax)}{1+x^2}\,dx = \dfrac{1}{2}\operatorname{arctg} a\,\ln(1+a^2).$ GK I 353(164)

18. $\int_0^1 \dfrac{\ln(1+ax)}{1+ax^2}\,dx = \dfrac{1}{2\sqrt{a}}\operatorname{arctg}\sqrt{a}\,\ln(1+a)\quad [a > 0].$ BH[114](21)

19. $\int_0^a \dfrac{\ln(ax+b)}{(1+x)^2}\,dx = \dfrac{1}{a-b}\left[\dfrac{1}{2}(a+b)\ln(a+b) - b\ln b - a\ln 2\right]$

$\qquad\qquad\qquad [a > 0,\ b > 0,\ a \neq b].$ BH[114](22)

20. $\int_0^\infty \dfrac{\ln(ax+b)}{(1+x)^2}\,dx = \dfrac{1}{a-b}[a\ln a - b\ln b]\quad [a > 0,\ b > 0].$ BH[139](8)

4.291

21. $\int_0^\infty \ln(a+x) \frac{x\,dx}{(b^2+x^2)^2} = \frac{1}{2(a^2+b^2)} \left(\ln b + \frac{a\pi}{2b} + \frac{a^2}{b^2} \ln a \right)$ $[a>0, b>0]$.

BH[139](9)

22. $\int_0^1 \ln(1+x) \frac{1+x^2}{(1+x)^4} dx = -\frac{1}{3} \ln 2 + \frac{23}{72}$.

Li[114](12)

23. $\int_0^1 \ln(1+x) \frac{1+x^2}{a^2+x^2} \cdot \frac{dx}{1+a^2 x^2} = \frac{1}{2a(1+a^2)} \left[\frac{\pi}{2} \ln(1+a^2) - 2 \operatorname{arctg} a \cdot \ln a \right]$

$[a>0]$.   Li[114](11)

24. $\int_0^1 \ln(1+x) \frac{1-x^2}{(ax+b)^2 (bx+a)^2} dx = \frac{1}{a^2-b^2} \left\{ \frac{1}{a-b} \left[\frac{a+b}{ab} \ln(a+b) - \right. \right.$

$\left. \left. -\frac{1}{a} \ln b - \frac{1}{b} \ln a \right] + \frac{4 \ln 2}{b^2-a^2} \right\}$ $[a>0, b>0, a^2 \neq b^2]$.

Li[114](13)

25. $\int_0^\infty \ln(1+x) \frac{1-x^2}{(ax+b)^2} \cdot \frac{dx}{(bx+a)^2} = \frac{1}{ab(a^2-b^2)} \ln \frac{b}{a}$ $[a>0, b>0]$.

Li[139](14)

26. $\int_0^1 \ln(1+ax) \frac{1-x^2}{(1+x^2)^2} dx = \frac{1}{2} \frac{(1+a)^2}{1+a^2} \ln(1+a) -$

$-\frac{1}{2} \cdot \frac{a}{1+a^2} \ln 2 - \frac{\pi}{4} \cdot \frac{a^2}{1+a^2}$ $[a > -1]$.

BH[114](23)

27. $\int_0^\infty \ln(a+x) \frac{b^2-x^2}{(b^2+x^2)^2} dx = \frac{1}{a^2+b^2} \left(a \ln \frac{b}{a} - \frac{b\pi}{2} \right)$ $[a>0, b>0]$.

BH[139](11)

28. $\int_0^\infty \ln(a-x)^2 \frac{b^2-x^2}{(b^2+x^2)^2} dx = \frac{2}{a^2+b^2} \left(a \ln \frac{a}{b} - \frac{b\pi}{2} \right)$ $[a>0, b>0]$.

BH[139](12)

29. $\int_0^\infty \ln(a-x)^2 \frac{x\,dx}{(b^2+x^2)^2} = \frac{1}{a^2+b^2} \left(\ln b - \frac{a\pi}{2b} + \frac{a^2}{b^2} \ln a \right)$ $[a>0, b>0]$.

BH[139](10)

4.292

1. $\int_0^1 \frac{\ln(1 \pm x)}{\sqrt{1-x^2}} dx = -\frac{\pi}{2} \ln 2 \pm 2G$.

GH[325](20)

2. $\int_0^1 \frac{x \ln(1 \pm x)}{\sqrt{1-x^2}} dx = -1 \pm \frac{\pi}{2}$.

GH[325](22c)

3. $\int_{-a}^{a} \frac{\ln(1+bx)}{\sqrt{a^2-x^2}} dx = \pi \ln \frac{1+\sqrt{1-a^2b^2}}{2} \quad \left[0 \leqslant |b| \leqslant \frac{1}{a}\right].$ 4.292

BH[145](16, 17)+, GH[325](21e)

4. $\int_{0}^{1} \frac{x \ln(1+ax)}{\sqrt{1-x^2}} dx = -1 + \frac{\pi}{2} \cdot \frac{1-\sqrt{1-a^2}}{a} + \frac{\sqrt{1-a^2}}{a} \arcsin a \, [|a| \leqslant 1];$

$$= -1 + \frac{\pi}{2a} + \frac{\sqrt{a^2-1}}{a} \ln(a + \sqrt{a^2-1}) \quad [a \geqslant 1].$$

GH[325](22)

5. $\int_{0}^{1} \frac{\ln(1+ax)}{x\sqrt{1-x^2}} dx = \frac{1}{2} \arcsin a (\pi - \arcsin a) = \frac{\pi^2}{8} - \frac{1}{2} (\arccos a)^2$

$[|a| \leqslant 1].$ BH[120](4), GH[325](21a)

1. $\int_{0}^{1} x^{\mu-1} \ln(1+x) dx = \frac{1}{\mu} [\ln 2 - \beta(\mu+1)] \quad [\operatorname{Re} \mu > -1].$ BH[106](4)+ 4.293

2. $\int_{1}^{\infty} x^{\mu-1} \ln(1+x) dx = \frac{1}{\mu} [\beta(-\mu) - \ln 2] \quad [\operatorname{Re} \mu < 0].$ Ba IV 315(17)

3. $\int_{0}^{\infty} x^{\mu-1} \ln(1+x) dx = \frac{\pi}{\mu \sin \mu\pi} \quad [-1 < \operatorname{Re} \mu < 0].$ GH[325](3)+

4. $\int_{0}^{1} x^{2n-1} \ln(1+x) dx = \frac{1}{2n} \sum_{k=1}^{2n} \frac{(-1)^{k-1}}{k}.$ GH[325](2b)

5. $\int_{0}^{1} x^{2n} \ln(1+x) dx = \frac{1}{2n+1} \left[\ln 4 + \sum_{k=1}^{2n+1} \frac{(-1)^k}{k}\right].$ GH[325](2c)

6. $\int_{0}^{1} x^{n-\frac{1}{2}} \ln(1+x) dx = \frac{2 \ln 2}{2n+1} + \frac{(-1)^n \cdot 4}{2n+1} \left[\pi - \sum_{k=0}^{n} \frac{(-1)^k}{2k+1}\right].$

GH[325](2f)

7. $\int_{0}^{\infty} x^{\mu-1} \ln|1-x| dx = \frac{\pi}{\mu} \operatorname{ctg}(\mu\pi) \quad [-1 < \operatorname{Re} \mu < 0].$

BH[134](4), Ba IV 315(18)

8. $\int_{0}^{1} x^{\mu-1} \ln(1-x) dx = -\frac{1}{\mu} [\psi(\mu+1) - \psi(1)] \quad [\operatorname{Re} \mu > -1].$ Ba IV 316(19)

9. $\int_{1}^{\infty} x^{\mu-1} \ln(x-1) dx = \frac{1}{\mu} [\pi \operatorname{ctg}(\mu\pi) + \psi(\mu+1) - \psi(1)] \quad [\operatorname{Re} \mu < 0].$

Ba IV 316(20)

4.293

10. $\int_0^\infty x^{\mu-1} \ln(1+\gamma x) dx = \dfrac{\pi}{\mu \gamma^\mu \sin \mu \pi}$ $\quad [-1 < \operatorname{Re} \mu < 0, \ |\arg \gamma| < \pi]$. BH[134](3)

11. $\int_0^\infty \dfrac{x^{\mu-1} \ln(1+x)}{1+x} dx = \dfrac{\pi}{\sin \mu \pi} [C + \psi(1-\mu)]$ $\quad [-1 < \operatorname{Re} \mu < 1]$.

Ba IV 316(21)

12. $\int_0^1 \dfrac{\ln(1+x)}{(1+x)^{\mu+1}} dx = \dfrac{-\ln 2}{2^\mu \mu} + \dfrac{2^\mu - 1}{2^\mu \mu^2}$. BH[114](6)

13. $\int_0^1 \dfrac{x^{\mu-1} \ln(1-x)}{(1-x)^{1-\nu}} dx = B(\mu, \nu)[\psi(\nu) - \psi(\mu+\nu)]$

$\quad [\operatorname{Re} \mu > 0, \ \operatorname{Re} \nu > 0]$. Ba IV 316(22)

14. $\int_0^\infty \dfrac{x^{\mu-1} \ln(\gamma+x)}{(\gamma+x)^\nu} dx = \gamma^{\mu-\nu} B(\mu, \nu-\mu)[\psi(\nu) - \psi(\nu-\mu) + \ln \gamma]$

$\quad [0 < \operatorname{Re} \mu < \operatorname{Re} \nu]$. Ba IV 316(23)

4.294

1. $\int_0^1 \ln(1+x) \dfrac{(p-1)x^{p-1} - px^{-p}}{x} dx = 2\ln 2 - \dfrac{\pi}{\sin p\pi}$ $\quad [0 < p < 1]$. BH[114](2)

2. $\int_0^1 \ln(1+x) \dfrac{1+x^{2n+1}}{1+x} dx = 2\ln 2 \sum_{k=0}^n \dfrac{1}{2k+1} - \sum_{j=1}^{2n+1} \dfrac{1}{j} \sum_{k=1}^j \dfrac{(-1)^{k-1}}{k}$. BH[114](7)

3. $\int_0^1 \ln(1+x) \dfrac{1-x^{2n}}{1+x} dx = 2\ln 2 \cdot \sum_{k=0}^{n-1} \dfrac{1}{2k+1} - \sum_{j=1}^{2n} \dfrac{1}{j} \sum_{k=1}^j \dfrac{(-1)^{k-1}}{k}$. BH[114](8)

4. $\int_0^1 \ln(1+x) \dfrac{1-x^{2n}}{1-x} dx = 2\ln 2 \cdot \sum_{k=0}^{n-1} \dfrac{1}{2k+1} + \sum_{j=1}^{2n} \dfrac{(-1)^j}{j} \sum_{k=1}^j \dfrac{(-1)^{k-1}}{k}$. BH[114](9)

5. $\int_0^1 \ln(1+x) \dfrac{1-x^{2n+1}}{1-x} dx = 2\ln 2 \sum_{k=0}^n \dfrac{1}{2k+1} + \sum_{j=1}^{2n+1} \dfrac{(-1)^j}{j} \sum_{k=1}^j \dfrac{(-1)^{k-1}}{k}$.

BH[114](10)

6. $\int_0^1 \ln(1-x) \dfrac{1-(-1)^n x^n}{1-x} dx = \sum_{j=1}^n \dfrac{(-1)^j}{j} \sum_{k=1}^j \dfrac{1}{k}$. BH[114](15)

7. $\int_0^1 \ln(1-x) \dfrac{1-x^n}{1-x} dx = -\sum_{j=1}^n \dfrac{1}{j} \sum_{k=1}^j \dfrac{1}{k}$. BH[114](16)

8. $\int_0^\infty \ln(1-x)^2 x^p dx = \dfrac{2\pi}{p+1} \operatorname{ctg} p\pi$ $\quad [-2 < p < -1]$. BH[134](13)+

9. $\int_0^1 [\ln(1+x)]^n (1+x)^r dx = (-1)^{n-1} \frac{n!}{(r+1)^{n+1}} +$

$$+ 2^{r+1} \sum_{k=0}^n \frac{(-1)^k n! (\ln 2)^{n-k}}{(n-k)!(r+1)^{k+1}}.$$ Li[106](34)+   4.294

10. $\int_0^1 [\ln(1-x)]^n (1-x)^r dx = (-1)^n \frac{n!}{(r+1)^{n+1}} \quad [r > -1].$ BH[106](35)+

11. $\int_0^1 \left(\ln \frac{1}{1-x^2}\right)^n x^{2q-1} dx = \frac{n!}{2} \zeta(n+1, q+1) \quad [-1 < q < 0].$ BH[311](15)+

12. $\int_0^1 (\ln x)^{2n} \ln(1-x^2) \frac{dx}{x} = -\frac{\pi^{2n+2}}{2(n+1)(2n+1)} |B_{2n+2}|.$ BH[309](5)+

1. $\int_0^\infty \ln(\mu x^2 + \beta) \frac{dx}{\gamma + x^2} = \frac{\pi}{\sqrt{\gamma}} \ln(\sqrt{\mu \gamma} + \sqrt{\beta})$ 4.295

$[\operatorname{Re}\beta > 0, \operatorname{Re}\mu > 0, \quad |\arg \gamma| < \pi].$ Ba V 218(27)

2. $\int_0^1 \ln(1+x^2) \frac{dx}{x^2} = \frac{\pi}{2} - \ln 2.$ GH[325](2g)

3. $\int_0^\infty \ln(1+x^2) \frac{dx}{x^2} = \pi.$ GH[325](4c)

4. $\int_0^\infty \ln(1+x^2) \frac{dx}{(a+x)^2} = \frac{2a}{1+a^2}\left(\frac{\pi}{2a} + \ln a\right) \quad [a > 0].$ BH[319](6)+

5. $\int_0^1 \ln(1+x^2) \frac{dx}{1+x^2} = \frac{\pi}{2} \ln 2 - G.$ BH[114](24)

6. $\int_1^\infty \ln(1+x^2) \frac{dx}{1+x^2} = \frac{\pi}{2} \ln 2 + G.$ BH[144](5)

7. $\int_0^\infty \ln(a^2 + b^2 x^2) \frac{dx}{c^2 + g^2 x^2} = \frac{\pi}{cg} \ln \frac{ag+bc}{g}$

$[a > 0, \ b > 0, \ c > 0, \ g > 0].$ BH[136](11, 12, 13, 14)+

8. $\int_0^\infty \ln(a^2 + b^2 x^2) \frac{dx}{c^2 - g^2 x^2} = -\frac{\pi}{cg} \operatorname{arctg} \frac{bc}{ag}$

$[a > 0, \ b > 0, \ c > 0, \ g > 0].$ BH[136](15)+

4.295

9. $\int_0^\infty \dfrac{\ln(1+p^2x^2) - \ln(1+q^2x^2)}{x^2}\,dx = \pi(p-q)\quad [p>0,\, q>0].$ \hfill F II 653

10. $\int_0^1 \ln\dfrac{1+a^2x^2}{1+a^2}\dfrac{dx}{1-x^2} = -(\operatorname{arctg} a)^2.$ \hfill BH[115](2)

11. $\int_0^1 \ln(1-x^2)\dfrac{dx}{x} = -\dfrac{\pi^2}{12}.$

12. $\int_0^\infty \ln(1-x^2)^2 \dfrac{dx}{x^2} = 0.$ \hfill BH[142](9)+

13. $\int_0^1 \ln(1-x^2)\dfrac{dx}{1+x^2} = \dfrac{\pi}{4}\ln 2 - \mathbf{G}.$ \hfill GH[325](17)

14. $\int_1^\infty \ln(x^2-1)\dfrac{dx}{1+x^2} = \dfrac{\pi}{4}\ln 2 + \mathbf{G}.$ \hfill BH[144](6)

15. $\int_0^\infty \ln(a^2-x^2)^2\dfrac{dx}{b^2+x^2} = \dfrac{\pi}{b}\ln(a^2+b^2)\quad [b>0].$ \hfill BH[136](16)

16. $\int_0^\infty \ln(a^2-x^2)^2\dfrac{b^2-x^2}{(b^2+x^2)^2}\,dx = -\dfrac{2b\pi}{a^2+b^2}\quad [b>0].$ \hfill BH[139](20)

17. $\int_0^1 \ln(1+x^2)\dfrac{dx}{x(1+x^2)} = \dfrac{1}{2}\left[\dfrac{\pi^2}{12} - \dfrac{1}{2}(\ln 2)^2\right].$ \hfill BH[114](25)

18. $\int_0^\infty \ln(1+x^2)\dfrac{dx}{x(1+x^2)} = \dfrac{\pi^2}{12}.$ \hfill BH[141](9)

19. $\int_0^1 \ln(\cos^2 t + x^2 \sin^2 t)\dfrac{dx}{1-x^2} = -t^2.$ \hfill BH[114](27)+

20. $\int_0^\infty \ln(a^2+b^2x^2)\dfrac{dx}{(c+gx)^2} = \dfrac{2\ln b}{cg} + \dfrac{b^2}{a^2g^2+b^2c^2}\left(\dfrac{a}{b}\pi + 2\dfrac{c}{g}\ln\dfrac{c}{g}\right.$

$\left. + 2\dfrac{a^2g}{b^2c}\ln\dfrac{a}{b}\right)\quad [a>0,\, b>0,\, c>0,\, g>0].$ \hfill BH[139](16)+

21. $\int_0^1 \ln(a^2 + b^2 x^2) \frac{dx}{(c+gx)^2} = \frac{2}{c(c+g)} \ln a + \frac{b^2}{a^2 g^2 + b^2 c^2} \left[\frac{2a}{b} \operatorname{arcctg} \frac{a}{b} + \right.$ 4.295

$\left. + \frac{cb^2 - ga^2}{b^2(c+g)} \ln \frac{a^2 + b^2}{a^2} - 2 \frac{c}{g} \ln \frac{c+g}{c} \right]$

$[a > 0, \ b > 0, \ c > 0, \ g > 0]$. BH[114](28)+

22. $\int_0^\infty \frac{\ln(1 + p^2 x^2)}{r^2 + q^2 x^2} dx = \frac{\pi}{qr} \ln \frac{q + pr}{q}$ $[qr > 0, p > 0]$.

F II 751+, BH[318](1)+, BH[318](4)+

23. $\int_0^\infty \frac{\ln(1 + a^2 x^2)}{b^2 + c^2 x^2} \frac{dx}{d^2 + g^2 x^2} = \frac{\pi}{b^2 g^2 - c^2 d^2} \left[\frac{g}{d} \ln\left(1 + \frac{ad}{g}\right) - \right.$

$\left. - \frac{c}{b} \ln\left(1 + \frac{ab}{c}\right) \right]$ $[a > 0, b > 0, c > 0, d > 0, g > 0, b^2 g^2 \neq c^2 d^2]$.

BH[141](10)

24. $\int_0^\infty \frac{\ln(1 + a^2 x^2)}{b^2 + c^2 x^2} \frac{x^2 dx}{d^2 + g^2 x^2} = \frac{\pi}{b^2 g^2 - c^2 d^2} \left[\frac{b}{c} \ln\left(1 + \frac{ab}{c}\right) - \right.$

$\left. - \frac{d}{g} \ln\left(1 + \frac{ad}{g}\right) \right]$ $[a > 0, b > 0, c > 0, d > 0, g > 0, b^2 g^2 \neq c^2 d^2]$.

BH[141](11)

25. $\int_0^\infty \ln(a^2 + b^2 x^2) \frac{dx}{(c^2 + g^2 x^2)^2} = \frac{\pi}{2c^3 g} \left(\ln \frac{ag + bc}{g} - \frac{bc}{ag + bc} \right)$

$[a > 0, \ b > 0, \ c > 0, \ g > 0]$. GH[325](18a)

26. $\int_0^\infty \ln(a^2 + b^2 x^2) \frac{x^2 dx}{(c^2 + g^2 x^2)^2} = \frac{\pi}{2cg^3} \left(\ln \frac{ag + bc}{g} + \frac{bc}{ag + bc} \right)$

$[a > 0, \ b > 0, \ c > 0, \ g > 0]$. GH[325](18b)

27. $\int_0^1 \ln(1 + ax^2) \sqrt{1 - x^2} dx = \frac{\pi}{2} \left\{ \ln \frac{1 + \sqrt{1+a}}{2} + \frac{1}{2} \frac{1 - \sqrt{1+a}}{1 + \sqrt{1+a}} \right\}$

$[a > 0]$. BH[117](6)

28. $\int_0^1 \ln(1 + a - ax^2) \sqrt{1 - x^2} dx = \frac{\pi}{2} \left\{ \ln \frac{1 + \sqrt{1+a}}{2} - \frac{1}{2} \frac{1 - \sqrt{1+a}}{1 + \sqrt{1+a}} \right\}$

$[a > 0]$. BH[117](7)

29. $\int_0^1 \ln(1 - a^2 x^2) \frac{dx}{\sqrt{1 - x^2}} = \pi \ln \frac{1 + \sqrt{1 - a^2}}{2}$ $[a^2 < 1]$. BH[119](1)

4.295

30. $$\int_0^1 \ln(1-a^2x^2)\frac{dx}{x\sqrt{1-x^2}} = \frac{\pi^2}{4} - (\arccos a)^2 \quad [a^2 < 1].$$
Li[120](11)

31. $$\int_0^1 \ln(1-x^2)\frac{dx}{\sqrt{(1-x^2)(1-k^2x^2)}} = \ln\frac{k'}{k}\boldsymbol{K}(k) - \frac{\pi}{2}\boldsymbol{K}(k').$$
BH[120](12)

32. $$\int_0^1 \ln(1 \pm kx^2)\frac{dx}{\sqrt{(1-x^2)(1-k^2x^2)}} = \frac{1}{2}\ln\frac{2\pm 2k}{\sqrt{k}}\boldsymbol{K}(k) - \frac{\pi}{8}\boldsymbol{K}(k').$$
BH[120](8), BH[120](14)

33. $$\int_0^1 \frac{\ln(1-k^2x^2)}{\sqrt{(1-x^2)(1-k^2x^2)}}\,dx = \ln k'\,\boldsymbol{K}(k).$$
BH[119](27)

34. $$\int_0^1 \ln(1-k^2x^2)\sqrt{\frac{1-k^2x^2}{1-x^2}}\,dx = (2-k^2)\boldsymbol{K}(k) - (2-\ln k')\boldsymbol{E}(k).$$
BH[119](3)

35. $$\int_0^1 \sqrt{\frac{1-x^2}{1-k^2x^2}}\ln(1-k^2x^2)\,dx = \frac{1}{k^2}(1+k'^2-k'^2\ln k')\boldsymbol{K}(k) -$$
$$- (2-\ln k')\boldsymbol{E}(k).$$
BH[119](7)

36. $$\int_{-1}^1 \ln(1-x^2)\frac{dx}{(a+bx)\sqrt{1-x^2}} = \frac{2\pi}{\sqrt{a^2-b^2}}\ln\frac{\sqrt{a^2-b^2}}{a+\sqrt{a^2-b^2}}$$
$$[a>0, b>0, a\neq b].$$
BH[145](15)

37. $$\int_0^1 \ln(1-x^2)(px^{p-1} - qx^{q-1})\,dx = \psi\left(\frac{p}{2}+1\right) - \psi\left(\frac{q}{2}+1\right)$$
$$[p>-2, q>-2].$$
BH[106](15)

38. $$\int_0^1 \ln(1+ax^2)\frac{dx}{\sqrt{1-x^2}} = \pi\ln\frac{1+\sqrt{1+a}}{2} \quad [a\geq -1].$$
GH[325](21b)

39. $$\int_0^1 \ln(1+x^2)x^{\mu-1}\,dx = \frac{1}{\mu}\left[\ln 2 - \beta\left(\frac{\mu}{2}+1\right)\right] \quad [\operatorname{Re}\mu > 2].$$
BH[106](12)

40. $$\int_0^\infty \ln(1+x^2)x^{\mu-1}\,dx = \frac{\pi}{\mu\sin\frac{\mu\pi}{2}} \quad [-2<\operatorname{Re}\mu<0].$$
BH[311](4)+, Ba IV 315(15)

41. $$\int_0^\infty \ln(1+x^2)\frac{x^{\mu-1}\,dx}{1+x} = \frac{\pi}{\sin\mu\pi}\left\{\ln 2 - (1-\mu)\sin\frac{\mu\pi}{2}\beta\left(\frac{1-\mu}{2}\right) - \right.$$
$$\left. - (2-\mu)\cos\frac{\mu\pi}{2}\beta\left(\frac{2-\mu}{2}\right)\right\} \quad [-2<\operatorname{Re}\mu<1].$$
Ba IV 316(25)

1. $$\int_0^1 \ln(1 + 2x\cos t + x^2)\frac{dx}{x} = \frac{\pi^2}{6} - \frac{t^2}{2}.$$ BH[114](34)   4.296

2. $$\int_{-\infty}^\infty \ln(a^2 - 2ax\cos t + x^2)\frac{dx}{1+x^2} = \pi \ln(1 + 2a\sin t + a^2).$$ BH[145](28)

3. $$\int_0^\infty \ln(1 + 2x\cos t + x^2)x^{\mu-1}\,dx = \frac{2\pi}{\mu}\frac{\cos\mu t}{\sin\mu\pi}$$
$$[|t| < \pi, \ -1 < \operatorname{Re}\mu < 0].$$ Ba IV 316(27)

1. $$\int_0^1 \ln\frac{ax+b}{bx+a}\frac{dx}{(1+x)^2} = \frac{1}{a-b}\left[(a+b)\ln\frac{a+b}{2} - a\ln a - b\ln b\right]$$
$$[a > 0, \ b > 0].$$ BH[115](16)   4.297

2. $$\int_0^\infty \ln\frac{ax+b}{bx+a}\frac{dx}{(1+x)^2} = 0 \quad [ab > 0].$$ BH[139](23)

3. $$\int_0^1 \ln\frac{1-x}{x}\frac{dx}{1+x^2} = \frac{\pi}{8}\ln 2.$$ BH[115](5)

4. $$\int_0^1 \ln\frac{1+x}{1-x}\frac{dx}{1+x^2} = G.$$ BH[115](17)

5. $$\int_0^\infty \ln\left(\frac{1+x}{1-x}\right)^2 \frac{dx}{x(1+x^2)} = \frac{\pi^2}{2}.$$ BH[141](13)

6. $$\int_u^v \ln\frac{v+x}{u+x}\frac{dx}{x} = \frac{1}{2}\left(\ln\frac{v}{u}\right)^2 \quad [uv > 0].$$ BH[145](33)

7. $$\int_0^\infty \frac{b\ln(1+ax) - a\ln(1+bx)}{x^2}\,dx = ab\ln\frac{b}{a} \quad [a>0,\ b>0].$$ F II 657

8. $$\int_0^1 \ln\frac{1+ax}{1-ax}\frac{dx}{x\sqrt{1-x^2}} = \pi \arcsin a \quad [|a| \leq 1].$$ GH[325](21c), BH[122](2)

9. $$\int_u^v \ln\left(\frac{1+ax}{1-ax}\right)\frac{dx}{\sqrt{(x^2-u^2)(v^2-x^2)}} = \frac{\pi}{v}F\left(\arcsin av, \frac{u}{v}\right) \quad [|av| < 1].$$ BH[145](35)

1. $$\int_0^\infty \ln\frac{1+x^2}{x}\frac{x^{2n-1}}{1+x}\,dx = \frac{\ln 2}{2n} + \frac{1}{4n^2} - \frac{1}{2n}\beta(2n+1).$$ BH[137](1)   4.298

4.298

2. $\int_0^\infty \ln\frac{1+x^2}{x} \frac{x^{2n}}{1+x} dx = \frac{\ln 2}{2n} + \frac{1}{4n^2} - \frac{1}{2n}\beta(2n+1).$ BH[137](3)

3. $\int_0^\infty \ln\frac{1+x^2}{x} \frac{x^{2n-1}}{1-x} dx = \frac{\ln 2}{2n} + \frac{1}{4n^2} - \frac{1}{2n}\beta(2n+1).$ BH[137](2)

4. $\int_0^\infty \ln\frac{1+x^2}{x} \frac{x^{2n}}{1-x} dx = -\frac{\ln 2}{2n} - \frac{1}{4n^2} + \frac{1}{2n}\beta(2n+1).$ BH[137](4)

5. $\int_0^\infty \ln\frac{1+x^2}{x} \frac{x^{2n-1}}{1+x^2} dx = \frac{\ln 2}{2n} + \frac{1}{4n^2} - \frac{1}{2n}\beta(2n+1).$ BH[137](10)

6. $\int_0^1 \ln\frac{1+x^2}{x} x^{2n} dx = \frac{1}{2n+1}\left\{(-1)^n\frac{\pi}{2} + \ln 2 - \frac{1}{2n+1} + 2\sum_{k=0}^{n-1}\frac{(-1)^k}{2n-2k-1}\right\}.$ BH[294](8)

7. $\int_0^1 \ln\frac{1+x^2}{x} x^{2n-1} dx = \frac{1}{2n}\left\{(-1)^{n+1}\ln 2 + \ln 2 - \frac{1}{2n} + (-1)^{n+1}\sum_{k=1}^{n-1}\frac{(-1)^k}{k}\right\}.$ BH[294](9)+

8. $\int_0^1 \ln\frac{1+x^2}{x} \frac{dx}{1+x^2} = \frac{\pi}{2}\ln 2.$ BH[115](7)

9. $\int_0^\infty \ln\frac{1+x^2}{x} \frac{dx}{1+x^2} = \pi \ln 2.$ BH[137](8)

10. $\int_0^\infty \ln\frac{1+x^2}{x} \frac{dx}{1-x^2} = 0.$ BH[137](9)

11. $\int_0^1 \ln\frac{1-x^2}{x} \frac{dx}{1+x^2} = \frac{\pi}{4}\ln 2.$ BH[115](9)

12. $\int_1^\infty \ln\frac{1+x^2}{x+1} \frac{dx}{1+x^2} = \frac{3\pi}{8}\ln 2.$ BH[144](8)

13. $\int_0^1 \ln\frac{1+x^2}{x+1} \frac{dx}{1+x^2} = \frac{3\pi}{8}\ln 2 - G.$ BH[115](18)

14. $\int_1^\infty \ln\frac{1+x^2}{x-1} \frac{dx}{1+x^2} = \frac{3\pi}{8}\ln 2 + G.$ BH[144](9)

15. $\int_0^1 \ln\frac{1+x^2}{1-x}\frac{dx}{1+x^2} = \frac{3\pi}{8}\ln 2.$ BH[115](19) 4.298

16. $\int_0^\infty \ln\frac{1+x^2}{x^2}\frac{x\,dx}{1+x^2} = \frac{\pi^2}{12}.$ BH[138](3)

17. $\int_0^\infty \ln\frac{a^2+b^2x^2}{x^2}\frac{dx}{c^2+g^2x^2} = \frac{\pi}{cg}\ln\frac{ag+bc}{c}$ $[a>0,\ b>0,\ c>0,\ g>0].$ BH[138](6, 7, 9, 10)+

18. $\int_0^\infty \ln\frac{a^2+b^2x^2}{x^2}\frac{dx}{c^2-g^2x^2} = \frac{1}{cg}\operatorname{arctg}\frac{ag}{bc}$ $[a>0,\ b>0,\ c>0,\ g>0].$ BH[138](8, 11)+

19. $\int_0^\infty \ln\frac{1+x^2}{x^2}\frac{x^2\,dx}{(1+x^2)^2} = \frac{\pi}{4}(\ln 4 - 1).$ BH[139](21)

20. $\int_0^1 \ln\left(\frac{1-x^2}{x^2}\right)^2 \sqrt{1-x^2}\,dx = \pi.$ F II 652+

21. $\int_0^1 \ln\frac{1+2x\cos t+x^2}{(1+x)^2}\frac{dx}{x} = \frac{1}{2}\int_0^\infty \ln\frac{1+2x\cos t+x^2}{(1+x)^2}\frac{dx}{x} = -\frac{t^2}{2}$ $[|t|<\pi].$ BH[115](23), BH[134](15)

22. $\int_0^\infty \ln\frac{1+2x\cos t+x^2}{(1+x)^2} x^{p-1}\,dx = -\frac{2\pi(1-\cos p\pi)}{p\sin p\pi}$ $[|p|<1,\ |t|<\pi].$ BH[134](17)

23. $\int_0^1 \ln\frac{1+x^2\sin t}{1-x^2\sin t}\frac{dx}{\sqrt{1-x^2}} = \pi\ln\operatorname{ctg}\left(\frac{\pi-t}{4}\right)$ $[|t|<\pi].$ GH[325](21d)

1. $\int_0^\infty \ln\frac{(x+1)(x+a^2)}{(x+a)^2}\frac{dx}{x} = (\ln a)^2\ [a>0].$ BH[134](14) 4.299

2. $\int_0^1 \ln\frac{(1-ax)(1+ax^2)}{(1-ax^2)^2}\frac{dx}{1+ax^2} = \frac{1}{2\sqrt{a}}\operatorname{arctg}\sqrt{a}\ln(1+a)\ [a>0].$ BH[115](25)

3. $\int_0^1 \ln\frac{(1-a^2x^2)(1+ax^2)}{(1-ax^2)^2}\frac{dx}{1+ax^2} = \frac{1}{\sqrt{a}}\operatorname{arctg}\sqrt{a}\ln(1+a)\ [a>0].$ BH[115](26)

4.299 4. $\int_0^1 \ln \dfrac{(x+1)(x+a^2)}{(x+a)^2} x^{\mu-1} dx = \dfrac{\pi(a^\mu - 1)^2}{\mu \sin \mu\pi}$ $[a > 0, \operatorname{Re} \mu > 0]$. BH[134](16)

4.311 1. $\int_0^\infty \ln(1+x^3) \dfrac{dx}{1-x+x^2} = \dfrac{2\pi}{\sqrt{3}} \ln 3$. Li[136](8)

2. $\int_0^\infty \ln(1+x^3) \dfrac{dx}{1+x^3} = \dfrac{\pi}{\sqrt{3}} \ln 3 - \dfrac{\pi^2}{9}$. Li[136](6)

3. $\int_0^\infty \ln(1+x^3) \dfrac{x\,dx}{1+x^3} = \dfrac{\pi}{\sqrt{3}} \ln 3 + \dfrac{\pi^2}{9}$. Li[136](7)

4. $\int_0^\infty \ln(1+x^3) \dfrac{1-x}{1+x^3} dx = -\dfrac{2}{9}\pi^2$. BH[136](9)

4.312 1. $\int_0^\infty \ln \dfrac{1+x^3}{x^3} \dfrac{dx}{1+x^3} = \dfrac{\pi}{\sqrt{3}} \ln 3 + \dfrac{\pi^2}{9}$. BH[138](12)

2. $\int_0^\infty \ln \dfrac{1+x^3}{x^3} \dfrac{x\,dx}{1+x^3} = \dfrac{\pi}{\sqrt{3}} \ln 3 - \dfrac{\pi^2}{9}$. BH[138](13)

4.313 1. $\int_0^\infty \ln x \ln(1+a^2 x^2) \dfrac{dx}{x^2} = \pi a(1 - \ln a)$ $[a > 0]$. BH[134](18)

2. $\int_0^\infty \ln(1+c^2 x^2) \ln(a^2 + b^2 x^2) \dfrac{dx}{x^2} =$

$= 2\pi\left[\left(c + \dfrac{b}{a}\right)\ln(b+ac) - \dfrac{b}{a}\ln b - c \ln c\right]$

$[a > 0, b > 0, c > 0]$. BH[134](20, 21)+

3. $\int_0^\infty \ln(1+c^2 x^2) \ln\left(a^2 + \dfrac{b^2}{x^2}\right) \dfrac{dx}{x^2} = 2\pi\left[\dfrac{a+bc}{b}\ln(a+bc) - \dfrac{a}{b}\ln a - c\right]$

$[a > 0, a + bc > 0]$. BH[134](22, 23)+

4. $\int_0^\infty \ln x \ln \dfrac{1+a^2 x^2}{1+b^2 x^2} \dfrac{dx}{x^2} = \pi(a-b) + \pi \ln \dfrac{b^b}{a^a}$ $[a > 0, b > 0]$. BH[134](24)

5. $\int_0^\infty \ln x \ln \dfrac{a^2 + 2bx + x^2}{a^2 - 2bx + x^2} \dfrac{dx}{x} = 2\pi \ln a \arcsin \dfrac{b}{a}$ $[a \geq |b|]$. BH[134](25)

6. $\displaystyle\int_0^\infty \ln(1+x) \frac{x \ln x - x - a}{(x+a)^2} \frac{dx}{x} = \frac{(\ln a)^2}{2(a-1)}$ $[a > 0]$.   BH[141](7)   4.313

7. $\displaystyle\int_0^\infty \ln(1-x)^2 \frac{x \ln x - x - a}{(x+a)^2} \frac{dx}{x} = \frac{\pi^2 + (\ln a)^2}{1+a}$ $[a > 0]$.   Li[141](8)

1. $\displaystyle\int_0^1 \ln(1+ax) \frac{x^{p-1} - x^{q-1}}{\ln x} dx = \sum_{k=1}^\infty \frac{a^k}{k} \ln \frac{p+k}{q+k} + \ln \frac{p}{q}$ 4.314

    $[a > 0,\ p > 0,\ q > 0]$.   BH[123](18)

2. $\displaystyle\int_0^\infty \left[\frac{(q-1)x}{(1+x)^2} - \frac{1}{x+1} + \frac{1}{(1+x)^q}\right] \frac{dx}{x \ln(1+x)} = \ln \Gamma(q)$ $[q > 0]$.   BH[143](7)

3. $\displaystyle\int_0^1 \frac{x \ln x + 1 - x}{x (\ln x)^2} \ln(1+x)\, dx = \ln \frac{4}{\pi}$.   BH[126](12)

4. $\displaystyle\int_0^1 \frac{\ln(1-x^2)\, dx}{x(q^2 + \ln^2 x)} = -\frac{\pi}{q} \ln \Gamma\left(\frac{q+\pi}{\pi}\right) + \frac{\pi}{2q} \ln 2q + \ln \frac{q}{\pi} - 1$ $[q > 0]$.

    Li[327](12)+

1. $\displaystyle\int_0^1 \ln(1+x)(\ln x)^{n-1} \frac{dx}{x} = (-1)^{n-1} (n-1)! \left(1 - \frac{1}{2^n}\right) \zeta(n+1)$.   BH[116](3)   4.315

2. $\displaystyle\int_0^1 \ln(1+x)(\ln x)^{2n} \frac{dx}{x} = \frac{2^{2n+1} - 1}{(2n+1)(2n+2)} \pi^{2n+2} |B_{2n+2}|$.   BH[116](1)

3. $\displaystyle\int_0^1 \ln(1-x)(\ln x)^{n-1} \frac{dx}{x} = (-1)^n (n-1)!\, \zeta(n+1)$.   BH[116](4)

4. $\displaystyle\int_0^1 \ln(1-x)(\ln x)^{2n} \frac{dx}{x} = -\frac{2^{2n}}{(n+1)(2n+1)} \pi^{2n+2} |B_{2n+2}|$.   BH[116](2)

1. $\displaystyle\int_0^1 \ln(1-ax^r)\left(\ln \frac{1}{x}\right)^p \frac{dx}{x} = -\frac{1}{r^{p+1}} \Gamma(p+1) \sum_{k=1}^\infty \frac{a^k}{k^{p+2}}$ 4.316

    $[p > -1,\ a < 1,\ r > 0]$.   BH[116](7)

2. $\displaystyle\int_0^1 \ln(1 - 2ax \cos t + a^2 x^2) \left(\ln \frac{1}{x}\right)^p \frac{dx}{x} = -2\Gamma(p+1) \sum_{k=1}^\infty \frac{a^k \cos kt}{k^{p+2}}$.   Li[116](8)

1. $\displaystyle\int_0^\infty \ln \frac{\sqrt{1+x^2} + a}{\sqrt{1+x^2} - a} \frac{dx}{\sqrt{1+x^2}} = \pi \arcsin a$ $[|a| < 1]$.   BH[142](11)   4.317

4.317

2. $\int_0^1 \ln \frac{\sqrt{1-a^2x^2} - x\sqrt{1-a^2}}{1-x} \frac{dx}{x} = \frac{1}{2}(\arcsin a)^2.$ BH[115](32)

3. $\int_0^1 \ln \frac{1+\cos t\sqrt{1-x^2}}{1-\cos t\sqrt{1-x^2}} \frac{dx}{x^2 + \operatorname{tg}^2 v} = \pi \operatorname{ctg} t \frac{\cos \frac{v-t}{2}}{\sin \frac{v+t}{2}}.$ BH[115](30)

4. $\int_0^1 \ln\left(\frac{x+\sqrt{1-x^2}}{x-\sqrt{1-x^2}}\right)^2 \frac{x\,dx}{1-x^2} = \frac{\pi^2}{2}.$ BH[115](31)

5. $\int_0^1 \ln\{\sqrt{1+kx} + \sqrt{1-kx}\} \frac{dx}{\sqrt{(1-x^2)(1-k^2x^2)}} =$

$= \frac{1}{4} \ln(4k)\,\boldsymbol{K}(k) + \frac{\pi}{8}\boldsymbol{K}(k').$ BH[121](8)

6. $\int_0^1 \ln\{\sqrt{1+kx} - \sqrt{1-kx}\} \frac{dx}{\sqrt{(1-x^2)(1-k^2x^2)}} =$

$= \frac{1}{4}\ln(4k)\,\boldsymbol{K}(k) + \frac{3}{8}\pi\,\boldsymbol{K}(k').$ BH[121](9)

7. $\int_0^1 \ln\{1 + \sqrt{1-k^2x^2}\} \frac{dx}{\sqrt{(1-x^2)(1-k^2x^2)}} =$

$= \frac{1}{2}\ln k\boldsymbol{K}(k) + \frac{\pi}{4}\boldsymbol{K}(k').$ BH[121](6)

8. $\int_0^1 \ln\{1 - \sqrt{1-k^2x^2}\} \frac{dx}{\sqrt{(1-x^2)(1-k^2x^2)}} =$

$= \frac{1}{2}\ln k\,\boldsymbol{K}(k) - \frac{3}{4}\pi\boldsymbol{K}(k').$ BH[121](7)

9. $\int_0^1 \ln \frac{1+p\sqrt{1-x^2}}{1-p\sqrt{1-x^2}} \frac{dx}{1-x} = \pi \arcsin p \quad [p^2 < 1].$ BH[115](29)

10. $\int_0^1 \ln \frac{1+q\sqrt{1-k^2x^2}}{1-q\sqrt{1-k^2x^2}} \frac{dx}{\sqrt{(1-x^2)(1-k^2x^2)}} = \pi F(\arcsin q, k') \quad [q^2 < 1].$ BH[122](15)

4.318

1. $\int_0^1 \frac{\ln(1-x^q)}{1+(\ln x)^2} \frac{dx}{x} = \pi \left[\ln \Gamma\left(\frac{q}{2\pi} + 1\right) - \frac{\ln q}{2} + \frac{q}{2\pi}\left(\ln \frac{q}{2\pi} - 1\right)\right] \quad [q > 0].$

BH[126](11)

2. $\int_0^\infty \ln(1+x^r) \left[\frac{(p-r)x^q - (q-r)x^q}{\ln x} + \frac{x^q - x^p}{(\ln x)^2} \right] \frac{dx}{x^{r+1}} =$ 4.318

$$= r \ln \left(\operatorname{tg} \frac{q\pi}{2r} \operatorname{ctg} \frac{p\pi}{2r} \right) \quad [p < r, \; q < r]. \qquad \text{BH[143](9)}$$

In Integralen, in denen $\ln(a + bx^r)$ vorkommt, substituiert man zweckmäßigerweise $x^r = t$ und sucht dann die sich ergebenden Integrale in den Tabellen auf. Beispielsweise ist

In integrals containing $\ln(a + bx^r)$ it is advantageous to substitute $x^r = t$; the resulting integrals can be found in these tables. For instance,

$$\int_0^\infty x^{p-1} \ln(1+x^r)\, dx = \frac{1}{r} \int_0^\infty t^{\frac{p}{r}-1} \ln(1+t)\, dt = \frac{\pi}{p \sin \frac{p\pi}{r}} \quad \text{(cf. 4.293 3.).}$$

1. $\int_0^\infty \ln(1 - e^{-2a\pi x}) \frac{dx}{1+x^2} = -\pi \left[\frac{1}{2} \ln 2a\pi + a(\ln a - 1) - \ln \Gamma(a+1) \right]$ 4.319

$[a > 0].$ BH[354](6)

2. $\int_0^\infty \ln(1 + e^{-2a\pi x}) \frac{dx}{1+x^2} = \pi \left[\ln \Gamma(2a) - \ln \Gamma(a) + \right.$

$$+ a(1 - \ln a) - \left(2a - \frac{1}{2}\right) \ln 2 \bigg] \quad [a > 0]. \qquad \text{BH[354](7)}$$

3. $\int_0^\infty \ln \frac{a + be^{-px}}{a + be^{-qx}} \frac{dx}{x} = \ln \frac{a}{a+b} \ln \frac{p}{q} \quad \left[\frac{b}{a} > -1, \; pq > 0 \right].$ F II 643, BH[354](1)

1. $\int_{-\infty}^\infty x \ln \operatorname{ch} x\, dx = 0.$ BH[358](2)+   4.321

2. $\int_0^\infty \ln \operatorname{ch} x \frac{dx}{1-x^2} = 0.$ BH[138](20)+

1. $\int_0^\pi \ln \sin x \cdot x\, dx = \frac{1}{2} \int_0^\pi \ln \cos^2 x \cdot x\, dx = -\frac{\pi^2}{2} \ln 2.$ BH[432](1, 2), F II 652   4.322

2. $\int_0^\infty \frac{\ln \sin^2 ax}{b^2 + x^2} dx = \frac{\pi}{b} \ln \frac{1 - e^{-2ab}}{2} \quad [a > 0, \; b > 0].$ GH[338](28b)

3. $\int_0^\infty \frac{\ln \cos^2 ax}{b^2 + x^2} dx = \frac{\pi}{b} \ln \frac{1 + e^{-2ab}}{2} \quad [a > 0, \; b > 0].$ GH[338](28a)

4.322   4. $\int_0^\infty \dfrac{\ln \sin^2 ax}{b^2 - x^2} dx = -\dfrac{\pi^2}{2b} + a\pi \quad [a > 0, \ b > 0].$ BH[418](1)

5. $\int_0^\infty \dfrac{\ln \cos^2 ax}{b^2 - x^2} dx = a\pi \quad [a > 0].$ BH[418](2)

6. $\int_0^\infty \dfrac{\ln \cos^2 x}{x^2} dx = -\pi.$ F II 695

7. $\int_0^{\pi/4} \ln \sin x \, x^{\mu-1} dx = -\dfrac{1}{2\mu}\left(\dfrac{\pi}{4}\right)^\mu \left[\ln 2 + \dfrac{2}{\mu} - \sum_{k=1}^\infty \dfrac{\zeta(2k)}{4^{2k-1}(\mu + 2k)}\right]$

$[\mathrm{Re}\,\mu > 0].$ Li[425](1)

8. $\int_0^{\pi/2} \ln \sin x \, x^{\mu-1} dx = -\dfrac{1}{\mu}\left(\dfrac{\pi}{2}\right)^\mu \left[\dfrac{1}{\mu} - \sum_{k=1}^\infty \dfrac{\zeta(2k)}{4^k (\mu + 2k)}\right]$

$[\mathrm{Re}\,\mu > 0].$ Li[430](1)

9. $\int_0^{\pi/2} \ln (1 - \cos x) x^{\mu-1} dx = \dfrac{-1}{\mu}\left(\dfrac{\pi}{2}\right)^\mu \left[\dfrac{2}{\mu} + \sum_{k=1}^\infty \dfrac{\zeta(2k)}{4^{2k-1}(\mu + 2k)}\right]$

$[\mathrm{Re}\,\mu > 0].$ Li[430](2)

10. $\int_0^\infty \ln (1 \pm 2p \cos \beta x + p^2) \dfrac{dx}{q^2 + x^2} = \dfrac{\pi}{q} \ln (1 \pm pe^{-\beta q}) \quad [p^2 < 1];$

$= \dfrac{\pi}{q} \ln (p \pm e^{-\beta q}) \quad [p^2 > 1].$ F II 724–725

4.323   1. $\int_0^\pi \ln \mathrm{tg}^2 x \, x \, dx = 0.$ BH[432](3)

2. $\int_0^\infty \dfrac{\ln \mathrm{tg}^2 ax}{b^2 + x^2} dx = \dfrac{\pi}{b} \ln \mathrm{th}\, ab \quad [a > 0, \ b > 0].$ GH[338](28c)

3. $\int_0^\infty \ln \left(\dfrac{1 + \mathrm{tg}\, x}{1 - \mathrm{tg}\, x}\right)^2 \dfrac{dx}{x} = \dfrac{\pi^2}{2}.$ GH[338](26)

4.324   1. $\int_0^\infty \ln \left(\dfrac{1 + \sin x}{1 - \sin x}\right)^2 \dfrac{dx}{x} = \pi^2.$ GH[338](25)

2. $\int_0^\infty \ln\frac{1+2a\cos px + a^2}{1+2a\cos qx + a^2}\frac{dx}{x} = \ln(1+a)\ln\frac{q^2}{p^2} \quad [-1 < a \leq 1];$ 4.324

$$= \ln\left(1+\frac{1}{a}\right)\ln\frac{q^2}{p^2} \quad \left[a < -1 \text{ oder } a \geq 1\right].$$
GH[338](27)

3. $\int_0^\infty \ln(a^2\sin^2 px + b^2\cos^2 px)\frac{dx}{c^2+x^2} =$

$$= \frac{\pi}{c}[\ln(a\,\text{sh}\,cp + b\,\text{ch}\,cp) - cp]$$
$[a > 0, b > 0, c > 0, p > 0].$ GH[338](29)

1. $\int_0^1 \ln\ln\left(\frac{1}{x}\right)\frac{dx}{1+x} = -C\ln 2 + \sum_{k=2}^\infty (-1)^k \frac{\ln k}{k} =$ 4.325

$$= -C\ln 2 + 0{,}159\,868\,905\ldots$$ GH[325](25a)

2. $\int_0^1 \ln\ln\left(\frac{1}{x}\right)\frac{dx}{x+e^{i\lambda}} = \sum_{k=1}^\infty \frac{(-1)^k}{k}e^{-ik\lambda}(C+\ln k).$ GH[325](26)

3. $\int_0^1 \ln\ln\left(\frac{1}{x}\right)\frac{dx}{(1+x)^2} = \int_1^\infty \ln\ln x\,\frac{dx}{(1+x)^2} =$

$$= \frac{1}{2}\left[\psi\left(\frac{1}{2}\right) + \ln 2\pi\right] = \frac{1}{2}\left(\ln\frac{\pi}{2} - C\right).$$ BH[147](7)

4. $\int_0^1 \ln\ln\left(\frac{1}{x}\right)\frac{dx}{1+x^2} = \int_1^\infty \ln\ln x\,\frac{dx}{1+x^2} = \frac{\pi}{2}\ln\frac{\sqrt{2\pi}\,\Gamma\left(\frac{3}{4}\right)}{\Gamma\left(\frac{1}{4}\right)}.$ BH[148](1)

5. $\int_0^1 \ln\ln\left(\frac{1}{x}\right)\frac{dx}{1+x+x^2} = \int_1^\infty \ln\ln x\,\frac{dx}{1+x+x^2} = \frac{\pi}{\sqrt{3}}\ln\frac{\sqrt[3]{2\pi}\,\Gamma\left(\frac{2}{3}\right)}{\Gamma\left(\frac{1}{3}\right)}.$ BH[148](2)

6. $\int_0^1 \ln\ln\left(\frac{1}{x}\right)\frac{dx}{1-x+x^2} = \int_1^\infty \ln\ln x\,\frac{dx}{1-x+x^2} =$

$$= \frac{2\pi}{\sqrt{3}}\left[\frac{5}{6}\ln 2\pi - \ln\Gamma\left(\frac{1}{6}\right)\right].$$ BH[148](5)

7. $\int_0^1 \ln\ln\left(\frac{1}{x}\right)\frac{dx}{1+2x\cos t + x^2} = \int_1^\infty \ln\ln x\,\frac{dx}{1+2x\cos t + x^2} =$

$$= \frac{\pi}{2\sin t}\ln\frac{(2\pi)^{t/\pi}\,\Gamma\left(\frac{1}{2}+\frac{t}{2\pi}\right)}{\Gamma\left(\frac{1}{2}-\frac{t}{2\pi}\right)}.$$ BH[147](9)

4.325

8. $\int_0^1 \ln \ln \frac{1}{x} x^{\mu-1} dx = -\frac{1}{\mu}(C + \ln \mu) \quad [\operatorname{Re} \mu > 0].$ BH[147](1)

9. $\int_1^\infty \ln \ln x \frac{x^{n-2} dx}{1 + x^2 + x^4 + \ldots + x^{2n-2}} =$

$$= \frac{\pi}{2n} \operatorname{tg} \frac{\pi}{2n} \ln 2\pi + \frac{\pi}{n} \sum_{k=1}^{n-1} (-1)^{k-1} \sin \frac{k\pi}{n} \ln \frac{\Gamma\left(\frac{n+k}{2n}\right)}{\Gamma\left(\frac{k}{2n}\right)} \quad \left[n \text{ geradzahlig} \atop \text{even}\right];$$

$$= \frac{\pi}{2n} \operatorname{tg} \frac{\pi}{2n} \ln \pi + \frac{\pi}{n} \sum_{k=1}^{\frac{n-1}{2}} (-1)^{k-1} \sin \frac{k\pi}{n} \ln \frac{\Gamma\left(\frac{n-k}{n}\right)}{\Gamma\left(\frac{k}{n}\right)} \quad \left[n \text{ ungeradzahlig} \atop \text{odd}\right].$$

BH[148](4)

10. $\int_0^1 \ln \ln \left(\frac{1}{x}\right) \frac{dx}{(1+x^2)\sqrt{\ln \frac{1}{x}}} = \int_1^\infty \ln \ln x \frac{dx}{(1+x^2)\sqrt{\ln x}} =$

$$= \sqrt{\pi} \sum_{k=0}^\infty \frac{(-1)^{k+1}}{\sqrt{2k+1}} [\ln(2k+1) + 2\ln 2 + C].$$ BH[147](4)

11. $\int_0^1 \ln \ln \left(\frac{1}{x}\right) \frac{x^{\mu-1} dx}{\sqrt{\ln \frac{1}{x}}} = -(C + \ln 4\mu) \sqrt{\frac{\pi}{\mu}} \quad [\operatorname{Re} \mu > 0].$ BH[147](3)

12. $\int_0^1 \ln \ln \left(\frac{1}{x}\right) \left(\ln \frac{1}{x}\right)^{\mu-1} x^{\nu-1} dx = \frac{1}{\nu^\mu} \Gamma(\mu) [\psi(\mu) - \ln(\nu)]$

$[\operatorname{Re} \mu > 0, \quad \operatorname{Re} \nu > 0].$ BH[147](2)

4.326

1. $\int_0^1 \ln(a - \ln x) x^{\mu-1} dx = \frac{1}{\mu} [\ln a - e^{a\mu} \operatorname{Ei}(-a\mu)] \quad [\operatorname{Re} \mu > 0, a > 0].$

BH[107](23)

2. $\int_0^{1/e} \ln \left(2 \ln \frac{1}{x} - 1\right) \frac{x^{2\mu-1}}{\ln x} dx = -\frac{1}{2} [\operatorname{Ei}(-\mu)]^2 \quad [\operatorname{Re} \mu > 0].$ BH[145](5)

4.327

1. $\int_0^1 \ln[a^2 + (\ln x)^2] \frac{dx}{1+x^2} = \pi \ln \frac{2\Gamma\left(\frac{2a+3\pi}{4\pi}\right)}{\Gamma\left(\frac{2a+\pi}{4\pi}\right)} + \frac{\pi}{2} \ln \frac{\pi}{2} \quad \left[a > -\frac{\pi}{2}\right].$

BH[147](10)

2. $\int_0^1 \ln[a^2 + 4(\ln x)^2] \frac{dx}{1+x^2} = \pi \ln \frac{2\Gamma\left(\frac{a+3\pi}{4\pi}\right)}{\Gamma\left(\frac{a+\pi}{4\pi}\right)} + \frac{\pi}{2} \ln \pi \quad [a > -\pi].$

BH[147](16)+

3. $\int_0^\infty \ln[a^2+(\ln x)^2] x^{\mu-1} dx = \frac{2}{\mu}[-\cos a\mu \operatorname{ci}(a\mu) - \sin a\mu \operatorname{si}(a\mu) + \ln\bar{}a]$ 4.337

$$[a > 0, \quad \operatorname{Re}\mu > 0]. \quad \text{GH[325](28)}$$

Enthält der Integrand einen Logarithmus, dessen Argument ebenfalls einen Logarithmus enthält (steht beispielsweise unter dem Integralzeichen $\ln\ln\frac{1}{x}$), so substituiert man zweckmäßigerweise $\ln x = t$ und sucht dann das transformierte Integral in den Tabellen auf.	If the integrand contains a logarithm whose argument also contains a logarithm, for instance, if the integrand contains $\ln\ln\frac{1}{x}$ the substitution $\ln x = t$ is used before looking up the integral.

Logarithmus und Exponentialfunktion
The logarithmic and the exponential function   4.33–4.34

1. $\int_0^\infty e^{-\mu x} \ln x \, dx = -\frac{1}{\mu}(C + \ln\mu) \quad [\operatorname{Re}\mu > 0].$ BH[256](2)   4.331

2. $\int_1^\infty e^{-\mu x} \ln x \, dx = -\frac{1}{\mu} \operatorname{Ei}(-\mu) \quad [\operatorname{Re}\mu > 0].$ BH[260](5)

3. $\int_0^1 e^{\mu x} \ln x \, dx = -\frac{1}{\mu}\int_0^1 \frac{e^{\mu x}-1}{x} dx \quad [\mu \neq 0].$ GH[324](81a)

1. $\int_0^\infty \frac{\ln x \, dx}{e^x + e^{-x} - 1} = \frac{2\pi}{\sqrt{3}}\left[\frac{5}{6}\ln 2\pi - \ln \Gamma\left(\frac{1}{6}\right)\right]$ (cf. 4.325 6.).   BH[257](6)   4.332

2. $\int_0^\infty \frac{\ln x \, dx}{e^x + e^{-x} + 1} = \frac{\pi}{\sqrt{3}} \ln\left[\frac{\Gamma\left(\frac{2}{3}\right)}{\Gamma\left(\frac{1}{3}\right)}\sqrt{2\pi}\right]$ (cf. 4.325 5.).

BH[257](7)+, Li[260](3)

$\int_0^\infty e^{-\mu x^2} \ln x \, dx = -\frac{1}{4}(C + \ln 4\mu)\sqrt{\frac{\pi}{\mu}} \quad [\operatorname{Re}\mu > 0].$ BH[256](8), F II 813+   4.333

$\int_0^\infty \frac{\ln x \, dx}{e^{x^2}+1+e^{-x^2}} = \frac{1}{2}\sqrt{\frac{\pi}{3}}\sum_{k=1}^\infty (-1)^k \frac{C+\ln 4k}{\sqrt{k}} \sin\frac{k\pi}{3}.$ BH[357](13)   4.334

1. $\int_0^\infty e^{-\mu x}(\ln x)^2 dx = \frac{1}{\mu}\left[\frac{\pi^2}{6} + (C+\ln\mu)^2\right] \quad [\operatorname{Re}\mu > 0].$ Ba IV 149(13)   4.335

4.335

2. $\int_0^\infty e^{-x^2}(\ln x)^2 \, dx = \dfrac{\sqrt{\pi}}{8}\left[(C + 2\ln 2)^2 + \dfrac{\pi^2}{2}\right].$ F II 813

3. $\int_0^\infty e^{-\mu x}(\ln x)^3 \, dx = -\dfrac{1}{\mu}\left[(C + \ln \mu)^3 + \dfrac{\pi^2}{2}(C + \ln \mu) - \psi''(1)\right].$ MHP 26

4.336

1. $\int_0^\infty \dfrac{e^{-x}}{\ln x} \, dx = 0.$ BH[260](9)

2. $\int_0^\infty \dfrac{e^{-\mu x} \, dx}{\pi^2 + (\ln x)^2} = v'(\mu) - e^\mu \quad [\operatorname{Re} \mu > 0].$ MHP 26

4.337

1. $\int_0^\infty e^{-\mu x} \ln(\beta + x) \, dx = \dfrac{1}{\mu}[\ln \beta - e^{\mu\beta} \operatorname{Ei}(-\beta\mu)]$

   $[|\arg \beta| < \pi, \operatorname{Re} \mu > 0].$ BH[256](3)

2. $\int_0^\infty e^{-\mu x} \ln(1 + \beta x) \, dx = -\dfrac{1}{\mu} e^{\mu/\beta} \operatorname{Ei}\left(-\dfrac{\mu}{\beta}\right) \quad [|\arg \beta| < \pi, \operatorname{Re} \mu > 0].$

   Ba IV 148(4)

3. $\int_0^\infty e^{-\mu x} \ln|a - x| \, dx = \dfrac{1}{\mu}[\ln a - e^{-a\mu}\overline{\operatorname{Ei}}(a\mu)] \quad [a > 0, \operatorname{Re} \mu > 0].$ BH[256](4)

4. $\int_0^\infty e^{-\mu x} \ln \dfrac{\beta}{\beta - x} \, dx = \dfrac{1}{\mu}[e^{-\beta\mu}\operatorname{Ei}(\beta\mu)]$ MHP 26

[β nicht positiv reell, $\operatorname{Re} \mu > 0$]. | [$\beta$ cannot be a positive real number, $\operatorname{Re} \mu > 0$].

4.338

1. $\int_0^\infty e^{-\mu x} \ln(\beta^2 + x^2) \, dx = \dfrac{2}{\mu}[\ln \beta - \operatorname{ci}(\beta\mu)\cos(\beta\mu) - \operatorname{si}(\beta\mu)\sin(\beta\mu)].$

   $[\operatorname{Re} \beta > 0, \operatorname{Re} \mu > 0].$ BH[256](6)

2. $\int_0^\infty e^{-\mu x} \ln(x^2 - \beta^2)^2 \, dx = \dfrac{2}{\mu}[\ln \beta^2 - e^{\beta\mu}\operatorname{Ei}(-\beta\mu) - e^{-\beta\mu}\operatorname{Ei}(\beta\mu)]$

   $[\operatorname{Im} \beta > 0, \operatorname{Re} \mu > 0].$ BH[256](5)

4.339 $\int_0^\infty e^{-\mu x} \ln\left|\dfrac{x+1}{x-1}\right| dx = \dfrac{1}{\mu}[e^{-\mu}\overline{\operatorname{Ei}}(\mu) - e^\mu \operatorname{Ei}(-\mu)] \quad [\operatorname{Re} \mu > 0].$ MHP 27

4.341 $\int_0^\infty e^{-\mu x} \ln \dfrac{\sqrt{x + ai} + \sqrt{x - ai}}{\sqrt{2a}} \, dx = \dfrac{\pi}{4\mu}[\mathbf{H}_0(a\mu) - N_0(a\mu)]$

   $[a > 0, \operatorname{Re} \mu > 0].$ Ba IV 149(20)

1. $\int_0^\infty e^{-2nx} \ln(\operatorname{sh} x)\, dx = \dfrac{1}{2n}\left[\dfrac{1}{n} + \ln 2 - 2\beta(2n+1)\right].$ BH[256](17)   4.342

2. $\int_0^\infty e^{-\mu x} \ln(\operatorname{ch} x)\, dx = \dfrac{1}{\mu}\left[\beta\left(\dfrac{\mu}{2}\right) - \dfrac{1}{\mu}\right]$ $[\operatorname{Re}\mu > 0]$.   Ba IV 165(32)

3. $\int_0^\infty e^{-\mu x}[\ln(\operatorname{sh} x) - \ln x]\, dx = \dfrac{1}{\mu}\left[\ln\dfrac{\mu}{2} - \dfrac{1}{2\mu} - \psi\left(\dfrac{\mu}{2}\right)\right]$ $[\operatorname{Re}\mu > 0]$.

   Ba IV 165(33)

$\int_0^\pi e^{\mu\cos x}[\ln(2\mu\sin^2 x) + C]\, dx = -\pi K_0(\mu).$ W 80(16)   4.343

Logarithmus, Exponentialfunktion und Potenzen   4.35–4.36
The logarithmic and the exponential function and powers

1. $\int_0^1 (1-x) e^{-x} \ln x\, dx = \dfrac{1-e}{e}.$ BH[352](1)   4.351

2. $\int_0^1 e^{\mu x}(\mu x^2 + 2x)\ln x\, dx = \dfrac{1}{\mu^2}\left[(1-\mu)e^\mu - 1\right].$ BH[352](2)

3. $\int_1^\infty \dfrac{e^{-\mu x}\ln x}{1+x}\, dx = \dfrac{1}{2} e^\mu [\operatorname{Ei}(-\mu)]^2$ $[\operatorname{Re}\mu > 0]$.   NI 32(10)

1. $\int_0^\infty x^{\nu-1} e^{-\mu x} \ln x\, dx = \dfrac{1}{\mu^\nu}\Gamma(\nu)[\psi(\nu) - \ln\mu]$ $[\operatorname{Re}\mu > 0,\ \operatorname{Re}\nu > 0]$.   4.352

   BH[353](3), Ba IV 315(10)+

2. $\int_0^\infty x^n e^{-\mu x} \ln x\, dx = \dfrac{n!}{\mu^{n+1}}\left[1 + \dfrac{1}{2} + \dfrac{1}{3} + \ldots + \dfrac{1}{n} - C - \ln\mu\right]$

   $[\operatorname{Re}\mu > 0]$.   Ba IV 148(7)

3. $\int_0^\infty x^{n-\frac{1}{2}} e^{-\mu x} \ln x\, dx = \sqrt{\pi}\,\dfrac{(2n-1)!!}{2^n \mu^{n+\frac{1}{2}}}\left[2\left(1 + \dfrac{1}{3} + \dfrac{1}{5} + \ldots + \dfrac{1}{2n-1}\right) - C - \ln 4\mu\right]$ $[\operatorname{Re}\mu > 0]$.   Ba IV 148(10)

4. $\int_0^\infty x^{\mu-1} e^{-x} \ln x\, dx = \Gamma'(\mu)$ $[\operatorname{Re}\mu > 0]$.   GH[324](83a)

4.353

1. $$\int_0^\infty (x-v)\, x^{v-1} e^{-x} \ln x\, dx = \Gamma(v) \quad [\text{Re}\, v > 0].$$
GH[324](84)

2. $$\int_0^\infty \left(\mu x - n - \frac{1}{2}\right) x^{n-\frac{1}{2}} e^{-\mu x} \ln x\, dx = \frac{(2n-1)!!}{(2\mu)^n} \sqrt{\frac{\pi}{\mu}} \quad [\text{Re}\, \mu > 0].$$
BH[357](2)

3. $$\int_0^1 (\mu x + n + 1)\, x^n e^{\mu x} \ln x\, dx = e^\mu \sum_{k=0}^n (-1)^{k-1} \frac{n!}{(n-k)!\, \mu^{k+1}} + (-1)^n \frac{n!}{\mu^{n+1}}$$
$$[\mu \neq 0].$$
GH[324](82)

4.354

1. $$\int_0^\infty \frac{x^{v-1} \ln x}{e^x + 1}\, dx = \Gamma(v) \sum_{k=1}^\infty \frac{(-1)^{k-1}}{k^v} [\psi(v) - \ln k] \quad [\text{Re}\, v > 0].$$
GH[324](86a)

2. $$\int_0^\infty \frac{x^{v-1} \ln x}{(e^x + 1)^2}\, dx = \Gamma(v) \sum_{k=2}^\infty \frac{(-1)^k (k-1)}{k^v} [\psi(v) - \ln k] \quad [\text{Re}\, v > 0].$$
GH[324](86b)

3. $$\int_0^\infty \frac{(x-v)\, e^x - v}{(e^x + 1)^2}\, x^{v-1} \ln x\, dx = \Gamma(v) \sum_{k=1}^\infty \frac{(-1)^{k-1}}{k^v} \quad [\text{Re}\, v > 0].$$
GH[324](87a)

4. $$\int_0^\infty \frac{(x-2n)\, e^x - 2n}{(e^x + 1)^2}\, x^{2n-1} \ln x\, dx = \frac{2^{2n-1} - 1}{2n} \pi^{2n} |B_{2n}|.$$
GH[324](87b)

5. $$\int_0^\infty \frac{x^{v-1} \ln x}{(e^x + 1)^n}\, dx = (-1)^n \frac{\Gamma(v)}{(n-1)!} \sum_{k=n}^\infty \frac{(-1)^k (k-1)!}{(k-n)!\, k^v} [\psi(v) - \ln k]$$
$$[\text{Re}\, v > 0].$$
GH[324](86c)

4.355

1. $$\int_0^\infty x^2 e^{-\mu x^2} \ln x\, dx = \frac{1}{8\mu} (2 - \ln 4\mu - C) \sqrt{\frac{\pi}{\mu}} \quad [\text{Re}\, \mu > 0].$$
BH[357](1)+

2. $$\int_0^\infty x(\mu x^2 - vx - 1) e^{-\mu x^2 + 2vx} \ln x\, dx = \frac{1}{4\mu} + \frac{v}{4\mu} \sqrt{\frac{\pi}{\mu}} \exp\left(\frac{v^2}{\mu}\right)\left[1 + \Phi\left(\frac{v}{\sqrt{\mu}}\right)\right]$$
$$[\text{Re}\, \mu > 0].$$
BH[358](1)

3. $$\int_0^\infty (\mu x^2 - n)\, x^{2n-1} e^{-\mu x^2} \ln x\, dx = \frac{(n-1)!}{4\mu^n} \quad [\text{Re}\, \mu > 0].$$
BH[353](4)

4. $$\int_0^\infty (2\mu x^2 - 2n - 1)\, x^{2n} e^{-\mu x^2} \ln x\, dx = \frac{(2n-1)!!}{2(2\mu)^n} \sqrt{\frac{\pi}{\mu}} \quad [\text{Re}\, \mu > 0].$$
BH[353](5)

1. $\int_0^\infty \exp\left[-\mu\left(\frac{x}{a}+\frac{a}{x}\right)\right] \ln x \, \frac{dx}{x} = 2 \ln a \, K_0(2\mu) \quad [a>0, \; \operatorname{Re}\mu > 0].$ 4.356

GH[324](91)

2. $\int_0^\infty \exp\left(-ax - \frac{b}{x}\right) \ln x \, [2ax^2 - (2n+1)x - 2b] x^{n-\frac{1}{2}} \, dx =$

$$= 2\left(\frac{b}{a}\right)^{\frac{n}{2}} \sqrt{\frac{\pi}{a}} e^{-2\sqrt{ab}} \sum_{k=0}^\infty \frac{(n+k)!}{(n-k)!(2k)!!(2\sqrt{ab})^k}$$
$[a>0, \; b>0].$ BH[357](4)

3. $\int_0^\infty \exp\left(-ax - \frac{b}{x}\right) \ln x \, [2ax^2 + (2n-1)x - 2b] \frac{dx}{x^{n+\frac{3}{2}}} =$

$$= 2\left(\frac{a}{b}\right)^{\frac{n}{2}} \sqrt{\frac{\pi}{a}} e^{-2\sqrt{ab}} \sum_{k=0}^\infty \frac{(n+k-1)!}{(n-k-1)!(2k)!!(2\sqrt{ab})^k}$$
$[a>0, \; b>0].$ BH[357](11)

Für / For $n = \frac{1}{2}$:

4. $\int_0^\infty \exp\left(-ax - \frac{b}{x}\right) \ln x \, \frac{ax^2 - b}{x^2} \, dx = 2K_0(2\sqrt{ab}) \quad [a>0, \; b>0].$

GH[324](92c)

Für / For $n = 0$:

5. $\int_0^\infty \exp\left(-ax - \frac{b}{x}\right) \ln x \, \frac{2ax^2 - x - 2b}{x\sqrt{x}} \, dx = 2\sqrt{\frac{\pi}{a}} e^{-2\sqrt{ab}}$

$[a>0, \; b>0].$ BH[357](7), GH[324](92a)

Für / For $n = -1$:

6. $\int_0^\infty \exp\left(-ax - \frac{b}{x}\right) \ln x \, \frac{2ax^2 - 3x - 2b}{\sqrt{x}} \, dx = \frac{1 + 2\sqrt{ab}}{a} \sqrt{\frac{\pi}{a}} e^{-2\sqrt{ab}}$

$[a>0, \; b>0].$ Li[357](6), GH[324](92b)

1. $\int_0^\infty \exp\left(-\frac{1+x^4}{2ax^2}\right) \ln x \, \frac{1 + ax^2 - x^4}{x^2} \, dx = -\frac{\sqrt{2a^3\pi}}{2\sqrt{e}} \quad [a>0].$ BH[357](8)   4.357

2. $\int_0^\infty \exp\left(-\frac{1+x^4}{2ax^2}\right) \ln x \, \frac{x^4 + ax^2 - 1}{x^4} \, dx = \frac{\sqrt{2a^3\pi}}{2\sqrt{e}} \quad [a>0].$ BH[357](9)

4.357 3. $\int_0^\infty \exp\left(-\frac{1+x^4}{2ax^2}\right) \ln x \cdot \frac{x^4 + 3ax - 1}{x^6} dx = \frac{(1+a)\sqrt{2a^3\pi}}{2\sqrt[a]{e}}$ $[a > 0]$.   BH[357](10)

4.358 1. $\int_1^\infty x^{\nu-1} e^{-\mu x} (\ln x)^m \, dx = \frac{1}{\mu} \frac{\partial^m}{\partial \nu^m} \{\mu^{1-\nu} \Gamma(\mu, \nu)\}$ $[\operatorname{Re} \mu > 0, \ \operatorname{Re} \nu > 0]$.

MHP 26

2. $\int_0^\infty x^{\nu-1} e^{-\mu x} (\ln x)^2 \, dx = \frac{\Gamma(\nu)}{\mu^\nu} \{[\psi(\nu) - \ln \mu]^2 + \zeta(2, \nu)\}$

$[\operatorname{Re} \mu > 0, \ \operatorname{Re} \nu > 0]$.   MHP 26

3. $\int_0^\infty x^{\nu-1} e^{\mu x} (\ln x)^3 \, dx = \frac{\Gamma(\nu)}{\mu^\nu} \{[\psi(\nu) - \ln \mu]^3 +$

$+ 3\zeta(2, \nu)[\psi(\nu) - \ln \mu] - 2\zeta(3, \nu)\}$

$[\operatorname{Re} \mu > 0, \ \operatorname{Re} \nu > 0]$.   MHP 26

4.359 1. $\int_0^\infty e^{-\mu x} \frac{x^{p-1} - x^{q-1}}{\ln x} dx = \frac{1}{\mu} [\lambda(\mu, p-1) - \lambda(\mu, q-1)]$

$[\operatorname{Re} \mu > 0, \ p > 0, \ q > 0]$.   MHP 27

2. $\int_0^1 e^{-\mu x} \frac{x^{p-1} - x^{q-1}}{\ln x} dx = \sum_{k=0}^\infty \frac{\mu^k}{k!} \ln \frac{p+k}{q+k}$ $[\operatorname{Re} \mu > 0, \ p > 0, \ q > 0]$.

BH[352](9)

4.361 1. $\int_0^\infty \frac{(x+1) e^{-\mu x}}{\pi^2 + (\ln x)^2} dx = \nu'(\mu) - \nu''(\mu)$ $[\operatorname{Re} \mu > 0]$.   MHP 27

2. $\int_0^\infty \frac{e^{-\mu x} \, dx}{x[\pi^2 + (\ln x)^2]} = e^\mu - \nu(\mu)$ $[\operatorname{Re} \mu > 0]$.   MHP 27

4.362 1. $\int_0^1 x e^x \ln(1-x) \, dx = 1 - e$.   BH[352](5)+

2. $\int_1^\infty e^{-\mu x} \ln(2x-1) \frac{dx}{x} = \frac{1}{2}\left[\operatorname{Ei}\left(-\frac{\mu}{2}\right)\right]^2$ $[\operatorname{Re} \mu > 0]$.   Ba IV 148(8)

4.363 1. $\int_0^\infty e^{-\mu x} \ln(a+x) \frac{\mu(x+a)\ln(x+a) - 2}{x+a} dx =$

$= \frac{1}{4} \int_0^\infty e^{-\mu x} \ln(a-x)^2 \frac{\mu(x-a)\ln(x-a)^2 - 4}{x-a} dx = (\ln a)^2$

$[\operatorname{Re} \mu > 0, \ a > 0]$.   BH[354](4, 5)

2. $\int_0^1 x(1-x)(2-x) e^{-(1-x)^2} \ln(1-x)\, dx = \dfrac{1-e}{4e}.$ BH[352](4)   4.363

1. $\int_0^\infty e^{-\mu x} \ln[(x+a)(x+b)] \dfrac{dx}{x+a+b} =$ 4.364

$= e^{(a+b)\mu}\{\mathrm{Ei}(-a\mu)\,\mathrm{Ei}(-b\mu) - \ln(ab)\,\mathrm{Ei}[-(a+b)\mu]\}$
$[a>0,\ b>0,\ \mathrm{Re}\,\mu>0].$ BH[354](11)

2. $\int_0^\infty e^{-\mu x} \ln(x+a+b)\left(\dfrac{1}{x+a}+\dfrac{1}{x+b}\right) dx =$

$= (1+\ln a\ln b)\ln(a+b) + e^{-(a+b)\mu}\{\mathrm{Ei}(-a\mu)\,\mathrm{Ei}(-b\mu) +$
$+ (1-\ln(ab))\,\mathrm{Ei}[-(a+b)\mu]\}$ $[a>0,\ b>0,\ \mathrm{Re}\,\mu>0].$

BH[354](12)

$\int_0^\infty \left[e^{-x} - \dfrac{x}{(1+x)^{p+1}\ln(1+x)}\right]\dfrac{ax}{x} = \ln p$ $[p>0].$ BH[354](15)   4.365

1. $\int_0^\infty e^{-\mu x} \ln\left(1+\dfrac{x^2}{a^2}\right)\dfrac{dx}{x} = [\mathrm{ci}(a\mu)]^2 + [\mathrm{si}(a\mu)]^2$ $[\mathrm{Re}\,\mu>0].$ NI 32(11)+   4.366

2. $\int_0^\infty e^{-\mu x} \ln\left|1-\dfrac{x^2}{a^2}\right|\dfrac{dx}{x} = \mathrm{Ei}(a\mu)\,\mathrm{Ei}(-a\mu)$ $[\mathrm{Re}\,\mu>0].$ MH 18

3. $\int_0^\infty x e^{-\mu x^2} \ln\left|\dfrac{1+x^2}{1-x^2}\right| dx = \dfrac{1}{\mu}[\mathrm{ch}\,\mu\,\mathrm{sh}\,i(\mu) - \mathrm{sh}\,\mu\,\mathrm{ch}\,i(\mu)]$

$[\mathrm{Re}\,\mu>0]$ (cf. 4.339).   MHP 27

$\int_0^\infty x e^{-\mu x^2} \ln \dfrac{x+\sqrt{x^2+2\beta}}{\sqrt{2\beta}}\, dx = \dfrac{e^{\beta\mu}}{4\mu} K_0(\beta\mu)$ $[|\arg\beta|<\pi,\ \mathrm{Re}\,\mu>0].$ 4.367

Ba IV 149(19)

$\int_0^{2u} e^{-\mu x^2} \ln \dfrac{x^2(4u^2-x^2)}{u^4}\dfrac{dx}{\sqrt{4u^2-x^2}} = \dfrac{\pi}{2} e^{-2u^2\mu}\left[\dfrac{\pi}{2} N_0(2iu^2\mu) -\right.$ 4.368

$\left. - (C-\ln 2) J_0(2iu^2\mu)\right]$ $[\mathrm{Re}\,\mu>0].$ Ba IV 149(21)+

1. $\int_0^\infty x^{\nu-1} e^{-\mu x}[\psi(\nu) - \ln x]\, dx = -\dfrac{\Gamma(\nu)\ln\mu}{\mu^\nu}$ $[\mathrm{Re}\,\nu>0].$ Ba IV 149(12)   4.369

2. $\int_0^\infty x^n e^{-\mu x}\left\{\left[\ln x - \dfrac{1}{2}\psi(n+1)\right]^2 - \dfrac{1}{2}\psi'(n+1)\right\} dx =$

$= \dfrac{n!}{\mu^{n+1}}\left\{\left[\ln\mu - \dfrac{1}{2}\psi(n+1)\right]^2 + \dfrac{1}{2}\psi'(n+1)\right\}$ $[\mathrm{Re}\,\mu>0].$

MHP 26

Logarithmus und Hyperbelfunktionen
The logarithmic and the hyperbolic function

4.37

4.371

1. $$\int_0^\infty \frac{\ln x}{\operatorname{ch} x}\, dx = \pi \ln \left[\frac{\sqrt{2\pi}\,\Gamma\left(\frac{3}{4}\right)}{\Gamma\left(\frac{1}{4}\right)}\right].$$ Li[260](1)+

2. $$\int_0^\infty \frac{\ln x\, dx}{\operatorname{ch} x + \cos t} = \frac{\pi}{\sin t} \ln \frac{(2\pi)^{\frac{t}{\pi}} \Gamma\left(\frac{\pi + t}{2\pi}\right)}{\Gamma\left(\frac{\pi - t}{2\pi}\right)} \qquad [t^2 < \pi^2].$$ BH[257](7)+

3. $$\int_0^\infty \frac{\ln x\, dx}{\operatorname{ch}^2 x} = \psi\left(\frac{1}{2}\right) + \ln \pi = \ln \pi - 2\ln 2 - C.$$ BH[257](4)+

4.372

1. $$\int_1^\infty \ln x\, \frac{\operatorname{sh} mx}{\operatorname{sh} nx}\, dx =$$

$$= \frac{\pi}{2n} \operatorname{tg} \frac{m\pi}{2n} \ln 2\pi + \frac{\pi}{n} \sum_{k=1}^{n-1} (-1)^{k-1} \sin \frac{km\pi}{n} \ln \frac{\Gamma\left(\frac{n+k}{2n}\right)}{\Gamma\left(\frac{k}{2n}\right)}$$

$$\left[m + n\ \begin{array}{c}\text{ungeradzahlig}\\\text{odd}\end{array}\right];$$

$$= \frac{\pi}{2n} \operatorname{tg} \frac{m\pi}{2n} \ln \pi + \frac{\pi}{n} \sum_{k=1}^{\frac{n-1}{2}} (-1)^{k-1} \sin \frac{km\pi}{n} \ln \frac{\Gamma\left(\frac{n-k}{n}\right)}{\Gamma\left(\frac{k}{n}\right)}$$

$$\left[m + n\ \begin{array}{c}\text{geradzahlig}\\\text{even}\end{array}\right].$$
BH[148](3)+

2. $$\int_1^\infty \ln x\, \frac{\operatorname{ch} mx}{\operatorname{ch} nx}\, dx =$$

$$= \frac{\pi}{2n} \frac{\ln 2\pi}{\cos \frac{m\pi}{2n}} + \frac{\pi}{n} \sum_{k=1}^n (-1)^{k-1} \cos \frac{(2k-1)m\pi}{2n} \ln \frac{\Gamma\left(\frac{2n+2k-1}{4n}\right)}{\Gamma\left(\frac{2k-1}{4n}\right)}$$

$$\left[m + n\ \begin{array}{c}\text{ungeradzahlig}\\\text{odd}\end{array}\right];$$

$$= \frac{\pi}{2n} \frac{\ln \pi}{\cos \frac{m\pi}{2n}} + \frac{\pi}{n} \sum_{k=1}^{\frac{n-1}{2}} (-1)^{k-1} \cos \frac{(2k-1) m\pi}{2n} \ln \frac{\Gamma\left(\frac{2n-2k+1}{2n}\right)}{\Gamma\left(\frac{2k-1}{2n}\right)} \quad 4.372$$

$$\left[m + n \; \begin{array}{l}\text{geradzahlig}\\ \text{even}\end{array}\right].$$

BH[148](6)+

1. $\displaystyle\int_0^\infty \frac{\ln(a^2 + x^2)}{\operatorname{ch} bx} dx = \frac{\pi}{b}\left[2\ln \frac{2\Gamma\left(\frac{2ab+3\pi}{4\pi}\right)}{\Gamma\left(\frac{2ab+\pi}{4\pi}\right)} - \ln \frac{2b}{\pi}\right] \quad \left[b>0, \; a>-\frac{\pi}{2b}\right].$ 4.373

BH[258](11)+

2. $\displaystyle\int_0^\infty \ln(1+x^2) \frac{dx}{\operatorname{ch}\frac{\pi x}{2}} = 2\ln\frac{4}{\pi}.$ BH[258](1)+

3. $\displaystyle\int_0^\infty \ln(a^2+x^2) \frac{\operatorname{sh}\left(\frac{2}{3}\pi x\right)}{\operatorname{sh}\pi x} dx = 2\sin\frac{\pi}{3}\ln\frac{6\Gamma\left(\frac{a+4}{6}\right)\Gamma\left(\frac{a+5}{6}\right)}{\Gamma\left(\frac{a+1}{6}\right)\Gamma\left(\frac{a+2}{6}\right)}$

$[a > -1].$ BH[258](12)

4. $\displaystyle\int_0^\infty \ln(1+x^2) \frac{dx}{\operatorname{sh}^2 ax} = \frac{2}{a}\left[\ln\frac{a}{\pi} + \frac{\pi}{2a} - \psi\left(\frac{\pi+a}{\pi}\right)\right] \quad [a>0].$

BH[258](5)

5. $\displaystyle\int_0^\infty \ln(1+x^2) \frac{\operatorname{ch}\frac{\pi}{2}x}{\operatorname{sh}^2\frac{\pi}{2}x} dx = \frac{2\pi-4}{\pi}.$ BH[258](3)

6. $\displaystyle\int_0^\infty \ln(1+x^2) \frac{\operatorname{ch}\frac{\pi}{4}x}{\operatorname{sh}^2\frac{\pi}{4}x} dx = 4\sqrt{2} - \frac{16}{\pi} + \frac{8\sqrt{2}}{\pi}\ln(\sqrt{2}+1).$ BH[258](2)

1. $\displaystyle\int_0^\infty \ln(\cos^2 t + e^{-2x}\sin^2 t)\frac{dx}{\operatorname{sh} x} = -2t^2.$ BH[259](10)+ 4.374

2. $\displaystyle\int_0^\infty \ln(a+be^{-2x})\frac{dx}{\operatorname{ch}^2 x} = \frac{2}{(b-a)}\left[\frac{a+b}{2}\ln(a+b) - a\ln a - b\ln 2\right]$

$[a>0, \; a+b>0].$ Li[259](14)

1. $\displaystyle\int_0^\infty \ln\operatorname{ch}\frac{x}{2}\frac{dx}{\operatorname{ch} x} = G - \frac{\pi}{4}\ln 2.$ BH[259](11) 4.375

4.375

2. $\int_0^\infty \ln \operatorname{cth} x \, \frac{dx}{\operatorname{ch} x} = \frac{\pi}{2} \ln 2.$ BH[259](16)

4.376

1. $\int_0^\infty \frac{\ln x}{\sqrt{x} \operatorname{ch} x} dx = 2\sqrt{\pi} \sum_{k=0}^\infty \frac{(-1)^{k+1}}{\sqrt{2k+1}} \{\ln(2k+1) + 2\ln 2 + C\}.$ BH[147](4)

2. $\int_0^\infty \ln x \, \frac{(\mu+1)\operatorname{ch} x - x \operatorname{sh} x}{\operatorname{ch}^2 x} x^\mu \, dx = 2\Gamma(\mu+1) \sum_{k=0}^\infty \frac{(-1)^{k+1}}{(2k+1)^{\mu+1}}$

$[\operatorname{Re} \mu > -1].$ BH[356](10)

3. $\int_0^\infty \ln x \, \frac{(n+1)\operatorname{ch} x - x \operatorname{sh} x}{\operatorname{ch}^2 x} x^n \, dx = \frac{(-1)^n}{2^n} \beta^{(n)}\left(\frac{1}{2}\right).$

4. $\int_0^\infty \ln 2x \, \frac{n \operatorname{sh} 2ax - ax}{\operatorname{sh}^2 ax} x^{2n-1} \, dx = -\frac{1}{n}\left(\frac{\pi}{a}\right)^{2n} |B_{2n}|.$ BH[356](9)+

5. $\int_0^\infty \ln x \, \frac{ax \operatorname{ch} ax - (2n+1)\operatorname{sh} ax}{\operatorname{sh}^2 ax} x^{2n} \, dx = 2 \frac{2^{2n+1} - 1}{(2a)^{2n+1}} (2n)! \, \zeta(2n+1).$

BH[356](14)

6. $\int_0^\infty \ln x \, \frac{ax \operatorname{ch} ax - 2n \operatorname{sh} ax}{\operatorname{sh}^2 ax} x^{2n-1} \, dx = \frac{2^{2n-1} - 1}{2n} |B_{2n}| \left(\frac{\pi}{a}\right)^{2n} \quad [a > 0].$

BH[356](15)

7. $\int_0^\infty \ln x \, \frac{(2n+1)\operatorname{ch} ax - ax \operatorname{sh} ax}{\operatorname{ch}^2 ax} x^{2n} dx = -\left(\frac{\pi}{2a}\right)^{2n+1} |E_{2n}| \quad [a > 0].$

BH[356](11)

8. $\int_0^\infty \ln x \, \frac{2ax \operatorname{sh} ax - (2n+1) \operatorname{ch} ax}{\operatorname{ch}^3 ax} x^{2n} dx = \frac{2}{a}(2^{2n-1} - 1)\left(\frac{\pi}{2a}\right)^{2n} |B_{2n}|$

$[a > 0].$ BH[356](2)

9. $\int_0^\infty \ln x \, \frac{2ax \operatorname{ch} ax - (2n+1)\operatorname{sh} ax}{\operatorname{sh}^3 ax} x^{2n} dx = \frac{1}{a}\left(\frac{\pi}{a}\right)^{2n} |B_{2n}|.$ BH[356](6)+

10. $\int_0^\infty \ln x \, \frac{x \operatorname{sh} x - 6 \operatorname{sh}^2\left(\frac{x}{2}\right) - 6\cos^2\frac{t}{2}}{(\operatorname{ch} x + \cos t)^2} x^2 \, dx = \frac{(\pi^2 - t^2)t}{3 \sin t} \quad [0 < t < \pi].$

BH[356](16)+

11. $\int_0^\infty \ln(1+x^2) \, \frac{\operatorname{ch} \pi x + \pi x \operatorname{sh} \pi x}{\operatorname{ch}^2 \pi x} \frac{dx}{x^2} = 4 - \pi.$ BH[356](12)

12. $\int_0^\infty \ln(1+4x^2) \dfrac{\operatorname{ch}\pi x + \pi x \operatorname{sh}\pi x}{\operatorname{ch}^2 \pi x} \dfrac{dx}{x^2} = 4\ln 2.$ BH[356](13) 4.376

$\int_0^\infty \ln 2x \, \dfrac{ax - n(1 - e^{-2ax})}{\operatorname{sh}^2 ax} x^{2n-1}\, dx = \dfrac{1}{2n}\left(\dfrac{\pi}{a}\right)^{2n}|B_{2n}|.$ Li[356](8)+ 4.377

Logarithmus und trigonometrische Funktionen
The logarithmic and the trigonometric function

4.38–4.41

1. $\int_0^1 \ln x \sin ax\, dx = -\dfrac{1}{a}[C + \ln a - \operatorname{ci}(a)] \quad [a > 0].$ GH[338](2a) 4.381

2. $\int_0^1 \ln x \cos ax\, dx = -\dfrac{1}{a}\left[\operatorname{si}(a) + \dfrac{\pi}{2}\right] \quad [a > 0].$ BH[284](2)

3. $\int_0^{2\pi} \ln x \sin nx\, dx = -\dfrac{1}{n}[C + \ln(2n\pi) - \operatorname{ci}(2n\pi)].$ GH[338](1a)

4. $\int_0^{2\pi} \ln x \cos nx\, dx = -\dfrac{1}{n}\left[\operatorname{si}(2n\pi) + \dfrac{\pi}{2}\right].$ GH[338](1b)

1. $\int_0^\infty \ln\left|\dfrac{x+a}{x-a}\right| \sin bx\, dx = \dfrac{\pi}{b}\sin ab \quad [a > 0,\ b > 0].$ Ba IV 77(11) 4.382

2. $\int_0^\infty \ln\left|\dfrac{x+a}{x-a}\right| \cos bx\, dx = \dfrac{2}{b}[\cos ab \operatorname{si}(ab) - \sin ab \operatorname{ci}(ab)]$

$$[a > 0,\ b > 0]. \quad \text{Ba IV 18(9)}$$

3. $\int_0^\infty \ln\dfrac{a^2+x^2}{b^2+x^2} \cos cx\, dx = \dfrac{\pi}{c}(e^{-bc} - e^{-ac}) \quad [a > 0,\ b > 0,\ c > 0].$

F III 526+, BH[337](5)

4. $\int_0^\infty \ln\dfrac{x^2+x+a^2}{x^2-x+a^2} \sin bx\, dx = \dfrac{2\pi}{b}\exp\left(-b\sqrt{a^2-\dfrac{1}{4}}\right)\sin\dfrac{b}{2}$

$$[b > 0]. \quad \text{Ba IV 77(12)}$$

5. $\int_0^\infty \ln\dfrac{(x+\beta)^2+\gamma^2}{(x-\beta)^2+\gamma^2} \sin bx\, dx = \dfrac{2\pi}{b} e^{-\gamma b}\sin\beta b$

$$[\operatorname{Re}\gamma > 0,\ |\operatorname{Im}\beta| \leq \operatorname{Re}\gamma,\ b > 0]. \quad \text{Ba IV 77(13)}$$

4.383

1. $\int_0^\infty \ln(1 + e^{-\beta x}) \cos bx \, dx = \dfrac{\beta}{2b^2} - \dfrac{\pi}{2b \, \mathrm{sh}\left(\dfrac{\pi b}{\beta}\right)}$ [Re $\beta > 0$, $b > 0$].

Ba IV 18(13)

2. $\int_0^\infty \ln(1 - e^{-\beta x}) \cos bx \, dx = \dfrac{\beta}{2b^2} - \dfrac{\pi}{2b} \mathrm{cth}\left(\dfrac{\pi b}{\beta}\right)$ [Re $\beta > 0$, $b > 0$].

Ba IV 18(14)

4.384

1. $\int_0^1 \ln(\sin \pi x) \sin 2n\pi x \, dx = 0$.

GH[338](3a)

2. $\int_0^1 \ln(\sin \pi x) \sin(2n+1)\pi x \, dx = 2 \int_0^{\frac{1}{2}} \ln(\sin \pi x) \sin(2n+1)\pi x \, dx =$

$= -\dfrac{1}{(2n+1)\pi}\left[2C + 2\ln 2 + \psi\left(\dfrac{1}{2} + n\right) + \psi\left(-\dfrac{1}{2} - n\right)\right] =$

$= \dfrac{2}{(2n+1)\pi}\left[\ln 2 - 2 - \dfrac{2}{3} - \ldots - \dfrac{2}{2n-1} - \dfrac{1}{2n+1}\right]$.

GH [338](3b)

3. $\int_0^1 \ln(\sin \pi x) \cos 2n\pi x \, dx = 2 \int_0^{\frac{1}{2}} \ln(\sin \pi x) \cos 2n\pi x \, dx =$

$= -\ln 2 \quad [n = 0]$;

$= -\dfrac{1}{2n} \quad [n > 0]$.

GH[338](3c)

4. $\int_0^1 \ln(\sin \pi x) \cos(2n+1)\pi x \, dx = 0$.

GH[338](3d)

5. $\int_0^{\pi/2} \ln \sin x \sin x \, dx = \ln 2 - 1$.

BH[305](4)

6. $\int_0^{\pi/2} \ln \sin x \cos x \, dx = -1$.

BH[305)](5)

7. $\int_0^{\pi/2} \ln \sin x \cos 2nx \, dx = -\dfrac{\pi}{4n}$.

Li[305](6)

8. $\int_0^\pi \ln \sin x \cos[2m(x - n)] \, dx = -\dfrac{\pi \cos 2mn}{2m}$.

Li[330](8)

9. $\displaystyle\int_0^{\pi/2} \ln\sin x \sin^2 x\, dx = \frac{\pi}{8}(1 - \ln 4).$ BH[305](7) 4.384

10. $\displaystyle\int_0^{\pi/2} \ln\sin x \cos^2 x\, dx = -\frac{\pi}{8}(1 + \ln 4).$ BH[305](8)

11. $\displaystyle\int_0^{\pi/2} \ln\sin x \sin x \cos^2 x\, dx = \frac{1}{9}(\ln 8 - 4).$ BH[305](9)

12. $\displaystyle\int_0^{\pi/2} \ln\sin x \, \text{tg}\, x\, dx = -\frac{\pi^2}{24}.$ BH[305](11)

13. $\displaystyle\int_0^{\pi/2} \ln\sin 2x \sin x\, dx = \int_0^{\pi/2} \ln\sin 2x \cos x\, dx = 2(\ln 2 - 1).$ BH[305](16, 17)

14. $\displaystyle\int_0^{\pi} \frac{\ln(1 + p\cos x)}{\cos x} dx = \pi \arcsin p \quad [p^2 < 1].$ F II 478–479

15. $\displaystyle\int_0^{\pi} \ln\sin x \frac{dx}{1 - 2a\cos x + a^2} = \frac{\pi}{1 - a^2} \ln \frac{1 - a^2}{2} \quad [a^2 < 1];$

$\displaystyle\qquad\qquad = \frac{\pi}{a^2 - 1} \ln \frac{a^2 - 1}{2a^2} \quad [a^2 > 1].$ BH[331](8)

16. $\displaystyle\int_0^{\pi} \ln\sin bx \frac{dx}{1 - 2a\cos x + a^2} = \frac{\pi}{1 - a^2} \ln \frac{1 - a^{2b}}{2} \quad [a^2 < 1].$ BH[331](10)

17. $\displaystyle\int_0^{\pi} \ln\cos bx \frac{dx}{1 - 2a\cos x + a^2} = \frac{\pi}{1 - a^2} \ln \frac{1 + a^{2b}}{2} \quad [a^2 < 1].$ BH[331](11)

18. $\displaystyle\int_0^{\pi/2} \ln\sin x \frac{dx}{1 - 2a\cos 2x + a^2} = \frac{1}{2}\int_0^{\pi} \ln\sin x \frac{dx}{1 - 2a\cos 2x + a^2} =$

$\displaystyle\qquad\qquad = \frac{\pi}{2(1 - a^2)} \ln \frac{1 - a}{2} \quad [a^2 < 1];$

$\displaystyle\qquad\qquad = \frac{\pi}{2(a^2 - 1)} \ln \frac{a - 1}{2a} \quad [a^2 > 1].$

BH[321](1), BH[331](13)

19. $\displaystyle\int_0^{\pi} \ln\sin bx \frac{dx}{1 - 2a\cos 2x + a^2} = \frac{\pi}{1 - a^2} \ln \frac{1 - a^b}{2} \quad [a^2 < 1].$ BH[331](18)

4.384

20. $\int_0^\pi \ln\cos bx \, \frac{dx}{1 - 2a\cos 2x + a^2} = \frac{\pi}{1-a^2} \ln\frac{1+a^b}{2}$ $[a^2 < 1]$. BH[331](21)

21. $\int_0^{\pi/2} \frac{\ln\cos x \, dx}{1 - 2p\cos 2x + p^2} = \frac{\pi}{2(1-p^2)} \ln\frac{1+p}{2}$ $[p^2 < 1]$;

$\qquad = \frac{\pi}{2(p^2-1)} \ln\frac{p+1}{2p}$ $[p^2 > 1]$. BH[321](8)

22. $\int_0^\pi \ln\sin x \, \frac{\cos x \, dx}{1 - 2a\cos x + a^2} = \frac{\pi}{2a} \frac{1+a^2}{1-a^2} \ln(1-a^2) - \frac{a\pi \ln 2}{1-a^2}$ $[a^2 < 1]$;

$\qquad = \frac{\pi}{2a} \frac{a^2+1}{a^2-1} \ln\frac{a^2-1}{a^2} - \frac{\pi \ln 2}{a(a^2-1)}$ $[a^2 > 1]$.

Li[331](9)

23. $\int_0^\pi \ln\sin bx \, \frac{\cos x \, dx}{1 - 2a\cos 2x + a^2} = \int_0^\pi \ln\cos bx \, \frac{\cos x \, dx}{1 - 2a\cos 2x + a^2} = 0$

$[0 < a < 1]$. BH[331](19, 22)

24. $\int_0^\pi \ln\sin x \, \frac{\cos^2 x \, dx}{1 - 2a\cos 2x + a^2} = \frac{\pi}{4a} \frac{1+a}{1-a} \ln(1-a) - \frac{\pi \ln 2}{2(1-a)}$ $[0 < a < 1]$;

$\qquad = \frac{\pi}{4a} \frac{a+1}{a-1} \ln\frac{a-1}{a} - \frac{\pi \ln 2}{2a(a-1)}$ $[a > 1]$.

BH[331](16)

25. $\int_0^{\pi/2} \ln\sin x \, \frac{\cos 2x \, dx}{1 - 2a\cos 2x + a^2} = \frac{1}{2} \int_0^\pi \ln\sin x \, \frac{\cos 2x \, dx}{1 - 2a\cos 2x + a^2} =$

$\qquad = \frac{\pi}{2a(1-a^2)} \left\{ \frac{1+a^2}{2} \ln(1-a) - a^2\ln 2 \right\}$ $[a^2 < 1]$;

$\qquad = \frac{\pi}{2a(a^2-1)} \left\{ \frac{1+a^2}{2} \ln\frac{a-1}{a} - \ln 2 \right\}$ $[a^2 > 1]$.

BH[321](2), BH[331](15), Li[321](2)

26. $\int_0^{\pi/2} \ln\cos x \, \frac{\cos 2x \, dx}{1 - 2a\cos 2x + a^2} =$

$\qquad = \frac{\pi}{2a(1-a^2)} \left\{ \frac{1+a^2}{2} \ln(1+a) - a^2\ln 2 \right\}$ $[a^2 < 1]$;

$\qquad = \frac{\pi}{2a(a^2-1)} \left\{ \frac{1+a^2}{2} \ln\frac{1+a}{a} - \ln 2 \right\}$ $[a^2 > 1]$. BH[321](9)

4.385

1. $\int_0^\pi \ln\sin x \, \frac{dx}{a + b\cos x} = \frac{\pi}{\sqrt{a^2-b^2}} \ln\frac{\sqrt{a^2-b^2}}{a+\sqrt{a^2-b^2}}$ $[a > 0, a > b]$.

BH[331](6)

2. $\int_0^{\pi/2} \ln \sin x \, \frac{dx}{(a \sin x \pm b \cos x)^2} = \int_0^{\pi/2} \ln \cos x \, \frac{dx}{(a \cos x \pm b \sin x)^2} =$ 4.385

$$= \frac{1}{b(a^2+b^2)} \left(\mp a \ln \frac{a}{b} - \frac{b\pi}{2} \right) \quad [a > 0, b > 0]. \quad \text{BH[319](1, 6)}+$$

3. $\int_0^{\pi/2} \frac{\ln \sin x \, dx}{a^2 \sin^2 x + b^2 \cos^2 x} = \int_0^{\pi/2} \frac{\ln \cos x \, dx}{b^2 \sin^2 x + a^2 \cos^2 x} = \frac{\pi}{2ab} \ln \frac{b}{a+b}$

$$[a > 0, b > 0]. \quad \text{BH[317](4, 10)}$$

4. $\int_0^{\pi/2} \ln \sin x \, \frac{\sin 2x \, dx}{(a \sin^2 x + b \cos^2 x)^2} =$

$$= \int_0^{\pi/2} \ln \cos x \, \frac{\sin 2x \, dx}{(b \sin^2 x + a \cos^2 x)^2} = \frac{1}{2b(b-a)} \ln \frac{a}{b}$$

$$[a > 0, b > 0]. \quad \text{BH[319](3, 7), Li[319](3)}$$

5. $\int_0^{\pi/2} \ln \sin x \, \frac{a^2 \sin^2 x - b^2 \cos^2 x}{(a^2 \sin^2 x + b^2 \cos^2 x)^2} \, dx =$

$$= \int_0^{\pi/2} \ln \cos x \, \frac{a^2 \cos^2 x - b^2 \sin^2 x}{(a^2 \cos^2 x + b^2 \sin^2 x)^2} \, dx = \frac{\pi}{2b(a+b)} \quad [a > 0, b > 0]. \quad \text{Li[319](2, 8)}$$

1. $\int_0^{\pi/2} \ln \sin x \, \frac{\sin x}{\sqrt{1 + \sin^2 x}} \, dx = \int_0^{\pi/2} \frac{\cos x \ln \cos x}{\sqrt{1 + \cos^2 x}} \, dx = -\frac{\pi}{8} \ln 2.$ BH[322](1, 6) 4.386

2. $\int_0^{\pi/2} \frac{\sin^3 x \ln \sin x}{\sqrt{1 + \sin^2 x}} \, dx = \int_0^{\pi/2} \frac{\cos^3 x \ln \cos x}{\sqrt{1 + \cos^2 x}} \, dx = \frac{\ln 2 - 1}{4}.$ BH[322](2, 7)

3. $\int_0^{\pi/2} \ln \sin x \, \frac{dx}{\sqrt{1 - k^2 \sin^2 x}} = -\frac{1}{2} K(k) \ln k - \frac{\pi}{4} K(k').$ BH[322](3)

4. $\int_0^{\pi/2} \frac{\ln \cos x \, dx}{\sqrt{1 - k^2 \sin^2 x}} = \frac{1}{2} K(k) \ln \frac{k'}{k} - \frac{\pi}{4} K(k').$ BH[322](9)

1. $\int_0^{\pi/2} \ln \sin x \sin^\mu x \cos^\nu x \, dx = \int_0^{\pi/2} \ln \cos x \cos^\mu x \sin^\nu x \, dx =$ 4.387

$$= \frac{1}{4} B\left(\frac{\mu+1}{2}, \frac{\nu+1}{2} \right) \left[\psi\left(\frac{\mu+1}{2} \right) - \psi\left(\frac{\mu+\nu+2}{2} \right) \right]$$

$$[\operatorname{Re} \mu > -1, \operatorname{Re} \nu > -1]. \quad \text{GH[338](6c)}$$

4.387

2. $\int_0^{\pi/2} \ln \sin x \, \sin^{\mu-1} x \, dx = \dfrac{\sqrt{\pi}\, \Gamma\left(\dfrac{\mu}{2}\right)}{4\Gamma\left(\dfrac{\mu+1}{2}\right)} \left[\psi\left(\dfrac{\mu}{2}\right) - \psi\left(\dfrac{\mu+1}{2}\right)\right]$ [Re $\mu > 0$].

GH[338](6a)

3. $\int_0^{\pi/2} \ln \sin x \, \cos^{\nu-1} x \, dx = \dfrac{\sqrt{\pi}\, \Gamma\left(\dfrac{\nu}{2}\right)}{4\Gamma\left(\dfrac{\nu+1}{2}\right)} \left[\psi\left(\dfrac{\nu}{2}\right) - \psi\left(\dfrac{\nu+1}{2}\right)\right]$ [Re $\nu > 0$].

GH[338](6b)

4. $\int_0^{\pi/2} \ln \sin x \, \sin^{2n} x \, dx = \dfrac{(2n-1)!!}{(2n)!!} \dfrac{\pi}{2} \left\{ \sum_{k=1}^{2n} \dfrac{(-1)^{k+1}}{k} - \ln 2 \right\}$.

F II 816

5. $\int_0^{\pi/2} \ln \sin x \, \sin^{2n+1} x \, dx = \dfrac{(2n)!!}{(2n+1)!!} \left\{ \sum_{k=1}^{2n+1} \dfrac{(-1)^k}{k} + \ln 2 \right\}$.

BH[305](13)

6. $\int_0^{\pi/2} \ln \sin x \, \cos^{2n} x \, dx = -\dfrac{(2n-1)!!}{(2n)!!} \dfrac{\pi}{4} \left[\sum_{k=1}^{n} \dfrac{1}{k} + \ln 4 \right] =$

$= -\dfrac{(2n-1)!!}{(2n)!!} \dfrac{\pi}{4} [C + \psi(n+1) + \ln 4]$.

BH[305](14)

7. $\int_0^{\pi/2} \ln \sin x \, \cos^{2n+1} x \, dx = -\dfrac{(2n)!!}{(2n+1)!!} \sum_{k=0}^{n} \dfrac{1}{2k+1} =$

$= -\dfrac{(2n)!!}{2(2n+1)!!} \left[\psi\left(n+\dfrac{3}{2}\right) - \psi\left(\dfrac{1}{2}\right) \right]$.

GH[338](7b)

8. $\int_0^{\pi/2} \ln \cos x \, \sin^{2n} x \, dx = -\dfrac{(2n-1)!!}{2^{n+1} \cdot n!} \dfrac{\pi}{2} \{C + 2\ln 2 + \psi(n+1)\}$.

BH[306](8)

9. $\int_0^{\pi/2} \ln \cos x \, \cos^{2n} x \, dx = -\dfrac{(2n-1)!!}{2^n n!} \dfrac{\pi}{2} \left\{ \ln 2 + \sum_{k=1}^{2n} \dfrac{(-1)^k}{k} \right\}$.

BH[306](10)

10. $\int_0^{\pi/2} \ln \cos x \, \cos^{2n-1} x \, dx = \dfrac{2^{n-1}(n-1)!}{(2n-1)!!} \left[\ln 2 + \sum_{k=1}^{2n-1} \dfrac{(-1)^k}{k} \right]$.

BH[306](9)

4.388

1. $\int_0^{\pi/4} \ln \sin x \, \dfrac{\sin^{2n} x}{\cos^{2n+2} x} \, dx = \dfrac{1}{2n+1} \left[\dfrac{1}{2} \ln 2 + (-1)^n \dfrac{\pi}{4} + \sum_{k=0}^{n-1} \dfrac{(-1)^k}{2n-2k-1} \right]$.

BH[288](1)

2. $\int_0^{\pi/4} \ln \sin x \, \frac{\sin^{2n-1} x}{\cos^{2n+1} x} \, dx = \frac{1}{4n} \left[-\ln 2 + (-1)^n \ln 2 + \sum_{k=1}^{n-1} \frac{(-1)^k}{n-k} \right].$ Li[288](2) 4.388

3. $\int_0^{\pi/4} \ln \cos x \, \frac{\sin^{2n} x}{\cos^{2n+2} x} \, dx = \frac{1}{2n+1} \left[-\frac{1}{2} \ln 2 + (-1)^{n+1} \frac{\pi}{4} + \sum_{k=0}^{n} \frac{(-1)^{k-1}}{2n-2k+1} \right].$

BH[288](10)

4. $\int_0^{\pi/4} \ln \cos x \, \frac{\sin^{2n-1} x}{\cos^{2n+1} x} \, dx = \frac{1}{4n} \left[-\ln 2 + (-1)^n \ln 2 + \sum_{k=0}^{n-1} \frac{(-1)^k}{n-k} \right].$ BH[288](11)

5. $\int_0^{\pi/2} \ln \sin x \, \frac{\sin^{p-1} x}{\cos^{p+1} x} \, dx = -\frac{\pi}{2p} \operatorname{cosec} \frac{p\pi}{2} \quad [0 < p < 2].$ BH[310](4)

6. $\int_0^{\pi/2} \ln \sin x \, \frac{dx}{\operatorname{tg}^{p-1} x \sin 2x} = \frac{1}{4} \frac{\pi}{p-1} \sec \frac{p\pi}{2} \quad [p^2 < 1].$ BH[310](3)

1. $\int_0^{\pi} \ln \sin x \, \sin^{2n} 2x \cos 2x \, dx = -\frac{(2n-1)!!}{(2n)!!} \frac{\pi}{4n+2}.$ BH[330](9) 4.389

2. $\int_0^{\pi/4} \ln \sin x \cdot \cos^n 2x \sin 2x \, dx = -\frac{1}{4(n+1)} \{ C + \psi(n+2) + \ln 2 \}.$ BH[285](2)

3. $\int_0^{\pi/4} \ln \cos x \, \cos^{\mu-1} 2x \operatorname{tg} 2x \, dx = \frac{1}{4(1-\mu)} \beta(\mu) \quad [\operatorname{Re} \mu > 0].$ BH[286](2)

4. $\int_0^{\pi/2} \ln \sin x \, \sin^{\mu-1} x \cos x \, dx = \int_0^{\pi/2} \ln \cos x \, \cos^{\mu-1} x \sin x \, dx = -\frac{1}{\mu^2}$

$[\operatorname{Re} \mu > 0].$ BH[306](11)

5. $\int_{-\pi/2}^{\pi/2} \ln \cos x \, \cos^p x \cos px \, dx = -\frac{\pi}{2^p} \ln 2 \quad [p > -1].$ BH[337](6)

6. $\int_0^{\pi/2} \ln \cos x \, \cos^{p-1} x \sin px \sin x \, dx = \frac{\pi}{2^{p+2}} \left[C + \psi(p) - \frac{1}{p} - 2\ln 2 \right]$

$[p > 0].$ BH[306](12)

4.391

1. $\displaystyle\int_0^{\pi/4} (\ln\cos 2x)^n \cos^{p-1} 2x \,\operatorname{tg} x \, dx =$

$\displaystyle = \int_0^{\pi/4} (\ln\sin 2x)^n \sin^{p-1} 2x \,\operatorname{tg}\left(\frac{\pi}{4} - x\right) dx = \frac{1}{2}\beta^{(n)}(p) \quad [p > 0].$

BH[286](10), BH[285](18)

2. $\displaystyle\int_0^{\pi/4} (\ln\sin 2x)^n \sin^{p-1} 2x \,\operatorname{tg}\left(\frac{\pi}{4} + x\right) dx = \frac{(-1)^n n!}{2}\zeta(n+1, p).$ BH[285](17)

3. $\displaystyle\int_0^{\pi/4} (\ln\cos 2x)^{2n-1} \operatorname{tg} x \, dx = \frac{1 - 2^{2n-1}}{4n}\pi^{2n}|B_{2n}|.$ BH[286](7)

4. $\displaystyle\int_0^{\pi/4} (\ln\cos 2x)^{2n} \operatorname{tg} x \, dx = \frac{2^{2n} - 1}{2^{2n+1}}(2n)!\,\zeta(2n+1).$ BH[286](8)

4.392

1. $\displaystyle\int_0^{\pi/4} \ln(\sin x \cos x) \frac{\sin^{2n} x}{\cos^{2n+2} x} dx =$

$\displaystyle = \frac{1}{2n+1}\left[(-1)^{n+1}\frac{\pi}{2} - \ln 2 + \frac{1}{2n+1} + 2\sum_{k=0}^{n-1}\frac{(-1)^{k-1}}{2n - 2k - 1}\right].$ BH[294](8)

2. $\displaystyle\int_0^{\pi/4} \ln(\sin x \cos x) \frac{\sin^{2n-1} x}{\cos^{2n+1} x} dx =$

$\displaystyle = \frac{1}{2n}\left[(-1)^n \ln 2 - \ln 2 + \frac{1}{2n} + (-1)^n \sum_{k=1}^{n-1}\frac{(-1)^k}{k}\right].$ BH[294](9)

4.393

1. $\displaystyle\int_0^{\pi/2} \ln\operatorname{tg} x \sin x \, dx = \ln 2.$ BH[307](3)

2. $\displaystyle\int_0^{\pi/2} \ln\operatorname{tg} x \cos x \, dx = -\ln 2.$ BH[307](4)

3. $\displaystyle\int_0^{\pi/2} \ln\operatorname{tg} x \sin^2 x \, dx = -\int_0^{\pi/2} \ln\operatorname{tg} x \cos^2 x \, dx = \frac{\pi}{4}.$ BH[307](5,6)

4. $\displaystyle\int_0^{\pi/4} \frac{\ln\operatorname{tg} x}{\cos 2x} dx = -\frac{\pi^2}{8}.$ GH[338](10b)+

5. $\displaystyle\int_0^{\pi/2} \sin x \ln\operatorname{ctg}\frac{x}{2} dx = \ln 2.$ Lo III 290

1.
$$\int_0^{\pi/2} \frac{\ln \tg x \, dx}{1 - 2a \cos 2x + a^2} = \frac{\pi}{2(1-a^2)} \ln \frac{1-a}{1+a} \quad [a^2 < 1];$$

$$= \frac{\pi}{2(a^2-1)} \ln \frac{a-1}{a+1} \quad [a^2 > 1]. \qquad \text{BH[321](15)} \quad 4.394$$

2.
$$\int_0^{\pi/2} \frac{\ln \tg x \cos 2x \, dx}{1 - 2a \cos 2x + a^2} = \frac{\pi}{4a} \frac{1+a^2}{1-a^2} \ln \frac{1-a}{1+a} \quad [a^2 < 1];$$

$$= \frac{\pi}{4a} \frac{a^2+1}{a^2-1} \ln \frac{a-1}{a+1} \quad [a^2 > 1]. \qquad \text{BH[321](16)}$$

3.
$$\int_0^{\pi} \frac{\ln \tg bx \, dx}{1 - 2a \cos 2x + a^2} = \frac{\pi}{1-a^2} \ln \frac{1-a^b}{1+a^b} \quad [0 < a < 1, b > 0]. \qquad \text{BH[331](24)}$$

4.
$$\int_0^{\pi} \frac{\ln \tg bx \cos x \, dx}{1 - 2a \cos 2x + a^2} = 0 \quad [0 < a < 1]. \qquad \text{BH[331](25)}$$

5.
$$\int_0^{\pi/4} \ln \tg x \, \frac{\cos 2x \, dx}{1 - a \sin 2x} = -\frac{\arcsin a}{4a} (\pi + \arcsin a) \quad [a^2 \leq 1]. \qquad \text{BH[291](2, 3)}$$

6.
$$\int_0^{\pi/4} \ln \tg x \, \frac{\cos 2x \, dx}{1 - a^2 \sin^2 2x} = -\frac{\pi}{4a} \arcsin a \quad [a^2 < 1]. \qquad \text{BH[291](9)}$$

7.
$$\int_0^{\pi/4} \ln \tg x \, \frac{\cos 2x \, dx}{1 + a^2 \sin^2 x} = -\frac{\pi}{4a} \Arsh a = -\frac{\pi}{4a} \ln (a + \sqrt{1+a^2})$$
$$[a^2 < 1]. \qquad \text{BH[291](10)}$$

8.
$$\int_0^u \frac{\sin x \ln \ctg \frac{x}{2}}{1 - \cos^2 \alpha \sin^2 x} dx =$$

$$= \cosec 2\alpha \left\{ \frac{\pi}{2} \ln 2 + L(\varphi - \alpha) - L(\varphi + \alpha) - L\left(\frac{\pi}{2} - 2\alpha\right) \right\}$$
$$[\tg \varphi = \ctg \alpha \cos u; \, 0 < u < \pi]. \qquad \text{Lo III 290}$$

9.
$$\int_0^{\pi/4} \frac{\ln \tg x \sin 2x \, dx}{1 - \cos^2 t \sin^2 2x} = \cosec 2t \left[L\left(\frac{\pi}{2} - t\right) - \left(\frac{\pi}{2} - t\right) \ln 2 \right]. \qquad \text{Lo III 290+}$$

1.
$$\int_0^{\pi/2} \frac{\ln \tg x \, dx}{\sqrt{1 - k^2 \sin^2 x}} = -\ln k' \mathbf{K}(k). \qquad \text{BH[322](11)} \quad 4.395$$

2.
$$\int_u^{\pi/4} \frac{\ln \tg x \sin 4x \, dx}{(\sin^2 u + \tg^2 v \sin^2 2x) \sqrt{\sin^2 2x - \sin^2 u}} =$$

$$= -\frac{\pi}{2} \frac{\cos^2 v}{\sin u \sin v} \ln \frac{\sin v + \sqrt{1 - \cos^2 u \cos^2 v}}{\sin u (1 + \sin v)}$$
$$\left[0 < u < \frac{\pi}{2}, 0 < v < \frac{\pi}{2} \right]. \qquad \text{Lo III 285+}$$

4.396

1. $$\int_0^{\frac{\pi}{2}} \ln(a\,\mathrm{tg}\,x)\sin^{\mu-1}2x\,dx = 2^{\mu-2}\ln a\,\frac{\left\{\Gamma\left(\frac{a}{2}\right)\right\}^2}{\Gamma(a)}$$
$$[a>0,\ \mathrm{Re}\,\mu>0].\qquad \text{Li[307](8)}$$

2. $$\int_0^{\frac{\pi}{2}} \ln\mathrm{tg}\,x\,\cos^{2(\mu-1)}x\,dx = -\frac{\sqrt{\pi}\,\Gamma\left(\mu-\frac{1}{2}\right)}{4\,\Gamma(\mu)}\left[C+\psi\left(\frac{2\mu-1}{2}\right)+\ln 4\right]$$
$$\left[\mathrm{Re}\,\mu>\frac{1}{2}\right].\qquad \text{BH[307](9)}$$

3. $$\int_0^{\frac{\pi}{2}} \ln\mathrm{tg}\,x\,\cos^{q-1}x\,\mathrm{ctg}\,x\,\sin[(q+1)x]\,dx = -\frac{\pi}{2}[C+\psi(q+1)]$$
$$[q>-1].\qquad \text{BH[307](11)}$$

4. $$\int_0^{\frac{\pi}{2}} \ln\mathrm{tg}\,x\,\cos^{q-1}x\,\cos[(q+1)x]\,dx = -\frac{\pi}{2q}\quad [q>0].\qquad \text{BH[307](10)}$$

5. $$\int_0^{\frac{\pi}{4}} (\ln\mathrm{tg}\,x)^n\,\mathrm{tg}^p\,x\,dx = \frac{1}{2^{n+1}}\beta^{(n)}\left(\frac{p+1}{2}\right)\quad [p>-1].\qquad \text{Li[286](22)}$$

6. $$\int_0^{\frac{\pi}{2}} (\ln\mathrm{tg}\,x)^{2n-1}\,\frac{dx}{\cos 2x} = \frac{1-2^{2n}}{2n}\pi^{2n}|B_{2n}|.\qquad \text{BH[312](6)}$$

7. $$\int_0^{\frac{\pi}{4}} \ln\mathrm{tg}\,x\,\mathrm{tg}^{2n+1}x\,dx = \frac{(-1)^{n+1}}{4}\left[\frac{\pi^2}{12}+\sum_{k=1}^{n}\frac{(-1)^k}{k^2}\right].\qquad \text{GH[338](8a)}$$

4.397

1. $$\int_0^{\frac{\pi}{2}} \ln(1+p\sin x)\,\frac{dx}{\sin x} = \frac{\pi^2}{8}-\frac{1}{2}(\arccos p)^2\quad [p^2<1].\qquad \text{BH[313](1)}$$

2. $$\int_0^{\frac{\pi}{2}} \ln(1+p\cos x)\,\frac{dx}{\cos x} = \frac{\pi^2}{8}-\frac{1}{2}(\arccos p)^2\quad [p^2<1].\qquad \text{BH[313](8)}$$

3. $$\int_0^{\pi} \ln(1+p\cos x)\,\frac{dx}{\cos x} = \pi\arcsin p\quad [p^2<1].\qquad \text{BH[331](1)}$$

4. $$\int_0^{\frac{\pi}{2}} \frac{\cos x\,\ln(1+\cos\alpha\cos x)}{1-\cos^2\alpha\cos^2 x}\,dx = \frac{L\left(\frac{\pi}{2}-\alpha\right)-\alpha\ln\sin\alpha}{\sin\alpha\cos\alpha}\quad \left[0<\alpha<\frac{\pi}{2}\right].\qquad \text{Lo III 291}$$

5. $$\int_0^{\frac{\pi}{2}} \frac{\cos x\,\ln(1-\cos\alpha\cos x)}{1-\cos^2\alpha\cos^2 x}\,dx = \frac{L\left(\frac{\pi}{2}-\alpha\right)+(\pi-\alpha)\ln\sin\alpha}{\sin\alpha\cos\alpha}\quad \left[0<\alpha<\frac{\pi}{2}\right].$$
$$\text{Lo III 291}$$

6. $\int_0^\pi \ln(1 - 2a\cos x + a^2)\cos nx\, dx = \frac{1}{2}\int_0^{2\pi} \ln(1 - 2a\cos x + a^2)\cos nx\, dx =$ 4.397

$$= -\frac{\pi}{n}a^n \quad [a^2 < 1]; \qquad\qquad \text{BH[330](11), BH[332](5)}$$

$$= -\frac{\pi}{na^n} \quad [a^2 > 1]. \qquad\qquad \text{GH[338](13a)}$$

7. $\int_0^\pi \ln(1 - 2a\cos x + a^2)\sin nx \sin x\, dx =$

$$= \frac{1}{2}\int_0^{2\pi}\ln(1 - 2a\cos x + a^2)\sin nx \sin x\, dx = \frac{\pi}{2}\left(\frac{a^{n+1}}{n+1} - \frac{a^{n-1}}{n-1}\right)$$

$$[a^2 < 1]. \qquad \text{BH[330](10), BH[332](4)}$$

8. $\int_0^\pi \ln(1 - 2a\cos x + a^2)\cos nx \cos x\, dx =$

$$= \frac{1}{2}\int_0^{2\pi}\ln(1 - 2a\cos x + a^2)\cos nx \cos x\, dx = -\frac{\pi}{2}\left(\frac{a^{n+1}}{n+1} + \frac{a^{n-1}}{n-1}\right)$$

$$[a^2 < 1]. \qquad \text{BH[330](12), BH[332](6)}$$

9. $\int_0^\pi \ln(1 - 2a\cos 2x + a^2)\cos(2n-1)x\, dx = 0 \quad [a^2 < 1]. \qquad \text{BH[330](15)}$

10. $\int_0^\pi \ln(1 - 2a\cos 2x + a^2)\sin 2nx \sin x\, dx = 0 \quad [a^2 < 1]. \qquad \text{BH[330](13)}$

11. $\int_0^\pi \ln(1 - 2a\cos 2x + a^2)\sin(2n-1)x \sin x\, dx =$

$$= \frac{\pi}{2}\left(\frac{a^n}{n} - \frac{a^{n-1}}{n-1}\right) \quad [a^2 < 1]. \qquad \text{BH[330](14)}$$

12. $\int_0^\pi \ln(1 - 2a\cos 2x + a^2)\cos 2nx \cos x\, dx = 0 \quad [a^2 < 1]. \qquad \text{BH[330](16)}$

13. $\int_0^\pi \ln(1 - 2a\cos 2x + a^2)\cos(2n-1)x \cos x\, dx =$

$$= -\frac{\pi}{2}\left(\frac{a^n}{n} + \frac{a^{n-1}}{n-1}\right) \quad [a^2 < 1]. \qquad \text{BH[330](17)}$$

14. $\int_0^{\pi/2}\ln(1 + 2a\cos 2x + a^2)\sin^2 x\, dx = -\frac{a\pi}{4} \quad [a^2 < 1];$

$$= \frac{\pi \ln a^2}{4} - \frac{\pi}{4a} \quad [a^2 > 1].$$

$$\text{BH[309](22), Li[309](22)}$$

4.397 15. $\displaystyle\int_0^{\pi/2} \ln(1 + 2a\cos 2x + a^2) \cos^2 x\, dx = \frac{a\pi}{4}$ $[a^2 < 1]$;

$$= \frac{\pi \ln a^2}{4} + \frac{\pi}{4a} \quad [a^2 > 1].$$
BH[309](23), Li[309](23)

16. $\displaystyle\int_0^\pi \frac{\ln(1 - 2a\cos x + a^2)}{1 - 2b\cos x + b^2}\, dx = \frac{2\pi \ln(1 - ab)}{1 - b^2}$ $[a^2 \leqslant 1,\, b^2 < 1]$. BH[331](26)

4.398 1. $\displaystyle\int_0^\pi \ln \frac{1 + 2a\cos x + a^2}{1 - 2a\cos x + a^2} \sin(2n+1)x\, dx = (-1)^n \frac{2\pi a^{2n+1}}{2n+1}$ $[a^2 < 1]$. BH[330](18)

2. $\displaystyle\int_0^{2\pi} \ln \frac{1 - 2a\cos x + a^2}{1 - 2a\cos nx + a^2} \cos mx\, dx = 2\pi\left(\frac{n}{m} a^{\frac{m}{n}} - \frac{a^m}{m}\right)$ $[a^2 \leqslant 1]$;

$$= 2\pi\left(\frac{n}{m} a^{-\frac{m}{n}} - \frac{a^{-m}}{m}\right) \quad [a^2 \geqslant 1].$$
BH[332](9)

3. $\displaystyle\int_0^\pi \ln \frac{1 + 2a\cos 2x + a^2}{1 + 2a\cos 2nx + a^2} \operatorname{ctg} x\, dx = 0.$ BH[331](5), Li[331](5)

4.399 1. $\displaystyle\int_0^{\pi/2} \ln(1 + a\sin^2 x) \sin^2 x\, dx = \frac{\pi}{2}\left(\ln \frac{1 + \sqrt{1+a}}{2} - \frac{1}{2}\frac{1 - \sqrt{1+a}}{1 + \sqrt{1+a}}\right)$
$[a > -1]$. BH[309](14)

2. $\displaystyle\int_0^{\pi/2} \ln(1 + a\sin^2 x) \cos^2 x\, dx =$

$$= \frac{\pi}{2}\left(\ln \frac{1 + \sqrt{1+a}}{2} + \frac{1}{2}\frac{1 - \sqrt{1+a}}{1 + \sqrt{1+a}}\right) \quad [a > -1].$$ BH[309](15)

3. $\displaystyle\int_0^{\pi/2} \frac{\ln(1 - \cos^2\beta \cos^2 x)}{1 - \cos^2\alpha \cos^2 x}\, dx = -\frac{\pi}{\sin\alpha} \ln \frac{1 + \sin\alpha}{\sin\alpha + \sin\beta}$

$$\left[0 < \beta < \frac{\pi}{2},\ 0 < \alpha < \frac{\pi}{2}\right].$$
Lo III 285

4.411 1. $\displaystyle\int_0^\pi \ln \frac{1 + \sin x}{1 + \cos\lambda \sin x} \frac{dx}{\sin x} = \lambda^2$ $[\lambda^2 < \pi^2]$. BH[331](2)

2. $\displaystyle\int_0^{\pi/2} \ln \frac{p + q\sin ax}{p - q\sin ax} \frac{dx}{\sin ax} = \int_0^{\pi/2} \ln \frac{p + q\cos ax}{p - q\cos ax} \frac{dx}{\cos ax} =$

$$= \int_0^{\pi/12} \ln \frac{p + q\operatorname{tg} ax}{p - q\operatorname{tg} ax} \frac{dx}{\operatorname{tg} ax} = \pi \arcsin \frac{q}{p} \quad [p > q > 0].$$
F II 703+, BH[315](5), BH[315](13), BH[315](17)+

3. $\displaystyle\int_0^{\pi/2} \frac{\cos x}{1-\cos^2\alpha\,\cos^2 x}\ln\frac{1+\cos\beta\,\cos x}{1-\cos\beta\,\cos x}\,dx = \frac{2\pi}{\sin 2\alpha}\ln\frac{\cos\frac{\alpha-\beta}{2}}{\sin\frac{\alpha+\beta}{2}}$ 4.411

$$\left[0 < \alpha \leqslant \beta < \frac{\pi}{2}\right].$$ Lo III 284

1. $\displaystyle\int_0^{\pi/4} \ln\,\mathrm{tg}\left(\frac{\pi}{4}\pm x\right)\frac{dx}{\sin 2x} = \pm\frac{\pi^2}{8}.$ BH[293](1)   4.412

2. $\displaystyle\int_0^{\pi/4} \ln\,\mathrm{tg}\left(\frac{\pi}{4}\pm x\right)\frac{dx}{\mathrm{tg}\,2x} = \pm\frac{\pi^2}{16}.$ BH[293](2)

3. $\displaystyle\int_0^{\pi/4} \ln\,\mathrm{tg}\left(\frac{\pi}{4}\pm x\right)(\ln\,\mathrm{tg}\,x)^{2n}\frac{dx}{\sin 2x} = \pm\frac{2^{2n+2}-1}{4(n+1)(2n+1)}\pi^{2n+2}|B_{2n+2}|.$ BH[294](24)

4. $\displaystyle\int_0^{\pi/4} \ln\,\mathrm{tg}\left(\frac{\pi}{4}\pm x\right)(\ln\,\mathrm{tg}\,x)^{2n-1}\frac{dx}{\sin 2x} = \pm\frac{1-2^{2n+1}}{2^{2n+2}\,n}(2n)!\,\zeta(2n+1).$

BH[294](25)

5. $\displaystyle\int_0^{\pi/4} \ln\,\mathrm{tg}\left(\frac{\pi}{4}\pm x\right)(\ln\,\sin 2x)^{n-1}\frac{dx}{\mathrm{tg}\,2x} = \frac{(-1)^{n-1}}{2}(n-1)!\,\zeta(n+1).$ Li[294](20)

1. $\displaystyle\int_0^{\pi/2} \ln(p^2+q^2\,\mathrm{tg}^2 x)\frac{dx}{a^2\sin^2 x + b^2\cos^2 x} = \frac{\pi}{ab}\ln\frac{ap+bq}{a}$ 4.413

$[a > 0,\ b > 0,\ p > 0,\ q > 0].$ BH[318](1-4)+

2. $\displaystyle\int_0^{\pi/2} \ln(1+q^2\,\mathrm{tg}^2 x)\frac{1}{p^2\sin^2 x + r^2\cos^2 x}\frac{dx}{s^2\sin^2 x + t^2\cos^2 x} =$

$$= \frac{\pi}{p^2 t^2 - s^2 r^2}\left\{\frac{p^2-r^2}{pr}\ln\left(1+\frac{qr}{p}\right) + \frac{t^2-s^2}{st}\ln\left(1+\frac{qt}{s}\right)\right\}$$

$[q > 0,\ p > 0,\ r > 0,\ s > 0,\ t > 0].$ BH[320](18)

3. $\displaystyle\int_0^{\pi/2} \ln(1+q^2\mathrm{tg}^2 x)\frac{\sin^2 x}{p^2\sin^2 x + r^2\cos^2 x}\frac{dx}{s^2\sin^2 x + t^2\cos^2 x} =$

$$= \frac{\pi}{p^2 t^2 - s^2 r^2}\left\{\frac{t}{s}\ln\left(1+\frac{qt}{s}\right) - \frac{r}{p}\ln\left(1+\frac{qr}{p}\right)\right\}$$

$[q > 0,\ p > 0,\ r > 0,\ s > 0,\ t > 0].$ BH[320](20)

4.413

4. $$\int_0^{\pi/2} \ln(1+q^2\operatorname{tg}^2 x) \frac{\cos^2 x}{p^2\sin^2 x + r^2\cos^2 x} \frac{dx}{s^2\sin^2 x + t^2\cos^2 x} =$$

$$= \frac{\pi}{p^2 t^2 - s^2 r^2} \left\{ \frac{p}{r} \ln\left(1 + \frac{qr}{p}\right) - \frac{s}{t} \ln\left(1 + \frac{qt}{s}\right) \right\}$$

$[q > 0,\ p > 0,\ r > 0,\ s > 0,\ t > 0]$. BH[320](21)

5. $$\int_0^{\pi} \frac{\ln \operatorname{tg} rx\, dx}{1 - 2p\cos x + p^2} = \frac{\pi}{1-p^2} \ln \frac{1-p^{2r}}{1+p^{2r}} \quad [p^2 < 1].$$ BH[331](12)

4.414

1. $$\int_0^{\pi/2} \ln(1 - k^2 \sin^2 x) \frac{dx}{\sqrt{1 - k^2 \sin^2 x}} = \ln k'\, \boldsymbol{K}(k).$$ BH[323](1)

2. $$\int_0^{\pi/2} \ln(1 - k^2 \sin^2 x) \frac{\sin^2 x\, dx}{\sqrt{1 - k^2 \sin^2 x}} = \frac{1}{k^2} \{(k^2 - 2 + \ln k')\, \boldsymbol{K}(k) +$$

$$+ (2 - \ln k')\boldsymbol{E}(k)\}.$$ BH[323](3)

3. $$\int_0^{\pi/2} \ln(1 - k^2 \sin^2 x) \frac{\cos^2 x\, dx}{\sqrt{1 - k^2 \sin^2 x}} = \frac{1}{k^2} [(1 + k'^2 - k'^2 \ln k')\, \boldsymbol{K}(k) -$$

$$- (2 - \ln k')\boldsymbol{E}(k)].$$ BH[323](6)

4. $$\int_0^{\pi/2} \ln(1 - k^2 \sin^2 x) \frac{dx}{\sqrt{(1 - k^2 \sin^2 x)^3}} = \frac{1}{k'^2} [(k^2 - 2)\, \boldsymbol{K}(k) +$$

$$+ (2 + \ln k')\, \boldsymbol{E}(k)].$$ BH[323](9)

5. $$\int_0^{\pi/2} \ln(1 - k^2 \sin^2 x) \frac{\sin^2 x\, dx}{\sqrt{(1 - k^2 \sin^2 x)^3}} = \frac{1}{k^2 k'^2} [(2 + \ln k')\boldsymbol{E}(k) -$$

$$- (1 + k'^2 + k'^2 \ln k')\, \boldsymbol{K}(k)].$$ BH[323](10)

6. $$\int_0^{\pi/2} \ln(1 - k^2 \sin^2 x) \frac{\cos^2 x\, dx}{\sqrt{(1 - k^2 \sin^2 x)^3}} = \frac{1}{k^2} [(1 + k'^2 + \ln k')\, \boldsymbol{K}(k) -$$

$$- (2 + \ln k')\, \boldsymbol{E}(k)].$$ BH[323](16)

7. $$\int_0^{\pi/2} \ln(1 - k^2 \sin^2 x) \sqrt{1 - k^2 \sin^2 x}\, dx = (1 + k'^2)\, \boldsymbol{K}(k) -$$

$$- (2 - \ln k')\, \boldsymbol{E}(k).$$ BH[324](18)

8. $$\int_0^{\pi/2} \ln(1 - k^2 \sin^2 x) \sin^2 x \sqrt{1 - k^2 \sin^2 x}\, dx =$$

$$= \frac{1}{9k^2} \{(-2 + 11k^2 - 6k^4 + 3k'^2 \ln k')\boldsymbol{K}(k) + [2 - 10k^2 -$$

$$- 3(1 - 2k^2) \ln k']\, \boldsymbol{E}(k)\}.$$ BH[324](20)

9. $\int_0^{\pi/2} \ln(1 - k^2 \sin^2 x) \cos^2 x \sqrt{1 - k^2 \sin^2 x}\, dx =$ 4.414

$= \dfrac{1}{9k^2}\{(2 + 7k^2 - 3k^4 - 3k'^2 \ln k')\mathbf{K}(k) - [2 + 8k^2 - 3(1 + k^2)\ln k']\mathbf{E}(k)\}.$

BH[324](21), Li[324](21)

10. $\int_0^{\pi/2} \ln(1 - k^2 \sin^2 x) \dfrac{\sin x \cos x\, dx}{\sqrt{(1 - k^2 \sin^2 x)^{2n+1}}} =$

$= \dfrac{2}{(2n-1)^2 k^2}\{[1 + (2n-1)\ln k']k'^{1-2n} - 1\}.$ BH[324](17)

1. $\int_0^\infty \ln x \sin ax^2\, dx = -\dfrac{1}{4}\sqrt{\dfrac{\pi}{2a}}\left(\ln 4a + C - \dfrac{\pi}{2}\right)$ $[a > 0]$. GH[338](19) 4.415

2. $\int_0^\infty \ln x \cos ax^2\, dx = -\dfrac{1}{4}\sqrt{\dfrac{\pi}{2a}}\left(\ln 4a + C + \dfrac{\pi}{2}\right)$ $[a > 0]$. GH[338](19)

1. $\int_0^{\pi/2} \dfrac{\cos x \ln(1 + \sqrt{\sin^2 \beta - \cos^2 \beta\, \mathrm{tg}^2 \alpha \sin^2 x})}{1 - \sin^2 \alpha \cos^2 x}\, dx =$ 4.416

$= \operatorname{cosec} 2\alpha\{(2\alpha + 2\gamma - \pi)\ln \cos \beta + 2L(\alpha) - 2L(\gamma) + L(\alpha + \gamma) - L(\alpha - \gamma)\}$

$\left[\cos \gamma = \dfrac{\sin \alpha}{\sin \beta}\,;\ 0 < \alpha < \beta < \dfrac{\pi}{2}\right].$ Lo III 291

2. $\int_0^{\pi/2} \dfrac{\cos x \ln(1 - \sqrt{\sin^2 \beta - \cos^2 \beta\, \mathrm{tg}^2 \alpha \sin^2 x})}{1 - \sin^2 \alpha \cos^2 x}\, dx =$

$= \operatorname{cosec} 2\alpha\{(\pi + 2\alpha - 2\gamma)\ln \cos \beta + 2L(\alpha) + 2L(\gamma) - L(\alpha + \gamma) + L(\alpha - \gamma)\}$

$\left[\cos \gamma = \dfrac{\sin \alpha}{\sin \beta}\,;\ 0 < \alpha < \beta < \dfrac{\pi}{2}\right].$ Lo III 291

3. $\int_\beta^{\pi/2} \dfrac{\ln(\sin x + \sqrt{\sin^2 x - \sin^2 \beta})}{1 - \cos^2 \alpha \cos^2 x}\, dx =$

$= -\operatorname{cosec} \alpha\left\{\operatorname{arctg}\left(\dfrac{\mathrm{tg}\,\beta}{\sin \alpha}\right)\ln \sin \beta + \dfrac{\pi}{2}\ln \dfrac{1 + \sin \alpha}{\sin \alpha + \sqrt{1 - \cos^2 \alpha \cos^2 \beta}}\right\}$

$\left[0 < \alpha < \pi,\ 0 < \beta < \dfrac{\pi}{2}\right].$ Lo III 285

4. $\int_0^{\pi/4} \ln \mathrm{tg}\, x\, (\ln \cos 2x)^{n-1} \mathrm{tg}\, 2x\, dx = (-1)^{n-1} \dfrac{(n-1)!}{4} \sum_{k=0}^\infty \dfrac{1}{(1+2k)^{n+1}} =$

$= (-1)^{n-1} \dfrac{(n-1)!}{2^{n+3}} \zeta\left(n+1, \dfrac{1}{2}\right).$ BH[287](20)

Logarithmus, trigonometrische Funktionen und Potenzen
The logarithmic and the trigonometric function and powers

4.42–4.43

4.421

1. $\int_0^\infty \ln x \sin ax \, \dfrac{dx}{x} = -\dfrac{\pi}{2}(C + \ln a) \quad [a > 0].$ F II 816+

2. $\int_0^\infty \ln ax \sin bx \, \dfrac{x \, dx}{\beta^2 + x^2} = \dfrac{\pi}{2} e^{-b\beta'} \ln (a\beta') -$

$\qquad - \dfrac{\pi}{4}[e^{b\beta'} \operatorname{Ei}(-b\beta') + e^{-b\beta'} \operatorname{Ei}(b\beta')] \quad [\beta' = \beta \operatorname{sign} \beta;\ a > 0,\ b > 0].$

Ba IV 76(5), NI 27(10)+

3. $\int_0^\infty \ln ax \cos bx \, \dfrac{\beta' \, dx}{\beta^2 + x^2} = \dfrac{\pi}{2} e^{-b\beta'} \ln (a\beta') +$

$\qquad + \dfrac{\pi}{4}[e^{b\beta'} \operatorname{Ei}(-b\beta') - e^{-b\beta'} \operatorname{Ei}(b\beta')] \quad [\beta' = \beta \operatorname{sign} \beta;\ a > 0,\ b > 0].$

Ba IV 17(3), NI 27(11)+

4. $\int_0^\infty \ln ax \sin bx \, \dfrac{x \, dx}{x^2 - c^2} = \dfrac{\pi}{2}\{-\operatorname{si}(bc)\sin bc +$

$\qquad + \cos bc [\ln ac - \operatorname{ci}(bc)]\} \quad [a > 0,\ b > 0,\ c > 0].$ BH[422](5)

5. $\int_0^\infty \ln ax \cos bx \, \dfrac{dx}{x^2 - c^2} = \dfrac{\pi}{2c}\{\sin bc [\operatorname{ci}(bc) - \ln ac] - \cos bc \operatorname{si}(bc)\}$

$\qquad [a > 0,\ b > 0,\ c > 0].$ BH[422](6)

4.422

1. $\int_0^\infty \ln x \sin ax \, x^{\mu-1} \, dx = \dfrac{\Gamma(\mu)}{a^\mu} \sin \dfrac{\mu\pi}{2} \left[\psi(\mu) - \ln a + \dfrac{\pi}{2} \operatorname{ctg} \dfrac{\mu\pi}{2} \right]$

$\qquad [a > 0,\ |\operatorname{Re}\mu| < 1].$ BH[411](5)

2. $\int_0^\infty \ln x \cos ax \, x^{\mu-1} \, dx = \dfrac{\Gamma(\mu)}{a^\mu} \cos \dfrac{\mu\pi}{2} \left[\psi(\mu) - \ln a - \dfrac{\pi}{2} \operatorname{tg} \dfrac{\mu\pi}{2} \right]$

$\qquad [a > 0,\ 0 < \operatorname{Re}\mu < 1].$ BH[411](6)

4.423

1. $\int_0^\infty \ln x \, \dfrac{\cos ax - \cos bx}{x} \, dx = \ln \dfrac{a}{b} \left(C + \dfrac{1}{2} \ln ab \right) \quad [a > 0, b > 0].$ GH[338](21a)

2. $\int_0^\infty \ln x \, \dfrac{\cos ax - \cos bx}{x^2} \, dx = \dfrac{\pi}{2}[(a-b)(C-1) + a\ln a - b\ln b]$

$\qquad [a > 0,\ b > 0].$ GH[338](21b)

3. $\int_0^\infty \ln x \, \dfrac{\sin^2 ax}{x^2} \, dx = -\dfrac{a\pi}{2}(C + \ln 2a - 1) \quad [a > 0].$ GH[338](20b)

4.424

1. $\int_0^\infty (\ln x)^2 \sin ax \, \frac{dx}{x} = \frac{\pi}{2} C^2 + \frac{\pi^3}{24} + \pi C \ln a + \frac{\pi}{2} (\ln a)^2 \quad [a > 0].$

Ba IV 77(9), F II 816+

2. $\int_0^\infty (\ln x)^2 \sin ax \, x^{\mu-1} dx = \frac{\Gamma(\mu)}{a^\mu} \sin \frac{\mu\pi}{2} \left[\psi'(\mu) + \psi^2(\mu) + \pi\psi(\mu) \operatorname{ctg} \frac{\mu\pi}{2} - \right.$

$\left. - 2\psi(\mu) \ln a - \pi \ln a \operatorname{ctg} \frac{\mu\pi}{2} + (\ln a)^2 - \pi^2 \right]$

$[a > 0, \ 0 < \operatorname{Re} \mu < 1].$ Ba IV 77(10)

4.425

1. $\int_0^\infty \ln(1+x) \cos ax \, \frac{dx}{x} = \frac{1}{2} \{[\operatorname{si}(a)]^2 + [\operatorname{ci}(a)]^2\} \quad [a > 0].$ Ba IV 18(8)

2. $\int_0^\infty \ln \left(\frac{b+x}{b-x} \right)^2 \cos ax \, \frac{dx}{x} = -2\pi \operatorname{si}(ab) \quad [a \geq 0, \ b > 0].$ Ba IV 18(11)

3. $\int_0^\infty \ln(1+b^2 x^2) \sin ax \, \frac{dx}{x} = -\pi \operatorname{Ei}\left(-\frac{a}{b}\right) \quad [a > 0, \ b > 0].$

GH[338](24), Ba IV 77(14)

4. $\int_0^1 \ln(1-x^2) \cos(p \ln x) \frac{dx}{x} = \frac{1}{2p^2} + \frac{\pi}{2p} \operatorname{cth} \frac{p\pi}{2}.$ Li[309](1)+

4.426

1. $\int_0^\infty \ln \frac{b^2+x^2}{c^2+x^2} \sin ax \, x \, dx = \frac{\pi}{a^2} [(1+ac) e^{-ac} - (1+ab) e^{-ab}]$

$[b \geq 0, \ c \geq 0, \ a > 0].$ GH[338](23)

2. $\int_0^\infty \ln \frac{b^2 x^2 + p^2}{c^2 x^2 + p^2} \sin ax \, \frac{dx}{x} = \pi \left[\operatorname{Ei}\left(-\frac{ap}{c}\right) - \operatorname{Ei}\left(-\frac{ap}{b}\right) \right]$

$[b > 0, \ c > 0, \ p > 0, \ a > 0].$ Ba IV 77(15)

4.427

$\int_0^\infty \ln(x + \sqrt{\beta^2 + x^2}) \frac{\sin ax}{\sqrt{\beta^2 + x^2}} dx = \frac{\pi}{2} K_0(a\beta) + \frac{\pi}{2} \ln(\beta) [I_0(a\beta) - L(a\beta)]$

$[\operatorname{Re} \beta > 0, \ a > 0].$ Ba IV 77(16)

4.428

1. $\int_0^\infty \ln \cos^2 ax \, \frac{\cos bx}{x^2} dx = \pi b \ln 2 - a\pi \quad [a > 0, \ b > 0].$ Ba IV 22(29)

2. $\int_0^\infty \ln(4 \cos^2 ax) \frac{\cos bx}{x^2 + c^2} dx = \frac{\pi}{c} \operatorname{ch}(bc) \ln(1 + e^{-2ac})$

$\left[0 < b < 2a < \frac{\pi}{c} \right].$ Ba IV 22(30)

3. $\int_0^\infty \ln \cos^2 ax \, \frac{\sin bx}{x(1+x^2)} dx = \pi \ln(1 + e^{-2a}) \operatorname{sh} b - \pi \ln 2(1 - e^{-b})$

$[a > 0, \ b > 0].$ Ba IV 82(36)

4.428   4.
$$\int_0^\infty \ln\cos^2 ax \, \frac{\cos bx}{x^2(1+x^2)} dx = -\pi \ln(1+e^{-2a}) \operatorname{ch} b + (b+e^{-b})\pi \ln 2 - a\pi$$
$[a > 0, \ b > 0]$.   Ba IV 22(31)

4.429
$$\int_0^1 \frac{(1+x)x}{\ln x} \sin(\ln x)\, dx = \frac{\pi}{4}.$$ BH[326](2)+

4.431   1.
$$\int_0^\infty \ln(2 \pm 2\cos x)\, \frac{\sin bx}{x^2+c^2} x\, dx = -\pi \operatorname{sh}(bc) \ln(1 \pm e^{-c})$$
$[b > 0, \ c > 0]$.   Ba IV 22(32)

2.
$$\int_0^\infty \ln(2 \pm 2\cos x)\, \frac{\cos bx}{x^2+c^2} dx = \frac{\pi}{c} \operatorname{ch}(bc) \ln(1 \pm e^{-c}) \quad [b > 0, \ c > 0].$$
Ba IV 22(32)

3.
$$\int_0^\infty \ln(1 + 2a\cos x + a^2)\, \frac{\sin bx}{x} dx = -\frac{\pi}{2} \sum_{k=1}^{E(b)} \frac{(-a)^k}{k}[1 + \operatorname{sign}(b-k)]$$
$[0 < a < 1, \ b > 0]$.   Ba IV 82(35)

4.
$$\int_0^\infty \ln(1 - 2a\cos x + a^2)\, \frac{\cos bx}{x^2+c^2} dx =$$
$$= \frac{\pi}{c} \ln(1 - ae^{-c}) \operatorname{ch}(bc) + \frac{\pi}{c} \sum_{k=1}^{E(b)} \frac{a^k}{k} \operatorname{sh}[c(b-k)]$$
$[|a| < 1, \ b > 0, \ c > 0]$.   Ba IV 22(33)

4.432   1.
$$\int_0^\infty \ln(1 - k^2 \sin^2 x)\, \frac{\sin x}{\sqrt{1-k^2\sin^2 x}} \frac{dx}{x} =$$
$$= \int_0^\infty \ln(1 - k^2 \cos^2 x)\, \frac{\sin x}{\sqrt{1-k^2\cos^2 x}} \frac{dx}{x} = \ln k' \mathbf{K}(k).$$ BH [412, 414](4)

2.
$$\int_0^{\pi/2} \ln(1 - k^2\sin^2 x)\, \frac{\sin x \cos x}{\sqrt{1-k^2\sin^2 x}} x\, dx =$$
$$= \frac{1}{k^2}\{\pi k'(1 - \ln k') + (2-k^2)\mathbf{K}(k) - (4 - \ln k')\mathbf{E}(k)\}.$$ BH[426](3)

3.
$$\int_0^{\pi/2} \ln(1 - k^2\cos^2 x)\, \frac{\sin x \cos x}{\sqrt{1-k^2\cos^2 x}} x\, dx =$$
$$= \frac{1}{k^2}\{-\pi - (2-k^2)\mathbf{K}(k) + (4 - \ln k')\mathbf{E}(k)\}.$$ BH[426](6)

4.
$$\int_0^\infty \ln(1 - k^2\sin^2 x)\, \frac{\sin x \cos x}{\sqrt{1-k^2\sin^2 x}} \frac{dx}{x} =$$
$$= \frac{1}{k^2}\{(2 - k^2 - k'^2 \ln k')\mathbf{K}(k) - (2 - \ln k')\mathbf{E}(k)\}.$$ BH[412](5)

4.432

5. $\displaystyle\int_0^\infty \ln(1-k^2\cos^2 x)\frac{\sin x \cos x}{\sqrt{1-k^2\cos^2 x}}\frac{dx}{x} =$

$\qquad = \dfrac{1}{k^2}\{(k^2-2+\ln k')\,\boldsymbol{K}(k)+(2-\ln k')\,\boldsymbol{E}(k)\}.$ BH[414](5)

6. $\displaystyle\int_0^\infty \ln(1\pm k\sin^2 x)\frac{\sin x}{\sqrt{1-k^2\sin^2 x}}\frac{dx}{x} =$

$\qquad = \displaystyle\int_0^\infty \ln(1\pm k\cos^2 x)\frac{\sin x}{\sqrt{1-k^2\cos^2 x}}\frac{dx}{x} =$

$\qquad = \displaystyle\int_0^\infty \ln(1\pm k\sin^2 x)\frac{\operatorname{tg} x}{\sqrt{1-k^2\sin^2 x}}\frac{dx}{x} =$

$\qquad = \displaystyle\int_0^\infty \ln(1\pm k\cos^2 x)\frac{\operatorname{tg} x}{\sqrt{1-k^2\cos^2 x}}\frac{dx}{x} =$

$\qquad = \displaystyle\int_0^\infty \ln(1\pm k\sin^2 2x)\frac{\operatorname{tg} x}{\sqrt{1-k^2\sin^2 2x}}\frac{dx}{x} =$

$\qquad = \displaystyle\int_0^\infty \ln(1\pm k^2\cos^2 2x)\frac{\operatorname{tg} x}{\sqrt{1-k^2\cos^2 2x}}\frac{dx}{x} =$

$\qquad = \dfrac{1}{2}\ln\dfrac{2(1\pm k)}{\sqrt{k}}\,\boldsymbol{K}(k)-\dfrac{\pi}{8}\boldsymbol{K}(k').$ BH[413](1–6), BH[415](1–6)

7. $\displaystyle\int_0^\infty \ln(1-k^2\sin^2 x)\frac{\sin^3 x}{\sqrt{1-k^2\sin^2 x}}\frac{dx}{x} =$

$\qquad = \dfrac{1}{k^2}\{(k^2-2+\ln k')\,\boldsymbol{K}(k)+(2-\ln k')\,\boldsymbol{E}(k)\}.$ BH[412](6)

8. $\displaystyle\int_0^\infty \ln(1-k^2\cos^2 x)\frac{\sin^3 x}{\sqrt{1-k^2\cos^2 x}}\frac{dx}{x} =$

$\qquad = \dfrac{1}{k^2}\{(2-k^2-k'^2\ln k')\,\boldsymbol{K}(k)-(2-\ln k')\,\boldsymbol{E}(k)\}.$ BH[414](6)+

9. $\displaystyle\int_0^\infty \ln(1-k^2\sin^2 x)\frac{\sin x\cos^2 x}{\sqrt{1-k^2\sin^2 x}}\frac{dx}{x} =$

$\qquad = \dfrac{1}{k^2}\{(2-k^2-k'^2\ln k')\,\boldsymbol{K}(k)-(2-\ln k')\boldsymbol{E}(k)\}.$ BH[412](7)

10. $\displaystyle\int_0^\infty \ln(1-k^2\cos^2 x)\frac{\sin x\cos^2 x}{\sqrt{1-k^2\cos^2 x}}\frac{dx}{x} =$

$\qquad = \dfrac{1}{k^2}\{(k^2-2+\ln k')\,\boldsymbol{K}(k)+(2-\ln k')\,\boldsymbol{E}(k)\}.$ BH[414](7)

4.432

11. $\int_0^\infty \ln(1 - k^2 \sin^2 x) \dfrac{\operatorname{tg} x}{\sqrt{1 - k^2 \sin^2 x}} \dfrac{dx}{x} =$

$= \int_0^\infty \ln(1 - k^2 \cos^2 x) \dfrac{\operatorname{tg} x}{\sqrt{1 - k^2 \cos^2 x}} \dfrac{dx}{x} = \ln k' \, \boldsymbol{K}(k).$ BH[412, 414](9)

12. $\int_0^\infty \ln(1 - k^2 \sin^2 x) \dfrac{\sin^2 x \, \operatorname{tg} x}{\sqrt{1 - k^2 \sin^2 x}} \dfrac{dx}{x} =$

$= \dfrac{1}{k^2} \{(k^2 - 2 + \ln k') \, \boldsymbol{K}(k) + (2 - \ln k') \, \boldsymbol{E}(k)\}.$ BH[412](8)

13. $\int_0^\infty \ln(1 - k^2 \cos^2 x) \dfrac{\sin^2 x \, \operatorname{tg} x}{\sqrt{1 - k^2 \cos^2 x}} \dfrac{dx}{x} =$

$= \dfrac{1}{k^2} \{(2 - k^2 - k'^2 \ln k') \, \boldsymbol{K}(k) - (2 - \ln k') \, \boldsymbol{E}(k)\}.$ BH[414](8)

14. $\int_0^\infty \ln(1 - k^2 \sin^2 x) \dfrac{\sin x}{\sqrt{(1 - k^2 \sin^2 x)^3}} \dfrac{dx}{x} =$

$= \int_0^\infty \ln(1 - k^2 \cos^2 x) \dfrac{\sin x}{\sqrt{(1 - k^2 \cos^2 x)^3}} \dfrac{dx}{x} =$

$= \dfrac{1}{k'^2} \{(k^2 - 2) \, \boldsymbol{K}(k) + (2 + \ln k') \, \boldsymbol{E}(k)\}.$ BH[412, 414](13)

15. $\int_0^{\pi/2} \ln(1 - k^2 \sin^2 x) \dfrac{\sin x \cos x}{\sqrt{(1 - k^2 \sin^2 x)^3}} x \, dx =$

$= \dfrac{1}{k^2} \left\{(1 + \ln k') \dfrac{\pi}{k'} - (2 + \ln k') \, \boldsymbol{K}(k)\right\}.$ BH[426](9)

16. $\int_0^{\pi/2} \ln(1 - k^2 \cos^2 x) \dfrac{\sin x \cos x}{\sqrt{(1 - k^2 \cos^2 x)^3}} x \, dx =$

$= \dfrac{1}{k^2} \{-\pi + (2 + \ln k') \, \boldsymbol{K}(k)\}.$ BH[426](15)

17. $\int_0^\infty \ln(1 - k^2 \sin^2 x) \dfrac{\sin x \cos x}{\sqrt{(1 - k^2 \sin^2 x)^3}} \dfrac{dx}{x} =$

$= \int_0^\infty \ln(1 - k^2 \cos^2 x) \dfrac{\sin^3 x}{\sqrt{(1 - k^2 \cos^2 x)^3}} \dfrac{dx}{x} =$

$= \dfrac{1}{k^2} \{(2 - k^2 + \ln k') \, \boldsymbol{K}(k) - (2 + \ln k') \, \boldsymbol{E}(k)\}.$

BH[412](14), BH[414](15)

18. $\int_0^\infty \ln(1 - k^2 \sin^2 x) \dfrac{\sin^3 x}{\sqrt{(1-k^2\sin^2 x)^3}} \dfrac{dx}{x} =$ 4.432

$$= \int_0^\infty \ln(1 - k^2 \cos^2 x) \dfrac{\sin x \cos x}{\sqrt{(1-k^2\cos^2 x)^3}} \dfrac{dx}{x} =$$

$$= \dfrac{1}{k^2 k'^2} \{(2 + \ln k') E(k) - (2 - k^2 + k'^2 \ln k') K(k)\}. \quad \text{BH[412](15), BH[414](14)}$$

19. $\int_0^\infty \ln(1 - k^2 \sin^2 x) \dfrac{\sin x \cos^2 x}{\sqrt{(1-k^2\sin^2 x)^3}} \dfrac{dx}{x} =$

$$= \int_0^\infty \ln(1 - k^2 \cos^2 x) \dfrac{\sin^2 x \, \operatorname{tg} x}{\sqrt{(1-k^2\cos^2 x)^3}} \dfrac{dx}{x} =$$

$$= \dfrac{1}{k^2} \{(2 - k^2 + \ln k') K(k) - (2 + \ln k') E(k)\}. \quad \text{BH[412](16), BH[414](17)}$$

20. $\int_0^\infty \ln(1 - k^2 \sin^2 x) \dfrac{\sin^2 x \, \operatorname{tg} x}{\sqrt{(1-k^2\sin^2 x)^3}} \dfrac{dx}{x} =$

$$= \int_0^\infty \ln(1 - k^2 \cos^2 x) \dfrac{\sin x \cos^2 x}{\sqrt{(1-k^2\cos^2 x)^3}} \dfrac{dx}{x} =$$

$$= \dfrac{1}{k^2 k'^2} \{(2 + \ln k') E(k) - (2 - k^2 + k'^2 \ln k') K(k)\}.$$
$$\text{BH[412](17), BH[414](16)}$$

21. $\int_0^\infty \ln(1 - k^2 \sin^2 x) \dfrac{\operatorname{tg} x}{\sqrt{(1-k^2\sin^2 x)^3}} \dfrac{dx}{x} =$

$$= \int_0^\infty \ln(1 - k^2 \cos^2 x) \dfrac{\operatorname{tg} x}{\sqrt{(1-k^2\cos^2 x)^3}} \dfrac{dx}{x} =$$

$$= \dfrac{1}{k'^2} \{(k^2 - 2) K(k) + (2 + \ln k') E(k)\}. \quad \text{BH[412, 414](18)}$$

22. $\int_0^\infty \ln(1 - k^2 \sin^2 x) \sqrt{1 - k^2 \sin^2 x} \, \sin x \, \dfrac{dx}{x} =$

$$= \int_0^\infty \ln(1 - k^2 \cos^2 x) \sqrt{1 - k^2 \cos^2 x} \, \sin x \, \dfrac{dx}{x} =$$

$$= (2 - k^2) K(k) - (2 - \ln k') E(k). \quad \text{BH[412, 414](1)}$$

4.432

23. $\int_0^{\pi/2} \ln(1 - k^2 \sin^2 x) \sqrt{1 - k^2 \sin^2 x} \sin x \cos x \cdot x \, dx =$

$= \dfrac{1}{27k^2} \{3\pi k'^3 (1 - 3\ln k') + (22k'^2 + 6k^4 - 3k'^2 \ln k') \boldsymbol{K}(k) -$
$- (2 - k^2)(14 - 6 \ln k') \boldsymbol{E}(k)\}.$ BH[426](1)

24. $\int_0^{\pi/2} \ln(1 - k^2 \cos^2 x) \sqrt{1 - k^2 \cos^2 x} \sin x \cos x \cdot x \, dx = \dfrac{1}{27k^2} \{-3\pi -$

$- (22k'^2 + 6k^4 - 3k'^2 \ln k') \boldsymbol{K}(k) + (2 - k^2)(14 - 6 \ln k') \boldsymbol{E}(k)\}.$ BH[426](2)

25. $\int_0^\infty \ln(1 - k^2 \sin^2 x) \sqrt{1 - k^2 \sin^2 x} \, \text{tg}\, x \, \dfrac{dx}{x} =$

$= \int_0^\infty \ln(1 - k^2 \cos^2 x) \sqrt{1 - k^2 \cos^2 x} \, \text{tg}\, x \, \dfrac{dx}{x} =$

$= (2 - k^2) \boldsymbol{K}(k) - (2 - \ln k') \boldsymbol{E}(k).$ BH[412, 414](2)

26. $\int_0^\infty \ln(\sin^2 x + k' \cos^2 x) \dfrac{\sin x}{\sqrt{1 - k^2 \cos^2 x}} \dfrac{dx}{x} =$

$= \int_0^\infty \ln(\sin^2 x + k' \cos^2 x) \dfrac{\text{tg}\, x}{\sqrt{1 - k^2 \cos^2 x}} \dfrac{dx}{x} =$

$= \int_0^\infty \ln(\sin^2 2x + k' \cos^2 2x) \dfrac{\text{tg}\, x}{\sqrt{1 - k^2 \cos^2 2x}} \dfrac{dx}{x} =$

$= \dfrac{1}{2} \ln\left[\dfrac{2(\sqrt{k'})^3}{1 + k'}\right] \boldsymbol{K}(k).$ BH[415](19–21)

Logarithmus, trigonometrische Funktionen und Exponentialfunktion
The logarithmic and the trigonometric function and exponentials

4.44

4.441

1. $\int_0^\infty e^{-qx} \sin px \ln x \, dx = \dfrac{1}{p^2 + q^2}\left[q \, \text{arctg}\, \dfrac{p}{q} - p\, C - \dfrac{p}{2} \ln(p^2 + q^2)\right]$

$[q > 0].$ BH[467](1)

2. $\int_0^\infty e^{-qx} \cos px \ln x \, dx = -\dfrac{1}{p^2 + q^2}\left[\dfrac{q}{2} \ln(p^2 + q^2) + p\, \text{arctg}\, \dfrac{p}{q} + qC\right]$

$[q > 0].$ BH[467](2)

4.442

$\int_0^{\pi/2} \dfrac{e^{-p\, \text{tg}\, x} \ln \cos x \, dx}{\sin x \cos x} = -\dfrac{1}{2}[\text{ci}(p)]^2 + \dfrac{1}{2}[\text{si}(p)]^2 \quad [\text{Re}\, p > 0].$ NI 32(11)

Zyklometrische Funktionen
Inverse Trigonometric Functions

4.5

Zyklometrische Funktionen
Inverse trigonometric functions

4.51

$$\int_0^\infty \operatorname{arcctg} px \operatorname{arcctg} qx \, dx =$$

4.511

$$= \frac{\pi}{2} \left\{ \frac{1}{p} \ln\left(1 + \frac{p}{q}\right) + \frac{1}{q} \ln\left(1 + \frac{q}{p}\right) \right\} \quad [p > 0, q > 0].$$ BH[77](8)

$$\int_0^\pi \operatorname{arctg}(\cos x) \, dx = 0.$$ BH[345](1)   4.512

Arkussinus, Arkuscosinus und Potenzen
Arcsines, arccosines, and powers

4.52

1. $$\int_0^1 \frac{\arcsin x}{x} \, dx = \frac{\pi}{2} \ln 2.$$ F II 624, F II 633   4.521

2. $$\int_0^1 \frac{\arccos x}{1 \pm x} \, dx = \mp \frac{\pi}{2} \ln 2 + 2G.$$ BH[231](7), BH[231](8)

3. $$\int_0^1 \arcsin x \, \frac{x}{1+qx^2} \, dx = \frac{\pi}{2q} \ln \frac{2\sqrt{1+q}}{1+\sqrt{1+q}} \quad [q > -1].$$ BH[231](1)

4. $$\int_0^1 \arcsin x \, \frac{x}{1-p^2 x^2} \, dx = \frac{\pi}{2p^2} \ln \frac{1+\sqrt{1-p^2}}{2\sqrt{1-p^2}} \quad [p^2 < 1].$$ Li[231](3)

5. $$\int_0^1 \arccos x \, \frac{dx}{\sin^2 \lambda - x^2} = 2 \operatorname{cosec} \lambda \sum_{k=0}^\infty \frac{\sin[(2k+1)\lambda]}{(2k+1)^2}.$$ BH[231](10)

6. $$\int_0^1 \arcsin x \, \frac{dx}{x(1+qx^2)} = \frac{\pi}{2} \ln \frac{1+\sqrt{1+q}}{\sqrt{1+q}} \quad [q > -1].$$ BH[235](10)

7. $$\int_0^1 \arcsin x \, \frac{x}{(1+qx^2)^2} \, dx = \frac{\pi}{4q} \frac{\sqrt{1+q}-1}{1+q} \quad [q > -1].$$ BH[234](2)

8. $$\int_0^1 \arccos x \, \frac{x}{(1+qx^2)^2} \, dx = \frac{\pi}{4q} \frac{\sqrt{1+q}-1}{\sqrt{1+q}} \quad [q > -1].$$ BH[234](4)

4.522

1. $\int_0^1 x\sqrt{1-k^2x^2}\,\arccos x\,dx = \dfrac{1}{9k^2}\left[\dfrac{3}{2}\pi + k'^2\,\boldsymbol{K}(k) - 2(1+k'^2)\,\boldsymbol{E}(k)\right].$

   BH[236](9)

2. $\int_0^1 x\sqrt{1-k^2x^2}\,\arcsin x\,dx =$

   $= \dfrac{1}{9k^2}\left[-\dfrac{3}{2}\pi k'^3 - k'^2\,\boldsymbol{K}(k) + 2(1+k'^2)\,\boldsymbol{E}(k)\right].$

   BH[236](1)

3. $\int_0^1 x\sqrt{k'^2+k^2x^2}\,\arcsin x\,dx = \dfrac{1}{9k^2}\left[\dfrac{3}{2}\pi + k'^2\,\boldsymbol{K}(k) - 2(1+k'^2)\,\boldsymbol{E}(k)\right].$

   BH[236](5)

4. $\int_0^1 \dfrac{x\,\arcsin x}{\sqrt{1-k^2x^2}}\,dx = \dfrac{1}{k^2}\left[-\dfrac{\pi}{2}k' + \boldsymbol{E}(k)\right].$

   BH[237](1)

5. $\int_0^1 \dfrac{x\,\arccos x}{\sqrt{1-k^2x^2}}\,dx = \dfrac{1}{k^2}\left[\dfrac{\pi}{2} - \boldsymbol{E}(k)\right].$

   BH[240](1)

6. $\int_0^1 \dfrac{x\,\arcsin x}{\sqrt{k'^2+k^2x^2}}\,dx = \dfrac{1}{k^2}\left[\dfrac{\pi}{2} - \boldsymbol{E}(k)\right].$

   BH[238](1)

7. $\int_0^1 \dfrac{x\,\arccos x}{\sqrt{k'^2+k^2x^2}}\,dx = \dfrac{1}{k^2}\left[-\dfrac{\pi}{2}k' + \boldsymbol{E}(k)\right].$

   BH[241](1)

8. $\int_0^1 \dfrac{x\,\arcsin x\,dx}{(x^2-\cos^2\lambda)\sqrt{1-x^2}} = \dfrac{2}{\sin\lambda}\sum_{k=0}^{\infty}\dfrac{\sin[(2k+1)\lambda]}{(2k+1)^2}.$

   BH[243](11)

9. $\int_0^1 \dfrac{x\,\arcsin kx}{\sqrt{(1-x^2)(1-k^2x^2)}}\,dx = -\dfrac{\pi}{2k}\ln k'.$

   BH[239](1)

10. $\int_0^1 \dfrac{x\,\arccos kx}{\sqrt{(1-x^2)(1-k^2x^2)}}\,dx = \dfrac{\pi}{2k}\ln(1+k).$

    BH[242](1)

4.523

1. $\int_0^1 x^{2n}\arcsin x\,dx = \dfrac{1}{2n+1}\left[\dfrac{\pi}{2} - \dfrac{2^n n!}{(2n+1)!!}\right].$

   BH[229](1)

2. $\int_0^1 x^{2n-1}\arcsin x\,dx = \dfrac{\pi}{4n}\left[1 - \dfrac{(2n-1)!!}{2^n n!}\right].$

   BH[229](2)

3. $\int_0^1 x^{2n}\arccos x\,dx = \dfrac{2^n n!}{(2n+1)(2n+1)!!}.$

   BH[229](4)

4. $\int_0^1 x^{2n-1} \arccos x \, dx = \dfrac{\pi}{4n} \dfrac{(2n-1)!!}{2^n n!}$. BH[229](5) 4.523

5. $\int_{-1}^1 (1-x^2)^n \arccos x \, dx = \pi \dfrac{2^n n!}{(2n+1)!!}$. BH[254](2)

6. $\int_{-1}^1 (1-x^2)^{n-\frac{1}{2}} \arccos x \, dx = \dfrac{\pi^2}{2} \dfrac{(2n-1)!!}{2^n n!}$. BH[254](3)

1. $\int_0^1 (\arcsin x)^2 \dfrac{dx}{x^2 \sqrt{1-x^2}} = \pi \ln 2$. BH[243](13) 4.524

2. $\int_0^1 (\arccos x)^2 \dfrac{dx}{(\sqrt{1-x^2})^3} = \pi \ln 2$. BH[244](9)

Arkustangens, Arkuscotangens und Potenzen
Arctangents, arccotangents, and powers 4.53–4.54

1. $\int_0^1 \dfrac{\arctg x}{x} dx = \int_1^\infty \dfrac{\arcctg x}{x} dx = \mathbf{G}$. F II 476, BH[253](8) 4.531

2. $\int_0^\infty \dfrac{\arcctg x}{1 \pm x} dx = \pm \dfrac{\pi}{4} \ln 2 + \mathbf{G}$. BH[248](6), BH[248](7)

3. $\int_0^1 \dfrac{\arctg x}{x(1+x)} dx = -\dfrac{\pi}{8} \ln 2 + \mathbf{G}$. BH[235](11)

4. $\int_0^\infty \dfrac{\arctg x}{1-x^2} dx = -\mathbf{G}$. BH[248](2)

5. $\int_0^1 \arctg qx \dfrac{dx}{(1+px)^2} = \dfrac{1}{2} \dfrac{q}{p^2+q^2} \ln \dfrac{(1+p)^2}{1+q^2} + \dfrac{q^2-p}{(1+p)(p^2+q^2)} \arctg q$

$[p > -1]$. BH[234](7)

6. $\int_0^1 \arcctg qx \dfrac{dx}{(1+px)^2} = \dfrac{1}{2} \dfrac{q}{p^2+q^2} \ln \dfrac{1+q^2}{(1+p)^2} +$

$+ \dfrac{p}{p^2+q^2} \arctg q + \dfrac{1}{1+p} \arcctg q \quad [p > -1]$. BH[234](10)

7. $\int_0^1 \dfrac{\arctg x}{x(1+x^2)} dx = \dfrac{\pi}{8} \ln 2 + \dfrac{1}{2} \mathbf{G}$. BH[235](12)

4.531

8.
$$\int_0^\infty \frac{x \operatorname{arctg} x}{1+x^4} \, dx = \frac{\pi^2}{16}.$$
BH[248](3)

9.
$$\int_0^\infty \frac{x \operatorname{arctg} x}{1-x^4} \, dx = -\frac{\pi}{8} \ln 2.$$
BH[248](4)

10.
$$\int_0^\infty \frac{x \operatorname{arcctg} x}{1-x^4} \, dx = \frac{\pi}{8} \ln 2.$$
BH[248](12)

11.
$$\int_0^\infty \frac{\operatorname{arctg} x}{x\sqrt{1+x^2}} \, dx = \int_0^\infty \frac{\operatorname{arcctg} x}{\sqrt{1+x^2}} \, dx = 2\mathbf{G}.$$
BH[251](3), BH[251](10)

12.
$$\int_0^1 \frac{\operatorname{arctg} x}{x\sqrt{1-x^2}} \, dx = \frac{\pi}{2} \ln(1+\sqrt{2}).$$
F II 697

13.
$$\int_0^1 \frac{x \operatorname{arctg} x \, dx}{\sqrt{(1+x^2)(1+k'^2 x^2)}} = \frac{1}{k^2}\left[F\left(\frac{\pi}{4}, k\right) - \frac{\pi}{2\sqrt{2(1+k'^2)}}\right].$$
BH[244](14)

4.532

1.
$$\int_0^1 x^p \operatorname{arctg} x \, dx = \frac{1}{2(p+1)}\left[\frac{\pi}{2} - \beta\left(\frac{p}{2}+1\right)\right] \quad [p > -2].$$
BH[229](7)

2.
$$\int_0^\infty x^p \operatorname{arctg} x \, dx = \frac{\pi}{2(p+1)} \operatorname{cosec} \frac{p\pi}{2} \quad [-1 > p > -2].$$
BH[246](1)

3.
$$\int_0^1 x^p \operatorname{arcctg} x \, dx = \frac{1}{2(p+1)}\left[\frac{\pi}{2} + \beta\left(\frac{p}{2}+1\right)\right] \quad [p > -1].$$
BH[229](8)

4.
$$\int_0^\infty x^p \operatorname{arcctg} x \, dx = -\frac{\pi}{2(p+1)} \operatorname{cosec} \frac{p\pi}{2} \quad [-1 < p < 0].$$
BH[246](2)

5.
$$\int_0^\infty \left(\frac{x^p}{1+x^{2p}}\right)^{2q} \operatorname{arctg} x \frac{dx}{x} = \frac{\sqrt{\pi^3}}{2^{2q+2} p} \frac{\Gamma(q)}{\Gamma\left(q+\frac{1}{2}\right)} \quad [q > 0].$$
BH[250](10)

4.533

1.
$$\int_0^\infty (1 - x \operatorname{arcctg} x) \, dx = \frac{\pi}{4}.$$
BH[246](3)

2.
$$\int_0^1 \left(\frac{\pi}{4} - \operatorname{arctg} x\right) \frac{dx}{1-x} = -\frac{\pi}{8} \ln 2 + \mathbf{G}.$$
BH[232](2)

3.
$$\int_0^1 \left(\frac{\pi}{4} - \operatorname{arctg} x\right) \frac{1+x}{1-x} \frac{dx}{1+x^2} = \frac{\pi}{8} \ln 2 + \frac{1}{2}\mathbf{G}.$$
BH[235](25)

4. $\displaystyle\int_0^1 \left(x \operatorname{arcctg} x - \frac{1}{x}\operatorname{arctg} x\right)\frac{dx}{1-x^2} = -\frac{\pi}{4}\ln 2.$ BH[232](1) 4.533

$\displaystyle\int_0^\infty (\operatorname{arctg} x)^2 \frac{dx}{x^2\sqrt{1+x^2}} = \int_0^\infty (\operatorname{arctg} x)^2 \frac{x\,dx}{\sqrt{1+x^2}} = -\frac{\pi^2}{4}+4G.$ 4.534

BH[251](9), BH[251](17)

1. $\displaystyle\int_0^1 \frac{\operatorname{arctg} px}{1+p^2x}\,dx = \frac{1}{2p^2}\operatorname{arctg} p \ln(1+p^2).$ BH[231](19) 4.535

2. $\displaystyle\int_0^1 \frac{\operatorname{arctg} px}{1+p^2x}\,dx = \frac{1}{p^2}\left\{\frac{\pi}{4}+\frac{1}{2}\operatorname{arcctg} p\right\}\ln(1+p^2)\quad [p>0].$ BH[231](24)

3. $\displaystyle\int_0^\infty \frac{\operatorname{arcctg} qx}{(p+x)^2}\,dx = -\frac{q}{1+p^2q^2}\left(\ln pq - \frac{\pi}{2}pq\right)\quad [p>0, q>0].$ BH[249](1)

4. $\displaystyle\int_0^\infty \frac{\operatorname{arcctg} qx}{(p+x)^2}\,dx = \frac{q}{1+p^2q^2}\left(\ln pq + \frac{\pi}{2pq}\right)\quad [p>0, q>0].$ BH[249](8)

5. $\displaystyle\int_0^\infty \frac{x\operatorname{arcctg} px}{q^2+x^2}\,dx = \frac{\pi}{2}\ln\frac{1+pq}{pq}\quad [p>0, q>0].$ BH[248](9)

6. $\displaystyle\int_0^\infty \frac{x\operatorname{arcctg} px\,dx}{x^2-q^2} = \frac{\pi}{4}\ln\frac{1+p^2q^2}{p^2q^2}\quad [p>0, q>0].$ BH[248](10)

7. $\displaystyle\int_0^\infty \frac{\operatorname{arctg} px}{x(1+x^2)}\,dx = \frac{\pi}{2}\ln(1+p)\quad [p\geqslant 0].$ F II 751

8. $\displaystyle\int_0^\infty \frac{\operatorname{arctg} px}{x(1-x^2)}\,dx = \frac{\pi}{4}\ln(1+p^2)\quad [p\geqslant 0].$ BH[250](6)

9. $\displaystyle\int_0^\infty \operatorname{arctg} qx \frac{dx}{x(p^2+x^2)} = \frac{\pi}{2p^2}\ln(1+pq)\quad [p>0, q\geqslant 0].$ BH[250](3)

10. $\displaystyle\int_0^\infty \operatorname{arctg} qx \frac{dx}{x(1-p^2x^2)} = \frac{\pi}{4}\ln\frac{p^2+q^2}{p^2}\quad [q\geqslant 0].$ BH[250](6)

11. $\displaystyle\int_0^\infty \frac{x\operatorname{arctg} qx}{(p^2+x^2)^2}\,dx = \frac{\pi q}{4p(1+pq)}\quad [p>0, q\geqslant 0].$ BH[252](12)+

12. $\displaystyle\int_0^\infty \frac{x\operatorname{arcctg} qx}{(p^2+x^2)^2}\,dx = \frac{\pi}{4p^2(1+pq)}\quad [p>0, q\geqslant 0].$ BH[252](20)+

4.535 13. $\int_0^1 \dfrac{\operatorname{arctg} qx}{x\sqrt{1-x^2}}\, dx = \dfrac{\pi}{2} \ln\left(q + \sqrt{1+q^2}\right).$ BH[244](11)

4.536 1. $\int_0^\infty \operatorname{arctg} qx \,\operatorname{arcsin} x \,\dfrac{dx}{x^2} = \dfrac{1}{2} q\pi \ln \dfrac{1+\sqrt{1+q^2}}{\sqrt{1+q^2}} +$

$+ \dfrac{\pi}{2} \ln\left(q + \sqrt{1+q^2}\right) - \dfrac{\pi}{2} \operatorname{arctg} q.$ BH[230](7)

2. $\int_0^\infty \dfrac{\operatorname{arctg} px - \operatorname{arctg} qx}{x}\, dx = \dfrac{\pi}{2} \ln \dfrac{p}{q} \quad [p>0,\, q>0].$ F II 644

3. $\int_0^\infty \dfrac{\operatorname{arctg} px \,\operatorname{arctg} qx}{x^2}\, dx = \dfrac{\pi}{2} \ln \dfrac{(p+q)^{p+q}}{p^p q^q} \quad [p>0,\, q>0].$ F II 751

4.537 1. $\int_0^1 \operatorname{arctg}(\sqrt{1-x^2}) \dfrac{dx}{1-x^2\cos^2\lambda} = \dfrac{\pi}{\cos\lambda} \ln\left[\cos\left(\dfrac{\pi-4\lambda}{8}\right) \operatorname{cosec}\left(\dfrac{\pi+4\lambda}{8}\right)\right].$

BH[245](9)

2. $\int_0^1 \operatorname{arctg}(p\sqrt{1-x^2}) \dfrac{dx}{1-x^2} = \dfrac{1}{2}\pi \ln\left(p + \sqrt{1+p^2}\right) \quad [p>0].$ BH[245](10)

3. $\int_0^1 \operatorname{arctg}(\operatorname{tg}\lambda \sqrt{1-k^2x^2}) \sqrt{\dfrac{1-x^2}{1-k^2x^2}}\, dx = \dfrac{\pi}{2k^2}[E(\lambda,k) - k'^2 F(\lambda,k)] -$

$- \dfrac{\pi}{2k^2} \operatorname{ctg}\lambda(1 - \sqrt{1-k^2\sin^2\lambda}).$ BH[245](12)

4. $\int_0^1 \operatorname{arctg}(\operatorname{tg}\lambda \sqrt{1-k^2x^2}) \sqrt{\dfrac{1-k^2x^2}{1-x^2}}\, dx =$

$= \dfrac{\pi}{2} E(\lambda,k) - \dfrac{\pi}{2} \operatorname{ctg}\lambda (1 - \sqrt{1-k^2\sin^2\lambda}).$ BH[245](11)

5. $\int_0^1 \dfrac{\operatorname{arctg}(\operatorname{tg}\lambda \sqrt{1-k^2x^2})}{\sqrt{(1-x^2)(1-k^2x^2)}}\, dx = \dfrac{\pi}{2} F(\lambda,k).$ BH[245](13)

4.538 1. $\int_0^\infty \operatorname{arctg} x^2 \dfrac{dx}{1+x^2} = \int_0^\infty \operatorname{arctg} x^3 \dfrac{dx}{1+x^2};$ BH[252](10), BH[252](11)

$= \int_0^\infty \operatorname{arcctg} x^2 \dfrac{dx}{1+x^2} = \int_0^\infty \operatorname{arcctg} x^3 \dfrac{dx}{1+x^2} = \dfrac{\pi^2}{8}.$

BH[252](18), BH[252](19)

2. $\int_0^1 \dfrac{1-x^2}{x^2} \operatorname{arctg} x^2\, dx = \dfrac{\pi}{2}(\sqrt{2}-1).$ BH[244](10)+

$$\int_0^\infty x^{s-1} \operatorname{arctg}(ae^{-x})\, dx = 2^{-s-1}\, \Gamma(s)\, a\Phi\left(-a^2, s+1, \frac{1}{2}\right).$$

Ba IV 222(47)   4.539

$$\int_0^\infty \operatorname{arctg}\left(\frac{p \sin qx}{1 + p \cos qx}\right) \frac{x\, dx}{1 + x^2} = \frac{\pi}{2} \ln(1 + pe^{-q}) \quad [p > -e^q].$$

BH[341](14)+   4.541

Zyklometrische Funktionen und Potenzen
Inverse trigonometric functions and exponentials   4.55

1. $$\int_0^1 (\arcsin x)e^{-bx}\, dx = \frac{\pi}{2b}[I_0(b) - \mathbf{L}_0(b)].$$

Ba IV 160(1)   4.551

2. $$\int_0^1 x(\arcsin x)e^{-bx}\, dx = \frac{\pi}{2b^2}[\mathbf{L}_0(b) - I_0(b) + b\mathbf{L}_1(b) - bI_1(b)] + \frac{1}{b}.$$

Ba IV 161(2)

3. $$\int_0^\infty \left(\operatorname{arctg} \frac{x}{a}\right) e^{-bx}\, dx = \frac{1}{b}[-\operatorname{ci}(ab)\sin(ab) - \operatorname{si}(ab)\cos(ab)]$$

$$[\operatorname{Re} b > 0]. \quad \text{Ba IV 161(3)}$$

4. $$\int_0^\infty \left(\operatorname{arcctg} \frac{x}{a}\right) e^{-bx}\, dx = \frac{1}{b}\left[\frac{\pi}{2} + \operatorname{ci}(ab)\sin(ab) + \operatorname{si}(ab)\cos(ab)\right]$$

$$[\operatorname{Re} b > 0]. \quad \text{Ba IV 161(4)}$$

$$\int_0^\infty \frac{\operatorname{arctg} \frac{x}{q}}{e^{2\pi x} - 1}\, dx = \frac{1}{2}\left[\ln \Gamma(q) - \left(q - \frac{1}{2}\right)\ln q + q - \frac{1}{2}\ln 2\pi\right] \quad [q > 0].$$

WW 251   4.552

$$\int_0^\infty \left(\frac{2}{\pi}\operatorname{arcctg} x - e^{-px}\right)\frac{dx}{x} = C + \ln p \quad [p > 0].$$

NI [66](12)   4.553

Arkustangens und Hyperbelfunktion
The arctangent and the hyperbolic function   4.56

$$\int_{-\infty}^\infty \frac{\operatorname{arctg} e^{-x}}{\operatorname{ch}^{2q} px}\, dx = \frac{1}{2}\int_{-\infty}^\infty \frac{\Pi(x)}{\operatorname{ch}^{2q} px}\, dx = \frac{\sqrt{\pi^3}}{4p}\frac{\Gamma(q)}{\Gamma\left(q + \frac{1}{2}\right)} \quad [q > 0].$$

Li [282](10)   4.561

Zyklometrische und trigonometrische Funktionen
Inverse and direct trigonometric functions   4.57

$$\int_0^{\pi/2} \arcsin(k \sin x) \frac{\sin x\, dx}{\sqrt{1 - k^2 \sin^2 x}} = -\frac{\pi}{2k} \ln k'.$$

BH[344](2)   4.571

4.572
$$\int_0^\infty \left(\frac{2}{\pi}\operatorname{arcctg} x - \cos px\right)dx = C + \ln p \quad [p > 0].$$
NI 66(12)

4.573
1.
$$\int_0^\infty \operatorname{arcctg} qx \sin px\, dx = \frac{\pi}{2p}\left(1 - e^{-\frac{p}{q}}\right) \quad [p > 0, q > 0].$$
BH[347](1)+

2.
$$\int_0^\infty \operatorname{arcctg} qx \cos px\, dx = \frac{1}{2p}\left[e^{-\frac{p}{q}}\operatorname{Ei}\left(\frac{p}{q}\right) - e^{\frac{p}{q}}\operatorname{Ei}\left(-\frac{p}{q}\right)\right]$$
$$[p > 0,\ q > 0].$$
BH[347](2)+

3.
$$\int_0^\infty \operatorname{arcctg} rx \frac{\sin px\, dx}{1 \pm 2q\cos px + q^2} =$$
$$= \pm \frac{\pi}{2pq}\ln\frac{1 \pm q}{1 \pm qe^{-\frac{p}{r}}} \quad [q^2 < 1,\ r > 0,\ p > 0];$$
$$= \pm \frac{\pi}{2pq}\ln\frac{q \pm 1}{q \pm e^{-\frac{p}{r}}} \quad [q^2 > 1,\ r > 0,\ p > 0].$$
BH[347](10)

4.
$$\int_0^\infty \operatorname{arcctg} px \frac{\operatorname{tg} x\, dx}{q^2\cos^2 x + r^2\sin^2 x} = \frac{\pi}{2r^2}\ln\left(1 + \frac{r}{q}\operatorname{th}\frac{1}{p}\right)$$
$$[p > 0,\ q > 0,\ r > 0].$$
BH[347](9)

4.574
1.
$$\int_0^\infty \operatorname{arctg}\left(\frac{2a}{x}\right)\sin(bx)\, dx = \frac{\pi}{b}e^{-ab}\operatorname{sh}(ab) \quad [\operatorname{Re} a > 0,\ b > 0].$$
Ba IV 87(8)

2.
$$\int_0^\infty \operatorname{arctg}\frac{a}{x}\cos(bx)\, dx = \frac{1}{2b}[e^{-ab}\overline{\operatorname{Ei}}(ab) - e^{ab}\operatorname{Ei}(-ab)]$$
$$[a > 0,\ b > 0].$$
Ba IV 29(7)

3.
$$\int_0^\infty \operatorname{arctg}\left[\frac{2ax}{x^2+c^2}\right]\sin(bx)\, dx = \frac{\pi}{b}e^{-b\sqrt{a^2+c^2}}\operatorname{sh}(ab) \quad [b > 0].$$
Ba IV 87(9)

4.
$$\int_0^\infty \operatorname{arctg}\left(\frac{2}{x^2}\right)\cos(bx)\, dx = \frac{\pi}{b}e^{-b}\sin b \quad [b > 0].$$
Ba IV 29(8)

4.575
1.
$$\int_0^\pi \operatorname{arctg}\frac{p\sin x}{1 - p\cos x}\sin nx\, dx = \frac{\pi}{2n}p^n \quad [p^2 < 1].$$
BH[345](4)

2.
$$\int_0^\pi \operatorname{arctg}\frac{p\sin x}{1 - p\cos x}\sin nx \cos x\, dx = \frac{\pi}{4}\left(\frac{p^{n+1}}{n+1} + \frac{p^{n-1}}{n-1}\right) \quad [p^2 < 1].$$
BH[345](5)

3. $\int_0^\pi \arctg \frac{p \sin x}{1 - p \cos x} \cos nx \sin x \, dx = \frac{\pi}{4}\left(\frac{p^{n+1}}{n+1} - \frac{p^{n-1}}{n-1}\right) \quad [p^2 < 1].$ 4.575

BH[345](6)

1. $\int_0^\pi \arctg \frac{p \sin x}{1 - p \cos x} \frac{dx}{\sin x} = \frac{\pi}{2} \ln \frac{1+p}{1-p} \quad [p^2 < 1].$ BH[346](1) 4.576

2. $\int_0^\pi \arctg \frac{p \sin x}{1 - p \cos x} \frac{dx}{\tg x} = -\frac{\pi}{2} \ln(1 - p^2) \quad [p^2 < 1].$ BH[346](3)

1. $\int_0^{\pi/2} \arctg(\tg \lambda \sqrt{1 - k^2 \sin^2 x}) \frac{\sin^2 x \, dx}{\sqrt{1 - k^2 \sin^2 x}} =$ 4.577

$= \frac{\pi}{2k^2}[F(\lambda, k) - E(\lambda, k) + \ctg \lambda (1 - \sqrt{1 - k^2 \sin^2 \lambda})].$ BH[344](4)

2. $\int_0^{\pi/2} \arctg(\tg \lambda \sqrt{1 - k^2 \sin^2 x}) \frac{\cos^2 x \, dx}{\sqrt{1 - k^2 \sin^2 x}} =$

$= \frac{\pi}{2k^2}[E(\lambda, k) - k'^2 F(\lambda, k) + \ctg \lambda (\sqrt{1 - k^2 \sin^2 \lambda} - 1)].$ BH[344](5)

Zyklometrische und trigonometrische Funktionen und Potenzen
An inverse and a direct trigonometric function and a power
4.58

$\int_0^\infty \arctg x \cos px \frac{dx}{x} = \int_0^\infty \arctg \frac{x}{p} \cos x \frac{dx}{x} = -\frac{\pi}{2} \Ei(-p) \quad [\Re(p) > 0].$ 4.581

F III 533, NI 25(13)

Zyklometrische Funktionen und Logarithmus
Inverse trigonometric functions and logarithms
4.59

1. $\int_0^1 \arcsin x \ln x \, dx = 2 - \ln 2 - \frac{1}{2}\pi.$ BH[339](1) 4.591

2. $\int_0^1 \arccos x \ln x \, dx = \ln 2 - 2.$ BH[339](2)

$\int_0^1 \arccos x \frac{dx}{\ln x} = -\sum_{k=0}^\infty \frac{(2k-1)!!}{2^k k!} \frac{\ln(2k+2)}{2k+1}.$ BH[339](8) 4.592

1. $\int_0^1 \arctg x \ln x \, dx = \frac{1}{2}\ln 2 - \frac{\pi}{4} + \frac{1}{48}\pi^2.$ BH[339](3) 4.593

4.593　2.　$\int_0^1 \operatorname{arcctg} x \ln x \, dx = -\frac{1}{48}\pi^2 - \frac{\pi}{4} - \frac{1}{2}\ln 2.$　BH[339](4)

4.594　$\int_0^1 \operatorname{arctg} x(\ln x)^{n-1} (\ln x + n) \, dx = \frac{n!}{(-2)^{n+1}} (2^{-n} - 1)\zeta(n+1).$　BH[339](7)

4.6 Mehrfache Integrale / Multiple Integrals

4.60 Variablentransformation in Mehrfachintegralen / Change of variables in multiple integrals

4.601　1.　$$\iint_{(\sigma)} f(x, y) \, dx \, dy = \iint_{(\sigma')} f[\varphi(u, v), \psi(u, v)] |\Delta| \, du \, dv,$$

wobei $x = \varphi(u, v)$, $y = \psi(u, v)$ ist und | where $x = \varphi(u, v)$ and $y = \psi(u, v)$, and

$$\Delta = \frac{\partial \varphi}{\partial u}\frac{\partial \psi}{\partial v} - \frac{\partial \psi}{\partial u}\frac{\partial \varphi}{\partial v} \equiv \frac{D(\varphi, \psi)}{D(u, v)}$$

die (JACOBIsche) Funktionaldeterminante der Funktionen φ und ψ bedeutet. | is the functional determinant, or JACOBIAN, of φ and ψ.

2.　$$\iiint_{(V)} f(x, y, z) \, dx \, dy \, dz =$$

$$= \iiint_{(V')} f[\varphi(u, v, w), \psi(u, v, w), \chi(u, v, w)] |\Delta| \, du \, dv \, dw,$$

wobei $x = \varphi(u, v, w)$, $y = \psi(u, v, w)$, $z = \chi(u, v, w)$ ist und | where $x = \varphi(u, v, w)$, $y = \psi(u, v, w)$ and $z = \chi(u, v, w)$, and

$$\Delta = \begin{vmatrix} \frac{\partial \varphi}{\partial u} & \frac{\partial \varphi}{\partial v} & \frac{\partial \varphi}{\partial w} \\ \frac{\partial \psi}{\partial u} & \frac{\partial \psi}{\partial v} & \frac{\partial \psi}{\partial w} \\ \frac{\partial \chi}{\partial u} & \frac{\partial \chi}{\partial v} & \frac{\partial \chi}{\partial w} \end{vmatrix} \equiv \frac{D(\varphi, \psi, \chi)}{D(u, v, w)}$$

die Funktionaldeterminante der Funktionen φ, ψ, χ bedeutet.
Dabei wird sowohl in 4.601 2. als auch in 4.601 1. folgendes vorausgesetzt:
a) Die Funktionen φ, ψ, χ sind nebst ihren ersten partiellen Ableitungen im Integrationsbereich stetig;
| is the functional determinant of φ, ψ, and χ.
Concerning 4.601 2. and 4.601 1. the following assumptions have been made:
(a) the functions φ, ψ, χ and their first partial derivatives are continuous throughout the domain of integration;

b) die Funktionaldeterminante ändert in diesem Bereich ihr Vorzeichen nicht;

(b) in this domain the functional determinant Δ does not change sign;

c) die Funktionalbeziehungen zwischen den ursprünglichen Variablen x, y, z und den neuen Variablen u, v, w sind im Integrationsbereich eineindeutig;

(c) in this domain there is a one-to-one relationship between the old and the new integration variables;

d) der Bereich V (bzw. σ) geht beim Übergang von den Variablen x, y, z zu den Variablen u, v, w in den Bereich V' (bzw. σ') über.

(d) V' corresponds to V and σ' corresponds to σ.

Übergang zu Polarkoordinaten.

Transformation to polar coordinates.   4.602

$$x = r\cos\varphi, \quad y = r\sin\varphi; \quad \frac{D(x,y)}{D(r,\varphi)} = r.$$

Übergang zu Kugelkoordinaten.

Transformation to spherical coordinates.   4.603

$$x = r\sin\theta\cos\varphi, \quad y = r\sin\theta\sin\varphi, \quad z = r\cos\theta, \quad \frac{D(x,y,z)}{D(r,\theta,\varphi)} = r^2\sin\theta.$$

Vertauschung der Integrationsfolge und Variablentransformation
Change of the order of integration and change of variables   4.61

1. $\displaystyle\int_0^\alpha dx \int_0^x f(x,y)\,dy = \int_0^\alpha dy \int_y^\alpha f(x,y)\,dx.$ 4.611

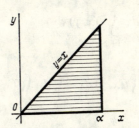

2. $\displaystyle\int_0^\alpha dx \int_0^{\frac{\beta}{\alpha}x} f(x,y)\,dy = \int_0^\beta dy \int_{\frac{\alpha}{\beta}y}^\alpha f(x,y)\,dx.$

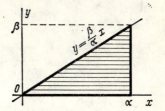

1. $\displaystyle\int_0^R dx \int_0^{\sqrt{R^2-x^2}} f(x,y)\,dy =$

$\displaystyle = \int_0^R dy \int_0^{\sqrt{R^2-y^2}} f(x,y)\,dx.$ 4.612

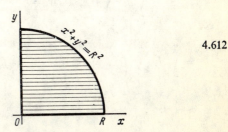

4.612 2. $\displaystyle\int_0^{2p} dx \int_0^{\frac{q}{p}\sqrt{2px-x^2}} f(x,y)\,dy =$

$\displaystyle = \int_0^q dy \int_{p\left[1-\sqrt{1-\left(\frac{y}{q}\right)^2}\right]}^{p\left[1+\sqrt{1-\left(\frac{y}{q}\right)^2}\right]} f(x,y)\,dx.$

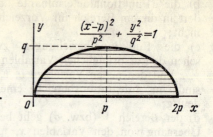

4.613 1. $\displaystyle\int_0^\alpha dx \int_0^{\frac{\beta}{\beta+x}} f(x,y)\,dy =$

$\displaystyle = \int_0^{\frac{\beta}{\beta+\alpha}} dy \int_0^\alpha f(x,y)\,dx +$

$\displaystyle + \int_{\frac{\beta}{\beta+\alpha}}^1 dy \int_0^{\frac{\beta}{y}(1-y)} f(x,y)\,dx.$

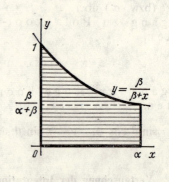

2. $\displaystyle\int_0^\alpha dx \int_{\beta x}^{\delta-\gamma x} f(x,y)\,dy =$

$\displaystyle = \int_0^{\alpha\beta} dy \int_0^{y/\beta} f(x,y)\,dx + \int_{\alpha\beta}^\delta dy \int_0^{\frac{\delta-y}{\gamma}} f(x,y)\,dx$

$\displaystyle \left[\alpha = \frac{\delta}{\beta+\gamma},\ \alpha>0,\ \beta>0,\ \gamma>0\right].$

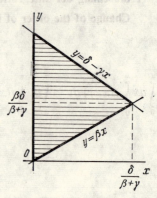

3. $\displaystyle\int_0^{2\alpha} dx \int_{\frac{x^2}{4\alpha}}^{3\alpha-x} f(x,y)\,dy =$

$\displaystyle = \int_0^\alpha dy \int_0^{2\sqrt{\alpha y}} f(x,y)\,dx + \int_\alpha^{3\alpha} dy \int_0^{3\alpha-y} f(x,y)\,dx.$

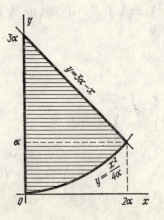

4. $\int_0^R dx \int_{\sqrt{R^2-x^2}}^{x+2R} f(x,y)\,dy = \int_0^R dy \int_{\sqrt{R^2-y^2}}^{R} f(x,y)\,dx +$

$+ \int_R^{2R} dy \int_0^R f(x,y)\,dx + \int_{2R}^{3R} dy \int_{y-2R}^{R} f(x,y)\,dx.$

4.613

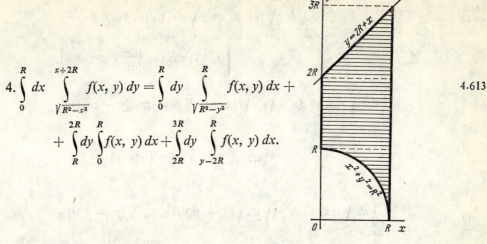

$\int_0^{\pi/2} d\varphi \int_0^{2R\cos\varphi} f(r,\varphi)\,dr = \int_0^{2R} dr \int_0^{\arccos\frac{r}{2R}} f(r,\varphi)\,d\varphi.$

4.614

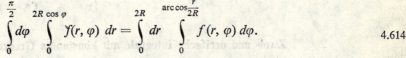

oder
or

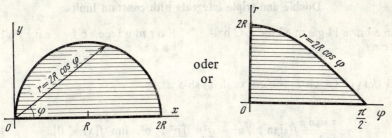

$\int_0^R dx \int_0^{\sqrt{R^2-x^2}} f(x,y)\,dy = \int_0^{\pi/2} d\varphi \int_0^R f(r\cos\varphi, r\sin\varphi)\,r\,dr.$

4.615

$\int_0^{2R} dx \int_0^{\sqrt{2Rx-x^2}} f(x,y)\,dy =$

$= \int_0^{\pi/2} d\varphi \int_0^{2R\cos\varphi} f(r\cos\varphi\, r, \sin\varphi)\,r\,dr.$

4.616

$\int_\alpha^\beta dx \int_{\varphi_1(x)}^{\varphi_2(x)} f(x,y)\,dy = \int_0^\beta dx \int_0^{\varphi_2(x)} f(x,y)\,dy - \int_0^\beta dx \int_0^{\varphi_1(x)} f(x,y)\,dy -$

4.617

4.617
$$-\int_0^\alpha dx \int_0^{\varphi_2(x)} f(x, y)\, dy + \int_0^\alpha dx \int_0^{\varphi_1(x)} f(x, y)\, dy$$

$$\left[\varphi_1(x) \leqslant \varphi_2(x) \ \text{für/for} \ \alpha \leqslant x \leqslant \beta\right].$$

4.618
$$\int_0^y dx \int_0^{\varphi(x)} f(x, y)\, dy = \int_0^y dx \int_0^1 f[x, z\varphi(x)]\, \varphi(x)\, dz \quad [y = z\varphi(x)];$$

$$= y \int_0^1 dz \int_0^{\varphi(yz)} f(yz, y)\, dy \quad [x = yz].$$

4.619
$$\int_{x_0}^{x_1} dx \int_{y_0}^{y_1} f(x, y)\, dy = \int_{x_0}^{x_1} dx \int_0^1 (y_1 - y_0)\, f[x, y_0 + (y_1 - y_0)t]\, dt$$

$$[y = y_0 + (y_1 - y_0)\, t].$$

4.62

Zwei- und dreifache Integrale mit konstanten Grenzen

Double and triple integrals with constant limits

4.620 Formeln allgemeinen Charakters. | Formulae of general character.

1. $\displaystyle\int_0^\pi d\omega \int_0^\infty f'(p\ \text{ch}\ x + q \cos \omega\ \text{sh}\ x)\ \text{sh}\ x\, dx =$

$$= -\frac{\pi\, \text{sign}\, p}{\sqrt{p^2 - q^2}} f(\text{sign}\, p\, \sqrt{p^2 - q^2}) \quad [p^2 > q^2,\ \lim_{x \to +\infty} f(x) = 0].$$

Lo III 389

2. $\displaystyle\int_0^{2\pi} d\omega \int_0^\infty f'[p\ \text{ch}\ x + (q \cos \omega + r \sin \omega)\ \text{sh}\ x]\ \text{sh}\ x\, dx =$

$$= -\frac{2\pi\, \text{sign}\, p}{\sqrt{p^2 - q^2 - r^2}} f(\text{sign}\, p\, \sqrt{p^2 - q^2 - r^2}) \quad [p^2 > q^2 + r^2,\ \lim_{x \to +\infty} f(x) = 0].$$

Lo III 390

3. $\displaystyle\int_0^\pi \int_0^\pi \frac{dx\, dy}{\sin x \sin^2 y} f'\left[\frac{p - q \cos x}{\sin x \sin y} + r \operatorname{ctg} y\right] =$

$$= -\frac{2\pi\, \text{sign}\, p}{\sqrt{p^2 - q^2 - r^2}} f(\text{sign}\, p\, \sqrt{p^2 - q^2 - r^2}) \quad [p^2 > q^2 + r^2,\ \lim_{x \to +\infty} f(x) = 0].$$

Lo III 280

4. $\displaystyle\int_{-\infty}^\infty dx \int_{-\infty}^\infty f'(p\ \text{ch}\ x\ \text{ch}\ y + q\ \text{sh}\ x\ \text{ch}\ y + r\ \text{sh}\ y)\ \text{ch}\ y\, dy =$

$$= -\frac{2\pi\, \text{sign}\, p}{\sqrt{p^2 - q^2 - r^2}} f(\text{sign}\, p\, \sqrt{p^2 - q^2 - r^2}) \quad [p^2 > q^2 + r^2,\ \lim_{x \to +\infty} f(x) = 0].$$

Lo III 390

5. $\int_0^\infty dx \int_0^\pi f(p\,\text{ch}\,x + q\cos\omega\,\text{sh}\,x)\,\text{sh}^2 x \sin\omega\,d\omega =$ 4.620

$$= 2\int_0^\infty f(\text{sign}\,p\,\sqrt{p^2 - q^2}\,\text{ch}\,x)\,\text{sh}^2 x\,dx \quad [\lim_{x\to+\infty} f(x) = 0].$$ Lo III 391

6. $\int_0^\infty dx \int_0^{2\pi} d\omega \int_0^\pi f[p\,\text{ch}\,x + (q\cos\omega + r\sin\omega)\sin\theta\,\text{sh}\,x]\,\text{sh}^2 x \sin\theta\,d\theta =$

$$= 4\int_0^\infty f(\text{sign}\,p\,\sqrt{p^2 - q^2 - r^2}\,\text{ch}\,x)\,\text{sh}^2 x\,dx$$

$$[p^2 > q^2 + r^2, \lim_{x\to+\infty} f(x) = 0].$$ Lo III 390

7. $\int_0^\infty dx \int_0^{2\pi} d\omega \int_0^\pi f\{p\,\text{ch}\,x + [(q\cos\omega + r\sin\omega)\sin\theta + s\,\text{ch}\,\theta]\,\text{sh}\,x\} \times$

$$\times \text{sh}^2 x \sin\theta\,d\theta = 4\pi\int_0^\infty f(\text{sign}\,p\,\sqrt{p^2 - q^2 - r^2 - s^2}\,\text{ch}\,x)\,\text{sh}^2 x\,dx$$

$$[p^2 > q^2 + r^2 + s^2, \lim_{x\to+\infty} f(x) = 0].$$ Lo III 391

1. $\int_0^{\pi/2}\int_0^{\pi/2} \dfrac{\sin y \sqrt{1 - k^2\sin^2 x \sin^2 y}}{1 - k^2\sin^2 y}\,dx\,dy = \dfrac{\pi}{2\sqrt{1 - k^2}}.$ Lo I 252(90) 4.621

2. $\int_0^{\pi/2}\int_0^{\pi/2} \dfrac{\cos y \sqrt{1 - k^2\sin^2 x \sin^2 y}}{1 - k^2\sin^2 y}\,dx\,dy = \boldsymbol{K}(k).$ Lo I 252(91)

3. $\int_0^{\pi/2}\int_0^{\pi/2} \dfrac{\sin\alpha \sin y\,dx\,dy}{\sqrt{1 - \sin^2\alpha \sin^2 x \sin^2 y}} = \dfrac{\pi\alpha}{2}.$ Lo I 253

1. $\int_0^\pi\int_0^\pi\int_0^\pi \dfrac{dx\,dy\,dz}{1 - \cos x \cos y \cos z} = 4\pi\,\boldsymbol{K}^2\!\left(\dfrac{\sqrt{2}}{2}\right).$ MO 137 4.622

2. $\int_0^\pi\int_0^\pi\int_0^\pi \dfrac{dx\,dy\,dz}{3 - \cos y \cos z - \cos x \cos z - \cos x \cos y} = \sqrt{3}\pi\,\boldsymbol{K}^2\!\left(\sin\dfrac{\pi}{12}\right).$ MO 137

3. $\int_0^\pi\int_0^\pi\int_0^\pi \dfrac{dx\,dy\,dz}{3 - \cos x - \cos y - \cos z} =$

$$= 4\pi[18 + 12\sqrt{2} - 10\sqrt{3} - 7\sqrt{6}]\,\boldsymbol{K}^2[(2 - \sqrt{3})(\sqrt{3} - \sqrt{2})].$$ MO 137

4.623
$$\int_0^\infty \int_0^\infty \varphi(a^2 x^2 + b^2 y^2)\, dx\, dy = \frac{\pi}{4ab} \int_0^\infty \varphi(x)\, x\, dx.$$

4.624
$$\int_0^\pi \int_0^{2\pi} f(\alpha \cos\theta + \beta \sin\theta \cos\psi + \gamma \sin\theta \sin\psi) \sin\theta\, d\theta\, d\psi =$$

$$= 2\pi \int_0^\pi f(R \cos p) \sin p\, dp = 2\pi \int_{-1}^1 f(Rt)\, dt \quad [R = \sqrt{\alpha^2 + \beta^2 + \gamma^2}].$$

Mehrfache Integrale

4.63–4.64   **Multiple integrals**

4.631
$$\int_p^x dt_{n-1} \int_p^{t_{n-1}} dt_{n-2} \cdots \int_p^{t_1} f(t)\, dt = \frac{1}{(n-1)!} \int_p^x (x-t)^{n-1} f(t)\, dt,$$

wobei die Funktion $f(t)$ im Intervall $[p, q]$ stetig und $p \leqslant x \leqslant q$ ist.

if $f(t)$ is continuous on $[p, q]$ and $p \leqslant x \leqslant q$.

F II 700   F II 700

4.632   1.
$$\iint\limits_{\substack{x_1 \geqslant 0,\, x_2 \geqslant 0,\ldots,\, x_n \geqslant 0 \\ x_1 + x_2 + \ldots + x_n \leqslant h}} \int dx_1\, dx_2 \ldots dx_n = \frac{h^n}{n!}$$

[Volumen des n-dimensionalen Simplexes].

[volume of n-dimensional simplex].

F III 386–387   F III 386–387

2.
$$\iint\limits_{x_1^2 + x_2^2 + \ldots + x_n^2 \leqslant R^2} \cdots \int dx_1\, dx_2 \ldots dx_n = \frac{\sqrt{\pi^n}}{\Gamma\left(\frac{n}{2} + 1\right)} R^n$$

[Volumen der n-dimensionalen Kugel].

[volume of n-dimensional sphere].

F III 387–388   F III 387–388

4.633
$$\iint\limits_{x_1^2 + x_2^2 + \ldots + x_n^2 \leqslant 1} \cdots \int \frac{dx_1\, dx_2 \ldots dx_n}{\sqrt{1 - x_1^2 - x_2^2 - \ldots - x_n^2}} = \frac{\pi^{\frac{n+1}{2}}}{\Gamma\left(\frac{n+1}{2}\right)} \quad [n > 1]$$

[halber Flächeninhalt der $(n+1)$-dimensionalen Sphäre $x_1^2 + x_2^2 + \ldots + x_{n+1}^2 = 1$].

[half the (hyper) surface of the $(n+1)$-dimensional sphere $x_1^2 + x_2^2 + \ldots + x_{n+1}^2 = 1$].

F III 388   F III 388

4.634
$$\iint\limits_{\substack{x_1 \geqslant 0,\, x_2 \geqslant 0,\ldots,\, x_n \geqslant 0 \\ \left(\frac{x_1}{q_1}\right)^{\alpha_1} + \left(\frac{x_2}{q_2}\right)^{\alpha_2} + \ldots + \left(\frac{x_n}{q_n}\right)^{\alpha_n} \leqslant 1}} \int x_1^{p_1-1} x_2^{p_2-1} \ldots x_n^{p_n-1}\, dx_1\, dx_2 \ldots dx_n =$$

$$= \frac{q_1^{p_1} q_2^{p_2} \cdots q_n^{p_n}}{\alpha_1 \alpha_2 \cdots \alpha_n} \frac{\Gamma\left(\frac{p_1}{\alpha_1}\right)\Gamma\left(\frac{p_2}{\alpha_2}\right)\cdots\Gamma\left(\frac{p_n}{\alpha_n}\right)}{\Gamma\left(\frac{p_1}{\alpha_1}+\frac{p_2}{\alpha_2}+\ldots+\frac{p_n}{\alpha_n}+1\right)} \quad 4.634$$

$[\alpha_i > 0, \, p_i > 0, \, q_i > 0, \, i = 1, 2, \ldots, n]$. F III 390

1. $\displaystyle\int\limits_{\substack{x_1 \geq 0,\, x_2 \geq 0,\ldots,x_n \geq 0 \\ \left(\frac{x_1}{q_1}\right)^{\alpha_1} + \left(\frac{x_2}{q_2}\right)^{\alpha_2} + \ldots + \left(\frac{x_n}{q_n}\right)^{\alpha_n} \geq 1}} \int \cdots \int f\left[\left(\frac{x_1}{q_1}\right)^{\alpha_1} + \left(\frac{x_2}{q_2}\right)^{\alpha_2} + \ldots + \left(\frac{x_n}{q_n}\right)^{\alpha_n}\right] \times$ 4.635

$$\times x_1^{p_1-1} x_2^{p_2-1} \cdots x_n^{p_n-1} \, dx_1 \, dx_2 \cdots dx_n =$$

$$= \frac{q_1^{p_1} q_1^{p_2} \cdots q_n^{p_n}}{\alpha_1 \alpha_2 \cdots \alpha_n} \frac{\Gamma\left(\frac{p_1}{\alpha_1}\right)\Gamma\left(\frac{p_2}{\alpha_2}\right)\cdots\Gamma\left(\frac{p_n}{\alpha_n}\right)}{\Gamma\left(\frac{p_1}{\alpha_1}+\frac{p_2}{\alpha_2}+\ldots+\frac{p_n}{\alpha_n}\right)} \int\limits_1^\infty f(x) x^{\frac{p_1}{\alpha_1}+\frac{p_2}{\alpha_2}+\ldots+\frac{p_n}{\alpha_n}-1} \, dx$$

vorausgesetzt, das auf der rechten Seite stehende Integral konvergiert absolut. | provided that the integral on the right converges absolutely.

F III 392

2. $\displaystyle\int\limits_{\substack{x_1 \geq 0,\, x_2 \geq 0,\ldots,x_n \geq 0 \\ \left(\frac{x_1}{q_1}\right)^{\alpha_1} + \left(\frac{x_2}{q_2}\right)^{\alpha_2} + \ldots + \left(\frac{x_n}{q_n}\right)^{\alpha_n} \leq 1}} \int \cdots \int f\left[\left(\frac{x_1}{q_1}\right)^{\alpha_1} + \left(\frac{x_2}{q_2}\right)^{\alpha_2} + \ldots + \left(\frac{x_n}{q_n}\right)^{\alpha_n}\right] \times$

$$\times x_1^{p_1-1} x_2^{p_2-1} \cdots x_n^{p_n-1} \, dx_1 \, dx_2 \cdots dx_n =$$

$$= \frac{q_1^{p_1} q_2^{p_2} \cdots q_n^{p_n}}{\alpha_1 \alpha_2 \cdots \alpha_n} \frac{\Gamma\left(\frac{p_1}{\alpha_1}\right)\Gamma\left(\frac{p_2}{\alpha_2}\right)\cdots\Gamma\left(\frac{p_n}{\alpha_n}\right)}{\Gamma\left(\frac{p_1}{\alpha_1}+\frac{p_2}{\alpha_2}+\ldots+\frac{p_n}{\alpha_n}\right)} \int\limits_0^1 f(x) x^{\frac{p_1}{\alpha_1}+\frac{p_2}{\alpha_2}+\ldots+\frac{p_n}{\alpha_n}-1} \, dx$$

vorausgesetzt, das auf der rechten Seite stehende einfache Integral konvergiert absolut; alle Zahlen q_i, α_i, p_i sind positiv. | provided that the integral on the right converges absolutely; q_i, α_i, p_i positive.

F III 392

Insbesondere gilt | In particular,

3. $\displaystyle\int\limits_{\substack{x_1 \geq 0,\, x_2 \geq 0,\ldots,x_n \geq 0 \\ x_1 + x_2 + \ldots + x_n \leq 1}} \int \cdots \int x_1^{p_1-1} x_2^{p_2-1} \cdots x_n^{p_n-1} e^{-q(x_1+x_2+\ldots+x_n)} \, dx_1 \, dx_2 \cdots dx_n =$

$$= \frac{\Gamma(p_1)\Gamma(p_2)\cdots\Gamma(p_n)}{\Gamma(p_1+p_2+\ldots+p_n)} \int\limits_0^1 x^{p_1+p_2+\ldots+p_n-1} e^{-qx} \, dx$$

$[n > 0, \, p_1 > 0, \, p_2 > 0, \, \ldots, \, p_n > 0]$.

4. $\displaystyle\int\limits_{\substack{x_1 \geq 0,\, x_2 \geq 0,\ldots,x_n \geq 0 \\ x_1^{\alpha_1} + x_2^{\alpha_2} + \ldots + x_n^{\alpha_n} \leq 1}} \int \cdots \int \frac{x_1^{p_1-1} x_2^{p_2-1} \cdots x_n^{p_n-1}}{(1 - x_1^{\alpha_1} - x_2^{\alpha_2} - \ldots - x_n^{\alpha_n})^\mu} \, dx_1 \, dx_2 \cdots dx_n =$

4.635
$$= \frac{1}{\alpha_1 \alpha_2 \ldots \alpha_n} \frac{\Gamma\left(\frac{p_1}{\alpha_1}\right) \Gamma\left(\frac{p_2}{\alpha_2}\right) \ldots \Gamma\left(\frac{p_n}{\alpha_n}\right)}{\Gamma\left(1 - \mu + \frac{p_1}{\alpha_1} + \frac{p_2}{\alpha_2} + \ldots + \frac{p_n}{\alpha_n}\right)} \Gamma(1 - \mu)$$

$$[p_1 > 0,\, p_2 > 0,\, \ldots,\, p_n > 0,\, \mu < 1].$$ F III 392

4.636 1.
$$\iint \cdots \int_{\substack{x_1 \geq 0,\, x_2 \geq 0, \ldots, x_n \geq 0 \\ x_1^{\alpha_1} + x_2^{\alpha_2} + \ldots + x_n^{\alpha_n} \geq 1}} \frac{x_1^{p_1-1} x_2^{p_2-1} \ldots x_n^{p_n-1}}{(x_1^{\alpha_1} + x_2^{\alpha_2} + \ldots + x_n^{\alpha_n})^\mu} dx_1\, dx_2 \ldots dx_n =$$

$$= \frac{1}{\alpha_1 \alpha_2 \ldots \alpha_n \left(\mu - \frac{p_1}{\alpha_1} - \frac{p_2}{\alpha_2} - \ldots - \frac{p_n}{\alpha_n}\right)} \frac{\Gamma\left(\frac{p_1}{\alpha_1}\right) \Gamma\left(\frac{p_2}{\alpha_2}\right) \ldots \Gamma\left(\frac{p_n}{\alpha_n}\right)}{\Gamma\left(\frac{p_1}{\alpha_1} + \frac{p_2}{\alpha_2} + \ldots + \frac{p_n}{\alpha_n}\right)}$$

$$\left[p_1 > 0,\, p_2 > 0,\, \ldots,\, p_n > 0;\, \mu > \frac{p_1}{\alpha_1} + \frac{p_2}{\alpha_2} + \ldots + \frac{p_n}{\alpha_n}\right].$$ F III 398–399

2.
$$\iint \cdots \int_{\substack{x_1 \geq 0,\, x_2 \geq 0, \ldots, x_n \geq 0 \\ x_1^{\alpha_1} + x_2^{\alpha_2} + \ldots + x_n^{\alpha_n} \leq 1}} \frac{x_1^{p_1-1} x_2^{p_2-1} \ldots x_n^{p_n-1}}{(x_1^{\alpha_1} + x_2^{\alpha_2} + \ldots + x_n^{\alpha_n})^\mu} dx_1\, dx_2 \ldots dx_n =$$

$$= \frac{1}{\alpha_1 \alpha_2 \ldots \alpha_n \left(\frac{p_1}{\alpha_1} + \frac{p_2}{\alpha_2} + \ldots + \frac{p_n}{\alpha_n} - \mu\right)} \frac{\Gamma\left(\frac{p_1}{\alpha_1}\right) \Gamma\left(\frac{p_2}{\alpha_2}\right) \ldots \Gamma\left(\frac{p_n}{\alpha_n}\right)}{\Gamma\left(\frac{p_1}{\alpha_1} + \frac{p_2}{\alpha_2} + \ldots + \frac{p_n}{\alpha_n}\right)}$$

$$\left[\mu < \frac{p_1}{\alpha_1} + \frac{p_2}{\alpha_2} + \ldots + \frac{p_n}{\alpha_n}\right].$$ F III 393

3.
$$\iint \cdots \int_{\substack{x_1 \geq 0,\, x_2 \geq 0, \ldots, x_n \geq 0 \\ x_1^{\alpha_1} + x_2^{\alpha_2} + \ldots + x_n^{\alpha_n} \leq 1}} x_1^{p_1-1} x_2^{p_2-1} \ldots x_n^{p_n-1} \sqrt{\frac{1 - x_1^{\alpha_1} - x_2^{\alpha_2} - \ldots - x_n^{\alpha_n}}{1 + x_1^{\alpha_1} + x_2^{\alpha_2} + \ldots + x_n^{\alpha_n}}} \times$$

$$\times dx_1\, dx_2 \ldots dx_n = \frac{\sqrt{\pi}}{2} \frac{\Gamma\left(\frac{p_1}{\alpha_1}\right) \Gamma\left(\frac{p_2}{\alpha_2}\right) \ldots \Gamma\left(\frac{p_n}{\alpha_n}\right)}{\alpha_1 \alpha_2 \ldots \alpha_n} \frac{1}{\Gamma(m)} \times$$

$$\times \left\{\frac{\Gamma\left(\frac{m}{2}\right)}{\Gamma\left(\frac{m+1}{2}\right)} - \frac{\Gamma\left(\frac{m+1}{2}\right)}{\Gamma\left(\frac{m+2}{2}\right)}\right\}$$

mit | with

$$m = \frac{p_1}{\alpha_1} + \frac{p_2}{\alpha_2} + \ldots + \frac{p_n}{\alpha_n}.$$ F III 393

4.637
$$\iint \cdots \int_{\substack{x_1 \geq 0,\, x_2 \geq 0, \ldots, x_n \geq 0 \\ x_1 + x_2 + \ldots + x_n \leq 1}} f(x_1 + x_2 + \ldots + x_n) \times$$

$$\times \frac{x_1^{p_1-1} x_2^{p_2-1} \ldots x_n^{p_n-1}\, dx_1\, dx_2 \ldots dx_n}{(q_1 x_1 + q_2 x_2 + \ldots + q_n x_n + r)^{p_1 + p_2 + \ldots + p_n}} =$$

$$= \frac{\Gamma(p_1) \Gamma(p_2) \ldots \Gamma(p_n)}{\Gamma(p_1 + p_2 + \ldots + p_n)} \int_0^1 f(x) \frac{x^{p_1 + p_2 + \ldots + p_n - 1}}{(q_1 x + r)^{p_1} (q_2 x + r)^{p_2} \ldots (q_n x + r)^{p_n}}\, dx,$$

wobei die Funktion $f(x)$ im Intervall $(0, 1)$ stetig ist

where $f(x)$ is continuous on $(0, 1)$

$$[q_1 \geq 0, q_2 \geq 0, \ldots, q_n \geq 0; r > 0].$$

1. $$\int_0^\infty \int_0^\infty \cdots \int_0^\infty \frac{x_1^{p_1-1} x_2^{p_2-1} \ldots x_n^{p_n-1} e^{-(q_1 x_1 + q_2 x_2 + \ldots + q_n x_n)}}{(r_0 + r_1 x_1 + r_2 x_2 + \ldots + r_n x_n)^s} dx_1 dx_2 \ldots dx_n =$$ 4.638

$$= \frac{\Gamma(p_1)\Gamma(p_2)\ldots\Gamma(p_n)}{\Gamma(s)} \int_0^\infty \frac{e^{-r_0 x} x^{s-1} dx}{(q_1 + r_1 x)^{p_1}(q_2 + r_2 x)^{p_2}\ldots(q_n + r_n x)^{p_n}}$$

für positive p_i, q_i, r_i, s. Unter der Bedingung $p_1 + p_2 + \ldots + p_n > s$ gilt dieses Resultat auch für $r_0 = 0$.

if p_i, q_i, r_i, s are positive; the formula also holds for $r_0 = 0$ provided $p_1 + p_2 + \ldots + p_n > s$.

2. $$\int_0^\infty \int_0^\infty \cdots \int_0^\infty \frac{x_1^{p_1-1} x_2^{p_2-1} \ldots x_n^{p_n-1}}{(r_0 + r_1 x_1 + r_2 x_2 + \ldots + r_n x_n)^s} dx_1 dx_2 \ldots dx_n =$$

$$= \frac{\Gamma(p_1)\Gamma(p_2)\ldots\Gamma(p_n)\Gamma(s-p_1-p_2-\ldots-p_n)}{r_1^{p_1} r_2^{p_2} \ldots r_n^{p_n} r_0^{s-p_1-p_2-\ldots-p_n} \Gamma(s)} \quad [p_i > 0, r_i > 0, s > 0].$$

3. $$\int_0^\infty \int_0^\infty \cdots \int_0^\infty \frac{x_1^{p_1-1} x_2^{p_2-1} \ldots x_n^{p_n-1}}{[1 + (r_1 x_1)^{q_1} + (r_2 x_2)^{q_2} + \ldots + (r_n x_n)^{q_n}]^s} dx_1 dx_2 \ldots dx_n =$$

$$= \frac{\Gamma\left(\frac{p_1}{q_1}\right)\Gamma\left(\frac{p_2}{q_2}\right)\ldots\Gamma\left(\frac{p_n}{q_n}\right)}{q_1 q_2 \ldots q_n r_1^{p_1 q_1} r_2^{p_2 q_2} \ldots r_n^{p_n q_n}} \cdot \frac{\Gamma\left(s - \frac{p_1}{q_1} - \frac{p_2}{q_2} - \ldots - \frac{p_n}{q_n}\right)}{\Gamma(s)}$$

$$[p_i > 0, q_i > 0, r_i > 0, s > 0].$$

1. $$\underset{x_1^2 + x_2^2 + \ldots + x_n^2 \leq 1}{\int\int \cdots \int} (p_1 x_1 + p_2 x_2 + \ldots + p_n x_n)^{2m} dx_1 dx_2 \ldots dx_n =$$ 4.639

$$= \frac{(2m-1)!!}{2^m} \frac{\sqrt{\pi^n}}{\Gamma\left(\frac{n}{2} + m + 1\right)} (p_1^2 + p_2^2 + \ldots + p_n^2)^m.$$ F III 394–395

2. $$\underset{x_1^2 + x_2^2 + \ldots + x_n^2 \leq 1}{\int\int \cdots \int} (p_1 x_1 + p_2 x_2 + \ldots + p_n x_n)^{2m+1} dx_1 dx_2 \ldots dx_n = 0.$$

F III 395

1. $$\underset{x_1^2 + x_2^2 + \ldots + x_n^2 \leq 1}{\int\int \cdots \int} e^{p_1 x_1 + p_2 x_2 + \ldots + p_n x_n} dx_1 dx_2 \ldots dx_n =$$ 4.641

$$= \sqrt{\pi^n} \sum_{k=0}^\infty \frac{1}{k!\, \Gamma\left(\frac{n}{2} + k + 1\right)} \left(\frac{p_1^2 + p_2^2 + \ldots + p_n^2}{4}\right)^k.$$ F III 395

4.641 2. $\iint \cdots \int\limits_{x_1^2+x_2^2+\ldots+x_{2n}^2 \leq 1} e^{p_1 x_1 + p_2 x_2 + \ldots + p_{2n} x_{2n}} dx_1 dx_2 \ldots dx_{2n} =$

$$= \frac{(2\pi)^n I_n(\sqrt{p_1^2 + p_2^2 + \ldots + p_{2n}^2})}{(p_1^2 + p_2^2 + \ldots + p_{2n}^2)^{n/2}}.$$

F III 395+

4.642 $\iint \cdots \int\limits_{x_1^2+x_2^2+\ldots+x_n^2 \leq R^2} f(\sqrt{x_1^2 + x_2^2 + \ldots + x_n^2}) dx_1 dx_2 \ldots dx_n =$

$$= \frac{2\sqrt{\pi^n}}{\Gamma\left(\frac{n}{2}\right)} \int_0^R x^{n-1} f(x) dx,$$

wobei die Funktion $f(x)$ im Intervall $(0, R)$ stetig ist. | where $f(x)$ is continuous on $(0, R)$.

F III 396–397

4.643 $\int_0^1 \int_0^1 \cdots \int_0^1 f(x_1 x_2 \ldots x_n)(1-x_1)^{p_1-1}(1-x_2)^{p_2-1} \ldots (1-x_n)^{p_n-1} \times$

$$\times x_2^{p_1} x_3^{p_1+p_2} \ldots x_n^{p_1+p_2+\ldots+p_{n-1}} dx_1 dx_2 \ldots dx_n =$$

$$= \frac{\Gamma(p_1) \Gamma(p_2) \ldots \Gamma(p_n)}{\Gamma(p_1 + p_2 + \ldots + p_n)} \int_0^1 f(x)(1-x)^{p_1+p_2+\ldots+p_n-1} dx$$

vorausgesetzt, das auf der rechten Seite stehende Integral konvergiert absolut. | provided the integral on the right converges absolutely.

F III 399

4.644 $\overbrace{\iint \cdots \int}^{n-1}_{x_1^2+x_2^2+\ldots+x_n^2=1} f(p_1 x_1 + p_2 x_2 + \ldots + p_n x_n) \frac{dx_1 dx_2 \ldots dx_{n-1}}{|x_n|} =$

$$= 2 \iint \cdots \int\limits_{x_1^2+x_2^2+\ldots+x_{n-1}^2 \leq 1} f(p_1 x_1 + p_2 x_2 + \ldots + p_n x_n) \frac{dx_1 dx_2 \ldots dx_{n-1}}{\sqrt{1-x_1^2-x_2^2-\ldots-x_{n-1}^2}} =$$

$$= \frac{2\sqrt{\pi^{n-1}}}{\Gamma\left(\frac{n-1}{2}\right)} \int_0^\pi f(\sqrt{p_1^2 + p_2^2 + \ldots + p_n^2} \cos x) \sin^{n-2} x \, dx \quad [n \geq 3],$$

wobei die Funktion $f(x)$ im Intervall | where $f(x)$ is continuous in

$$[-\sqrt{p_1^2 + p_2^2 + \ldots + p_n^2}, \sqrt{p_1^2 + p_2^2 + \ldots + p_n^2}]$$

F III 399–401

stetig ist.

4.645 Die Funktionen $f(x_1, x_2, \ldots, x_n)$ und $g(x_1, x_2, \ldots, x_n)$ seien im abgeschlossenen und beschränkten Bereich (D) stetig. Ferner sei m der kleinste und M der größte Wert, den die Funktion g im Bereich (D) annimmt. Mit $\varphi(u)$ werde eine für $m \leq u \leq M$ stetige Funktion bezeichnet. Schließlich sei | Let he functions $f(x_1, x_2, \ldots, x_n)$ and $g(x_1, x_2, \ldots, x_n)$ be continuous on a bounded closed domain (D); let m and M be the minimum and the maximum, respectively, of g in (D); let $\varphi(u)$ be continuous in the interval $m \leq u \leq M$; finally, let $\psi(u)$ be defined as

1. $$\psi(u) = \iint\limits_{m \leq g(x_1, x_2, \ldots, x_n) \leq u} \cdots \int f(x_1, x_2, \ldots, x_n)\, dx_1\, dx_2 \ldots dx_n,$$ 4.645

wobei das Integral über denjenigen Teil des Bereichs (D) erstreckt wird, in dem die Ungleichung $m \leq g(x_1, x_2, \ldots, x_n) \leq u$ erfüllt ist. Dann gilt	where the integral is taken over the part of (D) in which the inequality $m \leq g(x_1, x_2, \ldots, x_n) \leq u$ is satisfied. Then

2. $$\iint\limits_{m \leq g(x_1, x_2, \ldots x_n) \leq M} \cdots \int f(x_1, x_2, \ldots, x_n)\, \varphi[g(x_1, x_2, \ldots, x_n)]\, dx_1\, dx_2 \ldots dx_n =$$

$$= (S)\int_m^M \varphi(u)\, d\psi(u) = (R)\int_m^M \varphi(u)\, \frac{d\psi(u)}{du}\, du,$$

wobei das mittlere Integral im Sinne von STIELTJES zu verstehen ist. Hat $\psi(u)$ eine stetige Ableitung $\frac{d\psi}{du}$, so existiert das rechts stehende RIEMANNsche Integral.	where the second integral is a STIELTJES integral; the RIEMANN integral on the right exists if $\psi(u)$ has a continuous derivative $\frac{d\psi}{du}$.
In der Formel 4.645 2. kann M auch gleich $+\infty$ sein; dann ist unter $\int_m^{+\infty}$ der $\lim\limits_{M \to +\infty}\int_m^M$ zu verstehen.	In 4.645 2. M can be $+\infty$ as well; then $\int_m^{+\infty}$ means $\lim\limits_{M \to +\infty}\int_m^M$.

$$\iint\limits_{\substack{x_1 \geq 0, x_2 \geq 0, \ldots, x_n \geq 0 \\ x_1 + x_2 + \ldots + x_n \leq 1}} \cdots \int \frac{x_1^{p_1-1} x_2^{p_2-1} \ldots x_n^{p_n-1}}{(q_1 x_1 + q_2 x_2 + \ldots + q_n x_n)^r}\, dx_1\, dx_2 \ldots dx_n =$$ 4.646

$$= \frac{\Gamma(p_1)\, \Gamma(p_2) \ldots \Gamma(p_n)}{\Gamma(p_1 + p_2 + \ldots + p_n - r + 1)\, \Gamma(r)} \int_0^\infty \frac{x^{r-1}\, dx}{(1 + q_1 x)^{p_1} (1 + q_2 x)^{p_2} \ldots (1 + q_n x)^{p_n}}$$

$$[p_1 > 0,\ p_2 > 0,\ \ldots,\ p_n > 0,\ q_1 > 0,\ q_2 > 0,\ \ldots,\ q_n > 0,$$
$$p_1 + p_2 + \ldots + p_n > r > 0].$$ F III 402–403

$$\iint\limits_{0 \leq x_1^2 + x_2^2 + \ldots + x_n^2 \leq 1} \cdots \int \exp\left\{\frac{p_1 x_1 + p_2 x_2 + \ldots + p_n x_n}{\sqrt{x_1^2 + x_2^2 + \ldots + x_n^2}}\right\} dx_1\, dx_2 \ldots dx_n =$$ 4.647

$$= \frac{2\sqrt{\pi^n}}{n(p_1^2 + p_2^2 + \ldots + p_n^2)^{\frac{n}{4} - \frac{1}{2}}} I_{\frac{n}{2}-1}\left(\sqrt{p_1^2 + p_2^2 + \ldots + p_n^2}\right).$$ F III 404–405

$$\int_0^\infty \int_0^\infty \cdots \int_0^\infty \exp\left[-\left(x_1 + x_2 + \ldots + x_n + \frac{\lambda^{n+1}}{x_1 x_2 \ldots x_n}\right)\right] \times$$ 4.648

$$\times x_1^{\frac{1}{n+1}-1} x_2^{\frac{2}{n+1}-1} \ldots x_n^{\frac{n}{n+1}-1}\, dx_1\, dx_2 \ldots dx_n =$$

$$= \frac{1}{\sqrt{n+1}} (2\pi)^{\frac{n}{2}} e^{-(n+1)\lambda}.$$ F III 405

Bezeichnungen
Notations

Sofern der Buchstabe k nicht als Summationsindex auftritt, bezeichnet er eine Zahl mit $0 \leqslant k \leqslant 1$. Auch in Integralen, die auf elliptische Integrale führen, wird diese Bezeichnung benutzt. Ferner ist $k' = \sqrt{1-k^2}$.

Unless the letter k is used as the summation index, it indicates a value $0 \leqslant k \leqslant 1$. This notation is used in integrals that lead to elliptic integrals. In the latter case, $k' = \sqrt{1-k^2}$.

$R(x)$ — Rationale Funktion. / A rational function

$\operatorname{Re} z \equiv x$ — Realteil von $z = x + iy$. / The real part of the complex number $z = x + iy$.

$\operatorname{Im} z \equiv y$ — Imaginärteil von $z = x + iy$. / The imaginary part of the complex number $z = x + iy$.

$\bar{z} = x - iy$ — Die zu $z = x + iy$ konjugiert komplexe Zahl. / The complex conjugate of $z = x + iy$.

$\arg z$ — Argument von $z = x + iy$. / The argument of $z = x + iy$.

$\operatorname{sign} x$ — Vorzeichen der reellen Zahl x; / The sign of the real number x;

$\operatorname{sign} x = +1$ für $x > 0$; / if $x > 0$;
$\operatorname{sign} x = -1$ für $x < 0$. / if $x < 0$.

$E(x)$ — größte ganze Zahl, die nicht größer als x ist (x reell). / The integral part of x (x real).

$\int_a^{(b+)}, \int_a^{(b-)}$ — Linienintegrale. Der Integrationsweg geht vom Punkt a aus, nähert sich dem Punkt b (auf einer Geraden, wenn nichts anderes vermerkt ist), umläuft auf einem kleinen Kreis in positivem (negativem) Umlaufsinn den Punkt b und kehrt zum Punkt a zurück, wobei der ursprüngliche Weg in umgekehrter Richtung durchlaufen wird. / Contour integral. The path of integration starts from a, approaches b (along a straight line, except when stated otherwise) passes around b along the smallest possible circle in the positive (negative) sense and returns to a along the original path in the opposite direction.

$\int_C$ — Linienintegral längs der Kurve C. / Line integral along the curve C.

$$n! = 1 \cdot 2 \cdot 3 \ldots n, \quad 0! = 1.$$
$$(2n+1)!! = 1 \cdot 3 \ldots (2n+1).$$
$$(2n)!! = 2 \cdot 4 \ldots (2n).$$
$$\binom{p}{n} = \frac{p(p-1)\ldots(p-n+1)}{1 \cdot 2 \ldots n}, \quad \binom{p}{0} = 1.$$
$$(a)_n = a(a+1)\ldots(a+n-1) = \frac{\Gamma(a+n)}{\Gamma(a)}.$$
$$\sum_{k=m}^{n} u_k = u_m + u_{m+1} + \ldots + u_n,$$
$$\sum_{k=m}^{n} u_k = 0, \quad \text{wenn} \atop \text{if} \quad n < m.$$

$\sum\limits_{n}'$, $\sum\limits_{m,n}'$	Die Summe ist über alle ganzzahligen n außer $n = 0$ bzw. über alle ganzzahligen m und n außer $m = n = 0$ zu erstrecken.	The term with $n = 0$ or with $n = m = 0$, respectively, has to be excluded.
$O(f(z))$	Ordnung von $f(z)$. Es sei $\lim z = z_0$. Existiert ein $M > 0$ derart, daß in einer hinreichend kleinen Umgebung von z_0 die Ungleichung $\|g(z)\| \leqslant M\|f(z)\|$ gilt, so schreibt man $g(z) = O(f(z))$.	Order of $f(z)$. Let $\lim z = z_0$. If there exists an $M > 0$ such that $\|g(z)\| \leqslant M\|f(z)\|$ in a sufficiently small neighbourhood of z_0 the inequality, then $g(z) = O(f(z))$.

Nicht allgemein übliche Bezeichnungen | **Standardbezeichnung**
Notation not in general use | **Standard notation**

Arch	Ar cosh
Arcth	Ar coth
Arsh	Ar sinh
Arth	Ar tanh
arcctg	arc cot
arctg	arc tan
ch	cosh
ctg	cot
cth	coth
sh	sinh
tg	tan
th	tanh
$E(x)$	$[x]$

Zitierte Literatur *
References *

A   Adams, E., Smithsonian Mathematical Formulae and Tables of Elliptic Functions. Washington 1922.

AK   Appell, P., et J. de Fériet Kampé, Fonctions hypergéométriques et hypersphériques; Polynomes D'Hermite. Paris 1926.

Ba I   Bateman Manuscript Project. Higher Transcendental Functions I (ed. A. Erdélyi). New York–Toronto–London 1953.

Ba II   Bateman Manuscript Project. Higher Transcendental Functions II (ed. A. Erdélyi). New York–Toronto–London 1953.

Ba III   Bateman Manuscript Project. Higher Transcendental Functions III (ed. A. Erdélyi). New York–Toronto–London 1955.

Ba IV   Bateman Manuscript Project. Tables of Integral Transforms I (ed. A. Erdélyi). New York–Toronto–London 1954.

Ba V   Bateman Manuscript Project. Tables of Integral Transforms II (ed. A. Erdélyi). New York–Toronto–London 1954.

Be   Bertrand, J., Traité de calcul différentiel et de calcul intégral, v. 2., Calcul intégral. Intégrales définies et indéfinies. Paris 1870.

BH   Bierens de Haan, D., Nouvelles tables d'intégrales définies. Amsterdam 1867.

Br_{08} Bromwich, T. J., An Introduction to the Theory of Infinite Series. London 1908.

Br_{26} Bromwich, T. J., An Introduction to the Theory of Infinite Series, 2nd ed. London 1926.

Bu   Buchholz, H., Die konfluente hypergeometrische Funktion mit besonderer Berücksichtigung ihrer Anwendungen. Berlin–Göttingen–Heidelberg 1953.

BF   Byrd, P. F., and M. D. Friedman, Handbook of Elliptic Integrals for Engineers and Physicists. Berlin–Göttingen–Heidelberg 1954.

C   Cesàro, E., Elementares Lehrbuch der Algebraischen Analysis und der Infinitesimalrechnung (deutsch von G. Kowalewski). Leipzig 1904.

CH   Courant, R., und D. Hilbert, Methoden der mathematischen Physik I. Berlin 1931.

D   Dwight, H. B., Tables of Integrals and Other Mathematical Data. New York 1934.

ED   Efros, A. M., A. M., Danilewski, Operazionnoe istschislenie i konturnye integraly. Charkow 1937.

F I   Fichtenholz, G. M., Differential- und Integralrechnung I., 9. Aufl. Berlin 1975 (Übersetzung aus dem Russischen — translated from the Russian).

F II   Fichtenholz, G. M., Differential- und Integralrechnung II., 6. Aufl. Berlin 1974 (Übersetzung aus dem Russischen — translated from the Russian).

F III   Fichtenholz, G. M., Differential- und Integralrechnung III., 8. Aufl. Berlin 1977 (Übersetzung aus dem Russischen — translated from the Russian).

Ga   Gauss, C. F., Werke, Bd. III. Göttingen 1876.

Ge   Gelfond, A. O., Differenzenrechnung. Berlin 1958 (Übersetzung aus dem Russischen — translated from the Russian).

Gou I   Goursat, E., Cours d'analyse mathématique, I., 5ème éd. Paris 1943.

Gou II   Goursat, E., Cours d'analyse mathématique, II., 7ème éd. Paris 1949.

GHH   Gröbner, W., M. Hofreiter, N. Hofreiter, J. Laub und E. Peschl, Integraltafeln, Teil I, Unbestimmte Integrale. Braunschweig 1944.

---

* Vgl. die Fußnote im Vorwort zur 3. russ. Auflage.   * See the Preface to the 3rd Russian edition.

GH	GRÖBNER, W., und N. HOFREITER, Integraltafeln, Teil II, Bestimmte Integrale. Wien—Innsbruck 1958.
GK J	GÜNTER, N. M., und R. O. KUSMIN, Aufgabensammlung zur höheren Mathematik, I., 9. Aufl. Berlin 1975 (Übersetzung aus dem Russischen — translated from the Russian).
GK II	GÜNTER, N. M., und R. O. KUSMIN, Aufgabensammlung zur höheren Mathematik, II., 6. Aufl. Berlin 1976 (Übersetzung aus dem Russischen — translated from the Russian).
H	HOBSON, E. W., The Theory of Spherical and Ellipsoidal Harmonics. Cambridge 1931.
JE	JAHNKE, E., und F. EMDE, Funktionentafeln mit Formeln und Kurven, 2. Aufl. Leipzig 1933.
J	JAMES, H. M., N. B. NICHOLS and R. S. PHILLIPS (editors), Theory of Servomechanisms. New York—London 1947.
Jo	JOLLEY, L., Summation of Series. London 1925.
Kn	KNOPP, K., Theorie und Anwendung der unendlichen Reihen, 4. Aufl. Berlin—Heidelberg 1947.
Kr	KRETSCHMAR, W. A., Sadatschnik po algebre, isd. 2-je. Moskwa—Leningrad 1950.
Ku	KUSMIN, R. O., Besselevy funkzii. Moskwa—Leningrad 1935.
La	LASKA, W., Sammlung von Formeln der reinen und angewandten Mathematik. Braunschweig 1888.
Le	LEGENDRE, A. M., Exercices calcul intégral. Paris 1811.
Li	LINDMAN, C. E., Examen des Nouvelles tables d'intégrales définies de M. BIERENS DE HAAN. Amsterdam 1867, Stockholm 1891.
Lo I	LOBATSCHEWSKI, N. I., Polnoe sobranie sotschinenij, I. Moskwa—Leningrad 1946.
Lo III	LOBATSCHEWSKI, N. I., Polnoe sobranie sotschinenij, III. Moskwa—Leningrad 1951.
Lo V	LOBATSCHEWSKI, N. I., Polnoe sobranie sotschinenij, V. Moskwa—Leningrad 1951.
M	MCLACHLAN, N. W., Theory and Application of MATHIEU Functions. Oxford 1947.
MH	MCLACHLAN, N. W., et P. HUMBERT, Formulaire pour le calcul symbolique, 2nde éd. Paris 1950.
MHF	MC.LACHLAN, N. W., P. HUMBERT et L. POLI, Supplément au formulaire pour le calcul symbolique. Paris 1950.
MO	MAGNUS, W., und F. OBERHETTINGER, Formeln und Sätze für die speziellen Funktionen der mathematischen Physik, 2. Aufl. Berlin—Göttingen—Heidelberg 1948.
MC	MEYER ZUR CAPELLEN, W., Integraltafeln, Sammlung unbestimmter Integrale elementarer Funktionen. Berlin—Göttingen—Heidelberg 1950.
Na	NATANSON, I. P., Konstruktive Funktionentheorie. Berlin 1955 (Übersetzung aus dem Russischen — translated from the Russian).
NG	NIELSEN, N., Handbuch der Theorie der Gammafunktion. Leipzig 1906.
NI	NIELSEN, N., Theorie des Integrallogarithmus und verwandter Transcendenten. Leipzig 1906.
No	NOWOSELOW, S. I., Obratnye trigonometritscheskie funkzii. Moskwa—Leningrad 1950.
P	PEIRCE, B. O., A Short Table of Integrals, 3rd ed. Boston 1929.
Si	SIKORSKI, J. S., Elementy teorii elliptitscheskich funkzij s priloshenijami k mechanike. Moskwa—Leningrad 1936.
Sm III$_2$	SMIRNOW, W. I., Lehrgang der höheren Mathematik, Teil III$_2$, 11. Aufl. Berlin 1976 (Übersetzung aus dem Russischen — translated from the Russian).
St	STRUTT, M. O., LAMÉsche, MATHIEUsche und verwandte Funktionen in Physik und Technik. Berlin 1932.
T	TIMOFEJEW, A. F., Integrirowanie funkzij, T. 1. Moskwa—Leningrad 1933.
W	WATSON, G. N., A Treatise on the Theory of Bessel Functions. Cambridge 1922.
WW	WHITTAKER, E. T., and G. N. WATSON, A Course of Modern Analysis, 4th ed. Cambridge 1952.
Z	ZURAWSKI, A. M., Sprawotschnik po elliptitscheskim funkzijam. Moskwa—Leningrad 1941.

Für Nachträge/Notes

Für Nachträge/Notes

Für Nachträge /Notes